MODERN METHODS OF PLANT ANALYSIS

FOUNDED BY

K. PAECH   M. V. TRACEY

VOLUME VII

REDACTION BY

H. F. LINSKENS · B. D. SANWAL · M. V. TRACEY

CONTRIBUTORS

D. I. ARNON - E. J. BARRON - F. CHAPEVILLE - D. D. DAVIES - G. A. DIN - R. J. ELLIS
D. S. FEINGOLD - P. FROMAGEOT - M. GIBBS - T. W. GOODWIN
D. P. HACKETT - W. Z. HASSID - E. J. HEWITT - T. HIGUCHI - H. JANECKE
I. KAWAMURA - P. S. KRISHNAN - M. LATA - M. LOSADA - S. MAHADEVAN
E. F. NEUFELD - D. J. D. NICHOLAS - R. ROHRINGER - B. D. SANWAL
P. K. STUMPF - A. L. TAPPEL - J. F. TURNER - D. WANG - E. R. WAYGOOD
G. WEBSTER - M. W. ZINK

WITH 64 FIGURES

SPRINGER-VERLAG
BERLIN · GÖTTINGEN · HEIDELBERG
1964

MODERNE METHODEN DER PFLANZENANALYSE

BEGRÜNDET VON

K. PAECH   M. V. TRACEY

7. BAND

REDIGIERT VON

H. F. LINSKENS · B. D. SANWAL · M. V. TRACEY

BEARBEITET VON

D. I. ARNON - E. J. BARRON - F. CHAPEVILLE - D. D. DAVIES - G. A. DIN - R. J. ELLIS
D. S. FEINGOLD - P. FROMAGEOT - M. GIBBS - T. W. GOODWIN
D. P. HACKETT - W. Z. HASSID - E. J. HEWITT - T. HIGUCHI - H. JANECKE
I. KAWAMURA - P. S. KRISHNAN - M. LATA - M. LOSADA - S. MAHADEVAN
E. F. NEUFELD - D. J. D. NICHOLAS - R. ROHRINGER - B. D. SANWAL
P. K. STUMPF - A. L. TAPPEL - J. F. TURNER - D. WANG - E. R. WAYGOOD
G. WEBSTER - M. W. ZINK

MIT 64 ABBILDUNGEN

SPRINGER-VERLAG
BERLIN · GÖTTINGEN · HEIDELBERG
1964

© by Springer-Verlag oHG. Berlin · Göttingen · Heidelberg 1964
Softcover reprint of the hardcover 1st edition 1964

Library of Congress Catalog Card Number A 55-6022

ISBN 978-3-642-48143-7   ISBN 978-3-642-48141-3 (eBook)
DOI 10.1007/ 978-3-642-48141-3

Inhaltsverzeichnis. — Contents.

Special Methods of Isolation and Purification.

Enzymes du Métabolisme du Soufre. Par F. CHAPEVILLE et P. FROMAGEOT. Avec 1 Figure   1

A. Enzymes d'activation et de transfert du sulfate   1
    I. Enzymes d'activation du sulfate.   3
        1. Mesure de l'activité enzymatique   3
           a) Système complet d'activation   3
           b) ATP-sulfurylase — ROBBINS et LIPMANN (1958a)   3
           c) ADP-sulfurylase — ROBBINS et LIPMANN (1958a)   4
           d) APS-kinase — ROBBINS et LIPMANN (1958a).   4
        2. Préparation de l'ATP-sulfurylase, de l'ADP-sulfurylase et de l'APS-kinase   4
    II. Enzymes de transfert du sulfate.   6
        1. Mesure de l'activité de la phénol-sulfokinase   6
        2. Préparation de la phénol-sulfokinase — GREGORY et LIPMANN (1957) . .   7

B. Systèmes réducteurs du sulfate   8
    1. Mesure de l'activité sulfato-réductrice   8
    2. Préparation du système réducteur du sulfate, HILZ, KITTLER et KNAPE
        (1959).   9

C. Sulfatases   10
    1. Arylsulfatases   11
        Préparation des arylsulfatases   11
    2. Chondrosulfatases.   12
    3. Myrosulfatases   12
    4. Glucosulfatases   12
    5. Stéroïdesulfatases   12
    6. Cholinesulfatase   12

D. Enzymes du métabolisme du thiosulfate, du sulfite et du sulfure   13
    I. Enzymes du métabolisme du thiosulfate   13
        1. Thiosulfate-réductase   13
        2. Rhodanèse.   14
    II. Enzymes du métabolisme du sulfite   14
        1. Formation du sulfite par désulfination   14
        2. Sulfite-réductase   15
        3. Sulfite-oxydase   16
    III. Enzymes du métabolisme du sulfure   16
        1. Formation d'hydrogène sulfuré à partir de la cystéine   16
        2. Sérine-sulfhydrase   18
        3. Oxydation du sulfure   19
Bibliographie   19

Enzymes of Phosphate Metabolism. By P. S. KRISHNAN.   20
    I. Special Features of Enzyme Isolation and Purification   20
        1. Choice of Starting Material.   20
        2. Conventional versus Modern Methods of Enzyme Purification.   20
           a) Automatic Device for Ammonium Sulfate Addition   22
           b) Dialysis of Enzyme Solutions   22
        3. Stability of Enzymes on Dilution   22
    II. Assay of Phosphatase Activity   22
        1. Colorimetric Estimation of Phosphorus   23
        2. Aspects of Enzyme Assay in Plants and Microorganisms   25
    III. Intracellular Localization of Enzymes by the Technique of Differential Cen-
        trifugation   29
        1. Nuclear Fraction   32

       2. Plastid Fraction 32
       3. Mitochondrial Fraction 33
       4. Cell Wall Fraction 33
       5. Association of Enzymes with Larger Particles 33
       6. Isolation of Plant Protoplasts 34
       7. Localization of Phosphatases 34
   IV. Phosphatases Acting on Hexose Phosphates 35
       1. The Hydrolysis of Glucose-1-Phosphate 35
       2. C$_1$ Diphosphatases 36
          a) Fructose Diphosphatase. 36
          b) Fructose Diphosphatase of Animal Tissue 36
          c) Fructose Diphosphatase of Plants 37
          d) Neutral Fructose Diphosphatase 39
          e) Acidic Fructose Diphosphatase. 39
          f) Acid C$_1$ Diphosphatase of *Escherichia coli* 39
          g) Role of Alkaline Fructose Diphosphatase in Green Plants . . . 39
       3. Sedoheptulose-1,7-Diphosphatase 40
       4. Glycerate-2,3-Diphosphatase 40
    V. Enzymes Acting on Condensed Phosphates 41
       1. Inorganic Pyrophosphatases 41
          a) Inorganic Pyrophosphatase of Microorganisms 41
          b) Inorganic Pyrophosphatases of Animal Tissue 43
       2. Tripolyphosphatases 44
       3. Tetrapolyphosphatases 45
       4. Trimetaphosphatases 45
       5. Tetrametaphosphatases 47
   VI. Enzymes Acting on Polyphosphates 47
       Type I Enzyme: Polyphosphate Depolymerases 47
       Type II Enzyme: Polyphosphatases 47
          a) Type I 48
          b) Type II 49
          c) Type III 49
       Type III Enzyme: Enzymes Transfering Phosphate Groups from Polyphos-
       phates to Suitable Acceptors 49
       Polyphosphate-ADP-phosphotransferase of *Escherichia coli* 50
  VII. General Phosphomonoesterases 50
       1. Acid Phosphatases 51
       2. Alkaline Phosphatases. 51
          a) Alkaline Phosphatases of Microorganisms 52
 VIII. Phosphoprotein Phosphatase 52
   IX. Ortho Phosphoserine Phosphatase 54
    X. Phosphatase Acting on Nucleotides and Nucleic Acids 55
       1. Nucleotide Pyrophosphatases. 55
          a) Nucleotide Pyrophosphatase of Potato 56
          b) Nucleotide Pyrophosphatase of Tobacco. 56
       2. 5'-Nucleotidases 57
          a) 5'-Nucleotidase of *Clostridium stricklandii* 57
       3. Ribonucleases 58
          a) Ribonucleases of Plants 58
          b) Endonucleases from Microorganisms 62
          c) Phosphodiesterases Acting on Polynucleotides 62

  References 63

Enzymes of Inorganic Nitrogen Metabolism. By E. J. HEWITT and D. J. D. NICHOLAS.
  With 3 Figures 67

  A. Nitrate Reductase 67
     I. Preparation. 68
        1. Fungi and Higher Plants 68
        2. Bacteria 76
           a) Methods of extraction 76
           b) Fractionation and Stability 78
    II. Measurement of Activity 83
        1. Assay Methods 83
        2. Sources of Error 89

Inhaltsverzeichnis. — Contents. VII

III. Electron Donors and Co-Factors. 91
   1. Fungi and Higher Plants 91
     a) Pyridine nucleotides and other electron donors 91
     b) Flavins 93
     c) Metal and Anion Requirements 94
   2. Bacteria 95
  IV. Properties and Mechanisms 96
   1. Fungi and Higher Plants 96
   2. Bacteria 101
  V. Physiological Factors 108
   1. Fungi and Higher Plants 108
   2. Bacteria 109
B. Other Nitrate Reduction and Nitro Reductase Enzymes 110
  1. Aldehyde and Xanthine Oxidases of Animal Origin 110
  2. Aldehyde Oxidase of Potato 110
  3. Reduction of Aromatic Nitro-Compounds 111
C. Nitrite, Nitric Oxide, Hyponitrite and Hydroxylamine Reductases and Related Enzymes 112
  I. Preparation 113
   1. Fungi and Higher Plants 113
   2. Bacteria 115
  II. Fractionation and Stability 115
  III. Measurement of Activity 121
   1. Interference and Errors 121
   2. Assay Methods 126
     a) Use of CONWAY Methods for NH₃ Determinations 126
     b) Hydroxylamine 128
     c) Oximes, YAMAFUJI and AKITA (1952) 129
     d) Hyponitrite 129
     e) Assay Methods Using Reduced Benzyl Viologen and Other Dyes under Anaerobic Conditions 130
     f) Manometric Assay of Nitrite Reductase 133
     g) Manometric Assay of Nitric Oxide Reductase 136
   3. Preparation of Hyponitrite as a Substrate 136
  IV. Electron Donors and Co-Factors and Inhibitors 137
   1. Fungi and Higher Plants 137
     a) Nitrite Reductase 137
     b) Hyponitrite Reductase 141
     c) Hydroxylamine Reductase 142
   2. Bacteria 142
     a) Nitrite Reductase 142
     b) Nitric Oxide Reductase 144
     c) Hydroxylamine Reductase 146
  V. Properties and Mechanisms 146
   1. Fungi and Higher Plants 146
   2. Bacteria 148
     a) Nitrite Reductase 148
     b) Nitric Oxide Reductase 153
     c) Hydroxylamine Reductase 155
  VI. Physiological Factors and Other Features 157
   1. Fungi and Higher Plants 157
     a) Nitrite Reductase 157
     b) Hydroxylamine Reductase 158
D. Glutamic Dehydrogenase 158
E. Oxidation and Other Reactions of Ammonia, Hydroxylamine and Nitrite . . . 158
  I. Fungi and Higher Plants 158
   Hydroxylamine "Oxidase" of Higher Plants 160
  II. Bacteria (Enzymes of Nitrification) 161
   1. Preparation 161
   2. Measurement of Activity 164
   3. Cofactors 164
   4. Properties and Mechanisms 166
   Reversal of NH₂OH reductase "Ammonia dehydrogenase" artefact . . 169
Literature 169

Enzymes of Vitamin Metabolism. By T. W. Goodwin. With 5 Figures 173

  A. Enzymes Concerned with Biosynthesis 173

      1. Thiamine 173

        a) General Preparation of Enzyme Fractions 174

        b) Pyrimidine Kinase 174

        c) Thiazole Kinase 175

        d) Thiamine Phosphate Synthetase 175

        e) Thiamine Phosphatase 175

      2. Riboflavin 175

      3. Nicotinic Acid 176

        a) Tryptophan Pyrrolase 177

        b) Formylase (Kynurenine Formamidase) 178

        c) Kynurenine Hydroxylase 178

        d) Kynureninase 179

      4. Folic Acid 179

      5. Pantothenic Acid 180

        Pantothenate Synthetase 180

  B. Activating Enzymes 180

      1. Thiamine 180

        a) Thiamine Pyrophosphokinase 180

        b) Thiamine Pyrophosphate Kinase 182

      2. Riboflavin 182

        a) Flavokinase 183

        b) FAD Pyrophosphorylase 185

      3. Pyridoxal (Vitamin B_6) 186

        Pyridoxal Phosphokinase 186

      4. Nicotinic Acid 187

        a) PRPP-Nicotinic Acid Transferase 188

        b) DPN-Pyrophosphorylase 188

        c) DPN-Synthetase 189

        d) DPN-Kinase 189

      5. Folic Acid 189

        a) Folic Acid Reductase 190

        b) Dihydrofolic Acid Reductase 190

      6. Pantothenic Acid 190

        a) Pantothenate Kinase 191

        b) Coupling Enzyme 192

        c) Phosphopantothenylcysteine Decarboxylase 192

        d) Dephospho-CoA Pyrophosphorylase 192

        e) Phosphotransacetylase 193

        f) Dephospho-CoA Kinase 194

  C. Degrading Enzymes 194

      1. Thiamine 194

        a) Thiaminase 194

        b) Thiamine Phosphatase 195

        c) Thiamine Pyrophosphatase 195

      2. Riboflavin 196

        a) FAD-ase (Nucleotide Pyrophosphatase) 196

        b) FMN-Phosphatase 196

      3. Nicotinic Acid 196

        a) Nucleotide Pyrophosphatase 196

        b) DPN-ase (Pyridine Transglycosidase) 197

        c) NMN-Phosphatase 197

      4. Folic Acid 198

      5. Pantothenic Acid 198

        a) "CoA-3'-Nucleotidase" 198

        b) "CoA Pyrophosphatase" 199

        c) "CoA-Peptidase" 199

      6. Biotin 199

        d-Biotin Oxidase 199

      7. Inositol 200

        Phytase 200

      8. Carotene 201

Inhaltsverzeichnis. — Contents.   IX

Carotene Oxidase (Lipoxidase) 201
References 202

Enzyme des L-Ascorbinsäure-Stoffwechsels. Von H. JANECKE. Mit 7 Abbildungen . . . 204
  I. Enzyme bei der Biosynthese der L-Ascorbinsäure 204
    1. Indirekte Umwandlung 205
      a) Aldolkondensation 205
      b) Acyloin-Reaktion 205
    2. Direkte Umwandlung 207
      a) L-Galaktonsäure-γ-laktondehydrogenase 210
      b) Enzympräparat nach MAPSON u. ISHERWOOD (1958) 211
      c) Enzympräparat nach NAKATANI 211
      d) L-Gulonsäuredehydrogenase nach ISHIKAWA u. NOGUCHI (1957) . . . 214
      e) Herstellung der gereinigten Laktonase I 214
  II. L-Ascorbinsäure oxydierende Enzymsysteme 216
    1. Peroxydase 216
    2. Cytochrom c — Cytochromoxydase 217
    3. Polyphenoloxydase 217
    4. Laccase 218
    5. Ascorbinsäureoxydase 218
      a) Methoden der Kupferbestimmung 220
      b) Eigenschaften der AS-Oxydase 221
      c) Gewinnung der Ascorbinsäure-Oxydase 223
      d) Aktivitätsbestimmungen der AS-Oxydase 224
      e) Vorkommen der AS-Oxydase 225
    6. Andere Ascorbinsäure oxydierende Enzyme 226
  III. L-Ascorbinsäure reduzierende Enzymsysteme 226
    1. Herstellung des gereinigten Enzympräparates nach JANECKE u. RAGAB . 228
    2. Darstellung der L-Dehydroascorbinsäure 229
    3. Verbreitung der DAS-Reduktase 233
Literatur 233

Enzymes Involved in the Synthesis and Breakdown of Indoleacetic Acid. By S. MAHADEVAN.
With 1 Figure 238
  I. General Methods 238
    Separations of IAA from its Precursors in the Reaction Mixture 239
  II. Enzymes Involved in the Synthesis of IAA 240
    1. Enzymes Involved in the Conversion of TTP to IAA 240
      a) Distribution 240
      b) Pathways for the Conversion of TTP to IAA 241
      c) Conversion of IPyA to IAc and IAA 241
      d) Preparation and Properties of Some Enzymes Converting TTP to IAA 242
    2. Enzymes Involved in the Conversion of IAc to IAA 243
    Preparation of the Enzymes 243
    3. Enzymes Involved in the Conversion of TNH₂ to IAc and IAA 244
    4. Enzymes Involved in the Conversion of IAN to IAA 246
  III. Enzymes Involved in the Breakdown of IAA 248
    1. General Properties of the IAA Oxidation Reaction 249
    2. Preparation and Properties of some IAA Oxidases 251
      a) Lupin Enzyme 251
      b) Pea Enzyme 253
      c) Wheat Leaf Enzyme 254
      d) Pineapple Enzyme 255
      e) *Omphalia* Enzyme 256
References 258

Enzymes of Aromatic Biosynthesis. By TAKAYOSHI HIGUCHI and ICHIJI KAWAMURA.
With 2 Figures 260
  I. Enzymes of Aromatic Biosynthesis from Non-Aromatic Compounds in Micro-
    organisms 260
      a) 2-Keto-3-Deoxy-D-Araboheptonic Acid-7-Phosphate (KDHP) Syn-
        thetase 262
      b) The Enzyme Converting KDHP to Dehydroquinate 265
      c) Dehydroshikimic (DHS) Dehydrase 265

d) 5-Dehydroquinase 267
e) Quinic Dehydrogenase 268
f) 5-Dehydroshikimic Reductase 270
g) ATP-Shikimic Acid Transphosphorylase. 272
h) The Enzymes Forming 3-Enolpyruvyl Shikimate-5-Phosphate (ESP)
   from 5-Phosphoshikimic Acid and Phosphoenolpyruvate 273
i) Prephenic Aromatase 274
k) Prephenic Dehydrogenase 274
II. The Shikimic Acid Pathway in Higher Plants 275
III. Conversion of Simple Phenylpropanoids to Lignin and Related Compounds 277
a) Tyrase 278
b) Phenylalanine deaminase 281
IV. Conversion of Coniferin and Coniferyl Alcohol to Coniferous Lignin . . 283
1. Dehydrogenation of Coniferyl Alcohol by Laccase and Peroxidase . . . 285
a) Mushroom Enzyme 285
b) The Enzyme of Spruce Cambial Sap 287
c) The Enzyme of Japanese Lacquer 286
2. Coupled Oxidation of Coniferyl Alcohol by Yellow Enzyme-Peroxidase
   Systems 287
References 288

Enzymes of Amino Acid Metabolism.
Part 1: Enzymes of Deamination, Decarboxylation, Transmethylation and Intermediary Metabolism. By B. D. Sanwal and Madhu Lata. With 7 Figures 290

A. Enzymes of Oxidative Deamination 291
I. Pyridine Nucleotide linked Dehydrogenases 291
1. L-Glutamic Dehydrogenase of Higher Plants 291
2. L-Glutamic Acid Dehydrogenase of Neurospora and Other Microorganisms 293
3. L-Alanine Dehydrogenase 294
4. L-Leucine Dehydrogenase 295
II. D- and L-Amino Acid Oxidases 296
1. D-Amino Acid Oxidase of Microorganisms 297
2. D-Glutamic Acid Oxidase 298
3. L-Amino Acid Oxidase of Microorganisms 298
III. Amine Oxidases 300
Amine Oxidase of Pea Seedlings 300

B. Enzymes of Amino Acid Decarboxylation 302
I. Amino Acid Decarboxylases of Higher Plants 302
Glutamic Decarboxylase 303
II. Amino Acid Decarboxylases of Microorganisms 304
1. Glutamic Acid Decarboxylase of Escherichia coli 305
2. Lysine Decarboxylase of E. coli 306
3. Arginine Decarboxylase of E. coli 307
4. Leucine Decarboxylase 307
5. L-Trytophan Decarboxylase 308
6. Diaminopimelic Acid Decarboxylase 308
7. Tyrosine Decarboxylase 309

C. Enzymes of Non-Oxidative Deamination 309
I. Amino Acid Deaminases 309
1. Aspartase 309
2. Histidase 310
II. Dehydrative Deaminases 310
1. Dehydrases 310
a) L-Threonine (and L-Serine) Dehydrase 310
b) D-Serine (and D-Threonine) Dehydrase 311
2. Desulfhydrases 312
a) L-Cysteine Desulfhydrase 312
b) D-Cysteine Desulfhydrase 313
c) Homocysteine Desulfhydrase 313
III. Amino Acid C—S Cleaving Enzymes 313
1. Methionine Dethiomethylase 314
2. Dimethylpropionthetin Dethiomethylase 314

3. Cystathionase 315
4. Alliinase 315
5. C—S-Lyase 316
IV. Amino Acid Reductases 317
1. Proline Reductase 318
2. Glycine Reductase System 318
D. Enzymes of Transmethylation 319
I. Enzymes of Metabolism of S-Adenosylmethionine and Thetins 321
1. Methionine Activating Enzyme 321
2. Cleavage of S-Adenosylmethionine 322
3. S-Methylmethionine-Homocysteine Transmethylase 322
4. Adenosylmethionine-Homocysteine Transmethylase 323
5. Adenosylmethionine-Nicotinic Acid Transmethylase (Nicotinic Acid Methylpherase) 324
E. Enzymes of Biosynthesis 325
I. Enzymes of L-Threonine Biosynthesis 325
1. β-Aspartokinase 325
2. Aspartic β-Semialdehyde Dehydrogenase 326
3. Homoserine Dehydrogenase 327
4. L-Homoserine Kinase 327
5. Threonine Synthetase (Homoserine Phosphate Mutaphosphatase) . . . 328
II. Enzymes of L-Histidine Biosynthesis 329
1. Imidazoleglycerol Phosphate Dehydrase 329
2. Imidazoleacetol Phosphate Transaminase 331
3. L-Histidinol Phosphate Phosphatase 332
4. L-Histidinol Dehydrogenase 332
III. Enzymes of L-Proline Biosynthesis 333
Δ'-Pyrroline-5-Carboxylate Reductase 334
IV. Enzymes of Ornithine Synthesis 335
1. Higher Plants, Molds and Animals 335
Ornithine δ-transaminase 335
2. *Escherichia coli* 336
a) Amino Acid Transacetylase 336
b) Acetylornithine δ-Transaminase 336
c) Acetylornithinase 336
V. Enzymes of the Urea Cycle and Related Compounds (synthetic) 337
1. Carbamyl Phosphate Synthesizing Enzymes 337
2. Carbamate Kinase 338
3. Ornithine Transcarbamylase 338
4. Argininosuccinate Synthetase 339
5. Argininosuccinase 340
6. Arginase 341
VI. Enzymes of Arginine Degradation 342
1. Arginine Desiminase 342
2. Citrullinase System (Citrulline Ureidase) 343
VII. Enzymes of Synthesis of Branched-chain Amino Acids (Leucine, Isoleucine, Valine) 343
1. Acetolactate Synthetase (Acetolactate Forming Enzyme) 345
2. α-Hydroxy-β-keto Acid Reductoisomerase 346
3. α, β-Dihydroxy Acid Dehydrase 346
4. Branched-Chain Amino Acid Transaminase 347
VIII. Degradation of the Branched Chain Amino Acids 347
IX. Formation of Succinic Acid from Glutamic Acid 347
1. Succinic Semialdehyde Dehydrogenases 348
2. γ-Hydroxybutyrate Dehydrogenase 348
X. Enzymes of Synthesis of Glycine and Serine 349
1. 3-Phosphoglycerate $\rightarrow$ Serine 349
2. Formate $\rightarrow$ Serine 350
a) Hydroxymethyltetrahydrofolate Dehydrogenase 350
b) Serine Aldolase (Serine Transhydroxymethylase) 351
3. Glycoldehyde $\rightarrow$ Glycine 352
XI. Enzymes of Synthesis of Tryptophan 352

          1. Indole-3-Glycerol Phosphate Synthetase 353
          2. Tryptophan Synthetase 354
     References 355

Part 2: **Transaminases and Racemases.** By B. D. SANWAL, M. W. ZINK and GEORGE DIN.
     With 1 Figure. 361

     A. Transaminases 361
          I. D-Amino Acid Transaminases 362
         II. L-Amino Acid Transaminases 363
             1. Glutamate-Aspartate Transaminase 363
                a) Assay of Glutamate-Aspartate Transaminase 363
                b) Purification of the Enzyme from Plants 366
                c) Properties of the Plant Enzyme 367
             2. Glutamate-Alanine Transaminase 368
             3. Cysteinesulfinate Transaminase 369
             4. β-Alanine (β-Aminoisobutyrate)-Glutamate Transaminase 371
             5. α-Alanine-Glycine Transaminase 371
             6. α-Alanine-β-Alanine Transaminase 372
             7. γ-Aminobutyrate-Glutamate Transaminase. 373
             8. Kynurenine Transaminase 374
             9. Glutamate-Phosphohistidinol Transaminase 375
            10. Ornithine Transaminase 376
            11. Serine-Alanine Transaminase 376
            12. Tyrosine-Glutamate Transaminase 377
            13. Glutamine and Asparagine Transaminases 377
            14. Alanine, Phenylalanine, Glutamate-Branched Chain Amino Acids . . . 378
            15. Other Transaminases 379
        III. Transamidination 380
             1. Assay 381
             2. Properties 381
     B. Racemases 381
             1. Alanine Racemase 381
             2. Glutamate Racemase 383
             3. Threonine Racemase 384
             4. Methionine Racemase 385
             5. Lysine Racemase 386
             6. Proline Racemase 387
             7. α-ε-Diaminopimelic Acid (DAP) Racemase 388
             8. Other Amino Acid Racemases 389
     References 389

Enzymes of Peptide and Protein Metabolism. By GEORGE WEBSTER. With 11 Figures   392

     A. Enzymes Concerned with Syntheses (Synthetases) 393
          I. Amide Synthesis 393
             1. Glutamine Synthetase 393
                a) Occurrence 393
                b) Assay 393
                c) Preparation 394
                d) Properties 397
             2. Asparagine Synthetase 400
                a) Occurrence 400
                b) Assay 400
                c) Preparation 401
                d) Properties 401
         II. Peptide Synthesis 401
             1. Glutamylcysteine Synthetase 401
                a) Occurrence 402
                b) Assay 402
                c) Preparation of Glutamylcysteine Synthetase 402
                d) Properties 403
             2. Glutathione Synthetase 403
                a) Occurrence 403
                b) Assay 403

c) Preparation 403
d) Properties 404
III. Protein Synthesis 405
   1. Amino Acid-Activating Enzymes 405
      a) Occurrence 406
      b) Assay 406
      c) Preparation 407
      d) Properties 407
   2. Enzymes Concerned with Protein Synthesis 408
      a) Occurrence 408
      b) Assay 408
      c) Preparation 409
      d) Properties 409
B. Enzymes concerned with Degradation (Amidases, Peptidases and Proteases) . . 412
  I. Amidases 413
   Glutaminase and Asparaginase 413
      a) Occurrence 413
      b) Assay 413
      c) Preparation 413
      d) Properties 414
  II. Peptidases 414
      a) Occurrence 415
      b) Assay 415
      c) Preparation and Properties 415
  III. Proteases 415
      a) Occurrence 416
      b) Assay 416
      c) Preparation 417
      d) Properties 417
References 418

Enzymes of Synthesis of Purine and Pyrimidine Nucleotides. By DALTON WANG and E. R. WAYGOOD. With 2 Figures 421

A. Enzymes of Synthesis of Purine Nucleotides 422
  I. 5-Phosphoribosylpyrophosphate Kinase 422
   1. Assay 422
   2. Enzyme Preparation 424
      a) Procedures of KORNBERG et al. (1955) 424
      b) Procedures of KORN et al. (1955) 425
  II. 5-Phosphoribosylpyrophosphate Amidotransferase 426
   1. Assay 426
   2. Enzyme Preparations 427
  III. Glycinamide Ribotide Kinosynthase 429
  IV. Glycinamide Ribotide Transformylase 432
  V. Formylglycinamidine Ribotide Kinosynthase 435
  VI. Enzyme for the Synthesis of 5-Aminoimidazole Ribotide 436
  VII. 5-Aminoimidazole Ribotide Carboxylase and Enzyme for the Synthesis of
     5-Amino-4-Imidazole-N-Succinocarboxamide Ribotide 436
  VIII. Enzymatic Cleavage of 5-Amino-4-Imidazole-N-Succinocarboxamide Ribotide
     (AISCAR Splitting Enzyme) 437
  IX. 5-Amino-4-Imidazolecarboxamide Ribotide Transformylase and Inosinicase . 438
B. Enzymes of Synthesis of Pyrimidine Nucleotides 439
  I. Carbamyl Phosphate Synthetase 439
  II. Carbamyl Phosphate-Aspartate Transcarbamylase 441
  III. Dihydroorotase 442
  IV. Dihydroorotic Acid Dehydrogenase 444
  V. Orotidine-5′-Phosphate Pyrophosphorylase 446
  VI. Orotidine-5′-Phosphate Decarboxylase 446
References 446

Enzymes of Fat Metabolism.

A. Plant Lipases. By EDWARD J. BARRON. With 1 Figure 448
  Assay Procedures 448

1. Lipase Assay by the Release of Fatty Acids 449
2. Lipase Assay by the Decrease in Ester Content 450
3. Other Assay Procedures 450
4. Activators and Inhibitors 450
5. pH Optima 451
6. Specificity of Attack. 451
7. Purification 451
References 452

B. Phospholipases. By EDWARD J. BARRON. With 1 Figure 454
     I. Phospholipase A 454
         Assay Procedure 455
    II. Phospholipase B (Lysophospholipase B) 456
         Assay Procedure 456
         a) Estimation of Activity by Measuring the Decrease in Ester Bond . . 456
         b) Measurement of Activity by Estimating GPC Formed. 457
         c) Other Assay Procedures. 457
            Substrate Specificity 457
            pH Optima and Stability 458
            Activators and Inhibitors 458
            Purification Procedures 459
   III. Phospholipase C. 459
         Assay Procedure 459
         Other Assay Procedures 460
         pH Optimum and Stability 460
         Activators 460
         Substrate Specificity 460
         Purification Procedures 460
         Purification of Cottonseed Enzyme 460
   IV. Phospholipase D 461
         Assay Procedure 461
         Other Assay Procedures 461
         pH Optima and Stability. 462
         Activators 462
         Substrate Specificity 462
         Purification Procedure 462
    V. Other Enzymes Attacking Phospholipids 462
        1. Phosphoinositide Phosphorylase 462
        2. Phosphatidic Acid Phosphatase. 462
        3. Lysolecithin Isomerase 463
References 463

C. β-Oxidation. By P. K. STUMPF 465
     I. Even Chain Fatty Acids 465
    II. Odd-Chain Fatty Acids 466
References 466

D. α-Oxidation. By P. K. STUMPF 467
References 468

E. Lipoxidase. By A. L. TAPPEL. 469
     1. Assay Method 469
     Procedure 469
         a) Manometric 469
         b) Spectrophotometric. 469
     2. Purification Procedure. 470
     3. Properties 470
References 471

F. Synthesis of Fatty Acids. By EDWARD J. BARRON 472
     1. Preparation of Particles 472
     2. Extraction of Acetone Powder 472
     3. Assay 473
     4. pH Optima and Stability 473
References 473

Enzymes of Carbohydrate Synthesis. By DAVID S. FEINGOLD, ELIZABETH F. NEUFELD, and W. Z. HASSID. With 1 Figure 474

A. Formation of Precursors of Complex Saccharides 478
    I. Preparation of Substrates 478
    II. Separation and Identification of Reaction Products 479
    III. Enzymes which Catalyze the Formation of Sugar 1-Phosphates 481
        1. D-Galactokinase from *Saccharomyces fragilis* 481
        2. D-Glucuronic Acid Kinase from *Phaseolus aureus* 483
        3. D-Galactokinase and L-Arabinokinase from *Phaseolus aureus* 484
    IV. Enzymes which Catalyze the Formation of Sugar Nucleotides 484
        1. Sugar Nucleotide Pyrophosphorylases from *Phaseolus aureus* 485
        2. UDP-D-Glucose Pyrophosphorylase from *Phaseolus aureus* 486
        3. UDP-D-Glucose Pyrophosphorylase from Brewer's Yeast 488
        4. UDP-N-Acetyl-D-Glucosamine Pyrophosphorylase from *Phaseolus aureus* 489
        5. UDP-N-Acetyl-D-Glucosamine Pyrophosphorylase from Baker's Yeast . . 490
        6. GDP-D-Mannose Pyrophosphorylase from Brewer's Yeast 491
        7. α-D-Galactose-1-Phosphate Uridyl Transferase from *Saccharomyces fragilis* 492
    V. Enzymes which Catalyze Transformations of Sugar Nucleotides 492
        1. 4-Epimerases 492
          a) UDP-D-Galactose 4-Epimerase from *Saccharomyces fragilis* 493
          b) 4-Epimerases from Higher Plants 494
        2. UDP-D-Glucose Dehydrogenase from Peas 495
        3. UDP-D-Glucuronic Acid Decarboxylase from Wheat Germ 497

B. Synthesis of Disaccharides 498
    I. Enzymes which Catalyze the Formation of Sucrose and Sucrose Phosphate . 498
        1. UDP-D-Glucose-D-Fructose Transglucosylase from Wheat Germ 499
        2. UDP-D-Glucose-D-Fructose 6-Phosphate Transglucosylase from Wheat Germ 500
    II. Enzyme which Catalyzes the Formation of Trehalose Phosphate (UDP-D-Glucose-D-Glucose 6-Phosphate Transglucosylase from Yeast) 501

C. Synthesis of Glycosides 504
    I. Enzyme which Catalyzes the Formation of Diphenol D-Glucosides (UDP-D-Glucose-Diphenol Transglucosylase from Wheat Germ) 504
    II. Enzyme which Catalyzes the Formation of Phenolic Gentiobiosides (UDP-D-Glucose-Phenol-D-Glucoside Transglucosylase from Wheat Germ) 506

D. Synthesis of Polysaccharides 507
    I. Enzyme which Catalyzes the Formation of Callose 507
    II. Enzyme which Catalyzes the Formation of Chitin 508
    III. Enzyme which Catalyzes the Formation of D-Xylodextrins 509
    IV. Enzymes which Catalyze the Formation of Starch 510
        1. Phosphorylase from Potatoes 511
          Procedure 1 511
          Procedure 2 512
        2. Q-Enzyme (Branching Enzyme) from Potatoes 514
        3. D-Enzyme from Potatoes 515
References 516

Enzymes of Glycolysis. By MARTIN GIBBS and JOHN F. TURNER. With 1 Figure 520
    I. Assay of Glycolytic System 520
        1. Preparation of Glycolytic System 521
        2. Properties 522
    II. Phosphorylase 522
        1. Assay Method 522
        2. Purification 523
        3. Properties 523
    III. Phosphoglucomutase 523
        1. Assay Method 524
        2. Preparation of Phosphoglucomutase Extracts from Plant Tissues . . . 524
        3. Properties 524
    IV. Hexokinase 525
        1. Assay Method 525
        2. Preparation of Hexokinase Extracts from Plant Tissues 525
        3. Properties 526

      V. Phosphoglucose Isomerase 526
         1. Assay Method 526
         2. Preparation of Phosphoglucose Isomerase Extracts from Plant Tissues . 526
         3. Properties 527
     VI. Phosphohexokinase 527
         1. Assay Method 527
         2. Preparation of Phosphohexokinase from Plant Tissue 528
         3. Properties 528
    VII. Fructose-1,6-Diphosphatase 528
         1. Assay Method 528
         2. Purification Procedure (RACKER and SCHROEDER, 1958) 529
         3. Properties 529
   VIII. Aldolase 530
         1. Assay Method 530
         2. Purification Procedure 532
         3. Properties 532
     IX. Triosephosphate Isomerase 533
         Assay Method 533
      X. TPN Triosephosphate Dehydrogenase 534
         1. Assay Method 534
         2. Purification Procedure (GIBBS, 1955) 535
         3. Properties 535
     XI. DPN Triosephosphate Dehydrogenase 535
         1. Assay Method 535
         2. Purification Procedure (HAGEMAN and ARNON, 1955) 536
         3. Properties 536
    XII. Triosephosphate — Phosphoglycerate Dehydrogenase 536
         1. Assay Method 536
         2. Purification Procedure (ROSENBERG and ARNON, 1955) 537
         3. Properties 537
   XIII. Phosphoglycerate Kinase 537
         1. Assay Method 537
         2. Preparation Procedure (AXELROD and BANDURSKI, 1953) 538
         3. Properties 539
    XIV. Phosphoglyceric Acid Mutase 539
         1. Assay Method 539
         2. Purification (ITO and GRISOLIA, 1959) 540
         3. Properties 540
     XV. Enolase, 2-Phosphoglycerate Dehydrase 541
         1. Assay Method 541
         2. Purification Procedure (BOSER, 1959) 541
         3. Properties 542
    XVI. Pyruvate Kinase 543
         1. Assay Method 543
         2. Purification Procedure (MILLER and EVANS, 1957) 543
         3. Properties (Data of MILLER and EVANS, 1957) 544
   References 544

Enzymes of the Pentose Phosphate Cycle. By E. R. WAYGOOD and R. ROHRINGER. With
1 Figure 546
      I. Hexokinase (Glucokinase and Fructokinase) 549
         1. Assay 550
         2. Purification 551
     II. Glucose-6-Phosphate Dehydrogenase (Zwischenferment). Lactonase and
         6-Phosphogluconic Dehydrogenase 552
         1. Assay 552
         2. Purification 554
         3. Lactonase (Assay and Purification) 556
    III. Phosphoriboisomerase 556
         1. Assay 556
         2. Purification 557
         3. Preparation of Isomerase Product 558
     IV. Phosphoketopentoepimerase and Transketolase 558

1. Assay (Transketolase) 559
2. Purification (Transketolase) 560
3. Assay (Phosphoketopentoepimerase). 562
4. Purification (Phosphoketopentoepimerase) 563
V. Transaldolase 564
1. Assay 564
2. Preparation of Sedoheptulose-7-Phosphate 565
3. Purification 565
VI. Phosphohexoisomerase 566
1. Assay 566
2. Purification 567
References 567

Enzyme Systems in Photosynthesis. By MANUEL LOSADA and DANIEL I. ARNON. With
5 Figures 569

A. The Photosynthetic Structures of Plants and Bacteria. 570
I. Chloroplasts 571
1. Isolation and Purification of Whole Chloroplasts 571
2. Preparation of "Broken" Chloroplasts and Chloroplast Extracts 572
II. Chromatophores. 572
1. Isolation and Purification of *Chromatium* Chromatophores 572
a) Culture of Bacteria 572
b) Preparation and Fractionation of Cell Extracts 573
2. Isolation and Purification of *Rhodospirillum rubrum* Chromatophores . . 574
a) Culture of Bacteria 574
b) Preparations of Cell-Free Extracts 574
c) Preparation of Purified Chromatophores. 574

B. Carbon Dioxide Assimilation in Photosynthesis 574
I. Separation of Light and Dark Phases in Photosynthesis 575
II. Characteristic Enzymes of the Reductive Carbohydrate Cycle 577
1. Phosphoribulokinase 577
a) Assay Method 577
b) Purification Procedure 578
c) Properties 578
2. Carboxylation Enzyme 579
a) Assay Method 579
b) Purification Procedure (Method of WEISSBACH et al., 1956) 579
c) Purification Procedure (Method of JAKOBY et al., 1956) 580
d) Properties. 581

C. Photosynthetic Phosphorylation 582
I. Cyclic Photophosphorylation 583
1. Procedure for Cyclic Photophosphorylation in Chloroplasts 584
2. Procedure for Cyclic Photophosphorylation in Chromatophores 585
II. Noncyclic Photophosphorylation 586
1. Procedure for Noncyclic Photophosphorylation in Chloroplasts 587
2. Procedure for Noncyclic Photophosphorylation in Chromatophores . . . 588

D. Isolated Protein Constituents of the Photosynthetic Apparatus 589
I. The TPN-Reducing System 589
1. Chloroplast Ferredoxin 592
a) Isolation and Purification 593
b) Crystallization of Parsley Ferredoxin 594
c) Crystallization of Spinach Ferredoxin 595
2. Ferredoxin-TPN Reductase 596
a) Isolation and Purification 597
b) Assay Method 598
c) Properties. 598
d) Crystallization Procedure 599
II. Cytochromes in Leaves and Algae 599
1. Extraction and Purification 600
a) Preparation of Cytochrome f from Parsley 600
b) Crystallization of *Porphyra tenera*-Cytochrome 553 600
2. Properties 601
III. Bacterial Cytochromes 602

       1. RHP and Cytochrome c_2 from *R. rubrum* 603
         a) Preparation 603
         b) Properties 603
       2. RHP and Cytochrome c_2 from *Chromatium*. 604
       3. Cytochrome-552 from *Rhodopseudomonas palustris* 605
  E. Quinone Constituents of the Photosynthetic Apparatus 605
     I. Extraction and Purification of Vitamin K_1 607
    II. Extraction and Purification of Plastoquinone 607
       1. Method of CRANE (1959b) 607
       2. Method of BISHOP (1958, 1959) 608
   III. Properties of Plastoquinone 609
   IV. Role of Quinones in Photosynthetic Reactions 610
       1. Photosynthetic Phosphorylation 610
       2. Noncyclic Electron Transport 610
       3. Bacterial Photophosphorylations 612
    V. Concluding Remarks 612
 References 613

Enzymes of the Krebs Cycle, the Glyoxalate Cycle and Related Enzymes. By D. D.
DAVIES and R. J. ELLIS. With 2 Figures 616
  A. Enzymes of the Krebs Cycle 616
     I. Condensing Enzyme 616
    II. Aconitase 618
   III. *Iso* Citric Dehydrogenase (TPN Specific) 620
   IV. *Iso* Citric Dehydrogenase (DPN Specific) 621
    V. α-Keto Acid Oxidases 621
   VI. Succinyl CoA Synthetase (P Enzyme) 623
  VII. Succinic Dehydrogenase 624
 VIII. Fumarase 627
   IX. Malic Dehydrogenase 628
  E. Enzyme Activities Related to the Krebs Cycle 629
     I. Tartaric Dehydrogenase 629
    II. Malease 629
   III. Pyruvic (De) Carboxylase 630
   IV. Lactic Dehydrogenase 632
    V. Phosphoenolpyruvic Carboxylase 634
   VI. Phosphoenolpyurvate Carboxykinase 635
  VII. Malic Enzyme 635
  C. The Glyoxalate Cycle 636
     I. Malate Synthetase 636
    II. Isocitritase 637
  D. Enzymes Related to the Glyoxalate Cycle 638
     I. Glycollic Acid Oxidase 638
    II. Glyoxalic Acid Reductase 641
   III. Glycolaldehyde Dehydrogenase 642
   IV. Glycine Oxidase 642
    V. Formic Dehydrogenase 643
 References 643

Enzymes of Terminal Respiration. By DAVID P. HACKETT. With 12 Figures 647
  A. Characterization of the Intact Respiratory Chain 647
     I. Measurement of Respiratory Rate 648
       1. Manometry 648
       2. Volumetry 648
       3. Polarography 648
       4. Spectrophotometry 649
    II. Effect of Oxygen Partial Pressure on Respiratory Rate 650
       1. General Considerations 650
       2. Methods 651
   III. Coupling to Phosphorylation 651

      1. Manometric Method 652
      2. Other Methods 652
  IV. The Use of Inhibitors 653
      1. Oxidase Inhibitors 653
      2. Inhibition within the Cytochrome System 654
      3. Inhibition in the Flavoprotein Region 655
   V. Spectrophotometric Methods 656
      1. Special Problems 657
      2. Instruments 658
      3. Procedures 659
      4. Analysis of Results 662
B. The Respiratory Chain Components 663
   I. DPNH Dehydrogenase 663
      1. Diaphorase 664
      2. Quinone Reductase 664
      3. Cytochrome c Reductase 665
      4. DPNH-Ferricyanide Reduction 667
  II. Transhydrogenase 667
 III. Cytochromes "b" 668
  IV. Cytochromes "c" 672
   V. Cytochrome Oxidase (a—a_3) 675
  VI. Lipid Components (Coenzyme Q) 677
C. Other Pathways to Oxygen 679
   I. From Pyridine Nucleotides to Oxygen 679
      1. General Considerations 679
      2. The Copper Oxidases 681
         a) Polyphenol Oxidase 681
         b) Laccase 683
         c) Ascorbic Acid Oxidase 684
      3. Peroxidases 684
      4. Glycolic Acid (α-Hydroxy Acid) Oxidase 686
  II. From other Substrates to Oxygen 688
      1. Glucose Oxidase 688
      2. Carbohydrate Oxidase 689
References 690

Summary of Recommendations on Enzyme Terminology. (By the Commission on
   Enzymes of the International Union of Biochemistry, 1961) 695
      Enzyme units 695
      Symbols of enzyme kinetics 695
      The nomenclature of coenzymes 696
      Classification and nomenclature of cytochromes 696
      Classification and nomenclature of enzymes 697
      The terminology of enzyme formation 698
      "Appendix B": Recommended Symbols for Enzyme Kinetics 699
      "Appendix C": List of Cytochromes 699
      "Appendix D": Key to Numbering and Classification of Enzymes . . . 700
Sachverzeichnis (Deutsch-Englisch) 702
Subject Index (English-German) 718
Table des Matières pour la Contribution: F. CHAPEVILLE et P. FROMAGEOT, Enzymes
   du Métabolisme du Soufre 734

Inhalt der übrigen Bände. — Contents of other Volumes.

1. Band. — Volume I.

Allgemeine Maßnahmen und Bestimmungen bei der Aufarbeitung von Pflanzenmaterial. Von K. PAECH, Tübingen, Germany.

General Methods for Separation: Making and Handling Extracts. By N. W. PIRIE, Rothamsted, Great Britain.

General Methods for Separation. Electrical-Transport Methods. By R. L. M. SYNGE, Bucksburn, Great Britain.

Multiplikative Verteilung. Von E. HECKER, Tübingen, Germany.

Die chromatographische Analyse in Säulen. Von G. BRAUNITZER, Tübingen, Germany.

Papierchromatographie. Von H. HELLMANN, Tübingen, Germany.

Colorimetric, Absorptimetric and Fluorimetric Methods. By J. GLOVER, Liverpool, Great Britain.

Refraktometrie und Interferometrie, Polarimetrie, Nephelometrie. Von G. KORTÜM und M. KORTÜM-SEILER, Tübingen, Germany.

Principles of Biological Assay. By M. J. R. HEALY, Rothamsted, Great Britain.

Methods Involving Labelled Atoms. By J. GLOVER, Liverpool, Great Britain.

Estimation of pH Values. (Living Tissues and Saps.) By J. SMALL, Belfast, Great Britain.

Oxidation-Reduction Potentials. By R. HILL, Cambridge, Great Britain.

Gasometric Analysis in Plant Investigation. By R. H. KENTEN, Rothamsted, Great Britain.

Cytochemical Methods. By F. R. WHATLEY, Berkeley, Calif., USA.

Mineral Components and Ash Analysis. By E. C. HUMPHRIES, Rothamsted, Great Britain.

2. Band. — Volume II.

Mono- and Oligosaccharides and Acidic Monosaccharide Derivatives. By D. J. BELL, Edinburgh, Great Britain.

Acyclic Sugar Alcohols. By S. A. BARKER, Birmingham, Great Britain.

Inosite und verwandte Naturstoffe. Von G. DANGSCHAT, Berlin-Frohnau, Germany.

Ascorbinsäure. Von W. FRANKE, Bonn, Germany.

Phosphorylated Sugars. By Dr. A. A. BENSON, Pennsylvania, USA.

Starch, Glycogen, Fructosans and Similar Polysaccharides. By W. J. WHELAN, Bangor, Caernarvonshire, Great Britain.

Cellulose and Hemicelluloses. By M. A. JERMYN, Melbourne, Australia.

Pektine. Von F. A. HENGLEIN, Karlsruhe, Germany.

Chitin. By M. V. TRACEY, Rothamsted, Great Britain.

The Analysis of Plant Gums and Mucilages. By E. L. HIRST, Edinburgh, Great Britain, and J. K. N. JONES, Kingston, Ontario, Canada.

Glycosides as a General Group. By Dr. A. R. TRIM, Trumpington, Cambs., Great Britain.

Fats and Other Lipids. By M. L. MEARA, Middleton, Manchester, Great Britain.

Volatile Alcohols, Aldehydes, Ketones and Esters. By D. F. MEIGH, Maidstone, Kent, Great Britain.

Volatile Acids. By R. SCARISBRICK, London, Great Britain.

Nichtflüchtige Mono-, Di- und Tricarbonsäuren. (Unter Ausschluß chromatographischer Methoden.) Von JOHANNES WOLF, Karlsruhe, Germany.

Non Volatile Mono-, Di- and Tricarboxylic Acids. (Chromatographic and Ion Exchange Methods). By S. L. RANSON, Newcastle-upon-Tyne, Great Britain.

Lactones. By L. J. HAYNES, Edinburgh, Great Britain.

3. Band. — Volume III.

Die niederen Terpene (ätherische Öle und Harze allgemein). Von O. MORITZ, Kiel, Germany.

Pyrethrins and Allied Compounds. By R. F. PHIPERS, Berkhamsted, Herts., Great Britain.

Triterpene und Triterpen-Saponine. Von M. STEINER und H. HOLTZEM, Bonn, Germany.

Phytosterine, Steroidsaponine und Herzglykoside. Von A. STOLL und E. JUCKER, Basel, Switzerland.

Carotenoids. By T. W. GOODWIN, Liverpool, Great Britain.

The Determination of Rubber and Gutta in Plants. By H. M. BENEDICT, Stanford, California, USA

Simple Benzene Derivatives. By D. D. CLARKE and F. F. NORD, New York, USA.

Natural Tropolones. By H. ERDTMAN, Stockholm, Sweden.

Ein- und zweikernige Chinone. Von O. HOFFMANN-OSTENHOF, Wien, Austria.

Natural Phenylpropane. Derivatives By GEORGE DE STEVENS and F. F. NORD, New York, N. Y., USA.

Lignans. By D. ERDTMAN, Stockholm. Sweden.

Anthocyanins, Chalcones, Flavones, and Related Water-Soluble Plant Pigments. By T. A. GEISSMAN, Los Angeles, USA.

Lignin. Von K. FREUDENBERG, Heidelberg, Germany.

Natürliche Gerbstoffe. Von OTTO TH. SCHMIDT, Heidelberg, Germany.

Anthraglykoside und Dianthrone. Von W. SCHMID, Tübingen, Germany.

Growth Substances in Higher Plants. By POUL LARSEN, Bergen, Norway.

Antibiotics. By F. A. SKINNER, Rothamsted, Great Britain.

4. Band. — Volume IV.

Peptides (Bound Amino Acids) and Free Amino Acids. By R. L. M. SYNGE, Bucksburn, Aberdeenshire, Great Britain.

Proteins. By N. W. PIRIE, Rothamsted, Great Britian.

Seed Proteins. By J. PACE, St. Albans, Herts., Great Britain.

Methods of Determining the Nutritive Value of Proteins. By J. DUCKWORTH, Bucksburn, Aberdeenshire, Great Britain.

Urea and Ureides. By M. V. TRACEY, Rothamsted, Great Britain.

Chlorophylls. By J. H. C. SMITH and A. BENITEZ, Stanford, California, USA.

Haematin Compounds. By E. F. HARTREE, Cambridge, Great Britain.

Nucleic Acids, their Components and Related Compounds. By R. MARKHAM, Cambridge, Great Britain.

Adenosine Diphosphate, Adenosine Triphosphate. By H. G. ALBAUM, Brooklyn, New York, USA.

Codehydrasen I und II (Diphospho-pyridin-nucleotid und Triphospho-pyridin-nucleotid). Von K. HASSE, Karlsruhe, Germany.

Thiamine and its Derivatives. By Sir R. A. PETERS, Babraham, Cambs, .Great Britain, and J. R. P. O'BRIEN, Oxford, Great Britain.

The Alkaloids. By B. T. CROMWELL, Hull, Great Britain.

Amine und Betaine. Von E. WERLE, München, Germany.

Pantothensäure und Coenzym A. Von E. WERLE, München, Germany.

Riboflavin, Folic Acid and Biotin. By F. M. STRONG, Madison, Wisc., USA.

Melanins. By M. THOMAS, Newcastle-upon-Tyne, Great Britain.

Blausäure-Verbindungen. Von P. SEIFERT, Heidelberg, Germany.

Senföle, Lauchöle und andere schwefelhaltige Pflanzenstoffe. Von A. STOLL und E. JUCKER, Basel, Switzerland.

5. Band. — Volume V.

Emission and Atomic Absorption Spectrochemical Methods. By D. J. DAVID, Canberra, A.C.T., Australia.

Mass Spectrometric Methods. By K. BIEMANN, Cambridge, Mass., USA.

Plant Spectra: Absorption and Action. By W. L. BUTLER and K. H. NORRIS, Beltsville, Maryland, USA.

Gefriertrocknung. Von H. Moor, Zürich, Switzerland.

Vapour Phase Chromatography. By S. P. Burg, Miami, Florida, USA.

Ion-Exchange Chromatography. By N. K. Boardman, Canberra, A.C.T., Australia.

Molecular Sieving other than Dialysis. By N. K. Boardman, Canberra, A.C.T., Australia.

Dünnschicht-Chromatographie. Von Egon Stahl, Saarbrücken, Germany.

Paper Chromatography on a Preparative Scale. By F. A. Hommes and H. F. Linskens, Nijmegen, The Netherlands.

Determination of Size, Shape and Homogeneity of Macromolecules in Solution. By I. J. O'Donnell and E. F. Woods, Parkville, Victoria, Australia.

Optical Rotatory Dispersion. Its Application to Protein Conformation. By E. F. Woods and I. J. Donnell, Parkville, Victoria, Australia.

Diffuse Röntgenkleinwinkelstreuung. Von O. Kratky, Graz, Austria.

Méthodes Calorimétriques pour l'Analyse des Végétaux. Par Henri Prat, Marseille, France.

Surface Factors Affecting the Penetration of Compounds into Plants. By A. E. Dimond, New Haven, Conn. USA.

Tissue and Single Cell Cultures of Higher Plants as a Basic Experimental Method. By A. C. Hildebrandt, Madison, Wisc., USA.

Immunological Methods. By J. A. van der Veken, D. H. M. van Slogteren and J. P. H. van der Want, Lisse, The Netherlands.

Polarography and Tensammetry. By B. Breyer, Lidcombe, N.S.W., Australia.

Fallout Contamination in Plants. By J. V. Possingham, Merbein, Victoria, Australia, and P. S. Davis, Lucas Heights, N.S.W., Australia.

6. Band. — Volume VI.

Siliciumverbindungen. Von W. Heinen, Madison, Wisc., USA.

Determination of Sulfhydryl Groups. By Herbert Stern, Urbana, Ill., USA.

Phosphatide und Glykolipide. Von Ulrich Beiss, Braunschweig, Germany.

Natürlich vorkommende Acetylenverbindungen. Von F. Bohlmann und W. Sucrow, Berlin-Charlottenburg, Germany.

Natürliche Chromone. Von M. Hesse und H. Schmid, Zürich, Switzerland.

Orchinol. Von Richard Braun, Amherst, Mass., USA.

Humulones, Lupulones and other Constituents of Hops. By J. R. Hudson, Nutfield, Redhill, Surrey, Great Britain.

Lichen Substances. By S. Shibata, Hongo, Tokyo, Japan.

Kinetin and Kinetin-Like Compounds. By Carlos O. Miller, Bloomington, Indiana, USA.

Gibberelline. Von Rüdiger Knapp, Gießen, Germany.

Pflanzliche Toxine. Von Richard Braun, Amherst, Mass., USA.

Phytagglutinine. Von Josef Tobiška, Brno, ČSSR.

Isolierung und Analyse von Bakterienzellwänden. Von F. W. Zilliken und R. Lambert, Nijmegen, The Netherlands.

General Methods of Enzyme Chemistry.

Der Nachweis enzymatischer Aktivität. Von W. Heinen, Madison, Wisc., USA.

Allgemeine Charakterisierung eines Enzyms. Von W. Heinen, Madison, Wisc., USA.

Thunberg-Technik. (Methylenblau-Methode) Von W. Heinen, Madison, Wisc., USA, und H. F. Linskens, Nijmegen, The Netherlands.

Interpretation of Results. By M. V. Tracey, North Ryde, Australia.

General Methods of Preparation. By B. D. Sanwal, Winnipeg, Man., Canada.

General Aspects of Enzyme Purification and Characterization. By Hans G. Boman and Walter Björk, Uppsala, Sweden.

Purification of Enzymes by Ion Exchange Chromatography. By Hans G. Boman, Uppsala, Sweden.

Die Analyse von Enzymen im Boden. Von Ed. Hofmann, Freising-Weihenstephan, Germany.

Inhibition and Activation of Enzymes. By Fay Bendall, Cambridge, Great Britain.

Enzymic Assays of Amino-Acids and Keto Acids. By B. D. Sanwal, Winnipeg, Man., Canada.

Enzymatische Bestimmung von Metaboliten. Von G. Pfleiderer, Frankfurt a. Main, Germany.

Mitarbeiter von Band 7.   Contributors to Volume VII.

DANIEL I. ARNON, Professor of Cell Physiology and Biochemist in the Experiment Station, 251 Hilgard Hall, University of California, Berkeley 4, California (USA).

EDWARD J. BARRON, Associate Director of Research, The Virginia Mason Foundation, 1118, 9th Avenue, Seattle 1, Washington (USA).

FRANCOIS CHAPEVILLE, Chercheur au Commissariat à l'Energie Atomique, Département de Biologie, B.P. n° 2, Gif-sur-Yvette (S. et Oi.) (France).

DAVID DAVIES, Dr., Reader in Plant Biochemistry, University of East Anglia, Norwich, Norfolk (Great Britain).

GEORGE ALLAN DIN, Department of Microbiology, University of Manitoba, Winnipeg, Manitoba (Canada).

REGINALD JOHN ELLIS, B. Sc., Ph. D., Department of Biochemistry, University of Oxford, Oxford (Great Britain).

DAVID S. FEINGOLD, Dr., Associate Professor of Bacteriology, Department of Biological Science, University of Pittsburgh, Pittsburgh 13, Pennsylvania (USA).

PIERRE FROMAGEOT, Chef du Service de Biochemie du Commissariat à l'Energie Atomique, B. P. n° 2, Gif-sur-Yvette (S. et O.) (France).

MARTIN GIBBS, Professor of Biochemistry, Department of Biochemistry, Cornell University, Ithaca, N. Y. (USA).

T. W. GOODWIN, Professor, Department of Agricultural Biochemistry, University College of Wales, Aberystwyth (Great Britain).

DAVID P. HACKETT, Associate Professor, Department of Biochemistry, University of California, Berkeley 4, California (USA).

WILLIAM Z. HASSID, Professor of Biochemistry, Department of Biochemistry, University of California, Berkeley 4, California (USA).

E. J. HEWITT, Dr., Senior Plant Physiologist, Long Ashton Research Station, Bristol (Great Britain).

TAKAYOSHI HIGUCHI, Associate Professor, Doctor of Agricultural Science, Laboratory of Wood Chemistry, Faculty of Agriculture, Gifu University, Naka, Gifu (Japan).

H. JANECKE, Professor Dr., Pharmazeutisches Institut der Universität Frankfurt, Frankfurt a. M. (Germany).

ICHIJI KAWAMURA, Professor, Doctor of Agricultural Science, Laboratory of Wood Chemistry, Faculty of Agriculture, Gifu University, Naka, Gifu (Japan).

P. S. KRISHNAN, Professor of Biochemistry, Lucknow University, Lucknow, U.P. (India).

MADHU LATA, Dr., Research Associate, Pharmacology Department, School of Medicine, University of Manitoba, Winnipeg (Canada).

MANUEL LOSADA, Dr., Head of the Sección de Bioquímica y Fisiología Celular, C.S.I.C., Centro de Investigaciones Biológicas, Madrid (Spain).

S. MAHADEVAN, C.S.I.R., Scientists Pool Officer, Department of Biochemistry, Indian Institute of Science, Bangalore 12 (India).

ELIZABETH F. NEUFELD, Biochemist, National Institute of Arthritis and Metabolic Diseases, National Institute of Health, Bethesda, Md. (USA).

D. J. D. NICHOLAS, Professor Dr., Head of Chemical Microbiology Department, University of Bristol, Research Station, Long Ashton, Bristol (Great Britain).

R. ROHRINGER, Dr., Plant Physiologist, Canada Department of Agriculture, Research Station, P. O. Box 6200, Winnipeg, Manitoba (Canada).

B. D. SANWAL, Dr., Associate Professor, Department of Microbiology, University of Manitoba, Winnipeg (Canada).

P. K. STUMPF, Professor of Biochemistry, University of California, Davis, California (USA).

A. L. TAPPEL, Professor, Department of Food Science and Technology, University of California, Davis, California (USA).

JOHN F. TURNER, Professor Dr., Department of Agriculture, University of Sydney, Sydney (Australia).

DALTON WANG, Dr., Biochemist, Boyce Thompson Institute, Yonkers, New York (USA).

E. R. WAYGOOD, Professor, Department of Botany, The University of Manitoba, Winnipeg (Canada).

GEORGE WEBSTER, Professor, Institute for Enzyme Research, University of Wisconsin, Madison, Wisconsin (USA).

MICHAEL W. ZINK, Department of Microbiology, University of Manitoba, Winnipeg, Manitoba (Canada).

Special Methods of Isolation and Purification.

Enzymes du Métabolisme du Soufre.

Par

F. Chapeville et P. Fromageot.

Avec 1 Figure.

A. Enzymes d'activation et de transfert du sulfate.

En 1952 BERNSTEIN et McGILVERY ont obtenu à partir d'un homogénat de foie de rat une préparation qui, en présence d'ATP[1] et d'ions magnésium, catalyse la synthèse de l'acide aminophénylsulfurique à partir d'aminophénol et de sulfate minéral. Si l'on incube un tel système avec le sulfate, il s'accumule dans le milieu un intermédiaire appelé «sulfate actif» qui permet une estérification plus rapide de l'aminophénol.

Etudiant l'estérification du phénol, DE MEIO, WIZERKANIUK et FABIANI (1953) ont montré que la synthèse du phénylsulfate est inhibée par la triphosphatase présente dans les microsomes et que deux enzymes au moins interviennent dans cette estérification: le premier agit en présence d'ATP et permet la formation du sulfate actif, le second assure le transfert du sulfate ainsi activé sur le phénol.

ROBBINS et LIPMANN (1957) ont isolé le sulfate actif et montré que sa structure correspond à celle du 3′-phosphoadénosine-5′-phosphosulfate

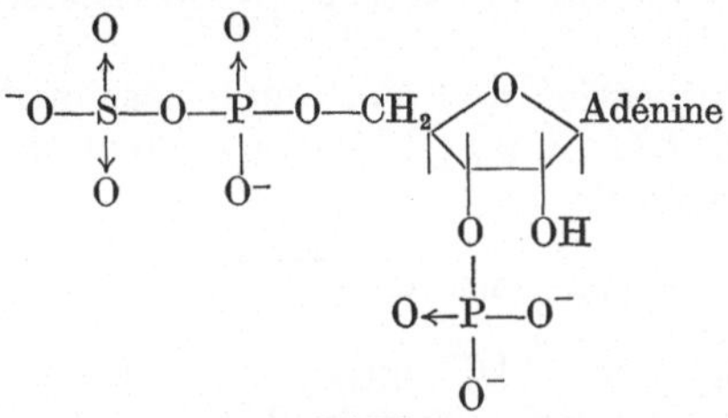

PAPS

Cet anhydride mixte phosphosulfurique se révéla être l'intermédiaire obligatoire commun dans la formation biologique de différents types d'esters sulfuriques (GREGORY et LIPMANN, 1957; NOSE et LIPMANN, 1958; D'ABRAMO et LIPMANN, 1957; KORN, 1959; SPENCER et HARADA, 1959). D'autres recherches sur le rôle biologique du PAPS ont montré que ce composé participe à la réduction biologique du sulfate (WILSON et BANDURSKI, 1958a; HILZ et KITTLER, 1958; HILZ, KITTLER et KNAPE, 1959; RAGLAND, 1959).

La synthèse du PAPS à partir du sulfate minéral a lieu seulement en présence de l'ATP et elle se fait en deux stades (BANDURSKI, WILSON et SQUIRES, 1956),

---

[1] AMP = acide adénosinemonophosphorique, ADP = acide adénosinediphosphorique, APS = adénosine-5′-phosphosulfate, ATP = acide adénosinetriphosphorique, EDTA = acide éthylènediaminetétra-acétique, p-NP = p-nitrophénol, p-NPS = p-nitrophénylsulfate, PAP = 3′-phosphoadénosine-5′-phosphate, PAPS = 3′-phosphoadénosine-5′-phosphosulfate, PP = pyrophosphate, Tris = trihydroxyméthylaminométhane.

le premier correspondant à la formation de l'adénosine-5′-phosphosulfate (APS) et le second à la phosphorylation de ce dernier en position 3′ (Robbins et Lipmann, 1956).

$$\text{I) ATP} + SO_4^{--} \quad \underset{}{\overset{\text{ATP-sulfurylase}}{\rightleftharpoons}} \quad \text{APS} + \text{Pyrophosphate}$$

$$\text{II) APS} + \text{ATP} \quad \overset{\text{APS-kinase}}{\longrightarrow} \quad \text{PAPS} + \text{ADP}$$

L'enzyme qui catalyse la première réaction a été appelé ATP-sulfurylase. Dans la deuxième réaction intervient une autre molécule d'ATP et un enzyme du type phosphokinase: l'adénosine-phosphosulfate-kinase (APS-kinase) (Robbins et Lipmann, 1958a). On a montré que la réaction catalysée par l'ATP-sulfurylase est réversible et qu'à l'équilibre la concentration en APS est très faible (Robbins et Lipmann, 1956; 1958a). Ces observations ont été mises à profit pour la mesure de l'activité enzymatique au cours de la purification de l'ATP-sulfurylase à partir de la levure de boulangerie par Robbins et Lipmann (1957, 1958a). En utilisant le même matériel biologique Wilson et Bandurski (1958b) ont montré que dans la réaction I le sulfate peut être remplacé par d'autres anions minéraux: le sulfite, le séléniate, le chromate, le tungstate et le molybdate.

En présence de ces anions l'ATP-sulfurylase permet la substitution du pyrophosphate de l'ATP de la façon suivante:

$$\text{III) ATP} + \text{Anion} \quad \underset{}{\overset{\text{Réaction enzymatique}}{\rightleftharpoons}} \quad 5'\text{-AMP-Anion} + \text{PP}$$

$$\text{IV) } 5'\text{-AMP-Anion} \quad \overset{\text{Réaction spontanée}}{\longrightarrow} \quad \text{AMP} + \text{Anion}$$

Les anhydrides mixtes autres que ceux qui contiennent le sulfate ou le séléniate sont très instables et leur existence n'a pu être mise en évidence que par échange du pyrophosphate (PP) de l'ATP avec le pyrophosphate ^{32}P introduit dans le milieu.

Au cours de la purification de l'ATP-sulfurylase et de l'APS-kinase à partir de la levure, Robbins et Lipmann (1958a) ont obtenu une fraction protéique qui catalyse la réaction suivante:

$$\text{V) APS} + \text{Phosphate minéral} \quad \overset{\text{ADP-sulfurylase}}{\longrightarrow} \quad \text{ADP} + SO_4^{--}$$

Le rôle physiologique de l'ADP-sulfurylase est inconnu.

Le transfert du groupement sulfate du PAPS sur les accepteurs possédant un groupement OH est assuré par des enzymes spécifiques appelés sulfokinases.

Suivant la nature de l'accepteur, on distingue les phénol-sulfokinases, les stéroïdes-sulfokinases, etc. En outre, une sulfokinase peut catalyser aussi le transfert du groupe sulfate d'un sulfoconjugué sur un autre accepteur. Ainsi Gregory et Lipmann (1957) ont obtenu du sulfate de phényle à partir de sulfate de p.nitrophényle, en présence de phénol, de 3′-phosphoadénosine-5′-phosphate (PAP) et de phénol-sulfokinase.

$$\text{p.nitrophénylsulfate} + \text{phénol} \quad \underset{\text{PAP}}{\overset{\text{sulfokinase}}{\longrightarrow}} \quad \text{p.nitrophénol} + \text{phénylsulfate}$$

le PAP intervient dans cette réaction par sa conversion en PAPS.

Les essais de purification des sulfokinases ont porté jusqu'à présent sur les enzymes d'origine animale. Les préparations purifiées suivantes ont été obtenues: phénol-sulfokinase par Gregory et Lipmann (1957), sulfokinase des stéroïdes par Nose et Lipmann (1958) et chondroitine-sulfokinase par d'Abramo et Lipmann (1957). Harada et Spencer (1960) ont montré que dans l'*Aspergillus orizae* le sulfate actif est l'intermédiaire au cours de la synthèse de l'ester sulfurique de la choline, ce qui suppose la présence dans ce champignon de la sulfokinase correspondante.

En absence de données sur la préparation des sulfokinases d'origine végétale et prenant en considération l'importance de ces enzymes dans le métabolisme du sulfate, nous décrirons un procédé de purification d'une sulfokinase animale.

I. Enzymes d'activation du sulfate.

1. Mesure de l'activité enzymatique.

a) Système complet d'activation.

Ce système contient l'ATP-sulfurylase et l'APS-kinase. Il permet la synthèse du PAPS à partir de l'ATP et du sulfate minéral. Deux méthodes ont été proposées pour la mesure de l'activité des préparations enzymatiques.

Dans la première, la formation du PAPS est mesurée par la quantité du p. nitrophénylsulfate (p. NPS) synthétisé à partir du p. nitrophénol après addition de phénol-sulfokinase.

La deuxième consiste à mesurer, en présence d'un excès de pyrophosphatase, le phosphate minéral formé à partir du pyrophosphate libéré au cours de la réaction I.

Méthode I Gregory et Lipmann (1957). Cette méthode est décrite plus loin (voir: purification de la phénol-sulfokinase).

Méthode II Robbins et Lipmann (1957). La préparation enzymatique utilisée est soumise à une dialyse préalable afin d'éliminer les ions phosphate et sulfate qu'elle peut contenir.

Le mélange réactionnel d'un volume total de 1 ml contient: tampon Tris-HCl 0,1 M à pH 8,5; ATP 0,01 M; $MgCl_2$ 0,01 M et la préparation enzymatique. Pour chaque détermination, on utilise deux tubes à essai contenant cette solution; dans un des tubes (A) on introduit K_2SO_4 à la concentration 0,02 M, l'autre (B) sert de témoin. Dans le cas des préparations enzymatiques purifiées qui peuvent être dépourvues de la pyrophosphatase on ajoute en outre 10 μg de cet enzyme cristallisé. Le mélange ainsi préparé est incubé à 37° pendant 30 minutes. La réaction est arrêtée par addition d'acide trichloroacétique. Le phosphate minéral formé est dosé par la méthode de Fiske et Subbarow (1925). La différence entre les quantités de phosphate trouvées dans les tubes A et B permet de connaître la quantité du PAPS synthétisée: 2 μmoles de phosphate correspondent à 1 μmole de PAPS.

b) ATP-sulfurylase — Robbins et Lipmann (1958a).

A l'équilibre, la concentration de l'APS en présence de cet enzyme étant extrêmement faible, il est plus facile pour mesurer l'activité de l'ATP-sulfurylase d'utiliser comme substrats l'APS et le pyrophosphate et de suivre la disparition de ce dernier au cours de la formation de l'ATP et du sulfate (réaction I).

Le milieu réactionnel contient: tampon Tris-HCl 0,1 M à pH 8,5; pyrophosphate: 0,003 M; APS: 0,003 M; $MgCl_2$ $2 \cdot 10^{-4}$ M et l'enzyme, dans un volume total de 0,2 ml. Lorsqu'on emploie des préparations enzymatiques brutes ou semi-purifiées, il est nécessaire de les soumettre préalablement à une dialyse et de rendre inactive la pyrophosphatase qu'elles contiennent par omission des ions Mg^{++}. Toutefois, en absence de ce métal, l'activité de l'ATP-sulfurylase est diminuée d'environ 40%. Dans le calcul de l'activité enzymatique, on tient compte de cette inhibition.

Le mélange réactionnel ci-dessus préparé dans des tubes jaugés à 10 ml, est incubé pendant 15 minutes à 37°. La réaction enzymatique est arrêtée par immersion des tubes dans un bain-marie bouillant, pendant 90 secondes. Après refroidissement dans l'eau glacée, le mélange est additionné de 0,02 ml de $MgCl_2$, 0,1 M et de 0,1 ml d'une solution (100 $\mu g/ml$) de pyrophosphatase. L'ensemble est de nouveau incubé à 37° pendant 5 minutes. Le dosage du phosphate présent dans le milieu est effectué par la méthode de Fiske et Subbarow (1925). L'utilisation des tubes jaugés facilite ce dosage.

La quantité de pyrophosphate utilisée par l'ATP-sulfurylase est déterminée par rapport à un essai témoin réalisé en absence d'APS. L'activité spécifique de l'enzyme est définie par le nombre de μmoles de pyrophosphate utilisées pour la synthèse de l'ATP, par mg de protéine et par heure.

c) ADP-sulfurylase — Robbins et Lipmann (1958a).

La réaction enzymatique est suivie par la disparition du phosphate au cours de la réaction V. Le mélange réactionnel contient: tampon Tris-HCl 0,1 M à pH 8,5; APS 0,003 M; phosphate 0,003 M et la préparation enzymatique. Le volume total est de 0,2 ml. L'incubation est effectuée à 25° pendant 15 minutes. La réaction est arrêtée par addition de 0,5 ml d'acide trichloroacétique à 5%. Les protéines sont éliminées par centrifugation et 0,5 ml du surnageant est utilisé pour le dosage du phosphate.

Dans les conditions ci-dessus, la vitesse de la réaction enzymatique est linéaire seulement pendant un temps court et diminue avant que 50% de l'APS et du phosphate aient réagi; aussi la durée de l'incubation ne doit pas être prolongée. Pour rendre inactive la pyrophosphatase éventuellement présente dans les préparations, on prend soin d'éviter la présence d'ions Mg^{++}. On remarquera qu'en leur absence l'activité de l'ADP-sulfurylase n'est pas diminuée.

d) APS-kinase — Robbins et Lipmann (1958a).

L'activité spécifique de cet enzyme est exprimée par la quantité en μmoles du PAPS formé à partir de l'APS par mg de protéine et par heure, dans les conditions standard. Le mélange réactionnel contient: tampon Tris-HCl 0,1 M à pH 8,5; ATP 0,005 M; $MgCl_2$ 0,001 M; APS (ne contenant pas de 3'-phospho-adénosine-5'-phosphate) 0,001 M et la préparation enzymatique. Le volume total est de 0,5 ml. Ce mélange est incubé à 37° pendant 20 minutes. La réaction est arrêtée par chauffage pendant 90 secondes dans un bain-marie bouillant. Les échantillons sont ensuite dilués et le PAPS formé est dosé en présence de phénol-sulfokinase et de p. nitrophénol, par la méthode de Gregory et Lipmann (1957) (Voir phénol-sulfokinase).

2. Préparation de l'ATP-sulfurylase, de l'ADP-sulfurylase et de l'APS-kinase.

Les trois enzymes sont préparés à partir d'un même extrait de levure de boulangerie. Robbins et Lipmann (1958a), (1958b).

Traitement de la levure. La levure fraiche est mélangée et agitée avec 1 partie d'éther et 1,5 partie de neige carbonique. On laisse reposer 30 minutes, on élimine l'excès d'éther et on étale la levure en couche mince sur un tissu. Un courant d'air chaud, au dessus de la préparation, permet une évaporation rapide du reste de l'éther. On recommence le mélange avec 1,5 partie de neige carbonique. On soumet la masse à une évaporation jusqu'à élimination complète de l'éther; la présence des traces de celui-ci inactive les enzymes au cours des traitements ultérieurs. On peut aussi disperser simplement la levure dans de l'azote liquide. La préparation de levure ainsi obtenue est conservée dans un congélateur.

A 3,6 kg de la préparation gelée, on ajoute 3,6 litres d'une solution de HK_2PO_4 0,5 M et on agite à la température du laboratoire jusqu'à obtention d'une suspension homogène. L'extraction par agitation se poursuit pendant 16 heures à 2°. Dans les stades suivants on travaille toujours entre 0 et 4°. Après l'extraction, les débris cellulaires sont éliminés par centrifugation. Le volume du surnageant recueilli (fraction I) est de 4,5 litres.

Acidification et précipitation par le chlorure de sodium. La fraction I est amenée à pH 5,0 par addition de 100 ml d'acide acétique 2 M. La suspension est clarifiée par centrifugation, le surnageant représente la fraction II. A 4,3 litres de la fraction II, on ajoute lentement, sous agitation 1,08 kg de NaCl. Après la dissolution du sel, on laisse le précipité se déposer pendant 16 h à 2°. Une partie du surnageant est siphonnée; le reste est centrifugé pendant 8 minutes.

Le précipité seul est conservé. Il est dissous dans 300 ml de Tris 0,02 M. Le pH est ensuite ajusté à 7 environ, à l'aide d'une solution de KOH 2 N. Cette fraction (III) est très stable et congelée elle se conserve pendant plusieurs mois.

La fraction III est dialysée pendant 22 heures contre 16 litres d'une solution de Tris 0,01 M — EDTA 10^{-3} M à pH 7,5. Après dialyse la préparation est acidifiée à pH 5,8 avec de l'acide acétique. Dans ces conditions l'ATP-sulfurylase précipite et l'APS-kinase reste en solution. Le précipité (fraction IV) est recueilli par centrifugation et le surnageant (fraction V) est conservé pour la purification de l'APS-kinase.

Purification de l'ATP-sulfurylase. *Fractionnement avec le sulfate d'ammonium.* Au cours de toutes les étapes, on utilise une solution saturée à la température du laboratoire de $(NH_4)_2SO_4$. Cette solution est additionnée de 2 mmoles de EDTA et de 40 mmoles de KOH par litre.

La fraction IV est dissoute dans 30 ml de tampon Tris 0,01 M à pH 8 contenant de l'ATP à la concentration de 10^{-3} M. Le volume est ajusté à 50 ml et additionné de 33 ml d'une solution saturée de $(NH_4)_2SO_4$. On centrifuge immédiatement et on élimine un lourd précipité brun. Au surnageant on ajoute de nouveau 16 ml de la solution saturée de $(NH_4)_2SO_4$. Après une heure un léger précipité est recueilli par centrifugation pendant 10 minutes. Il est dissous dans 1 ml de tampon Tris à pH 7,5 et dialysé une nuit contre Tris-EDTA (fraction VI).

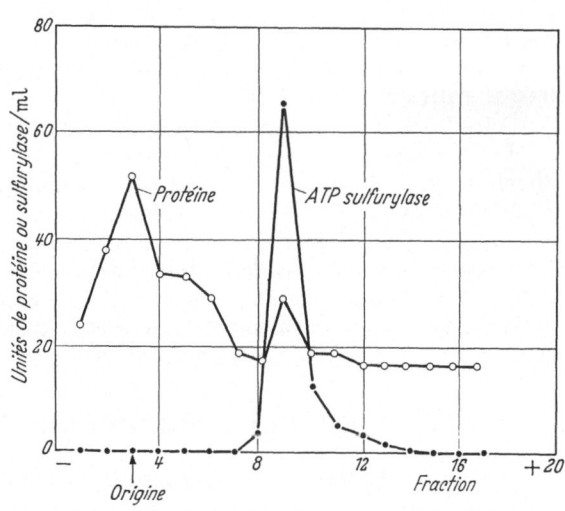

Fig. 1. Purification de l'ATP-sulfurylase par électrophorèse sur Géon 426 (ROBBINS et LIPMANN 1958 b).

Electrophorèse sur résine. La résine (Géon 426, B. F. Goodrich Company) est lavée plusieurs fois à l'eau, puis 2 fois avec une solution: Tris 0,1 M + NaCl 0,003 M à pH 8. La résine est disposée en une couche: 30 cm . 4 cm . 1 cm. On place une partie de la fraction VI dans la résine à 10 cm de l'extrémité reliée à la cathode. L'électrophorèse est effectuée sous 175 volts (30 à 35 milliampères)

Tableau 1. *Séparation de l'ATP-sulfurylase, de l'APS-kinase et de l'ADP-sulfurylase de la levure* (Robbins et Lipmann 1958a).

Fraction	ATP-sulfurylase (Réaction inverse). Activité spécifique	APS-kinase Activité spécifique	ADP-sulfurylase Activité spécifique
I — Extrait brut	5,6	2,0	—
II — Surnageant acide . . .	8,1	1,6	—
III — Précipité NaCl.	32	4,6	0,37
IV — Précipité acide	47	1,2	0,15
VI — $(NH_4)_2SO_4$ 40 à 50%/$_0$ . .	2.600	0	2,0
VII — Electrophorèse	7.100	0	0
VIII — Précipité alcoolique . .	0	6,3	0,94
IX — Fraction $(NH_4)_2SO_4$ . .	—	—	3,0

pendant 24 heures à 2°. Dans ces conditions, l'ATP-sulfurylase migre de 9 à 12 cm vers l'anode. La résine est découpée en fragments de 1,5 cm et chaque fragment est suspendu dans 5 ml d'eau. Après filtration, la résine est lavée encore avec 5 ml d'eau. Les fractions aqueuses réunies sont centrifugées pour éliminer complètement la résine (fraction VII). On détermine la concentration en protéines de chaque fraction et mesure leur activité (Fig. 1 et Tableau 1). La fraction VII se montre homogène à l'ultracentrifugation et son activité spécifique est environ 1400 fois supérieure à celle de l'extrait brut de levure. Le rendement est de 10%.

Purification de l'APS-kinase. *Précipitation par l'éthanol.* La fraction V est acidifiée à pH 5,2 avec de l'acide acétique 0,1 M et clarifiée par centrifugation. Le précipité est rejeté. Le surnageant refroidi dans un bain de glace et de sel est agité et additionné lentement de 0,1 volume d'éthanol à 95% refroidi. La température de la solution est maintenue entre −1° et −3°. Le précipité qui se forme est recueilli par centrifugation à −2°. Il est dissous dans 25 ml de Tris 0,02 M à pH 8 (fraction VIII).

Cette fraction contenant l'APS-kinase est assez stable.

Fractionnement par le sulfate d'ammonium. La fraction VIII est amenée à 30 ml et est additionnée de 20 ml d'une solution saturée de $(NH_4)_2SO_4$. Après 15 minutes, la suspension est centrifugée (Servall SS-1) et le surnageant recueilli est additionné de 10 ml de la solution saturée de $(NH_4)_2SO_4$. Une heure après on centrifuge de nouveau pendant 10 minutes et recueille le précipité qui est dissous dans 5 ml de tampon Tris 0,02 M à pH 7,5 (fraction IX). Les préparations ainsi obtenues ne sont pas également stables et souvent elles perdent leur activité au cours de la conservation dans un congélateur.

II. Enzymes de transfert du sulfate.

1. Mesure de l'activité de la phénol-sulfokinase.

Deux méthodes ont été utilisées pour mesurer l'activité des préparations enzymatiques.

Méthode I. Robbins et Lipmann (1957). Mesure de la vitesse du transfert du groupement sulfate du PAPS sur le p-nitrophénol (p-NP). En milieu alcalin, le p-NP étant jaune et le p-NPS incolore, la diminution de la concentration du p-NP se traduit par la diminution de la densité optique.

Le milieu d'incubation contient: chlorhydrate d'imidazol: 100 μmoles à pH 7; cystéine: 10 μmoles; p-NP: 1 μmole; PAPS: 0,5 μmole; préparation enzymatique, volume total 1 ml. Incubation: 30 minutes à 37°. Les protéines sont éliminées après l'incubation par addition de 2 ml d'alcool et centrifugation. A 2,5 ml du surnageant recueilli, on ajoute 0,5 ml de KOH 0,1 N et on mesure la densité optique à 400 mμ.

Méthode II. Gregory et Lipmann (1957). Mesure de la vitesse du transfert du groupement sulfate du p-nitrophénylsulfate (p-NPS) sur le phénol en présence du PAP.

Le milieu d'incubation contient: Tampon Tris-HCl pH 7,8: 100 μmoles; p-NPS: 0,5 μmole; phénol: 0,5 μmole; PAP: 0,001 μmole; volume total: 0,9 ml. Le mélange introduit dans une cuve du spectrophotomètre est additionné de 0,1 ml de solution enzymatique contenant du glutathion à la concentration 0,1 M. On suit aussitôt après l'addition de l'enzyme la variation à 400 mμ de la densité optique.

La quantité d'enzyme présente est directement proportionelle à l'augmentation de la densité optique jusqu'à une valeur de 0,100 par minute en cuve de 10 mm d'épaisseur.

Une unité d'activité est définie par la quantité d'enzyme nécessaire pour obtenir un changement de densité optique de 0,100 par minute.

2. Préparation de la phénol-sulfokinase — Gregory et Lipmann (1957).

On utilise le foie de lapin; cet organe est refroidi et broyé pendant 30 secondes en présence de 3 volumes de tampon phosphate 0,05 M à pH 7,4 contenant 0,01 M de EDTA. La suspension ainsi obtenue est centrifugée 30 minutes à 25.000 g environ. Le surnageant laiteux recueilli (12 unités d'enzyme par ml; 0,4 unité par mg de protéine) est additionné de 10% de son volume d'une solution de sulfate de protamine à 2%. La suspension est clarifiée par centrifugation pendant 20 minutes. A cette solution (9 unités par ml, 0,5 unité par mg), on ajoute 33% de son volume d'une suspension d'un gel d'alumine Cγ (suspension: 13 mg de poids sec par ml), afin d'éliminer les enzymes qui interviennent dans l'activation du sulfate; une partie de l'activité transférante est perdue au cours de ce traitement.

La solution rougeâtre ainsi obtenue est additionnée de $(NH_4)_2SO_4$ jusqu'à 50% de saturation; le précipité recueilli est suspendu dans un petit volume de tampon Tris ou phosphate à pH 7 et est dialysé contre le même tampon. Après centrifugation, le surnageant limpide contient 45 unités par ml et 1,0 unité par mg, soit 86% de l'activité initiale.

Une telle préparation enzymatique donne des témoins légèrement colorés en absence de PAP. Ces témoins sont complètement incolores à 400 mμ si l'on prépare l'extrait brut de foie suivant la méthode de Hilz et Lipmann (1955). Cet extrait est traité par 1/3 de son volume de gel d'alumine Cγ et concentré soit par lyophylisation, soit par précipitation avec du sulfate d'ammonium à 85% de saturation.

L'activité enzymatique présente dans le foie est variable selon les animaux, sans qu'il apparaisse de corrélation avec l'âge ou le sexe.

Cette phénol-sulfokinase possède une activité maximale entre les pH 7 et 8 (Tris, triéthanolamine, ou phosphate) et une très faible activité à pH 6 (citrate). Ni les ions Mg^{++} ou Mn^{++}, ni l'EDTA, le fluorure, le pyrophosphate, le molybdate ou le tungstate à la concentration 0,01 M ne modifient son activité. L'addition d'un composé sulfhydrylé est nécessaire pour obtenir l'activité maximale de l'enzyme. Dans ce but, au moins 10 minutes avant chaque essai, on dilue la préparation enzymatique dans une solution 0,1 M de glutathion réduit.

B. Systèmes réducteurs du sulfate.

La plupart des microorganismes et des végétaux utilisent le sulfate minéral pour la synthèse des composés soufrés organiques réduits. On a montré que parallèlement ces mêmes organismes réduisent le sulfate en sulfite et en sulfure (Fromageot et Chapeville 1958).

Sous l'influence des enzymes spécifiques, ces deux formes du soufre réagissent avec le carbone des molécules organiques pour former des sulfonates ou des thiols correspondants (Chapeville et Fromageot 1957; Schlossmann et Lynen 1957).

Le rôle du sulfite comme intermédiaire au cours de la réduction du sulfate en sulfure a été démontré avec *Desulfovibrio desulfuricans* (Millet 1954) et avec la levure (Hilz, Kittler et Knape 1959).

L'étude du mécanisme de la réduction du sulfate en sulfite et en sulfure se révèle difficile du fait que jusqu'à ces dernières années il a été impossible d'obtenir des préparations acellulaires actives.

De telles préparations ont été obtenues pour la première fois en 1958 par Wilson et Bandurski (1958a), Hilz et Kittler (1958), Hilz, Kittler et Knape (1959), à partir de la levure, puis par Ishimoto (1959), Ishimoto et Fujimoto (1959), Peck (1960), à partir de bactéries sulfato-réductrices.

Dans les laboratoires qui ont abordé l'étude de la réduction du sulfate, on a montré que l'intermédiaire obligatoire est, chez la levure, le sulfate actif (PAPS), et chez les bactéries sulfato-réductrice, l'adénosine-5'-phosphosulfate (APS).

S'il en est ainsi, les préparations enzymatiques obtenues jusqu'à présent sont très complexes, et contiennent l'ATP-sulfurylase, l'APS-kinase et un système réducteur du PAPS ou de l'APS.

1. Mesure de l'activité sulfato-réductrice.

Dosage du sulfite. Le sulfite n'apparaissant au cours de la réduction du sulfate qu'en faibles quantités, il est nécessaire d'utiliser pour son dosage des méthodes très sensibles. Hilz, Kittler et Knape (1959) ont employé avec succès la technique décrite par Grant (1947) qui utilise la réaction de la fuchsine avec le formol. Hilz (1960) procède ainsi: A 0,40 ml du mélange réactionnel on ajoute 0,60 ml d'une solution saturée de chlorure mercurique, pour déprotéiniser, et on centrifuge. A une partie aliquote du surnageant on ajoute 4 ml du réactif et, après 8 minutes exactement, on mesure la densité optique à 578 mμ.

Le réactif, préparé le jour de son emploi, comporte: 60 ml d'eau distillée, 3 ml d'acide sulfurique concentré, 0,25 ml d'une solution à 3 $^0/_0$ de fuchsine dans de l'alcool à 80 $^0/_0$, et enfin 0,25 ml d'une solution de formol à 40 $^0/_0$.

Isolement du sulfite radioactif. La réduction du sulfate a été étudiée à l'aide de sulfate ^{35}S, par Wilson et Bandurski (1958a); Hilz, Kittler et Knape (1959); Chapeville et Fromageot (1957); Fromageot et Perez-Milan (1959).

Hilz, Kittler et Knape (1959) isolent le sulfite ^{35}S de la façon suivante: Le milieu réactionnel contenant $Na_2^{35}SO_4$ est incubé dans un récipient du type flacon laveur, afin de pouvoir réaliser par barbotage, un entrainement des gaz dissous dans le liquide. Durant les 10 premières minutes d'incubation, on fait passer dans le milieu un courant d'azote. A la fin de l'incubation, on introduit dans le milieu 10 à 15 μmoles de sulfite entraineur, une ou deux gouttes d'octanol, et, après agitation, 1 ml de H_2SO_4 2 N. Un nouveau passage d'azote permet l'entrainement du SO_2 formé. Celui-ci est recueilli à l'état de $SrSO_3$ dans une solution neutre de nitrate de strontium à 7 $^0/_0$. Le précipité de $SrSO_3$ est lavé plusieurs fois à l'eau, et sa radioactivité est mesurée, soit directement, soit après sa conversion en $SrSO_4$ par oxydation avec du perhydrol et précipitation du sulfate sous forme de $BaSO_4$.

Le sulfite peut être recueilli aussi à l'état de K_2SO_3, en particulier lorsque l'incubation a été faite dans des cellules de WARBURG. Dans ce but, on utilise des cellules avec deux tubulures latérales; l'une contient le sulfite entraineur, et l'autre H_2SO_4. La cupule centrale contient 0,2 ml de KOH et un rouleau de papier filtre. Après l'incubation, on verse dans la solution le sulfite, puis après homogénéisation, l'acide sulfurique, et on poursuit l'agitation à la température de l'incubation pendant 60 minutes. La totalité du SO_2 qui s'est formé est captée dans la solution alcaline. Celle-ci est recueillie et est additionnée de 0,15 ml de perhydrol. Le sulfate est isolé à l'état de sel de baryum.

Dosage du sulfure. Celui-ci est effectué par les micro-méthodes classiques.

Isolement du sulfure radioactif. On opère comme pour l'isolement du sulfite, en introduisant à la fin de l'incubation, comme entraineur, 5 à 10 μmoles de HKS non radioactif. Après acidification, l'hydrogène sulfuré dégagé est capté dans 10 ml d'une solution d'acétate de cadmium à 1%. Si l'incubation est effectuée dans des cellules de WARBURG, l'hydrogène sulfuré formé est capté dans une solution de KOH 4 N, puis précipité à l'état de CdS par addition de 1 ml d'acétate de cadmium à 10% et de 1 ml d'acide acétique pur. Après plusieurs lavages à l'eau on mesure la radioactivité du CdS recueilli.

2. Préparation du système réducteur du sulfate.
HILZ, KITTLER et KNAPE (1959).

Extrait. On introduit dans le désintégrateur de Braun (Fa. B. Braun, Melsungen) 10 g de levure de boulangerie et 10 ml de tampon Tris (0,2 M à pH 8) avec 50 g de billes de verre ($\varnothing = 0,35$ mm) et on agite à froid pendant 4 fois 30 secondes. Dix préparations ainsi obtenues sont réunies et séparées des billes par filtration sur verre fritté G-3. Si l'extrait est utilisé directement pour l'incubation, il est clarifié par centrifugation.

Fractionnement par l'acétone. A 170 ml de l'extrait non centrifugé on ajoute 170 ml d'eau et 119 ml d'acétone à $-5°$ (concentration finale 26%). Après centrifugation on rejette le précipité; la solution obtenue sert à la préparation des fractions suivantes: *Ac.45:* Aux 366 ml du surnageant, on ajoute de nouveau 163 ml d'acétone. Après centrifugation, le précipité recueilli est dissous dans 50 ml de tampon Tris 0,02 M à pH 7,5 et centrifugé de nouveau. Le surnageant est dialysé pendant 90 minutes contre l'eau distillée; sa concentration en protéines est de 70 mg/ml. *Ac.55:* Le surnageant acétonique Ac.45 est additionné de 216 ml d'acétone. Le précipité est dissous dans 25 ml de tampon Tris et traité comme celui de la fraction Ac.45. La concentration en protéines est de 7 mg/ml. *Ac75:* A 600 ml du surnageant Ac.55, on ajoute 965 ml d'acétone et recueille le précipité. Après dissolution dans le tampon Tris et centrifugation suivie de dialyse, on obtient une fraction qui contient 0,6 mg de protéines par ml.

Fraction NaCl. Au mélange de 10 ml de Ac. 45 + Ac. 55, en parties égales on ajoute 2,5 g de NaCl et recueille par centrifugation le précipité formé. Il est dissous dans 4 ml de tampon Tris 0,02 M à pH 7,5 (protéines: 20 mg/ml).

L'activité des différentes fractions obtenues est indiquée dans le tableau 2.

En utilisant la préparation acellulaire (Extrait) dans les mêmes conditions que ci-dessus mais additionnée en plus de glucose-6-phosphate (G-6-P), de TPN et d'acide lipoïque, on obtient la réduction du sulfate jusqu'au sulfure. Le tableau 3 montre l'importance des différents cofacteurs pour cette réaction.

Tableau 2. *Réduction du sulfate en sulfite par les différentes fractions protéiques précipitées par l'acétone* (HILZ, KITTLER et KNAPE 1959).

Fraction	Méthode isotopique		Détermination directe du sulfite
	SO_3H^- ipm au total	SO_3H^- ipm par μmole	$m\mu$ moles de SO_3H^- formé
Extrait	4.380	129	0
Ac.45	2.370	59	0
Ac.55	17.000	930	0
Ac.75	3.500	46	0
Ac.45 + Ac.55	157.000	3.080	216
Ac.45 + Ac.75	6.700	142	0
Ac.55 + Ac.75	7.050	184	0
Ac.45 + Ac.55 + Ac.75	238.000	5.600	337

Composition: a) méthode isotopique: 100 μmoles du tampon phosphate pH 7,4; DPN: 0,6 μmole; ATP: 15 μmoles; $MgCl_2$: 7 μmoles; Fructose diphosphate: 0,2 μmole; EDTA: 3 μmoles; Nicotinamide: 50 μmoles; Na_2SO_3: 7,5 μmoles; $Na_2^{35}SO_4$: 4 μC et la préparation enzymatique (0,75ml Extrait; 0,25 ml Ac.45; 0,50 ml Ac.55; 0,50 ml Ac.75). Volume total: 1,50 ml. Après l'incubation à 30° C pendant 60 minutes, on isole le sulfite marqué. b) Détermination directe du sulfite: L'incubation est effectuée dans les mêmes conditions que dans a) mais en absence du sulfite entraineur et le sulfate radioactif est remplacé par 10 μmoles de K_2SO_4.

Le système réducteur du sulfate à partir de *Desulfovibrio* (ISHIMOTO, 1959 et ISHIMOTO et FUJIMOTO, 1959) a été obtenu après broyage des cellules en présence d'alumine ou leur désintégration par les ultrasons. L'adénosine-5'-phosphosulfate-réductase a été purifiée par ISHIMOTO et FUJI-MOTO (1961). Des enzymes impliqués dans la réduction du 3'-phosphoadénosine-5'-phosphosulfate ont été obtenus de la levure par ASAHI, BANDURSKI et WILSON (1961) et par WILSON, ASAHI et BANDURSKI (1961).

Tableau 3. *Importance des différents co-facteurs sur la réduction du sulfate en sulfure* (HILZ, KITTLER et KNAPE 1959).

Mélange réactionnel	CdS: Ipm/min
complet	6.230
— ATP	2.320
— G-6-P	2.300
— DPN	2.980
— TPN	2.800
— DPN-TPN \\	
— ATP-Ac. Lipoïque /	920

L'activité de la préparation est mesurée dans l'appareil de WARBURG par la consommation d'hydrogène moléculaire en présence du méthylviologène comme transporteur d'électrons, l'hydrogène sulfuré formé étant capté par de la potasse.

Le mélange réactionnel de volume total 3 ml contient: Tampon phosphate 250 μmoles; pH 6,8; K_2SO_4: 10 μmoles; $MgCl_2$: 50 μmoles; ATP: 7,5 μmoles; G-6-P: 10 μmoles; DPN: 0,75 μmole; TPN: 0,4 μmole; Ac. lipoïque S_2: 0,5 μmole; HKS: 5 μmoles; Extrait de levure: 1,50 ml et $Na_2^{35}SO_4$: 3 μC. Incubation en anaérobiose pendant 3 heures à 30°.

C. Sulfatases.

Ces enzymes hydrolysent les esters organiques de l'acide sulfurique suivant la réaction:

$$R—O—SO_3^- + H_2O \longrightarrow ROH + SO_4^{--} + H^+$$

Les sulfatases sont largement répandues dans le règne animal et végétal et d'après la nature du substrat hydrolysé, on les divise en: Arylsulfatases, chondrosulfatases, myrosulfatases, glucosulfatases, stéroidesulfatases et cholinesulfatase.

Mesure de l'activité enzymatique. Après incubation de l'ester sulfurique en présence d'une préparation enzymatique, on dose soit le sulfate minéral libéré soit le produit organique issu de l'hydrolyse.

Plusieurs techniques ont été décrites par DODGSON et SPENCER (1953a, b), DODGSON, MELVILLE, SPENCER et WILLIAMS (1954), ROY (1953), FROMAGEOT (1955), HILZ et LIPMANN (1955), et par SPENCER (1960).

Le sulfate est dosé à l'état de sels de benzidine ou de baryum. Dans le cas des composés cycliques dont le spectre d'absorption est différent lorsqu'ils se trouvent à l'état d'esters sulfuriques, on suit au spectrophotomètre la quantité de $R-O-SO_3^-$ disparue ou celle de $R-OH$ apparue. Certains phénols enfin peuvent être dosés spécifiquement par voie chimique (ABBOT 1947).

1. Arylsulfatases.

Ces enzymes hydrolysent les esters sulfuriques des phénols. Ils ont été mis en évidence dans de nombreux organismes: mammifères, insectes, mollusques, champignons, et bactéries.

Parmi 160 souches différentes de *Staphylococcus pyogenes* examinées, deux seulement contiennent une arylsulfatase (BARBER, BROOKSBANK et KUPER 1951). Une arylsulfatase a été mise en évidence par les mêmes auteurs dans *Salmonella paratyphi*. YOUNG, MORRISSON et WHITEHEAD (1952) ont examiné 212 souches des bactéries et n'ont détecté une arylsulfatase que dans certaines Salmonelles et Mycobactéries.

Dans la vase découverte par la marée, DODGSON, MELVILLE, SPENCER et WILLIAMS (1954) et DODGSON, SPENCER et WILLIAMS (1955) ont mis en évidence deux microorganismes assez riches en arylsulfatase: *Trichosporon cutaneum*(levure) et *Alcaligenes metalcaligenes* (bactérie). La sulfatase de *A. metalcaligenes* se distingue des autres arylsulfatases connues par son activité à des pH allant jusqu'à 11.

L'action de cette sulfatase sur divers substrats est donnée par DODGSON, SPENCER et WILLIAMS (1956).

Préparation des arylsulfatases.

FROMAGEOT (1955) a décrit la préparation des arylsulfatases A et B à partir de foie de mammifères.

Préparation de l'arylsulfatase d'alcaligenes metalcaligenes. DODGSON, MELVILLE, SPENCER et WILLIAMS (1954) et DODGSON, SPENCER et WILLIAMS (1955). *Traitement par l'acétone.* Le microorganisme est cultivé dans des récipients de 16 litres contenant le milieu nutritif ajusté à pH 7,4 à 7,6. Les cultures aérées sont maintenues à la température de 25°. Après 6 jours, les cellules sont recueillies, lavées à l'eau à 0° et suspendues dans l'acétone froide. Recueillies par filtration, elles sont lavées avec de l'acétone. L'acétone est éliminée sous vide. 5 g de poudre (A) ainsi obtenue sont suspendus dans 75 ml d'eau froide pendant 90 minutes. La suspension est ensuite centrifugée à 0° et les débris cellulaires lavés avec 60 ml d'eau. Les surnageants réunis sont additionnés d'acétone froide jusqu'à la concentration de 85%. Le précipité qui se forme est séparé, lavé à l'acétone froide et séché sous vide (poudre B). Cette préparation maintenue à 0° conserve son activité enzymatique.

Electrophorèse. On dissous 0,1 g de la poudre B dans 0,6 ml de tampon phosphate 0,1 M à pH 8,0. On fractionne 0,1 ml de cette solution par électrophorèse sur papier, pendant 18 heures sous 110 volts en présence de tampon phosphate 0,1 M à pH 8,0. Après l'électrophorèse, le papier (Whatman No. 100) est séché à la température du laboratoire et découpé en fragments de 1 cm. Les protéines sont localisées avec du bleu de bromophénol et leur activité est mesurée en incubant un fragment du papier d'électrophorèse avec 0,5 ml de p-nitrophénylsulfate 0,0015 M dans une solution de phosphate 0,1 M à pH 8,5. L'activité enzymatique

est exprimée par la quantité en μg de p-nitrophénol formé, par mg d'azote protéique. Activité enzymatique des différentes fractions isolées: Cellules entières:
53; Poudre A: 370; Poudre B 1190; Fraction séparée par électrophorèse 1650.

2. Chondrosulfatases.

Ces enzymes qui libèrent le sulfate de l'acide chondroitine-sulfurique ont
été découvertes par NEUBERG et RUBIN (1914) dans les bactéries putréfiantes,
puis mises en évidence dans d'autres microorganismes: *Pseudomonas fluorescens
non liquefaciens, Proteus vulgaris, Bact. pyogenes* (NEUBERG et HOFMANN 1931a, b);
Penicillium spinulosum (PINCUS 1950) et dans la flore orale des animaux (PINCUS
1949).

Une préparation purifiée soluble de la chondrosulfatase a été obtenue par
DODGSON, LLOYD et SPENCER (1957) à partir de *Proteus vulgaris* traité par l'acétone. La fraction protéique finale est 30 fois plus active que le matériel de départ.
Dans les conditions optimales elle libère de l'acide chondroitine-sulfurique, 65 μg
de SO_4^{--} par heure et par mg d'enzyme.

3. Myrosulfatases.

La myrosulfatase a été découverte par NEUBERG et WAGNER (1926) dans les
graines de *Sinapis alba*. Le même type d'enzyme a été aussi mis en évidence dans
d'autres crucifères où ils existent à côté de leurs substrats qui sont les sulfates de
glucosides de sénévol: sinigrine, sinalbine, glycochéroline etc. La présence de
myrosulfatase dans les bactéries et dans les tissus animaux a été signalée par
NEUBERG et al. (1927, 1931a).

La plupart des préparations obtenues à partir des crucifères possèdent à côté
de l'activité sulfatasique l'activité thioglucosidasique. Ce point de vue, soutenu par
NEUBERG et VON SCHÖNEBECK (1933), ISHIMOTO et YAMASHINA (1949) a été discuté
par ETTLINGER et LUNDEEN (1957) et par REESE, CLAPP et MANDELS (1958) qui
ont considéré le détachement de l'ion sulfate comme consécutif au détachement du
glucose sous l'effet d'une β-thioglucosidase suivi d'un réarrangement de LOSSEN.
GAINES et GOERING (1960) ont applani ces divergences en montrant que l'on peut
isoler des graines de *Brassica juncea* une β-thioglucosidase et une sulfatase qui
catalysent la coupure respectivement des liaisons thioglucoside et ester sulfurique.

4. Glucosulfatases.

Une glucosulfatase a été mise en évidence par SODA et HATTORI (1931) d'abord
dans l'escargot commun du Japon, puis dans de nombreux mollusques marins
(SODA 1936; DODGSON et SPENCER, 1954) dans des bactéries (TANKO 1932) et
dans des champignons (YAMASHINA 1951). Cet enzyme hydrolyse le glucose-6-
sulfate. Des préparations enzymatiques solubles ont été obtenues par DODGSON
et SPENCER (1954) à partir de *Littorina littorea*. Les conditions optimales de
l'hydrolyse enzymatique du glucose-6-sulfate ont été examinées par les mêmes
auteurs.

5. Stéroïdesulfatases.

Cette sulfatase a été découverte par HENRY, THEVENET et JARRIGE (1952)
dans les sucs intestinaux des mollusques et étudiée chez ces derniers particulière-
ment par ROY (1956). Des préparations purifiées ont été obtenues par ce même
auteur qui en a déterminé l'affinité à l'égard de 15 sulfates de stéroïdes différents.

6. Cholinesulfatase

Une cholinesulfatase a été isolée de *Pseudomonas nitroreducens* par TAKEBE
(1961).

D. Enzymes du métabolisme du thiosulfate, du sulfite et du sulfure.

Ces trois composés minéraux du soufre sont des intermédiaires dans la transformation oxydative en sulfate du soufre organique réduit. On a démontré en outre le rôle du sulfure et du sulfite comme précurseurs immédiats des molécules soufrées organiques (SCHLOSSMANN et LYNEN, 1957; CHAPEVILLE et FROMAGEOT, 1957).

Jusqu'à présent on possède peu de données relatives à l'isolement et à la purification des enzymes qui interviennent dans le métabolisme de ces trois formes du soufre. En général, les réactions qu'ils catalysent ont été étudiées soit sur des organismes entiers, soit à l'aide de préparations peu purifiées.

I. Enzymes du métabolisme du thiosulfate.

1. Thiosulfate-réductase.

La réduction du thiosulfate par les bactéries a été démontrée par BEIJERINCK en 1895. CLARKE (1953) a examiné la formation de H_2S à partir du thiosulfate dans un grand nombre des bactéries appartenant à des groupes divers. Sur 235 souches examinées, 178 réduisent le thiosulfate. MITSUHASHI et MATSUO (1950) en utilisant le *Proteus vulgaris*, et ISHIMOTO, KOYAMA, et NAGAI (1955) des extraits acellulaires de bactéries sulfatoréductrices, ont montré que le thiosulfate est réduit de la façon suivante: $S_2O_3^{--} + H_2 \rightarrow SO_3^{--} + H_2S$. Suivant les organismes et les conditions, les donateurs d'hydrogène sont soit des composés organiques comme le lactate, le pyruvate, le formiate etc. soit l'hydrogène moléculaire. ISHIMOTO et al. (1957) et POSTGATE (1956a) ont montré que le cytochrome c_3 agit comme transporteur d'électrons au cours de la réduction. Les préparations enzymatiques purifiées par ISHIMOTO et KOYAMA (1957) sont dépourvues de l'hydrogènase, mais elles utilisent comme donateur d'hydrogène le méthyl viologène réduit.

Mesure de l'activité de la thiosulfate-réductase. ISHIMOTO et KOYAMA (1957). La réduction du thiosulfate est suivie par la mesure de l'oxydation du méthyl viologène réduit. La réaction s'effectue en anaérobiose dans des tubes de THUNBERG. Le mélange réactionnel contient: 0,5 ml du tampon phosphate 0,067 M à pH 7,0; 1,0 ml de la préparation enzymatique et 0,5 ml d'une solution 0,002 M de méthylviologène réduit dans le tampon phosphate 0,013 M à pH 7. La tubulure latérale contient 0,5 ml d'une solution de $Na_2S_2O_3$ 0,1 M. Lorsque l'anaérobiose est obtenue, on mélange les solutions et on suit à 30° la décoloration au photomètre.

Le méthylviologène réduit est préparé en chauffant sous atmosphère d'hydrogène, une solution de méthylviologène dans du tampon phosphate avec de la poudre de zinc (POSTGATE 1956b).

Préparation de la thiosulfate-réductase de Desulfovibrio desulfuricans. ISHIMOTO et KOYAMA (1957). Le microorganisme est cultivé en anaérobiose stricte dans le milieu liquide suivant: peptone: 5 g; lactate de sodium: 5 g; $MgSO_4$ 7 H_2O: 1,5 g; Na_2SO_4: 1,5 g; KH_2PO_4: 0,1 g; eau de robinet à pH 7: 1000 ml. Le milieu est stérilisé, ajusté à pH 7, ensemencé et incubé à 37° de 48 à 72 heures. Les cellules sont recueillies par centrifugation et lavées 2 fois dans une solution de NaCl à 0,8%. Un extrait aqueux est obtenu par broyage des bactéries en présence d'alumine et centrifugation à 4000 tours pendant 40 minutes. Le surnageant est centrifugé de nouveau à 16.000 tours pendant 20 minutes. Le précipité gris recueilli par cette dernière centrifugation est lavé avec du tampon phosphate

0,067 M à pH 7, puis mis en suspension dans le même tampon (P). Le surnageant (S) brun foncé est additionné de 1/20ème de son volume de $MnCl_2$ 2 M pour éliminer les acides nucléiques. Le précipité recueilli après 5 minutes de centrifugation à 1000 tours est extrait avec du tampon phosphate (P_{Mn}). Le surnageant (S_{Mn}) est traité par le $(NH_4)_2SO_4$. On recueille 2 fractions: la première qui précipite à 33% de saturation (P_{33}) et la seconde entre 33 et 67% (P_{67}). Elles sont dissoutes dans le tampon phosphate 0,067 M à pH 7. Rapportée à la même quantité d'azote, la quantité en μg de méthyl viologène oxydé en 1 heure à 30° par les diverses fractions est la suivante: Extrait initial: 60; S: 74; P: 11; S_{Mn}: 74; S_{33}: 110; P_{33}:32; S_{67}: 5; P_{67}: 140.

2. Rhodanese.

Le thiosulfate réagit avec l'acide cyanhydrique de la façon suivante: $Na_2S_2O_3 + HCN \rightarrow HSCN + Na_2SO_3$. Cette réaction est catalysée par un enzyme, la rhodanese, mise en évidence dans différents tissus des animaux (Lang 1933 a, b). Sörbo (1955) a cristallisé la rhodanese du foie de boeuf. Dans le règne végétal cet enzyme a été peu étudié; McChesney (1958) l'a détecté dans les cultures de certains thiobacilles.

II. Enzymes du métabolisme du sulfite.

Le sulfite formé chez les êtres vivants peut avoir trois origines:
1) Rupture de la liaison C—S d'une molécule organique sulfinée.
2) Réduction du sulfate et du thiosulfate
3) Oxydation du sulfure.
Il a été démontré que les voies métaboliques possibles de son utilisation sont les suivantes:
1) Réduction en sulfure
2) Oxydation en sulfate
3) Combinaison avec une molécule organique et formation d'une liaison C—S.

1. Formation du sulfite par désulfination.

La formation de sulfite à partir de l'acide cystéinesulfinique a été mise en évidence d'abord dans les tissus des animaux (Fromageot, Chatagner et Bergeret 1948), puis dans les bactéries (Kerney et Singer 1952) et dans les végétaux (Perez-Milan, Schliack et Fromageot 1959). Cette dégradation correspond aux réactions suivantes:
1) Acide cystéinesulfinique + acide α-cétonique $\rightarrow$ acide β-sulfinylpyruvique + acide aminé
2) Acide β-sulfinylpyruvique + H_2O $\rightarrow$ sulfite + acide pyruvique.
L'acide α-cétonique impliqué peut être l'acide α-cétoglutarique, l'acide oxaloacétique, l'acide pyruvique.
Seule la première réaction (transamination) est enzymatique. L'acide β-sulfinyl-pyruvique par suite de son instabilité n'a jamais pu être isolé; il se décompose rapidement en acide pyruvique et en sulfite. Ellis et Davies (1961) ont montré que l'enzyme catalysant la transamination entre l'acide cystéinesulfinique et les acides α-cétoglutarique ou oxaloacétique est l'acide aspartique-glutamique-transaminase.

Mesure de l'activité enzymatique. Le sulfite formé à partir de l'acide cystéine-sulfinique est dosé à l'état de sulfate après oxydation ou directement dans le milieu réactionnel par la méthode polarographique (Fromageot et Patino-Bun 1961). Cette dernière méthode permet des mesures continues.

Dans une cellule pour polarographie maintenue à 37° on introduit les solutions suivantes préalablement ajustées à pH 7,8: acide α-cétoglutarique: 20 μmoles, 0,5 ml; tampon phosphate 0,06 M, 5 ml; préparation enzymatique. On chasse l'air pendant 10 mn par un courant d'azote puis on maintient la cellule sous azote. On introduit alors l'acide cystéinesulfinique: 20 μmoles, 0,5 ml. Volume final 8 ml. Dans la cellule plonge l'électrode à goutte de mercure et une électrode à calonel-KCl saturé. L'ensemble est connecté à un polarographe. L'électrode de mercure est rendue positive et on lui applique une tension de $+ 0,1$ volt. Avant chaque expérience, on étalonne l'appareil par addition de quantités connues de sulfite au mélange réactionnel complet mais privé d'acide cystéinesulfinique. La courbe de réponse reliant la déflection du galvanomètre à la quantité de sulfite présente est linéaire jusqu'à 10—12 μmoles de sulfite. A des concentrations supérieures à $1 \cdot 10^{-4}$ M la cystéine gêne les mesures.

2. Sulfite-réductase.

La réduction du sulfite en sulfure a été démontrée chez un grand nombre de microorganismes. BUTLIN et POSTGATE (1956) ont recueilli les principales données bibliographiques sur cette question.

Mesure de l'activité sulfito-réductrice. ISHIMOTO et YAGI (1961). *Méthode manométrique:* On mesure la consommation d'hydrogène en présence d'hydrogénase, de méthylviologène comme transporteur d'électrons, de sulfite de sodium et de la préparation enzymatique. On prend: tampon phosphate pH 6,2: 75 μmoles; Na_2SO_3: 4 μmoles; méthylviologène: 1 μmole; hydrogénase: 10 mg de protéine, et la préparation de sulfite-réductase. Le volume final est 1 ml. L'atmosphère est de l'hydrogène pur. Température 30°. Les solutions de sulfite et de méthylviologène sont placées dans des tubulures latérales. On ne considère que la période pendant laquelle la consommation d'hydrogène croit linéairement en fonction du temps. L'unité enzymatique est la quantité d'enzyme catalysant l'absorption de 1 μmole d'hydrogène par heure. La vitesse est proportionnelle à la quantité d'enzyme dans l'intervalle 1—15 unités.

Méthode photométrique: On mesure le temps de décoloration du méthylviologène réduit en présence de sulfite dans des tubes de Thunberg. Le milieu comporte: tampon phosphate pH 6,0: 100 μmoles; Na_2SO_3: 10 μmoles; méthylviologène réduit: 4 μmoles et la préparation enzymatique. Volume final: 5 ml. L'atmosphère est constituée de CO à la pression de 0,5 atmosphère. Température 30°. L'unité de sulfite-réductase est la quantité d'enzyme qui oxyde 2 μmoles de méthylviologène réduit en une heure. Avec des préparations partiellement purifiées, la vitesse d'oxydation du méthylviologène est inversement proportionnelle à la quantité d'enzyme dans l'intervalle 1—10 unités. La présence de CO est destinée à inhiber l'hydrogénase qui peut être présente et qui catalyse, à pH 6, la réoxydation du méthylviologène réduit. A pH 7, cet effet est très faible, ce qui explique qu'il ne soit pas nécessaire d'introduire du CO dans les tubes destinés à la mesure de la thiosulfate-réductase.

Préparation du système réducteur du sulfite à partir de Desulfovibrio desulfuricans (ISHIMOTO et YAGI 1961). Les bactéries sont cultivées dans les mêmes conditions que celles utilisées pour la préparation de la thiosulfate-réductase. Les cellules, 33 g dans 50 ml d'eau, sont broyées dans un appareil à ultrasons pendant 10 mn, sous 560 Kc. Les débris cellulaires sont éliminés par une centrifugation de 20 mn à 6000 g. On centrifuge le surnageant pendant 20 mn à 18.000 g. Le culot contient l'hydrogénase. Au surnageant, on ajoute une solution saturée de sulfate d'ammonium et on recueille le précipité formé entre 33% et

67% de la saturation. Ce précipité est privé de sulfate d'ammonium par dialyse contre de l'eau. On centrifuge. A 50,5 ml du surnageant on ajoute trois fois du gel de phosphate de calcium à raison de 0,6 g par g de protéine, agite, attend 15 mn et centrifuge. Au surnageant on ajoute 0,51 g de gel de phosphate de calcium. Ce gel est recueilli et élué par du tampon phosphate pH 7,0 successivement 0,013 M, 0,067 M et 0,25 M. L'éluat qui correspond au tampon phosphate 0,067 M contient la plus grande partie de la sulfite-réductase qui peut être concentrée par précipitation par le sulfate d'ammonium entre 30% et 70% de la saturation. La thiosulfate-réductase n'est pas adsorbée sur le gel dans ces conditions. L'hydrogénase se trouve en partie dans la fraction précipitée par centrifugation à 18.000 g, en partie dans la préparation de sulfite-réductase.

3. Sulfite-oxydase.

L'oxydation du sulfite en sulfate a été étudiée par Handler et Fridovich (1956) avec des préparations de foie de mammifères, par Tager et Rautanen (1956) et par Fromageot, Vaillant et Perez-Milan (1960) avec des préparations de feuilles et de racines d'avoine. Les enzymes responsables de l'oxydation du sulfite chez les végétaux n'ont pas été purifiés : la principale difficulté de la mesure de l'oxydation enzymatique du sulfite réside dans l'obtention de préparations dépourvues d'ions métalliques libres. Ces derniers, en effet, catalysent l'oxydation rapide du sulfite. Les métaux peuvent être complexés par l'addition de polyalcools, ou de certaines protéines pures, telles que la sérumalbumine.

III. Enzymes du métabolisme du sulfure.

Le sulfure formé par les être vivants provient soit de la réduction du sulfate ou autres formes du soufre minéral moins oxydé, soit de la rupture de la liaison C—S des composés soufrés organiques réduits tels que la cystéine ou l'homocystéine (Fromageot et Desnuelle 1942). Ainsi formé il est soit oxydé soit utilisé pour la synthèse des molécules organiques du type R—SH. On examinera ici deux types de systèmes enzymatiques : ceux qui permettent la rupture de la liaison C—S et ceux qui, par contre, catalysent la formation d'une telle liaison.

Les systèmes enzymatiques qui permettent la formation de H_2S à partir d'un thiol organique sont désignés sous le nom de désulfhydrases bien que cette formation ne corresponde pas toujours à une libération directe de H_2S mais puisse résulter de plusieurs réactions successives. Cette question a fait l'objet d'une revue (Hanson 1956).

1. Formation d'hydrogène sulfuré à partir de la cystéine.

Trois types de réactions ont été mis en évidence soit chez les bactéries, soit chez les animaux :

1) Transamination entre la cystéine et un acide α-cétonique avec formation de l'acide β-mercaptopyruvique et désulfuration de ce dernier.

2) Désamination oxydative de la cystéine avec formation de NH_3 et de l'acide β mercaptopyruvique (Tamiya 1954).

3) Désulfhydration de la cystéine avec formation de l'acide α-aminoacrylique qui, spontanément, se décompose en acide pyruvique et NH_3, ou avec formation de sérine.

Les réactions 1 et 2 sont catalysées par deux enzymes différents : une transaminase ou une désaminase de la cystéine et une désulfurase de l'acide β-mercaptopyruvique. En présence de la cystéine, le soufre libéré est réduit non enzymatiquement en H_2S. La réaction 3 correspond à la désulfhydration directe de la cystéine et elle est catalysée par un seul enzyme.

La transamination de la cystéine avec l'acide α-cétoglutarique a été démontrée par CHATAGNER et SAURET-IGNAZI (1956) et la désulfuration de l'acide β-mercapto-pyruvique a été étudiée par HANSON et MANTEL (1953), MEISTER, FRASER et TICE (1954) avec des préparations bactériennes et par KUN et FANSHIER (1959) avec des préparations purifiées d'organes animaux.

La désulfhydration directe de la cystéine (réaction 3) a été étudiée par DESNUELLE et FROMAGEOT (1939), par SMYTHE (1945), par CHAPEVILLE et FROMAGEOT (1958) et par BRÜGGEMANN et WALDSCHMIDT (1962). CHAPEVILLE et FROMAGEOT (1958) ont obtenu à partir du sac vitellin de l'embryon d'oiseau, une préparation purifiée qui catalyse soit l'échange entre le soufre du groupe thiol de la L-cystéine et celui du sulfure minéral, soit la substitution de ce groupe thiol par un groupe sulfonate formé à partir du sulfite. Une autre désulfhydrase a été isolée du foie de chat par MADLO (1960).

Des préparations brutes qui dégradent la L-cystéine en H_2S, acide pyruvique et NH_3 ont été obtenues à partir de différentes bactéries par KALLIO et PORTER (1950), KALLIO (1951), DELWICHE (1951), OHIGASHI et al. (1952), METAXAS et DELWICHE (1955) et SUDA et al. (1954). Puis SAZ et BROWNELL (1954) ont obtenu à partir d'un mutant d'*Escherichia coli* un extrait acellulaire actif vis-à-vis de la D-cystéine seulement. La plupart des enzymes étudiés sont actifs vis-à-vis de la cystéine et ne transforment pas les dérivés de substitution de cet amino-acide. Cependant GMELIN, HASENMAIER et STRAUSS (1957) ont montré que dans les graines d'*Albizzia lophanta* (mimosacées) il existe un enzyme, une C—S-lyase qui attaque non seulement la L-cystéine mais aussi ses dérivés S-substitués et les sulfoxydes correspondants. Cet enzyme a été obtenu sous une forme purifiée par SCHWIMMER et KJAER (1960).

Détermination de l'activité enzymatique de la C-S-lyase des graines d'Albizzia lophanta. HANSEN, KJAER et SCHWIMMER (1959). La S-(2.4.dinitrophényl)cystéine est scindée sous l'influence de l'enzyme en 2.4.dinitrothiophénol, acide pyruvique et ammoniaque. A pH 4 le substrat a un maximum d'absorption à 337 mμ, et le produit un maximum à 400 mμ. On met à profit cette différence pour mesurer, par la variation de la densité optique à 400 mμ l'activité de la préparation enzymatique. Avec la S-(2.4.dinitrophényl)cystéine obtenue selon ZAHN et TRAUMANN (1954), on prépare une solution stock du substrat par dissolution de 28,7 g de celui-ci dans 20 ml d'éthanol absolu. On ajoute 1 ml de HCl N et on dilue le tout à 50 ml avec de l'eau distillée. Une telle solution est stable au moins 2 mois à la température ambiante. On prend 0,3 ml de cette solution stock, 1,5 ml de tampon phosphate 0,2 M, pH 6,5 et une solution de l'enzyme. Le volume final est de 3 ml. Une variation de la densité optique à 400 mμ de 0,054 par minute à 23—25° correspond à une unité d'activité enzymatique. Il est important que le pH du milieu réactionnel soit de 6,5. A des pH supérieurs, on assiste, même en l'absence d'enzyme, à une variation de la densité optique à 400 mμ, par suite probablement d'une transarylation conduisant de la S-(2.4.dinitrophényl)cystéine à la N-(2.4.dinitrophényl)cystéine. On remarquera que le pH 6,5 est éloigné du pH d'activité optimale de l'enzyme qui est de 8,5.

Purification de la C-S lyase. On concasse 200 g de graines d'*Albizzia lophanta* dans un broyeur à cylindres (type moulin à café). Par flottation sur du trichloréthylène, on sépare l'endosperme de la cuticule plus dense. On élimine le trichloréthylène par séchage à l'air. On broie l'endosperme et on l'extrait par de l'eau (360 ml) par agitation pendant 1 heure à la température ordinaire. Il est bon de procéder à cette opération sous une hotte. La suspension obtenue est filtrée sur une gaze. On obtient 290 ml d'un extrait laiteux (fraction A). Celui-ci est additionné de 6 ml de HCl N, ce qui amène le pH à 5,5. On centrifuge pendant 30 minutes à 2.100 g. Le surnageant (210 ml) constitue la fraction B. Il est neutralisé par addition de NaOH N et chauffé à 65 ± 3° pendant 10 minutes. Après refroidissement et élimination du précipité par centrifugation à 2.100 g on obtient la fraction C (200 ml). A celle-ci on ajoute 60 g de sulfate d'ammonium. Le précipité recueilli par centrifugation est dissous dans la quantité suffisante d'eau pour obtenir 25 ml (fraction D). La fraction D est précipitée à nouveau par le sulfate d'ammonium à 42% de la saturation, par addition d'une solution saturée de ce sel à pH 6,54. Le précipité est dissous dans de l'eau de façon à obtenir un volume total de 15 ml (fraction E). Par dialyse de cette solution à 5° C contre de l'eau distillée pendant 48 heures, on recueille un précipité qui est lyophiliseé. Poids 150 mg. C'est la fraction F.

On parvient à une purification plus poussée en traitant la fraction D par de l'alumine pour chromatographie (Sigma). Dans ce but on ajoute à 25 ml de la fraction D 1,6 ml d'alumine. Après agitation on recueille l'alumine, on la lave par 10 ml d'eau distillée. On élue l'alumine par trois fois 10 ml de solutions de sulfate d'ammonium respectivement à 5,0, 9,3 et 13,8%. Les deux derniers éluats constituent la fraction G.

L'activité enzymatique totale ainsi que l'activité exprimée en unités par mg d'azote protéique sont données dans le tableau 4.

Tableau 4. *Purification de la C—S-lyase d'Albizzia lophanta* (Schwimmer et Kjaer 1960).

Fraction	Volume ml	Activité enzymatique	
		Unités totales	Unités par mg N
A	290	208.000	3
B	210	70.100	12
C	200	66.300	19
D	25	47.200	79
E	15	43.900	145
F	—	35.000	210
G	20	5.000	680

2. Sérine-sulfhydrase.

Cet enzyme, découvert par Schlossmann et Lynen (1957) catalyse la fixation du soufre du sulfure sur le carbone β de la L-sérine avec formation de la L-cystéine suivant la réaction:

$$HOCH_2—CH(NH_2)—COOH + H_2S \rightleftarrows HS—CH_2—CH(NH_2)—COOH + H_2O$$

L'enzyme a été mis en évidence dans la levure de boulangerie, dans *E. Coli*, dans la feuille d'épinard et chez d'autres microorganismes utilisant le soufre minéral pour la synthèse des acides aminés soufrés (Brüggemann et al. 1962). Il possède le phosphate de pyridoxal comme coenzyme. Une préparation 50 fois plus active que l'extrait brut de la levure a été obtenue. Dans ce but, on désintègre les cellules de la levure par agitation avec des billes de verre et traite l'extrait acellulaire avec une solution de protamine à 2%, puis précipite la protéine active par le sulfate d'ammonium entre 40 et 65% de saturation. L'adsorption ultérieure sur le gel d'alumine à pH 5,7 et l'élution avec le tampon Tris 0,10 M à pH 8,0 permettent d'obtenir la fraction

Tableau 5. *Spécificité de la sérine-sulfhydrase.* (Schlossmann et Lynen 1957).

	Réactifs ajoutés			Substances trouvées	
	L-Sérine μmoles	D-Sérine μmoles	Na$_2$S μmoles	cystéine μmoles	Sérine μmoles
a)	2,6	—	17,2	2,1	—
	1,3	1,3	17,2	1,1	—
	1,3	—	17,2	1,2	—
b)	10,5	—	17,2	6,8	3,0
	5,25	5,25	17,2	3,2	6,7
	10,5	—	—	—	10,0

Le mélange réactionnel contient: tampon Tris 0,15 M à pH 8; EDTA 4 μmoles; phosphate de pyridoxal 25 μg; préparation enzymatique 0,25 mg dans les essais a) et 1,0 mg dans les essais b)

active qui se conserve pendant plusieurs mois à − 15° C mais n'est active qu'en présence du phosphate de pyridoxal. Une préparation 2000 fois plus active que l'extrait brut à été obtenue de la levure par LORCH (1960).

Le tableau 5 montre la spécificité de l'enzyme pour la L-sérine et indique les conditions expérimentales dans lesquelles sont effectuées les incubations.

3. Oxydation du sulfure.

L'oxydation du sulfure a été étudiée in vivo (FROMAGEOT et ROYER 1943) et in vitro (BAXTER, VAN REEN, PEARSON et ROSENBERG 1958), avec des homogenats de tissus animaux. Il est vraisemblable que le sulfite intervient comme intermédiaire au cours de cette oxydation. Il en est probablement de même chez les bactéries autotrophes où l'oxydation du sulfure est couplée avec la réduction du gaz carbonique (LEES 1955).

Bibliographie.

ABBOT, L. D.: Arch. Biochem. Biophys. 15, 205 (1947). — D'ABRAMO, F., and F. LIPMANN: Biochim. biophys. Acta 25, 211 (1957). — ASAHI, T., R. S. BANDURSKI and L. G. WILSON: J. biol. Chem. 236, 1830 (1961).

BANDURSKI, R. S., L. G. WILSON and C. L. SQUIRES: J. Am. chem. Soc. 78, 6408 (1956). — BARBER, M. B. W. L., BROOKSBANK et S. W. A. KUPER: J. Path. Bact. 63, 57 (1951). — BAXTER, C. F., R. VAN REEN, P. B. PEARSON and C. ROSENBERG: Biochim. biophys. Acta 27, 584 (1958). — BEIJERINCK, M. W.: Zbl. Bakt. (2 Abt.) 1, 1 (1895). — BERNSTEIN, S., and R. W. McGILVERY: J. biol. Chem. 198, 195 (1952). — BRÜGGEMANN, J., K. SCHLOSSMANN, M. MERKENSCHLAGER and M. WALDSCHMIDT: Biochem. Z. 335, 392 (1962). — BRÜGGEMANN, J., and M. WALDSCHMIDT: Biochem. Z. 335, 408 (1962). — BUTLIN, K. R., and J. POSTGATE: Colloque sur la Biochimie du Soufre, Roscoff C.N.R.S. Paris (1956).

CHAPEVILLE, F., and P. FROMAGEOT: Biochim. biophys. Acta 26, 538 (1957); — Bull. Soc. Chim. biol. (Paris) 40, 1965 (1958). — CHATAGNER, F., and G. SAURET-IGNAZI: Bull. Soc. Chim. biol. (Paris) 38, 415 (1956). — CLARKE, P. H.: J. gen. Microbiol. 8, 397 (1953).

DELWICHE, E. A.: J. Bact. 62, 717 (1951). — DE MEIO, R. H., M. WIZERKANIUK and E. FABIANI: J. biol. Chem. 203, 257 (1953). — DESNUELLE, P., and CL. FROMAGEOT: Enzymologia 6, 80 (1939). — DODGSON, K. S., A. G. LLOYD and B. SPENCER: Biochem. J. 65, 131 (1957). — DODGSON, K. S., T. H. MELVILLE, B. SPENCER and K. WILLIAMS: Biochem. J. 58, 182 (1954). — DODGSON, K. S., and B. SPENCER: Biochem. J. 53, 444 (1953a); 55, 436 (1953b); 57, 310 (1954). — DODGSON, K. S., B. SPENCER and K. WILLIAMS: Biochem. J. 61, 374 (1955); 64, 216 (1956).

ELLIS, R. J., and D. D. DAVIES: Biochem. J. 78, 615 (1961). — ETTLINGER, M. G., and A. J. LUNDEEN: J. Amer. chem. Soc. 79, 1764 (1957).

FISKE, C. H., and Y. SUBBAROW: J. biol. Chem. 66, 375 (1925). — FROMAGEOT, CL.: Methods in enzymology. COLOWICK, S. P., and N. O. KAPLAN, Vol. II p. 324. New York: Academic Press 1955. — FROMAGEOT, CL., and P. DESNUELLE: Bull. Soc. Chim. biol. (Paris) 24, 1269 (1942). — FROMAGEOT, P., and F. CHAPEVILLE: IVth Intern. Congress. Biochem.13, 69 (1958). — FROMAGEOT, CL., F. CHATAGNER and B. BERGERET: Biochim. biophys. Acta 2, 294 (1948). — FROMAGEOT, CL., et A. ROYER: Enzymologia 11, 361 (1943/45). — FROMAGEOT, P., and U. PATINO-BUN: Biochim. biophys. Acta 46, 533 (1961). — FROMAGEOT, P., and H. PEREZ-MILAN: Biochim. biophys. Acta 32, 457 (1959). — FROMAGEOT, P., R. VAILLANT and H. PEREZ-MILAN: Biochim. biophys. Acta 44, 77 (1960).

GAINES, R. D., and K. J. GOERING: Biochem. Biophys. Res. Comm. 2, 207 (1960). — GMELIN, R. G., G. HASENMAIER and G. STRAUSS: Z. Naturforsch. 12b, 687 (1957). — GRANT, W. M.: Analyt. Chem. 19, 345 (1947). — GREGORY, J. D., and F. LIPMANN: J. biol. Chem. 229, 1081 (1957).

HANDLER, P., et I. FRIDOVICH: Colloque sur la Biochimie du Soufre-Roscoff, C.N.R.S. p. 83 (1956). — HANSEN, S. E., A. KJAER and S. SCHWIMMER: C. R. Lab. Carlsberg 31, 193 (1959). — HANSON, H.: Colloque sur la Biochimie du Soufre-Roscoff, C.N.R.S. Paris (1956). — HANSON, H., and E. MANTEL: Hoppe-Seylers Z. physiol. Chem. 295, 141 (1953). — HARADA, T., and B. SPENCER: J. gen. Microbiol. 22, 520 (1960). — HENRY, R., M. THEVENET and P. JARRIGE: Bull. Soc. Chim. biol. (Paris) 34, 897 (1952). — HILZ, H.: Thèse. Hambourg (1960). — HILZ, H., and M. KITTLER: Bioch. biophys. Acta 30, 650 (1958). — HILZ, H., M. KITTLER and G. KNAPE: Biochem. Z. 332, 151 (1959). — HILZ, H., and F. LIPMANN: Proc. nat. Acad. Sci. (Wash.) 41, 880 (1955).

Ishimoto, M.: J. Biochem. (Tokyo) 46, 105 (1959). — Ishimoto, M., and D. Fujimoto: Proc. Japan Acad. 35, 243 (1959). — Ishimoto, M., and D. Fujimoto: J. Biochem. (Tokyo) 50, 299 (1961). — Ishimoto, M., and J. Koyama: J. Biochem. (Tokyo) 44, 233 (1957). — Ishimoto, M., J. Koyama and Y. Nagai: J. Biochem. (Tokyo) 42, 41 (1955). — Ishimoto, M., J. Koyama, T. Yagi and M. Shiraki: J. Biochem. (Tokyo) 44, 413 (1957). — Ishimoto, M., and T. Yagi: J. Biochem. (Tokyo) 49, 103 (1961). — Ishimoto, M., et I. Yamashina: Symposia on Enzyme Chem. (Japan) 2, 36 (1949).

Kallio, R. E.: J. biol. Chem. 192, 371 (1951). — Kallio, R. E., and J. R. Porter: J. Bact. 60, 607 (1950). — Korn, E. D.: J. biol. Chem. 234, 1647 (1959). — Kearney, E. B., and T. P. Singer: Biochim. biophys. Acta 8, 698 (1952). — Kun, E., and D. A. Fanshier: Biochim. biophys. Acta 32, 338 (1959).

Lang, K.: Biochem. Z. 259, 243 (1933a); 263, 262 (1933b). — Lees, H.: Biochemistry of autotrophic bacteria. Butterworths Scientific Publications, London (1955). — Lorch, E.: Dissertation. Munich (1960).

Madlo, Z.: Coll. Czechoslov. Chem. Comm. 25, 729 (1960). — McChesney, C. A.: Nature (Lond.) 181, 347 (1958). — Meister, A , P. E. Fraser and S. V. Tice: J. biol. Chem. 206, 561 (1954). — Metaxas, M. A., and E. A. Delwiche: J. Bact. 70, 735 (1955). — Millet, J.: Compt. rend. 238, 408 (1954); — Mitsuhashi, S., and Y. Matsuo: Jap. J. exp. Med. 20, 729 (1950).

Neuberg, C., and E. Hofmann: Biochem. Z. 234, 345 (1931a); — Naturwissenschaften 19, 484 (1931b). — Neuberg, C., and D. Rubin: Biochem. Z. 67, 82 (1914). — Neuberg, C., and O. von Schönebeck: Biochem. Z. 265, 223 (1933). — Neuberg, C., and J. Wagner: Biochem. Z. 174, 457 (1926); — Z. exp. Med. (Moskau) 56, 334 (1927). — Nose, Y., and F. Lipmann: J. biol. Chem. 233, 1348 (1958).

Ohigashi, K., A. Tsunetoshi, M. Uchida and K. Ichihara: J. Biochem. (Tokyo) 39, 211 (1952).

Peck, H. D.: J. biol. Chem. 235, 2734 (1960). — Perez-Milan, H., J. Schliack and P. Fromageot: Biochim. biophys. Acta 36, 73 (1959). — Pincus, P.: Brit. Dent. J. 86, 226 (1949); — Nature (Lond.) 66, 187 (1950). — Postgate, J.: J. gen. Microbiol. 14, 545 (1956a); 15, 186 (1956b).

Ragland, J. B.: Arch. Biochem. Biophys. 84, 541 (1959). — Reese, E. T., R. C. Clapp and M. Mandels: Arch. Bioch. Biophys. 75, 228 (1958). — Robbins, P. W., and F. Lipmann: J. Amer. chem. Soc. 78, 6409 (1956); — J. biol. Chem. 229, 837 (1957); 233, 681 (1958a); 233, 686 (1958b). — Roy, A. B.: Biochem. J. 53, 12 (1953); 62, 41 (1956).

Saz, A. K., and L. W. Brownell: Arch. Biochem. Biophys. 52, 291 (1954). — Schloss-mann, K., and F. Lynen: Biochem. Z. 328, 591 (1957). — Schwimmer, S., and A. Kjaer: Biochim. biophys. Acta 42, 316 (1960). — Smythe, C. V.: Advanc. Enzymol. 5, 235 (1945). — Soda, T.: J. Fac. Sci. Tokyo Univ. 3, 150 (1936). — Soda, T., and C. Hattori: Bull. chem. Soc. Japan 6, 258 (1931). — Sörbo, B.: In Methods in enzymology. Colowick, S. P., and N. O. Kaplan. Vol. II p. 334 New York, Acad. Press 1955. — Spencer, B.: Biochem. J. 75, 435 (1960). — Spencer, B., and T. Harada: Biochem. J. 73, 34 (1959). — Suda, M., T. Saigo and K. Ichihara: Med. J. Osaka Univ. 5, 127 (1954).

Tager, J. M., and N. Rautanen: Biochim. biophys. Acta 18, 111 (1956). — Takebe, I.: J. Biochem. (Tokyo) 50, 245 (1961). — Tamiya, N.: J. Biochem. (Tokyó) 41, 199, 287 (1954). — Tanko, B.: Biochem. Z. 247, 486 (1932).

Wilson, L. G., T. Asahi and R. S. Bandurski: J. biol. Chem. 236, 1822 (1961). — Wilson, L. G., and R. S. Bandurski: J. Amer. chem. Soc. 80, 5576 (1958a); — J. biol. Chem. 233, 975 (1958b).

Yamashina, I.: J. chem. Soc. Japan 72, 124 (1951). — Young, L., A. R. Morrison and J. E. M. Whitehead: Nature (Lond.) 169, 711 (1952).

Zahn, H., and K. Traumann: Z. Naturforsch. 9b, 518 (1954).

Enzymes of Phosphate Metabolism.

By

P. S. Krishnan.

I. Special Features of Enzyme Isolation and Purification.

1. Choice of Starting Material.

Plant tissues vary considerably in their content of protein and it may, therefore, be anticipated that some tissues, in, reference to others, may lend themselves
more readily to enzyme isolation and purification. The leaves of the tapioca plant
contain 20—36% of crude protein (ROGERS, 1959), whereas the phylloclades of
cactus have only about 7%, both on a dry weight basis (SANWAL, 1960). The
nature of the nonprotein material of the starting material also influences the
course of enzyme purification. The presence of mucilaginous material in tissues
such as cactus interferes with conventional techniques of enzyme enrichment.
Whereas, in general, the investigator is concerned with the enzyme-make up of a
given tissue and preliminary screening does not find much scope, it is sometimes
possible to choose a special tissue where the given enzyme is present in optimum
concentration. In the animal tissue, high alkaline phosphatase activity seems to be
associated with intestinal mucosa, acid phosphatase with kidney and spleen and,
even more so, with prostate and 5'-nucelotidase activity with testis. Germinating
seeds constitute a good sources of acid phosphatases (NEWMARK and WENGER,
1960). Plant tissue and microorganisms are, in general, poor sources of alkaline
phosphatase, but GAREN and LEVINTHAL (1960) showed that when *Escherichia
coli* was grown in a medium containing limiting amounts of orthophosphate, as
much as 6% of the total protein synthesized was alkaline phosphatase. The
availability of such a concentrated source facilitated the isolation of the phosphatase and study of its properties. The diurnal variation observed in some
enzymes (SANWAL and KRISHNAN, 1960) may have considerable significance from
the point of view of enzyme isolation; in order to have a source material with
maximum enzyme activity, the harvesting may have to be carried out at a definite
hour of the day or night.

2. Conventional versus Modern Methods of Enzyme Purification.

Conventional methods, such as ammonium sulfate fractionation, acetone precipitation,
adsorption on tricalcium phosphate and aluminum hydroxide gels, were introduced during
the early phase of research in enzyme isolation. These techniques still find wide application, but
are being increasingly supplemented by modern techniques such as chromatography and
electrophoresis. By combining ammonium sulfate precipitation with chromatography on
DEAE cellulose and hydroxy apatite and zone electrophoresis on starch column, ROGERS and
REITHEL (1960) succeeded in demonstrating that the acid phosphatase of *Escherichia coli*
can be separated into five different activities. TOMLINSON and WARREN (1960) showed that the
acid phosphatase activity of ling cod muscle could be separated into five fractions, each
possessing acid phosphomonoesterase activity, by initial precipitation of extracts with
ammonium sulfate, followed by stepwise chromatography on DEAE cellulose in the basic
⁻m.

a) Automatic Device for Ammonium Sulfate Addition.

The most important single step in enzyme purification is, probably, precipitation with ammonium sulfate. RACKER and SCHROEDER (1958) described an automatic device for the slow addition of solid ammonium sulfate. Nomograms representing the amount of salt to be added to obtain a desired concentration are available (GREEN and HUGHES, 1955).

b) Dialysis of Enzyme Solutions.

The use of Visking tubes for dialysis seems to be a standard practice in many laboratories. KLOTZ and HUGHES (1956) suggested that these tubes be previously cleaned by boiling with water and soaking. To ensure thorough mixing of the contents during dialysis glass beads may be enclosed in the dialysis sacs. Dialysis has been sought to be expedited by stirring of the contents of the bag, or the fluid outside the bag, or both. OGSTON (1960) investigated the effect of stirring on the rate of dialysis of sodium chloride and acetic acid from Visking tubes. Under no condition did stirring increase the rate of dialysis more that 2-fold. He concluded that dialysis of salt is a rapid process and stirring has less effect than appears to be usually thought.

3. Stability of Enzymes on Dilution.

Quite often enzyme solutions suffer diminution in activity on dilution with water or buffers, especially after a certain stage of purification has been achieved. KUNITZ (1952) made use of gelatin solution for dilution in order to preserve the activity of yeast pyrophosphatase. JEFFREE (1955, 1956) observed that the loss in activity of diluted seminal plasma was prevented by the addition of gelatin or of long chain amines. REINER, TSUBOI and HUDSON (1955) and TSUBOI and HUDSON (1956) found that losses in activity of prostatic, erythrocyte and yeast phosphatases consequent on dilution or surface denaturation could be prevented by the presence of a surface active agent in the reagent solution. NEWMARK and WENGER (1960) demonstrated that incorporation of Triton X-100 in the diluting medium served to fortify the enzyme and that the effect of Triton was supplemented to a small degree by the addition of EDTA.

II. Assay of Phosphatase Activity.

The simplest technique consists in the colorimetric estimation of the orthophosphate liberated as the result of the enzyme action. Methods for estimating orthophosphate are discussed below. Quite often it is possible to have independent estimation of the organic (or inorganic) component entering into esterification with phosphate. When phenyl phosphate is used as substrate, the phenol formed can be estimated coveniently by the method of HOCKENHULL, ASHTON, FANTES and WHITEHEAD (1954) or KIND and KING (1954), or according to FISHMAN and LERNER (1953). The use of para nitrophenyl phosphate as substrate permits easy estimation of para nitrophenol by noting the increase in optical density at 400 mμ in an alkaline medium (BESSEY, LOWRY and BROCK, 1946; TORRIANI, 1960). Salicylic acid formed from orthocarboxyphenylphosphate can be estimated by the method of HOFSTEE (1954). Phenolphthalein liberated from phenolphthalein phosphate may be estimated according to SUOMALAINEN, LINKO and OURA (1960). NIGAM, DAVIDSON and FISHMAN (1959) have described the assay of glycerophosphatase activity by estimating the glycerol liberated. Investigating the kinetics of the hydrolysis of orthophosphate monoesters of phenol, para nitrophenol and glycerol by prostatic acid phosphatase, these authors were able to show that there was equimolar liberation of inorganic orthophosphate and the organic component. The ammonia split off from phosphoramidate may be assayed by a suitable method such as CONWAY's microdiffusion technique (SINGER and FRUTON, 1957). 5′-nucleotidase activity can be estimated by determination of adenosine formed, by the decrease in absorbency at 265 mμ in the presence of adenosine desaminase (SEGAL and BRENNER, 1960). Acting on polynucleotides, the phosphatase activity could be determined by carrying out absorption measurements on the perchloric acid-soluble fraction (ANDERSON and HEPPEL, 1960). Manometric estimation of nucleoside pyrophosphatases is possible, by measuring the CO_2 evolved when the nucleotide is broken down in a bicarbonate medium (CLAYTON and HANSELMAN, 1960). DPN or TPN-pyrophosphatase can be estimated by allowing the enzyme to act on the coenzyme and determining the residual material spectrophotometrically after reduction with a specific dehydrogenase (CLAYTON and HANSELMAN, 1960).

Working with crude preparations, it is often found that the activity of a specific enzyme is interfered with by unspecific phosphatase action. A suitable correction can be applied in many cases. When adenylic acid is used as substrate the amount of orthophosphate liberated is due both to specific 5′-nucleotidase activity and to unspecific phosphatase activity. The nonspecific activity could be determined separately using phenyl phosphate as substrate or by making use of the ability of Ni in millimolar concentration to inhibit 5′-nucleotidase activity completely without affecting nonspecific phosphatase activity (AHMED and REIS, 1958). Formaldehyde inhibition can be made use of to distinguish between the acid phosphatase of red cells and that of the prostate: the former is inactivated, whereas the latter is stable (AHMED and KING, 1959). L-tartrate inhibits prostatic but not erythrocyte acid phosphatase (ABDUL-FADL and KING, 1949). FISHMAN and LERNER (1953) described a method of estimating acid phosphatase of prostatic origin, based on the ability of L-tartrate to inhibit the prostatic component.

Phosphatases fall into two classes, those that are specific hydrolases, in that water is the only phosphate acceptor, and those which are transferases as well as hydrolases, in that other hydroxy compounds in addition to water may act as phosphate acceptors. MORTON (1955b) has reviewed the early literature on the phosphoryl transferring activity of phosphatases. Phosphotransferase activity is conveniently measured using phenyl phosphate as the phosphate donor and glycerol or methanol as the acceptor (TSUBOI and HUDSON, 1953). At the end of a suitable period of incubation, the enzyme digest is deproteinized and analysed for phenol and orthophosphate. The difference, in micromoles, between phenol and phosphate liberated represents the extent of phosphoryl transfer. NIGAM and FISHMAN (1959) found that β-glycerophosphate was the best donor and n-butanol the acceptor, using prostatic acid phosphatase. AGREN (1959) used dephosphorylated phosphopeptone as the phosphate acceptor. It may be mentioned in this connection that TUNIS and CHARGAFF (1960a, b) have claimed the nonidentity of phosphatase and nucleoside phosphotransferase activity of carrot. By Celite adsorption and paper electrophoresis it was possible to obtain a phosphatase fraction completely devoid of transferase activity and a transferase fraction which always retained slight hydrolytic activity. Moreover Cu^{++} and Zn^{++} were found to have a specific inhibitory effect on phosphatase activity, without affecting the nucleoside phosphotransferase activity.

1. Colorimetric Estimation of Phosphorus.

1. The method of FISKE and SUBBAROW. The most popular method of estimating phosphorus colorimetrically is that of FISKE and SUBBAROW (1925). Many minor

modifications have been suggested from time to time, in view of the lability of some phosphate esters towards acid and molybdate, and the possible interference by naturally occurring compounds. JOHNSON and JOHNSON (1959), who were assaying pyrophosphatase activity, carried out the color development at pH 4.1 using 1 M acetate buffer. RAFTER (1958) recommended that pyrophosphate be added to the standard orthophosphate solution also, since there was some inhibition by pyrophosphate of orthophosphate estimation. TSUBOI and HUDSON (1953) showed that β-glycerophosphate present in high concentration will interfere with the estimation of orthophosphate. DRYER, TAMMES and ROUTH (1957) found that the presence of trichloroacetic acid gave elevated values for color and therefore incorporated the acid in the orthophosphate samples used as standard. The last mentioned authors used semidine as the reducing agent and showed that the reagent was so sensitive and the final color stable, that it could be used advantageously in estimating low concentrations of phosphate, which cannot be accurately determined by the method of FISKE and SUBBAROW. The presence of excess of chlorophyll in enzyme preparations sometimes gives rise to spurious color for phosphorus after TCA-inactivation of the enzyme digest. Such interference can be eliminated by preliminary treatment with *iso*butanol before adding the acid-molybdate reagent (HEWITT and NOTTON, 1960).

2. **The method of LOWRY and LOPEZ.** In view of the labile nature of some phosphorylated compounds in the presence of sulphuric acid and molybdate, LOWRY and LOPEZ (1946) devised a method of color development in acetate buffer of pH 4.0, employing ascorbic acid as the reducing agent. This method has met with extensive application in phosphate analysis in enzyme digests. The presence of interfering material in tissue extracts was noted by LOWRY and LOPEZ (1946). Powerful inhibition has since been shown to be associated with liver, brain and bacterial extracts (PEEL and LOUGHMAN, 1957). TEWARI and KRISHNAN (1960) found that the linear condensed phosphates inhibit the development of color of orthophosphate by the LOWRY-LOPEZ method. BRUEMMER and O'DELL (1956) and PEEL and LOUGHMAN (1957) independently and simultaneously discovered that the addition of minute quantities of copper ions helped to overcome the interference due to naturally occurring material. For maximal color development in the presence of 0.2 mM EDTA a higher concentration of copper (0.35 mM) was required (NEWMARK and WENGER, 1960). According to TEWARI and KRISHNAN (1960), the effect of copper is not specific; low concentrations of nickel were also effective in overcoming the inhibition due to polyphosphates.

3. **The iso-butanol extraction method of BERENBLUM and CHAIN.** The present author is firmly convinced that the most accurate method of assay of orthophosphate in the presence of interfering material is by preliminary extraction of the phosphomolybdate complex into *iso*-butanol, as originally recommended by BERENBLUM and CHAIN (1938), or into *iso*-butanol-benzene mixture, as in the modification by WEIL-MALHERBE (1953). The manipulative details are certainly cumbersome, compared to direct development of color, but with some experience the method can be employed for routine analysis.

4. **Bioassay of orthophosphate.** The usual colorimetric methods fail when the phosphate concentrations are very low. Under such conditions a microbiological method can be advantageously used (TORRIANI, 1960). The solution to be assayed is used as a phosphate source for the growth of *Escherichia coli* W wild type. After 24 hours the optical density is measured and phosphorus concentration read on a standard calibration curve.

2. Aspects of Enzyme Assay in Plants and Microorganisms.

1. The pH of substrate solution. The importance of the prior adjustment of the pH of the substrate solution to correspond with that of the buffer cannot be over-emphasized. Solutions of sodium β-glycerophosphate and sodium pyrophosphate, for example, have high buffering capacity and materially alter the pH values of the buffer systems usually employed in the acid and near neutral range. When the pH-activity relationship for an enzymatic reaction is to be determined, the substrate solution employed should be separately adjusted to the pH values of the various buffers used in the experiment.

2. Choice of buffers. The absolute value for enzyme activity is sometimes dependent upon the nature of buffer solution. ZITTLE and DELLA MONICA (1950) reported that tetraborate competitively inhibited alkaline phosphatase from the intestine. GRAN and EEG-LARSEN (1960) made a comparative study of the activity of purified intestinal phosphatase in 4 different buffers, veronal, ethanolamine, glycine and tetraborate, all of 0.068 M concentration and pH 10.0. The activity of the enzyme was the same in the first 3 buffers. With tetraborate buffer, however, the rate of hydrolysis was only 29% of the rate obtained with the other systems. Also, the addition of phosphorylated vitamin D-2 was without effect on the enzyme activity assayed using veronal, ethanolamine or glycine buffers, but the initial activity in borate buffer was increased several fold. The inhibition of enzyme activity in tetraborate buffer was, however, much lower when purified bone phosphatase was used and was not apparent when crude extracts of rat kidney were assayed. CLAYTON and HANSELMAN (1960) observed that when phosphate buffer was substituted for Tris buffer, nucleoside pyrophosphatase activity of tobacco roots towards DPN was reduced by about 30%. On the other hand, some authors have stated that the nature of buffer employed has practically no influence on the pH optimum of an enzyme. POGELL and McGILVERY (1954) reported that the use of barbital, serine or borate buffer yielded equivalent values for the activity of fructose diphosphatase of liver tissue.

SANWAL (1960) studied the phosphatase activity against β-glycerophosphate of whole homogenates of cactus tissue using veronal, acetate and borate buffers. Veronal and acetate buffers elicited practically the same activity in the acid range. However, the use of borate buffers in the near-neutral, neutral and alkaline ranges led to lowered activity compared to that with veronal buffer. It was also found that the addition of Mg^{++} activated the phosphatase activity when borate buffer was employed. The use of borate buffer for pH values 6 and 7, preceded by acetate or veronal buffers for pH 4 and 5 and succeeded by veronal buffer for the alkaline range, would, thus, yield values suggesting an apparent peak in the alkaline range.

Reports are available that the shape of the pH-activity curve might vary with the type of buffer employed. RACKER and SCHROEDER (1958) reported that the pH optimum of purified fructose diphosphatase was 8.5 using Tris buffer, but 8.2 when veronal buffer was employed. According to TORRIANI (1960) the pH optima for the acid and alkaline phosphatase activity of *Escherichia coli* against *para*-nitrophenyl phosphate varied from 3 to 5 and from 8.8 to 9.8, respectively, depending on the nature of the buffer employed.

Activation by metals may also be affected by the type and the concentration of buffer used in the assay system. OGINSKY and RUMBAUGH (1955) reported that the Co-activated acid pyrophosphatase of *Streptococcus faecalis* showed considerably higher activity in histidine buffer than in acetate, phosphate or borate buffers. The requirement for histidine could not be considered to be specific, since it could be replaced with almost equal effectiveness by cysteine; lysine was also active to some extent. GAREN and LEVINTHAL (1960) found that the purified alkaline phosphatase of *Escherichia coli* did not require Mg^{++} when the assays were conducted in 1.0 M Tris buffer at pH 8.0; however, when the concentration of Tris buffer was lowered to 0.01 M, full activity was obtained only after addition of Mg^{++}.

VISWANATHAN (1961) has observed that the degree of activation of alkaline C-1 fructose diphosphatase of tapioca leaves by EDTA was considerably higher in Tris buffer than in veronal buffer.

It seems advisable to specify the composition of the buffers employed when reporting the kinetics of an enzymatic reaction.

3. Manner of adding metallic ions. The addition of Mg^{++} to an assay system for pyrophosphatase activity results in the formation of insoluble precipitate; whereas the enzymatic splitting of pyrophosphate takes place, considerable uncertainty exists as to the effective concentration of the metal and substrate. BLOCH-FRANKELTHAL (1954) reported that precipitation began when the molal ratio of $MgCl_2$ to sodium pyrophosphate attained the value of 1. At lower ratios a soluble complex was obtained. (An effect of the complex formation is the decrease in pH of the mixture and allowance for this must be made in the assay system). The author recommended that pyrophosphate should always be added immediately before the enzyme, which was put in last, so as to avoid the ageing of the $MgCl_2$-pyrophosphate mixtures before their coming in contact with the enzyme. A similar difficulty arises when oxalate and Ca^{++} are present side by side in enzyme assay systems. The addition of fluoride, as inhibitor, also causes precipitation of Mg^{++}. Hexokinase assays in crude preparations are usually performed in the presence of added fluoride, which serves to inhibit phosphatase action. In order to avoid precipitation in hexokinase assays caused by the interaction of fluoride, ATP and Mg^{++} in concentrated solution, LONG (1952) recommended that the tissue be homogenized in a fluoride-phosphate medium and the enzyme preparation added to the rest of reaction mixture containing ATP and Mg^{++}. Where it is not possible to homogenize tissue in fluoride medium, the inhibitor may be added separately to the hexokinase assay system after ATP and Mg^{++} have been mixed, but immediately before enzyme addition.

In passing, it may be pointed out that the ratio of metal to substrate may affect the pH optimum for enzyme action. KUNITZ (1952) obtained pH optima of about 6.3 and 7.2 for Mg^{++}-activated pyrophosphatase, the different values being obtained at different metal-substrate ratios.

4. Control experiments. The common practice of running the enzyme control is by acidifying with trichloroacetic acid or perchloric acid before the addition of substrate. Heat-treated enzyme preparations are sometimes employed as controls. A word of caution may not be out of place regarding the use of heat-treated enzyme as control. Phosphatases, in general, are fairly resistant to short term heating at 50—60°; an extreme example is the phosphatase of *Azotobacter agilis*, which is remarkably stable to heat-treatment (JOHNSON and JOHNSON, 1959). During recent years a few isolated cases have been reported, where an enzymic activity in crude preparations could be demonstrated only after heating, or an enzyme was apparently more active after a short period of heat-treatment. SWARTZ, KAPLAN and FRECH (1956) found that DPN-pyrophosphatase and 5'-nucleotidase were present in sonicates of *Proteus vulgaris* in an inhibited state, and that significant activity could be demonstrated only after the extracts were placed in boiling water for several minutes. The maximum activity was obtained after two minutes of boiling and even after 15 minutes about 50% of the maximum remained. The authors attributed these results to the presence in sonicates of excessive amounts of a heat-labile inhibitor, probably of protein nature, along with comparatively heat-stable enzymes. Assaying thyroxine deiodination by liver tissue, STANBURY, MORRIS, CORRIGAN and LASSITER (1960) found that when whole liver homogenate was preheated for purpose of control, the rate at which thyroxine was deiodinated was in fact sharply accelerated as compared to unheated samples. The authors postulated that the mechanism of heat-activation consisted in the release of iron in the ferrous form, which was essential for the enzymatic reaction.

5. Protein determination. Protein estimation is necessary to assess the enrichment in activity of an enzyme as a result of the particular purification method employed. It also enables a decision to be made whether the total mass of a given homogenate has been recovered in the individual fractions separated during differential centrifugation, an important criterion in enzyme localization. Methods for protein determination have been summarised by LAYNE (1957) and may be classified under two general headings: one, the estimation of total nitrogen by the micro-KJELDAHL method and the calculation of protein by multiplication with a suitable factor (usually 6.25), and the second, a direct determination based on spectrophotometry, turbidimetry or colorimetry.

A few remarks may not be out of place regarding colorimetric methods which find extensive use in enzyme purification, especially under conditions where the spectrophotometric assay is inapplicable. The method of LOWRY, ROSENBROUGH, FARR and RANDALL (1951), employing the FOLIN-CIOCALTEAU phenol reagent, has found wide application because of its simplicity and high degree of sensitivity. CHOW and GOLDSTEIN (1960) made a critical study of the molecular basis of color production with the phenol reagent and concluded that whereas the color intensity of completely hydrolysed protein could be accounted for almost entirely by its tyrosine and tryptophan content, the higher intensity obtained with whole proteins has to be attributed to chromogenic amino acid sequences. Even admitting that chromogenicity is not confined to a few specific groups, but is common to a number of different sequences, it may be anticipated that different proteins may give rise to different color intensities per unit of total nitrogen. The use of crystalline albumin as a common standard for all proteins may, therefore, lead to significant error. On theoretical grounds it appears that an assay based on the peptide linkage would be the method of choice in protein assay. The biuret method of GORNALL, BARDAWILL and DAVID (1949) has much to commend it. Whether protein is estimated by KJELDAHL-nitrogen or colorimetry, a preliminary precipitation is advantageous, especially when dealing with crude extracts. No ideal protein-precipitant exists, instances being known where some proteins escape quantitative precipitation (LINEWEAVER and MURRAY, 1947; JANSEN, JANG and BALLS, 1952; SCANDRETT, 1960). The presence of chlorophyll may interfere with the final development of color; this difficulty may be overcome by ether alcohol (3:1) extractions of the TCA-precipitate (BRAWERMAN and KONIGSBERG, 1960).

6. An evaluation of the homogenate technique of assay. The homogenate technique undoubtedly offers several advantages (POTTER, 1957), but there are some important limitations which should not be lost sight of. The activity observed for a given enzyme in the homogenate may not be the optimum possible *in vitro*. That some of the enzymes may be in a latent condition in homogenates is clear from the researches of DE DUVE and his school on animal tissue (cf. BENDALL and DE DUVE, 1960). BONTING and ROSENTHAL (1960) made a comparative study of the enzymatic activity of liver tissue prepared for analysis by various methods, inclusive of homogenates; according to these authors, the highest enzyme activities were obtained in frozen-dried sections.

Another disadvantage of working with whole homogenates, instead of the separated fractions, is the error arising from nonspecific enzymatic action. To illustrate from liver assays, if glucose-6-phosphatase estimations are not performed on the isolated microsomal fraction, the results are likely to be affected by nonspecific phosphatase action. C_1 fructose diphosphatase activity of plant homogenates is likely to be elevated due to nonspecific phosphatase action on fructose-6-phosphate; an approximate correction may be applied by running simultaneous assays with fructose-6-phosphate as substrate.

SEGAL and BRENNER (1960) found that the 5′-nucleotidase activity of isolated liver microsomes exhibited a pH optimum in the range 6.5 to 8.0 and was unaffected by added Mg^{++}. When the effect of pH and Mg^{++} was tested with the whole liver homogenate, the 5′-nucleotidase activity rose with pH upto 9.5 and there was marked stimulation by Mg^{++} at the higher pH values. This Mg^{++}-stimulated activity manifested at high pH values was presumed to be caused by the 5′-nucleotidase activity of nuclei.

The presence of inhibitors can lead to decreased activity in the homogenates compared to isolated fractions. SEGAL and BRENNER (1960) found that considerably less orthophosphate was obtained when 5′-AMP was incubated with whole homogenate of rat liver than with the supernatant fluid from a 10-minutes' centrifugation at $5000 \times g$, or with the microsomes separated from the supernatant. The authors believed that an endogenous inhibitor was present in the particles sedimentable at $5000 \times g$. This inhibitor was heat-stable. It was also observed that this inhibition tended to disappear when subjected to treatment such as freezing and thawing, grinding in a POTTER-ELVEHJEM homogenizer, or ageing at $0°$ C for several hours.

Still another complicating factor in the assay of enzymes in whole homogenates and in partially purified preparations is the possibility of masking or interference due to other enzymatic reactions taking place simultaneously. TEWARI and KRISHNAN (1960) have shown that when hexokinase is assayed by the method of LONG (1952), the net amount of glucose which disappears from the system is the resultant of two opposing actions, namely, utilization by hexokinase reaction and formation by some unexplained reaction. TURNER and TURNER (1960) have pointed out that phosphatase action on glucose-1-phosphate may be masked under certain conditions by mutase activity, if the assay system contains significant amounts of orthophosphate.

Yet another error associated with the homogenate technique has been brought to light recently by HAGEMAN and FLESHER (1960). These authors found higher values for aldolase activity with acetone powder than with fresh homogenates of corn seedlings. They hypothesized that certain enzymes may be selectively adsorbed by fat and discarded in the preparation of homogenates.

The K_m value obtained using the whole homogenate may not be the same as the value with the various separated fractions, or the final purified or solubilized enzyme preparation. TRIANTAPHYLLAPOULOS and TUBA (1959) compared the kinetics of alkaline phosphatase of rat small intestines in whole homogenates and in separated nuclear, mitochondrial, microsomal and supernatant fractions. Over three fourths of the enzyme activity was present in the microsomal fraction, which had also by far the highest specific activity; the rest was almost equally distributed between the other 3 fractions. Using β-glycerophosphate as substrate, the K_m values were as follows:

Unfractionated homogenate . . . . 0.0041 M
Nuclear fraction . . . . 0.0034 M
Mitochondrial fraction . . . . . . 0.0030 M
Microsomal fraction . . . . 0.0025 M
Supernatant fraction . . . . 0.0023 M

The slight differences observed in the K_m values were attributed by the authors to the difference in the availability of the substrate for the enzyme in the individual fractions. In the case of nuclei and mitochondria the substrate has to pass through membranes in order to react with enzyme and this may have a retarding effect on the rate of formation of enzyme-substrate complex. The enzyme in the supernatant is mostly in the soluble readily accessible form and consequently the K_m value of this fraction may be expected to be the lowest. The authors also stated that the degree of homogenization had some influence on the K_m value of whole homogenate. Michaelis constant determined on crude homogenates was found to be slightly higher than that for the same homogenates after they had been subjected to further treatment in a glass homogenizer.

It should also be borne in mind that the experimental values for the various enzymes by the homogenate and other techniques of *in vitro* analysis represent the maximal rates attained under optimal conditions of pH and substrate concentration and offer little information on the actual velocities at any given moment under steady state conditions *in vivo*. Since our knowledge of the intracellular organization of plant tissue is very meagre, it may not be advisable to assume that *in vitro* results are applicable, *in toto*, *in vivo*.

7. Influence of orthophosphate on phosphatase assay. The effect of orthophosphate on phosphatase action deserves special mention, in view of the fact that it is invariably present in whole homogenates of tissue. The inhibition of phosphatase action by orthophosphate was observed by JACOBSEN (1932, 1933), who concluded that the rate of hydrolysis of glycerophosphate by alkaline kidney phosphatase depended on the ratio of substrate to orthophosphate. PFANKUCH (1936) reported the inhibition by orthophosphate on the hydrolysis of α- and β-glycerophosphates, fructose diphosphate, phosphoglycerate and phytate by potato phosphatase. MORTON (1955a) found that phosphate inhibited the alkaline phosphatase of calf intestinal mucosa. TURNER and TURNER (1960) reported that where as glucose was without effect, inorganic orthophosphate in low concentration inhibited the breakdown of both glucose-1-phosphate and glucose-6-phosphate by pea seed phosphatase. The authors showed that the inhibition was of the competitive type. The degree of inhibition of phosphatase activity by orthophosphate depends upon whether the enzyme is of the "acid" or "alkaline" type. TORRIANI (1960) found that whereas the alkaline phosphatase activity of Escherichia coli was strongly inhibited by orthophosphate, its acid phosphatase activity was less affected. Arsenate also brought about powerful inhibition of alkaline phosphatase activity (GAREN and LEVINTHAL, 1960). NORBERG (1960) found that whereas phosphoprotein phosphatase was markedly inhibited by comparatively large quantities of orthophosphate, produced either as a result of enzyme action or present at the outset in the system, a certain minimal concentration of phosphate was required to obtain the greatest possible activity. A further activation through detergents was also shown to occur and the experiments appeared to indicate that this effect may consist in a loosening and breaking of covalent phosphorus linkage in the substrate.

III. Intracellular Localization of Enzymes by the Technique of Differential Centrifugation.

Studies based on histochemical analysis and centrifugation of whole organisms have played a definite role in elucidating the intracellular location of enzymes, but the most powerful tool is undoubtedly the technique of differential centrifugation of suspensions of cell-free preparations. The criteria which must be rigorously satisfied before the localization of an enzyme may be accepted have been reemphasized by JOHNSON and JOHNSON (1959) in their investigation on the heat resistant inorganic pyrophosphatase from Azotobacter agilis. Data on pyrophosphatase distribution in sonicates of A. agilis are given in Table 1.

Table 1. Pyrophosphatase distribution in sonicates of Azotobacter agilis.

Fraction	Total orthophosphate liberated, μg	Protein, μg	Specific activity	Activity recovered, %	Enrichment
Crude extract . .	618	52	125	(100)	(1.0)
45 p 30	6.0	6	11	0.9	0.09
144 p 60	4.0	3	14	0.7	0.11
144 s 60	587	38	164	95.0	1.31
% recovery . . .	96.6	90.4		96.6	

Of the 618 μg of inorganic orthophosphate produced by the pyrophosphatase of the crude extracts, approximately 95% was located in the 144 s 60 fraction, that is, the supernatant obtained on centrifugation at 144,000 $\times g$ for 60 minutes. The enzyme was, therefore, located totally in the nonsedimentable cytoplasmic constituents of A. agilis. There was 90.4% recovery of the total mass of the crude extract in the sum total of the fractions. There was an enrichment in the supernatant by a factor of 1.31 compared to the activity of the crude extract. The same pattern of distribution was obtained when the cells were disrupted by a totally different method, namely, grinding with alumina. All the criteria were thus fulfilled for establishing a particular locus as the site for enzyme activity.

The colloquium on the intracellular localization of enzymes held recently (DE DUVE, 1960) highlights the rapid advances being continuously made in the field of enzymes of animal tissue. It has, in addition, emphasized the utmost importance of the close cooperation between electron microscopists, histochemists and biochemists. Whereas the data from animal tissue, liver in particular, are, for the most part, unequivocal, considerable uncertainty attends the location of the enzymes of plant tissue. A satisfactory fractionation procedure for the whole plant cell does not exist, because of the larger variety of particulate components and the difficulty experienced in their morphological indentification. It is not uncommon to find that results obtained by the technique of differential centrifugation are at variance with data from histochemical studies. Yin (1948), Yin and Tung (1948), Yin and Sun (1949) and Paech and Krech (1952) concluded from histochemical studies that phosphorylase was localized in chloroplasts. However, Stocking (1952) could not confirm the above result by differential centrifugation. In spite of the fact that rat liver has been the best studied tissue, staining techniques and differential centrifugation of homogenates have yielded contradictory results for the location of esterase, the former technique pointing to lysosome association and the latter the soluble fraction (De Duve, 1960).

Among the factors which influence the intracellular distribution of plant enzymes, over which experimental control has been difficult, are the mechanical aspects involved in the liberation of the cell contents and the composition of the suspending medium.

i. Grinding device. The plant cell wall presents a rather formidable barrier to the release of protoplasm, necessitating high shearing forces for its rupture (Crook, 1946; Wildman and Bonner, 1947). The mildest form of disintegration involves the use of a mortar and pestle usually with sand or glass powder as abrasive, or of a glass homogenizer. Drastic treatments, such as the use of Waring blendor, or Virtis homogenizer, aimed at cell-free preparation from plant material may cause excessive damage to the particulate components of cells, and the possibility of artifacts must always be kept in mind. Saltman (1953) reported that the partition of hexokinase depended on the method of preparation of homogenates; when potatoes were ground with sand in a mortar all the hexokinase activity was found in the particulates; when homogenized in a Waring blendor, over 2/3 of the activity was in soluble form. In the case of wheat germ the proportion of soluble to particulate-bound hexokinase depended on the treatment of tissue prior to homogenization. Newcomb (1951) reported that ascorbic acid oxidase, which was essentially localized on the wall fragments of tobacco pith in homogenates effected in Potter-Elvehjem glass homogenizer, tended to pass into the supernatant when a Waring blendor was used.

On the other hand, some plant tissues may require comparatively drastic treatment for cell rupture. Grinding in a Waring blendor for 30 seconds has been advocated by some laboratories. The experience of the present author has been that leaves of the tapioca plant, and *Bougainvillaea* and the phylloclades of cactus require about 2—3 minutes of grinding in a Waring blendor for yielding cellfree preparations. Reference should be made here to the artifacts resulting from incomplete homogenization of tissue. In such cases, the nuclear fraction sedimenting at low speeds contains an abnormal proportion of essentially cytoplasmic enzymes (De Duve and Berthet, 1954).

Comar (1942) states that freezing of plant tissue should be avoided, as nonplastid cytoplasm is also precipitated with plastid. Opinion on this point does not seem unanimous (Clayton and Hanselman, 1960). Since acetone treatment disrupts particulate structures it cannot be employed in localization studies (Mazelis and Vennesland, 1957).

ii. Suspending Medium.

(a) Aqueous media. On homogenization the intracellular components are released into an unnatural environment, and so it is logical to choose a suspension medium approximating to the soluble phase of the cytoplasm, so that the cell particulates remain morphologically, structurally, and functionally intact. A wide variety of aqueous extraction media, ranging from water, sucrose, mannitol, sorbitol, to high molecular weight polyethylene glycols, has been used by investigators in the field.

The vacuole occupies a large portion of the total volume of the living plant cell. In succulents and many other plant tissues the contents of the vacuoles are quite acidic, the degree of acidity being a function of the physiological age-status of the tissue. On the other hand, the cytoplasmic pH is generally thought to be close to neutrality. While the vacuolar reservoir of acidity is effctively prevented by the tonoplast from making contact with the cytoplasm, one of the consequences of homogenization is to dilute enormously all of the protoplasmic constituents into an acid environment. While the effects of dilution are not assessable, the effect of acid on the cytoplasm may be to cause the rapid and irreversible aggregation of the cytoplasmic proteins (SINGER, EGGMAN, CAMBELL and WILDMAN, 1952). Chloroplast inactivation by acids occurs from pH 4.2 downwards (CLENDENNING, BROWN and WALLDOV, 1955). The inhibiting action of acidic leaf sap upon "active" chloroplasts has been sought to be eliminated by prior neutralization. Use of strongly buffered grinding fluids may provide little protection, since the sensitive chloroplasts are exposed to the acids during the intermingling of cell contents that precedes their release. KINZEL and URL (1954) attempted to prevent the inactivation of the enzymes during their isolation from acidic leaves by soaking the leaves beforehand in dilute ammonia. COLES and WAYGOOD (1957) introduced the technique of ammonia infiltration of succulent leaves before maceration. WALKER and BROWN (1957) effected instantaneous neutralization during homogenization of succulents by adding a predetermined amount of Tris buffer. It has become a common practice to incorporate neutral or slightly alkaline buffers of low molarity in aqueous grinding fluids, but some authors have failed to specify the final pH of the homogenate. PRICE and THIMANN (1954) used 0.2 M sucrose in 0.03 M phosphate buffer of pH 7.0 for grinding apical etiolated pea stems and the coleoptiles of etiolated oat seedlings. MAZELIS (1954), in studies on particulate adenylyic kinase, used 0.5 M sucrose in 0.2 M Tris buffer at pH 7.4 for grinding spinach and tobacco leaves in a WARING blendor.

The experience of the writer of this review has been that whereas 0.05 M phosphate buffer of pH 7.2 is sufficient to yield sucrose homogenates of pH 6.8—7.0 in the case of nonacidic leaves like those of the tapioca plant and Bougainvillaea, as high a concentration of buffer as 0.5 M may be needed to yield a neutral homogenate of a succulent tissue such as the phylloclade of the cactus. The presence of such high concentration of salts in the final homogenates may tend to bring about agglutination of the cell particles. The use of neutral buffers of high molarity may be avoided by employing 0.05 M buffer of pH 8—10 in preparing sucrose solutions but the uncertainty exists that cell contents may instantaneously get exposed during the shearing process to a highly alkaline environment, leading to the inactivation of sensitive enzymes.

(b) Non-aqueous media. Water-soluble nuclear-enzymes are known to diffuse easily out of the nucleus during isolation in aqueous media, so that these enzymes appear as artifacts in extranuclear fractions. Also, soluble cytoplasmic enzymes may diffuse into the nucleus and get indicated as true nuclear enzymes. These sources of error are sought to be eliminated by homogenization in non-aqueous media. STERN and MIRSKY (1952—53) isolated wheat germ nuclei by using cyclo-hexane-carbon tetrachloride mixture as the suspension fluid. Recent investigations go to show that enzymes involved in photosynthesis may be lost from the chloro-plasts during the isolation of these particles in aqueous media. Ribulose di-phosphate carboxylase, a vital enzyme in photosynthesis, was found to be present only in small proportion in pea chloroplasts isolated in aqueous media (SMILLIE and FULLER, 1959). RACKER and SCHROEDER (1958) used 0.35 M sodium chloride for grinding spinach leaves and arrived at the conclusion that the specific alkaline C-1 fructose-1:6-diphosphatase is essentially a soluble enzyme and that as such it does not play a part in photosynthetic fixation of CO_2. SMILLIE (1960) found that whereas the major part of alkaline fructose-1:6-diphosphatase was present in the soluble fraction when chloroplasts were isolated in aqueous media, its specific activity in the plastid fraction was greater than in the other fractions. Fractionating pea leaves and *Euglena* cells in nonaqueous media by the procedure of STOCKING (1959), SMILLIE (1960) found that the alkaline fructose diphosphatase was actually localized in the chloroplasts.

Nevertheless the use of organic media for homogenization and differential centrifugation does not seem to have found wide application. Many enzymes are sensitive to organic solvents: also, mitochondria, which are rich in lipid material, may be expected to undergo drastic structural alterations as a result of contact with fat solvents. Although nuclei and nuclear fragments remain coherent in non-aqueous media, final purification of the nuclear fraction is difficult. Even with wheat germ, which is rich in nuclei, the nuclear fraction cannot be freed from considerable cytoplasmic contamination.

Presence of activators and inhibitors in tissue. The error likely to arise in localization studies from the presence of activators and inhibitors has not been taken notice of by many workers in the field. It is not uncommon to find that plants contain activators and inhibitors for a number of enzymatic reactions. Orthophosphate which is present in all plant tissue acts as a powerful inhibitor for some phosphatases. NIGAM, DAVIDSON and FISHMAN (1959) showed that the hydro-lysis of β-glycerophosphate by prostatic acid phosphatase was inhibited by oxalate, pyruvate, maleate, glutamate, malonate, glucuronate, saccharate and L-tartrate. Some of these acids have wide distribution in plant tissue especially in succulents. The usual practice in localization experiments has been to suspend the isolated particulate fractions in the specific grinding fluid employed, before assays are conducted. A soluble activator or inhibitor will get concentrated in the super-natant fraction during the process of differential centrifugation of whole homo-genates and consequently the data for the recovery of activity may be misleading. It is advisable to have independent assays conducted after suspending the separated particles in filtrates from heat-treated whole homogenates, or to mix the various separated fractions and redetermine activity on the mixture.

1. Nuclear Fraction.

JOHNSTON, ASATIR and STERN (1957) reported the isolation of nuclei from wheat embryos, using 2 M sucrose in 0.001 M calcium chloride for disintegration. The high concentration of sucrose helps to prevent rupture of the free nuclei. Calcium chloride is introduced to reduce further the breakage of nuclei.

2. Plastid Fraction.

The isolation of plastids by differential centrifugation has been reviewed by WIER and STOCKING (1952). The popular suspending medium for the preparation of intact chloroplasts has been 0.5 M sucrose, with or without a small proportion of buffer to maintain neutrality. ALLEN, ARNON, CAPINDALE, WHATLEY and DURMAN (1955) have used 0.35 M sodium chloride for the isolation of spinach chloroplasts. Osmotic pressure of leaf sap varies diurnally, seasonally, with the position of leaf on the plant, with the species and habitat and with all other factors influencing intracellular supplies of water, solutes and colloids. For these reasons, the term "isotonic" rarely can be applied with precision to artificial media for leaf chloroplasts. In comparison with the chloroplast's fluctuating natural environment, the composition and osmotic pressure of standard leaf grinding fluids (e.g. 0.5 M sucrose) have been remarkably uniform. GOOD (1960) has reported the composition of five different solutions made up of KCl or sucrose with pH values varying from 6.3 to 7.8 which he used in the isolation of the chloroplasts from pea leaves.

3. Mitochondrial Fraction.

HACKETT (1955) has reviewed the various techniques of isolation of plant mitochondria. Although a few workers have used water alone, most have employed solutions of sugar (0.2—0.5 M), or inorganic salts. A number of special substances such as serum albumin, nicotinamide, or cysteine, may be used to stabilize the mitochondria. As a typical example of the isolation of mitochondria, the experiments of NEUFELD, FEINGOLD and HASSID (1959) may be cited. Mung bean seedlings were ground with acid-washed sand in a chilled mortar in medium consisting of 0.4 M sucrose, 0.01 M sodium-potassium phosphate buffer of pH 7.0 and 0.01 M mercaptoethanol. The ground suspension was passed through cheese cloth and subjected to differential centrifugation. Most of the studies of plant mitochondria have been carried out with etiolated or non-pigmented tissue, very few having been done on chlorophyllous tissue, as the presence of plastids and vacuolar contents makes isolation and purification difficult. PIERPOINT (1959) described the preparation of mitochondria from the leaf of the tobacco plant by grinding with buffered sucrose medium in a chilled mortar. PIERPOINT (1960) pointed out that whereas a sucrose solution was suitable for the isolation of functional mitochondria from animal tissue, it was not sufficient in itself for mitochondrial isolation from leaves. Even after incorporation of EDTA, phosphate, Mg^{++} and citrate into the sucrose solution, the mitochondria isolated from tobacco leaves were somewhat functionally damaged and were heavily contaminated with chloroplast fragments. He devised a modified medium for working with tobacco leaves; the medium was satisfactory also for the leaves of garden lupine and yam, but failed in the case of leaves of French beans.

4. Cell Wall Fraction.

In a few cases, the cell wall fraction has been isolated and examined for enzymatic activity. The best example is, probably, ascorbic acid oxidase of barly roots; HONDA (1955) prepared homogenates of the roots by grinding in either water or media initially of 0.8 M tonicity and adjusted to 0.4 M final tonicity.

5. Association of Enzymes with Larger Particles.

As would be evident from the foregoing sections, the isolation of morphologically uniform cellular components from plant tissue is attended with considerable uncertainty. Some workers have obtained particulate fractions from homogenates in water or sucrose solution by centrifugation at speeds in the vicinity of $2000 \times g$, after removal of unbroken cells and starch granules by centrifugation at 100 to $200 \times g$. The particulate material sedimenting between 200 and $2000 \times g$, which the present author prefers to designate as "Larger particles", is not homogeneous, being made up of cell-wall fraction, nuclei, nuclear fragments, plastids, plastid fragments and any agglutinated mitochondria. A number of enzymes have been shown to be "associated" with these larger particles.

A word of caution may be necessary against the usual practice of rejecting the starch fraction obtained by centrifugation at very low speeds. RONGINE DE FEKETE, LELOIR and CARDINI (1960) could obtain definite evidence for uridine-diphosphateglucose mediation in starch synthesis in plants only by using as enzyme source the starch granules obtained from dwarf string beans.

6. Isolation of Plant Protoplasts.

Several aspects of the metabolism of microorganisms have been elucidated by the use of protoplasts. The use of enzymes to digest the cell walls of bacteria (WEIBULL, 1958) and fungi (EDDY and WILLIAMSON, 1957; BACHMANN and BONNER, 1959) led COCKING (1960) to use fungal cellulase to obtain protoplasts from root tips of tomato seedlings. It would appear that the availability of plant protoplasts would permit the study in future of aspects of plant metabolism which were not possible by the conventional techniques.

7. Localization of Phosphatases.

Surprisingly few accounts are available relating to the localization of the different phosphatases in a given tissue and these few studies have been confined to animal tissue. The data reported by AHMED and KING (1959) for placenta and by NOVIKOFF, PODBER and RYAN (1950) for rat liver are reproduced in the following table.

Table 2. *Distribution of phosphatases in animal tissue (units/g tissue).*

	Acid phosphatase		Alkaline phosphatase			Pyrophosphatase			5'-Nucleotidase	
	Units	%	Units	%	+ Mg units	Units	%	+ Mg units	Units	%
Human placenta										
Total	2.8	100	24	100	30.8	2.2	100	5.2	4.5	100
Residue	0.2	7	2.0	8	2.4	0.2	9	0.4	0.8	18
(Cell debris)										
Nuclear fraction	0	0	1.6	7	1.7	0	0	0	0.1	2
Mitochondria	0.4	14	3.0	13	3.2	0.2	9	0.7	1.0	22
Microsomes	0.4	14	8.0	33	9.6	0.8	36	1.6	1.6	35
Cytoplasm	1.6	57	8.8	37	10.8	0.8	36	1.9	1.0	22
Recovery %		92		98			90			99
Rat liver										
Total	90—100		85—105						85—105	
Nuclear fraction	5—10		10—18						40—50	
Mitochondria	35—40		17—20						40—45	
Microsomes	5—10		0—10						5—10	
"Supernatant"	35—50		55—70						10—15	

In placenta, the greater part of the acid phosphatase was in the cytoplasmic fractions. The alkaline phosphatase and pyrophosphatase were distributed mainly between the cytoplasmic and microsomal fractions. 5'-nucleotidase activity was distributed between cell debris, mitochondria, microsomes and cytoplasm. In rat liver, 5'-nucleotidase activity occurred mostly in the nuclear and mitochondrial fractions, acid phosphatase in the mitochondrial and supernatant fractions and alkaline phosphatase mostly in the supernatant.

The location in baker's yeast of the acid phosphatase acting at pH optimum of 4.3 and of alkaline phosphatases with optima at pH 7.9 and 8.9 was arrived conducting the assays on intact cells and preparations obtained after cell damage (SUOMALAINEN, LINKO and OURA, 1960). Using intact yeast, either in the fresh condition or after drying, the acid phosphatase activity was the same. There was, however, no demonstrable alkaline phosphatase activity under these conditions. Cytolysates of yeast, on the other hand, possessed alkaline phosphatase activity. It was apparent that acid phosphatase was localized on the surface, whereas the alkaline phosphatases were exclusively intracellular. Since the acid phosphatase activity was of higher order after cytolysis, or after freezing and thawing, it was also concluded that a certain part of the acid phosphatase was localized inside the cell membrane.

IV. Phosphatases Acting on Hexose Phosphates.

1. The Hydrolysis of Glucose-1-Phosphate.

The discovery of a specific glucose-6-phosphatase localized in the microsomal fraction of rat liver (HERS and DE DUVE, 1950) raises the question whether glucose-1-phosphate is hydrolysed directly or after conversion to glucose-6-phosphate through the activity of phosphoglucomutase. Evidence from both animal and plant tissue has been conflicting. The direct hydrolysis of glucose-1-phosphate from silk worm blood was reported by FAULKNER (1955). On the other hand, BROH-KAHN and MIRSKY (1948) and GOODLAD and MILLS (1957) believed that the main route of hydrolysis of glucose-1-phosphate in rat liver was through glucose-6-phosphate. MORTON (1955a) found that purified alkaline phosphatase from cow's milk and calf intestinal mucosa hydrolysed both glucose-1-phosphate and glucose-6-phosphate. AXELROD (1947) found that glucose-1-phosphate was not hydrolysed by the acid phosphatase of citrus fruit, although other esters such as fructose diphosphate were hydrolysed. Glucose-1-phosphate was hydrolysed by extracts from potato tubers (PORTER, 1953), tomato leaves (SPENCER, 1954) and wheat leaves (ROBERTS, 1956), but it was not certain whether the hydrolysis was direct, or through glucose-6-phosphate.

TURNER and TURNER (1960) have recently conducted detailed studies on the hydrolysis of hexosephosphates by a 20-fold purified enzyme preparation from pea seeds; the authors came to the conclusion that, whereas it was doubtful whether two separate enzymes are involved, both glucose-1-phosphate and glucose-6-phosphate were hydrolysed directly.

Enzyme preparation (TURNER and TURNER, 1960). Immature pea seeds were blended with water. The enzyme was precipitated from the extract with ammonium sulfate (55—75% saturation), inert material removed by acidification, and the enzyme further purified by adsorption on tricalcium phosphate gel and elution.

Enzyme assay. A suitable aliquot of the enzyme preparation (containing approximately 0.25 mg of protein) was incubated with Tris (0.75 mmole)-acetic acid buffer, pH 5.4, in a volume of 14.5 ml and the reaction was started by the addition of 7.5 μmoles of substrate in 0.5 ml of water. The reaction time was 60 minutes with glucose-1-phosphate as substrate and 10 minutes with glucose-6-phosphate. The amount of enzyme preparation added was adjusted to give not more than 10% hydrolysis of the substrate, under which condition the rate of reaction was approximately constant during the incubation period. The reaction mixture was inactivated by the addition of 0.1 volume of 40% (w/v) trichloroacetic acid and aliquots utilized for the colorimetric estimation of inorganic phosphate. The activity was expressed in terms of μ moles of phosphate liberated per mg of protein per minute.

Properties of the enzyme. The pH-activity relationship for the hydrolysis of glucose-1-phosphate and glucose-6-phosphate was similar, with pH optimum at 5.4 for the former and 5.4—5.7 for the latter. Mg^{++} was without effect. The addition of EDTA in final concentration of 10 mM stimulated the enzyme by 20%. The phosphatase activity was powerfully inhibited by iodoacetate, mercuric chloride, fluoride and molybdate. A wide range of substrates was attacked; the hydrolysis of ATP, p-nitrophenyl phosphate, 3-phosphoglycerate and fructose-1,6-diphosphate was even more rapid than that of the glucose phosphates. Of the two glucose phosphates, the 6-phosphate was hydrolysed faster than the 1-phosphate.

That glucose-1-phosphate was hydrolysed directly by the enzyme preparation was based on the following evidence. Chromatographic examination of the enzyme digest gave no indication of glucose-6-phosphate. There was no decrease in the sum of acid-labile phosphate + orthophosphate during the course of hydrolysis. These two observations showed that glucose-6-phosphate did not accumulate in significant amounts during enzymatic action on glucose-1-phosphate. It was quite possible that the 6-phosphate might be getting hydrolysed as soon as formed, so that there would be no accumulation. However, the involvement of phosphoglucomutase action was improbable since the mutase enzyme of pea seed had a pH optimum at 7.5, being only a tenth as active at pH 5.4; also the addition of magnesium chloride, which markedly stimulated mutase activity, was without effect on phosphatase action on glucose-1-phosphate.

The enzymic activity towards glucose-1-phosphate closely resembled that towards glucose-6-phosphate in pH optimum and in inhibition by various agents. Also there was no indication of any separation of the two activities during the purification procedure.

SANWAL and KRISHNAN (1961) studied the hydrolysis of sugar phosphates by cactus phosphatase purified about 30-fold on the basis protein and over 150-fold on dry weight basis. The activity against fructose diphosphate seemed to be distinct from that towards hexose monophosphates in having a single pH optimum at 6, as compared to two optima each towards glucose-1-phosphate (pH 5 and 7) and glucose-6-phosphate (pH 5.5 and 7). A further point of difference was the marked activation of fructose diphosphatase activity by Mg^{++} at pH 5.0, contrasting with the inhibition of glucosemonophosphatase activity at this pH. It was difficult to decide whether glucose-1-phosphate was hydrolysed directly or through glucose-6-phosphate. In support of the former view was the experimental observation that the addition of Mg^{++} did not have a marked effect on the hydrolysis of glucose-1-phosphate at pH 7.0. It was possible to show a distinction between the two monophosphatase activities; in the presence of added Mg^{++} glucose-6-phosphatase activity had an additional peak of activity at pH 6; also, Mg^{++} inhibited glucose-6-phosphatase activity at pH 7.0, but activated glucose-1-phosphatase activity.

NEWMARK and WENGER (1960) obtained a 1000-fold purified preparation of lupine phosphatase. The enzyme hydrolysed glucose-6-phosphate, but had negligible activity towards glucose-1-phosphate.

2. C_1 Diphosphatases.

a) Fructose Diphosphatase.

Fructose-1,6-diphosphate + H_2O → fructose-6-phosphate + orthophosphate

The splitting of fructose diphosphate to fructose-6-phosphate by a phosphatase is an important step in the pentose phosphate reductive pathway of carbohydrate metabolism and in photosynthesis (RACKER, 1955, 1956). This enzyme is also a link in gluconeogenesis and in the transformation of fructose to glucose (LEUTHHARDT, TESTA and WOLF, 1953; GINSBERG and HERS, 1960).

b) Fructose Diphosphatase of Animal Tissue.

A specific alkaline C_1 fructose diphosphatase was discovered in animal tissue by GOMORI (1943). This enzyme was studied in detail by POGELL and McGILVERY

(1952, 1954), MOKRASCH and McGILVERY (1956) and MOKRASCH, DAVIDSON and McGILVERY (1956). POGELL and McGILVERY (1952) reported that there was a 3 to 4-fold increase in the total fructose diphosphatase activity of liver on incubating extracts for 8 hours at 38° at pH 4.5. This was due to proteolytic activity of particles. Of the commercial enzymes studied, only papain was able to bring about activation comparable to liver residue. The enzyme had a pH optimum at 9.3—9.5 and required metal for activation. The optimum activity of the enzyme was achieved by employing a mixture of three activators, Mg^{++}, Mn^{++} and cysteine. Assay of the enzyme in crude tissue is rendered difficult because of the presence of non-specific phosphatases. Most of the interfering enzymes are destroyed by homogenizing the tissue in lactate buffer of pH 3.5—4.0. A rough correction for contaminating activity at pH 9.5 can be made by substituting fructose-6-phosphate in the assay and subtracting the value thus obtained from the fructose diphosphatase assay.

Assay of fructose diphosphatase activity of liver. The method described is that of FREEDLAND and HARPER (1959). Liver tissue was homogenized at 0—4° in 40 volumes of 0.05 M sodium lactate buffer, pH 5.0. After filtering through cheese cloth it was used for both direct and activation assays.

Direct assay. 0.1 ml homogenate was added to 0.9 ml substrate mixture, both of which had been incubated at 37.5° for 5 minutes. The substrate mixture contained sufficient substrate and cofactors to provide the following concentrations in the final incubation mixture: 0.0025 M magnesium sulfate, 0.005 M manganese sulfate, 0.03 M cysteine, 0.005 M fructose diphosphate, 0.02 M serine buffer; final pH 9.4. After incubating at 37.5° for 30 minutes the reaction was stopped by adding 1 ml of 10% trichloroacetic acid.

Activation assay after proteolysis. 0.5 ml of 1% solution of papain was added to 4.5 ml of homogenate and the mixture incubated at 37.5° for 20 minutes. 0.1 ml of the papain-treated enzyme was added to 0.9 ml of substrate mixture, which had been incubated for 5 minutes at 37.5°. The reaction was terminated after 15 minutes of incubation. In the final calculation the necessary correction was made for dilution of the homogenate due to addition of the papain solution. Inorganic phosphate liberated was estimated colorimetrically. One unit of enzyme was defined as the amount which caused the release of 1 μmole of orthophosphate in 1 minute at 37.5° and pH 9.4.

The amount of the enzyme in liver is potentially a controlling factor in the rate of gluco-neogensis (MORKASCH, DAVIDSON and McGILVERY, 1956). Evidence was obtained that the increase in fructose diphosphatase activity of liver following glucogenic stress of the animal was partly due to alteration of the inactive precursor to the active form, since there was a fall in the level to which the enzyme was activated on proteolysis *in vitro*. FREEDLAND and HARPER (1959) concluded that the adaptation in fructose diphosphatase in rat was due to protein synthesis and not due to activation of a precursor already present; regardless of whether fructose diphosphatase increased or decreased, the percentage activation on proteolysis *in vitro* was constant.

c) Fructose Diphosphatase of Plants.

(a) Alkaline fructose diphosphatase. GIBBS and HORECKER (1954) obtained evidence for fructose diphosphate splitting in plants. RACKER (1955) reported an absolutely specific, alkaline, Mg^{++} requiring, fructose diphosphatase in spinach leaves. RACKER and SCHROEDER (1958) purified this enzyme about 2000-fold and studied its properties.

(i) Occurrence of alkaline fructose diphosphatase. The enzyme is widely distributed in the photosynthetic tissue of plants and bacteria. Its presence has been demonstrated in pea leaves (GIBBS and HORECKER, 1954; FULLER, 1959),

spinach leaves (RACKER and SCHROEDER, 1958) and in cell-free extracts of photosynthetic tissue from barely, alfalfa, *Spirodela*, the fern *Pteris gautherii* the algae
Euglena gracilis and *Chlamydomonas reinhardii* and the photosynthetic bacteria
Rhodospirillum rubrum and *Chromatium* (SMILLIE, 1960). Conversely, the enzyme
was not present in extracts of pea root or an apoplastidic (streptomycin-bleached)
strain of *Euglena gracilis*; the activity of the enzyme in etiolated pea leaves was
less than 5% of the maximum activity found in green pea leaves (SMILLIE, 1960).
VISWANATHAN (1961) found that the leaves of the tapioca plant, an excellent
starch bearer, contain powerful alkaline fructose diphosphatase activity, side by
side with activity on the neutral and acid range.

(ii) **Assay of alkaline fructose diphosphatase activity.** RACKER and SCHROEDER
(1958) used a spectrophotometric method of assay for following the course of
purification of the enzyme. In other experiments, assays were done by estimating
inorganic phosphate liberated. Proportionality between activity and enzyme
concentration was linear between 0.0015 and 0.015 μmole hexosediphosphate
cleaved per minute in the spectrophotometric method and between 0.2 and 2 μmoles
hexosediphosphate cleaved in 10 minutes in the colorimetric assay.

Spectrophotometric assay. The assay is based on the formation of fructose-
6-phosphate, which is measured by reduction of TPN in the presence of
glucose-6-phosphate isomerase and glucose-6-phosphate dehydrogenase. In a
final volume of 1 ml, the following reagents were pipetted into a quartz micro cell
with a light path of 10 mm:100 μmoles of Tris buffer pH 8.8, 1 μmole of fructose
diphosphate, 600 μg EDTA, 5 μmoles MgCl$_2$, 160 μg TPN, 0.3 unit of glucose-
6-phosphate dehydrogenase and 0.15 units of glucose-6-phosphate isomerase. The
enzyme preparation, suitably diluted with distilled water, was added to the reaction
mixture. Density readings were taken in a BECKMAN DU spectrophotometer at
340 mμ at 30 seconds intervals.

Colorimetric assay. In a final volume of 1 ml the following reagents were
added: 100 μmoles of Tris buffer pH 8.8, 5 μmoles of fructose diphosphate,
600 μg, EDTA, 5 μmoles of magnesium chloride and the enzyme. The final pH
was 8.5. After 10 minutes at 25° the solution was deproteinized (5% TCA) and
orthophosphate determined in the supernatant.

One *unit* of enzyme was defined as that amount which catalysed the turnover
of 1 μmole substrate per minute under the conditions of assay. Calculations were
based on the assumption that 1 μmole of TPNH gave an optical density of 6.22 in
these tests. The specific activity was defined as enzyme units per mg protein.

(iii) **Enzyme isolation** (RACKER and SCHROEDER, 1958). The purification
procedure consisted in the preparation of the juice from the leaves, ammonium
sulfate fractionation, heat treatment, adsorption on calcium phosphate gel and
elution, followed by ammonium sulfate refractionation.

(iv) **Properties of the enzyme.** The purified enzyme was specific for fructose
diphosphate. There was no hydrolytic activity against mono and other diphosphate
esters, including fructose-1-phosphate, fructose-6-phosphate, ribulose-1,5-di-
phosphate and sedoheptulose-1,7-diphosphate. The pH optimum was 8.5 using
Tris buffer and 8.2 using veronal buffer. The K_m value was 3×10^{-4}. The enzyme
was completely inactive in the absence of Mg^{++}. Mn^{++} in low concentrations was
even more stimulatory than Mg^{++}, but in higher concentration Mn^{++} was inhibitory. A remarkable feature of the enzyme activity was that incorporation of
EDTA gave considerable stimulation. Sodium fluoride and p-chloromercuri-
benzoate were powerful inhibitors. The spinach alkaline fructose diphosphatase
is in many respects similar to the liver enzyme but differs from it in being almost
devoid of activity at pH 7, at which pH the liver enzyme is appreciably active.

d) Neutral Fructose Diphosphatase.

In addition to the specific alkaline C_1 phosphatase, crude preparations of spinach leaves contain an enzyme hydrolysing fructose diphosphate at neutral pH. Separation between the neutral enzyme and the alkaline enzyme was possible because the former was precipitated between 40 and 60% saturation with ammonium sulfate, whereas the latter remained in the supernatant (RACKER and SCHROEDER, 1958). The neutral enzyme differed from the alkaline enzyme in having a pH optimum at 6.9, in hydrolysing sedoheptulose-1,7-diphosphate even more rapidly than fructose-1,6-diphosphate and in not requiring Mg^{++} and EDTA.

e) Acidic Fructose Diphosphatase.

CHAKRAVORTY, CHAKRABORTTY and BURMA (1959) discovered and purified 17-fold an acidic C_1 diphosphatase from spinach leaves. The starting material was the acetone powder of the leaves, which was extracted with 0.1 M sodium acetate, pH 7. The extract was subjected to fractionation with acetone and ammonium sulfate and the enzyme further purified by adsorption on calcium phosphate gel and elution. The enzyme had its pH optimum between 5 and 6. Mg^{++} was without effect. It acted on both fructose diphosphate and ribulose diphosphate. The activity towards fructose-6-phosphate was only 4% of that towards the diphosphates. The K_m value was 1.1×10^{-3} using ribulose diphosphate and 1.04×10^{-3} using fructose diphosphate. The similarity of the K_m values, the constancy of the ratio of the activities towards both substrates during various stages of purification and data from competitive inhibition studies, indicated that the same enzyme was acting on both ribulose diphosphate and fructose diphosphate. Fluoride, Be^{++}, mercuric chloride, EDTA and p-chloromercuribenzoate caused inhibition of the enzyme activity.

f) Acid C_1 Diphosphatase of *Escherichia coli*.

ROGERS and REITHEL (1960) succeeded in preparing a hexose phosphatase fraction from *Escherichia coli* free from nucleoside monophosphatase activity, by employing ammonium sulfate fractionation and chromatography on diethyl aminoethyl (DEAE) cellulose. Further separation of this fraction into three subfractions was possible by chromatography on hydroxyapatite and rechromatography on DEAE cellulose. The final purification achieved was of the order of 180-fold. The hexose phosphatase fraction and subfractions obtained therefrom were found to hydrolyse fructose-1,6-diphosphate, glucose-1-phosphate, fructose-6-phosphate and lactose-1-phosphate; in addition, β-glycerophosphate and ortho carboxyphenyl phosphate were split. The hydrolysis of fructose-1,6-diphosphate was characterised by the fact that the pH optima were 2.6 and 4.3, with the possibility of a third optimum at pH 5.5. Although each of the three subfractions derived from the hexose phosphatase fraction hydrolysed hexose phosphates, each could be distinguished by its relative activity against these substrates and by its pH optimum for activity. In all, four different hexose phosphatase activities were recognized. The authors suggested that these hexose phosphatases represent complexes of enzymes or multiple forms of enzyme.

g) Role of Alkaline Fructose Diphosphatase in Green Plants.

RACKER and SCHROEDER (1958) believed that alkaline fructose diphosphatase did not participate in photosynthesis, being apparently absent from plastids; the neutral enzyme was implicated in the process. A survey by SMILLIE (1960) showed that the alkaline enzyme was widely distributed in photosynthetic tissues of plants and bacteria. On the other hand, the enzyme was either absent, or present in very low concentration, in plant tissue devoid of chlorophyll. Measurement of the activity of the enzyme in growing leaves indicated a pattern similar to that of a number of enzymes definitely known to be involved in photosynthesis. No evidence was obtained for the participation of the neutral enzyme in photosynthesis, and this enzyme was found absent in extracts of *Chromatium*. These lines of evidence, considered along with the experimental observation that the enzyme was localized in chloroplasts isolated using non-aqueous media, led Smillie (1960) to conclude that alkaline fructosediphosphatase constitutes an integral link in the photosynthetic fixation of CO_2.

3. Sedoheptulose-1,7-Diphosphatase.

Racker and Schroeder (1958) prepared a specific C_1 sedoheptulose-1,7-diphosphatase from yeast by autolysis and fractionation with ammonium sulfate. Contamination with transketolase was got rid of by dialysis against 0.6% EDTA at pH 7.8. The final preparation was 8-fold purified and contained almost the entire activity of the starting material. The activity was assayed by incubating at 37° for 30 minutes 0.01 ml 1 M Tris buffer pH 7.2, 0.03 ml of 0.007 M sedoheptulose-1,7-diphosphate and the enzyme solution added to give a final volume of 0.1 ml; the mixture was deproteinized with 0.1 ml of 10% TCA and orthophosphate estimated in the supernatant. The enzyme had its optimum activity at pH 7.2 and did not have a requirement for Mg^{++}. It specifically cleaved $\check{C}_1$ phosphate from sedoheptulose-1,7-diphosphate and had no effect on fructose diphosphate or sedoheptulose-7-phosphate. Fluoride, *p*-chloromercuribenzoate, Zn^{++} and Cu^{++} were found to have inhibitory effect on the enzyme.

Use of hexose phosphatases in plant analysis. The spectrophotometric method of assay of biological intermediates has overshadowed the conventional enzymic and non-enzymic methods of assay of hexose phosphates, in view of the specificity, ease of operation and wide application. Nevertheless, the availability of specific enzymes such as C_1 fructose-1,6-diphosphatase, C_1 sedoheptulose-1,7-diphosphatase, glucose-6-phosphatase and lactose-1-phosphatase may have considerable significance in the colorimetric assay of the respective substrates by hydrolytic splitting of inorganic orthophosphate. These enzymes may come in handy also in the removal of small quantities of the substrates from assay systems.

4. Glycerate-2,3-Diphosphatase.

A hydrolytic enzyme apparently specific for glycerate-2,3-diphosphate has wide distribution in animal tissue, where it has been shown to be present in high concentration in beef heart, brain, muscle, liver and kidney and in rabbit brain, heart and muscle and in chicken breast muscle. The liver and kidney of rabbit have lower activities. The enzyme occurs in high concentration also in yeast. Joyce and Grisolia (1958) reported the purification of the enzyme starting from bakers yeast and acetonized chicken breast muscle. The most notable feature of the purification procedure was the use of mercuric acetate to precipitate contaminants. The crude preparations were activated by Hg^{++} but the purified preparations were not affected. The enzyme was inactive towards ATP, β-glycerophosphate, 3-phosphoglycerate, 2-phosphoglycerate and fructose-1, 6-diphosphate. Although the muscle enzyme and yeast enzyme resembled each other in many respects, they differed in some properties. The pH optimum for the muscle enzyme was in the neighbourhood of 7, while that of the yeast enzyme was about 6; the

K_m value for the former was 1.8×10^{-3} M and for the latter 3.3×10^{-4} M. Also, the muscle enzyme was inhibited to a higher degree by 3-phosphoglycerate than the yeast enzyme. When either of the enzymes acted on 2,3-diphosphoglycerate, a mixture of 2- and 3-monophosphoglycerates was obtained, probably because the enzyme preparations contained high mutase activity.

V. Enzymes Acting on Condensed Phosphates.

1. Inorganic Pyrophosphatases.

Pyrophosphatases are extensively distributed in animal tissue, plants and microorganisms. As for the specific role played by these enzymes, a few hypotheses have been put forward. Pyrophosphate is formed in many ATP-mediated biosynthetic reactions (AXELROD, 1956); these are equilibrium reactions and if pyrophosphate is removed as soon as formed, the synthesis may be expected to proceed to completion. Pyrophosphate binds metal ions and may thus inhibit enzymatic reactions requiring metal for full activity (NASHKOV and NASHKOVA, 1960). Removal of pyrophosphate would release the enzymes from inhibition. The orthophosphate formed from pyrophosphate may be delivered at specific sites needed, as for example to the mitochondria for purposes of formation of high energy phosphate bonds. LEVINSON (1957) has postulated that pyrophosphatases play a role in spore germination by hydrolysing phosphate compounds on spore coats, thereby rendering the spores more permeable. In addition to hydrolytic removal, one may visualize that pyrophosphate may enter into transfer reactions, in which case a portion of the pyrophosphate energy will be preserved in the new phosphate ester.

Pyrophosphatases display isodynamic properties, three types having been recognized on the basis of the pH optima (ROCHE, 1950).

1. pH optimum in the range 7.2 to 8.2: strongly activated by Mg^{++} (JENNER and KAY, 1931, NAGANNA and MENON, 1948); abundant in animal tissues, for example, erythrocytes (KAY, 1928), muscle (LOHMANN, 1928), nervous tissue (GORDON, 1950; SEAL and BINKLEY, 1957). The yeast enzyme also falls into this category (KUNITZ, 1952).

2. pH optimum in the range 5 to 5.5: not significantly activated by Mg^{++}; very strongly inhibited by fluoride; always associated with alkaline phosphatase in animal tissue and abundant especially in liver (ROCHE and BAUDOIN, 1943); encountered also in plants and microorganisms. These enzymes have not been purified sufficiently.

3. pH optimum in the region 3.2 to 4.0: These enzymes also have not been characterized properly.

a) Inorganic Pyrophosphatase of Microorganisms.

1. Yeast pyrophosphatase. Of all the pyrophosphatases the yeast enzyme has been the best studied. KUNITZ (1952) crystallized the enzyme and it has since been available for the specific estimation of pyrophosphate in biological systems.

Assay.

The enzymic activity is expressed in terms of the rate of liberation of orthophosphate phosphorus from a solution of sodium pyrophosphate. The assay system is made up of 4 ml 0.1 M veronal-HCl buffer, pH 7.2; 1 ml 0.01 M $Na_4P_2O_7$; 1 ml 0.01 M $MgCl_2$ and 1 ml enzyme in 0.02 M buffer pH 7.2 or 1 ml buffer without enzyme for blank. Mixtures are incubated at 30° C for 15 minutes, at the end of which suitable aliquots are withdrawn for orthophosphate estimation.

One enzyme unit is defined as the amount of enzyme liberating 1 mg of orthophosphate phosphorus per minute at 30° C and pH 7.2. Specific activity is expressed in terms of units of activity per mg protein.

Enzyme isolation. The following is the procedure adopted by KUNITZ (1952). Baker's yeast was plasmolysed with toluene and extracted with water. The solution was brought to 0.5 saturation with ammonium sulfate and the precipitate filtered off and rejected. The enzyme was precipitated from the filtrate by adding ammonium sulfate to 0.7 saturation. The precipitate was taken up in water, the pH adjusted to 5.35 and inert material removed by slow autolysis at 5° C, and the enzyme fractionated with ammonium sulfate. At this stage contaminants were removed by adsorption on tricalcium phosphate gel. The process of precipitation with ammonium sulfate and treatment with gel was repeated. Ammonium sulfate was removed by dialysis, and the enzyme crystallized by the cautious addition of alcohol at controlled pH. The enzyme was purified by recrystallization and dried with acetone. The yield was 10 to 15 mg per pound of yeast.

Properties. Crystalline pyrophosphatase of yeast is a soluble colorless protein of the albumin type. Its molecular weight is of the order of 100,000. The enzyme requires the presence of Mg^{++} for its catalytic activity, the concentration being equimolar to, or slightly more than that of pyrophosphate at the optimum pH of 7.2. Co^{++} and Mn^{++} also activate when Mg^{++} is absent, but they exert an inhibitory effect on the activation by Mg^{++}. Ca^{++} acts as an inhibitor. The kinetics of the enzymic hydrolysis of pyrophosphate in the presence of optimal Mg^{++} concentration follow the zero order reaction. The hydrolysis proceeds to completion and is irreversible.

Specificity. KUNITZ (1952) reported that the several times recrystallized enzyme was absolutely specific for inorganic pyrophosphate. Inorganic polyphosphates and organic pyrophosphate esters such as ATP and ADP were not split to any significant degree. SCHLESINGER and COON (1960) made the surprising discovery that when used in conjunction with Zn^{++}, Mn^{++} or Co^{++}, instead of Mg^{++}, the crystalline pyrophosphatase of yeast was able to hydrolyse a number of nucleoside tri- and diphosphates. In the presence of Zn^{++}, ATP, GTP, UTP, CTP and inosine triphosphate were attacked with great ease; the corresponding diphosphates were hydrolysed less readily. The pH optimum for ATP-ase action was 6.0, the cleavage taking place between the β and γ-phosphate groups to yield ADP, which was then slowly converted to AMP. The author considered that the nucleotide pyrophosphatase activity of yeast inorganic pyrophosphatase may have physiologic significance.

2. **The pyrophosphatase of Streptococcus faecalis.** OGINSKY and RUMBAUCH (1955) showed that cell-free extracts of *S. faecalis* hydrolyse pyrophosphate both in the alkaline and acid range and that these reactions have an absolute requirement for divalent metals. Mg^{++} was most active in the alkaline range and Co^{++} in the acid range. A separation could not be effected between the two types of activities by fractionation procedures. The alkaline enzyme had a pH optimum at 8. Ca^{++}, Fe^{++} and Mn^{++} also stimulated the activity, but they were far less effective than Mg^{++}, resembling in their action the metal activation of pyrophosphatases of animal tissues. The acid pyrophosphatase had a pH optimum of 5.3 to 5.4 and its pattern of activation was quite different from that of any heretofore described. The enzyme was activated primarily by cobaltous ions and more strongly so in the presence of histidine. Present alone, histidine did not stimulate the acid phosphatase activity. The requirement for histidine was not specific, since it could be replaced with almost equal effectiveness by cysteine. Mn^{++} and Zn^{++} also activated the enzyme, but to a less extent than Co^{++}. The

authors also showed that the addition of Co^{++} to the culture medium for growth of *S. faecalis* increased the level of acid pyrophosphatase activity, as measured in the presence of Co^{++} and histidine. There was, however, very little activity in the absence of added Co^{++} in the test system, indicating that the presence of Co^{++} in culture medium resulted in a marked increase in the "activatable" or "potential" enzyme.

1×10^{-3} or 1×10^{-2} M fluoride did not inhibit the acid pyrophosphatase.

3. Pyrophosphatase of Azotobacter agilis. JOHNSON and JOHNSON (1959) reported that the pyrophosphatase of *A. agilis* was located totally in the nonsedimentable cytoplasmic constituents when the cells were disrupted by sonic vibration or by grinding with alumina. The kinetics of the enzymatic reaction were studied in the supernatant obtained on centrifugation of sonicates at $144,000 \times g$ for 60 minutes. The enzyme had absolute requirement for Mg^{++}; with Co^{++} there was 10% of the activity obtained with Mg^{++}; Mn^{++} and Ca^{++} were without effect. An outstanding property of the enzyme was its unusual heat-stability, heating at 75° C for two hours not affecting the enzyme activity. This is quite surprising considering the fact that the organism is a nonspore-forming mesophile; in contrast, MARSH and MILITZER (1956) reported that the pyrophosphatase of a spore-forming thermophile was resistant for less than 1 hour to heating at 75° C. With dialysed preparations of the *Azotobacter* enzyme this heat-stability could be demonstrated only in the presence of added Mg^{++}.

4. Pyrophosphatases of spores.

(a) Bacterial spores. LEVINSON, SLOAN and HYATT (1958) reported the presence in the spores of *Bacillus megaterium* of a pyrophosphatase with properties quantitatively and qualitatively different from the vegetative cell enzyme. Cell-free preparations were obtained by rupturing untreated and heat-shocked spores in a MICKLE disintegrator. The extracts contained as the major component a Mn^{++}-stimulated pyrophosphatase active at neutral pH and a Co^{++}-activated acid phosphatase as a minor component. The crude extracts of spores showed surprising specificity in that there was no activity towards glycerophosphate, ATP, glucose-1-phosphate and tri*meta*- and tetra*meta*phosphates. On germination there was a decrease in the neutral, Mg^{++}-stimulated, pyrophosphatase activity. Extracts of vegetative cells were characterized by the fact that there was no Mn^{++}-activated enzyme, but increased Co^{++}-activated pyrophosphatase.

(b) Mold spores. Pyrophosphatase of the spores of *Aspergillus niger*.

BHATNAGAR and KRISHNAN (1960b) found that freshly collected spores of *A. niger* did not have phosphatase activity, but that when germination set in phosphatase activity appeared and progressively increased. The properties of the enzyme were studied in broken cell preparations. The enzyme activity was heat labile, unlike catalase of ungerminated spores (BHATNAGAR and KRISHNAN, 1960a). In general properties, the enzyme closely resembled the unspecific acid phosphatases of plants. Two other substrates tested, β-glycerophosphate and phenyl phosphate, were also hydrolysed. The enzyme was not activated by Mn^{++}, Co^{++} or Zn^{++} and was inhibited by fluoride, molybdate, L-tartrate and formaldehyde.

b) Inorganic Pyrophosphatases of Animal Tissue.

In general, animal pyrophosphatases have not been as well characterized as microbiological pyrophosphatases.

1. Pyrophosphatase of swine brain. An inorganic pyrophosphatase of swine brain was purified about 165-fold (Seal and Binkley, 1957). The enzyme had a pH optimum of 7.6 to 7.8 and required Mg^{++}. It was activated by EDTA, cysteine, glutathione and thioglycolic acid; on the other hand, it was inhibited by *para*-chloromercuribenzoate and by a variety of polyvalent metals. Although EDTA used in low concentrations activated the enzyme, presumably by complexing with inhibitory metals, higher concentrations were found to have definite inhibitory action on the enzyme activity.

2. Pyrophosphatase of Liver. Pyrophosphatases acting optimally at different pH values have been described in liver homogenates (Bamann and Gall, 1937; Norberg, 1950). Rafter (1958) demonstrated that pyrophosphatase activity was concentrated in the mitochondrial fraction from mouse liver. The pH optimum was 5.3 in 0.08 M sodium acetate. The particulate preparations had some activity also towards tripolyphosphate and ATP and much less of activity towards β-glycerophosphate and trimetaphosphate. The enzyme was inhibited by fluoride and by iodoacetate and *para*-chloromercuribenzoate. In contrast to the usual requirement for added Mg^{++}, the mouse particles were fully active without added Mg^{++}. Nevertheless, Mg^{++} was believed to be involved, since fluoride led to marked inhibition. A notable feature of the pyrophosphatase activity of the particles was the rapid loss of activity in the absence of substrate at the pH value for optimal pyrophosphatase activity.

Rafter (1960) was able to show that whereas mouse liver mitochondria metabolised pyrophosphate largely by hydrolysis to inorganic phosphate, there was simultaneous phosphoryl transfer to glucose to give rise to glucose-6-phosphate, which was estimated spectrophotometrically. The pH optimum for phosphotransferase activity was 5.1. Except for the lack of inhibition by phosphate and arsenate of glucose-6-phosphate formation, the hydrolytic and nonhydrolytic cleavage of inorganic pyrophosphate in liver mitochondria showed many properties in common. The author concluded that one and the same enzyme brought about both the reactions and postulated the following mechanism.

Enzyme + pyrophosphate $\rightleftharpoons$ Enzyme — P + orthophosphate
(a) Enzyme-P + HOH $\rightleftharpoons$ Enzyme + orthophosphate
(b) Enzyme-P + ROH $\rightleftharpoons$ Enzyme + ROP

where Enzyme-P is the hypothetical intermediate, ROH a hydroxyl containing acceptor and ROP the corresponding esterified compound.

The use of pyrophosphate in the direct phosphorylation of glucose is probably unique, since in the case of other condensed phosphates such transfer of phosphate is mediated through adenine nucleotides.

2. Tripolyphosphatases.

(Linear) tripolyphosphate has been identified in yeast extracts (Kornberg, 1956). A probable mechanism of its formation was apparent when Lieberman (1956) reported its synthesis by muscle adenylate kinase.

$$ADP + PP \rightleftharpoons 5' — AMP + PPP$$

Incidentally, the reversal of the reaction would enable the utilization of the energy of tripolyphosphate for the synthesis of ADP from AMP. Such a reaction finds a counterpart in the polyphosphate-AMP-phosphotransferase of *Myco-bacterium* (Winder and Denneny, 1957). A second mechanism of formation of triphosphate is by enzymatic cleavage of deoxyguanosine triphosphate (Kornberg, Lehman, Bessman, Simms and Kornberg, 1958).

Deoxyguanosine triphosphate $\rightarrow$ Deoxyguanosine + tripolyphosphate

This enzymatic reaction is a unique type of cleavage of nucleoside triphosphate. The enzyme has been purified and shown to be highly specific towards deoxyguanosine triphosphate and in having very great affinity for it ($K_m = 2.5 \times 10^{-6}$ M).

The enzyme triphosphatase was discovered by Neuberg and Fisher (1937, 1938) in aqueous extracts of Taka-diastase. Acting on tripolyphosphate the enzyme gave rise to pyrophosphate and orthophosphate. Mattenheimer (1956a, b) demonstrated the occurrence of a specific tripolyphosphatase in yeast along with enzymes acting on other condensed phosphates. Tripolyphosphatase and pyrophosphatase were separated from the rest by precipitation at pH 4.1 from extracts of brewer's yeast and further purified by precipitation with salts. A separation between pyrophosphatase and tripolyphosphatse was achieved by paper electrophoresis. The pH optimum in barbital buffer was in the neighbourhood of 8 and Mg^{++} was needed for maximum activity. Paper chromatographic examination revealed that tripolyphosphate was degraded through the intermediate stage of pyrophosphate (MATTENHEIMER, 1956c). KORNBERG (1956) reported the purification of a triphosphatase from yeast. An extract of yeast obtained with the aid of sonic vibration was freed from contaminating material by treatment with protamine solution. The enzyme was adsorbed on alumina gel and eluted with 0.02 M tripolyphosphate at pH 9.4 and the polyphosphate removed by passing through a column of Dowex 1. The enzyme preparation was free from metaphosphatase activity, but still contained pyrophosphatase activity. When the enzyme acted on tripolyphosphate, the latter was completely degraded to orthophosphate. Whether pyrophosphate was formed as intermediate was not certain since the enzyme preparation contained pyrophosphatase also. In fact, the assay system for tripolyphosphatase activity contained added pyrophosphatase. The enzyme had pH optimum at 6.1 and required Mg^{++} for full activity.

Tripolyphosphatases occur also in plant tissue. The purified phosphoesterase of pea seedlings (PIERPOINT, 1957a, b) was found to hydrolyse tripolyphosphate. Chromatography of the reaction product showed that pyrophosphate was formed as a result of enzyme action, even though the enzyme preparation by itself had powerful pyrophosphatase activity. The enzyme assays were carried out at pH 6.0 using 0.075 M citrate buffer. The pea enzyme differs from the yeast enzyme in its ability to hydrolyse a wide range of esters including trimetaphosphate, and in not requiring Mg^{++} for activity.

The occurrence of isodynamic, Mg^{++} activated, tripolyphosphatases in ox and human sera was reported by EBEL and MEHR (1957).

3. Tetrapolyphosphatases.

MATTENHEIMER (1956a, b) reported the presence in yeast of a specific tetrapolyphosphatase. The pH optimum in barbital buffer was in the neighbourhood of 8.0. Mg^{++} was needed for optimal action. Tetrapoly- and the other polyphosphatases were precipitated between 40 and 60% saturation with ammonium sulfate, whereas the bulk of the metaphosphatases was precipitated between 20 and 40%. Tetrapolyphosphatase could be distinguished from other enzymes by the different degrees of adsorption on alumina and calcium bentonite. The enzymatic degradation of tetrapolyphosphate was shown to proceed by way of tripolyphosphate and pyrophosphate (MATTENHEIMER, 1956c). RAFTER (1959) prepared a polyphosphatase fraction from yeast, which split tetrapolyphosphate at pH 8.0 in the presence of Mg^{++}; however the preparation was more active towards tripolyphosphate and especially pyrophosphate. The formation of tetrapolyphosphate in cell-free systems has not been demonstrated unequivocally. RAFTER (1959) believed that tetrapolyphosphate was formed from trimetaphosphate in the presence of an enzyme fraction from yeast and inorganic orthophosphate.

4. Trimetaphosphatases.

Of the (cyclic) metaphosphates only trimetaphosphate has been reported in microorganisms. KORNBERG (1956) isolated both trimetaphosphate and tripolyphosphate from extracts of yeast prepared with hot water or cold TCA. The existence of the compound in animal tissue seems doubtful. The mechanism of formation of trimetaphosphate has not yet been elucidated, since unlike pyrophosphate, tripolyphosphate and the polyphosphates, the formation of trimetaphosphate could not be demonstrated in cell-free systems.

Enzymes hydrolysing trimetaphosphate have wide distribution in microorganisms, plants and animal tissues. Kitosato (1928 a, b), Meyerhof, Shatas and Kaplan (1953) and Mann (1944 a, b) could show only orthophosphate as a result of trimetaphosphate hydrolysis. Rothenbach and Hinkelman (1954) showed that the metaphosphatase in extracts of germinated and ungerminated barley liberated orthophosphate from trimetaphosphate.

The enzyme preparation of Meyerhof, Shatas and Kaplan (1953) was obtained by precipitation from sonic extracts of baker's yeast with ammonium sulfate between 30 to 60% saturation and dialysis. The pH optimum was about 7. The enzyme was activated by Mg^{++} and less so by Mn^{++}, Co^{++} and Zn^{++}. Fluoride and cyanide inhibited. Pyrophosphate was hydrolysed with double the speed of trimetaphosphate; hexametaphosphate was also hydrolysed, but at 1/3 the speed. The authors could not obtain any evidence for the intermediate formation of pyrophosphate when the enzyme degraded trimetaphosphate to orthophosphate.

Kornberg (1956) reported the preparation of an apparently specific trimetaphosphatase from yeast. Extracts prepared by subjecting yeast to sonic vibrations were fractionated with ammonium sulfate, followed by alcohol. A 30-fold purified enzyme preparation was obtained which was almost free of tripolyphosphatase activity. The enzyme system for activity determination consisted, besides the enzyme, of succinate buffer, pH 6.1, Mg Cl_2 and tripolyphosphatase. The orthophosphate formed at the end of the incubation period was estimated. When the tripolyphosphatase was left out from the assay system, it was found that only small amounts of orthophosphate were formed, indicating that the product of hydrolysis of trimetaphosphate was tripolyphosphate. The formation of tripolyphosphate was actually confirmed by ion exchange chromatographic studies. The yeast enzyme is thus a cyclophosphatase having phosphodiesterase, but not phosphomonoesterase activity.

Mattenheimer (1956 a, b) obtained evidence for a specific trimetaphosphatase in maceration extracts of brewer's yeast, with a pH optimum of about 7 and activated by Mg^{++}. The cleavage of trimetaphosphate gave rise to tripolyphosphate and then to pyrophosphate and orthophosphate (Mattenheimer, 1956 c).

The pea enzyme of Pierpoint (1957 a, b) acting on trimetaphosphate at pH 6 gave rise exclusively to orthophosphate; neither tripolyphosphate, nor pyrophosphate could be identified. This is not surprising when one considers the fact that the enzyme preparation possessed also tripolyphosphatase and pyrophosphatase activity.

Rafter (1959) reported a unique reaction catalysed by extracts of yeast prepared by shaking with glass beads. At pH 9.0, and in the presence of orthophosphate the enzyme preparation brought about the disappearance of trimetaphosphate. Mg^{++} was not necessary; in high concentrations it actually inhibited the reaction. The product was not identified with certainty, but was considered to be tetrapolyphosphate. Granting that tetrapolyphosphate was the product formed from trimetaphosphate, the reaction is unique from two points of view (a) a cyclic phosphodiester bond was cleaved by an agent other than water, namely, orthophosphate, and a new P—O—P linkage was formed (b) inorganic phosphate was directly incorporated into trimetaphosphate without the mediation of adenosine polyphosphates.

5. Tetrametaphosphatases.

MATTENHEIMER (1956a, b) reported a specific tetrametaphosphatase in yeast with pH optimum near 7 in barbital buffer, and requiring Mg^{++} for full activity. The author also showed that autoproteolysis led to the destruction of tetrametaphosphatase, along with tetrapolyphosphatase and polyphosphatase. Tetrametaphosphate was shown to be degraded to orthophosphate by way of tetrapolyphosphate, tripolyphosphate, and pyrophosphate (MATTENHEIMER, 1956c). The enzyme was therefore a cyclophosphatase, rupturing the phosphodiester bond. The pea enzyme preparation of PIERPOINT (1957) did not act on tetrametaphosphate. EBEL and MEHR (1957) demonstrated the occurrence of isodynamic tetrametaphosphatases in human and ox serum.

VI. Enzymes Acting on Polyphosphates.

Polyphosphates find wide distribution in microorganisms and have attracted considerable attention during recent years as potential phosphagens (DREWS, 1960). The demonstration of the biosynthesis of high molecular polyphosphates in cell-free systems has been a triumph of the enzymologist (YOSHIDA and YAMATKA, 1953; KORNBERG, KORNBERG and SIMMS, 1956). Three types of enzymes are known to act on polyphosphates:

Type I Enzyme: Polyphosphate Depolymerases.

Enzymes capable of depolymerising high molecular polyphosphates without the formation of inorganic phosphate have been described by INGELMAN and MALMGREN in yeast, and the mycelia of *Aspergillus niger* and *Penicillium* (1948, 1949a, b). The enzymes are assayed by the drop in viscosity when they act on solutions of highly polymerized polyphosphates. The final scission product was believed to be tetra- or pentaphosphate. The significance of the occurrence of these enzymes is unknown, unless it be that the energy liberated during the rupture of the high energy bonds may be harnessed for biosynthetic reactions (HOFFMANN-OSTENHOF and WEIGERT, 1952).

Type II Enzyme: Polyphosphatases.

These are hydrolytic in nature, and lead to the liberation of orthophosphate. The biological role of these enzymes, which occur also in animal tissues which do not contain the substrate, is uncertain. It has been postulated that they prevent the accumulation in tissues of the polyphosphates, which are toxic because of their protein denaturing and metal chelating properties. According to ROTHSTEIN and MEIR (1954) polyphosphates in small amounts constitute an integral component of the yeast cell membrane, serving to bind metallic ions and are thereby concerned with the enzymatic membrane transfer of electrolytes. The cytoplasmic polyphosphatases may then be thought to function by regulating the level of the membrane-bound polyphosphates.

MANN (1944a, b) reported the presence in aqueous extracts of the mycelium of *Aspergillus niger* of an enzyme with powerful hydrolytic action against "hexametaphosphate". Pyrophosphate, metaphosphate, α-phosphoglycerol, β-phosphoglycerol, glucose-1-phosphate, fructose-6-phosphate, fructose-1,6-diphosphate, 3-phosphoglycerate, adenosine triphosphate (all three atoms of phosphorus), muscle adenylate, yeast adenylate, aneurin diphosphate and phytate were also split. MATTENHEIMER (1956a, b) reported the occurrence in yeast of a specific polyphosphatase, besides the enzymes acting on the lower linear and the cyclic condensed phosphates. Evidence has accumulated to show that the polyphosphatases are not confined to microorganisms, but are present also in animal and plant tissue. Ebel and Mehr (1957) showed that ox and human sera possess hydrolytic activity against polyphosphates. Four isodynamic polyphosphatases were shown to be present with pH optima 3.8 to 4.0, 5.6 to 6.6, 7.4 and 8.4. Mg^{++} activated the polyphosphatase action. Significant changes in the activities of the enzyme were found to be associated with the stage of growth and with pathological conditions.

Depending on the range of molecular complexity of the polyphosphate molecules acted upon, three types of polyphosphates can be distinguished.

a) Type I.

The first type consists of a number of polyphosphatases isolated from various sources which attack only relatively short chain molecules. An example of this type is the enzyme isolated from pea seedlings by Pierpoint (1957a, b) which splits "commercial hexametaphosphate". Muhammed, Rodgers and Hughes (1959) reported that Pierpoint's enzyme did not mineralize bacterial or synthetic polyphosphates of high molecular weight. The enzyme preparation of Mattenheimer (1956 b) should also be included in this class.

Pierpoint (1957a) reported the purification of the pea enzyme. Pea seedlings were extracted with citrate solution, clarified at pH 4.5 and the enzyme precipitated with ammonium sulfate between 50 and 80% saturation. The precipitate was dissolved in citrate buffer and dialysed. The enzyme was further fractionated by the pH-gradient elution method using Dowex 50. Adsorbing the enzyme at pH 4.5 and eluting with buffer of increasing pH values, two distinct components were obtained, but the two fractions had similar, though not identical, properties.

Commercial "hexametaphosphate" was acted on by the enzyme, liberating orthophosphate. The enzyme was, however, not specific for "hexametaphosphate". The pea enzyme was characterized by the close association of phosphomonoesterase and phosphodiesterase activities. Diphenylphosphate, the typical substrate for phosphodiesterase action, was not hydrolysed, but the enzyme acted on trimetaphosphate, a reaction involving the splitting of diphosphoester and monophosphoester bonds. However, the closely related cyclic phosphate, tetrametaphosphate, was not hydrolysed. In addition, the enzyme was active against pyrophosphate, tripolyphosphate, and β-glycerophosphate. Acting on ribonucleic acid, about 7% of the phosphorus was liberated in the ortho form, due primarily to phosphomonoesterase activity and not to diesterase action.

The pH optimum for the purified enzyme acting on the various substrates was 6 using citrate buffer; but assays with crude preparation were usually conducted at pH 5. Metal was not necessary for the enzymic activity and EDTA did not have any effect. Sulfhydryl groups were not essential for enzymatic action, since iodoacetate and para-chloromercuribenzoate did not inhibit the enzyme action.

Mattenheimer (1956a, b) obtained evidence for a specific polyphosphatase acting on high molecular polyphosphate, as distinct from tetrapolyphosphatase, tripolyphosphatase, pyrophosphatase, tetrametaphosphatase and trimetaphosphatase. It was possible to distinguish the enzyme from the other phosphatases on the basis of precipitation with ammonium sulfate, and adsorption on carriers like alumina and calcium bentonite. The enzymic hydrolysis of Graham's salt (polymerization grade 27 to 30) showed a sharp maximum at pH 7.7 in barbital buffer. Mg^{++} was needed for full activity. The enzymatic degradation of Graham's salt to orthophosphate was shown to proceed with the formation of intermediate and low polymers (MATTENHEIMER, 1956c).

Polyphosphate → polyphosphate + orthophosphate
(n 27 to 30)   (n 10 ?)
Polyphosphate → oligophosphate + orthophosphate
(n 10 ?)   (n 4—10)
Oligophosphate → tripolyphosphate + orthophosphate
(n 4 to 10)

b) Type II

The second type of polyphosphatase attacks long chain and short chain poly-phosphates, as well as ring phosphates. This type is exemplified by the unspecific phosphatase isolated from tapioca leaves by AGRAWAL and KRISHNAN (1959). The purification of this enzyme was not taken sufficiently far to pronounce a definite opinion on homogeneity and substrate specificity.

c) Type III

The third type which attacks only long chain polyphosphates has been purified from *Corynebacterium xerosis* (MUHAMMED, RODGERS and HUGHES, 1959). The enzyme had no action on trimetaphosphate, tetrametaphosphate, pyrophosphate, ATP and glucose-6-phosphate. It had a pH optimum at 7.0 with very little activity at pH 6.0. Mg^{++} activated the enzyme activity. The enzyme was purely hydrolytic in character, without ability to transfer polyphosphate phosphorus to ADP or AMP. The hydrolysis of polyphosphate was shown to be irreversible. By carrying out the reaction in a dialysing chamber the authors were able to demonstrate that when polyphosphates were broken down no dialysable intermediates were formed other than orthophosphate. The mechanism of action was thus clearly different from that envisaged by MATTENHEIMER (1956c) for the yeast enzyme. Since TEWARI (1960) had shown that polyphosphate molecules with end group molecular weights below 6000 are dialysable, it would appear that the *Corynebacterium* enzyme is able to act only on polyphosphates with end group molecular weights in excess of 6000.

Type III Enzyme: Enzymes Transfering Phosphate Groups from Polyphosphates to Suitable Acceptors.

Enzymes transferring phosphate from polyphosphate to AMP or ADP have been shown to occur in cell-free preparations of *Mycobacterium tuberculosis* (WINDER and DENNENY, 1957), *Escherichia coli* and *Corynebacterium diphtheriae* (KORNBERG, KORNBERG and SIMMS, 1956; KORNBERG, 1957). Since ATP so formed can be made to phosphorylate glycerol or glucose in the presence of the corresponding kinase it would appear as if polyphosphates can function as phosphagens.

Polyphosphate-AMP-phosphotransferase of Mycobacterium. WINDER and DENNENY (1957) prepared cell-free extracts of *Mycobacteria* by grinding with alumina. Assay of the enzyme was based on the transformation of AMP to ADP, which in the presence of glycerol-kinase, also present in the extract, brought about the phosphorylation of added glycerol. The assay system was made up of 1 ml of 0.5% (w/v) $MgSO_4 \cdot 7H_2O$, 1 ml 2% (w/v) glycerol in water, 1 ml 0.5 M succinate buffer, pH 6.3, 1 ml of 0.2% (w/v) 5'-AMP, 1 ml of 1% sodium poly-phosphate in 0.5% K_2SO_4 and 3 ml of cell extract. Control tubes lacking poly-phosphate and nucleotide were also set up. 2 ml samples were pipetted into 2 ml volume of 10% TCA at zero time. After incubation for 3 to 5 hours, further samples were withdrawn for analysis. The sample tubes were capped, heated for 10 minutes at 90° C to extract polyphosphate from precipitated protein, centrifuged and the supernatant fluid used for the determination of the sum of orthophosphate and acid-labile phosphate. The decrease in this figure as a result of incubation was a measure of the amount of stable phosphate ester (glycerophosphate) and hence a measure of the formation of ATP and ADP from AMP. By carrying out the enzymatic reaction in the absence of added glycerol and by ion exchange chromatography of the products formed, it was shown that ADP and not ATP was formed. The pH optimum for the reaction was 6.3, there being very little activity at pH 7. Mg^{++} activated the reaction.

Polyphosphate-ADP-phosphotransferase of Escherichia coli

$$(PO_3^-)n + n\,ADP \rightleftharpoons n\,ATP$$

Assay of enzymic activity. The formation of ATP from polyphosphate and ADP was measured by determining the acid soluble P^{32} released from labeled polyphosphate as a result of the enzymatic action. The assay system contained: glycyl glycine buffer, pH 7.0, 12.5 μmoles; $MgCl_2$ 1 μmole; ADP 0.17 μmole; ammonium sulfate 10.5 μmoles; labeled polyphosphate 0.056 μmole and enzyme in a total volume of 0.25 ml.

The ADP was omitted in the control. After 15 minutes of incubation at 37° C, 0.25 ml of 7% $HClO_4$ was added and then 0.50 ml of a 0.15% serum albumin. The ATP formed was estimated in the supernatant isotopically. Mg^{++} was essential for the reaction. The addition of ammonium sulfate was found to have considerable stimulating effect.

Purification. KORNBERG, KORNBERG and SIMMS (1956) purified the enzyme starting from *E. coli*. Extracts in glycylglycine buffer, 0.02 M, pH 7.0, were prepared by the use of sonic vibrations. The enzyme was precipitated with streptomycin sulfate and eluted with phosphate buffer of pH 7.4 and the eluate fractionated with ammonium sulfate. Streptomycin precipitation and elution were repeated, the enzyme isolated as a pellet by high speed centrifugation and suspended in glycylglycine buffer. Mg^{++} was absolutely essential for the reaction.

VII. General Phosphomonoesterases.

Many living cells contain at least two types of phosphomonoesterases, the "acid" and the "alkaline" phosphatases, with a wide range of specificity. ROCHE (1950) classified the isodynamic phosphomonoesterases into four types based essentially on their pH optima, and outlined their principal sources and chief characteristics. Plant tissue, in contrast to animal tissue, contains mainly phosphomonoesterases acting in the acid range. The plant phosphatases fall mostly into type 2 of ROCHE (1950), that is, they have pH optima in the range 5.0 to 5.5, are not affected by Mg^{++} and are inhibited by Fe^{++}. Less common in plants are those with pH optima in the region 3.4 to 4.2, which are activated by metals. ROBERTS (1958) made a detailed study of the effect of Mg^{++}, Mn^{++}, Fe^{++}, Zn^{++}, Co^{++} and Ni^{++} on the acid phosphatase activity of dialysed wheat leaf juice against 16 substrates, and concluded that the phosphatase activity resulted from a group of enzymes, some of which require a metal for full activity. In general, the phosphatases were more influenced by metal at pH 3.9 than at 5.7. The author also emphasised that in reporting the activity of plant phosphomonoesterases, it is necessary to specify the pH at which the reaction was carried out, the substrate employed in the assay and the source of tissue used in the experiment.

1. Acid Phosphatases.

The phosphatase of white lupine seedlings (NEWMARK and WENGER, 1960) may be considered as a typical acid phosphatase.

Assay. The substrate reagent consisted of 0.016 M *para*-nitrophenylphosphate, 0.01 M $MgCl_2$, 0.067 M acetate buffer and 0.025% Triton X-100, adjusted to pH 5.3 and stored frozen. Three volumes of the substrate reagent were added to one volume of the enzyme solution to be assayed, which had been diluted with 0.05% Triton X-100 in 0.01 M EDTA. The phosphate liberated in 15 or 30 minutes at 37° C was estimated colorimetrically and corrected for enzyme blank.

One unit of acid phosphatase activity was defined as that amount which catalysed the formation of 1 μmole of *para*-nitrophenol from nitrophenylphosphate in 30 minutes at 37°. The specific activity was expressed as units per mg protein.

Purification of the enzyme. Cotyledons from 5 days old etiolated lupine seedlings were extracted with pH 5 acetate buffer and the extracts clarified. The purification of the enzyme was achieved by ammonium sulfate fractionation, heat denaturation of inert protein, ethanol fractionation, and ammonium sulfate refractionation, resulting in about 1000-fold enrichment.

Properties. The enzyme had pH optimum in the rage 5.2 to 5.5 in 0.05 M acetate buffer. The K_m value for *para*-nitrophenylphosphate in acetate buffer at pH 5.3 was 3.2×10^{-4} M in the presence of Mg^{++}. Tested at the optimum pH, Mg^{++} and Mn^{++} were only mildly stimulatory, and the surprising observation was made that EDTA augments the stimulation by Mg^{++}, while counteracting that by Mn^{++}. Fluoride was found to exert a noncompetitive inhibition on the enzyme activity. The effect of *para* chloromercuribenzoate was either not apparent, or was of a *stimulatory* nature depending on both its concentration and that of the substrate.

Substrate specificity. The lupine enzyme behaved as a typical plant acid phosphatase in having a broad range of activity. Most of the monophosphate esters and carbohydrate esters tested were hydrolysed, though with varying speeds. However, the activity towards glucose-1-phosphate was of a low order compared to that towards glucose-6-phosphate. The enzyme was active also against inorganic pyrophosphate, ATP and ADP, but not against DPN. It has powerful 3'-nucleotidase activity, but only a low order of activity against 5'-nucleotides, differing in this respect from many other plant phosphomonoesterases. There was no diesterase activity tested against diphenylphosphate, but trimetaphosphate was split at about a third of the speed of *para*-nitrophenylphosphate, indicating cyclophosphatase activity. Judged on the basis of substrate specificity, the lupine enzyme resembles other plant phosphatases (ITO, KONDO and WATANABE, 1955) in that, in addition to phosphomonoesters, terminal P—O—P bonds are also hydrolysed. In contrast, the prostatic acid phosphatase, the best studied acid phosphatase of animal origin (TSUBOI and HUDSON, 1955; REINER, TSUBOI and HUDSON 1955), is a true phosphomonoesterase.

2. Alkaline Phosphatases.

These enzymes occur in high concentration in intestinal mucosa, in kidney and in placenta (AHMED and KING, 1959). Their exact physiological role is unknown at the moment, but it may be pointed out that all these tissues are concerned with active transfers. On the basis of extensive studies of the development patterns of alkaline and acid phosphatases of avian embryos, MOOG (1958) suggested that acid phosphatase which remains at relatively constant levels of activity during the growth of the chick embryo, functions as a constitutive enzyme, whereas alkaline phosphatase, which shows marked fluctuations in level and locus of activity, functions as an adaptive enzyme.

a) Alkaline Phosphatases of Microorganisms.

Alkaline phosphatase of yeast. The Fe^{++}-dependent alkaline phosphatase of yeast (STADTMAN, 1959) deserves special reference. The author showed that the conventional steps for alcohol dehydrogenase preparation can result in the purification to a considerable degree and presumably the coprecipitation of a very similar protein possessing phosphatase activity. A separation between the two enzymes was possible either by selective adsorption of the alcohol dehydrogenase on hydroxyapatite, or by ageing at 0°, whereby the dehydrogenase decayed more rapidly. The pH optimum for the hydrolysis of para-nitrophenylphosphate was 7.5 to 8.0, using Tris buffer. Fe^{++} was absolutely essential for the enzyme activity. An interesting feature of the enzyme reaction was the marked stimulation attendant on the addition of pyrophosphate, or various nucleotides such as ATP, ADP and AMP. The routine assay system for the enzyme contained added AMP. The enzyme had a rather limited substrate specificity. Phosphoramidate was split even more rapidly than para nitrophenylphosphate; however, creatine phosphate, acetylphosphate, phosphoenolpyruvate, glucose-1-phosphate, glucose-6-phosphate, ATP and AMP were not hydrolysed.

The alkaline phosphatase of Escherichia coli. GAREN and LEVINTHAL (1960) showed that when E. coli was grown under condition of phosphate deprivation alkaline phosphatase activity made its appearance in considerable amounts. An enzyme preparation was obtained as an apparently homogeneous protein by extracting the enriched cells with 0.1 M Tris buffer of pH 7.4 at 82° for 15 minutes, dialysing, and repeated adsorption and elution from DEAE cellulose column.

The pH optimum against para-nitrophenylphosphate was 8.0 using Tris buffer. All the phosphomonoesters tested were hydrolysed inclusive of 5'- and 3'-nucleotides, and hexose phosphate, indicating that the organic moieties of the phosphomonoester substrates had only a small effect on the rate of enzyme activity. There was no diesterase activity, since bis-nitrophenyl phosphate was hydrolysed only to a limited extent. Alkaline phosphatase from cow's milk and calf intestinal mucosa has also been found to act as general phosphomonoesterase (MORTON, 1955a). The bacterial enzyme did not liberate orthophosphate from creatine phosphate, pyrophosphate or ATP. The alkaline phosphatase from cow's milk and calf intestinal mucosa was reported to be active against creatine phosphate (MORTON, 1955a). The K_m value for para-nitrophenyl phosphate was 1.2×10^{-5} at pH 8.0 and temperature 25°. The turnover number was 2700 molecules of para-nitrophenyl phosphate cleaved per minute at 25°, a value which is considerably lower than that of mammalian alkaline phosphatases. Metal chelating agents inhibited the enzyme activity leading to the assumption that a metal was involved at the active site. In this respect the bacterial enzyme resembled the alkaline phosphatase from swine kidney which was believed to contain Zn^{++} (MATHIES, 1958). The bacterial alkaline phosphatase was strongly inhibited by phosphate and arsenate, whereas the acid phosphatase from the same organism was less affected (TORRIANI, 1960). 0.01 M sodium fluoride, which inhibited the acid phosphatase of E. coli by 85%, was without effect on the alkaline phosphatase activity.

VIII. Phosphoprotein Phosphatase.

Phosphoproteins have assumed considerable significance during recent years. There is a rapid incorporation of administered P^{32} into the phosphoprotein fraction, indicating high metabolic acitivity for protein bound phosphorus. It has been postulated that the primary step in oxidative phosphorylation results in the formation of a phosphorylated protein, with subsequent transfer of the high energy phosphate to adenine nucleotides. The reversible phosphate transfer between yolk phosphoprotein and ATP was studied by RABINOWITZ and LIPMANN (1960).

The existence of enzymes catalysing the dephosphorylation of casein and phosvitin in animal tissue was reported by HARRIS (1946) and FEINSTEIN and VOLK (1949) and in plant tissue by SAMPATHKUMAR, SUNDERARAJAN and SARMA (1957). SUNDERARAJAN and SARMA (1954, 1957, 1959) reported the purification of beef spleen phosphoprotein phosphatase. These authors concluded that the spleen enzyme is distinct from simple nonspecific phosphomonoesterases. SINGER and FRUTON (1957) purified phosphoramidase from beef spleen some 50-fold and showed that the same enzyme brought about the hydrolysis of phosvitin and *ortho*-carboxyphenylphosphate. Surprisingly enough, the enzyme did not act on β-glycerophosphate, although phenylphosphate, another commonly used substrate for phosphomonoesterase activity was hydrolysed. Starting with the preparation of SINGER and FRUTON (1957), GLOMSET (1959) took the stage of purification further using chromatography on DEAE cellulose column and preparative electrophoresis and obtained a preparation which moved as single peak on electrophoresis in acetate buffer pH 5.6 and had a blue violet color. REVEL and RACKER (1960) carried out extensive studies on a preparation of beef spleen phosphoprotein phosphatase which they prepared by a modification of the procedure of SUNDERARAJAN and SARMA (1954, 1957).

Assay of phosphoprotein phosphatase. The enzyme was assayed by REVEL and RACKER (1960) in a test system containing 8 μmoles of casein-bound phosphate, 100 μmoles of acetate buffer of pH 5.8, 5 μmoles of neutralised ascorbic acid and enzyme in a total volume of 1 ml. After incubation for 15 or 30 minutes at 37°, the reaction was stopped by adding 1 ml ice cold 20% TCA and the amount of inorganic phosphate liberated was determined on an aliquot of supernatant solution after centrifugation. One unit was defined as the liberation of 1 μmole of phosphorus per minute.

Enzyme localization. PAIGEN (1959) reported that the enzyme was associated with cytoplasmic particles.

Enzyme purification. REVEL and RACKER (1960) extracted beef spleen with 0.5 M sodium chloride-0.2 M acetate solution of pH 5.0 in a WARING blendor and fractionated the extract with ammonium sulfate. The enzyme separated during subsequent dialysis but was solubilized by extracting the residue with 0.5 M sodium chloride-0.2 M acetate solution of pH 5.0. Inert protein was removed by heat treatment and nucleic acid contamination with protamine sulfate and the enzyme concentrated by ammonium sulfate reprecipitation. The preparation was dialysed against sodium chloride-acetate solution of pH 6.0, during which the enzyme remained in solution. Fractionation on DEAE-cellulose column yielded a 300-fold purified enzyme. The procedure adopted by SUNDERARAJAN and SARMA (1959) yielded a preparation which had a specific activity twice that obtained by REVEL and RACKER (1960).

Substrate specificity. SUNDERARAJAN and SARMA (1959) and REVEL and RACKER (1960) observed that the enzyme was able to hydrolyse a wide range of phosphate esters tested in the pH range 4.5 to 5.8. ATP and inorganic phosphate and aromatic monophosphate esters were split with ease. The two terminal groups were split from ATP during enzymatic action. During dephosphorylation of casein there was very little release of acid soluble nitrogenous material. In addition, the enzyme was shown to act on phosphorylenolpyruvate, phosphoamide, 3-phosphoglyceric acid, fructose-1,6-diphosphate, and acetyl phosphate. Diphenyl phosphate was not split.

Properties. The enzyme required the presence of reducing agents for full activity, ascorbic acid being the most efficient. These reducing agents could be replaced almost completely by Fe^{++}, which was as effective as ascorbic acid. No other metal could replace Fe^{++}. It may be pointed out that the enzyme preparation of SINGER and FRUTON (1957), in contrast to that of REVEL and RACKER (1960), was less dependent on reducing agents and was actually inhibited by Fe^{++}. REVEL and RACKER (1960) believed that the enzyme was an iron-protein complex, and that the metal had to be in the reduced state for eliciting enzyme activity. The stability properties of the spleen enzyme were rather unusual. While activity of the purified enzyme was preserved after exposure to 70° C for long periods, the enzyme was rapidly inactivated, even at 37° C, in the presence of ascorbic acid or thioglycolic acid. Fe^{++} as well as substrate protected the enzyme against inactivation in the presence of reducing agents. The enzyme was not inhibited by L-tartrate, which was a powerful inhibitor for the acid phosphatase of spleen.

IX. Ortho Phosphoserine Phosphatase.

A phosphatase with either absolute or marked specificity towards ortho-phosphoserine has been reported in *Escherichia coli* (SMITH, SHUSTER, ZIMMERMAN and GUNSALUS, 1956), in baker's yeast (SCHRAMM, 1958) and chicken and rat liver (NEUHAUS and BYRNE, 1959a, b, 1960; NEMER, WISE, WASHINGTON and ELWYN, 1960). The enzyme is distinguished by the fact that simultaneously with de-phosphorylation of phosphoserine, it catalyses the exchange of L-serine with the serine moiety of L-phosphoserine (NEUHAUS and BYRNE, 1959a; BORKENHAGEN and KENNEDY, 1958); however, the enzyme was unable to bring about an exchange of P^{32} labeled orthophosphate with the phosphate moiety of serine phosphate (NEUHAUS and BYRNE, 1959b). In addition to the exchange reaction between L-phosphoserine and L-serine, the enzyme can catalyse the transfer of a phosphoryl group from D-phosphoserine to L-serine. Examples of phosphatases which also carry out transfer or exchange reactions are well known, but the exchange reaction catalysed by phosphoserine phosphatase has several novel features. Usually, rapid exchange or transfer is catalysed only by nonspecific phosphatases, whereas phosphoserine phosphatase is highly specific. Furthermore very high concentrations of acceptor are usually needed to compete with water in the hydrolysis, whereas low concentrations of L-serine are effective with phosphoserine phosphatase. If serine is biosynthesized by a route involving the formation of serine phosphate, this phosphatase may be expected to play an important role in the final step of dephosphorylation.

Assay of phosphatase activity. The phosphatase activity can be assayed by the rate of liberation of inorganic phosphate from phosphoserine. The assay system used by NEUHAUS and BYRNE (1959b) consisted of 0.01 M $MgCl_2$, 0.05 M succinate-acetate buffer of pH 5.90 and 0.01 M DL-phosphoserine. After tempera-ture equilibration at 38° the reaction was started by adding the enzyme pre-paration. 0.5 ml aliquots were withdrawn immediately and again after 10 minutes and added to 1.0 ml of 10% TCA. Orthophosphate estimations were carried out on the centrifuged supernatant. A unit of enzyme was defined as that amount which caused the liberation of 1.0 μmole of orthophosphate per ml of incubation mixture in 10 minutes. Exchange assay was carried out by estimating the amount of L-serine-C^{14} incorporated into a phosphoserine pool.

Enzyme localization: According to NEMER, WISE, WASHINGTON and ELWYN (1960) the enzyme occurs in the supernatant fraction of rat liver homogenates.

Enzyme purification. SCHRAMM (1958) obtained the phosphatase from baker's yeast. The dried material was extracted with bicarbonate solution, the enzyme fractionated with ammonium sulfate, inert material removed by alumina, and fractionated on DEAE cellulose. About 30-fold enrichment of activity was achieved by this procedure. NEUHAUS and BYRNE (1959b) started from acetonized chicken liver. A 20-fold enriched preparation was obtained by extraction with 0.001 M EDTA at pH 7.4, ammonium sulfate fractionation, acid and heat treatments, ammonium sulfate refractionation, adsorption on alumina Cγ and elution with 0.1 M DL-phosphoserine, pH 7.4 and final acetone fractionation. BROKEN-HAGEN and KENNEDY (1958) used rat liver as the starting material. By a series of steps involving extraction with buffer, acid and heat treatment and fractionation on calcium phosphate gel the enzyme was obtained with an overall purification of 25—40-fold.

Properties. The pH optimum for hydrolysis was 5.9 to 6.6 and for the exchange reaction 6.9 to 7.3. Mg^{++} activated both the reactions. The following K_m values for phosphatase action were obtained: L-serine, 5.8×10^{-5} M; D-serine, 4.2×10^{-3} M; DL-serine, 1.1×10^{-4}. The K_m value for exchange reaction with L-serine was 5.7×10^{-5} M, which was practically the same as for phosphatase action. *Para*-chloromercuribenzoate inhibited the reaction, but neither L nor D tartrate inhibited. The hydrolytic reaction was inhibited by L- and D-serine and by L-alanine and glycine. Whereas the phosphoserine phosphatase preparations from all sources investigated resembled one another in hydrolytic and exchange reactions and in inhibition by serine, there was quantitative difference in the sensitivity to serine inhibition. The enzyme preparations from chicken and rat liver were inhibited by low concentrations of L-serine, whereas relatively high concentrations were required to inhibit the enzymes from baker's yeast and *E. coli*.

Specificity. The phosphate esters of both L- and D-serine were hydrolysed. NEUHAUS and BYRNE (1959b) reported that *para*-nitrophenylphosphate was split with about half the speed of phosphoserine. The enzyme acting on the phenol ester was in all probability a contaminating phosphatase distinct from orthoserine phosphatase. The purified preparation obtained by NEMER, WISE, WASHINGTON and ELWYN (1960) was found to split also phosphohydroxypyruvate, but at a slower rate than that of phosphoserine. The authors considered that phospho-hydroxypyruvate may be a natural substrate for the enzyme. As for the exchange reaction, several phosphorylated compounds tested could not replace phospho-L-serine in the conversion of free labeled serine to phosphoserine. Also, there was reason to believe that serine was the specific acceptor of the phosphate group of phospho-L-serine.

X. Phosphatase Acting on Nucleotides and Nucleic Acids.

1. Nucleotide Pyrophosphatases.

These are specific enzymes with wide distribution, bringing about the hydrolytic rupture of the internal pyrophosphate bond of nucleotides such as DPN, TPN and FAD. DPN-pyrophosphatase splits DPN to AMP and NMN and has to be distinguished from DPN-nucleosidase which catalyses the cleavage of the glycosidic bond between nicotinamide and ribose (HANDLER and KLEIN, 1942). The two enzymic activities can be distinguished by the fact that NMN specifically inhibits the nucleosidase activity. The availability of this specific enzyme has enabled the preparation of some rare mononucleotides and has thrown light on the structure of some biologically occurring compounds.

a) Nucleotide Pyrophosphatase of Potato.

KORNBERG and PRICER (1950) carried out a detailed study of the nucleotide pyrophosphatase of potato.

Assay of DPN splitting activity. The assay was based on the spectrophotometric determination of residual DPN by reduction with alcohol dehydrogenase in the presence of alcohol. An incubation mixture of 1.0 ml contained 0.1 ml DPN (0.02 M, pH 6) 0.2 ml of potassium phosphate buffer (0.5 M, pH 7.0) and enzyme solution containing 1 to 5 units. After incubation for 20 minutes 0.1 ml of the incubation mixture was added to each of two absorbtion cells (1 cm) containing 2.8 ml of alcohol-pyrophosphate buffer mixture. After an initial reading, 0.1 ml of water was added to one cell (blank) and 0.1 ml of alcohol dehydrogenase (5 to 10 μg) to the other. The optical density at 340 mμ was read off after 5 minutes.

A unit of enzyme activity was defined as the amount causing the splitting of 1 μM of substrate per hour. Specific activity was given by enzyme units per mg of protein.

Purification. Potatoes were extracted with 0.40 saturated ammonium sulfate in a WARING blendor. The enzyme was precipitated with ammonium sulfate, taken up in water, dialysed, fractionated with alcohol, adsorbed on tricalcium phosphate gel, eluted with ammonium sulfate solution adjusted to pH 7.5 with ammonia, reprecipitated with ammonium sulfate and purified further by adsorbtion on tricalcium phosphate gel and elution. A 750-fold purified preparation was obtained with an overall yield of 11%.

Properties. There was a broad optimum of activity between pH 6.5 and 8.5. Borate and pyrophosphate inhibited strongly at pH 8.5 but not at pH 7.4, suggesting that the quaternary pyrophosphate and tertiary borate ions were the inhibitory agents. The inhibition by NMN was feeble, but AMP inhibited powerfully. DPN splitting was not stimulated by Mg^{++} or Ca^{++} and was inhibited by fluoride only in the presence of phosphate.

In addition to DPN the following were split: $DPNH_2$, TPN, FAD, ATP, thiamine pyrophosphate.

b) Nucleotide Pyrophosphatase of Tobacco.

The pyrophosphatase of tobacco roots (CLAYTON and HANSELMAN, 1960) is associated with both soluble and particulate fractions, in contrast to potato, where it is in soluble form and rabbit kidney, where it is associated with mitochondria (KORNBERG and LINDERBERG, 1948). CLAYTON and HANSELMAN (1960) were able to show that ATP and DPN splitting activities by root preparations are catalysed by different enzymes.

Localization of enzyme. In general fresh roots were employed, but the authors stated that storage at $-20°$ C did not affect the intracellular distribution. Fairly drastic grinding was needed, namely, initial grinding in a WARING blendor for 60 seconds, followed by homogenization in a hand-operated TEN BROECK apparatus, and reextraction of residue obtained on passing through cheese cloth. The grinding fluid consisted of 0.5 M sucrose in 0.1 M Tris, pH 7.5, with added 0.01 M cysteine to prevent darkening during homogenization. The nuclear fraction was obtained as a sediment following 5 minutes' centrifugation at 750 $\times$g. The supernatant was then separated into mitochondrial and soluble fractions by centrifugation for 15 minutes at 20,000 $\times$g. Recombination of the nuclear, mitochondrial and supernatant fractions did not result in significant stimulation or inhibition of the activities. With DPN as test substrate there was an overall recovery of activity of 93%. The mitochondrial fraction contained about 50% of the total activity, the rest being distributed between the nuclear (27%) and the supernatant fractions (16%). The specific activity of the supernatant was only a fourth of that of the whole extract, that of the nuclear fraction was almost the same as that of the initial extract, while that of the mitochondrial fraction was almost 3-fold higher.

Neither the soluble enzyme nor the mitochondrial enzyme could be purified. Most studies were conducted with the mitochondrial fraction and a few with the soluble fraction. The pH optimum was in the range of 7.5 to 8.0 using Tris buffers of varying concentration. Ca^{++}, Mg^{++}, NMN and fluoride were without effect. Rare earth chlorides were found to be extremely potent inhibitors of nucleotide pyrophosphatase activity.

2. 5'-Nucleotidases.

Phosphatases exhibiting a marked or absolute specificity for 5'-nucleotides have been shown to occur in many tissue. Detailed studies have been made of the enzymes from snake venom (HEPPEL and HILMOE, 1951), seminal plasma (HEPPEL and HILMOE, 1951) and Clostridium stricklandii (HERMAN and WRIGHT, 1959).

Localization. DE LAMINARNDE, ALLARD and CANTERO (1958) obtained the following data for the intracellular distribution of 5'-nucleotidase in rat liver by biochemical analysis of cell fractions obtained on differential centrifugation.

Table 3. *Distribution of 5'-nucleotidase in rat liver.*

Fraction	Percent recovery	Specific activity
Whole homogenate	(100)	0.35
Nuclear fraction	40.2	0.65
Mitochondrial fraction	13.7	0.21
Microsomal fraction	37.9	0.67
Supernatant	9.2	0.08
Recovery	101	

The above results show that 5'-nucleotidase activity in liver was especially concentrated in the nuclear and microsomal fractions. NOVIKOFF, PODBER, RYAN and NOE (1953) had earlier reported that 5'-AMP splitting activity was present in both the "heavy" and the "light" microsomes of liver, unlike a number of other enzymes, which were present only in one or the other of these fractions.

a) 5'-Nucleotidase of Clostridium stricklandii.

Assay. The assay system used by HERMAN and WRIGHT (1959) consisted, in a total volume of 0.4 ml, of the following components: AMP 2 to 3 μmoles; $FeSO_4$ 2 μmoles; triethanolamine, pH 7.35, 40 μmoles and the enzyme solution.

Each incubation tube was thoroughly flushed with helium, stoppered immediately and incubated at 38° for 20 minutes. The reaction was stopped by the addition of 0.10 ml of 10% TCA and suitable aliquots of the supernatant used for orthophosphate determination.

A unit of activity corresponded to the liberation of 1 μmole of orthophosphate per minute.

Purification. Cell-free extracts of *C. stricklandii* prepared by sonic vibrations were dialysed against 0.01 M Tris, pH 7.4 and then against 0.01 M triethanolamine, pH 7.35. Nucleic acid was then removed by precipitation with protamine. Conventional methods of enzyme purification failed to increase the specific activity of the enzyme. However, adsorption on DEAE cellulose and elution resulted in 7-fold enrichment.

Properties. The pH optimum lay in the range of 7.0 to 8.5. The stability behaviour towards heat was unusual. When buffered at pH 7.35 with triethanol-amine (final concentration 0.1 M) approximately half the activity remained after 10 minutes at 100°. When tested in distilled water without buffer there was a rapid destruction of enzyme activity. The enzyme was completely dependent on certain metals, notably ferrous iron, for activity. The cleavage of AMP in the presence of Fe^{++} was some 200 times faster than in its absence. The authors believed that ferrous iron served both as a reductant and a heavy metal activator for the enzyme. A determination of K_m was not possible because of apparent saturation of enzyme at low levels of substrate. Segal and Brenner (1960) reported K_m value of 1.0 to 1.2×10^{-5} M for the microsomal fraction of rat liver and Heppel and Hilmoe (1951) $< 10^{-4}$ M for seminal plasma enzyme, using AMP as substrate. In contrast to the seminal plasma enzyme, the liver microsomal 5'-nucleotidase had maximal activity at neutral pH and had no requirement for added Mg^{++}.

Substrate specificity. The enzyme preparation from *C. stricklandii* brought about the hydrolysis of 5'-mononucleotides of adenine, cytosine, hypoxathine and uracil, and also of ADP, ATP and DPN. All these reactions were dependent completely on the presence of Fe^{++}. There was no cleavage of 3'-AMP, 2'-AMP, TPN, inorganic pyrophosphate, glucose-6-phosphate and ribose-5-phosphate.

3. Ribonucleases.

Ribonucleases may be defined as those enzymes which degrade RNA so that it no longer shows the typical macromolecular properties. Every ribonuclease is a phosphodiesterase because it is able to split the phosphodiester link of RNA.

a) Ribonucleases of Plants.

Ribonuclease was detected in plants for the first time by Jono (1930), who by following the release of inorganic phosphate from yeast RNA, found the highest activity in germinating soy beans. Schlamowitz and Graner (1946) extended this work and reported the separation of nuclease from phosphomonoesterase. Axelrod (1947) pointed out the heat stability of ribonuclease activity present in citrus phosphatase preparations. Ribonucleases have since been studied in detail in many plant tissues, including sprouted castor beans (Bheemeshwar and Sreenivasayya, 1953), sprouted barely and rye grass (Shuster and Kaplan, 1953), pea leaves (Holden and Pirie, 1955a) (Markham and Strominger, 1956), tobacco leaves (Frisch-Niggemeyer and Reddi, 1957), spinach (Tuve and Anfinsen, 1960) and carnivorous plants (Mathews, 1960). The enzyme was obtained in a high state of purity by Holden and Pirie (1955a), Pierpoint (1956), Frisch-Niggemeyer and Reddi (1957), Shuster (1957) and by Tuve and Anfinsen (1960). The properties of the enzyme were studied by these workers as well as by Reddi (1958) and Shuster, Khorana and Heppel (1959).

Assay of the enzyme. The following assay system was used by Tuve and Anfinsen (1960) for the spinach enzyme and was based on the procedure of Anfinsen, Redfield, Choate, Page and Carrol (1954). 1 ml of 0.4% RNA in 0.1 M acetate buffer, pH 5.0, the enzyme and acetate buffer in a total volume of 2.5 ml were incubated in a 12 ml centrifuge tube at 24° for 25 minutes. 0.5 ml of 0.75% uranyl acetate in 25% perchloric acid was added to stop the reaction. After centrifugation for 4 minutes, 0.1 ml aliquots were diluted to 3.1 ml and optical density determined at 260 mμ in silica cells. Corrections were made for appropriate blanks containing no enzyme, or, in case of highly coloured crude enzyme preparations, zero time blanks containing enzyme. One enzyme unit corresponded to an increase in optical density of 0.010.

Products formed during enzymatic degradation may be identified by chromatography, ultra violet spectral analysis and by treatment with purified nucleotidases.

Distribution in plants. In sprouting wheat grain the highest activity was found on the 6th day, the activity being higher in the sprouts than in the endosperms (SISKYAN, VASILENA and KOSHLOYANTS, 1952). Ribonuclease activity was found in the "sap" from the leaves of broad bean, french bean, tomato, potato and many other plants (HOLDEN and PIRIE 1955a). Measurements have been made by these authors of the distribution of ribonuclease in leaf fractions of pea and tomato. Extracts of the washed fibre contained 0.33 as much enzyme as the original sap and washings. Of the activity present in the sap only less than 20% was sedimented after 30 minutes centrifugation at $100,000 \times g$. During germination of pea seed for 17 days, ribonuclease activity expressed as units per g wet weight, increased between 2nd and 6th days, then remained constant till the 10th day and then gradually fell. In the initial stages of germination the greater part of activity of ribonuclease was in the cotyledons, but as germination progressed the % of the total activity in the cotyledons decreased, while that in the shoot increased. Roots contained only a small proportion of the total activity; the activity per g fresh weight was lower in root than in shoot. SHUSTER (1957) reported that the ribonuclease activity of barley increased 3-fold during germination, while the 3'-nucleotidase activity increased 10—15-fold.

Localization of the enzyme. PIRIE (1957) reported ribonuclease activity in microsomes from healthy tobacco leaves. MATSUSHITA and IBUKI (1960) found the presence of ribonuclease in the microsomes from young wheat roots and from pea seedlings. Whereas the ribonuclease activity of the microsomes from young wheat roots was inhibited by EDTA, the enzyme from pea seedlings was not inhibited; in fact, it was rather activated by EDTA. When the microsomes of pea seedlings were fractionated into endoplasmic reticulum and ribonucleoprotein particles, different ribonucleases were found in the two fractions, differing in pH optimum and the effect of added EDTA. Three day old pea seedlings were homogenized in 0.5 M sucrose and fractionated by centrifugation at $16,000 \times g$ for 15 minutes to remove mitochondrial sediment. The supernatant was centrifuged at $50,000 \times g$ for 30 minutes to sediment microsomes consisting of endoplasmic reticulum and a part of ribonucleoproteins. The resulting supernatant was further centrifuged at $100,000 \times g$ for 120 minutes to sediment ribonucleoprotein particles. The washed microsomes were suspended in water and 0.2 vol. butanol added and subjected to sonic oscillation for 10 minute. NaCl was added to 0.15 M and sonic oscillation continued for 10 minutes. The suspension was centrifuged at $5000 \times g$ for 20 minutes and the clear water phase separated from butanol phase, and dialysed. Tested in 0.1 M acetate buffer pH 6.0, which was the pH optimum, 2.5×10^{-3} M EDTA was found to inactivate the ribonuclease activity. The ribonuclease associated with the ribonucleoprotein particles was solubilized by metal binding compounds, such as EDTA, deoxycholate, citrate or phosphate. The enzyme had an optimum pH of 5.6 and the activity was not inhibited by EDTA, indicating that metal ions were not needed for activity.

The ribonuclease of endoplasmic reticulum of pea seedlings corresponded to the enzyme of wheat roots. The ribonuclease of the ribonucleoprotein particles was similar to ribonuclease of rat liver microsomes (TASHIRO, 1958), of mouse pancreas microsomes (DICKMANN and TRUPIN, 1958) and in the ribonucleoprotein particles of *E. coli* (Elson, 1958). But these enzymes were all latent and were rendered active only when the microsomes or ribonucleoprotein particles were treated with EDTA or urea.

Purification and properties of plant ribonucleases.

A. Pea leaf enzyme (Holden and Pirie, 1955 a, b). The leaves harvested from plants grown for about 3 weeks were minced and the sap squeezed by hand pressure. The fibre was extracted with citrate. The sap and the fibre extract were freed from inert protein by adjustment of pH to 4.5. The enzyme was fractionated with ammonium sulfate, dialysed and refractionated with ammonium sulfate. A 230-fold purified preparation was obtained which did not act on DNA, glycerophosphate or diphenyl phosphate. The pH for optimum activity was 5.5. The leaf enzyme was much less stable to heat treatment then pancreatic ribonuclease. Compared to the pancreatic enzyme, the leaf enzyme was much more easily affected by several inhibitors.

Pierpoint (1956) was able to achieve separation of contaminating metaphosphatase and phosphomonoesterase activities from the pea leaf ribonuclease preparation of Holden and Pirie (1955a) by the use of ion exchange chromatography. The purified enzyme was not affected by 5×10^{-2} M azide, fluoride and arsenate. Molybdate, which inhibits some plant phosphatases, had little or no effect on the ribonuclease activity when tested in citrate buffer. However, molybdate inhibited the enzyme in buffers which did not form complexes with metals as readily as citrate. Thus, in maleate buffer, 2.2×10^{-3} M molybdate inhibited the enzyme by 50%. Structural analogues of purines and pyrimidines had no effect in concentrations up to 1.25×10^{-3} M. Chloramphenicol or terramycin, tested in concentrations which inhibit protein synthesis, had no effect on the leaf ribonuclease.

B. Tobacco leaf ribonuclease. The procedure adopted by Frisch-Niggemeyer and Reddi (1957) consisted in freezing and grinding the leaves, centrifugation, removal of precipitate formed at pH 4.5, collection of precipitate formed between 40—85% ammonium sulfate saturation, dialysis and alumina chromatography, followed by β-glycerophosphate elution.

C. Rye grass ribonuclease. Shuster and Kaplan (1953) found that all preparations of "b nucleotidase" from germinating barley and rye grass degraded RNA completely, without the release of more than about 6% of the total phosphate as inorganic phosphate, similar to the tobacco leaf ribonuclease of Pirie (1950). Shuster (1957) was able to effect a separation between 3'-nucleotidase, ribonuclease and deoxyribonuclease in germinating rye grass. Shuster, Khorana and Heppel (1959) studied in detail the mechanism of action of purified rye grass ribonuclease.

Rye grass seeds, after two days of germination, were ground in a Waring blendor and the homogenate centrifuged to remove cell debris and starch. The precipitate obtained on 90% saturation with ammonium sulfate was dissolved in water, dialyzed and adsorbed on alumina C_γ gel and eluted with 1.0 M NaHCO$_3$ and dialyzed. It was then adsorbed on calcium phosphate gel and eluted with 1.0 M ammonium sulfate. A second treatment with alumina gel resulted in adsorption of 80% of the 3'-nucleotidase, leaving behind the ribonuclease in solution. The supernatant was dialyzed and the enzyme further purified by two more treatments with alumina gel. During the different steps in the purification the ratios of the activities of the three enzymes, ribonuclease, deoxyribonuclease and 3'-nucleotidase, did not remain constant. Ribonuclease preparation was enriched 58-fold, deoxyribonuclease preparation 29-fold and 3'-nucleotidase preparation 10-fold.

The optimum pH for ribonuclease action was 4.8, whereas that for deoxyribonuclease was 5.5 and for 3'-nucleotidase 7.5. The K_m value was 2.5×10^{-3} M total RNA-phosphate. The ribonuclease was found considerably more heat-stable than 3'-nucleotidase and deoxyribonuclease. During heat-treatment ribonuclease activity was most stable between pH 7 and pH 9, whereas 3'-nucleotidase had maximum heat stability in 0.05 NHCl. During purification the heat-stability of ribonuclease and 3'-nucleotidase increased, but the addition of small amounts of crude enzyme made the purified 3'-nucleotidase heat-labile, whereas it had no effect on heat-stability of ribonuclease. The factor causing heat-lability of 3'-nucleotidase was found to be non-dialyzable, precipitable with acetone, stable to heating for 10 minutes at 100° C in 1 N HCl and destroyed by ashing.

Mg^{++}, in concentration of 1×10^{-2} M, which was a potent inhibitor of pancreatic ribonuclease, gave rise to only 30% inhibition with rye grass ribonuclease and none with deoxyribonuclease. KCN and cysteine did not inhibit at pH 4.5, but use of higher pH values resulted in marked inhibition.

D. *Spinach ribonuclease*. The preparation obtained by TUVE and ANFINSEN (1960) from spinach is the purest plant ribonuclease obtained so far. Spinach leaves were put through a coarse meat grinder and repeatedly frozen and thawed. The mince was homogenized with 0.1 M potassium phosphate buffer pH 5.7 and after stirring for 2 hours at 4° the extract was separated by centrifugation. Inert protein was now precipitated and removed by adjusting pH to 5.1. The enzyme was adsorbed on Amberlite resin XE—64 and eluted with molar potassium phosphate buffer pH 6.0. Further purification was achieved by ammonium sulfate fractionations and chromatography on a DEAE-cellulose column. Over 2500-fold purification was achieved by the above procedure.

The activity of the purified enzyme preparation reached a maximum in 0.1 M buffers at pH 5.0 to 6.0 with a distinct double hump. In 0.1 M citrate or citrate-phosphate buffer the activities were slightly higher, and the humps exaggerated. It was not certain whether the enzyme preparation from spinach leaves contained 2 ribonucleases with differing pH optima. There was no activity at alkaline pHs either in crude or purified preparations. p-Chloromercuribenzoate, mercaptoethanol and EDTA, all 0.1 M, and 0.001 M Mg^{++} or Ca^{++} had no effect on the activity of the purified enzyme.

Comparison of the various plant nucleases. All have a pH optimum in the range 4.5 to 6.0. The pea leaf enzyme has an optimum at pH 5.5 (HOLDEN and PIRIE, 1955 b), tobacco leaf enzyme at 5.1 (FRISCH-NIGGEMEYER and REDDI, 1957) and rye grass enzyme at 4.5 (SHUSTER, 1957).

Tobacco leaf ribonuclease appears to be heat-stable, as is the pancreatic enzyme, whereas rye grass and spinach ribonuclease appear to be intermediate in heat stability, and pea leaf ribonuclease is heat-labile.

In contrast to pancreatic ribonuclease, which gives rise only to pyrimidine mononucleoside cyclic phosphates, all of the ribonucleases obtained from plant sources in purified form are capable of completely hydrolysing RNA to mononucleoside 2'-3'-cyclic phosphates. In every case guanosine 2'-3'-cyclic phosphate is the earliest detectable mononucleotide. In addition, spinach and rye grass ribonucleases further catalyse the hydrolysis of purine and pyrimidine nucleoside cyclic phosphate to nucleoside 3' phosphates, whereas the ribonucleases of tobacco and pea leaves hydrolyse only the purine nucleoside cyclic phosphates.

b) Endonucleases from Microorganisms.

The plant nucleases described above, as well as pancreatic ribonuclease, cannot be strictly designated as *endonucleases* since mononucleotides can be detected early during the enzyme action. Their action proceeds in two distinct stages; the first is an intramolecular transphosphorylation to form products with 2′-3′-cyclic phosphodiester end groups; this is followed by a hydrolysis of all or some of the 2′-3′-cyclic phosphate groups. Two endonucleases of microbial origin have been described which, by contrast, bring about a direct hydrolytic attack on the phosphodiester bonds of RNA and polynucleotides.

Nuclease of Micrococcus pyrogenes var. aureus. The enzyme occurring in culture filtrates was purified by Reddi (1959) by a procedure involving heat-treatment, fractionation with ammonium sulfate and isolation by precipitation at pH 4.5. The activity of the enzyme was assayed by spectrophotometric measurement of the absorption of the breakdown products which were not precipitated by acid. The enzyme had a pH optimum of 8.8, using borate buffer. Ca^{++} activated the enzyme action. Both RNA and DNA were hydrolysed by the enzyme, resulting in the formation of oligonucleotides with 3′-phosphomonoester end groups, as well as 3′-mononucleotides.

Nuclease of Azotobacter agilis. This enzyme (Stevens and Hilmoe, 1960a, b) resembles the nuclease from *Micrococcus pyogenes* and differs from other endonucleases in bringing about direct hydrolysis of the phosphodiester linkage without preliminary intramolecular transphosphorylation. The *Azotobacter* enzyme is, however, unique in that acting on RNA, DNA, or polynucleotide, the products formed are oligonucleotides with 5′-phosphomonoester end groups, with no detectable mononucleotides.

The enzyme was isolated from cell-free extracts of cells by ammonium sulfate fractionation, followed by treatment with protamine sulfate, aluminum hydroxide C_γ and chromatography on carboxymethyl cellulose.

c) Phosphodiesterases Acting on Polynucleotides.

In addition to the nucleases described in the foregoing section phosphodiesterases also bring about the splitting of polynucleotides. Of the phosphodiesterases which hydrolyse nucleic acids and polynucleotides to nucleoside monophosphates, two have been extensively studied: snake venom phosphodiesterase yielding nucleoside 5′-phosphates and spleen phosphodiesterase giving rise to nucleoside 3′-phosphates. Both of these enzymes also hydrolyse simple esters of appropriate mononucleotides, such as benzoyl adenosine 5′-phosphate in the case of venom enzyme, and benzoyl adenosine 3′-phosphate in the case of the spleen enzyme. Anderson and Heppel (1960) have reported the purification of a third type of phosphodiesterase, occurring in ascites cells, an enzyme whose specificity resembles that of the snake venom enzyme in that when acting on polynucleotides and (very slowly on) RNA nucleoside 5′ phosphates are liberated. But the enzyme from leukemic cells differs from the venom enzyme (Razzell and Khorana, 1959) in that the substrate must have at least two nucleoside residues, simple esters such as benzoyl adenosine 5′ phosphate and p-nitrophenyl thymidine 5′-phosphate not being split. Lehman (1960) reported a phosphodiesterase in *Escherichia coli*, rapidly hydrolysing DNA which has undergone prior degradation.

The Phosphodiesterase of Escherichia coli.

Assay is based on the formation of acid-soluble product as a result of enzymic degradation. The incubation mixture (0.30 ml) contained 20 μmoles of glycine buffer pH 9.2, 2 μmoles of $MgCl_2$, 10 mμmoles of partially degraded p^{32}-DNA and a suitable quantity of enzyme preparation. The mixture was incubated at 37° C for 30 min. 0.2 ml of a solution of calf thymus DNA (2.5 mg per ml) was added as carrier and then 0.5 ml of cold 0.5 N $HClO_4$. After 5 minutes at 0°, the precipitate was removed by centrifugation and aliquots of supernatant taken for radio-activity measurement.

A unit of enzyme was the amount causing the production of 10 mμmoles of acid soluble P^{32} in 30 minutes.

Purification. A 140-fold purified preparation was obtained by precipitating the enzyme from sonic extract of cells with protamine sulfate, elution with phosphate buffer, concentration of eluate with ammonium sulfate and fractionation with the salt, followed by DEAE-cellulose chromatography.

Mode of action and specificity. Acting on partially degraded DNA, the enzyme brought about a step wise hydrolysis starting from the 3' hydroxyl end, giving rise to 5' mononucleotides. This *exonuclease* mode of action was thus similar to that of venom diesterase, but different from the *endonuclease* action of desoxyribo-nuclease which attacked DNA in a random manner. Free dinucleotides were not hydrolysed by the enzyme, so that dinucleotides were formed along with mono-nucleotide as a result of enzyme action. The 5'-terminal dinucleotide of a poly-deoxynucleotide chain also was not cleaved. However, in contrat to venom phos-phodiesterase, the *E. coli* enzyme was able to degrade bacteriophage DNA bearing glucosylated hydroxymethyl cytosine. RNA was hydrolysed very slowly. The enzyme showed only slight activity towards intact double-stranded DNA. The conventional diesterase substrate Ca-bis [(p-nitrophenyl)-phosphate]$_2$ was not split. Using 0.07 M glycine buffer, the pH optimum was 9.2 to 9.8. Mg^{++} was needed for maximum activity. Phosphate ions stimulated the enzyme activity.

References.

ABDUL-FADL, M. A. M., and E. J. KING: Biochem. J. **45**, 51 (1949). — AGRAWAL, M., and P. S. KRISHNAN: Enzymologia **21**, 18 (1959). — AGREN, G.: Acta Chem. Scand. **13**, 1048 (1959). — AHMED, Z., and E. J. KING: Biochim. Biophys. Acta **34**, 313 (1959). — AHMED, Z., and J. L. REIS: Biochem. J. **69**, 386 (1958). — ALLEN, M. B., D. I. ARNON, J. B. CAPINDALE, F. R. WHATLEY and L. J. DURMAN: J. Am. Chem. Soc. **77**, 414 (1955). — ANDERSON, E. P., and L. A. HEPPEL: Biochim. Biophys. Acta **43**, 79 (1960). — ANFINSEN, C. B., R. R. RED-FIELD, W. L. CHOATE, J. PAGE and W. R. CARROL: J. biol. Chem. **207**, 201 (1954). — AXEL-ROD, B.: Advances in Enzymology **17**, 159 (1956); — J. biol. Chem. **167**, 57 (1947).

BACHMANN, B. J., and D. M. BONNER: J. Bacteriol **78**, 550 (1959). — BAMANN, E., and H. GALL: Biochem. Z. **293**, 1 (1937). — BENDALL, D. S., and C. DEDUVE: Biochem. J. **74**, 444 (1960). — BHEEMESHWAR, B., and M. SREENIVASAYYA: J. Sci. Ind. Res. India **12**B, 529 (1953). — BERENBLUM, J., and E. CHAIN: Biochem. J. **32**, 295 (1938). — BESSEY, O. A., O. H. LOWRY and M. J. BROCK: J. biol. Chem. **164**, 321 (1946). — BHATNAGAR, G. M., and P. S. KRISHNAN: Arch. Mikrobiol. (a) **36**, 131 (1960); (b) **36**, 161 (1960). — BLOCH FRANKELTHAL, L.: Biochem. J. **57**, 87 (1954). — BONTING, S. L., and I. M. ROSENTHAL: Nature (Lond.) **185**, 686 (1960). — BORKENHAGEN, L. F., and E. P. KENNEDY: Biochim. Biophys. Acta **28**, 222 (1958); — J. biol. Chem. **234**, 849 (1960). — BRAWERMAN, G., and N. KONIGSBERG: Biochim. Biophys. Acta **43**, 374 (1960). — BROH-KAHN, R. H., and J. A. MIRSKY: Arch. Biochem. **16**, 87 (1948). — BRUEMMER, J. H., and B. L. O'DELL: J. biol. Chem. **219**, 283 (1956).

CHAKRAVORTY, M., H. C. CHAKRABORTTY and D. P. BURMA: Arch. Biochem. Biophys. **82**, 21 (1959). — CHOW, S., and A. GOLDSTEIN: Biochem. J. **75**, 109 (1960). — CLAYTON, R. O., and L. M. HANSELMAN: Arch. Biochem. Biophys. **87**, 161 (1960). — COCKING, E. C.: Nature (Lond.) **187**, 962 (1960). — COLES, C. H., and E. R. WAYGOOD: Canad. J. Bot. **35**, 25 (1957). — COMAR, C. L.: Bot. Gaz. **104**, 122 (1942). — CROOK, E. M.: Biochem. J. **40**, 197 (1946). — CLENDENNING, K. A., T. E. BROWN and E. E. WALLDOV: Gatlingburg conference 1955. GAFFRON, H. (Ed.) Interscience Publishers, New York, 1957.

DICKMAN, S. R., and K. M.TRUPIN: Biochim. Biophys. Acta **30**, 200 (1958). — DE DUVE, C.: Nature (Lond.) **187**, 836 (1960). — DE DUVE, C., and J. BERTHET: Int. Rev. Cytol. **3**, 223

(1954). — De Lamirande, G., C. Allard and A. Cantero: J. Biophys. Biochem. Cytol. **41**, 373 (1958). — Drews, G.: Arch. Mikrobiol. **36**, 387 (1960). — Dryer, R. L., A. R. Tammes and J. I. Routh: J. biol. Chem. **225**, 177 (1957).

Ebel, J. P., and E. Mehr: Bull. Soc. Chem. Biol. **39**, 1535 (1957). — Eddy, A. A., and D. H. Williamson: Nature (Lond.) **179**, 1252 (1957). — Elson, D.: Biochim. Biophys. Acta **27**, 216 (1958).

Faulkner, P.: Biochem. J. **60**, 590 (1955). — Feinstein, R. N., and M. E. Volk: J. biol. Chem. **177**, 339 (1949). — Fishman, W. H., and F. Lerner: J. biol. Chem. **200**, 89 (1953). — Fiske, C. H., and Subbarow Y.: J. biol. Chem. **66**, 375 (1925). — Fleury, P., and J. Courtois: Enzymologia **1**, 377 (1937); **5**, 254 (1938). — Frankenthal, L., J. S. Roberts and C. Neuberg: Expt. Med. Surgery **1**, 386 (1943). — Freedland, R. A., and A. E. Harper: J. biol. Chem. **234**, 1350 (1959). — Frisch-Niggemeyer, W., and K. K. Reddi: Biochim. Biophys. Acta **26**, 40 (1957). — Fuller, R. C.: Plant. Physiol. **34**, 651 (1959).

Garen, A., and C. Levinthal: Biochim. Biophys. Acta **38**, 470 (1960). — Gibbs, M., and B. L. Horecker: J. Biol. Chem. **208**, 813 (1954). — Ginsberg, V., and H. G. Hers: Biochim. Biophys. Acta **36**, 227 (1953). — Glomset, J. A.: Biochim. Biophys. Acta **32**, 349 (1959). — Good, N. E.: Biochim. Biophys. Acta **40**, 502 (1960). — Goodland, G. A. J., and G. T. Mills: Biochem. J. **66**, 354 (1957). — Gomori, G.: J. biol. Chem. **148**, 139 (1943). — Gordon, J. J.: Biochem. J. **46**, 96 (1950). — Gornall, A. G., C. J. Bardawill and M. M. David: J. biol. Chem. **177**, 751 (1949). — Gran, F. C., and N. Eeg-Larsen: J. Nutrition **71**, 137 (1960). — Green, A. A., and W. L. Hughes: Methods in enzymology Vol. I, p. 67 Colowick, S. P., and N. O. Kaplan (Ed.) New York: Academic Press Inc. 1955. — Green, A., and C. Levinthal: Biochim. Biophys. Acta **38**, 470 (1960). — Greenstein, J. P.: in Biochemistry of cancer. 2nd Ed. p. 431. New York: Academic Press Inc. 1954.

Hackett, D. P.: Int. Rev. Cytol. **4**, 143 (1955). — Hageman, R. H., and D. Flesher: Arch. Biochem. Biophys. **87**, 203 (1960). — Handler, P., and J. R. Klein: J. biol. Chem. **143**, 49 (1942). — Harris, D. I.: J. biol. Chem. **165**, 541 (1946). — Heppel, L. A., and R. J. Hilmoe: J. biol. Chem. **186**, 665 (1951). — Herman, E. C., and B. E. Wright: J. biol. Chem. **234**, 122 (1959). — Hers, H. G., and C. De Duve: Bull. Soc. Chim. Biol. **32**, 20 (1950). — Hewitt, E. J., and B. A. Notton: Chem. and Ind. **1960**, 1046. — Hockenhull, D. J., G. C. Ashton, K. H. Fantes and B. K. Whitehead: Biochem. J. **57**, 93 (1954). — Hoffman-Ostenhof, O., and W. Weigert: Naturwissenschaften **13**, 303 (1952). — Hofstee, B. H. J.: Arch. Biochem. Biophys. **51**, 139 (1954). — Holden, M., and N. W. Pirie: (a) Biochem. J. **60**, 39 (1955a); **60**, 53 (1955b). — Honda, S.: Plant Physiol. **30**, 174 (1955).

Ingelman, B., and H. Malmgren: Acta Chem. Scand. **2**, 365 (1948); **3**, 157 (1949a); **3**, 1331 (1949b). — Ito, E., S. Kondo and S. Watanabe: J. Biochem (Japan) **42**, 793 (1955).

Jacobson, E.: Biochem. Z. **249**, 21 (1932); **267**, 89 (1933). — Jansen, E. F., R. Jang and A. K. Balls: J. biol. Chem. **196**, 247 (1952). — Jeffree, G. M.: Biochim. Biophys. Acta **20**, 50 (1956); — Nature (Lond.) **175**, 509 (1955). — Jenner, H. D., and H. D. Kay: J. biol. Chem. **93**, 733 (1951). — Johnson, E. J., and M. K. Johnson: J. Bacteriol. **78**, 792 (1959). — Johnston, F. B., M. N. Asatir and H. Stern: Plant Physiol. **32**, 124 (1957). — Jono, Y.: Acta Schol. Med. Univ. Imp. Kioto. **13**, 162 (1930). — Joyce, B. K., and S. Grisolia: J. biol. Chem. **233**, 350 (1958).

Kay, H. D.: Biochem. J. **22**, 1446 (1928). — Kind, P. R. N., and E. J. King: J. Clin. Pathol. **7**, 322 (1954). — Kinzel, H., and W. Url: Physiol. Plant. **7**, 835 (1954). — Kitosato, T.: Biochem. Z. **197**, 257 (1928a); **201**, 206 (1928b). — Klotz, I. M., and T. R. Hughes: Meth. Biochem. Analysis **3**, 265 (1956). — Kornberg, A.: J. biol. Chem. **182**, 779 (1950); — Kornberg, S. R.: J. biol. Chem. **218**, 23 (1956). — Kornberg, A., S. R. Kornberg and E. S. Simms: Biochim. Biophys. Acta **20**, 215 (1956). — Kornberg, A., and O. Linderberg: J. biol. Chem. **176**, 665 (1948). — Kornberg, A., and W. E. Pricer: J. biol. Chem. **182**, 763 (1950). — Kornberg, S. R.: Biochim. Biophys. Acta **26**, 294 (1957). — Kornberg, S. R., J. R. Lehman, M. J. Bessman, E. S. Simms and A. Kornberg: J. biol. Chem. **233**, 159 (1958). — Kunitz, M.: J. gen. Physiol. **35**, 423 (1952). — Kurata, K.: J. Biochem. (Japan) **17**, 343 (1933).

Layne, E.: Method in enzymology, Vol. III, 447, Colowick, S. P., and N. O. Kaplan (Editors). New York: Academic Press Inc. 1957. — Lehman, I. R.: J. biol. Chem. **235**, 1479 (1960). — Leloir, L., and S. H. Goldenberg: J. biol. Chem. **235**, 919 (1960). — Leuthhardt, F., E. Testa and H. P. Wolf: Helv. Chim. Acta **36**, 227 (1953). — Levinson, H. S.: in Symposium on spores, p. 130. Published by American Institute of Biological Sciences 1957. — Levinson, H. S., J. D. Sloan and M. T. Hyatt: J. Bacteriol. **75**, 291 (1958). — Lieberman, J.: J. biol. Chem. **219**, 307 (1956). — Lineweaver, H., and C. W. Murray: J. biol. Chem. **171**, 565 (1947). — Lohmann, K.: Biochem. Z. **203**, 172 (1928). — Long, C.: Biochem. J. **50**, 407 (1952). — Lowry, O. H., and J. A. Lopez: J. biol. Chem. **162**, 421 (1946). — Lowry, O. H., N. J. Rosenbrough, A. L. Farr and R. J. Randall: J. biol. Chem. **193**, 265 (1951).

Mann, T.: Biochem. J. (a) **38**, 339 (1944); (b) **38**, 345 (1944). — Markham, R., and J. L. Strominger: Biochem. J. **64**, 46p. (1956). — Marsh, C., and W. Militzer: Arch. Biochem.

Biophys. **60**, 439 (1956). — MATHEWS, R. E. F.: Biochim. Biophys. Acta **38**, 552 (1960). — MATHIES, J. C.: J. biol. Chem. **233**, 1121 (1958). — MATSUSHITA, S., and F. IBUKI: Biochim. Biophys. Acta **40**, 358 (1960). — MATTENHEIMER, H.: Z. physiol. Chem. (a) **303**, 107 (1956); (b) **303**, 115 (1956); (c) **303**, 126 (1956). — MAZELIS, M.: Plant Physiol. **29**, 113 (1954). — MAZELIS, M., and B. VENNESLAND: Plant Physiol. **32**, 591 (1957). — MEYERHOFF, O., R. SHATAS and A. KAPLAN: Biochim. Biophys. Acta **12**, 121 (1953). — MOKRASCH, L. C., W. D. DAVIDSON and R. W. McGILVERY: J. biol. Chem. **222**, 174 (1956). — MOKRASCH, L. C., and R. W. McGILVERY: J. biol. Chem. **221**, 909 (1956). — MOOG, F.: Symposium Soc. Study Develop. Growth **17**, 121 (1958). — MORTON, R. K.: Biochem. J. **61**, 232 (1955a); — Discuss. Faraday Soc. **20**, 149 (1955b). — MUHAMMED, A., A. RODGERS and D. E. HUGHES: J. gen. Microbiol. **20**, 482 (1959). — MUNEMURA, S.: J. Biochem. (Japan) **17**, 343 (1933).

NAGANNA, B., and V. K. N. MENON: J. biol. Chem. **174**, 501 (1948). — NASHKOV, D., and O. NASHKOVA: Enzymologia **21**, 380 (1960). — NEMER, M. J., E. M. WISE, F. M. WASHINGTON and D. ELWYN: J. biol. Chem. **235**, 2063 (1960). — NEUBERG, C., and H. A. FISHER: Enzymologia **2**, 241 (1937); — Enzymologia **2**, 360 (1938). — NEUBERG, C., A. GRAUER and I. MANDL: Enzymologia **14**, 157 (1950). — NEUFELD, E. F., D. S. FEINGOLD and W. J. HASSID: Arch. Biochem. Biophys. **83**, 96 (1959). — NEUHAUS, F. C., and W. L. BYRNE: (a) J. biol. Chem. **234**, 109 (1959); (b) **234**, 113 (1959); **225**, 2019 (1960). — NEWCOMBE, H.: Proc. Soc. expt. Biol. Med. **76**, 504 (1951). — NEWMARK, M. A., and B. S. WENGER: Arch. Biochem. Biophys. **89**, 110 (1960). — NIGAM, V. N., H. M. DAVIDSON and W. H. FISHMAN: J. biol. Chem. **234**, 1550 (1959). — NIGAM, V. N., and W. H. FISHMAN: J. biol. Chem. **234**, 2394 (1959). — NORBERG, B.: Acta Chem. Scand. **4**, 601 (1950); **14**, 650 (1960). — NOVIKOFF, A. B., E. PODBER and J. RYAN: Federation Proc. **9**, 210 (1950). — NOVIKOFF, A. B., E. PODBER, J. RYAN and E. NOE: J. Histochem. Cytochem. **1**, 20 (1953).

OGINSKY, E. L., and H. L. RUMBAUGH: J. Bacteriol. **70**, 92 (1955). — OGSTON, A. G.: Arch. Biochem. Biophys. **89**, 181 (1960).

PAECH, K., and E. KRECH: Planta **41**, 391 (1952). — PAIGEN, K.: J. biol. Chem. **233**, 388 (1958). — PEEL, J. J., and B. C. LOUGHMAN: Biochem. J. **65**, 709 (1957). — PFANKUCH, E.: Z. physiol. Chem. **241**, 34 (1936). — PIERPOINT, W. S.: Biochim. Biophys. Acta **21**, 136 (1956); — Biochem. J. (a) **65**, 67 (1957); (b) **67**, 466 (1957); **71**, 519 (1959); **75**, 501 (1960). — PIRIE, N. W.: Biokhimiya. **22**, 140 (1957); — Biochem. J. **47**, 614 (1950). — POGELL, B. M., and R. W. McGILVERY: J. biol. Chem. **197**, 293 (1952); **208**, 149 (1954). — PORTER, H. K.: J. exp. Bot. **4**, 44 (1953). — POTTER, V. R.: in Manometric techniques, page 170. UMBREITT, W. W., R. H. BURRIS and J. F. STAUFFER (Editors). Minneapolis, Minnesota: Burgess Publishing Company 1957. — PRICE, C. A., and E. V. THIMANN: Plant Physiol. **29**, 113 (1954).

RABINOWITZ, M., and F. LIPMANN: J. biol. Chem. **235**, 1043 (1960). — RACKER, E.: Nature (Lond.) **175**, 249 (1955); — Harvey Lectures, p. 143. 1955/56. — RACKER, E., and E. A. R. SCHROEDER: Arch. Biochem. Biophys. **74**, 326 (1958). — RAFTER, G. W.: J. biol. Chem. **230**, 643 (1958); — Arch. Biochem. Biophys. **81**, 238 (1959); — J. biol. Chem. **235**, 2475 (1960). — RAZZEL, W. E., and KHORANA, H. G.: J. biol. Chem. **234**, 2105 (1959). — REDDI, K. K.: Biochim. Biophys. Acta **28**, 386 (1957); **36**, 123 (1959). — REINER, J. M., K. K. TSUBOI and P. B. HUDSON: Arch. Biochem. Biophys. **55**, 191 (1955). — REVEL, H. R., and E. RACKER: Biochim. Biophys. Acta **43**, 465 (1960). — ROBERTS, D. W. A.: J. biol. Chem. **219**, 711 (1956); **230**, 213 (1958). — ROCHE, J.: (a) Biochem. J. **25**, 1724 (1931); — (b) Bull. Soc. Chim. Biol. **13**, 841 (1931); — ROCHE, J., and J. BAUDOIN: C. R. Soc. Biol. **137**, 245 (1943). — ROCHE, J.: in The enzymes, Vol. I, Part I, p. 473. SUMNER, J. B., and V. MYERBACK (Ed.). New York: Academic Press Inc. 1950. — ROGERS, D. J.: Economic Bot. **13**, 261 (1959). — ROGERS, D., and F. S. REITHEL: Arch. Biochem. Biophys. **89**, 97 (1960). — RONGINE DE FEKETE, M. A., L. F. LELOIR and C. E. CARDINI: Nature (Lond.) **187**, 918 (1960). — ROTHENBACH, E. F., and S. T. HINKELMAN: Naturwissenschaften **41**, 555 (1954).— ROTHSTEIN, A., and R. MEIER: J. Cell. Comp. Physiol. **38**, 245 (1954).

SALTMAN, P.: J. biol. Chem. **200**, 145 (1953). — SAMPATHKUMAR, K. S. V., T. A. SUNDERARAJAN and P. S. SARMA: Enzymologia **18**, 228 (1957). — SANWAL, G. G.: Ph. D. Thesis. Lucknow University (1960). — SANWAL, G. G., and P. S. KRISHNAN: Nature (Lond). **188**, 664 (1960); — Enzymologia (1961 in Press). — SCANDRETT, E. J.: Nature (Lond.) **186**, 558 (1960). — SCHAFFNER, A., and F. KRUMEY: Z. physiol. Chem. **243**, 149 (1936). — SCHLAMOWITZ, M., and R. L. GRANER: J. biol. Chem. **163**, 487 (1946). — SCHLESSINGER, M. J., and M. J. COON: Biochim. Biophys. Acta **41**, 30 (1960). — SCHRAMM, M.: J. biol. Chem. **233**, 1169 (1958). — SCHUSTER, L., H. G. KHORANA and L. A. HEPPEL: Biochim. Biophys. Acta **33**, 452 (1954). — SEAL, U. S., and F. BINKLEY: J. biol. Chem. **228**, 193 (1957). — SEGAL, H. L., and B. M. BRENNER: J. biol. Chem. **235**, 471 (1960). — SHUSTER, L.: J. biol. Chem. **229**, 289 (1957). — SHUSTER, L., and N. O. KAPLAN: J. biol. Chem. **201**, 535 (1953). — SHUSTER, L., H. G. KHORANA and L. A. HEPPEL: Biochim. Biophys. Acta **33**, 452 (1959). — SINGER, S. J., L. EGGMAN, J. M. CAMBELL and S. G. WILDMAN: J. biol. Chem. **197**, 233 (1952). — SINGER, M. F., and J. S.

FRUTON: J. biol. Chem. **229**, 111 (1957). — SISAKYAN, N. M., N. A. VASILENA and N. N. KOSHLOYANTS: Dokl. Akad. Nauk (U.S.S.R.) **112**, 300 (1952). — SMILLIE, R. M.: Nature (Lond.) **187**, 1024 (1960). — SMILIE, R. M., and FULLER R. C.: Plant Physiol. **34**, 651 (1959). — SMITH, R. A., C. W. SHUSTER, S. ZIMMERMAN and I. C. GUNSALUS: Bacteriological Proceedings, Society of American Bacteriologists Baltimore 1956, p. 107. — SOUMALAINEN, H., M. LINKO and E. OURA: Biochem. Biophys. Acta **37**, 482 (1960). — SPENCER, D.: Aust. biol. Sci. **7**, 151 (1954). — STADTMAN, T.: Biochim. Biophys. Acta **32**, 95 (1959). — STANBURY, J. B., M. L. MORRIS, H. J. CORRIGAN, W. E. LASSITER: Endocrinology **67**, 353 (1960). — STERN, H., and A. E. MIRSKY: J. gen. Physiol. **36**, 181 (1952—53). — STEVENS, A., and R. J. HILMOE: (a) J. biol. Chem. **235**, 3016 (1960); (b) **235**, 3023 (1960). — STOCKING, C. R.: Am. J. Bot. **39**, 282 (1952); — Plant. Physiol. **34**, 56 (1959). — SUNDERARAJAN, J. A., and P. S. SARMA: Biochem. J. **56**, 125 (1954); **65**, 261 (1957); **71**, 537 (1959). — SWARTZ, M. N., N. O. KAPLAN and M. E. FRECH: Science **123**, 50 (1956).

TASHIRO, Y.: J. Biochem. (Japan) **45**, 937 (1958). — TEWARI, K. K., and P. S. KRISHNAN: Anal. Chem. Acta **22**, 111 (1960). — TEWARI, K. K.: Ph. D. Thesis, Luckow University (1960). — THOAI, N. V.: Bull. Soc. Chim. Biol. **34**, 264 (1941). — TOMLINSON, N., and R. A. J. WARREN: Canad. J. Biochem. Physiol. **38**, 605 (1960). — TORRIANI, A.: Biochim. Biophys. Acta **38**, 460 (1960). — TRIANTAPHYLLAPOULOS, E., and J. TUBA: Canad. J. Biochem. Physiol. **37**, 699 (1959). — TSUBOI, K. K., and P. B. HUDSON: Arch. Biochem. Biophys. **43**, 339 (1953); **55**, 206 (1955); **61**, 197 (1956). — TUNIS, M., and E. CHARGAFF: (a) Biochem. Biophys. Acta **37**, 257 (1960); (b) **37**, 267 (1960). — TURNER, D. H., and J. F. TURNER: Biochem. J. **74**, 486 (1960). — TUVE, T. W., and C. B. ANFINSEN: J. biol. Chem. **235**, 3437 (1960).

VISWANATHAN, P. N.: Unpublished results (1961).

WALKER, D. A., and J. M. A. BROWN: Biochem. J. **67**, 79 (1957). — WEIBULL, C.: Ann. Rev. Microbiol. **12**, 1 (1958). — WEIL-MALHERBE, H.: Biochem. J. **55**, 741 (1953). — WIER, T. E., and C. R. STOCKING: Bot. Rev. 18, 14 (1952). — WILDMAN, S. G., and J. BONNER: Arch. Biochem. **14**, 381 (1947). — WINDER, F. G., and J. M. DENNENY: J. gen. Microbiol. **17**, 573 (1957).

YIN, H. C.: Nature (Lond.) **162**, 928 (1948). — YIN, H. C., and Y. T. TUNG: Science **108**, 87 (1948); — Plant. Physiol. **24**, 103 (1949). — YIN, H. C., and C. N. SUN: Plant. Physiol. **24**, 103 (1949). — YOSHIDA, A., and A. YAMATKA: J. Biochem (Japan) **40**, 85 (1953). ZITTLE, C. A., and E. S. DELLA MONICA: Arch. Biochem. **26**, 112 (1950).

Enzymes of Inorganic Nitrogen Metabolism.

By

E. J. Hewitt and D. J. D. Nicholas

With 3 figures.

The enzymes considered in this chapter include nitrate reductase, nitrite reductase, hyponitrite reductase, hydroxylamine reductase, enzymes of denitrification, and some other enzymes such as aldehyde oxidase with related properties. The properties of the various enzymes are described in relation to the different plants and micro-organisms from which they have been obtained[1].

Nitrate reductase has been studied more intensively and in a wider range of species than the other enzymes considered here and it is therefore described first in detail. Many features of the other systems are conveniently considered against this background. A brief historical introduction is followed by details of extraction, assay, substrate and co-factor specificity, and other characteristics.

A. Nitrate Reductase.

Nitrate reduction in higher plants has been the subject of several investigations since the time of SCHIMPER (1888, 1890) who first observed the dependence of nitrate reduction on light *in vivo*. The relationship has been observed in several studies since, notably by DITTRICH (1930), ECKERSON (1932) EISENMENGER (1933), and BURSTRÖM (1943) and is considered again later.

Attempts to demonstrate nitrate reduction *in vitro* were first made by ECKERSON (1924, 1931) who observed slow disappearance of nitrate when tomato leaf extracts were incubated with glucose. ANDERSON (1924) and SOMER (1936) were unable to confirm ECKERSON's conclusions and doubted whether the production of nitrite in plant tissues when extracts were incubated in the presence of nitrate, glucose and toluene was due to a catalytic factor without light. The amounts produced were very small and not markedly dependent on the period of incubation over several hours.

Bacteria utilise nitrate nitrogen and the overall process whereby nitrate is reduced to ammonia with the subsequent formation of amino acids, protein and other nitrogenous cell constituents is known as *"nitrate assimilation"*.

---

[1] The following abbreviations are used in this chapter: ATP adenosine triphosphate, DPN DPNH oxidised and reduced diphosphopyridine nucleotides, TPN TPNH oxidised and reduced triphosphopyridine nucleotides, FMN FMNH$_2$ oxidised and reduced riboflavin phosphate, FAD FADH$_2$ oxidised and reduced flavin adenine dinucleotide, p CMB parachloromercuric benzoate, PMA phenyl mercury acetate, EDTA ethylenediamine tetraacetic acid, TRIS 2-amino-2-(hydroxyethyl)-1:3-propanediol. Mb Methylene blue, PMS Phenazine methosulphate, Phen S Phenazine sulphate, BV Benzyl viologen, MV Methylviologen, Py Pyocyanine, Tol. B. Toluylene blue. —H —H$_2$ the reduced forms of these dyes. Cyt. Cytochrome (b, c etc. as shown). (DPN, DPNH, TPN, TPNH have how been renamed NAD, NADH, NADP, NADEH respectively).

GAYON and DUPETIT (1886) suggested that nitrite was the first product of nitrate reduction in bacteria. QUASTEL, STEPHENSON and WHETHAM (1925) found that *B. coli* produced nitrite when grown anaerobically; subsequently nitrite has been identified as a product of nitrate reduction in a range of bacteria (WOODS 1938; VERHOEVEN, 1956). Nitrate reductase, the enzyme responsible for the reduction of nitrate has been studied in some detail in bacteria. Under certain conditions, however, several micro-organisms can use nitrate or some of its reduction products as terminal hydrogen acceptors in place of oxygen. This process has been termed *"nitrate respiration"* by SATO (1956) or *"dissimilatory nitrate reduction"* by VERHOEVEN (1952). A similar process has been described by EGAMI et al. (1957) in higher plants.

Many classifications have been proposed for the various types of nitrate reduction but none has been entirely satisfactory. JENSEN (1904) suggested five categories according to the products of the reaction whereas VERHOEVEN (1952, 1956) differentiated between three types of nitrate reduction thus: (a) *assimilation* in which nitrate is reduced only for the elaboration of nitrogenous cell materials, (b) *incidental dissimilation* in which nitrate acts as a non-essential hydrogen acceptor, (c) *true dissimilation* in which nitrate acts, under certain conditions as the essential hydrogen acceptor which enables the organism to grow. The separation of incidental and true nitrate dissimilation does not seem to be of great value and has only a teleological significance; organisms which carry out incidental dissimilation are those that have a fermentative respiration when both oxygen and nitrate are limiting factors. Thus VERHOEVEN (1952, 1956) showed that his strain of *Escherichia coli* produced large amounts of lactic and succinic acids under anaerobic conditions but in the presence of either nitrate or oxygen complete oxidation to carbon dioxide occurred. Japanese workers (1956, 1957) have classified nitrate reducing organisms on the basis of the behaviour of their cytochromes towards nitrate. This is not now satisfactory since in *Neurospora crassa* (WALKER andNICHOLAS, 1961a) and *Escherichia coli* (TANIGU CHI et al., 1953, 1956, 1957) an iron system appears to be required for the dissimilatory nitrate reductase (FEWSON and NICHOLAS, 1960a, 1961 a, b, c) but not for the assimilatory enzyme. FEWSON and NICHOLAS (1961) have proposed that nitrate reduction is best considered under two headings: (1) *nitrate assimilation* in which the end products are nitrogenous cell constituents; (2) *nitrate respiration* or *nitrate dissimilation* in which nitrate is used instead of oxygen, usually under anaerobic conditions, as the terminal hydrogen acceptor. Denitrification may be regarded as a special type of nitrate respiration and is defined as the production of nitrogen gas or the oxides of nitrogen by the reduction of nitrate, nitrite or any other suitable intermediate compound.

I. Preparation.

1. Fungi and Higher Plants.

NASON and EVANS (1952, 1953) isolated nitrate reductase for the first time from the fungus *Neurospora crassa* (wild-type strain). Nitrate reductase was first obtained from higher plants in extracts of soybean *(Glycine max)* leaves and potato, barley, tomato, melon and maize roots or leaves by EVANS and NASON (1953), and its activity was observed in extracts of many plants by others since referred to below.

a) **Methods of extraction** (Table 1). (1) *Fungi.* NASON and EVANS (1953) extracted the enzyme from *Neurospora* after growing it for 4 to 5 days in darkness on FRIES's basal medium at 28—30° C with forced aeration. Mycelia were washed in glass distilled water, frozen 1 to 3 hours at —15° C, ground in a TEN BROECK homogeniser with 3 weights of 0.1 M K_2HPO_4 buffer and centrifuged at 20,000 g. About 85 per cent of the total activity was obtained in the supernatant extract.

NICHOLAS, NASON and McELROY (1954) used the same extraction procedure for *Neurospora* and *Aspergillus niger*. These fungi were grown for 5 days at 25° C on the medium used by NICHOLAS (1952) with intermittant shaking to prevent sporulation as the spore pigment interferes with the assay. KINSKY and McELROY (1958) found that mycelia grown at 37° C had relatively little activity and used room temperature (about 20°). Mycelia were stored below 0° and activity increased during the first week of storage, possibly due to loss of nitrite reductase. Frozen mycelia were broken with a hammer and enzyme was extracted by grinding in a Servall Omnimixer with 4 weights of ice cold 0.1 M potassium phosphate buffer at pH 7, to 7.5, and centrifuging at 10,000 g. About 90 per cent of the total activity was extracted.

(ii) *Higher plants*. EVANS and NASON (1953) extracted the enzyme from soybean by grinding the first trifoliate leaves of plants aged 8 to 10 days with 3 times their weight of 0.1 M K_2HPO_4 buffer pH 9.0 and two parts by weight of alumina powder (Alcoa A-301) for two minutes in a WARING blendor followed by grinding for three minutes in a TEN BROECK homogeniser all at 0—4° C. The extract was centrifuged at 4° C at 20,000 g and the activity was assayed in the supernatant solution which contained only about 10 per cent of the total activity associated with the crude homogenate. Extracts prepared for subsequent fractionation were made in 0.1 M K_2HPO_4 buffer at pH 7.0 by grinding first in a chilled mortar and then in a TEN BROECK homogeniser.

NICHOLAS and NASON (1955a) omitted the alumina powder and added cysteine and EDTA each at final concentrations of 10^{-4} M. This improved the stability and yield of enzyme by protecting it against inactivation by heavy metals or oxidation of sulphydryl groups. Extraction from cauliflower (*Brassica oleracea* var. *Botrytis*) leaf used by HEWITT, FISHER and CANDELA (1955), CANDELA, FISHER and HEWITT (1957) was carried out a pH 7.0 with 10^{-4} M cysteine and EDTA in 1954, and at pH 8.8 with 10^{-3} M cysteine and EDTA in 1955 by grinding by hand in a chilled mortar with acid-extracted silica sand. CANDELA et al. (1957) found that grinding in a mechanically-driven closely fitting glass pestle and mortar caused severe losses of activity even when immersed at —5° C. Leaves were frozen at —18° C before extraction in some experiments. Unpublished work by HEWITT with M. M. R. K. AFRIDI has shown that preliminary freezing for periods of 30 minutes to a few hours results in increased activity in extracts of the order of 20 to 50 per cent.

SPENCER (1959) separated wheat embryos from the endosperm after 48 hours germination in dilute nitrate solution, and ground them in an ice cold mortar at 0—2° C with about 10 weights of 0.1 M K_2HPO_4/KH_2PO_4 pH 7.5 containing 10^{-3} M glutathione, which slightly increased activity when included, and centrifuged at 5,000 g.

CRESSWELL (unpublished work) at Long Ashton obtained good results with leaves of wheat *(Triticum vulgaris)*, water cress *(Nasturtium officinale)* and vegetable marrow *(Cucurbita pepo)*, by grinding in a mortar chilled to —18° C with three weights of 0.1 M sodium phosphate pH 8.0 containing cysteine and EDTA at 10^{-3} M. The last-named species has yielded higher activities than most others tested at Long Ashton or recorded by other workers and would appear to be a particularly convenient source of the enzyme from higher plants. Values of

Table 1. *Methods used for extracting nitrate reductase systems from fungi and higher plants.*

Source	Extracting solution[1]	Vol./Wt.	Method of Extraction[2]	Reference
Neurospora crassa and *Aspergillus niger*	0.1 M K_2HPO_4 (pH 7.5)	3:1	Frozen mycelia (−15° 3 hours). Ground at 0—4° in Ten Broeck homogeniser. Centrifuge 20,000 g. 85% activity in supernatant fluid.	Nason and Evans (1952, 1953); Nicholas, Nason and McElroy (1954)
N. crassa	0.1 M K_2HPO_4[1] pH 7—7.5	4:1	Frozen, homog. in Servall omnimixer 0—4°. Centrifuge 10,000 g. 90% activity in supernatant fluid	Kinsky and McElroy (1958)
Soybean (*Glycine max*) leaves	(a) 1st trifoliate 0.1 M K_2HPO_4 (pH 9.0) for fractionation	3:1 + 2 wts Alcoa A301 alumina powder	Waring blendor 2 min 0—4° followed by Ten Broeck homog. Centrif. 20,000 g 10% activity in supernatant fluid.	Evans and Nason (1953)
Soybean (*Glycine max*) leaves	(b) 0.1 M K_2HPO_4, pH 7.0 for comparative assays		Ground in chilled mortar and Ten Broeck homogeniser	Evans and Nason (1953)
Soybean leaves	0.1 M K_2HPO_4 (pH 9.0) with 10^{-4} M EDTA and cysteine	3:1	As Evans and Nason (1953) but omitting alumina.	Nicholas and Nason (1955a)
Cauliflower (*Brassica oleracea* var. *Botrytis*) leaves	0.1 M Na_2HPO_4 (pH 8.8) 10^{-3} M EDTA and cysteine	3:1	Leaves frozen and ground by hand in chilled mortar at 0° with neutral acid extracted silica sand. Centrif. 20,000 g.	Hewitt, Fisher and Candela (1955)
Cauliflower (*Brassica oleracea* var. *Botrytis*) leaves	0.1 M Na/HPO_4 pH 7.0 10^{-4} M EDTA and cysteine	3:1	As above	Candela, Fisher and Hewitt (1957)
Wheat leaf. 12 days age	0.025 M K_2HPO_4 and 5×10^{-4} M EDTA	3:1	90 sec in Blendor, followed by 3 min homog. in Potter-Elvehjem macerator. 4°.	Anacker and Stoy (1958)
Wheat (*Triticum vulgaris*) embryo	0.1 M K/HPO_4 pH 7.5 10^{-3} M glutathione	10:1	Ground by hand in chilled mortar at 0—2°. Centrif. 5,000 g.	Spencer (1959)
Maize (*Zea mays*) leaves	0.1 M Tris pH 7.3—7.8 to produce final pH of 7.0 10^{-2} M cysteine 3×10^{-4} M EDTA	4:1	Ground in omnimixer at 2°, centrifuged at 30,000 g.	Hageman and Flesher (1960)

[1] Footnote to all tables: K/HPO_4 or Na/HPO_4 indicates mixture of potassium or sodium phosphates of molarity shown adjusted to pH value.
[2] Temperatures ° Centigrade (C).

3,000 mμmoles nitrite/g fresh wt. hour at 25° (20 mμmoles NO_2/mg/min) have been frequently obtained, compared with values of 1,200 from soybean (EVANS and NASON, 1953), or of 1,500 from cauliflower.

HAGEMAN and FLESHER (1960) used a much higher level of cysteine (0.01 M) than usual for extraction of the enzyme from maize (Zea mays). It was found (HAGEMAN, unpublished work) that little or no activity could be detected when lower concentrations were used, or at any time if the extract showed a brown pigment. The EDTA concentration was 3×10^{-4} M. TRIS buffer 0.1 M pH 7.3 to 7.8 was used in a ratio of 4 ml per g tissue. The initial pH was adjusted to about 7.6 so that the pH of the extract was about 7.0. Extraction was carried out at 2° C with a Servall Omnimixer at maximum speed. This method was originally worked out by HAGEMAN and WAYGOOD (1959) for the extraction of triose-phosphate dehydrogenase from maize and resulted in an outstanding increase in initial activity.

CRESSWELL, HAGEMAN and HEWITT (unpublished work) have investigated the cysteine requirements necessary for the extraction of the enzyme from different species. It was found that high concentrations of 10^{-2} or 5×10^{-2} M were required with maize; somewhat lower concentrations around 10^{-3} M were adequate for vegetable marrow at certain stages of development whilst at other stages concentrations up to 10^{-2} M were required for maximum activity. For cauliflower, where natural thiol compounds are abundant, concentrations of 5×10^{-4} M were generally adequate, and higher concentrations were sometimes inhibitory. It is thus necessary to establish the optimum concentration of cysteine to be used before proceeding to routine work.

Extracts of bean leaves (Phaseolus vulgaris) and tomato leaves (Lycopersicum esculentum) are often without any activity, are brown in colour, and inhibit activity in extracts from other plants (CRESSWELL, unpublished work). Inclusion of cysteine up to 5×10^{-2} M during extraction of bean or tomato leaves decreases or abolishes the inhibitory effect. Activity may be detected in tomato leaf extracts prepared with high cysteine concentrations, but has not been found in bean leaf extracts. Any tendency to develop a brown colour leads to loss of activity.

CRESSWELL, HAGEMAN and HEWITT (unpublished) did not observe the in-activation of the enzyme when extracted by a high speed blendor, as found on many occasions by CANDELA et al. (1957). The difference may be due to excessive surface denaturation from the use of longer periods of maceration by CANDELA et al., since the present method is to blend for only 60 to 90 seconds with the vessel immersed in ice. The use of hand grinding in a mortar with or without sand therefore appears unjustified and is laborious.

b) **Fractionation and stability.** In order to obtain maximal specific activity and good yields, the inclusion of a thiol compound such as cysteine and a chelating agent such as EDTA are necessary, as described above. Preliminary tests must be made to determine the most satisfactory concentration for each in relation to the source of the enzyme. It is advantageous to maintain the presence of these protective agents during subsequent stages of fractionation, but generally at lower concentrations. Nitrate reductase enzymes are very sensitive to damage by heavy metals, e.g., copper at 10^{-4} M causes 64% inhibition of the Neurospora enzyme (NICHOLAS and NASON, 1954a). Special precautions are therefore necessary to protect the enzyme during isolation. Ammonium sulphate is a notable source of heavy metal impurities, including lead, zinc and copper, and these may be present in amounts equivalent to about 5×10^{-4} M heavy metals in saturated ammonium sulphate used to precipitate protein fractions. In the work of HAGEMAN, CRESSWELL and HEWITT (1962) saturated ammonium sulphate is extracted exhaustively with diphenylthiocarbazone in carbon tetrachloride at pH 6—7.5 and then with 8-hydroxyquinoline in chloroform until colourless. Chloroform is removed with carbon tetrachloride and the solution is allowed to separate for a few days before use. The pH is then adjusted to about 7.5 with strong ammonia solution. Ammonium sulphate solutions prepared thus are free of iron, manganese, zinc, copper, lead and several other heavy metals. As an added precaution 10^{-3} M concentration of cysteine hydrochloride with correction of pH to 7.5 by addition of an equivalent amount of normal potassium hydroxide, is always included immediately before use. Ammonium sulphate solutions may also be relatively acid and usually require neutralisation to pH 7—8 by careful addition of strong ammonium hydroxide until a faint smell of ammonia is noticeable.

Dialysis sacs should be prepared by soaking in 5×10^{-3} or 10^{-2} M glutathione or cysteine for at least 24 hours before required and should be washed in water before use. NICHOLAS and NASON (1954b) found that nitrate reductase was extremely sensitive to effects of metals or oxidising compounds in untreated dialysis sacs, whereas dialysis itself was not especially harmful. The marrow leaf enzyme studied by CRESSWELL can be dialysed exhaustively without harm when these precautions are adopted.

The examples given below illustrate stages in the fractionation of the enzyme from different sources and are given in an abbreviated note form. Units are specified in Table 2A.

(i) *Fungi.* NASON and EVANS (1953) *(Neurospora crassa):* All steps carried out at 0—4° C. Saturated ammonium sulphate solution neutralised to pH 7—7.5 but not purified. Crude extracts (60 units/mg) brought to 43% saturation by ammonium sulphate. Precipitate retained and dissolved in 40 ml phosphate buffer, 0.1 M pH 7.0 (121 units/mg with 50% recovery). Ammonium sulphate solution added to 24% saturation, precipitate discarded, and saturation raised to 46%. Precipitate dissolved in 16 ml phosphate buffer (417 units/mg with 20% of initial total activity). Calcium phosphate gel (7.2 ml prepared by the method of KEILIN and HARTREE, 1938) aged for 9 months was added. Collected by centrifugation after 15 min washed twice with 5 ml 0.1 M phosphate buffer pH 7.5 and eluted with 4ml 0.1 M pyrophosphate buffer pH 7.0 (2690 units/mg; 10% of initial activity). Eluate brought to 60% saturation with ammonium sulphate (4,000 units per mg 9.7% of initial activity).

NICHOLAS and SCAWIN (1956) *(Neurospora crassa):* The first three steps were similar to those described by NASON and EVANS (1953) except that the calcium phosphate gel was eluted with 0.2 M pyrophosphate (pH 7.5). The subsequent ammonium sulphate precipitation was carried out with solid ammonium sulphate added to 100% saturation and the precipitate was washed 4 times with saturated ammonium sulphate. The precipitate (195 units/mg) was redissolved in water and treated with a suspension of Darco G60 charcoal at the rate of 30 mg per ml of enzyme solution. The suspension was centrifuged after 15 min and the supernatant contained 0.8% of the original activity with a specific activity of 1,540 units per mg. This fractionation allowed the demonstration of a phosphate requirement for maximal activity.

KINSKY and MCELROY (1958) *(Neurospora crassa):* The procedure involved only the use of ammonium sulphate precipitations with the exception of the precipitation of nucleic acids by protamine. A novel feature was the inclusion of glutathione at 10^{-3} M in the buffers used to redissolve precipitates as this was found to enhance stability of the enzyme. In accordance with other procedures the enzyme was mostly precipitated by 46% saturation ammonium sulphate. The ammonium sulphate fractionation in combination with use of maleate instead of phosphate buffers was effective in revealing a marked effect of phosphate.

(ii) *Higher plants.* EVANS and NASON (1953) (Soybean): 60 g primary leaves yielded 190 ml crude extract containing 1.8 *Neurospora* units/mg protein. Enzyme was adsorbed during 10 minutes on 1,000 mg calcium phosphate gel contained in 95 ml. The gel was centrifuged, washed by suspension for one minute in 95 ml 0.1 M phosphate buffer pH 7.5 and eluted with 38 ml 0.1 M pyrophosphate buffer pH 7.0 (42 units/mg; 46% recovery of initial activity). The eluate was brought to 40% saturation with ammonium sulphate and the precipitate was dissolved in 15 ml 0.1 M phosphate buffer pH 7.0 (39 units/mg; 23% recovery). A second adsorption on 80 mg of calcium phosphate gel followed by washing with 0.1 M phosphate buffer pH 7.5 and elution with 3 ml 0.1 M pyrophosphate buffer pH 7.0 yielded 121 units/mg with a recovery of 7% of the initial activity.

SPENCER (1959) (wheat embryo): Centrifugation at 105,000 g indicated that the enzyme was present in the soluble cytoplasmic fraction after extraction. The extent of purification achieved was only an increase of 7 fold in specific activity with 58% recovery of total activity. The procedure depended on fractionation with solid ammonium sulphate only and all precipitations were carried out in 0.1 M potassium phosphate pH 7.5 containing 10^{-3} M glutathione. The enzyme was precipitated between 28 and 35% saturation ammonium sulphate; a somewhat lower concentration than that found effective for the *Neurospora* enzyme.

CRESSWELL, HAGEMAN and HEWITT (vegetable marrow). Immediately after extraction 2 volumes saturated ammonium sulphate solution containing 10^{-3} M cysteine at 0° and purified as described above, are added very gradually to the extract (50 to 300 ml) at 0° and left 30 min before centrifuging. The protein is redissolved in 0.2 initial volumes of potassium phosphate 0.1 M pH 7.5 and containing 10^{-3} M cysteine. At this stage G 50 Sephadex (Pharmacia, Uppsala Sweden) which has been finely ground may be added in the proportion of 1 g per 5 ml to reduced volume and ammonium sulphate content. The protein solution, with or without Sephadex, is centrifuged and dialysed for 16 hours at 2° C in 3—5 l 0.03 M potassium phosphate buffer containing 10^{-3} cysteine at pH 7.5. Such dialysed and ammonium sulphate precipitated preparations do not show any response in terms of increased activity resulting from added FAD or FMN.

Thermal inactivation data for nitrate reductase are given in Table 11. It will be seen that these enzymes are very thermolabile, but can be stored at —18° C for several weeks without loss of activity if isolated in the presence of cysteine or glutathione; a conclusion confirmed by CRESSWELL, HAGEMAN and HEWITT (unpublished) for the marrow leaf enzyme after dialysis. The stability of the enzymes during dialysis appears to vary. NICHOLAS and NASON (1954a) found glutathione was necessary to protect the enzyme obtained from *Neurospora* during dialysis. At Long Ashton, CRESSWELL (unpublished) found that the enzyme from marrow leaves was stable to dialysis without added cysteine if the dialysis sac was pretreated as described. SPENCER (1959) reported that the wheat embryo enzyme was also entirely stable to dialysis without the addition of gluta-thione. On the other hand SPENCER (1959) found that the wheat embryo enzyme was apparently very sensitive to surface denaturation effects since bubbling nitrogen gas through a solution of the enzyme resulted in total inactivation; air or carbon monoxide had similar effects.

ANACKER and STOY (1958). (Wheat leaf.) Leaves were harvested at 10—12 days extracted in 0.025 M K_2HPO_4 and the extract diluted to 0.01 M K_2HPO_4. 80 ml was applied to a column of calcium phosphate. This was prepared as follows. 0.1 mole $Ca(H_2PO_4)_2$ was dissolved in 5 l water and 1 l of 1.5 M K_2HPO_4 added

Table 2. *Methods used for extracting nitrate reductase systems from bacteria.*

Source	Extracting solution	Vol./wt.	Method of extraction	Reference
Escherichia coli	0.1 M phosphate (pH 7.5)	not specified	Ultrasonic breakage of the cells.	SATO and NIWA (1952)
Escherichia coli (B strain)	0.1 M phosphate (pH 7.0) + 10^{-4} M Na-EDTA	3:1	Cells harvested in a Sharples centrifuge at 30,000 g, washed with 1% w/v saline to remove nitrite; frozen at $-15°$ for 3 hr before homogenizing in a cold pestle and mortar using an equal weight of alumina (Alcoa A-301). After grinding for 10 min cold buffer added and grinding continued for another 5 min. Centrifuge 15 min at 4° at 6,000 g. Supernatant solution used as enzyme source.	NICHOLAS and NASON (1955 b)
Escherichia coli (YAMAGUTCHI strain)	0.1 M phosphate (pH 6.8)	not specified	Ultrasonic disruption of cells (550 k cycles).	TANIGUCHI, ASANO, IIDA, KONO, OHMACHI, EGAMI (1957)
Escherichia coli (B strain)	0.1 M phosphate (pH 7.5)	4:1	Cells collected by centrifuging, washed twice in 0.9% w/v NaCl. Frozen 16 hr. $-15°$. Ground with 2 parts alumina (Alcoa A-310) for 10 min. Then cold buffer added and grinding continued for another 10 min at 4°. Centrifuged at 8000 g for 10 min. Activity measured in supernatant solution.	MEDINA and HEREDIA (1958); HEREDIA and MEDINA (1960)
Escherichia coli (YAMAGUTCHI strain)	0.1 M phosphate (pH 6.8) and 10^{-3} M Na-EDTA	3:1	Cells collected in a Sharples centrifuge, washed with cold distilled water. Frozen cells ground in cold mortar for 40 min with 1.5 wt. alumina powder (Wako No. 800) and for further 20 min with $3 \times$ wt. of cold 0.1 M phosphate (pH 6.8) containing 10^{-3} M Na-EDTA. Centrifuged at 2,000 g for 20 min to remove alumina and cell debris. Supernatant solution centrifuged at 20,000 g for 40 min and 107,000 g for 1 hr and particles resuspended in buffer solution.	ITAGAKI and TANIGUCHI (1951)
Escherichia coli (YAMAGUTCHI strain)	0.1 M phosphate (pH 7.1)	50:1	Grinding cell paste with twice their weight of alumina powder for 30 min, then for another 10 min with $25 \times$ their weight of buffer. Centrifuged at 2000 g for 20 min for cell-free extract (A). Particulate preparation obtained by centrifuging extract A for a further 30 min at 15,000 g. Particulate enzyme solubilized in 5 mM Na_2HPO_4 (pH 8.3) containing 0.1 m M KNO_3, heated to 60° for 5 min. Heated suspension cooled to 4° and kept for about 15—20 hr. Soluble enzyme obtained after centrifuging at 50,000 g in a Sharples centrifuge.	TANIGUCHI and ITAGAKI (1960)

Pseudomonas aeruginosa (NCIB 8104)	0.1 M phosphate (pH 7.4)	3:1	Cells centrifuged in a Sharples super centrifuge washed with 0.85% w/v NaCl(1 litre/10 g cells). Ultrasonic treatment of cells for 15 min with an ultrasonic probe 20 k cycles/sec at 4°. Centrifuged at 20,000 g for 30 min to prepare cell-free extracts.	FEWSON and NICHOLAS (1960a, 1961a)
Achromobacter fischeri = Photobacterium fischeri	Water	15:1	Cells collected by centrifugation washed with 3% w/v NaCl. Lysis of cells in water. Cell-free particulate fraction collected after centrifugation.	SADANA and McELROY (1957)
Hansenula anomala Strain 317	0.2 M phosphate (pH 7.5)	Thick cream paste (amounts not specified)	Cells collected by centrifuging at 4° and washed with 0.9% w/v saline until no nitrite detected. Packed cells resuspended in 0.2 M phosphate (pH 7.5) to make a thick cream. 50 ml aliquots and 5 g 200 mesh powdered pyrex glass mixed and put in 9-kc Raytheon sonic oscillator; cells disrupted for 40 min in cold. Cell debris removed by low speed centrifugation and supernatant then centrifuged at 107,000 g for 50 min 90% activity in supernatant solution.	SILVER (1957)
Azotobacter vinelandii (O) strain	0.2% w/v KCl	5:1	Cells harvested, after 20 hr. growth at 30° with vigorous aeration, and washed 3 × with 0.2% w/v KCl and centrifuged to remove nitrite. Cell yield 5 to 10 g/litre. Cell-free extracts prepared by grinding cell-paste with 1.5 × wt of alumina powder (Wako 800) for 30 min in a cold mortar. Then for another 10 min with cold 0.2% w/v KCl. Centrifuged 2000 g 20 min and at 105,000 g for 60 min. Three fractions collected, (a) large particles (b) small particles (c) supernatant solution. Particles suspended in 0.005 M phosphate (pH 6).	EVANS (1954)
Rhizobium japonicum from soybean root nodules	0.1 M phosphate (pH 7) + 10^{-4} M Na-EDTA	3:1	Nodules ground in a cold mortar. Extracts centrifuged at 20,000 g. Residue resuspended and centrifuged at 600 g. Supernatant solution centrifuged at 20,000 g to give whole cells of Rhizobia.	CHENIAE and EVANS (1956)
Glycine max. Merr. variety Ogden				

slowly with continuous stirring over 2 hours. To this solution was added in a trickle 1 l of 1.4 M $CaCl_2$ with stirring. The precipitate of $CaHPO_4 2 H_2O$ was allowed to settle and the supernatant solution decanted. The precipitate was washed twice with 1 l water on a Büchner funnel. The solid was resuspended in 2.5 l 0.05 M NaOH and maintained at 40° with occasional stirring for about 1 hour until the pH fell to 8—9. The solution was decanted and replaced with fresh 0.05 M NaOH and the resuspended solid left for 24 hours. This was repeated a further three times. The solid was collected by decantation and washed on a Büchner funnel with 0.005 M NaH_2PO_4 until the pH was 6—7. The solid was washed with ethanol and acetone and dried in air at room temperature.

The adsorption column was 35 mm diameter and contained 5 g of solid added in 30 ml. The enzyme was eluted by increasing concentration of K_2HPO_4 pH 7.0 and the most active fraction obtained by elution with 0.05 M buffer. Rechromatography and gradient elution with sodium pyrophosphate 0.001—0.1 M, pH 7.0 resulted in an increased specific activity of 100 times. The enzyme was then precipitated by ammonium sulphate 0—35% saturation, with a further two-fold increase in specific activity. The preparation was not homogeneous in the ultracentrifuge, but the activity was associated with protein having a sedimentation constant of 19 S and a molecular weight of 500,000—600,000. At this stage the molybdenum content was 0.2 $\mu g/mg$ with an activity of 200 units/mg.

Stability was not improved by cysteine but was increased by a short period of electrodialysis at 5 volts/cm. The purified enzyme lost 50 per cent of activity in 180 and 36 hours at —15° and 0° respectively.

2. Bacteria.

Early investigators were concerned with determining nitrite produced from nitrate in the culture medium and in later work washed cell suspensions were used to follow nitrite production. Cells have been collected by centrifugation in an ordinary centrifuge, in a Sharples superspeed centrifuge and more recently in a Servall centrifuge fitted with a continuous flow head. A few investigators only collect the bacteria in the cold but this is not essential. The cells are usually washed with saline (in most instances between 0.85 and 1% w/v NaCl) to remove the free nitrite which would interfere with the subsequent assay of nitrate reductase.

a) Methods of extraction (Table 2).

The cells are frequently frozen in a deep-freeze cabinet for times varying from 1 to 24 hr before they are disrupted. Nicholas and Nason (1955b) added an equal weight of alumina (Alcoa A-301) to the cell paste of *E. coli* in a cold mortar and pestle, grinding for 10 min before adding 0.1 M phosphate containing 10^{-4} M Na-EDTA (pH 7.0) when grinding was continued for another 5 min. The addition of the Na-EDTA prevented Cu^{++} ions from inactivating the enzyme. The extract was centrifuged at 4,000 g for 15 min and over 75% of the nitrate reductase was in the supernatant solution. The Japanese workers have also used alumina (Wako 800) to disrupt the bacterial cells and have also included Na-EDTA in their extracting solutions as in the procedure of Nicholas and Nason (1955a). A more efficient and less laborious method has been the use of either sonic or ultrasonic equipment to disrupt cells. [Sato and Niwa (1952), Taniguchi et al. (1957), Nicholas and Fewson (1960a, 1961a,b,c.)] The M.S.E. Mullard ultrasonic titanium probe (20k/cycles) has been successfully used to extract nitrate reductase from a range of denitrifying bacteria (Fewson and Nicholas, 1960a, 1961a, b, c). It is essential to keep the bacterial suspension cold during the ultrasonic treatment and this is achieved by circulating iced water through a double walled glass vessel containing the bacterial suspension shown in Fig. 1. A comparison of the various methods for extracting the enzyme from *Ps. aeruginosa* given in Table 3 shows that the ultrasonic treatment was the most effective technique. It is preferable to use a titanium probe since those made of stainless steel give rise to free radicles which could denature the enzyme. The sonic treatments using the Raytheon 9 kc or 10 kc are not usually as effective in releasing the enzyme from bacterial cells as are the ultrasonic techniques.

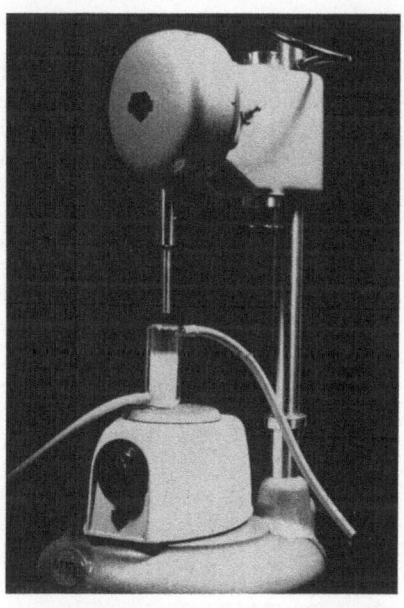

Fig. 1. Apparatus for disrupting bacteria using the M.S.E./Mullard ultrasonic titanium probe (20 k/cycles) (NICHOLAS and FISHER, 1960). — The bacterial suspension is put in a double-walled glass vessel of 30 ml capacity. The cell suspension is agitated by a magnetic stirrer and cooled by circulating ice-cold water by means of an electric pump through the outer jacket. The tip of the probe, immersed 1/8th in. into the suspension is operated at maximum frequency for 15 min.

A simple procedure is used for disrupting salt water bacteria, namely suspending them in distilled water when lysis occurs within a few minutes. Thus SADANA and McELROY (1957) suspended 1 g wet weight of *Achromobacter fischeri*/10 ml distilled water and the cell-free particulate fraction was collected by high speed centrifugation in a Servall centrifuge for 20 min. A crude nitrate reductase preparation was obtained by resuspending the sediment in distilled water (1 g wet weight/15 ml water) and recentrifuging as described previously. Although most of the proteins were extracted by the first water lysis the supernatant solution contained little nitrate reductase activity. The first crude water extract contained most of the DPNH-cytochrome reductase activity. The

Table 3. *A comparison of various methods for extracting nitrate reductase from Ps. aeruginosa* (FEWSON and NICHOLAS, unpublished).

Method	% of total N in the supernatant solution after centrifuging at 6,000 g for 20 min.	Nitrate reductase activity in the homogenate as a % of the activity in the original cells.	Nitrate reductase in the supernatant solution after centrifuging at 18,000 g for 45 min as a % of the activity in the original cells.
1. Autolysis in 1% w/v NaF . . .	51	49	10
2. Repeated freezing and thawing	28	88	9
3. Lysozyme treatment	37	113	34
4. Grinding with alumina powder[1]	54	—	17
5. Grinding with alumina (ALCOA) powder[2]	29	—	15
6. Grinding with silicon carbide[1] .	27	—	14
7. HUGHES Press (—20° C)	66	88	43
8. MICKLE disintegrator (Ballotini No. 12 beads)	52	89	27
9. M.S.E. Ultrasonic probe (20 k cycles/sec)	78	72	55

[1] Universal Grinding Wheel Co. Ltd. (Stafford) Grit size 1200.
[2] Aluminium Company of America (Pittsburgh) Size: A-301.

supernatant solution left after the second lysis was used as the source of the enzyme. It should be noted that the genus *Achromobacter* is synonymous with *Photobacterium* and the latter is now the accepted name.

b) Fractionation and Stability.

As mentioned in the introduction there are two types of nitrate reductases in bacteria. The one associated with the dissimilation of nitrate is usually particulate. The other is the assimilatory enzyme which is usually found in the soluble parts of the cells, i.e. in the supernatant solution left after centrifuging the extracts at 30,000 *g*. for 30 min and even at 144,000 *g* in an ultracentrifuge for 1 hr.

1. Dissimilatory Enzyme. The Japanese school has worked extensively on the dissimilatory or respiratory nitrate reductase. Attempts have been made to solubilize the particulate enzyme in *E. coli* using deoxycholate or iso-butanol (IIDA and TANIGUCHI, 1959). The heat treatment method used to solubilize the enzyme by TANIGUCHI and IIDA (1960) was as follows: The frozen bacterial cells were ground in a mortar for 30 min with twice their weight of alumina powder (Wako No. 800), for another 10 min with five times their weight of cold 0.1 M phosphate buffer (pH 7.1) and then centrifuged at 2000 *g* for 20 min to obtain the cell-free extract. The latter was further centrifuged at 15,000 *g* for 30 min and the particulate fraction so obtained was washed twice with cold distilled water. The particulate preparation (containing 0.2—0.6 mg N/ml) in 5 mM Na_2HPO_4 (pH 8.3) containing 0.1 mM KNO_3 was heated to 60° for 5 min and then cooled to 4° and kept at this temperature for about 15—20 hr. The nitrate reductase solubilized in this way was centrifuged in a Sharples centrifuge at 50,000 *g* for 10 min and the enzyme was then fractionated as shown in Table 4.

To the solubilized enzyme was added 1 M acetate buffer (pH 5.0) to produce a final concentration of 0.02 M and then 70—150 mg calcium phosphate gel/100 ml supernatant solution and the solution stirred for 10 min and allowed to stand for another 10 min. The gel was centrifuged and washed twice with distilled water and once with 2—10 mM phosphate buffer (pH 7.1). A large amount of the adsorbed enzyme was eluted from the gel by 0.1 M phosphate buffer (pH 7.1) containing 0.1 mM glutathione. The combined eluates (80 ml from 100 g cell paste) were then fractionated further with solid ammonium sulphate. The fraction between 38—53% saturation was collected and dissolved in the phosphate/glutathione solution. The fraction between 30—40% saturation was prepared from this first ammonium sulphate preparation. The precipitate formed was centrifuged and dissolved in the phosphate/glutathione buffer. This fraction was

Table 4. *Solubilization and purification of E. coli nitrate reductase* (TANIGUCHI and ITAGAKI, 1960).

Step No.	Fraction	Total N (mg)	Units ($\times 10^{-4}$)	Specific activity ($\times 10^{-3}$)
1	Cell paste	(270 g wet weight)		0.18
2	Particulate preparation	2500	135	0.54
3	Supernatant solubilized after heat treatment	580	100	1.72
4	Calcium phosphate gel eluate	227	90	3.95
5	$(NH_4)_2SO_4$ 1st ppt. (38—53% sat.)	55.2	80	14.5
6	$(NH_4)_2SO_4$ 2nd ppt. (30—40% sat.)	11.2	50	56.2
7	Spinco L 1st pellet	3.04	40	132
9	Spinco L 3rd pellet	1.50	28	186

then centrifuged in a Spinco Model L at 110,000 g for 3—4 hr. The pellet containing the active enzyme was washed three times with the phosphate/glutathione buffer in order to remove the inactive lighter components. The final preparation can be stored for one week at —15° with less than 30—40% loss of activity.

Other methods used to solubilize the particulate preparation include treatments with dodecyl sulphate, steapsin or chymotrypsin.

To the particulate preparation (3—5 mg N/ml) suspended in 0.05 M phosphate buffer (pH 7.1) was added 10% w/v dodecyl sulphate solution to give a final concentration of 0.5 to 1%. After standing for 2 hr at 30°, a reddish brown supernatant solution was obtained after centrifuging at 20,000 g for 40 min and this was used as the dodecyl sulphate solubilized preparation.

The steapsin method used by TANIGUCHI and ITAGAKI (1959) is as follows: To the heated preparation immediately cooled to 30° in ice water was added 0.15% w/v (final concentration) of "steapsin" from pig pancreas. After incubating for 45 hr at 30° the soluble nitrate reductase was in the supernatant solution after centrifuging for 40 min at 20,000 g. The supernatant solution was centrifuged at 107,000 g in the preparative Spinco for 1 hr and the particle free solution so obtained contained between 95 to 100% of the original activity of the particles. The solubilized preparation of nitrate reductase was free from the system transferring electrons from the natural electron donors, reduced pyridine nucleotides and formate which were actively utilized by the original particulate system. Only reduced dyes, especially methyl viologen, were effective donors.

Nitrate reductase was extracted from actively denitrifying cells of *Pseudomonas aeruginosa* by means of the ultrasonic probe (FEWSON and NICHOLAS, 1960a, 1961a, b, c). After 30 min treatment 89% of the cell nitrogen was extracted into the supernatant solution indicating an efficient disruption of the cells. Although the cell suspension was well stirred and kept at 4° about half the enzyme activity was lost after the 30 min treatment. The proportion of enzyme activity in the supernatant solution (after 6,000 g/20 min) increased only slightly between 15 and 30 min because of the progressive denaturation of the enzyme during ultrasonic treatment. The maximum amount of enzyme (55% of activity in original cells) was found in the supernatant solution after the 15 min period.

Table 5. *Distribution of nitrate reductase in extracts of Ps. aeruginosa* (FEWSON and NICHOLAS, 1960a, 1961a, b ,c).

A cell homogenate was prepared with an ultrasonic probe. Enzyme activity was determined in the supernatant solutions obtained after centrifuging the homogenate at 4° C at the g values and times indicated.

Centrifugation		Enzyme activity of the supernatant solution as % of the activity in the homogenates
g	min	
1,000	10	100
6,000	20	80
18,000	30	71
50,000	30	55
139,000	60	39

The distribution of the nitrate reductase in cell free extracts was determined by centrifuging homogenates prepared as described above, at various g values. The results given in Table 5 show that most of the activity, i.e., 71% was found in the supernatant solution after centrifuging at 18,000 g for 30 min whereas only 39% was left after centrifuging at 140,000 g for 1 hr. After the ultrasonic disruption of the cells nitrate reductase activity was not associated with particles of any particular size since the enzyme was distributed between the various fractions collected. When any one of the precipitates resuspended in 0.1 M phosphate buffer (pH 7.4) was subjected to further ultrasonic treatment for 10 min and then centrifuged under the same conditions, more enzyme activity

Table 6. *Purification of nitrate reductase from Ps. aeruginosa* (Fewson and Nicholas, 1960a, 1961a, b, c).

15 g frozen cells were suspended in 15 ml 0.1 M phosphate buffer (pH 7.4) and disintegrated by means of the ultrasonic probe (20 k/c) for 15min at 4° C. The homogenate was diluted to 70 ml with the same buffer and centrifuged at 18,000 g for 45 min. The supernatant solution was used as the source of the crude enzyme.

Fraction	Procedure	Volume ml	Total protein mg	Protein mg/ml	Total units m M.NO$_2$/ 10 min	Specific activity m M.NO$_2$/10 min/mg protein	Recovery %	Purification
1	Supernatant solution	60	510	8.50	109,000	214	—	—
2	Precipitate from 0—43% ammonium sulphate saturation of fraction 1 was centrifuged at 5,000 g for 15 min and dissolved in 0.1 M phosphate buffer (pH 7.4)	58	139	2.40	118,000	830	108	3.8
3	Protamine sulphate (100 mg) was added to fraction 2. Centrifuged at 5,000 g for 5 min. Activity in the supernatant solution.	75	131	1.75	109,000	850	100	4.0
4	Precipitate from 0—43% ammonium sulphate saturation of fraction 3 was centrifuged at 5,000 g for 15 min and dissolved in 0.1 M phosphate buffer (pH 7.4)	55	68.8	1.25	101,000	1,470	93	6.9
5	Calcium phosphate gel (50 mg) added to fraction 4. Centrifuged at 3,000 g for 5 min. Activity in the supernatant solution	54	63.7	1.18	95,900	1,500	88	7.0
6	Calcium phosphate gel (500 mg) added to fraction 5 and centrifuged at 3,000 g for 5 min. Gel washed with 0.25 M phosphate buffer and the enzyme eluted with 0.2 M sodium pyrophosphate (pH 7.4)	40	15.2	0.38	53,700	3,520	49	16.5
7	Precipitate from 0—55% ammonium sulphate saturation of fraction 6 was centrifuged at 5,000 g for 15 min and dissolved in 0.1 M phosphate buffer (pH 7.4)	33	7.6	0.23	39,100	5,150	36	24.1
8	Fraction 7 was purified by electrophoresis on granular starch and the enzyme eluted with 0.1 M phosphate buffer (pH 7.4)	5	0.6	0.12	14,400	24,000	13	115

was released into the supernatant solution. Thus when the precipitate collected by centrifuging at 18,000 g for 30 min was treated with an ultrasonic probe and then recentrifuged at 18,000 g for 30 min, 30% of the activity in the residue was released into the supernatant solution.

The enzyme was fractionated as shown in Table 6. About 15 g frozen cells suspended in 15 ml 0.1 M phosphate (pH 7.4) were disrupted with an ultrasonic probe for 15 min at 4° C. The cell homogenate was diluted to 70 ml with the same buffer and centrifuged at 18,000 g for 45 min. The supernatant solution (fraction I) was used as the source of the crude enzyme and was purified by means of ammonium sulphate precipitation, calcium phosphate gel treatments, and by electrophoresis on granular starch. Details of the fractionation are as follows:

An aliquot (45 ml) of the saturated ammonium sulphate solution (adjusted to pH 7.4 with 0.1 N NaOH) was added to 60 ml of fraction I. The precipitate collected by centrifuging at 5,000 g for 15 min was dissolved in 0.1 M phosphate buffer (pH 7.4) (fraction 2). Protamine sulphate (100 mg in 20 ml phosphate buffer) was added to fraction 2 and the precipitate collected by centrifuging at 5,000 g for 5 min was discarded. The supernatant solution (fraction 3) was brought to 43% saturation by adding 57 ml saturated ammonium sulphate (pH 7.4). The precipitate centrifuged at 5,000 g for 15 min was dissolved in 50 ml 0.1 M phosphate buffer (pH 7.4) (fraction 4).

Calcium phosphate gel (50 mg) was added to fraction 4 and after thorough mixing was collected by centrifuging at 3,000 g for 5 min and discarded. A further aliquot of gel (500 mg) was added to the supernatant solution (fraction 5) and after standing for 15 min with occasional stirring, it was centrifuged at 3,000 g for 5 min. The gel was washed with 20 ml 0.25 M phosphate buffer (pH 7.4) and then thoroughly mixed with 40 ml 0.2 M sodium pyrophosphate (adjusted to pH 7.4 with N HCl) in a TEN-BROECK glass homogeniser. After standing at 4° C for 20 min the gel was removed by centrifuging at 3,000 g for 5 min. The nitrate reductase activity was in the supernatant solution (fraction 6) and was precipitated by adding 48 ml saturated ammonium sulphate solution (pH 7.4). The precipitate collected by centrifugation at 5,000 g for 15 min was dissolved in 30 ml 0.1 N phosphate buffer (pH 7.4) (Fraction 7).

Fraction 7 contained 36% of the activity of the original supernatant solution (Fraction 1) and represented a 24-fold purification of the enzyme. Further purification was achieved by electrophoresis on granular starch.

Washed starch powder (120 g) suspended in 100 ml 0.1 M borate buffer (pH 8.4) was poured into a glass trough ($1.5 \times 2.0 \times 50$ cm) to form a layer 1 cm deep. After about one hour the excess buffer was removed with filter paper strips placed on the surface of the starch. Connections between the ends of the starch bed and the reservoirs containing the buffer (0.1 M borate, pH 8.4) with the platinum electrodes, were made with thin wicks (10 cm long) passing through narrow glass tubes filled with saturated KCl and 2% agar. The platinum electrodes were connected to a constant voltage D.C. output transformer (Shandon) and a potential of 250 volts applied. Fluctuations in the current usually occurred when the apparatus was first switched on, so that the current was passed for about 5 hr to stabilise the unit before the enzyme was applied. The starch was cooled by passing cold water through a jacket surrounding the glass trough. The nitrate reductase activity of Fraction 7 was precipitated by adding an equal volume of saturated ammonium sulphate solution (pH 7.4). The precipitate, centrifuged at 5,000 g for 15 min, was dissolved in a minimal amount (3 ml) of 0.1 M borate buffer (pH 8.4) and mixed with dry starch powder (2 g) to give a "putty-like" consistency. A small section (1 cm) was cut out of the middle of the starch bed and replaced by the enzyme-starch mixture. The surface of the starch was carefully levelled and a current of 5 mA at 250 volts applied for 36 hr. Sections (1·5 cm) of the starch bed were then removed and eluted with 5.0 ml aliquots of 0.1 M phosphate buffer (pH 7.4). During electrophoresis most of the protein moved quite rapidly towards the cathode but the nitrate reductase moved only slightly from the origin. The eluate from the starch fraction represented a 115-fold purification of the crude enzyme and contained 13% of the original activity.

The purified enzyme was stable when stored at $-17°$ C and retained about 65% of its activity after 9 months. At 4° C, however, all the activity was lost after 72 hr and this effect could not be offset by additions of glutathione (5×10^{-3} M) or Na-EDTA (10^{-4} M) either alone or in combination.

Nitrate reductase has been purified from *Achromobacter fischeri* by SADANA and MCELROY (1957).

Details of the method are as follows: In the purification procedure, the activity of nitrate reductase was followed by determining the amount of nitrite formed using reduced benzyl viologen as an electron donor. Unless otherwise stated, the temperature of the solutions during purification was maintained between 2 and 5° C. Specific activity was increased $1.5 \times$ with 88 % recovery.

Step 1. The pH of the clear supernatant of the second extract was adjusted to 4.5 with 0.2 M acetic acid. The resulting active precipitate was removed by centrifugation and resuspended in about one third of the original volume of 0.05 M phosphate buffer, pH 6.5.

Step 2. Protamine treatment. The isoelectric precipitate contained large amounts of nucleic acids and approximately 6—8 mg protein/ml. The pH was adjusted to 6.0 by the addition of 0.2 M acetic acid, and protamine sulphate (15 mg/ml, pH 5.0) was added with vigorous stirring. The precipitate was removed periodically by centrifugation, and the ratio of optical density at 280 mμ/260 mμ was determined with a Beckman spectrophotometer. The crude fraction had a 280 mμ/260 mμ ratio of about 0.3, and protamine sulphate addition was continued until no more precipitate was obtained and the ratio reached about 1.0. Usually 200 ml of protamine sulphate was required to remove most of the nucleic acids present in the extract obtained from 100 g (wet weight) of bacterial cells. Specific activity was doubled with little loss of total activity.

Step 3. Adsorption and elution from calcium phosphate gel. Seventy-five milliliters (15 mg/ml) of calcium phosphate gel was added to 100 ml of the supernatant from step 2; the solution was adjusted to pH 6 and stirred for 15 min. After centrifugation in the cold, the supernatant fluid was discarded and the gel washed once with distilled water and twice with 0.05 phosphate buffer, pH 7.5. The enzyme was eluted by 4—5 successive treatments with 20 ml portions of 0.4 M phosphate buffer, pH 8.5. Specific activity was again doubled and 68 % of the initial activity was retained.

Step 4. Fractionation with ammonium sulphate at pH 7.0. The combined eluates from step 3 were brought to 0.5 saturation with solid ammonium sulphate, and the pH was adjusted to 7. The suspension was stirred for 15 min at 4° C and centrifuged for 15 min at 12,000 r.p.m. in a Servall. The concentration of the supernatant solution was then raised to 0.7 saturation with solid ammonium sulphate (pH7.0). The suspension was stirred and centrifuged as before. The precipitate was dissolved in 50 ml of 0.05 M phosphate buffer, pH 7.5, and dialyzed overnight against the same buffer. Specific activity was approximately doubled again at this stage with 56 % of total activity.

Step 5. Selective heat denaturation of inactive proteins. The dialyzed supernatant from step 4 was heated in a water bath at 55° C for 10 min, with continuous shaking, and then immediately cooled in an ice bath. The impurities were removed by centrifugation. The heat treatment completely inactivated the residual DPNH-cytochrome reductase, but resulted in no loss of nitrate reductase.

Step 6. Further purification was achieved by repeated ammonium sulphate fractionations, the fraction between 0.57 and 0.62 saturation being most active. No measurable nitrite reductase could be detected in these preparations. The yield of final purified preparation was about 0.5—0.6 mg protein from 100 g of bacterial cells (wet weight), and on a protein basis represents a 250-fold purification. The recovery of initial activity was only 5 per cent. The final enzyme preparation was pink in colour and was stable for several weeks when stored in deep freeze. It was not sedimented by centrifugation for 3 hr. at 100,000 g. Ammonium sulphate fractionation resulted in progressive loss of about 11 % of initial activity at each of the first three steps and in increased specifi activity fifteen times.

2. Assimilatory enzyme. NICHOLAS and NASON (1955b) extracted a soluble enzyme from *E. coli* which was concerned with the assimilation of nitrate. The frozen cells were homogenized in a cold pestle and mortar using an equal weight of alumina powder (Alcoa A-301). After grinding for 10 min, 3 × their weight of cold 0.1 M K_2HPO_4 (pH 7.0) containing 10^{-4} M Na-EDTA was added slowly and grinding continued for another 5 min when the material was centrifuged for 15 min at 4° C at 5,000 g. The supernatant solution contained 75% of the nitrate reductase activity present in the crude extract. The 0—70% and 0—50% ammonium sulphate fractions were prepared in the usual way. After dissolving the latter in the phosphate/versene buffer it was adsorbed on to an equal volume of calcium phosphate gel. The gel was washed twice with 0.1 M phosphate (pH 7.0) and the enzyme eluted with 0.2 M pyrophosphate (pH 7.0). This resulted in a 15-fold purification of the enzyme with 33% recovery of the units in the crude extract.

The purified enzyme stores well at —15° at pH 7.8 but loses more than 90% of its activity after 5 min at 65°. The addition of glutathione (10^{-3} M) enhanced the storage of the enzyme and the use of glutathione treated sacs improved the stability of the enzyme during dialysis (NICHOLAS and NASON, 1954a).

II. Measurement of Activity.

1. Assay Methods.

Three methods of measuring activity have been used, Tables 7 and 8. These are the colorimetric measurement of net nitrite production, manometric estimation of nitrite, and spectrophotometric measurement of oxidation of reduced pyridine nucleotides. Estimation of nitrate loss (d) is difficult because high concentrations are needed to saturate the enzymes and nitrite formed interferes in the estimation of nitrate.

a) The GRIESS-ILOSVAY colorimetric method. This depends on the diazotisation of an aromatic amino compound by nitrite in acid solution and coupling with a suitable reagent to give an intense red-purple colour. The most satisfactory procedure is that of SNELL and SNELL (1949) which was used by EVANS and NASON (1953) and in subsequent work, and is slightly modified here.

1 ml of 1% sulphanilamide in N HCl is added with mixing, to a suitable aliquot of the reaction mixture or a solution derived from it. 1 ml of 0.01% N-(1-naphthyl)-ethylenediamine hydrochloride in water is added and mixed immediately. The colour develops rapidly and reaches maximum intensity in about 30 minutes, but is stable for hours. Aliquots containing 10—100 mμmoles nitrite in a final volume of 6 ml will produce extinctions between about 0.08 and 0.8 at 540 mμ. The BEER-LAMBERT law is obeyed up to extinctions near 1.0. Volumes can be adjusted to meet particular requirements.

Units of activity are variously given in Tables 2 and 8 on the basis of mμmoles nitrite (net) produced in a given period per mg protein. Units of activity are not always comparable as different flavins and electron donors have been used and temperatures also vary. The conditions must therefore be fully specified.

Assay conditions vary considerably and are summarised in Tables 7 and 8. When reduced pyridine nucleotides and FAD are used, their expense means that reaction volumes should be kept to a minimum in order to economise in the amounts of these reagents, whilst maintaining adequate concentrations for saturation according to the dissociation constants and optimum concentrations summarised in Tables 9 and 10. A reaction mixture of 1 or 2 ml is often convenient, and provides 3 or 4 ml after addition of the reagents for nitrite estimation. A small point of occasional importance is that if large aliquots of heavily buffered solutions are taken, additional precautions may be needed to ensure sufficient acid is present to obtain rapid diazotisation and for this reason HEWITT et al. (1955), CANDELA et al. (1957) increased to 1 N the strength of acid used to prepare

Table 7. *Conditions used for colorimetric assay of nitrate reductase systems in fungi and higher plants.*

Assay	Time (min)	Temp. °C	Notes	Reference
0.26 ml 0.2 M pyrophosphate pH 7.0 0.05 ml 1.5 10^{-5} M FAD 0.04 ml 2×10^{-3} M TPNH 0.1 ml 0.1 M KNO_3 0.05 ml extract	5	23 or 28	Started by adding enzyme. Control assays omitting TPNH. 1 Unit = quantity to produce 10^{-3} μM NO_2/5 min.	NASON and EVANS (1953)
0.24 ml 0.2 M pyrophosphate pH 7.0 0.05 ml 10^{-3} M FMN 0.04 ml 2×10^{-3} M TPNH 0.02 ml 10^{-3} M KCN 0.1 ml 0.1 M KNO_3 0.05 ml extract	10	25	Assay started and controls as above. 1 Unit = quantity to produce 10^{-3} μmole NO_2/10 min.	NICHOLAS, NASON and McELROY (1954)
0.24 ml 0.2 M orthophosphate pH 7.0 0.05 ml 10^{-3} M FMN 0.04 ml 2×10^{-3} M TPNH 0.02 ml 10^{-3} M KCN 0.1 ml 0.1 M KNO_3 0.05 ml extract	10	Room temp.	Assay started by adding KNO_3. Controls omitting TPNH. 1 Unit = quantity to produce 10^{-3} μmole in 10 min.	SILVER and McELROY (1954)
0.23 ml 0.1 M TRIS buffer pH 7.5 0.02 ml boiled TRIS extract of pig heart acetone-powder 0.05 ml 2×10^{-3} M TPNH 0.1 ml 0.1 M KNO_3 0.1 ml enzyme solution	10	25	Controls omitting TPNH, units as NICHOLAS, NASON and McELROY (1954).	NICHOLAS and SCAWIN (1956)
0.20—0.29 ml 0.1 M K_2HPO_4, pH 7—7.5 0.05 ml 10^{-5} M FAD 0.05 ml 2×10^{-3} M TPNH 0.1 ml 0.1 M $NaNO_3$ 0.1—1.0 ml extract	10	Room temp.	Controls as above. 1 Unit = quantity to produce increase over control of 0.19 mμM NO_2	KINSKY and McELROY (1958)
0.11 ml 0.5 M phosphate pH 7.0 0.05 ml 2.6×10^{-5} M FAD 0.04 ml 2×10^{-3} M TPNH 0.1 ml 0.1 M KNO_3 0.2 ml extract 0.5 μM NH_2OH added in assay of homogenates, see text (sources of error)	10	28	Started with enzyme extract. Controls omitting TPNH. Units of activity as mμM NO_2 produced/mg dry wt. or per mg protein.	EVANS and NASON (1953)

0.1—0.25 ml K$_2$HPO$_4$/KH$_2$PO$_4$ pH 7.5 0.1 ml 0.1 M KNO$_3$ 0.05 ml 10^{-5} M FAD 0.05 ml 4×10^{-3} M DPNH 0.05—0.02 ml of solution containing 100—250 μg protein. Assay of crude extract carried out in presence of 0.05 ml 2×10^{-3} M DPNH 0.05 ml 3 M Ethanol and 0.05 ml (40 μg) cryst. alcohol dehydrogenase	30	30	Started by adding extract. Stopped by removal of DPNH (see text, sources of error).	Spencer (1959)
1.0 ml 0.1 M K$_2$HPO$_4$ pH 7.5 0.5 ml 1.36×10^{-3} M DPNH 0.2 ml 0.1 M KNO$_3$ 0.2 ml extract Water to 2.0 ml	15	27	Started by adding DPNH followed by extract. Stopped by adding sulphanilic acid solution. Controls omitting DPNH.	Hageman and Flesher (1960)

the sulphanilamide solution, and KINSKY and McELROY (1958) used 2.5N acid. The conditions used to avoid errors due to reaction between nitrite and reduced pyridine nucleotides are described later under "Sources of Error."

Assays may be started adding enzyme or electron donor, usually TPNH or DPNH. The reaction is immediately stopped by adding the sulphanilamide or by other methods outlined later.

The problems of control experiments require comment. Crude extracts may contain some nitrite or may produce some by chemical reduction. Usually, the reaction occurs to a negligible extent if DPNH or TPNH or alternatively enzyme are omitted. If endogenous nitrite is very low, either procedure may provide a satisfactory control measurement, subject to the limitations stated for "Sources of Error" described below. It is necessary to establish that assays are linear with time over the period selected (see "Sources of Error"). The method of measuring "zero time" and "end of assay" values has the merit of keeping all factors the same except for some utilisation of electron donor, which should be small in relation to the total amount present. The precision of this method is greater for longer assay times and depends upon quick working for the zero time, especially with very active systems. A 1 min and $T + 1$ min assay might be advantageous.

b) Manometric estimation of nitrite. RO-BINSON (1954) described a method which depended on the manometric estimation of nitrite production by measurement of nitrogen liberated on addition of sulphamic acid in acid solution at the end of the assay period: $HONO + C_6H_5SO_2NH_2 \rightarrow N_2 + C_6H_5SO_2OH + H_2O$. One μmole nitrite yields 22.4 μl. This method is therefore less sensitive than the colorimetric estimation which readily detects 10 mμmole nitrite: equivalent to only 0.224 μl of nitrogen. It has merits where whole cells are used or where colorimetric methods are not applicable, when activities are high.

The manometric method for estimation of nitrate reduction was modified by HEWITT and HALLAS (1959) in a study of azide inhibition of aldehyde oxidase of potato. The procedure is applicable to any system analogous to nitrate reductase, where nitrite is the reaction product or substrate. In theory nitrite reacts with azide in acid solution as follows:

$$NO_2^- + N_3^- + 2H^+ \rightarrow N_2O + N_2 + H_2O$$

One μmole nitrite yields 44.8 μl of gas. As N$_2$O is considerably more soluble than nitrogen the volume of gas above the liquid phase is less than would be obtained if the whole solution comprised nitrogen. HEWITT and HALLAS observed experimentally the evolution of 82—86 μl of gas from 2 μM nitrite in a

Table 8. *Conditions used for assay of nitrate reductase systems in bacteria.*

Method	Assay	Time (min)	Temp. (°C)	Notes	Reference
Colorimetric test for nitrite	0.1 M phosphate (pH 7.4) 0.02 M Na formate 10^{-5} M methylene blue 0.004 M KNO_3 1 mg cell-free enzyme preparation Total volume 5 ml	50	35	Assays in THUNBERG tubes under anaerobic conditions.	SATO and NIWA (1952)
Colorimetric test for nitrite	0.31 ml 0.1 M phosphate (pH 7.0) 0.04 ml 4×10^{-3} μM DPNH 0.05 ml 10^{-4} M FMN or boiled pig heart 0.1 ml 0.1 M KNO_3 0.1 ml enzyme	10	25	Assays in air.	NASON and NICHOLAS (1955)
Colorimetric test for nitrite	0.028 M phosphate (pH 6.8) 0.15 μmoles DPNH 20 μmoles KNO_3 0.25 ml enzyme Total volume 5 ml	20	15	Assays in THUNBERG tubes under anaerobic conditions.	TANIGUCHI, ASANO, IIDA, KONO, OHMACHI, EGAMI (1957)
Colorimetric test for nitrite	15 μmoles phosphate (pH 7.5) 0.4 μmoles DPNH 0.1 μmole vitamin K_3 Total volume 1 ml	10	25	Enzyme system coupled to menadione reductase. Electrons transferred via vitamin K_3 to nitrate. Assays done anaerobically and others in air.	MEDINA and HEREDIA (1958) HEREDIA and MEDINA (1960)
Colorimetric test for nitrite and DPNH oxidation at 340 mμ	0.05 M citrate-phosphate (pH 6.4) 2.5 μmoles reduced phenosafranine (reduced with palladium + H_2) 5 μmoles KNO_3 enzyme preparation Final volume 2.5 ml 5 μmoles methyl viologen (reduced with $Na_2S_2O_4$) or 20 μmoles sodium formate or 0.5 μmoles DPNH used as alternative donors to reduced phenosafranine	10—20	30	Assays in THUNBERG tubes. After-incubation time, neutralized $CdSO_4$ solution added and nitrite determined in deproteinized supernatant solution. When DPNH used tubes shaken vigorously to oxidise DPNH by its oxidase at end of incubation period.	ITAGAKI and TANIGUCHI (1959)

Assay	Reaction mixture			Procedure	Reference
Colorimetric test for nitrite or spectrophotometric measurement of oxidation of MBH_2 at 590 mμ	250 μmoles phosphate (pH 7.1) 0.05 μmoles methyl viologen (MVH) or methylene blue (MB) 50 μmoles KNO_3 0.1 ml enzyme Total volume 0.5 ml After 5 min preincubation reaction started by adding < 1 mg $Na_2S_2O_4$	5	30	After reaction proceeded 5 min, tube shaken to reoxidise remaining MVH before testing for nitrite.	TANIGUCHI and ITAGAKI (1960)
Colorimetric test for nitrite.	0.7 ml 0.1 M phosphate (pH 7.4) 0.1 ml DPNH (250 mμM) 0.1 ml 10^{-2} M KNO_3 0.1 ml enzyme	10	30	After incubation, 0.1 ml 2 M Zn acetate and 1.9 ml 95% w/v ethanol added and the mixture thoroughly shaken. Nitrite determined in a suitable aliquot (0.2 to 1.0 ml) of the supernatant solution after centrifuging at 3000 g for 3 min.	FEWSON and NICHOLAS (1960a, 1961a, b, c)
Colorimetric test for nitrite	0.5 ml 0.2 M phosphate (pH 7.5) 1—1.5 mg benzyl viologen (BV) 0.05 ml 0.2 M KNO_3 3—6 μg enzyme protein Final volume 1 ml 2 ml $Na_2S_2O_4$ (0.2 mg/ml) in sidearm of THUNBERG tube	10		Assay in THUNBERG tubes. Reaction started by tipping in 0.2—0.3 ml of hydrosulphite solution from the side-arm of the tube. Care taken not to add too much $Na_2S_2O_4$ at once. More added from time to time as colour of BVH disappeared.	SADANA and McELROY (1957)
Colorimetric test for nitrite	0.2 M phosphate (pH 7.5) 0.13—0.20 μmoles DPNH 0.05 ml boiled extract of pig heart 0.1 ml 0.1 M KNO_3 0.01 to 0.05 ml enzyme Final volume 0.5 ml	10	Laboratory temp.	Assays in open tubes. Reaction stopped by adding the nitrite test reagents.	SILVER (1957)
Colorimetric test for nitrite and DPNH oxidation followed at 340 mμ	0.05 M phosphate (pH 6.7) 1.5 μmoles DPNH 30 μmoles KNO_3 0.2—0.6 mg N (enzyme) Final volume 3 ml Reduced Nile green (5 μmoles) also used instead of DPNH in the above assay. Dye reduced with H_2 gas and palladium or with $Na_2S_2O_4$	10—30	30	Assay in THUNBERG tubes. Enzyme preparation in sidearm. After incubation, neutralized $CdSO_4$ solution added to deproteinize the extract. Nitrite measured in clear supernatant solution.	TANIGUCHI and OEMACHI (1960)
Colorimetric test for nitrite	10 μmoles phosphate buffer (pH 7.5) 10 μmoles Na succinate (pH 7.5) 10 μmoles $NaNO_3$ 0.1 ml bacterial cell suspension Total volume 0.5 ml	10	30	Assays in air.	EVANS (1954) CHENIAE and EVANS (1956)

fluid volume of 4.5 ml at 27° C in an atmosphere of nitrogen. The application of this manometric method to the measurement of azide inhibition of nitrate reductase systems is described later (Properties and Mechanisms).

c) **Spectrophotometric measurement of DPNH or TPNH oxidation.** NASON and EVANS (1953) measured the changes in light absorption at 340 mμ associated with oxidation of TPNH or DPNH to follow the reaction in nitrate reductase systems in *Neurospora*.

The assay system illustrated by NASON and EVANS (1953) comprised the following components in 3.6 ml of 0.03 M pyrophosphate/0.1 M phosphate buffer pH 7.0: 0.83 μmole TPNH, 100 μmole KNO$_3$, 0.3 μmole FMN and 424 units of enzyme protein as defined by nitrite production in the section on assay.

The complete system produced a change of extinction at 340 mμ of 0.06 per minute over 15 minutes, but was not linear with time, since the initial rate over 3 minutes was about 0.1 per minute. EVANS and NASON (1953) assayed the soybean enzyme eluted from calcium phosphate gel in the presence of 0.63 μmole TPNH, 0.01 μmole FAD, 100 μmole KNO$_3$ and 45 units of enzyme in 0.17 M phosphate buffer at pH 7.0. Extinction at 340 mμ decreased at the rate of 0.03/min.

This method of assay is of use with relatively very active preparations obtained after partial purification. When initial activities in crude preparations, of the order of 1 to 10 mμmole NO$_2$/mg/min are expected, the method is not sufficiently sensitive in view of possible sources of interference such as TPNH and DPNH oxidase systems. These are especially active in plant roots (CRESSWELL, unpublished; VAIDYANATHAN and STREET, 1959) and may completely mask any evidence for nitrate reductase activity assayed by this means. Even if such interfering systems are not present it may be calculated from stoichiometry that as 1 mμmole NO$_2$ is equivalent to 1 mμmole TPNH, i.e. (0.74 μg) the change in extinction due to oxidation of 10 mμmole NO$_2$ using the molecular extinction coefficient of 6.22×10^3 M^{-1} cm^2 is about 0.02 and too small for reliable and accurate observation, if produced over a period of several minutes. EVANS and NASON (1953) combined spectrophotometric observations at 340 mμ with estimations of nitrite formation to establish the stoichiometric nature of the reaction with the soybean enzyme. There was a rapid endogenous oxidation of TPNH in the absence of added nitrate. The rate was further increased when nitrate was added and the difference in TPNH oxidised under the two sets of circumstances was equivalent to the nitrite produced.

NICHOLAS and NASON (1954b, c) and KINSKY and McELROY (1958) used a spectrophotometric assay at 340 mμ under anaerobic conditions, in order to investigate the electron transfer step between TPNH and FAD in the *Neurospora* enzyme. The reaction mixture was contained in cuvettes attached to a modified THUNBERG tube which was evacuated before tipping the enzyme. The assay mixture comprised 4 μmoles TPNH and 0.2 μmoles FAD in phosphate buffer, at pH 7.5. TPNH was omitted from the reference cuvettes. The next step from FMNH$_2$ (used in place FADH$_2$) to molybdenum and nitrate was studied by measurement of nitrite production under anaerobic conditions.

d) **Estimation of nitrate loss.** When concentrations of the order of 10^{-3} M NO$_3$ are used it is possible to measure 10—20 per cent loss of nitrate. The best method is that of WOOLLEY, HICKS and HAGEMAN (1960). This is a modification of the procedure of NELSON, KURTZ and BRAY (1954) where nitrate is reduced to nitrite by zinc in the presence of citric acid, diazotised, and coupled — all in the same reaction mixture. The reagent specified by WOOLLEY et al. consists of 100 g barium sulphate (analytical reagent grade) 75 g citric acid, 10 g manganese sulphate dihydrate, 4 g sulphanilic acid, 2 g zinc dust and 2 g 1-naphthylamine finely ground to an intimate mixture. 0.8 g is used in each estimation and added to the sample (1 ml) in 9 ml of 20% w/v acetic acid containing 0.2 ppm Cu^{2+}. The mixture is shaken on a shaker at 25° C for 3 min and then centrifuged to stabilise the end-point. The colour is measured at 540 mμ for nitrite. The reagents used must be initially free of copper as revealed by diphenylthiocarbazone. The recovery of nitrate is linear and reproducible for the amount added, but corresponds to almost exactly one-third of the value given by equimolar concentrations of nitrite. When estimating nitrate loss in the presence of nitrite formation the latter must be determined separately and a correction made for the nitrite value which would be obtained if that amount were added to a mixture estimated by the above method. When the method is used in the presence of benzyl viologen the column separation described under hydroxylamine assay must be adopted. It is important to control the conditions of assay with regard to time, temperature and concentration of all reagents for reproducible results.

2. Sources of Error.

The assays described here are subject to several possible sources of error or interference including simultaneous loss of nitrite by nitrite reductase, reoxidation of nitrite to nitrate, reaction between nitrite and components of the reaction mixture, adsorption on precipitated protein. Turbidity due to precipitated protein, non-linear response to protein added or with time and destruction of TPNH or DPNH by other oxidative systems may also occur.

a) **Enzymic interference.** Loss of nitrite by enzymic reduction was not regarded as significant by EVANS and NASON (1953) except during prolonged incubation with homogenates. At Long Ashton nitrite reductase activity has not caused losses of nitrite formed by nitrate reductase action during standard 10—30 minute assays with extracts of cauliflower, marrow, soybean, or wheat, when TPNH and DPNH and flavin co-factors have been used under anaerobic or aerobic conditions. EVANS and NASON (1953) found that 10^{-3} M concentration hydroxylamine hydrochloride effectively inhibited the slight nitrite reductase action in soybean extracts but also caused 12 per cent inhibition of nitrate reductase. Hydroxylamine is not normally used at Long Ashton. HAGEMAN and FLESHER (1960) omitted the use of hydroxylamine owing to its inhibitory effect. As EVANS and NASON (1953) observed a unit stoichiometric relationship between TPNH oxidation and nitrite production in the soybean preparation, it is evident that simultaneous loss of nitrite must have been negligible under these conditions. SILVER and MCELROY (1954) included cyanide at about 3×10^{-5} M in order to inhibit nitrite reductase differentially from nitrate reductase. NICHOLAS, NASON and MCELROY (1954) similarly used cyanide of a low concentration (4×10^{-5} M) to produce differential inhibition of nitrite and nitrate reductases.

Oxidation of nitrite to nitrate occurs in the presence of catalase or peroxidase and endogenous hydrogen peroxide (THURLOW, 1925), but has not been reported to be involved in assays with plant extracts, but recent work by CRESSWELL (unpublished) now suggests this source of error may be important.

Other sources of interference in nitrate reductase assay arise from non-linearity due to the loss of DPNH or TPNH as the result of simultaneous oxidase activity. Reduced pyridine nucleotide oxidases are extremely active in roots VAIDYANATHAN and STREET (1959) (CRESSWELL, unpublished) TPNH and DPNH-quinone reductase enzymes are also widespread and especially active in roots (WOSILAIT and NASON, 1954; CRESSWELL, unpublished). Evacuation and gassing with nitrogen decreases the rate of endogenous oxidation but is sometimes ineffective in preventing it altogether, as found by EVANS and NASON (1953) and

CRESSWELL (unpublished). The latter observed that in some preparations practically the whole of the normal level of DPNH added to nitrate reductase assay is destroyed by the oxidase system in 1 minute. For these reasons it is essential to establish that linearity obtains over a sufficient time period adequately to cover the duration of the assay. Another source of error in that connection is the loss of reduced pyridine nucleotide as a result of nitrate reduction. In very active systems this may proceed to an extent which results after a period in insufficient reduced pyridine nucleotide remaining to saturate the enzyme. Autoxidation of the flavin yields hydrogen peroxide which may inactive the enzyme, and DPNH cytochrome-c reductase can cause loss of DPNH according to SADANA and McELROY (1957).

b) **Chemical and physical errors.** A serious source of error which was not fully appreciated in the original work by EVANS and NASON (1953) and NASON and EVANS (1953) with plants and fungi is caused by rapid chemical reaction between nitrite and reduced pyridine nucleotides in acid solution produced during diazotization.

This interference was studied by the writers in separate experiments at Long Ashton namely on nitrate reductase (HEWITT, Long Ashton Ann. Reports 1956, p. 40; 1957, p. 41) and during investigations on nitrite reductase by MEDINA and NICHOLAS (1957), on nitrate reductase by HEWITT and AFRIDI (unpublished work) and independently by SADANA and McELROY (1957) and SPENCER (1959).

DPNH and to a lesser extent TPNH react with the diazotised product of nitrite and sulphanilamide in acid solution. This was shown by the effects on the apparent nitrite content present at a standard concentration, of adding DPNH, nitrite, sulphanilamide and the naphthyl reagent in different sequences. It is supposed that DPNH in acid solution reduces diazotised sulphanilamide to a phenylhydrazine derivative which is not able to couple with the naphthyl reagent. The extent of the interference over a range of nitrite concentrations is almost constant for a given amount of DPNH. 50 mμmoles DPNH decrease colour by 24 per cent and 500 mμmoles DPNH decrease colour by 66 per cent. TPNH interferes less than DPNH as equivalent concentrations DPN, TPN and nicotinamide do not interfere. There is no deamination of the adenine part of the molecule and no reaction between DPNH and nitrite at physiological pH values.

This source of interference can be overcome by removal of residual DPNH and TPNH at the end of the reaction. This can be achieved either by precipitation of the barium compounds (MEDINA and NICHOLAS, 1957), or as zinc salts or by enzymic oxidation.

In the barium or zinc precipitation methods 0.1 ml of M barium or zinc acetate are added followed by 5 volumes of cold ethanol, thorough mixing and centrifuging after 10 minutes standing at 0° C (MEDINA and NICHOLAS, 1957). This procedure also precipitates protein, clarifies the solution for colorimetric determination of nitrite, and prevents adsorption of nitrite on protein. The final volume which is increased may cause loss of sensitivity when low concentrations of nitrite are present. The calibration for extinction at 540 mμ is preferably made in the presence of comparable amounts of ethanol.

Oxidation of DPNH with acetaldehyde and alcohol dehydrogenase is readily achieved at pH values at or below 7.5 by adding 0.1 ml of M acetaldehyde followed by 0.1 ml of a solution containing at least 1,000 units (RACKER, 1950) of crystalline alcohol dehydrogenase, SPENCER (1959), NICHOLAS (1959), AFRIDI and HEWITT (unpublished). This oxidises more than 95 per cent of the total DPNH in 30 seconds and can be used to terminate the reaction; SADANA and McELROY (1957) and

SPENCER (1959) found it unnecessary to remove residual reduced pyridine nucleotides if these were being generated continuously at only catalytic concentrations by lactic or alcohol dehydrogenases, but removal is nevertheless preferable under these circumstances.

The enzymic removal of TPNH can be accomplished by the use of a quinone reductase (WOSILAIT and NASON, 1954) as used by WALKER and NICHOLAS (1961 a).

Another source of error in the assay by nitrite estimation is the loss of the red dye or of nitrite before diazotisation by adsorption on precipitated protein. This appears to occur to a variable extent and depends upon the amount of protein present and other factors which are not at present understood. This source of interference is conveniently avoided by adding 0.2 ml of saturated solution of sodium dodecyl sulphate before the addition of the acid sulphanilamide reagent, and after the addition of acetaldehyde and alcohol dehydrogenase (AFRIDI and HEWITT, unpublished). The sodium dodecyl sulphate should be filtered after standing at 0—4° C overnight. When unusually large amounts of protein are present, the volume of sodium dodecyl sulphate may require to be increased. Addition of ethanol also prevents adsorption of the red dye (CRESSWELL, unpublished).

Ascorbic acid present in some plant extracts reacts with nitrite on acidification (EVANS and McAULIFFE, 1956). This interference is usually small when endogenous ascorbic acid oxidation is rapid, but can be eliminated by the addition of partially purified cucumber ascorbic acid oxidase (HEWITT and DICKES, 1961).

III. Electron Donors and Co-Factors.

1. Fungi and Higher Plants.

The requirements and specificity for various electron donors, flavins, metals and anions are given in Table 9 and in scheme on page 100.

a) Pyridine nucleotides and other electron donors.

In *N. crassa* and *A. niger* the nitrate reductase is essentially TPNH specific (NICHOLAS, NASON and McELROY, 1954), but in many higher plants, e.g. soybean, cauliflower, vegetable marrow, DPNH and TPNH function equally well (EVANS and NASON, 1953; HEWITT and others, unpublished). The wheat embryo system (SPENCER, 1959) and maize leaf system (HAGEMAN and FLESHER, 1960) are DPNH specific, but the wheat leaf system also functions with DPNH (ANACKER and STOY, 1958) or TPNH (SPENCER, 1959), and the activity in corn roots was obtained with TPNH or DPNH in the work of EVANS and NASON (1953).

The saturation concentrations and MICHAELIS constants also vary considerably as seen from Table 3A. The optimum concentration of DPNH for the wheat embryo system appeared to be exceptionally high for higher plants.

NASON and EVANS (1953) were unable to replace TPNH by succinate, cysteine, glutathione or ascorbic acid as electron donors for nitrate reductase from *Neurospora*. VAIDYANATHAN and STREET (1959) however obtained slow reduction of nitrate by a crude preparation from excised tomato roots cultured in sterile media, when ascorbate and ferrous iron (7×10^{-4} M) were present.

In order to maintain an adequate concentration of reduced pyridine nucleotides especially in crude preparations containing oxidase systems, the activity of catalytic concentrations of DPNH or TPNH can be maintained by linking nitrate reductase to an appropriate pyridine nucleotide dependant dehydrogenase system.

Table 9. *Substrate, electron donor and co-factor requirements of nitrate reductase systems in fungi and higher plants.*

Source	Nitrate	Reduced pyridine nucleotide or other electron donors	Flavin	Metals and other co-factors	pH opt.	PO$_4$ requirements or other anions	Reference
Neurospora crassa	K_m 1.4×10^{-3} Opt. 4×10^{-3} M	TPNH K_m 7×10^{-5} M Opt. 3×10^{-4} M DPNH K_m 1.4×10^{-4} M Opt. 7×10^{-4} M TPNH[1] 20×DPNH	FAD (Natural) K_m 3.2×10^{-7} M Opt. 2.5×10^{-6} M FMN K_m 3×10^{-6} M Opt. 8×10^{-6} M FAD[1] 2×FMN		7.0 sharp in pyro-phosphate		NASON and EVANS (1953)
N. crassa				Mo			NICHOLAS, NASON and McELROY (1954)
N. crassa				Mo			NICHOLAS and NASON (1954a, b)
Soybean leaves				Mo			NICHOLAS and NASON (1955a)
N. crassa						PO$_4^{3-}$ opt. >5×10^{-3} M pH 7.5. Phosphate replacable by tellurate or selenate and partially so by arsenate, silicate or sulphate	NICHOLAS and SCAWIN (1956)
N. crassa		TPNH K_m 1.3×10^{-5} TPNH[1] 10×DPNH	FAD K_m 10^{-7} M			PO$_4^{3-}$ opt. 7×10^{-2} M K_m approx.10^{-3} M at pH7.0 Phosphate replacable by arsenate or tungstate, but not by silicate or sulphate	KINSKY and McELROY (1958)
Soybean leaves	K_m 7.5×10^{-3} M Opt. 2×10^{-2} M	TPNH K_m 2.3×10^{-5} M Opt. 8×10^{-5} DPNH K_m 3.2×10^{-5} Opt. 8×10^{-5} M TPNH[1] = DPNH DPNH Opt. 10^{-3} M	FAD (Natural) K_m 10^{-7} FMN K_m 3.7×10^{-6} Opt. 1.5×10^{-5} M FAD[1] 2×FMN FAD (Natural)		6.0		EVANS and NASON (1953)
Wheat leaf	—		FAD (Natural)	Mo	—	—	ANACKER and STOY (1958)
Wheat embryo	K_m 3.8×10^{-4} M Opt. 5×10^{-3} M	DPNH specific K_m 8×10^{-6} M Opt. 4×10^{-4} M DPNH Specific	FAD Specific K_m approx.10^{-7} M		7.4 sharp	PO$_4^{3-}$ 10^{-2} M approx. causes 30% stimulation	SPENCER (1959)
Maize			No detectable response to added flavins				HAGEMAN and FLESHER (1960)

[1] Indicates relative activities at optima.

SPENCER (1959) used crystalline alcohol dehydrogenase (RACKER, 1950) at pH 7.5 in the presence of 0.3 M concentration ethanol and 2×10^{-4} M concentration DPNH for assays on crude wheat embryo extracts where 0.05 ml of enzyme (0.75 mg/ml) was used. The procedure involving excess acetaldehyde to destroy residual DPNH was of course not employed. HAGEMAN, CRESSWELL and HEWITT (1962) linked glucose-6-phosphate and its dehydrogenase (Zwischenferment) and catalytic amounts of TPN to nitrate reductase obtained from vegetable marrow.

EVANS and NASON (1953) were able to link nitrate reductase from soybean with photochemically driven reduction of TPNH, brought about by illumination of chloroplast grana. The system caused the reduction of 65 mμmoles nitrate in 30 minutes in the presence of 200 mμmoles TPN and 1400 lumens/sq. ft. illumination.

Following observations by STOY (1955) on the action spectra for photosynthesis and nitrate reduction, STOY (1956) has shown that photochemical reduction of riboflavin can be linked directly to nitrate reductase action without the addition of a pyridine nucleotide. In the presence of EDTA (2×10^{-3} M) riboflavin (2.7×10^{-4} M) phosphate buffer, pH 7.0, KNO_3 (6×10^{-3} M) and 28 Neurospora units of a partially purified wheat nitrate reductase and a high pressure mercury vapour light intensity of 5×10^4 ergs/cm²/sec under anaerobic conditions, 30 mμmoles of nitrite were produced in 10 min at 28°. Under aerobic conditions only 0.5 mμmoles nitrite were produced. The activity with irradiated riboflavin was nearly double that observed with DPNH in the usual assay system.

HAGEMAN, CRESSWELL and HEWITT (1962) have shown that nitrate reductase obtained from vegetable marrow leaves can function with reduced benzyl viologen in place of pyridine nucleotides as the primary electron donor. Under these conditions activity may be increased approximately 4 to 5 times that observed with DPNH or TPNH. The ratio of reduced to oxidised dye is important and determines whether the reaction stops at nitrite as the end product or whether nitrite reduction to ammonia also occurs. This point is referred to in more detail in connection with nitrite and hydroxylamine reductases. For nitrate reductase assays with benzyl viologen, however, anaerobic conditions are necessary and dye is "titrated" into an evacuated THUNBERG tube (containing the other components) in 0.1 ml aliquots to maintain a faint blue colour. The concentration of dye is 1.5 mg/ml (3.65×10^{-3} M) of which between 25 and 40 per cent is in the reduced state (see nitrite reductase). HEWITT (unpublished) found similarly enhanced activities in tomato leaf extracts when benzyl viologen was used and values were similar ot those obtained for marrow.

b) Flavins.

The fungal enzymes appear to be relatively specific for FAD as shown by the low value of the MICHAELIS constant and the ratio of activities at optimal concentrations for FAD and FMN. A similar conclusion applies to the soybean enzyme described by EVANS and NASON (1953). In both Neurospora and soybean enzymes FAD was shown to be the natural flavin after dissociation by ammonium sulphate, and measurement of flavin level by D-amino acid oxidase and fluorescence tests (EVANS and NASON, 1953; NASON and EVANS, 1953). Although FMN can replace FAD at half activity in soybean, in the wheat embryo system (SPENCER, 1959) the FAD requirement is specific.

Flavins may be supplied as such or in boiled pig heart acetone powder, but at Long Ashton (HEWITT and others, unpublished) and in the work of HAGEMAN and FLESHER (1960) no response to added flavins has been obtained with crude preparations of nitrate reductase from higher plants. It is probable that the flavin and protein are not easily dissociated as was found for the enzyme from *E. coli* by NICHOLAS and NASON (1955 b).

c) Metal and Anion Requirements.

NICHOLAS, NASON and McELROY (1954), NICHOLAS and NASON (1954a, b, c, 1955a) showed conclusively that nitrate reductases are metallo-proteins containing molybdenum in preparations from *N. crassa, A. niger* and soybean. The metal is firmly bound to the protein and is only dissociated by dialysis against cyanide (NICHOLAS and NASON, 1954a, b, 1955a). After removal of cyanide the metal can be returned to the apo-enzyme and activity is then restored.

Protein extracted from molybdenum deficient fungi (NICHOLAS and NASON, 1954a) or higher plants (HEWITT, FISHER and CANDELA, 1955; CANDELA, FISHER and HEWITT, 1957) is inactive as a nitrate reductase and no reactivation occurs on adding molybdenum to the cell-free extract. Activity is restored if the metal is added before cell rupture and a further period of incubation of a few hours is allowed (NICHOLAS and NASON, 1954a; CANDELA, FISHER and HEWITT, 1957; HEWITT, 1957; HEWITT and AFRIDI, 1959). This process is inhibited by certain antimetabolites, lack of oxygen or low temperature (HEWITT and AFRIDI, 1959, and unpublished). EVANS and HALL (1955) showed that the distribution in protein fractions of molybdenum given as ^{99}Mo to soybean was similar to the specific activity of nitrate reductase. In the best preparations of ANACKER and STOY (1958) the proteins of molecular weight about 500,000 to 600,000 contained 0.2 μg Mo/mg of protein. This would correspond to 1 atom Mo per molecule of enzyme if pure, or higher ratios if not pure. A light absorption peak at 680 mμ was not constant, and indicated some impurity.

In addition to molybdenum, iron may also be required for the *Neurospora* enzyme at early stages of growth when grown under conditions of low oxygen tension where nitrate serves as a terminal oxidase electron acceptor (WALKER and NICHOLAS, 1961a). In older mycelia which are no longer submerged and where rapid oxygen exchange normally occurs the iron requirement is no longer observed. The need for iron may be related to the increased cytochrome content produced under low oxygen tension conditions and the fact that nitrate acts as inducer for both the nitrate reductase and associated TPNH cytochrome c reductase properties of the nitrate reductase complex (KINSKY and McELROY, 1957).

Anion requirements for nitrate reductase have been observed for the *Neurospora* enzyme by NICHOLAS and SCAWIN (1956) and KINSKY and McELROY (1956, 1958) and stimulation by phosphate was reported by SPENCER (1959) for the wheat embryo nitrate reductase. NICHOLAS and SCAWIN (1956) found that phosphate stimulation was optimal at about 5×10^{-3} M at pH 7.5 and that activity was increased by approximately 50 per cent at this concentration. KINSKY and McELROY (1958) observed a somewhat greater stimulation in similar preparations with an optimum concentration around 7×10^{-2} M. NICHOLAS and SCAWIN (1956) found that tellurate, selenate, arsenate and silicate, but not tungstate or vanadate could wholly or partially replace phosphate, whilst KINSKY and McELROY (1958) observed replacement by arsenate or tungstate but not by silicate. From a practical standpoint, it is desirable to include phosphate at 10^{-2} M or higher concentration in assay systems, especially when TRIS buffer has been used during previous stages of preparation. NASON and EVANS (1953) found that pyrophosphate stimulated activity of the *Neurospora* enzyme after dialysis.

2. Bacteria.

A variety of hydrogen or electron donors have been used for nitrate reductase in bacteria. Thus GREEN, STICKLAND and TARR (1934) showed that cell-free extracts of *Bacterium coli (Escherichia coli)* reduced nitrate to nitrite in the presence of lactate or formate or reduced dyes, e.g. methylene blue or benzyl viologen. The Japanese school, EGAMI and SATO (1947, 1948, 1949), SATO and EGAMI (1949), and SATO (1956), who have made an extensive study of the nitrate reductase system in *Escherichia coli,* have often used as substrate formate or lactate linked through its dehydrogenase and a suitable dye such as methylene blue for nitrate reduction. They have implicated haem proteins (cytochrome b in particular) as an important link between substrate and the nitrate acceptor. SATO and EGAMI (1949) suggested that cytochrome b_1 might be identical with nitrate reductase since they found that CO inhibited the enzyme and this was claimed to be reversed by light. This suggestion was withdrawn later by SATO and NIWA (1952) when they found that thiourea inhibited nitrate reduction but did not affect the aerobic oxidation of cytochrome b_1. JOKLIK (1950) was moreover unable to repeat the CO inhibition reported by SATO and EGAMI (1949). These authors, however, still believe that this cytochrome plays a key role in nitrate reduction. AUBEL, LUBICHINSKY and PROVOST (1953) studied nitrate reductase in *E. coli* and showed that the enzyme did not contain cytochrome b_1.

NICHOLAS and NASON (1955b) purified a soluble assimilatory nitrate reductase from *E. coli* (strain B) and this utilized DPNH as an electron donor and it required FAD and Mo which function as electron carriers. Cytochromes were not involved in this system.

Nitrate reductase from actively denitrifying cells of *Pseudomonas aeruginosa* will also utilize DPNH and to a lesser extent TPNH as an electron donor (FEWSON and NICHOLAS, 1960a, 1961a, b, c). A comparison between various electron donors for reduction of nitrate by the purified enzyme is shown in Table 10. It is clear that reduced forms of benzyl viologen or methyl viologen or methylene blue were even more effective donors than were DPNH or TPNH. Reduced flavin, ferro-cyto-chrome c or Mo^{5+} were also effective donors for the purified enzyme. In *Pseudomonas stutzeri,* however, it is necessary to regenerate DPNH by the alcohol dehydrogenase enzyme or TPNH by the isocitric acid dehydrogenase enzyme since there is a very active DPNH and TPNH oxidases present in extracts of the bacteria (CHUNG and NAJJAR, 1956a, b).

Many of the purified enzymes especially those concerned with the dissimilation of nitrate do not utilize DPNH or TPNH as electron donors. Thus TANIGUCHI and ITAGAKI (1959) found that reduced dyes only, e.g. benzyl viologen, were effective donors for nitrate reduction and under these conditions flavin did not appear to be essential but a metal requirement was suggested from inhibition studies. Recently TANIGUCHI and ITAGAKI (1960) prepared a protein containing Mo and Fe. They claim that the Mo functions as an electron carrier as suggested earlier by NICHOLAS and STEVENS (1955) for the *Neurospora* enzyme. In *Achromobacter fischeri* SADANA and McELROY (1957) used reduced benzyl viologen as an electron donor for nitrate reductase. The dye was reduced by tipping in 0.2 ml 0.3% w/v $Na_2S_2O_4$ from a sidearm of a THUNBERG tube into the well containing substrate, enzyme and dye. Care was taken to avoid excessive addition of the dithionite since it inactivates the enzyme as well as reducing nitrite formed in the reaction mixture. This technique is difficult to operate and it is preferable to reduce the benzyl viologen with palladised asbestos and hydrogen in specially designed titration vessels (Fig. 2) or a special THUNBERG tube apparatus (Fig. 3) to be described under nitrite reductase (p. 131). These procedures do not inactivate the enzyme nor is the nitrite reduced by dye reduced by hydrogen and the palladised asbestos if reduction is properly controlled.

A large number of bacteria and algae utilize hydrogen gas for the reduction of nitrate: *Aerobacter aerogenes* (LASCELLES and STILL, 1944), *Ankistrodesmus braunii* (KESSLER, 1957), *Azotobacter vinelandii* (PHELPS and WILSON, 1941), *Bacterium formicum* (STEPHENSON and STICKLAND, 1931), *Clostridium welchii* (WOODS, 1938), *Micrococcus denitrificens* (VERHOEVEN, KOSTER and VAN NIEVELT, 1954; FEWSON and NICHOLAS, 1961), *Proteus vulgaris* (BACK, LASCELLES and STILL, 1946) and *Scenedesmus obliquus* (KESSLER, 1957). Although cell free extracts of some of these organisms have been prepared the enzymes concerned have not been fully characterized.

A porphyrin requirement for nitrate reduction has been demonstrated in *Haemophilus influenzae* (GRANICK and GILDER, 1946) and in *Staphylococcus aureus* (LASCELLES, 1956). Cytochromes have been suggested to be required for nitrate reduction by VERHOEVEN and TAKEDA (1956) and SATO (1956). The cytochrome c contents of *Pseudomonas aeruginosa, Ps. stutzeri* and *M. denitrificans* were shown to increase during denitrification and the addition of nitrate to whole cells caused an oxidation of the cytochromes. These findings with whole cells, however, do not necessarily imply that cytochromes are components of the nitrate reductase enzyme since they might be the result of (a) the haem compounds acting at some point in the electron transfer chain before the terminal nitrate reductase enzyme or (b) the cytochromes being involved in a reduction of inorganic nitrogen compounds at a lower oxidation state than nitrate.

IV. Properties and Mechanisms.

1. Fungi and Higher Plants.

a) Stability. Most nitrate reductase preparations are relatively thermolabile (Tables 11 and 12) and are easily denatured. The enzyme from soybean obtained by EVANS and NASON (1953) was about 46 per cent inactivated in 5 minutes at 30° and lost 28 per cent of activity at 20° during a similar period. This lability is obviously a factor to be considered in the choice of assay conditions. When dialysed at 0—2° in the presence of cysteine in suitably treated dialysis sacs, the enzymes may remain active when stored for prolonged periods at low temperatures.

b) Effects of inhibitors. (i) *Sulphydryl reagents.* Compounds such as p-chloromercuribenzoate (p-CMB), phenylmercury acetate, iodosobenzoate, and heavy metals strongly inhibit nitrate reductase. The probable site of action is at the sulphydryl groups, but no titrations with purified proteins have been reported and the number of these groups which are involved in enzyme function is not known. Inhibition by p-CMB is reversible by incubation with comparable or higher concentrations of cysteine or glutathione. The soybean and *Neurospora* enzymes appear to be more sensitive than the wheat embryo enzyme as seen from some data in Table 11.

(ii) *Metal prosthetic groups — and other inhibitors.* Nitrate reductase enzymes are inhibited by cyanide (Table 11) at varying concentrations depending upon the source of the enzyme. That obtained from wheat embryo appeared to be more sensitive than those prepared from soybean or *Neurospora*.

The effect of vanadate reported by SPENCER (1959) is interesting as this appeared highly inhibitory to the wheat embryo enzyme but was not reported as such for the *Neurospora* enzyme when tested by NICHOLAS and SCAWIN (1956) as a possible replacement for phosphate in relation to the anion requirement noted earlier. Tungstate was not inhibitory to the enzymes from either source.

Azide at concentrations between 10^{-5} and 10^{-3} M has been widely reported as inhibitory to nitrate reductase enzymes. It is probable as with aldehyde oxidase of potato studied by HEWITT and HALLAS (1959) that such reports of inhibition are valid. It is necessary, however, to point out that this effect where reported elsewhere has not been shown unequivocally, since the fact that azide and nitrite react together rapidly at a low pH produced when the acid sulphanilamide reagent is added, has been overlooked. In each case the inhibition produced by azide had been measured by measuring the decrease in nitrate production colorimetrically following incubation and assay of the enzyme with or without the presence of azide. As noted in the section on manometric methods of assay, HEWITT and HALLAS (1959) found that azide at concentrations of the order of those reported to inhibit nitrate reductase reacts directly with nitrite in the acid medium used for diazotisation with consequent loss of nitrite as nitrogen and nitrous oxide. Azide at 10^{-3} M concentration results in almost total loss of nitrite present at 10^{-5} to 10^{-4} M concentration, i.e., 10 to 100 mμmoles per ml. It is therefore necessary to measure azide inhibition by the method described by HEWITT and HALLAS (1959) where the difference in gas evolution is observed in assay systems carried out in the presence of azide added either before the reaction or at the end of the reaction period. VILLANEUVA (1959c) independently noted the interference by azide.

The system suggested by HEWITT and HALLAS to test for azide inhibition in aldehyde oxidase of potato was as follows. The main compartment of a WARBURG manometer flask contained 3.5 ml of 0.13 M phosphate buffer pH 6.0 and enzyme. Other pH values would be chosen for nitrate reductase systems. When azide inhibition was to be tested various concentrations ranging from 10^{-4} M to 5×10^{-3} M were also included in the reaction mixture. The reaction was started by adding 0.1 ml of M acetaldehyde contained in a small cup hooked on to the centre well and dislodged by a jolt. At the end of the assay period 0.5 ml 10^{-2} M sodium azide was tipped from one side arm followed by 0.4 ml of 2 N hydrochloric acid from a second side arm. The rapid gas evolution was measured in the usual manner. Later work indicated that when 10^{-3} M concentration of azide was present as an inhibitor it was unnecessary to add more for gas production. If the concentration of azide is too high, blank values due to decomposition of azide in acid solutions become higher than desirable for precise measurements; the concentration used in the side arm should therefore be reduced to 5×10^{-3} M (HEWITT and HALLAS, unpublished). In order to use the manometric assay a minimum of about 0.5 to 1 μmole nitrite require to be produced.

c) **Mechanism of action.** The enzymes prepared from *Neurospora crassa* and soybean were shown by NICHOLAS and NASON (1954a, b, c) and NICHOLAS and STEVENS (1956a, b) to contain molybdenum as stated above. The behaviour of the enzymes when separated from the metal shows that its presence is essential for the final step in electron donation to nitrate. Reduction of flavin by reduced pyridine nucleotide occurs in the absence of the metal. As restoration of pentavalent molybdenum (Mo^{5+}) to the apo-enzyme allows perceptible nitrate reduction in the absence of reduced flavin or reduced pyridine nucleotide it is concluded that the pentavalent metal is as effective an electron donor as the reduced flavin or pyridine nucleotide. Electron donation from the pentavalent molybdenum is regarded as the site of reversible cyanide inhibition. The metal-free apo-enzyme is also able to transfer electrons from reduced pyridine nucleotides via flavins to dyes such as 2,3,6-trichlorophenol-indophenol; that is functioning as a diaphorase system. Spin resonance measurements have confirmed that Mo^{5+} is formed during

Table 10. *Substrate, electron donor cofactor requirements*

Source	Nitrate	Reduced pyridine nucleotide or other electron donors*
Escherichia coli		Reduced Methylene blue (MBH$_2$) Formate
Escherichia coli (B strain)	K_m 10^{-4} M	K_m DPNH 2×10^{-4} M
Escherichia coli (Yamagutchi strain)		MBH$_2$ Formate DPNH
Escherichia coli (B strain)	K_m 10^{-4} M	DPNH TPNH
Escherichia coli (Yamagutchi strain)	K_m 7.5×10^{-5} M with Phenosafrani- ne-H$_2$ as donor	*Optimum* MBH$_2$ 1.7×10^{-3} M: 0.17 MVH 1.7×10^{-3} M: 0.85 Phenosafranine-H$_2$ 1.7×10^{-3} M: 0.56 2,6:dichlorophenolindo- phenol H$_2$ 1.7×10^{-3} M: 0.11 Nile blue H$_2$ 1.7×10^{-3} M: 0.48 Thionine H$_2$ 1.7×10^{-3} M: 0.15
Escherichia coli (Yamagutchi strain)	K_m 5.1×10^{-4} M	*Optimum* Reduc. Methyl viologen 2×10^{-3} M: 100 Reduc. Benzyl viologen 2×10^{-3} M: 100 Phenosafranine-H$_2$ (PSH$_2$) 2×10^{-3} M: 58 Pyocyanine-H 2×10^{-3} M: 55 MBH$_2$ 2×10^{-3} M: 5.3 Phenazine methosulph. H 2×10^{-3} M: 2.7 FMNH$_2$ 0.2×10^{-4} M: 31 FADH$_2$ 0.2×10^{-4} M: 29 Formate 5.0×10^{-3} M: 16 DPNH 0.2×10^{-4} M: 6.8 TPNH 0.2×10^{-4} M: 2
Pseudomonas aeruginosa (NCIB 8704)	K_m 1.6×10^{-5} M	DPNH K_m 4.2×10^{-5} *Optimum* DPNH 2×10^{-4} M: 100 TPNH 2×10^{-4} M: 8 Ferrocytochrome c 2.5×10^{-4} M: 70 FADH$_2$ 2.5×10^{-4} M: 35 MoCl$_5$(Mo^{5+}) 1×10^{-3} M: 25 FeSO$_4$(Fe^{2+}) 1×10^{-3} M: 4 BVH 2.5×10^{-4} M: 340 MVH 2.5×10^{-4} M: 360 MBH$_2$ 2.5×10^{-4} M: 410 Xanthine 1×10^{-4} M: 6 Hypoxanthine 1.5×10^{-3} M: 34 Acetaldehyde 3×10^{-3} M: 14
Achromobacter fischeri = *Photobacterium fischeri*	K_m 6.5×10^{-5} M with BVH or DPNH as donor	BVH MVH DPNH coupled cyt. c reductase

* Figures show relative activity with different electron donors at optimum concentrations.

of nitrate reductase systems in bacteria.

Flavin*	Metals and other cofactors	pH optimum	Reference
not detected	Fe cytochrome b (later retracted)	7.4	SATO and NIWA (1952)
FAD K_m 0.5×10^{-6} M FMN	Mo	7—8 50% loss at 5 or at 9.5	NICHOLAS and NASON (1955)
FAD	Fe cytochrome b_1 and unidentified soluble factor	7.4	TANIGUCHI, ASANO, IIDA, KONO, OHMACHI, EGAMI (1957)
Anaerobically DPNH-system flavin dependent (FAD)	Fe cytochrome b vitamin K_3 Transfer of electrons from DPNH or TPNH to vitamin K_3 catalyzed by menadione reductase.	7.5	MEDINA and HEREDIA (1958) HEREDIA and MEDINA (1960)
Flavin	Menadione Fe^{++} cytochrome b_1 as intermediary electron carriers	6.4 activated by citrate or pyrophosphate	ITAGAKI and TANIGUCHI (1959)
no bound flavin	1 atom Mo/mole enzyme 40 atoms Fe/mole enzyme cytochrome b_1	8.3	TANIGUCHI and ITAGAKI (1960)
FAD K_m 2×10^{-7} M opt. 1×10^{-6} M: 100 FMN K_m 1×10^{-7} M opt. 8×10^{-7} M: 65	Mo: 0.15 μg/mg protein Fe: 5 μg/mg protein cytochrome c	7.4	FEWSON and NICHOLAS (1960a) (1961a, b)
DPNH cyt. reductase coupled to nitrate requires FMN or FAD K_m 1.6×10^{-5} M		7—8.5 acetate, phosphate TRIS or glycine buffer	SADANA and McELROY (1957)

Table 10

Source	Nitrate	Reduced pyridine nucleotide or other electron donors
Hansenula anomala Strain 317	$K_m\ 3.2 \times 10^{-4}$ M	DPNH $K_m\ 0.2 \times 10^{-4}$ M TPNH $K_m\ 0.3 \times 10^{-4}$ M
Azotobacter vinelandii (O)	$K_m\ 5.3 \times 10^{-4}$ M	2,6, Dichlorophenol indophenol H_2 0.08 Thionine H_2 0.10 MB H_2 0.10 Toluylene blue H_2 0.38 Nile blue H_2 0.44 Phenosafranine H_2 0.40 DPNH 0.20
Rhizobium japonicum from soybean nodules	no record	*Optimum* DPNH 1.6 10^{-4} M: 100 TPNH 1.6 10^{-4} M: 10 Na Succinate (no record)

nitrate reductase action in *Ps. aeruginosa* (Fewson and Nicholas, 1961 a, b, c). The sulphydryl groups were considered by Nicholas and Nason (1954b, c) to function in binding the pyridine nucleotide to the protein.

The phosphate requirement was considered by Mahler (1956), Kinsky and McElroy (1958), Hewitt (1959) to be related to the determination of the appropriate oxidation-reduction potential of hexavalent molybdenum in molybdoflavoproteins. Kinsky and McElroy showed that phosphate stimulated the transfer of electrons from reduced flavin to nitrate but had no effect on reduction of flavin by the TPNH. The mechanism of the anion effect is not clear however since the role of those anions that can form hetero-poly acid complexes which facilitate reduction of Mo^{6+} is neither consistent nor similar. The extent to which the anion may serve to bind molybdenum to the protein as suggested by Nicholas and Scawin (1956) is also uncertain in view of the ease of removal of phosphate compared with the closely associated molybdenum-protein complex. It is possible that a clearer idea of the role of the anion may be obtained when more highly purified preparations of the enzymes are available. The phosphate requirement, replacable by arsenate is also observed for the xanthine oxidase and aldehyde oxidase systems as described by Mackler, Mahler and Green (1954) and Mahler, Mackler, Green and Bock (1954). It is nevertheless quite clear that in nitrate reductase and related enzymes, the anion is involved in the final electron transfer stage.

The mechanism of action of nitrate reductase in bacteria, fungi and plants may therefore be represented by the following scheme, based on the work of Nicholas

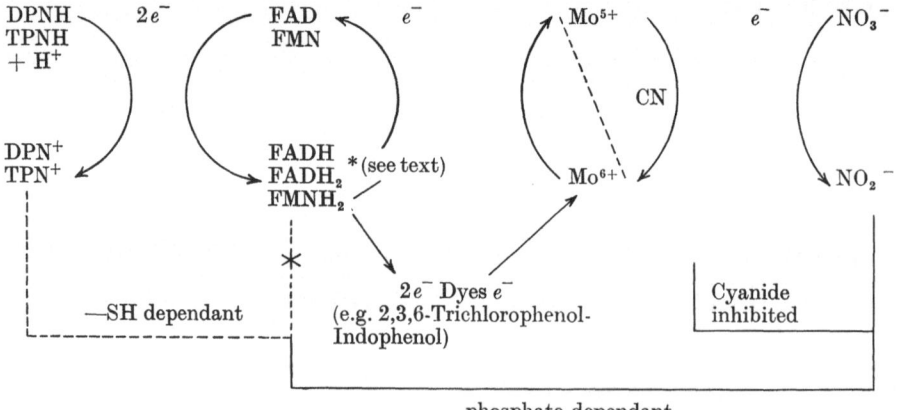

phosphate dependant

(Continued).

Flavin	Metals and other cofactors	pH optimum	Reference
FAD FMN DPNH system stimulated 2 fold by FAD or FMN. Reduced nile blue; no flavin requirement	Mo CO insensitive: heavy metal — not identified	7—8 6.7	SILVER (1957) TANIGUCHI and OHMACHI (1960)
FAD	Fe Mo (deficiency studies only)	6.5—7.5	EVANS (1954) CHENIAE and EVANS (1960)

and NASON; NICHOLAS and STEVENS; KINSKY and MCELROY; FEWSON and NICHOLAS.

The question of whether a semi-quinone corresponding to a single electron change is involved at the flavin level is not established, but is likely in the view of MAHLER and GLENN (1956). They suggested in a general thesis on metallo-flavoproteins that the metal and the flavin formed a resonance pair able to stabilise the existence of a semi-quinone one electron reduction state in the flavin. It is also likely that two electron changes can occur at the flavin level in accordance with the diaphorase activity shown by ability to reduce two electron acceptor dyes, e.g. 2,3,6-trichloro-indophenol, and in some analogous enzymes, e.g. xanthine oxidase, with the formation of hydrogen peroxide during substrate oxidation with oxygen. It is of special interest that whereas the dyes normally reduced by flavins are two electron acceptors, benzyl viologen which will serve as a primary electron donor (HAGEMAN, CRESSWELL and HEWITT, 1961) undergoes only a single electron change (MICHAELIS and HILL, 1933).

It has been pointed out elsewhere (HEWITT, 1958) that as two electrons are involved in the reduction of nitrate to nitrite some additional mechanism must be postulated to explain the final step of the reaction. It is possible that the ability of flavins to function as two electron carriers and also (in the view of MAHLER and GLENN, 1956) to form stabilised metal-semi-quinone resonating structures may explain the overall mechanism more satisfactorily thus:

$$\text{TPNH(DPNH)} + \text{H}^+ + \text{FAD(FMN)} \longrightarrow \text{TPN}^+(\text{DPN}^+) + \text{FADH}_2(\text{FMNH}_2) \qquad 1.$$

$$\text{FADH}_2 + \text{Mo}^{6+} \longrightarrow \text{FAD}\bigg\langle{}^{\text{H}}_{\text{Mo}^{5+}} + \text{H}^+ \qquad 2.$$

(stabilised metal-semi-quinone state)

$$\text{FAD}\bigg\langle{}^{\text{H}}_{\text{Mo}^{5+}} + \text{NO}_3^- + \text{H}^+ \longrightarrow \text{FAD} + \text{Mo}^{6+} + \text{NO}_2^- + \text{H}_2\text{O} \qquad 3.$$

$$\text{Sum: TPNH} + \text{H}^+ + \text{NO}_3^- \longrightarrow \text{TPN}^+ + \text{NO}_2^- + \text{H}_2\text{O}$$

The alternative is to postulate that one or two flavin molecules donate two electrons to two Mo^{5+} sites. Further purification of the enzyme may show the ratio of Mo/flavin/enzyme and decide this point.

2. Bacteria.

Similar molybdo-flavoprotein systems to those occurring in fungi have been described in *E. coli* (NICHOLAS and NASON, 1955 b) and in yeast *Hansenula anomala* (SILVER, 1957). The effects of inhibitors on these enzymes are shown in Table 12.

Table 11. *Stability of, and effects of inhibitors on, nitrate reductase preparations from fungi and higher plants.*

Source	Thermal denaturation and stability[1]	Inhibition by —SH reagents[2] or flavin analogues	Other inhibitors[2]	Reference
N. crassa	0—43% $(NH_4)_2SO_4$ fraction 10—20/month —15°; 24—46% $(NH_4)_2SO_4$ fraction 50/day 4° or —15°; phosphate gel eluate 50/week —15°; 50/5 min 40° pH 7; 100/5 min 50°	p. CMB 5×10^{-6} M: 43; 6×10^{-5} M: 94; Reversed by cysteine or glutathione	KCN 10^{-3} M:76; 10^{-4} M: 32; NaN₃ 10^{-3} M: 84; 10^{-4} M: 68; $CuSO_4$ 10^{-3} M: 88; 10^{-4} M: 64; K Ethyl Xanthate 5×10^{-3} M: 56; 5×10^{-4} M: 32; o-Phenanthroline 5×10^{-3} M: 64; 5×10^{-4} M: 14; EDTA 5×10^{-3} M: 11; 5×10^{-4} M: NIL; Thiourea 5×10^{-3} M: 44; 5×10^{-4} M: Nil; NaF 5×10^{-3} M: Nil	NASON and EVANS (1953)
N. crassa isolated in presence of glutathione	Stable at all stages in 0.1 M K/HPO₄ pH 7.5 <50 *p. annum* —17°	p. CMB 10^{-5} M: 100; 10^{-7} M: 9 (both NO₃ reductase and TPNH-cytochrome-c reductase)	NaN₃ 10^{-3} M: 98; KCN 10^{-3} M: 92; (NO₃ reductase activity only)	KINSKY and McELROY (1958)
Soybean leaves	Fraction IV: Phosphate gel, $(NH_4)_2SO_4$ 0—40% and phosphate gel again 50—70/week —15°; 28/5 min 20°; 46/5 min 30°; 71/5 min 40°; 100/5 min 50°	p. CMB 10^{-3} M: 94; 10^{-4} M: 90; 10^{-5} M: 80; 10^{-6} M: 51; Reversible by 10^{-3} M cysteine	KCN 10^{-3} M: 83; 10^{-4} M: 27; NaN₃ 10^{-3} M: 98; 10^{-4} M: 39; $CuSO_4$ 10^{-4} M: 32; NH₂OH 10^{-3} M:12; CO: Nil; o-Phenanthroline 2×10^{-3} M: 10; α-α-Dipyridyl 2×10^{-3} M: 10; Thiourea 2×10^{-3} M: 56; NaF: Nil; K Ethyl xanthate 2×10^{-3} M: 56; 8-Hydroxyquinoline 2×10^{-3} M: 11	EVANS and NASON (1953)
Wheat embryo	50/week —15°; 38/5 min 30°; 96/5 min 45°; 100/5 min 60°	p. CMB 2×10^{-4} M: 100; 2×10^{-5} M: 81; 2×10^{-6} M: 0	KCN 10^{-4} M: 100; 10^{-5} M: 84; 10^{-6} M: 38; NaN₃ 10^{-4} M: 87; 10^{-5} M: 35; Salicylaldoxime 2×10^{-3} M: 47; Na VO₃ 10^{-3} M: 100; 10^{-4} M: 77; 10^{-5} M: 32; 10^{-6} M: 13; NaWO₄ 10^{-3} M: Nil; α-α-Dipyridyl and o-Phenanthroline 2×10^{-3} M: Nil; 8-Hydroxyquinoline 2×10^{-3} M: 8	SPENCER (1959)

[1] Values are as per cent inactivation for times and temperatures shown.
[2] Values are as per cent inhibition for concentrations shown.

Both molybdenum and cytochrome have been shown to participate in nitrate reduction by denitrifying bacteria (FEWSON and NICHOLAS, 1960a, 1961a,b,c). The sequence of electron transfer to the alternative hydrogen acceptors oxygen and nitrate in *Pseudomonas aeruginosa* is as follows:

DPNH $\longrightarrow$ FAD $\longrightarrow$ cytochrome c $\longrightarrow$ nitrate reductase $\longrightarrow$ NO_3^-
(Mo)

cytochrome oxidase

O_2

Preliminary evidence that both Fe and Mo are essential for the dissimilatory enzyme was obtained in nutritional deficiency and inhibitor studies (Table 12) and by the concentration of these metals in the purified enzyme. At least half of the iron was present as cytochrome c. Electron spin resonance measurements showed that during nitrate reduction molybdenum undergoes a valency change probably $Mo^{6+} \rightarrow Mo^{5+}$ and an FAD semiquinone is formed (FEWSON and NICHOLAS, 1961a). The sequence of electron transfer was confirmed by the fact that added mammalian or bacterial cytochrome c was reduced enzymically by either DPNH or reduced FAD and re-oxidised by either Mo^{6+} or nitrate in the presence of the enzyme only. The reduction of nitrate by reduced cytochrome c was inhibited by metal chelating agents indicating that cytochrome precedes Mo in the electron transfer sequence. Thus the first part of the system is a DPNH-dependent cytochrome c reductase which can reduce endogenous cytochrome c and is probably the pen-ultimate electron sequence of nitrate reductase. The latter enzyme is probably a Mo containing protein only. The inhibition of nitrate reduction by oxygen in cell-homogenates was facilitated by cytochrome oxidase. A similar effect of oxygen on nitrate reduction in *E. coli* has been observed by STICKLAND (1931).

FEWSON and NICHOLAS (1961b) have also shown that a similar scheme operates in *Micrococcus denitrificans* but in addition DPN can be reduced by hydrogenase in this organism:

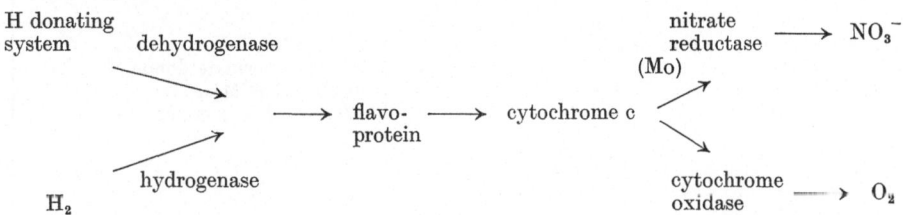

SADANA and McELROY (1957) proposed a similar scheme for nitrate reduction in *Achromobacter fischeri* (= *Photobacterium fischeri*) but had no evidence that Mo was involved:

DPNH $\longrightarrow$ FAD $\longrightarrow$ Fe^{+++} $\longrightarrow$ Bacterial cytochrome c $\longrightarrow$ O_2

Reduced Benzyl viologen $\longrightarrow$ nitrate reductase

NO_3^-

Table 12. *Stability of, and effects of inhibitors*

Source	Thermal denaturation and stability[1]	Inhibition by —SH reagents or flavin analogues[2]
Escherichia coli	Total loss of activity at 60°	not tried
Escherichia coli (B strain)	90/5 min/65°	pCMB 5×10^{-4} M: 60 reversed completely by GSH.10^{-3} M
Escherichia coli (YAMAGUTCHI strain)	Activity halved at 50°	not tried
Escherichia coli (B strain)		
Escherichia coli (YAMAGUTCHI strain)	10^{-4} M cysteine stabilizes pure enzyme when stored for 2 weeks (—15°). No loss of activity. Stable to short term dialysis. Cysteine or GSH (1 to 5×10^{-4} M) essential for prolonged dialysis. 100/5 min/80°.	When DPNH electron donor: 10^{-4} M pCMB: 98 Alloxan 10^{-4} M: 65 10^{-3} M: 100 Reversed by 10^{-3} M cysteine
Escherichia coli (YAMAGUTCHI strain)	no record	5×10^{-4} M pCMB: 60 when formate donor 5×10^{-4} M pCMB: 20 when MVH donor
Pseudomonas aeruginosa (NCIB 8704)	60/5 min/55°	10^{-2} M pCMB competitively inhibited the activation of the apo-enzyme by FAD. Effect reversed completely with 10^{-3} M glutathione

[1] Values are as per cent inactivation for times and temperatures shown.
[2] Values are as per cent inhibition for concentrations shown.

on, the nitrate reductase systems from bacteria.

Other inhibitors[2]	Reference
KCN 10^{-3} M: 100 Thiourea.0.2 M: 100	Sato and Niwa (1952)
KCN.10^{-3} M: 80 NaN$_3$.10^{-3} M: 90 8-Hydroxyquinoline 10^{-3} M: 40	Nicholas and Nason (1955 b)
2-nHeptyl-4-hydroxyquinoline-N-oxide.(HOQNO). 10^{-6} M: 70 KCN.10^{-3} M: 100	Taniguchi, Asano, Iida, Kono, Ohmachi and Egami (1957)
KCN.10^{-3} M: 100 α-α -Dipyridyl 6×10^{-3} M: 67 o-Phenanthroline 5×10^{-3} M: 90 Na Diethyldithiocarbamate.10^{-3} M: 29 Salicylic acid 10^{-3} M 32 8-Hydroxyquinoline 10^{-3} M 29 Thiourea 10^{-3} M: 41 2-nHeptyl-4 hydroxyquinoline N-oxide (HOQNO) 4×10^{-4} M 95	Medina and Heredia (1958) Heredia and Medina (1960)

	DPNH	PSH$_2$	MVH	
KCN 10^{-2} M	100	90	100	Itagaki and Taniguchi (1959)
NaN$_3$ 10^{-3} M	100	100	100	
CO dark (1 atm.)	0	0	0	
8-Hydroxyquinoline.10^{-2} M	30	0	0	
Thiourea.10^{-2} M	85	45	46	
Na-EDTA 10^{-2} M	50	25	15	
Amytal 10^{-3} M	40	0	0	
Quinine 10^{-3} M	30	0	5	
Dicoumarol.	80	8	0	
HOQNO.10^{-4} M	77	0	0	

		Donor		
		Formate	MVH	
Amytal	2×10^{-3} M	75	20	Taniguchi and Itagaki (1960)
Menadione	5×10^{-3} M	40	0	
Dicoumarol	3×10^{-5} M	70	25	
Dicoumarol } Menadione } +	3×10^{-5} M } 5×10^{-5} M }	70	—	
HOQNO	10^{-5} M	60	20	
Phenazine methosulphate	3.3×10^{-4} M	160	5	

Other inhibitors	Reference
KCN. 5×10^{-3} M: 10 0;5×10^{-4} M: 60; 5×10^{-5} M: 33; 10^{-5} M: 6. Na-EDTA, 5×10^{-5} M: 33. Na-Diethyldithiocarbamate. 5×10^{-5} M: 70; 5×10^{-4} M: 34. 8-Hydroxyquinoline 10^{-3} M: 32 o-Phenanthroline 5×10^{-3} M: 85; 10^{-3} M: 48; 5 10^{-4} M: 25; 5×10^{-5} M: 24. α-α -Dipyridyl 5×10^{-3} M: 47; 5×10^{-4} M: 14. KCNS 5×10^{-3} M: 39; 5×10^{-4} M: 24; 5×10^{-5} M: 14. Dithiol 5×10^{-3} M: 76; 5×10^{-4} M: 60; 5×10^{-5} M: 24 Mepacrine hydrochloride 5×10^{-3} M: 70; 10^{-3} M: 37 Quinine hydrochloride. 5×10^{-3} M: 45 CuSO$_4$ 5×10^{-3} M: 75; 10^{-3} M: 65; 5×10^{-5} M: 16.	Fewson and Nicholas (1960 a) (1961 a, b)

Table 12

Source	Thermal denaturation and stability[1]	Inhibition by —SH reagents or flavin analogues[a]
Achromobacter fischeri ≡ *Photobacterium fischeri*	Stored for several months. Stable to prolonged dialysis. No loss of activity. 10 min/50° 10—15/10 min/60° 100/10 min/70°	pCMB 3×10^{-3} M: 100. Reversed completely with 10^{-3} GSH
Hansenula anomala (strain 317)	Stores for 6 weeks —60° losing 50% of its activity. GSH.10^{-3} M, enhances storage at 5° C 100/5 min/68°	5×10^{-3} M pCMB: 98 5×10^{-4} M pCMB: 83 5×10^{-5} M pCMB: 58 5×10^{-6} M pCMB: 35 Reversed by 10^{-3} M GSH or cysteine
Azotobacter vinelandii (O)	no record	5×10^{-4} M pCMB: 75 5×10^{-4} M pCMB $+ 10^{-3}$ M GSH: 20 Inhibition reversed by GSH
Rhizobium japonicum from soybean plants. (*Glycine max.* Merr. variety Ogden)	no record	no record

Although they grew the bacteria in peptone medium without the addition of molybdenum it is unlikely that their cultures were in fact deficient in the micronutrient since no purification methods were used. More recent work (Nicholas, unpublished) has shown that Mo is required for nitrate reduction in *Photobacterium sepia*. The different effects of inhibitors on the systems using DPNH or reduced benzyl viologen observed by Sadana and McElroy (Table 12) are however as yet unexplained.

Many schemes have been postulated for nitrate reduction in *E. coli* and Taniguchi and Itagaki (1960) have recently obtained a homogeneous Mo-containing protein which reduces nitrate with cytochrome b_1 as the donor. Thus the Japanese school have now confirmed the work of Nicholas and Nason (1954), Nicholas and Stevens (1955, 1956a, b) that Mo is a constituent of nitrate reductase. Thus all enzyme preparations have probably all contained the same terminal nitrate reductase, i.e. a molybdenum containing protein with either succinic or lactic dehydrogenases (Sato and Niwa, 1952), DPNH-flavin reductase (Nicholas and Nason, 1955b) menadione reductase (Wainwright, 1955), or quinone reductase (Heredia and Medina, 1960) as the penultimate hydrogen donating system.

(Cont.)

Other inhibitors[2]				Reference
		BVH	DPNH	
KCN	5×10^{-2} M	0	50	SADANA and McELROY (1957)
	2×10^{-2} M	0	95—98	
NaN₃	10^{-3} M	50		
	3×10^{-3} M	99	100	
Na₂S	5×10^{-4} M	100	100	
CO (dark)	100%	0	100	
CO (light)	100%	0	4	
α-α -Dipyridyl	5×10^{-3} M	0	50	
	2×10^{-2} M	0	88	
o-Phenanthroline	5×10^{-4} M	0	48	
	10^{-3} M	0	97	
Thiourea	2×10^{-1} M	0	52	
Dimethylglyoxime	2.5×10^{-2} M	0	80	
Na-Versenate	10^{-2} M	0	0	
Antimycin A	50 μg	0	0	
KCN	10^{-3} M: 98; 10^{-4} M: 20			SILVER (1957)
NaN₃	10^{-3} M: 99; 10^{-4} M: 66			
Salicylaldoxime	10^{-2} M: 13; 10^{-4} M: 9			
o-Phenanthroline	10^{-3} M: 10; 10^{-4} M: 7			
CuCl₂	10^{-3} M: 94; 10^{-4} M: 16			
NaF	10^{-2} M: 4			
CO	100% no inhibition			
		DPNH	Reduced Nile blue	
KCN	10^{-3} M	90	85	TANIGUCHI and OHMACHI (1960)
	10^{-4} M		20	
NaN₃	10^{-3} M		90	
	10^{-4} M	75	80	
CO (dark)	1 atm.	0	0	
o-Phenanthroline	10^{-3} M	60	0	
Na-EDTA	10^{-2} M	40	0	
Amytal	10^{-3} M	55	0	
Dicoumarol	10^{-3} M	0	0	
KCN	10^{-2} M: 83; 10^{-3} M: 4			EVANS (1954)
NaN₃	10^{-3} M: 81; 10^{-4} M: 38			CHENIAE and EVANS (1960)
CuSO₄	10^{-4} M: 41			

[1] Values are as per cent inactivation for times and temperatures shown.
[2] Values are as per cent inhibition for concentrations shown.

When nitrates are dissimilated, cytochromes are required in the penultimate electron donating system but during the slower process of assimilation haem compounds are not necessary for nitrate reduction.

A generalised scheme for electron transport during microbial nitrate reduction is as follows:

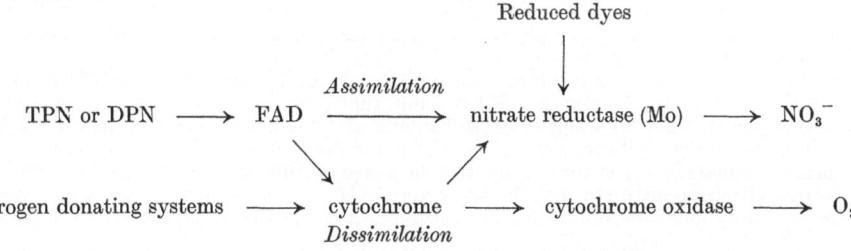

Scheme I. *Alternative pathways of electron transfer to nitrate and oxygen in micro-organisms.*

V. Physiological Factors.

1. Fungi and Higher Plants.

a) **Adaptive features.** The nitrate enzymes obtained from fungi including *Neurospora* by Nason and Evans (1953), Kinsky and McElroy (1958), and *Aspergillus niger* by Nicholas, Nason and McElroy (1954) and various higher plants by Hewitt, Fisher and Candela (1955), (Hewitt 1957), Tang and Wu (1957), Hewitt and Afridi (1959, and unpublished), Rijven (1958), Mulder, Boxma and Van Veen (1959), and Hageman and Flesher (1960) are inducible by, or adaptive to the substrate, nitrate. It follows therefore that the nitrogen nutrition of the organism or plant is of the greatest importance in determining the level of activity that can be extracted. Unpublished work by Afridi and Hewitt with cauliflower has shown that a daily supply of nitrate at an optimal concentration of 12—15 mM is necessary to maintain maximum nitrate reductase activity. Suppression of the nitrate supply resulted in an appreciable decrease in activity after 24 hours and activity was almost negligible after 3 or 4 days without nitrate. Restoration of nitrogen as ammonia did not suffice to restore activity and response to nitrate was inhibited by several antimetabolites.

Hageman and Flesher (1960) have shown that for maize, nitrate reductase activity is dependent on the concentration of nitrate in the nutrient medium over the range from below 2 to above 15 mM NO_3. Plants should be sampled about 6 hours after application of nitrate nutrient which is preferably given three times daily.

The presence of ammonia in the nutrient medium may cause suppression of enzyme activity in some organisms, e.g., in the fungus *Scopulariopsis brevicaulis* even when nitrate is also given (Morton, 1956). This effect occurs *in vivo* and not *in vitro*. It is necessary therefore to specify and examine the effects of the composition of the nutrient medium before assessing the significance of qualitative measurements of activity. The addition of molybdenum to extracts prepared from molybdenum deficient fungi (Nicholas et al., 1954), or plants (Candela et al., 1957) does not result in any restoration of the enzyme activity. Introduction of molybdenum by addition to culture media (Nicholas et al., 1954) or by infiltration into excised tissues (Hewitt et al., 1955; Hewitt, 1957; Candela et al., 1957; Hewitt and Afridi, 1959) results in the production of enzyme activity after a period of between one and several hours. This response is inhibited by antimetabolites known to inhibit protein synthesis and by low temperature (0—5°) or under nitrogen (Hewitt and Afridi, 1959, and unpublished). The synthesis of the enzyme therefore depends upon the presence of the prosthetic metal *in vivo*, as well as on the substrate.

b) **Environmental factors, age and species.** Light has a direct and large effect on nitrate reductase activity present in leaf extracts. Several workers including Schimper (1888, 1890), Dittrich (1930), Éckerson (1932), Eisenmenger (1933), Burström (1943) observéd that light is essential for nitrate assimilation by intact plants. Evans and Nason (1953) showed that photochemical reduction of triphosphopyridine nucleotide could be linked to nitrate reductase activity but this observation did not seem to provide a complete explanation of the effect of light, since adequate carbohydrate reserves might still serve to provide a supply of reduced pyridine nucleotides in darkness. Hewitt, Fisher and Candela (1955), Candela, Fisher and Hewitt (1957), Hageman and Flesher (1960) independently found that when cauliflower or maize plants are transferred to darkness, nitrate reductase activity in extracts steadily falls. It decreases to nil in cauliflower after 5—6 days and is negligible in maize after 2 days. Restoration of light results in a rapid reappearance of activity which reaches normal levels in about 8 to 12 hours. Partial shading was found by Hageman and Flesher (1960) to result in marked decreases in nitrate reductase activity in maize; 25 per cent shading for 8 days or 14 days decreased activity to 65 or 30 per cent of the normal levels respectively. It is therefore desirable to harvest plants about 8 hours after a period of maximum illumination. Values on dull days or from self shaded leaves may be appreciably less than those observed on sunny days.

An adequate water supply is also essential for high nitrate reductase activity. It has been found by Cresswell (unpublished work) that wilting causes rapid loss of activity and this is not restored until several hours after the plants have recovered. Eckerson (1924, 1931) noted that immediate extraction of excised foliage was necessary to avoid loss of reductase activity associated with wilting. Anaerobic conditions also cause loss of the enzyme from plant tissues (Hewitt et al., 1955; Candela et al., 1957) and from certain fungi (Morton, 1956). Walker and Nicholas (1961a) found that nitrate reductase activity in *Neurospora* was dependent on iron as well as molybdenum during the first two or three days growth when the mycelium was still submerged. This was attributed to effects of a semi-anaerobic environment. Decreased oxygen tension resulted in increased nitrate reductase activity which converted nitrate to nitrite so that the latter accumulated. The response to anaerobic conditions as shown by increased production of nitrate reductase has been termed nitrate dissimilation and may be analogous to nitrate reductase in denitrification in *Pseudomonas aeruginosa* described already.

The age and leaf position number significantly affect apparent nitrate reductase activity in plant extracts. HEWITT et al. (1955) and CANDELA et al. (1957) showed that in cauliflower the activity per unit of protein increased and then decreased in successive leaves along the stem axis, whilst activity of a different enzyme, acid phosphatase, increased progressively towards the plant apex. Nitrate reductase activity on a protein basis was also very high in the leaf petioles, though total activity was low. CRESSWELL and HEWITT (unpublished) showed that age of vegetable marrow plants has a large effect on the activity in extracts of cotyledons which decreases as the plants grow, due to the production of inhibitory or denaturing substances. The interaction between leaf position, plant age, species, cultural environment and cysteine content of the extracting medium, which is now under investigation (CRESSWELL, HAGEMAN and HEWITT, unpublished), is complex. It is however clear that each factor must be taken into account when quantitative comparisons of activity are required, and preliminary trials are necessary to establish the conditions which will yield maximum activities *in vitro*, especially with regard to the optimum concentration of cysteine in the extracting medium.

EVANS and NASON (1953) concluded that nitrate reductase activity is associated with younger tissues. This was based on the decrease observed with ageing of the primary leaves of soybean. The work of CRESSWELL (unpublished) has confirmed this observation but when cysteine is added to the extracting medium activity is partially restored for a much longer period of growth in vegetable marrow. Embryonic and more mature tissues also appear to produce enzymes with different specificity as noted for wheat, SPENCER (1959). EGAMI, OHMACHI, LIDA and TANIGUCHI (1957) reported that anaerobic and aerobic nitrate reductase systems occured during the first two days growth of cotyledons of *Vigna sesquipedalis*.

Activity varies in extracts from different species even when inhibitory factors are eliminated as far as possible by the presence of cysteine. Thus high activities of the order of 15,000 mμmoles nitrite produced/hour/g fresh weight are observed with marrow and maize, intermediate values of about 2,000 are found with cauliflower. The lower values of about 1,500 calculated from the data of EVANS and NASON for soybean were obtained without the use of cysteine during extraction. The activity of root extracts is certainly very low as in work of CRESSWELL and HEWITT. Whether unidentified factors other than those discussed in section I B mask its presence or whether a distinct system as inferred by VAIDYANATHAN and STREET (1959) is involved, remains to be determined. The high values reported by EVANS and NASON (1953) for roots of several plants do not accord with the experience of CRESSWELL and HEWITT (unpublished) or VAIDYANATHAN and STREET (1959).

2. Bacteria.

a) **Effect of oxygen tension.** The respiratory or dissimilatory nitrate reductase is usually more active when the bacteria are grown under anaerobic conditions when nitrate is the obligatory terminal acceptor. This effect is well illustrated for nitrate reductase and nitrite reductase from *Pseudomonas aeruginosa* (FEWSON and NICHOLAS, 1960 a, 1961 a,b,c; WALKER and NICHOLAS, 1960, 1961 a,b). The activities of both enzymes were reduced by increasing oxygen tension in the medium, and this is the usual pattern for the denitrifying enzymes as well as the dissimilatory ones from *Pseudomonas aeruginosa*, and *Photobacterium fischeri* (NICHOLAS, unpublished results). The assimilatory nitrate reductase systems, however, are not as markedly affected by changes in oxygen tension in the medium.

b) **Composition of the culture medium.** Changes in the constituents of the culture medium can affect the activities of nitrate reductase in bacteria. Thus when *E. coli* was grown in an inorganic medium containing nitrate as sole souce of N with a sugar as C source, the nitrate reductase was mainly of the assimilatory type and was easily extracted in phosphate buffer (NICHOLAS and NASON, 1955b). When the bacteria were grown in a more complex medium containing peptone then the nitrate reductase appeared to be associated with cell particles and was not readily extracted. It is therefore of prime importance to standardize and define the medium in which the bacteria are grown before valid comparisons can be made between the results of various workers. Unfortunately many investigators have not paid sufficient attention to the constitution of the basal culture media. It has been observed that ionic copper is very inhibitory to the formation of nitrate reductase in *Pseudomonas aeruginosa* (FEWSON and NICHOLAS, 1960a, 1961 a, b, c). There is enough copper as a contaminant in the basal culture medium to depress production of the enzyme in the cells since removal of Cu^{2+} from the culture solution produced cells with a much higher nitrate reductase activity (NICHOLAS, unpublished results). Thus it is advisable to check that the copper content of the culture media used is not at an inhibitory level for nitrate reduction.

c) **Temperature.** The effect of temperature during growth, on nitrate reduction in bacteria is best illustrated in the denitrifying ones. Results of typical experiments with *Pseudomonas* *aeruginosa* show that the optimum temperature for growth is between 30 and 37° C. Nitrate and nitrite reductase activities paralleled each other closely over a range of temperatures and were maximal in cells grown at 37° C. Temperatures above 37° C had less effect on enzyme activity than on cell yield and at 43° C the enzymes were still 60 per cent of the maximum activity at 37° C whilst growth was markedly decreased. The effect of temperature during growth on the activity of the dissimilatory nitrate reductase closely parallels bacterial growth since under these conditions nitrate reduction is the rate limiting step for growth. This may also be the case for the assimilatory enzyme when nitrate is the sole source of nitrogen but is not necessarily so when alternative nitrogenous sources are present in the medium.

d) **Length of growing period.** It is important to collect bacteria during the late exponential phase of growth since nitrate reductase falls off rapidly in non-dividing cells. The effects of various growing periods on cell yields and denitrifying enzymes of *Pseudomonas aeruginosa* showed that growth was rapid during the first 24 hr but then cell yield increased by only 15% during the next 40 hr. Both nitrate and nitrite reductases were relatively low during the early stage of growth but increased rapidly and reached a maximum at the end of the exponential phase (31 hr). Subsequently nitrate reductase activity decreased by about 15% during the next 30 hr. Nitrite reductase activity decreased more markedly and 65 hr after inoculation it was only 50% of that at the end of the logarithmic phase. Thus cells should be routinely collected between 25 and 30 hr after inoculation.

B. Other Nitrate Reduction and Nitro Reductase Enzymes.

1. Aldehyde and Xanthine Oxidases of Animal Origin.

Aldehyde oxidase of liver from several animals and xanthine oxidase from milk and from liver are able to mediate electron transfer from aldehydes or DPNH to nitrate, cytochrome c, ferricyanide, or oxygen, with various degrees of specificity (GREEN and BEINERT, 1953; MACKLER, MAHLER and GREEN, 1954; MAHLER et al., 1954; AVIS, BERGEL and BRAY, 1956; MAHLER and GLENN, 1956). These reactions appear to depend on the presence of a flavin and the single electron transfer reactions with nitrate, cytochrome c and ferricyanide also depend on the presence of molybdenum, and on the presence of inorganic phosphate, silicate or arsenate where this has been tested. Aldehyde oxidase of pig liver also appears to contain iron combined with a porphyrin according to MAHLER, MACKLER, GREEN and BOCK (1954). The properties of this group of enzymes has been reviewed elsewhere (HEWITT, 1958). They do not appear to be represented in plants or fungi.

2. Aldehyde Oxidase of Potato.

Finally mention must be made of aldehyde oxidase of potato tubers *(Solanum tuberosum)* described by BERNHEIM (1928). This enzyme has been studied also at Long Ashton (HEWITT and HALLAS, unpublished work). The properties of the enzyme include the capacity to transfer electrons from several aldehydes including acetaldehyde to nitrate to produce nitrite, without the addition of reduced pyridine nucleotides. Anaerobic conditions are necessary for the reaction as in the presence of oxygen, hydrogen peroxide is produced and causes irreversible inactivation of the enzyme. The enzyme obtained by HEWITT and HALLAS was not able to reduce methylene blue. It was irreversibly inhibited by cyanide and reversibly inhibited by 10^{-4} M azide (HEWITT and HALLAS, 1959). 8-Hydroxyquinoline, $\alpha\alpha$-dipyridyl, EDTA, 2:2-'diquinolyl, diethyldithiocarbamate and several other metal chelating reagents produced little or no inhibition. No evidence for a dissociable flavin could be obtained and inhibitors of flavin enzymes were without effect. No combination of several metals including iron and molybdenum restored activity after dialysis against cyanide in the usual manner. A metal component is inferred to be present but has not been identified. The enzyme was inhibited by *p*-chloromercuribenzoate and this was reversed by glutathione. It is extracted from acetone powder of potato tubers by 0.1 M sodium chloride at pH 6.0 and chromatographic adsorption and elution on a calcium phosphate gel column increases specific activity about ten times. Further fractionation is obtained by heating for 5 min at 55°, precipitation by ammonium sulphate between 40 and 52 per cent saturation and precipitation by 30 per cent ethanol at 5°. The best preparations representing about 100 fold increase in specific activity showed no evidence of any absorption bands corresponding to flavin or cytochrome components.

The interest in this enzyme lies in the fact that it is distinct from nitrate reductase and may represent either another molybdo-protein or else a different mechanism for nitrate reduction in one part of one species, viz. *Solanum tuberosum*.

3. Reduction of Aromatic Nitro-Compounds.

It is possible that alternative or simultaneous mechanisms for nitrate reduction might involve aromatic nitro-compounds although these have frequently been found to inhibit the growth of plants and bacteria. It is of interest that bacteria resistant to chloramphenicol may reduce it to the inactive arylamine (VILLANUEVA, 1959a, b).

DIXON (1926) found that xanthine oxidase could reduce *m*-dinitrobenzene, and BUEDING and JOLLIFE (1946) found that the same enzyme reduced trinitrotoluene to hydroxylaminodinitrotoluene. Reduction of aryl nitro-compounds was later noted in bacteria (SMITH and WORREL, 1949; EGAMI, EBATO and SATO, 1951) and fungi (ZUCKER and NASON, 1956).

EGAMI, EBATO and SATO (1950) reported that nitrite inhibited the reduction of chloramphenicol, and *vice versa*, in *Strep. hemolyticus* and similar findings were reported in *Pseudomonas* spp. and *Nocardia* spp. (CARTWRIGHT and CAIN, 1959b). SAZ and SLIE (1954), however, were unable to demonstrate this effect in *E. coli*. SAZ and MARMUR, 1953; SAZ and SLIE, 1954a, b observed that cell-free extracts of *E. coli* reduced the nitro-groups of chloramphenicol, *p*-nitrobenzoic acid and other nitro-compounds to the corresponding arylamines in the presence of L-cysteine. DPNH served as an electron donor and it was suggested that nitroso- and hydroxylamino-compounds were intermediates in the reduction.

ZUCKER and NASON (1956) partially purified an enzyme system from *N. crassa* which catalyzed the reduction of *m*-dinitrobenzene to *m*-nitroaniline. A metal requirement was indicated by cyanide and azide inhibition and FAD and sulphydryl groups were essential for enzyme activity. The following sequence of reduction was suggested:

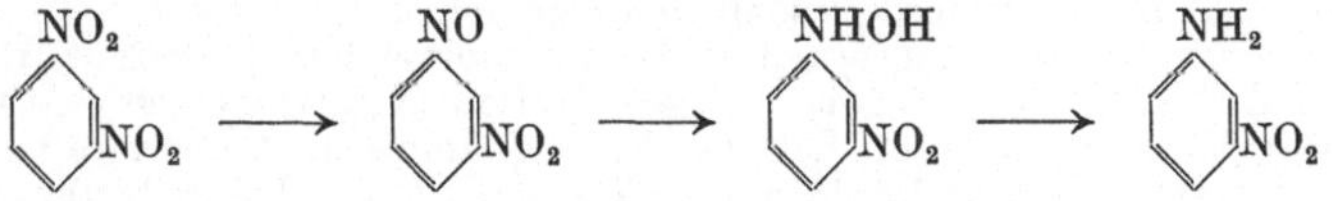

WESTERFELD, RICHERT and HIGGINS (1957) found that *A. niger* reduced *p*-nitrobenzene sulphonamide to *p*-aminobenzene sulphonamide and suggested that an iron-flavoprotein enzyme might be involved. CARTWRIGHT and CAIN (1959b) were unable to separate enzymes responsible for individual steps of the reduction of nitrobenzoic acids, but reported a stimulation by FAD and Fe^{++}. These authors concluded that reduction was not the primary mode of attack on the nitro groups of aromatic compounds and that aminobenzoic acids do not lie on the normal direct oxidative pathway of nitrobenzoic acid metabolism in *Pseudomonas* and *Nocardia* (CARTWRIGHT and CAIN, 1959a). VILLANEUVA (1959a, and b) obtained a 200-fold purification of a nitro-reductase from a *Nocardia* sp. and demonstrated a requirement for FAD and sulphhydryl groups. He was unable to show a metal participation in enzyme action. DPNH was about 10 times as effective as TPNH.

NICHOLAS (1959a,b) concluded that the reduction of aromatic compounds by *N. crassa* is not important in nitrate assimilation but is merely a detoxication mechanism. It is probable that this conclusion also applies to other micro-organisms.

C. Nitrite, Nitric Oxide, Hyponitrite and Hydroxylamine Reductases and Related Enzymes.

In contrast with the well documented information on nitrate reductase systems, the identity and properties of nitrite reductase, hydroxylamine reductase and other associated enzymes are not clearly characterised for higher plants. More information is available regarding corresponding enzymes obtained from fungi and some bacteria but the systems often appear to be complex and incompletely understood. Some general survey of bacterial enzymes precedes the next section.

YAMAGATA (1939) found a nitrite reductase enzyme in cell-free extracts of *Bacillus pyocyaneus (Ps. aeruginosa)* and TANIGUCHI, MITSUI, TOYADA, TAMADA and EGAMI (1953) found that extracts of *B. pumulis* reduced nitrite when reduced methylene blue was the electron donor. Nitrite reduction also occurs in *Azotobacter species* (ROBERTS and AZIM, 1956; SPENCER, TAKAHASHI and NASON, 1957), *Debarymyces spp.* (WICKERAM, 1957), *Desulfovibrio desulfuricans* (SENEZ and PICHINOTY, 1958; KESSLER et al. (1957), *E. coli* (LASCELLES and STILL, 1946) and in algae (KESSLER, 1953, 1955, 1959).

CHUNG and NAJJAR (1956a, b) identified nitric oxide as the product of nitrite reduction in *Ps. stutzeri*. IWASAKI, MATSUBAYASHI and MORI (1956) reported that extracts of an unnamed bacterium reduced nitrite to nitric oxide which in turn was converted to nitrogen gas. Similar observations were made with whole cells of *Thiobacillus denitrificans* by BAALSRUD and BAALSRUD (1954). WALKER and NICHOLAS (1960, 1961b) have characterised a nitrite reductase from *Pseudomonas aeruginosa*.

FEWSON and NICHOLAS (1960a, b) have shown that nitric oxide is readily utilized by extracts of bacteria or algae previously grown on nitrate, but that the uptake of the gas was much reduced when the micro-organisms were grown on ammonia as the sole source of nitrogen. Another interesting feature was the uptake of NO in extracts of bacteria or blue gree algae that fix atmospheric nitrogen. FEWSON and NICHOLAS (1960a, b; 1961a, b, c) prepared a nitric oxide reductase from actively denitrifying cells of *Ps. aeruginosa*.

Hyponitrite was suggested as an intermediate by KLUYVER and DONKER (1926) and by LLOYD and CRANSTON (1930). ALLEN and VAN NIEL (1952) using *Ps. stutzeri* and KLUYVER and VERHOEVEN (1954), FEWSON and NICHOLAS (1960b, 1961d) with *Ps. aeruginosa* and *M. denitrificans* were unable to demonstrate utilization of hyponitrite. McNALL and ATKINSON (1957) showed that *E. coli* reduced hyponitrite to ammonia.

Hydroxylamine has been suggested to be an intermediate both in nitrate reduction (VIRTANEN and CSÁKY, 1948) and in nitrogen fixation (VIRTANEN and RAUTENEN, 1951; WILSON, 1958). VERHOEVEN (1952) detected hydroxylamine in the culture medium during dissimilatory nitrate reduction by *Denitrobacillus licheniformis* but reduction of hydroxylamine to ammonia was not demonstrated in *Azotobacter agile* (SPENCER, TAKAHASHI and NASON, 1957). *Clostridium welchii* (WOODS, 1938), *E. coli* (LASCELLES and STILL, 1946; McNALL and ATKINSON, 1956, 1957) and algae (KESSLER, 1957) produced ammonia.

TANIGUCHI et al. (1956, 1957) found that hydroxylamine was reduced to ammonia in cell-free extracts of a halotolerant bacterium when reduced methylene blue was the hydrogen donor. SPENCER et al. (1957) using *Azotobacter agile* and WALKER and NICHOLAS (1960, 1961c) using *Pseudomonas aeruginosa* characterised hydroxylamine reductase enzymes prepared from these bacteria. KLAUSMEYER and BARD (1954) reported on an ammonium dehydrogenase from *B. subtilis* which apparently catalysed the reversible formation of ammonia from hydroxylamine but ROUSSOS, TAKAHASHI and NASON (1957) showed that this effect was non-enzymatic. A non-enzymatic reduction of hydroxylamine to ammonia has also been shown to occur in extracts of *Desulfovibrio desulfuricans* (SENEZ and PICHINOTY, 1958a, b).

I. Preparation.

1. Fungi and Higher Plants (Table 13).

(i) Nitrite reductase. NASON, ABRAHAM and AVERBACH (1954) reported the presence of a nitrite reductase in extracts of *Neurospora crassa* prepared as described by NASON and EVANS (1953) for nitrate reductase. In spite of the assay method — measurement of nitrite production — used for nitrate reductase, the nitrite reductase of NASON et al. (1954) catalysed ammonia production from nitrite. This paradox has not been resolved by later work. Fractionation by ammonium sulphate of unspecified concentration and calcium phosphate gel resulted in a 10-fold increase in specific activity.

MEDINA and NICHOLAS (1957a, c), NICHOLAS, MEDINA and JONES (1960) prepared a similar enzyme for *N. crassa*. NICHOLAS (1959b) reported that the enzyme may be associated with cell particles as the enzyme activity was increased by the use of deoxycholate and was stabilised by 10% w/v sucrose in the extracting medium. A 50-fold increase in specific activity was obtained by the fractionation procedure shown in Table 14.

NASON et al. (1954) reported also that soybean leaves yielded an enzyme which catalysed the production of ammonia from nitrite in the presence also of nitrate reductase. The enzyme was increased in specific activity by 15 times when adsorbed on and eluted from alumina c γ gel. ROUSSOS and NASON (1960) described an enzyme system which caused the oxidation of DPNH on addition of nitrite, which did not disappear. This system is discussed separately later, in relation to the oxidation of nitrite and hydroxylamine. VAIDYANATHAN and STREET (1959) reported the presence of nitrite reductase in water extracts of excised sterile tomato

Table 13. *Methods used for extracting enzymes concerning nitrite, hyponitrite and hydroxylamine metabolism from fungi and higher plants*

Source	Extracting Solution[1]	Vol/wt.	Methods of Extraction	References
Neurospora crassa (wild type)	not used	—	Washed 4—5 days old mycelia, frozen 1—3 hours —15° powdered while frozen. Homogenised in TEN BROECK homogeniser at 0°. Centrifuged 3,000 g.	ZUCKER and NASON (1955) Hydroxylamine reductase
N. crassa	0.1 M PO$_4$ buffer[1] pH 7.5	4:1	Mycelia frozen —17° 12 hours ground in mortar and TEN BROECK homogeniser at 0° with buffer. Centrifuged at 3,000 g. 80% activity in supernatant fluid	MEDINA and NICHOLAS (1957 a, c) Nitrite, hyponitrite and hydroxylamine reductases
N. crassa cultured in presence of 50 μg/L pyridoxine	as above	4:1	as above	NICHOLAS, MEDINA and JONES (1960). Nitrite reductase
Soybean	0.05 M PO$_4$ pH 7.7[1] 10^{-3} M glutathione and 10^{-4} M EDTA	3:1	2—3 min at 4° in WARING blendor strained and centrifuged at 4,000 g.	ROUSSOS and NASON (1960) "Pyridine nucleotide Nitrite and hydroxylamine enzymes"
Vegetable marrow (*Cucurbita pepo*) leaves	0.1 M Tris pH 7.5 to 7.8 5×10^{-3} M cysteine 3—5×10^{-4} M EDTA	3:1	Leaves chilled in cold room, cut, macerated 1.5 min 0° C in WARING blendor, strained and centrifuged 15 min 0° C 20,000 g. Activity in supernatant fluid.	HAGEMAN, CRESSWELL and HEWITT (1962) Nitrite and hydroxylamine reductases

[1] Phosphate buffer compositions as indicated in Table 1. Cation not given unless specified.

Table 14. *Purification of Nitrite Reductase from Neurospora crassa.*
Mycelia weighing 60 g (fresh wt.) from about 20 flasks were used for purification of the enzyme. The mycelium was homogenised in 0.01 M phosphate (pH 7.4) and the supernatant solution left after centrifuging the homogenate at 4000 g was fractionated as described in the text. The activity of the enzyme was measured in the usual reaction mixture (NICHOLAS, MEDINA and JONES, 1957 b).

	Total Protein (mg)	Total Units	Protein (mg/ml)	Specific Activity mμmoles NO₂ reduced/mg protein/10 min	Percentage Recovery
1. Supernatant solution	1050	10,800	10.5	10.3	100
2. Calcium phosphate gel added to (1) (25/100 mg protein). Stand for 5 min at 0°. Centrifuge at 4000 g. Gel washed twice with 0.01 M phosphate (pH 7.4) and eluted with 0.05 M phosphate (pH 7.4). Activity of supernatant solution	104	9,000	1.3	86.3	83
3. Protamine sulphate added to (2) (15 mg/10 mg protein). Stand for 15 min. Centrifuge at 4,000 g. Activity of supernatant solution	56.5	9,000	0.70	159	83
4. (NH₄)₂SO₄ precipitate of (3), between 40 and 60% saturation, dissolved in 0.01 M phosphate (pH 7.4)	43.5	8,100	0.60	186	75
5. Calcium phosphate gel added to (4) (25 mg/100 mg protein). Stand 15 min, centrifuged at 4,000 g. Gel washed twice with 0.01 M phosphate (pH 7.4) and eluted with 0.05 M phosphate (pH 7.4). Activity of eluate	14.4	3,200	0.24	220	30
6. (NH₄)₂SO₄ precipitate of (5), between 50 and 60% saturation, dissolved in 0.1 M phosphate (pH 7.4)	4.4	2,300	0.11	523	21

roots grown with nitrate and which were then centrifuged at 600 g. The enzyme rapidly converted nitrite to a compound resembling hyponitrite which accumulated.

HAGEMAN, CRESSWELL and HEWITT (1962) have however obtained definite evidence of an active nitrite reductase from leaves of vegetable marrow and maize. The enzyme is readily extracted by TRIS buffer (0.1 M pH 7.6 containing 3×10^{-4} M EDTA and 5×10^{-3} M cysteine) and is precipitated by 66 per cent saturation ammonium sulphate. It is stable in glutathione-treated dialysis sacs to dialysis against 0.03 M phosphate pH 7.5 in the presence of 10^{-3} M cysteine. Ammonium sulphate can be removed and the enzyme solution concentrated by passing through a column of Sephadex G50 (Pharmacia Uppsala, Sweden) and addition of Sephadex to the solution followed by centrifugation. The enzyme can be adsorbed on calcium phosphate gel from 0.05 M phosphate pH 7.5 and eluted with 0.2 M buffer pH 8.5. It is inactivated at pH 4.0 but is stable for several weeks at pH 7.4 when stored at $-18°$. Activities ranged from $5-15$ μmoles/hr/g fresh wt. in terms of ammonia production and were therefore about 1000 times as great as those reported as ammonia formation by nitrite or hyponitrite reductases of tomato by VAIDYNATHAN and STREET (1959)[1].

(ii) **Hyponitrite reductase.** MEDINA and NICHOLAS (1957a, b) described the properties of an enzyme system obtained from *N. crassa* which catalysed reduction of hyponitrite to ammonia. The enzyme was present in the same phosphate buffer extracts from which nitrite and hydroxylamine reductases were obtained (Table 13) but further fractionation has not so far been reported.

The existence of pyridine nucleotide-linked hyponitrite reductase in preparations from higher plants was reported by VAIDYANATHAN and STREET (1959). Activity was very weak in extracts prepared as described for nitrite reductase.

---

[1] HUZISIGE, H., and K. SATOH, Botan. Mag. Tokyo **74**, 178 (1961) reported the partial purification of a nitrite enzyme from spinach leaves which was activated by light and grana. Activity was similar to preparations of HAGEMAN et al. but ammonia production was not reported. K_m for nitrite was 3×10^{-4} M.

(iii) **Hydroxylamine reductase.** ZUCKER and NASON (1954, 1955) obtained nearly 4-fold enrichment of hydroxylamine reductase specific activity present in phosphate buffer extracts of frozen *Neurospora* mycelia when precipitated with neutralised ammonium sulphate between 45 and 65 per cent saturation. Adsorption on calcium phosphate gel from water and elution by 0.5 M K_2HPO_4 produced a further two fold increase in specific activity. Formation of ammonia was shown to occur in preparations which were dialysed for 2 hours at 4° C in a large volume of metal-free water containing 3×10^{-3} M pyrophosphate pH 8.0 and 10^{-3} M cysteine hydrochloride. Under these circumstances half the activity was lost. Reaction also appeared to cease after 30 minutes due to enzyme inactivation. Endogenous ammonia production was observed to occur non-enzymically from hydroxylamine as reported also by NICHOLAS et al. (1960) and from the enzyme solution as well. Each produced up to 0.5 μmole, namely 8.5 μg. The enzymic ammonia production over a similar period attained 14 μg in 15 min and 16 μg in 30 min.

MEDINA and NICHOLAS (1957c) found that hydroxylamine reductase was present in the extracts used to obtain nitrite and hyponitrite reductase from *N. crassa*. The hydroxylamine reductase was especially active. The ratio of activities as ammonia formation from nitrite, hyponitrite and hydroxylamine was 2:1:8 approximately.

VAIDYANATHAN and STREET (1959) found that hydroxylamine was converted to ammonia by the tomato root extracts described for nitrite reductase. The activity of the hydroxylamine system was relatively high under aerobic conditions (10 μmoles/hour/g fresh wt. approx.) but the yield of ammonia was only 30—35% of the hydroxylamine lost, possibly because of peroxidation in air as shown by CRESSWELL and HEWITT (1960).

HAGEMAN, CRESSWELL and HEWITT (1962) found a hydroxylamine reductase in the extracts of vegetable marrow leaves used to study nitrite reductase. The enzyme was similar to nitrite reductase with respect to ammonium sulphate precipitation, dialysis and storage. Activity was comparable to that reported for tomato roots. Unlike the tomato roots however, the activity of marrow hydroxylamine reductase assayed anaerobically with reduced benzyl viologen was unrelated to and often less than half the nitrite reductase which was highly active (5—15 μmoles/hr/g fresh wt.) whilst yields of ammonia ranged from nil to 100 per cent in preparations that contained nitrite reductase yielding 85—100 per cent as ammonia. No ammonia was produced under aerobic conditions. Activity was linear with time for over 30 min (CRESSWELL, unpublished work).

2. Bacteria.

The collection of cells and methods for extraction of nitrite, nitric oxide and hydroxylamine reductases are similar to those already described for nitrate reductase (Tables 15 and 16).

II. Fractionation and Stability.

Nitrite reductases have been prepared from bacteria assimilating nitrate and also from denitrifying bacteria. They are usually less stable than nitrate reductase. Nitric oxide reductases have been isolated from denitrifying bacteria only although they are also present when nitrates are assimilated. TANIGUCHI et al. (1958) extracted nitrite and hydroxylamine reductases from a halo-tolerant *Bacillus pumilis* using ultrasonic treatment. They found that extracts of this bacterium

Table 15. *Methods used for extracting nitrite reductase systems from bacteria*

Source	Extracting solution	Vol./wt.	Method of extraction	Reference
Pseudomonas stutzeri	water	1 : 1	Cells collected at 0—4°C in a Sharples centrifuge. Cells washed with 0.85% w/v saline then mixed with alumina powder (1557 AB Buehler Ltd., Chicago). Ground cells extracted 3 × with equal volume of water and then centrifuged 15—20 min at 2000 g. Precipitate discarded. Finer cell particles separated by centrifuging at 20,000 g for 30—40 min were washed 3 × in cold water. Clear supernatant solution fractionated.	Najjar and Chung (1956)
Halotolerant Bacillus (Strain 203)	0.05 M phosphate (pH 7.5)	3 : 1	Cells collected by centrifugation. Disrupted by ultrasound (550kc)/30 min in the cold. Cell debris centrifuged at 500 g for 30 min supernatant solution centrifuged at 22,000 g/30 min and particles collected.	Taniguchi, Sato and Egami (1956)
Azotobacter agile (No. 9104 Amer. Type Collection)	0.05 M tris-(hydroxymethyl) amino methane (pH 7.1) + 10^{-3} M glutathione	3 : 1	Cells harvested at 2000 g washed twice with 0.2% w/v KCl to remove nitrite. Suspended cells disrupted in a Raytheon (10 kc) sonic oscillator at 0.94 amp output and centrifuged at 25,000 g for 20 to 30 min. Supernatant solution used as cell-free extract. Grinding cell paste with alumina powder (Alcoa A-301) also gave active extracts.	Spencer, Takahashi and Nason (1957)
Micrococcus (strain 203)	Deionised water	3 : 1	Lyopholized cells disrupted in deionised water, centrifuged 20,000 g/30 min at 0°. To the supernatant solution was added phosphate buffer (pH 6.8) and NaCl to a final concⁿ. of 0.02 M and 0.3 M respectively and an equal volume of cold acetone. Precipitate centrifuged and dissolved in 0.02 M phosphate (pH 6.8). Alternative method, lyopholized cells given ultrasonic treatment (490 kc), 30 min. Cell debris centrifuged 20,000 g/30 min. To supernatant solution, phosphate buffer (pH 6.5) and NaCl added to final concentration 0.01 M and 0.05 M respectively and precipitate obtained between 50 and 70 volume-% cold acetone collected and dissolved in 0.02 M phosphate (pH 6.8).	Asano (1959)
Pseudomonas aeruginosa (NCIB 8704)	0.05 M phosphate (pH 7.1)	1 : 1	Cells collected in a Sharples centrifuge, washed with cold 0.85% w/v saline. 15 g cells suspended in 15 ml 0.05 M phosphate (pH 7.1) disrupted with an ultrasonic probe (20 k cycles/sec) at 4°/15 min. Centrifuged 25,000 g/40 min. Supernatant solution, source of enzyme.	Walker and Nicholas (1960) (1961 b)

Table 16. *Methods used for extracting nitric oxide reductase and hydroxylamine reductase systems from bacteria*

Source	Extracting solution	Vol./wt.	Type of extraction	Reference
Pseudomonas stutzeri	Iced water	1 : 1	Cells collected in a Sharples centrifuge at room temperature, washed with 2 litre cold water. Cell paste mixed with 3 × weight of alumina powder (No. 1557 A. B. Buehler Ltd., Chicago) ground at 4°C. Ground cells extracted 3 × with equal volume of iced water, centrifuged 30—40 min at 20,000 g to eliminate cell debris and cell particles. Sediment washed 3 × with cold water.	CHUNG and NAJJAR (1956 b) (NO reductase).
Pseudomonas aeruginosa (NCIB 8704)	0.1 M phosphate (pH 8.0)	1 : 1	Cells collected in a Sharples centrifuge washed with cold 0.85% w/v NaCl to remove nitrite. Cells disrupted with an ultrasonic-probe (20 k cycles/sec) for 15 min in 4°. Centrifuged at 18,000 g for 45 min. Supernatant solution used as source of enzyme.	FEWSON and NICHOLAS (1960 b) (NO reductase).
Clostridium welchii (NCIB 273)	Washed cells used in assay	10:1	Cells centrifuged and washed 2 × with 0.02 M phosphate (pH 7.1). Cells suspended in water or buffer; 10 mg dry weight/ml.	WOODS (1938) (hydroxylamine reductase)
Bact. coli (2 strains) *Azotobacter vinelandii* (O)	0.05 M tris-(hydroxy-methyl) amino- ethane (pH 7.1) + 10⁻³ M GSH.	3 : 1	Cells collected by centrifuging at 2,000 g, washed twice in 0.2% KCl to remove nitrite. Cells suspended in 3 × their weight of buffer subjected to sonic breakage in 10 kc Raytheon oscillator 0.94-amp. Centrifuged at 20,000—25,000 g/20—30 ml. Supernatant solutin used as cell-free extract. Extracts can also be prepared by grinding cells with alumina powder (Alcoa A 301).	SPENCER, TAKAHASHI and NASON (1957) (hydroxylamine reductase)
Pseudomonas aeruginosa (NCIB 8704)	0.1 M phosphate (pH 7.4)	3 : 1	Cells collected in a Sharples centrifuge, washed with 0.85% w/v NaCl and stored at —17°. Cells suspended in the buffer disrupted by an MSE Mullard ultrasonic probe (20 kc/sec) 15 min at 4° in a double walled glass vessel through which iced water was circulated.	WALKER and NICHOLAS (1961 c) (hydroxylamine reductase)
Desulfovibrio desulfuricans	2% w/v NaCl and 0.5% w/v Na₂SO₄	50 : 1	Cells collected in a Sharples centrifuge washed twice with distilled water and then with 2% w/v NaCl. 50 ml suspension of the bacteria given ultrasonic treatment (420 k cycles) for 45 min in an atmosphere of H₂ with 0.5% w/v Na₂SO₄ at 20°. Alternatively aqueous extracts prepared from acetone powder of the cells by grinding with powdered glass at 0°. Extracts centrifuged at 26,000 g at 4°/15 min.	SENEZ and PICHINOTY (1958 b) (hydroxylamine reductase)

would sometimes reduce $NO_2 \rightarrow NH_3$ but on other occasions appreciable amounts of N_2 gas were formed by denitrification when formate or reduced methylene blue was the electron donor. Thus far the factors affecting this alternative way of utilizing nitrite have not been determined.

SPENCER, TAKAHASHI and NASON (1957) found that nitrite and hydroxylamine reductases from *Azotobacter agile* were located in the cytoplasmic fraction since considerable activity was left in the supernatant solution after centrifuging the cell-free extracts prepared with alumina powder and buffered hypertonic sucrose solution in an ultracentrifuge at 144,000 g. The enzymes were purified 2 or 3 fold only as shown in Table 17. The cell-free extract was treated with half its volume of alumina C_γ gel (12.1 to 16.5 mg dry weight per ml) for 15 to 30 min with intermittent stirring. The gel was collected by centrifugation at 3,000 g for 5 min washed twice with 5 to 10 times its volume of 0.05 M tris buffer (pH 7.1) containing glutathione (10^{-3} M) and eluted twice with 0.1 M phosphate buffer (pH 7.5) containing glutathione. For each elution carried out for 30 min with occasional stirring, the buffer volume was one quarter the volume of the cell-free extract. The two eluates were combined and this fraction was used for enzyme studies. In some cases this fraction was centrifuged at 144,000 g and the active supernatant solution collected. This step was effective in removing most of the TPNH oxidase activity which was mainly associated with the cell particles. Their attempts to purify the nitrite reductase enzyme further with ammonium sulphate treatment resulted in a complete loss of nitrite reductase activity. Hydroxylamine reductase, however, was precipitated between 30 and 50% ammonium sulphate saturation but this resulted in only a two fold increase in specific activity of the enzyme and the yield was poor. Partially purified nitrite and hydroxylamine reductases were stable for several weeks at $-15°$ when stored in 0.1 M phosphate buffer (pH 7.5) containing 10^{-3} M glutathione. The two enzymes store at 0 to 2° C for about 1 week without loss of activity. The crude cell-free extracts, however, are relatively unstable. Both enzymes lose all their activities after 5 min at 50° C but are stable to dialysis for 6 hr against 0.1 M K_2HPO_4 and 10^{-3} M glutathione provided the dialysis membrane is soaked for 30 min in the same solution prior to use as described by NICHOLAS and NASON (1954a). The activity and stability of nitrite reductase was greatly enhanced when both tris and phosphate buffers used for purification were treated with 8-hydroxyquinoline in chloroform in order to remove heavy metals. The metallic inhibitors were not identified.

Nitrite reductase from denitrifying organisms first reported in cell-free extracts of *Pseudomonas aeruginosa* has been partially purified from *Ps. stutzeri* by CHUNG

Table 17. *Summary of Purification of Nitrite and Hydroxylamine Reductases from Azotobacter agile* (SPENCER, TAKAHASHI and NASON, 1957).

Fraction	Nitrite Reductase				Hydroxylamine Reductase			
	Total units	Total protein mg	Specific activity units/mg protein	Recovery %	Total units	Total protein mg	Specific activity units/mg protein	Recovery %
1. Crude extract	5,300	376	14.2		8,900	727	12.2	
2. Supernatant solution after high speed centrifugation	3,600	264	13.8	67.5				
3. Al Cγ gel eluate	4,850	114	42.6	91.0	5,300	240	22.1	59.5

and NAJJAR (1956a, b) who used a single ammonium sulphate precipitation only, resulting in a two fold purification. They showed that the enzyme was associated with cell particles and that its product was nitric oxide. WALKER and NICHOLAS (1960, 1961b) purified nitrite reductase from *Ps. aeruginosa* more than 600 fold as shown in Table 18. Crude extracts (fraction 2) contained an inhibitor which reduced the activity to 10% of that in the 0 to 100% ammonium sulphate precipitate of this fraction. The inhibitor retained in the supernatant solution was heat labile, unstable to dialysis and it probably affected —SH groups since the inclusion of 10^{-3} M glutathione in the assay markedly increased enzyme activity of fraction 2, whereas activity of fraction 3 which did not contain the inhibitor was unaffected. The optimum enzyme activity was obtained at 37° after 10 min incubation at pH 7.2.

GAYON and DUPETIT (1886) described the conversion of nitrate to nitrite, nitric oxide, nitrous oxide, nitrogen and to ammonia by soil bacteria. Since then nitric oxide has been identified as a product of nitrite reduction in *Ps. stutzeri*, *Bacillus subtilis* (NAJJAR and ALLEN, 1953, 1954), *Thiobacillus denitrificans* (BAALSRUD and BAALSRUD, 1954) and in mixed cultures of denitrifying bacteria (WIJLER and DELWICHE, 1954) and in a species of *Micrococcus* (IWASAKI, MATSU-BAYASHI, MORI, 1956). In addition IWASAKI et al. (1956) showed nitric oxide was utilized by growing cells of *E. coli* (Bn) a non-denitrifying organism although the products of the reaction were not identified (McNALL and ATKINSON, 1956).

CHUNG and NAJJAR (1956a, b) isolated an enzyme from *Ps. stutzeri* which converted nitric oxide to nitrogen gas. Subsequently FEWSON and NICHOLAS (1960a, b) purified the enzyme (25 fold) from *Ps. aeruginosa* as shown in Table 19. The enzyme is stabilized by adding to it 10^{-3} M glutathione and it stores for about one week at $-17°$ C.

Table 18. *Purification of Nitrite Reductase from Pseudomonas aeruginosa*
(WALKER and NICHOLAS, 1961b).
15 g frozen cells suspended in 15 ml 0.05 M phosphate (pH 7.1) were disrupted with an ultra-sonic probe and then diluted to 60 ml with the same buffer. The enzyme was assayed using 0.30 μM PyH$_2$ as an electron donor, and protein determined by the FOLIN-CIOCALTEAU reagent (LOWRY et al., 1951).

	Volume (ml)	Enzyme Activity (μM NO$_2$/10 min)	% Recovery of enzyme	Protein (mg)	Specific Activity (μM NO$_2$/10 min/mg protein)	Purification
Homogenate	60	—	—	—	—	—
Supernatant solution after centrifuging homogenate at 25,000 g for 40 min	50	130	100	1562	0.08	1
Precipitate collected between 44 to 60% saturation with ammonium sulphate after centrifuging at 20,000 g for 5 min: dissolved in 50 ml 0.01 M phosphate (pH 7.1)	50	421	325	478	0.88	11
500 mg Ca$_3$ (PO$_4$)$_2$ gel stirred into fraction 3. After standing for 15 min centrifuged at 5,000 g. for 3 min. Enzyme eluted with 50 ml 0.1 M phosphate (pH 8.1) containing 176 g/l ammonium sulphate	50	286	219	28.1	10.2	127
Precipitate collected between 60 to 90% saturation with ammonium sulphate after centrifuging at 5,000 g for 20 min: dissolved in 20 ml 0.05 M phosphate (pH 7.1)	20	193	149	4.0	48.2	602

Table 19. *Purification of Nitric Oxide Reductase from Pseudomonas aeruginosa*
(FEWSON and NICHOLAS, 1960 b, 1961 d).

Frac-tion	Procedure	Volume (ml)	Total protein (mgm.)	Protein (mgm./ml)	Total units (μl NO/hr.)	Specific activity (μl NO/mgm. protein/hr.)	Recovery (per cent)	Purifi-cation
1	Supernatant solution after centrifuging at 18,000 *g*.	17.5	301	17.2	1,350	4.5	—	—
2	Precipitate from 0—43 per cent ammonium sulphate saturation of fraction 1, centrifuged at 5,000 *g*. for 5 min dissolved in 0.005 M phosphate buffer (pH 8.0)	17.5	155	8.86	1,260	8.1	93	1.8
3	Fraction 2 was added to a diethyl-aminoethyl cellulose column. The column was eluted with 20 ml aliquots of 0.01 M phosphate buffer (pH 8.0). Activity of the fourth aliquot	22.0	8.6	0.39	960	113.5	71	25

Actively denitrifying cells of *P. aeruginosa* (NCIB 8704) were grown in a nitrate-peptone medium (ref. 17) under anaerobic conditions for 24 hr. at 37° C. The harvested cells were washed and stored at —15° C. 5 g frozen cells were suspended in 5 ml 0.1 M phosphate buffer (pH 8.0) and disrupted with an ultrasonic probe (20 kc./sec) for 15 min at 4° C. The homogenate was diluted to 20 ml with the same buffer and centrifuged at 18,000 *g* for 45 min. The supernatant solution was used as the source of the crude enzyme. The enzyme activity was assayed using the following WARBURG reaction mixture. Main vessel, 1.0 ml enzyme, 0.1 M phosphate buffer (pH 8.0), 0.1 ml 10^{-3} M pyocyanine; side-arm, 0.1 mg. crystalline alcohol dehydrogenase, 0.1 ml 2 per cent v/v ethyl alcohol, 0.1 ml (300 mμM) reduced diphosphorpyridine nucleotide; centre well, 0.2 ml 20 per cent w/v potassium hydroxide. Gas phase, 25 per cent NO/75 per cent nitrogen. Temp. 30° C. Boiled enzyme controls were included.

Table 20. *Purification of Hydroxylamine Reductase from P. aeruginosa*
(WALKER and NICHOLAS, 1961 c).

15 g frozen cells suspended in 15 ml 0.05 M phosphate (pH 7) were disrupted with an ultrasonic probe and then diluted to 60 ml with the same buffer. — The enzyme was assayed using 0.3 μM PyH$_2$ as an electron donor, and protein determined by the FOLIN-CIOCALTEAU reagent.

	Volume (ml)	Enzyme activity (μmoles NH$_2$OH/10 min)	Recovery of enzyme (%)	Protein	Specific activity (μmoles NH$_2$OH 10 min/mg protein)	Purifi-cation ratio
I Homogenate	60	896	100	—	—	—
II Supernatant solution after centrifuging homogenate at 25,000 *g* for 40 min	50	808	90	1329	0.61	—
III Precipitate collected between 41 to 55% saturation with ammonium sulphate by centrifuging at 20,000 *g* for 5 min was dissolved in 25 ml 0.05 M phosphate (pH 7.1)	25	584	65	588	1.0	1.6
IV 270 mg Ca$_3$(PO$_4$)$_2$ gel stirred into fraction III. After standing 15 min and centrifuging at 5000 *g* for 3 min the enzyme was eluted with 50 ml 0.1 M phosphate (pH 8.0) containing 176 g/l ammonium sulphate and 10^{-3} M reduced glutathione . .	50	372	42	34.9	10.7	17.5
V Precipitate collected between 30 to 60% saturation with ammonium sulphate by centrifuging at 5000 *g* for 20 min and dissolved in 20 ml 0.05 M phosphate (pH 7.0)	20	281	31	9.0	28.7	47

Hydroxylamine reductase occurs in denitrifying bacteria. FEWSON and NICHO-LAS (1961a—d) have shown that even in actively denitrifying cells of *Pseudomonas aeruginosa* grown anaerobically in a peptone and nitrate medium some (^{15}N) nitrate was assimilated. Thus 95% was lost as gas and the rest assimilated but this 5% $^{15}NO_0$ was equivalent to 15% of the total cell nitrogen. This then accounts for the induction of hydroxylamine reductase in actively denitrifying cells.

Hydroxylamine reductase was purified 47-fold from cell-free extracts of *Ps. aeruginosa* by WALKER and NICHOLAS (1961c). The fractionation was achieved using a combination of ammonium sulphate and calcium phosphate gel treatments as shown in Table 20. The purified enzyme was pale pink in colour and contained a cytochrome of the c type. Fraction V was free from nitrite and nitric oxide reductases but it did contain a weak catalase activity. The latter, however, did not parallel hydroxylamine reductase and the ratio of catalase/hydroxylamine reductase in crude extracts (fraction II) was 10:1 and in the purified enzyme 0.2:1. Optimum enzyme activity occurred at 37° during 10 min incubation at pH 7.8.

III. Measurement of Activity.

1. Interference and Errors.

This subject is considered first as certain conclusions reached in recent work are now seen to be dependent on the method of assay and conditions under which it was used and several points merit note in this context. The sources of error already discussed in relation to nitrate reductase apply equally to the enzymes in this section, where comparable methods are involved. Unlike nitrate reductase where the assay is usually carried out by uncomplicated measurement of nitrite production under conditions such that no further loss of nitrite occurs, the assay of the enzymes considered here presents problems. Although measurement of loss of substrate would appear a simple approach, this procedure is not wholly acceptable unless certain conditions are observed. Each of the substrates for the enzymes under consideration is liable to oxidation as well as reduction. Assay by measurement of substrate loss is therefore strictly valid only under anaerobic conditions or where oxygen uptake is unequivocally excluded. Nitrite and hypo-nitrite may each be reduced by two or more different possible routes yielding unidentified products, oxides or nitrogen, hyponitrite and hydroxylamine as discussed by FEWSON and NICHOLAS (1961c). It is likely therefore in some instances at least for some doubt to exist as to which, or how many, enzyme systems are being assayed. Each of the potential first products of reductive reactions may be further metabolised and may not therefore accumulate. The same electron donors may be involved in these later reactions as in the first step. Stoichiometric comparisons between electron donor requirements and substrate loss may be inaccurate for this reason. The conclusions to be drawn from particular assay methods must therefore be qualified in a number of respects.

The assay of enzymes in the group under consideration is preferably carried out by showing that substrate loss and end-product formation are related in a stoichiometric manner. The significance of this point is revealed in the work of HAGEMAN, CRESSWELL and HEWITT (1962) where it was shown that nitrite loss and ammonia production were, within limits of probable experimental error, stoichiometrically related for the marrow leaf preparation under specified conditions of assay but may not show this relationship under other conditions whilst the slower reduction of hydroxylamine often produced much lower yields of ammonia with identical preparations, later, stoichiometric results occurred.

Table 21. *Conditions used for assay of enzymes concerned in nitrite, hyponitrite and hydroxylamine metabolism in fungi and higher plants.*

Method	Assay Mixture	Time/Temp.	Notes	References
Oxidn. of DPNH measured at 340 mμ	2.6 ml 0.1 M pyrophosphate pH 8 0.05 ml 1.5 × 10⁻⁵ M FAD 0.05 ml 6 × 10⁻³ DPNH 0.1 ml 0.4 M NH₂OH HCl 0.2 ml enzyme	Extinction measured at 15 or 30 sec intervals for 2 or 3 min, Room temperature	Reaction started by adding NH₂OH. Endogenous oxidn. of DPNH measured initially for 3 min.	ZUCKER and NASON (1955) Hydroxylamine reductase *N. crassa*
Est. of NH₃ produced in Conway cells by phenol-hypochlorite. Assay performed in Conway cell, alternative assay by decrease in extinction at 245 mμ	*Outer compartment* 0.15 ml 0.2 M pyrophosphate pH 7.5 0.05 ml boiled pig heart extract 0.1 ml 10⁻² M DPNH 0.1 ml 4 × 10⁻³ M Na₂·N₂O₂ 0.2 ml enzyme (8 mg/ml) *Inner compartment* 1.5 ml 0.01 N HCl	20 min	All reagents boiled to remove ammonia	MEDINA and NICHOLAS (1957 a, b, c). Hyponitrite reductase *N. crassa*
Est. of NO₂ loss by GRIESS-ILOSWAY method (see text)	1.5 × 10⁻³ M PO₄ pH 7.5 0.1 ml Boiled Pig Heart acetone powder 8 × 10⁻⁴ M DPNH 2 × 10⁻⁵ M NaNO₂ 0.1 ml enzyme 14 mg protein/ml Total vol. 1.0 ml	10 min 25°	Reaction stopped by adding 0.1 ml 1% v/v acetaldehyde and cryst. alcohol dehydrogenase Followed by 0.1 ml M zinc acetate and 3 ml ethanol to remove DPNH	NICHOLAS, MEDINA and JONES (1960). Nitrite reductase *N. crassa*
Est. of NH₃ produced in Conway cells by alkaline phenate-hypochlorite or oxidation of DPNH measured at 340 mμ	0.8 ml 0.1 M pyrophosphate pH 7.5 0.2 ml 2 × 10⁻⁵ M FAD 0.3 ml 9 × 10⁻³ M DPNH 0.2 ml 10⁻³ M Mn (omitted from *Neurospora*) 0.5 ml 10⁻² M NaNO₂ (*Neurospora*) 0.2 ml 5 × 10⁻³ M NaNO₂ (soybean) 0.2 ml enzyme	15—40 min (*Neurospora*) 4—8 min (Soybean)	Reaction stopped by adding 0.2 ml 10⁻² M KCN	NASON, ABRAHAM and AVERBACH (1954). Nitrite reductase *N. crassa* and soybean
Oxidn. of DPNH or TPNH measured at 340 mμ under aerobic conditions	1.27 ml 0.1 M pyrophosphate pH 7.0 0.3 ml 2.1 × 10⁻³ M DPNH (or TPNH) 0.3 ml 10⁻³ M NaNO₂ or NH₂OH HCl 0.03 ml 10⁻² M MnCl₂ 1.2 ml crude boiled extract diluted 10 fold with buffer 0.2 ml enzyme	2—10 min read at 30—120 sec intervals. Room temperature	Started by adding DPNH. Endogenous oxidation of DPNH measured for 1 to 2 min before adding NaNO₂ or NH₂OH	ROUSSOS and NASON (1960). "Pyridine nucleotide — nitrite and hydroxylamine enzymes": Soybean

Disappearance of NO₂ or NH₂OH and simultaneous production of ammonia using CONWAY cells and alkaline phenate-hypochlorite method under *anaerobic* conditions

1.0 ml 0.1 M K/HPO₄ pH 7.5
0.2 ml 0.02 M NaNO₂
or 0.02 M NH₂OH HCl
0.2 or 0.4 ml enzyme
6 ml benzyl viologen solution
1.5 mg/ml total concentration; approx. one-half in reduced state (see text)
Alternative assay using
0.2 ml 3×10⁻³M TPNH
ard 0.2—0.3 ml oxidised benzyl viologen solution or 0.2 ml 10⁻³M TPN
0.5 ml 3×10⁻²M glucose-6-phosphate
0.1 ml 10⁻²M MgSO₄
0.1 ml glucose-6-phosphate dehydrogenase and 0.2—0.3 ml benzyl viologen solution

20—30 min 21—27°

Reaction started by anaerobic addition of benzyl viologen. Reaction stopped by admission of air

Reaction started by tipping TPNH from side arm of THUNBERG tube

Reaction stopped by adding aliquot to 0.1 ml M Zn acetate followed by 3 ml ethanol

HAGEMAN, CRESSWELL and HEWITT (1962). Vegetable marrow nitrite and hydroxylamine reductase

The simplest procedure for the assay of nitrite reductase is to follow nitrite disappearance under conditions summarised in Tables 21, 22 and 24. Where this method is shown to be valid it still has the disadvantage, owing to the extreme sensitivity of the colorimetric method, that low substrate concentrations of nitrite are often used. There is then a real danger of failure to saturate the enzyme either initially or during the course of reaction, although the saturation constants (Tables 25, 27) are generally much smaller than for nitrate. A serious source of error which was not generally appreciated in earlier work is the reaction between reduced pyridine nucleotides, especially DPNH, and nitrite in an acid medium (Table 23). Prevention of this interference has already been considered in detail in connection with nitrate reductase. One point requires note. The effect of DPNH on the results of the assay depends on the method used and the enzymes concerned. In nitrate reductase measurements the failure to remove excess DPNH causes low values which are practically independent of whether DPNH or enzyme are omitted from blanks or whether boiled enzyme or zero time measurements are used as nitrite is absent or negligible in all these control systems. In nitrite reductase assays however, where nitrite is present initially, the omission of DPNH from control assays will give totally different results from those obtained if boiled enzyme, no enzyme or zero time controls are used in the presence of DPNH, if this compound is not removed before nitrite estimation. Thus, inclusion of DPNH for zero time and assay time systems will give low but comparable values for different assays whilst omission of DPNH in zero time systems will give inaccurate results that are very high and not comparable for different assays. Non-enzymic loss of nitrite occurs with presence of copper with production of oxides of nitrogen (NICHOLAS, unpublished work). Assay methods based on spectrophotometric measurements of concurrent oxidation of reduced pyridine nucleotides suffer from other sources of error already noted.

Endogenous oxidation of DPNH and TPNH may often be relatively rapid, and as found by ZUCKER and NASON (1955), ROUSSOS and NASON (1960), SPENCER, TAKAHASHI and NASON

Table 22. *Conditions used for assay of nitrite reductase systems from bacteria*

Method	Assay	Time (min)	Temp. (°C)	Notes	References
WARBURG assay — production of NO and N₂ gases. Nitrite utilized by colorimetric test	*Warburg* / *main compartment* / 0.08 M phosphate (pH 6.8) / $10\ \mu M$ $MgCl_2$ / 1 mg isocitric dehydrogenase / 30 mg enzyme / *1st sidearm:* / $20\ \mu M$ glucose-6-phosphate / $10\ \mu M$ sodium isocitrate / $10\ \mu M$ $NaNO_2$ / *2nd sidearm:* / $0.1\ \mu M$ TPN / Total volume 3 ml / 0.2 ml 20% w/v KOH in centre well. / Gas phase N_2.	180	30	All substrates tipped in at zero time.	NAJJAR and CHUNG (1956)
Colorimetric test for nitrite, DPNH oxidation at 340 mμ.	0.033 M phosphate (pH 6.8) / $0.24\ \mu M$ MbH_2 / $2.5\times10^{-3}\ \mu M$ FAD / $10\ \mu M$ $NaNO_2$ / 0.1 ml enzyme / Total volume 3 ml	15	Room temp.	Assays in THUNBERG tubes under anaerobic conditions.	TANIGUCHI, SATO and EGAMI (1956)
Colorimetric test for nitrite	0.25 ml 0.1 M TRIS (pH 7.1) / 0.03 ml 2×10^{-3}M TPNH / 0.05 ml boiled pig heart or 1.2×10^{-4} M FAD / 0.15 ml 2×10^{-4} M. $NaNO_2$ / 0.02 ml enzyme	10—15	Room temp.	After incubation 1.5 ml H_2O, 0.5 ml sulfanilamide (1% w/v 3 N HCl) and 0.5 ml N-1-naphthylethylenediamine hydrochloride (0.02% w/v) added. A similar reaction mixture at zero time or one from which TPNH was omitted was used as a control.	SPENCER, TAKAHASHI and NASON (1957)
Colorimetric test for nitrite	*Thunberg method* / 0.05 M phosphate buffer / $5\ \mu$moles PyH_2 / or MbH_2 or MVH / or $PhMSH_2$ or Tol BH_2 or. / $215\ \mu$moles glucose or succinate / 0.1 ml enzyme (in sidearm) / Final volume 2.5 ml	20	37	Reaction stopped by adding $CdSO_4$ (pH 4.2). Kaolin or active charcoal was used for decolourising the extract. Electron donors prepared with H_2 and palladised asbestos. 2:6-dichlorophenol indophenol and phenazine methosulphate reduced with ascorbic acid. Methyl viologen reduced with Mg powder.	ASANO (1959)

Process			Assay composition	Notes	Reference
Production of NO and N₂ gases in a Warburg apparatus	60	37	*Warburg assay* — *Main compartment*: 0.08 M phosphate buffer (pH 6.8); 11.5 μmoles MBH$_2$ or 30 μmoles sodium formate. *Sidearm*: 0.1 ml enzyme. Total volume 2 ml. 0.1 ml 20% w/v KOH in centre well. Gas phase — N$_2$. When sodium formate electron donor NaCl was added (0.5 M final concentration)	Gas output measured at 10 min intervals over 60 min period. NaCl requirement for these enzymic reactions.	
Colorimetric test for nitrite; Production of NO and N₂ gases in a Warburg apparatus	10	37	*Well of Thunberg Tube*: 1.3 ml 5×10^{-2}M phosphate (pH 7.1); 0.1 ml 2×10^{-3}M pyocyanine; 0.1 mg 5% w/v palladised asbestos. *Sidearm*: 0.3 ml 5×10^{-2}M phosphate (pH 7.1); 0.1 ml 2×10^{-3}M NaNO$_2$. Pyocyanine can be replaced by MB FAD, FMN or riboflavin	THUNBERG tubes evacuated and filled with H$_2$ gas purified by passing it through a 'De-oxo' catalytic de-oxygenator. Tubes evacuated with water pump and gassed six times and incubated at 30° until Py was fully reduced, as judged by eye. After 10 min nitrite determined in 0.5 ml aliquots.	WALKER and NICHOLAS (1960b)

(1957), may contribute up to 50 per cent of the total rate observed in this way. Measurement of steady state rates before and after substrate addition allow correction for the endogenous rate. The presence of quinones and the difficulties of producing entirely anaerobic conditions or of excluding traces of endogenous hydrogen peroxide may make the elimination of the endogenous reaction a source of difficulty. Thus a change in extinction at 340 mμ of 0.5 in 3 ml volume corresponds to an oxygen requirement of only 2.8 μl for a DPNH oxidase system. Measurements of oxygen uptake in systems appropriate to spectrophotometric methods, may therefore be insensitive and not detected.

Estimation of ammonia as a product of nitrite reductase depends on the uninhibited activity of at least one, perhaps two, or more, other enzymes whose activity is not limiting to the rate of nitrite reduction being studied.

The assay of enzymes involved in metabolism of hydroxylamine and nitric oxide also presents problems which are similar to those described above, except that the saturation by hydroxylamine occurs at high concentrations which may be inhibitory (Tables 25, 28). Conditions are shown in Tables 21 and 24. The estimation of hydroxylamine, by either CSÁKY (1948) method where hydroxylamine is oxidised to nitrite by iodine in acetic acid or by 8-hydroxyquinoline by the method of FREAR and BURRELL (1955), is very sensitive. For this reason substrate concentration sometimes tends to be kept low or may become limiting during the reaction.

Estimation of ammonia production by the Conway (1957) method is subject to the error caused by chemical decomposition of hydroxylamine in the presence of potassium carbonate to produce ammonia non-enzymically (Zucker and Nason, 1955; Medina and Nicholas, 1957a, c). This leads to apparently high values. The error is mainly compensated by measuring ammonia production chemically with boiled enzyme and zero time assays and allowing for this in the assay. (Hageman, Cresswell and Hewitt, 1962) or by utilisation of all hydroxylamine as substrate (Medina and Nicholas, 1957c).

In experiments with purified hydroxylamine reductase from *Pseudomonas aeruginosa* to determine the stoichiometry of the reaction, Walker and Nicholas (1960, 1961b) found it necessary to continue the enzyme assays until hydroxylamine was fully reduced so that the chemical formation of ammonia cannot occur. Non-enzymic disappearance of hydroxylamine may occur in the presence of copper or cobalt according to Spencer et al. (1957).

Table 23. *Effect of Various Mole Ratios of DPNH/NO₂ and TPNH/NO₂ on the Nitrite Test* (Medina and Nicholas, 1957b). (Values found as % of NO₂ added.)

DPNH/NO₂	0	2.5	5	7.5	10
	100	50	45	31	25
TPNH/NO₂	0	2	4.1	6.2	8.3
	100	79	75	56	49

Ammonium sulphate remaining after its use for protein precipitation is an obvious source of error. Much can be removed by dialysis when the enzyme is not inactivated by this procedure. The amounts produced in assay systems of the order of 5—100 μg are similar in magnitude to those likely to remain in 0.2 ml of enzyme solution after normal dialysis unless this is exhaustive in large volumes with stirring or unless further fractionations or Sephadex treatment are adopted.

2. Assay Methods.

The detailed methods are used at Long Ashton for determination of substrates and products concerned in this section are described below.

a) Use of Conway Methods for NH₃ Determinations, Conway (1957), Russell (1944).

Ammonia is determined in Conway units by distilling it into 0.01 N HCl followed by the colorimetric determination of ammonia with alkaline phenate and hypochlorite (Conway, 1957). All reagents used are dispensed in water that has been boiled to remove ammonia.

The following reaction mixture was used by Walker and Nicholas (1960, 1961a, b) to follow ammonia production from hydroxylamine: *Outer compartment:* 1 ml enzyme; 0.5 ml (1 μM) NH₂OH.HCl; 0.5 ml 0.1 M phosphate buffer (pH 7.4). *Inner compartment:* 1.5 ml 0.01 N HCl. After incubating for 2 hour 1 ml saturated K₂CO₃ was added and well mixed with the contents of the outer well and the greased lid replaced rapidly. After incubating for 90 min the ammonia contents of aliquots from the centre well were determined by the Russell phenate method (described below; Medina and Nicholas, 1957a and c).

Hageman, Cresswell and Hewitt (1962) determined ammonia formation by hydroxylamine reductase of marrow leaf in the presence of hydroxylamine by estimating ammonia at zero time and after the assay period of 10, 20 or 30 mins. 1 ml aliquots of the reaction mixture containing initially in about 7.5 ml of phosphate buffer .05 M pH 7.5, 2 μmoles NH₂OH and 20 μmoles of oxidised benzyl viologen (produced on opening the THUNBERG tube) were introduced into CONWAY cells prepared as described below. Cells were washed in paraffin after wiping to remove grease, boiled in water with Teepol detergent which was flooded away to carry off the supernatant paraffin. The cells were soaked for 2 hours in N HCl, rinsed in copious amounts of water and oven-dried. The rims of the inner compartments were also greased before adding 1.5 ml 0.01 N HCl. 1.5 ml of saturated K_2CO_3 was added rapidly after the introduction of the reaction mixture sample, with the lid partially removed and the cell tilted on a block of 3 mm polythene to keep the reaction mixture under the covered area. The lid was immediately drawn over the whole area and the contents gently mixed by swirling. The cell was placed on a rocker tray and kept at 25° C for 4 hours or 16 hours before final estimation which gave 90 and 100 per cent recoveries of the ammonia present. Quantities of fluid are designed to cover the whole surface area of each compartment with minimum depth.

The RUSSELL (1945) method for determining ammonia by phenate-hypochlorite method is as follows: *Reagents* 1. Alkaline-phenate solution containing 25% w/v phenol in 2.7 N NaOH; prepared by adding 25 g of crystalline phenol to 100 ml water, 54 ml 5 N NaOH is then stirred into the solution which is stored in a dark bottle in a refrigerator. 2. Hypochlorite solution; 25 g calcium hypochlorite (bleaching powder) is ground and then dissolved, as far as possible, in 300 ml hot water. Next 135 ml K_2CO_3 (20 g anhydrous salt/100 ml solution previously boiled to free it of ammonia) is mixed into the solution which is then heated briefly to about 90 C° and after cooling is made to 500 ml. A small portion of the mixture is filtered to test for Ca ions. Should the test be positive more carbonate is added to the mixture until a negative test is obtained. The solution is then filtered and stored in a refrigerator (this solution should be water clear and contain between 1.3 and 1.4 g of free Cl/100 ml. 3. Manganese solution; 0.003 M manganese chloride or sulphate.

An aliquot (0.5 to 1 ml) of the HCl solution in the CONWAY unit containing 0.5 to 5 μg NH_3—N, is put in a test tube followed by 0.05 ml manganese chloride solution and 1 ml of the alkaline phenate reagent and 0.5 ml of the hypochlorite solution. The two last named reagents must be cold when added to the sample. Losses of ammonia from the alkaline solution may be reduced further by keeping the sample tubes in an ice bath during their preparation. The contents of the tubes are mixed by gentle rotation and put into a briskly boiling water bath for 5 min then cooled, shaken and made to a convenient volume (6 or 10 ml). The blue colour formed is measured in photoelectric colorimeter using a filter having an absorption near 625 mμ or preferably in a spectrophotometer.

This test is very sensitive for ammonia since as little as 0.1 μg NH_3—N gives a detectable colour and 0.5 to 0.6 μg NH_3—N may be determined accurately. If distillation of water occurs from the centre well, the whole contents are removed with a fine graduated pipette followed by washings to make 2 ml. The sample should be either neutral or with an acidity not exceeding 0.01 or 0.02 N.

Other workers have used the NESSLER reagent to determine micro-amounts of NH_3—N in CONWAY units (SPENCER, TAHASHI and NASON, 1957). The NESSLER reagent is prepared by grinding 10 g mercuric iodide (HgI_2) in a mortar with water (100 ml) in which are dissolved 5 g KI and 20 g NaOH. The precipitate settles out completely after 24 hr so that the clear solution can be decanted. The reagent must be kept in a well-stoppered bottle in the cold and in the dark. The NESSLER reagent covers a similar range of NH_3—N concentrations as does the hpenate-hypochloritere agent. Measurements are made at 430 mμ.

Since micro-amounts of ammonia are determined in the CONWAY units it is imperative that the last traces of ammonia are removed from reagents and glass-ware. Thus all reagents must be freshly prepared in solutions that are boiled to remove ammonia. Enzymes must also be dissolved in buffers similarly treated. When ammonium sulphate is used to fractionate enzymes then the purified enzyme must be dialysed exhaustively as described already.

b) Hydroxylamine.

(i) CSÁKY (1958) method. The modified CSÁKY method used by WALKER and NICHOLAS (1960, 1961 c) for determining hydroxylamine is as follows: A suitable aliquot was diluted to 3.5 ml with deionised water and 1 ml 1% w/v sulphanil-amide in 25% w/v acetic acid and 0.5 ml 1.3% w/v iodine in glacial acetic acid added. After 4 min 1 ml 2% w/v sodium arsenite in water was added and the tubes shaken until all the iodine was reduced when 1 ml 0.02% w/v aqueous naphthyl-amine was added. After standing for 20 min the intensity of the red azo dye was determined in a spectrophotometer. It is important to use freshly prepared reagents since the iodine solution in particular is readily reduced when exposed to air. The time periods fixed for each step must be standardised to obtain reproducible results. The amount of enzyme used was such that it reduced 100 mμM/NH$_2$OH since at this level, the substrate was not rate limiting. CRESS-WELL (unpublished) found improved reproducibility if the sulphanilic acid reagent of CSÁKY was added after the arsenite. The method has been modified by E. J. HEWITT (unpublished) so that oxidation occurs under weakly alkaline conditions similar to those described for oximes by YAMAFUGI and AKITA (1952). To the sample containing up to 250 μm Mol NH$_2$OH in 6 ml of 0.016 M phosphate buffer pH 8.5, are added in sequence immediately with rapid mixing by inversion at each stage, the following: 1 ml iodine reagent, prepared by dissolving 2.5 g iodine in 40 ml 18.5% w/v potassium iodide, diluting to 200 ml for storage and further diluting 15 ml to 100 ml before use, 1 ml 0.8% w/v sodium arsenite, 1 ml 1.0% w/v sulphanilamide in 1N hydrochloric acid, 1 ml 0.1% N-(1-naphthyl)-ethylenediamine dihydrochloride in a mixture of 45% v/v acetic acid and 0.15 N hydrochloric acid. Full colour intensity is obtained in 45 min. The method is reproducible and Beer's Law is obeyed up to 250 μm Mol. The recovery of hydro-xylamine as nitrite is 90.5%. The modification described here was found necessary because of severe interference in the usual CSÁKY method when using benzyl viologen (HAGEMAN, CRESSWELL and HEWITT, 1962). When carried out as described the recovery of hydroxylamine is 88 and 83.5% as nitrite in the presence of 8 and 12.5 molar excess ratios respectively of benzyl viologen. Beer's Law is obeyed up to 70 μm Mol, above which the extinction coefficient decreases. The recovery is also decreased to reproducible extents when nitrite is also present as well as benzyl viologen. The evidence for this is that the values for nitrite alone determined by this method or by the Griess method described earlier are similar but the difference between the two methods in the presence of hydroxyl-amine is less than that recorded for hydroxylamine alone by the method just described. The effect of nitrite increases as the nitrite concentration increases and also increases with increasing dye concentration although recovery of varying concentrations of hydroxylamine is also practically constant for a given nitrite concentration. As nitrite can be determined alone in the presence of hydroxylamine it is possible to determine with satisfactory precision the changes in hydroxyl-amine concentration in the presence of nitrite.

(ii) 8-Hydroxyquinoline method (FREAR and BURRELL, 1955). The method is based on the reaction between 8-hydroxyquinoline and hydroxylamine to produce 5,8-quinoline-5-(8-hydroxy-5-quinolylimide). This has a light absorption maximum at 705 mμ which conforms to BEER's Law up to 10^{-2} M. Reagents: 1.0 M sodium carbonate, 1% w/v 8-hydroxyquinoline, 12% w/v trichloracetic acid. Standard curves are prepared as follows: 1 ml hydroxylamine solution containing up to 0.25 μmole is transferred to a 15 ×125 mm tube followed by 1 ml 0.05 N phosphate buffer pH 6.8 and water to make 2.8 ml. 0.2 ml trichloracetic acid is followed by 1 ml 8-hydroxyquinoline solution with gentle mixing and 1 ml 1.0 M sodium carbonate. The mixture is shaken vigorously to aerate, stoppered and placed in a boiling water bath for 1 min, cooled and measured after 15 min. $\varepsilon = 1.5 \times 10^4$ M^{-1} cm^2. A blank mixture is also determined.

(iii) *o*-Phenanthroline method (MEDINA and NICHOLAS, 1957a). The method used for determining hydroxylamine by the Fe^{3+} *o*-phenanthroline method is as follows: aliquots of reaction mixture corresponding to concentrations between 0 and 5 μg NH$_2$OH HCl in 1 ml distilled water are put in test tubes followed by 0.3 ml FeCl$_3$ (200 μg Fe^{3+} ml) 0.3 ml 0.1% w/v *o*-phenanthroline and the volume made to 5 ml with distilled water. The red colour is determined in a SPEKKER absorbtiometer 0.5 cm cell using green filters or preferably in a spectrophotometer. The effective range is from 0 to 5 μg hydroxylamine.

(iv) Nitrobenzene method (WOOD, 1953). This method has only been described in outline but merits attention. It is reported to be specific for hydroxylamine. It is based on condensation of hydroxylamine and nitrobenzene in alkaline solution to produce the diazonium derivative which is coupled to α-napthol as in the GRIESS-ILOSWAY reaction.

c) Oximes, YAMAFUJI and AKITA (1952).

Mild acid hydrolysis of oximes regenerates hydroxylamine but according to YAMAFUJI and AKITA (1952) the yield may be only 60 per cent of theoretical. These workers have described a new method using oxidation by iodine in mildly alkaline solution to oxidise oximes to nitrite. The conditions as described are as follows: 5 ml of oxime solution (10^{-4} M) are added to a mixture containing 1 ml 20% w/v sodium salt of sulphanilic acid, and 0.3 ml 0.1 N iodine solution. After 2 minutes at room temperature the excess iodine is removed by additions of 0.3 ml 0.1 N sodium thiosulphate followed by 1 ml 30% w/v acetic acid, 3 ml 0.5 M hydrochloric acid, and 1 ml 0.01% w/v α-naphthylamine. Extinction is measured at 540 mμ as nitrite. Oximes as pyruvic oxime yield to 98—100 per cent whilst free hydroxylamine yields 93 per cent of the expected nitrite production. When tested with diacetyl monoxime (HEWITT, unpublished) negligible nitrite formation occurred. The modified method for hydroxylamine described above was tested in the presence of excess pyruvate which interfered seriously. When the solution with buffer was heated to 80° *after* adding the 1 ml iodine reagent and 1 drop 1N potassium hydroxide, and left to cool to room temperature before adding the other reagents recoveries of 90 and 100 per cent as nitrite were obtained in the absence and presence respectively of excess pyruvate (HEWITT, unpublished). Diacetyl monoxime gave about 50 per cent recovery by this method.

d) Hyponitrite.

The estimation of hyponitrite is complicated by its instability especially below pH 9. This subject is well reviewed by FREAR and BURRELL (1958). YAMAFUJI and OSAJIMA (1961) have described a method for its determination in the presence of hydroxylamine and nitrite. The method is based on the fact that both hydroxylamine and hyponitrite produce an oxime in the method of Yamafuji and Akita (1952), but unlike hydroxylamine, hyponitrite is decomposed on boiling. Application of the oxime method on boiled and unboiled aliquots gives the value for hyponitrite by difference. Estimation of nitrite separately allows determination of the three compounds in a mixture. Spectrophotometric estimation of hyponitrite in the presence of nitrite and nitrate is described by Addison, Gamlen and Thompson (1952). It might be noted that other procedures e.g. that of Csáky (1948) also include both hyponitrite and hydroxylamine and can therefore be similarly applied for the differential determination of the two before and after boiling. Nicholas (Medina and Nicholas, 1957) has also described a method for hydroxylamine determination in the presence of hyponitrite or oximes which is based on the fact that hydroxylamine will reduce ferric to ferrous iron whilst hyponitrite and oximes do not. The ferrous iron is determined with o-phenathroline and 0.1 μg hydroxylamine can be detected as described above. Hyponitrite absorbs light in the ultraviolet region with an absorption maximum at 248 mμ; $\varepsilon = 3.98 \times 10^3$ M^{-1} cm^2 (Addison, Gamlen and Thompson, 1952). This property has been used by Nicholas (1959) to follow hyponitrite disappearance in the presence of the *Neurospora* enzyme systems. α-oxyhyponitrite (H$_2$N$_2$O$_3$) has an extinction of $\varepsilon = 8.31 \times 10^3$ at 248 mμ (Addison et al., 1952).

e) Assay Methods Using Reduced Benzyl Viologen and Other Dyes under Anaerobic Conditions.

(i) Hageman, Cresswell and Hewitt (1962). Benzyl viologen (BDH England 4,4'-dipyridylium-1,1'-dibenzylchloride) (mol. wt. 409) is reduced by hydrogen in the presence of palladised asbestos Fig. 2. The hydrogen is high purity oxygen-free and is passed through a deoxygenating cartridge, reducing valve and 3 cm length of thermometer capillary. Palladised asbestos is washed to remove fine material and is prevented from entering the storage reservoir when dye is transferred after reduction by a filter of Tygan gauze filter cloth around the outlet from the reduction vessel. The hydrogen in the reduction vessel and the storage vessel escapes through Bunsen valves and maintains a constant pressure of about 3—5 lb. per sq. in. The storage vessel is flushed with hydrogen for 30 min before transfer of the dye and any dye used previously is left in the storage vessel until the next supply is prepared. This arrangement ensures that traces of oxygen from minor leakages are absorbed and do not affect the reduction of the freshly prepared dye. Reduced dye can be stored for use for several days in this apparatus. The benzyl viologen solution is initially 1.5 mg/ml (3.66×10^{-3} M approx.); with a purity of about 90 per cent this value may be only 3.3×10^{-3} M. Reduction is controlled by titration of the benzyl viologen into 0.5 ml of 2×10^{-3} M potassium permanganate in N sulphuric acid. The end-point is a change from trace pink to trace blue through colourless. Titration is carried out by placing the permanganate solution in a Thunberg tube which is attached to the benzyl viologen reservoir and burette system by a three-way stopcock. This is used for evacuation and admission of the benzyl viologen from the burette. The upper end of the burette is connected to the upper part of the storage vessel and is under hydrogen gas. The burette is supplied by a three way stopcock which permits dye to be conveyed from reducing vessel to storage or in the reverse direction, from storage vessel to the burette or from reducing vessel to burette for initial titration. Dye is transferred by hydrogen pressure from reducing to storage vessel by appropriate control of the stopcocks and valves. 0.5 ml of 2×10^{-3} M KMnO$_4$ is equivalent to 1 ml of 5×10^{-3} M reduced dye, which is a single electron donor, and therefore to 1.37 ml of dye of the concentration stated and to 1.37×100/per cent purity. The observed ratio 1.37 to volume of dye equivalent to 0.5 ml $KMnO_4 \times 100$ gives the percentage reduction. Values of about 60 per cent reduction give good results in nitrite reductase assay. Over-reduction of dye can occur as noted by MICHAELIS and HILL (1933) and in this condition nitrite is lost chemically without ammonia production (HAGEMAN, CRESSWELL and HEWITT, 1962). Over reduction is indicated

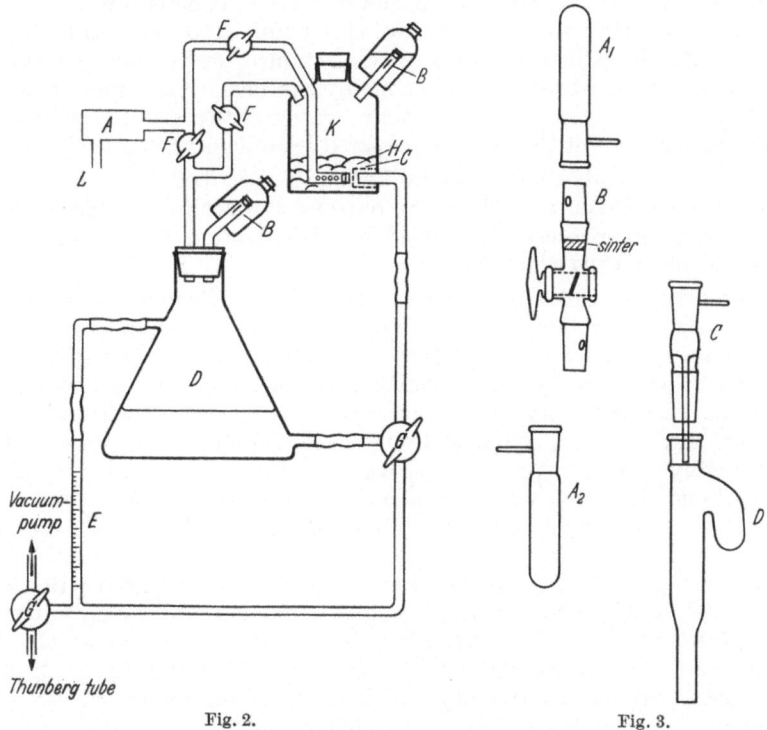

Fig. 2.   Fig. 3.

Fig. 2. Apparatus for preparation titration and volumetric introduction of reduced benzyl viologen under anaerobic conditions. Key: A = de-oygenator, B = Bunsen valve, C = filter gauze. D = dye reservoir, E = burette (2.4 or 10ml.), F = 1-way stopcock, G = 3-way stopcock, H = Pd. asbestos, K = reduction vessel, L = hydrogen inlet (from cylinder). (HAGEMAN, CRESSWELL and HEWITT, 1962).

Fig. 3. Apparatus for the transfer of reduced electron donors under anaerobic conditions (WALKER and NICHOLAS, 1961 b, c). THUNBERG tubes A_1 and A_2 are 6 ml capacity; D is a 1 cm glass cuvette fused to a THUNBERG top. All the joints are "Quickfit" B14. The method of operation is given in the text.

by titration values approaching 80 per cent reduction or below the theoretical for a single electron change. Accurate control of dye reduction and addition permit calculation of stoichiometric data.

The method of assay for nitrite reductase and for hydroxylamine reductase used by HAGEMAN, CRESSWELL and HEWITT (1962) consists of introduction of 4—6 ml of partially reduced dye under hydrogen into a thoroughly evacuated cold THUNBERG tube containing the other components (buffer, substrate, enzyme etc.) of the system. Assay times between 5 and 30 min are suitable at 20—27°. The assay is terminated by opening the tube and shaking when autoxidation of dye occurs. Oxidised dye does not interfere in nitrite estimation.

Since bacterial nitrite reductase has to be determined under anaerobic conditions special apparatus is used and precautions taken to ensure the complete absence of air from the reaction mixtures. Thus WALKER and NICHOLAS (1960, 1961b, c) used special THUNBERG tubes as shown in Fig. 3, to add electron donors free from palladised asbestos to the enzyme and its substrate under anaerobic conditions. The oxidised donor and palladised asbestos was put in Tube A, which was attached to assembly B. The tap was closed. Tube A, evacuated and filled with hydrogen (six times) was incubated at 30° C until the electron donor was fully reduced. Either A_2 or cuvette D (fitted with adaptor C) containing the enzyme and nitrite, was then attached to B and evacuated exhaustively. The electron donor was then drawn from tube A_1 into either tube A_2 or cuvette D by opening the tap in B. The palladised asbestos was retained on the sintered disc (No. 2 porosity). The reduced form of either methylene blue, pyocyanine or FMN was used as an electron donor. A rigorous removal of oxygen is essential since even traces of the gas inhibit the enzyme prepared from denitrifying bacteria.

The methods of dye reduction by hydrogen described here avoid the toxic effects of using dithionite employed by Sadana and McElroy (1957) whilst dye reduction can be precisely determined by the titration method of Hageman, Cresswell and Hewitt (1962).

Oxidised benzyl viologen produces a yellow precipitate in the Csáky (1948) hydroxylamine estimation, which slowly disappears on standing for about one hour (Cresswell, Hageman and Hewitt, unpublished). Observed values however are low. This interference is conveniently avoided in unpublished work by Cresswell, Hageman and Hewitt by separating dye from hydroxylamine (and nitrite and nitrate also if required) by passing a solution of the mixture through carboxymethyl cellulose powder (Whatman CM70). The powder is washed in 0.1 N HCl and then with water. Columns 0.6 cm diam. by 2.5 cm long retain all dye from a 1 ml sample containing about 1 mg dye. The cellulose powder is poured from a measure as a thin suspension into the column when full of water and allowed to settle rapidly as the outlet drains. 1 ml sample of reaction mixture chilled to 0° in an ice bath when the tube was opened to minimise peroxidase action (Cresswell and Hewitt, 1960), is immediately added to the drained column, allowed to drain and washed through with water up to 25 ml. 5 ml samples are used for determination of hydroxylamine by the Csáky method. Interference between dye and hydroxylamine estimations is minimised by taking 0.1 ml aliquots of the reaction mixture and making to 6.0 ml with water before application of the Csáky procedure. Recently (Hewitt unpublished) satisfactory results have been obtained with the method of Yamafuji and Akita (1952) using alkaline iodine oxidation, with the modification described on pages 128 and 129.

Walker and Nicholas (1960, 1961 a, b) found that maximum hydroxylamine reductase activity in crude extracts of *Ps. aeruginosa* was obtained with 10^{-3} M hydroxylamine and the Michaelis-Menten constant was approximately 4×10^{-4} for NH_2OH (Table 28). There was, however, also a non-enzymatic reduction of hydroxylamine by the reduced pyocyanine used as the electron donor but at 2×10^{-4} M which was used in the assay, this was negligible. Although this concentration was below saturation, when not more than 0.12 μmole NH_2OH was reduced activity was proportional to enzyme concentration and also to the time of incubation. They also found that assays had to be done under strict anaerobic conditions when it was found that for every μmole of electron donor oxidised one μmole of hydroxylamine was reduced.

The Japanese workers (Taniguchi et al., 1958) have used Thunberg techniques to measure nitrite reductase activity with either glucose or succinate or another suitable substrate, linked through its dehydrogenase to nitrite or a reduced dye, as an electron donor. The reaction was stopped by adding cadmium sulphate at pH 4.2. Kaolin or active charcoal was also used to clarify the solution. This method, however, is not satisfactory since these substances adsorb nitrite and even small losses result in serious errors since only 2.5 μmole NO_2^- is used in the assay mixture. Methylene blue, toluylene blue and pyocyanine were reduced by ascorbic acid and the methyl viologen by Mg powder. Since ascorbic acid can reduce nitrite chemically great care must be taken that it is not present in excess in the assay mixture. The correct stoichiometric amount required for the reduction of the dye must be added. When a dehydrogenase system is used to link its substrate to nitrate or nitrite it is important that conditions are such that this enzyme is not the rate limiting reaction, otherwise the results for nitrate or nitrite reductases will be erroneous.

f) Manometric Assay of Nitrite Reductase. Table 22.

CHUNG and NAJJAR (1956a and b) assayed nitrite reductase in the WARBURG anaerobically under nitrogen. They calculated the amount of nitric oxide and nitrogen produced from nitrite by means of two simultaneous equations; Table 24:

$$N_2 + NO = G \text{ (total gas measured)}$$
$$2N_2 + NO = NO_2^- \text{ (utilised in assay)}$$

$G = \mu l$ gas output in the absence of alkaline sulphite.

NO_2^- is expressed, for purposes of calculation, as the volume in μl equivalent to the fraction of the gram molecular volume corresponding to the amount of nitrite reduced, assuming it to be represented by a monatomic gaseous compound of nitrogen. The amount of nitrogen produced can be determined independently by including alkaline sulphite in one sidearm of a double sidearm WARBURG flask. The alkaline sulphite absorbs the nitric oxide formed. The Japanese workers (TANIGUCHI et al., 1958) have also used the WARBURG method to determine N_2 production by nitrite reductase from a *Micrococcus*.

WALKER and NICHOLAS (1960, 1961a) also used a WARBURG method to measure NO and N_2 production by nitrite reductase from *Ps. aeruginosa*. They regenerated catalytic amounts of the electron donor, e.g. reduced forms of pyocyanine, methylene blue, by reducing them with DPNH. They found it necessary to flush the WARBURG flask and contents for 1 hr. with hydrogen gas which was freed from oxygen by passing it through a "De-oxo" catalytic de-oxygenator (Baker Platinum Division, Englehard Industries Ltd., 52 High Holborn, London, W.C.1) and then through two DRESCHEL bottles each containing 10^{-5} M pyocyanine and approximately 0.2 g palladised asbestos. The reaction was started when the dyes had been fully reduced, as judged by eye. The contents of the sidearm were added to the flasks and gas production measured at intervals until it ceased (1—2 hr.). It is very important to remove the last traces of oxygen since it markedly inhibits the nitrite reductase enzyme from denitrifying bacteria. The assimilatory nitrite reductases, however, can be assayed in air. Since reduced forms of methylene blue, pyocyanine and flavin inhibited at the higher concentrations required when large aliquots were used in the WARBURG assay, a method was devised whereby about 100 mμmoles of the donor was continually regenerated chemically with DPNH (WALKER and NICHOLAS, 1960, 1961a). After 20 min incubation the following acceptors oxidised the following amounts of DPNH (μmole): Mb, 0.192; Py 0.135 and FAD or FMN 0.070. The oxidation of DPNH started slowly as soon as the acceptor was added but the reduction of the acceptor was delayed presumably until the last traces of oxygen had been eliminated from the system. The amount of acceptor reduced, between 20 and 40 min, agreed closely with the DPNH oxidised. It was at this stage that the assay was started.

Table 24. Assay system for nitric oxide reductase and hydroxylamine reductase systems from bacteria

Method	Assay	Time min	Temp. °C	Notes	References
Colorimetric test for nitrite and uptake of NO measured in WARBURG apparatus	WARBURG assay — NO and N₂ production *Main compartment* 0.08 M phosphate (pH 6.8) 0.1 ml enzyme Final volume 2.4 to 3.0 ml *1st sidearm* 20 μmoles NaNO₂ 4 mg Difco yeast extract 0.46 μmoles DPN *2nd sidearm* 0.2 ml 5% w/v sodium sulphite in 0.1 N NaOH or 0.2 ml H₂O 0.2 ml 20% w/v KOH in centre well Gas phase pure N₂ (oxygen-free)	180	30	Total gas formed represented N_2 and NO. The latter could be absorbed into alkaline sulphite placed in one sidearm. Simultaneous measurement of the total volume of gas formed, NO_2 reduced, NO uptake. Two simultaneous equations used to obtain exact values of NO and N_2. $N_2 + NO =$ gas observed $2\,N_2 + NO = NO_2^-$ reduced (see text page 133) Therefore: — $N_2 = NO_2^-$ — (gas observed) and $NO = 2 \times$ (gas observed) — $(NO_2^-$ reduced)	CHUNG and NAJJAR (1956a, b) NO reductase
	WARBURG assay — NO uptake *Main compartment* 0.08 M TRIS buffer (pH 7.9) 1 mg d-isocitric dehydrogenase 2.5 to 5 mg enzyme *1st sidearm* 0.1 μmole TPN or DPN 0.1 μmole FMN or FAD 10 μmoles MgCl₂ *2nd sidearm* 20 μmoles glucose-6-phosphate 10 μmoles d-isocitrate Gas phase. 25% NO in N₂ gas (oxygen-free)	120	30	NO generated by adding 5 N HNO_3 to copper turnings under N_2 gas. Commercially available gas also used. All manometers flushed with O_2-free N_2 for 35 min prior to introduction of NO. 25% NO in 75% N_2	
NO uptake	WARBURG assay *Main compartment* 1.1 ml 0.1 M phosphate 0.1 ml 2.7×10^{-3} M pyocyanine 1.0 ml enzyme *Sidearm* 0.1 ml (500 μmoles) DPNH 0.1 ml 5% w/v ethanol 0.1 ml (0.1 mg) crystalline alcohol dehydrogenase *Centre well* 0.2 ml 20% w/v KOH *Gas phase* 4.5% NO/95.5% N_2	60	30	WARBURG flushed with O_2-free N_2 for 40 min and then with 20% NO in N_2 gas until damp indicator paper, held in the exit gas stream of the sidearm stopper, showed acid. The apparatus was equilibrated at 30° and the DPNH generating system added to start the reaction. Nitric oxide generated in N_2 gas by adding 6 N.HNO_3 to Cu-turnings. $3\,Cu + 8\,HNO_3 \rightarrow 2\,NO + 3\,Cu\,(NO_3)_2 + 4\,H_2O$	FEWSON and NICHOLAS (1960b) NO reductase

Method	Reagents	Time (min)	Temp. (°C)	Notes	Reference
Oxidation of hydroxylamine with I_2 and determination of nitrite formed by a GRIESS-ILOSVAY reagent. Hydrogen uptake in a WARBURG apparatus. NESSLER reagent to determine NH_3 formed	*Centre well* 1 ml 0.2 M phosphate (pH 7.1) 1 ml bacterial suspension 3 ml H_2O *Sidearm* 0.1 ml 0.02 M NH_2OH	130	30	WARBURG readings taken at 10 min intervals over a period of 130 min. Residual hydroxylamine and ammonia determined at the end of the reaction	WOODS (1938) Hydroxylamine reductase
Oxidation of hydroxylamine with I_2 and nitrite formed determined colorimetrically	0.26 ml 0.1 M TRIS buffer (pH 7) 0.02 ml 2×10^{-3} M TPNH 0.01 ml 10^{-4} M FMN or FAD 0.15 ml 4×10^{-4} M $NH_2OH.HCl$ 0.02 ml 0.2 M glucose-6-phosphate 0.02 ml 0.2 M $MnCl_2$ 0.02 ml enzyme Final volume 0.5 ml	10—20	Room temp.	CZÁKY method used to oxidise residual hydroxylamine to nitrite	SPENCER, TAKAHASHI and NASON (1957) Hydroxylamine reductase
Oxidation of hydroxylamine with I_2 and nitrite formed determined colorimetrically. Ammonia formation determined in CONWAY units	THUNBERG *assay* *Tube* 1.2 ml 5×10^{-2} M phosphate (pH 7.0) 0.2 ml 2×10^{-3} M pyocyanine 0.1 mg 5% w/v palladised asbestos 0.1 ml enzyme *Sidearm* 0.3 ml 5×10^{-2} M phosphate (pH 7.0) 0.2 ml 2×10^{-3} M $NH_2OH.HCl$ CONWAY *Units* *Main compartment* 0.2 ml 10^{-2} M phosphate (pH 7) 0.1 ml enzyme *Centre well* 1 ml 0.1 N HCl After incubation saturated K_2CO_3 added and NH_3 distilled into HCl	15 20	30 Room temp.	Ammonia determined in an aliquot of the HCl using the phenate-hypochlorite reagents	WALKER and NICHOLAS (1961c) Hydroxylamine reductase
WARBURG assay H_2 uptake. Residual NH_2OH oxidised to nitrite determined colorimetrically. Ammonia determined in CONWAY units	*Centre Well* 1 ml 0·06 M phosphate (pH 7.4) 20 μmoles cysteine hydrochloride 20 μM benzyl viologen 1.05 mg dry wt. extract 3.3×10^{-3} M $NH_2OH.HCl$ *Sidearm* *Gas phase:* H_2 (purified over red-hot Cu).	180	37	H_2 uptake measured at 30 min intervals over a period of 200 min	SENEZ and PICHINOTY (1958b) Hydroxylamine reductase

g) Manometric Assay of Nitric Oxide Reductase (Table 24).

Nitric oxide gas was prepared by adding 6 N HNO_3 to copper turnings according to the following equation:

$$3Cu + 8HNO_3 = 3Cu(NO_3)_2 + 2NO + 4H_2O$$

The method used by FEWSON and NICHOLAS (1960b, 1961d) is as follows: copper turnings were put in a 500 ml "Quickfit" round bottomed flask fitted with a dropping funnel containing 6 N HNO_3. In addition the flask was fitted with two outlets. Oxygen-free nitrogen was flushed through the apparatus for 10 min to remove oxygen and then the reaction was started by adding nitric acid slowly from the dropping funnel. The NO evolved was displaced by a slow stream of N_2 (oxygen-free) and after passing through a DRESCHEL bottle containing water (degassed previously by boiling), the gas mixture was collected over water in a 10 litre glass aspirator.

The nitric oxide content of the gas mixture was determined in a gas burette connected to a reservoir containing saturated pyrogallol in 20% w/v KOH. The volume of gas was measured and then oxygen added from a cylinder. After vigorous shaking the contraction in gas volume due to the reaction of NO and O_2 to give nitrogen dioxide was measured. The latter and the oxygen dissolved readily in the alkaline pyrogallol. The gas mixture usually contained from 20 to 30% nitric oxide in nitrogen (FEWSON and NICHOLAS, 1960b, 1961d).

The rate of nitric oxide uptake was measured in the WARBURG apparatus using DPNH as the donor which was continuously regenerated with crystalline alcohol dehydrogenase and alcohol. The WARBURG apparatus was flushed with oxygen-free nitrogen for 40 min to displace all the air and then with 20% nitric oxide in nitrogen, until damp indicator paper showed an acid reaction when held in the exit gas stream from the side-arm stopper indicating that flushing with NO is complete. The apparatus was equilibrated at 30° C and then a DPNH generating system was added to start the reaction. Great care must be taken to eliminate the last traces of oxygen, otherwise nitrogen dioxide is formed which will dissolve to give a mixture of nitrous and nitric acids and thus invalidate the assay. On no account should the WARBURG apparatus be evacuated for the purpose of removing the last traces of air since leaks develop under pressure that result in air re-entering *via* rubber tubing connections of the WARBURG gassing assembly (FEWSON and NICHOLAS, 1960a, 1961a).

3. Preparation of Hyponitrite as a Substrate.

As hyponitrite is an unstable compound some reference to its preparation seems appropriate. JONES and SCOTT (1924) and STEVENS (1958) have described procedures both of which have been employed at Long Ashton by MEDINA and NICHOLAS (1957a, 1957c). The methods as used are outlined below. ADDISON et al. (1952) have also described a simple modified method for the preparation of pure sodium hyponitrite from nitrate based on the early description of DIVERS *(loc. cit.)* and PARTINGTON and SHAH (1931).

Sodium hyponitrite was prepared (MEDINA and NICHOLAS 1957) by allowing 5 gm of hydroxylamine sulphate in 12 ml distilled water to react with 3 gm of sodium nitrite in 8 ml distilled water at 0° C. 5 ml of 0.5 N silver nitrate solution with 20 ml of distilled water was added to the solution and the yellow precipitate of silver hyponitrite was allowed to settle out in the dark for 15 min before filtering it through a No. 42 Whatman paper in a BÜCHNER funnel. The precipitate well washed with cold distilled water, was transferred to a specimen tube containing 2 ml of saturated sodium chloride and filtered. The filtrate contained sodium hyponitrite. The antipyrene test[1] on aliquots of the filtrate, which is specific for

---

[1] Acetic acid solution of antipyrene gives a green colour with nitrite; to the sample is added an equal volume of 1 % antipyrene in 10 % acetic acid, SNELL and SNELL (1949).

nitrite was negative and hydroxylamine was not detected using a method in which ferric iron is reduced to ferrous iron, the latter determined by o-phen-anthroline. This method is specific for hydroxylamine, detecting as little as 0.1 μg in the presence of hyponitrite. The amount of hyponitrite in solution was determined quantitatively by oxidising the hyponitrite to nitrite with iodine as in the CsÁKY method and the nitrite was determined by the sulphanilamide and N-(1-naphthyl) ethylenediamine reagents. The amount of hyponitrite present was 5/2 of the nitrite formed, calculated for the reaction:

$$5H_2N_2O_2 + O_2 \rightarrow 2HNO_2 + 4H_2O + 4N_2O$$

Sodium hyponitrite can also be prepared by a modification of the method of ADDISON et al. (1952) described by STEVENS (1959). This preparation has been used by MEDINA and NICHOLAS (1957) in their studies with hyponitrite reductase in *Neurospora crassa*. Sodium amalgam was prepared by adding small pieces of sodium metal to warm mercury in a silica crucible until the hot liquid began to solidify. The temperature of the amalgam was then raised until it was fused and a scum of molten sodium hydroxide, carbonate etc. formed on the surface. Most of this adhered to the walls of the crucible when the amalgam was agitated. The remaining bright molten metal was poured into petroleum ether and broken up into pieces in a mortar. The resulting metal amalgam (which was only slowly acted upon by water) was heated with a little distilled water to 80° under nitrogen for a short time and then washed in quick succession with cold distilled water, ethanol, acetone and petroleum ether. It was stored under petroleum ether until required for use. By this procedure a source of carbonate contamination in the sodium hyponitrite was eliminated.

Sodium nitrite (20 g) was dissolved in water (60 ml) and added drop wise *via* a tap-funnel to the purified sodium amalgam in a conical flask. Provision was made for a thermometer to dip into the reacting mixture and the bung was also fitted with entry and exit tubes for nitrogen to displace air from the flask. The temperature of the reacting mixture was kept at 60—80° as, according to MELLOR (1928), a maximum yield of hyponitrite is obtained below 100°.

The addition of sodium nitrite was stopped just before the amalgam had completely liquified, and the reaction was allowed to proceed for a further 5—10 min. The mixture was then cooled and the remaining liquid amalgam was separated. The aqueous portion was filtered rapidly by suction through a No. 541 Whatman paper and poured into a 20-fold excess of ethanol under nitrogen. After allowing this mixture to stand 1 hr, the precipitated sodium hyponitrite was filtered off by suction (No. 541 paper), washed with ethanol and ether and dried *in vacuo* over P_2O_5 in a desiccator for 24 hr. The product appeared to be of a high degree of purity because: (1) silver nitrate produced a bright yellow flocculent precipitate with a fresh solution of the salt, (2) nitrite was not detected when a concentrated solution of the salt was analysed by paper chromatography, (3) the material was shown to be free from carbonate by boiling it with water for a few minutes to destroy hyponitrite followed by the addition of a few drops of fresh baryta solution. No turbidity was observed.

The compound, when dry, could be kept indefinitely in a stoppered container.

IV. Electron Donors and Co-Factors and Inhibitors.

1. Fungi and Higher Plants (Table 25).

a) Nitrite Reductase.

The *Neurospora* enzyme studied by NASON et al. (1954) utilised DPNH or TPNH and the latter was 1.5 times the more efficient. The addition of FAD

Table 25. *Substrate, electron donor and co-factor requirements of enzymes concerned*

Source	Substrate	Pyridine nucleotide	Flavin
Neurospora crassa	NH_2OH K_m 3.8×10^{-3} (toxic conc.)	DPNH K_m 0.7×10^{-4} M TPNH K_m 10^{-4} M DPNH $2 \times$ rate of TPNH	FAD specific K_m 3×10^{-8} (approx.) Opt. 5×10^{-7} M
N. crassa	$Na_2N_2O_2$	DPNH	Flavin
N. crassa	NH_2OH	DPNH	FAD or FMN
N. crassa	NO_2 K_m 1.1×10^{-4}	DPNH > TPNH Opt. 1.2×10^{-3} M	FAD (natural) FMN K_m (both) 5×10^{-5}
Tomato (*Lycopersicon esculentum*) excised, sterile cultured roots	$NaNO_2$ $Na_2N_2O_2$ NH_2OH	DPNH No data for TPNH	FMN
Soybean	NO_2 $K_m = 7 \times 10^{-6}$ Opt. 8×10^{-5} M NH_2OH $K_m = 4 \times 10^{-7}$ Opt. 4×10^{-6} M	DPNH $1.5 \times$ TPNH K_m 10^{-4} M Opt. 2×10^{-4} M DPNH $1.5 \times$ TPNH K_m 10^{-4} M Opt. 2×10^{-4} M	
Vegetable Marrow	NO_2 (to ammonia) K_m 4.8 NH_2OH K_m 4—5	TPNH specific in presence of catalytic amounts of Benzyl viologen	No response with NO_2 FMN or FAD opt. 5×10^{-4} M with NH_2OH

increased activity four times whilst FMN was stated to have no effect. Cyanide, 8-hydroxyquinoline and salicylaldoxime between 10^{-3} and 5×10^{-6} M inhibited activity between 100 and 50 per cent. A metal co-factor was inferred but not identified.

MEDINA and NICHOLAS (1957c) and NICHOLAS, MEDINA and JONES (1960) described in more detail the properties of a system from *Neurospora crassa* which led to nitrite disappearance in the presence of a complex requirement for several cofactors (Table 25). Flavin requirements were equally satisfied by FMN or FAD and the enzyme was inhibited by atabrine, the effect of which was reversed to the extent of 50 per cent by one-hundredth part of FAD, which was shown to be the natural flavin by paper chromatography, electrophoresis and by D-amino acid oxidase activity. DPNH was a more effective electron donor than TPNH. Inhibition by p-chloromercuribenzoate was reversed by reduced glutathione. Iodosobenzoate inhibition was not reversed by glutathione, possibly because disulphide bonds were formed by the reaction with this reagent.

Deficiencies of copper, iron or magnesium during growth depressed enzyme activity. The addition of copper at 1 μg per mg and of iron at 0.2 μg per mg protein in the crude extract produced reactivation *in vitro*. Copper and iron increased in content in direct proportion to specific activity as the enzyme was purified (NICHOLAS, 1959). Magnesium had no reactivating effect *in vitro* and did not accumulate in the purified enzyme. Pyridoxine, pyridoxal or pyridoxal phosphate stimulated the activity of crude but not of the most purified preparations. Pyridoxine also led to a rapid increase in activity if added to deficient mycelia before extraction. Removal of hydroxylamine as an oxime when produced in secondary reactions was considered to be a possible explanation of the effect of pyridoxal etc. added to crude extracts. There was no evidence for any reduction of oximes by *Neurospora* enzymes. There is also a specific phosphate requirement *in nitrite hyponitrite and hydroxylamine metabolism in fungi and higher plants.*

Other cofactors	Metals	pH opt.	Reference
Phosphate or arsenate (NICHOLAS, 1959)	Unidentified but inferred	7.5—8.5	ZUCKER and NASON (1955) Hydroxylamine reductase
Phosphate 3.3×10^{-3} M or arsenate or pyrophosphate (specific phosphate reqn 0.2 M opt.) pyridoxine (see text)	Cu and Fe Mn and Mg during growth only Cu and Fe *in vitro* Mg during growth	7.5 7.6	MEDINA and NICHOLAS (1957a), Hyponitrite reductase MEDINA and NICHOLAS (1957b), NICHOLAS (1959 a, b) NICHOLAS, MEDINA and JONES (1960) Nitrite reductase
	Mn ? ?		VAIDYANATHAN and STREET (1959). Nitrite, hyponitrite and hydroxylamine reductases
(Unidentified thermostable co-factor)	Mn K_m 6×10^{-6} Opt. 4×10^{-5} M Mn K_m 1.4×10^{-5} M Opt. 4×10^{-5} M	7.0 (TRIS) 6.0 ($P_2O_7{}^{4-}$) 7.0 (TRIS) 6.7 (P_2O^{4-})	ROUSSOS and NASON (1960) "Pyridine nucleotide — nitrite and hydroxylamine enzymes"
Unidentified co-factors can be replaced by benzyl viologen	No response to Mn, unidentified metal ?	> 7.0	HAGEMAN et al (1962), CRESSWELL et al (1962). Nitrite and hydroxylamine reductases

(NICHOLAS, 1959 a, b) which cannot be replaced by other anions in contrast to nitrate and hydroxylamine reductases. The enzyme appears to be inhibited about 60 per cent by 1.5×10^{-4} M concentration of hyponitrite but not appreciably by hydroxylamine when assayed in terms of nitrite loss. NICHOLAS (1959a, b) concluded the copper functions in the final stages of electron transfer to nitrite since cuprous copper did not reduce added flavin and the removal of flavin did not depress activity observed when cuprous copper was added to the enzyme in the absence of DPNH. The function of the iron in the system has not been determined. Manganese deficiency during growth also depressed activity, possibly by inhibiting hydroxylamine reductase.

Several metal chelating compounds inhibit the enzyme (Table 26) including cyanide, 8-hydroxyquinoline, diethyldithiocarbamate, 2,2'-diquinolyl, o-phenanthroline, $\alpha\alpha$-dipyridyl, salicylic acid, and EDTA. 2 *n*-heptyl-4-hydroxyquinoline-N-oxide and hydrazine also inhibit powerfully. A dialysable inhibitor also develops in zinc deficient mycelia.

According to VAIDYANATHAN and STREET (1959) manganese at 3.3×10^{-5} M is essential for nitrite reduction to hyponitrite by tomato root crude extracts. DPNH was found to function as an electron donor but could be replaced by succinate, formate or lactate, and DPNH was also rapidly oxidised by DPNH oxidase. In the soybean preparation of NASON et al. (1954) DPNH or TPNH and FAD or FMN were equally active. Unlike the *Neurospora* enzyme that from soybean was dependent on manganese for full activity as shown by ammonia production.

The natural electron donors and cofactors have not been elucidated for the marrow leaf nitrite reductase enzyme studied by HAGEMAN, CRESSWELL and HEWITT (1962). TPNH (10^{-3} M) or TPN and glucose-6-phosphate together with its dehydrogenase will produce full activity in the presence of 10^{-4} M oxidised benzyl viologen which is reduced by a TPNH specific diaphorase enzyme

Table 26. *Stability of, and effects of inhibitors on, enzymes concerned in nitrite, hyponitrite and hydroxylamine metabolism in fungi and higher plants*

Source	Thermal denaturas and stability[1]	Inhibition by —SH reagents or flavin analogues	Other Inhibitions[3]	References and Enzymes
Neurospora crassa	Stable at pH 7.5—9.0 —15° 100 at 4° overnight 100 at 50° 5 min 50 during dialysis in 10^{-3} M cysteine and 3×10^{-3} M pyrophosphate		KCN 10^{-5} M : 100 NaN$_3$ 10^{-3} M : 65 o-Phenanthroline 10^{-7} M : 30 8-Hydroxyquinoline 10^{-3} M : 30 Dipyridyl 10^{-3} M : 10 Sulphite 10^{-3} M : 100	Zucker and Nason (1955) Hydroxylamine reductase
N. crassa	50/16 hours —17° Inactivated by dialysis	Atabrine 2×10^{-3} M : 60 Reversed by FAD 2×10^{-5} M pCMB inhibition reversed by glutathione Iodosobenzoate inhibition irreversible	KCN $(10^{-2}$M) 8-hydroxyquinoline $(10^{-2}$ M) EDTA $(10^{-2}$ M) diethyl dithiocarbamate (10^{-1}) dipyridyl $(10^{-2}$ M) o-phenanthroline $(10^{-3}$ M) 2-Heptyl-4-hydroxyquinoline N-oxide $(10^{-3}$ M) naphthoquinone $(10^{-4}$ M) and urethane $(10^{-1}$M) inhibit between 60 and 100 2:4 Dinitrophenol $(10^{-3}$ M) 2:2'-Diquinolyl $(10^{-4}$ M) inhibit 50. All inhibitors tested by 10 min pre-incubation at concentration (M) given before dilution in assay system	Nicholas (1959 a, b), Nicholas, Medina and Jones (1960) Nitrite reductases
Soybean	100/5 min 50° 60—90/month at —20° Stable 1 week at 4° after fractionation on diethylaminoethyl cellulose	10^{-4} M p CMB NO$_2$ reaction 0 NH$_2$OH reaction 100 10^{-3} M p CMB NO$_2$ reaction 75 NH$_2$OH reaction 100 NH$_2$OH partly reversed by GSH NO$_2$ not reversed Atabrine 3×10^{-4} M : 29 for NO$_2$ 0 for NH$_2$OH[2]	KCN 10^{-4} M : 10 for both KCN 10^{-6} M : 26 for NO$_2$ 60 for NH$_2$OH NaN$_3$ 10^{-5} M : 87 for NO$_2$: 90 for NH$_2$OH 10^{-6} M : 43 for NO$_2$: 43 for NH$_2$OH Na diethyl dithiocarbamate 10^{-3} M :54 for NO$_2$; 100 for NH$_2$OH o-Phenanthroline 10^{-3} M : 33 for NO$_2$ 100 for NH$_2$OH	Roussos and Nason (1959) "pyridine nucleotide-nitrite and hydroxylamine enzymes"

Vegetable marrow *Cucurbita pepo*			
Stable at —18° several weeks (<10/week/—18°) Stable to dialysis in PO$_4$/10⁻³ M cysteine buffer pH 7.5 at 0°	Similar stability to nitrite reductases	p CMB 10⁻³ M and PMA 10⁻³ M : 80	No data	KCN 5×10⁻³ M : 100 5×10⁻⁴ : 85 NaN$_3$ 2 × 10⁻⁴ M:Nil o-Phenanthroline 10⁻³ M: Nil αα Dipyridyl 10⁻³ M: Nil Diethyl dithiocarbamate 10⁻³ M : Nil 8-Hydroxyquinoline 10⁻³ M : Nil Hydrazine and pyruvate 5×10⁻³: Nil all when assayed with benzyl viologen	KCN 2.2 × 10⁻³ M:80 5 × 10⁻⁴ M:70 NaN$_3$ 7.3 × 10⁻⁴ M:Nil Hydrazine 5 × 10³ M:Nil NaNO$_2$ 10⁻⁴ M:60—80	HAGEMAN, CRESSWELL and HEWITT (1962) Nitrite reductase and unpublished data.	HAGEMAN et al (1962) CRESSWELL et al (1962) Hydroxylaminereductase

¹ Values are as per cent inactivation for times and temperatures shown. — ² Values are as per cent inhibition for concentrations shown.

present in the system. Reduced benzyl viologen alone present at 10⁻³ M with a ratio of not less than 25 per cent reduction will cause nitrite reduction to ammonia. TPNH or DPNH alone at 10⁻³ M are unable to drive the system either under aerobic or anaerobic conditions. Addition of FAD or FMN or boiled extract have no effect unless glucose-6-phosphate is also present in addition to TPNH (HEWITT, unpublished). The state of reduction of the dye is critical. At ratios of less than 15 per cent reduced dye nitrite reductase activity is negligible but nitrate reductase will reduce nitrate to nitrite in the presence of 10⁻⁴ M concentration of dye of which approximately 10—20 per cent is reduced. When dye reduction is increased nitrite reductase activity appears at about 15 per cent reduction and increases steadily to about 60 per cent reduction beyond which there is little further increase in activity whilst non-enzymic nitrite loss without ammonia formation is observed. A redox potential effect may be inferred from this relationship, CRESSWELL, HAGEMAN and HEWITT (1962)[1].

The enzyme is inhibited 80 per cent by p-CMB and phenylmercuriacetate at 10⁻³ M and 100 per cent by cyanide at 10⁻³ M but not by azide at 2 ×10⁻⁴ M. Specific activities are not decreased by deficiencies of zinc, or copper, the last actually increasing specific activity by 50 per cent. Deficiencies of iron or manganese have increased and decreased activities, CRESSWELL et al. (1962). Addition of manganese had no effect on the activity *in vitro*, and there was no inhibition by o-phenanthroline, 8-hydroxyquinoline or diethyldithiocarbamate at 10⁻³ M. *Neurospora* nitrite reductase was also relatively insensitive to these compounds, Table 26. It would appear that the metal requirements for the *Neurospora*, tomato root and marrow leaf enzymes differ quantitatively and possibly qualitatively also.

b) Hyponitrite Reductase.

DPNH functions as electron donor according to NICHOLAS (1959a, b) for the *Neurospora* enzyme. There was no change in extinction at

---

[1] LOSADA, M., A. PANEQUE, J. M. RAMIREZ and F. F. DEL CAMPO: Biochem. biophys. Res. Commun. 10, 298 (1963) have now shown that ferredoxin prepared from spinach will function as the direct electron donor in the presence of illuminated chloroplast fragments and G. BEST and E. J. HEWITT (unpublished) have obtained a similar result for vegetable marrow and have also used TPNH and ferredoxin in place of benzyl viologen.

$245\,\mathrm{m}\mu$ unless DPNH was included and the reaction could be blocked at the stage of hydroxylamine formation in media due to oxime formation, especially when deficient in manganese (NICHOLAS, 1959a).

Either FAD or FMN fulfilled the flavin requirements. Inhibition by atabrine was reversed by FAD. Evidence for the role of sulphydryl groups was obtained from inhibition by p-CMB which was reversible by glutathione but hyponitrite reductase was less sensitive than either nitrite or hydroxylamine reductases to this reagent. Deficiencies of iron, copper or manganese reduced activity after extraction.

VAIDYANATHAN and STREET (1959) found that hyponitrite reductase activity was extremely low in tomato root crude extracts and no certain data regarding electron donors are available. As with the *Neurospora* enzyme systems hyponitrite reductase appeared to be the limiting factor in nitrite conversion to ammonia.

In the marrow leaf system studied by HAGEMAN, CRESSWELL and HEWITT (1962), however, hyponitrite did not accumulate or limit the rate of reduction of nitrite to ammonia when reduced benzyl viologen was used as an electron donor or as a cofactor in the presence of TPNH. On the other hand the activity of marrow root preparations was found to be relatively low in terms of ammonia production from nitrite.

c) Hydroxylamine Reductase.

The enzyme described by ZUCKER and NASON (1954, 1955) from *Neurospora* (Table 25) utilised DPNH in preference to TPNH which gave half the activity of DPNH. There was stoichiometric production of 1 mole ammonia for the oxidation of 1 mole of DPNH and for the loss of mole of hydroxylamine. The enzyme was thought to be FAD specific.

The enzyme was especially sensitive to cyanide which caused 50 per cent inhibition at 5×10^{-6} M. A metal component was inferred but not identified, and attempts to reverse the inhibition by cyanide with several metals were unsuccessful.

MEDINA and NICHOLAS (1957c) and NICHOLAS (1959a, b) found that DPNH was the effective electron donor for the *Neurospora* enzyme, but obtained comparable responses to either FAD or FMN. Although manganese and magnesium deficiencies caused decreases in the activity of extracts, there was no response to either metal *in vitro*. It is possible that metallo-enzymes are concerned as recent reviews (HEWITT, 1957, 1958, 1959) have indicated that metallo-enzymes are not formed during growth of the metal-deficient organism and are not reactivated by the metal *in vitro* after extraction.

VAIDYANATHAN and STREET (1959) found that succinate, formate or lactate functioned as electron donors in the presence of DPN for the relatively active hydroxylamine reductase in tomato root crude extracts. Nicotinamide was added to inhibit DPN-ase activity. HAGEMAN, CRESSWELL and HEWITT (1962) used reduced benzyl viologen as the primary electron donor for the marrow leaf system as described for nitrite reductase. Dye reduction appeared to be less critical between 25 and 60 per cent than for nitrite reductase. TPNH alone was inactive but there was a response to added flavins at 10^{-4} M in the presence of TPN and glucose-6-phosphate. Cyanide inhibited 70 per cent at 5×10^{-4} M but azide at 7×10^{-4} M did not inhibit. Nitrite inhibits 60—80 per cent at 10^{-4} M, but hydrazine did not HEWITT (unpublished) CRESSWELL et al. (1962)[1].

2. Bacteria.

a) Nitrite Reductase.

Nitrite reductase from *Bacillus pumilis* was shown by TANIGUCHI, MITSUI, TOYODA, YAMADA and EGAMI (1953) to be utilized by reduced methylene blue

---

[1] LOSADA et al. (1963) (p. 141) found that ferredoxin will function as electron donor in photochemical hydroxylamine reduction by spinach chloroplasts.

Table 27. *Substrate, electron donor and cofactor requirements of nitrite reductase systems from bacteria*

Source	Substrate NO₂	Reduced pyridine nucleotide or other electron donors	Flavin	Metal and other cofactors	pH optimum	Reference
Pseudomonas stutzeri	*Optimum* 20 μM NaNO₂ Total vol 3.0 ml	DPNH regenerated with glucose phosphate dehydrogenase from *Leuconostoc mesenteroides.* TPNH regenerated with isocitric dehydrogenase from pig heart. Because of highly active DPNH and TPNH oxidase K_m values could not be determined	FAD or FMN	Cu and Fe	phosphate 6.8	CHUNG and NAJJAR (1956 a, b)
Halotolerant *Bacillus* (strain 203)	no record	MbH₂ 1 μM 100, FADH₂ 1 μM 90		metallo-protein		TANIGUCHI, SATO and EGAMI (1956)
Azotobacter agile (No. 9104 Amer. Type culture)	K_m 6.3 × 10⁻⁵ M	DPNH regenerated with glucose-6-phosphate dehydrogenase TPNH regenerated with isocitric dehydrogenase	FAD; FAD 4 × 10⁻⁶ M	Metal component not identified	TRIS or pyrophosphate or phosphate } 7.1	SPENCER, TAKAHASHI and NASON (1957)
Micrococcus (strain 203)	K_m 5.6 × 10⁻⁵ M	MbH₂: 100, Tol BH₂: 40, FADH₂: 35, Phen S.H₂: (phenazine sulph) 40, Phen MSH: phenazine methosulphate) 80	FAD required in presence of DPNH or glucose	cytochrome b₄	phosphate buffer + NaCl } 6.8	ASANO (1959)
Pseudomonas aeruginosa (NCIB 8704)	K_m 3.1 × 10⁻⁵ M	PyH₂ K_m 7 × 10⁻⁸ M, MbH₂ K_m 3 × 10⁻⁷ M, FMNH₂ K_m 8 × 10⁻⁸ M, FADH₂ K_m 8 × 10⁻⁸ M	When PyH₂ donor, FAD, 10⁻⁷M increased enzyme activity by 100% and FMN, 10⁻⁸M, by 30%	Cu and Fe	phosphate 7.2	WALKER and NICHOLAS (1960, 1961b)

either alone or with formate as electron donors. The enzyme, from a *Micrococcus* strain utilised DPNH as a donor but not reduced FAD (Taniguchi, Sato and Egami, 1956). When reduced methylene blue was used, nitrite reduced by the enzyme to hydroxylamine was less than was expected from the amount of nitrite reduced but no gas was formed in the reaction mixture. The enzyme apparently contained a metal component but it was not identified. Nitrite reductase purified 2 to 4 fold from *Azotobacter agile* utilised either DPNH or TPNH and required FAD for maximal activity (Spencer, Takahashi and Nason, 1957). Extracts of *Desulfovibrio desulfuricans* reduced nitrite to ammonia with either hydrogen gas or pyruvate as the hydrogen donor (Senez, Pichinoty and Konovalchikoff-Mazoyer, 1956; Senez and Pichinoty 1958a, b). The reduced cytochrome c_3 from this bacterium reduced nitrite non-enzymatically.

The nitrite reductases from denitrifying organisms have been studied in some detail (Table 27). Thus the enzyme from *Pseudomonas aeruginosa* utilized reduced methylene blue as donor (Yamagata, 1939). Chung and Najjar (1956a, b) showed that in cell-free extracts of *Pseudomonas stutzeri* and *B. subtilis* the continuous regeneration of either DPNH or TPNH by its dehydrogenase was necessary before nitrite reductase could be demonstrated and even then the enzyme activity was very low. They found that FAD stimulated enzyme activity. They claimed that Fe and Cu were required for its action since the enzyme dialysed against salicylaldoxime and then distilled water, was reactivated by adding either Cu or Fe (5×10^{-4} M) whereas other metals were either without effect or inhibitory. Walker and Nicholas (1960, 1961a) purified a nitrite reductase 600-fold from *Ps. aeruginosa* and showed that it reduced nitrite to nitric oxide when either reduced flavins (riboflavin H_2, $FMNH_2$, $FADH_2$) reduced pyocyanine or reduced methylene blue were electron donors (Table 27) but DPNH, TPNH or reduced cytochrome c were relatively ineffective donors. The enzyme contained 1.5 mμmoles FAD/mg protein and FAD was required for maximum enzyme activity when reduced pyocyanine was the donor but not when reduced methylene blue was used. In the purified form there was a light absorption maximum between 630 and 635 mμ which was reduced by the electron donors and re-oxidised specifically by nitrite. Iron and copper concentrated in purified enzyme fractions.

Taniguchi et al. (1958) and Asano (1959) showed that reduced methylene blue was an effective donor for nitrite reductase from a halotolerant *Micrococcus*. Extracts contained cytochrome b_4 which was reduced anaerobically with ascorbic acid and re-oxidised by nitrite. The enzyme was separated into a soluble factor (supernatant solution after centrifuging at 100,000 g) and a particulate fraction collected between 10,000 and 20,000 g (Asano, 1959). There was little enzyme activity when either reduced methylene blue or reduced phenazine methosulphate were donors unless the soluble and particulate fractions were combined.

b) Nitric Oxide Reductase.

Chung and Najjar (1956a, b) showed that nitric oxide reductase from *Ps. stutzeri* utilized DPNH or TPNH, when regenerated by dehydrogenases required because of active oxidases in their crude extracts, and that either FMN or FAD stimulated activity. They claimed that both Cu and Fe were required for enzyme activity. The dialysis techniques used by these workers to establish metal requirements are open to question since Fe, Cu and, to a lesser extent, Zn reactivated the enzyme after it had been dialysed against salicylaldoxime. The metal effects could therefore have resulted from the removal of residual chelate bound to the enzyme since the reversal of the inhibition was in the order of the affinities of

Table 28. *Substrate, electron donor and cofactor requirements of nitric oxide reductase and hydroxylamine reductase systems from bacteria*

Source	Substrate	Reduced pyridine nucleotide or other electron donors	Flavin	Other cofactors including metals	pH optimum	Reference
Pseudomonas stutzeri	NO	DPNH or TPNH generated continuously with glucose-6-phosphate dehydrogenase from *Leuconostoc mesenteroides*. High DPNH and TPNH oxidase activity prevented any K_m determinations for the electron donors.	FAD or FMN	Cu, Fe	0.08 M TRIS and 0.1 M phosphate 8.0	CHUNG and NAJJAR (1956 a, b) NO reductase
Pseudomonas aeruginosa (NCIB 8704)	NO	DPNH generated with alcohol dehydrogenase and ethanol. TPNH generated with isocitric dehydrogenase and d-isocitrate.	FAD or FMN K_m 4×10^{-5} M	Fe	0.1 M TRIS or phosphate or carbonate buffers 8.0	FEWSON and NICHOLAS (1960 b, 1961 d) NO reductase
Clostridium welchii (NCIB 273) *Bact. coli* (2 strains)	$NH_2OH \cdot HCl$ 0.01 M *optimum* NH_2OH	Hydrogen	no record	no record	Rate of reduction increased steadily with rising pH throughout the range of phosphate buffers.	WOODS (1938) Hydroxylamine reductase
Azotobacter agile (No. 9104 Amer. Type culture)	$NH_2OH \cdot HCl$ K_m 4.8×10^{-5} M	DPNH generated with glucose-6-phosphate dehydrogenase. TPNH generated with isocitric acid dehydrogenase.	*optimum* FAD 8×10^{-6} M FMN 9×10^{-6} M	Mn K_m 4×10^{-5} M	6.5 to 8.0 in TRIS buffer	SPENCER, TAKAHASHI and NASON (1957) Hydroxylamine reductase
Pseudomonas aeruginosa (NCIB 8704)	$NH_2OH \cdot HCl$ K_m 4×10^{-4} M	PyH_2 K_m 1.2×10^{-8} M MbH_2 K_m 4.4×10^{-8} M *Relative efficiency of Donors (percentage)* PyH_2 100 MbH_2 100 1.4 Naphthoquinone 150 p. Hydroquinone 10 Benzyl viologen K_m 5×10^{-5} M	$FADH_2$ K_m 10^{-8} M 0.9 mμmole/mg protein. When MbH_2 or PyH_2 donor, FAD was required but not FMN.	Mn (Co, 10% as effective as Mn)	7.8 in phosphate buffer	WALKER and NICHOLAS (1960) Hydroxylamine reductase
Desulfovibrio desulfuricans	K_m NH_2OH HCl 10^{-3} M		no record	cytochrome c_3	6.8—8.5 phosphate buffer	SENEZ and PICHINOTY (1958 a, b) Hydroxylamine reductase

the metals for salicylaldoxime. FEWSON and NICHOLAS (1960a) showed that nitric oxide reductase from *Ps. aeruginosa* was a flavoprotein dependent on Fe but *not* Cu for its activity (Table 28).

c) Hydroxylamine Reductase.

WOODS (1958) showed that hydroxylamine was reduced to ammonia by *Clostridium welchii* when hydrogen gas was the donor and TANIGUCHI, SATO and EGAMI (1956) demonstrated that reduced methylene blue was an effective donor for hydroxylamine reduction in extracts of a halotolerant bacteria. They later purified the enzyme 200-fold and showed that dialysis against demineralized water resulted in 70—80% loss of activity but the addition of Mn^{++} almost completely reactivated the enzyme. TANIGUCHI et al. found that the purified enzyme contained a cytochrome of the c type (λ max. oxidised 405 and 635 mμ, reduced 420, 521, 554 mμ). The purified enzyme contained 0.029% Fe and 0.03%Mn. They found that DPNH and succinate were suitable donors when the enzyme was assayed anaerobically and that FAD was required for activity when the former was the donor. SPENCER, TAKAHASHI and NASON (1957) used TPNH continuously regenerated with glucose-6-phosphate dehydrogenase, because a strong oxidase was present in their extracts for hydroxylamine reductase purified 2—3 fold from *Azotobacter agile*.

WALKER and NICHOLAS (1960, 1961b) purified the enzyme (50-fold) from *Ps. aeruginosa* and showed either reduced pyocyanine, reduced methylene blue or $FADH_2$ were effective donors at low concentrations (5×10^{-8} M) so that it was necessary to regenerate catalytic amounts of donors continuously with hydrogen gas and palladised asbestos in special THUNBERG tubes. They observed that about 50 mμmoles NH_2OH were reduced by Pd-H_2 in the absence of other donors and about 100 mμmoles from a total of 800 mμmoles present was reduced non-enzymatically by $FADH_2$. The MICHAELIS-MENTEN constants (Table 28) for the electron donor enzyme complexes were very low: $FADH_2$, K_m 10^{-8}; PyH_2 K_m, 1.2×10^{-8}; MbH_2, K_m 4.4×10^{-8}. The purified enzyme contained FAD which could be partly removed by ammonium sulphate precipitation and the enzyme was reactivated by FAD but not FMN. Manganese (2.5×10^{-7} M) reactivated the enzyme after it had been dialysed against 10^{-3} M sodium diethyldithiocarbamate and then exhaustively against phosphate buffer.

Thus hydroxylamine reductases so far isolated from bacteria are flavoproteins that are dependent on Mn^{++} for maximum activity and resemble the *Neurospora* enzymes. In some bacteria although DPNH or TPNH function as electron donors in crude extracts for reduction of hydroxylamine this probably results from the enzymatic reduction of endogenous flavin by active DPNH or TPNH dependent reductases which then reduce hydroxylamine. When reduced forms of donors such as either pyocyanine or methylene blue were used, flavin stimulated the enzyme (WALKER and NICHOLAS, 1960, 1961b).

V. Properties and Mechanisms.

1. Fungi and Higher Plants.

Little is known at present with regard to the possible mechanism of action of enzymes from plants and fungi catalysing ammonia formation from nitrite and less oxidised compounds. Thus nitrite reductase in *Neurospora* is not known to yield hyponitrite (NICHOLAS, 1959b) although hyponitrite reductase was identified by MEDINA and NICHOLAS (1957a), whereas according to VAIDYANTHAN and

STREET (1959) nitrite reductase of excised tomato roots causes accumulation of a compound resembling hyponitrite which is only slowly metabolised. This compound is reported to inhibit nitrite reductase activity in both systems. Whereas hyponitrite production in the tomato root enzyme is highly dependent on manganese, the *Neurospora* enzyme requires copper and iron whilst that obtained by HAGEMAN et al. (1962) from marrow leaves was not affected by manganese supply *in vitro* or by copper deficiency (CRESSWELL et al. 1962).

MEDINA and NICHOLAS (1957 c) concluded that hydroxylamine reductase was the most active and hyponitrite reductase the least active of the three *Neurospora* enzymes under comparison. VAIDYANTHAN and STREET (1959) reached a similar conclusion for the excised tomato root enzymes. On the other hand HAGEMAN, CRESSWELL and HEWITT (1962) found quite clearly that hydroxylamine reductase was less active than nitrite reductase from marrow leaf. They concluded that free hydroxylamine is unlikely to be a natural intermediate in nitrite reduction to ammonia, but the significance of bound forms of hydroxylamine cannot be assessed at the present time. Unlike the *Neurospora* system (ZUCKER and NASON, 1955), no evidence for a manganese requirement has been observed with the marrow leaf enzyme by HAGEMAN et al. (1962). FEWSON and NICHOLAS (1960 b) found that homogenates of plant and fungal tissues were able to catalyse the uptake of nitric oxide under anaerobic conditions in the presence of DPNH, and WALKER (unpublished) found that nitrite is reduced to nitric oxide by the *Neurospora* nitrite reductase preparations. These properties were adaptive to the presence of nitrate during growth. FEWSON and NICHOLAS (1961 c) considered that a compound resembling nitroxyl (NOH) is the natural intermediate in nitrite reduction and is in equilibrium with hyponitrite as a side reaction. Their scheme does not however at present resolve the problem raised by the work of HAGEMAN et al. (1962) regarding the more rapid metabolism of nitrite than of hydroxylamine to ammonia by the marrow leaf enzyme. It is of course feasible that a bound form of hydroxylamine could be metabolized at a faster rate than an external supply of the compound. Nitrite also inhibits hydroxylamine reduction; CRESSWELL et al. (1962).

An alternative scheme proposed for utilising hydroxylamine is by chemical reaction with α-keto acids and subsequent reduction to the corresponding amino acids (VIRTANEN and SARIS, 1955, 1956). Thus pyruvic, α-ketoglutaric, oxaloacetic and glyoxalic oximes have been found in *Azotobacter* and in *Torulopsis* (SARIS and VIRTANEN, 1957).

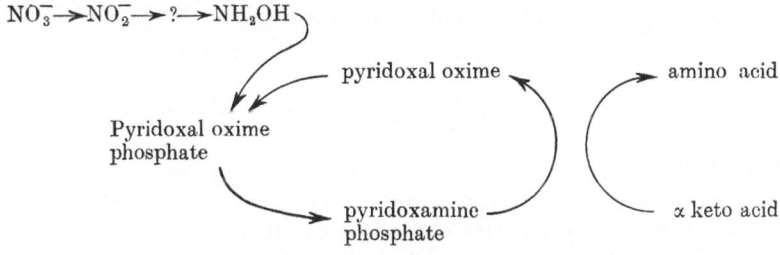

Scheme after SILVER and MCELROY (1954).

SILVER and MCELROY (1954) as a result of their work with mutants of *Neurospora crassa* proposed that an oxime of pyridoxine was involved. They found that pyridoxine-requiring mutants accumulated nitrite when the vitamin was deficient, suggesting that the vitamin is required for nitrite reduction. They consider that pyridoxine may act by condensing with hydroxylamine to form an oxime which could then be reduced to an amine with subsequent formation

10*

of amino acids by transamination. They were unable to demonstrate, however, that pyridoxal oxime phosphate was utilized by *N. crassa*.

Nicholas (1957) found that another pyridoxine requiring mutant, however, had no increased requirement for pyridoxine when grown on nitrate instead of ammonia as sole N source. He further showed that ^{14}C labelled pyruvic oxime, α-ketoglutaric oxime and pyridoxal oxime were not reduced to amino acids in extracts of *N. crassa* since all the ^{14}C was recovered in the corresponding α-keto acids, apparently after release of hydroxylamine for subsequent reduction to ammonia. It was suggested, therefore, that oximes are important only in detoxication mechanisms for hydroxylamine.

2. Bacteria.

a) Nitrite Reductase.

Since the time Yamagata (1939) first reported nitrite reductase in cell-free extracts of *Ps. aeruginosa* this enzyme has been characterised in a range of bacteria (Taniguchi et al., 1952, 1956; Chung and Najjar, 1954, 1956 a, b; Senez et al., 1956; Pichinoty and Senez, 1958; Spencer et al. 1957; Asano, 1959 and Walker and Nicholas, 1960, 1961 a).

Nitrite reductases have been isolated from bacteria assimilating nitrate to ammonia. Thus the enzyme was demonstrated in cell-free extracts of a halo-tolerant *Bacillus pumilis* when reduced methylene blue or formate was the electron-donor (Taniguchi et al.). The nitrite reductase from *Micrococcus* (strain 203) (Asano, 1959) utilised DPNH or $FADH_2$ as an electron donor. When MbH_2 was the donor, nitrite was reduced to ammonia but the yield was lower than expected from the amount of nitrite reduced. This could mean that an intermediate product had accumulated in the reaction mixture although no gas was detected. The compound was unlikely to be hyponitrite since crude extracts capable of reducing nitrite to ammonia did not reduce hyponitrite. The enzyme was inhibited by cyanide, azide, hydroxylamine (10^{-4} M) and by carbon monoxide (Table 29). Metal chelating agents such as $\alpha\alpha'$-dipyridyl, o-phenanthroline and sodium diethyldithiocarbamate also reduced enzyme activity. By adding Mg^{++} and Mn^{++} to the culture medium the bacterium changed from assimilating nitrate to the process of denitrification (Yamata and Virtanen, 1956). The following scheme for electron transfer has been proposed for nitrite reductase from the *Micrococcus*.

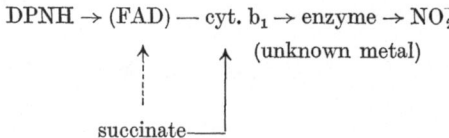

$$DPNH \rightarrow (FAD) \text{—} cyt.\ b_1 \rightarrow enzyme \rightarrow NO_2^-$$

When DPNH was the donor the flavin was obligatory whereas succinate reduced cytochrome b_1 without mediation of flavin. They claim that cytochrome b_1 is involved in electron transfer but this is based solely on spectral changes on addition of nitrite to the enzyme. The terminal nitrite reductase is thought to contain a metal but this has not been identified (Taniguchi et al., 1958).

An assimilatory TPNH- or DPNH-dependent nitrite reductase, stimulated by flavin, was prepared from *Azotobacter agile*. This enzyme, however, has not been purified sufficiently (2 fold only) for a mechanism of action to be proposed for it (Spencer et al., 1957). A novel mechanism has been suggested for nitrite reductase from *Desulfovibrio desulfuricans* which reduced nitrite to ammonia with either hydrogen gas or pyruvate as the hydrogen donor (Senez, Pichinoty and KONAVALCHIKOFF-MAZOYER, 1956; PICHINOTY and SENEZ, 1958). It is claimed that the reduced cytochrome c_3 involved in the electron transfer sequence, reduced nitrite non-enzymatically.

The denitrifying nitrite reductase enzymes have been studied by numerous workers. Thus CHUNG and NAJJAR (1956a, b) proposed that nitrite reductase from *Ps. stutzeri* was a flavoprotein dependent on Cu and Fe for its activity. They suggested the following electron flow scheme although their evidence for it is meagre:

$$\begin{matrix} \text{TPNH} & & \text{FAD} & & \text{Cu}^{++} \\ \text{or} & + \text{H}^+ \to & \text{or} & \to & \text{and} & \to \text{cytochrome} \to \text{NO}_2^- \\ \text{DPNH} & & \text{FMN} & & \text{Fe}^{+++} \end{matrix}$$

They suggested that the metals might function by forming a metallo-protein complex so as to co-ordinate with the substrate.

TANIGUCHI et al. (1958) and ASANO (1959) have suggested the following mechanism of action for a denitrifying *Micrococcus* (strain 203):

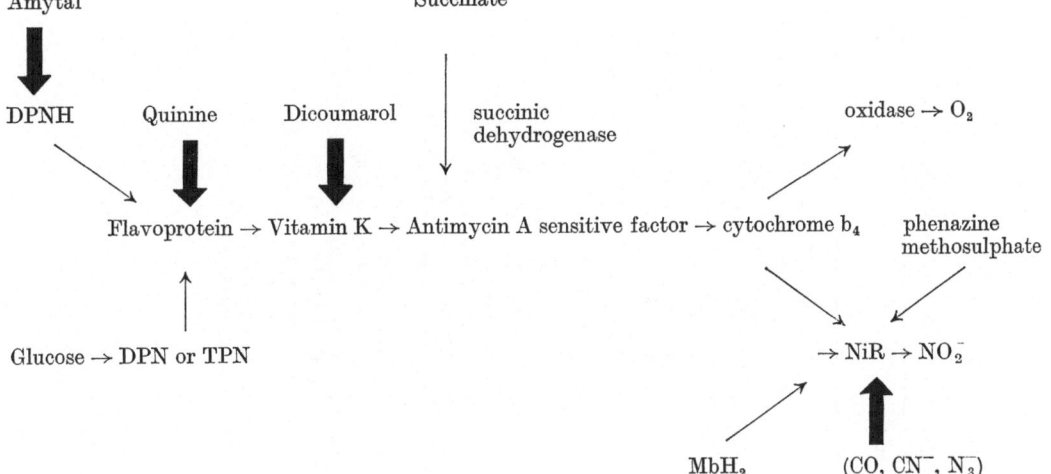

Scheme for nitrite reductase action in *Micrococcus*.
Heavy arrows indicate inhibitions.

The inhibition by amytal suggests the participation of a pyridine nucleotide in the glucose-NiR system. The mode of action of quinine inhibition was not clarified. Although it is known to inhibit flavin function the reversal of this effect by FAD was not tried. The dicoumarol effect was taken as evidence for a vitamin K requirement. None of these compounds, however, affected the succinate-NiR system. The cytochrome b_4 was located in the particles. Antimycin A at relatively high concentrations (0.5 to 1 μg/mg dry weight) inhibited the electron transport system. Nitrite reductase activity was reduced by CO, CN$^-$, N$_3^-$ and by metal complexing agents when reduced phenazine methosulphate was the donor. The enzyme was resolved into a soluble and a particulate fraction; both were required for enzyme activity. Interesting features of several bacterial enzyme systems are the differing effects of inhibitors according to the nature of the electron donors (Table 29).

A quite different mechanism for reduction of nitrite to nitrogen gas has been suggested in a soil bacterium (IWASAKI and MORI, 1958) in which nitrite is reduced

Table 29. *Stability of and effect of inhibitors on the nitrite reductase systems from bacteria*

Source	Thermal denaturation and stability	Inhibition by —SH reagents[1] or flavin analogues	Other Inhibitors[1]	References
Pseudomonas stutzeri	No loss of activity after dialysis for 5 to 18 hr against distilled water	no record	KCN 5×10^{-2}M: 90; 10^{-2}: 40; 10^{-3}: 0 Na diethyldithiocarbamate 1.10^{-2}M: 80 o-Phenanthroline 2×10^{-3}M: 0 8-Hydroxyquinoline 10^{-3}M: 0 Na-versenate 10^{-3}M: 0 Salicylaldoxime 5×10^{-2}M: 80 Potassium ethylxanthate 5×10^{-2}M: 75	CHUNG and NAJJAR (1956a, b)
Halotolerant Bacillus (strain 203)	no record	no record	MbH$_2$ as donor KCN 10^{-2}M: 100; 10^{-3}: 35 NaN$_3$ 10^{-2}M: 100; 10^{-3}: 90; 10^{-4}: 15 NH$_2$OH 10^{-4} M 100 CO (dark) 1 atm: 80 CO (light) 1 atm: 80 $\alpha\alpha'$-Dipyridyl 4×10^{-3} M: 15 o-Phenanthroline 2×10^{-3} M: 30 8-Hydroxyquinoline 10^{-3} M: 0 Alkylthiourea 10^{-3} M: 30 Salicylaldoxime 10^{-3} M: 25 Sodium diethyldithiocarbamate 2×10^{-4} M: 60 Dithizone 5×10^{-4} M: 35 Na-EDTA 2×10^{-3} M: 30	TANIGUCHI, SATO and EGAMI (1956)
Azotobacter agile (No. 9104 Amer. Type culture)	Purified enzyme stable for several weeks at —15°C in phosphate (pH 7.5) containing 10^{-3} M glutathione. Crude extracts unstable. Loss of activity on dialysis for 6 hr against 0.1 M K$_2$HPO$_4$ and 10^{-3}M glutathione. 100/5 min/50°	10^{-3} M *p*CMB inhibited 100% 50% reversal with 10^{-3}M glutathione	KCN 10^{-5} M: 100; 10^{-6} M: 70 Na-EDTA 10^{-3} M: 0 8-Hydroxyquinoline 10^{-4} M: 29 Thiourea 10^{-3} M: 7 Hydrazine 10^{-3} M: 0 Alkylthiourea 10^{-4} M: 0	SPENCER, TAKAHASHI and NASON (1957)

Microccocus (strain 203) — Asano (1959)

no record

Donor: MbH₂ PMSH (phenazine methosulphate)
4×10^{-3}M pCMB 20 10

		Donor		
		MbH₂	PMSH	Formate-MbH₂
KCN	10^{-2}M	100	100	65
	10^{-3}M	35	80	
NaN₃	10^{-2}M	100		
	10^{-3}M	90	15	45
	10^{-4}M	15		
o-Phenanthroline	2×10^{-3}M	30	35	45
α α'Dipyridyl	2×10^{-3}M	20	10	35
Sodium diethyldithio-carbamate	2×10^{-3}M	60	15	15
Salicylaldoxime	10^{-3}M	25	30	
Alkylthiourea	10^{-3}M	30	0	
Na-EDTA	2×10^{-3}M	30		
Dithizone	5×10^{-3}M	35		
NH₂OH	10^{-3}M	100	5	50
	10^{-4}M	70	0	
CO (light)	1 atm	70	70	60
dark	1 atm	70	70	60
NO	0.1 atm	100	45	

Pseudomonas aeruginosa (NCIB 8704) — Walker and Nicholas (1960)

0/30 min/30°
50/30 min/37°
100/20 min/41°

Donor: PyH₂ FMNH₂ MbH₂
10^{-4}M pCMB 59 62 37
10^{-4} pCMB + 5×10^{-3}M glutathione 0 0 0
Glutathione completely reversed inhibition

	PyH₂	FMNH₂	MbH₂
KCN 2.5×10^{-4}M	27	16	0
10^{-3}M	90	84	9
o-Phenanthroline 5×10^{-3}M	49	43	0
α-α'Dipyridyl 5×10^{-3}M	35	0	0
Neo-cuproine (satd.)	34	46	0
2,4 Dinitrophenol	44	41	52

[1] Values are per cent inhibition for concentration or treatment shown.

to hydroxylamine, the latter then reacts with another molecule of nitrite to produce either N_2O or N_2 gas as follows:

$$HNO_2 \xrightarrow{\ +\ 4e^-\ } NH_2OH \xrightarrow{\ +HNO_2\ } N_2O$$

$$N_2 + HNO_2 + 2e^-$$

Nicholas (unpublished) has not substantiated this finding in *Ps. aeruginosa*; indeed the high concentration of hydroxylamine (10^{-2} M) used by the Japanese workers inhibited nitrite reduction.

Walker and Nicholas (1960, 1961a) purified a nitrite reductase, more than 600-fold, from extracts of *Ps. aeruginosa*. The enzyme reduced nitrite to nitric oxide when either a reduced flavin (riboflavin-H_2, $FMNH_2$, $FADH_2$), reduced pyocyanine, reduced methylene blue, or reduced 1,4-naphthoquinone was the electron donor. DPNH, TPNH, or reduced cytochrome c were ineffective. FAD was required for maximum activity when reduced pyocyanine was the donor but not when reduced methylene blue was used. The enzyme has a cytochrome c type spectrum and an additional light absorption maximum between 630 and 635 mμ which is probably associated with copper. The latter was reduced by electron donors that were effective for the purified enzyme and re-oxidised specifically by nitrite. Metal deficiency studies and the concentration of metals in the purified enzyme showed that Fe and Cu are required for nitrite reduction. The enzyme contains —SH groups and either phosphate or sulphate was essential for maximum activity. The mechanism of action proposed was as follows:

$$Cu^{1+} \rightleftarrows Cu^{2+}$$
$$PyH_2 \rightarrow FAD \rightarrow Cu\text{-protein} \rightarrow NO_2^-$$

Flavins are known to act as electron carriers during nitrite reduction in many organisms when reduced pyridine nucleotides are electron donors. In the present work with nitrite reductase from *P. aeruginosa*, although the purified enzyme contained FAD, which functioned when reduced pyocyanine was the electron donor, neither DPNH or TPNH were effective donors. Since flavin was reduced enzymically by DPNH and TPNH in crude extracts and reduced flavin acts as an effective donor for the purified enzyme it seems likely that flavin is a natural carrier for nitrite reduction in the bacteria. In this connection it is of interest that a-DPNH or a-TPNH-dependent flavin reductase was precipitated in an earlier ammonium sulphate fraction and the DPNH-flavin system was an effective donor for the purified nitrite reductase.

A copper requirement for the enzyme was suggested by the reduced activity of nitrite reductase in extracts of cells deficient in copper and by accumulation of the metal in purified fractions. The purified enzyme had a λ-max. 630—635 mμ which was diminished on reduction with either PyH_2, flavin H_2 or MbH_2 and restored specifically by nitrite. This maximum was removed after dialysis against cyanide and reconstituted specifically with copper (2 mμmoles/mg protein). Enzyme activity, which was lost during cyanide dialysis was not restored by returning the metal. It is significant, however, that the amount of copper (2 mμmoles/mg protein) required to restore the λ-max. 630 mμ after cyanide dialysis, was similar to the copper content of the purified enzyme viz. 1—1.7 mμmoles/mg protein. A blue copper-containing protein has been crystallized from

P. aeruginosa, having a similar maximum to the λ-max. 630—635 mμ of nitrite reductase, which was reduced with lactate and yeast lactic dehydrogenase (Hono, 1957). It was suggested that it functioned as an electron carrier between cytochromes of the c type and cytochrome oxidase. It is possible, however, that this blue copper protein is similar to the nitrite reductase described by Walker and Nicholas (1960, 1961a).

b) Nitric Oxide Reductase.

Chung and Najjar (1956a, b) suggested that this enzyme from *Ps. stutzeri* is a flavoprotein requiring either DPNH or TPNH and stimulated by either FAD or FMN. The fact that the enzyme was inhibited by cyanide, salicylaldoxime, sodium diethyldithiocarbamate (Table 30) suggested a metal constituent. After dialysis against salicylaldoxime and then distilled water, the enzyme was reactivated by Cu and Fe. Although they had no direct experimental evidence they suggested a scheme of electron transfer similarto that proposed by them for nitrite reductase:

$$\begin{matrix} TPNH & & FAD & & Cu_2 \\ or & \rightarrow & or & \rightarrow & and & \rightarrow cytochrome \rightarrow NO \rightarrow N_2 \\ DPNH & & FMN & & Fe_3 \end{matrix}$$

They postulated that nitrite and nitric oxide reductases are similar enzymes. Fewson and Nicholas (1960b) have shown that nitrite and nitric oxide reductases from *Ps. aeruginosa* are separate enzymes. The nitric oxide reductase was shown to be an iron-dependent flavoprotein and was not dependent on copper for its activity. The dialysis techniques used by Chung and Najjar (1956a, b) to establish metal requirements are open to question since Fe, Cu and, to a smaller extent, Zn reactivated the enzyme after it had been subjected to prolonged dialysis against salicylaldoxime. The metal effects could, therefore, have resulted from the removal of the residual chelate bound to the enzyme since the reversal of the inhibition was in the order of the affinities of the metals for salicylaldoxime.

Nitric oxide reductase purified 25-fold from *P. aeruginosa*, as shown in Table 19, is a metallo-flavoprotein. The activity of the purified enzyme, which utilized reduced pyocyanine but not reduced pyridine nucleotides as the hydrogen donor, was maximal at pH 8.0. Sulphydryl groups appear to be required for the enzyme activity since glutathione partially offset the inhibition of *p*-chloromercuribenzoate. A flavin requirement for the enzyme was indicated by the reversal of mepacrine inhibition with flavin mononucleotide (FMN) or flavin adenine dinucleotide (FAD) and confirmed by precipitating the enzyme three times with ammonium sulphate (pH 7.0), when its activity, diminished by about 70 per cent, was reconstituted by FMN or FAD (Michaelis-Menten constant, $K_m = 4 \times 10^{-5}$). At a concentration of 2×10^{-3} M the iron and copper chelating agents, $\alpha\alpha'$-dipyridyl, *o*-phenanthroline, potassium ethyl xanthate and 2:9-dimethyl-1:10-phenanthroline, inhibited the enzyme by more than 40 per cent. Salicylaldoxime, α-benzoin oxime and sodium diethyldithiocarbamate, however, which chelate copper but not iron between pH 7 and 8 (Hallaway, 1959), reduced the enzyme activity by less than 10 per cent, Table 30. There is some evidence for the concentration of iron, but not copper, in partially purified preparations of the enzyme. Extracts of cells deficient in iron contained decreased nitric oxide reductase activity (25 per cent of normal cells) whereas that of cells deficient in copper, molybdenum, zinc, or manganese was unaffected. It seems unlikely, therefore, that copper is required for nitric oxide reductase activity in this organism.

Table 30. *Stability and effects of inhibitors on the nitric oxide reductase and hydroxylamine reductase systems in bacteria* (Values are per cent inhibition for concentration or treatment shown).

Source	Thermal denaturation and stability	Inhibition by —SH reagents and flavin analogues	Other inhibitors	Reference
Pseudomonas stutzeri	no record	no record	Na diethyldithiocarbamate 10^{-2} M: 48; 5×10^{-3} M: 30; Salicylaldoxime 10^{-2} M: 45; 5×10^{-3} M: 35	CHUNG and NAJJAR (1956a, b) NO reductase
Pseudomonas aeruginosa (NCIB 8704)	68/10 min/60°	pCMB 2.10^{-3} M: 40; pCMB 2.10^{-3} M: $+$ 0.5 μmole GSH: } 15	KCN 5×10^{-4} M: 100; Na-EDTA 2×10^{-3} M: 33; Na-diethyldithiocarbamate 2×10^{-3} M: 0; αα'-Dipyridyl 2×10^{-3} M: 59; o-Phenanthroline 2×10^{-3} M: 73; K-ethyl xanthate 2×10^{-3} M: 40; Neocuproine 2×10^{-3} M: 54; Salicylaldoxime 2×10^{-3} M: 10; α-Benzoin oxime saturated 0	FEWSON and NICHOLAS (1960b, 1961d) NO reductase
Clostridium welchii (NCIB 273) Bact. coli (2 strains)	no record	no record	no record	WOODS (1938) Hydroxylamine reductase
Azotobacter agile (No. 9104 Amer. Type Culture)	Purified enzyme stable for several weeks at —15° when stored in 0.1 M phosphate (pH 7.5)+10^{-3}M GSH. Stores at 2° for 1 week without marked loss of activity. Crude extracts unstable.	10^{-3} M pCMB 15% inhibition	KCN, 10^{-2} M: 100; 2×10^{-3}: 13; 10^{-3} M: 7; Na-EDTA 10^{-3} M: 64; 8-hydroxyquinoline 10^{-3} M: 45	SPENCER, TAKAHASHI and NASON (1957) Hydroxylamine reductase

			PyH₂	MbH₂	
Pseudomonas aeruginosa (NCIB8704)	When PyH₂ donor enzyme stable 37° for 60 min 50/30 min/40° 100/5 min/42°	10⁻³ M *p*CMB competitively inhibited the activation of the enzyme by Mn. PyH₂ as donor 10⁻⁴ M *p*CMB: 96% inhibition. MbH₂ as donor 10⁻⁴ M *p*CMB: 91% inhibition. Reversed completely by 10⁻⁴ M glutathione	KCN 2.5×10^{-4} M 18; 1×10^{-3} M 100 Na-EDTA 5×10^{-3} M 20 o-Phenanthroline 5×10^{-3} M 82 α,α-Dipyridyl 5×10^{-3} M 73 Neocuproine satd. 56 Na-diethyldithiocarbamate 1×10^{-3} M 60; 5×10^{-3} M 100 2-n′-Heptyl-4-hydroxyquinoline-N-oxide satd. (HOQNO) 0 Na arsenite 5×10^{-3} M 80 2,4 Dinitrophenol 5×10^{-3} M 81	21 90 — 78 59 0 72 100 0 0 79	WALKER and NICHOLAS (1960c) Hydroxylamine reductase
Desulfovibrio desulfuricans	no record	10⁻² M Iodoacetate: 29 10⁻³ M Hg⁺⁺: 100 10⁻⁴ M Hg⁺⁺: 84 10⁻⁵ M Hg⁺⁺: 12	KCN 10⁻³ M: 73; 10⁻⁴ M: 57; 10⁻⁵ M: 36 Na-arsenite 10⁻² M: 49; 10⁻³ M: 22 NaN₃ 10⁻² M: 79; 10⁻³ M: 24 Hydrazine 10⁻² M: 65; 10⁻³ M: 23 Phenyl hydrazine 10⁻² M: 80; 10⁻³ M: 57; 10⁻⁴ M: 34 Phenyl-urethane 1/3rd satn: 50; 1/30th satn.: 12 Cu⁺⁺ 10⁻³ M: 100; 10⁻⁴ M: 0 Zn⁺⁺ 10⁻³ M: 0 K₂CrO₄ 10⁻⁴ M: 100; 10⁻⁵ M: 12 K₂Cr₂O₇ 10⁻⁴ M: 100; 10⁻⁵ M: 19 KIO₄ 2×5.10^{-3} M: 100; 2.5×10^{-5} M: 15		SENEZ and PICHINOTY (1958a, b) Hydroxylamine reductase

c) Hydroxylamine Reductase.

Hydroxylamine has been suggested to be an intermediate in the reduction of nitrate to ammonia in micro-organisms (VERHOEVEN, 1952; EGAMI, YAMADA and TANIGUCHI, 1952; NASON, ABRAHAM and AVERBACH, 1954) while organo-hydroxylamine compounds occur in yeasts (VIRTANEN and CSÁKY, 1948; VIRTANEN and SARIS, 1955, 1956). WOODS (1938) first demonstrated an enzymatic reduction of hydroxylamine to ammonia in *Clostridium welchii* and similar reductions have been reported in *E. coli* (BACK et al., 1946; GROSSOWITZ and LICHENSTEIN, 1946; and McNALL and ATKINSON, 1957).

Hydroxylamine reduction was first reported in a denitrifying *Micrococcus* by TANIGUCHI et al. (1958). The suggestion by TANIGUCHI et al. (1958) that hydroxylamine reductase has a physiological function in a halotolerant *Micrococcus* other than for reducing hydroxylamine to ammonia is probably incorrect since NICHOLAS (unpublished) has shown *Ps. aeruginosa*, grown in peptone medium simultaneously denitrified and assimilated ¹⁵N labelled nitrate into the cell. About 5% of the nitrate in the medium was incorporated into the bacterium which was equivalent to 14% of the total N and 95% was lost as N₂ gas.

The following electron transport sequence has been suggested by the Japanese school (TANIGUCHI et al., 1958) for hydroxylamine reduction (see scheme p. 156).

The evidence supporting this scheme is as follows: A DPNH-cytochrome b_4 reductase activity was present in crude extracts when 1 M sodium chloride was added. In the presence of hydroxylamine reductase the reduced form of purified cytochrome b_4, chemically prepared with ascorbate, was rapidly reoxidised by adding hydroxylamine. The ultrasonic extracts of the cells contained an active succinic dehydrogenase as well as hydroxylamine reductase and cytochrome b_4. The addition of succinate to these extracts caused the reduction of cytochrome b_4 but the reduced cytochrome was immediately re-oxidised by adding hydroxylamine. Hydroxylamine reductase was inhibited by several chelating agents (Table 30). The purified enzyme lost 80% of its activity when dialyzed against demineralized water. Manganese reactivated the dialysed enzyme. The haem protein nature of the enzyme was suggested by the inhibition by CO and its reversal by light and also by spectral investigations. The purified enzyme showed light absorption maxima when oxidised (λmax. 405 and 635 mμ) and when reduced (λmax. 435, 521 and 554 mμ with a shoulder at 548 mμ). These are, however, distinct from cytochrome b_4. The reduced haemochromogen of the enzyme was identical with that of mammalian cytochrome c. The purified enzyme contained 0.029% Fe and 0.03% Mn. Iron was not removed by dialysis. The Soret peak (420 mμ) showed rapid oxidation by air or by hydroxylamine, but the oxidation was faster in air. Manganese accelerated both reactions. They suggest that the haem-iron that combines with CO is the active centre of the enzyme and they favour the idea that haem iron undergoes a reversible valency change during the reaction. The possibility that the CO inhibition might affect a new type of terminal oxidase other than cytochrome a_3 was considered possible.

SPENCER et al. (1957) found that a hydroxylamine reductase from *Azotobacter agile* utilized DPNH and was flavin- and Mn-dependent. Since their preparations were purified 2 to 4 fold only they were unable to suggest a mechanism for enzyme action. The reductase nature of this system is not fully established.

WALKER and NICHOLAS (1961 c) characterized hydroxylamine reductase from denitrifying cells of *Ps. aeruginosa*. The enzyme purified 50 fold utilized reduced pyocyanine, reduced methylene blue, reduced flavin (riboflavin-H_2, $FMNH_2$ or $FADH_2$) (Table 28). Neither TPNH nor DPNH functioned. The purified enzyme contained FAD (0.9 μm mole/mg protein) and when PyH_2 or MbH_2 was the donor there was a flavin requirement. Metal deficiency studies showed that the enzyme activity was markedly reduced in extracts of felts deficient in manganese. A deficiency of iron also reduced activity by 30%. Although purified pre-

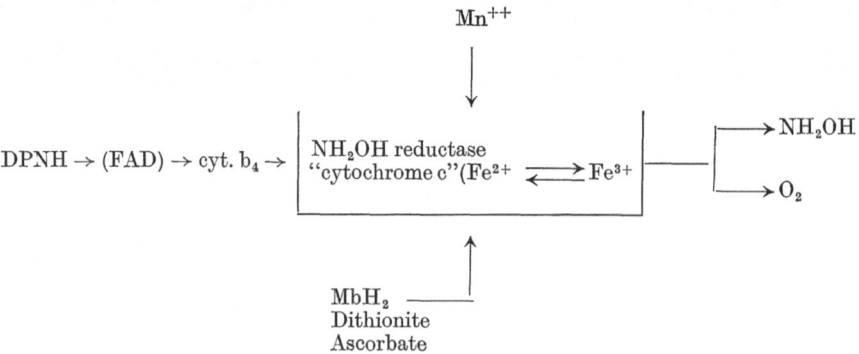

Scheme for Hydroxylamine reductase from a denitrifying *Micrococcus*.

parations contained a cytochrome of the c type there was no evidence that it functioned in electron transfer. The dialysed enzyme was activated three fold by manganese (0.2 μM) after dialysing it first against 1 mM sodium diethyl-dithiocarbamate in 10 μM phosphate (pH 7) and then against phosphate alone for another 18 hr. to remove the chelating agent. Although Co (0.5 μM) sub-stituted for Mn it was only one-tenth as effective. The activation of the dialysed enzyme by Mn was inhibited competitively by p-chloromercuribenzoate, suggesting that —SH groups may be involved in binding the metal to the enzyme. Effects of several inhibitors were similar when methylene blue or pyocyanine were electron donors except for those of neocuproine and arsenite (Table 30). A possible sequence of electron transfer in the purified enzyme was suggested as follows:

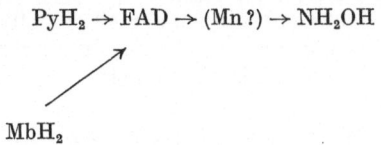

$$PyH_2 \rightarrow FAD \rightarrow (Mn\,?) \rightarrow NH_2OH$$

MbH$_2$

It is not known, however, whether Mn undergoes valency changes during hydroxylamine reduction or whether it simply orientates the substrate, to facilitate reaction with the electron donor and enzyme. The addition of Mn to extracts of bacteria deficient in the micronutrient activated hydroxylamine reductase so that the metal appears to be easily dissociated from the enzyme and does not appear to be required for the formation of the apoenzyme. They suggest that the reduced enzyme activity in extracts of cells deficient in iron does not necessarily imply that the metal is directly required for enzyme action. It is known that enzymes in the nitrate assimilation sequence are induced by their substrates and the iron effect could be caused by a depression in nitrate, nitrite and nitric oxide reductases which are iron-dependent there byresulting in decreased hydroxylamine production. A similar mechanism has been suggested to account for the effect of Mo deficiency in depressing glutamic dehydrogenase in *Neurospora crassa* (NICHO-LAS and MABEY, 1957).

VI. Physiological Factors and Other Features.

1. Fungi and Higher Plants.

a) Nitrite Reductase.

The importance of a supply of pyridoxal, pyridoxal phosphate or pyridoxine during growth of *Neurospora* for nitrite reductase activity has been noted. NICHO-LAS (1959) reported that 50 μg/l pyridoxine was optimal for this requirement. The question of whether pyridoxine yields an oxime with hydroxylamine has also been discussed. The need for magnesium during growth to obtain enzyme activity is difficult to interpret at present as this metal has so many functions. The enzyme is inducible in relation to nitrite or nitrate in the medium.

VAIDYANATHAN and STREET (1959) considered that nitrite reductase which resulted in hyponitrite formation is inducible by nitrate or nitrite during growth of excised tomato roots in sterile cultures.

CRESSWELL, HAGEMAN and HEWITT (unpublished) have found that activity of nitrite reductase obtained from marrow leaves decreases with leaf age in a manner which closely parallels that of nitrate reductase. Activities are highest in young rapidly expanding leaves. Nitrite reduction *in vivo* as measured by disappearance of nitrite infiltrated into intact marrow leaf discs is directly dependent on light, and proceeds only very slowly when the leaves are kept in the dark. Rates of 3000 mμ moles/hour/g fresh wt. occur in illuminated leaf tissues (Cresswell and Hewitt, unpublished work) and are commensurate with the enzymic rates observed *in vitro* (Hageman, Cresswell and Hewitt, 1962).

b) Hydroxylamine Reductase.

Zucker and Nason (1955) concluded that hydroxylamine reductase of *N. crassa* is inducible since activity was lost if the fungus was grown with alanine or with ammonium chloride instead of ammonium nitrate. Maximum activities were produced in mycelia grown with nitrate. The presence of nitrite suppressed activity but ammonia did not have this effect.

Vaidyanathan and Street (1959) concluded that hydroxylamine reductase of excised tomato roots is induced by nitrate or nitrite. Activity was absent from roots grown with ammonia or glutamic acid.

D. Glutamic Dehydrogenase.

There is general agreement that ammonia is probably the final inorganic product of nitrate assimilation in plants and fungi (Nason, 1956; McElroy and Spencer, 1956; Evans, 1956; Nicholas, 1957, 1959) and that it combines with α-keto glutaric acid to form glutamic acid. Ammonia has also been found in culture media during dissimilatory nitrate reduction by spore-forming bacteria (Klaeser, 1914; Verhoeven, 1952) and during dissimilation of nitrate in *Bacillus pumilis* (Taniguchi et al., 1953), in a halotolerant *Micrococcus* (Taniguchi et al., 1958) and in *E. coli* (Bn) (McNall and Atkinson, 1957). Ammonia has been suggested to be the "key intermediate" in nitrogen fixation, i.e. the final inorganic product of fixation before incorporation into organic compounds (Wilson, 1958). This key reaction is controlled by glutamic dehydrogenase. This enzyme catalyses the reversible incorporation of ammonia into glutamic acid:

$$\underset{\alpha \text{ ketoglutarate}}{RCO \cdot COOH} + NH_3 + \underset{TPNH}{DPNH} + H^+ \rightleftarrows \underset{\text{L} + \text{glutamate}}{RCH \cdot NH_2 \cdot COOH} + \underset{TPN^+}{DPN^+} + H_2O$$

It therefore controls the last step in the assimilation of inorganic nitrogen from nitrate and its intermediate reduction products.

The occurrence assay, purification and properties of the L-glutamic dehydrogenase of higher plants are described by Sanwal and Lata on page 291f. of this volume and those of *Neurospora* and other micro-organisms on page 293f.

E. Oxidation and Other Reactions of Ammonia, Hydroxylamine and Nitrite.

I. Fungi and Higher Plants.

There is now good reason to believe that hydroxylamine, nitrite and possibly hyponitrite and ammonia undergo oxidation as well as reduction *in vivo* in plants and fungi. Frear and Burrell (1958) found that ^{15}N-labelled hyponitrite when infiltrated into leaves of soybean plants was converted to nitrate. Raistrick and Stössl (1958) found that when ammonia was supplied to *Penicillium atrovenetum* 60 per cent was recovered as β nitropropionic acid, whilst only 10 per cent of added nitrate appeared in this form. β nitropropionic acid comprised the major nitrogenous constituent of the fungus. Its occurrence in plants has been reported by Carrie (1934), Carter (1943), Carter and McChesney (1949) and Morris,

PAGAN and WARMKE (1954), and in *Aspergillus flavus* by BUSH, TONSTER and BROCKMAN (1951). The enzymic systems involved in these reactions have not been identified.

The oxidation of hydroxylamine is naturally involved in bacterial nitrification where this compound occurs as an intermediate and is described later. A reaction involving the peroxidation of hydroxylamine to mainly unknown products, but including a small proportion of nitrite has now been described by CRESSWELL and HEWITT (1960) for higher plants. This system is described below, but before doing so, reference in some detail to the work of ROUSSOS and NASON (1960) is necessary.

ROUSSOS and NASON (1960) described a non-particulate system obtained from phosphate buffer extracts of soybean leaves which they called "pyridine nucleotide — nitrite and hydroxylamine enzymes," (Tables 13, 21, 25 and 26). This system caused the catalytic oxidation of DPNH on the addition of substrate concentrations of nitrite or hydroxylamine. Hydroxylamine disappeared but no loss of nitrite was observed, and no ammonia formation occurred with either "substrate." Hydroxylamine disappeared in amounts apparently equivalent to the oxidation of DPNH above that corresponding to the endogenous rate.

The first point of importance to note in this work is that no products of reduction of nitrite or hydroxylamine when tested in this system by ROUSSOS and NASON could be identified. An unidentified, heat-stable and alkali-labile cofactor, without which no reaction occurred, was also inferred to be required for the substrate-dependent oxidation of DPNH. No evidence was obtained for the participation of flavins. Cytochrome c, menadione, vitamin K_1, 1,4-benzoquinone, α-lipoic acid, pyridoxal phosphate, ATP, phosphopyruvic acid, and several other compounds failed to replace the unidentified cofactor. The rate of reaction as measured by oxidation of DPNH in the presence of added nitrite or hydroxylamine was increased 5 times with nitrite and 35 times with hydroxylamine when manganese at 10^{-4} M was added. The endogenous rate of DPNH oxidation was stated to be unaffected by manganese and was quite high. The addition of copper (10^{-4} M) also stimulated DPNH oxidation in the presence of hydroxylamine but not in the presence of nitrite and for the former the effects of the two metals were additive. It was stated (without experimental details) that under aerobic conditions no oxygen uptake was observed in the nitrite or hydroxylamine-stimulated oxidations or in the endogenous systems, and that DPNH oxidation in the presence of nitrite or hydroxylamine was still observed under anaerobic conditions. No ammonia, amino acids or identifiable products of these reactions could be detected.

It is relevant to note that although ROUSSOS and NASON (1960) claimed a stoichiometric correspondence between disappearance of DPNH and of hydroxylamine in their soybean system they did not show that hydroxylamine disappearance was dependent on the presence of DPNH. SPENCER et al. (1957) reported that the "hydroxylamine reductase" of *Azotobacter agile* was "stimulated" 25—55 per cent under aerobic as compared with anaerobic conditions and was also stimulated by manganese. The observation was unexplained but it is likely that hydroxylamine oxidation was involved, as ammonia formation was negligible.

In spite of their failure to observe loss of nitrite or to obtain ammonia production from hydroxylamine, ROUSSOS and NASON (1960) were nevertheless of the opinion that reductase mechanisms were involved. FREAR and BURRELL (1955) had already inferred that the disappearance of hydroxylamine in the presence of extracts of soybean leaves which was observed in aerobic systems was due to a hydroxylamine reductase. The system was dependent upon the addition of manganese, but the formation, if any, of ammonia was not ascertained. In the opinion of Cresswell and Hewitt as stated elsewhere by Hageman, Cresswell and Hewitt (1962) the systems described by Roussos and Nason (1960) do not constitute satisfactory evidence of nitrite and hydroxylamine reductase in plants. An alternative explanation representing the views of Cresswell and Hewitt is presented below.

Hydroxylamine "Oxidase" of Higher Plants.

Much of the confusion and uncertainty regarding metabolism of hydroxylamine so far described for systems from higher plants by Roussos and Nason (1960) and Frear and Burrell (1955), has been resolved by the recent work of Cresswell and Hewitt (1960) and unpublished. This has shown that hydroxylamine as estimated by the Frear and Burrell (1955) method is rapidly oxidised in air by protein fractions present in extracts from vegetable marrow leaves or by highly purified horse radish peroxidase in the presence of manganese, hydrogen peroxide and a phenolic co-factor. The evidence is wholly consistent with the conclusion that the activity of marrow leaf extracts or of partially purified preparations isolated by a calcium phosphate gel adsorption and ammonium sulphate precipitation can be explained by the activity of a manganese and phenol-dependent peroxidation mechanism of the type described by Kenten and Mann (1949, 1950, 1952, 1953), Kenten (1955). These workers showed that carboxylic acids, indole acetic acid and other substrates are rapidly oxidised by peroxidase and hydrogen peroxide in the presence of manganese and a mono-phenol cofactor.

In the presence of hydrogen peroxide and partially purified enzyme obtained from marrow leaves by Cresswell and Hewitt (1960) activity was equally great under aerobic and anaerobic conditions. In the absence of peroxide, oxygen was essential for activity which was not dependent on DPNH or TPNH and in fact DPNH partially inhibited the activity, probably by competing for peroxide or mono-phenol or trivalent manganese in the highly complex system. In crude preparations the reaction sometimes proceeded without the addition of mono-phenol or manganese. This was considered to be due to the presence of appreciable concentrations of one or other of these components even after adsorption and elution of the active protein from calcium phosphate gel and to the possibility of endogenous formation of hydrogen peroxide. The role of oxygen in the reaction is not yet elucidated. It is possible that hydrogen peroxide is generated in crude extracts by oxidase systems such as amine oxidase, by oxidation of substrates such as reduced flavins, or by copper-catalysed autoxidation of ascorbic acid. On the other hand the mechanism proposed by Chance (1952) for the manganese-catalysed peroxidation of dihydroxyfumaric acid, namely an aerobic reaction generating hydrogen peroxide and subsequent peroxidation may apply to the oxidation of hydroxylamine described by Cresswell and Hewitt. In unpublished work, Cresswell and Hewitt have used glucose and glucose oxidase to yield peroxide.

The assay conditions used to investigate the system were as follows: Hydroxylamine hydrochloride 2.5×10^{-4} to 10^{-3} M; 2,4-dichlorophenol, p-cresol or resorcinol, 5×10^{-5} to 10^{-4} M; hydrogen peroxide 10^{-4} to 10^{-3} M, or glucose 10^{-2} M and notatin; manganese chloride 10^{-6} to 10^{-3} M; TRIS or phosphate buffers 0.05 M, pH 8.0.

The optimum concentrations of phenol, manganese, or peroxide depend upon the source of enzyme and its purity, and on concentrations of each of the remaining components of this complex system.

The results obtained by Cresswell and Hewitt (1960) probably explain the need for, and nature of, the unidentified cofactor of Roussos and Nason (1960) in terms of a phenolic compound and also explain the importance of manganese. The failure to observe ammonia production or nitrite formation from hydroxylamine by ROUSSOS and NASON is also consistent with the results of CRESSWELL and HEWITT. The apparently stoichiometric relationship between DPNH oxidation and hydroxylamine loss might be the result of their comparable rates of reaction in the same system as independent substrates.

Other points that require further elucidation as being apparently inconsistent with the work reported here are:

(a) that in the soybean system the active protein was reported to be adsorbed on alumina $C\gamma$ gel whilst the marrow peroxidase is not,

(b) that hydroxylamine-, and nitrite-stimulated oxidation of DPNH were reported not to involve oxygen uptake and to occur under anaerobic conditions.

As no experimental details were given in relation to these last statements, their significance is hard to evaluate but their findings have not been confirmed in unpublished work of CRESSWELL and HEWITT. Endogenous hydrogen peroxide formation also appears to be likely from the experience of CRESSWELL and HEWITT.

It has been suggested by HAGEMAN, CRESSWELL and HEWITT (1962) that the hydroxylamine-stimulated oxidation of DPNH might be analogous to the sulphite stimulated oxidation of DPNH produced in the presence of peroxidase, manganese and monophenol in a system studied by KLEBANOFF (1961) a similar reaction has since been observed by CRESSWELL and HEWITT. Nitrite-catalysed oxidation of DPNH would also occur in the presence of peroxidase and endogenous hydrogen peroxide (THURLOW, 1925) and nitrate reductase. Under these circumstances there might be no net change in nitrite concentration.

II. Bacteria (Enzymes of Nitrification).

The chemoautotrophic bacteria *Nitrosomonas* and *Nitrobacter* derive their energy for growth by oxidising NH_4^+ and NO_2^- respectively: $NH_4^+ + 1^1/_2O_2 = 2H^+ + +H_2O + NO_2^- + 66$ K cal; $NO_2^- + {}^1/_2O_2 = NO_3^- + 17$ K cal. WINOGRADSKY (1890) showed that these were autotrophic bacteria since they grew in simple media without organic supplements. LEES and QUASTEL (1945, 1955) have studied the nitrification process in soil using their ingenious percolation techniques. The biochemistry of whole cells of *Nitrobacter* and *Nitrosomonas* has been reviewed by LEES (1955) but since that time notable advances have been made both with culturing these bacteria in sufficient quantity for biochemical studies and with cell-free extracts prepared from them.

1. Preparation.

a) **Cultural conditions.** Since great difficulty has been experienced with culturing *Nitrosomonas* and *Nitrobacter*, it is appropriate to consider briefly the problems involved. MEIKLEJOHN (1950) isolated the bacteria on silica gel plates whereas JENSEN (1950) did so on agar films. LEES (1952) has grown them in liquid culture using forced aeration. ENGEL and ALEXANDER (1958a, b) were the first to grow *Nitrosomonas europaea* in a clear medium without added calcium carbonate. They maintained the pH at 8.0 by the continuous addition of sterile potassium carbonate solution. Their cultures contained less than 1% of viable heterotrophic contaminants that were able to grow on nutrient agar. After 60 hr. growth in the logarithmic phase their culture contained $1 \cdot 9 \times 10^8$ viable cells/ml yielding 73 mg dry weight per litre. This work showed conclusively that calcium carbonate in the medium was not essential for growth. All that is required is a control of pH which was achieved by substituting a solution of potassium carbonate for solid calcium carbonate. Nicholas and Jones (1960) grew *Nitrosomonas europaea* in batch culture (300 litres) in the following medium: (g/litre glass distilled water); $(NH_4)_2SO_4$ 3; K_2HPO_4 0.13; $MgSO_4.7H_2O$ 0.24; $FeSO_4.7H_2O$ 0.5; adjusted with 5% w/v K_2CO_3 to pH 8.3, glass distilled water. The pH of the medium was maintained constant by titrating it continuously with sterile 10% w/v K_2CO_3. The phosphates were autoclaved separately and added to the rest of the medium when cool. In later work, Nicholas (unpublished) grew the bacteria in 50 litre batches in glass carboys; the pH was kept constant by automatically titrating sterile potassium carbonate into the medium by a solenoid valve which was controlled by a pH alarm relay connected to a pH meter. Sterile electrodes were immersed into the culture medium which was well aerated with sterile air through two glass sparges from a suitable pump. A 20% inoculum was used and the bacteria grown at 25° C for 3 days. Skinner and Walker (1961) devised a semicontinuous culture apparatus for culturing *Nitrosomonas*. The culture medium used was as follows: (g/litre); $(NH_4)_2SO_4$, 1.0—8.0; KH_2PO_4, 0.4; $MgSO_4.7H_2O$, 0.05; $CaCl_2.2H_2O$, 0.05; and when necessary 0.1 to 0.6 mg Fe (equimolar mixture of ferrous sulphate and Na-EDTA).

Nitrobacter agilis (ATCC 9842) was grown in 10 litre serum bottles containing 8 litre of the following medium: (g/litre); KNO_2, 2.4; $MgSO_4.7H_2O$, 1.5; NaCl, 1.5; $CaCl_2.2H_2O$, 0.1; K_2HPO_4, 4; KH_2PO_4, 4; $FeSO_4.7H_2O$, 0.08 and $KHCO_3$, 1.5. The phosphates were autoclaved separately and the bicarbonate and $FeSO_4$ were sterilized individually by passing them through a Seitz filter. This ensured a clear medium. Sterilized air was dispersed in the medium by means of two glass sparges (Aleem and Alexander, 1958). Since high nitrite concentrations prolong the lag phase of growth, the original medium contained only a low level of energy substrate. Its growth proceeded and nitrite was oxidised. Additional increments were added until a final concentration of 1.5 mg nitrite N/ml was metabolized. This procedure allows for initiation of growth with no significant lag and a relatively large cell yield.

b) **Methods of extraction** (Table 32). Nicholas and Jones (1960) collected *Nitrosomonas europaea* in a Sharples centrifuge but more recently Nicholas (unpublished) has employed the continuous flow Servall centrifuge fitted with the Szent Gyorgy-Blum continuous flow head. The latter procedure is preferable when handling small amounts of cells. The bacteria were suspended in 0.01 M phosphate and 0.05 M sodium borate buffer (pH 8.3) and then disintegrated by means of the ultrasonic probe for 15 min at 4° C in a double-walled glass vessel through which iced waterwas circulated (Fig. 1). The homogenate was centrifuged at 20,000 g for 20 min at 4° C and the supernatant solution was used as the source of the crude enzyme for subsequent fractionation.

Aleem and Alexander (1958) cooled the culture of *Nitrobacter* to 5° C during the active phase of growth and harvested the cells in the Sharples centrifuge. The cell paste was washed twice with cold distilled water and suspended in 25 ml 0.05 M K_2HPO_4. Cell extracts were prepared by treating the cell suspension in a 10 k/cycles Raytheon magneto restrictive oscillator for 15 min at 10° C when a deep viscous liquid formed. Residual intact cells and debris were removed by centrifuging at 6000 g for 20 min at 3° C. Extracts contained 1.7 to 3.3 mg N/ml.

Aleem and Alexander (1958) extracted hydroxylamine reductase from *Nitrosomonas* using a similar technique but the sonic treatment of the cells lasted 30 min. Residual intact cells were removed by centrifugation at 3500 g. for 30 min. The extract was bright red, resulting from high cytochrome content and containing

0.5 to 1.1 mg N/ml. ALEEM and NASON (1960) centrifuged sonic extracts of *Nitrobacter* at 10,000 g for 30 min. The resulting cell-free opalescent solution containing nitrite reductase activity was then fractionated by successive centrifugation for 30 min intervals at 27,000 g, 58,000 g, 95,000 g and 144,000 g respectively. The reddish-brown pellets obtained after each centrifugation were suspended separately in 0.1 M tris (hydroxymethyl) aminomethane buffer (pH 7.5) equal in volume to 1/3rd of that of the starting supernatant extract. All the particulate fractions contained the nitrite oxidase activity.

c) **Fractionation and stability.** The hydroxylamine oxidase from *Nitrosomonas* has been fractionated 50-fold as shown in Table 33 (NICHOLAS and JONES, 1960).

Fraction (2) collected between 80 and 90 per cent saturation with ammonium sulphate dissolved in 0.01 M phosphate and 0.005 M borate buffer (pH 8.4) resulted in a 27 fold purification with 44 per cent recovery of the enzyme. Aliquots (5 ml) of fraction 2 were put on DEAE-cellulose columns (15 cm × 1.2 cm) and the enzyme was washed in turn with 10 ml volumes of the following buffers (pH 8.5): 0.05 M, 0.1 M phosphate borate: 0.25 M phosphate and finally 0.2 M pyrophosphate. The latter fraction contained the bulk of the activity resulting in a 44-fold purification. The DEAE-cellulose varies markedly with different batches; lately using recent material, which is presumably better quality, the enzyme is eluted from the column in 0.25 M phosphate buffer. Thus it is important to check the enzyme activity of the various eluants especially when different batches of the cellulose are used. ENGEL and ALEXANDER (1959) used crude extracts of *Nitrosomonas* to study the oxidation of hydroxylamine but no fractionations were attempted.

Table 32. *Methods used for extracting hydroxylamine oxidase and nitrite oxidase systems from nitrifying bacteria*

Source	Extracting solution	Vol./Wt.	Type of extraction	Reference
Nitrosomonas europaea	0.01 M phosphate } pH 8.0 0.05 M borate	3:1	Bacteria collected in a SERVALL centrifuge fitted with a SZENT-GEORGY continuous flow head. Bacteria suspended in the buffer and disintegrated by means of a MULLARD ultrasonic probe 20 kc/15 min at 4°. Homogenate centrifuged 20,000 g/20 min/4°.	NICHOLAS and JONES (1960) Hydroxylamine oxidase
Nitrobacter agilis (Amer. Type Culture No. 9482)	0.05 M phosphate (pH 7.5)	5:1	Bacteria collected in a SHARPLES centrifuge at 5°, washed 2 × with cold distilled water and suspended in 25 ml 0.05 M phosphate. Cell extracts prepared by treating suspension in a 10 kc RAYTHEON magneto-restrictive oscillator/30 min. Residual intact cells centrifuged at 3,500 g/30 min. Supernatant solution used as enzyme source.	ALEEM and ALEXANDER (1958) Nitrite oxidase
Nitrobacter agilis (ATCC 9482)	0.05 M phosphate (pH 7.5)	5:1	Bacteria collected in a SHARPLES centrifuge at 5° C, washed 2 × with cold distilled water and suspended in 25 ml 0.05 M phosphate + 10^{-3} M glutathione. Sonic oscillation as described by ALEEM and ALEXANDER (1958) centrifuged 10,000 g/30 min. Supernatant solution recentrifuged 144,000 g/1 hr. Red pellet contains 50% of activity.	ALEEM and NASON (1959, 1960) Nitrite oxidase

11*

Table 33. *Fractionation of the Enzyme System in Cell-free Extracts of Nitrosomonas europaea that Oxidizes Hydroxylamine to Nitrite Using Phenazine Methosulphate as the Acceptor* (Nicholas and Jones, 1960).

Fraction	Volume (ml)	Total units (mμM nitrite/20 min)	Total protein (mg.)	Specific activity (mμM nitrite/20 min/ mg. protein)	Purification (fold)	Percentage recovery of enzyme
(1) Crude extract after centrifuging at 20,000 g for 20 min	20	6,500	37.4	192		
(2) Fraction collected between 80 and 90 per cent saturation with ammonium sulphate dissolved in 0.01 M phosphate and 0.005 M borate buffer (pH 8.4)	10	2,850	1.1	5,200	27	44
(3) 5 ml aliquots of (2) put on a DEAE-cellulose column (15 cm $\times$ $\times$1.2 cm). Column washed in turn with 10 ml aliquots of the following buffers (pH 8.5): 0.05 M, 0.1 M phosphate-borate; 0.25 M, 0.5 M phosphate and finally 0.2 M pyrophosphate . . . . Activity of pyrophosphate eluate	5	510	0.2	8,500	44	8

Aleem and Alexander (1958) first prepared extracts of *Nitrobacter agile* that oxidised nitrite to nitrate and later Aleem and Nason (1960) showed that the nitrite oxidase activity was associated with cell-particles. Solubilization and fractionation of nitrite oxidase, however, has not been attempted. Malavolta, Delwiche and Burge (1960) have studied CO_2 fixation and phosphorylation in extracts of *Nitrobacter agile*.

2. Measurement of Activity.

The oxidation of ammonia or hydroxylamine to nitrite by *Nitrosomonas* and the oxidation of nitrite to nitrate by *Nitrobacter* have been followed by the well known colorimetric test for nitrite. The Warburg method can also be used to follow oxygen uptake by whole cells or in extracts prepared from them. Hydroxylamine can be determined either by the Csáky or 8-hydroxylquinoline methods discussed earlier. Hydroxylamine oxidase can also be measured by following the reduction of the added acceptor, e.g. mammalian or bacterial cytochrome c at 551 mμ or methylene blue at 625 mμ (Nicholas and Jones, 1960) or 2,4,5-triphenyltetrazolium chloride at 485 mμ (Engel and Alexander, 1959) under anaerobic conditions in Thunberg tubes. The oxidation of hydroxylamine is also measured under aerobic conditions when some of the hydroxylamine is oxidised to nitrite. Lees and Simpson (1957) reported that in cell suspensions of *Nitrobacter* cytochrome bands at λ max. 589, 551 and 520—525 mμ were reduced on adding nitrite. Aleem and Nason (1960) determined the difference spectra for steady state and oxidised conditions by means of the Cary recording spectrophotometer with cell-free fractions prepared from *Nitrobacter*. The various methods and conditions adopted are summarised in Table 34.

3. Cofactors.

Nicholas and Jones (1960) showed that the addition of 0.2 μmole of either phenazine methosulphate or mammalian or bacterial cytochrome c activated the oxidation of hydroxylamine in 0.1 ml extracts (1 mg protein/ml) of *Nitrosomonas*. Phenazine methosulphate was the best acceptor and pyocyanine, methylene blue, benzyl viologen or cytochrome c were usually less active whereas ferricyanide, 2,3,6-trichloroindophenol dye, DPN, TPN or glutathione had little effect (Table 35). When cytochrome c was used, phosphate was required, *viz.* 10 μmoles PO_4/ml enzyme but at higher concentrations the anion was inhibitory. The phosphate

Table 34. *Conditions used for assay of hydroxylamine oxidase and nitrite oxidase systems from nitrifying bacteria*

Method	Assay	Time min	Temp.°C	Notes	Reference
Colorimetric test for nitrite formed from hydroxylamine. Oxygen uptake measured in the WARBURG apparatus	1 ml 0.01 M phosphate }pH 7.5 or 8.0 0.05 M borate 0.2 μmole cytochrome c in buffer or 0.2 μmole phenazine methosulphate in buffer 0.1 μmole $NH_2OH.HCl$ Final volume 2 ml WARBURG assay as for colorimetric assay but final volume 3 ml. hydroxylamine in sidearm	15—20	30	After incubation period 0.5 ml 1% w/v sulphanilamide in N.HCl and 1 ml 0.01% w/v N-(1-naphthyl) ethylenediamine dihydrochloride added and the volume made to 5 ml with deionised water	NICHOLAS and JONES (1960) Hydroxylamine oxidase
Colorimetric test for nitrite left after oxidation to nitrate. Oxygen uptake in WARBURG	Colorimetric test 100 μmoles phosphate (pH 7.4) 5 μmoles $FeCl_3$ 60 μmoles $NaNO_2$ 0.5 ml cell-free preparation WARBURG assay *Main compartment* 50 μmoles phosphate (pH 7.4) 5 μmoles $FeCl_3$ 100 μmoles KNO_2 Total volume 3 ml *Sidearm* 0.1 ml extract in sidearm 0.24 20% w/v KOH in centre well O_2 uptake in air.	60 Aliquots tested at intervals up to 90 min	30 30	Nitrite determined by the GRIESS-ILOSVAY test and nitrate formed by the phenoldisulphonic acid method	ALEEM and ALEXANDER (1958) Nitrite oxidase
Colorimetric test for residual nitrite	Assay as described by ALEEM and ALEXANDER (1958) Colorimetric assay WARBURG assay Spectrophotometric measurements of C_1 (max.λ 415 and 550 mμ) and A_1 (max. λ 585 and 438 mμ) on adding NO_2^-	15—20 60	Room temp.	Cary recording spectrophotometer used to follow cytochrome changes	ALEEM and ALEXANDER (1959,1960) Nitrite oxidase

inhibition which was competitive for cytochrome c did not occur with phenazine methosulphate. Citrate, selenate, arsenate, tungstate, phosphate and pyrophosphate inhibited hydroxylamine oxidase (Table 36). Engel and Alexander (1959) found that methylene blue was reduced by extracts of *Nitrosomonas* when hydroxylamine was added to them under anaerobic conditions in Thunberg tubes. The use of 2,4,5-triphenyl tetrazolium chloride as hydrogen acceptor gave results analogous to methylene blue. The dye was readily reduced yielding the red formazan which could be measured at 485 mμ.

The oxidation of nitrite by cell-free extracts of *Nitrobacter* is stimulated by adding iron salts (Aleem and Alexander, 1958). This was confirmed by Aleem and Nason (1960) who showed, however, that this was a variable effect. Substrate and cofactor concentrations are shown in Table 35.

4. Properties and Mechanisms.

Hydroxylamine oxidase prepared from *Nitrosomonas* has a broad pH optimum between 7 and 8.6. Hydrazine, a well known inhibitor of nitrite formation from hydroxylamine in washed cells of *Nitrosomonas* (Lees, 1955), was found by Nicholas and Jones (1960) to have a similar effect on the cell free extracts. They showed that there was an active cytochrome oxidase in the extracts and the subsequent addition of hydroxylamine resulted in the immediate enzymatic reduction of cytochrome c with concomitant formation of nitrite. The addition of hydrazine instead of hydroxylamine produced an enzymatic reduction of cytochrome c but nitrite was not produced. An increase in hydroxylamine concentration decreased the inhibition of nitrite production by hydrazine. Thus hydrazine inhibition may be due to a competition with hydroxylamine for a common acceptor such as cytochrome c which is indispensable for the oxidation of hydroxylamine. Cyanide inhibited when a range of acceptors was used. Carbon monoxide depressed cytochrome oxidase activity and also nitrite formation. Neither hyponitrite nor ammonia alone was oxidised by these extracts but addition of ammonia stimulated the oxidation of hydroxylamine by about 60 per cent. They also found that nitrite formed from hydroxylamine in these extracts using cytochrome c or phenazine methosulphate as acceptor, was only 40—73 per cent of the theoretical values expected from oxygen uptake results obtained in the Warburg apparatus. The hydroxylamine was completely oxidised in agreement with the oxygen uptake values. There is evidence that the first step is a dehydrogenation resulting in the production of N_2O, NO and N_2 and the second step an oxidation involving cytochrome c and its oxidase (Nicholas unpublished results). Engel and Alexander (1959) who coupled the oxidation of hydroxylamine to oxidation-reduction dyes found that it was inhibited by cyanide. The reduction of methylene blue was related to the soluble portion of cell extracts after removing the particulate components by centrifuging at 144,000 g.

Aleem et al. (1962) have shown that washed living cells of *Nitrosomonas* oxidise nitrohydroxylamine to nitrite with yields of about 75 per cent and almost stoichiometric oxygen uptake. They suggest that nitrohydroxylamine (prepared as described by Angeli 1896, 1897) is a natural intermediate in nitrification of ammonia and hydroxylamine to nitrite and stated that rates of nitrification of the three compounds were similar.

Falcone, A. B., A. L. Shug, and D. J. D. Nicholas. (Biochem. Biophys. Res Comm. **9**, 126 (1962) recently reported that components of the respiratory chain present in cell particles are reduced by hydroxylamine and reoxidised by oxygen via cytochrome oxidase. The particles contain flavin, cytochromes b and a as well as cytochrome oxidase. Copper has been detected in these particles using election paramagnetic resonance methods (Nicholas D. J. D., P. W. Wilson, W. Heinen, G. Palmer, and H. Beinert. Nature (Lond.) **196**, 1 (1962).

LEES and SIMPSON (1957) observed the formation of cytochrome absorption bands at λ max. 589, 551 and 520—525 mμ when nitrite was added to cell sus- pensions of *Nitrobacter*. ALEEM and ALEXANDER (1958) were the first to prepare cell-free extracts of *Nitrobacter* that oxidised nitrite to nitrate. They showed that these extracts required iron for activity and a pH optimum between 7.5 and 8. The enzyme which was associated with cell-particles and was inhibited by cyanide, *p.*-chloromercuribenzoate, 2,4-dinitrophenol and iodoacetate (Table 36). ALEEM and NASON (1960) examined various particulate fractions of *Nitrobacter* and showed that addition of nitrite to them resulted in the appearance of light absorption bands λ max. 550, 520 mμ (the alpha and beta bands of cytochrome c) and at 585—590 mμ (the alpha and gamma bands of cytochrome a$_1$). Additions of dithionite or cyanide resulted in similar bands which were more intense and in addition the λ 415 mμ band corresponding to the gamma band of cytochrome c. They tentatively ascribed the effect of cyanide to an inhibition of the cytochrome a$_1$-portion of the electron transport scheme. They suggested that cytochrome c and cytochrome oxidase orientated in the particle are necessary for

Table 35. *Substrate, electron acceptor, cofactor requirements of hydroxylamine oxidase and nitrite oxidase systems from nitrifying bacteria.*

Source	Substrate	Acceptors	Flavin	Metals, cofactors	pH optimum	Reference
Nitrosomonas europaea	optimum. NH$_2$OH 0.1 μmole	optimum concentration per 0.1 ml crude extract 0.2 μmole phenazine metho-sulphate 0.2 μmole cytochrome c 0.2 μmole pyocyanine 0.2 μmole methylene blue 0.2 μmole benzyl viologen 5 μmoles FeCl$_3$	not detected	cytochrome c	7.0—8.6	NICHOLAS and JONES (1960). Hydroxylamine oxidase
Nitrobacter agilis Amer. Type Culture (ATCC 9482)	optimum. 100 μmoles KNO$_2$ in reaction mixture		not detected	Fe	7.5 to 8.0 in 0.02 M phosphate 0.2 M TRIS 0.2 M carbonate/bicarbonate	ALEEM and ALEXANDER (1958) Nitrite oxidase
Nitrobacter agilis Amer. Type Culture (ATCC 9482)	no record	no record	10^{-3} M atabrine dihydro-chloride, 10^{-3} M quinine sulphate inhibited the enzyme	Fe^{++} or Fe^{+++}	7.5	ALEEM and NASON (1959, 1960) Nitrite oxidase
Nitrosomonas washed cells	Nitrohydroxyl-mine	Oxygen				ALEEM et. al. (1962)

Table 36. *Stability of and effects of inhibitors on hydroxylamine oxidase and nitrite oxidase systems from nitrifying bacteria*

Source	Thermal denaturation and stability*	Inhibition by —SH reagents or flavin analogues*	Other inhibitors*	Reference
Nitrosomonas europaea	40: 15 min/50°	no pCMB inhibition	50 μmoles Citrate: 93 50 μmoles Selenate: 93 50 μmoles Arsenate: 85 50 μmoles Tungstate: 95 50 μmoles Phosphate: 84 50 μmoles Pyrophosphate: 90 10^{-4} M Hydrazine: 90 10^{-3} M KCN: 80 } cytochrome c as acceptor (final volume: 1.0 ml)	Nicholas and Jones (1960) Hydroxylamine oxidase
Nitrobacter agilis Amer. Type culture (ATCC 9482)	no record	10^{-3} M pCMB: 34 10^{-4} M pCMB: 21 10^{-2} M Iodoacetate: 18	KCN 10^{-4} M: 76; 10^{-5} M: 3 2.4 dinitrophenol 10^{-2} M: 84	Aleem and Alexander (1958) Nitrite oxidase
Nitrobacter agilis Amer. Type culture (ATCC 9482) 15—20 μg	Aged preparations unchanged nitrite oxidase activity but decreased rates of phosphorylation	no record	2.4. dinitrophenol . . . . 5×10^{-4} M: 100 Dicoumarol . . . . 5×10^{-5} M: 100 Thyroxine . . . . 5×10^{-5} M: 100 Antimycin A . . . 15 to 20 μg/ml: 50—100 2-n-heptyl-4-hydroxyquinoline-N-oxide (HOQNO) . . . . 50 μg/ml: 70	Aleem and Nason (1959, 1960) Nitrite oxidase

* Values are per cent inhibition for concentration or treatment shown.

nitrite oxidase activity. Nitrite is specific as a substrate for nitrite oxidase since neither succinate, DPNH nor lactate reduced the cytochrome components. The particulate fraction collected between 10,000 and 144,000 g was inactivated after dialysis against cyanide and glutathione, as described by Nicholas and Nason (1954). The addition of Fe^{3+} or Fe^{2+} (4×10^{-3} M) reactivated the dialysed enzyme but magnesium, copper, zinc, manganese, cobalt, nickel, tungstate, vanadate or borate ions were without effect. Tungstate and copper increased the non-enzymatic reduction of nitrite. The cyanide-dialysed fraction showed reduced cytochrome bands on adding nitrite but the addition of $FeSO_4$ (5×10^{-3} M final concentration) together with nitrite enhanced the formation of the reduced bands of cytochrome c and a_3. This effect they attributed to the non-enzymatic reduction of cytochrome c by Fe^{2+} as shown to occur by Weber, Lenhoff and Kaplan (1956) and by Nicholas et al. (1960). The role of Fe in the enzyme is not clear since the addition of Fe^{3+} plus nitrite did not restore the steady state difference spectrum of the cyanide-dialyzed enzyme. Thus far there is no evidence for an enzymatic reduction of Fe^{3+} by nitrite either aerobically or anaerobically. They were unable to detect a flavin component or show a requirement for it in their crude preparations. Atabrine dihydrochloride at 10^{-3} M, inhibited the oxidation of nitrite by intact cells and in particles collected at 144,000 g by 60 and 25 per cent respectively and quinine sulphate also resulted in a 30 per cent inhibition. The inhibition data, however, do not provide unequivocal evidence for a flavin requirement since these inhibitors are not specific as flavin antagonists. The mechanism of nitrite oxidase action is suggested to involve the transfer of electrons from nitrite to molecular oxygen via cytochrome c and cytochrome a_1-like components as follows:

$$NO_2^- \rightarrow \text{cytochrome } c \rightarrow \text{cytochrome } a_1 \rightarrow O_2$$

They suggest that iron might operate between nitrite and cytochrome c. The exact site and role of iron in this scheme has not been determined.

ALEEM and NASON (1960) showed that the particulate nitrite oxidase enzyme was coupled to phosphorylation but the P/O ratios were small varying between 0.03 and 0.2. Dinitrophenol, thyroxine or dicoumarol even at 5×10^{-6} M did not, however, uncouple the system. Relatively high concentrations of antimycin A and 2-n-heptyl-4-hydroxyquinoline-N-oxide inhibited the oxidation of nitrite (Table 36).

Reversal of NH_2OH reductase: "Ammonia dehydrogenase" artefact.

KLAUSMEYER and BARD (1954) reported that extracts of *Bacillus subtilis* contained an enzyme which catalysed the reduction of DPN on the addition of NH_4OH. The reaction was reversible and enzymic. They concluded that their system comprised a reversible DPN hydroxylamine reductase-ammonia dehydrogenase. ZUCKER and NASON (1955) calculated that the equilibrium constant for the system

$$NH_2OH + DPNH + H^+ \rightleftharpoons NH_3 + DPN^+ + H_2O$$

to be 10^{35} and concluded that reversibility was highly unlikely, and could not be demonstrated experimentally. ROUSSOS, TAKAHASHI and NASON (1957) reinvestigated the problem and observed an apparently enzymic reduction of DPN, but not of TPN on the addition of NH_4OH to extracts prepared from *B. subtillis* as described by KLAUSMEYER and BARD (1954). This activity was not observed when NH_4Cl or $(NH_4)_2SO_4$ were used. It was however produced when other alkali reagents such as NaOH or KOH were added. The reduction of DPN was reversed on the addition of alcohol dehydrogenase and acetaldehyde. However, alcohol dehydrogenase was absent from the preparations and the enzyme systems involved in the reduction were not identified. Several DPN-specific reversible dehydrogenases would however behave thus with an endogenous substrate. It was concluded that the original view regarding the identity of ammonia dehydrogenase was based on an artefact and that such a system does not exist.

Literature.

ADDISON, C. C., G. A. GAMLEN and R. THOMPSON: J. chem. Soc. **1952**, 338. — ADELSTEIN, S. J., and B. L. VALLEE: J. biol. Chem. **233**, 589 (1958). — ADLER, E., N. B. DAS, H. VON EULER and U. HEYMAN: C. R. Lab. Carlsberg Sér. chim. **22**, 15 (1938). — ADLER, E., G. GUNTHER u. J. E. EVERETT: Hoppe-Seylers Z. physiol. Chem. **255**, 27 (1938). — ADLER, E., V. HELLMSTROM, G. GUNTHER u. H. VON EULER: Hoppe-Seylers Z. physiol. Chem. **255**, 14 (1938). — ALEEM, M. I. H., and M. ALEXANDER: J. Bact. **76**, 510 (1958). — ALEEM, M. I. H., H. LEES, R. LYRIC and D. WEISS: Biochem. biophys. Res. Comm. **7**, 126(1962). — ALEEM, M. I. H., and A. NASON: Biochem. biophys. Res. Commun. **1**, 323 (1959); — Proc. nat. Acad. Sci (Wash.) **46**, 763 (1960). — ALLEN, M. B., and C. B. VAN NIEL: J. Bact. **64**, 397 (1952). — ANACKER, W. F., and V. STOY: Biochem. Z. **330**, 141 (1958). — ANDERSON, V. L.: Ann. Bot. (Lond.) **38**, 699(1924). — ANGELI, A.: Gazetta Chimica Italiana **26** (II) 17 (1896). — ANGELI, A.: Gazetta Chimica Italiana **27** (II) 35 (1897). — ASANO, A.: J. Biochem. (Japan) **46**, 781 (1959). — AUBEL, E., B. LUBO-CHINOKY et A. PROUVOST: C. R. Acad. Sci. (Paris) **236**, 145 (1953). — AVIS, P. G., F. BERGEL and R. C. BRAY: J. chem. Soc. **1956**, 1219.

BAALSRUD, K., and K. S. BAALSRUD: Arch. Mikrobiol. **20**, 34 (1954). — BACK, K. J. C., J. LASCELLES and J. L. STILL: Aust. J. Sci. Res. **9**, 25 (1946). — BARBAN, S.: J. Bact. **68**, 493 (1954). — BEIJERINCK, M. W.: Zbl. Bakt., II. Abt. **7**, 561 (1901); — Folia Mikrobiol. (Delft) **3**, 91 (1914). — BERGER, J., and G. S. AVERY, jr.: Amer. J. Bot. **30**, 290 (1943); **31**, 11 (1944). —

BERNHEIM, F.: Biochem. J. 22, 344 (1928). — BONNER, J., and S. G. WILDMAN: Arch. Biochem. 10, 497 (1946). — BUEDING, B., and N. JOLLIFE: J. Pharmacol. 88, 300 (1946). — BURSTROM, H.: Ann. Roy. Agr. Coll. (Sweden) 11, 1 (1943). — BUSH, M. T., O. TONSTER and J. E. BROCKMAN: J. biol. Chem. 188, 685 (1951).

CAIN, R. B.: J. gen. Microbiol. 19, 1 (1958). — CANDELA, M., E. G. FISHER and E. J. HEWITT: Plant Physiol. 32, 280 (1957). — CARRIE, M. S.: J. Soc. chem. Ind. (Lond.) 53, 288 (1934). — CARTER, C. L.: J. Soc. chem. Ind. (Lond.) 62, 238 (1943). — CARTER, C. L., and W. J. McCHESNEY: Nature (Lond.) 164, 575 (1949). — CARTWRIGHT, N. J., and R. B. CAIN: Biochem. J. 71, 248 (1959); 73, 305 (1959). — CHANCE, B.: J. biol. Chem. 197, 577 (1952). — CHENIAE, G., and H. J. EVANS: In Inorganic Nitrogen Metabolism, p. 184 (W. D. McELROY and B. GLASS, eds.). Baltimore: Johns Hopkins Univ. Press 1956; — Biochim. biophys. Acta 26, 654 (1957); — Plant Physiol. 35, 454 (1960). — CHUNG, C. W., and V. A. NAJJAR: J. biol. Chem. 218, 617 (1956a); 218, 627 (1956b). — CONWAY, E. J.: Microdiffusion Analysis and Volumetric Error. 4th Edn. London: Crosby Lockwood 1957. — CRESSWELL, C. F., and E. J. HEWITT: Biochem. biophys. Res. Commun. 3, 544 (1960). — CRESSWELL, C. F., R. H. HAGEMAN and E. J. HEWITT: Biochemic. J. 83, 38 P. (1962). — CSÁKY, T. Z.: Acta chem. scand. 2, 450 (1948).

DAMODARAN, M., and K. R. NAIR: Biochem. J. 32, 1064 (1938). — DITTRICH, W.: Planta 12, 69 (1930). — DIXON, M.: Biochem. J. 20, 703 (1920).

ECKERSON, S. H.: Bot. Gaz. 77, 377 (1924); — Contr. Boyce Thompson Inst. 3, 405 (1931); 4, 119 (1932). — EGAMI, F., and M. ITAHASHI: Igaku Seibutsugaku (Med. and Biol.) 19, 292; Cited in Chem. Abstr. 1951, 10278. — EGAMI, F., K. OHMACHI, K. IIDA and S. TANIGUCHI: Biokhimiya 22, 115 (Transln. Consultants Bureau Inc. New York) (1957a); 22, 122 (1957b). — EGAMI, F., T. YAMADA and S. TANIGUCHI: Cong. intern. biochem. Resumes Communs 2nd Congr. Paris, 78 (1952). — EISENMENGER, W. S.: J. Agr. Res. 46, 255 (1933). — ENGEL, M. S., and M. ALEXANDER: Nature (Lond.) 181, 136 (1958a); — J. Bact. 76, 217 (1958b); 78, 796 (1959). — EULER, H. VON, E. ADLER, G. GUNTHER and L. ELLIOTT: Enzymologia 6, 337 (1938). — EVANS, H. J.: Plant Physiol. 29, 298 (1954); — Soil Sci. 81, 199 (1956). — EVANS, H. J., and A. NASON: Plant Physiol. 28, 233 (1953). — EVANS, H. J., and N. S. HALL: Science 122, 922 (1955). — EVANS, H. J., and C. MCAULIFFE: In Inorganic Nitrogen Metabolism, p. 189 (W. D. McELROY and B. GLASS, eds.). Baltimore: Johns Hopkins Univ. Press 1956.

FEWSON, C. A., and D. J. D. NICHOLAS: Biochem. J. 77, 3 (1960a); — Nature (Lond.) 188, 794 (1960b); — Biochim. biophys. Acta 49, 335 (1961a). — Biochim. biophys. Acta 48, 208 (1961b); — Nature (Lond.) 190, 2 (1961c); — Biochem. J. 78, 9 P. (1961d). — FINCHAM, J. R. S.: J. gen. Microbiol. 5, 793 (1951). — FREAR, D. S., and R. C. P. BURRELL: Anal. Chem. 27, 1664 (1955); — Plant Physiol. 33, 105 (1958).

GAYON, U., and G. DUPETIT: Soc. Sci. Phys. Nat. (Bordeaux) 3e Ser. 11, 201 (1886). — GRANICK, S., and H. GILDER: J. gen. Physiol. 30, 1 (1946). — GREEN, D. E., and H. BEINERT: Biochim. biophys. Acta 11, 599 (1953). — GREEN, E. D., L. H. STICKLAND and H. L. A. TARR: Biochem. J. 28, 1812 (1934). — GROSSWICZ, N., and Y. LICHTENSTEIN: Congr. Intern. Biochem. Résumés Communs. p. 98, 3rd Congress Brussels (1955).

HAGEMAN, R. H., C. F. CRESSWELL and E. J. HEWITT: Nature (Lond.) 193, 247 (1962). — HAGEMAN, R. H., and D. FLESHER: Plant Physiol. 35, 700 (1960). — HAGEMAN, R. H., and E. R. WAYGOOD: Plant Physiol. 34, 396 (1959). — HALLAWAY, M.: In Data for Biochemical Research, p. 154 (DAWSON, R. M. C., ELLIOTT, D. C., ELLIOTT, W. H. and JONES, K. M. eds.). Oxford: Clarendon Press 1959. — HEREDIA, C. F., and A. MEDINA: Biochem. J. 77, 24 (1960). — HEWITT, E. J.: Nature (Lond.) 180, 1020 (1957); — In Encyclopaedia of Plant Physiology, Vol. 4, p. 427 (W. RUHLAND ed.). Berlin-Göttingen-Heidelberg: Springer 1958; — Biol. Rev. 34, 333 (1959). — HEWITT, E. J., and AFRIDI M. M. R. K.: Nature (Lond.) 183, 57 (1959). — HEWITT, E. J., and G. J. DICKES: Biochem. J. 78, 384 (1961). — HEWITT, E. J., E. G. FISHER and M. CANDELA: Long Ashton Agr. Hort. Res. St. Ann. Rep. p. 202 (1955). — HEWITT, E. J., and D. G. HALLAS: Nature (Lond.) 184, 1485 (1959). — HOFMAN, T., and H. LEES: Biochem. J. 52, 140 (1952); 54, 579 (1953). — HOLZER, H., and S. SCHNEIDER: Biochem. Z. 329, 361 (1957).

IIDA, K., and S. TANIGUCHI: J. Biochem. (Tokyo) 46, 1041 (1959). — ITAGAKI, E., and S. TANIGUCHI: J. Biochem. (Tokyo) 43, 295 (1956). — IWASAKI, H., R. MATSUBAYASHI and T. MORI: J. Biochem. (Tokyo) 43, 295 (1956). — IWASAKI, H., and T. MORI: J. Biochem. (Tokyo) 45, 133 (1938).

JACOBI, G.: Naturwissenschaften 44, 265 (1957). — JENSEN, H. L.: Handb. der Tech. Mykol. Bd. III, p. 182. H. Lafar (1904); — Tidsskr. Planteavl. 54, 62 (1950). — JOKLIK, W.: Aust. J. Sci. Res. Ser. B. 3, 28 (1950). — JONES, L. W., and A. W. SCOTT: Amer. chem. Soc. 46, 2172 (1924).

KEILIN, D., and E. F. HARTREE: Proc. roy. Soc. London B 124, 397 (1938). — KENTEN, R. H.: Biochem. J. 59, 110 (1955). — KENTEN, R. H., and P. J. G. MANN: Biochem. J. 45,

255 (1949); **46**, 67 (1950); **52**, 125 (1952); **53**,.498 (1953). — KESSLER, E.: Flora (Jena) **140**, 1 (1953); — Nature (Lond.) **176**, 1069 (1955); — In Utilization of Nitrogen and its Compounds in Plants. Soc. exp. Biol. Sympos. XIII, p. 87 (H. K. PORTER ed.), Cambridge Univ. Press (1959). — KESSLER, E., W. ARTHUR and J. E. BRUGGER: Arch. biochem. Biophys. **71**, 326 (1957). — KINSKY, S. C., and W. D. MCELROY: In Inorganic Nitrogen Metabolism P. (W. D. MCELROY and B. GLASS, eds.). Baltimore: Johns Hopkins Univ. Press 1956; — Arch. Biochem. **73**, 466 (1958). — KLAESER, M.: Zbl. Bakt., II. Abt. **41**, 365 (1914). — KLAUSMEYER, R., and R. BARD: J. Bact. **68**, 129 (1954). — KLEBANOFF, S. J.: Biochim. biophys. Acta **48**, 93 (1961). — KLINKHAMMER, F.: Arch. Mikrobiol. **33**, 357 (1959). — KLUYVER, A. J., and R. J. L. DONKER: Chem. Zelle u. Gewebe **13**, 134 (1926). — KLUYVER, A. J., and C. B. VAN NIEL: The Microbe's Contribution to Biology. Harvard, Mass.: Harvard Univ. Press 1956. — KLUYVER, A. J., and W. VERHOEVEN: Microbiol. Serol. **20**, 241 (1954).

LASCELLES, J.: J. gen. Microbiol. **15**, 404 (1956). — LASCELLES, T., and J. L. STILL: Aust. J. Sci. Res. **7**, 93 (944); — Aust. J. exp. Biol. med. Sci. **24**, 159 (1946). — LEES, H.: Biochem. J. **42**, 534 (1948); — Plant and Soil **1**, 221 (1949); — Nature (Lond.) **167**, 355 (1951); — Biochem. J. **52**, 134 (1952); — Biochemistry of Autotrophic Bacteria. London: Butterworths Scientific Publications 1955. — LEES, H., and J. MEIKLEJOHN: Nature (Lond.) **161**, 398 (1948). — LESS, H., and J. H. QUASTEL: Nature (Lond.) **155**, 276 (1945). — LESS, H., and J. R. SIMPSON: Biochem. J. **65**, 297 (1957). — LENHOFF, H. M., D. J. D. NICHOLAS and N. O. KAPLAN: J. biol. Chem. **220**, 983 (1956). — LLOYD, M., and J. A. CRANSTON: Biochem. J. **24**, 529 (1930).

MACKLER, B., H. R. MAHLER and D. E. GREEN: J. biol. Chem. **210**, 149 (1954). — MAHLER, H. R.: Advanc. Enzymol. **17**, 233 (1956). — MAHLER, H. R., and J. L. GLENN: In Inorganic Nitrogen Metabolism, p. 575 (W. D. MCELROY and B. GLASS eds.) Baltimore: Johns Hopkins Univ. Press 1956. — MAHLER, H. R., B. MACKLER, D. E. GREEN and R. M. BOCK: J. biol. Chem. **210**, 465 (1954). — MALAVOLTA, E., C. C. DELWICHE and W. D. BURGE: Biochem. biophys. Res. Comm. **2**, 445 (1960). — MCELROY, W. D., and D. SPENCER: In Inorganic Nitrogen Metabolism, p. 137 (W. D. MCELROY and B. GLASS eds.) Baltimore: Johns Hopkins Univ. Press 1956. — MCNALL, E. G., and D. E. ATKINSON: J. Bact. **72**, 226 (1956); **74**, 60 (1957). — MEDINA, R., E. F. HEREDIA: Biochim. biophys. Acta **28**, 452 (1958). — MEDINA, A., and D. J. D. NICHOLAS: Nature (Lond.) **179**, 533 (1957a); — Biochim. biophys. Acta **23**, 440 (1957b); **25**, 138 (1957c). — MEIKLEJOHN, J.: J. gen. Microbiol. **8**, 58 (1953); — In Autotrophic Micro-organisms. 4th Symposium Soc. Gen. Microbiol. p. 68 (B. A. FRY and J. L. PEEL, eds.) Cambridge University Press 1954. — MELLOR, J. W.: A Comprehensive Treatise on Inorganic and Theoretical Chemistry, Vol. 8, p. 404. London: Longmans Green & Co. Ltd. 1928. — MICHAELIS, L., and E. S. HILL: J. gen Physiol. **16**, 859 (1933). — MORRIS, M. P., C. PAGAN and H. E. WARMKE: Science **119**, 322 (1954). — MORTON, A. G.: J. exp. Bot. **7**, 97 (1956). — MULDER, E. G., R. BOXMA and W. L. VAN VEEN: Plant and Soil **10**, 335 (1959).

NAJJAR, V. A., and M. B. ALLEN: J. biol. Chem. **206**, 209 (1954). — NASON, A.: In Inorganic Nitrogen Metabolism p. 109. (W. D. MCELROY and B. GLASS, eds.) Baltimore: Johns Hopkins Univ. Press 1956. — NASON, A., and H. J. EVANS: Arch. Biochem. **39**, 234 (1952); — J. biol. Chem. **202**, 655 (1953). — NASON, A., R. G. ABRAHAM and B. C. AVERBACH: Biochim. biophys. Acta **15**, 160 (1954). — NICHOLAS, D. J. D.: Analyst **77**, 629 (1952); — Nature (Lond.) **179**, 800 (1957); — In Proceedings IV International Congress Biochemistry Vol. XIII — Colloquia p. 473 Pergamon Press 1959a; — In 13th Symposium Soc. exp. Biol. p. 1 Cambridge Univ. Press (1959b); — Minor Mineral Elements. Ann. Rev. Plant Phys. **12**, 63 (1961). — NICHOLAS, D. J. D., and O. T. G. JONES: Nature (Lond.) **185**, 512 (1960). — NICHOLAS, D. J. D., and MABEY G. L.: J. gen. Microbiol. **22**, 184 (1960). — NICHOLAS, D. J. D., and A. NASON: J. biol. Chem. **207**, 353 (1954a); — Arch. Biochem. **51**, 311 (1954b); — J. biol. Chem. **211**, 183 (1954c); — Plant Physiol. **30**, 135 (1955a); — J. Bact. **69**, 580 (1955b). — NICHOLAS, D. J. D., A. NASON and W. D. MCELROY: J. biol. Chem. **207**, 341 (1954). — NICHOLAS, D. J. D., A. MEDINA and O. T. G. JONES: Biochim. biophys. Acta **37**, 468 (1960). — NICHOLAS, D. J. D., and J. H. SCAWIN: Nature (Lond.) **178**, 1474 (1956). — NICHOLAS, D. J. D., and H. M. STEVENS: Nature (Lond.) **176**, 1066 (1955); — In Inorganic Nitrogen Metabolism p. 178 (W. D. MCELROY and B. GLASS eds.) Baltimore: Johns Hopkins Univ. Press 1956. — NISMAN, B.: Bact. Rev. **18**, 16 (1954).

OKUNUKI, K.: Bot. Mag. (Tokyo) **51**, 270 (1939).

PARTINGTON, J. R., and C. C. SHAH: J. chem. Soc. **1931**, 2071. — PHELPS, A. S., and P. W. WILSON: Proc. Soc. exp. Biol. (N.Y.) **47**, 473 (1941). — POLLOCK, M. R.: Brit. exp. Path. **27**, 419 (1946).

QUASTEL, J. H., M. STEPHENSON and M. D. WHETHAM: Biochem. J. **19**, 304 (1925). RACKER, E.: J. biol. Chem. **184**, 313 (1950). — RAISTRICK, H., and A. STÖSSL: Biochem. J. **68**, 647 (1958). — RIJVEN, A. H. G. C.: Aust. J. biol. Sci. **11**, 142 (1958). — ROBERTS, W. R., and M. A. AZIM: In Inorganic Nitrogen Metabolism p.176 (W. D. MCELROY and B. GLASS eds.) Baltimore: Johns Hopkins Press 1956. — ROBINSON, J.: Arch. Biochem. **52**, 148 (1954). —

Roussos, G. G.: Fed. Proc. **18**, 1239 (1959). — Roussos, G. G., and A. Nason: J. biol. Chem. **235**, 2997 (1960). — Roussos, G. G., H. Takahashi and A. Nason: J. Bact. **73**, 594 (1957). — Russel, J. A.: J. biol. Chem. **156**, 457 (1945).

Sacks, L. E., and H. A. Barker: J. Bact. **58**, 11 (1949). — Sadana, J. C., and W. D. McElroy: Arch. Biochem. **67**, 16 (1957). — Sanwal, B. D.: Arch. Biochem. **93**, 377 (1961). — Sato, R.: In Inorganic Nitrogen Metabolism p. 163 (W. D. McElroy and B. Glass eds.) Baltimore: Johns Hopkins Press University 1956. — Sato, R., and F. Egami: Bull. Chem. Soc. (Japan) **22**, 137 (1949). — Sato, R., and M. Niwa: Bull. Chem. Soc. (Japan) **25**, 202 (1952). — Saz, A. R., and J. Marmur: Proc. Soc. exp. Biol. (N.Y.) **82**, 783 (1953). — Saz, A. R., and L. M. Martinez: J. biol. Chem. **223**, 285 (1956). — Saz, A. R., and R. B. Slie: J. Am. chem. Soc. **75**, 4626 (1953); — J. biol. Chem. **210**, 407 (1954). — Schimper, A. F. W., Bot. Ztg. **46**, pp. 65, 81, 97, 113, 129, 145 (1888); — Flora **73**, 207 (1890). — Senez, J. C., and F. Pichinoty: Biochem. biophys. Acta **27**, 569 (1958a); **28**, 355 (1958b). — Senez, J. C., F. Pichinoty and M. Konovalchikoff-Mazoyer: C. R. Acad. Sci. (Paris) **242**, 570 (1956). — Silver, W. S.: J. Bact. **73**, 241 (1957). — Silver, W. S., and W. D. McElroy: Arch. Biochem. **51**, 379 (1954). — Simpson, J. K., and W. C. Evans: Biochem. J. **55**, XXIV (1953). — Skinner, F. A., and N. Walker: Arch. Mikrobiol. **38**, 339 (1961). — Smith, G. M., and C. S. Worrel: Arch. Biochem. **24**, 216 (1949). — Snell, F. D., and C. T. Snell: In Colorimetric Methods of Analysis 3rd Edn. New York: Van Nostrand 1949. — Sommer, A. L.: Plant Physiol. **11**, 429 (1936). — Spencer, D.: Aust. J. biol. Sci. **12**, 181 (1959). — Spencer, D., H. Takahashi and A. Nason: J. Bact. **73**, 553 (1957). — Stevens, H. M.: Anal. chim. Acta **21**, 456 (1959). — Stickland, L. H.: Biochem. J. **25**, 1543 (1931). — Stoy, V.: Physiol. Plantarum **8**, 963 (1955); — Biochim. biophys. Acta **21**, 395 (1956). — Susuki, N., and S. Susuki: Sci. Rep. Tohuku Univ., 4th Ser. **20**, 195 (1954).

Takahashi, H., and A. Nason: Biochim. biophys. Acta **23**, 433 (1957). — Taniguchi, S., and K. Ohmachi: J. Biochem. (Tokyo) **48**, 50 (1960). — Taniguchi, S., A. Akira, I. Katsuhira, K. Masakiyo, O. Kazuchiyo and F. Egami: Proc. Int. Symposium Enz. Chem. Tokyo p. 238 (1958). — Taniguchi, S., A. Asano, K. Iida, M. Kono, K. Ohmachi and F. Egami: Proc. Int. Symposium Enzyme Chem. Tokyo and Kyoto, p. 238 (1957). — Taniguchi, S., and K. Iida: Biochem. biophys. Acta **44**, 253 (1960). — Taniguchi, A., and E. Itagaki: Biochem. biophys. Acta **31**, 294 (1959); **44**, 263 (1960). — Taniguchi, S., H. Mitsui, J. Toyoda, T. Yamada and F. Egami: J. Biochem. (Japan) **40**, 175 (1953). — Taniguchi, S., R. Sato and F. Egami: In Inorganic Nitrogen Metabolism p. 87 (W. D. McElroy and B. Glass eds.) Baltimore: Johns Hopkins Univ. Press 1956. — Tang, Pei-Sung, and Hsiung-Yü Wu: Nature (Lond.) **179**, 1355 (1957). — Thurlow, S.: Biochem. J. **19**, 175 (1925).

Vaidyanathan, C. S., and H. E. Street: Nature (Lond.) **184**, 531 (1959). — Vallee, B.: In The Enzymes Vol. 3, p. 225 (P. D. Boyer, H. Lardy and K. Myrbäck, eds.) New York: Academic Press 1960. — Vallee, B. L., and F. L. Hoch: J. Amer. Chem. Soc. **77**, 821 (1955). — Verhoeven, W.: Aerobic Spore Forming Nitrate Reducing Bacteria. Thesis Delft (1952); — In Inorganic Nitrogen Metabolism p. 61 (W. D. McElroy and B. Glass, eds.) Baltimore: Johns Hopkins Univ. Press 1956. — Verhoeven, W., A. L. Kooter and M. C. A. Van Niebelt: Antonie v. Leeuwenhoek **20**, 273 (1954). — Verhoeven, W., and Y. Takeda: In Inorganic Nitrogen Metabolism p. 159. (W. D. McElroy and B. Glass eds.) Baltimore: Johns Hopkins Univ. Press 1956. — Villanueva, J. R.: Biochem. J. **72**, 36P (1959a); — J. gen. Microbiol. **20**, VI (1959b); — Nature (Lond.) **184**, 549 (1959c). — Virtanen, A. I., and T. Z. Csàky: Nature (Lond.) **161**, 814 (1948). — Virtanen, A. I., and T. Laine: Biochem. J. **33**, 412 (1939). — Virtanen, A. I., and N. Rautenen: In The Enzymes p. 1089. (J. B. Sumner and K. Myrbäck eds.) New York: Academic Press 1951. — Virtanen, A. I., and N. E. Saris: Acta chem. scand. **9**, 337 (1955); **10**, 483 (1956).

Wachsman, J. T.: J. biol. Chem. **223**, 19 (1956). — Wainwright, S. D.: Biochim. biophys. Acta **18**, 583 (1955). — Walker, G. C., and D. J. D. Nicholas: Biochem. J. **77**, 4P (1960); — Walker, G. C. and D. J. D. Nicholas: Nature (Lond.) **189**, 141 (1961a). — Biochim. biophys. Acta **49**, 350 (1961b); **49**, 361 (1961c). — Webber, M. W., Lenhoff H. M. and N. O. Kaplan: J. biol. Chem. **220**, 93 (1956). — Westerfeld, W. W., D. A. Richert and E. S. Higgins: J. biol. Chem. **227**, 379 (1957). — Wickeram, L. J.: J. Bact. **74**, 837 (1957). — Wijler, J., and C. C. Delwiche: Plant and Soil **5**, 155 (1954). — Wilson, P. W.: In Handbuch der Pflanzenphysiologie Bd. VIII p. 9. (K. Mothes ed.) Berlin-Göttingen-Heidelberg: Springer 1958. — Wood, J. G.: Ann. Rev. Plant Physiol. **4**, 1 (1953). — Woods, D.: Biochem. J. **32**, 2000 (1938). — Wosilait, W. D., and A. Nason: J. biol. Chem. **206**, 255 (1954).

Yamata, T., and A. I. Virtanen: Acta chem. scand. **10**, 20 (1956). — Yamagata, S.: Acta phytochim. (Tokyo) **11**, 145 (1939). — Yamafuji, K., and T. Akita: Enzymologia **15**, 313 (1952). — Yamafugi, K., and Y. Osajima: Nature (Lond.) **190**, 534 (1961).

Zucker, M., and A. Nason: Fed. Proc. **13**, 328 (1954); — J. biol. Chem. **213**, 463 (1955); — In Methods of Enzymology vol. II p. 406 (S. P. Colowick and N. O. Kaplan eds.) New York: Academic Press 1956.

Enzymes of Vitamin Metabolism.

By

T. W. Goodwin

With 5 Figures.

In the generally accepted sense of the term, vitamins are complex organic chemicals which are essential for the normal functioning of the metabolic processes of an animal and have to be supplied in the diet. Some micro-organisms, bacteria, fungi and algae, also depend on an external supply of such compounds which, in this context, are termed *growth factors*. Many other micro-organisms and all higher plants are, on the other hand, entirely independent of an external source of vitamins. These organisms utilize vitamins in their metabolism but they have evolved the necessary enzymic apparatus for the synthesis of all necessary growth factors.

Plants are, therefore, potentially good sources of enzymes carrying out the biosynthesis of vitamins, but although considerable advances have recently been made in studying the mechanisms involved (GOODWIN, 1963) very few of the investigations have reached the stage at which individual enzymes have been isolated.

The first section of this chapter *Enzymes concerned with biosynthesis* will therefore be comparatively short.

The techniques which were evolved for the extraction and purification of vitamins invariably led to the isolation of what might be called the *basic vitamin*, that is the organic structure, for example thiamine, which an animal cannot synthesize. It soon became apparent, however, that these organic molecules did not represent the metabolically active forms of the vitamins and that they needed to be *activated* in some way before they were effective; if we continue with thiamine as our example, it has to be converted into thiamine pyrophosphate before it is active as an enzyme prosthetic group. The second section of this chapter is, therefore, taken up with a discussion of enzymes dealing, in the broadest sense, with the *activation of vitamins*.

The final section of the chapter is concerned with plant enzymes which destroy vitamins as such, or *break down the active forms to the free vitamins*.

Enzymes which are concerned with reversible changes which vitamins, as prosthetic groups of enzymes, undergo (e.g. $DPN^+ \rightleftharpoons DPNH$) are not discussed here; they are more appropriately considered in the sections dealing with general metabolism.

Apart from ascorbic acid, which is the subject of a separate chapter (p. 204 ff), it can be assumed that if a vitamin is not mentioned in this article, insufficient is currently known about its enzymology for it to be considered here.

A. Enzymes Concerned with Biosynthesis.

1. Thiamine.

Recently CAMIENER and BROWN (1960) have shown that the following sequence of reactions represents the final steps in the biosynthesis of thiamine in yeast:

The functional form of thiamine, thiamine pyrophosphate (TPP) is synthesized from thiamine by pyrophosphate transfer and not from TMP by orthophosphate transfer (see p. 180f).

2-Methyl-4-amino-5-hydro-
xymethylpyrimidine
(thiamine-pyrimidine)

$$\xrightarrow[\text{pyrimidine kinase}]{\text{ATP, Mg}^{2+}}$$

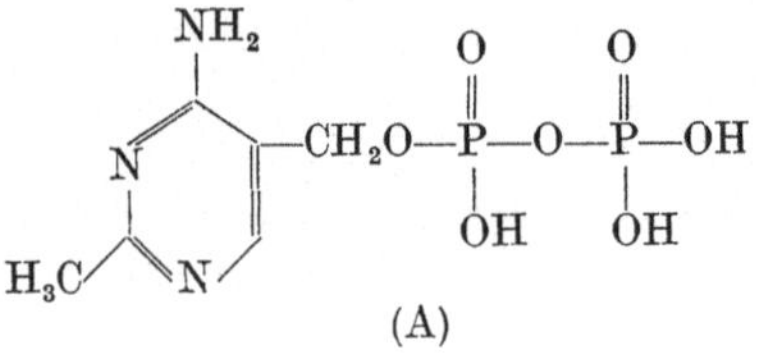

(A)

$$\text{(1)}$$

4-Methyl-5-(β-hydroxyethyl)
thiazole
(thiamine-thiazole)

$$\xrightarrow[\text{thiazole kinase}]{\text{ATP, Mg}^{2+} \quad \text{ADP (?)}}$$

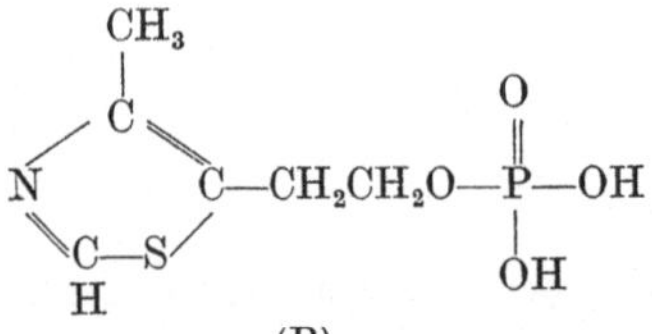

(B)

$$\text{(2)}$$

A + B

$$\xrightarrow[\substack{\text{P—P (?)}\\\text{Thiamine phosphase}\\\text{synthetase}}]{\text{Mg}^{2+}}$$

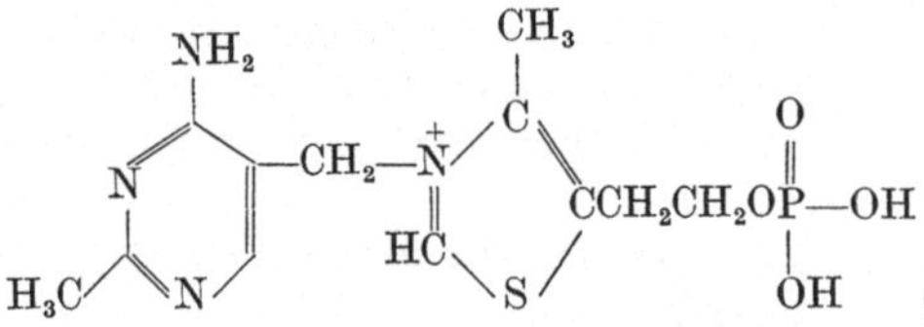

Thiamine monophosphate (TMP)

$$\text{(3)}$$

Thiamine monophosphate
TMP

$$\xrightarrow[\substack{\text{Thiamine}\\\text{monophosphatase}}]{\text{P}}$$

Thiamine

$$\text{(4)}$$

a) General Preparation of Enzyme Fractions.

Baker's yeast is autolysed by standard methods (McDonald, 1955) in the presence of toluene, but 0.02 M phosphate buffer (pH 6.9) is used instead of water. Solid $(NH_4)_2SO_4$ is added at 4° and the fraction precipitating between 58—65% saturation collected. The precipitate is dissolved in 0.02 M phosphate buffer (pH 7.0) and dialysed first for 18 hr. against 40 volumes of 0.005 M phosphate buffer, pH 7.0 containing 0.005 M EDTA and 0.02 M cysteine (buffer A), the buffer being changed after 6 hours, and then for 12 hr. against 0.0044 M phosphate buffer, pH 7.0, containing 0.0044 M EDTA and 0.0044 M cysteine. The solution is then fractionated at 4° on a column (2.2 ×35 cm) of 18 g DEAE-cellulose slurry. The column is washed with 500 ml of buffer A, the dialysed enzyme (about 1.5 g protein in 44 ml) applied, and a further 200 ml of buffer A run through. The elution is then carried out using a NaCl/cysteine logarithmic gradient. The mixing chamber contains initially 1 l of 0.005 M NaCl in buffer A and the reservoir 1500 ml of 0.130 M NaCl in buffer A; 10 ml fractions are collected and 0.2 ml of 1.0 M cysteine (pH 7.0) added to each fraction to preserve enzymic activity. Tubes 14—120 are combined as fraction 1 and tubes 124—185 as fraction 2. Fraction 1 contains two enzymes, pyrimidine kinase (reaction 1) and thiamine pyrophosphokinase; fraction 2 contains thiazole kinase (reaction 2) and thiamine phosphate synthetase (reaction 3); thiamine monophosphatase (reaction 4) which occurs in the crude yeast extract is lost during purification.

b) Pyrimidine Kinase.

It has not yet been possible to decide whether the pyrimidine pyrophosphate is formed via the orthophosphate, which can be detected as a metabolite in yeast, or by direct pyrophosphate transfer. If the latter were true, then the enzyme would be more correctly named pyrimidine pyrophosphokinase.

Assay. (CAMIENER and BROWN, 1960). A complete assay has not yet been described but the formation of the pyrimidine pyrophosphate can be determined by paper chromatography (isobutyric acid, NH_4OH, water, 198:3:99) and the spots located by ultra-violet light or by bioautography. The latter technique involves placing the developed chromatogram in contact with the surface of a solid medium containing per litre, K_2HPO_4 21 g, KH_2PO_4 9 g, Na citrate, $2H_2O$, 1.0 g, $MgSO_4$, 0.1 g, $(NH_4)_2SO_4$, 2.0 g, mannitol, 20 g, thiamine-free yeast extract 10 g, pantothenic acid 1.0 mg, nicotinamide, 1.0 mg, riboflavin 1.0 mg, *p*-aminobenzoic acid, 2.0 mg, pyridoxine, 4.0 mg, folic acid, 0.2 mg, biotin, 0.02 mg and agar, 20 g) in a 22×30 cm sterile petri dish as a 5 mm layer and seeded with *Salmonella typhimurium* Ath 4 (a mutant requiring either thiamine or the pyrimidine portion of thiamine). After 5 min, the chromatogram is removed and the plate covered and incubated for 16 hr. at 37°. The resulting growth zones correspond to the areas of the chromatogram which contain the thiamine-pyrimidine and its phosphate esters.

The incubation mixture contains 58 mμmoles of hydroxymethyl pyrimidine, 10 μmoles ATP, 10 μmoles $MgCl_2$, 0.1 ml of enzyme (fraction 1) in a final volume of 1.0 ml of 0.1 M phosphate buffer, pH 6.9. Incubation is for 3 hr. at 37° and the reaction is stopped by heating the mixture in a boiling water bath.

c) Thiazole Kinase.

Formation of thiazole phosphate can be demonstrated by incubating thiazole, 20 mμmoles; ATP 10 μmoles, $MgCl_2$, 10 μmoles and enzyme (fraction 2) in a final volume of 1.3 ml of 0.08 M glycylglycine buffer pH 7.4, for 3 hr. at 37° (CAMIENER and BROWN, 1960).

The thiazole phosphate produced is detected chromatographically on paper with the solvent system described in the previous section, and confirmed by bioautography using the thiazole-requiring mutant *E. coli* 26−43 growing on DAVIS's (1949) medium solidified by the addition of 2% agar.

d) Thiamine Phosphate Synthetase.

Formation of "thiamine" is demonstrated with the incubation mixture described in the previous section with the addition of pyrimidine pyrophosphate (5 mμmoles). The "thiamine" produced is determined microbiologically using *Lactobacillus viridescens* (ATCC No. 12706) (CAMIENER and BROWN, 1960) which responds to thiamine, thiamine phosphate and thiamine pyrophosphate; all three may be present if crude enzyme extracts are used.

If the incubate is first treated with takadiastase, to hydrolyse any thiamine phosphates to free thiamine, then the vitamin can be determined fluorimetrically after oxidation to thiochrome (PETERS and O'BRIEN, 1955; LEDER, 1959).

e) Thiamine Phosphatase.

The yeast enzyme has not yet been examined in any detail (see CAMIENER and BROWN, 1960) but this hydrolysis is carried out by takadiastase preparations and by seminal phosphatase (see p. 195).

2. Riboflavin.

The outline of the general pathway of riboflavin biosynthesis is now fairly clearly established (Fig. 1) (see e.g. GOODWIN, 1960), but detailed enzymological studies have been carried out only on the last step — the conversion of 5,6-di-methyl-8-ribityllumazine (compound G) into riboflavin. In this reaction it would appear that one molecule of G is broken down to provide an active 4-C unit (or possibly two 2-C units) which then condenses with a further molecule of G to form riboflavin (Plaut, 1960; Goodwin and Horton, 1961) (reaction 5).

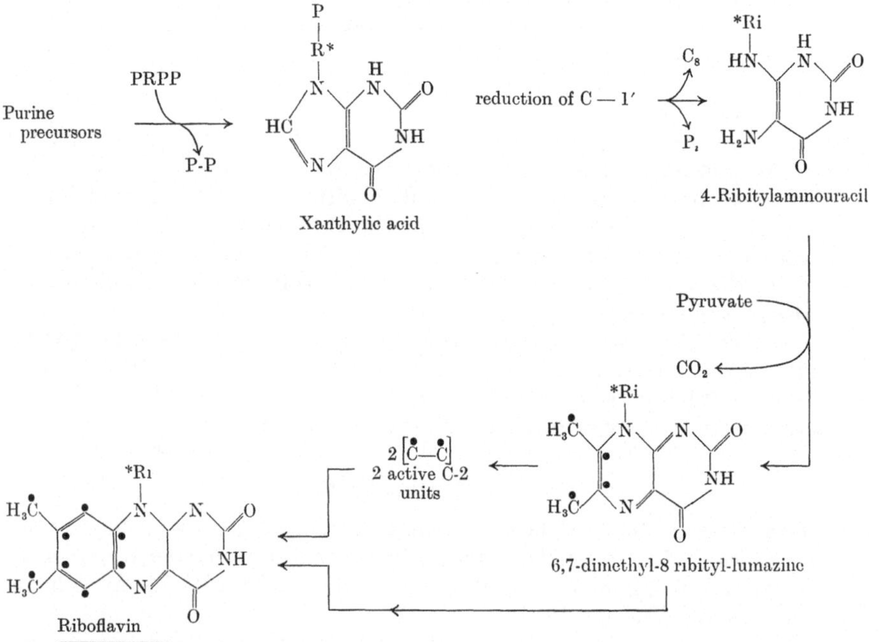

Xanthylic acid

4-Ribitylaminouracil

6,7-dimethyl-8-ribityl-lumazine

Riboflavin

* R = Ribose; Ri = Ribitol.

Fig. 1.

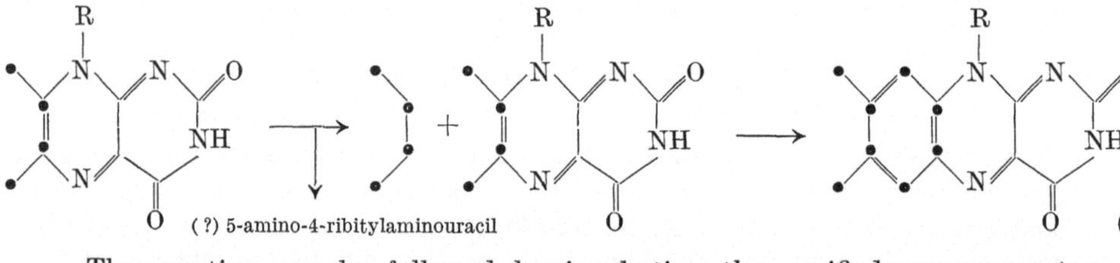

(?) 5-amino-4-ribitylaminouracil   (5)

The reaction can be followed by incubating the purified enzyme system (2.9 mg of material with specific activity 35 mμmoles of riboflavin formed per mg protein her hour at 37°) with 50 μmoles of 6,7-dimethyl-8-ribityllumazine in a final volume of 3 ml phosphate buffer pH 6.9 at 37° for 1 hr. The reaction is then stopped by the addition of 1.5 ml of 15% trichloroacetic acid and the protein removed by centrifugation. The riboflavin in the supernatant is separated from 6,7-dimethyl-8-ribityllumazine by column (Magnesol) chromatography (Plaut, 1960) or by paper chromatography using distilled water as solvent (Goodwin and Horton, 1961), and determined spectrophotometrically or fluorimetrically by standard procedures (Strong, 1955).

3. Nicotinic Acid.

The general pathway for the synthesis of nicotinic acid from tryptophan in fungi (and animals) is now well established (Fig. 2) (see Goodwin, 1960). In higher plants and bacteria the pathway is still obscure, the clearest piece of information being that tryptophan is not the source of nicotinic acid.

The enzymes which have been purified are those concerned with the first four steps (Fig. 2) and have been named tryptophan pyrrolase, formylase, kynurenine 3-hydroxylase and kynureninase, respectively.

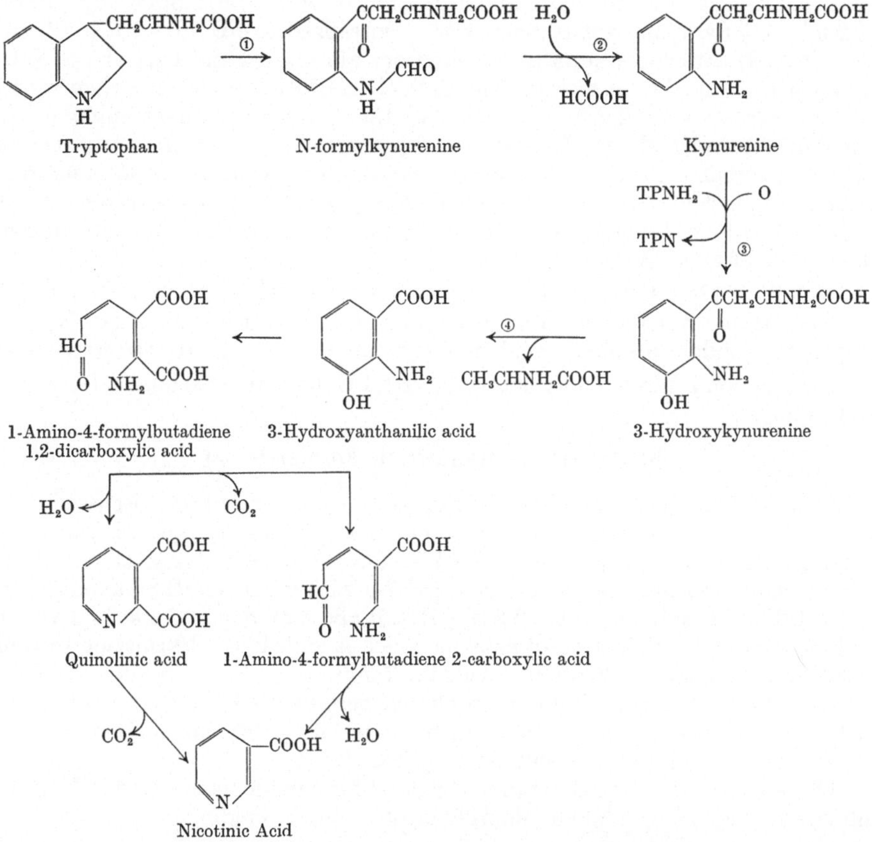

Fig. 2. Pathway of nicotinic acid biosynthesis in animals and fungi.

a) Tryptophan Pyrrolase.

The enzyme, which is an iron-porphyrin enzyme, has only been studied in detail when extracted from liver and *Pseudomonas* spp. It catalyses reaction 1 (Fig. 2) and functions as an oxygen transferase because both oxygen atoms which are utilized per molecule of tryptophan oxidized appear in each molecule of N-formyl kynurenine produced.

Assay (KNOX, 1955). The principal of the assay is to hydrolyse the resulting formylkynurenine to kynurenine in the presence of formylase (kynurenine form-amidase). The amount of kynurenine present is then determined spectrophoto-metrically at 365 mμ. The extinction coefficient of kynurenine at 365 mμ, and pH 6.5 is 4.53×10^3.

When crude liver homogenates are used excess formylase is always present; if plant tissues are to be examined then the presence of this enzyme in the extract must be demonstrated. If it is not present then it must be added to the reaction mixture.

Four cups (two concentrations of enzyme and two non-tryptophan blanks) are equilibrated in a Dubnoff shaker at 37° in an atmosphere of oxygen. Two of the cups contain 0.3 ml L-tryptophan (0.03 M) and all cups contain 1 ml NaH_2PO_4/ Na_2HPO_4 buffer (0.2 M, pH 7.0), a suitable volume of enzyme and water to a final volume of 4 ml. The incubation is continued for 60 min, stopped by addition of 2 ml 15% (w/w) metaphosphoric acid (neutralized to pH 7.0—7.5 by 2 ml of 1.0 N NaOH). After five minutes the solutions are filtered and 1 ml of 1 N NaOH is added to 3 ml of each filtrate. The extinction of each solution is then read at 365 mμ, care being taken to keep the solution away from direct sunlight. The differences between the readings of the experimentals and blanks are a measure of the kynurenine formed. Decrease in activity of preparations on standing is often due to lack of peroxidases producing H_2O_2 *in situ*. Activity is restored by the addition of 15 μmoles of glucose and sufficient glucose oxidase (notatin) to cause the uptake of 20 μl O_2/10 min.

Activity is expressed as μmoles kynurenine produced per ml enzyme per hour.

The enzyme has no action on D-tryptophan, DL-acetyltryptophan, N-acetyl tryptophan, indole propionic acid or tryptamine; K_m is 4×10^{-3} M tryptophan.

The reaction is inhibited by catalase, Cu^{2+}, CO (light-reversible), azide, cyanide and sulphide.

b) Formylase (Kynurenine Formamidase).

Assay (Knox, 1955). The basis of the assay has already been discussed in the previous section. The reaction is carried out in a 1 cm absorption cell, which contains 0.2 ml of formylkynurenine (0.01 M, formed by dissolving 12 mg of free acid in water and quickly adjusting to pH 7.0 with 0.1 N NaOH to avoid hydrolysis), 1.0 ml phosphate buffer (0.2 M, pH 7.5) and water to make a final volume of 3 ml after addition of 1 ml of enzyme which is added last. Extinction readings at 365 mμ are taken at 30 sec intervals for 3 min.

Activity is expressed as the density change per min produced by 1 ml of enzyme in the above system. The method can be used with crude enzyme extracts provided they give solutions of sufficient optical clarity.

The enzyme will attack many aromatic formamido compounds. It is not inhibited by cyanide, sulphide, fluoride or other metal-binding agents.

c) Kynurenine Hydroxylase.

The conversion of kynurenine into 3-hydroxykynurenine has up to the present been demonstrated only in intact liver mitochondria from rats and cats. The conversion specifically requires $TPNH_2$ and the reaction is typically that of a mixed function oxidase; one atom of oxygen from each molecule consumed appears on each molecule of 3-hydroxykynurenine synthesized, (see Goodwin, 1960).

No assay method has yet been described, but using the incubation mixture described below the formation of 3-hydroxykynurenine and disappearance of kynurenine can be demonstrated by paper chromatography. An adequate solvent system is methanol, butanol, benzene, water (2:1:1:1) containing either 1% acetic acid or 1% 15 N NH_4OH (de Castro, Price and Brown, 1956).

Washed mitochondria (200—400 mg wet wt.) are incubated aerobically at 37° for 2 hr. in 1.5 ml of 0.05 M phosphate buffer (pH 7.4) containing 4.0 μmoles

TPNH$_2$, 3.3 μmoles nicotinamide and 2.5 μmoles of L-kynurenine. The chromato-gram can be spotted directly without deproteinization. TPN + citrate can replace TPNH$_2$.

The enzyme is specific for L-kynurenine; no action is observed with D-kynu-renine, N-acetyl-L-kynurenine, N-acetyl-D-kynurenine, kynurenic acid or anthra-nilic acid.

d) Kynureninase.

This enzyme has been purified from *Neurospora crassa* (Em 5256 A) (JAKOBY and BONNER, 1953; SARAN, 1958) (it catalyses reaction 4, Fig. 2). The amount present is increased by 600 times when the mould is cultured in the presence of tryptophan. It is a pyridoxal phosphate-dependent enzyme and is activated by Mg^{2+}; L-kynurenine, and N-formyl-L-kynurenine are also substrates. Carbonyl reagents and *p*-chloromercuribenzoate inhibit the reaction.

Purification (JAKOBY and BONNER, 1953). Lyophilized mycelium is shaken with sand or ballotini (120 mesh) in a MICKLE shaker in 0.1 M phosphate buffer, pH 8.1. The suspension is strained through cheese cloth and then centrifuged for 40 min at 18,000 g at 4°. The enzyme is precipitated from the supernatant with (NH$_4$)$_2$SO$_4$ in the fraction precipitating between 42 and 60% saturation, but the salt must be added in steps of 48, 54 and 60% saturation. The combined precipi-tates are dissolved in 0.1 M phosphate buffer pH 8.1 and treated with protamine (30 mg/100 mg protein). The precipitate is removed by centrifugation and the supernatant dialysed for 4 hr. against two changes of distilled water. The pre-cipitate is discarded and ethanol (− 30°) is added to the supernatant at − 2° over a period of ten minutes to a final concentration 20% (v/v). During this period the temperature of the mixture should reach − 14°. The concentration of ethanol is then further increased to 55% (v/v) during the next 10 min. After stirring for 10 min the precipitate is discarded by centrifugation at 15° (20000 g for 15 min), and then the ethanol concentration brought to 70% (v/v) over 5 minutes; after stirring for 5 min the precipitate is collected by centrifugation, dissolved in 0.05 M phosphate buffer (pH 8.1), dialysed against two changes of distilled water for 3 hr. each, and centrifuged. The supernatant is treated with calcium phosphate gel [45 mg/ml, pH 6.5 (acetic acid)], the mixture stirred for 20 min in an ice bath and the sediment discarded.

Assay (JAKOBY and BONNER, 1953). The reaction mixture, in a fluorimeter cuvette, contains 1.2 μg pyridoxal phosphate, 2.4 μmoles magnesium sulphate, 0.12 μmole of 3-hydroxy-L-kynurenine and the enzyme in 1.2 ml of 0.1 M phos-phate buffer (pH 8.1). The mixture is incubated at 25°, and the increase in fluores-cence, due to the formation of hydroxyanthranilic acid, measured at one minute intervals, with 50 mμmoles of hydroxyanthranilic acid as standard. The blank value is determined by extrapolating to zero time.

4. Folic Acid.

(I)

The pteridine residue of folic acid (I) clearly arises in a similar way as do rings B and C of riboflavin; that is, from a purine precursor. Furthermore, work with

12*

extracts from *Mycobact. avium* indicate that the complete molecule is built up by the condensation of a pteridine (the exact compound is still uncertain but it is possibly 2-amino-4-hydroxypteridine-6-aldehyde) with *p*-aminobenzoyl glutamate. However, no enzymic studies have yet been developed to any significant extent (see Goodwin, 1960).

5. Pantothenic Acid.

The general pattern of biosynthesis of pantothenic acid is outlined in Fig. 3. Most of the evidence leading to this scheme is nutritional, although cell-free systems have been obtained which carry out the hydroxymethylation of keto-valine and the condensation of pantoate and β-alanine (see Maas, 1959). It is with the enzyme catalysing the latter reaction that we are now concerned.

Pantothenate Synthetase.

The enzyme which carries out reaction 4 (Fig. 3) has been found in both bacteria and fungi (Maas, 1959) but only the bacterial enzyme has been purified

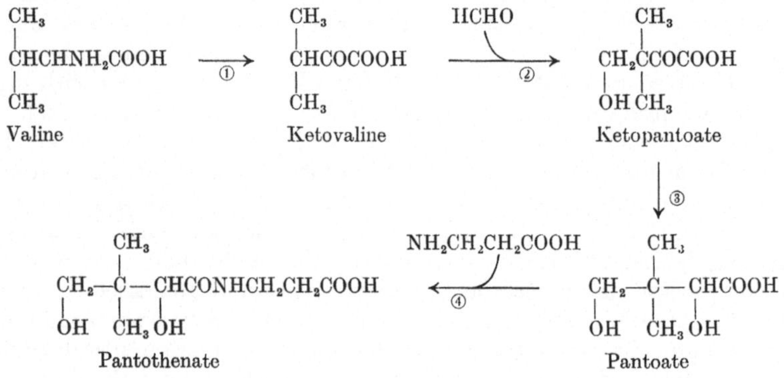

Fig. 3. The probable pathway of biosynthesis of pantothenic acid.

and examined in detail. ATP is specifically required as an energy source and pantoyladenylate is probably an intermediate in the reaction. Additionally Mg^{2+} or Mn^{2+}, and K^+ or NH_4^+ are required as activators.

Assay (Novelli, 1955a). Tubes are set up containing per ml 10 μmoles of ATP, 100 μmoles of KCl, 20 μmoles of β-alanine, 20 μmoles of K pantoate, 10 μmoles of $MgSO_4$, 100 μmoles Tris buffer (pH 8.5) and varying amounts of enzyme solution. The tubes are incubated at 25° for 30 min and the reaction stopped by heating in a boiling water bath for 3 min. The pantothenic acid formed is assayed microbiologically with a pantothenate-requiring mutant of *E. coli* (Maas and Davis, 1950).

A unit of activity is the amount which causes the synthesis of 1 μmole of pantothenate under the conditions just described.

B. Activating Enzymes.

1. Thiamine.

a) Thiamine Pyrophosphokinase.

The metabolically functional form of thiamine is thiamine pyrophosphate (TPP) (also designated co-carboxylase, thiamine diphosphate) and it is formed from thiamine and ATP in the presence of thiamine pyrophosphokinase and Mg^{2+} (reaction 6) (WESTENBRINK, 1959). The mechanism involves the transfer of an intact pyrophosphate residue from ATP; this was demonstrated using [^{32}P]-ATP (FORSANDER, 1956); furthermore thiamine monophosphate is not a substrate for the enzyme (CAMIENER and BROWN, 1959, 1960; NOSE, UEDA and KANASAKI, 1959).

$$\text{Thiamine} + \text{ATP} \xrightarrow{Mg^{2+}} \text{TPP} + \text{AMP} \qquad (6)$$

The enzyme needs Mg^{2+} as activator; Mn^{2+} is also effective but at concentrations higher than 2×10^{-3} M it is inhibitory. The reaction is also stimulated by low concentrations of orthophosphate (2×10^{-3} M), but completely inhibited by molar concentrations. The enzyme has a broad pH optimum stretching between 6 and 7.5 (STEYN-PARVÉ, 1952). Yeast thiamine pyrophosphokinase is inhibited by substances present in a crude extract of baker's yeast, by the pyrimidine residue of thiamine (WESTENBRINK, STEYN-PARVÉ and VEDLMAN, 1947) and by oxythiamine (II) (EUSEBI and CERECEDO, 1950), but not by pyrithiamine (III) (EICH and CERECEDO, 1954); the enzyme from animal sources responds differently to these inhibitors. Intact yeast cells phosphorylate considerable amounts of thiamine, baker's yeast being more effective than brewer's yeast; phosphorylation is stimulated by the presence of glucose and the pH optimum varies between 6 and 3, so the reaction cannot be directly related to that of isolated thiamine kinase (WESTENBRINK et al., 1947).

(II)   (III)

Preparation (WESTENBRINK, 1955). Baker's yeast (10 g) is frozen at $-70°$ for 15 min and then thawed. The process is repeated twice. Solid KCl is added to a concentration 0.5 M and the mixture incubated at 37° with shaking for 3 hr. and then kept in the refrigerator overnight. The resulting material is centrifuged and the slightly opalescent light-brown supernatant is fractionated with $(NH_4)_2SO_4$; the fraction precipitating between 45 and 60 vol.-% contains the enzyme. This increases the specific activity about ten fold. Reprecipitation [$55-65$ vol.-% $(NH_4)_2SO_4$] removes residual phosphatase activity but does not increase the specific activity. Further purification using other methods have not been reported.

The specific activity of the crude extract can be increased fivefold by dialysis against 0.02 M KCl; this is presumably owing to removal of natural inhibitors.

Brewer's yeast is treated slightly differently in the early stages; 200 g are plasmolysed with $20-30$ g NaCl and then 1 l of 0.2 M phosphate, pH 10.1, is added. After 10 min the mixture is centrifuged and the supernatant discarded. The residue is suspended in 200 ml distilled water, incubated for $2-3$ hr. at 37° and centrifuged. The supernatant contains the enzyme, which can then be purified as described in the previous paragraph.

Assay (WESTENBRINK, 1955). The reaction mixture (5 ml) contains thiamine-HCl, 5 mg; 0.02 M ATP (K salt), 0.5 ml; 0.1 mMg SO_4 0.5 ml; enzyme solution (varying vol.); 0.1 M phosphate buffer, pH 7.0, to volume. The high thiamine concentration is required to inhibit phosphatase activity. The mixture is incubated at 27° for 1 hr. and the reaction stopped by transferring 1 ml of incubate into

5 ml HCl (0.05 N) and boiling for 1 min. The TPP formed is determined as co-carboxylase (Westenbrink and Steyn-Parvé, 1950). Sodium pyruvate (0.2 ml of 2.5% solution) in 0.1 M phosphate buffer, pH 6.2 is placed in the side arm of a Warburg vessel. Into the main compartment are added 1 ml of a solution containing TPP (0.05—0.2 μg) and 0.5 ml of a suspension of alkaline-washed yeast, both being at pH 6.2. The CO_2 evolved is then measured conventionally, together with those from a blank (boiled enzyme) and three standard TPP concentrations. A standard curve is constructed and the unknown TPP concentration read off.

b) Thiamine Pyrophosphate Kinase.

When incubated with thiamine, baker's yeast accumulates thiamine triphosphate (TTP) as well as TPP (Kiessling, 1956, 1957, 1959). Experiments with $^{32}P_i$ and with $[\gamma\text{-}^{32}P]$-ATP indicate that TTP is formed by the transfer of

$$TPP + ATP \rightleftharpoons TTP + ADP \qquad (7)$$

P_i from ATP to TPP (equation 7). The metabolic significance of TPP is not yet known.

Purification. This has been achieved by Greiling (1958); details have not yet been published.

Assay (Kiessling, 1959). Baker's yeast (20 g) is suspended in 0.1 M succinate buffer pH 3.7, 0.2 g thiamine added and the suspension incubated at 27° with oxygen as the gas phase. Samples of the suspension are taken at various times, centrifuged in the cold and the supernatant discarded. The cells are suspended in cold 40% trichloroacetic acid to a final concentration of 8%. After 15 min at —10°, the suspension was stirred and centrifuged. The supernatant is extracted with ether to remove the trichloroacetic acid, and a measured volume chromatographed on Whatman 52 paper (previously washed in aqueous EDTA) using isobutyric acid, N NH_4OH, and 0.1 M EDTA (100:60:1.6) as solvent system. The TTP is located by running a second separation on the same paper and spraying with alkaline potassium ferricyanide; the resulting thiochrome derivative is then detected by its fluorescence. The unsprayed TTP is eluted with water and hydrolysed to thiamine by a yeast phosphatase (Westenbrink et al., 1950; see also p. 196). The thiamine is then determined fluorimetrically following its conversion to thiochrome. For full details see Peters and O'Brien (1955).

2. Riboflavin.

The functional forms of riboflavin are riboflavin 5′-phosphate (loosely called flavinmononucleotide, FMN (IV)] and flavin adenine dinucleotide [FAD (V)].

(IV)

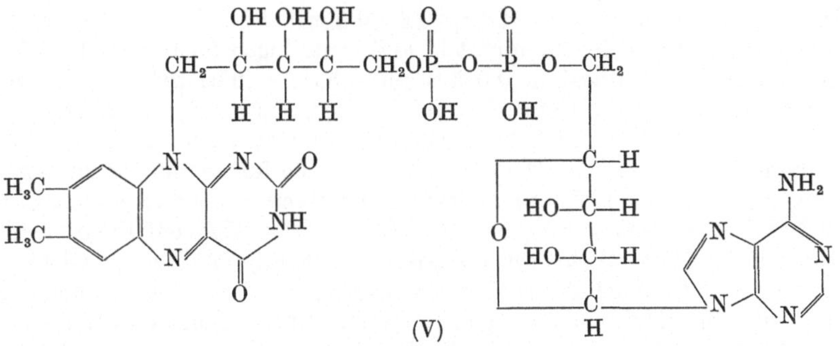

(V)

a) Flavokinase.

FMN is synthesized from riboflavin and ATP (equation 8) in the presence of *flavokinase*; the reaction is essentially irreversible.

The optimum pH for the yeast enzyme is 7.8—8.5 and Zn^{2+}, Mn^{2+} and Co^{2+} can substitute for Mg^{2+} as activator. AMP is a competitive inhibitor whilst lumiflavin (VI) and Co^{2+}, in high concentration, are also inhibitory (KEARNEY and ENGLARD, 1951). Apart from riboflavin, dichloroflavin and arabitylflavin are substrates; isoriboflavin (VII) galacto-, sorbityl- and dulcitylflavin are neither substrates nor inhibitors (KEARNEY and ENGLARD, 1951). The enzyme follows a zero-order reaction with K_m values for riboflavin and ATP 1.0×10^{-5} M and 1.7×10^{-5} M respectively (KEARNEY, 1955). The optimum pH for the enzyme from higher plants is 8.6 and the optimum temperature 55°, Mg^{2+}, Zn^{2+} and Mn^{2+} are activators, although the last two are inhibitory at higher concentrations; CN^-, Hg^{2+}, Fe^{2+}, Cu^{2+}, MnO_4^- and hydroxylamine are inhibitory. K_m for riboflavin is $1.5 \times \times 10^{-5}$ M (GIRI, KRISHNASWAMY and RAO, 1958).

$$\text{Riboflavin} + \text{ATP} \xrightarrow{\text{Mg}^{2+}} \text{FMN} + \text{ADP} \tag{8}$$

Preparation from yeast (KEARNEY, 1955). Brewer's yeast (500 g) is autolysed in 1 l of water at 36° for 2 hr. with efficient stirring. Water (500 ml) is then added and the mixture rapidly cooled to 30°; stirring with cooling is continued for 15 min by which time the yeast has reached 10°. It is then centrifuged (4000 to 5000 rpm) for 30—45 min. The activity is precipitated from the supernatant by 40% saturation with $(NH_4)_2SO_4$ (71.5 g/100 ml = 100% saturation). The precipitate is collected by centrifugation (5,000 rpm for 1 hr.), dissolved in 30—40 ml

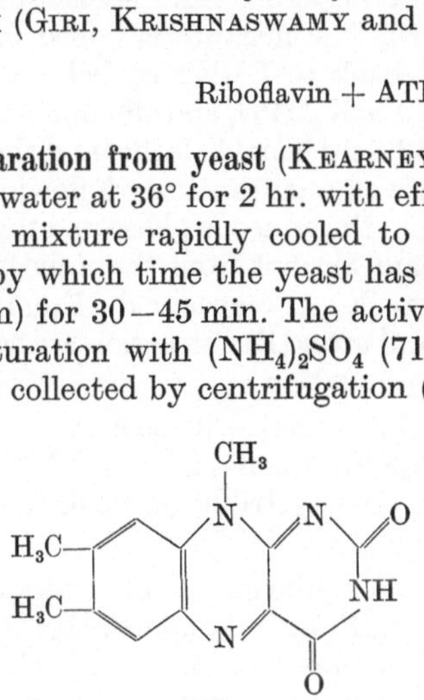

(VI)

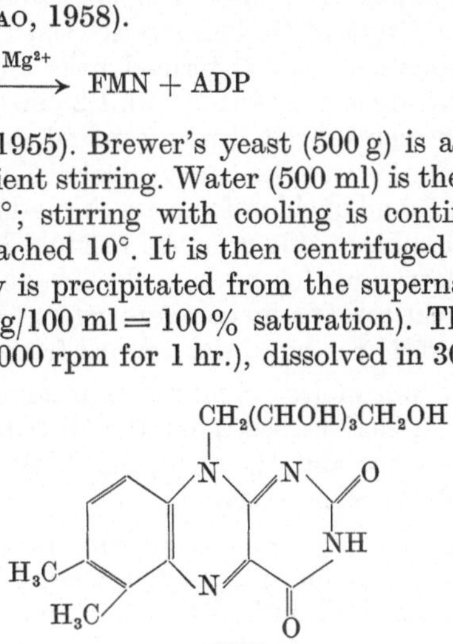

(VII)

water and dialysed in the cold against running water for 16—18 hr. The precipitate is centrifuged off and discarded. The dialysed enzyme solution (8—10 mg protein per ml) is brought to pH 5.0 at 0° with 1 N acetic acid (0.2—0.3 ml), aluminium hydroxide gel C_γ (60—80 mg/mg protein) added, and the mixture stirred for 15 min and then centrifuged for 30 min at 10,000 rpm. The supernatant is adjusted to pH 6.0 at 0° with 0.5 M K_2HPO_4 (0.4–0.5 ml), brought to 42% saturation with $(NH_4)_2SO_4$, stirred for 30 min at 0° and centrifuged for 15 min (10,000 rpm). The precipitate is dissolved in 0.015 M phosphate buffer, pH 7.2, the volume being one half that of the solution from which it was precipitated. The solution is centrifuged if not clear, and the $(NH_4)_2SO_4$ concentration raised to 42% saturation by addition of solid $(NH_4)_2SO_4$. After stirring for 15 min the suspension is centrifuged (10,000 rpm). The precipitate is dissolved in a small volume of water and dialysed at 0° against running water for 2–3 hr. The centrifuged residue is freeze-dried. The final stage, the dissolution of the lyophilized powder, should only be carried out immediately before use because solutions of the purified enzyme are very unstable. The powder is gently suspended in water and 0.2 M succinate buffer pH 6.0 added to give a molarity of 0.05 M and 5–6 mg enzyme/ml. After standing at 0° for 45 min any denatured protein is removed by centrifugation at 0° (18,000 rpm). This purification increases the specific activity from 0.57 unit per mg to 290 units/mg, with a loss of total activity of about 50%. One unit is the amount of enzyme which catalyses the synthesis of 1 mμmole of FMN in two hours at 30° per 5 ml reaction mixture under standard reaction conditions. Specific activity is units/mg protein. All manipulation must be carried out in subdued light.

Preparation from seeds (Giri et al., 1958). Finely powdered seeds (100 g) are extracted with water (300 ml) for 60 min at 0–5° with intermittent shaking, and the extract centrifuged at 1000 g. The supernatant is then treated in essentially the same way as that described by Kearney (1955). A recovery of 38% with a 75 fold purification can be achieved.

Assay (A) (Kearney, 1955). The reaction mixture (5 ml) contains 7.5×10^{-2} M Tris buffer, 1×10^{-4} M riboflavin, 1×10^{-3} M ATP, 1×10^{-3} M $MgSO_4$ and 50–100 units of enzyme. The mixture is incubated for 2 hr. at 30° and the reaction stopped by the addition of trichloroacetic acid (2 ml). The mixture is boiled for 10 min (to hydrolyze any FAD formed from FMN back to FMN), cooled and filtered; 5 ml of filtrate is neutralized with 1.25 ml of 2.4 M K_2HPO_4 and 450 mμ E measured. 5 ml of the neutralized filtrate is mixed with 12.5 ml benzyl alcohol and the mixture bubbled for 30 sec. The benzyl alcohol is removed quantitatively following centrifugation; 5.0 ml of chloroform are then added to the aqueous phase and the mixture further bubbled; this removes residual benzyl alcohol from the aqueous phase, which after centrifuging is collected and its $E_{450m\mu}$ measured. From the two readings the total flavin contents before (A) and after (B) extraction are calculated, and the FMN produced determined from the equation:

FMN (mμ moles/5.0 ml reaction mixture) $= 1.25\,B - 0.125\,A$.

The method is based on the distribution of riboflavin and FMN between benzyl alcohol and the neutralized filtrate, the distribution coefficients being 3.6 and 0.044 respectively.

Assay (B) (Giri et al., 1958). The reaction mixture (2 ml) contains 0.1 ml M riboflavin, 0.2 ml, mM ATP, 0.1 ml, 0.1 M NaF, 0.1 ml, 0.3 mM $MgSO_4$, 0.1 ml, 0.1 M-veronal-HCl buffer pH 8.6, 0.7 ml, enzyme preparation 0.8 ml. The mixture is incubated at 55° for 60 min and the reaction stopped by the addition of 0.8 ml of 17.5% trichloroacetic acid. The reaction tube is placed in a boiling water bath for 5 min (to hydrolyse any FAD formed to FMN). The solution is cooled, centrifuged and portions submitted to circular paper chromatography (Giri and Krishnaswamy, 1956) with butanol-acetic acid water (4:1:5, v/v) as solvent. The FMN formed is located with the aid of an ultra violet lamp, eluted with glass-distilled water and determined fluorimetrically (see e.g. Bessey, Lowry and Love, 1949).

b) FAD Pyrophosphorylase.

FAD is synthesized from FMN and ATP in the presence of FAD pyrophosphorylase by the reversible reaction indicated in equation (9). Mg^{2+} is required as activator, ADP does not replace ATP, and metaphosphate and inorganic orthophosphate do not replace inorganic pyrophosphate (P-P). Activity may be missed

$$FMN + ATP \rightleftharpoons FAD + P—P \qquad (9)$$

in crude extracts because of the presence of FAD-hydrolysing enzymes. The enzyme has only been purified from brewer's yeast but it is present in extracts of higher plants (GIRI, 1957; GIRI et al., 1958). The optimum pH of the enzyme is about 7.5 and K_m for FMN, ATP and FAD are 1.4×10^{-6} M, 1.2×10^{-5} M and 5.3×10^{-6} M, respectively (KORNBERG, 1955).

Purification (KORNBERG, 1955). Brewer's yeast (100 g) is autolysed with 300 ml of 0.1 M $NaHCO_3$ (saturated with a mixture of 95% N_2 and 5% CO_2) for 24 hr. at 23°. The remaining steps are, unless stated otherwise, carried out at 3°. After centrifugation the autolysate is diluted with H_2O to 324 ml and 108 g of $(NH_4)_2SO_4$ added; the resulting precipitate is removed by centrifuging, dissolved in 60 ml water and dialysed against running demineralized water for 1 hr. The dialysed solution is diluted to 75 ml with water and mixed with 75 ml of 0.1 M Na acetate (pH 5.0). After 5 min 95% ethanol (12 ml) is added dropwise with constant stirring at $-1°$ and the mixture centrifuged. The supernatant is treated similarly at $-2°$ with 23 ml of 95% ethanol and the precipitate collected by centrifugation. It is dissolved in 48 ml H_2O and sufficient 0.1 N NaOH to give a neutral solution; the pH is then adjusted to pH 5.85 with 0.02 N acetic acid and 13.8 ml of aluminium hydroxide gel C_γ (15.5 g dry wt./l) added. After 10 min the suspension is centrifuged and the residue is washed with 0.02 M acetate buffer pH 6.0 (13 ml) and then eluted with 3×14 ml 0.02 M phosphate buffer pH 6.0. The combined eluates are diluted to 48 ml with water and 1.0 N acetic acid (1.0 ml) added at 0° with constant stirring, this is followed by 95% ethanol (6.9 ml) at $-1°$ to $-2°$. The precipitate is removed by centrifugation, dissolved in 20 ml H_2O and sufficient 0.1 NaOH to give a neutral solution; the pH is then adjusted to pH 6.0 with 0.02 N acetic acid and the solution treated with 21.4 ml of calcium phosphate gel (8.2 g dry wt./l). After 10 min the suspension is centrifuged and the residue washed with 0.02 M Na acetate buffer, pH 6.0 (17 ml) and eluted four times with 0.01 M phosphate buffer (pH 7.7). This final solution contains about $1.3 - 1.9$ mg protein/ml and represents an overall yield of 45% and a 91 times purification; at 3° it loses $20 - 30\%$ its activity in 4 days. It contains 83 units/mg protein (one unit is the amount catalysing the synthesis of 1 mμmole FAD per hour).

Assay (KORNBERG, 1955a). The incubation mixture (1 ml) contains 0.15 M $MgCl_2$, 0.05 ml; 0.02 M ATP, 0.1 ml; 2×10^{-4} M FMN, 0.1 ml; 0.25 M KH_2PO_4/ K_2HPO_4 buffer (pH 7.5), 0.1 ml, enzyme $3 - 10$ units, water to volume. The mixture is incubated for $6 - 15$ min at 37° and the reaction stopped by immersion in boiling water for 3 min. After cooling and centrifuging the supernatant is assayed for FAD by activation of D-amino acid apooxidase (BURTON, 1955).

In this assay the oxygen consumption (equation 10) is measured manometrically in a conventional WARBURG apparatus. The main compartment of each WARBURG flask contains 0.1 M pyrophosphate buffer pH 8.3, 1 ml; catalase[1]

$$R\,CHNH_2COOH + O_2 + H_2O \longrightarrow R\,COCOOH + NH_3 + H_2O_2 \qquad (10)$$

---

[1] Red blood cells can be used as a source of catalase, which is added to ensure complete destruction of the H_2O_2 formed.

0.5 μg; 10^{-4} M FAD[1], 0.1 ml; D-aminoacid oxidase; final volume 2.3 ml. DL-Alanine (5%) (0.2 ml) is in the side arm and the centre well contains 0.2 ml 2 N NaOH and filter paper to absorb CO_2. After equilibration, readings are taken at 5 min intervals and the rate should be constant until at least 100 μl O_2 are consumed. The range of the assay is $1-5$ units, one unit being the amount of enzyme catalysing the consumption 1 μl of O_2 per minute.

3. Pyridoxal (Vitamin B_6).

Pyridoxal functions metabolically as pyridoxal 5-phosphate (VIII).

CHO
HO—⟨ring⟩—CH_2OP—OH, H_3C—N, O, OH
(VIII)

CH_2OH
HO—⟨ring⟩—CH_2OH, H_3C—N
(IX)

CH_2NH_2
HO—⟨ring⟩—CH_2OH, H_3C—N
(X)

Pyridoxal Phosphokinase.

Pyridoxal phosphate is formed from pyridoxal and ATP in the presence of pyridoxal phosphokinase (equation 11). The enzyme will also phosphorylate

$$\text{Pyridoxal} + \text{ATP} \rightarrow \text{Pyridoxal 5-phosphate} + \text{ADP} \tag{11}$$

pyridoxine (IX) and pyridoxamine (X) at approximately the same rate as pyridoxal; it is specific for ATP and inhibited by adenine, adenosine, AMP, ADP, F' (0.1 M), pyrophosphate and inorganic phosphate at high concentrations. The optimum pH is 6.9. The enzyme is activated by Mg^{2+}, Fe^{2+} or Co^{2+} but the last inhibits at high concentrations. K_m values for ATP, pyridoxal and Mg^{2+} are 3.3×10^{-4} M, 1.4×10^{-4} M and 4.7×10^{-4} M, respectively (HURWITZ, 1953; GUNSALUS and RAZZELL, 1955; BAUM and GUNSALUS, 1961).

Purification (HURWITZ, 1953). Dried brewers' yeast (500 g) is autolysed in 1 l of water as described under "Flavokinase" (p. 183). The enzyme is precipitated from the autolysate by the addition of 28 g $(NH_4)_2SO_4$ per 100 ml of extract. This and all subsequent manipulations are carried out at 0°. The precipitate is recovered by centrifugation, suspended in 60 ml of water and dialysed against distilled water for 15 hr. The solution is then centrifuged and the clear supernatant is adjusted to pH 5 with N acetic acid and stirred for 30 min. The precipitate is removed by centrifugation and $1/30$ volume of 3 M acetate buffer, pH 5, is added to the supernatant followed by ethanol, cooled to $-10°$ slowly over 15 min, to give a final volume of 12% (v/v). The precipitate is recovered by centrifugation, and dried in a vacuum desiccator. This material represents a recovery of 230% (owing to removal of phosphatases) and a 440-fold purification. The enzyme can be further purified by a second $(NH_4)_2SO_4$ precipitation followed by heating at 55° for 5 min. The resultant represents a recovery of $40-50\%$ and a 1000-fold purification.

Assay (HURWITZ, 1953). The assay is based on the activation of tyrosine apodecarboxylase. The reaction mixture (2 ml) contains $10-100$ units pyridoxal kinase in 1.2 ml (one unit is 1 mμmole of pyridoxal phosphate formed in 2 hr. at 30°), 0.3 M phosphate buffer, pH 6.85, 0.2 ml; 0.05 M Na_4ATP, 0.2 ml; 0.05 M pyridoxal hydrochloride neutralized to pH 6.85, 0.2 ml; 0.01 M $MgSO_4$, 0.2 ml and water to final vol. The mixture is incubated for 2 hr. at 30° and the reaction

---

[1] Or the unknown material suitably diluted.

stopped by immersion of the tube in a boiling water bath for 3 min. The suspension is diluted to 5 ml and filtered (solution A).

The residual ATP, which might react with pyridoxal in the next step in the presence of crude tyrosine apodecarboxylase, is removed by treatment with potato apyrase (KRISHNAN, 1955). The reaction mixture contains 0.8 ml of A (containing about 1.6 μmoles of ATP), 0.1 M succinate buffer, pH 6.5, 0.5 ml; 0.045 M $CaCl_2$, 0.2 ml; apyrase 0.5 ml (200 units)[1]; and water to 2 ml. The mixture is incubated for 30 min at 30° and the reaction stopped by chilling in an ice bath. The solution is then diluted to 5 ml (solution B). The pyridoxal phosphate produced is assayed manometrically in a WARBURG apparatus. Not more than 1.8 ml of B are placed in the main compartment containing 1 M Na acetate (pH 5.5) 0.2 ml, and tyrosine apodecarboxylase (730 units)[2] 0.5 ml, and water to 2.5 ml; in the side arm is placed 0.03 M tyrosine suspension, pH 5.5 (0.5 ml). After equilibration at 30° for 10 min the CO_2 released is measured during the first 10 minutes following tipping.

Potato apyrase. Peeled potato (1 kg) is homogenized in a WARING blendor with 1 l of neutralized 0.01 M KCN for 5 min, squeezed through cheese cloth, and the extract centrifuged in the cold. The enzyme is precipitated from the supernatant by addition of $(NH_4)_2SO_4$ (450 g for every litre) with constant stirring. The precipitate is filtered off (Whatman No. 1) and allowed to drain overnight. The precipitate is dissolved in water and dialysed in the cold for 24 hr. against running water. The resultant precipitate is removed by centrifugation and the $(NH_4)_2SO_4$ precipitation and dialysis repeated on the superntant. To the resulting solution is added $(NH_4)_2SO_4$ (30 g per 100 ml) and the precipitate removed by filtration. Further $(NH_4)_2SO_4$ (15 g/100 ml) is added to the filtrate and the enzyme is precipitated. This is finally purified by dialysis and centrifugation.

Tyrosine apodecarboxylase (crude) is prepared from cells of *Streptococcus faecalis* strain R grown for 15 ∓ 3 hr. at 37° (BELLAMY and GUNSALUS, 1945). The cells are pipetted with stirring into 10 vol. of acetone at − 20°. After 15 min the powder is filtered off, and washed once with acetone and once with ether at − 20°. The powder (20 mg/ml) is suspended in 0.02 M phosphate buffer pH 5.5 and incubated at 37° for 20 hr. The residue is centrifuged off, and the supernatant freeze-dried and stored in a desiccator at 0°. Under these conditions it is stable indefinitely.

4. Nicotinic Acid.

The functional forms of nicotinic acid are DPN (XI) and TPN (XII). The overall mechanisms involved in their formation from nicotinic acid are indicated in equations (12 − 15) (HANDLER, 1959).

(XI)

---

[1] The apyrase unit is the amount of enzyme required to liberate 1 μg of P_i in 30 min at 30° in the presence of 300 μg or more of ATP-P.

[2] The tyrosine decarboxylase unit is the amount of enzyme required to liberate 1 μl of CO_2 in 60 min at 30° under the conditions specified above.

(XII)

Nicotinic acid + phosphoribosylpyrophosphate (PRPP) → desamido-NMN + P-P   (12)

Desamido-NMN + ATP → Desamido-DPN + P-P   (13)

Desamido-DPN + ATP + glutamine → DPN + AMP + glutamate + P-P   (14)

DPN + ATP ⇌ TPN + ADP   (15)

The enzymes involved are PRRP-nicotinic acid transferase (reaction 12), DPN pyrophosphorylase (reaction 13), DPN-synthetase (reaction 14) and DPN kinase (reaction 15).

a) PRPP-Nicotinic Acid Transferase.

Assay (Preiss and Handler, 1958). The incubation mixture consists of 0.37 μmole [^{14}C]nicotinic acid, 0.33 μmole PRPP, 5 μmoles MgCl$_2$, 20 μmoles P$_i$, pH 7.4, and an appropriate amount of enzyme in a final volume of 1 ml. The mixture is incubated at 37° for 6 hr. The reaction is stopped by heating and the denatured protein removed by centrifugation. Samples (0.02 ml) of the filtrate are then chromatographed on paper using a solvent system containing 7 parts 7 95% ethanol to 3 parts of 1 M ammonium acetate, adjusted to pH 5·0 with HCl.

The amount of desamido-NMN formed is then determined from the radio-activity of the appropriate spot.

b) DPN-Pyrophosphorylase.

This enzyme, which catalyses reaction (13) has been found in red cells, the cytoplasmic fraction of yeast and nuclei of rat liver. Nicotinamide mononucleotide (NMN) is also a substrate for the enzyme, but is much less effective (Handler, 1959). Neither 2,4-DNP (10^{-4} M) nor F$^-$ (0.05 M) inhibits the enzyme (Kornberg, 1955a). The enzyme is activated by Mg^{2+} and has been purified from liver much more extensively than from yeast.

Assay (Kornberg, 1955b) **(DPN-formation).** The incubation mixture contains 0.1 ml ATP (0.02 M), 0.05 ml desamido-NMN or NMN (0.05 M), 0.2 ml glycyl-glycine buffer (0.25 M, pH 7.4), 0.1 ml MgCl$_2$, enzyme (1 unit or less) and water to a final volume of 1.0 ml. It is necessary to add 0.1 ml nicotinamide (2 M) when as saying crude extracts to inhibit DPN-nucleosidase. After incubation for 20 min at 38°, the solution is deproteinized with 1 ml of 10% trichloroacetic acid, neutralized with 2 N NaOH, and the DPN formed assayed spectrophotometrically as DPNH$_2$ by using the alcohol dehydrogenase method for reduction and assumin g the extinction coefficient of DPNH$_2$ to be 6.22×10^6 cm^2/mole at 340 mμ.

Assay (Preiss and H andler, 1958) **(desamido-DPN formation).** The incubation mixture contains 0.2 μmole [^{14}C]desamido-NMN, 1 μmole ATP, 15 μmoles Tris buffer (pH 7.4), 5 μmoles MgCl$_2$ and an appropriate amount of enzyme in a final volume of 0.5 ml; after incubation for 20 min at 37° the mixture is heated as described under A above, and the amount of desamido-DPN determined from the radioactivity of the appropriate spot on the chromatogram.

c) DPN-Synthetase.

Assay (PREISS and HANDLER, 1957, 1958). The incubation mixture consists of desamido-DPN (1.7 μmoles), glutamine (20 μmoles), ATP (2 μmoles), phosphate buffer pH 7.4 (40 μmoles) and 0.4 ml enzyme (dialysed yeast autolysate) in a final volume 1 ml. After incubation for 100 min at 37° 1 ml of water was added and the mixture placed in a boiling water bath for 1.5 min and cleared by centrifugation. Portions of the mixture (1 ml) are taken for DPN assay by the alcohol dehydrogenase method. The enzyme is strongly inhibited by azaserine.

d) DPN-Kinase.

This enzyme, catalysing reaction (15), was observed in extracts of yeast in 1940, but its further study and purification have been carried out only with pigeon liver (WANG and KAPLAN, 1954). It is assayed by measuring the activation of a TPN-specific isocitric dehydrogenase from pig heart. DPN-kinase is specific for DPN and although it will phosphorylate the 3-acetylpyridine analogue of DPN, it exhibits no activity for desamido-DPN, DPNH$_2$ or DP-CoA (dephospho-CoA, see p. 192); indeed DPNH$_2$ is an effective inhibitor. The enzyme is also specific for ATP as phosphate donor. K_m for DPN is 6×10^{-4} M.

Assay (WANG and KAPLAN, 1954). The incubation mixture contains 0.1 ml of each of the following: DPN (25 mg/ml H$_2$O), ATP (0.04 M), MgCl$_2$ (0.05 M), Tris (0.05 M, pH 7.5) and DPN-kinase. After incubation for 60 min at 37°, 0.5 ml H$_2$O is added and the mixture heated in boiling water for 2 min. After centrifugation, a portion of the supernatant is transferred to a quartz spectrophotometer cell containing enough Tris (pH 7.5) to make the final concentration of buffer 0.05 M and the final volume 2.9 ml; 0.05 M isocitrate (0.05 ml) and 0.1 ml of isocitrate dehydrogenase (prepared according to the method of GRAFFLIN and OCHOA, 1950) are added and 340 mμ E measured every 30 sec. The reaction should be complete in 2 min and the increase in E$_{340}$ mμ measures the conversion of TPN into TPNH$_2$.

The activity of the DPN-kinase is expressed as μmoles TPN formed per hour.

This assay can be used with crude animal tissue extracts provided that sufficient nicotinamide is added to inhibit DPN-ase (HANDLER and KLEIN, 1942); 50% inhibition is achieved with 1.5×10^{-3} M nicotinamide (ZATMAN et al., 1953). However, DPN-ase from *Neurospora* is inhibited only at very high nicotinamide concentrations (0.1 M).

5. Folic Acid.

It is now known that the active metabolic form of folic acid (I), is 5,6,7,8-tetrahydrofolic acid (XIII). The reduction of (I) to (XIII) has been observed in crude

(XIII)

extracts of soya bean leaves (IWAI and YOSHIDA, 1954) but the enzymes involved
have not been purified. However, work with purified enzymes from bacterial and
animal sources have demonstrated that the reduction takes place in two steps
via a-dihydrofolic acid (see GOODWIN, 1960). The first step is catalysed by folic
acid reductase and the second step by dihydrofolic acid reductase. The latter
enzyme has been purified from both chicken and sheep liver; the chicken enzyme
requires $TPNH_2$ specifically. whilst the sheep enzyme requires $DPNH_2$ specifically.
However, both are inhibited by low levels of (about $1-2 \times 10^{-9}$ M) aminopterin
(4-aminofolic acid) and amethopterin (4-amino-10-methylfolic acid); sulphydryl
inhibitors do not inhibit dihydrofolic reductase.

a) Folic Acid Reductase.

The enzyme catalyses reaction (16) and is specific for $TPNH_2$; it has an

$$\text{Folic acid} + TPNH_2 \rightleftharpoons \text{Dihydrofolic acid} + TPN \qquad (16)$$

optimum pH of 5.0 (liver enzyme) and is strongly inhibited by aminopterin
(1.2×10^{-9} M). Teropterin (diglutamylfolic acid) is a substrate, but pteroyl-
aspartic acid is not.

Assay (PETERS and GREENBERG, 1959). The enzyme preparation (1.0 ml),
0.1 M Na acetate buffer, pH 5.0 (1.0 ml), 0.25 mg $TPNH_2$ and folic acid (the
amount required is that which will yield a zero-order reaction, as determined in
preliminary experiments) in a final volume of 2.4 ml are incubated at 37° for
30 min. The reaction is stopped by the addition of 1.0 ml of trichloroacetic acid
(15%). The deproteinized solution is assayed by measuring the non-enzymic
decomposition of dihydrofolic acid.

The unit of activity is the amount which will reduce 1 mμmole of folic acid in
30 min under the conditions just defined.

b) Dihydrofolic Acid Reductase.

This enzyme catalyses reaction (17).

$$\text{Dihydrofolic acid} + TPNH_2 \rightarrow \text{5,6,7,8-tetrahydrofolic acid} + TPN \qquad (17)$$

Assay (MISRA, HUMPHREYS, FREIDKIN, GOLDIN and CRAWFORD, 1961). The
reaction mixture (3 ml) consists of 0.05 M K phosphate buffer, pH 7.5, 0.01 M
mercaptoethanol, 0.1 μmole dihydrofolic acid, 0.2 μmole $TPNH_2$ and 50 μl of
enzyme. Incubation is for 5 min at 28°. The amount of dihydrofolate reduced
(mμmoles) is calculated from $- \Delta E$ at 340 mμ, assuming $- \Delta \varepsilon = 11,270$, resulting
from the combined decrease, is due to TPN oxidized and dihydrofolate reduced.

Dihydrofolic acid is prepared by suspending 20 mg of folic acid in 2 ml H_2O
and adding N KOH dropwise until the folic acid dissolves; 5 ml of potassium
ascorbate (100 mg/ml, pH 6.0) and 200 mg sodium dithionite are added. After
standing for 5 min at room temperature the solution is cooled to 0° and 2 N HCl
added dropwise until the pH is 2.8 (thymol blue). After further standing at 0° the
precipitate is collected by centrifugation, resuspended in 5 ml potassium ascorbate
and reprecipitated with HCl as just described. The precipitate is washed with
4×10 ml of cold 0.005 N HCl and stored at 0° as a suspension in 0.005 N HCl.

6. Pantothenic Acid.

The metabolically active form of pantothenic acid is coenzyme A and steps
involved in the main pathway of synthesis of coenzyme A are indicated in Fig. 4
(BROWN, 1959). The first step has been observed in yeast and the third in both

yeast and *Neurospora crassa*; however the enzymes have not been isolated from these sources so only their assay will be described here.

An alternative pathway of synthesis in which pantothenic acid is converted into pantetheine before phosphorylation, is probably limited to *Lactobacillus arabinosus* (BROWN, 1959) and does not exist in animal tissues as was previously suggested (NOVELLI, 1959).

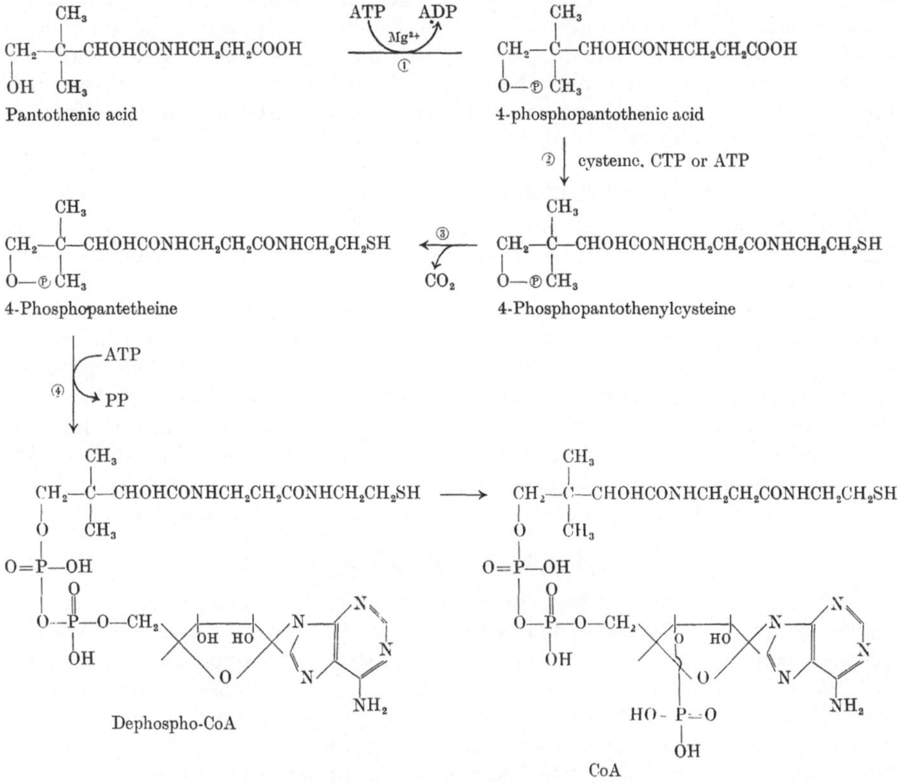

$$CH_3$$
$$CH_2-C-CHOHCONHCH_2CH_2COOH$$
$$OH \quad CH_3$$
Pantothenic acid

ATP ADP
$$\xrightarrow[\text{Mg}^{2+}]{\textcircled{1}}$$

$$CH_3$$
$$CH_2-C-CHOHCONHCH_2CH_2COOH$$
$$O-\textcircled{P} \quad CH_3$$
4-phosphopantothenic acid

$$\textcircled{2} \downarrow \text{ cysteine, CTP or ATP}$$

$$CH_3$$
$$CH_2-C-CHOHCONHCH_2CH_2CONHCH_2CH_2SH$$
$$O-\textcircled{P} \quad CH_3$$
4-Phosphopantetheine

$$\xleftarrow{\textcircled{3}} \quad CO_2$$

$$CH_3$$
$$CH_2-C-CHOHCONHCH_2CH_2CONHCH_2CH_2SH$$
$$O-\textcircled{P} \quad CH_3$$
4-Phosphopantothenylcysteine

$$\textcircled{4} \quad \text{ATP} \searrow \text{PP}$$

Dephospho-CoA

$$\longrightarrow$$

CoA

Fig. 4. Biosynthesis of coenzyme A (main pathway).

a) Pantothenate Kinase.

This enzyme catalyses reaction (1) (Fig. 4); its purification from animal and bacterial sources has been described by BROWN (1959).

Assay (BROWN, 1959). The incubation mixture contains 0.083 μmole pantothenic acid; 20 μmoles ATP[1]; 10 μmoles MgCl$_2$; 0.2 ml enzyme extract in a final volume of 2 ml. Tris buffer (pH 7.4, 0.05 M). The mixture is incubated at 37° for 3 hours and then placed in a boiling water bath for 5 min. Denatured protein is removed by centrifugation and portions of the supernatant are treated with 0.1 ml of a suspension of intestinal phosphatase (40 mg/ml) 0.1 ml Na bicarbonate buffer (pH 8.5, M) and water to 1 ml, and incubated at 37° for 2 hours. The reaction mixture is assayed microbiologically for pantothenic acid before and after incubation with intestinal phosphatase. The organism used is *Saccharomyces carlsbergensis* strain 4228 (ATKIN, WILLIAMS, SCHULTZ and FREY, 1944) which responds to phosphopantothenic acid (KING and STRONG, 1951), CoA and pantetheine (CRAIG and SNELL, 1951), and pantothenylcysteine (BROWN and SNELL, 1953).

b) Coupling Enzyme.

This enzyme catalyses the condensation of 4-phosphopantothenate and cysteine to yield 4-phosphopantothenylcysteine (reaction 2, Fig. 4); CTP is an essential co-factor[1]. Its purification from mammalian and bacterial sources has been described (BROWN, 1959).

Assay (BROWN, 1959). The reaction mixture contains 4-phosphopantothenic acid (0.083 μmole); cysteine (20 μmoles); $MgCl_2$ (10 μmoles), CTP (5 μmole), and enzyme in 2 ml Tris buffer pH 7.4 (0.04 M). Incubation is carried out for 3 hours at 37°. The reaction is stopped by heating in a boiling water bath and the mixture incubated with intestinal phosphatase as described under "Pantothenic acid kinase". The pantothenylcysteine produced is determined microbiologically using *L. helveticus* strain 80 PC (CRAIG and SNELL, 1951).

c) Phosphopantothenylcysteine Decarboxylase.

The enzyme, which has not been fully purified, catalyses reaction 3 (Fig. 4) and has been demonstrated in both yeast and *N. crassa*. It has no action on pantothenylcysteine.

Assay (BROWN, 1959). The reaction mixture contains 0.0165 μmole phosphopantothenylcysteine; 10 μmoles cysteine (as reducing agent), 0.2 ml of enzyme extract, in a final volume of 1 ml of Tris buffer (pH 7.4, 0.04 M). The mixture is incubated at 37° for 3 hours and then heated in a boiling water bath for 5 min. After removal of denatured protein portions of the supernatant are heated with intestinal phosphatase as described under "Pantothenic acid kinase" (p. 191). The reaction mixture is then assayed microbiologically for pantetheine (*L. helveticus* strain 80) before and after treatment with intestinal phosphatase.

d) Dephospho-CoA Pyrophosphorylase.

This enzyme carries out reaction 4 (Fig. 4) which is the only reversible reaction in the series of steps concerned with CoA biosynthesis. Although it has been purified considerably it has not been separated from dephospho-CoA kinase (p. 194) which converts one product of the reaction, dephospho-CoA, into CoA (see below). Because of this the activity is assayed by measuring either the CoA or pyrophosphate produced under standard conditions. In crude extracts CoA formation is the most accurate because of the presence of inorganic pyrophosphatase. However, in purified extracts, from which pyrophosphatase has been eliminated, pyrophosphate determination is also an accurate index of activity.

Assay (NOVELLI, 1955b). Tubes are prepared containing 0.1 ml 4-phosphopanthetheine (reduced, 0.02 M), 0.1 ml ATP (0.1 M), 0.05 ml $MgCl_2$ (0.1 M), 0.1 ml cysteine HCl (0.1 M), 0.05 ml Tris buffer (M, pH 7.7), and enzyme 10−20 units of CoA when CoA is measured, or 1−2 μmoles CoA when pyrophosphate is measured) to a final volume of 1 ml. The tubes are incubated at 37° for 60 min and the reaction stopped by boiling the tubes for 3 min. The CoA produced is measured by the phosphotransacetylase reaction (STADTMAN, NOVELLI and LIPMANN, 1951) (see below) or, alternatively, the pyrophosphate is determined by the method of FLYNN, JONES and LIPMANN (1954) (see below).

A unit of activity is the amount of enzyme producing one unit of CoA or liberating 0.0025 μmole of inorganic pyrophosphate in 1 hr. under the conditions just described, and with the protein content of the preparation measured turbidimetrically (BÜCHER, 1947).

---

[1] Although any other nucleoside trisphosphate can substitute for ATP in the animal system (rat liver and kidney), CTP is specific for the bacterial system (*E. coli, Proteus morganii*).

e) Phosphotransacetylase.

The enzyme catalyses reaction (18).

$$\text{Acetyl phosphate} + \text{CoA} \rightleftharpoons \text{Acetyl CoA} + \text{P}_i \qquad (18)$$

However in the presence of arsenate, acetyl CoA acetylates the arsenate (reaction 19) and the resulting unstable acetyl arsenate spontaneously hydrolyses to acetate and arsenate (reaction 20).

$$\text{Acetyl-CoA} + \text{arsenate} \rightarrow \text{Acetyl arsenate} + \text{CoA} \qquad (19)$$
$$\text{Acetyl arsenate} + \text{H}_2\text{O} \rightarrow \text{Acetate} + \text{arsenate} \qquad (20)$$
$$\text{Acetyl phosphate} + \text{H}_2\text{O} \rightarrow \text{Acetate} + \text{P}_i \qquad (21)$$

The sum of these reactions is indicated in reaction 21, and as the rate of this reaction, as measured by the disappearance of acetyl phosphate, is proportional to the amount of enzyme present under the prescribed conditions, the enzyme can be assayed quantitatively.

Assay (STADTMAN, 1955). Tubes are prepared containing 0.4 ml water and 0.1 ml each of Tris buffer (0.1 M, pH 8.0), dilithium acetyl phosphate (0.06 M), coenzyme A (1.6×10^{-4} M), cysteine HCl (0.1 M) and enzyme (50 units/ml). The tubes are incubated for 5 min at 28°, 0.1 ml K arsenate (0.5 M, pH 8.0) added and the incubation continued for a further 15 min. The residual acetyl phosphate is then determined by the hydroxamic acid method (LIPMANN and TUTTLE, 1945). A unit of enzyme is the amount required to catalyse arsenolysis of 1 μmole of acetyl phosphate in 15 min under the conditions first defined.

Pyrophosphate determination (FLINN et al., 1954). A sample containing $0.1-0.5$ μmole of inorganic pyrophosphate and preferably less than 1 μmole inorganic orthophosphate is placed in a 10 ml volumetric tube (for a KLETT-SUMMERSON colorimeter) or flask, and water added to approximately 7 ml; then are added 1 ml of molybdate reagent (2.5% ammonium molybdate in 5 N H_2SO_4), 0.4 ml (70 μmoles) cysteine HCl (0.175 M), and 0.4 ml of the aminonaphthol-sulphonic acid solution (29 g $NaHSO_3$ and 1 g $NaSO_3$ are dissolved in 200 ml H_2O; 0.5 g of 1,2,4-aminonaphtholsulphonic acid is ground in a mortar with a little of the sulphite-bisulphite solution; the remainder of the solution is then added, and the mixture filtered) and the contents mixed immediately. After standing for 7 min the sample is read using a suitable colorimeter or spectro-photometer (the method described used filter S 66 with the KLETT-SUMMERSON colorimeter); a further reading is made after 90 min and then a correction applied which on average amounted to 11.9% over a range $0.258-0.774$ μmole P_i. Exact timing is essential. A correction is also required if adenosine polyphosphates are present, and high concentrations of versene interfere with the colour develop-ment with pyrophosphate.

Acetyl phosphate determination (hydroxamic acid method) (LIPMANN and TUTTLE, 1945). To 2 ml of test solution containing $0.5-5$ μmoles of acetyl phosphate is added 1.0 ml of neutralized hydroxylamine (4 M $NH_2OH.HCl$ mixed with an equal volume of 3.5 M NaOH just before use). The mixture is allowed to stand for 10 min at room temperature and then 3.0 ml of the ferric chloride reagent are added (prepared by mixing equal volume of 5% $FeCl_3$ in 0.1 N NaOH, 12% trichloroacetic acid and 3 N HCl). The solution is centrifuged if necessary and the absorption at 540 mμ compared with a standard curve. Proportionality is achieved over the range $0-1$ μmole of acetyl phosphate per ml final solution.

f) Dephospho-CoA Kinase.

This enzyme catalyses reaction 5 (Fig. 4). It is specific for ATP, requires cysteine for maximum activity, and is not inhibited by adenosine or adenosine $2'$-, $3'$- or $5'$-phosphate. Its specificity with respect to dephospho-CoA has not yet been examined because of lack of suitable assay methods for CoA analogues. K_m for dephospho-CoA is 3×10^{-6} M.

Assay (Wang, 1955). Tubes are prepared containing 20 units of dephospho-CoA, 2 μmoles ATP (0.04 M), 5 μmoles MgCl$_2$ (0.05 M), 3 μmoles cysteine HCl (0.1 M), 40 μmoles Tris buffer (pH 8.2, 0.5 M) and 0.05 mg enzyme. The tubes are incubated at 37° for 30 min and the reaction stopped by immersion in boiling water; this step also denatures any acetyl phosphatase which might be present. The mixture is then assayed for CoA by the phosphotransacetylase method (p. 193).

A unit of activity is the amount of enzyme catalysing the formation of one unit of CoA in 30 min under the conditions defined above.

C. Degrading Enzymes.

1. Thiamine.

a) Thiaminase.

Thiaminase catalyses the fission of the methylene quaternary nitrogen bond of thiamine with the transfer of the pyridimine residue (Py) to an amine acceptor; the free thiazole (Th) residue being liberated (reaction 22) (Fujita, 1954, 1955).

$$Py - CH_2 - \overset{+}{Th} + RNH_2 \rightarrow Py - CH_2NHR + Th + H^+ \qquad (22)$$

The enzyme is widespread in aquatic animals but is rare in higher plants, algae and fungi excluding yeasts, and widely distributed in yeasts and pteridophyta. In plants the enzyme from bracken has been most fully investigated (Kenten, 1957, 1958).

Purification (Kenten, 1957, 1958). The pinnae of bracken fronds mixed with 1.5 vol. of Na$_2$HPO$_4$ (0.033 M) are minced in a domestic mincer and squeezed by hand through strong cotton cloth. Air-dried bracken can also be used as starting material. The extracts are filtered through fluted Whatman No. 1 filter paper at 0−4° for 2 days. Solid (NH$_4$)$_2$SO$_4$ is then added at 0° with stirring until a concentration of 650 g/l is reached; after stirring for one hour the precipitate is collected by centrifuging at 1000 ×g, suspended in water and dialysed overnight at 0°. Sufficient 0.5 M phosphate, pH 7.5, is added to the dialysed material to give a final concentration of 0.01 M. The suspension is frozen for 24 hr., thawed and filtered through a pad of Hyflo Super-Cel which is finally washed with a small volume of 0.01 M phosphate pH 7.5. Solid (NH$_4$)$_2$SO$_4$ is added to the filtrate at 0° to a final concentration of 500 g/l. The precipitate is collected by centrifugation at 10,000 g, suspended in the minimum volume of 0.01 M phosphate buffer, pH 7.5, and dialysed against several changes of distilled water at 0−4° for 24 hr.

The preparation, which should contain 35−50% of the original extract and represent a ten fold purification, will keep for at least two months if frozen at − 18°.

Assay. The activity of the enzyme can be determined by (a) measuring the thiamine disappearing fluorimetrically as thiochrome (p. 175), (b) by measuring the heteropyrithiamine formed when pyridine is the amine acceptor (KENTEN, 1957) and (c) manometrically in a HCO_3'/CO_2 buffer by measuring the CO_2 produced accompanying the release of H^+ (KENTEN, 1958). The first is probably the least effective because of the presence in plants of heat-stable factors, possibly flavonoids, which combine with thiamine to yield products which cannot be oxidized to thiochrome (HASEGAWA et al., 1957).

Heteropyrithiamine determination (KENTEN, 1957). The reaction mixture contains 5.9 μmoles thiamine hydrochloride, 1 ml pyridine (0.1 M), 0.5 ml phosphate buffer (pH 7.5, 0.5 M), enzyme and water to 10 ml, and is incubated at 37° for 30 min. A sample (2 ml) is taken at zero time and pipetted into 1 ml HPO_3 (20% w/v); a second sample is taken after 30 min and treated similarly. These samples are centrifuged and the supernatant incubated with 1 ml NaOH (20% w/v) at 37° for 1 hr. This treatment reduces the absorption at 386 mμ due to thiamine after oxidation with $K_3Fe(CN)_6$ but has no effect on that due to heteropyrithiamine. Then 0.5 ml of oxidizing reagent [4 ml $K_3Fe(CN)_6$ (1% w/v) + 6 ml NaOH (20% w/v)] is added and the mixture allowed to stand at room temperature for 15 min. The excess $K_3Fe(CN)_6$ is removed by adding 1 ml H_2O_2 (0.05% w/v) and allowed to stand for 20 min with occasional shaking. The absorption at 386 mμ of the experimental and blank solutions is measured using, in the compensating cell, samples of the mixture from which thiamine and the enzyme have been omitted but which have been treated with the oxidizing mixture. The amount of heteropyrithiamine formed is calculated using $\varepsilon = 8000$ for the oxidation product. A correction has to be made to the absorption at 386 mμ for the contribution of oxidized thiamine remaining in the reaction mixture. This can be calculated on the basis of the absorption of the incubation mixture minus thiamine oxidized, in the usual way.

The range of optimal accuracy is the production of about $0.7-1$ μmole of heteropyrithiamine.

Manometric assay (KENTEN, 1958). The WARBURG flasks contain 0.005 M thiamine, 0.05 M acceptor amine (pyridine or trimethylamine), enzyme, 0.05 M $NaHCO_3$, gas phase $CO_2 + N_2$ (5:95, v/v), pH 7.5. The incubation temperature is 37°, and the amount of enzyme is adjusted so that the CO_2 evolution does not exceed 350 μl/hr. Under these conditions the rate of evolution of CO_2 is directly proportional to the enzyme concentration.

The unit of activity is the amount of enzyme which catalyses the formation of 1 μmole of heteropyrithiamine in 1 hr. under the conditions described under "heteropyrithiamine determination".

b) Thiamine Phosphatase.

An unstable enzyme in yeast which splits orthophosphate out of thiamine monophosphate is discussed on p. 174.

Aspergillus oryzae is a good source of an enzyme which rapidly hydrolyses thiamine phosphate. Other plant preparations with similar activity towards thiamine phosphate are also on the market (Association of Vitamin Chemists, 1947).

c) Thiamine Pyrophosphatase.

There are indications that nucleotide pyrophosphatase (p. 196) has slight activity in hydrolysing thiamine pyrophosphate to thiamine monophosphate; but this may be due to contamination of the preparation.

2. Riboflavin.

a) FAD-ase (Nucleotide Pyrophosphatase).

The enzyme which splits FAD to FMN and AMP is very probably the same enzyme as that which hydrolyses DPN and TPN in a similar manner. It is discussed in detail under "nicotinic acid" on this page.

b) FMN-Phosphatase.

Very little is known about the hydrolysis of FMN in living tissues, but takadiastase and similar preparations are probably active.

3. Nicotinic Acid.

a) Nucleotide Pyrophosphatase.

This enzyme catalyses the hydrolysis of the pyrophosphate link in FAD, TPN, DPN and Coenzyme A (Kornberg and Pricer, 1950) (reactions 23, 24, 25, 26).

$$DPN + H_2O \rightarrow NMN + AMP \text{ (adenosine 5'-phosphate)} \tag{23}$$
$$TPN + H_2O \rightarrow NMN + ADP \text{ (adenosine 2',5'-diphosphate)} \tag{24}$$
$$FAD + H_2O \rightarrow FMN + AMP \text{ (adenosine 5'-phosphate)} \tag{25}$$
$$CoA + H_2O \rightarrow \text{Pantetheine-4 phosphate} + ADP \text{ (adenosine 3',5'-diphosphate)} \tag{26}$$

Purification. 200 g peeled potatoes are blended with 400 ml of 0.40 saturated $(NH_4)_2SO_4$ for 90 seconds, and the extract filtered at 2° on fluted paper. The precipitate is dissolved in 550 ml water and dialysed against running water (8−18°) for 90 min.

The dialysed solution is brought to pH 4.4 with acetic acid (1 M), and cooled to −0.5°; 95% ethanol is then added with stirring. The precipitate is removed by centrifugation at 0°. This ethanol fractionation can be repeated (Kornberg, 1955c). The supernatant is diluted with water to give a protein concentration of 1.5 mg/ml and calcium phosphate gel (dry wt. 7.9 mg/ml and aged for 2 months) added and the mixture stirred for 5 min at room temperature. The precipitate is collected by centrifugation, washed with 4×100 ml of 0.1 M potassium phosphate buffer pH 7.4, and the enzyme eluted with 3×100 ml of 0.20 saturated $(NH_4)_2SO_4$ adjusted to pH 7.5 with ammonia water. The eluates can be concentrated by adding solid $(NH_4)_2SO_4$, collecting the precipitate and redissolving in a small volume of water (protein concentration 2 mg/ml). To obtain increased specific activity a second calcium phosphate adsorption can be carried out.

Assay (Kornberg, 1955c). A mixture containing 0.1 ml DPN (0.02 M, pH 6), 0.2 ml $KH_2PO_4 - K_2HPO_4$ buffer (0.5 M, pH 7.0)[1], 1−5 enzyme units and water to a final vol. of 1 ml is incubated for 20 min at 38°. A portion (0.10 ml) of the mixture is transferred to an absorption cell containing, in 2.8 ml, 0.5 M ethanol, 0.02 M nicotinamide, and 0.1 M glycine-NaOH buffer (pH 9.5). Readings taken at 340 mμ before and after (5 min) the addition of crystalline yeast alcohol dehydrogenase. From these readings the amount of residual DPN can be calculated.

The unit of enzyme activity is that amount which splits one micromole of substrate (in this case DPN) per hour. FAD can be assayed by the D-amino acid oxidase method (p. 185), TPN by the isocitric dehydrogenasemethod; and pantetheine 4-phosphate microbiologically (see p. 192).

---

[1] With CoA, the optimum pH is 5.0 (Novelli, 1959).

b) DPN-ase (Pyridine Transglycosidase).

This enzyme catalyses (a) the hydrolysis of DPN with the liberation of free nicotinamide by cleavage of the nicotinamide-ribose linkage (reaction 27), (b) the transfer of adenosine diphosphate ribose from one pyridine group to another as in reaction (28), where X is a pyridine compound related to nicotinamide. The enzyme will also hydrolyse TPN but will not attack $DPNH_2$ or $TPNH_2$.

$$DPN + H_2O \rightarrow Nicotinamide + R—P—P—R—A + H^+ \qquad (27)$$
$$\text{(Adenosine diphosphate ribose)}$$

$$DPN + X \rightarrow Nicotinamide + X^+R—P—P—R—A \qquad (28)$$
$$(DPN = \overset{+}{N}—R—P—P—R—A)$$

Assay (KAPLAN, 1955). The reaction mixture which contains 0.3 ml of phosphate buffer (0.1 M, pH 7.2), 0.5 μmole DPN and enzyme (about 1 unit) in a final volume of 0.6 ml, is incubated for 7.5 min at 37°. At the end of this period 3 ml of M KCN is added and the absorption at 325 mμ measured (325 mμ is the maximum of the DPN-cyanide complex). The difference between this reading and that obtained at zero time indicates the amount of enzymic activity.

The enzyme can also be assayed using yeast alcohol dehydrogenase. This however only indicates loss of DPN; the cyanide method measures the disappearance of the quaternary N atom present when the nicotinamide occurs in a riboside linkage.

The unit of activity is the amount of enzyme which hydrolyses 0.01 μmole of DPN under the described conditions and with protein measured by the method of LOWRY, ROSENBROUGH, FARR and RANDALL (1951).

Purification. The enzyme has been obtained from mycelia of *Neurospora* grown on a zinc-deficient medium (NASON, KAPLAN and COLOWICK, 1951); these contain about 50 times more enzyme than do mats grown on a normal medium. The mats are harvested, washed with triple-distilled water, frozen for 1—3 hr., homogenized in a glass homogenizer with three times their weight of 0.1 M phosphate buffer (pH 7.5) and centrifuged at 13,000 rpm (Servall) at 4° for 10 min. The supernatant which contains virtually all the enzyme is adjusted to pH 5 with HCl and the resulting precipitate discarded. Acetone is then added to the supernatant at 0° to a concentration of 35% and the precipitate discarded; the concentration of acetone is increased to 60% and the resulting precipitate dissolved in 0.1 M K_2HPO_4. The pH of this solution is brought to pH 2.7 at 0° with HCl, and the enzyme reprecipitated by adding acetone to a final concentration of 60%. The precipitate is centrifuged off at 0° and triturated with 5 ml 0.1 M phosphate buffer (pH 7.5). The resulting solution represents a 30% yield and a 30-fold purification.

c) NMN-Phosphatase.

It has been reported that the 5'-nucleotidase of seminal plasma will remove the phosphate group from NMN (HEPPEL and HILMOE, 1955), but no report of similar activity in plant tissues has yet appeared.

Assay (HEPPEL and HILMOE, 1955). The incubation mixture contains 0.1 ml of 1 M glycine/NaOH buffer, pH 8.5, 0.1 ml $MgCl_2$ (0.1 M), 3 μmoles NMN and enzyme in a final volume of 1.24 ml. The mixture is incubated at 37° for 15 min and the reaction stopped by adding trichloroacetic acid to a final concentration of 5%. Samples and substrate blanks are then analysed for inorganic orthophosphate by standard procedures.

The unit of activity is the amount of enzyme liberating 1 μmole of inorganic orthophosphate per hour.

4. Folic Acid.

There are various reports in animals and bacteria of enzyme systems which split folic acid (p. 179) at the C-9 and N-10 bond (see Woods, 1959) but only one report exists of such activity in higher plants. Braganca, Krishnamurthi and Ghanekar (1959) found that extracts of pea seedlings made in 0.5 M phosphate buffer at pH 5.5 will carry out this cleavage. The product of the reaction is 2-amino-4-hydroxy-6-formylpteridine. The enzyme is inhibited by dialysis, or by the addition of catalase, CN^- or azide; dialysed extracts are reactivated by traces of H_2O_2. The reaction appears to be peroxidatic. Non-germinated seedlings are inactive but activity appears within 20 hr. of germination.

Assay (Kenkare and Braganca, 1958). The reaction is carried out at 37° for 60 min by incubating 150 μg folic acid in 2 ml of a mixture of 0.06 M phosphate buffer, pH 5.5, 1.5×10^{-3} M ATP, 1.0×10^{-3} M $MnCl_2$, 1.0×10^{-3} M Na citrate, 1.0×10^{-4} M DPN, and the appropriate amount of enzyme.

Enzyme activity is determined by measuring the amount of a diazotizable amine liberated by means of the Bratton and Marshall (1939) method.

5. Pantothenic Acid.

Various enzymes are known which degrade coenzyme A but little is known of the breakdown of pantothenic acid itself (Novelli, 1959). The general reactions observed to date are indicated in Fig. 5.

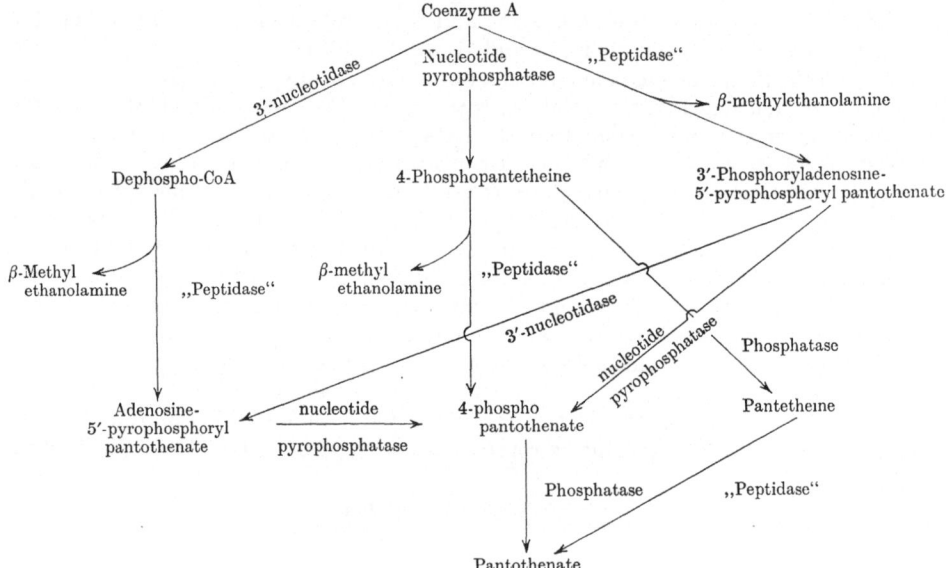

Fig. 5. Some major routes of coenzyme A degradation.

a) „CoA-3′-Nucleotidase".

This enzyme which is present in ryegrass seeds specifically removes phosphate from 3′-nucleotides including Coenzyme A (Fig. 5) (Shuster and Kaplan, 1955; Cohen and Goodwin, 1961). The phosphate monoesterase from human prostate gland will also carry out this reaction but is less active (Novelli, 1959).

The ryegrass enzyme will not appreciably attack other phosphate monoesters. It is inhibited at pH 7.5 by 10^{-3} M cysteine, 5×10^{-5} M glutathione and 5×10^{-5} M

KCN; no inhibition is observed with sulphydryl compounds below pH 6.0, so that when reduced Coenzyme A (CoASH) is the substrate, a pH of 5.0—5.5 is used.

Purification. Ryegrass seed (500 g) is soaked in water for 2—3 days and then homogenized in batches with cold water (1 l) for 3 minutes at a time. The extract is squeezed through cheese cloth, centrifuged and precipitated by 90% saturation with ammonium sulphate. The floating precipitate is collected, dissolved in 150 ml cold water and dialysed overnight against cold water. The enzyme is adsorbed on to alumina C_γ-gel (22 mg/ml, 100—150 ml), the mixture held at 8° for 10 min and then centrifuged. The gel is washed with 2 ×150 ml water and then with 2 ×150 ml 0.5 M ammonium sulphate; the enzyme is then eluted with 4 ×150 ml 1.0 M $NaHCO_3$ and the eluate dialysed overnight against cold tap water. The enzyme is again adsorbed, this time on to calcium phosphate gel (13 mg/ml, 150 ml) and after standing at 8° for 10 min the mixture is centrifuged in the cold. The gel is washed with 2 ×150 ml distilled water and the enzyme eluted with 5 ×50 ml 1.0 M ammonium sulphate. The eluate can be either dialysed overnight against water and stored at 4° or 18°, or further purified by readsorbing on aluminia C_γ gel and calcium phosphate gel.

Procedure (SHUSTER and KAPLAN, 1955). The incubation mixture consists of 0.1 ml 0.04 M coenzyme A, 0.1 M Tris-HCl buffer (pH 5.5)[1] and enzyme to a final volume of 2.0 ml. The mixture is incubated for 15 min at 37° and the reaction stopped by the addition of 1.0 ml of 20% trichloroacetic acid; if necessary any protein thrown out of solution is centrifuged off, and samples taken for inorganic orthophosphate analysis by standard methods.

A unit of enzyme activity is the amount releasing 1 μmole of inorganic phosphate per hour under the conditions just defined.

b) "CoA Pyrophosphatase".

Three enzymes are known which split the pyrophosphate bridge in CoA (Fig. 5). (A) Nucleotide pyrophosphatase: this enzyme, which is obtained from potatoes, has already been discussed on p. 196; (B) intestinal phosphatase, which cleaves the molecule to yield adenosine 3′, 5′-diphosphate and pantetheine 4′-phosphate, but the presence of monoesterase activity in intestinal phosphatase as usually prepared results in the final products being adenosine, pantetheine and inorganic orthophosphate (NOVELLI, 1959); (C) dephospho-CoA pyrophosphorylase which reversibly splits dephospho-CoA to ATP and pantetheine 4-phosphate (see p. 192).

c) "CoA-Peptidase".

According to NOVELLI (1959) enzymes which can hydrolyse the peptide bond between β-mercaptoethylamine and pantothenic acid (Fig. 5) are widespread but have not yet been investigated in full detail.

6. Biotin.

d-Biotin Oxidase.

An enzyme which oxidizes d-biotin (XIV) with the liberation of CO_2 from the carboxyl group is present in some but not all animal tissues tested (QUASTEL,

---

[1] pH 7.5 is used with substrates other than CoA.

1955). It has never been looked for in plant tissues. The reaction is almost completely inhibited, under anaerobic conditions, by 0.01 M azide, malonate, nor-biotin, bis-homobiotin, desthiobiotin and *l*-biotin. It is stimulated by fumarate.

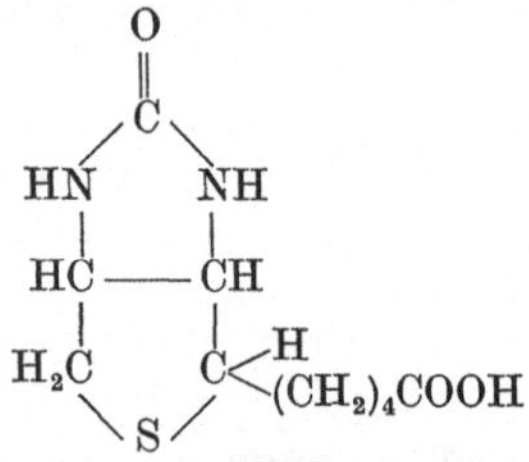

$$
\begin{array}{c}
O \\
\parallel \\
C \\
HN \qquad NH \\
HC\text{---}CH \\
H_2C \qquad C{<}^{H}_{(CH_2)_4COOH} \\
S
\end{array}
$$

(XIV)

Assay (Quastel, 1955). Tissue slices and [^{14}C-carboxyl]biotin in $3.0-3.2$ ml Ringer-phosphate are placed in the main compartments of Warburg flasks. The centre wells contain 0.2 ml 20% NaOH and the side arms 0.2 ml 8 N H_2SO_4, the gas phase is air. After shaking at 37° for 3 hr. the acid is tipped into the main compartment to stop enzyme activity and to drive off any CO_2 in solution. After 20 min standing at 37° the alkali is removed from the centre well and about 5 mg Na_2CO_3 added. The carbonate is then precipitated as barium carbonate, filtered off, washed, weighed and assayed for radioactivity.

7. Inositol.

Phytase.

Phytase is an enzyme with a wide distribution in plants, (including fungi), bacteria and animals. It is particularly abundant in cereal seeds which are also rich in the substrate for the enzyme, phytic acid — inositol hexaphosphoric acid (XV).

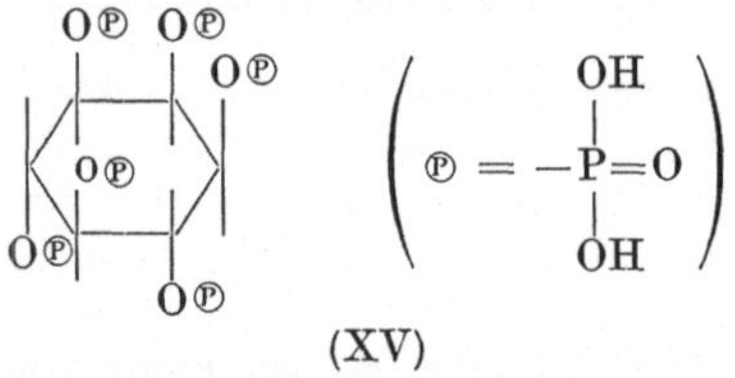

$$
\begin{array}{c}
O\textcircled{P} \quad O\textcircled{P} \\
O\textcircled{P} \\
O\textcircled{P} \\
O\textcircled{P} \\
O\textcircled{P}
\end{array}
\qquad
\left(
\textcircled{P} = -\overset{OH}{\underset{OH}{P}}=O
\right)
$$

(XV)

Preparation and purification (Peers, 1953). Wheat wholemeal (30 g) is extracted with water (175 ml) at $0-4°$ for 6 hr. The residue is centrifuged off and the supernatant filtered. The filtrate is poured with vigorous stirring into ice-cold acetone (600 ml) and allowed to stand 10 min; the supernatant is then decanted. The precipitate is collected on a Büchner funnel, washed with acetone, acetone/ether and finally acetone. The residue is extracted with water (75 ml) at $0-4°$ for 2 hr. and the extract saturated with ammonium sulphate. After 2 hr. the precipitate is collected, dried *in vacuo*, dissolved in water (20 ml) and dialysed for 48 hr. at $0-4°$ against 3×2 l distilled water. The precipitate is removed by filtration and the solution used as the source of the enzyme.

The enzyme is activated by Mg^{2+}; the inhibition caused by heavy metal salts is mainly due to the precipitation of insoluble phytates.

Assay (Peers, 1953). The incubation mixture contains enzyme ($200-1000$ units), 5 ml of 0.2 M acetate buffer, pH 5.15, containing 0.004 M $MgSO_4$, prewarmed to 55°, and the appropriate amount of phytate solution (at 55°) to bring the final volume to 10 ml and the final substrate phytate concentration to about 1.6×10^{-3} M (i.e. about 300 μg/ml phytate-P). The mixture is incubated for 60 min at 55° and 2 ml samples are taken for inorganic phosphate analysis. These values obtained are corrected for blank values on samples taken at zero time.

A unit of activity is the amount releasing 1 μg inorganic phosphate under the conditions just described. Commercial sodium phytate is often unsuitable as a substrate because of the large amount of inorganic orthophosphate present. A satisfactory preparation can be made by dissolving phytin (10 g) in water (50 ml) with the aid of the minimum amount of concentrated HCl. Ferric phytate is precipitated by the addition of 20% (w/v) $FeCl_3$ in 2 N HCl, collected by centrifugation, washed with 3×50 ml N HCl followed by 5×50 ml H_2O, evenly suspended in water and dissolved by adding 10 N NaOH with vigorous stirring. The solution is centrifuged, the pH adjusted to 5.15 with HCl and the volume made up to 200 ml for use.

8. Carotene.

a) Carotene Oxidase (Lipoxidase).

A plant enzyme which has been studied in considerable detail is lipoxidase or carotene oxidase. It is widely distributed but is present in high amount in soya beans and carries out the coupled oxidation of carotenoids and certain unsaturated fatty acids, in particular linoleic acid (see e.g. GOODWIN, 1952).

The crystalline enzyme contains no iron or detectable prosthetic group, has a molecular weight of 102,000 and an isoelectric point at pH 5.4 (THEORELL, HOLMAN and ÅKESON, 1947).

Preparation (THEORELL et al., 1947; KUNKEL, 1951). Defatted soya bean meal is extracted with acetate buffer (pH 4.5). The extract is then brought to pH 6.75 and barium acetate, basic lead acetate and acetone added to precipitate inactive material. The active fraction is then precipitated with $(NH_4)_2SO_4$ (400 g/kg solution). The precipitate is dissolved in a minimum of water and further inactive material removed by heating to 63° for 5 min. An equimolar mixture (35.7 g/100 ml enzyme solution) of K_2HPO_4 and KH_2PO_4 is added to the supernatant and the precipitate removed by centrifugation. A further 5 g of the mixture is added and the precipitate centrifuging at 400 g for 10 min discarded. Recentrifugation of the supernatant at 2000 g for 1 hr. produces an active precipitate which is dissolved in water, filtered, dialysed for 10 hr. against distilled water, and used as required.

Assay (TOOKEY, WILSON, LOHMAN and DUTTON, 1958). The substrate used contains linoleic acid (0−1070 μmoles) plus β-carotene (0−7.73 μmoles) dissolved in 6 ml of peroxide-free diethylether by gentle warming. 15 ml of warm ethanol are then added and the mixture reduced to 9 ml on a water bath. The solution is quickly added with stirring to 171 ml of 0.05 M borate buffer (pH 9.0 and presaturated with oxygen). The resulting dispersion is used and is stable for several hours.

The substrate (45 ml) is placed in an ERLENMEYER flask held at 20° in a constant temperature bath, oxygen is then passed over the surface for 5 min and then 5 ml of the enzyme solution is added; the flask is shaken and 2 ml samples are taken at intervals of 20 or 30 seconds. The samples are immediately poured into a mixture of 4 ml cold absolute ethanol and 8 ml hexane. The carotene passes into the hexane layer which is separated off, washed with water, dried with anhydrous sodium sulphate and made up to 10 ml. The extinction of this solution at 436 mμ is measured in a spectrophotometer. The molecular extinction coefficient of β-carotene at 436 mμ is 106,300.

Carotene oxidation is approximately linear with time up to 50% oxidation.

References.

ADLER, E., S. ELLIOT and L. ELLIOTT: Enzymologia 8, 80 (1940). — *Association of Vitamin Chemists Inc.*: Methods of Vitamin Assay New York: Interscience 1947. — ATKIN, L., W. L. WILLIAMS, A. S. SCHULTZ and C. N. FREY: Ind. Eng. Chem. Anal. Ed. 16, 67 (1944).

BAUM, R. H., and I. C. GUNSALUS: In Biochemist's Handbook, p. 407. London: Ed. C. Long 1961. — BELLAMY, W. D., and I. C. GUNSALUS: J. Bact. 50, 95 (1945). — BESSEY, O. A., O. H. LOWRY and R. H. LOVE: J. biol. Chem. 180, 755 (1949). — BRAGANCA, B. M., V. KRISH-NAMURTHI and D. S. GHANEKAR: Proc. 4th Int. Congr. Biochem. 11, 109 (1959). — BRATTON, A. C., and E. K. MARSHALL: J. biol. Chem. 128, 537 (1939). — BROWN, G. M.: Fed. Proc. 17, 197 (1958); — J. biol. Chem. 234, 370 (1959). — BROWN, G. M., and E. E. SNELL: J. Amer. Chem. Soc. 75, 2782 (1953). — BÜCHER, T.: Biochim. Biophys. Acta 1, 292 (1947). — BURTON, K.: In Meth. Enzymol. 2, 199. Ed. COLOWICK, S. P., and N. O. KAPLAN. New York: Academic Press 1955.

CAMIENER, G. W., and G. M. BROWN: J. Amer. Chem. Soc. 81, 3800 (1959); — J. biol. Chem. 235, 2404, 2411 (1960). — COHEN, R. Z., and T. W. GOODWIN: Phytochemistry 1, 47 (1961). — CRAIG, J. A., and E. E. SNELL: J. Bact. 61, 283 (1951).

DAVIS, B. D.: Proc. Nat. Acad. Sci. U.S.A. 35, 1 (1949). — DE CASTRO, E. T., J. M. PRICE and R. P. BROWN: J. Amer. Chem. Soc. 78, 2904 (1956).

EICH, S., and L. R. CERECEDO: J. biol. Chem. 207, 295 (1954). — EUSEBI, A. J., and L. R. CERECEDO: Fed. Proc. 9, 169 (1950).

FLYNN, R. M., M. E. JONES and F. LIPMANN: J. biol. Chem. 211, 791 (1954). — FOR-SANDER, O.: Soc. Sci. Fenn. Comm. Phys. Math. 19, No. 22 (1956). — FUJITA, A.: Adv. in Enzymol. 15, 389 (1954); — In Meth. Enzymol. 2, 622. Ed. COLOWICK, S. P., and N. O. KAPLAN: New York: Academic Press 1955. — FUTTERMAN, S.: J. biol. Chem. 228, 1031 (1957).

GIRI, K. V.: Abst. Intern. Conf. Flavin Enzymes, Nagoya (Japan) p. 4 (1957). — GIRI, K. V., and P. R. KRISHNASWAMY: J. Indian Inst. Sci. 38, 232 (1956). — GIRI, K. V., P. R. KRISHNASWAMY and N. A. RAO: Biochem. J. 70, 66 (1958). — GOODWIN, T. W.: Compar-ative Biochemstry of the Carotenoids. London: Chapman & Hall 1952; — Recent Advan-ces in Biochemistry. London: Churchill 1960; — The Biosynthesis of Vitamins. London: Academic Press 1963. — GOODWIN, T. W., and A. A. HORTON: Nature 191, 772 (1961). — GRAFFLIN, A., and S. OCHOA: Biochim. biophys. Acta 4, 205 (1950). — GREILING, H.: Proc. Comm. 4th Int. Cong. Biochem. p. 46 (1958). — GUNSALUS, I. C., and W. E. RAZZELL: In Methods in Enzymology 2, 646. Ed. COLOWICK, S. P., and N. O. KAPLAN. New York: Aca-demic Press 1955.

HANDLER, P.: Proc. 4th Int. Cong. Biochemistry 11, 39 (1959). — HANDLER, P., and J. R. KLEIN: J. biol. Chem. 143, 49 (1942). — HASEGAWA, E., T. TAMAKI, T. TANAKA and A. FUJITA: J. Vitaminol. (Japan) 3, 30 (1957). — HEPPEL, L. A., and R. J. HILMOE: In Methods in Enzymology 2, 546. Ed. S. P. COLOWICK and N. O. KAPLAN. New York: Academic Press 1955. — HURWITZ, J.: J. biol. Chem. 205, 935 (1953).

IWAI, K., and T. YOSHIDA: Bull. Res. Inst. Food Sci. Kyoto Univ, No. 15, p. 115 (1954). JAKOBY, W. B., and D. M. BONNER: J. biol. Chem. 205, 699 (1953).

KAPLAN, N. O.: In Methods in Enzymology 2, 664. Ed. S. P. COLOWICK and N. O. KAPLAN. New York: Academic Press 1955. — KEARNEY, E. B.: In Methods in Enzymology 2, 640. Ed. COLOWICK, S. P., and N. O. KAPLAN. New York: Academic Press 1955. — KEARNEY, E. B., and S. ENGLARD: J. biol. Chem. 193, 821 (1951). — KENKARE, U. W., and B. M. BRA-GANCA: Nature (Lond.) 181, 548 (1958). — KENTEN, R. H.: Biochem. J. 67, 25 (1957); 69, 439 (1958). — KIESSLING, K. H.: Ark. Kemi 10, 279 (1956); — Acta Chem. Scand. 11, 97 (1957); 13, 1358 (1959). — KING, T. E., and F. M. STRONG: J. biol. Chem. 189, 315 (1951). — KNOX, W. E.: In Methods in Enzymology 2, 242. Ed. S. P. COLOWICK and N. O. KAPLAN. New York: Academic Press 1955. — KORNBERG, A.: In Methods in Enzymology 2, 673. Ed. COLOWICK, S. P., and N. O. KAPLAN. New York: Academic Press 1955a; — In Methods in Enzymology 2, 670. Ed. S. P. COLOWICK and N. O. KAPLAN. New York: Academic Press 1955b; — In Methods in Enzymology 2, 655. Ed. S. P. COLOWICK and N. O. KAPLAN. New York: Academic Press 1955c. — KORNBERG, A., and W. E. PRICER jr.: J. biol. Chem. 182, 763 (1950). — KRISH-NAN, P. S.: In Methods in Enzymology 2, 591. Ed. COLOWICK, S. P., and N. O. KAPLAN. New York: Academic Press 1955. — KUNKEL, H. O.: Arch. Biochem. 30, 306 (1951).

LEDER, I.: Biochem. Biophys. Res. Comm. 1, 63 (1959). — LIPMANN, F., and L. C. TUTTLE: J. biol. Chem. 158, 505 (1945). — LOWRY, O. H., N. J. ROSEBROUGH, A. L. FARR and R. J. RANDALL: J. biol. Chem. 193, 265 (1951).

MAAS, W. K.: In Proc. Symp. 4th Int. Cong. Biochem. 11, 161 (1959). — MAAS, W. K. and B. D. DAVIS: J. Bact. 60, 733 (1950). — McDONALD, M. R.: In Methods in Enzymol-ogy 1, 269. Ed. S. P. COLOWICK and N. O. KAPLAN. New York: Academic Press 1955. —

Misra, D. K., S. R. Humphreys, M. Friedkin, A. Goldin and E. J. Crawford: Nature 189, 39 (1961).
Nason, A., N. O. Kaplan and S. P. Colowick: J. biol. Chem. 188, 397 (1951). — Nose, Y., K. Ueda and T. Kawasaki: Biochim. biophys. Acta 34, 277 (1959). — Novelli, G. D.: In Methods in Enzymology 2, 619. Ed. S. P. Colowick and N. O. Kaplan. New York: Academic Press 1955a; — In Methods in Enzymology 2, 667. Ed. Colowick, S. P., and N. O. Kaplan. New York: Academic Press 1955b; — Proc. 4th Int. Cong. Biochem. 11, 169 (1959).
Peers, F. G.: Biochem. J. 53, 102 (1953). — Peters, J. M., and D. M. Greenberg: Biochim. Biophys. Acta 32, 273 (1959). — Peters, R. A., and J. R. P. O'Brien: In Mod. Meth. Plant Analys. (Ed. K. Paech and M. V. Tracey) 4, 435 (1955). — Plaut, G. W. E.: J. biol. Chem. 235, PC 41 (1960). — Preiss, J., and P. Handler: J. Amer. Chem. Soc. 79, 1514, 4246 (1957); — J. biol. Chem. 233, 488 (1958).
Quastel, J. H.: In Methods in Enzymology 2, 631. Ed. S. P. Colowick and N. O. Kaplan. New York: Academic Press 1955.
Saran, A.: Biochem. J. 70, 182 (1958). — Shuster, L., and N. O. Kaplan: In Methods in Enzymology 2, 551. Ed. S. P. Colowick and N. O. Kaplan. New York: Academic Press 1955. — Stadtman, E. E.: In Methods in Enzymology 1, 596. Ed. S. P. Colowick and N. O. Kaplan. New York: Academic Press 1955. — Stadtman, E. R., G. D. Novelli and F. Lipmann: J. biol. Chem. 191, 365 (1951). — Steyn-Parvé, E. P.: Biochim. biophys. Acta 8, 310 (1952). — Strong, F. M.: Mod. Meth. Plant Analys. (Ed. K. Paech and M. V. Tracey) 4, 643 (1955).
Theorell, H., R. T. Holman and A. Åkeson: Acta Chem. Scand. 1, 571 (1947). — Tookey, H. L., R. G. Wilson, R. L. Lohman and H. J. Dutton: J. biol. Chem. 230, 65 (1958).
Wang, T. P.: In Methods in Enzymology 2, 649. Ed. Colowick, S. P., and N. O. Kaplan. New York: Academic Press 1955. — Wang, T. P., and N. O. Kaplan: J. biol. Chem. 206, 311 (1954). — Ward, G. B., G. M. Brown and E. E. Snell: J. biol. Chem. 213, 869 (1955). — Westenbrink, H. G.: Proc. 4th Int. Cong. Biochem. Vienna, 11, 73 (1959). — Westenbrink, H. G. K., E. P. Steyn-Parvé and H. Vedlman: Biochim. biophys. Acta 1, 154 (1947). — Westenbrink, H. G. K., and E. P. Steyn-Parvé: Intern. Rev. Vitamin Research 21, 461 (1950). — Woods, D. D.: Proc. 4th Int. Cong. Biochem. Vienna, 11, 89 (1959).
Zatman, L. J., N. O. Kaplan and S. P. Colowick: J. biol. Chem. 200, 197 (1953).

Enzyme des L-Ascorbinsäure-Stoffwechsels.

Von

H. Janecke.

Mit 7 Abbildungen.

Die L-Ascorbinsäure nimmt eine besondere Stellung im Stoffwechsel der Pflanzenzelle ein. Ihre spezielle Funktion als Wasserstoffüberträger bei Atmungsvorgängen konnte in vivo nachgewiesen werden (z. B. FRANKE, 1958). Es erfolgt hierbei ihre Umwandlung in L-Dehydroascorbinsäure. An keimenden Erbsen wurde gefunden, daß das System L-Ascorbinsäure-L-Dehydroascorbinsäure in die Endatmung eingeschaltet ist, d. h. in den letzten Oxydationsschritt, bei dem der aus dem Substrat stammende Wasserstoff zu Wasser oxydiert wird (MAPSON u. MOUSTAFA, 1956).

I. Enzyme bei der Biosynthese der L-Ascorbinsäure.

Es handelt sich bei der L-Ascorbinsäure um eine Verbindung, die den Kohlenhydraten nahesteht. Sie kann als die Enolform des 3-Keto-L-gulonsäure-γ-laktons angesehen werden[1].

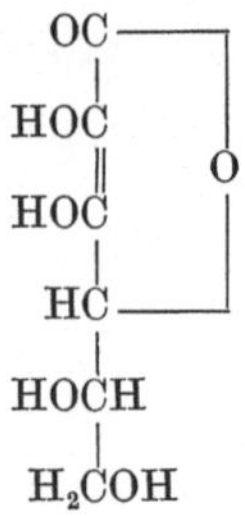

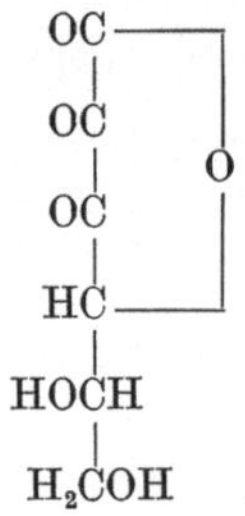

$$
\begin{array}{cc}
\text{OC} & \text{OC} \\
\text{HOC} & \text{OC} \\
\text{HOC} \quad \text{O} & \text{OC} \quad \text{O} \\
\text{HC} & \text{HC} \\
\text{HOCH} & \text{HOCH} \\
\text{H}_2\text{COH} & \text{H}_2\text{COH}
\end{array}
$$

L-Ascorbinsäure (AS) L-Dehydroascorbinsäure (DAS)

Auf Grund der nicht möglichen Reduktion der Carbonylgruppe (mit $LiAlH_4$ in absol. Äther) der AS nehmen PETUELY u. BAUER (1952) an, daß die funktionelle Gruppe ein mesomeres System darstellt, in dem die π-Elektronen relativ gleichmäßig verteilt sind und das daher nur schwer polarisierbar ist.

Die AS kann gebunden und in freier Form in pflanzlichen und tierischen Geweben vorkommen (Literatur z. B. HERRMANN u. ZOBEL, 1962)[1].

Im Text häufig wiederkehrende Abkürzungen: AS L-Ascorbinsäure, DAS L-Dehydroascorbinsäure, DKS 2,3-Diketo-L-gulonsäure, GSH Glutathion (red.), GSSG Glutathion (oxyd.), TPN Triphospho-pyridin-nucleotid (oxyd.), TPNH Triphospho-pyridin-nucleotid (red.), DPN Diphospho-pyridin-nucleotid (oxyd.), DPNH Diphospho-pyridin-nucleotid (red.), FMN Flavin-mononucleotid (oxyd.), FMNH Flavin-mononucleotid (red.), FAD Flavin-adenin-dinucleotid (oxyd.), FADH Flavin-adenin-dinucleotid (red.), UTP Uridintriphosphat, UDPG Uridindiphosphatgalaktose, UDP-Glucuronsäure Uridindiphosphatglucuronsäure, UDP-Galaktose Uridindiphosphatgalaktose, DIP 2,6-Dichlorphenolindophenol.

[1] Bestimmungsmethoden siehe dieses Handbuch Bd. 2, S. 95; Synthese, Physiologie siehe z. B. SEBRELL u. HARRIS (1954).

Unsere Kenntnisse über die Biosynthese gehen auf Ray (1934) zurück, der mit Erbsen-embryonen zeigen konnte, daß auf synthetischem Medium Hexosen bei der Bildung der AS eine Rolle spielen. Mit Pentosen und L-Zuckern wurden dagegen nichtbefriedigende Ergebnisse erzielt (Mapson, Cruickshank u. Chen, 1949; Åberg, 1949).

Es war naheliegend, die AS-Bildung mit der Photosynthese in Zusammenhang zu bringen. Eingehende Studien an Kartoffeln von Franke (1952, 1954, 1955, 1958) führten zu dem Ergebnis, daß der AS-Gehalt abhängig von der Atmung steigt und daß somit die AS-Synthese ein dissimilatorischer, mit dem Stoffabbau und -umbau verbundener Vorgang ist, siehe auch Mirimanoff (1956). Die Bildung kann daher prinzipiell in allen zur Atmung befähigten Organen erfolgen, wie die Erfahrung auch bisher gezeigt hat.

Nach dem Stand unserer Kenntnisse (Zusammenfassungen: Franke, 1958; Mapson, 1955, 1960; Gäumann u. Renner, 1955; Bünning u. Gäumann, 1960; Hollmann, 1961; Sobotka u. Stewart, 1961) sind zwei Möglichkeiten der AS-Bildung in Betracht zu ziehen:

1. Indirekte Umwandlung von Kohlenstoffverbindungen,
2. Direkte Umwandlung von Zuckern ohne Aufspaltung der C_6-Kette.

1. Indirekte Umwandlung.

Indirekte Umwandlung kann durch Aldolkondensation zwischen L-Glycerin-aldehyd und Hydroxybrenztraubensäure erfolgen (Hough u. Jones, 1951, 1952; Isherwood, Chen u. Mapson, 1954; Mapson, 1955). Ein der Aldolase (Meyerhoff, Lohmann u. Schuster, 1936) ähnliches Enzym wurde von Stumpf (1948) aus Erbsen gewonnen und auf das 92fache angereichert.

a) Aldolkondensation.

Herstellung des Enzyms nach Stumpf *(1948).* 25 g über Nacht bei 2° C ein-geweichte Erbsen wurden in 10 Vol. kalter 0,1%iger Kaliumcarbonatlösung sus-pendiert. Nach dem Abseihen erfolgte Ammoniumsulfatfällung (70% Sättigung), Dialyse 5 Std, nochmalige Ammoniumsulfatfällung (61%ige Sättigung), Dialyse 12 Std, Fällung am I.P. (pH 5,5) bei 0° C, Dialyse und nochmalige Fällung am I.P. Das Enzym katalysiert die reversible Reaktion (schwächer als tierische Aldolase):

Fructosediphosphat $\leftrightarrows$ Dihydroxyacetonphosphat + D-3-Glycerinaldehydphosphat

Das Gleichgewicht der Reaktion wird nach etwa 10 min erreicht. Die Aktivität wurde nach der Methode von Herbert u. Mitarb. (1940) bei pH 8,5 bestimmt, bei der das gebildete Triosephosphat mit Cyanid zum Cyanhydrin umgesetzt wird. Die Bestimmung des Triosephosphat-P erfolgte nach Hydrolyse mit starkem Alkali (2 n-NaOH, 10 min, 20° C) colorimetrisch mit Ammoniummolybdat nach Fiske u. Subbarow (1925).

b) Acyloin-Reaktion.

Eine andere Möglichkeit wird in der Acyloin-Reaktion zwischen Tartron-semialdehyd und L-Glycerinaldehyd unter Bildung von 3-Keto-L-gulonsäure gesehen. Nach Cox, Hirst u. Reynolds (1932) steht die 3-Keto-gulonsäure mit der AS in einem tautomeren Gleichgewicht, das zugunsten der AS verschoben ist. Roy u. Mitarb. (1946) sowie Smythe u. King (1942) nehmen auch die Bildung aus Triose-Einheiten an. Smith (1952) verfolgte die AS-Biosynthese in Kartoffel-scheiben bei Gegenwart von Fluorid und von Jodacetat und fand eine Hemmung der AS-Synthese. Dieses Ergebnis deutete er in der Weise, daß bei der Spaltung von Fructose-1,6-diphosphat aus L-Glycerinphosphat L-Glycerinaldehyd-3-phos-phat gebildet würde, das sich mit Dihydroxyaceton zu L-Sorbose-1,6-diphosphat kondensiert. Über Oxydation dieser Verbindung zu 2-Keto-gulonsäure soll dann Bildung von AS erfolgen, siehe Abb. 1. Isherwood, Chen u. Mapson (1954)

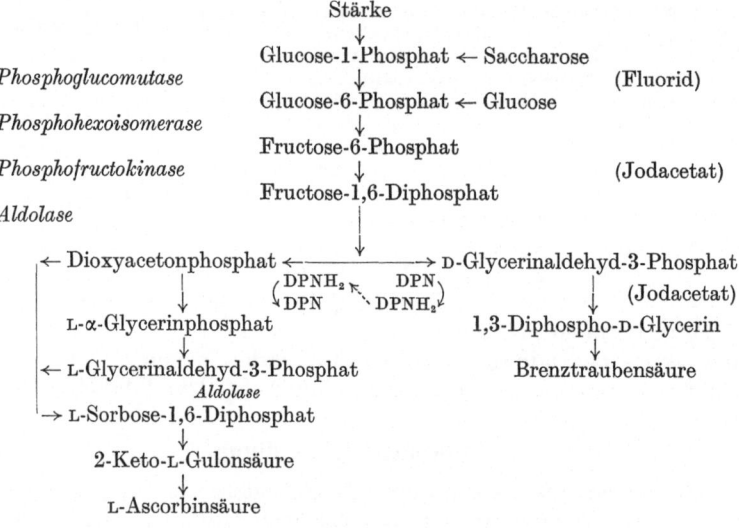

Abb. 1. Ascorbinsäurebildung nach F. G. SMITH (1952).

verabfolgten DL- und D-Glycerinaldehyd an Kressesamen und fanden keine Beeinflussung der AS-Synthese mit D-Glycerinaldehyd. Mit DL-Glycerinaldehyd
wurde Hemmung beobachtet. D-Glycerinsäure tritt in keimenden Kressesamen auf.
Die Ergebnisse von ISHERWOOD, CHEN u. MAPSON (1954a) zeigen jedoch, daß die
Bildung von AS und D-Glycerinsäure parallel verläuft und daß D-Glycerinsäure
nicht als Vorläufer der AS angesehen werden kann. Diese Annahme wird noch
gestützt durch die Beobachtung, daß D-Glucuronsäure-γ-lakton und L-Gulon-
säure-γ-lakton, die als Vorläufer der AS gelten, die Bildung von D-Glycerinsäure
in Kressesamen hemmen.

NATH, CHITALE u. BELAVADY (1952) nehmen an, daß ein Kondensations-
produkt aus Acetessigsäure und Glucose (Glucose-cycloacetoacetat) der Vorläufer der AS ist. Es bildet in größerer Ausbeute AS als mit Glucose und Acetessigsäure allein. Glucose und Acetessigsäure können sich in vitro kondensieren
(WEST, 1927; GONZALES, 1934); es bildet sich 2-Tetrahydroxybutyl-5-methyl-
4-carbäthoxyfuran, das eine gewisse formale Ähnlichkeit mit der AS besitzt:

$$HC\!-\!\!-\!\!-C \cdot COOC_2H_5$$

2-Tetrahydroxybutyl-5-methyl-4-carbäthoxyfuran

L-Ascorbinsäure

Darstellung der 2-Tetrahydroxybutyl-5-methyl-4-carboxyfuran (NATH u. Mitarb., 1952).
100 g Glucose (feinst gepulvert) werden mit 100 g geschmolzenem Zinkchlorid gut vermischt.
In einem runden Pyrex-Kolben, der in kochendes Wasser eintaucht, wird diese Mischung mit
50 g Äthylacetoacetat und 50 ml absol. Alkohol versetzt. Nach etwa 25 min hat sich daraus
unter intensivem Umrühren ein gelber Sirup gebildet. In dieser Zeit muß nochmals die gleiche
Menge an absol. Alkohol (50 ml) hinzugegeben werden. Die Sirupmasse wird in 150 ml Wasser
gegossen. Im Kühlschrank bilden sich über Nacht feine, nadelförmige Kristalle, die gesammelt

und mit eiskaltem Wasser gewaschen werden. Durch mehrfaches Umkristallisieren aus heißem Wasser werden sie gereinigt. Ausbeute: 25g, Schmelzpunkt: 153° C (korr.), α_D^{20}- -19,00° (Wasser); schwer wasserlöslich, daher heiß lösen.

Die Untersuchungen wurden an Bohnen *(Phaseolus mungo)* in wäßrigem Milieu bei p_H 7,4 durchgeführt. Daß AS gebildet wird, wurde papierchromatographisch (Rundfiltermethode) nach dem Verfahren von NATH u. BHATTATHIRY (1956) bestimmt. Es gelang, Glucose-cycloacetoacetat in keimenden Samen nachzuweisen, so daß dieser Syntheseweg durchaus möglich ist (NATH u. BELKHODE, 1958). Thiamin (Cocarboxylase) und Pantothensäure (Coenzym A) wirkten als Co-Faktoren. Die Antivitamine Pyrithiamin und Pantoyltaurin hemmten die Biosynthese. Im tierischen Bereich besteht auch die Möglichkeit zu einem ähnlichen Ablauf der AS-Biosynthese (NATH, CHAKRABORTY, HATWALNE u. LAUE, 1948).

2. Direkte Umwandlung.

LOEWUS, JANG u. SEEGEMILLER (1956) nehmen auf Grund ihrer Isotopenversuche mit Glucose-1-C^{14} an reifenden Erdbeeren an, daß Hexosemonophosphat zu 6-Phosphogluconsäure (oder ihrem Lakton) oxydiert wird; es erfolgt dann Oxydation am C_3 zur entsprechenden Ketosäure und Seitentausch am C_5-Atom. Die Ausbildung der Endiolstruktur und des Laktonringes, wenn das Zwischenprodukt als freie Säure vorliegt, soll spontan oder enzymatisch erfolgen. Die enzymatische Oxydation am C_3 durch eine Dehydrogenase, die spezifisch auf C_3-L und C_3-D eingestellt ist, würde die Beobachtungen von ISHERWOOD, CHEN u. MAPSON (1954), MAPSON, ISHERWOOD u. CHEN (1954), JACKSON, WOOD u. PROSSER (1961) erklären, daß D-Altronsäure-γ-lakton und L-Galaktonsäure-γ-lakton von Kressesamen und D-Mannonsäure-γ-lakton und L-Gulonsäure-γ-lakton von Ratten zu AS verarbeitet wurden. Die Ergebnisse von LOEWUS u. Mitarb. (1956) lassen aber noch die Möglichkeit der AS-Synthese auf einem anderen Wege zu, denn nur 65 bis 70% der zugeführten Menge an C^{14} wurden an derselben Stelle (C_1) gefunden; daneben beobachteten sie noch 14—18 % in C_6 und 7—18% in C_3. Das Hexosemonophosphat soll Glucose-6-phosphat sein (LOEWUS u. JANG, 1958). Galaktose-1-C^{14} (Übergang in Glucose kann erfolgen) war ein ausgezeichneter Vorläufer für AS und Galakturonsäure, während Xylose-1-C^{14} und Arabinose-1-C^{14} besser AS als Galakturonsäure bildeten. Untersuchungen an etiolierten Kressesamen brachten ein ähnliches Ergebnis (LOEWUS u. JANG, 1957; LOEWUS, FINKLE u. Jang, 1958). LOEWUS, JANG u.SEEGEMILLER (1958) bringen die Biosynthese der AS in Beziehung zum Pentosestoffwechsel und betrachten 3-Keto-D-gluconsäure-6-phosphat von HORECKER (1951, 1958) als unmittelbarer Vorläufer des Ribulose-5-phosphates angesehen, als mögliches Zwischenprodukt. Weiterhin gelang es LOEWUS u. KELLY (1959), die Umwandlung von D-Glucuronsäurelakton in L-Gulonsäure (und dessen Lakton) nachzuweisen. Damit erfolgte die Bestätigung eines wichtigen Bezirkes ihres Schemas, das 2 Möglichkeiten der AS-Biosynthese vorsieht.

Die mit C^{14} gekennzeichnete Säure erschien in der Fraktion aus Erdbeeren, die sich in 70%igem Äthanol löste. Die Auftrennung gelang an Dowex 1 (formate) (FINKLE u. LOEWUS, 1960). Nach dem Konzentrieren erfolgte die papierchromatographische Auftrennung durch Auftragen von breiten Zonen mit Äthylacetat-Eisessig-Wasser (3:1:3, obere Phase) 20 Std lang, absteigend (RAO, BERI u. RAO, 1951); Trocknung und Kontaktaufnahme mit Kodak-No-Screen X-ray film (12 Tage lang) in Kassetten schlossen sich an. Es hatten sich L-Gulonsäure und ihr Lakton gebildet. Das L-Gulonsäure-γ-lakton wurde mit synthetischem Material verglichen, hergestellt durch Reduktion von Natrium-D-gluconat mit $NaBH_4$, nach WOLFRAM u. ANNO (1952). Die weitere Charakterisierung gelang als Amid (HUDSON u. KOMATSU, 1919), als Phenylhydrazid (FISCHER u. CURTIS, 1892) und mit Hilfe der Perjodatoxydation (SEEGMILLER, AXELROD u. McCREADY, 1955). Ein Übergang von C^{14} in andere Stellungen wurde praktisch nicht beobachtet.

Es erfolgt wohl zunächst eine enzymatische Aufspaltung des Laktonringes, dann eine Isomerisierung zu D-Fructuronsäure (Wahba, Hickman u. Ashwell, 1958) und schließlich eine stereospezifische Reduktion zur Gulonsäure.

Von anderen Autoren wird die 2-Keto-l-gulonsäure als Vorläufer der AS angesehen. Sie konnten nachweisen, daß durch *Aspergillus niger* (Galli, 1946) in stark saurem Medium und durch Hefe (Monzini, 1954) unter mehr physiologischen Bedingungen diese Säure in AS übergeführt werden kann. Es scheint aber, daß die freie Säure kein Zwischenprodukt ist (Burns u. Evans, 1956); sie konnte die AS-Synthese in Kressekeimen und in Petersilienblättern nicht beeinflussen (Åberg, 1953; Chen, 1950; s. hierzu auch Nakatani, 1958). Es ist möglich, daß Derivate, z. B. Ester oder Laktone, die die Zellwand besser durchdringen, andere Ergebnisse zeigen (Isherwood, Chen u. Mapson, 1954).

Das Verdienst des Arbeitskreises um Mapson und Isherwood ist es, erkannt zu haben, daß die Vorstellung einer Inversion des Moleküls der Glucose oder Galaktose um 180° das Verständnis der AS-Biosynthese wesentlich erleichtert (vgl. auch die Arbeiten von Loewus u. Mitarb.).

Nur vier Zuckerderivate waren fähig, die AS-Bildung in Keimlingen zu fördern oder die Exkretion von AS nach Injektion bei Ratten zu erhöhen. Es waren dieses die l-Gulonsäure- und l-Galaktonsäure-γ-laktone sowie D-Glucuronsäure-γ-lakton, D-Galakturonsäurelakton und D-Galakturonsäuremethylester. Damit Umbildung zu AS erfolgen kann, sind an die Ausgangsstoffe allgemein folgende Anforderungen zu stellen (Mapson, 1960):

1. Hydroxyl am C_2 und C_5 in l-Konfiguration,

2. Hydroxyl am C_4 in D-Konfiguration.

Diese Bedingungen werden von den l-Gulonsäure- und l-Galaktonsäure-γ-laktonen erfüllt. Zeigt das Hydroxyl am C_5-Atom D-Konfiguration, so entsteht D-Araboascorbinsäure. Besonders leicht soll die Umwandlung in Pflanzen erfolgen, wenn die OH-Gruppe am C_3 in D-Konfiguration vorliegt: l-Galaktonsäurestruktur; siehe Abb. 2.

Das Glucose- oder Galaktosemolekül soll zunächst durch Oxydation in die D-Glucuronsäure bzw. D-Galakturonsäure übergehen. Unter Inversion des Moleküls um 180° und anschließender Reduktion entstehen l-Gulonsäure-γ-lakton bzw. Galaktonsäure-γ-lakton, die schließlich unter nochmaliger Oxydation in AS übergehen. Die Ergebnisse, die zunächst an intakten Pflanzen erhalten wurden, konnten mit Extrakten und mit teilweise gereinigten Enzymen weiter verfolgt werden. Mapson u. Mitarb. (1954) zeigten als erste die in vitro-Bildung von AS mit Extrakten aus Erbsen. Die Enzyme befinden sich in den Mitochondrien. Die Umwandlung von l-Galaktonsäure-γ-lakton zu AS (60% Ausbeute) verläuft wesentlich schneller als diejenige von l-Gulonsäure-γ-lakton. Eine Oxydation der Galaktonsäure und anderer Zuckersäuren wurde nicht beobachtet.

An der Oxydation scheint das Cytochrom-System beteiligt zu sein (Mapson, 1960). Die Reaktion wird durch Cyanid, Azid und CO (nur im Dunkeln) gehemmt. Energiereiche Phosphorverbindungen scheinen für die Umwandlung nicht erforderlich zu sein. Dinitrophenol, das bei der oxydativen Phosphorylierung hemmt, war ohne Wirkung. ATP hemmt. Es gelang, die l-Galaktonsäure-γ-laktondehydrogenase gereinigt aus Mitochondrien von Blumenkohl zu gewinnen und einige ihrer Eigenschaften kennenzulernen (Mapson u. Breslow, 1958). Das Enzym ist ein Flavoprotein (Absorptionsspektren, Hemmung durch Riboflavin), benötigt SH-Gruppen für seine Aktivität und besitzt eine hohe Spezifität für das l-Galaktonsäure-γ-lakton. Es vermag z. B. nicht D-Mannonsäure-γ-lakton, l- und D-Gulon-

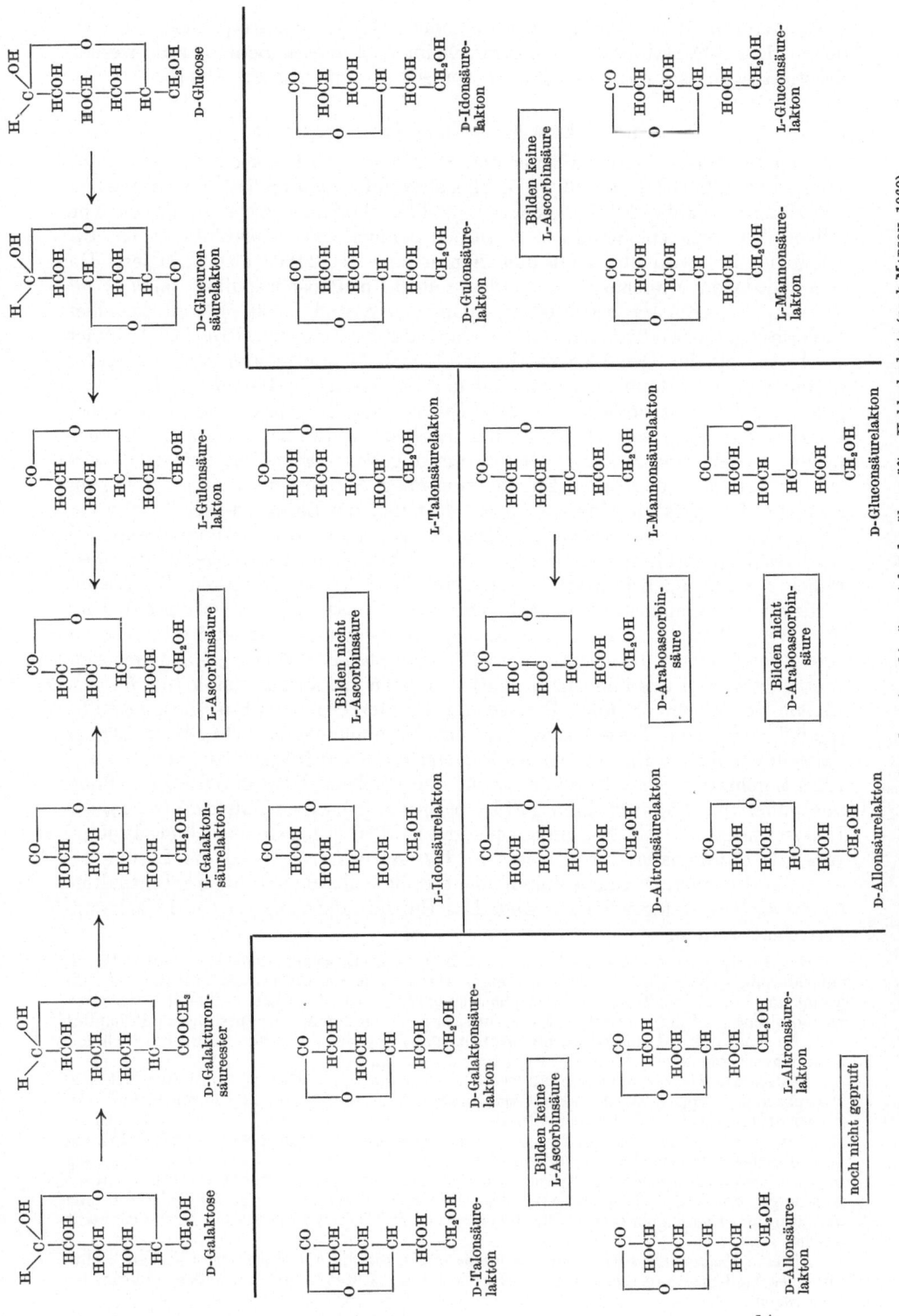

Abb. 2. Formelbilder der in Beziehung zur Biosynthese der L-Ascorbinsäure stehenden überprüften Kohlenhydrate (nach MAPSON, 1960).

säure-γ-lakton, d-Gluconsäure-γ-lakton oder d-Galaktonsäure-γ-lakton zu oxydieren. Das l-Galaktonsäure-γ-lakton scheint nach diesen Befunden die wesentlichste Ausgangssubstanz zur Bildung von AS zu sein (Mapson, 1960).

a) l-Galaktonsäure-γ-laktondehydrogenase.

1 kg Blumenkohl (ohne Blätter und Stiel) wird mit 1 l Saccharose-Phosphatlösung (0,4 m-Saccharose und 0,1 m-Phosphatpuffer vom p_H 7,4) innig verrieben. Der Mörser soll auf —20° C, die Saccharose-Phosphatlösung auf 0° C vor dem Verreiben, das höchstens 30 min in Anspruch nehmen soll, abgekühlt werden. Der erhaltene Brei wird abgeseiht und 20 min lang bei 20000 g zentrifugiert. Die abgeschiedenen Mitochondrien werden mit 0,1 m-Phosphatpuffer (p_H 7,4) gewaschen, zentrifugiert bei 20000 g, 20 min lang, und in 50—100 ml derselben Pufferlösung suspendiert, auf —20° C abgekühlt und mit dem 10fachen Volumen an Aceton gefällt. Der Niederschlag wird nach 15 min gesammelt, das Aceton teilweise unter Vakuum entfernt und der Rückstand in 150 ml des Phosphatpuffers (0,1 m) suspendiert. Es folgt Dialyse der Suspension gegen denselben Puffer (3 Std lang) bei + 1° C, um den Rest an Aceton zu entfernen. Die klare Lösung (gegebenenfalls zentrifugieren, um ungelöstes Eiweiß zu entfernen) wird auf 40%ige Sättigung durch Zugabe von 243 g Ammoniumsulfat pro 1 Lösung gebracht. Das gefällte Eiweiß wird entfernt und die Lösung durch Zugabe von 205 g Ammoniumsulfat pro l auf 70%ige Sättigung gebracht. Es folgt Lösung des Niederschlages in Phosphatpuffer (p_H 6,8) und 2 Std lange Dialyse gegen denselben Puffer. Die folgenden Reinigungsstufen sind bei 1° C durchzuführen. Die Lösung wird mit Calciumphosphatgel nach Keilin u. Hartree (1938) (s. unten), 1 mg für 1,5 mg Protein, versetzt, 15 min lang stehen gelassen, dann 5 min lang bei 1500 g zentrifugiert. Das Gel wird mit Phosphatpuffer (0,05 m, p_H 6,8) gewaschen, zentrifugiert und aus ihm durch Elution mit 0,5 m-Phosphatpuffer (p_H 7,4) das Enzym gewonnen. Es folgt Dialyse des Eluates gegen 0,01 m-Phosphatpuffer (p_H 6,8), 3 Std lang. Diese Lösung wird mit Ammoniumsulfat (336 g/l) auf 53%ige Sättigung und nach Entfernung des Niederschlages auf 80%ige Sättigung (weitere 192 g/l) gebracht. Die Lösung des gebildeten Niederschlages in 5 ml 0,1 m-Phosphatpuffer (p_H 6,8) wird dann 2 Std lang gegen denselben Puffer dialysiert, das Dialysat nochmals mit Calciumphosphatgel bei 1° C (1 mg Gel für 1,5 mg Protein) versetzt, 15 min lang stehen gelassen, das Gel dekantiert und die Lösung bei —3° C aufbewahrt. Diese Lösung soll nicht weniger als 10 mg Protein/ml enthalten, da schwächere weniger haltbar sind. Das Enzym konnte so auf das 12fache angereichert werden.

Herstellung des Calciumphosphatgels nach Keilin u. Hartree (1938). 150 ml Calciumchloridlösung (132 g $CaCl_2 \cdot 6H_2O$ im Liter) werden mit Leitungswasser auf 1600 ml verdünnt und mit 150 ml Trinatriumphosphatlösung (152 g $Na_3PO_4 \cdot 12H_2O$ im Liter) versetzt. Die Mischung wird mit verdünnter Essigsäure auf p_H 7,4 gebracht. Das entstandene Präcipitat wird 3—4mal durch Dekantieren mit großen Wassermengen (15—20 l) und schließlich mit dest. Wasser gewaschen. Ausbeute: 9,1 g Calciumphosphat.

Versuchsansatz. In ein Thunberg-Gefäß werden die Enzymlösung, 1 ml 0,01 m-KCN in Phosphatpuffer (p_H 7,4), 0,5 ml l-Galaktonsäure-γ-lakton (24 mg/ml) gegeben und mit Phosphatpuffer auf 5 ml gebracht. In das Seitengefäß werden 1 ml Phenazinmethosulfat [1 mg/ml in Phosphatpuffer (p_H 7,4) gelöst] und 0,08 ml 2,6-Dichlorphenolindophenol (1 mg pro ml) eingefüllt. Das Gefäß wird evakuiert, mehrmals mit N_2 durchspült und 10 min lang bei 37° C gehalten. Die Farbstofflösung wird dann in den Reaktionsraum überführt und die Ablesungen am EEL-Colorimeter mit Filter EEL 607 in Abständen von 2 min 12 min lang durchgeführt. Das Ergebnis eines Blindversuches ohne Enzymzusatz muß in Abzug gebracht werden.

Als eine *Enzymeinheit* wurde die Menge definiert, die die Abnahme der optischen Dichte um einen Skalenteil pro min bewirkte. Das Verfahren läßt sich auch auf andere Colorimeter übertragen.

Phenazinmethosulfat wurde durch Zugabe von Dimethylsulfat zu Phenazin, gelöst in Nitrobenzol, hergestellt (KEHRMANN, 1913; HILLEMANN, 1938).

Neben dieser Dehydrogenase fanden MAPSON u. ISHERWOOD (1958) in Erbsen ein weiteres Enzymsystem (s. folgende Herstellungsvorschrift), das die Reaktion zwischen TPNH und D-Galakturonsäureestern katalysiert. Ein Reaktionsprodukt soll das L-Galaktonsäure-γ-lakton sein. Die freie Säure wurde nicht reduziert, und TPNH ließ sich nicht durch DPNH ersetzen. Die γ-Laktone der D-Glucuronsäure und D-Mannuronsäure wurden mit geringerer Geschwindigkeit reduziert.

b) Enzympräparat nach MAPSON u. ISHERWOOD (1958).

Erbsen, die, um Infektionen zu vermeiden, 1 min lang in 1%ige Sublimatlösung getaucht und anschließend sorgfältig abgespült wurden, wurden 24 Std lang in Wasser (20° C) eingeweicht und dann 24 Std lang keimen gelassen. Die so teilweise gekeimten Samen (30 g) wurden mit Seesand und 40 ml einer Lösung von 0,4 m-Saccharose-0,1 m-Phosphat (p_H 7,4) und 4 m-Mg^{2+} bei 0° C verrieben. Der Brei wurde zentrifugiert (500 g, 5 min), die überstehende Flüssigkeit weiter durch Zentrifugieren bei 20000 g, 20 min, fraktioniert. Das Enzym befand sich in dem löslichen Anteil. Die Enzymreaktion konnte nach ihrem Abklingen mit Mitochondrien in bisher nicht geklärter Weise wieder in Gang gebracht werden. Für einige Versuche wurden Präparate verwendet, die noch einer Fällung mit Aceton (30 und 70% V/V) bei niederer Temperatur (ASKONAS, 1951) oder einer Ammoniumsulfatsättigung (50 und 70%, bei p_H 7,0) unterworfen worden waren.

Die Eingliederung der beiden Enzyme in ein Schema der AS-Biosynthese in Pflanzen zeigt die folgende Abbildung:

$$TPN$$
$$\Updownarrow$$
$$\text{D-Galakturonsäureverbindungen} + TPNH \rightarrow \text{L-Galaktonsäureverbindungen} + TPN$$
$$\downarrow$$
$$\text{L-Galaktonsäure-}\gamma\text{-lakton}$$
$$|$$
$$(\text{L-Galaktonsäure-}\gamma\text{-laktondehydrogenase})$$
$$\downarrow$$
$$\text{L-Ascorbinsäure}$$

NAKATANI (1958) konnte mit einer löslichen Enzymzubereitung aus grünen Erbsen D-Glucuronsäure-γ-lakton sowie D-Galakturonsäure-γ-lakton und Diaceton-2-keto-L-gulonsäure in AS umwandeln (siehe hierzu MAPSON, ISHERWOOD u. CHEN, 1954). Er nimmt folgenden Reaktionsablauf an:

$$\text{D-Glucuronsäure-}\gamma\text{-lakton} \rightarrow \text{L-Gulonsäure} \rightarrow \text{2-Keto-gulonsäure}$$
$$\text{(aus Diaceton-2-keto-1-gulonsäure)}$$
$$\downarrow$$
$$\text{L-Ascorbinsäure}$$

Dinitrophenol war ohne Wirkung auf die Synthese von AS. Die freie Carboxylgruppe der 2-Keto-gulonsäure scheint nach NAKATANI erforderlich zu sein, da ihr Methylester nicht in AS übergeführt werden konnte. Der optimale p_H-Wert der Reaktion liegt bei 5,8—6,2 (Phosphatpuffer), bei 33° C. Das Enzym ist bei 50° C relativ stabil, bei 100° C wurde es schnell inaktiviert. Hemmung erfolgte mit Jodacetat, Kupferchlorid und Sublimat; Kaliumcyanid und Natriumfluorid hemmten in Konzentrationen von 10^{-3} m, Dialyse machte das rohe Enzym unwirksam.

c) Enzympräparat nach NAKATANI.

Grüne Erbsen wurden 1 min lang in Sublimatlösung (1%) getaucht, über Nacht mit laufendem Wasser behandelt und anschließend mit sterilem abgespült. Die Erbsen ließ man dann auf mit 10%iger Salzsäure gewaschenem und mit sterilem Wasser gespültem Seesand keimen. Die 6 Tage alten Keimlinge erwiesen sich als am wirksamsten; sie wurden mit dem gleichen Volumen an 0,1 m-Phosphatpuffer (pH 6,0), bei 0—5° C, im Mixer homogenisiert. Das Homogenisat wurde bei 1200 g, 20 min lang, bei 0° C zentrifugiert; die überstehende Flüssigkeit diente als Rohenzym. Sie konnte mit Hilfe der Ultrazentrifuge in weitere Fraktionen unterteilt werden. Die aktive Fraktion zeigte eine Sedimentationskonstante von $s_t = 2,2 \times 10^{-13}$, was auf ein Protein mit niederem Molekulargewicht hinweist.

Neben diesen Untersuchungen an Pflanzenmaterial wurden von verschiedenen Forschergruppen auch solche an *tierischem* durchgeführt. Die Mitochondrien-Mikrosomenfraktion, (Rattenleber) war allein verantwortlich für die Umwandlung der l-Gulonsäure- oder l-Galaktonsäure-γ-laktone in AS, zit. nach Mapson (1960). Von diesen erwiesen sich zunächst die Mitochondrien als wirksamer (Burns, Peyser u. Moltz, 1957). Eine Oxydation der freien Säure fand nicht statt. Bei Einsatz des gesamten Homogenisates wurden auch d-Glucuronsäure-γ-lakton und Ester der d-Galakturonsäure zu AS übergeführt. Chatterjee, Ghosh, Ghosh, Roy u. Guha (1957) fanden in den Lebern von Maus, Kaninchen und Ziege (nicht von Meerschweinchen, Taube oder Huhn) ein Enzymsystem, das auch d-Glucuronsäure-γ-lakton bei pH 7,0 in AS umzuwandeln vermag. Eine Anreicherung wurde mit Ammoniumsulfat (30—50% Sättigung) erreicht. Natriumgluconat, Natriumgalakturonat und Natriumglucuronat erwiesen sich als Ausgangssubstanzen zur AS-Synthese als ungeeignet. Das Enzymsystem erfordert nicht DPN, Mg^{2+} und Coenzym A. ATP in einer Konzentration von 0,0016 m wirkte hemmend. Hemmung wurde nicht mit ADP, AMP und Adenosin oder Adenin beobachtet. Diese Ergebnisse wurden durch weitere Veröffentlichungen ergänzt (Chatterjee u. Mitarb., 1958). Die Wirkung konnte nur mit Mikrosomen (Ziegenleber) in Gegenwart von CN^-, die unbedingt erforderlich zu sein scheinen, erreicht werden. Die Mikrosomenfraktion vermochte auch l-Gulonsäure-γ-lakton in AS umzuwandeln; CN^- oder ein Co-Faktor waren hierbei nicht erforderlich. Chatterjee, Chatterjee, Ghosh, Ghosh u. Guha (1959) konnten mit Natriumdesoxycholat oder mit Schlangengift ein Enzym aus der Mikrosomenfraktion in Lösung bringen, das auch die Umwandlung von d-Glucuronsäure-γ-lakton in AS katalysierte. Bei der Oxydation des l-Gulonsäure-γ-laktons in AS, einer Zwischenstufe der vorgenannten Reaktion, in 2-Keto-gulonsäure ist ein Co-Faktor erforderlich, der dem System mit Petroläther oder Äther entzogen werden kann (Chatterjee, Chatterjee, Ghosh, Ghosh u. Guha, 1959a). Die volle Wirksamkeit wird mit α-Tokopherol oder Vitamin K_1 und teilweise mit Mn^{2+} wiederhergestellt (Chatterjee, Kar, Ghosh u. Guha, 1960). Carpenter, Kitabchi, McCay u. Caputto (1959) fanden, daß die in Vitamin E-Mangel-Ratten gehemmte AS-Synthese durch Tokopherol, Dithizon, Diphenyl-p-phenylendiamin, Äthylendiaminotetraessigsäure und durch Kationen (Co^{2+}, Mn^{2+}, Ce^{3+}) aktiviert wird.

Andere Forschergruppen fanden in der löslichen Fraktion des Homogenisates der Rattenleber ein Enzym, das die Reaktion zwischen DPN und l-Gulonat katalysiert. Dabei entsteht ein Oxydationsprcdukt, das als 3-Keto-l-gulonsäure (unsicher) angesehen wird (Hassan u. Lehninger, 1956; Grollman u. Lehninger, 1957; Bublitz, Grollman u. Lehninger,

Tabelle 1 (nach Franke, 1958, gekürzt). *Verteilung der an der AS-Synthese beteiligten Enzyme in Geweben verschiedener Tierarten.*

Art und Gewebe	Enzyme		
	I	II	III
Ratte			
Leber	2,34	9,6	1,12
Niere	83,5	142,0	0,0
Herz	0,0	0,0	0,0
Milz	3,27	4,83	0,0
Gehirn	0,0	0,0	0,0
Maus			
Leber	4,4	10,2	0,60
Niere	14,5	28,5	0,0
Milz	0,0	0,0	0,0
Herz	0,0	0,0	0,0
Kaninchen			
Leber	10,7	23,4	0,78
Niere	32,0	20,7	0,0
Gehirn	0,0	0,0	0,0
Herz	0,0	1,0	0,0
Nebenniere . .	0,0	0,0	0,0
Hund			
Leber	10,9	35,7	0,46
Niere	4,75	57,3	0,0
Herz	0,0	0,0	0,0
Milz	0,0	0,0	0,0
Nebenniere . .	0,0	0,0	0,0
Taube			
Leber	6,24	45,0	0,0
Niere	7,65	25,8	1,22
Herz	0,0	10,5	0,0
Milz	0,0	0,0	0,0

Aktivität der Enzyme in μmol Substrat oder Produkt/g Frischgewicht/h bei 37° C.

1958). LEHNINGER u. Mitarb. fanden weiter im löslichen Teil des Homogenisates ein Enzym, das die Reduktion von D-Glucuronsäure-γ-lakton bei Gegenwart von TPNH fördert; es scheint in seiner Wirkung dem von MAPSON und ISHERWOOD (1958) in Erbsen gefundenen ähnlich zu sein. Aus diesen Ergebnissen ergibt sich folgender Reaktionsablauf, zit. nach MAPSON (1960):

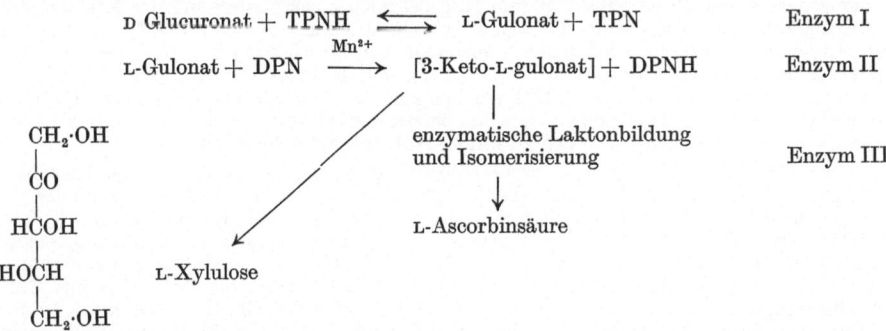

Mit diesem Reaktionsschema liefern GROLLMAN u. LEHNINGER (1957) einen wichtigen Beitrag zu der Frage, wie es zu erklären ist, daß der Mensch und einige Tiere (z. B. Affe, Meerschweinchen, Präriehund, Reh) nicht die Fähigkeit besitzen, AS zu synthetisieren (FRANKE, 1958; Diskussionsbemerkung von GUHA in MAPSON, 1960; FELIX, 1958). Sie untersuchten die als AS-haltig bekannten Organe einiger Tiere und konnten feststellen, daß die Gesamtenzymausstattung (der Säugetiere) jeweils nur in der Leber zu finden ist, die somit für diese Tiere als einziges Organ der AS-Bildung anzusehen ist. Bei den Arten, die keine AS synthetisieren können, fehlt das Enzym III, was mit dem Verlust oder der Mutation eines Gens zusammenhängen kann. Es ist möglich, daß die von LEHNINGER u. Mitarb. (1957) als Intermediärprodukt angesehene 3-Keto-L-gulonsäure in einer Zwischenreaktion unter Abspaltung von CO_2 in die L-Xylulose übergeführt wird. Es ist bekannt, daß Homogenisate aus der menschlichen Leber und Niere an Glucuronsäure diese Umwandlung vollziehen können. Aus den Tab. 1 u. 2 ist die Verteilung der drei Enzyme in verschiedenen Tieren zuersehen (FRANKE, 1958). Bei Amphibien, Reptilien und einigen Vögeln erfolgt die AS-Synthese vornehmlich in der Niere (GROLLMAN u. LEHNINGER, 1957; ROY u. GUHA, 1958).

Die Untersuchungen von KANFER, BURNS u. ASHWELL (1959) bringen eine

Tabelle 2 (nach FRANKE, 1958, gekürzt). *Verteilung der Enzyme in Geweben von Tieren, die zur AS-Synthese nicht befähigt sind.*

Art und Gewebe	Enzyme		
	I	II	III
Meerschweinchen			
Leber	3,33	1,91	0,0
Niere	6,03	5,72	0,0
Nebenniere . .	0,0	0,0	0,0
Herz	2,52	2,82	0,0
Milz	0,0	0,0	0,0
Cynomolgus-Affe			
Leber	11,8	14,8	0,0
Niere	10,5	47,5	0,0
Herz	0,0	0,0	0,0
Milz	0,0	0,0	0,0
Gehirn	0,0	0,0	0,0
Rhesus-Affe			
Leber	5,0	35,0	0,0
Menschenleber			
Patient M . . .	14,0	30,1	0,0
Patient R . . .	10,0	78,0	0,0

Einheiten wie in Tab. 1.

Erweiterung der Vorstellungen von LEHNINGER u. Mitarb. Sie konnten durch Untersuchungen mit gereinigten Leber- und Nierenpräparaten (hergestellt durch Zentrifugieren, Ultraschallbehandlung, Dialyse, Fällung mit Ammoniumsulfat) feststellen, daß der Laktonring für die Bildung von AS erforderlich ist (bei Gegenwart von freier Säure entsteht L-Xylulose) und daß die AS-Bildung im Gegensatz zu der der L-Xylulose Pyridinnucleotid-unabhängig ist und vielleicht 2-Keto-L-gulonsäure-γ-lakton als Zwischenprodukt hat. Bei L-Xylulose tritt 3-Keto-L-gulonsäure als Intermediärprodukt auf.

$$\text{L-Gulonsäurelakton} \xrightarrow{\text{O}_2} [\text{2-Keto-L-}\gamma\text{-gulonsäurelakton}] \longrightarrow \text{L-Ascorbinsäure}$$

$$\text{L-Gulonsäure} \xrightarrow{\text{DPN}} \text{3-Keto-L-gulonsäure} \longrightarrow \text{L-Xylulose}$$

ISHIKAWA u. NOGUCHI (1957) konnten aus Meerschweinchenleber- und -nierenpräparaten DPN- und TPN-spezifische L-Gulonsäuredehydrogenasen als rohe Enzymzubereitungen gewinnen. Das DPN-System bildet L-Xylulose, während im TPN-System Glucuronsäure entsteht.

d) l-Gulonsäuredehydrogenase nach Ishikawa u. Noguchi (1957).

Junge Meerschweinchen (etwa 300 g) wurden unter Äthernarkose getötet. Nach dem Ausbluten wurden die Niere und Leber mit 2,5 Volumen kalter isotonischer KCl-Lösung im Mixer homogenisiert, das Homogenisat 10 min lang bei 8500 g zur Entfernung der Mitochondrien zentrifugiert; die überstehende Flüssigkeit wurde anschließend bei 100000 g, 30 min lang, zentrifugiert. Die Proteinfraktion der 2. überstehenden Flüssigkeit wurde nach der Ausfällung mit Ammoniumsulfat (30—50 Gew./Vol.-%), p_H 7,4, in kaltem redest. Wasser gelöst und diente als Enzymmaterial bei den Untersuchungen. Die DPN-spezifische Gulonsäuredehydrogenase war nicht sehr stabil; sie verlor schon bei der Aufbewahrung bei —10° C ihre Aktivität. Die Untersuchungen wurden in der Warburg-Apparatur durchgeführt.

Auch Ashwell, Kanfer u. Burns (1959) stellten (aus Schweinenieren) eine (35fach) gereinigte l-Gulonsäuredehydrogenase (DPN-Enzym) her, die (auch) nicht sehr stabil ist. Das Enzym benötigt SH-Gruppen für seine Wirkung. Der Nachweis der gebildeten l-Xylulose erfolgte durch die von Ashwell u. Hickman (1957) modifizierte Cystein-Carbazol-Reaktion nach Dische. Ishikawa u. Noguchi (1957) bestimmten die l-Xylulose papierchromatographisch (Benzylalkohol-Eisessig-Wasser 3 :3:1, R_f-Wert 0,60). Sie färbt sich mit Orcin-Trichloressigsäure purpurrot, fluoresciert unter UV-Licht weiß-gelb, Klevstrand u. Nordal (1950). Die Vergleichssubstanz (Xylulose) wurde nach Schmidt u. Treiber (1933) synthetisiert.

Die Angaben von Ishikawa (1959) lassen erkennen, daß im tierischen Gewebe die Möglichkeit des Überganges von d-Glucuronsäurelakton und l-Gulonsäurelakton zu den entsprechenden freien Säuren gegeben sein muß. Yamada (1959) fand zwei Laktonasen (bezeichnet als I und II), die den Ring der beiden angegebenen Laktone aufspalten. Die Laktonasen unterscheiden sich in ihrer intercellularen Verteilung — Laktonase I im löslichen Anteil des Homogenisates, Laktonase II in der Mitochondrienfraktion —, Substratspezifität und Alkaliempfindlichkeit. Laktonase I wurde in der Leber der Ratte, des Rindes, des Kaninchens und in der Niere der Taube gefunden. Sie fehlt in der Leber des Menschen und des Affen sowie in der Niere, im Gehirn, in der Lunge und im Herz der Ratte. Aus diesen Befunden wird ihre Beziehung zur AS-Synthese deutlich. Laktonase II war in allen untersuchten Species anwesend.

e) Herstellung der gereinigten Laktonase I.

Frische, zerkleinerte Rinderleber wird mit 5 Vol. Aceton im Mixer zerkleinert, mit Trockeneis auf —20° C gebracht und dann zermahlen. Nach der Filtration wird der Filterkuchen mehrmals mit Aceton ausgespült und schließlich fein gerieben (6 Monate haltbar). Es folgt die Extraktion des Acetonpulvers mit eiskaltem Wasser (8 Vol.) unter Umrühren und anschließend Enteiweißung bei 63° C, 3 min. Zu 53 ml der Lösung werden 21 g festes Ammoniumsulfat (0,52-Sättigung) bei 0° C gegeben, nach 30 min langem Stehen wird im Kühlraum abzentrifugiert, 4000 g, und die überstehende Flüssigkeit mit festem Ammoniumsulfat auf 0,67-Sättigung gebracht. Die Präcipitate der 0,52- und 0,67-Sättigung werden in 20 ml 0,01 m-Tris-Puffer (p_H 7,35) gelöst und 2 Tage lang gegen kaltes redest. Wasser dialysiert. Die dialysierte Lösung (auf p_H 6,8 gebracht und die Cl⁻-Konzentration durch Verdünnen mit dest. Wasser auf 0,01 m eingestellt) wird so lange mit Aceton (auf —10° C abgekühlt) unter Umrühren versetzt, bis die Endkonzentration 40 Vol.-% beträgt, Temperatur —5 bis —2° C. Das Präcipitat wird bei der gleichen Temperatur abzentrifugiert, 4000 g, 10 min. Die Flüssigkeit wird dann mit Aceton bis zu einer Endkonzentration von 55 Vol.-% versetzt. Die beiden Fällungen werden in 70 ml Tris-Puffer[1] (p_H 7,75) gelöst und auf eine DEAE-Cellulosesäule[1] gegeben. Diese wird mit 0,01 m-Tris-Puffer (p_H 7,75) und anschließend mit einem solchen vom p_H 7,35 + 0,02 m-NaCl gewaschen. Die Waschflüssigkeit wird fraktioniert aufgefangen; sie enthält in einer Fraktion das Enzym in etwa 21facher Anreicherung.

Winkelman u. Lehninger (1958) gelang die 60fache Anreicherung der Aldonsäurelaktonase.

---

[1] Tris ist Tris-(hydroxymethyl)aminomethan. DEAE-Cellulose kann mit 2-Chlortriäthylaminhydrochlorid nach der Methode von Petersen u. Sober (1956) hergestellt werden. Bezugsquelle für DEAE-Cellulose in Deutschland z. B.: Machery, Nagel & Co, Düren, Mikro-Technik GmbH, Miltenberg a. Main.
Die für die Untersuchung erforderlichen Substrate wurden nach folgenden Verfahren gewonnen: d-Glucuronsäurelakton aus Gluconsäure nach Prager, Jacobsen u. Richter (1953); d-Galaktonsäurelakton aus Galaktose durch Oxydation mit Brom nach Clowes u. Tollnes (1899) und Levene u. Meyer (1921); l-Gulonsäure-γ-lakton durch Reduktion von d-Glucuronsäure mit Natriumborhydrazid nach Wolfrom u. Anno (1952); l-Gluconsäure-γ-lakton aus Xylose nach Hudson (1951).

Für die Verfolgung der Biosynthese der AS ist die Kenntnis über die enzymatische Bildung des L-*Gulonsäure-γ-laktons* aus der L-Gulonsäure von höchstem Interesse. Es gelang YAMADA (1959a) eine Laktonase gereinigt herzustellen, die diese Reaktion katalysiert. Er ging von Ochsenleber aus und stellte daraus ein Trockenpulver her (Herstellung erfolgt wie diejenige der Laktonase I, aber ohne die Reinigungsstufe mit Aceton bei pH 6,8, siehe S. 214). Die Laktonbildung verfolgte er mit Hydroxylaminhydrochlorid und bestimmte das gebildete Hydroxamat colorimetrisch:

2 m-Hydroxylaminhydrochlorid wurde täglich frisch hergestellt und mit 4 m-NaOH zu 4 m-NH$_2$OH auf den jeweiligen pH-Wert der Reaktion eingestellt; die Verdünnung auf 2 m erfolgte mit Wasser.

Beispiel eines Reaktionsansatzes. 0,05 m Tris-Maleat-Puffer (pH 6,1) 4 ml, 0,1 m Magnesium-chlorid 0,2 ml, 2 m-Hydroxylaminhydrochlorid (pH 6,1) 0,6 ml, 0,01 m-Glutathion (red.) 0,1 ml, Enzym-Lösung 0,5 ml, Substrat (Natriumgulonat 2 µM oder Natriumglucuronat 120 µM) bis zum Totalvolumen von 6,0 ml, Temperatur 37° C. Nach der Inkubation wurden

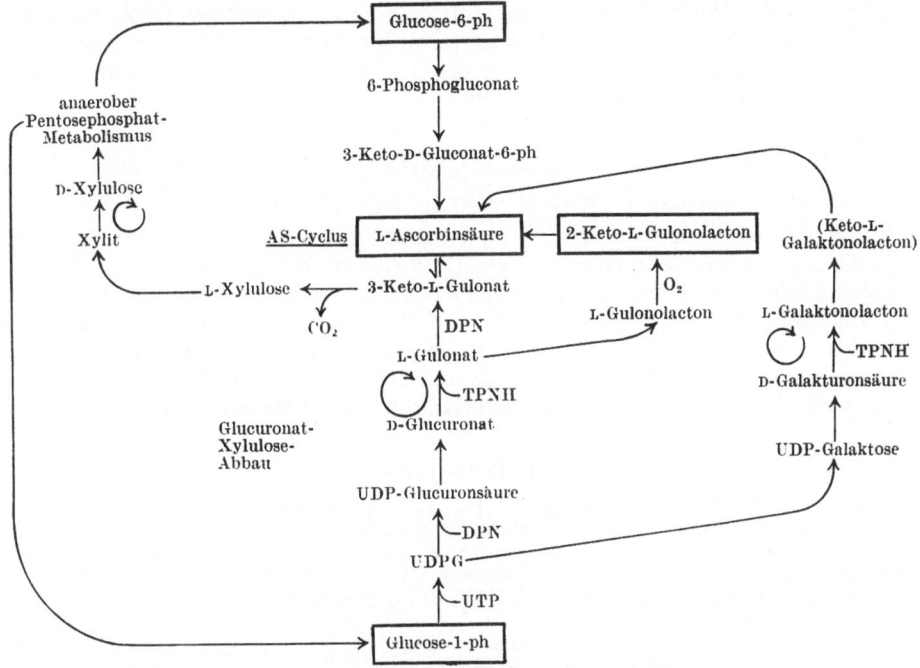

Abb. 3. Schema des Ascorbinsäure-Metabolismus (nach MAPSON, GIBBS, LOEWUS, KANFER u. a.).

aus der Reaktionsflüssigkeit 2 ml entnommen und mit 1 ml Reagens versetzt, das aus konz. Salzsäure (24%), Trichloressigsäure (6%) und FeCl$_3$ · 6H$_2$O (11%) besteht. Es folgt 5 min langes Zentrifugieren. Nach 60 min wird die gebildete Farbe der überstehenden Flüssigkeit bei 540 mµ in einem Spektralphotometer gemessen.

Das Enzym wird durch Schwermetalle und SH-blockierende Verbindungen gehemmt. Mg^{2+} (10^{-2} m) und Mn^{2+} (10^{-5} m) wirken stimulierend.

Es hat den Anschein, daß die Lakton-hydrolysierende und Lakton-bildende Wirkung von einem Enzym verursacht wird, dessen pH-Optimum bei 6,1—6,3 liegt. Gulonsäure wird in dem beschriebenen Versuchsansatz nur enzymatisch laktonisiert, während Glucuronsäure hauptsächlich nichtenzymatisch zum entsprechenden Lakton umgewandelt wird. Die Biosynthese verläuft nach YAMADA (1961) über L-Gulonsäure → L-Gulonolakton → AS und nicht über das D-Glucuronolakton → L-Gulonolakton → AS.

Abschließend kann festgestellt werden, daß wohl mehrere Wege dem pflanzlichen und tierischen Organismus zur Biosynthese der AS zur Verfügung stehen. Interessant ist die Ein-beziehung des Pentosephosphat-Cyclus.

Die Abb. 3 zeigt eine Zusammenfassung der bisherigen Befunde (entnommen aus BÜNNING u. GÄUMANN, 1960).

II. l-Ascorbinsäure oxydierende Enzymsysteme.

Aus der Beobachtung, daß die Ascorbinsäure (AS) in pflanzlichen und tierischen Geweben mit erhöhtem Stoffwechsel in oxydierter und reduzierter Form vorkommt, hat man geschlossen, daß diese beiden Verbindungen eine wichtige Rolle spielen. Als erster konnte Szent-Györgyi (1931) ein Enzym in Kohlblättern feststellen, das eine direkte Reaktion zwischen molekularem Sauerstoff und AS vermittelt. Die Pflanze besitzt mehrere Oxydasen, die direkt oder indirekt AS zu oxydieren vermögen. Sie sind Schwermetallproteide, die in ihren prosthetischen Gruppen Kupfer oder Eisen enthalten. Aber auch eine nicht-enzymatisch bedingte Oxydation von AS kann stattfinden. Glutathion (red.), Chinone, Aminosäuren, Proteine u. a. wirken oxydationshemmend, u. a. Barron, De Meio u. Klemperer (1936), Schulze u. Morgan (1946), Potterman (1949), Lund, Shaw u. Drinker (1921), v. Oettingen (1935), Rudra (1939), Boyer, Shaw u. Phillips (1942), Skinner u. McHargue (1946), Lindow, Elvehjem u. Petersen (1929), De Caro (1938), Lesne, Polonowski u. Briskas (1943), Hochberg, Melnick u. Oser (1945).

Die Oxydation verläuft in 2 Stufen:

1. AS $\rightleftharpoons$ DAS $\rightarrow$ DKS
2. DKS $\rightarrow$ unter Abbau $\rightarrow$ l-Threonsäure, Lyxonsäure, Xylonsäure, Oxalsäure

(z. B. Herbert, Hirst, Perceival, Reynolds u. Smith, 1933).

Die Umwandlung von DKS ist p_H-abhängig; sie findet in neutralem und alkalischem Milieu sehr schnell statt (Penney u. Zilva, 1943). Die Enzymsysteme führen die Oxydation im allgemeinen bis zur DAS. Es konnte aber auch in Pflanzensäften eine Umwandlung zur DKS und deren Abbau beobachtet werden (Janecke u. Müller). In Weizen hat Waygood (1950) einen Faktor beobachtet, der die Bildung von DKS aus DAS katalysiert. Borat und Cyanid öffnen den Laktonring der DAS schnell (Huelin, 1949; Mapson u. Ingram, 1951).

1. Peroxydase.

Gereinigte Peroxydasezubereitungen ($+ H_2O_2$) katalysieren erst nach Zugabe von Pflanzenextrakten die Oxydation von AS (Mapson u. Ingram, 1951). Die in den Säften vorhandenen Chinon-bildenden Polyphenole sind die Ursache zu diesem Verhalten (Szent-Györgyi, 1928; Huzak, 1937). DAS kann durch Glutathion oder Cystein wieder reduziert werden (Borsook u. Jeffries, 1936; Hopkins u. Morgan, 1936; Pfankuch, 1934):

$$\text{Phenol-verbindung} + \frac{\text{Peroxydase}}{+ H_2O_2} \rightarrow \text{Chinon}$$

$$\text{Chinon} + \text{AS} \rightarrow \text{DAS} + \text{Phenol-verbindung}$$

Die Reaktionsgeschwindigkeit hängt von der Art des Phenols ab, Tannin und Gallussäure reagieren nur langsam (Tauber, 1938).

Diemair und Häusser (1941) benutzten AS und Leuko-DIP zur *Peroxysadebestimmung*.

Prinzip. Durch Peroxydase erfolgt bei Gegenwart von Wasserstoffperoxyd eine Oxydation von Leuko-DIP (hergestellt durch Reduktion von DIP mit AS), die gemessen werden kann und ein Maß für die anwesende Peroxydase darstellt.

Reagentien. 0,05%ige AS-Lösung mit Zusatz von 1 ml einer 20%igen Essigsäurelösung in 100 ml 0,001 n-2,6-Dichlorphenol-Indophenollösung (DIP). 0,5%ige Wasserstoffperoxydlösung, genau eingestellt aus Perhydrol (Merck), gesättigte Natriumacetatlösung, Äther, dem auf 100 ml 0,1 ml 20%ige Essigsäurelösung zugesetzt wurde.

Methode. In einem 100 ml fassenden Scheidetrichter (s. Abb. 4) werden 4 ml 0,05%ige AS-Lösung gegeben; dazu fügt man 0,5 ml gesättigte Natriumacetatlösung und titriert die AS-Lösung mit 0,001 n-2,6-Dichlorphenol-Indolphenollösung bis zum Auftreten eines mausgrauen Farbtones. Es ist auf einen stets gleichmäßigen Umschlagfarbton zu achten. Dazu gibt man die ebenfalls bis zum

Auftreten des schmutziggrauen Farbtones mit DIP austitrierte Peroxydaselösung, füllt auf 50 ml im Scheidetrichter mit Wasser auf und versetzt die Lösung nach kräftigem Umschütteln mit 2 ml 0,5%igen H_2O_2. Nach 30 sec (mit Stoppuhr messen) wird das Reaktionsgemisch mit 50 ml des vorbereiteten Äthers überschichtet und nach weiteren 7 sec kräftig durchgeschüttelt, bis der gesamte carminrote Farbstoff sich im Äther gelöst hat. Die ätherische Farbstofflösung wird in ein 50 ml Meßkölbchen gebracht und hier auf 50 ml ergänzt. Die Ablesung erfolgt im Stufenphotometer mit Filter S 53. Durch Multiplikation des ermittelten Extinktionskoeffizienten mit dem Faktor 8,76 ergibt sich die Anzahl ml 0,001 n-2,6-Dichlorphenol-Indophenollösung (DIP) in 50 ml Äther. Als Vergleichslösung dient die vorbereitete Ätherlösung. Der so ermittelte Blindwert wird von der gemessenen Farbstoffmenge abgezogen; die peroxydatische Wirksamkeit wird als Farbstoffzahl (F.Z.) angegeben.

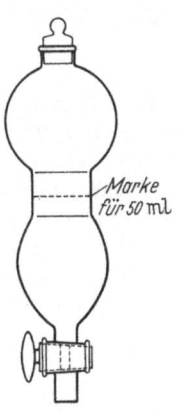

2. Cytochrom c — Cytochromoxydase.

Diese Oxydase soll auch im pflanzlichen Material vorkommen (HILL u. BHAGVAT, 1939; BHAGVAT u. HILL, 1950; MARSH u. GODDARD, 1939; GODDARD, 1939 u. 1944; THOMPSON, 1950). Sie wird durch HCN, H_2S und Azid, im Dunkeln auch durch CO reversibel, gehemmt (STOTZ, HARRIS, SCHULTZE u. KING, 1938). Cytochromoxydase oxydiert bei Abwesenheit von Cytochrom AS nicht; auch hier liegt eine indirekte Oxydation vor.

Abb. 4. Scheidetrichter

3. Polyphenoloxydase.

Dabei handelt es sich um ein Kupferproteid, das in Pflanzen weit verbreitet ist und das bei der Verfärbung (Bräunung) von pflanzlichen Stoffen, z. B. von Lebensmitteln (Äpfeln, Birnen, Kartoffeln, Pfirsichen u. a.), neben der nichtenzymatisch bedingten, eine Rolle spielt (DAWSON u. MALETTE, 1945; KUBOWITZ, 1937; DALTON u. NELSON, 1938; KEILIN u. MANN, 1939). Die weite Verbreitung dieses Enzyms läßt vermuten, daß es neben der Peroxydase, Cytochromoxydase und AS-Oxydase eine wichtige Rolle als terminale Oxydase spielt (JAMES, 1953, 1953 a); in Kartoffeln soll es etwa $^2/_3$ der gesamten Atmung beherrschen (BOWELL u. WHITING, 1938; LEVY u. SCHADE, 1948). Nach RUBIN und CHETVERIKOWA (1951) erfolgt in reifem Kohl die Atmung ausschließlich über Metallenzyme. Viele Pflanzen, die keine AS-Oxydase besitzen, haben einen hohen Gehalt an Polyphenoloxydase. Die indirekte Oxydation der AS mit Polyphenoloxydase soll nach folgendem Schema verlaufen:

O_2+ Phenolverbindungen (o- o. p-) → o- oder p-Chinon
Chinon (o- oder p-) + AS → DAS + Phenolverbindungen

Ihre Aktivität kann z. B. mit Hilfe von spektrophotometrischen und colorimetrischen Methoden bestimmt werden (EIGER u. DAWSON, 1949; KEILIN u. MANN, 1938; PONTING u. JOSLYN, 1947). Auf die *chronometrische*, die das obige Reaktionsschema zur Grundlage hat, wird besonders verwiesen (MILLER, MALETTE, ROTH u. DAWSON, 1944). Das Verfahren soll den manometrischen überlegen sein:

Prinzip. Die durch die Polyphenoloxydase (Catecholase) aus Brenzkatechin gebildete Menge an o-Benzochinon ist der Menge AS äquivalent, die hinzugegeben wurde. So lange noch AS vorhanden ist, wird das entstandene o-Benzochinon reduziert. Der Endpunkt wird von Jodid-Stärke (als Indicator) angezeigt. Es wird diejenige Zeit bestimmt, die bei einer gegebenen Menge an Enzym zur Bildung einer bestimmten Menge an o-Benzochinon, begrenzt durch die vorgelegte Menge an AS, erforderlich ist. Die „spezifische Aktivität" wird in Einheiten pro mg Trockengewicht angegeben; Bestimmung des Trockengewichtes nach Malette u. Mitarb. (1948) siehe dieses Handbuch Bd. 7, S. 225; vergl. auch Bd. 6, S. 312. Eine Einheit ist diejenige Menge an Enzym, die die Sauerstoffaufnahme von μl pro min katalysiert; diese entspricht einer o-Benzochinonbildung von $1,49 \times 10^{-8}$ Mol/sec (Gregg u. Nelson, 1940). Die Herstellung einer gereinigten Polyphenoloxydase beschreiben z. B. Malette u. Dawson (1949).

4. Laccase.

Dieses Enzym ähnelt in seiner Wirkung sehr der Polyphenoloxydase. Es vermag jedoch Tyrosin und p-Kresol zu oxydieren. Mit seiner Reinheit nimmt aber auch seine Fähigkeit ab, AS zu oxydieren. Durch Zugabe von p-Phenylendiamin (aber nicht von Brenzkatechin) kann seine Aktivität wieder hergestellt werden. Die Oxydation von AS erfolgt indirekt. Die Laccase, die in Rhus-Arten vorkommt, z. B. Redfield (1950), ist ein Kupferproteid. Ein sich ähnlich verhaltendes Enzym hat man auch in Tieren gefunden, Cadden u. Dill (1942). Der optimale pH-Wert der AS-Oxydation durch Laccase liegt bei 6,7, Bertrand (1945).

5. Ascorbinsäureoxydase.

Während die beschriebenen Enzyme indirekt die Oxydation von AS katalysieren, erfolgt diese mit der Ascorbinsäureoxydase (AS-Oxydase), die in der Literatur auch als Ascorbinase, Ascorbicase, Ascorbase und Ascorbinsäuredehydrase bezeichnet wird, direkt. Es findet eine Oxydation von AS zu DAS statt. Die chemische Oxydation führt über das DAS weiter zu Lyxonsäure, Xylonsäure, Threonsäure, Glykolaldehyd und Oxalsäure (Strohecker u. Matt, 1951). Die AS-Oxydase ist (ebenfalls) ein Kupferproteid, das im Pflanzenreich weit verbreitet vorkommt.

Szent-Györgyi (1930) beobachtete als erster, daß die von ihm gefundene „Hexuronsäure" (l-Ascorbinsäure) nicht nur indirekt, sondern auch direkt durch ein spezifisches Enzym, das in Kohlblättern vorkommt, oxydiert werden kann.

Ähnliche Enzyme wurden in der Folge u. a. von Zilva (1934), Tauber u. Kleiner (1935), die dem Enzym den Namen gaben, sowie von Hopkins u. Morgan (1936), Scrinivasan (1935, 1936), Wacholder u. Hamel (1937), Meiklejohn u. Stewart (1941), Armentano u. Bartok (1942), Diemair u. Zerban (1942, 1943), Wacholder (1942), Somogyi (1945), Lovett-Janison u. Nelson (1940), Malette, Lewis, Ames, Nelson u. Dawson (1948), Dunn u. Dawson (1951) beschrieben und untersucht.

Das Enzym oxydiert mit unterschiedlicher Geschwindigkeit auch Analoge der AS: d-Araboascorbinsäure, l-Glucoascorbinsäure und l-Galaktoascorbinsäure (Johnson u. Zilva, 1937). Reduktone und Reduktinsäure werden auch oxydativ verändert (Snow u. Zilva, 1938). Die Spezifität des Enzyms scheint sich auf die Dienol-Gruppierung zu beziehen. Es bildet nicht Wasserstoffperoxyd (Steinmann u. Dawson, 1942), wie früher angenommen wurde (Szent-Györgyi, 1939; Roberts, 1939, Jagle, 1939, Ebihara, 1939). Zum Unterschied hierzu wurde bei der Cu^{2+}-Oxydation der AS H_2O_2-Bildung beobachtet (Lyman, Schultze u. King, 1937; Dekker u. Dickinson, 1940; Hand u. Greisen, 1942; Steinmann u. Dawson, 1942; Silverblatt, Robinson u. King, 1943; Peterson u. Walton, 1943; Weissberger, Lu Valle u. Thomas, 1943; Weissberger u. Lu Valle, 1944).

Mehrmals wurde die Existenz eines spezifischen Enzyms in Frage gestellt und die beobachtete AS-Oxydation den in Pflanzen vorkommenden Cu^{2+} zugeschrieben (Barron u. Mitarb., 1936). Stotz u. Mitarb., 1937 konnten mit Kupfer-Eiweißmodellen zeigen, daß diese ähnliche Eigenschaften aufweisen wie die AS-Oxydasen aus Kürbis, Gurke, Blumenkohl und Kohl. Ihre Ergebnisse wurden von Somogyi (1945) bestätigt. Ähnliche Beispiele siehe Dawson (1950). Ramasarma, Datta u. Doktor (1940) konnten mit einem Präparat eine besonders hohe Aktivität erzielen, die sich nur mit der Annahme der Existenz eines Enzymes erklären läßt. LOVETT-JANISON u. NELSON (1940) fanden in einer AS-Oxydase-Zubereitung aus Kürbissaft eine 1100 mal höhere Aktivität als sie Cu^{2+} allein besitzen; das gleiche Präparat war 13000 mal aktiver als Kupferalbumin und 4100 mal wirksamer als Kupfergelatine.

Nach SOMOGYI (1945) zeigt die Cu^{2+}-Katalyse einen anderen Verlauf wie diejenige mit AS-Oxydase, wie in Abb. 5 dargestellt ist.

Versuchsansatz. 5 ml AS-Lösung (15 mg-%) werden mit 5 ml Kohlextrakt (Herstellung siehe unten) bzw. mit 5 ml Kupfersulfatlösung (1 mg-%) versetzt. Die Einstellung des p_H-Wertes erfolgte durch Zugabe von 0,03—0,3 ml 0,25 n-HCl bzw. von 0,03—0,8 ml 0,10 n-$NaHCO_3$-Lösung (von p_H 3,0—8,0). Die Verwendung eines Puffergemisches hatte auf den Ablauf des Versuches keinen wesentlichen Einfluß. Die Versuchszeit betrug 5 min. Die Kupfersulfatlösung wurde so gewählt, daß sie bei p_H 6,0 eine 15 mg-%ige AS-Lösung innerhalb von 15 min vollständig oxydiert.

Der Kohlextrakt wurde auf einen bestimmten Gehalt an AS-Oxydase eingestellt.

Herstellung des Kohlextraktes: 10 g Kohlblätter werden mit etwas Seesand und 50 ml tridest. Wasser etwa 5—8 min lang verrieben, 3—5 min stehen gelassen, 10 min lang zentrifugiert (3000 U/min) und dann filtriert. In einem Vorversuch wird die Aktivität des Extraktes bestimmt. Hierzu stellt man fest, in wieviel Minuten eine 15 mg-%ige AS-Lösung bei p_H 6,0 (Phosphatpuffer) vollständig oxydiert wird. Der Pflanzenextrakt wird dann mit tridest. Wasser so weit verdünnt, daß dieser eine 15 mg-%ige AS-Lösung innerhalb von 15 min vollständig zur DAS oxydiert.

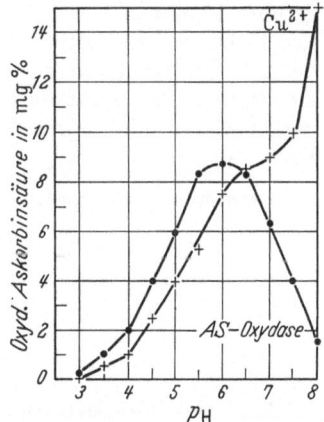

Abb. 5. Oxydation der L-Ascorbinsäure bei verschiedenen pH-Werten innerhalb von 5 min mit AS-Oxydase aus Kohl bzw. mit Cu^{2+} ($CuSO_4$).

HOCHBERG, MELNICK u. OSER (1945), SCHULTE u. SCHILLINGER (1952) und auch BERNER (1947) halten die AS-Oxydase für ein spezifisches Enzym und nicht für einen losen Komplex von Kupfer mit mehr oder weniger spezifischen Proteinen. GIRI u. SESHAGIRIRAO (1946) prüften den Einfluß von Adenin, Guanin, Xanthin, Harnsäure, Theophyllin und Kreatinin auf die Cu^{2+}-katalysierte und die durch AS-Oxydase beschleunigte Oxydation der AS. Sie fanden ein unterschiedliches Verhalten. Die angeführten Verbindungen hemmen die Cu-Katalyse (p_H 7,2) bei Gegenwart von Luft, aber nicht die AS-Oxydasen aus Cucurbita maxima und aus Tricosanthus anguina; siehe auch METZNER (1957).

Die AS-Oxydase verliert auf der Kationenaustauschersäule, Amberlite IR 100 AG (0,1 m-Acetatpuffer, p_H 5,6), nicht das enzymgebundene Metall, das nicht dissoziierbar ist, JOSELOW, DAWSON u. LEWIS (1951), DAWSON (1950). Cu^{64} tauscht das Kupfer im inaktivierten Enzym aus, nicht im ruhenden, DAWSON (1950), KIRSCHENBAUM (1956). RAMASARMA u. Mitarb. (1940) haben als erste die Anwesenheit von Cu festgestellt. Sie fanden, daß Aktivität und Kupfergehalt bei der Enzymreinigung zunehmen. Ihre besten Zubereitungen enthielten 0,03% Cu. In der Zwischenzeit wurden von anderen Untersuchern hiervon abweichende Kupfergehalte angegeben: z. B. 0,25% STOTZ (1940); 0,15% LOVETT-JANISON u. NELSON (1940); 0,0009% DIEMAIR u. ZERBAN (1944); 0,27% KIRSCHENBAUM (1956). Auch andere Autoren fanden eine Beziehung zwischen Cu-Gehalt und Enzymaktivität, GRAUBARD (1939), LOVETT-JANISON u. NELSON (1940), MEIKLEJOHN u. STEWART (1941), DIEMAIR u. ZERBAN (1944), während SPRUYT u. VOGELSANG (1938) sowie ARMENTANO u. BARTOK (1942) eine solche Relation ablehnen.

a) Methoden der Kupferbestimmung.

Die Kupferbestimmung nach Warburg (1927), Warburg u. Krebs (1927), benutzt die Oxydation von Cysteinlösungen zum Nachweis von Schwermetallen (Fe, Mn, Cu). Pyrophosphat hemmt die Wirkung des Eisens und des Mangans, nicht aber diejenige des Kupfers. Aus den Oxydationsgeschwindigkeiten kann der Gehalt an Cu berechnet werden. Die Untersuchungen werden in Warburg-Gefäßen (15 ml) mit Birnenansatz bei 20° C (Thermostat) unter Schütteln durchgeführt. 10 min nach dem Einhängen in den Thermostaten überführt man den Inhalt der Birne in den Hauptraum, 5 min später (durch Austreiben der Kohlensäure entsteht zunächst ein geringer positiver Druck) beginnt die Messung der O_2-Absorption. Der Gasraum enthält Luft. Die Ausschläge sollen nicht größer sein als 3 mm pro min; durch Variation der Schüttelgeschwindigkeit ist zu kontrollieren, ob sie unabhängig davon sind.[1]

Versuchsansatz (Beispiel)

	Gefäß 1	Gefäß 2	Gefäß 3
Hauptraum . . .	2 ml	2 ml	2 ml 0,2 m-Pyrophosphat
Birne	6 mg	6 mg	6 mg Cystein-HCl
	0,1 ml	0,1 ml	0,1 ml Aqua bidest.
		0,1 ml	0,1 ml Untersuchungslösung
			0,1 ml Kupferlösung mit
			2×10^{-4} mg Cu

Natriumpyrophosphatlösung. 180 g Natriumpyrophosphat p.a. werden unter Erwärmen in 100 ml Wasser gelöst, filtriert und auf Zimmertemperatur abgekühlt, wobei 90 g Salz ausfallen. Man saugt ab, wäscht mit eiskaltem Wasser, trocknet bei 100° C und stellt daraus eine 0,2 m Lösung her. Die Aufbewahrung der Lösung erfolgt in glasierten Porzellanflaschen, die einige Tage vorher schon mit Pyrophosphatlösung gereinigt worden waren. Zum Gebrauch setzt man soviel eisenfreie Salzsäure (zu 10 ml 1 ml n-HCl) hinzu bis der p_H-Wert 7,88 beträgt. Dieser erniedrigt sich bei Zusatz der Cysteinhydrochlorid-Lösung auf 7,64.

Standardkupferlösung. Auflösen von reinem Kupfervitriol in 0,01 n-HCl, so daß eine Lösung von 2×10^{-4} mg Cu in 0,1 ml entsteht.

Reinigung der Gläser. Diese erfolgt mit Chromschwefelsäure, dann mit dest. Wasser und Trocknen im Trockenschrank. Leitungswasser ist zu vermeiden.

Berechnung der Cu-Menge (unter Berücksichtigung der angegebenen Beispieles). Die Anfangsgeschwindigkeit in den 3 Gefäßen seien: a_1 in Gefäß 1, a_2 in Gefäß 2 und a_3 in Gefäß 3, dann ist die Kupfermenge x in 0,1 ml Untersuchungslösung

$$ x = \frac{a_2 - a_1}{a_3 - a_2} \cdot 2 \times 10^{-4} \text{ mg} . $$

Diemair u. Zerban (1944) bestimmten den Cu-Gehalt ihrer AS-Oxydase-zubereitungen mit Dithizon nach Schwaibold, Bleyer u. Nagel (1938) (weitere Literatur zur Dithizon-Methode in Iwantscheff, 1958).

Reagentien. Schwefelsäure, reinst, Dithizon (Merck), Ammoniak, reinst. Die Reagentien sollen möglichst frei von Schwermetallen sein. Es empfiehlt sich, den Blindwert in einem analog der Bestimmungsmethode anzustellenden Versuch zu ermitteln und diesen in Abzug zu bringen.

Arbeitsvorschrift. 100 ml der zu untersuchenden, fast neutralen Lösung werden mit 5 ml 10%iger Schwefelsäure angesäuert und solange mit kleinen Mengen an Dithizonlösung (6 mg in 100 ml CCl_4) ausgeschüttelt, bis die grüne Färbung dieser Lösung unverändert bleibt. Die vereinigten Auszüge werden mit CCl_4 auf 30 ml aufgefüllt und nach dem Waschen mit 10 ml bidest. Wasser. Wasser so oft wie 10 ml einer Ammoniaklösung (1:1000) ausgewaschen, bis die Waschflüssigkeit nicht oder kaum mehr gefärbt wird. Nach kurzem Durchschütteln mit 1%iger Schwefelsäure kann nun die Messung der CCl_4-Lösung mit dem Filter S 53 im Stufenphotometer vorgenommen werden. Hg wird gemeinsam mit Cu extrahiert. Durch Waschen mit 10 ml 5%iger Jodkali-Lösung wird der Hg-Komplex zersetzt. Nach dem Waschen mit

---

[1] Die Methode von Warburg (1927) wurde von Dalton u. Nelson (1939), Dills u. Nelson (1942), Lovett-Janison u. Nelson (1940), Ludwig u. Nelson (1939), Parkinson jr. u. Nelson (1940), Powers, Lewis u. Dawson (1944) angewendet.

Wasser wird das überschüssige Dithizon mit verdünntem Ammoniak entfernt und die Cu-Bestimmung, wie beschrieben, durchgeführt. (Die nach der Abtrennung des Cu verbliebene Flüssigkeit kann zur Pb- und Zn-Bestimmung herangezogen werden.)

Herstellung der Auswertungskurve. Durch Verdünnen einer Standardlösung von $CuSO_4 \cdot 5H_2O$ werden Kupferlösungen mit 5—60 γ Cu hergestellt, die nach Zusatz von etwa 5% einer 10%igen Schwefelsäurelösung mit einer Dithizonlösung (6 mg in 100 ml CCl_4) in Anteilen von einigen ml erschöpfend, d. h. bis keine Farbänderung der Dithizonlösung mehr auftritt, im Schütteltrichter extrahiert werden. Die CCl_4-Lösung wird nach dem Auffüllen auf 30 ml durch kurzdauerndes Durchschütteln mit verdünntem Ammoniak (1:100) so oft gewaschen, bis dieser farblos bleibt. Die Messung erfolgt z. B. im Stufenphotometer mit dem Filter S 53 (1 cm Cuvette).

AMES u. DAWSON (1945) beschreiben eine geeignete polarographische Methode zur Bestimmung von Cu in Kupferproteiden.

b) Eigenschaften der AS-Oxydase.

Nähere Angaben über die AS-Oxydase verdanken wir DAWSON (1950). Nach ihm besitzt sie ein MG von 150000 und enthält 6 Atome Cu. Es wird bei der Katalyse schnell inaktiviert; das Cu-Proteid wird hierbei jedoch nicht gespalten (POWERS u. DAWSON, 1944). TADOKARA u. TAKASUGI (1939) berichten über die Gewinnung einer kristallinen AS-Oxydase, deren Aktivität aber geringer war als diejenige von nichtkristallinem Material, so daß Zweifel an dem kristallinen Zustand geäußert wurden (Kritik von DAWSON u. MALETTE, 1945). KIRSCHENBAUM (1956) untersuchte die Zusammensetzung der AS-Oxydase aus *Cucurbita pepo condensa*, hergestellt nach dem Verfahren von DAWSON u. MAGEE (1955), und fand nach saurer Hydrolyse papierchromatographisch als Bestandteile des Eiweißanteiles: Glycin, Alanin, Valin, Leucin, Isoleucin, Serin, Threonin, Asparaginsäure, Glutaminsäure, Phenylalanin, Tyrosin, Lysin, Arginin, Histidin, Methionin und Prolin. Die Anwesenheit von Cystein ist zweifelhaft, während Cystin vorzuliegen scheint. Das UV-Absorptionsspektrum der AS-Oxydase ähnelt dem eines Proteins, das Tyrosin und Tryptophan enthält und gibt keinen Anhalt für die Anwesenheit von Nucleotiden, wie TADOKARA u. TAKASUGI u. Mitarb. (1939, 1940, 1941) annehmen. Die AS-Oxydase besitzt keine endständige Aminogruppe, wie mit den Methoden von SANGER (1945, 1952) (weitere Literatur siehe u. a. FRAENKEL-CONRAT, HARRIS u. LEVY, 1955; PORTER, 1950; SCHRAMM u. BRAUNITZER, 1953; ROVERY u. DESNUELLE, 1954; ACKER u. LAURILLA, 1953; BRENNER, 1953; KHORANA, 1952; BRAUNITZER, 1957) und von EDMAN (1950) (siehe auch FRAENKEL-CONRAT, HARRIS u. LEVY, 1955) festgestellt wurde. Die IR-Spektroskopie ergab eine Kurve, die Eiweißkurven sehr ähnlich ist. Die Entfernung des Kupfers aus dem Molekül der AS-Oxydase läßt die Absorptionsbande bei 8,1 μ verschwinden. Die Entfernung des Cu gelingt leicht durch Dialyse gegen Cyanidlösung oder gegen eine Lösung mit einem pH-Wert unter 5. Das Absorptionsspektrum im sichtbaren Bereich zeigt zwei Maxima: 6060 Å und 4125 Å. Die blaue Farbe des Enzyms steht in Beziehung zu der Absorption bei 6060 Å. Bei der Reaktion des Enzyms mit AS verschwindet die blaue Farbe sofort und eine goldgelbe bleibt. Spektralanalytisch zeigt sich diese Farbänderung in dem Verlust der 6060 Å-Bande, die Bande bei 4125 Å bleibt. Bei Oxydation (Lufteinblasen) kehrt die blaue Farbe zurück. KIRSCHENBAUM (1956) hat von der AS-Oxydase folgende Vorstellung: Das Kupfer liegt in dem ruhenden Enzym in der Cu^{1+}- und Cu^{2+}-Form vor. Der Cu^{2+}-Anteil ist über 4 Histidinreste gebunden und verleiht dem Enzym die blaue Farbe (6060 Å). Dieser Cu-Anteil kann während der Wirkung des Enzyms (auf AS) durch radioaktives Kupfer ersetzt werden. Der Cu^{1+}-Anteil liegt in noch unbekannter Bindungsform vor. Er ist verantwortlich für die Absorption bei 4125 Å und die gelbe Farbe, die auftritt, wenn das Enzym mit AS in Reaktion tritt. Williams (1953) fand, daß Cu^{1+}-Komplexe niemals blau, sondern farblos oder blaßgelb gefärbt sind. Nach Dunn u. Dawson (1951) besitzt eine weitgehend gereinigte AS-Oxydase die Eigenschaften eines Globulins und zeigt eine Aktivität von 2000 Einheiten pro mg und von 750 Einheiten pro μg Cu. Die Aktivität ist über 1000mal größer als diejenige einer äquivalenten Menge ionisierten Kupfers, Dawson u. Magee (1955). Das gereinigte Enzym verbraucht für ein Mol AS ein Grammatom Sauerstoff. Wasserstoffperoxydbildung findet nicht statt. Cu^{2+} bewirken eine größere Sauerstoffaufnahme und bilden H_2O_2, Steinmann u. Dawson (1942), Dekker u. Dickenson (1940), Hand u. Greisen (1942), Lyman, Schultze u. King (1937).

Für die Verfolgung der Aktivität und Charakterisierung eines Enzyms ist die Kenntnis des optimalen pH-Wertes, der günstigsten Reaktionstemperatur und der Hemmstoffe erforderlich.

In der Literatur wird ein ziemlich breiter Bereich für den *optimalen pH-Wert* der AS-Oxydase angegeben. Dieser liegt nach Lardy (1950) zwischen pH 4,6 und 6,6 und scheint von der Reinheit des Präparates abhängig zu sein. Powers u. Mitarb. (1944) fanden einen solchen von 5,6 (Citratpuffer) in ihrer höchstgereinigten Enzymzubereitung. Diemair u. Zerban (1944) bestimmten in rohen Preßsäften aus Gurken und Kürbis unterschiedliche Werte. In ihren gereinigten Enzympräparaten lag der optimale pH-Wert scharf bei pH 6,0. In diesem Zusammenhang sei an den Verlauf der AS-Oxydation mit Cu^{2+} erinnert, Somogyi (1945), siehe dieses Handbuch Bd. 7, S. 219. Nach Bergner (1947) und Schmidt-Nielsen u. Spilling (1942) nimmt die Stabilität der AS gegenüber der AS-Oxydase mit fallendem pH-Wert zu. Die *optimale Temperatur* liegt nach Diemair u. Zerban (1944), Ranguekar, De u. Subrahmanyan (1948) und anderen bei 37—38° C. Die ersten Autoren fanden bei einer Temperaturerhöhung um 10° C eine Zunahme der Reaktionsgeschwindigkeit um das Zwei- bis Zweieinhalbfache. Ponting u. Joslyn (1948) ermittelten als optimale Temperatur 43° C. Durch Erhitzen auf höhere Temperatur (z. B. 70° C, 10 min lang) erfolgt Inaktivierung, Matukawa (1940). Diese Beobachtung hat lebensmitteltechnologische Bedeutung. Zahlreich sind die Angaben in der Literatur über das Verhalten der AS-Oxydase beim Blanchierprozeß (z. B. Dhopeshwarka u. Magar, 1952; Nomura u. Matsunaga, 1953; Hartzler u. Guerrant, 1952; Yamaguchi u. Joslyn, 1951; Schulte u. Schillinger, 1952). Nach Günther (1948) zerstört ein Erhitzen auf 100° C die AS-Oxydase, eine Abkühlung auf —20° C hat hemmende Wirkung. Valle (1951) fand bei —25° C und bei —45° C noch keine Zerstörung der AS-Oxydase, wenn sie in Citronensaft aufbewahrt wurde.

Zahlreich sind die Angaben, die die *Hemmung* der AS-Oxydase betreffen. Man fand, daß Cyanid, Sulfid, Kohlenmonoxyd, Natriumdiäthyldithiocarbamat, Phenylmerkurichlorid, 8-Hydroxychinolin, Wasserstoffperoxyd, Thioharnstoff, Carotin u. a. die Aktivität des Enzyms hemmen oder ganz aufheben (Konzentrationsabhängigkeit), Lit. z. B. Lardy (1950), Srinivasan (1936), Diemair u. Zerban (1944), Stotz u. Mitarb. (1937), Ranguekar, De u. Subrahmanyan (1948), Russell (1954), Jones u. Garton (1952), Mapson u. Moustafa (1956), Balakhovskič, Drozdova u. Fedorova (1953) u. a. Antioxydantien, wie Tocopherol, Propylgallat und Nordihydroguajaretsäure (NDGA), setzen die Aktivität in gleicher Weise herab, Tappel u. Marr (1954). Nach Frieden (1952) hemmen auch Cu^{2+}. Der natürliche Inhibitor in Erdbeeren soll eine Cr-Verbindung sein; Tomatensaft hemmt durch eine flüchtige, instabile Verbindung, Inagaki, Fukuba u. Matsushita (1954, 1955). Die gleichen Autoren (1955a) beschreiben noch Nasunin (Glucosid aus Delphinidin, 2 Mol. Glucose und p-Cumarsäure), Anthocyanin und bestimmte Proteinfraktionen aus Pflanzenmaterial als Inhibitoren.

Interessant ist die Beobachtung von RUBIN u. CHETVERIKOVA (1955), daß ein Toxin aus Botrytis cinerea durch Proteolyse die AS-Oxydase inaktiviert.

c) Gewinnung der Ascorbinsäure-Oxydase.

Die Vorschriften von FUJITA u. SAKAMOTO (1938), SOMOGYI (1945) und von TAUBER u. KLEINER (1935) wurden in Bd. 2 dieses Handbuches, S. 109, aufgeführt.

DIEMAIR u. ZERBAN (1944) beschreiben die Herstellung von 5 verschieden aktiven Enzymlösungen. Durch fraktionierte Ausfällung mit Ammoniumsulfat, Dialyse gegen 0,001 n-Acetatpuffer (p_H 6,0) bei 0° C, Ultrafiltration unter Stickstoff bei 5 Atm. Druck (Ultrafeinfilter „fein" der Membranfiltergesellschaft Sartorius-Werke, Göttingen) wurde schließlich nach Adsorption an Filtrol-Neutrol und Elution mit Glycerin (mit 10 ml Phosphatpuffer auf p_H 6,8 gebracht) eine 130 fache Aktivitätssteigerung der AS-Oxydase aus Cucurbita pepo L. erreicht. Der N-Gehalt der Enzymsubstanz betrug 14,6%. Bei kräftigem Schütteln der Enzymlösung trat Schüttelinaktivierung ein. Die Aktivität wurde mit Hilfe der Titration mit 2,6-Dichlorphenol-Indophenol bei p_H 6,0 bestimmt.

Eine *Enzymeinheit* nach DIEMAIR u. ZERBAN ist diejenige Menge, durch die 1 mg AS bei 20° C und einem p_H von 6,0 (Acetatpuffer) oxydiert wird. Gemessen wird nur die Anfangsgeschwindigkeit der Reaktion (nach genau einer Minute). Oxydiert eine Enzymlösung mehr als 50% der vorgelegten AS, 10 mg-%, so wird diese verdünnt und die Bestimmung wiederholt.

Die *Enzymaktivität* gibt an, wieviel mg AS unter den beschriebenen Bedingungen (siehe Enzymeinheit) von 1 mg Enzymtrockenmasse nach Abzug der noch enthaltenen Ammoniumsulfatmenge oxydiert werden.

Die nach dem Verfahren von DAWSON u. MAGEE (1955) hergestellte AS-Oxydase soll eine etwa 80%ige Reinheit besitzen:

Herstellung des Rohextraktes aus *Cucurbita pepo condensa (Yellow summer squash)*. Man schält die äußere Schicht ab, zerkleinert sie und preßt den Saft ab.

Fraktionierung mit Ammoniumsulfat. Der Saft (etwa 50 l) wird mit Borax auf p_H 7,6 eingestellt, mit m-Bariumacetat (10 ml/l Saft) versetzt und durch Zugabe von festem Ammoniumsulfat bei Zimmertemperatur auf 1,6 m (0,3 Sättigung) gebracht. Die überstehende Flüssigkeit wird abfiltriert und mit der gleichen Menge an Ammoniumsulfat versetzt. Die Aufarbeitung des Präcipitates erfolgt dann durch *weitere Fraktionierung mit Ammoniumsulfat.* Hierzu verteilt man das Präcipitat in 2,5 l kaltem Wasser unter Rühren, läßt es absetzen, filtriert (unter Zusatz von Celite) und dialysiert die anfallende Lösung gegen dest. Wasser bis sich gerade ein gelbes Präcipitat zu bilden beginnt. Dieses wird abgeschieden und zu der Lösung wird ein Fünftel ihres Volumens an Aluminiumgel nach WILLSTÄTTER (1928) (s. unten) hinzugegeben. Das Aluminiumgel mit dem adsorbierten Enzym muß sofort abfiltriert werden; seine Elution erfolgt mit 0,2 m-Na_2HPO_4. Das Filtrat wird dann im Kühlraum wieder der Dialyse gegen dest. Wasser unterworfen.

Fällung mit Aceton. Die dialysierte Enzymlösung versetzt man nun mit etwa einem Fünftel ihres Volumens an Aceton (5° C) und filtriert sofort. Das Filtrat wird nun etwa 7 mal immer mit der gleichen Menge an Aceton versetzt. Bei der etwa 4. bis 6. Acetonbehandlung fällt je ein grüner oder blauer Niederschlag an. Die einzelnen Präcipitate werden getrennt in 150—200 ml 0,2 m-Na_2HPO_4 gelöst und gegen bidest. Wasser dialysiert. Aus den konzentrierten Lösungen fällt das Enzym als blaues, amorphes Protein an. Die Ausbeute beträgt etwa 18% der ursprünglichen Aktivität.

Herstellung von Aluminiumhydroxydgel nach WILLSTÄTTER (1928), mod. Man löst 170 g $Al_2(SO_4)_3 \cdot 18 H_2O$ in 2 l warmen Wasser (60° C) in einer etwa 12 l fassenden Flasche und fügt zu der Lösung 2 l einer warmen, wäßrigen Lösung von 50 g Ammoniumsulfat [$(NH_4)_2SO_4$]. Unter Umrühren werden etwa 150 ml Ammoniak (konz.) hinzugegeben. Die Flasche wird nun mit heißem Leitungswasser aufgefüllt, ihr Inhalt umgerührt und nach kurzer Ruhe (30—40 Min) erfolgt erneute Zugabe einer kleinen Menge an Ammoniak. Die überstehende Flüssigkeit wird nach kurzem Stehenlassen vorsichtig abgesaugt und mit warmem Wasser wieder aufgefüllt. Ungefähr 12 Waschungen sind erforderlich, den p_H-Wert des Waschwassers auf 8,0 zu bringen. Es folgen dann Waschungen mit dest. Wasser bis der p_H-Wert 7,0 beträgt. Schließlich wird das Präcipitat in soviel dest. Wasser verteilt bis die Suspension 2 l beträgt (Trockenmasse in 10 ml 0,144 g).

d) Aktivitätsbestimmungen der AS-Oxydase.

Zur Bestimmung der Aktivität der Ascorbinsäureoxydase eignet sich im allgemeinen jede quantitative AS-Bestimmung. Es empfiehlt sich, neben dem eigentlichen Versuchsansatz, AS und AS-Oxydase bei p_H 5,7—6,0, noch Ansätze ohne AS-Oxydaselösung (dafür die gleiche Menge an bidest. Wasser), sowie mit durch Hitze inaktivierter AS-Oxydase mitlaufen zu lassen.

Zur Bestimmung der AS in schwach gefärbten Lösungen bietet sich das Verfahren von Barakat u. Mitarb. (1955, 1956), die Titration mit N-Bromsuccinimid, an.

Prinzip: Die AS wird mit N-Bromsuccinimid zur DAS oxydiert. Nach Zugabe von Jodkali und Stärkelösung kann der Endpunkt scharf bestimmt werden.

Es stören *nicht* Kohlenhydrate, wie Glucose, Lactose, Saccharose und Stärke; ferner nicht die folgenden Verbindungen Diketogulonsäure, Reduktion, Reduktinsäure, Harnsäure und Kreatinin, Alkohole, Formaldehyd und Aceton, Äthylacetat, Riboflavin, Thiaminhydrochlorid, Oxalate, Tartrate und Citrate, Aminosäuren, wie Glycin, Alanin, Valin und Isoleucin, ferner Ferro-, Ferri- und Kupfersalze. Als die Reaktion störend wurden bisher Sulfide, Sulfite, Thiosulfate und Thioharnstoff gefunden.

Reagentien. Jodkalilösung (4%), Essigsäure (3%), Stärkelösung (1%), N-Bromsuccinimidlösung (0,01%).

Ausführung. Die Extrakte werden wie üblich mit Metaphosphorsäurelösung (5%) hergestellt. Die Titration eines bekannten Volumens einer klaren Lösung wird mit möglichst frisch bereiteter, wäßriger 0,01%iger N-Bromsuccinimidlösung (genau einwiegen!) ausgeführt. Bei höheren Gehalten ist eine 0,1%ige Lösung zu verwenden. Zur Erkennung des Endpunktes gibt man 5 ml einer 4%igen KJ-Lösung, 2 ml einer 3%igen Essigsäure und etwas Stärkelösung hinzu.

Berechnung. AS-Gehalt $= V \cdot C \cdot \dfrac{176}{178}$

$V =$ Volumen der verbrauchten Titerlösung;
$C =$ Konzentration der Titerlösung in mg oder μg pro ml.

Berechnungsbeispiel:

Verbrauch an N-Bromsuccinimidlösung (0,01%)

Hauptversuch:	2,60 ml
Blindversuch:	—0,09 ml
Endverbrauch:	2,51 ml

$$AS = 2,51 \cdot \frac{0,01}{100} \cdot 0,9888$$
$$= 0,2482 \text{ mg/ml, abrunden auf}$$
$$= 0,25 \text{ mg/ml}$$

Dawson u. Magee (1955) beschreiben eine *manometrische Methode* im Warburg-Apparat zur Festlegung der Aktivität der AS-Oxydase.

Prinzip: Die Methode beruht darauf, daß in einem bestimmten Konzentrationsbereich die Sauerstoffaufnahme der AS der vorhandenen Enzymmenge proportional ist.

Reagentien. Das verwendete dest. Wasser soll einen Cu-Gehalt unter 0,05 γ/ml haben. Die AS-Lösung (0,028 m) wird durch Lösen von 250 mg AS in 50 ml dest. Wasser, das 50 mg Metaphosphorsäure enthält, hergestellt; sie ist frisch zu bereiten. 0,2 m-Na_2HPO_4-0,1m-Citronensäurepuffer (pH 5,7). Gelatinelösung: 750 mg Gelatine werden in 150 ml warmen Wassers unter Umschwenken gelöst, 2 mg Thymol dienen als Konservierungsmittel; bei 5° C aufbewahren, brauchbar etwa eine Woche. Enzymlösung: Die Stammlösung wird soweit verdünnt, bis ein ml zwischen 1—2,5 Einheiten enthält. Diese Konzentration erreicht man durch eine Serie von Verdünnungen mit eiskaltem Wasser. Die endgültige Verdünnungsstufe muß 2 ml Gelatinelösung in 10 ml enthalten. Bildet sich hierbei ein Präcipitat, so wird ein neuer Ansatz mit eiskalter 0,1 m-Acetatpufferlösung (pH 5,6) hergestellt.

Methode. Es werden 40 ml fassende WARBURG-Gefäße verwendet. Vor dem Verdünnen der Enzymlösung wird der Reaktionsansatz, bestehend aus 4 ml Phosphat-Citratpuffer, 1 ml Gelatinelösung, 1 ml AS-Lösung und 3 ml dest. Wasser in den Hauptraum des Gefäßes gefüllt; dann erfolgt die Verdünnung der Enzymlösung, von der in das Seitengefäß 1 ml gemessen wird, und die Temperierung im Thermostaten auf 25° C ($\pm 0,01°$ C). Durch Kippen des Gefäßes wird die Reaktion eingeleitet. Die Ablesungen erfolgen alle 2 min 10 min lang bei 120 Schüttelbewegungen pro min. Der Sauerstoffverbrauch wird auf eine min berechnet. Es ist erforderlich, daß ein Blindversuch ohne AS-Oxydase, die durch die gleiche Menge an Wasser ersetzt wird, mitläuft.

Definition der Einheit. Eine AS-Oxydase-Einheit ist diejenige Menge an Enzym, die eine anfängliche O_2-Aufnahme von 10 μl/min bewirkt.

Die *spezifische Aktivität* wird in Einheiten bezogen auf mg Trockengewicht ausgedrückt. Die Trockengewichtsbestimmung erfolgt nach MALETTE u. Mitarb. (1948). Hochgereinigte Enzyme werden vorher gegen 0,1 m-Acetatpuffer (pH 5,7) dialysiert. Eine Korrektion für das Trockengewicht des Puffers ist bei der Berechnung anzubringen.

Trockengewichtsbestimmung nach MALETTE *u.* Mitarb. (1948).

Abgemessene Anteile der Enzymlösung werden durch 3—5tägige Dialyse (Cellophanbeutel) gegen kupferfreies Wasser (10—15maliger Wechsel) salzfrei gemacht und auf ein bestimmtes Volumen aufgefüllt. Von diesem wird ein aliquoter Teil entnommen und dieser wird in einem kleinem Kolben (5 ml aus Pyrexglas) bei 100° C unter Einblasen eines schwachen Luftstromes (Capillare) getrocknet. Nach 20 min, wenn keine Feuchtigkeit mehr entweicht, wird das Erwärmen unterbrochen und die Kolben (Doppelbestimmung) werden über Calciumchlorid abgekühlt. Sie werden dann auf 0,01 mg genau auf einer Halbmikrowaage gewogen. Die Werte der Doppelbestimmungen sollen nicht mehr als $\pm 0,02$ mg differieren.

e) Vorkommen der AS-Oxydase.

Man hat bisher das Vorkommen der AS-Oxydase in einer Vielzahl von Pflanzen festgestellt: Äpfeln, Kohl, Kürbis, Gerste, Gurke, Banane, Steckrüben, Petersilie, Kartoffeln, Melone, Lattich, Blumenkohl, Pfirsichen, Kresse, Erbsen, Bohnen, Soja, Mais, Karotten, Spinat u. a. GONTARSKI (1948) fand sie in der Honigbiene. Ihr Vorkommen in anderen tierischen Geweben ist bisher nicht mit Sicherheit festgestellt worden, obwohl eine Umwandlung von AS zu DAS in diesen beobachtet wurde. Es sollen hierzu nach TRAVIA u. SANTAGATI (1952) noch energiereiche Phosphorsäureradikale erforderlich sein. BEZSSONOFF u. LEROUX (1944) fanden eine Oxydation von AS in der Spinalflüssigkeit, im Plasma und im Urin noch über die DAS hinaus. MICHELAZZI (1948) verfolgte die AS-Oxydation bei pH 7,3 manometrisch mit Filtraten von Mikroorganismen (*Cholera, Typhus* und *Saccharomycetes*); Phosphat wirkte oxydationsbeschleunigend. Bei der Prüfung von Pflanzenmaterial auf AS-Oxydase sind neben dem Wachstumszustand der Pflanze durch Varietäten bedingte Unterschiede und durch Lichtwechsel verursachte Einflüsse zu berücksichtigen, siehe u. a. MORRIS, WEART u. LINEWEAVER (1946), CHATTOPADHYAY u. BANERJEE (1952), TOMITA (1950), IVANOVA (1952), RANGUEKAR, DE u. SUBRAHMANYAN (1948), LABARRE u. PFEFFER (1946), SRINIVASAN u. WANDREKAR (1950, 1950a), YAMAGUCHI u. JOSLYN (1951), JAMES u. BOULTER (1955), SISAKYAN u. VASELEVA (1954).

6. Andere Ascorbinsäure oxydierende Enzyme.

Außer der beschriebenen kupferhaltigen AS-Oxydase und den vorangestellten verschiedenen Oxydasen, die auch Metallenzyme darstellen, gibt es noch einige andere Enzymsysteme, die auch AS oxydieren, sich aber von den erwähnten unterscheiden.

SRINIVASAN u. WANDREKAR (1950) fanden in Extrakten aus 4—5 Tage alten Keimlingen, die durch Dialyse von Cu befreit worden waren, eine fünfmal größere Aktivität als bei der Anwesenheit von Kupfer. Der Extrakt wirkte sogar hemmend auf die durch Cu-katalysierte AS-Oxydation. CLAYSON u. BEESLEY (1951) versuchen die unbedingte Notwendigkeit von Cu für die Wirkung der AS-Oxydase zu widerlegen: 2800 ml Kürbissaft wurden mit soviel Aceton versetzt bis der Gehalt daran 40% (v/v) betrug; der gebildete Niederschlag wurde dann in 250 ml Phosphatpuffer (pH 6,0) verteilt. Längere Lagerung setzte den Cu-Gehalt auf 0,013 mg-% herab, der dann vollständig durch Ionenaustauscher Zeo-Carb H1, beladen mit Kaliumionen, pH 6,0, entfernt werden konnte. Diese Behandlung hatte einen beträchtlichen Verlust der Gesamtaktivität zur Folge (25—60%). Die Zubereitung war jetzt aber kupferfrei wie mit Hilfe der Dithizon- und der manometrischen Methode nach WARBURG (1927) (s. dieses Hdbch. Bd. 7, S. 220) (vgl. auch Bd. 1, S. 415) ermittelt wurde, zeigte aber noch eine beachtliche AS-Oxydaseaktivität. Die Zugabe von kleinen Mengen an Kupfersulfat setzte die Aktivität herab.

MANDELS (1951, 1953, 1953a) fand ein Enzym auf der Oberfläche der Sporen von *Myrothecium verrucaria*, das AS oxydiert. Eine 20 min lange Behandlung mit 0,1 n-HCl zeigte keinen besonderen Einfluß auf die Lebenstätigkeit der Pilzsporen; das Enzym wurde bereits nach 30 sec vollständig inaktiviert. Die Säure ist bei dieser Behandlung nicht bis in das Innere der Zelle eingedrungen, da ihre Atmungs- und Assimilationsenzyme hierdurch nicht geschädigt wurden. [Über das Vorkommen von Enzymen auf der Oberfläche von Hefezellen siehe den Überblick von DOUNCE (1950).] Das Enzymsystem ist hitzelabil und nicht dialysierbar. Es nimmt pro Mol AS ein halbes Mol O_2 auf, was auch für eine enzymkatalysierte Oxydation spricht. Bei der Schwermetallkatalyse wird mehr Sauerstoff verbraucht (WEISSBERGER u. LU VALLE, 1944). Die Reaktion verläuft nach der Gleichung 1.Ordnung. Das Enzym besitzt ein breites pH-Optimum, von pH 4—8 (und höher), das aber von der Art des Puffers anhängig ist. Es erwies sich als resistent gegenüber Cyanid, Azid, Diäthyldithiocarbamat, Phenylthioharnstoff und gegen 8-Hydroxychinolin. SH-gruppengifte, wie Chlormerkuribenzoat und p-Chinon hemmen. Jodacetat wirkt leicht stimulierend. Isoascorbinsäure und Glucoascorbinsäure hemmen. Die Untersuchungen wurden mit der üblichen WARBURG-Technik durchgeführt, bei pH 6,25 (Phosphatpuffer) mit 4—5 mg Sporen oder einem Extrakt, der 10 mg Sporen entspricht; im Ansatz des Reaktionsgefäßes wurde das Substrat (3—4 mg AS) untergebracht. Die Angabe der Ergebnisse in μl O_2/mg Trockengewicht an Sporen pro Std errechnet man nach dem Ergebnis der ersten 10 min. In einigen Untersuchungen wurde auch GSH hinzugegeben, das aber keinen meßbaren Einfluß zeigte.

Es wird hier nochmals an die Arbeiten von TADOKARA u. Mitarb. (1939, 1940) erinnert, die als wesentlichen Bestandteil der AS-Oxydase ein Mononucleotid sehen, eine Beobachtung, der DUNN u. DAWSON (1951) nicht zustimmen.

Zum Abschluß sei noch auf Beobachtungen von WARD (1954) hingewiesen, der in *Physarum polycephalium* eine atypische AS-Oxydase feststellte. Die Oxydation von AS war von Cyanid, Azid, Sulfid, CO und Chelatbildern nicht beeinflußbar. Diäthyldithiocarbamat wirkt auf ihren Ablauf katalysierend. Nach seinen Untersuchungen ist es durchaus möglich, daß Wasserstoffperoxyd hierbei entsteht. Der von ihm verwendete rohe Enzymextrakt wird durch Dialyse inaktiviert. Er nimmt die Beteiligung von Flavinenzymen an.

Diese Zusammenstellung zeigt, daß in Pflanzen neben den metallhaltigen Oxydasen noch andere Enzymsysteme vorkommen, die eine Oxydation der AS bewirken.

III. L-Ascorbinsäure reduzierende Enzymsysteme.

Es ist wohl als gesichert anzunehmen, daß sowohl pflanzliches als auch tierisches Gewebe die Fähigkeit besitzt, DAS zu AS zu reduzieren. ZILVA, KIDD u. WEST (1938) beobachteten, daß beim Fortschreiten der Fruchtreife die Konzentration der DAS abnimmt und diejenige der AS zunimmt. Eine ähnliche Feststellung machte MAPSON (1954) an Keimen. Im allgemeinen ist der Anteil der DAS in Pflanzen wesentlich kleiner als derjenige der AS (SCHEUNERT u. THEILE, 1952; KOCH u. BRETTHAUER, 1956). Es scheint ein Gleichgewicht zu bestehen, das durch den Zellstoffwechsel aufrechterhalten wird. Bei künstlicher Störung, z. B. durch

Zerstörung der Zelle (SZENT-GYÖRGYI, 1931; STONE, 1937), Hemmung mit Narcoticis und anderen Enzymgiften (MAPSON u. BARKER, 1954) kommt es zur schnellen Bildung von DAS. Man nimmt an, daß das Oxydasesystem jetzt enger mit seinem Substrat (AS) in Berührung kommt und daß das System, welches die Reduktion der DAS vollzieht, empfindlicher ist als das Oxydasesystem (SE-BRELL u. HARRIS, 1954). Das Schema von WAYGOOD (1950) gibt ein Beispiel dafür, wie vielleicht der Wasserstofftransport unter Einschaltung des AS-DAS-Systems abläuft:

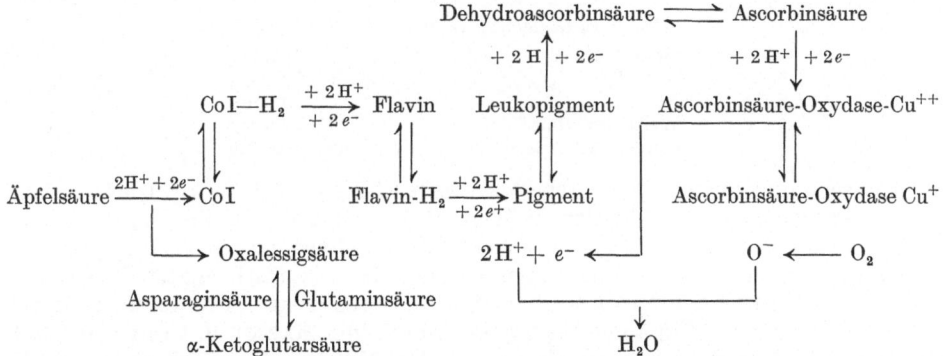

Abb. 6. Einschaltung des Ascorbinsäure-Dehydroascorbinsäure-Systems in den enzymatischen Wasserstofftransport beim Weizen (nach WAYGOOD, 1950, umgezeichnet nach METZNER, 1957).

Die Natur des zwischengeschalteten Pigmentes ist nicht bekannt. MAPSON (1954) denkt an PALLADINS (1910) Atmungschromogen Synergin, an ROBINSONS (1935) Leukoanthocyanin, an das Hermidin aus Mercurialis (HAAS u. HILL, 1925), an Vitamin P (BENTSATH, RUSZNAK u. SZENT-GYÖRGYI, 1937) und auch an das Oxydations- und Reduktionsenzym von WAVRA u. WEBB (1942).

JAMES u. Mitarb. (1943, 1944) nehmen an, daß in der Gerste ein Dehydrogenase-System vorkommt, das unter Einschluß von α-Hydroxysäuren DAS zu AS reduziert:

$$\text{Milchsäure} \leftrightarrows \text{Brenztraubensäure}$$
$$R \cdot CH \cdot OH \cdot COOH + DAS \leftrightarrows R \cdot CO \cdot COOH + AS$$

Mehrere Arbeiten zeigen die Beteiligung von DPN an der Umwandlung von DAS in AS (JAMES u. Mitarb., 1943, 1944; DAVISON, 1949; MATTHEWS, 1951). Der Wasserstoffacceptor mit Erbsenenzym soll nach MATTHEWS nicht DAS, sondern Monodehydroascorbinsäure sein, da DAS sich als unwirksam erwies.

Schutzwirkung von GSH. Es besteht eine enge Beziehung zwischen Glutathion (red.) und AS im pflanzlichen und tierischen Gewebe (PETT, 1936; HOPKINS u. MORGAN,1943). In keimenden Samen und treibenden Kartoffelknollen erscheinen beide Substanzen zu gleicher Zeit. Die Schutzwirkung von GSH gegen die Oxydation von AS wurde von mehreren Untersuchern beobachtet. Diese wurde auf die Fähigkeit von GSH, Kupfer zu binden, zurückgeführt und fand ihre Erklärung weiter darin, daß GSH DAS zu AS auch ohne Zugabe von Katalysatoren, wenn auch nur langsam, zu reduzieren vermag (DE CARO u. GIANNI, 1934; CROOK u. HOPKINS, 1938; MAWSON, 1935; BERSIN u. Mitarb., 1935).

PFANKUCH (1934, 1934a) fand als erster die enzymatisch katalysierte Reduktion von DAS. Er wies im Kartoffelpreßsaft eine hitzelabile Substanz nach, die mit Eiweißfällungsreagentien abgeschieden werden kann und die natürliche und zugesetzte DAS zu reduzieren vermag. HOPKINS u. Mitarb. (1936, 1938, 1941, 1943) sowie KOHMANN u. SANBORN (1937) beschreiben eine DAS-Reduktase aus Blumenkohl, die die Reduktion von DAS bei Gegenwart von GSH katalysiert

$$2\ GSH + DAS \leftrightarrows GSSG + AS.$$

15*

Das Enzym wurde aus Blumenkohlsaft mit Ammoniumsulfat gefällt. GSH ließ sich, wenn auch mit schlechteren Ergebnissen, durch Cystein und Thiomilchsäure ersetzen (Mapson, 1959). Die Bedenken von Barron (1939), daß verhältnismäßig wenig GSH in Pflanzen vorkommt, konnte durch die Untersuchungen von Hopkins widerlegt werden; GSH und AS liegen etwa in einem Verhältnis von 2:1 vor (zit. nach Mapson in Sebrell u. Harris, 1954).

Nach der Zusammenstellung von Isherwood (1959) wurden in pflanzlichen Geweben folgende Glutathionwerte gefunden:

Gehalt an Glutathion in 100 g Frischsubstanz.

Hefe . . . . 128,0 mg
Kartoffel . . . . 10,8 mg
Äpfel 0   mg
Weizenkeimlinge . . . . 12,6 mg
gequollene Erbsen . . . . . . etwa 60,0 mg

Tierische und pflanzliche Gewebe vermögen GSSG → GSH zu reduzieren. Conn u. Vennesland (1951) fanden in Weizenkeimextrakten ein Enzym, die Glutathionreduktase, das bei Gegenwart von TPN und Glucose-6-phosphat (pH 6,5) auf GSSG reduzierend wirkt. Das Enzym ist auf GSSG spezifisch eingestellt; Cystein, Homocystin und Glutamylcystin werden nicht reduziert. TPN war nicht durch DPN zu ersetzen. Ein ähnliches System wurde von Mapson u. Goddard (1951, 1951a) in Erbsen gefunden. Es vermochte GSSG in Anwesenheit von TPN, Mn^{2++} und von Mg^{2++}, sowie von Isocitronensäure oder von dl-Äpfelsäure zu reduzieren. Auch hier konnte TPN nicht durch GSSG ersetzt werden. Die rückläufige Reaktion konnte experimentell nicht festgestellt werden. Janecke u. Ragab (1956) fanden in Blumenkohlsaft ein Enzymsystem, das die Reduktion von DAS zu As katalysiert. Nach ihren Untersuchungen eignete sich GSH als Wasserstoffdonator am besten. Auch der Einsatz von GSSG führte zu ähnlichen Ergebnissen, da dieses durch ein im Saft vorhandenes System zunächst zu GSH reduziert wird. Sie nehmen an, daß TPN und ein Flavin-Enzym an der Reaktionskette beteiligt sind. Als Flavinkomponente verwendeten sie FMN, da ihnen FAD, das vielleicht bessere Dienste getan hätte, nicht zur Verfügung stand.

$$
\begin{array}{lcccc}
 & & \text{TPN} & & \text{TPNH} \\
\text{1. Stufe:} & 2\,\text{GSH} + & \text{oder} & \rightleftharpoons \text{GSSG} + & \text{oder} \\
 & & \text{DPN(?)} & & \text{DPNH}
\end{array}
$$

$$
\begin{array}{lcccc}
 & \text{TPNH} & \text{FMN(?)} & \text{TPN} & \text{FMNH} \\
\text{2. Stufe:} & \text{oder} + & \text{oder} & \rightleftharpoons \text{oder} + & \text{oder} \\
 & \text{DPNH} & \text{FAD} & \text{DPN} & \text{FADH}
\end{array}
$$

$$
\begin{array}{lccc}
 & \text{FMNH} & & \text{FMN} \\
\text{3. Stufe:} & \text{oder} + \text{DAS} & \rightleftharpoons & \text{oder} + \text{AS} \\
 & \text{FADH} & & \text{FAD}
\end{array}
$$

1. Herstellung des gereinigten Enzympräparates nach Janecke u. Ragab.

Zur Gewinnung des Rohsaftes wurden die Blumenkohlblütenstände mit ihren Strunken herausgeschnitten, sorgfältig gereinigt, mit Trockeneis und Aceton zum Gefrieren gebracht und dann sofort im Starmix vermahlen. Der Brei wurde auf 0° C abgekühlt und schnell ausgepreßt. Die Klärung des so erhaltenen Saftes erfolgte durch Zentrifugieren. Hierbei stieg die Temperatur nur wenig über 0° C an. Auf diese Weise wurden 1,1 l eines Blumenkohlsaftes erhalten, der mit der gleichen Menge gesättigter, auf 0° C abgekühlter Ammoniumsulfatlösung versetzt wurde. Unter Einhaltung der niederen Temperatur wurde das Gemisch auf pH 7,0 eingestellt. Nach 5 min langem Stehen hatte sich ein Niederschlag gebildet, der durch Zentrifugieren entfernt wurde. Die überstehende Flüssigkeit (2 l) wurde der durch Zusatz von 3,15 l gesättigter Ammoniumsulfatlösung zur 80%igen Sättigung gebracht. Es bildete sich wieder ein Niederschlag, der in einer kleinen Menge Pufferlösung (pH 7,0) gelöst wurde und dann einer 50%igen und anschließend einer 80%igen Ammoniumsulfatsättigung unterworfen wurde. Mit dem zuletzt entstandenen Niederschlag wurde nach seiner Lösung in Pufferlösung (pH 7,0) eine Dialyse (0° C, 8 Std) gegen Pufferlösung durchgeführt. Es folgte nochmals eine Ammoniumsulfatbehandlung und Dialyse zunächst gegen Pufferlösung, anschließend gegen bidest. Wasser. Durch Gefriertrocknung wurde aus dieser Lösung ein feines, leichtes, schwach gelbliches Pulver erhalten. Durch den Gefrierprozeß hatte eine Wirkungseinbuße nicht stattgefunden. Das Trockenpräparat behielt bei einer Temperatur von —15° C über $4^1/_2$ Monate seine volle Wirksamkeit. Auf die Trockensubstanz bezogen wurde eine 70fache Anreicherung erzielt. Das angereicherte Präparat hatte sein pH-Optimum bei pH 7,0. Yamaguchi u. Joslyn (1951) fanden 6,7—6,9, Crook (1941) 7,0.

Versuchsansatz (als Beispiel). 0,2 ml des *nicht gereinigten* Enzympräparates und 7,0 mg GSH wurden mit 2,0 mg DAS, gelöst in 2,0 ml 0,066 m-Phosphatpufferlösung (pH 7,0), nach dem Temperieren auf 25° C unter Luftausschluß in einem Thunberg-Röhrchen miteinander gemischt. Nach 15 min erfolgte nach Wegnahme des Vakuums die Zugabe von 20%iger Trichloressigsäure bis eine Endkonzentration von etwa 2% erreicht war. Die gebildete AS wurde mit DIP (0,001 n) nach Tillmans (siehe dieses Handbuch Bd. 2, S. 98, 100) bestimmt.

Bei den Ansätzen mit *gereinigtem* Enzym wurden die folgenden Mengen gewählt: 0,1 mg gereinigtes Enzympräparat, 7,0 mg GSH, 2,0 mg DAS, 2,5 mg TPN, 0,05 mg FMN, pH 7,0, Temp. 25° C, Reaktionszeit 25 min. Die nichtenzymatische Reduktion ist in einem parallel laufenden Versuch zu ermitteln und ist von dem Ergebnis im Versuch mit Enzymzusatz abzuziehen.

2. Darstellung der L-Dehydroascorbinsäure.

Wasserfreie Dehydro-L-Ascorbinsäure[1] wurde nach dem Verfahren von B. Pecherer hergestellt. In einen 3-Liter-Dreihalskolben, der mit einem Rührer, Thermometer, Gaseinleitungs- und Gasableitungsrohr ausgerüstet ist, gibt man 176 g AS (1 Mol), 520 g neutrales Bleicarbonat (1,96 Mol) und 1200 ml trockenen Methylalkohol. Der Inhalt des Kolbens wird auf —10° C abgekühlt und dann wird bei guter äußerer Kühlung in die gerührte Suspension langsam Chlor-Gas eingeleitet, so daß die innere Temperatur —6° C nicht übersteigt. Um ein Mol Chlor einzuleiten, benötigt man annähernd $1^1/_2$ Std. Man erhält eine farblose Suspension und ermittelt den Endpunkt entweder visuell oder durch Verwendung von Jodstärke-Papier. Ein Überschuß an Chlor wird durch Zugabe der erforderlichen Menge AS entfernt. Das Rühren wird noch ungefähr $1^1/_2$ Std lang fortgesetzt, nachdem der Endpunkt erreicht ist. Dabei nimmt die Entwicklung von CO_2 langsam ab. Zu dieser Zeit soll der pH-Wert der Lösung etwa 5—6 betragen. Nun entfernt man den Kolben aus dem Kältebad und filtriert die suspendierten Salze über eine dünne Schicht aus Filterschnitzeln ab. Es wird dreimal mit 150 ml CH_3OH gewaschen, die auf —5° C gekühlt sind. — In die klare farblose Lösung leitet man nicht länger als 5 sec lang H_2S ein. Der Überschuß an H_2S wird durch Durchsaugen eines kräftigen Luftstromes entfernt. In die erhaltene Suspension von PbS gibt man 5 g Filterschnitzel, schüttelt kräftig durch und saugt über eine dünne Lage Filterschnitzel ab. Es wird mit gekühltem CH_3OH gewaschen. Das klare, farblose Filtrat wird in einen Rundkolben eingebracht und Methanol bei einer Wasserbadtemperatur von 40° C unter Wasserstrahlvakuum abdestilliert.

Wenn sich der Sirup grün färbt, ersetzt man den Destillieraufsatz und Kühler durch ein Zwischenstück und schließt direkt an die Wasserstrahlpumpe an, ferner erhöht man die Badtemperatur im Verlauf von 30 min auf 100° C. Die grüne Farbe verschwindet und es resultiert

[1] Über Methoden zur Herstellung von DAS berichten noch Antener (1937), Crook (1948), Crook u. Morgan (1944), Karrer, Salomon, Morf u. Schöpp (1933), Kenyon u. Munro (1948), Knobloch (1948), Mapson u. Ingram (1951), Moll u. Wieters (1936), Pohloudek-Fabini u. Fürtig (1959), Ulmann (1954), Wieters (1935), Worker u. Antener (1937). Zur Formulierung der Dehydroascorbinsäure siehe: Eistert (1954), v. Euler u. Eistert (1957) und Albers u. Müller (1959).

ein blaß bernsteinfarbener Sirup. An diesem Punkt wird die Wasserstrahlpumpe durch eine Ölpumpe ersetzt und der Rückstand für eine Stunde lang in diesem Vakuum belassen. Der Kolben wird dabei direkt auf dem Dampfbad erhitzt. Bei diesem Vorgang tritt ein mehr oder weniger starkes Schäumen auf. Nach dieser Behandlung wird der schaumige, spröde Rückstand im Vakuum abgekühlt, in 500 ml abs. Äthanol gelöst und 18 Std lang bei 0° C aufbewahrt. Es bildet sich ein dichter mikrokristalliner Niederschlag. Dieser wird abfiltriert, mehrmals mit kleinen Anteilen von abs. Äthanol gewaschen und im Vakuum über CaCl$_2$ getrocknet, was etwa 2—3 Tage in Anspruch nimmt.

Aus den vereinigten Filtraten und Waschwassern, die schwach gelbbraun gefärbt sind, können ein 2. und ein 3. Mal durch Behandlung mit Holzkohle, nachfolgendes Eindampfen und Wiederholung des Trocknungsprozesses weitere Anteile erhalten werden. Es wurden insgesamt Ausbeuten von 30—39% erhalten.

Der Schmelzpunkt variiert mit der Art des Erhitzens und der Temperatur, bei der die Probe in den Schmelzblock eingebracht wird. Ein Schmelzpunkt von 220—225° C (u. Zers.) ist zufriedenstellend, und ein solches Material enthält 95—100% DAS. Frisch dargestellte, wasserfreie DAS ist weiß, und besitzt einen Schmelzpunkt von 225° C (Zers.); nach 6 Monaten bei gewöhnlicher Temperatur aufbewahrt, nimmt sie eine gelbliche bis braune Färbung an und der Schmelzpunkt sinkt auf 204—205° C (Zers.) ab. Eine Reinigung durch Umkristallisation ist nicht möglich, da sie in allen gebräuchlichen Lösungsmitteln unlöslich ist. Sie löst sich schnell in basischen Lösungsmitteln wie Pyridin, kann aber aus solchen Lösungen nicht zurückerhalten werden. Vollständige und schnelle Löslichkeit in Wasser kann nur beim Erwärmen auf 60° C erreicht werden; beim Abkühlen erscheint aber kein festes Produkt mehr.

Nach der von Pohloudek u. Fabini (1959) angegebenen Methode wird die AS mit Selendioxyd in absol. Methanol oxydiert. Es entsteht DHA neben Selen, das abfiltriert wird. Nach der Vorschrift wird das selenfreie Filtrat i. V. eingedampft, bis eine schaumige, farblose Masse zurückbleibt, die in wenig Äthanol aufgenommen wird. Hierbei färbt sich die Masse oft braun und bildet mit dem Äthanol eine sirupöse Lösung, aus der keine DHA ausfällt. Wird aber die AS-Lösung in kleinen Mengen in die Selendioxydlösung eingetragen, die Fällung jedesmal abfiltriert und dieses so lange fortgesetzt, bis keine Fällung mehr auftritt (über Nacht stehenlassen), so erhält man nach dem vorsichtigen Eindampfen der Lösung eine weiße, schwammige Substanz, aus der sich mit absol. Äthanol DHA abscheiden läßt [(nach Janecke u. Osthoff 1962)]. Die Identifizierung kann nach Strohecker, Heimann u. Matt (1955) erfolgen.

Die im Versuch entstandene AS kann z. B. nach der Methode von Barakat, Abd El-Wahab (1955, 1956, s. dieses Hdb. Bd. 7, S. 224) oder einer anderen bestimmt werden. Mit dem Verfahren nach Roe u. Mitarb. (1944—1948) erfaßt man die beiden Reaktionspartner. Es gibt darüber hinaus noch einen Anhalt, inwieweit Verluste an DAS (zu DKS umgewandelt) bei den Versuchen eingetreten sind.

Der *2,4-Dinitrophenylhydrazin-Methode* von J. H. Roe und C. H. Kuether (1944—1954) liegt folgendes *Prinzip* zugrunde: Die Ascorbinsäure (AS) wird mit milden Oxydationsmitteln zur Dehydroascorbinsäure (DAS) oxydiert, die leicht einer spontanen Umlagerung zur 2,3-Diketogulonsäure (DKS) unterliegt. Die Reaktion verläuft langsam in schwach saurer Lösung und sehr schnell in Lösungen mit p$_H$-Werten unter 1,0 sowie in neutralem oder alkalischem Medium[1]. Bei Behandlung mit 2,4-Dinitrophenylhydrazin bilden DAS und DKS Derivate, die bis-Dinitrophenylhydrazone. Die Reaktionsgeschwindigkeit ist bei der DKS größer als bei der DAS, was darauf hinweist, daß die Reaktion nur mit der DKS stattfindet, und die DAS erst nach Umlagerung in die 2,3-Diketogulonsäure reagiert. Wird das gebildete bis-2,4-Dinitrophenylhydrazon mit 85%iger H$_2$SO$_4$ behandelt, so bildet sich durch eine molekulare Neuordnung ein sehr stabiles, rötlich braunes Produkt, dessen Absorptionsmaxima bei 500—550 mμ und 350—380 mμ liegen. Die auf diese Weise erhaltene Farbe kann photometrisch gemessen werden.

Als Oxydationsmittel dienen Aktivkohle oder Brom. Erstere hat den Vorteil, daß sie sowohl ein oxydierendes als auch ein entfärbendes Reagens ist. Brom kann gut bei ungefärbten Extrakten verwendet werden und hat den Vorzug, daß sein Überschuß durch Einleiten eines Luftstromes einfach entfernt werden kann. Will man die drei Säuren AS, DAS und DKS einzeln bestimmen, so gelingt dies durch zusätzliche Verwendung von H$_2$S. In einem aliquoten Teil des Extraktes wird ohne Vorbehandlung das bis-2,4-Dinitrophenylhydrazon entwickelt und dieser liefert somit den Wert für die bereits vorhandene DAS und DKS. Ein anderer aliquoter Teil wird mit H$_2$S reduziert, wodurch die Dehydroascorbinsäure in AS übergeht, während die DKS nicht verändert wird. Die sich anschließende Bestimmung liefert somit den Wert für die 2,3-Diketogulonsäure. Dann erfolgt in einem dritten aliquoten Teil die Oxydation mit

---

[1] Die DAS besitzt ihre größte Stabilität bei p$_H$ 2,0 (Huelin, 1949, 1951).

anschließender Kupplung mit 2,4-Dinitrophenylhydrazin, und das ergibt den Wert für die Summe der drei Säuren. Durch Subtraktion erhält man dann die Einzelwerte für jede Säure.

In pflanzlichen Extrakten verhindert der Zusatz von Thioharnstoff zur HPO$_3$ die Oxydation der AS, dagegen ist dieser bei tierischen Geweben wegen der beträchtlichen Oxydationskraft des Oxyhämoglobins unwirksam und muß durch Zinn-II-chlorid ersetzt werden.

Spezifität. Diese Methode besitzt eine hohe Spezifität; eine ähnliche Färbung wird von den 2,4-Dinitrophenylhydrazonen der Hexosen, Pentosen und diesen ähnlichen Verbindungen gegeben. Die Reaktionsgeschwindigkeit der Zucker mit dem Reagens ist im Vergleich zu den Oxydationsprodukten der AS jedoch sehr langsam. Außerdem zersetzen sich die 2,4-Dinitrophenylhydrazone der Zucker in H$_2$SO$_4$ der angewendeten hohen Konzentration. Daher wartet man nach Entwicklung der Farbe noch 30 min lang, ehe die Werte photometrisch bestimmt werden. Die Kupplungsreaktion wird in Gegenwart von Thioharnstoff oder SnCl$_2$ ausgeführt, um eine Störung durch andere Farbstoffe zu verhindern. Störungen können auftreten bei behandelten Gemüsen, in denen Reduktone gebildet wurden, und bei Gemüsen und Obstsorten mit einem sehr hohen Zuckergehalt. Die Absorptionsmaxima der störenden, auch gefärbten Produkte liegen bei kürzeren Wellenlängen. Durch Verwendung eines 540 mμ Filters im Colorimeter, oder bei Verwendung eines Schmalspaltspektrophotometers eingestellt auf 540 mμ wird der Fehler beträchtlich, wenn auch nicht vollständig, verhindert.

Die bei der Titration nach J. TILLMANS (siehe dieses Handbuch Bd. 2, S. 98, 100) störenden Metallionen beeinträchtigen diese Methode *nicht.*

Reagentien. Noritkohle, Aktivkohle Carboraffin 101, Farbenfabriken Bayer Leverkusen, oder akt. Tierkohle, Merck (n. DAUBENMERKL, 1949, 1950); 5%ige Metaphosphorsäure-Lösung; 5%ige HPO$_3$-10%ige Essigsäure-Lösung; 2,4-Dinitrophenylhydrazin-Thioharnstoff-Reagens; 85%ige Schwefelsäure.

Die 5%ige Metaphosphorsäure-Lösung und die 5%ige HPO$_3$-10%ige Essigsäure-Lösung sollen wöchentlich frisch hergestellt werden. Für die letztere löst man 50 g reiner HPO$_3$ in ungefähr 800 ml dest. Wasser, fügt dann 100 ml Eisessig hinzu und füllt mit dest. Wasser auf 1000 ml auf. Die Säurelösungen werden im Kühlschrank aufbewahrt.

Für das 2,4-Dinitrophenylhydrazin-Thioharnstoff-Reagens löst man 2 g 2,4-Dinitrophenylhydrazin in 100 ml annähernd 9 n-H$_2$SO$_4$ (3 Vol. Tl. H$_2$O:1 Vol. Tl. konz. H$_2$SO$_4$), fügt 4 g Thioharnstoff hinzu, schüttelt bis zur vollständigen Lösung und filtriert. Das Reagens wird im Kühlschrank aufbewahrt und jeweils vor Gebrauch filtriert. Diese Lösung ist wenigstens einmal im Monat frisch herzustellen. Von Zeit zu Zeit prüfe man die Reduktionskraft. Zu diesem Zweck gebe man 2 ml einer 1%igen Sublimat-Lösung in ein Reagenzglas und füge tropfenweise das zu prüfende Reagens hinzu. Wenn noch genügend Thioharnstoff vorhanden ist, verursacht die Zugabe von 3—5 Tropfen einen reichlichen Niederschlag von HgCl.

Die Aktivkohle muß vor Gebrauch mit Säure gewaschen werden: Man gibt 200 g Aktivkohle in einen großen Kolben und erhitzt nach Zugabe von 1000 ml 10%iger HCl zum Sieden und saugt anschließend ab. Den Filterkuchen gibt man in ein großes Becherglas, fügt 1000 ml dest. Wasser hinzu, rührt sorgfältig auf und saugt erneut ab. Der gleiche Vorgang wird noch ein bis zweimal wiederholt. Die Kohle wird dann über Nacht bei 110—120° C getrocknet.

Zur Herstellung der Lösungen soll nur dest. Wasser benutzt werden.

Ausführung. Man zerkleinert das Material unter Zugabe einer 10%igen Suspension von SnCl$_2$ in 5%iger Metaphosphorsäure-Lösung und verdünnt dann auf eine Endkonzentration von 0,5% SnCl$_2$. Der geringe Niederschlag, der sich bei der Vermischung des SnCl$_2$ mit der HPO$_3$-Lösung bildet, ist für die Bestimmung ohne Bedeutung. Am besten verwendet man eine Verdünnung, bei der das Filtrat pro ml 1—12 γ Vitamin C enthält. Wird ein Mixer benutzt, so ist durch die Extraktionsflüssigkeit Kohlendioxyd zu leiten, um den Sauerstoff zu vertreiben, und auch über der Flüssigkeit muß man eine inerte Gasatmosphäre herstellen. Zu diesem Zwecke wird nach den ersten Sekunden des Zerkleinerns der Deckel des Mixers leicht geöffnet, um einen Schlauch hindurchzuführen, durch den das CO$_2$ über das Homogenisat geleitet wird. Nach dem Auffüllen auf das entsprechende Volumen wird durch ein schnell laufendes Faltenfilter filtriert.

Bei der Bestimmung in Obstpreßsäften werden die jeweiligen Früchte ausgepreßt und der Saft in 5%iger HPO$_3$-Lösung aufgefangen. Danach wird unter Zugabe von SnCl$_2$ auf die nötige Konzentration verdünnt. Für die Oxydation mit Kohle wird mit 5%iger HPO$_3$-10%iger Essigsäure-Lösung ohne Zugabe von SnCl$_2$ aufgefüllt.

Zur *Bestimmung von DKS und DAS* pipettiert man je 4 ml der SnCl$_2$-HPO$_3$-Verdünnung in 3 Colorimetergefäße und gibt in 2 davon je 1 ml der 2,4-Dinitrophenylhydrazin-Thioharnstoff-Lösung; das dritte Röhrchen dient als Blindprobe (ohne Zusatz des Reagenses). Alle drei Röhrchen stellt man dann in ein Wasserbad von 37° C und beläßt sie dort für 5 Std. Die Temperatur darf um höchstens ± 2° C schwanken. Nach dieser Zeit nimmt man die Röhrchen heraus und stellt sie umgehend in ein Becherglas mit Eiswasser, das noch große Mengen Eis enthält. Während des Stehens in dem Eisbad gibt man zu jeder der drei Proben 5 ml 85%ige

H_2SO_4 aus einer Bürette tropfenweise hinzu und schüttelt sie gut durch; dieser Schritt benötigt annähernd drei Minuten. Zum Schmieren des Bürettenhahnes nehme man konz. H_2SO_4 und *nicht Hahnfett*. Zu der Blindprobe gibt man erst jetzt 1 ml des 2,4-Dinitrophenylhydrazin-Reagenses. Sind die Röhrchen genügend abgekühlt (12—15° C), entfernt man sie aus dem Eisbad und führt 30 min nach Zugabe der Schwefelsäure die colorimetrische Bestimmung durch. Die Blindprobe dient zur Einstellung auf 100% Durchlässigkeit, oder auf Null bei einem Nullpunktinstrument. Das geeignetste Filter ist das mit einer maximalen Durchlässigkeit bei 540 mμ. Der erhaltene Wert gibt die Menge der vorhandenen DKS + DAS an.

Für die *Bestimmung der DKS* gibt man 75—100 ml der $SnCl_2$-HPO_3-Verdünnung in eine Gaswaschflasche, die am Gaseinleitungsrohr mit einer Glasfritte von etwa 20 mm Durchmesser ausgerüstet ist. Es wird nun 15 min lang H_2S durch die Lösung geleitet, wodurch das Sn als schwarzes SnS ausfällt. Es ist darauf zu achten, daß jegliche Volumenänderung und damit eine Änderung der Konzentration vermieden wird. Ist die Reduktion beendet, gibt man zu 50 ml der H_2S-gesättigten Lösung 0,5 g Thioharnstoff, schüttelt bis zur vollständigen Lösung und filtriert. Zur Vertreibung des überschüssigen Schwefelwasserstoffs leitet man noch 5 min lang CO_2 durch das Filtrat und verfährt danach wie oben bereits beschrieben wurde. Der erhaltene Wert gibt die Menge der vorhandenen DKS an, und man erhält durch Subtraktion von dem vorangegangenen Wert die Menge der vorhandenen DAS.

Zur *Bestimmung von AS, DAS und DKS* werden 50 ml des mit 5%iger HPO_3-10%iger Essigsäure-Lösung verdünnten Extraktes mit 1 g Aktivkohle versetzt. Die Essigsäure hat den Zweck, eine Adsorption der Ascorbinsäure an die Kohle zu verhindern und die Oxydation zu fördern. Sie wird bevorzugt an der Aktivkohle adsorbiert und verdrängt so den aktiven Sauerstoff in einer Menge, die zur schnellen Oxydation ausreicht. Nach dem Zusatz der Kohle schüttelt man kräftig um und filtriert frühestens nach 5 min durch ein schnell filtrierendes Faltenfilter. Anschließend gibt man soviel Thioharnstoff hinzu, daß eine 1%ige Lösung entsteht. Dann pipettiert man dreimal je 4 ml ab und verfährt wie vorher beschrieben. — Von dem erhaltenen Wert zieht man den aus der Bestimmung der DKS + DAS ab und erhält damit den Gehalt an Ascorbinsäure.

Für die *Aufstellung der Eichkurve* löst man 100 mg AS p.A. (Merck) in 100 ml der entsprechenden Säurelösung und oxydiert, wie bereits beschrieben, mit Aktivkohle. 10 ml dieser DAS-Lösung werden in einen 500 ml-Meßkolben pipettiert und mit der gleichen Säurelösung zur Marke aufgefüllt. Nun bereitet man sich eine Serie von Standardlösungen, indem man 5, 10, 20, 30, 40, 50 und 60 ml abpipettiert, 1% Thioharnstoff hinzusetzt und jeweils auf 100 ml verdünnt. Je 4 ml dieser Standardlösungen werden dann in die Röhrchen eingebracht und wie beschrieben behandelt. Die Eichkurve erhält man, indem man auf semilogarithmischem Papier auf der Ordinate die prozentuale Durchlässigkeit und auf der Abszisse die Konzentration in γ/ml einträgt. Man kann direkt die zugehörige Konzentration ablesen und erhält durch entsprechende Multiplikation mit dem Verdünnungsfaktor die Konzentration der Ausgangslösung.

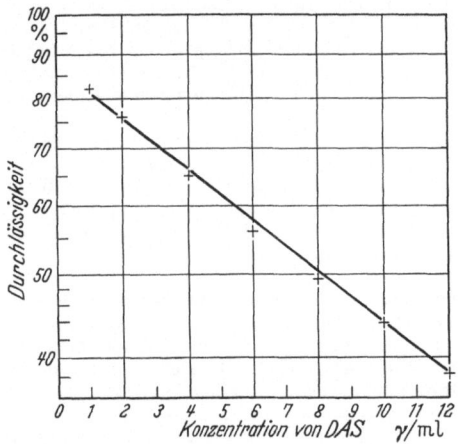

Abb. 7   Eichkurve

Bei der oft angewendeten Methode von EMMERIE und VAN EEKELEN (1934, 1936, 1937, 1938) zur Bestimmung von AS und DAS nebeneinander, siehe dieses Handbuch Bd. 2, S. 106, führt ein zu langes Stehen der Lösung (über Nacht) in H_2S-Atmosphäre zu beträchtlichen Verlusten. Die besten Ergebnisse wurden nach 2stündigem Stehenlassen erzielt; jedoch ist dann der Fehler noch größer als bei der Methode nach ROE und KUETHER (dieses Handbuch Bd. 7. S.230). Merkuriacetat verursacht einen kleineren Fehler als zu langes Stehenlassen unter Schwefelwasserstoff (MÜLLER, 1959).

DKS reduziert DIP in saurer Lösung nicht. Aus einer mit H_2S behandelten DKS-Lösung — die Reduktion erfolgt sehr langsam und nur unvollkommen — entstand bei der papierchromatographischen Untersuchung (Besprühen mit alkalischer 3,4-Dinitrobenzoesäurelösung[1]) neben dem DKS-Fleck, der einen Ansatz zu weiterer Auftrennung zeigte, in der Höhe

---

[1] Herstellung der 3,4-Dinitrobenzoesäurelösung: 1 g 3,4-Dinitrobenzoesäure wird in 100 ml 2 n-Sodalösung gelöst; die Flecke werden durch Erhitzen auf 105° C im Trockenschrank sichtbar gemacht.

der AS ein Fleck, der mit DIP und mit 3,4-Dinitrobenzoesäure reagierte. Es kann die Substanz vorliegen, von der Penney und Zilva (1945) feststellten, daß sie wohl reduziere, aber keine antiskorbutische Wirkung besäße. Es handelt sich nicht um Threonsäure, die mit 3,4-Dinitrobenzoesäure nicht reagiert (Müller, 1959). Ferner wurde eine weitere Substanz gefunden, die einen kleineren R_f-Wert besitzt und die auch reduziert. Hamburger und Joslyn (1941) empfehlen, H_2S nur 10 min lang einwirken zu lassen, da sich sonst reduzierende Verbindungen bilden, die mit DIP reagieren, aber nicht AS sind. Nach Ribeiro, Bonoldi und Ribeiro (1942) ist die H_2S-Reduktion von DAS bei 40° C in 30 min oder bei 50° C in 15min vollständig. Nach Krauze und Bozyk (1959) wird DAS bei p_H 4,7 oder 6,2 in 15 min vollständig durch H_2S reduziert. Es entstehen Nebenprodukte (Polysulfide), die die von ihnen angewendete polarographische Methode stören. Sie kommen zu dem Ergebnis, daß DAS in Gemüse und Obst erst bei der Untersuchung entsteht. Levenson, Rosen und Hitchings (1951) untersuchten die Kinetik der Reduktion der DAS mit Schwefelwasserstoff und fanden eine reversible (zu AS) und auch eine irreversible Reduktion zu einem unbekannten Stoff. Es besteht eine Abhängigkeit von der Temperatur und von der Konzentration des H_2S. Die irreversible ist bei p_H 3,5 nur gering. Mit diesen Angaben sollte gezeigt werden, daß sich DKS mit H_2S nur schwer reduzieren läßt (mit HJ läuft die Reduktion schneller ab) und daß die Reduktion von DAS nicht einheitlich verläuft.

Versuche mit Extrakten aus Blumenkohl, DKS zu reduzieren, waren bisher negativ (Janecke u. Müller, 1959).

3. Verbreitung der DAS-Reduktase.

Das Enzymsystem der DAS-Reduktase scheint nicht allgemein in allen Pflanzenfamilien vorzukommen. Es wurde bei den *Cruciferae, Solanaceae, Leguminosae, Cucurbitaceae, Umbelliferae* und *Liliaceae* mit guter Aktivität angetroffen, während es in *Rosaceae, Vitaceae* und *Rutaceae* zu fehlen scheint. In Citronen konnte nur eine geringe Aktivität festgestellt werden. Parrot und Dambrine (1956) fanden in diesen Früchten ein wesentlich wirksameres Enzymsystem. Es ist jedoch nicht auszuschließen, daß seine Aktivität oder gar seine Existenz von den Entwicklungs-, Wachstums- und Reifungsvorgängen abhängig ist. Hierfür spricht, daß es im jüngeren Gewebe des Spargels in aktiverer Form zu finden ist als im leicht verholzten Teil (Janecke u. Ragab, 1956, 1957). Diese Beobachtung stimmt auch mit den Befunden von Yamaguchi und Joslyn (1951) überein, die die höchste Aktivität in den keimenden Samen und in jungen Pflanzen fanden. Nach einer neueren Arbeit von Ragab (1962) ist das Vorkommen dieses Enzymsystems artbedingt. Er fand es bei der Untersuchung von 5 Citrusarten nur in dem Saft von Süßorangen und von Süßzitronen.

Literatur.

Åberg, B.: Physiol. Plant 2, 164 (1949). — Acker, R., and U. R. Laurilla: Bull Soc. Chem. Biol. 35, 413 (1953). — Ahlers, H., u. E. Müller: Naturwissenschaften 46, 75 (1959). — Ames, S. R., and C. R. Dawson: Ind. Eng. Chem. 17, 249 (1945). — Antener, J.: Helv. Chim. Acta 20, 742 (1937). — Armentano, L., u. H. A. Bartok: Biochem. Z. 311, 418 (1942). — Arnon, D. J.: In Symposium on Copper metabolism; S. 99. Herausgegeb. von M. D. McElroy u. B. Glass. Baltimore: John Hopkins Press 1950. — Ashwell, G., and J. Hickman: J. biol. Chem. 226, 65 (1957). — Ashwell, G., J. Kanfer and J. J. Burns: J. biol. Chem. 234, 472 (1959). — Askonas, B. A.: Biochem. J. 48, 42 (1951).

Babber, J.: J. Med. Research 38, 263 (1950). — Balakhovskic, S. D., N. N. Drozdowa u. V. N. Fedorova: Biokhimija 18, 112 (1953) (C.A. 1953, 8132). — Barakat, M. J., M. F. Abd El-Wahab u. M. M. El-Sadr: Anal. Chem. 27, 536; Z. analyt. Chemie 152, 227 (1955). — Barron, E. S. G.: Cold Spring Harbor Symposia Quant. Biol. 7, 145 (1939). — Barron, E. S. G., A. G. Barron u. F. Klemperer: J. biol. Chem. 116, 563 (1936). — Barron, E. S. G., R. H. De Meio u. F. Klemperer: J. biol. Chem. 112, 625 (1936). — Bentsath, A., St. Rusznak and A. Szent-Györgyi: Nature (Lond.) 139, 326 (1937). — Bergner, K. G.: Dtsch. Lebensm. Rundschau 43, 95 (1947). — Bersin, Th., H. Köster u. J. Jusatz: Hoppe-Seylers Z. physiol. Chem. 235, 12 (1935). — Bertrand, D.: Bull. soc. chim. biol. 27, 396 (1945). — Bezssonoff, N., u. H. Leroux: Z. Vitaminforsch. 14, 270 (1944). — Bhagvat, K., u. R, Hill: New Phytologist 50, 112 (1950). — Borsook, H., and C. E. P. Jeffries: Science 83.

398 (1936). — BOWELL, J. G., and G. C. WHITING: Ann. Bot. n. s. 2, 847 (1938). — BOYER, P. D.
J. H. SHAW u. P. H. PHILLIPS: J. biol. Chem. 143, 417 (1942). — BRAUNITZER, G.: Angew.
Chemie 69, 189 (1957). — BRENNER, M.: Chimia 7, 198 (1953). — BUBLITZ, C., A. P. GROLL-
MAN and A. L. LEHNINGER: Biochim. Biophys. Acta 27, 221 (1958). — BÜNNING, E., u. E.
GÄUMANN: Fortschr. Botan. 22, 229 (1960). — BURNS, J. J., and C. EVANS: J. biol. Chem.
223, 897 (1956). — BURNS, J. J., P. PEYSER and A. MOLTZ: Science 124, 1148 (1957).
    CADDEN, J. F., and L. V. DILL: J. biol. Chem. 143, 105 (1942). — CARPENTER, M. P.,
A. E. KITABCHI, P. B. McCAY and R. CAPUTTO: J. biol. Chem. 234, 2814 (1959). — CHATTER-
JEE, J. B., G. C. CHATTERJEE, N. C. GHOSH, J. J. GOSH and B. C. GUHA: Sci. and Culture
(Calcutta) 24, 534 (1959); — Naturwissenschaften 46, 475 (1959a). — CHATTERJEE, J. B.,
N. C. GHOSH, J. J. GHOSH, R. N. ROY and B. C. GUHA: Sci. and Culture (Calcutta) 23, 50
(1957). — CHATTERJEE, J. B., N. C. GHOSH, J. J. GHOSH and B. C. GUHA: Sci. and Culture
(Calcutta) 23, 382 (1958). — CHATTERJEE, J. B., N. C. KAR, N. C. GHOSH and B. C. GUHA:
Arch. Biochem. 86, 154 (1960). — CHATTOPADHYAY, H., and S. BANERJEE: Indian J. Med.
Research 40, 439 (1952) (C.A. 1953, 8193). — CHEN, Y-T.: Ph. D. Thesis. Cambridge University
(England) (1950). — CLAYSON, D. H. F., and J. A. BEESLEY: Chem. & Ind. (London) 191
(1951). — CLOWES, G. H. A., and B. TOLLENS: Ann. 310, 166 (1899). — CONN, E. E., and
B. VENNESLAND: Nature (Lond.) 167, 976 (1951). — COX, E. G., E. L. HIRST and R. J. W.
REYNOLDS: Nature (Lond.) 130, 88 (1932). — CROOK, E. M.: Biochem. J. 35, 226 (1941). —
CROOK, E. M., and F. G. HOPKINS: Biochem. J. 32, 1356 (1938). — CROOK, E. M.,and E. J.
MORGAN: Biochem. J. 38, 10 (1944).
    DALTON, F. R., and J. M. NELSON: J. Am. Chem. Soc. 60, 3085 (1938); 61, 2946 (1939). —
DAUBENMERKL, W. D.: Acta Pharmacol. 5, 270 (1949); — Acta Med. Scand. 138, Suppl.
Bd. 239, 1950; zit. in Z. analyt. Chemie 132, 469 (1951). — DAVISON, D. C.: Proc. Linnean
Soc. N. S. Wales 74, 37 (1949); zit. nach W. H. SEBRELL and R. S. HARRIS: The Vitamins. Vol. I.
New York: Academic Press 1954. — DAWSON, C. R.: In Symposium on Copper metabolism,
S. 18. 1950. Herausgegeb. von M. D. McELROY u. B. GLASS. Baltimore: J. Hopkins Press
1950. — DAWSON, C. R., and R. J. MAGEE: In S. P. COLOWICK u. N. O. KAPLAN Methods in
Enzymology Bd. II, S. 833. New York: Academic Press 1955. — DAWSON, C. R., and M. F.
MALETTE: Advances in Protein Chem. 2, 179 (1945). — DeCARO, L., u. M. GIANNI: Hoppe-
Seyler's Z. physiol. Chem. 228, 13 (1934); Boll. soc. ital. biol. sper. 13, 727 (1938). — DEKKER,
A. O., and R. G. DICKINSON: J. Am. Chem. Soc. 62, 2165 (1940). — DIEMAIR, W., u. K.
ZERBAN: Dtsch. Lebensm.-Rundschau 10, 60 (1942); — Biochem. Z. 316, 189, 335 (1944). —
DIEMAIR, W., u. H. HÄUSSER: Z. analyt. Chemie 122, 12,160 (1941). — DILLS, W. L., and J. M.
NELSON: J. Am. Chem. Soc. 64, 1616 (1942). — DHOPESHWARKA, G. A., and N. G. MAGAR:
J. Sci. Ind. Res. (India) 11A, 264 (1952); — C.A. 1953, 7694. — DOUNCE, A. L.: In J. B.
SUMNER and K. MYRBÄCK: The Enzymes, Vol. I, 1 S. 187. New York: Academic Press Inc.
1950. — DUNN, F. J., and C. R. DAWSON: J. biol. Chem. 189, 485 (1951).
    EBIHARA, T.: J. Biochem. (Tokio) 29, 199 (1939). — EDMAN, P.: Acta Chem. Scand. 4,
277, 283 (1950). — EIGER, I. J., and DAWSON C. R.: Arch. Biochem. 21, 181 (1949). — EISTERT,
B.: Angew. Chemie 66, 160 (1954). — EMMERIE, A., u. M. VAN EEKELEN: Biochem. Z. 28,
1151 (1934); 30, 25 (1936); — Z. Vitaminforsch. 6, 150 (1937); 7, 254 (1938). — EULER, H. v.,
u. B. EISTERT: Chemie und Biochemie der Reduktone und Reduktonate. S. 221. Stuttgart:
F. ENKE Verlag 1957. — EULER, H. v., K. MYRBÄCK u. H. LARSON: Hoppe-Seylers Z. physiol.
Chemie 245, 217 (1933).
    FELIX, K.: In Neuere Ergebnisse aus Chemie und Stoffwechsel der Kohlenhydrate, S. 56.
8. Colloquium der Ges. physiologische Chemie, 2./4. Mai 1957 Mosbach. Berlin, Göttingen,
Heidelberg: Springer Verlag 1958. — FINKLE, B. J., ST. KELLY u. F. A. LOEWUS: Biochim.
Biophys. Acta (Amsterdam) 38, 332 (1960). — FISCHER, E., u. R. S. CURTIUS: Ber. 25, 1025
(1892). — FISKE, C. H., u. Y. SUBBAROW: J. biol. Chem. 66, 375 (1925). — FRAENKEL-CONRAT,
H., J. J. HARRIS and A. L. LEVY: In Methods of Biochemical Analysis Bd. II, 359; herausgegeb. von D. GLICK. New York: Interscience Pub. 1955. — FRANKE, W.: Planta 41, 197
(1952); 44, 437 (1954); 45, 166 (1955); — Klin. Wochschr. 36, 17, 789 (1958). — FRIEDEN, E.:
Biochim. Biophys. Acta 9, 696 (1952). — FRIEDEN, E., and B. NAILE: Biochim. Biophys.
Acta 48, 448 (1954). — FUJITA, A., and T. SAKAMOTO: Biochem. Z. 297, 10 (1938).
    GALLI, A.: Ber. schweiz. Botan. Ges. 56, 113 (1948) (C.A. 1948, 2639d). — GÄUMANN, E.,
u. O. RENNER: Fortschr. Botan. 17, 586. Berlin, Göttingen, Heidelberg: Springer Verlag
1955. — GIRI, K. V., and P. SESHAGIRIRAO: Proc. Indian Acad. Sci. 24B, 264 (1946). — GOD-
DARD, D. R.: Am. J. Bot. 31, 270 (1944). — GONTARSKI, H.: Z. Naturforsch. 3b, 245 (1948). —
GONZALES, F. G.: Anal. Soc. espan. fis. quim. 32, 815 (1934). — GRAUBARD, M.: Enzymologia
5, 332 (1939). — GRAUBARD, M., u. J. M. NELSON: J. biol. Chem. 111, 757 (1935). — GREGG,
D. C., and J. M. NELSON: J. Am. Chem. Soc. 62, 2500 (1940). — GROLLMAN, A. P., u. A. L.
LEHNINGER: Arch. Biochem. 69, 458 (1957). — GÜNTHER, E.: Pharmazie 3, 158 (1948).
    HAAS, P., u. T. G. HILL: Ann. Bot. 39, 861 (1925). — HAMBURGER, J. J., and M. A.
JOSLYN: Food Research 6, 599 (1941). — HAND, D. B., and E. GREISEN: J. Am. Chem. Soc.

64, 358 (1942). — HANSL, N. R.: Österr. Akad. Wiss. Math.-naturw. Kl. Sitzber. Abt. I **164,** 25 (1955). — HARTZLER, E., and N. B. GUERRANT: Food Research **17,** 15 (1952). — HASSAN, M., u. A. L. LEHNINGER: J. biol. Chem. **223,** 123 (1956). — HERBERT, R. W., E. L. HIRST, G. V. PERCEIVAL, R. REYNOLDS and F. SMITH: J. Chem. Soc. **52,** 1270 (1933). — HERBERT, D., H. GORDON, V. SUBRAHMANYAN and D. E. GREEN: Biochem. J. **34,** 1108 (1940). — HERRMANN, J., u. M. ZOBEL: Z. Lebensm.-Unters. u. Forsch. **117,** 189 (1962). — HILL, R., and K. BHAGVAT: Nature (Lond.) **143,** 726 (1939). — HILLMANN, H.: Ber. dtsch. chem. Ges. **71,** 34 (1938). — HOCHBERG, M., D. MELNICK and B. L. OSER: J. Nutrition **30,** 225 (1945). — HOLLMANN, S.: Nicht)glykolytische Stoffwechselwege der Glucose. Stuttgart: Thieme 1961. — HONDA, S. G.: Plant Physiol. **30,** 174 (1955). — HOPKINS, F. G., and E. J. MORGAN: Nature (Lond.) **152,** 288 (1943); — Biochem. J. **30,** 1446 (1936). — HORECKER, B. L.: In W. D. MCELROY u. B. GLASS, Phosphorous metabolism, Vol. I. Baltimore: S. 117. J. Hopkins Press 1951; — Pentosephosphate und Heptulosephosphat im Kohlenhydratstoffwechsel in Neuere Ergebnisse aus Chemie und Stoffwechsel der Kohlenhydrate, 8. Colloquium Ges. physiol. Chemie 2./4. Mai 1957, Mosbach/Baden, S. 29. Berlin, Göttingen, Heidelberg: Springer Verlag 1958. — HOUGH, L., and J. K. JONES: Nature **167,** 180 (1951); — J. chem. Soc. **1952,** 4052. — HUDSON, C. S.: J. Am. Chem. Soc. **73,** 4498 (1951). — HUDSON, C. S., and S. KOMATSU: J. Am. Chem. Soc. **41,** 1141 (1919). — HUELIN, F. E.: Australian J. Sci. Research **2,** 346 (1949); — Z. analyt. Chemie **132,** 232 (1951) (Ref.). — HUZAK, S.: Hoppe-Seylers Z. physiol. Chemie **243,** 239 (1937).

INAGAKI, CH., H. FUKUBA and A. MATSUSHITA: Nat. Sci. Rept. Ochanomizu **5,** 92 (1955) (C.A. 1955, 9052); — Nat. Sci. Rept. Ochanomizu **4,** 235 (1954) (C.A. 1955, 5718); — Nat. Sci. Rept. Ochanomizu **5,** 313 (1955a) (C.A. 1956, 7171). — ISHERWOOD, F. A., Y-T. CHEN and L. W. MAPSON: Biochem. J. **56,** 1 (1954); **56,** 15 (1954a). — ISHERWOOD, F. A.: In Biochemical Symposia Nr. 17, herausgegeb. von E. M. CROOK. Cambridge: University Press 1959. — ISHIKAWA, S., and K. NOGUCHI: J. Biochem. (Tokio) **44,** 465 (1957). — IVANOVA, T. M.: Doklady Akad. Nauk S.S.S.R. **86,** 373 (1952). — IWANTSCHEFF, G.: Dithizon und seine Anwendung in der Mikro- und Spurenanalyse. Weinheim (Bergstraße): Verlag Chemie 1958.

Jackson, G. A., R. B. Wood u. M. V. Prosser: Nature (Lond.) **191,** 282 (1961). — JAGLE, M. F.: Bull. Soc. chim. biol. **21,** 41 (1939). — JAMES, W. O.: Proc. Roy. Soc. B **141,** 289 (1953); — Biol. Revs. Cambridge Phil. Soc. **28,** 245 (1953a). — JAMES, W. O., and D. BOULTER: New Physiologist **54,** 1 (1955). — JAMES, W. O., and J. M. CRAGG: New Physiologist **42,** 28 (1943). — JAMES, W. O., C. R. C. HEARD and G. M. JAMES: New Physiologists **43,** 62 (1944). — JANECKE, H., u. R. MÜLLER: nicht veröffentlicht. — JANECKE, H., u. M. H. H. RAGAB: Intern. Z. Vitaminforsch. **27,** 140, 203 (1956); — Z. Lebensm.-Untersuch. Forsch. **107,** 132 (1957) — Planta med. **5,** 116 (1957a). — JANECKE, H., u. D. OSTHOFF: nicht veröffentlicht (1962). — JOHNSON, S. W., and S. S. ZILVA: Biochem. J. **31,** 1366 (1937). — JONES, W. O., and N. GARTON: J. exptl. Botany **3,** 310 (1952). — JOSELOW, M., C. R. DAWSON and ST. LEWIS: J. biol. Chem. **191,** 1 (1951).

KANFER, J., J. J. BURNS and G. ASHWELL: Biochim. Biophys. Acta **31,** 556 (1959). — KANFER, J., G. ASHWELL u. J. J. BURNS: J. biol. Chem. **235,** 2518 (1960). KARRER, P., H. SALOMON, R. MORF u. K. SCHÖPP: Biochem. Z. **258,** 4 (1933). — KEILIN, D., and E. F. HARTREE: Proc. Roy. Soc. B **124,** 397 (1938). — KEILIN, D., and T. MANN: Proc. Roy. Soc. B **125,** 187 (1938); — Nature (Lond.) **143,** 23 (1939). — KENYON, J., and N. MUNRO: J. chem. Soc. (Lond.) **1948,** 158. — KEHRMANN, F.: Ber. dtsch. chem. Ges. **46,** 341 (1913). — KHORANA, H. G.: Quater. Rev. Chem. Soc. (Lond.) **6,** 340 (1952). — KIRSCHENBAUM, D. M.: Ph.D. Dissertation, Columbia University (Doctoral Diss. Ser. Publ. Nr. 17061, (1956). — KLEVSTRAND, R., u. A. NORDAL: Acta Chem. Scand. **4,** 1320 (1950). — KNOBLOCH, H.: Pharmazie **3,** 70 (1948). — KOCH, J., u. G. BRETTHAUER: Landwirt. Forsch. **9,** 51 (1956). — KOHMANN, E. F., u. N. H. SANBORN: Ind. Eng. Chem. **29,** 189, 1195 (1937). — KRAUZE, S., u. Z. BOZYK: Mitt. Gebiete Lebensm. u. Hyg. **50,** 228 (1959). — KUBOWITZ, F.: Biochem. Z. **292,** 221 (1937).

LABARRE, J., and S. PFEFFER: Rev. Can. biol. **5,** 233 (1946). — LARDY, H. A.: Respiratory Enzymes. Minneapolis: S. 15 Burgess Publ. Comp. 1950. — LESNE, E., M. POLONOWSKI and S. BRISKAS: Compt. rend. **217,** 406 (1943). — LEVENE, A., u. G. MEYER: J. biol. Chem. **46,** 307 (1921). — LEVENSON, ST. M., H. ROSEN and G. H. HITCHINGS: Arch. Biochem. Biophys. **33,** 50 (1951). — LEVY, H., and A. SCHADE: Arch. Biochem. Biophys. **19,** 273 (1948). — LINDOW, C. W., C. A. ELVEHJEM u. W. H. PETERSEN: J. biol. Chem. **82,** 465 (1929). — LOEWUS, F. A., B. J. FINKLE and R. JANG: Biochem. Biophys. Acta **30,** 629 (1958). — LOEWUS, F. A., and R. JANG: Biochem. Biophys. Acta **23,** 205 (1957); — J. biol. Chem. **232,** 505 (1958). — LOEWUS, F. A., R. JANG, u. C. G. SEEGEMILLER: J. biol. Chem. **222,** 649 (1956); **232,** 521, 541 (1958). — LOEWUS, F. A., and ST. KELLY: Biochem. Biophys. Res. Commun. **1,** 143 (1959). — LOVETT-JANISON, P. L., and J. M. NELSON: J. Am. Chem. Soc. **62,** 1409 (1940). — LUDWIG, B. J., and J. M. NELSON: J. Am. Chem. Soc. **61,** 2601 (1939). — LUND, C. C., L. A. SHAW and C. DRINKER: J. exptl. Med. **33,** 231 (1921).

MALETTE, M. F., S. R. AMES, J. M. NELSON and C. R. DAWSON: Arch. Biochem. 16, 283 (1948). — MALETTE, M. F., and C. R. DAWSON: J. Am. Chem. Soc. 69, 466 (1947); — Arch. Biochem. 23, 29 (1949). — MALETTE, M. F., S. LEWIS, S. R. AMES, J. M. NELSON and C. R. DAWSON: Arch. Biochem. 16, 283 (1948). — MANDELS, G. R.: Am. J. Botany 38, 213 (1951); — Exptl. Cell Res. 5, 48 (1953) (C 1954, 7670); — Arch. Biochem. Biophys. 42, 164, 362 (1953). — MAPSON, L. W.: Biochem. J. 39, 228 (1945); — Vitamins and Hormones XIII, 701 (1955); — in W. H. SEBRELL and R. S. HARRIS: The Vitamins, Vol. I, New York: Academic Press 1954; — Glutathione, Biochem. Symposium Nr. 17, Cambridge: University Press 1959; — Proc. of the Fourth Congr. of Biochemistry, Vienna 1958, Symposium XI, S. 1, Vitamin Metabolism. London, New York, Paris, Los Angeles: Pergamon Press 1960. — MAPSON, L. W., and J. BAKER: zit. in H. SEBRELL and R. S. HARRIS: The Vitamins, Vol. I. New York: Academic Press 1954. — MAPSON, L. W., F. A. ISHERWOOD and Y-T. CHEN: Biochem. J. 56, 21 (1954). — MAPSON, L. W., and E. BRESLOW: Biochem. J. 68, 395 (1958). — MAPSON, L. W., E. M. CRUICKSHANK and Y-T. CHEN: Biochem. J. 45, 171 (1949). — MAPSON, L. W., and D. R. GODDARD: Nature (Lond.) 167, 975 (1951); — Biochem. J. 49, 592 (1951 a). — MAPSON, L. W., and M. INGRAMM: Biochem. J. 48, 551 (1951); 29, 108 (1951 a). — MAPSON, L. W., and F. A. ISHERWOOD: Biochem. J. 64, 13 (1958). — MAPSON, L. W., and E. W. MOUSTAFA: Biochem. J. 62, 248 (1956). — MARSH, P. B., and D. R. GODDARD: Am. J. Botany 26, 724 (1939). — MATTHEWS, B.: J. biol. Chem. 189, 695 (1951). — MATUKAWA, D.: J. Biochem. (Japan) 32, 265 (1940). — MAWSON, C. A.: Biochem. J. 29, 569 (1935). — MEIKLEJOHN, G. T., and C. P. STEWART: Biochem. J. 35, 755 (1941). — METZNER, H.: Die Askorbinsäure in der Pflanzenzelle, in Protoplasma. Handbuch der Protoplasmaforschung Bd. II, B 2 b α. Wien: Springer Verlag 1957. — MEYERHOFF, O., K. LOHMANN u. P. SCHUSTER: Biochem. 286, 301 (1936). — MICHELAZZI, L.: Boll. soc. ital. biol. sper. 24, 1283 (1948). — MILLER, W. H., M. F. MALETTE, L. J. ROTH and C. R. DAWSON: J. Am. Chem. 66, 514 (1944). — MILLS, M. B., u. J. H. ROE: J. biol. Chem. 170, 159 (1947). — MIRIMANOFF, A.: J. suisse pharm. 94, 401 (1956). — MOLL, TH., u. H. WIETERS: Mercks Jahresber. 1935, 93; 1936, 65. — MONZINI, B.: Chimica (Milano) 9, 55 (1954). — MORRIS, H. J., C. A. WEAST and H. LINEWEAVER: Bot. Gaz. 107, 362 (1946). — MÜLLER, R.: Diss. Universität, Frankfurt a. M. 1959.

NAKATONI, M.: Bull. agric. chem. Soc. Japan 22, 261 (1958). — NATH, M. C., and M. L. BELKHODE: Proc. Soc. exp. Biol. (N. Y.) 99, 544 (1958); — Nature (Lond.) 183, 1258 (1959). — NATH, M. C., and E. P. M. BHAKATHIRY: Nature (Lond.) 178, 1233 (1956). — NATH, M. C., C. H. CHATHRABORTY, V. G. HATWALNE and K. S. LAUE: Nature (Lond.) 162, 660 (1948). — NATH, M. C., R. P. CHITALE and B. BELAVADY: Nature (Lond.) 170, 545 (1952). — NOMURA, D., and M. MATSUNAGA: Fermentation Technol. (Japan) 31, 174 (1953); C. A. 1953, 7694.

OETTINGEN, W. F. VON: Physiol. Revs. 15, 175 (1935).

PALLADIN, W.: Biochem. Z. 27, 442 (1910); C 1910, 4, 1308. — PARKINSON jr., G. G., and J. M. NELSON: J. Am. Chem. Soc. 62, 1693 (1940). — PARROT, J. L., et M. DAMBRINA: Bull. soc. chim. biol. (Paris) 38, 1355 (1956). — PECHERER, B.: J. Am. Chem. Soc. 73, 3827 (1951). — PENNY, J. R., and S. ZILVA: Biochem. J. 37, 39 (1943); 39, 1 (1945). — PETERSON, E. A., and H. A. SOBER: J. Am. Chem. Soc. 78, 751 (1956). — PETERSEN, R. W., and J. H. WALTER: J. Am. Chem. Soc. 65, 1212 (1943). — PETT, L. B.: Biochem. J. 30, 1228 (1936). — PETUELY, F., u. H. F. BAUER: Monatsh. Chem. 83, 758 (1952). — PFANKUCH, E.: Naturwissenschaften 22, 821 (1934); — Biochem. Z. 279, 115 (1934). — POHLOUDEK-FABINI, R., u. W. FÜRTIG: Arch. Pharm. 292/64, 350 (1959). — PONTING, J. D., and M. A. JOSLYN: Arch. Biochem. 19, 47 (1948). — PORTER, R. R.: Methods in Med. Research 3, 256 (1950). — POTTERMAN, L. E.: Acta Med. Scand. 134, Suppl. 134, 230 (1949). — POWERS, W. H., and C. R. DAWSON: J. Gen. Physiol. 27, 181 (1944). — POWERS, W. H., S. LEWIS and C. R. DAWSON: J. Gen. Physiol. 27, 167 (1944). — PRAGER, B., P. JACOBSON u. F. RICHTER: Beilsteins Handbuch der org. Chemie, Deutsche Chem. Ges. Berlin, 4. Aufl. 18. Bd. 203 (1953).

RAGAB, M. H.H.: Dissertation Universität, Frankfurt a. M. 1956; — Z. Lebensm. Untersuch. u. Forsch. 116, 397 (1962). — RAMASARMA, G. B., N. C. DATTA and N. S. DOKTOR: Enzymologia 8, 108 (1940). — RANGUEKAR, Y. B., S. S. DE and V. SUBRAHMANYAN: Indian J. Med. Res. 36, 361 (1948); (C.A. 1949, 8452). — RAO, P/ S., R. M. BERI and P. R. RAO: Proc. Ind. Acad. Sci. 34 A, 236 (1951). — RAY, S. N.: Biochem. J. 28, 996 (1934). — REDFIELD, A. C.: In Symposium on Copper Metabolism, S. 178; herausgegeb. W. D. McELROY u. B. GLASS. Baltimore: J. Hopkins Press 1950. — RIBEIRO, R. F., V. BONOLDI e O. F. RIBEIRO: Rev. fac. med. vet. Univ. São Paulo 2, 29 (1942) (C.A. 1944, 4714). — ROBERTS, E. A. H.: Biochem. J. 33, 836 (1939). — ROBINSON, G. M., and R. ROBINSON: J. Chem. Soc. 1935, 747. — ROE, J. H.: In Methods of Biochemical Analysis, Vol. I S. 115. New York: Interscience Publ. Inc. 1954. — ROE, J. H., u. C. H. KUETHER: J. biol. Chem. 165, 377 (1946) — ROE, J. H., M. B. MILLS, M. J. OESTERLING u. C. M. DAMRON: J. biol. Chem. 174, 201 (1948). — ROE, J. H., u. M. J. OESTERLING: J. biol. Chem. 152, 511 (1944). — ROVERY, M., and P. DESNUELLE: Bull. Soc. Chem. Biol. 36, 107 (1954). — ROY, R. N., and B. C. GUHA: Nature (Lond.) 182, 319 (1958). — ROY, S. C., S. K. ROY and B. C.

GUHA: Nature (Lond.) 158, 238 (1946). — RUBIN, B. A., and E. P. CHETVERIKOVA: Isvest. Akad. Nauk S.S.S.R. Ser. Biol. 1951, 120; — Biokhim. Plodov i Ovoshchei, Akad. Nauk S.S.S.R. Inst. Biokhim. Sbornik 3, 43 (1955) (C.A. 1956, 2754). — RUDRA, M. N.: Nature (Lond.) 144, 868 (1939). — RUSSEL, R. S.: Symp. Soc. Exptl. Biol. 8, 343 (1954).

SANGER, F.: Biochem. J. 39, 507 (1945); — Advances of Protein Chem. 7, 2 (1952). — SCHEUNERT, A., u. E. THEILE: Pharmazie 7, 776 (1952). — SCHMIDT-NIELSEN, S.: Kgl. Norske Videnskap. Selskap Forh. 15, 49 (1942). — SCHMIDT, O. T., u. R. TREIBER: Ber. 66, 1765 (1933). — SCHRAMM, G., u. G. BRAUNITZER: Z. Naturforsch. 8b, 65 (1953). — SCHULTE, K. E., u. A. SCHILLINGER: Z. Lebensm. Untersuch. u. Forsch. 94, 309 (1952). — SCHULZE, H. V., and A. F. MORGAN: J. Diseases Children 71, 593 (1946). — SCHWAIBOLD, J., B. BLEYER u. G. NAGEL: Biochem. Z. 297, 324 (1938). — SEBRELL, W. H. jr., and R. S. HARRIS: The Vitamins, Vol. I. New York: Academic Press 1954. — SEEGEMILLER, C. G., B. AXELROD and R. M. McCREADY: J. biol. Chem. 217, 765 (1955). — SILVERBLATT, E., A. L. ROBINSON and C. G. KING: J. Am. Chem. Soc. 65, 137 (1943). — SISAKYAN, N. M., and N. A. VASILEVA: Biokhimia 19, 730 (1954). — SKINNER, J. T., and J. S. McHARGUE: Am. J. Physiol. 145, 566 (1946). — SMITH, F. G.: Plant Physiol. 27, 736 (1952). — SMYTHE, C. V., u. C. G. KING: J. biol. Chem. 142, 529 (1942). — SNOW, G. A., and S. S. ZILVA: Biochem. J. 32, 1926 (1938). — SOBOTKA, H., and C. P. STEWART: Adv. in Clin. Chem. 4, 134 (1961) New York-London: Academic Press. SOMOGYI, J. C.: Z. Vitaminforsch. 16, 134 (1945). — SPRUYT, J. P., u. G. M. VOGELSANG: Arch. neerl. Physiol. 23, 424 (1938). — SRINIVASAN, M.: Biochem. J. 30, 2077 (1936); — Current 4, 407 (1935). — SRINIVASAN, A., and S. D. WANDREKAR: Nature (Lond.) 165, 765 (1950); — Proc. Indian Acad. Sci. 32 B, 321 (1950a). — STEINMANN, H., and C. R. DAWSON: J. Am. Chem. Soc. 64, 1212 (1942). — STONE, W.: Biochem. J. 31, 508 (1937). — STOTZ, E.: J. biol. Chem. 133 C (1944). — STOTZ, E., C. J. HARRIS and C. G. KING: J. biol. Chem. 119, 511 (1937). — STOTZ, E., C. J. HARRIS, M. O. SCHULTE and C. G. KING: J. biol. Chem. 124, 745 (1938). — STROHECKER, R., u. F. MATT: Z. analyt. Chemie 133, 343 (1951). — STROHECKER, R., W. HEIMANN u. F. MATT: Z. anal. Chem. 145, 401 (1955). — STUMPF, P. K.: J. biol. Chem. 176, 233 (1948). — SUMNER, J. B., and K. MYRBÄCK: The Enzymes, Vol. II, 1. New York: Academic Press Publ. 1951. — SZENT-GYÖRGYI, A.: Biochem. J. 22, 1387 (1928); — Science 72, 125 (1930); — J. biol. Chem. 90, 385 (1931); — On Oxidation, Fermentation, Vitamins, Health and Disease. Baltimore: Williams & Wilkins Co 1939.

TADOKARA, T., and N. TAKASUGI J. Chem. Soc. (Japan) 60, 188, 929, 1112, 1255 (1941); 61, 234, 356, 495, 703 (1940). — TADOKARA, T., N. TAKASUGI and T. SAITO: J. Chem. Soc. (Japan) 62, 119, 271, 419, 757 (1941) (C.A. 36,6556). — TAPPEL, A. L., and A. G. MARR: J. Agr. Food Chem. 2, 554 (1954). — TAUBER, H.: Ergebn. Enzymforsch. 7, 305 (1938). — TAUBER, H., and I. S. KLEINER: Proc. Soc. Exptl. Biol. Med. 32, 577 (1935); — J. biol. Chem. 110, 559 (1935). — THOMPSON, J. B.: In Copper Metabolism. Symposium, S. 141. Baltimore: J. Hopkins Press 1950. — TOMITA, K.: J. Fermentation Technol. (Japan) 28, 432, 437, 440 (1950). — TRAVIA, C. L., e U. SANTAGATI: Acta Vitaminol. 6, 97 (1952).

ULMANN, M.: Pharmazie 9, 523 (1954).

VALLE, T.: Ann. sper. agr. (Rom) 5, 221 (1951).

WACHOLDER, K.: Biochem. Z. 312, 394 (1942). — WACHOLDER, K., u. P. HAMEL: Klin. Wschr. 16, 10 (1937). — WAHBA, G., J. W. HICKMAN and G. ASHWELL: J. Am. Chem. Soc. 80, 2594 (1958). — WARBURG, O.: Biochem. Z. 187, 256 (1927); — Schwermetalle als Wirkungsgruppen von Enzymen. Berlin: Dr. W. Saenger 1946. — WARBURG, O., u. H. A. KREBS: Biochem. Z. 190, 143 (1927). — WARD, J. M.: Diss. Abstr. 14, 754 (1954). — WAWRA, C. Z., and J. L. WELB: Science 96, 302 (1942). — WAYGOOD, E. R.: Canad. J. Res. Section C 28, 7 (1950). — WEISSBERGER, A., and J. E. LU VALLE: J. Am. Chem. Soc. 66, 700 (1944). — WEISSBERGER, A., J. E. LU VALLE and D. S. THOMAS: J. Am. Chem. Soc. 65, 1934 (1944). — WEST, E. S.: J. biol. Chem. 74, 561 (1927). — WILLIAMS, R. J. P.: Biol. Rev. 28, 398 (1953). — WILLSTÄTTER, R.: Untersuchungen über Enzyme. S. 575. Berlin: Springer 1928. — WINKELMAN, J., and A. L. LEHNINGER: J. biol. Chem. 233, 794 (1958). — WOLFRAM, M. L., and K. ANNO: J. Am. Chem. Soc. 74, 5583 (1952). — WORKER, G., u. J. ANTENER: Helv. Chim. Acta 20, 144, 732 (1937).

YAMADA, K.: J. Biochem. (Japan) 46, 361, 529 (1959). — YAMADA, K., K. SUZUKI, Y. MANO u. N. SHIMAZONO: J. Biochem. (Japan) 50, 374 (1961). — YAMAGUCHI, M., and M. A. JOSLYN: Plant Physiol. 26, 757 (1951).

ZILVA, S. S.: Biochem. J. 28, 663 (1934). — ZILVA, S. S., F. KIDD and C. WEST: New Phytologist 37, 345 (1938).

Enzymes Involved in the Synthesis and Breakdown of Indoleacetic Acid.

By

S. Mahadevan.

With 1 Figure.

I. General Methods.

Detailed methods for the identification, purification, chromatography, bioassay and colorimetric methods of estimation of indoleacetic acid (IAA) have been dealt with by LARSEN (1955) in Modern Methods of Plant Analysis, Vol. III, to which the reader is referred.

Bioassay of IAA. The *Avena* curvature test has been generally used for the estimation of the minute quantities of IAA (usually fractions of a microgram), formed from such precursors as L-tryptophan (TTP), 3-indoleacetaldehyde (IAc), and tryptamine (TNH$_2$) upon incubation with plant extracts or crude enzyme preparations. This test, though most sensitive for IAA, is not specific for it, and two other indoles, IAc and IAN, are active in the test (LARSEN, 1949, 1955; BENTLEY and BICKLE. 1952), probably by being converted to IAA therein. TTP and TNH$_2$ can also give some delayed curvature though at high concentrations (SKOOG, 1937). IAA has therefore to be separated from other compounds in the reaction mixture prior to its estimation (see below); a precaution not taken in some earlier investigations. The drawbacks of bioassay methods are that they are time consuming, and that their day to day variation is troublesome for routine enzymological work.

Colorimetric methods of assay. The colorimetric estimation of IAA with the modified SALKOWSKI reagent require about 2 micrograms or more of IAA for a reliable quantitative estimation. The principle consists of the colorimetric or spectrophotometric estimation of a pink colour, which is proportional to IAA concentration within limits, developed by the action of a solution of ferric chloride and a strong mineral acid as H$_2$SO$_4$ or HClO$_4$ on IAA.

TANG and BONNER Method (1947).

Reagent. A mixture of 15 ml of 0.5 M FeCl$_3$, 500 ml distilled water, and 300 ml of H$_2$SO$_4$ sp. gr. 1.84. 8 ml of this reagent was added to 2 ml of aliquot containing 5—70 μgm. IAA, and the colour read after 30 mins. in a colorimeter with a green filter (No.54). GORDON and WEBER (1951) recommended a 1.5:1 ratio of the reagent to aliquot, with colour measurement after 40 mins.

Specificity. No appreciable colour is obtained with TTP, indolecarboxylic acid (ICA) and indole. IAM produced the same colour as IAA.

Abbreviations used: IAA = 3-indoleacetic acid; IAc = 3-indoleacetaldehyde; IAld = 3-indolealdehyde; IAM = 3-indoleacetamide; IAN = 3-indoleacetonitrile; IBA = 3-indolebutyric acid; ICA = 3-indolecarboxylic acid; IPA = 3-indolepropionic acid; IPyA = 3-indolepyruvic acid; NAA = α-naphthaleneacetic acid; NAc = α-naphthaleneacetaldehyde; TNH$_2$ = tryptamine; TTP = L-tryptophan; DCP = 2,4-dichlorophenol; MH = maleic hydrazide.

GORDON and WEBER Method (1951).

Reagent. 1 ml of 0.5 M FeCl₃ in 50 ml of 35% HClO₄.

2 ml of the reagent was added to 1 ml of the aqueous aliquot containing 0.2—45 μgm IAA and the colour read after 1 hr. at 530 mμ in a spectrophotometer or with a green filter in a colorimeter (No. 54). The colour is stable for 3 hrs.

GORDON and PALEG (1957) recommend the use of an ether aliquot of IAA with the above reagent, which yields a more intense colour in 35 mins. than the aqueous solution, especially if heated for 30 seconds in a boiling water bath, after colour development. A methanolic aliquot of IAA, though less sensitive than the aqueous solution, was useful for the direct assay of eluates of chromatograms with this solvent (see below for the procedure).

Sources of error. Bright light bleaches the pink colour developed with IAA, and the colour development should be done in darkness or dim light. The aliquot to be tested should be free of reducing agents as ascorbic acid or cysteine and poly-phenols e.g. catechol, as all these reduce the final colour intensity produced (PLATT and THIMANN, 1956; GORDON and PALEG, 1957). Some phenolic compounds themselves give colours with the reagent (STOWE, THIMANN and KEFFORD, 1955). Storage of the reagent in darkness or in coloured bottles (PLATT and THIMANN, 1956) and colour development at a constant temperature, to minimise fluctuations in sensitivity, are recommended (GOLDACRE, 1951).

Separations of IAA from its Precursors in the Reaction Mixture.

Solvent partitioning and chromatographic methods have been used for the separation of IAA from the other compounds in the reaction mixture prior to its biological or colorimetric assay (cf. LARSEN, 1955 for partitioning methods).

Partitioning methods. a) The reaction mixture, after acidification to pH 2.8—3 with N HCl or N H₃PO₄ is extracted thrice with half to equal volumes of ether[1]. The ether fraction (20—30 ml) containing acidic (IAA) and neutral compounds is extracted twice with 5 ml volumes of 1% NaHCO₃. The aqueous phase containing IAA is adjusted to pH 3 and IAA re-extracted into ether as above. The method separates IAA from TTP, IAc, TNH₂ and IAN. IPyA is unstable under alkaline conditions. The method is unreliable for quantities of a μgm. or less.

b) IAc may be separated from IAA by adjusting the reaction mixture to pH 8—9 with NaHCO₃ and extracting with ether as above to remove IAc. IAA is then extracted from the aqueous phase after acidification to pH 3 (GORDON and NIEVA, 1949; LARSEN, 1949). The ether extracts are evaporated and the IAA in the acidic fraction estimated by the *Avena* test or colorimetrically.

Sources of error. Acid fractions of crude plant extracts obtained by partitioning methods often contain compounds besides IAA which give chromophores like IAA with the SALKOWSKI reagent (GORDON and PALEG, 1957) and erroneous results are obtained especially when dealing with one to a few micrograms of IAA. Chromatographic resolution of this fraction is essential.

Chromatographic method (GORDON and PALEG, 1957). Ether extracts of the acidified (pH 3) reaction mixture are evaporated and the residue taken up in a known volume (about 50 μl.) of purified methanol[2]. Two 25 μl. aliquots are spotted 1 inch apart on Whatman No. 1 paper (2 × 14 inches) under a stream of nitrogen and chromatographed in isopropanol, ammonia 28%, water (10 : 1 : 1) or preferably n-butanol, ammonia, water (10 : 1 : 1). The IAA spot on one half of the chromatogram (cut lengthwise) is located by spraying with a dilute SALKOWSKI reagent and the corresponding area of the unsprayed half is cut out and the IAA

---

[1] Ether should be made peroxide free by freshly distilling over CaO and FeSO₄ according to the method of GARBARINI (1909) (cf. LARSEN, 1955).

[2] Methanol should be purified by distillation from powdered Zn and KOH (GORDON and PALEG, 1957).

eluted, within 1—2 hrs., by 3 five-minute rinses with 2 ml. portions of methanol. The methanol extract is evaporated on a water bath and the residue taken up in 1 ml of ether and the IAA estimated as given before, with the GORDON and WEBER reagent. Aliquots of the methanolic extract may also be directly estimated for IAA.

About 85 % of the IAA (10 μgm) added. to the plant extract containing 42 mg. solid matter may be recovered by this procedure. The method also resolves catechol and IAA.

II. Enzymes Involved in the Synthesis of IAA.

The enzymatic steps involved in the biosynthesis of IAA in plants are still not known (cf. GORDON, 1956) owing to the very low capacity of healthy tissues to synthesise this biologically active compound. However IAA is formed from several naturally occurring precursors by the action of various plant extracts and purified enzymes. These reactions, being heat inactivable, are regarded as enzymatic.

Of the potential precursors of IAA, TTP, IAN and TNH$_2$ have been isolated and IAc convincingly shown to exist in plants (cf. STOWE, 1959). The existence of 3-indolepyruvic acid (IPyA) in plants has not been conclusively settled, but it almost certainly serves as an intermediate in the conversion of TTP to IAA (STOWE, 1955; KAPER and VELDSTRA, 1958); it may well occur in plant material (STOWE and THIMANN, 1954; VLITOS and MEUDT, 1954). The occurrence and distribution of these precursors in plants has been reviewed by BENTLEY (1958) and STOWE (1959).

Indoleacetamide (IAM), esters of IAA (SEELEY, FAWCETT, WAIN and WIGHTMAN, 1956) and higher homologues of IAA such as indolebutyric acid (IBA) (FAWCETT, WAIN and WIGHTMAN, 1958) are converted to IAA in vivo, by certain plants, but these compounds have not been demonstrated in plants (STOWE, 1959).

1. Enzymes Involved in the Conversion of TTP to IAA.

Enzymatic conversion of TTP to an auxin, active in the *Avena* test, was first shown with a spinach leaf enzyme (WILDMAN, FERRI and BONNER, 1947), and has since been demonstrated with several plant extracts and in a few cases the product identified as IAA by paper chromatography and colour reactions. In general, the conversions have been only a small fraction of one percent of the added TTP.

a) Distribution.

Higher plants. Conversion of TTP to IAA has been shown by extracts of the following plant tissues: (i) by chromatography, colour reactions, bioassay and/or colorimetric assay — maize embryo (YAMAKI and NAKAMURA, 1952), mung seedlings (GORDON, 1956, 1958), rice kernel sap (MURAKAMI and HAYASHI, 1957); (ii) by colorimetry or bioassay — sunflower crown gall tumour, callus and normal stem tissues (HENDERSON and BONNER, 1952); (iii) by bioassay alone — *Avena* coleoptiles (WILDMAN and BONNER, 1948), pineapple leaf base (GORDON and NIEVA, 1949), tobacco ovaries (WILDMAN and MUIR, 1949), tobacco pollen tubes, style and pollinated ovaries (LUND, 1956).

Water-melon tissue slices and *Avena* coleoptiles fed with TTP-2-C^{14} (side chain carbon labelled) gave rise to numerous labelled indole compounds including IAA (DANNENBURG and LIVERMAN, 1957; LIVERMAN and DANNENBURG, 1957). Similarly cabbage tissues fed with TTP-β-C^{14} produced IAA, IAN, indolealdehyde (IAld) and indolecarboxylic acid (ICA) (WIGHTMAN and NEISH, 1960) [cf. GORDON (1954) for the list of other tissues converting TTP to IAA].

Lower plants. Several fungi produce IAA in the culture medium either only, or to an increased extent in the presence of TTP in the free or bound form. A definite identification of the product as IAA has been made in the following cases by isolation or chromatography and colour reactions: *Rhizopus suinus* (THIMANN, 1935); *Ustilago zeae* and *Gymnosporangium juniperivirginianae* (WOLF, 1952, 1956). In addition several species of *Taphrina*, *Ustilago* and *Exobasidium* and *Fusarium oxysporum* f. *vasinfectum* (HIRATA, 1957; 1958) probably produce IAA from TTP. No enzymatic work has been done with any of the fungi (see GRUEN [1959] for an extensive list). *Phycomyces* similarly forms IAA within its mycelium when TTP is in the medium (GRUEN, 1956).

Several bacteria have been shown to produce IAA when TTP was present in the medium and a partial list is given: *Escherichia coli* (HOPKINS and COLE, 1903; STOWE, 1955), *Agrobacterium tumefaciens* (KAPER and VELDSTRA, 1958; STOWE, 1955); *Pseudomonas fluorescens* (STOWE, 1955); several species of *Rhizobium* including *R. trifolii* (KEFFORD, BROCKWELL and ZWAR, 1960); *R. melioti*, *R. leguminosarum* and *Bact. radiobacter* (THIMANN, 1939; GEORGI and BEGUIN, 1939). [cf. FRIEBER (1921) and BOYSEN-JENSEN (1931) for the conversion of TTP to IAA by other bacteria].

Cell free preparations of *E. coli*, *A. tumefaciens* and *Ps. fluorescens* transaminate TTP with keto-acids in the presence of pyridoxal phosphate with the ultimate production of IAA (STOWE, 1955).

b) Pathways for the Conversion of TTP to IAA.

IPyA pathway. TTP may be either oxidatively deaminated (1) (THIMANN, 1935; WILDMAN et al., 1947) or transaminated (STOWE, 1955; MURAKAMI and HAYASHI, 1957) to IPyA, and the latter oxidatively decarboxylated (IV) to IAA (WILDMAN et al., 1947) or decarboxylated (II) to IAc (GORDON and NIEVA, 1949). IAc may be further oxidised (or dehydrogenated) to IAA (LARSEN, 1949; GORDON and NIEVA, 1949) or two moles of IAc dismuted to IAA and tryptophol (LARSEN, 1949).

TNH$_2$ pathway. TTP may be decarboxylated (V) to TNH$_2$ and the latter further oxidised (VI) by an amine oxidase to IAc. IAc undergoes subsequent conversion as above (SKOOG, 1937; GORDON and NIEVA, 1949).

Decarboxylation of TTP to TNH$_2$ has however not been demonstrated in plants or microorganisms (cf. STOWE, 1959). Recently WEISSBACH et al. (1959) have shown the production of TNH$_2$ from TTP by intestinal bacteria.

c) Conversion of IPyA to IAc and IAA.

Owing to its spontaneous decomposition, under neutral and alkaline aqueous conditions, to IAA, IAld, and several other indolic compounds (KAPER and

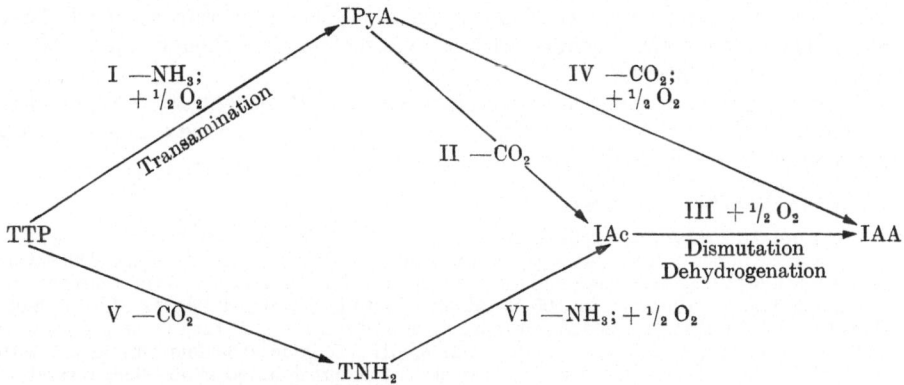

Fig. 1. Postulated pathways for the conversion of TTP to IAA.

VELDSTRA, 1958; SCHWARZ and BITANCOURT, 1957), it has so far not been possible to demonstrate unequivocally any enzyme mediated conversion of IPyA to either IAc or IAA (reactions II and IV in Fig. 1) (WILDMAN et al., 1947; GORDON and NIEVA, 1949; STOWE, 1955, 1959).

d) Preparation and Properties of Some Enzymes Converting TTP to IAA.

i) Mung seedling enzyme (GORDON, 1956a; 1958).

Preparation: Primary leaves and buds of light grown, 7—9 day old mung seedlings *(Phaseolus aureus)* were ground with sand in equal volume of cold 0.2 M phosphate buffer, pH 7.0 (with 0.3 M sucrose in experiments where intracellular organelles were separated) and centrifuged at $100 \times g$ for 10 minutes to remove cell debris.

3 ml aliquots of the brei and 3 ml of 0.85% TTP in 0.2 M phosphate buffer, final pH 7.4 were shaken in the dark at 30° C for 120 mins., when IAA production was linear with time. The reaction was stopped by boiling for 1 minute. Under these conditions, 3.3—4.7 μgm. of IAA were produced (colorimetric estimation) and the specific activity was 9—13 mμgm. IAA per min. per mgm. nitrogen.

Assay methods. The products were chromatographically resolved and IAA assayed either colorimetrically or by *Avena* test as given in "general methods" p. 238. IAA values obtained by the *Avena* test were 30—100% higher than those obtained by colorimetry.

Properties of the enzyme. The enzyme was "soluble", the activity not being sedimented by centrifugation at $105{,}000 \times g$ for 30 mins. On further centrifugation the activity moved as a heterogeneous fraction with a sedimentation constant (S'_{20}) of 0—4 (GORDON, 1958).

Amine oxidase inhibitors such as $\alpha\alpha$-dipyridyl at 10^{-3} M, and marsalid at 10^{-2} M, did not affect the activity, nor was TNH$_2$ oxidised by the enzyme to IAA. The conversion of TTP to IAA was therefore probably via the IPyA pathway.

ii) Sunflower tumour tissue enzyme (HENDERSON and BONNER, 1952).

Preparation. Crown gall tissue (primary strain, P-III) cultured on GAUTHERET's medium with 2% sucrose, mineral salts and 1% agar, was rapidly frozen after harvest, lyophilised and the powder suspended in buffer and used as the enzyme.

10 mgm of the lyophilised powder in 10 ml of 0.2 M phosphate buffer at pH 7.0 with 50 mgms. of TTP was incubated at 25° C for 5 hrs., when 0.8—1.6 μgm. of IAA was produced.

Assay methods. IAA was separated by the solvent partitioning method (a) given under "general methods", p. 238f. and was either assayed colorimetrically (TANG and BONNER method) or by the *Avena* test.

Values for IAA estimated by these two methods agreed within 10—20%.

Properties of the enzyme. The pH optimum was 8 in phosphate buffer and the optimum 60° C for a 5 hrs. reaction time; a rapid decline in rate occurred above 80° C. The reaction rate declined with time after 30 mins., though some activity was present even after 16 hrs.

The reaction rate increased with an increase in TTP concentration from 0.01 to 5 mgm. per ml, which was saturating. A natural inhibitor for the enzyme activity was present in lyophilised normal stem tissue.

iii) Escherichia coli enzyme (STOWE, 1955; unpub.).

Preparation. Freeze-dried cells of *E. coli* were homogenised in a bead type homogeniser with water or buffer and the homogenate centrifuged at $10{,}000 \times g$ for 20 mins. The supernatant was dialysed against water and used as enzyme or further purified. To the crude enzyme containing 3 mgm. protein/ml in 0.05 M sodium acetate buffer at pH 4.9 was added calcium phosphate gel (40 mgm./ml) to dilute it to a 20% mixture. Elution of the centrifuged gel with half volume of 0.02 M calcium oxalate solution, pH 4.5, gave a 30 fold enrichment with 10—20% increase in total activity. The enzyme may be precipitated at 40—60% saturation of (NH$_4$)$_2$SO$_4$ at pH 4.9.

20 μgm. enzyme, 2.5 μM L-tryptophan, 5 μM α-ketoglutaric acid, 1 μM pyridoxal-5-phosphate in 1 ml of 0.05 M sodium-Tris buffer, pH 7.9 incubated at 25° C for 15 min. produced about 0,07 μM (13 μgm.) equivalents of IAA; the product was probably largely IPyA, which gives the same colour as IAA with the SALKOWSKI reagent after 1 hour (Author's addition).

Assay method. IAA (IPyA) was assayed by adding directly SALKOWSKI reagent (GORDON and WEBER, 1951) to the reaction mixture as given in "general methods" (p. 238).

Properties of the enzyme. The pH optimum was between 7 and 8 in Tris buffer. Oxalacetic and α-ketoglutaric acids were specific for the reaction and could not be replaced by pyruvic, glyoxylic or α-ketobutyric acids. Glutamic acid was detected by chromatography when α-ketoglutaric acid was added, indicating a normal transamination reaction.

Pyridoxal-5-phosphate or pyridoxamine phosphate was essential for the activity of the purified enzyme; the activity was linear with the log. concentration of pyridoxal phosphate up to 1 μM.

IAA was identified as the product by chromatography and colour reactions, and no indole was detected. Owing to its spontaneous degradation, the formation of IPyA could not be demonstrated; nor is it known whether its decomposition to IAA was furthered by bacterial enzymes.

Intermediates in the reaction. Neither IPyA nor IAc have been unequivocally demonstrated as the intermediates in the conversion of TTP to IAA. However, based on the similarity between the pattern of spots, obtained by the chromatography (isopropanol, ammonia, water 10:1:1) of an acid fraction of a TTP-containing culture medium of *A. tumefaciens*, and the characteristic breakdown pattern of spots of IPyA under similar chromatographic conditions, the formation of IPyA as an intermediate has been suggested (KAPER and VELDSTRA, 1958).

A neutral compound was formed along with IAA when either the pineapple leaf enzyme (GORDON and NIEVA, 1949; see p. 244) or extracts of mung seedlings irradiated by ionising radiation (GORDON, 1956; see p. 244) was incubated with TTP. This compound was believed to be IAc because i) it formed adducts with dimedon or bisulphite, being released from the latter by the addition of alkali and ii) it was converted to an acid auxin, active in the *Avena* test, by treatment with soil or *Avena* coleptile juice.

Sources of error. Tryptophan samples often contain traces of IAA, which may seriously affect results, especially when very low amounts of IAA are being produced enzymatically. When TTP solutions are autoclaved IAA is formed, even in neutral solutions or pure water (KULESCHA, 1951). All TTP solutions should be freshly prepared and ether extracted (2 to 3 times with half volume of ether) after acidification of the solution to pH 3 to remove all traces of IAA. Recrystallisation of TTP is also recommended.

2. Enzymes Involved in the Conversion of IAc to IAA.

Enzymatic conversion of IAc to IAA has been obtained with extracts of the following tissues: pineapple leaf base (GORDON and NIEVA, 1949); *Avena* coleoptile (LARSEN, 1949); *Artemesia* roots (ASHBY, 1951), maize embryo (YAMAKI and NAKAMURA, 1952); mung bean seedling (GORDON, 1956).

IAc used in these studies was either a crude natural product from pineapple or *Taraxacum* roots (GORDON and NIEVA, 1949), or an impure, chemical preparation, obtained by the action of ninhydrin or isatin on TTP, and containing about 2% IAc but no IAA (LARSEN, 1944; GRAY, 1959). The preparation of pure crystalline IAc, as its bisulphite adduct, has been described by GRAY (1959).

Preparation of the Enzymes.

LARSEN (1955) has described the conversion of IAc to IAA by *Avena* coleoptile juice. Preparation of a purified enzyme from *Avena* by acetone precipitation and $(NH_4)_2So_4$ fractionation has been reported recently, but the details have not yet been published (SHIGEMURA and GORDON, 1960).

i) **Pineapple leaf enzyme** (GORDON and NIEVA, 1949). Frozen young leaf bases of pine-apple were blended in about 3 times their weight of cold 0.1 M phosphate buffer, pH 7 and centrifuged. $(NH_4)_2SO_4$, at 68 gm/100 ml brei, was added to the supernatant, and the precipi-tate removed by centrifugation after overnight flocculation at 0° C. The precipitate was resuspended in water and dialysed for 30 hrs. against distilled water in the cold, and used as the enzyme. The solid content was 54 mg./ml.

1 ml of the enzyme dispersion in 50 ml of phosphate buffer at pH 6.8 containing 6 μgm. equivalents of IAc produced 0.35 μgm. IAA in 4 hrs. at 27° C.

ii) **Maize embryo enzyme** (YAMAKI and NAKAMURA, 1952). Etiolated maize embryos were ground with an equal weight of cold 0.01 M phosphate buffer, pH 6.8 and the brei used after filtering through sintered glass.

1 ml extract and two ml of the buffer containing 0.1—1.5 mgm. equivalents of the impure, chemically prepared IAc, produced 25.3 μgm. of IAA in 4 hrs. at 22° C.

Assay methods. The method for the separation of neutral (IAc) and acidic (IAA) compounds has been given under "general methods", p. 238.

IAA, produced by the maize extract, was separated by paper chromatography with 70% ethanol as solvent; IAc has a R_f of 0.42—0.48 and IAA a R_f of 0.85—0.9 in this system. The region of the paper strip containing IAA was extracted with ether in a micro-Soxhlet apparatus for three hours, giving quantitative recovery of IAA (YAMAKI and NAKAMURA, 1952). IAA was estimated by the *Avena* curvature test.

Properties of the enzymes. The effects of pH, substrate concentration, inhibitors etc. have not been studied; the reactions have been conducted at pH 6—7.

In addition to IAc, naphthaleneacetaldehyde (NAc) was converted to an acid auxin, naphthaleneacetic acid (NAA), by *Avena* coleoptile juice and *Artemisia* root sap (LARSEN, 1951; ASHBY, 1951).

About 2 moles of IAc or NAc disappeared per mole of IAA or NAA formed. A dismutation reaction involving the simultaneous production of 1 mole IAA (or NAA) and 1 mole tryptophol (or α-naphthalene ethanol) for every 2 moles of IAc (or NAc) destroyed has been suggested (LARSEN, 1949; 1951; 1955). How-ever, this has been questioned (ASHBY, 1951) and tryptophol formation has not yet been demonstrated (cf. LARSEN, 1955).

The *Avena* coleoptile and mung bean enzymes (cf. p. 242) were sensitive to low doses of ionising radiation; the conversion of IAc to IAA by cell free extracts of these plants, prepared immediately after their exposure to 25 and 500 roentgen radiation, was inhibited by 20 and 40—60% respectively. The low levels of IAA present in irradiated plants were therefore attributed to the inhibition of this enzyme, which may catalyze the last step of the conversion of TTP to IAA (cf. Fig. 1, reaction III).

Indeed, extracts of irradiated plants incubated with TTP show a decreased IAA formation and a corresponding accumulation of a neutral compound, which gives bisulphite and dimedon adducts, and which is converted to an acid auxin (IAA) by coleoptile juice and hence believed to be IAc. Such inhibitions were overcome in 14 days following radiation doses up to 500 r (GORDON, 1956; 1956a).

Sources of error. Several plants extracts can enzymically degrade phenyl-acetaldehyde to the lower homologue benzaldehyde, and IAc may be similarly oxidised (KENTEN, 1953). Oxidation of IAc to products other than IAA occurred with the pea extract (CLARKE and MANN, 1957).

3. Enzymes Involved in the Conversion of TNH₂ to IAc and IAA.

The enzymatic conversion of TNH_2 to a neutral (IAc) and an acid auxin (IAA) was demonstrated with a crude pineapple leaf enzyme (GORDON and NIEVA, 1949) using the *Avena* test for IAA assay. TNH_2 was converted to IAA by maize embryo juice, the IAA being identified by chromatography and bioassay (YAMAKI and NAKAMURA, 1952). The conversions were less than 0.01 and 2% respectively.

The reaction pathway was probably via IAc, for the latter was also converted to IAA by these enzyme preparations (Fig. 1; reactions VI and III).

A highly purified plant amine oxidase from pea seedlings, oxidising several mono- and diamines to the respective aldehydes, also oxidised TNH_2 to IAc (KENTEN and MANN, 1952; MANN, 1955; CLARKE and MANN, 1957). Similar amine oxidases have been demonstrated in lavender leaves, red clover leaves and seedlings of lupin (KENTEN and MANN, 1952).

WERLE and ROEWER (1950) have demonstrated a plant monoamine oxidase in *Salvia uliginosa* leaf extracts oxidising TNH_2 to unknown products; it was insensitive to 10^{-3} M cyanide. The enzyme was also present in other plants.

Nature of the reaction.

a) $\text{Ind—CH}_2\text{—CH}_2\text{—NH}_2 + \text{O}_2 + \text{H}_2\text{O} \xrightarrow{\text{amine oxidase}} \text{Ind—CH}_2\text{—CHO} + \text{NH}_3 + \text{H}_2\text{O}_2$

b) $\text{Ind—CH}_2\text{—CH}_2\text{—NH}_2 + \text{O}_2 + \text{H}_2\text{O} + \text{catalase} \xrightarrow{\text{amine oxidase}} \text{Ind—CH}_2\text{—CHO} + \text{NH}_3 + \text{H}_2\text{O} + \frac{1}{2}\text{O}_2 + \text{catalase}$

Ind = indole.

The preparation of the enzyme has been described elsewhere (see, page 300 ff).

Unit of activity. The amount of oxidase, which at 28° C causes an uptake of 1 μl/hr. O_2 in the presence of 0.01 M 1:4-diaminobutane; catalase[1]; 0.067 M phosphate buffer pH 7, in a total volume of 3 ml.

Assay methods.

i) *Manometry.* O_2 uptake measured in a WARBURG's manometer at 28° C with a 3 ml reaction mixture and 0.2 ml of 5 N KOH in the centre well.

ii) *Assay of IAc.* The IAc formed is oxidised to IAA with silver oxide and the IAA estimated spectrophotometrically with the SALKOWSKI reagent.

To 0.5 ml of the reaction mixture containing 100—300 μgm. IAc, was added 0.5 ml water, 1 ml 0.1 M AgNO₃, 1 ml 0.1 N NaOH followed immediately by 1 ml of 0.2 M phosphate pH 7.0 in that order. The solution was thoroughly mixed after each addition and finally shaken for 5 mins. in the dark and filtered through Whatman No. 44 paper. 1 ml of filtrate was estimated for IAA by the addition of 9 ml water and 15 ml of a modified SALKOWSKI reagent (5 ml 0.5 M FeCl₃, 250 ml H₂O and 150 ml conc. H₂SO₄) and the colour read after 1 hr. at 530 mμ. Blank contained everything except TNH₂.

Precaution. Silver oxide had to be precipitated in the reaction mixture as addition of freshly prepared silver oxide gave erratic results.

iii) Ammonia estimation was done by diffusion in Conway dishes at room temperature.

Properties of the enzyme. The pH optimum, with TNH_2 as substrate, was about 8.1 in phosphate borate buffer. TNH_2 at 4×10^{-3} M gave maximal O_2 uptake, being inhibitory above 2×10^{-2} M. With the addition of catalase, 0.5 mole O_2 was rapidly taken up per mol TNH_2 oxidised and 1 mole of NH_3 was released. Thereafter on standing a slow oxidation took place to give a ratio of about 0.6 mole O_2/mole TNH_2 due to secondary oxidation, with the purified enzyme. With the crude enzyme this secondary oxidation proceeded to a greater extent.

IAc was identified as the reaction product by chromatography of the vacuum dried reaction mixture in isopropanol, ammonia, water (10 : 1 : 1) solvent and spraying with EHRLICH's reagent, when a yellow brown streak (R_f 0.9) of IAc resulted. On silver oxide oxidation of the products, a pink IAA spot (R_f 0.44)

---

[1] As shown in equation (a), H_2O_2 was a product of the reaction and this inhibited the enzyme on accumulation in the presence of the substrate (KENTEN and MANN, 1952; MANN, 1955). Addition of catalase (Katalase Fähigkeit = 27000) at 50 μgm per 3 ml reaction mixture decomposed the H_2O_2 formed (CLARKE and MANN, 1957).

was obtained on spraying with the reagent. IAc was also isolated as its 2,4-dinitro-phenylhydrazone and dimedone adduct (CLARKE and MANN, 1957).

Cyanide at 10^{-2} M and semicarbazide at 10^{-5} M caused a 70—100% inhibition while copper reagents such as Na diethyldithiocarbamate and potassium ethylxanthate at 10^{-3} or salicylaldoxime at 10^{-2} M caused a 40—90% inhibition with 1 :4 diaminobutane, β-phenylethylamine or ethanolamine as substrates (MANN, 1955). The effect of these inhibitors on TNH_2 oxidation has not been tried.

The enzyme can oxidise the following mono- and diamines: 5-OH tryptamine, tyramine, L-lysine and the above mentioned amines (KENTEN and MANN, 1952; MANN, 1955), the diamines being oxidised more rapidly than monoamines.

Sources of error. With the crude extract, IAc formed could not be detected due to secondary oxidations. TNH_2 was also peroxidatively destroyed by the crude enzyme, but these secondary reactions were eliminated on purification (CLARKE and MANN, 1957).

4. Enzymes Involved in the Conversion of IAN to IAA.

The conversion of IAN to IAA was first demonstrated in the *Avena* coleoptile tissue and the product identified as IAA by paper chromatography, activity in the pea curvature test and colour reaction (THIMANN, 1953; STOWE and THIMANN, 1954). The enzymatic nature of the conversion was shown with an enzyme from oat and barley leaves (THIMANN and MAHADEVAN, 1958).

Distribution of the ability to convert IAN to IAA in plants. Of 29 species of plants from 20 families, tested (including Pteridophytes, Gymnosperms and Angiosperms), the following tissues have been shown to convert IAN to IAA when infiltrated with IAN: Crucifers including cabbage leaves, cauliflower inflorescence, kohl-rabi leaves, and Brussel sprouts (MAHADEVAN, 1960; MAHADEVAN and THIMANN, unpub.; LIBBERT and BALLIN, 1959) and radish seedlings (MICHEL, unpub.); members of the Gramineae including oat coleoptiles and leaves, barley leaves, stem, roots and coleoptiles (MAHADEVAN, 1960), wheat and maize coleoptiles (SEELEY et al., 1956); and two members of the Musaceae, banana *(Musa paradisiaca)* and *Strelitzia reginae* (MAHADEVAN, 1960).

In addition mycelial mats of the following species of *Fusarium: f. oxysporum* f. *cubense, F.o.* f. *dianthi, F. roseum* and *Fusarium* isolates from carrot and Easter lily bulbs were capable of a rapid conversion of IAN to IAA (MAHADEVAN, 1960).

Method for the demonstration of the conversion of IAN to IAA in tissues. About 3 gms of tissue were vacuum infiltrated in about 10 ml of 0.05 M, pH 7 phosphate buffer containing 5×10^{-3} M KCN (to prevent any IAA oxidase activity) and 25 μgm./ml IAN. After incubation 12 hrs. at 25° C, the tissue and ambient solution were acidified to pH 3 and ether extracted. The ether extract was separated into neutral and acid fractions (cf. "general methods", p. 239) and the acid fraction chromatographed (isopropanol, NH_4OH, water (8:1:1) and sprayed with SALKOWSKI reagent. A pink spot at the R_f of IAA (0.3—0.35) indicates conversion.

Nature of the reaction.

$$\text{Ind—CH}_2\text{—CN} + 2\,\text{H}_2\text{O} \xrightarrow{\text{enzyme}} \text{Ind—CH}_2\text{—COOH} + \text{NH}_3$$

Preparation of barley leaf nitrilase. Frozen green barley leaves were blended with four times their weight of cold distilled water and the slurry strained through cloth. The brei was centrifuged at 10,000 × g for 1 hr. and the supernatant brought to 0.5 saturation with $(NH_4)_2SO_4$ and centrifuged. The precipitate was taken up in 0.1 M, pH 7 phosphate buffer and dialysed overnight against 0.005 M, pH 7 buffer in the cold. After removal of any precipitate by centrifugation the supernatant was brought to 0.15 saturation with respect to $(NH_4)_2SO_4$ and the precipitate removed by centrifugation. The supernatant was then brought to 0.3 saturation with $(NH_4)_2SO_4$. The precipitate formed was centrifuged, taken up in 0.1 M phosphate buffer pH 7, and dialysed against 0.005 M buffer, pH 7 for 24 hrs. in the cold with two changes of buffer. This enzyme (I) could be stored frozen for weeks at —10° and was six fold purified.

Further purification of the enzyme could be achieved with calcium phosphate gel adsorption (31 mgm./ml gel per 1.7 mgm/ml protein) and elution (with 0.05 M, pH 6.4 phosphate buffer) to give 28 fold purification.

0.1 ml enzyme (I) (equivalent to 3—4 gm. barley leaves) in 1 ml 0.05 M phosphate buffer, pH 7 and containing 100—350 μgm. IAN could form 20—30 μgm. IAA in 1 hr. at 35° C. Enzyme reaction was stopped by boiling for half a minute.

Assay methods.

i) *IAA estimation*. The reaction mixture was diluted and acidified to pH 3 with N H_3PO_4 and the IAA and IAN extracted three times with equal volumes of peroxide free ether. The ether extract was layered over 1 ml. of alkaline water (pH 9) in a separatory funnel having a 2 ml mark etched on it and the ether removed with a stream of air. The aqueous layer was made up to 2 ml with alkaline water (pH 9) and extracted three times with 5 ml volumes of CCl_4 which completely removed the IAN. The IAA in the aqueous layer was estimated colorimetrically after addition of 4 ml of SALKOWSKI reagent (GORDON and WEBER, 1951; see "general methods" p. 238). Quantitative separation of 10 μgm. IAA and 500 μgm. IAN could be achieved by this method.

ii) *Ammonia estimation*. Ammonia was estimated by diffusion on Conway microdiffusion dishes at room temperature followed by Nesslerisation.

Sources of error in assay. In the presence of a large amount of proteins and colouring matter in crude enzyme preparations a quantitative recovery of added IAA was not always achieved by the above method. Use of the purified enzyme (I) eliminated these errors.

Properties of the enzyme nitrilase. The enzyme had a pH optimum range between 5.5 and 8 in phosphate buffer at 25° C. A K_m of 5.1×10^{-5} M was obtained with IAN as substrate at 35° C. At high substrate levels a slight deviation from linearity in a LINEWEAVER-BURK plot occurred. There was no inhibition of activity at these concentrations, however.

The Q_{10} values for the reaction between 5 and 15°; 15 and 25°; or 25 and 35° C ranged between 2.06 to 2.1, and the energy of activation (5—35 °C) was calculated to be 14,000 cals./mole. One mole of ammonia was released per mole IAA formed; indoleacetamide could neither be detected as an intermediate nor was it hydrolysed by the enzyme at a rate fast enough to be considered as an intermediate.

Cyanide at 10^{-3} M, and iodoacetamide at 10^{-3} M had no effect on activity but p-chloromercuribenzoate at 10^{-4} M was inhibitory which was reversed (67%) by glutathione; it is therefore a sulphydryl enzyme. Heavy metal groups are not involved in the activity of the enzyme. The reaction is hydrolytic, proceeding equally well in air and *in vacuo*.

No cofactor requirements were found for the enzyme.

Several ring nitriles including 2-indoleacetonitrile, various para- und meta-substituted benzonitriles, 2-, 3-, and 4-cyanopyridines, α-naphthaleneacetonitrile etc. and aliphatic nitriles as propionitrile and β-Cl, β-methoxy and β-OH propionitriles were hydrolysed by the enzyme; these reactions were followed by estimating the ammonia produced.

The relatives rates of hydrolysis of the m- and p-substituted and the unsubstituted benzonitriles indicated that electron withdrawing groups on the ring such as Cl, F, NO_2 enhanced, while electron donating groups as OH, OCH_3 and NH_2 reduced the rate of hydrolysis compared to the unsubstituted benzonitrile. The mechanism of action of the enzymatic hydrolysis resembles alkali hydrolysis of nitriles, where a similar effect of substituents has been found (OGATA and OKANO, 1949), and probably involves an initial nucleophilic hydroxyl attack on the carbon atom of the nitrile group.

III. Enzymes Involved in the Breakdown of IAA.

The first systematic study of enzymatic IAA oxidation was made by TANG and BONNER (1947) with an enzyme from etiolated pea epicotyls; such a destruction was first shown by THIMANN (1934) with *Helianthus* leaf brei and later by LARSEN (1940) with *Phaseolus* extracts. Several plant extracts oxidise IAA enzymically and a representative list is given in Table 1. The enzyme is widely distributed in the higher plants, especially in the Leguminosae, and in some fungi. Moreover highly purified horse radish peroxidase and turnip peroxidase can destroy IAA oxidatively (KENTEN, 1955; RAY, 1956; YAMAZAKI and SOUZU, 1960).

Table 1. *Distribution of IAA oxidase enzyme in plants.*

Plant	Plant part where enzyme studied and distribution in the plant	Reference
Higher plants:		
Cabbage	Leaves	TANG and BONNER (1948) MELCHIOR (1958)
Lentils	Roots	PILET (1955; 1957)
Lupin	Green leaves and stems; lower activity in etiolated plants	STUTZ (1957)
Oats	Coleoptiles; lower activity in roots	TANG and BONNER (1948)
Osmunda cinnamomea	Youngest fronds; lower activity in older parts	BRIGGS et al. (1955)
Peas	Etiolated epicotyls, roots and cotyledons; lower activity in green plants. No activity in buds	TANG and BONNER (1947; 1948)
Pineapple	Vegetative stem tip; lower activity in leaves and roots	GORTNER and KENT (1953)
Quackgrass	Rhizomes	MUDD et al. (1959)
Spinach	Roots; no activity in leaves	TANG and BONNER (1948)
Wax bean	Roots	WAGENKNECHT and BURRIS (1950)
Wheat	Green leaves	WAYGOOD et al. (1956a and b)
Fungi:		
Lactarius spp.	Fruiting body	LEGRAND (1957)
Melampsora lini	Mycelium	OAKS and SHAW (1960)
Omphalia flavida (*Mycena citricolor*)	Extracellular enzyme in culture	SEQUIERA and STEEVES (1954) RAY and THIMANN (1956)
Polyporus versicolor	Extracellular enzyme in culture medium	TONHAZY and PELCZAR (1954)

Crown-gall tissue cultures of *Parthenocissus tricuspidata* (LIPETZ and GALSTON, 1959) and tissue cultures of *Picea glauca* (REINERT, SCHRAUDOLF and TAZAWA, 1957) secrete an IAA oxidising enzyme into the media; *Parthenocissus* cultures, both normal and crowngall, contain the enzyme in comparable amounts (PLATT, 1954).

Methods.

1. *Colorimetric.* The disappearance of IAA has been followed by the reduction in the pink colour developed by IAA with a modified SALKOWSKI reagent (see Section I; p. 238). The enzymatic oxidation products of IAA give very slight or no colour with these reagents. The reagent is diretly added to suitable aliquots of the reaction mixture containing 5—50 μgm. residual IAA and the colour read at 530 mμ in a spectrophotometer or with a green filter (No. 54) in a colorimeter.

2. *Manometric.* Conventional WARBURG manometric techniques (UMBREIT et al., 1949) for measuring O_2 uptake with KOH in the centre well, and CO_2

liberation by the "direct method" have been used, with either IAA or its sodium salt, or the enzyme in the side arm.

3. *Spectrophotometric.* This method has been used only with the *Omphalia* enzyme (see below) and horse radish peroxidase (RAY, 1956) but may be of general applicability. The principle consists of following spectrophotometrically at 261 mμ, the optical density (O.D.) changes which are directly proportional to the conversion of IAA to a labile intermediate. This intermediate decomposes secondarily (non-enzymatic) to the final products, without further change at 261 mμ. The secondary non-enzymatic reaction may be followed spectrophotometrically by O.D. changes at 272 mμ.

Method. Diluted *Omphalia* enzyme (0.1 ml in 10 ml reaction mixture) and 6×10^{-5} M IAA in citric acid buffer (0.005 M) pH 3.7 are placed in a 1 cm light path quartz cuvette and at zero time about 0.02 μmole of H$_2$O$_2$ is added as a starter for the reaction, stirred and the O.D. changes at 261 mμ and 272 mμ followed in a spectrophotometer with a hydrogen discharge lamp as an UV source. IAA in buffer serves as a blank and the initial O.D. adjusted to 0.1 for obtaining maximum sensitivity. The enzyme itself has negligible absorption above 230 mμ and IAA is not destroyed by these exposures to UV.

Precaution. The method is not suitable in the presence of Mn^{2+} in the reaction mixture, for with Mn^{2+} the changes in the O.D. at 261 mμ continue even after all IAA is used up.

Peroxidase activity assay. The peroxidase activity of IAA oxidase may be assayed by any one of the modifications of WILLSTÄTTER and STOLL's (1918) method for the estimation of purpurogallin, formed by reaction with pyrogallol and H$_2$O$_2$. (cf. KEILIN and HARTREE, 1951).

Method of STUTZ (1957). A sample of lupin enzyme (see below) 1.25 ml of 0.4 M phosphate buffer, pH 6.3 and 2.5 ml of 0.5% fresh pyrogallol were made up to 4.5 ml. with water and equilibrated at 20° C. 0.5 ml of 0.06% H$_2$O$_2$ was added at zero time and the reaction stopped after 10 mins. with 0.2 ml of 2 N H$_2$SO$_4$. The yellow purpurogallin colour was read at 420 mμ. Standard purpurogallin solution was made in acetone. Linearity is obtained up to saturation with purpurogallin. The values are expressed as modified Purpurogallin-Zahl (PZ') after calculating the values to a 500 ml basis.

$$PZ' = \frac{\text{mgm. purpurogallin in 500 ml}}{\text{volume of enzyme (ml)} \times \text{mgm. of protein}} \cdot$$

1. General Properties of the IAA Oxidation Reaction.

The overall reaction of enzymatic IAA oxidation may be written as:

$$IAA + O_2 \xrightarrow[\text{(activators)}]{\text{enzyme}} \text{Products} + CO_2 \qquad \text{(Eq. 1)}$$

where a mole of oxygen is taken up and a mole of CO$_2$ released per mole IAA destroyed. No IAA destruction occurs in the absence of oxygen.

The above stoichiometry has been obtained with the enzymes of pea (TANG and BONNER, 1947), bean root (WAGENKNECHT and BURRIS, 1950), pineapple (GORTNER and KENT, 1953), wheat leaf (WAYGOOD, OAKS and MACLACHLAN, 1956b), *Omphalia* (Ray and THIMANN, 1956), *Polyporus* (TONHAZY and PELCZAR, 1954), and horse radish peroxidase (KENTEN, 1955).

The carboxyl group of IAA is lost as CO$_2$ during enzymic oxidation; with the lupin enzyme negligible radioactivity is retained in the products when C^{14} OOH-labelled IAA is used (STUTZ, 1957); with the *Omphalia* enzyme C^{14}O$_2$ comes off rapidly and almost stoichiometrically (RAY and THIMANN, 1956). A free carboxyl group also appears to be essential since neither the ethyl ester of IAA (*Omphalia*, RAY and THIMANN, 1956) nor the amide (pea, TANG and BONNER, 1947; *Polyporus*, TONHAZY and PELCZAR, 1954; *Omphalia*, RAY and THIMANN, 1956) is oxidized.

Table 2. *Effect of inhibitors on IAA oxidising enzymes.*

Inhibitor	Lupin[a]	Pea[b-f]	Wheat[j]	Pineapple[k,l]	Omphalia[m,n]
Cyanide	$5 \cdot 10^{-5}$ (50)	10^{-3} (77)[b] 10^{-2} (100)	10^{-2} (100)	10^{-3} (93)[k]	10^{-5} (80)[m] 10^{-4} (100)
Azide	10^{-3} (50)	5×10^{-3} (76)[c]	10^{-2} (100)		
Hydroxylamine . .	5×10^{-4} (50)	5×10^{-3} (76)[c]	10^{-2} (100)		
Semicarbazide . .	5×10^{-5} (50)		10^{-2} (inb)		
Diethyldithio-carbamate . . .		10^{-3} (81)[c]		(inb)[k]	
Potassium ethyl-xanthate		10^{-3} (98)[c]			
8-hydroxyquinoline		5×10^{-3} (77)[c]		10^{-2}—10^{-3} [k] (no effect or promotion)	
Cu^{2+}	10^{-3} (inb)	10^{-3} (inb)[d]		4×10^{-9} (50)[l]	
Co^{2+}	10^{-3} (no effect)	10^{-3} (55)[e]			
Fe^{3+}	10^{-3} (no effect)	10^{-3} (inb)[d]			
Pyrogallol	10^{-4} (comp. inb)		5×10^{-4} (97) lag	1.5×10^{-6} (inb)[l]	7×10^{-6}(50)[n] 3.3×10^{-5} (95)
Catechol	10^{-4} (comp. inb)		5×10^{-4} (95) lag	1.5×10^{-6} (inb)[l]	10^{-5} (inb)[n]
Guaiacol	10^{-4} (comp. inb)	5×10^{-6} (inb)[f]		4×10^{-6} (inb)[l] 2×10^{-6} (prom)	
Hydroquinone . .	10^{-4} (comp. inb)		5×10^{-4} (92) no lag	2.5×10^{-6} (inb)[l]	10^{-5} (inb)[n]
Chlorogenic acid .		2.8×10^{-6} (57)[g] 4.9×10^{-6} (100)		3×10^{-6} (50)[k]	
Caffeic acid	10^{-4} (comp. inb)	2.8×10^{-6} (57)[g] 5.5×10^{-6} (100)		10^{-6} (inb)[l]	
Scopoletin		2.5×10^{-5} (60—70)[h]	(inb)	2×10^{-4} (prom)[l] 10^{-3} (inb)	
α- and β-naphthol .		10^{-5} (100)[i]		10^{-5} (prom)[l]	
Catalase	(no effect)	(inb)[d,h]	(no effect)		

Reference:

[a] STUTZ (1957)
[b] TANG and BONNER (1947)
[c] WAGENKNECHT and BURRIS (1950)
[d] GALSTON (1953)
[e] GALSTON and SIEGEL (1954)
[f] GOLDACRE (1951)
[g] RABIN and KLEIN (1957)

[h] ANDREAE and ANDREAE (1953)
[i] GOLDACRE et al. (1953)
[j] WAYGOOD et al. (1956)
[k] GORTNER and KENT (1953)
[l] GORTNER et al. (1958)
[m] RAY and THIMANN (1956)
[n] RAY (1960)

Abbreviations:
  inb — inhibition.
  comp. inb. — complete inhibition.
  prom. — promotion.
  Open figures indicate concentration in moles/litre.
  Figures in parentheses indicate percent inhibition.
  Small alphabets indicate reference.

The reaction products are complex, probably being formed non-enzymatically from a labile primary product (RAY, 1956). They include neutral (pea — MANNING and GALSTON, 1955; lupin — STUTZ, 1957) and polymeric compounds (lupin — STUTZ, 1957; *Polyporus* — TONHAZY and PELCZAR, 1954). The major component appears to be 3-methyl dioxindole or a nearly related compound (STOWE, RAY and THIMANN, 1957). Indolealdehyde, which also satisfies the above stoichiometry, is not a major product with the pea, *Omphalia*, lupin or *Polyporus* enzymes.

IAA oxidase preparations of pea, lupin, *Omphalia* and quackgrass (MUDD, JOHNSON, BURRIS and BUCHHOLTZ, 1959) can peroxidatively destroy pyrogallol with H_2O_2, while the *Polyporus* and *Lactarius* (LEGRAND, 1957a) enzymes do not and have a polyphenolase activity. With lupin and *Omphalia* the IAA oxidase and peroxidase activities are attributed to the same enzyme, for in the former the activities migrate together on electrophoresis (STUTZ, 1957), while with the latter a constant ratio of activities is obtained during the various purification steps or during the thermal inactivation of the enzyme (RAY, 1960).

All IAA oxidases are inhibited by cyanide and several heavy metal (Fe^{2+} and/or Cu^{2+}) inhibitors (cf. Table 2), and the *Omphalia* enzyme (RAY, 1960) undergoes an inhibition by carbon monoxide in the dark which is reversed by light like an iron oxidase. Polyphenols, especially pyrogallol, powerfully inhibit all IAA oxidase with peroxidase activity (cf. Table 2).

Manganous ions are in some cases essential for, or in other cases promote the activity of IAA oxidase under certain conditions depending on the source, stage of purification of the enzyme and the status of natural or added phenolic or other cofactors in the medium. Mn^{2+} cannot be replaced by Cu^{2+}, Fe^{2+}, Zn^{2+}, Mg^{2+} or Co^{2+} (WAGENKNECHT and BURRIS, 1950; WAYGOOD et al., 1956b; STUTZ, 1957). Mn^{2+} can also protect the enzyme during IAA oxidation and prevent the decline in rate of horse radish peroxidase, lupin and *Omphalia* enzymes. Phenolic cofactors as 2,4-dichlorophenol (DCP) may or may not be required for IAA oxidase action depending on the source and stage of purification of the enzyme, level of endogenous cofactors and Mn^{2+} status of the reaction mixture. The effects of Mn^{2+} and cofactors are given for each of the enzymes described.

The mechanism of enzymatic IAA oxidation is not understood and schemes involving flavoproteins-peroxidase, Mn^{2+}-phenol-peroxidase systems have been postulated (GALSTON, BONNER and BAKER, 1953; MACLACHLAN and WAYGOOD, 1956). An autocatalytic cyclical series of reactions involving a peroxidase has been envisaged, with the formation of labile free radical intermediates which would account for the multiplicity of the products and the powerful inhibition of the reaction by free radical inhibitors as polyphenols (MACLACHLAN and WAYGOOD, 1956; RAY, 1960; YAMAZAKI and SOUZU, 1960). A critical appraisal of the several postulated mechanisms and the role of Mn^{2+}, phenols and H_2O_2 in IAA oxidation is given by RAY (1958).

2. Preparation and Properties of some IAA Oxidases.

The preparation and properties of lupin, pea, wheat, pineapple and *Omphalia* IAA oxidases are described. In addition the substrate specifity of some others are included in Table 3.

a) Lupin Enzyme (STUTZ, 1957).

Preparation. The enzyme concentration is higher in the green than in the etiolated tissues. Fresh or frozen green lupin tissues (stem and leaves) were ground with sand and equal volume of water, the macerate expressed through cloth, centrifuged at $20,000 \times g$ for 10 mins., and dialysed first against tap and then against distilled water for 18 to 36 hrs. to remove inhibitors. Any precipitate formed was filtered off, and the supernatant kept frozen or used as the enzyme.

Table 3. *Specificity of the IAA oxidising enzymes.*
(values in parentheses indicate destruction as percent of IAA).

Enzyme	Compounds destroyed other than IAA	Compounds not destroyed	Reference
Lupin crude	IPA; IBA; 2-COOH-IAA		Stutz (1957) Stutz (1958)
Lupin purified	IPA (14); IGlyA (60); IiBA (137); 1-Me-IAA (45)	IAN; IBA; ICA; TTP 5-OH-IAA; IAld; 2-COOH-IAA; NAA Indoleglyoxylic acid	
Pea	IPA (55); IBA (22)* IPA (44); IBA (15)*	IAM; IPA; IBA; TTP; ICA** IAld; OFA***	*Wagenknecht and Burris (1950) **Tang and Bonner (1947) ***Manning and Galston (1955)
Wheat		IPA; IBA; NAA 2.4-D	Waygood et al. (1956b)
Omphalia	IiBA (150—200); 5-OH-IAA (slight); 7-OH-IAA (slight) IPA (slight with Mn²⁺)	TTP; IAN; IAM EtIA; skatole IAld; OFA; IPA	Ray and Thimann (1956)
Wax bean	IPA (30; IBA (15)		Wagenknecht and Burris (1950)
Poliporus	IBA (44); IPA (18); IAc (18)	TTP; indole skatole; IAM	Tonhazy and Pelczar (1954)
Horse-radish peroxidase	IPA; IBA (only with Mn²⁺)		Kenten (1955a)

Abbreviations:

IGlyA	3-indoleglycolic acid
IiBA	3-indoleisobutyric acid
OFA	o-formylaminoacetophenone
1-Me-IAA	1-methyl, indoleacetic acid
2-COOH-IAA	2-carboxyl, 3-indoleacetic acid
5-OH-IAA	5-hydroxy indoleacetic acid
7-OH-IAA	7-hydroxy indoleacetic acid
EtIA	ethyl indoleacetate

Other abbreviations as given before in Section 1.

For further purification, $(NH_4)_2SO_4$ was added to 80% to 100% saturation, the precipitate centrifuged and both the supernatant and the precipitate (taken up in water) dialysed separately. Both fractions may contain the enzyme depending on the tissue (etiolated or green, leaf or stem). The supernatant was concentrated at low temperature and both fractions (supernatant and precipitate) lyophilised.

Ten to fifty mgm. lots of each fraction were electrophoretically purified at 5° C (method and apparatus of Kunkel and Slater, 1952) on a starch bed of 5 cm² cross section saturated with 0.025 M phosphate, pH 7 or acetate buffer, pH 5.5. Electrophoresis was done by passing a 300 volts, 12—15 milliamps. current for 18 hrs. The starch bed was cut into 1 cm blocks, eluted with buffer and IAA oxidase and peroxidase activities estimated in each block. Peaks of both IAA oxidase and peroxidase activities coincide and are attributed to the same enzyme. Migration characteristics of enzyme were the same at pH 5.5 and 8.5. A thousand-fold purification was achieved by electrophoresis.

The incubation mixture consisted of 1 ml of dialysed or purified enzyme, 1 ml of 0.4 M phosphate buffer, pH 6.3; 10^{-2} M sodium salt of IAA, 10^{-3} M $MnCl_2$, and 10^{-4} to 10^{-3} M DCP in a total volume of 3 ml. The electrophoretically purified enzyme from green leaves had a maximal Q_{02} (N) of 2×10^5 μl. O_2/mgm. protein/min with IAA, and a PZ' of 1100 with pyrogallol.

Colorimetry for IAA was not done as the reaction products gave a positive SALKOWSKI reaction.

Properties of the enzyme. Dialysed or electrophoretically purified enzyme had a pH optimum of 6.3—6.5 in phosphate buffer.

A phenolic cofactor such as 2,4-dichlorophenol (optimal conc. 10^{-3} M) was essential for the activity and the further addition of 10^{-3} M Mn^{2+} increased O_2 uptake 4 to 5 times over that due to DCP alone. Mn^{2+} alone at 10^{-4} to 10^{-2} M had no effect on O_2 uptake. DCP may be replaced by a dialysable, heat-stable, ether-soluble natural activator present in the extract. The identity of this activator is not known but it gives a $FeCl_3$ test and has the IR spectrum of a phenol.

H_2O_2 at 10^{-2} M had no effect on the reaction.

The effect of inhibitors is given in Table 2. Citrate buffer inhibited the Mn^{2+} enhanced oxidation alone, probably by chelation of Mn^{2+}. Pyrophosphate completely inhibited IAA oxidase action but decreased the peroxidase activity by only 30%. A natural, dialysable ether-soluble inhibitor in the crude extract caused a lag in the O_2 uptake by 90—120 mins.

The substrate specificity of the enzyme is given in Table 3. Indoleglycolic acid (IGlyA), but not indoleglyoxylic acid, was oxidised to IAld and hence indole-glyoxylic acid is not an intermediate in the oxidation of IGlyA (STUTZ, 1957). Neither one can be an intermediate in the oxidation of IAA since IAA does not yield IAld. The methylene carbon of the side chain of IAA is retained in the products, as indicated by methylene-C^{14} labelled IAA experiments (STUTZ, 1957).

The nature of the products of the reaction is unknown. They include dark ether insoluble polymer (probably formed during extraction and separation), four acidic and three neutral-basic, ether-soluble compounds.

The ether-soluble compounds were chromatographed (isopropanol, ammonia, water 10:1:1) and sprayed with SALKOWSKI reagent or examined under UV or their spectrum taken after elution. Two neutral-basic (R_f 0.8 and 0.9) and one acidic compound (R_f 0.45) had absorbancy peaks at 280—290 mμ and gave a pink colour with SALKOWSKI reagent; they were probably indolic (for details see STUTZ, 1957).

b) Pea Enzyme (cf. LARSEN, 1955).

Preparation. Epicotyls of dark grown, 6—9 day old Alaska pea plants have generally been used. The epicotyls were homogenised with cold 0.017 M phosphate buffer at pH 6.6, strained through cloth, centrifuged at $2000 \times g$ and the crude extract used as enzyme either directly (GOLDACRE, GALSTON and WEINTRAUB, 1953), or after dialysis for 24—48 hrs. against 0.1 M phosphate buffer at 3° C.

An acetone precipitated preparation, free of natural inhibitors, was prepared by the addition of 40 ml of cold acetone to every 100 ml of cold centrifuged pea epicotyl brei in water. The resulting precipitate was washed twice with 40% acetone in water and finally taken up in 0.066 M phosphate or citrate-phosphate buffer, pH 6.6 (TANG and BONNER, 1947; GOLDACRE, 1951). Saturation of the centrifuged pea brei with $(NH_4)_2SO_4$ did not precipitate all the activity, the bulk of it being retained in the supernatant (MANNING and GALSTON, 1955).

Reaction mixtures with 1 ml of the enzyme (crude, dialysed or acetone precipitated) in 0.01—0.025 M phosphate or citrate-phosphate buffer, pH 6.1—6.6 and containing 2 to 3×10^{-4} M (about 35—50 μgm./ml) IAA have been used, with or without addenda and incubations were at 25° C. One gm. tissue equivalent of the enzyme converts 5—15 μgm. IAA per hour at 25° C.

Properties of the enzyme. The pH optimum for the undialysed enzyme was between 6.2 and 6.7 in citrate-phosphate buffer (TANG and BONNER, 1947) and 5.5 to 5.6 for the acetone precipitated enzyme (RABIN and KLEIN, 1957) or the dialysed (24—48 hrs.) enzyme in phosphate buffer, with a cofactor and Mn^{2+} (SHARPENSTEEN and GALSTON, 1959). Addition of Mn^{2+} (1.7×10^{-3} M) shifted the pH optimum of a dialysed enzyme in buffer from pH 5 to pH 4 (REINERT, SCHRAUDOLF and REINERT, 1957). Maximal enzyme activity occurred with 2.8×10^{-4} M

IAA and the Q_{10} of the reaction between $1-11°$ and $11-25°$ C was 1.6 (TANG and BONNER, 1947).

Mn^{2+} at 5×10^{-3} M replaced a third of the activity lost by 12 hrs. of dialysis (WAGENKNECHT and BURRIS, 1950). Almost all activity was lost on 24—48 hrs. dialysis of the enzyme; the activity was partially increased by the addition of a natural cofactor or DCP alone and fully increased with a cofactor plus Mn^{2+} at 5×10^{-5} M. Mn^{2+} also abolished a lag period of 20—25 mins. present under these conditions (SHARPENSTEEN and GALSTON, 1959). With the undialysed enzyme, Mn^{2+} at 10^{-5} M was about 75% inhibitory in the presence of 10^{-6} M DCP but enhanced the rate by 20 to 30% with 10^{-5} M and 10^{-4} M DCP over the DCP controls (HILLMAN and GALSTON, 1956).

DCP at an optimal concentration of $2-3 \times 10^{-5}$ M enhanced the activity of the undialysed extract by 20 to 1200%, the activity varying with the preparation; it was inhibitory above 2×10^{-4} M. Phenol, several halo phenols, 4-carboxyphenol, resorcinol were all activators (see GOLDACRE et al., 1953 for complete list). Maleic hydrazide (MH) at 1.25×10^{-3} M and 4-methylumbelliferone at 2.5×10^{-5} M enhanced the activity of the acetone-precipitated enzyme about two fold (ANDREAE and ANDREAE, 1953). The activity of the undialysed enzyme was increased by 80% with 8.8×10^{-4} M H_2O_2 in the dark (GALSTON et al. 1953); with an acetone precipitated enzyme, H_2O_2 at 1.3×10^{-4} M had no effect (ANDREAE and ANDREAE, 1953).

A natural, dialysable, alkali-unstable, acid-stable cofactor was obtained from etiolated pea bud tissues. This cofactor, which was insoluble in ether but soluble in 80% ethanol, promoted activity at all concentrations tried and its identity is unknown (SHARPENSTEEN and GALSTON, 1959).

The effect of inhibitors on the enzyme and the substrate specificity are given in Tables 2 and 3 respectively.

The nature of the products of the reaction are unknown, but indolealdehyde, o-aminoacetophenone, o-formamidoacetophenone and 8-OH quinoline could not be detected (MANNING and GALSTON, 1955).

Two neutral compounds which were extracted into $CHCl_3$ at neutral and acid pH's, appeared to be the major products. On chromatography (isopropanol, ammonia, water 2:1:1), they had R_f values of 0.94 and 0.91, and gave a yellow orange EHRLICH colour deepening to red, and a pink colour with $FeCl_3$. A small (5%) amount of indolealdehyde, which gave a 2,4-dinitrophenylhydrazone derivative with the reagent, was claimed to be produced (RACUSEN, 1955).

c) Wheat Leaf Enzyme

(WAYGOOD et al. 1956 a and b, MACLACHLAN and WAYGOOD, 1956).

Preparation. Leaves of light grown, 12—15 days old wheat plants were ground with sand in the cold and expressed. The press juice was centrifuged at $20,000 \times g$ for 30 mins. at 2° C and dialysed against several changes of distilled water for three days at 4° C. Intermittent freezing and thawing gave a clear preparation containing little or no green matter.

3 ml of the reaction mixture containing 0.5 ml enzyme (0.3 mgm. protein N); 0.05 M phosphate buffer, pH 6; 10^{-3} M $MnCl_2$; 5×10^{-4} M resorcinol and 2.22×10^{-3} M IAA were incubated at 30° C. The reaction went to completion in 40—50 mins.

Properties of the enzyme. The pH optimum was about 5 in phosphate buffer with a marked reduction in the rate above pH 6 and below pH 4 with either DCP or resorcinol as cofactor.

The curve relating enzyme concentration to activity was linear at low enzyme concentration (up to 0.25 ml under the above reaction conditions) but fell off at higher concentrations (0.5—1.0 ml). The amount of enzyme required for the completion of the reaction depended on IAA, Mn^{2+} and cofactor concentrations, and 0.5 ml enzyme was optimal under the above conditions. All the added IAA was not destroyed with lower (0.25 ml) or higher (1.0 ml) amounts of enzyme.

The K_m for the enzyme averaged $0.7-1.1 \times 10^{-3}$ M IAA, and the reaction rate was maximal at 1.7×10^{-3} M IAA. The Q_{10} (20—30° C) was 1.5 and the average energies of activation were 7160 cals./mol. between 20—30° C and 11700 cals./mol. between 10 and 20° C with DCP, resorcinol or maleic hydrazide (MH) as cofactors (MACLACHLAN and WAYGOOD, 1956). With resorcinol as cofactor, thermal inactivation of the reaction occurred above 30° C.

Mn^{2+} at 10^{-3} M and 10^{-2} M were optimal for the reaction with resorcinol and DCP as cofactors respectively; no O_2 uptake occurred with the dialysed enzyme in the absence of both Mn^{2+} and a cofactor, and only very slight O_2 uptake in absence of either one. DCP at 5×10^{-4} M and resorcinol at 10^{-3} M were optimal for activity; higher concentrations were inhibitory but the inhibition was reversible by added IAA. The ratios of IAA and the cofactors for optimal activity were IAA/resorcinol = 2.2 and IAA/DCP = 4.4.

Maleic hydrazide showed no optimal concentration and enhanced the rate even at 10^{-1} M.

A natural dialysable activator, capable of replacing resorcinol, could be extracted with ether from the concentrated dialysate (concentrated *in vacuo* at 45° C) after acidification to pH 2 with phosphoric acid. Its identity is unknown. Natural inhibitors in the leaf extracts were removed by dialysis. They were probably phenolic, as indicated by a blue $FeCl_3$ test.

The effect of inhibitors is given in Table 2. Citrate and pyrophosphate at 8.3×10^{-3} M completely inhibited, while below 0.83×10^{-3} M they extended the lag period, possibly by Mn^{2+} chelation.

A lag period in O_2 uptake, ranging from 2—8 mins. for the dialysed enzyme from winter grown plants and 10—20 mins. for the enzyme from summer grown plants, was observed.

This lag could be abolished greatly (80—90%) by (i) exposing the reaction mixture to 60 foot candles white or blue but not red light; (ii) using O_2 instead of air in the reaction chamber; (iii) allowing acetaldehyde to volatilize from the side-arm of the vessel; this reportedly produces H_2O_2 by the aldehyde-aldehyde oxidase system in the enzyme preparation and serves in IAA oxidation. The lag period may be photo-irreversibly but O_2-reversibly extended (up to 3 hrs.) by the addition of 5×10^{-6} M catechol. The lag period in CO_2 evolution was much less than that of O_2 uptake, indicating decarboxylation prior to oxidation. A build up of manganic ions, believed to be the primary electron acceptor, was supposed to occur during the lag or induction period (MACLACHLAN and WAYGOOD, 1956).

d) Pineapple Enzyme (GORTNER and KENT, 1953, 1958).

Preparation. 20—50 g samples of vegetative stem tips of pineapple *(Ananas comosus)* were blended with 1.5 times their weight of cold distilled water, and the brei strained and centrifuged. The milky extract was used directly. An inhibitor of the enzyme in the crude extract could be destroyed by photolysis for $3^1/_4$ hrs. in bright sunlight or removed by 23 hrs. dialysis. Dialysis partially precipitates the enzyme but the precipitate can be resuspended in 1% NaCl. To prepare an enzyme free of cofactors and inhibitors, 100 parts of the aqueous extract was treated twice with 5 parts of 0.1 M H_2O_2 with a 15 minutes interval between treatments, and the mixture kept cold for 1 hr. for the native catalase to destroy all excess H_2O_2. This enzyme had little or no activity without added phenolic cofactors.

The reaction mixture consisted of 0.05 to 0.1 ml of the enzyme extract in 1 ml water, 1 ml IAA (1.12 μmole), 0.2 ml of 1 M phosphate buffer, pH 4 and containing 5×10^{-3} M $MnCl_2$. Reaction was stopped after 10 mins. at 35° C by adding 3 ml of 1:4 $HClO_4$. Under these conditions 1 gm. tissue equivalent of enzyme destroyed 4.5 mgm. IAA.

Properties of the enzyme. The pH optimum for the enzyme was about 4 in phosphate buffer with essentially no activity between pH 6 and 7 (GORTNER and KENT, 1958). Increasing the concentration of the crude, undialysed enzyme to 0.5 ml in the reaction mixture completely inhibited IAA oxidation due to the presence of natural inhibitors which could be removed by dialysis or photolysis (see above).

Mn^{2+} at an optimal concentration of 5×10^{-3} M was essential for the activity of the crude, dialysed or H_2O_2-treated enzyme. A phenolic cofactor such as DCP did not enhance the activity of the crude or dialysed enzyme but was essential for the H_2O_2-treated enzyme. DCP at 1.5×10^{-4} M, p-coumaric acid at 10^{-5} M, the naturally occurring depside of p-coumaric and quinic acids, α- and β-naphthols at 10^{-5} M were all active as cofactors (GORTNER, KENT and SUTHERLAND, 1958; GORTNER and KENT, 1958 for complete list). p-Coumaric acid at higher concentrations acts as a competetive inhibitor in IAA oxidation. The effect of inhibitors is given in Table 2.

A natural heat-stable, dialysable photosensitive inhibitor of unknown identity was present in the crude extracts. It could be precipitated from the dialysate with basic lead acetate and recovered from it by the addition of 0.06 N H_2SO_4 followed by ethyl acetate extraction. The compound has an absorbancy maximum at 310—315 mμ and could be phenolic in nature.

The pineapple enzyme is the most active IAA oxidising enzyme so far studied and 4.7 μgm. of the H_2O_2 treated enzyme (equivalent to 1.2 mgm. fresh weight tissue) oxidises 150 μgm. (0.85 μmole) IAA in 10 mins. in the presence of p-coumaric acid.

e) Omphalia Enzyme (RAY and THIMANN, 1956; RAY, 1956, 1960).

Preparation. About 50 stilboids of the fungus *Omphalia flavida* were cultured for three weeks at 26° C in the dark, in 500 ml of a medium of the following composition:

Glucose	20 gms.
Ammonium tartrate . . .	9.2 gms.
KH_2PO_4	5.8 gms.
$MgSO_4$	2.5 gms.
$FeCl_3$ (1% solution) . . .	0.6 ml
Thiamine HCl	100 μgms.
Distilled water	1000 ml

The medium was decanted and fresh medium added to the mat for a further crop of the enzyme; enzyme production was increased after a few such transfers of the mat.

The decanted medium was filtered through paper, dialysed with agitation, against 10—20 vols. of glass distilled water for 4 days at 3° C, with 5 to 6 changes of water, filtered through sintered funnel and used or stored in a deep freeze where it is stable for at least 4 weeks. A typical preparation had a Qo_2 (N) of 560 μl O_2/hr./mgm. nitrogen and contained 66μgm. N/ml.

Partial purification (2—4 fold) was obtained by adding solid $(NH_4)_2SO_4$ to 90 to 100% saturation, centrifuging, and resuspending the active yellow-brown pellet in cold distilled water and dialysing. The enzyme was further purified four to five fold by adsorption on calcium phosphate gel and elution with 1.5×10^{-3} M Na_2HPO_4 followed by dialysis. Activity was 3.37 μmole IAA destroyed/mgm. dry wt./min. (values calculated from RAY, 1960).

1 to 2 ml of the dialysed medium in 0.05 M citric acid buffer, pH 3.7 in a total of about 3 ml have been used. In manometric experiments about 3×10^{-3} M IAA (sodium salt) and in spectrophotometric experiments about 6×10^{-5} M IAA were employed. Catalytic amounts of H_2O_2 (about 3.3×10^{-6} M) were added to overcome the induction period. Incubations were at 26° C.

Addition of 1% digitonin prevented the crystallisation of IAA at low pH, when the concentration was above 3×10^{-3} M, without affecting the rate.

Assay. To correct for the faint pink colour developed by the IAA oxidation products with the SALKOWSKI reagent, which also interfered with the IAA colour development, an IAA calibration curve was prepared with samples made by mixing 0.9 μmoles IAA in 10 ml buffer containing 0.04 μmole H_2O_2 with a similar solution, where the enzyme had been previously added and all IAA oxidised. Such a curve has some absorbance at zero IAA concentration and a slight upward concavity.

Because pyrogallol peroxidation products of *Omphalia* enzyme had an absorbancy maximum at 380 mμ instead of the 420 mμ for purpurogallin (see "methods", p. 249) the reaction was followed by measuring the O.D. changes at 380 mμ.

Peroxidase activity was assayed at 25° C with 3 ml reaction mixture containing 15 μmoles Na citrate (pH 3.7), 0.6 μmole pyrogallol, and 0.9 μmol H_2O_2. The reaction was stopped after 6 mins. with 0.1 ml 3.75% H_2SO_4 and the O.D. at 380 mμ determined.

A method for following IAA oxidation and pyrogallol peroxidation simultaneously is given by RAY (1960).

Properties of the enzyme. The pH optimum for the enzyme was 3.7 in citric acid buffer, with negligible activity between pH 6 and 7. The same optimum was found for the peroxidase activity of the enzyme. A lag or induction period in IAA oxidation, with a well dialysed sample of the enzyme, was overcome by the addition of catalytic amounts (about 3.3×10^{-6} M) of H_2O_2. In the absence of oxygen, no IAA oxidation occurred even with 4×10^{-5} M H_2O_2; thus the peroxidase action is not exerted on IAA, but only on polyphenols.

Mn^{2+} at 10^{-5} M increased the initial rate of O_2 uptake by only 20—30% and it depressed the decline in the rate of the reaction with time; at 10^{-3} M the initial rate was depressed, but the total O_2 consumption (after 17 hrs.) exceeded that of the controls, and 1.5 moles of O_2 instead of 1 mole O_2 per mole IAA was taken up. DCP at 10^{-5} to 10^{-3} M concentrations had no effect on O_2 uptake.

The effect of inhibitors is given in Table 2. Pyrogallol inhibition of IAA oxidation only affected the rate and not the lag period. A mixture of 95% CO/ 5% O_2 inhibited the reaction up to 50%, the controls being in 95% N_2/5% O_2 in the dark; this inhibition was completely reversed by light. However, increasing the oxygen concentration did not reverse the inhibition sufficiently to suggest that O_2 and CO competed for the same site on the enzyme (RAY, 1960). The substrate specificity of the enzyme is given in Table 3.

IAA at 5×10^{-6} M inhibited the peroxidase activity of the enzyme on pyrogallol by 50%, and 97% at 1.8×10^{-4} M. Two other substrates for the enzyme, 5-OH-IAA and IiBA (see substrate specificity, Table 3) at 1.8×10^{-4} M caused 65% and 54% inhibitions respectively of pyrogallol peroxidation.

Moreover, IAA oxidation was itself inhibited by 95% when pyrogallol peroxidation was inhibited 85—90% by H_2O_2; the inhibition therefore does not depend on competition for the enzyme or H_2O_2, but more probably on the formation of an inactive form of the enzyme in the presence of both substrates (RAY, 1960).

A labile intermediate was produced enzymatically during IAA oxidation following the uptake of O_2 and the evolution of CO_2. This reaction was cyanide-sensitive. The intermediate gave a stable complex with $NaHSO_3$ (3×10^{-3} M) and was not released from it by adding alkali; hence it may not be a simple carbonyl adduct. Although the intermediate had an indole type spectrum it is not IAld or an indolenine type of compound (RAY, 1956) and its identity is unknown. It decomposed spontaneously, or on the addition of acids, to the final neutral products.

A complex of neutral products, retaining quantitatively the 2-carbon of the indole ring, were extracted at pH 3.5 or 10 by $CHCl_3$ or ether from the reaction mixture. No pure products have been isolated. Indolealdehyde, o-aminoacetophenone or quinoline are not the major products as indicated by UV spectrum of the products.

The IR and UV spectra of a white compound, which precipitated from a cold $CHCl_3$ solution of the products, resembled those of 3-methyldioxindole but the compound did not have quite its solubility properties, and may be polymeric in nature (see section a above). A red compound also occurred in the $CHCl_3$ extract in major amounts (STOWE et al., 1957).

Chromatography of the products in benzene, n-butanol, methanol, water, 2:2:4:1 v/v, a solvent which separated neutral compounds without streaking, gave four spots (R_F 0.78, 0.69, 0.66, 0.62) of which two gave a pink SALKOWSKI colour (R_F 0.78 and 0.66).

Acknowledgment: The author wishes to thank Prof. K. V. THIMANN, Biological Laboratories, Harvard University, for kindly reading the manuscript and for his several suggestions.

References.

ANDREAE, W. A., and S. R. ANDREAE: Can. J. Bot. **31**, 426 (1953). — ASHBY, W. C.: Bot. Gaz. **112**, 237 (1951).

BENTLEY, J. A., and A. S. BICKLE: J. exptl. Bot. **3**, 406 (1952). — BRIGGS, W. R., G. MOREL, T. A. STEEVES, I. M. SUSSEX and R. H. WETMORE: Plant Physiol. **30**, 143 (1955). — BOYSEN-JENSEN, P.: Biochem. Z. **236**, 205 (1931).

CLARKE, A. J., and P. J. G. MANN: Biochem. J. **65**, 763 (1957).

DANNENBURG, W. N., and J. L. LIVERMAN: Plant Physiol. **32**, 263 (1957).

FAWCETT, C. H., R. L. WAIN and F. WIGHTMAN: Nature (Lond.) **181**, 1387 (1958). — FRIEBER, W.: Centr. Bakteriol. Parasitenk. **87**, 254 (1921).

GALSTON, A. W., J. BONNER and R. S. BAKER: Arch. Biochem. Biophys. **42**, 456 (1953). — GALSTON, A. W., and S. M. SIEGEL: Science **120**, 1070 (1954). — GEORGI, C. E., and A. E. BEGUIN: Nature (Lond.) **143**, 25 (1939). — GOLDACRE, P. L.: Australian J. Sci. Res. (B) **4**, 293 (1951). — GOLDACRE, P. L., A. W. GALSTON. and R. L. WEINTRAUB: Arch. Biochem. Biophys. **43**, 456 (1953). — GORDON, S. A.: Ann. Rev. Plant Physiol. **5**, 341 (1954); — In: R. L. WAIN and F. WIGHTMAN, "The Chemistry and Mode of Action of Plant Growth Substances", p. 65. London: Butterworths 1956; — Proceedings of the International Conference on the Peaceful uses of Atomic Energy. Vol. 11, p. 283. United Nations, New York 1956a; — Plant Physiol. **33**, 23 (1958). — GORDON, S. A., and F. S. NIEVA: Arch. Biochem. **20**, 367 (1949). — GORDON, S. A., and L. G. PALEG: Physiol. Plant. **10**, 39 (1957). — GORDON, S. A., and R. P. WEBER: Plant Physiol. **26**, 192 (1951). — GORTNER, W. A., and M. KENT: J. biol. Chem. **204**, 593 (1953); **233**, 731 (1958). — GORTNER, W. A., M. KENT and G. K. SUTHERLAND: Nature (Lond.) **181**, 630 (1958). — GRAY, R. A.: Arch. Biochem. Biophys. **81**, 480 (1959). — GRUEN, H. E.: Ann. Rev. Plant Physiol. **10**, 405 (1959).

HENDERSON, J. H. M., and J. BONNER: Amer. J. Bot. **39**, 444 (1952). — HILLMAN, W. S., and A. W. GALSTON: Physiol. Plant. **9**, 230 (1956). — HIRATA, S.: Ann. Phytopathol. Soc. Japan **22**, 153 (1957) [Quoted from GRUEN, H. E.: Ann. Rev. Plant Physiol. **10**, 405 (1959)]; — Miyazaki Daigaku Nôgakubu Kenkyû Jihô **4**, 119 (1958). — HOPKINS, F. G., and S. W. COLE J. Physiol. **29**, 451 (1903).

KAPER, J. M., and H. VELDSTRA: Biochim. Biophys. Acta **30**, 401 (1958). — KEFFORD, N. P., J. BROCKWELL and J. A. ZWAR: Australian J. biol. Sci. **13**, 456 (1960). — KEILIN, D., and E. F. HARTREE: Biochem. J. **49**, 88 (1951). — KENTEN, R. H., and P. J. G. MANN: Biochem. J. **50**, 360 (1952). — KENTEN, R. H.: Biochem. J. **55**, 350 (1953); **59**, 110 (1955); **61**, 353 (1955a). — KULESCHA, Z.: Thesis Paris, 114 pp. (1951). — KUNKEL, H. G., and R. J. SLATER: Proc. Soc. exptl. Biol. Med. **80**, 42 (1952).

LARSEN, P.: Planta **30**, 673 (1940); — Am. J. Bot. **36**, 32 (1949); — Plant Physiol. **26**, 697 (1951); — In: K. PAECH and M. V. TRACEY, "Modern Methods of Plant Analysis", Vol. III, p. 565. Berlin: Springer-Verlag 1955. — LEGRAND, G.: Compt. rend. soc. biol. **151**, 921 (1957); — Bull. soc. chim. biol. **39**, 1289 (1957a). — LIBBERT, E., and G. BALLIN: Naturwissenschaften **46**, 532 (1959). — LIPETZ, J., and A. W. GALSTON: Am. J. Bot. **46**, 193 (1959). — LIVERMAN, J. L., and W. L. DANNENBURG: Plant Physiol. **32**, supplement xviii (1957). — LUND, H. A.: Plant Physiol. **31**, 334 (1956).

MACLACHLAN, G. A., and E. R. WAYGOOD: Can. J. Biochem. and Physiol. **34**, 1233 (1956). — MAHADEVAN, S.: "The Enzymatic Hydrolysis of Nitriles", Ph.D. Thesis, Harvard Univ., Cambridge, Mass. (U.S.A.) (1960). — MAHADEVAN, S., and K. V. THIMANN (in preparation) for publication in Arch. Biochem. Biophys. — MANN, P. J. G.: Biochem. J. **59**, 609 (1955). — MANNING, D. T., and A. W. GALSTON: Plant Physiol. **30**, 225 (1955). — MELCHIOR, G. H.: Planta **50**, 557 (1958). — MICHEL, B. E.: (Personal communication). — MUDD, J. B., B. G. JOHNSON, R. H. BURRIS and K. P. BUCHHOLTZ: Plant Physiol. **34**, 144 (1959). — MURAKAMI, Y., and Y. HAYASHI: J. Agric. Chem. Soc. Japan **31** (7), 468 (1957).

OAKS, A., and M. SHAW: Can. J. Bot. **38**, 761 (1960). — OGATA, Y., and M. OKANO: J. Chem. Soc. Japan (Pure Chem. Section) **70**, 32 (1949).

PILET, P. E.: Act. Soc. helv. sci. nat. **135**, 133 (1955); — Experientia (Basel) **13**, 35 (1957). — PLATT, R. S. jr.: Ann. Biol. **30**, 349 (1954). — PLATT, R. S. jr., and K. V. THIMANN: Science **123**, 105 (1956).

RABIN, R. S., and R. M. KLEIN: Arch. Biochem. Biophys. **70**, 11 (1957). — RACUSEN, D.: Arch. Biochem. Biophys. **58**, 508 (1955). — RAY, P. M.: Arch. Biochem. Biophys. **64**, 193

(1956); — Ann Rev. Plant Physiol. **9**, 81 (1958); — Arch. Biochem. Biophys. **87**, 19 (1960). — RAY, P. M., and K. V. THIMANN: Arch. Biochem. Biophys. **64**, 175 (1956). — REINERT, J., H. SCHRAUDOLF and U. REINERT: Z. Naturforsch. **126**, 569 (1957). — REINERT, J., H. SCHRAU-DOLF and M. TAZAWA: Naturwissenschaften **44**, 588 (1957).

SCHWARZ, K., and A. A. BITANCOURT: Science **126**, 607 (1957). — SEQUEIRA, L., and T. A. STEEVES: Plant Physiol. **29**, 11 (1954). — SEELEY, R. C., C. H. FAWCETT, R. L. WAIN and F. WIGHTMAN: In: R. L. WAIN and F. WIGHTMAN, "Chemistry and Mode of Action of Plant Growth Substances", p. 234. London: Butterworths 1956. — SHARPENSTEEN, H. A., and A. W. GALSTON: Physiol. Plant. **12**, 465 (1959). — SHIGEMURA, T., and S. A. GORDON: Plant Physiol. **35**, Suppl. xxviii (1960). — SKOOG, F.: J. gen. Physiol. **20**, 311 (1937). — STOWE, B. B.: Biochem. J. **61**, (ix) (1955); — (unpublished report, 1955; personal communication); — Fortschr. Chem. org. Naturstoffe **17**, 248 (1959). — STOWE, B. B., and K. V. THIMANN: Arch. Biochem. Biophys. **51**, 499 (1954). — STOWE, B. B., K. V. THIMANN and N. P. KEFFORD: Plant Physiol. **31**, 162 (1956). — STOWE, B. B., P. M. RAY and K. V. THIMANN: Rapports et Comm. VIIIe Congres Int. de Botanique (Paris 1954) **11** suppl., 135 (1957). — STUTZ, R. E.: Plant Physiol. **32**, 31 (1957).

TANG, Y. W., and J. BONNER: Arch. Biochem. **13**, 11 (1947); — Amer. J. Bot. **35**, 570 (1948). — THIMANN, K. V.: J. gen. Physiol. **18**, 23 (1934); — J. biol. Chem. **109**, 279 (1935); — Proc. Internat. Congr. Soil Sci. 3rd. Comm. A, 24 (1939); — Arch. Biochem. Biophys. **44**, 242 (1953). — THIMANN, K. V., and S. MAHADEVAN: Nature (Lond.) **181**, 1466 (1958). — TON-HAZY, N. E., and M. J. PELCZAR: Science **120**, 141 (1954).

UMBREIT, W. W., R. H. BURRIS and J. H. STAUFFER: "Manometric techniques and Tissue Metabolism". Minneapolis: Burgess Publ. Co. 1949.

VLITOS, A. J., and W. MEUDT: Contrib. Boyce Thompson Inst. **174** B, 417 (1954).

WAGENKNECHT, A. C., and R. H. BURRIS: Arch. Biochem. **25**, 30 (1950). — WAYGOOD, E. R., A. OAKS and G. A. MACHLACHLAN: Can. J. Bot. **34**, 54 (1956a); **34**, 905 (1956b). — WEISSBACH, H., W. KING, A. SJOERDSMA and S. UDENFRIEND: J. biol. Chem. **234**, 81 (1959). WERLE, E., and F. ROEWER: Biochem. Z. **320**, 298 (1950). — WIGHTMAN, F., and A. C. NEISH read in First Annual meeting of the Canadian (1960).

Enzymes of Aromatic Biosynthesis.

By

Takayoshi Higuchi and Ichiji Kawamura.

With 2 Figures.

The higher plants form many kinds of aromatic compounds, such as aromatic amino acids, lignins, flavonoids, phenolic glycosides, coumarins, and alkaloids. Some of these, like the aromatic amino acids, are universally distributed in nature and others are found in a rather restricted number of plants. The biosynthetic pathways of these compounds in plants have been actively investigated recently by tracer experiments using C^{14}-labelled compounds. It has been assumed that the universally occurring aromatic amino acids are made from carbohydrates following essentially the same pathway — "DAVIS' scheme of aromatic biosynthesis" — as in microorganisms, and that the aromatic compounds restricted to certain species might be derived secondarily from the fundamental aromatic amino acids or intermediates involved in their biosynthesis (NEISH, 1960a). The investigations of aromatic biosynthesis in plants have now reached the stage where hypotheses about biochemical pathways obtained by the tracer experiments have to be confirmed by the series of biochemical reactions based on the enzymes or enzyme systems, as has been done in aromatic biosynthesis of microorganisms. However, few investigations on the enzymes of aromatic biosynthesis have hitherto been carried out.

I. Enzymes of Aromatic Biosynthesis from Non-Aromatic Compounds in Microorganisms.

The series of auxotrophic and enzymatic investigations of DAVIS (1955, 1958) with biochemical mutants of *E. coli* established a biosynthetic pathway of the aromatic amino acids from sugars. The biosynthetic pathway of phenylalanine and tyrosine published by these investigators is shown in Fig. 1. In the scheme, dehydroquinic acid, dehydroshikimic acid and shikimic acid were confirmed as intermediates in the formation of all the aromatic compounds required by *E. coli* for growth, and quinic acid was also confirmed as a growth factor for a mutant of *Aerobacter aerogenes*. It has further been proved that these bacterial extracts contain the enzymes promoting the reactions of Fig. 1, and that 5-dehydroquinic acid is formed by combination of D-erythrose-4-phosphate with phosphoenol pyruvate, the latter being formed by glycolysis. The mechanism of the formation of dehydroquinic acid was further elucidated recently by an enzymatic study, and 2-keto-3-deoxy-D-araboheptonic acid-7-phosphate was identified as an intermediate in this reaction (SRINIVASAN and SPRINSON, 1959). Prephenic acid plays a role as a branching compound in the formation of either phenylpyruvic acid or *p*-hydroxyphenylpyruvic acid, and it has been believed to be derived from compound "Z1", to which the structure 3-enolpyruvyl shikimate has been recently assigned. However, the enzymatic study recently done (LEVIN and SPRINSON,

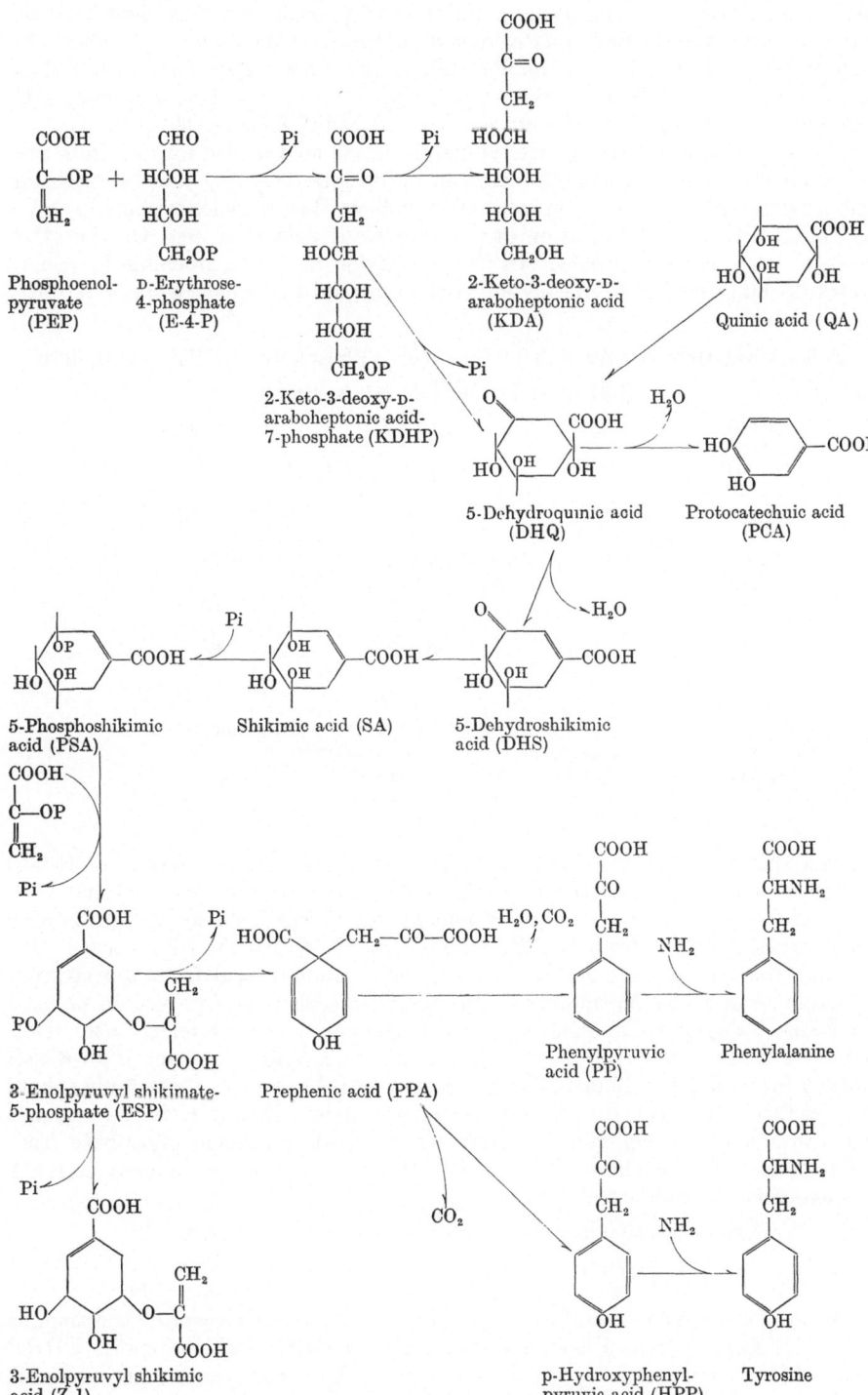

Fig. 1. Biosynthesis of phenylalanine and tyrosine in microorganisms.

1960), has shown that 3-enolpyruvyl shikimate-5-phosphate, rather than Z1 is the active intermediate in prephenate formation and that 3-enolpyruvyl shikimate-5-phosphate is formed by the condensation of phosphoenolpyruvate with 5-phosphoshikimic acid and then it is dephosphorylated to 3-enolpyruvyl shikimate (Z1). Phenylpyruvic acid is derived from prephenic acid by the catalysis of the enzyme prephenic aromatase and p-hydroxyphenylpyruvic acid is also formed from prephenic acid by the action of another enzyme "prephenic dehydrogenase" (Schwinck and Adams, 1959). There is no reason to believe that p-hydroxyphenylpyruvic acid is formed by hydroxylation of phenylpyruvic acid in $E.$ $coli$. On the other hand, the enzyme which catalyzes the conversion of 5-dehydroshikimic acid to protocatechuic acid has been isolated from a mutant of *Neurospora crassa* recently (Gross, 1958).

a) 2-Keto-3-Deoxy-D-Araboheptonic Acid-7-Phosphate (KDHP) Synthetase

(Srinivasan and Sprinson, 1959).

Phosphoenol-pyruvate (PEP)	Erythrose-4-phosphate (E-4-P)	2-Keto-3-deoxy-D-araboheptonic acid-7-phosphate (KDHP) + Pi	

The enzyme catalyzes the condensation of D-erythrose-4-phosphate (E-4-P) and phosphoenolpyruvate (PEP) to form 2-keto-3-deoxy-D-araboheptonic acid-7-phosphate (KDHP) and inorganic phosphate; it was isolated by Srinivasan and Sprinson (1959) from the $E.$ $coli$ mutant 83-24. By the discovery of this enzyme, the quantitative formation of 5-dehydroquinic acid from D-erythrose-4-phosphate and phosphoenolpyruvate, observed earlier, was proved to proceed via 2-keto-3-deoxy-D-araboheptonic acid-7-phosphate as an intermediate. Hurwitz and Weissbach (1959) have also shown that a cell-free extract of an $E.$ $coli$ mutant forms a 2-keto-3-deoxyheptonic acid (KDA) from D-ribose-5-phosphate or D-erythrose-4-phosphate plus phosphoenolpyruvate. However, as the enzymes also contained a phosphatase activity which yields inorganic phosphate from 2-keto-3-deoxy-7-phosphoheptonic acid (KDPA) the synthetic pathway of KDA was explained by following equations:

Phosphoenolpyruvate + O-erythrose-4-phosphate → KDPA + Pi

KDA + Pi

Assay. The activity of the enzyme can be determined by measuring the amount of KDHP formed from E-4-P and PEP. By the action of periodate, KDHP forms β-formylpyruvic acid, which reacts with thiobarbituric acid giving an intense pink color with an absorption maximum at 549 mμ; hence the activity can be assayed through the colorimetric determination of KDHP formed.

Procedure. The incubation mixture contains 100 μmoles of potassium phosphate buffer (pH 6.4), 0.5 μmole each of E-4-P and PEP, and enzyme fraction in a total volume of 1.0 ml. The reaction was started by the addition of enzyme after pre-incubation of the other constituents at 37° for 10 minutes. After 5 minutes at 37° the reaction was stopped by the addition of 0.4 ml of 10 per cent trichloroacetic acid and the mixture was centrifuged.

KDHP (0.1 to 0.05 μmoles) in 0.25 ml of solution was treated with 0.25 ml of 0.025 M periodic acid in 0.125 N H_2SO_4. After 45 minutes at room temperature, 0.5 ml of 2 per cent sodium arsenite in 0.5 N HCl was added to destroy the excess periodate (2 minutes at room temperature). Thiobarbituric acid solution (2 ml, 0.3%) was added and the tubes placed in a boiling water bath for 5 minutes. After cooling in a water bath at 40° the pink color was measured immediately at 549 mμ in a spectrophotometer. A unit of enzyme was defined as that amount of enzyme which will form 0.1 μmol of KDHP in 5 minutes under the above-mentioned conditions.

Paper chromatographic identification of KDHP. A solution equivalent to 3 to 5 μg of KDHP was spotted on acid-washed Whatman No. 1 paper and developed for 24 hours by descending chromatography with a mixture of tertiary amyl alcohol-formic acid (98 per cent)-water (3:3:1). The paper was dried at room temperature and sprayed with 0.1 M periodic acid in 0.125 N H_2SO_4. After 20 minutes at room temperature the paper was sprayed with 10 per cent sodium arsenite in 0.5 N HCl. The paper became colored by liberated iodine which was removed by further reaction with arsenite. This was repeated until further application of arsenite did not produce the color of iodine. The paper was allowed to dry at room temperature, sprayed with 0.6 per cent thiobarbituric acid (pH 2.0) and dried in an oven at 90° for 5 minutes. Both KDHP and 2-keto-3-deoxy-galactonic acid appeared as pink spots by this treatment. The R_f values for these two compounds were 0.47 and 0.55, respectively.

Preparation of enzyme. *E. coli* mutant 83−24 was grown for 18 hours with aeration at 37° in medium A [K_2HPO_4, 7.0 g; KH_2PO_4, 3.0 g; Na-citrate · 3H_2O, 0.5 g; $MgSO_4$· 7H_2O, 0.1 g; $(NH_4)_2SO_4$, 1.0 g; glucose, 2.0 g; H_2O, 1000 ml; pH 7.0] supplemented with 0.2 per cent yeast extract (Difco) and 0.2 per cent casein hydrolysate. The cells were harvested by centrifugation at 2°, washed with cold water, suspended in M/30 phosphate buffer, pH 7.4 (20 ml/5.0 g of wet bacteria), and disrupted by means of sonic oscillation for 30 minutes in a 9-Kc Raytheon oscillator cooled with circulating ice water. Centrifugation at 13,000 ×g in a Spinco preparative centrifuge yielded a clear greenish yellow solution containing 16 mg of protein per ml. All subsequent operations were carried out at 2°.

Two hundred ml of the cell free extract was treated with 28 ml of 2 per cent protamine sulfate solution and the precipitate was removed by centrifugation. To 218 ml of the supernatant solution were added slowly with stirring 54 g of $(NH_4)_2SO_4$ and, after being stirred for another 20 minutes, the precipitate was removed by centrifugation and discarded. The procedure was repeated with 21 g of $(NH_4)_2SO_4$ on the supernatant solution. The precipitate was removed by centrifugation, dissolved in 45 ml of M/30 potassium phosphate buffer, pH 7.4, and dialyzed against the same buffer.

The rest of the fractionation procedure was best carried out on a smaller scale. Twenty ml of the dialysate was adjusted to pH 5.4 with 1 N acetic acid and the precipitate was removed by centrifugation. The supernatant solution (cooled by an ice bath) was treated with 17 ml of acetone (pre-cooled to −15°) slowly with stirring and the precipitate was removed by centrifugation and discarded. The

17a

supernatant solution was treated similarly with 12 ml of acetone, and the precipitate was removed by centrifugation, dissolved in 5 ml of M/30 potassium phosphate buffer, pH 7.6 and dialyzed against 2 liters of 0.01 M phosphate buffer, pH 6.8, as described previously.

The diethylaminoethyl (DEAE)-cellulose column used in the next step was prepared as follows: To 3 g of DEAE-cellulose suspended in 100 ml of H_2O at room temperature, 1.5 ml of 1.0 M KH_2PO_4 was added with stirring. (The pH of the solution was 6.8). The cellulose was removed by filtration on a Büchner funnel, washed several times with 0.01 M potassium phosphate buffer, pH 6.8 and suspended in 70 ml of the same buffer. Ten ml of this suspension was placed on a chromatographic column (13 mm diameter) and packed by mild suction (length 38 mm). The column was transferred to a cold room at 2° and equilibrated with cold buffer by allowing approximately 50 ml to pass through the column (2 hours).

One ml of the dialyzed acetone fraction was placed on the column and eluted successively with 5.0 ml each of phosphate buffer, pH 6.8, of the following molar concentrations (flow rate 5 ml in 20 minutes): 0.01, 0.02, 0.04, 0.06, 0.08, 0.10, 0.10, 0.16 and 0.20. The desired enzyme activity was present in the two 0.1 M fractions. Four such fractions (from two column operations) were combined (3.8 mg of protein), and immediately concentrated by precipitation with $(NH_4)_2SO_4$ (80 per cent saturation). The precipitate was dissolved in 0.8 ml M/30 potassium phosphate buffer, pH 7.4, dialyzed against the same buffer, and clarified by centrifugation. Recovery of protein was 1.8 mg. A summary of the enzyme purification is given in following table. The $(NH_4)_2SO_4$ and acetone fractions were stable for at least several weeks at —15°.

Purification of KDHP synthetase

Fraction	Volume ml	Protein concentration mg/ml	Total activity units	Specific activity units/mg
Crude extract	200	16	1080	0.34
Ammonium sulfate (40—55%) .	45	15	790	1.20
Acetone (47—59%)	11	7.3	440	5.6
DEAE cellulose (0.1 M phosphate eluate)	110	0.19	420	20

Properties. *Specificity.* Erythrose-4-phosphate cannot be replaced by D-erythrose, D-glyceraldehyde-3-phosphate, ribose-5-phosphate, glucose-6-phosphate, glucosamine-6-phosphate or N-acetylglucosamine-6-phosphate. Pyruvate or pyruvate plus ATP cannot substitute for phosphoenolpyruvate.

Inhibitors. With the acetone fraction, ethylendiaminetetraacetate (EDTA) (4×10^{-4} M) did not inhibit the reaction. Co^{++}, Zn^{++}, Mg^{++} at a concentration of 2×10^{-3} M had no effect on the condensation. Fluoride, sodium arsenite, azide and iodoacetate did not affect the rate of formation of KDHP. p-Chloromercuribenzoate (2×10^{-5} M) inhibited the reaction completely and this inhibition could be reversed by cysteine. The enzyme therefore seems to have a thiol group as a prosthetic group. Sedoheptulose-1,7-diphosphate and several other phosphorylated carbohydrates were inhibitory to this reaction.

Effect of pH. The pH optimum is at 6.4 in potassium phosphate buffer. With Tris-maleate buffer, the H^+ ion concentration had little effect on the reaction between pH 6.4 and 7.4.

MICHAELIS *constant.* MICHAELIS constants for erythrose-4-phosphate and phosphoenolpyruvate, determined at pH 6.4 (potassium phosphate buffer) were 1.2×10^{-3} M and 3.5×10^{-3} M, respectively.

Reversibility. All attempts to reverse the condensation reaction were unsuccessful.

b) The Enzyme Converting KDHP to Dehydroquinate
(SRINIVASAN and SPRINSON, 1959).

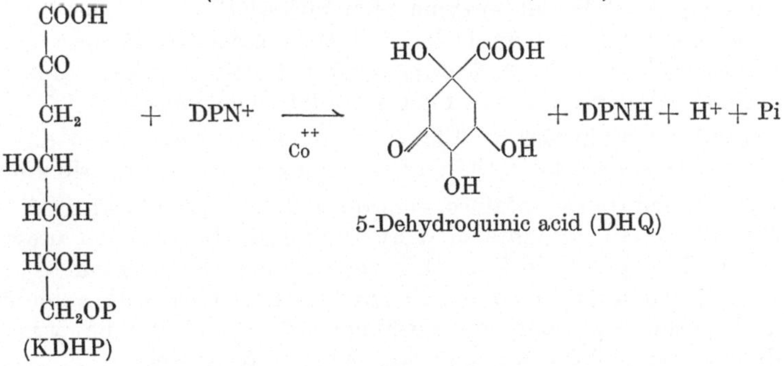

$$
\begin{array}{l}
\text{COOH} \\
|\\
\text{CO} \\
|\\
\text{CH}_2 \\
|\\
\text{HOCH} \\
|\\
\text{HCOH} \\
|\\
\text{HCOH} \\
|\\
\text{CH}_2\text{OP}
\end{array}
\quad + \quad \text{DPN}^+ \quad \xrightarrow[\text{Co}^{++}]{} \quad
\text{5-Dehydroquinic acid (DHQ)} \quad + \text{DPNH} + \text{H}^+ + \text{Pi}
$$

(KDHP)

The enzyme catalyzing above reaction was contained in the crude enzyme preparation of KDHP synthetase. However, the enzyme has not yet been purified.

Assay. The activity of the enzyme can be determined by the microbiological assay of the amount of 5-dehydroquinic acid (DHQ) formed and also by measuring the optical density at 340 mμ of DPNH under following procedure.

Procedure. The reaction mixture contained KDHP, 50 μmoles of potassium phosphate buffer, pH 7.4, 0.25 μmole of DPN, 1.0 μmole of Co^{++}, and 0.1 ml of enzyme (1.5 mg of protein) in a total volume of 1.0 ml. After incubation at 37° for 1 hour, 0.02 ml of 6 N HCl was added, proteins were removed by centrifugation, and 0.2 ml of the clear supernatant solution was used for the microbiological assay of the products with *E. coli* mutant A170-143 Sl (DAVIS and WEISS, 1953).

Preparation of the enzyme. The crude enzymes before the purification of KDHP synthetase by (NH$_4$)$_2$SO$_4$ fractionation contain this enzyme.

Properties. The enzyme required DPN as the electron acceptor.

Inhibitors. The enzyme required Co^{++} as cofactor and the reaction was completely inhibited by EDTA.

c) Dehydroshikimic (DHS) Dehydrase
(GROSS, 1958).

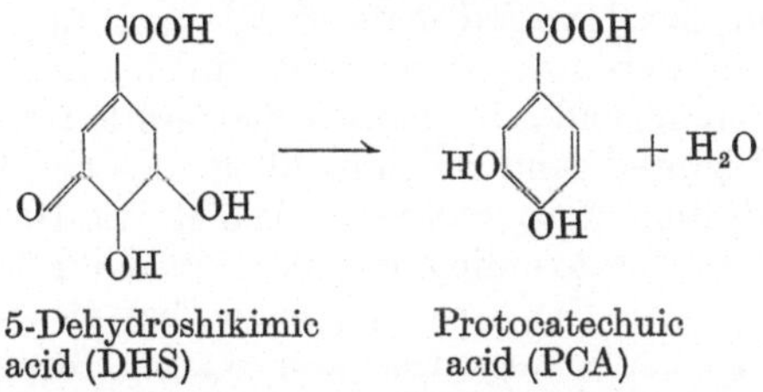

5-Dehydroshikimic acid (DHS) $\longrightarrow$ Protocatechuic acid (PCA) $+ H_2O$

This enzyme was isolated by GROSS from a *Neurospora crassa* mutant which accumulated 5-dehydroshikimic acid. The enzyme is an inducible enzyme and catalyzes the conversion of 5-dehydroshikimic acid (DHS) to protocatechuic acid (PCA). In this reaction 5-dehydroshikimic acid is converted to protocatechuic acid by dehydration across positions 2 and 3 of the enol form of 5-dehydroshikimic acid[1].

---

[1] A similar enzyme was also found in tea leaves recently (ISHIKAWA, 1960).

Assay. Dehydroshikimic dehydrase activity can be estimated by following the reduction of optical density at 234 mμ with the use of DHS as substrate (ε 234 = 1.1×10^4, pH 7.2, in 0.05 M Tris-HCl buffer). Some of the decrease of optical density is the result of the conversion of DHS to 5-dehydroquinic acid by dehydroquinase present in the enzyme preparation. However, the equilibrium of this reaction strongly favors DHS, and the extensive decrease in optical density observed is largely due to the conversion of DHS to β-ketoadipic acid by dehydroshikimic dehydrase and the PCA-degradative enzymes.

PCA concentrations in the reaction mixture can be determined directly by measuring the optical density of the ferric complex of the acid at 640 mμ.

Procedure. Preparations obtained contained large amounts of PCA oxidase and usually only small amounts of dehydroquinase. In order to prevent the oxidation of the PCA produced from DHS the reaction was run in an oxygen-free nitrogen atmosphere in 8 ml modified THUNBERG tubes filled with a double stop-cocked head permitting complete flushing with gas. Nitrogen was passed through alkaline pyrogallol before entering the reaction vessels.

The assay mixture usually contained 2.4 ml of 0.05 M Tris-HCl buffer, pH 7.4, and 0.5 ml of enzyme preparation containing 10 to 20 units of activity. A unit of activity is defined as that amount of enzyme which catalyzes the conversion of 10^{-9} moles of DHS to PCA per minute under the conditions described. DHS (3 μmoles for maximal initial velocity) in 0.1 ml of Tris-HCl buffer was added to the side arm and the tubes flushed with nitrogen for 1 hour at 0°. After a 10 minute equilibration period at 30° the reaction was initiated by tipping the contents of the side arm into the main compartment. The reaction was stopped by incubating the reaction vessels for 10 minutes at 60°. Tris-maleate buffer (2 ml of 0.5 M, pH 5.2) was then added and the reaction mixture heated for an additional 5 minutes at 60°. The flocculent precipitate was removed by centrifugation at 30,000×g for 15 minutes and the supernatant solution assayed for PCA.

Assays of PCA. At pH 5.3 to 5.5 PCA gave a consistent extinction coefficient at 640 mμ. In general, 2 ml of 0.5 M Tris-maleate buffer, pH 5.2 was added to 3 ml of reaction mixture (usually buffered at pH 7.4 with 0.05 M Tris-HCl) to bring the pH to 5.3 to 5.4. The optical density of either 2.9 ml of the above or of a sample diluted with water was determined after the addition of 0.1 ml of a 1.0 per cent aqueous FeCl$_3$ solution. The optical density varies proportionately with PCA content, between 3 and 20 μg per ml.

Preparation of enzyme. Mycelial mats of *N. crassa* mutant Y7655a obtained from 400 ml of medium were filtered and washed three times with distilled water. The mycelia were then ground with mortar and pestle for 5 minutes in the presence of washed and ignited sand weighing about one-half the wet weight of the mycelia. After the addition of about 1 ml of 0.05 M Tris-HCl, pH 7.4, buffer per g wet weight of mycelia, the grinding was continued for 5 minutes. The slurry produced was diluted with the same buffer so that the wet weight-to-volume ratio was 1:10, and it was homogenized in a Waring Blendor intermittently for 2 minutes at top speed. After an extraction period of from 6 to 10 hours, the debris was removed by centrifugation at about 1000×g. All operations subsequent to and including grinding were carried out at 4° to 5°. The crude extract was adjusted to pH 6.5 with 1 M acetic acid, and protamine sulfate was added until precipitation was complete (about 30 mg per 100 ml of extract). The pH was held constant during the operation by dropwise addition of a 1 M solution of sodium bicarbonate. After removal of the precipitate by centrifugation (10,000×g for 10 minutes), the supernatant solution was brought to 60 per cent saturation with (NH$_4$)$_2$SO$_4$.

The precipitate thus obtained, after collection by centrifugation as above, contained all of the original enzyme activity. The precipitate was then dissolved in 0.05 M Tris-HCl buffer, pH 7.4, and solid $(NH_4)_2SO_4$ was added slowly until 32 per cent saturation was obtained. After centrifugation enough $(NH_4)_2SO_4$ was added to the supernatant solution to yield 50 per cent saturation. The precipitate obtained between 32 and 50 per cent saturation was dissolved in 0.05 M Tris-HCl buffer and used. All operations were carried out at 4°. The specific activity of the enzyme solution thus obtained represented a 3-fold increase but the separation of the enzyme from PCA oxidase was unsuccessful.

Properties. *Inhibitors.* The enzyme is completely inhibited by *p*-chloromercuribenzoate (10^{-4} M). This inhibition is reversed by reduced glutathione (5×10^{-4} M) added either before or simultaneously with the initiation of the reaction. The enzyme therefore seems to have a thiol group which is necessary for activity.

Effect of pH. The pH optimum is between 7.4 and 7.6.

MICHAELIS *constant.* The MICHAELIS constant for the substrate is 6.0×10^{-4} M at pH 7.4.

d) 5-Dehydroquinase
(MITSUHASHI and DAVIS, 1954a).

This enzyme was isolated by MITSUHASHI and DAVIS from *E. coli*. The enzyme catalyzes the conversion of 5-dehydroquinic acid (DHQ) to 5-dehydroshikimic acid (DHS) and is found in *E. coli*, *Aerobacter aerogenes*, yeast, peas and spinach leaves.

Assay. The assay method is based on the fact that DHS has an ultra-violet absorption peak at 234 mμ, with a molar extinction coefficient of 11,900, whereas DHQ does not absorb significantly at this wavelength.

Procedure. 0.1 ml of 1 M phosphate buffer, pH 7.4, the enzyme sample, and enough water to make a final volume of 3.0 ml were placed in a silica cell having a 1 cm light path. No change in OD_{234} should take place over the next 2 to 3 minutes. 0.1 ml of 3 mM DHQ was added and the optical density at 234 mμ read at 30 second intervals (room temperature, ca. 25°).

A unit of enzyme was defined as the amount that catalyzes the formation of 1 μmole of DHS per minute under the above conditions. Thus 0.01 unit per 3.0 ml would cause an OD_{234} increment of 0.040 per minute. Specific activity was expressed as units per miligram of protein.

Extracts of *E. coli* or *Aerobacter*, prepared as described below, usually contained 0.2 unit of enzyme per milliliter and could be assayed directly. Although these extracts contained the enzyme for converting DHS to shikimic acid, this reaction did not proceed significantly in the absence of an added source of TPNH.

Preparation of enzyme. The extraction of the enzyme from *E. coli*, treatment with $MnCl_2$, and treatment with 0.2 vol. of calcium phosphate gel were identical with those described for DHS reductase, except that freshly harvested wet cells were used, and $MnCl_2$ was allowed to act for 30 minutes before centrifugation. The subsequent adsorption on calcium phosphate gel was performed with 0.8 to 1.0 vol. The gel was washed with distilled water and stirred for 20 minutes with

M/10 phosphate buffer, pH 6.1, in a volume one-twentieth that of the original extract. The supernatant, which contained most of the DHS reductase, was discarded, and the dehydroquinase was eluted from the gel by a similar treatment with buffer at pH 7.0.

After removal of the gel by centrifugation, the supernatant was subjected to $(NH_4)_2SO_4$ fractionation at pH 7.0, as described for DHS reductase. The dehydroquinase was collected in the fraction between 50 and 60% saturation. The precipitate was dissolved in M/30 Tris buffer (pH 7.4) and dialyzed against the same buffer. A typical purification is summarized in the accompanying table.

Summary of purification procedure

Fraction	Protein mg/ml	Specific activity units	Total activity units	Recovery %
Initial extract	21.4	0.011	23.6	100
$MnCl_2$ supernatant	16.4	0.014	19.6	80
Supernatant from first gel . . .	6.2	0.020	13.0	55
Eluate from second gel.	2.4	0.055	6.2	26
$(NH_4)_2SO_4$ fraction 0.50—0.60 .	2.2	0.089	3.0	12

The purified enzyme showed no significant loss of activity on storage at $-15°$ for 6 months and also was stable over a period of several days at $0°$. The enzyme obtained by this procedure had been freed of DHS reductase. With *Aerobacter* extracts there was no significant separation from quinic dehydrogenase, which catalyzes a DPN-linked interconversion of dehydroquinic acid and quinic acid. *E. coli*, however, lacks quinic dehydrogenase.

Properties. Dehydroquinase, like fumarase and aconitase, converts an α-hydroxy acid to the corresponding α,β-unsaturated acid. The preparation described above, however, could not be shown to react with high concentration of DL-malate, DL-isocitrate, or citrate. Furthermore, a purified preparation of fumarase in high concentration did not react with DHQ. Purified dehydroquinase did not convert quinic acid to shikimic acid.

Inhibitors. The enzyme does not appear to have any metal ion requirement, since it is not inhibited by 0.05 M EDTA in Tris buffer. The enzyme was 50 to 80% inhibited by 0.005 M $FeCl_3$, $ZnCl_2$, $CuSO_4$, or iodoacetamide. At the same concentration, Mg^{++}, Mn^{++}, arsenate, and azide were without effect, and Co^{++}, Fe^{++}, CN^- and hydroxylamine were slightly inhibitory.

Effect of pH. The pH optimum lies at about 8.0 with 90% of optimal activity at pH 9.0, and 80% at 7.0.

MICHAELIS *constant.* The dissociation constant for dehydroquinic acid is 4.4×10^{-5} M.

The dehydroquinase of cauliflower buds (BALINSKY and DAVIES, 1961 b) was purified 7.6-fold by using ammonium sulfate fractionation. The enzyme has some similar properties to the dehydroquinase of *E. coli*. But the plant enzyme differs considerably in its stability, being markedly inactivated on overnight storage at $-15°$. The MICHAELIS constant of the cauliflower enzyme was 61 μM at pH 7.4 for dehydroquinic acid. It is active between pH 6.5 and 9.0 and does not catalyse the dehydration of malate or citrate. There is no inhibition by metal-chelating agents.

e) Quinic Dehydrogenase

(MITSUHASHI and DAVIS, 1954 b).

Quinic acid (QA)   DHQ

This enzyme was isolated by MITSUHASHI and DAVIS (1954b) from *Aerobacter* mutant A 170-143 Sl. The enzyme catalyzes the conversion of quinic acid to 5-dehydroquinic acid and is supposed to be present in plants.

Assay. The assay method is based on the increase in absorption at 340 mμ that results from the conversion of DPN to DPNH. Neither quinic acid nor DHQ absorbs at this wave length. Cyanide is added to prevent re-oxidation of DPNH by DPNH oxidase.

Procedure. One-hundredth its volume of M KCN was added to the enzyme solution and allowed to stand at 0° for at least 5 minutes. (The cyanide treated enzyme is stable for at least 12 hours.) 0.1 ml of 1 M potassium carbonate-bicarbonate buffer, pH 9.4, 0.1 ml of 0.01 M DPN, enzyme and enough water to make a final volume of 3.0 ml were added in a silica cell having a 1 cm light path. No change in OD_{340} should take place over the next 2 to 3 minutes. 0.1 ml of 0.1 M quinic acid was added and the optical density at 340 mμ read at 30-second intervals (room temperature, ca. 25°).

A unit of enzyme was defined as the amount that catalyzes the formation of 1 μmole of DHQ (and of DPNH) per minute under the above conditions. If the molar extinction of DPNH at 340 mμ is taken as 6.22×10^3, 0.1 unit of enzyme per 3.0 ml would cause an OD_{340} increment of 0.207 per minute.

Specific activity was expressed as units per milligram of protein. The enzyme extract and purified preparations also contain high concentrations of dehydroquinase, but this enzyme has no interfering effect on the present assay method. However, for a crude preparation the bioassay method which responds to the DHS and shikimic acid secondarily formed is generally more sensitive and reliable. In the bioassay, enzyme (0.002—0.02 unit), DPN (0.5 μmole), phosphate buffer (pH 7.4, final concentration 0.1 M), and quinic acid (1 μmole) were incubated in a volume of 1.0 ml for 10 to 30 minutes. One ml of 0.2 N HCl was added, the precipitated protein was centrifuged off, and 0.1- to 0.4-ml portions of the supernatant were pasteurized (65°, 10 minutes) and assayed for DHS plus shikimic acid with *E. coli* mutant 83-1. The enzyme assays thus obtained on purified preparations were in satisfactory agreement with those obtained spectrophotometrically; and the bioassay showed extracts of wild-type *Aerobacter* to contain about one-fiftieth as much quinic dehydrogenase as extracts of mutant A 170-143 Sl, although the spectrophotometric assay failed to reveal any of this enzyme in the wild-type extracts.

Preparation of enzyme. The enzyme was obtained from *Aerobacter* mutant A 170-143 Sl, which is blocked before DHQ and in addition has a secondary mutation that improves its growth response to quinic acid or DHQ. The organisms were grown with aeration of Medium A (cf. p. 263) supplemented with 0.2% yeast extract or 0.2% casein hydrolyzate and harvested and washed with water after 24 hours. All further operations were carried out at 0°. Extracts were prepared by grinding the wet bacteria with three times their weight of glass powder for 5 minutes, and then for an additional 5 minutes with the gradual addition of ten times their weight of M/30 Tris buffer, pH 7.4. Centrifugation at $5000 \times g$ for 10 minutes yielded a clear supernatant containing about 10 mg/ml of protein. This supernatant was treated with 8% of its volume of 1 M $MnCl_2$, twice with calcium phosphate gel, and with $(NH_4)_2SO_4$ as described for dehydroquinase. A typical purification is summarized in the accompanying table.

The purified enzyme had been freed of DHS reductase and TPN and so could not link quinic acid oxidation with DHS reduction. It had also been freed of DPNH oxidase and therefore did not require the presence of CN^- in the spectrophotometric assay. The purified enzyme showed negligible loss of activity in storage at —15° for several weeks, and 30% loss in 6 months.

Summary of purification procedure

Fraction	Protein mg/ml	Specific activity units	Total activity units	Recovery %
Initial extract	10.8	0.25	270	100
MnCl$_2$ supernatant	7.1	0.37	254	94
Supernatant from first gel . . .	2.2	0.64	160	59
Eluate from second gel.	1.1	1.57	83	30
(NH$_4$)$_2$SO$_4$ fraction 0.50—0.60 .	2.0	2.15	22	8

Properties. With the purified enzyme TPN could not replace DPN as the electron acceptor. With initial extracts TPN showed considerable activity which could be accounted for by the combined action of the nucleotide transhydrogenase and the traces of DPN that were present in these extracts.

Inhibitors. There appears to be no metal ion requirement, since the enzyme is not inhibited by 0.05 M EDTA.

Effect of pH. The pH optimum is 9.8. About 50% of maximal activity is obtained at pH 8.5.

MICHAELIS *constant.* The dissociation constants at pH 9.4 for quinic acid and DPN are 4.9×10^{-4} M and 1.4×10^{-5} M, respectively.

f) 5-Dehydroshikimic Reductase

(YANIV and GILVARG, 1955).

This enzyme was isolated from *E. coli* by YANIV and GILVARG and later found in yeast, peas, spinach leaves and wheat. The enzyme catalyzes the conversion of 5-dehydroshikimic to shikimic acid.

Assay. Although the reaction is shifted to the side of shikimic acid (SA), DHS and TPNH are less readily available reagents than SA and TPN and therefore it is more convenient to follow the reverse reaction, measuring TPNH at 340 mμ. The unfavorable equilibrium, however, limits the range of linearity with respect to enzyme concentration in this assay. A more satisfactory assay was therefore developed by coupling reaction (1) with the essentially irreversible action of GSSG reductase (2), giving the over-all reaction (3). The activity can be assayed through the determination of SH groups released.

$$\text{Shikimic acid} + \text{TPN}^+ \rightleftharpoons \text{DHS} + \text{TPNH} + \text{H}^+ \qquad (1)$$
$$\text{GSSG} + \text{TPNH} + \text{H}^+ \rightarrow 2\,\text{GSH} + \text{TPN}^+ \qquad (2)$$

$$\text{Shikimic acid} + \text{GSSG} \rightarrow \text{DHS} + 2\,\text{GSH} \qquad (3)$$

Procedure. In a test tube were placed 0.04 ml of 1 M Tris-HCl buffer, pH 8.0, 0.04 ml of 0.1 M EDTA-KOH, pH 8.0, 0.02 ml of 0.002 M TPN, 0.04 ml of 0.025 M GSSG, 0.10 ml of 0.01 M shikimic acid, and suitable amounts of GSSG reductase. Water was added to adjust the final volume to 0.60 ml, and the DHS reductase was added last. After incubation at 30° for 20 minutes 0.60 ml of 6% metaphosphoric acid was added. After centrifugation a 0.80-ml portion of the supernatant was analyzed for SH content. As controls, SH should be determined at zero time and also after incubation without substrate.

A unit of enzyme was defined as the amount that catalyzes the formation of 1 μmole of DHS per minute under the above conditions. Specific activity was expressed as units per milligram of protein.

Preparation of enzyme. *E. coli* was grown with aeration on minimal medium A (cf. p. 263) supplement with 0.2% Difco yeast extract, harvested after 24 hours, and lyophilized. All further operations were carried out at 0°. Extracts were prepared by grinding the bacteria with three times their dry weight of glass powder for 5 minutes and then for an additional 10 minutes with gradual addition of M/30 Tris buffer, pH 8.0 (20 ml/g dry weight of bacteria). Centrifugation for 20 minutes at 5000×g yielded a clear yellow supernatant; this was adjusted to a protein concentration of 20 mg/ml. In order to reduce the nucleic acid content of the extract, one-twentieth its volume of 1 M $MnCl_2$ was added over a period of 10 minutes with occasional stirring. The resulting voluminous precipitate was removed by centrifugation at 5000×g for 20 minutes. The supernatant was adjusted to pH 5.5 with 0.1 M acetic acid, 0.2 vol. of calcium phosphate gel was added, and the whole was stirred for 20 minutes and then centrifuged. The precipitate was discarded. To the supernatant 1.2 vol. of gel was added, and the procedure was repeated except that the gel was retained. It was washed by being suspended in several volumes of water and centrifuged. The washed gel was eluted by being stirred for 20 minutes in a volume of M/10 phosphate buffer, pH 6.0 equal to the volume of the $MnCl_2$ supernatant. The eluted gel was removed by centrifugation. The supernatant was subjected to $(NH_4)_2SO_4$ fractionation at pH 7.0. The salt was added as the solid over a period of 20 minutes, stirring was continued for an additional 20 minutes and the precipitate was removed by centrifugation at 5000×g for 20 minutes. The material precipitating between 32 and 45 per cent saturation was dissolved in M/30 Tris buffer, pH 8.0, and dialyzed overnight against the same buffer. A typical fractionation was summarized in the accompanying table. The purified enzyme was free of dehydroquinase and retained 80% of its activity after storage at −15° for 4 months.

Summary of purification procedure

Fraction	Protein mg/ml	Specific activity units/mg	Total activity units	Recovery %
Initial extract	20	0.0082	7.9	100
$MnCl_2$ supernatant	8.4	0.0183	6.0	76
Calcium phosphate gel eluate . .	2.2	0.025	2.21	28
$(NH_4)_2SO_4$ fraction 0.32—0.45 .	2.9	0.078	1.18	15

Properties. The purified enzyme required TPN as the electron acceptor, DPN being inactive. Dihydroshikimic acid, 5-epishikimic acid, 3-phosphoshikimic acid, 5-phosphoshikimic acid, and quinic acid were not oxidized by the enzyme plus TPN.

Inhibitors. There appeared to be no metal ion requirement, since the enzyme was not inhibited by 0.006 M EDTA.

Effect of pH. The pH optimum was 8.5. About 50% of maximal activity was obtained at pH 10.5 or 7.5.

MICHAELIS *constant*. The MICHAELIS constants for TPN and shikimic acid, determined at pH 8.0, were 3.1×10^{-5} M and 5.5×10^{-5} M, respectively.

The dehydroshikimic reductase of etiolated pea epicotyls (BALINSKY and DAVIES, 1961a) was purified 78-fold by ammonium sulfate fractionation, heat treatment, ammonium citrate fractionation and column chromatography on calcium phosphate-cellulose. The enzyme has several properties in common with the corresponding enzyme of *E. coli*. Both enzymes are TPN-specific and the MICHAELIS constants for shikimic acid and TPN for two enzymes are of the same order of magnitude. MICHAELIS constants for the plant enzyme at pH 8.0 are 55 μM for shikimic acid and 31 μM for TPN. The enzyme does not require a multivalent-ion. However, pH optimum was somewhat different, the plant enzyme showed a pH optimum of 10.0, compared with pH 8.5 for the *E. coli* enzyme. The plant enzyme was inhibited by p-chloromercuribenzoate with reversal of inhibition by cysteine.

The dehydroshikimic reductase of mung bean *(Phaseoulus aureus)* seedlings (NANDY and GANGULI, 1961) was partially purified by heat treatment and acetone fractionation. The enzyme is TPN-specific and its optimum pH is 8.0. The MICHAELIS constants for shikimic acid and TPN at pH 8.0 are 8.7×10^{-5} M and 1.17×10^{-4} M, respectively. The enzyme was not inhibited by metal-chelating agents but was completely inhibited by p-chloromercuribenzoate with partial reversal by cysteine.

g) ATP-Shikimic Acid Transphosphorylase

(FEWSTER, 1959).

SA   5-Phosphoshikimic acid (PSA)

This enzyme was isolated from *E. coli* 518 by FEWSTER. The enzyme catalyzes the transphosphorylation from ATP to shikimic acid and is distributed in *Acetobacter suboxydans*, *Aerobacter aerogenes*, *Arthrobacter globiformis*, *Corynebacterium erythrogenes* and *Pseudomonas ovalis*.

Assay. The phosphorylation can be assayed by following the disappearance of free shikimic acid and also by measuring the decrease in ATP.

Procedure. The reaction mixture, containing suitable amounts of enzyme, shikimic acid and ATP, was treated with barium hydroxide and zinc sulfate to precipitate the phosphate ester after deproteinization, and shikimic acid was estimated by the method of GOITONDE and GORDON (1958). Further, the decrease of ATP in the reaction mixture was determined by the method of BERENBLUM and CHAIN (1938).

5-Phosphoshikimic acid in the reaction mixture could be identified by paper chromatography.

Preparation of enzyme. The cells of *E. coli* 518 were treated with ultrasonic oscillation, giving a crude extract; which was centrifuged at $100,000 \times g$. The supernatant fraction contained all the enzyme activity. Crude extracts catalyze the disappearance of shikimic acid at a rate of 140 μmoles/mg protein/hr. at the pH optimum of 6.5, and the disappearance of the acid is also accompanied by an equivalent decrease in ATP.

Properties. The enzyme requires Mg^{++} or Mn^{++} as cofactors. The enzyme is not inhibited by 2×10^{-4} M phenylmercuric acetate.

h) The Enzymes Forming 3-Enolpyruvyl Shikimate-5-Phosphate (ESP) from 5-Phosphoshikimic Acid and Phosphoenolpyruvate

(LEVIN and SPRINSON, 1960).

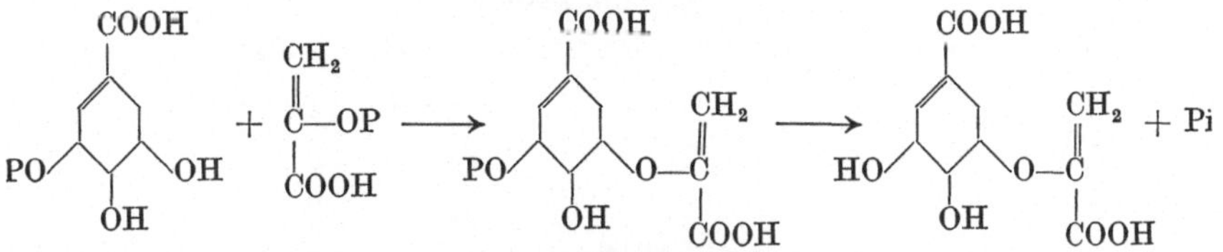

PSA   PEP   3-Enolpyruvyl shikimate-   Z 1
                     5-phosphate (ESP)

The enzymes were isolated from an *E. coli* mutant by LEVIN and SPRINSON. The enzymes catalyze the reaction shown above and the reaction product was suggested to be 3-enolpyruvyl shikimate-5-phosphate (ESP) which is then dephosphorylated to the Z 1 compound in DAVIS' scheme.

Assay. Unreacted PSA can be determined according to the shikimate assay after treatment by phosphatase; Z 1 and ESP can be also estimated by the determination of additional shikimate produced after acid hydrolysis of the reaction mixture and of the phosphatase-treated reaction mixture.

Procedure. The reaction mixture contained 1 μmole of PSA, 1 μmole of phosphoenol-pyruvate, 50 μmoles of Tris buffer, pH 8.2, and 0.05 ml of extract (1 mg of protein), in a final volume of 1.4 ml. It was incubated at 37° for 18 hours; proteins were precipitated by the addition of 0.8 ml of 25% trichloroacetic acid, and removed by centrifugation and filtration. A 0.8 ml aliquot of the reaction mixture was further incubated for 1 hour at 37° with 50 μmoles of Tris buffer, pH 8.2, 10 μmoles of MgCl$_2$, and 50 μg of intestinal alkaline phosphatase, in a final volume of 1.2 ml. An aliquot was used directly for shikimate assay (PSA). Another aliquot was acidified with 6 N HCl to give a concentration of 0.1 N, heated in a boiling water bath for 10 minutes, and assayed for additional shikimate produced (ESP).

Preparation of the enzymes. *E. coli* K-12 mutant 58-278, a phenylalanine auxotroph which accumulates prephenic acid, was grown with aeration for 18 hours at 35° on Medium A (cf. p. 263) enriched with 0.2% yeast extract and 0.2% casein hydrolysate. The cells were harvested by centrifugation at 3°, washed twice with 1/30 M phosphate buffer, pH 7.2, suspended in 0.01 M Tris buffer, pH 8.2, (20 ml per 5 g of wet bacteria), and subjected to sonic oscillation at 9 kc, with cooling, for 30 minutes. All subsequent operations were carried out at 2°. The clear supernatant solution, obtained by centrifugation, was treated with 0.5 ml of 2% protamine sulfate solution per 70 mg of protein, and the precipitate was removed by centrifugation and discarded. The supernatant solution was treated with solid (NH$_4$)$_2$SO$_4$ in order to obtain a fraction precipitating between 0.4 and 0.6 saturated (NH$_4$)$_2$SO$_4$. The precipitate was dissolved in 0.01 M Tris buffer, and was dialyzed extensively against this buffer.

Properties. The extracts contain the enzyme forming 3-enolpyruvylshikimate-5-phosphate (ESP) and the phosphatase dephosphorylating ESP to Z 1 (ES). Fluoride strongly inhibited the formation of Z 1 from 3-enolpyruvylshikimate-5-phosphate.

Attempts to demonstrate the conversion of Z 1 to prephenate in extracts of *E. coli* (K-12 mutant 58-278, or wild type) were unsuccessful and thus 3-enol-pyruvylshikimate-5-phosphate, rather than Z1 was believed to be the active intermediate.

For the determination of Z1, a 0.2 ml aliquot of the filtrate was heated in a boiling water bath for 10 minutes, and used for the assay of shikimate.

i) Prephenic Aromatase
(Weiss, Gilvarg, Mingoli and Davis, 1954).

$$\text{Prephenic acid (PPA)} \longrightarrow \text{Phenylpyruvic acid} + CO_2 + H_2O$$

The enzyme was found by Weiss et al. (1954) in the extracts of wild-type *E. coli* but not in the extracts of *E. coli* mutant 83-5, which is blocked between prephenic acid and phenylpyruvic acid. The enzyme catalyzes the conversion of prephenic acid to phenylpyruvic acid and was named prephenic aromatase.

Assay. The activity can be assayed by the spectrophotometric determination of phenylpyruvic acid formed in alkaline medium at 320 mμ.

Properties. As the purification of the enzyme has not yet been carried out, there is no description of the properties of the enzyme.

k) Prephenic Dehydrogenase
(Schwinck and Adams, 1959).

$$\text{PPA} + DPN^+ \longrightarrow \text{p-Hydroxyphenylpyruvic acid (HPP)} + DPNH + H^+ + CO_2$$

This is a soluble enzyme from an *E. coli* mutant isolated by Schwinck and Adams and it catalyzes an aromatization reaction which converts prephenic acid (PPA) to *p*-hydroxyphenylpyruvic acid (HPP). The enzyme is absent in a tyrosine auxotroph, indicating its essential function in tyrosine biosynthesis. Early observation on the formation of phenyllactic acid (HPL) from prephenic acid was shown to be based on the secondary product from p-hydroxyphenylpyruvic acid.

Assay. Prephenic dehydrogenase was assayed by following the formation of the Millon-reacting products.

Procedure. For routine assays incubation mixtures contained per ml: barium PPA, 2.7 μmoles; potassium phosphate, pH 6.5 and 7.5, 50 μmoles; DPN, 1 μmole; and enzyme. Incubations were carried out for 1 hour at 37° in a total vol. of 0.5—2.5 ml in stoppered tubes, or in small beakers in an incubator. For the Millon determination, the reaction was stopped by adding H_2SO_4 and color was measured in 10-fold diluted aliquots. It should be emphasized that the color yield in the MILLON-determination is identical for the 3 phenolic compounds, tyrosine, HPP and HPL. Therefore, this method does not distinguish among a number of possible enzymic products of PPA.

Identification of the phenolic reaction products. There are at least 3 plausible reaction products — HPP, HPL and tyrosine — which show identical MILLON color yields and so could not be distinguished on this basis. The following observations support HPP as the major reaction product: Spectrophotometric determination of the products. — HPP has strong absorption in alkaline solution with a maximum at 330 mμ and negligible absorption in neutral solution. In the enzyme incubation at low pH (6.5), however, quantitative interpretation of the resultant O.D. at 330 mμ was made difficult by the spontaneous conversion of PPA to phenylpyruvate, which has an absorption peak at 320 mμ in alkali. At pH values above 7.5, this spontaneous reaction is not significant and, under these conditions, the yield of HPP (calculated from the value ε 330 = 21.9 $\times 10^6$ for authentic HPP) accounted for between 60% and 85% of the total MILLON-positive products. Paper chromatographic identification. — R_f values in most solvents are quite similar for HPP and HPL. However, the two compounds can be easily distinguished by their different colors with diazotized sulfanilic acid and diazotized p-nitroaniline. n-Butanol-acetic acid-water (4:1:5 upper layer) is useful as a solvent.

Preparation of enzyme. Cells of mutant 83-5 were the source of enzyme. The growth medium contains K_2HPO_4 1 g, KH_2PO_4 1.5 g, $MgSO_4$ 0.1 g, $(NH_4)_2SO_4$ 1 g, L-phenylalanine 20 mg, and sodium DL-lactate 5 g per liter of water. Cells were grown with shaking in 1 l Erlenmeyer flasks for 20—22 hours at 37°, harvested by centrifugation in the cold and washed twice with 0.05 M potassium phosphate at pH 7.5. Washed cells were stored at —15° and provided a stable source of enzyme for at least 9 months. Sonic extracts were made by 30-min oscillation (Raytheon, 9 Kc) using 20 ml of 0.05 M potassium phosphate at pH 7.5/3.5 g of wet cells; supernatants obtained by 30-min centrifugation in the cold at 25,000 $\times g$ contained all the enzyme activity. The sonicate supernatant was treated with 0.2 vol. of 1% protamine sulfate and after 30 min at 0° the precipitate was removed by centrifugation. The protamine supernatants retained activity for several weeks at —15°.

Properties. Prephenic dehydrogenase requires DPN as a cofactor. When the protamine supernatant was treated with charcoal (Norit A), the reaction in the absence of DPN fell to negligible or low levels, but could be restored by adding DPN.

Effect of pH. The pH optimum of prephenic dehydrogenase is rather broad, and lies between 7.0 and 8.0.

MICHAELIS *constant.* The MICHAELIS constant for the substrate was about 2.0 $\times 10^{-3}$ M at pH 6.5 and 7.5.

Enzyme stability. In the complete incubation system containing both PPA and DPN, enzyme fractions appeared stable up to 3 hours at 25°, 30° or 35°. Preincubation of the enzyme in buffer for 90 min before the addition of PPA and DPN, however, resulted in almost total inactivation.

II. The Shikimic Acid Pathway in Higher Plants.

The recent review by NEISH (1961) has shown many definite reasons for believing that the shikimic acid pathway does function in higher plants. Shikimic acid is of widespread occurrence in higher plants including *Bryophyta*, *Pteridophyta*, *Gymnospermae* and *Angiospermae*, and is found in the meristematic and cambial tissues as well as in the leaves of the plants (Hasegawa et al., 1957, Higuchi, 1958 c).

Recent experiments using the tracer technique have revealed that shikimic acid is readily converted to a variety of aromatic compounds in a number of plant species (Neish, 1961).

Brown and Neish (1955) proved that C^{14}-labelled shikimic acid was incorporated readily into the aromatic nuclei of the lignins of wheat and maple plants, and Eberhardt and Schubert (1956) further found that when shikimic acid labelled in carbons 1 and 6 was fed to sugar cane, the vanillin subsequently isolated from the lignin was labelled in the same way without rearrangement of the carbon skeleton. Direct conversion of the cyclohexene ring of shikimic acid and dehydroshikimic acid to the aromatic nucleus has been confirmed in microorganisms (Davis, 1955; Gross, 1958).

McCalla and Neish (1959 a, b) and Gamborg and Neish (1959) further found that generally labelled shikimic acid was converted to phenylalanine, tyrosine and cinnamic acids of *Salvia splendens, Triticum vulgare* and *Fagopyrum tataricum*, and proved by degradation studies that these compounds were labelled only in the ring. The β-carbon was not labelled, as would be expected from the biosynthetic mechanism of phenylpropanoid compounds from prephenic acid in microorganisms.

Moreover, the tracer studies recently carried out have shown that shikimic acid is a good precursor of quercetin, 3,4-dihydroxyacetophenone, *o*-coumaric acid, tryptophan etc. (Neish, 1961).

As shown in Fig. 1 shikimic acid is made from phosphoenolpyruvate and D-erythrose-4-phosphate via dehydroquinic acid in microorganisms. In higher plants, it is believed that phosphoenolpyruvate is produced via the Embden-Meyerhof pathway of sugar metabolism. D-erythrose-4-phosphate can be probably formed through the transketolase and transaldolase reactions from D-fructose-6-phosphate and D-glyceraldehyde-3-phosphate during the respiration through the Embden-Meyerhof and pentose phosphate pathways, and photosynthesis. Neish (1961) explained the net formation of D-erythrose-4-phosphate by the cyclic processes among the sugar phosphates:

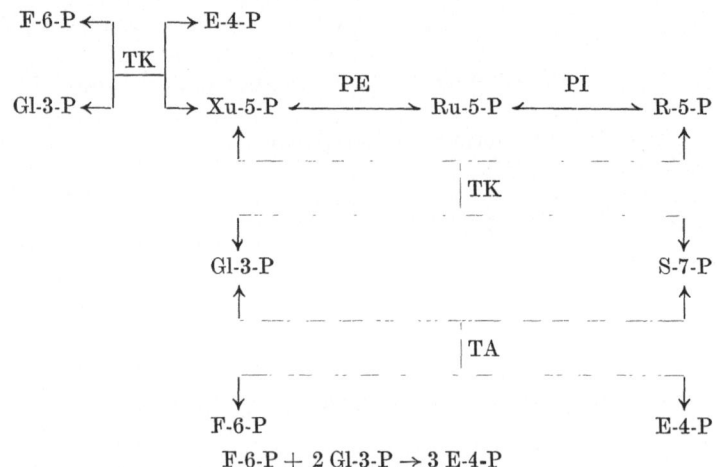

F-6-P + 2 Gl-3-P → 3 E-4-P

F-6-P = D-fructose-6-phosphate; Gl-3-P = D-glyceraldehyde-3-phosphate; Xu-5-P = D-xylulose-5-phosphate; Ru-5-P = D-ribulose-5-phosphate; S-7-P=sedoheptulose-7-phosphate E-4-P = D-erythrose-4-phosphate; TK = transketolase; TA = transaldolase; PE = phospho-ribulo-epimerase; PI = phosphoriboisomerase.

In fact, KRATZL and FAIGLE (1959) and ACERBO, SCHUBERT and NORD (1960) showed recently that glucose-1-C^{14} and -6-C^{14} were well incorporated into lignin. The distributions of C^{14} in vanillin molecules obtained from the lignins were very similar to those of C^{14} in tyrosine and phenylalanine formed from the glucose labelled in the same way by microorganisms. HASEGAWA and HIGUCHI (1960) found also that glucose-C^{14} was appreciably incorporated into shikimic acid as well as into lignin of *Eucalyptus nitens* and that the incorporation of glucose-1-C^{14} into these compounds was about the same as that of uniformly labelled glucose. This might suggest that the EMBDEN-MEYERHOF pathway plays an important role in the formation of D-erythrose-4-phosphate required for the shikimic acid synthesis in this plant under the experimental conditions used.

Furthermore, the distribution of quinic acid in plants is also wide, and in many cases, coexistence of the acid with shikimic acid has been found (HASEGAWA et al., 1954; MANSKAJA and KODINA, 1958). Some plants contain much quinic acid and less shikimic acid, and others contain both acids in about the same proportion or in the reverse proportion. WEINSTEIN, PORTER and LAURENCOT (1959a, b) found recently that quinic acid-C^{14} was converted to shikimic acid, phenylalanine and tyrosine in young rose blooms. This suggests that higher plants are also able to convert quinic acid to shikimic acid as has been shown for *A. aerogenes*.

As explained above, there may be a close resemblance between the biosynthetic pathway of aromatic compound in microorganisms and that in plants. However, few enzymatic studies on aromatic biosynthesis in plants have hitherto been done and only 5-dehydroshikimic reductase, dehydroshikimic dehydrase and 5-dehydroquinase have been obtained as crude extracts.

The descriptions of these enzyme preparations are not given here because the preparation methods are almost the same as those in microorganisms.

III. Conversion of Simple Phenylpropanoids to Lignin and Related Compounds.

A few aromatic compounds, such as protocatechuic acid, phenylpyruvic acid, and *p*-hydroxyphenylpyruvic acid are formed from carbohydrates via the shikimic acid pathway as explained above. Other aromatic secondary growth substances might be formed by a complex series of interconversions of these simple phenyl compounds.

There are many groups of phenylpropanoid derivatives, such as lignins, lignans, flavonoids, phenolic glycosides and coumarins in plants, and present evidence favors the view that phenylpyruvic acid and *p*-hydroxyphenylpyruvic acid formed via the shikimic acid pathway are the aromatic precursors of phenylpropanoid structures in these compounds.

BROWN, WRIGHT and NEISH (1959) have carried out a series of comparative studies on the efficiency of several labelled compounds as precursors of lignin and other aromatic compounds, and from the results of numerous feeding experiments, they reached the conclusion that lignin and related compounds are formed through the interconversions among phenylpropanoid acids shown in Fig. 2.

All compounds shown as intermediates in Fig. 2 have been fed to excised shoots of various plants and found to be readily converted to lignin, pungenin aglycone, the C_6(ring B)-C_3 portion of flavonoid molecule, and coumarins. However, an interesting taxonomic difference was noted. Out of the 11 species representing 10 plant families, only two could readily convert tyrosine to lignin, whereas all species readily utilized phenylalanine as a lignin precursor. The species using tyrosine were members of the family Gramineae, and the plants of this family also could utilize *p*-hydroxyphenylpyruvic acid and *p*-hydroxyphenyllactic acid as lignin precursors, although these compounds could not be utilized by other plants.

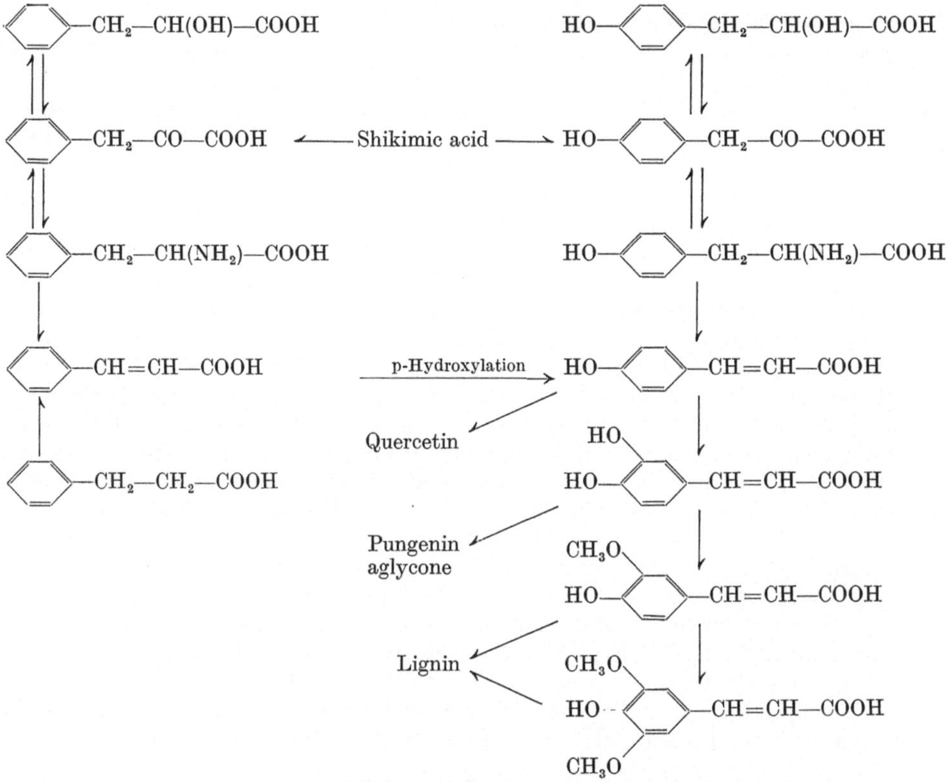

Fig. 2. Scheme for formation of lignin and related compounds from shikimic acid.

BROWN et al. (1959) have suggested that the failure of many species to use tyrosine for synthesizing lignin and related compounds is due to their inability to carry out reactions amounting to a dehydration of p-hydroxyphenyllactic acid. Recently, the enzymes phenylalanine deaminase (KOUKOL and CONN, 1960, 1961) and tyrase (NEISH, 1961) which catalyze the conversion of phenylalanine and tyrosine to the corresponding cinnamic acids and ammonia have been isolated. It is very interesting in connection with the ability to utilize tyrosine that phenylalanine deaminase has been found in many plants including species of the *Gramineae*, whereas tyrase has been found in significant amount only in species of the *Gramineae*.

The scheme in Fig. 2 is supported mainly by tracer experiments with living plants. Further studies on isolated enzyme systems are needed.

a) Tyrase
(NEISH, 1960).

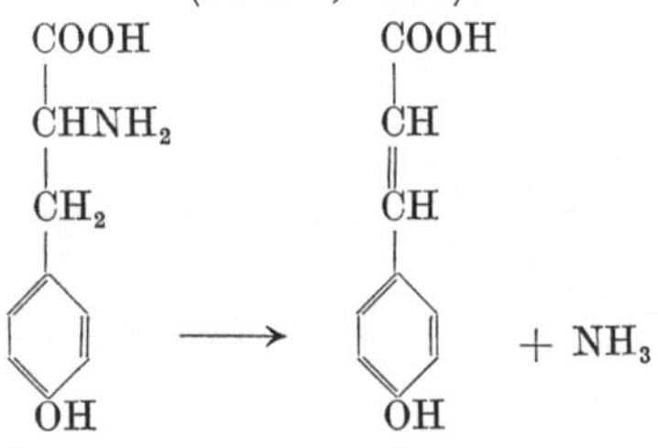

L-Tyrosine   p-Coumaric acid

This enzyme was isolated by NEISH from barley stems and rice seedlings. The enzyme catalyzes the deamination of L-tyrosine, giving equimolar amounts of *trans-p*-coumaric acid and ammonia, and has been named tyrase. The enzyme has been detected in sorghum, barley, rice, wheat, oat, corn and sugar cane plants; but not in pea, lupine, alfalfa or white sweet clover plants or yeast. In view of the distribution and role of tyrase it may be easily understandable that, as explained already, grasses can easily convert tyrosine to lignin while legumes can not.

Assay. Tyrase activity was estimated by measuring either fluorometrically or by the absorption at 333 mμ, the rate of formation of *p*-coumaric acid in the presence of excess L-tyrosine.

Procedure. Usually 1.4 ml of reaction mixture contained 1.4 mg of L-tyrosine, 100 μmoles of sodium borate at pH 8.8, and the enzyme being tested. This mixture was incubated in a glass-stoppered tube at 40° for 30—60 minutes, then acidified with 2.3 ml of 0.15 M hydrochloric acid, cooled in an ice bath and shaken manually with 2.0 ml of ice-cold ether for 1 minute. The phases were allowed to separate at 0°, 1.0 ml of the ether layer was pipetted into an 18×150 mm culture tube and the ether evaporated by an air stream at room temperature. The dry residue was dissolved in a suitable volume of 0.05 N sodium hydroxide. *p*-Coumarate was measured in this solution fluorometrically or by its absorption at 333 mμ, at a concentration of less than 2 μg per ml. A stock solution of *p*-coumaric acid (100 μg per ml) in dilute sodium bicarbonate (1 mM) was stored in a refrigerator. Aliquots of this were diluted 100-fold with 0.05 N sodium hydroxide to give the working standard.

When acetone powders were assayed, 0.05—0.20 g of the powder was suspended in 3 ml of a 0.1% solution of L-tyrosine in 0.1 M sodium borate (pH 8.8—9.0) in an 18×150 mm culture tube and incubated. Water (7 ml) was added and the debris removed by filtration with suction through a 0.7 cm disc of Whatman No. 1 paper. The residue was washed with 2—3 ml of water, the filtrate acidified by 0.2 ml of 5 N hydrochloric acid, then shaken with diethyl ether (10 ml) in a glass-stoppered tube. The ether extract was poured into a 20 ml beaker, evaporated to dryness, the residue dissolved in a suitable volume (5—50 ml) of 0.05 N sodium hydroxide and the *p*-coumarate in this solution estimated fluorometrically.

Preparation of the enzyme. Fresh barley stems were cut into convenient lengths with scissors, covered with acetone at −20° and blended for 1—1.5 minutes at full speed in a Waring blendor. The homogenates were filtered by suction and washed three times with acetone at −20°. After drying for about 5 minutes on the BÜCHNER funnel, the powder was spread on paper, air-dried in a fume hood for about 15 minutes and finally dried *in vacuo* at room temperature for at least 1 hour. The powders were stored at 4—6° in containers with tight caps.

The acetone powder was mixed with about 15 times its weight of 0.10 M sodium borate, pH 8.8, at room temperature. The mixture was cooled in an ice-bath and stirred manually from time to time during 30 minutes, then filtered through a double layer of cheese cloth. The filtrate was centrifuged at 8000×g and 0° for 15 minutes. The supernatant fluid contained much of the tyrase. All subsequent steps were carried out at 0—4°.

The supernatant was cooled in ice and solid $(NH_4)_2SO_4$ to give 70% saturation was added gradually, with stirring, during 5—10 minutes, then the precipitate was collected by centrifugation at 8000×g for 15 minutes. The precipitate was taken up in 0.05 M sodium borate at pH 8.8, using a volume about one-twentieth the original extract. This was dialzyed overnight against 50 volumes of the same buffer.

The dialyzed solution was titrated to pH 5.9—6.1 (glass electrode) using M acetic acid at 0°. A heavy precipitate formed and was removed by centrifugation at 11,000 $\times g$. The supernatant fluid contained practically all of the tyrase; it was titrated to pH 8.6 by N sodium hydroxide. A clear solution was obtained.

Solid $(NH_4)_2SO_4$ was added carefully to the clear solution at 0°. Tyrase was always concentrated in the fractions at 40—60% saturation. The precipitates in these fractions were collected by centrifugation at 8000 $\times g$, and dissolved in 0.02 M potassium phosphate at pH 6.8, to give a protein concentration between 1 and 2%. The solution was then dialyzed overnight against 50—100 volumes of the same buffer at 0°. The dialyzed solution was ready for chromatography on DEAE-cellulose.

The DEAE-cellulose was a commercial preparation; i. e. Biorad Cellex-D. It was washed with 0.1 N sodium hydroxide, suspended in deionized water, titrated to pH 6,8 by M phosphoric acid, filtered on a BÜCHNER funnel and washed with 0.02 M potassium phosphate at pH 6.8.

Two sizes of columns were employed. The small size (1.1×10 cm) was loaded with about 20 mg of protein, the large size (2.2×17 cm) was loaded with about 100 mg. The small column was operated at 0°, in an ice bucket, using a flow rate of about 30 ml per hour; the fractions (3 ml) were collected manually. The large column was operated at 4°, in a cold room, using a flow rate of about 90 ml per hour; the fractions (4 ml) were collected automatically. Columns were packed in Pyrex tubes and after washing the packing thoroughly with 0.02 M potassium phosphate (pH 6.8) the enzyme solution was put on and wahsed in. The column was developed using a linear gradient between equal volumes of 0.02 M potassium phosphate at pH 6.8 and 0.05 M potassium phosphate at the same pH but containing 0.4 M potassium chloride. The total volume in the gradient elution apparatus was 80 ml for the small column and 400 ml for the large one.

The tyrase content of fractions eluted from DEAE-cellulose columns was determined by a rapid fluorometric procedure. An aliquot (0.1—0.2 ml) of the eluate was mixed with 0.5 ml of 0.1% tyrosine in 0.15 M sodium borate (pH 9.0). This mixture was incubated for 30 minutes at 40°, then 1.0 ml of 0.13 N sodium hydroxide was added and the mixture poured into a round quartz cuvette for measurement of its fluorescence. All fractions had a fluorescence, but this was augmented 3—5 times in tubes containing appreciable amounts of tyrase. This simple, rapid method showed which fractions could be recombined. The enzyme was thus obtained in about 0.035 M potassium phosphate at pH 6.8, 0.2 M with respect to KCl. The protein concentration was usually 0.3—0.4 mg per ml. The active fractions were stored at −20°.

Properties. The purified enzymes also catalyze the deamination of L-phenylalanine and DL-*m*-tyrosine, giving *trans*-cinnamic acid and *trans-m*-coumaric acid as well as ammonia, respectively. However, the ratio of these reactions varies with the pH and also with the enzyme preparation. Thus it seems likely that two enzymes are involved, and their relative concentration changes as the plant matures.

Inhibitors and activators Tyrase was strongly inhibited by *p*-chloromercuribenzoate and by cyanide but not by iodoacetamide, EDTA and fluoride. Pyridoxal phosphate, folic acid, ATP and CoA had no effect. No evidence could be obtained for participation of a divalent cation in the tyrase reaction.

Effect of pH. The pH optimum is between 8.8 and 9.0.

MICHAELIS *constant.* The MICHAELIS constant for L-tyrosine is $0.74—1.0 \times 10^{-4}$ and that for *m*-tyrosine is $1.3—2.3 \times 10^{-3}$ at pH 9.0.

Tyrosine inhibits the formation of m-coumaric acid from m-tyrosine but p-coumaric formation is not inhibited by either L-phenylalanine or DL-m-tyrosine. *Reversibility.* Attempts to reverse the reaction were unsuccessful.

b) Phenylalanine deaminase
(KOUKOL and CONN, 1960, 1961).

L-Phenylalanine   Cinnamic acid

This enzyme was detected by KOUKOL and CONN from sweet clover. The enzyme catalyzes the formation of cinnamic acid and ammonia from phenylalanine and has been named phenylalanine deaminase. The enzyme is found in members of the Leguminosease and of the Gramineae. It seems likely that the enzyme plays an important role in the formation of the precursors of lignin and related phenylproponoid compounds.

Assay. Phenylalanine deaminase activity was estimated by measuring the C^{14} activity of the reaction product and also by absorption of cinnamic acid at 268 mμ in alkaline solution.

Procedure. a) The reaction mixture contained enzyme, 20 μmoles of L-phenylalanine and 100 μmoles of borate buffer, pH 8.8 in a final volume of 2.0 ml. The reaction mixture was incubated for 1 hour in an unstoppered test tube without shaking at 40°. The reaction was stopped by the addition of 0.1 ml of 5 N HCl and the volume was adjusted to 5.0 ml. With enzyme preparations of low specific activity, the acidified reaction mixture was heated in a boiling water bath for 10 minutes and the coagulated protein removed by centrifugation. The removal of the protein was omitted when DEAE-cellulose chromatography fractions were being assayed. The acidified reaction mixture was extracted once with 5 ml of ether, an aliquot of the ether phase was removed and the ether was evaporated under a stream of air. The residue was dissolved in 0.05 M NaOH and the absorbancy at 268 mμ determined.

b) The second procedure was based on the conversion of radioactive L-phenylalanine to cinnamic acid. The reaction mixture contained enzyme, 25 μmoles of L-phenylalanine, 0.45 μmoles of DL-phenylalanine-3-C^{14} of specific activity 1.3 μc/mole, and 200 μmoles of borate buffer, pH 8.8 in a final volume of 5.0 ml. The reaction mixture was incubated for 1 hour in a covered test tube without shaking at 40°. The reaction was stopped by the addition of 0.5 ml of 50% trichloracetic acid, immediately preceded by the addition of 0.5 ml of a 0.1% solution of unlabelled *trans*-cinnamic acid in 0.05 M NaOH. After 10 minutes, the reaction mixture was centrifuged to remove the protein, if necessary. The acidified reaction mixture was then extracted once with 10 ml of toluene. Following centrifugation at 500 $\times g$ for several minutes at room temperature, a 5 ml aliquot of the toluene phase was transfered to a Tri-carb vial containing 5 ml of counting mixture in toluene. The vial was covered, shaken, and counted at $-5°$ in a Packard Automatic Tri-carb Liquid Scintillation spectrometer. The counts per minute of the blank were subtracted from the total c.p.m. of each sample. Since the efficiency of counting and the specific activity of the L-phenylalanine were known, it was possible to calculate the quantity of cinnamic acid formed from the radioactivity in the toluene phase.

The unit of phenylalanine deaminase activity is defined as that quantity of the enzyme which catalyzes the formation of 1 μg of cinnamic acid per hour under the assay condition. 10 to 100 units of enzyme of specific activity 352 to 970 units per mg were usually employed in the assay.

Preparation of the enzyme. The acetone powder was prepared from barley stems in the pre-head stage. The yield of the powder expressed as the percent of the fresh weight was 10.5. The purifications reported below was performed on 40 g of acetone powder. All procedures of purification were carried out at 0—4°. The pH of the solution being fractionated with $(NH_4)_2SO_4$ was intermittently checked and adjusted with 5 M NaOH, if necessary, so that the pH never went below 7.0.

40 g of barley stem acetone powder was extracted for 1 hour with 1 liter of 0.1 M borate buffer, pH 8.5. The crude preparation was strained through cheese cloth and the filtrate was centrifuged at $4000 \times g$ to clarify the solution. The volume of the supernatant solution was 835 ml. 555 ml of a neutralized solution of saturated $(NH_4)_2SO_4$ was added to the supernatant solution from the last step to form a solution which was 40% saturated with respect to the salt. The precipitate formed was removed by centrifugation at $15,000 \times g$ for 15 minutes and discarded. Solid $(NH_4)_2SO_4$ (165 g) was added to the supernatant solution (1300 ml) to form a solution which was 60% saturated with respect to the salt. The precipitate formed was removed by centrifugation at $15,000 \times g$ for 15 minutes and dissolved in 30 ml of 0.02 M potassium phosphate buffer, pH 6.8. This solution was dialyzed overnight with stirring against 1 liter of the same buffer. The pH of the dialzyed solution (38 ml) was carefully adjusted to 6.0 with 1 M acetic acid. The precipitate formed was removed by centrifugation at $15,000 \times g$ for 15 minutes and discarded. The pH of the supernatant solution was immediately adjusted to 7.0 with 1 M NaOH and applied on a DEAE-cellulose column of the following description:

On the day before use, 20 g DEAE-cellulose was washed by suspension in 0.1 M NaOH, followed by neutralization to pH 6.8 with 1 M H_3PO_4. The DEAE-cellulose was collected on a BÜCHNER funnel and washed repeatedly by filtration with 0.02 M potassium phosphate buffer, pH 6.8. The DEAE-cellulose was resuspended in more buffer and packed so that the dimensions of the column were 2.2×16 cm. The column was placed in a 4° room overnight. After addition of the enzyme, the column was washed with 5 ml of the buffer used to equilibrate the column. The enzyme was eluted from the column by using a linear gradient between equal volumes (130 ml each) of 0.02 M potassium phosphate buffer, pH 6.8, and 0.5 M potassium phosphate buffer, pH 6.8 containing 0.4 M KCl. The column was eluted at the rate of 50 ml/hour and fractions (4 ml) were collected until the enzyme had been completely eluted.

With the above procedure, the enzyme is purified about 28-fold and the recovery is approximately 19%.

Properties. The enzyme is specific for L-phenylalanine. Cinnamic acid is not formed when phenylalanine deaminase is incubated with D-phenylalanine. The DEAE-cellulose fraction also catalyzes the deamination of L-tyrosine and DL-m-tyrosine to p-coumaric acid and m-coumaric acid, respectively. These conversions are probably due to the presence of tyrase. However, the ability of the DEAE-cellulose fractions to catalyze these conversions is completely lost on storage of the fractions for 3 months at $-10°$ although 75% of the original phenylalanine deaminase activity still remains under these conditions.

Inhibitors and activators. Phenylalanine deaminase is completely inhibited by 10^{-5}M p-chloromercuriphenylsulfonic acid. Iodoacetamide at a higher concentration causes 67% inhibition of the enzyme. This evidence clearly indicates that the enzyme is a sulfydryl enzyme.

10^{-2} M EDTA and 10^{-3} M α,α'-dipyridyl, do not affect the reaction. 10^{-3} M KCN inhibits the reaction 85%. L-tyrosine and both m- and p-coumaric acids markedly inhibit the enzyme reaction. Cinnamic acid is also inhibitory.

No evidence could be obtained for the participation of a divalent cation in the phenylalanine deaminase reaction. Pyridoxal phosphate and DPNH has no effect.

Effect of pH. The optimum pH is 8.8—9.2. The enzyme is active over the pH range 8—10.6.

MICHAELIS *constant.* The MICHAELIS constant for L-phenylalanine is 1.7×10^{-3}M at pH 8.8.

Reversibility. Attempts to reverse the reaction have been unsuccessful.

Stability of the Enzyme. The stability of the enzyme depends upon the age of the plant material used for preparing the acetone powder. Acetone powders of the ARAVAT variety of barley in the pre-head stage are completely stable for at least 3 months at 6°. However, if the acetone powder is prepared from the ARAVAT variety when the heads have emerged, the powder loses most of its activity before three months. Acetone powders prepared from the MARIOUT variety are also stable in storage, but are unstable to fractionation.

IV. Conversion of Coniferin and Coniferyl Alcohol to Coniferous Lignin.

TIEMANN and MENDELSOHN (1875) established the chemical structure of the glucoside coniferin, which is present in the cambial sap of conifers, and he suggested that lignin and coniferin are related genetically. Later, ERDTMAN (1933a, b) showed that the oxidation product of isoeugenol by mushroom phenoloxidases was a phenylcoumaran derivative which is claimed to occur in lignin, and he pointed out that lignin might be a dehydrogenation product of guaiacylpropane derivatives with an oxidized side-chain.

During the past 10 years FREUDENBERG et al. (1958) have been studying the biosynthesis of lignin from coniferyl alcohol. They found that when air is bubbled through a very dilute aqueous solution of coniferyl alcohol in the presence of mushroom phenoloxidases, a light-colored grey-brown precipitate with properties remarkably similar to those of spruce milled wood lignin is formed. FREUDENBERG further found that, when all the coniferyl alcohol in the reaction mixture has been used up and only very little lignin has been formed, large amounts of the following intermediates of low molecular weight are obtained:

(A) dehydro-diconiferyl alcohol (20%); (B) DL-pinoresinol (20%); (C) guaiacyl-glycerol-β-coniferyl ether (30%); (D) coniferyl aldehyde (1%); (E) the aldehyde of A (4%); (F) the aldehyde of C (3%); (G) dehydro-di-pinoresinol (4%); (H) guaiacylglycerol-bis-coniferyl ether (15%); (I) guaiacylglycerol-α-dehydrodiconiferyl-β-coniferyl ether; (J) Bis-dehydroconiferylalcohol; (K) DL-epipinoresinol.

The yield of each substance has its individual maximum at a different stage of formation of lignin.

By the continued action of the enzymes, these substances were dehydrogenated further and condensed with each other or with coniferyl alcohol to form the lignin polymolecule. Moreover, all of these intermediates and synthesized lignins were optically inactive like lignins isolated from woods, although several asymmetric carbons are present.

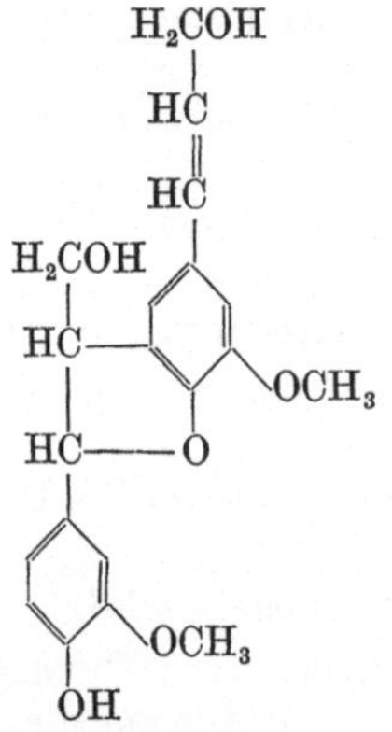

Dehydrodiconiferyl
alcohol (A)

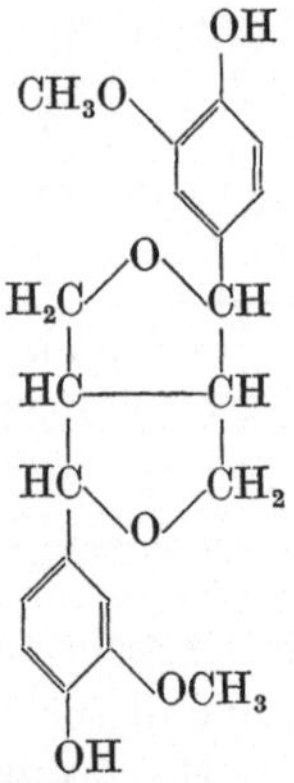

DL-pinoresinol
(B)
DL-epipinoresinol
(K)

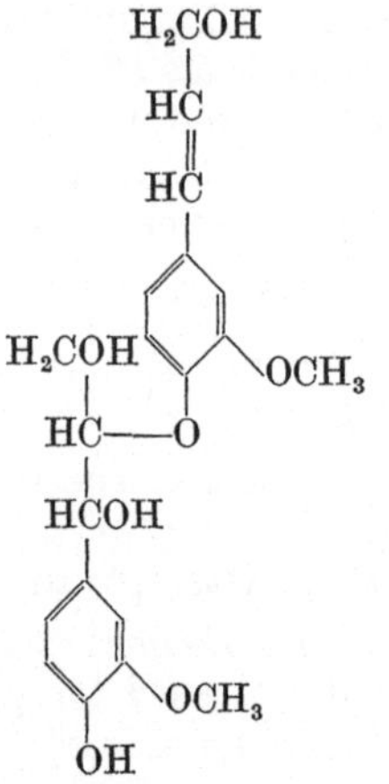

Guaiacylglycerol-
β-coniferyl ether (C)

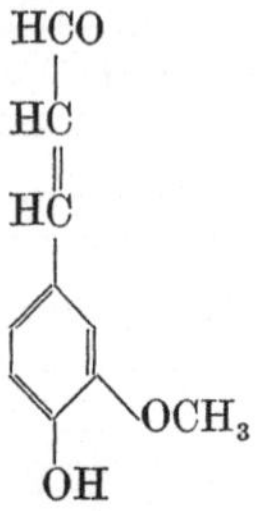

Coniferyl
aldehyde
(D)

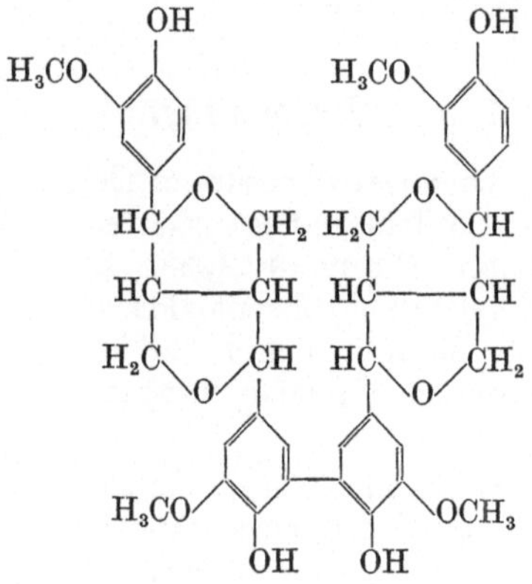

Dehydrodipinoresinol (G)

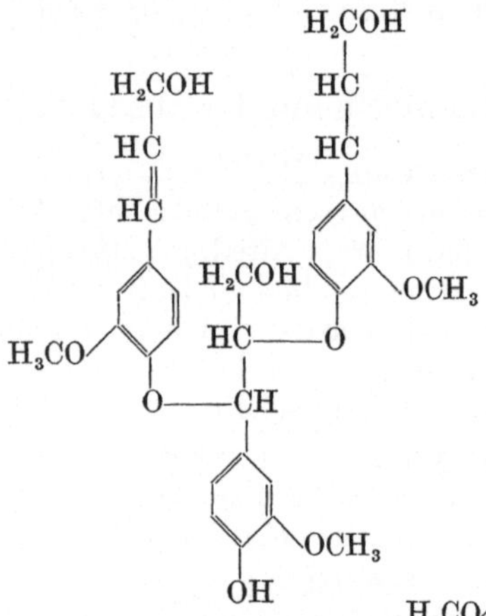

Guaiacylglycerol-bis-
coniferyl ether (H)

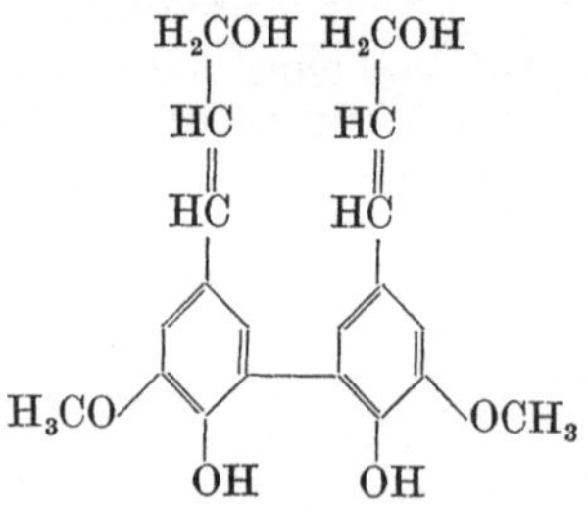

Bis-dehydro-
coniferylalcohol (J)

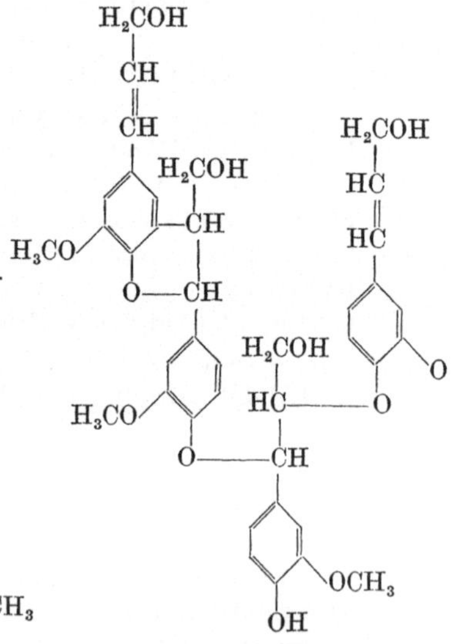

Guaiacylglycerol-α-
dehydrodiconiferyl-β-
coniferyl ether (I)

FREUDENBERG (1956), and ADLER (1958) discussed the mechanism of this dehydrogenative polymerization. They explained that the action of the enzymes is restricted to dehydrogenation of the phenolic group, and all subsequent reactions, such as the formation of quinone methides by the interaction of the radicals, the addition reactions of the quinone methides and possibly polymerization reactions, proceed automatically without further regulation by the enzymes.

HIGUCHI et al. (1953) investigated the properties of the phenoloxidase responsible for the lignification of bamboo-shoot and suggested that the enzyme is probably a laccase, and subsequently HIGUCHI (1955, 1957 and 1958a, b, c) confirmed that several laccases and peroxidases but not tyrosinase, can catalyze the dehydrogenative oxidation of coniferyl alcohol. FREUDENBERG et al. (1958) also confirmed later that the enzyme is a laccase.

FREUDENBERG (1959a, b) and KRATZL et al. (1956a, b; 1957; 1958a, b, c) found that radioactive coniferin or coniferyl alcohol is incorporated into spruce lignin without rearrangement of the carbon skeleton, and that radioactive phenylalanine is transformed into radioactive coniferin and subsequently incorporated into the lignin. They also found that during the vegetative period the tissues of the cambium and adjacent cells of spruce contain large amounts of coniferin, laccase and peroxidase, and that β-glucosidase is present between the fresh cells and the mature wood.

From these results FREUDENBERG reached the conclusion that coniferin produced from phenylpropanoid acids diffuses into the newly formed cells, where it is hydrolyzed to glucose and coniferyl alcohol, and the latter, coming into contact with laccase and peroxidase, is transformed into lignin.

FREUDENBERG further suggested that the angiosperm lignin is formed by copolymerization of coniferyl alcohol and sinapyl alcohol. In fact, a mixture of sinapyl alcohol and coniferyl alcohol gave a dehydrogenation polymer with properties similar to those of angiosperm lignin in the presence of laccase. KRATZL and BUCHTELA (1958) prepared this mixed polymer using laccase and showed that the polymer gives on ethanolysis syringoyl methyl ketone, which is obtainable from angiosperm lignin and is known to be derived from β-aryl ether, an important structural element of lignin.

1. Dehydrogenation of Coniferyl Alcohol by Laccase and Peroxidase.

a) Mushroom Enzyme.

CH$_2$OH   CH$_2$OH

CH   CH

CH $-(H^+ + e)$ CH

$\longrightarrow$

OH   OCH$_3$ O   OCH$_3$

Coniferyl alcohol

FREUDENBERG (1952) found that the phenoloxidases of mushroom catalyze the dehydrogenative polymerization of coniferyl alcohol to coniferous lignin, and later the mushroom enzyme catalyzing the dehydrogenative polymerization of guaiacyl and syringyl compounds including coniferyl alcohol was identified as a laccase by HIGUCHI (1958a, c). Further, HIGUCHI and ITO (1958) and FREUDENBERG et al. (1958) confirmed that peroxidase also catalyzes the dehydrogenative polymerization of coniferyl alcohol.

Assay. The enzyme activity can be estimated by measuring manometrically the oxygen uptake by coniferyl alcohol. The solubility of coniferyl alcohol in aqueous solution is, however, very low and dihydroferulic acid can be used as a good substitute for coniferyl alcohol.

Procedure. The following system, contained in manometer flasks of the WARBURG type, was usually used: main compartment; 20 μmoles of substrates dissolved in 1 ml phosphate buffer (1/15 M pH 6.8) and distilled water 1 ml. Side arm; Enzyme solution 0.3 ml. Center well; 15% KOH 0.2 ml. Total volume of fluid 3.0 ml.

The experiments were run at 30° with shaking at 120 oscillations per minute. The influence of CO on the enzyme activity was tested with respirometers filled with a gas mixture of 90% CO + 10% O_2, using one of 90% N_2 + 10% O_2 as control.

Preparation of enzyme (FREUDENBERG et al., 1958). Two kilograms of fresh mushrooms *(Psalliota campestris)* were homogenized with a small amount of phosphate buffer (pH 6.5, 0.05 M) and subsequently ground with 800 g of sea sand and pressed through muslin under 450 atmospheres pressure. The residue was extracted twice with each 250 ml of phosphate buffer and pressed again.

The extracts were combined and cooled to 0° and cold methanol was added dropwise until the volumetric ratio of water to methanol became 3 : 2 under vigorous stirring (2 hours). The precipitate was removed by centrifugation and cold methanol was added further to the well-cooled supernatant (−10°) until the ratio of water to methanol became 1 : 3. The grayish precipitate was collected by centrifugation and the supernatant discarded. The precipitate was extracted for 1 hour with well-cooled buffer solution (200 ml) under vigorous shaking at 0°. The suspension was centrifuged and the residue was extracted again in the same way. The extracts were combined and cold methanol was added until the ratio of water to methanol became 3 : 2. A small amount of precipitate that formed was centrifuged off and cold methanol was added further until ratio became 1 : 3. The light-colored grayish precipitate formed was extracted twice using 200 ml of cold buffer solution each time, as described above. The enzyme extract thus obtained was kept at −25°. The yield of the enzyme was about 25 to 35% of the total activity of the enzyme in crude extracts. The enzyme solution was stable for a long time at −25°.

In further purification, the enzyme solution was dialyzed against distilled water until the solution was phosphate-free, and a suitable amount of basic lead acetate solution (the minimum amount necessary to decolorize the solution but not to precipitate the enzyme) was added with stirring. The solution was kept overnight in a refrigerator and the precipitate was centrifuged off. To this supernatant calcium triphosphate gel or aluminium hydroxide gel was added until almost all the enzyme was adsorbed, the adsorbed gel was collected by centrifugation and, after washing with a small amount of water, was well eluted with 0.5 M phosphate buffer pH 6.5. By this procedure, about two-thirds of the total activity was lost but the specific activity of the enzyme obtained was raised 5 times. The Q_{O_2}* for hydroquinone using this enzyme was 20,000 at 25° and pH 6.5.

b) The Enzyme of Spruce Cambial Sap.

Three twenty-five-year-old spruce trees were cut down at the end of May and the trees were cut into one meter lengths. The bark was removed by a plane and the cambial tissues, including young woody tissues, were collected by a scraper.

---

* Q_{O_2} means the μl of oxygen used up per mg of dry enzyme per hour.

The debarked trees were washed with diluted buffer solution (pH 6.7) and the cambial tissues collected were extracted with the above washing (buffer solution) and pressed through muslin under 250 atmospheres pressure. These procedures were repeated once. The extracts (about 15 l.) were combined and well cooled, and purified with methanol as described above. $Q_{O_2} = 5000$ (hydroquinone). The residual tissues were ground with dry ice in a mortar and homogenized for several minutes in a Waring blendor. The homogenates were kept in 0.05 M phosphate buffer, pH 6.5, for 48 hours at 3° and filtered. The Q_{O_2} for hydroquinone, using the enzyme solution thus obtained, was 580 at pH 6.5 and 25°.

The rate of oxygen consumption during the oxidation of coniferyl alcohol and dihydroferulic acid is proportional to the amount of enzyme present.

c) The Enzyme of Japanese Lacquer
(HIGUCHI, 1958a).

Two liters of cold acetone was added with stirring to about 800 g of commercial Japanese lacquer. The brown-colored acetone solution was filtered, and the sediment was well washed with cold acetone until the filtrate became colorless. The sediment was extracted overnight with 1.5 liters of distilled water in a refrigerator. The extract containing the laccase was filtered on a BÜCHNER funnel, and 750 g of solid $(NH_4)_2SO_4$ was added to the filtrate (about 1.5 liter, blue-colored solution) under continuous stirring. The precipitate was centrifuged off, and then solid $(NH_4)_2SO_4$ was added to the supernatant at 40° until saturated. The blue precipitate was collected by centrifugation and was dissolved in small amounts of distilled water. This can be used as an enzyme solution after dialysis.

Properties. The enzyme, which is a copper enzyme, catalyzes the oxidation of guaiacyl- and syringyl compounds as well as mono- and ortho-phenols; nevertheless, the oxidation of guaiacyl- and syringyl compounds cannot be catalyzed by potato tyrosinase.

Inhibitors. Substances known to complex with copper, such as potassium cyanide, diethyldithiocarbamate, hydrogen sulfide, sodium azide, and salicylaldoxime inhibit the enzyme but it is not inhibited by carbon monoxide.

For the preparation of the dehydrogenation polymer of coniferyl alcohol, the description by FREUDENBERG in Vol. 4 of this series can be consulted.

2. Coupled Oxidation of Coniferyl Alcohol by Yellow Enzyme-Peroxidase Systems.

LYR (1957) found that the oxygen consumption by coniferyl alcohol in the coupled system of glucose oxidase and peroxidase-coniferyl alcohol is equivalent to about 0.5 atom per molecule as found in the oxidation by laccase. HIGUCHI (1958c) also confirmed the same phenomenon by the systems of mould glucose oxidase, D-amino acid oxidase from pig kidney and radish peroxidase. In the D-amino acid oxidase system, however, the keto acid produced is oxidized by peroxidase and therefore it is difficult to use for quantitative measurement.

Assay. Glucose oxidase catalyzes the reaction of glucose to gluconic acid and hydrogen peroxide following the equation:

$$C_6H_{12}O_6 + O_2 + H_2O \rightarrow C_6H_{12}O_7 + H_2O_2$$

Hydrogen peroxide thus produced is used in the coupled oxidation of coniferyl alcohol by peroxidase. Hence, the difference between O_2 uptake by the system of glucose, glucose oxidase, catalase, peroxidase and coniferyl alcohol (1) and that by the system of glucose, glucose oxidase, catalase and coniferyl alcohol (2) should correspond to the oxygen consumption by coniferyl alcohol.

Procedure. Oxygen-uptake was measured with a WARBURG respirometer in the following systems:

1. Main compartment; 20 μmoles of coniferyl alcohol and 25 μmoles of glucose in 2 ml phosphate buffer (1/15 M, pH 6.8).

Side arm 1; 0.3 ml each of peroxidase and catalase solution.

Side arm 2; glucose oxidase, 0.3 ml.

Center well; 15% KOH, 0.2 ml.

2. The same system, except that peroxidase solution was omitted. Total volume, 3.5 ml.

The experiments were run at 30°, shaking at 120 oscillations per minute.

As stated above, coniferyl alcohol was oxidized by both peroxidase and laccase and the oxygen consumption was equivalent to about 0.5 atom per molecule in both cases.

HIGUCHI (1958b, c) confirmed that the dehydrogenation polymer of coniferyl alcohol formed by the action of peroxidase is very similar to that formed by laccase and produced vanillin and well-known guaiacylpropanes after alkaline nitrobenzene oxidation, ethanolysis and hydrolysis.

References.

ACERBO, S. N., W. J. SCHUBERT and F. F. NORD: J. Amer. chem. Soc. **82**, 735 (1960). — ADLER, E.: Proc. Int. Congr. Biochem. 4th Congr. **2**, 137 (1958).

BALINSKY, D., and D. D. DAVIES: Biochem. J. **80**, 292 (1961a); — **80**, 300 (1961b). — BERENBLUM, I., and E. CHAIN: Biochem. J. **32**, 295 (1938). — BROWN, S. A., and A. C. NEISH: Nature (Lond.) **175**, 688 (1955). — BROWN, S. A., D. WRIGHT and A. C. NEISH: Canad. J. Biochem. Physiol. **37**, 25 (1959).

DAVIS, B. D.: Advanc. Enzymol. **16**, 247 (1955); — Arch. Biochem. Biophys. **78**, 497 (1958). — DAVIS, B. D., and U. WEISS: Arch. exp. Path. Pharmak. **220**, 1 (1953).

ERDTMAN, H.: Biochem. Z. **258**, 172 (1933a); — Liebigs Ann. **503**, 283 (1933). — EBERHARDT, G., and W. J. SCHUBERT: J. Amer. chem. Soc. **78**, 2835 (1956).

FEWSTER, J. A.: Biochem. J. **73**, 14 (1959). — FREUDENBERG, K.: Holz als Roh- und Werkstoff **10**, 339 (1952). — Angew. Chem. **68**,84 (1956); — Nature (Lond.) **183**, 1152 (1959a); — Chem. Ber. **92**, 89 (1959b). — FREUDENBERG, K., J. M. HARKIN, M. REICHERT and T. FUKUZUMI: Chem. Ber. **91**, 581 (1958).

GAMBORG, O. L., and A. C. NEISH: Canad. J. Biochem. Physiol. **37**, 1277 (1959). — GOITONDE, M. K., and M. W. GORDON: J. biol. Chem. **230**, 1043 (1958). — GROSS, S. R.: J. biol. Chem. **233**, 1146 (1958).

HASEGAWA, M., and T. HIGUCHI: J. Jap. For. Soc. **42**, 305 (1960). — HIGUCHI, T., and Y. ITO: J. Biochem. (Japan) **45**, 575 (1958). — HASEGAWA,M., T. NAKAGAWA and S. YOSHIDA: J. Jap. For. Soc. **39**, 159 (1957). — HASEGAWA, M., S. YOSHIDA and T. NAKAGAWA: Kagaku **24**, 421 (1954). — HIGUCHI, T., I. KAWAMURA and H. ISHIKAWA: J. Jap. For. Soc. **35**, 258 (1953). — HIGUCHI, T., I. KAWAMURA and I. MORIMOTO: J. Jap. For. Soc. **37**, 446 (1955). — HIGUCHI, T.: Physiol. Plant. **10**, 364 (1957); — J. Biochem. (Tokyo) **45**, 515 (1958a); **45**, 575 (1958b); — Proc. Int. Congr. Biochem. 4th Congr. **2**, 161 (1958c). — HURWITZ, J., and A. WEISSBACH: J. biol. Chem. **234**, 710 (1959).

ISHIKAWA, H.: Private communication (1960).

KOUKOL, J., and E. E. CONN: Abstr., Pacific Slope Biochem. Conference, Davis, California, Sept. 1960; — J. Biol. Chem. **236**, 2692 (1961). — KRATZL, K., and G. BILLEK: Holzforschung **10**, 161 (1956a). — KRATZL, K., G. BILLEK, E. KLEIN and K. BUCHTELA: Mh. Chem. **88**, 721 (1957). — KRATZL, K., and K. BUCHTELA: Mh. Chem. **90**, 1 (1958c). — KRATZL, K., and H. FAIGLE: Mh. Chem. **89**, 708 (1958b). — KRATZL, K., and H. FAIGLE: Mh. Chem. **90**, 768 (1959). — KRATZL, K., and G. HOFBAUER: Mh. Chem. **87**, 617 (1956b). — KRATZL, K., and G. HOFBAUER: Mh. Chem. **89**, 96 (1958a).

LEVIN, J. G., and D. B. SPRINSON: Biochem. Biopyhs. Res. Comm. **3**, 157 (1960). — LYR, H.: Naturwissenschaften **44**, 235 (1957).

MANSKAJA, S. M., and L. A. KODINA: Dokl. Akad. Nauk U.S.S.R. **123**, 4 (1958). — MCCALLA, D. R., and A. C. NEISH: Canad. J. Biochem. Physiol. **37**, 531 (1959a); — **37**, 537 (1959h). — MITSUHASHI, S., and B. D. DAVIS: Biochim. biophys. Acta **15**, 54 (1954a); — Biochim. biophys. Acta **15**, 268 (1954b).

NANDY, M., and N. C. GANGULI: Arch. Biochem. Biophys. **92**, 399 (1961). — NEISH, A. C.: Proc. Int. Congr. Biochem., 4th Congr. **2**, 82 (1958); — Annu. Rev. Pl. Physiol. **11**, 55 (1960); — Phytochemistry **1**, 1 (1961).

SCHWINCK, I., and E. ADAMS: Biochim. biophys. Acta **36**, 102 (1959). — SRINIVASAN, P. R., and D. B. SPRINSON: J. biol. Chem. **234**, 716 (1959).

TIEMANN, F., and B. MENDELSOHN: Ber. dtsch. chem. Ges. **8**, 1136, 1139 (1875).

WEINSTEIN, L. H., C. A. PORTER and H. J. LAURENCOT: Nature (Lond.) **183**, 326 (1959a); — Contr. Boyce Thompson Inst. **20**, 121 (1959b). — WEISS, U., C. GILVARG, E. S. MINGIOLI and B. D. DAVIS: Science **119**, 774 (1954).

YANIV, H., and C. GILVARG: J. biol. Chem. **213**, 787 (1955).

Enzymes of Amino Acid Metabolism.

Part 1.

Enzymes of Deamination, Decarboxylation, Transmethylation and Intermediary Metabolism.

By

B. D. Sanwal and Madhu Lata.

With 7 Figures.

Plants are autotrophic with regard to the amino acids and manufacture them from inorganic sources of nitrogen. Obviously then, they must possess all the necessary enzymes bringing about the synthesis and interconversion of the various amino acids; yet very little is so far known about the enzymes of higher plants involved in the metabolism of these compounds. Numerous tracer studies have been made using labeled carbon compounds (cf. WEBSTER, 1959) and in each case the results have pointed to the presence of reactions and pathways similar to those of microorganisms and, in some instances, of animal tissues. Tracer techniques, however, do not yield information regarding the nature of the enzymes bringing about the synthesis of a particular intermediate. From the viewpoint of comparative biochemistry it would, for instance, be interesting to know the nature of the enzyme synthesizing carbamyl phosphate in higher plants and fungi. It is well known that this compound is synthesized in animals by means of an irreversible reaction mediated by carbamyl synthetase; in bacteria, however, this compound is produced by means of a reversible reaction catalyzed by carbamate kinase. Similarly, in the synthesis of ornithine, a dichotomy of pathways is observed with different organisms. In gram positive bacteria and animal tissues, ornithine probably arises from glutamate *via* glutamic γ-semialdehyde, but in gram negative bacteria, like *E. coli*, ornithine is produced through a pathway involving N-acetylated derivatives of glutamic acid (VOGEL and BONNER, 1959). The manner in which ornithine is synthesized in plants is not known. Some plants possess along with ornithine, N-acetylornithine (VIRTANEN and LINKO, 1955), and it seems probable that the latter is produced by means of enzyme systems similar to those of *E. coli*.

Abbreviations: ADP, adenosine diphosphate; AMP, adenosine-5-phosphate; ASA, aspartic β-semialdehyde; ATP, adenosine triphosphate; BAP, β-aspartyl phosphate; CoA, coenzyme A; DCIP, dichlorophenol indophenol; DEAE-cellulose, diethylaminoethyl cellulose; DMPT, dimethylpropionthetin; DNAse, deoxyribonuclease; DPN and DPNH, oxidized and reduced diphosphopyridine nucleotide; EDTA, ethylenediaminetetraacetic acid; FAD, flavinadenine dinucleotide; GSH, reduced glutathione; pCMBA, para-chloromercuribenzoate; P_i, inorganic orthophosphate; PP, pyrophosphate; S-AMe, S-adenosylmethionine; S-MMe, S-methyl methionine; TCA, trichloracetic acid; THFA, tetrahydrofolic acid; TPN and TPNH, oxidized and reduced triphosphopyridine nucleotide, respectively; TPP, thiamine pyrophosphate; Tris, tris(-hydroxymethyl)aminomethane.

In the following pages, we have pointed out, wherever possible, the need for further investigations with plant tissues, especially from the viewpoint of comparative biochemistry. This review primarily deals with the enzymes of higher plants, fungi, algae, and bacteria. Enzymes which are known to occur excusively in animal tissues (as, for instance, the thetin transmethylases) have not been discussed, but references have been made to places where descriptions of the enzymes are to be found. Emphasis has been placed on the assay of enzymes, and detailed methods of isolation are given only for enzymes of the higher plants.

A. Enzymes of Oxidative Deamination.

I. Pyridine Nucleotide linked Dehydrogenases.

1. L-Glutamic Dehydrogenase of Higher Plants.

The enzyme catalyzes the reversible reaction:

$$\text{glutamate} + \text{DPN}^+ + \text{H}_2\text{O} \rightleftharpoons \alpha\text{-ketoglutarate} + \text{NH}_4^+ + \text{DPNH} + \text{H}^+$$

Occurrence. It has been reported to occur in pea seeds, cabbage and carrot roots, celery, radish, and white cabbage (ADLER, DAS, EULER and HEYMAN, 1938), bean and pea seedlings (DAMODARAN and NAIR, 1938; DAVIES, 1956; BONE, 1959a), oats (BERGER and AVERY, 1943, 1944; RAUTANEN and TAGER, 1955), spinach leaves (BONNER and WILDMAN, 1946), and cucumber seeds (EULER, ADLER, GÜNTHER and ELLIOT, 1939). In most cases only crude extracts or plant macerates have been used. BULEN (1956) purified the enzyme 45-fold from corn leaves and studied its characteristics. Since the enzyme is of foremost importance in converting ammonia to amino groups, it is possible that it occurs in all higher plants.

Intracellular localization. The enzyme seems to be localized in the mitochondria of the pea plant (DAVIES, 1956), oats (RAUTANEN and TAGER, 1955) and mung bean (BONE, 1959b). Treatment of the isolated mitochondria with 0.1% non-ionic detergent, O.P.C. 45 (Petrochemicals Ltd., London) results in the solubilization of the mitochondrial enzyme (BONE, 1959b). It is of interest to note that glutamic dehydrogenase from animal liver is also localized in the mitochondria.

Assay. GDH has been assayed by measuring dye reduction in THUNBERG tubes, uptake of oxygen in WARBURG manometric apparatus, or reduction of DPN at 340 mμ in a spectrophotometer. The previous two methods are inconvenient and the assay utilizing THUNBERG technique is also inaccurate. With crude plant preparations interference is caused in the spectrophotometric assay by turbidity and the green colouring matter of the extracts. It is, therefore, recommended that the assay be performed either by following the reduction of cytochrome C at 550 mμ in the presence of purified DPNH cytochrome C reductase (BENDALL and DE DUVE, 1960) or the reduction of DCIP in the presence of excess diaphorase at 600 mμ (AMES and GARRY, 1959).

1. Assay by cytochrome C reduction. The assay system consists of 0.02 M glycylglycine-NaOH buffer, pH 7.8; 0.03 M nicotinamide; 0.4 mM sodium cyanide; 1 mM EDTA; 0.7 mM DPN; 0.013 M potassium-L-glutamate; 80 μM cytochrome C; 6 units cytochrome C reductase (prepared according to the method of MAHLER, SARKAR, VERNON and ALBERTY, 1952), and 0.05 or 0.1 ml of enzyme in a total volume of 1.8 ml. The reaction is started by the addition of enzyme and the increase in extinction at 550 mμ is followed in a spectrophotometer with the help of a reference cell containing all the components of the assay system except glutamate.

2. Assay by dye reduction (AMES and GARRY, 1959). The assay cuvette contains 0.5 ml of 0.4 mM sodium dichlorophenolindophenol; 0.4 ml of 0.2 M triethanolamine buffer (pH 8.1); 0.005 ml of 50 mM DPN (brought to pH 5.0 with KOH); and excess of diaphorase (available from Worthington Biochemical Corporation, Freehold, N. J.), and dialyzed enzyme extract. After the endogenous oxidation ceases and dye is not reduced further, 0.05 ml of 0.5 M L-glutamate is added and the rate of dye bleaching is determined.

3. *Assay by DPNH oxidation.* With not very turbid solutions, assays can be conveniently performed by measuring DPNH oxidation in the reductive amination assay (Bulen, 1956), which is at least 5 times as sensitive as the oxidative de-amination assay. The system consists of 0.5 ml of 10^{-3} M DPNH solution; 0.2 ml of 0.2 M potassium α-ketoglutarate; 0.2 ml of 1.5 M $(NH_4)_2SO_4$, and 2.0 ml of 0.2 M Tris buffer, pH 8.15. At zero time, 0.1 ml of properly diluted enzyme is added and extinction readings at 340 mμ are taken every minute for 6 minutes in a spectrophotometer. The reference cuvette contains all of the reagents except α-ketoglutarate. The rate of reaction is linear for about 5 minutes.

One *unit* of GDH is defined as the amount of enzyme producing an optical density decrement of 1.0 (Bulen, 1956).

Purification of the enzyme from corn leaves (Bulen, 1956). The sap from 45-day old chilled leaves (1.7 kg) is obtained by grinding in a "Corona" corn mill (Landers, Frary and Clark, New Britain, Connecticut) and straining through 2 layers of cheesecloth. Five ml of a 0.2 M KH_2PO_4 solution containing 1% cysteine is added to each 30 ml of sap immediately. The sap is adjusted to pH 5.5 with K_2HPO_4 and after 15 minutes, the precipitate is removed by centrifugation. The protein precipitating between 0.29 and 0.48 $(NH_4)_2SO_4$ saturation is dissolved in 200 ml of 0.2 M K_2HPO_4, pH 8.0. Sufficient 0.2 M KH_2PO_4 is added to the solution to reduce the pH to 6.6, and water is added to a final volume of 450 ml. This solution is fractionated with $(NH_4)_2SO_4$ and the fraction precipitating between 0.42 and 0.49 saturation is dissolved in 50 ml of K_2HPO_4 buffer, pH 8.0. The pH of this solution is adjusted to pH 5.6 with 0.1 N H_2SO_4, stirred for 15 minutes with 3 g of aged calcium phosphate gel, and centrifuged. The enzyme is eluted from the gel with 20 ml of 0.2 M potassium phosphate, pH 6.4 and dialyzed against 0.05 M phosphate, pH 8.0 for 4 hours. The purified enzyme can be stored at $-25°$ C for periods up to one year.

Properties of the plant enzyme (Bulen, 1956). The pH optimum of the reductive amination reaction is 8.1. The initial rate in Tris is 11% higher than that observed in phosphate. The enzyme is DPN specific. It is very sensitive to pCMBA but not to o-iodosobenzoate, iodoacetate or cyanide. The pCMBA inhibition is reversed by glutathione. D-glutamic acid is neither a substrate nor an inhibitor. This point requires re-examination because it has been shown that GDH from animal tissues (Olson and Anfinsen, 1953) is inhibited by D-glutamate. Animal GDH is a metalloprotein containing $2-4$ g atoms of Zn per mole of protein (Adelstein and Vallee, 1958).

The K_m values for DPNH, α-ketoglutarate and ammonia are 3.65×10^{-5} M, 1.51×10^{-3} M and 0.101 M respectively.

The substrate specificity of plant GDH is not well known. It has been demonstrated that GDH from chicken liver also catalyzes the oxidative deamination of L-norvaline, L-α-aminobutyric acid, L-leucine, L-valine, DL-norleucine, L-isoleucine (Struck and Sizer, 1960) and L-alanine (Tomkins, Yielding and Curran, 1961). Treatment of crystalline GDH from animal sources with diethylstilbesterol, deaggregates the enzyme into smaller units and the latter are capable of oxidizing L-alanine at an increased rate (Tomkins et al., 1961).

2. L-Glutamic Acid Dehydrogenase of Neurospora and Other Microorganisms.

There are two types of glutamic dehydrogenases in microorganisms differentiated on the basis of their coenzyme specificity. The GDH of the green alga, *Ulva lactuca* is non-specific and can function with either DPN or TPN as electron acceptors in the deamination of L-glutamic acid (JACOBI, 1957). In this regard it resembles the GDH of animal liver. In bacteria, it is TPN-specific (ADLER, HELLSTRÖM, GUNTHER and VON EULER, 1938) or DPN-specific (BARBAN, 1954; NISMAN, 1954). The fungi so far investigated, like brewers' and bakers' yeast (HOLZER and SCHNEIDER, 1957), *Fusarium* (SANWAL, 1961), and *Neurospora crassa* (SANWAL and LATA, 1961) possess two different enzymes, one specific for TPN and another for DPN. The only exception seems to be *Allomyces arbuscula* (KLINKHAMMER, 1959) which possesses only one enzyme specific for DPN.

Assay. Both TPN- and DPN-dependent glutamic acid dehydrogenases can be assayed by following either the disappearance of the reduced coenzymes in the reductive deamination or their appearance in the oxidative deamination assay at 340 mμ. Since the equilibrium of the reaction favours the synthesis of glutamic acid, the reductive amination assay is more sensitive. DPNH oxidase, if present in crude extracts, interferes with the assay. The assay mixture in quartz cuvettes of 1 cm light path consists of, 0.1 ml of 0.2 M α-ketoglutarate; 0.1 ml of 1.2 M $(NH_4)_2SO_4$; 0.05 ml of 4.0×10^{-3} M TPNH (or 4.6×10^{-3} M DPNH, depending upon the enzyme being assayed); 0.01 $-$ 0.1 ml of enzyme preparation to produce extinction decrement of not more than 0.08 per minute, and enough 0.1 M Tris of the desired pH (see, enzyme properties below) to make 3.0 ml. The enzyme solution is added at zero time from an addermixer of the type described by BOYER and SEGAL (1954). The reference cuvette contains all the components of the assay mixture except α-ketoglutarate. The decrease in extinction is measured at 15 seconds interval for 60 seconds and the first two 15 second readings are used to calculate the activity of the enzymes.

One *unit* of the enzyme is defined as the amount which causes a decrease of 0.001 in optical density in one minute.

Purification of the TPN-dependent GDH of *Neurospora* (SANWAL and LATA, 1961). The fungus (any wild-type strain) is grown in the liquid medium N of VOGEL (1956) for 24 hours with shaking. The mycelium is harvested on two layers of muslin, washed with distilled water and excess water squeezed off. The cells are frozen at $-20°$ C for $2-3$ hours before use. The frozen cells are mixed with twice their weight of levigated alumina and crushed vigorously in a mortar till the whole mass acquires a "sticky" consistency. The disrupted cells are extracted with a buffer containing 0.1 M Tris containing 5×10^{-4} M β-mercaptoethanol, pH 8.5. The alumina and cell debris is removed by centrifugation. The pH of the supernatant solution is adjusted to 6.8 with 10% acetic acid and the precipitate is discarded. The solution is heated in a water bath to 55° C and held at that temperature for 5 minutes. The DPN-specific enzyme is labile to heat and is consequently denatured. The precipitate is centrifuged off and the supernatant is fractionated with solid $(NH_4)_2$ SO_4. The precipitate obtained between $0.40-0.50$ saturation is dissolved in a small volume of 0.02 M phosphate, pH 6.5, containing 5×10^{-4} M β-mercaptoethanol. The solution is dialyzed for 8 hours against the same buffer with frequent changes. To the dialyzed solution C$_\gamma$-aluminium hydroxide gel is added at the rate of 0.2 mg gel/mg protein and after 15 minutes, the enzyme is eluted with the same quantity of 0.2 M phosphate, pH 8.4, containing 5×10^{-4} M β-mercaptoethanol. The gel absorption is continued till all the enzyme has been adsorbed from the solution. The fractions showing the highest specific activity are pooled, lyophilized and stored at $-20°$ C for further use. The enzyme is stable for about 7 days.

Purification of the DPN-dependent GDH of *Neurospora* (Sanwal and Lata, 1961). The cell free extract is obtained in the same way as with the TPN-dependent GDH. The precipitate obtained after adjusting the pH to 7.0 is discarded. In ammonium sulphate fractionation, the precipitate obtained between $0.35-0.45$ saturation is dissolved in a small volume of 0.02 M phosphate buffer, pH 6.5, containing 5×10^{-4} M β-mercaptoethanol, and the solution is dialyzed against the same buffer for 4 hours. The enzyme is adsorbed and eluted from C_γ-aluminium hydroxide gel in the same way as the TPN-specific enzyme. Nearly all of the DPN-specific enzyme is adsorbed in the first two treatments. The first two eluates are pooled and enough ammoniacal $(NH_4)_2SO_4$ [made by adding 5 ml of concentrated NH_4OH to $0°$ C saturated $(NH_4)_2SO_4$] is added to 0.35 saturation. The precipitate is dissolved in a small volume of 0.02 M Tris, pH 7.5, and the solution is dialyzed for 6 hours against the same buffer. The dialyzed solution can be stored after lyophilization at $-20°$ C for approximately 24 hours.

Properties of DPN- and TPN-specific enzymes. The pH optimum of the DPN- and TPN-specific enzymes is 8.2, and 7.5 respectively. Both of the enzymes are inhibited by pCMBA and EDTA (Nicholas and Mabey, 1960). The enzymes obtained from *Fusarium* (Sanwal, 1961) are also inhibited by glutaric acid, the dicarboxylic acid analogue of glutamic acid. DPN is ineffective as an electron acceptor in the deamination of L-glutamic acid by the TPN-specific enzyme and TPN is ineffective for the DPN-specific enzyme. L-glutamine, L-aspartate, L-asparagine, L-alanine, L-valine, L-proline, L-threonine in the oxidative amination assay, and α-ketobutyrate, α-ketoisocaproate and α-ketoisovalerate in the reductive amination assay are ineffective with both enzymes.

The K_m values for the DPN-enzyme are 5.5×10^{-4} M for DPNH, 1.7×10^{-2} M for NH_4^+, 4.6×10^{-3} M for α-ketoglutarate, 3.3×10^{-4} M for DPN, and 5.5×10^{-3} M for L-glutamate. For the TPN-specific enzyme the corresponding K_m values are, 1.25×10^{-4} M for TPNH, 10^{-2} M for NH_4^+, 5.3×10^{-3} M for α-ketoglutarate, 0.5×10^{-4} M for TPN, and 45.0×10^{-3} M for L-glutamate.

3. L-Alanine Dehydrogenase.

The enzyme catalyzes the reversible reaction:

$$\text{pyruvate} + NH_4^+ + DPNH + H^+ \rightleftharpoons \text{L-alanine} + DPN^+ + H_2O$$

Occurrence. This enzyme has so far been found in bacteria only. Its occurrence in homogenates of ovaries of *Ascaris lumbricoides* reported by Pollak and Fairbairn (1955) is doubtful. The enzyme is of universal occurrence in members of the genus *Bacillus* (Wiame, Pierard and Ramos, 1961; Hong, Shen and Braunstein, 1959; O'Connor and Halvorson, 1960) and some species of *Arthrobacter*, *Micrococcus*, *Sarcina*, *Sporosarcina* (Wiame et al., 1961), *Rhizobium* (Jordan, 1959), and *Mycobacterium tuberculosis* (Goldman, 1959).

Assay. The enzyme is conveniently assayed by following the reductive amination of pyruvate by DPNH and NH_3. The mixture in a quartz cuvette of 1 cm light path contains, 0.2 ml of 0.16 M sodium pyruvate; 0.05 ml of 1.5×10^{-2} M DPNH; 0.15 ml of 3 M NH_4Cl; enzyme, and enough 0.1 M carbonate-bicarbonate buffer, pH 9.4 to make 3.0 ml (Wiame et al., 1961; O'Connor and Halvorson, 1960). Reference cuvette contains all the components of the mixture except pyruvate. After addition of the enzyme, decrease in extinction is measured at 340 mμ every 15 seconds. The results of 2 minute readings are plotted on paper and the rate of the reaction is calculated by extrapolation.

Different workers on this enzyme have used different definitions of a *unit*. Following the system used for other amino acid dehydrogenases, one unit of alanine dehydrogenase is defined as the amount which causes a decrease in extinction of 0.001 per minute.

Purification. Alanine dehydrogenase has been purified 19-fold from *Mycobacterium tuberculosis* var. *hominis* (GOLDMAN, 1959), about 60-fold from aerobic spores of *Bacillus cereus* (O'CONNOR and HALVORSON, 1960; 1961) and 9-fold from vegetative cells of *B. subtilis* (WIAME et al., 1961). It is easy to grow large quantities of cells of bacilli in a short time, and provided they are cultivated in low concentrations of glucose (approximately 0.2%), high yields of the enzyme are generally obtained.

The enzyme may be prepared from the vegetative cells of *B. subtilis* according to the procedure of WIAME et al. (1961).

Properties. The pH optimum for oxidative deamination is about 10.0. The enzyme is inhibited by $-SH$ group inhibitors, specially pCMBA. L-cysteine completely reverses the inhibition (WIAME et al., 1961). Fluorene carcinogens, 2-amino-fluorene and 2-acetylamino-fluorene, are slightly inhibitory (GOLDMAN, 1959). The K_m values are 3.5×10^{-3} M for alanine, 2.6×10^{-4} M for DPN, 8.8×10^{-4} M for pyruvate, 4.1×10^{-5} M for DPNH, and 0.36 M for NH_4^+. The equilibrium constant:

$$K' eq = \frac{(\text{pyruvate}) (\text{DPNH}) (NH_4^+) (H^+)}{(\text{alanine}) (DPN^+)}$$

is 9 ± 1.10^{-14} (WIAME et al., 1961).

Specificity of the reactants. DPN is specific as the coenzyme. As in the case of GDH, alanine dehydrogenase has a broad substrate specificity. Aliphatic amino acids with chain lengths of less than 7 carbons are metabolized by the enzyme. The V_{max} relative to alanine of α-aminobutyric acid is 31.6%, L-norvaline 16.7%, L-leucine 4.4%, L-isoleucine 4.0%, L-valine 2.9% and DL-serine 1.5% (O'CONNOR and HALVORSON, 1961). D-isomers of these substrates, glycine, and sarcosine are competitive inhibitors. Compounds with substitutions in the $-COOH$ group, the $-NH_2$ group or the $-H$ of α-carbon of L-alanine do not serve as the substrates of the enzyme.

4. L-Leucine Dehydrogenase.

The enzyme catalyzes the reversible reaction:

$$\alpha\text{-ketoisocaproate} + DPNH + NH_4^+ + H^+ \rightleftharpoons \text{L-leucine} + DPN^+ + H_2O$$

So far it has been demonstrated to occur in a common laboratory strain of *B. cereus*, but is, perhaps, widely distributed in other members of the genus *Bacillus* (SANWAL and ZINK, 1961).

Assay. The enzyme is assayed, depending upon requirements, by either the increase in absorbance in the oxidative deamination assay or the decrease in the reductive amination assay at 340 mμ. The latter assay is more sensitive but there is some interference by the DPNH oxidase and lactic dehydrogenase present in crude extracts. In the oxidative deamination assay, quartz cuvettes of 1 cm light path are used. The reaction mixture consists of 0.1 ml of 0.2 M L-leucine; 0.05 ml of 7.6×10^{-3} M DPN, and 2.8 ml of 0.1 M Glycine-NaOH buffer, pH 11.0. The reaction is started by adding 0.05 ml of a properly diluted enzyme solution and extinction measurements are made every 30 seconds for 2 minutes. In calculating the rate of reaction, the first two 30 second figures are used.

One *unit* of the enzyme is defined as the amount causing an increase of 0.001 in extinction per minute in the oxidative deamination assay.

Purification of the enzyme (SANWAL and ZINK, 1961). A common laboratory strain of *Bacillus cereus* is grown for 36—48 hours at 25° C in Medium E of VOGEL and BONNER (1956b) with a supplement of 0.2% glucose, 0.1% yeast extract and 0.1% casamino acids. Nutrient broth can also be used. The cells are harvested on a continuous flow centrifuge and can be stored, if not required for immediate use, at — 20° C for over 30 days. Cell free extracts are made by mixing 20 g frozen cells with 40 g acid-washed levigated alumina and crushing vigorously in a mortar. The whole mass is extracted with 100 ml of 0.1 M Tris, containing 10^{-3} M β-mercaptoethanol, pH 8.5. The cell debris and alumina is removed by centrifugation and the pH of the supernatant solution is adjusted to 7.0 and the precipitate formed is discarded. The solution is fractionated by solid $(NH_4)_2SO_4$ and the precipitate obtained between 0.40—0.65 saturation is dissolved in 40 ml of 0.1 M Tris, pH 8.5. While the solution is stirred, 28 ml of ammoniacal ammonium sulphate [made by adding 5 ml of concentrated NH_4OH to 100 ml of 0° C saturated $(NH_4)_2SO_4$] is slowly added and the precipitate discarded after centrifugation. To the supernatant solution, a further 26 ml of ammoniacal ammonium sulphate is gradually added. The precipitate is recovered by centrifugation and dissolved in 20 ml of 0.05 M Tris, pH 7.0. The enzyme solution is quickly brought to 60° C in a water bath and after heating for 5 minutes, the heavy precipitate is discarded. The supernatant solution is dialyzed for 12 hours against 3 l of 0.05 M Tris, pH 7.0. The dialyzed solution is fractionated by aged C_{γ}-aluminium hydroxide gel according to the method of OCHOA, MEHLER and KORNBERG (1948). The gel is added at the rate of 0.2 mg/ml protein and after adsorption for 15 minutes, it is eluted with 5 ml of 0.2 M K_2HPO_4, pH 8.7, containing 10^{-3} M β-mercaptoethanol. The absorption and elution process is repeated and the fractions showing the highest specific activity are pooled. The enzyme is stored at — 25° C where it is stable for over a month. With the above procedure a 30-fold increase in purity is obtained.

Properties. The pH optimum of the oxidative deamination is 11.3. The enzyme is very sensitive to sulphide and pCMBA. The inhibition by the latter is completely reversed by L-cysteine. Chelating agents do not affect the enzyme activity at all. The K_m values for different substrates are, 6.2×10^{-3} M for α-ketoisocaproate, 10^{-4} M for DPNH, 1.6×10^{-4} M for DPN$^+$ and 1.3×10^{-2} M for NH$_4^+$. The equilibrium constant,

$$K'eq = \frac{(\alpha\text{-ketoisocaproate) (DPNH) (NH}_4^+\text{) (H}^+)}{\text{(leucine) (DPN}^+)} = 11.1 \pm 1.10^{-14}.$$

Substrate specificity. The enzyme reacts with L-leucine, L-valine, L-isoleucine, and DL-norvaline. Using V_{max} for L-leucine as 100, the V_{max} of valine is 80, that of isoleucine is 65 and that of DL-norvaline is 25. DPN is absolutely specific for the enzyme.

II. D-and L-Amino Acid Oxidases.

The D- and L-amino acid oxidases are known to occur in animal tissues, fungi and bacteria. To the best of our knowledge such oxidases have never been purified from plants. From time to time suggestions have been made that D-amino acid oxidases occur in green plants. Thus, SUNDARAM and SARMA (1953) ascribed the utilization of DL-tryptophan to the presence of a D-amino acid oxidase in *Phaseolus aureus*. Similarly, the green alga, *Chlorella vulgaris* can utilize sulphate or the D- or L-isomer of methionine as the only source of sulfur. Use of the D-isomer suggests the presence of a D-amino acid oxidase (SHRIFT, 1954). In crude rye seedling homogenates, KRETOVICH and DROZDOVA (1948) demonstrated a rapid oxidation of L-amino acids. In the presence of polyphenol oxidase and pyrocatechin, the rate of oxidation increased 3—4 fold in the preparations of sunflower seedlings. However, the enzymes have never been purified. The oxidation of tryptophan in pea seedling tissue has been ascribed to an oxidase by WILTSHIRE (1953), and ROGERS (1955) has shown that crude extracts of acorn squash oxidize some L-amino acids (Table 1) aerobically at an insignificant rate.

1. D-Amino Acid Oxidase of Microorganisms.

The enzyme catalyzes the oxidation of some D-amino acids as follows:

$$R.CHNH_2COOH + O_2 + H_2O \rightarrow R.CO.COOH + NH_3 + H_2O_2$$
(D-amino acid)

Occurrence. The enzyme is known to occur in several species of *Aspergillus*, *Penicillium* (EMERSON, PUZISS and KNIGHT, 1950) *Neurospora* (BENDER and KREBS, 1950), and *Proteus morgani* (STUMPF and GREEN, 1946; CLARKE, 1958). For details of occurrence in other organisms, the review of KREBS (1951) should be consulted. MASSEY, PALMER and BENNETT (1961) have obtained the enzyme in a crystalline form from pig kidneys. The claim that the flavoprotein enzyme contains iron (KUBO et al., 1958) has been shown by these authors to be erroneous. The reversibility of the reaction under certain conditions has been demonstrated by RADHAKRISHNAN and MEISTER (1958).

Table 1. *Specificity of amino acid oxidases of Neurospora and of squash* (from WEBSTER, 1959).

Amino acid	Microliters of O_2 consumption/hour/mg protein	
	Neurospora	Squash
Alanine	102	0.6
Arginine	179	0.5
Aspartate	0	0.2
Cysteine	272	2.5
Glycine	0	0
Histidine	448	0
Leucine	320	0
Methionine	170	0
Phenylalanine . . .	339	0.3
Proline	0	0.3
Serine	0	0
Threonine	0	0
Tryptophan	138	0
Tyrosine	390	0
Valine	61	0

Assay. The enzyme is easily assayed by measuring the O_2 uptake in WARBURG manometric apparatus in the presence of FAD and excess catalase. The main compartment contains a suitable quantity of the enzyme (giving a measurable gas exchange), 20 μg FAD, 0.5 μg purified catalase (available commercially) and enough 0.03 M pyrophosphate buffer, pH 8.2 to make 2.6 ml. The side arm contains 0.2 ml of 0.06 M DL-methionine (or 0.03 M D-isomer). The central well contains a wick of filter paper with 0.2 ml of 2 N NaOH. Gas phase is O_2. After equilibration of the reaction mixture at 28.6° C, the substrate is tipped and readings are taken every 5 minutes for 20—30 minutes. The rate is calculated from either the 5 and 10 minute reading or 10 and 20 minute reading, whichever gives the higher value. This procedure is necessary for crude extracts owing to the existence of a lag which delays the development of maximum rates (BENDER and KREBS, 1950).

If crude extracts are being assayed, the addition of catalase and FAD are not necessary but controls are used to account for the autorespiration of crude extracts.

Preparation of the enzyme. The enzyme from the moulds or bacteria has not been purified to any great extent so far. All of the strains of *Neurospora crassa* do not produce the enzyme. The time at which maximum enzyme activity is obtained after growth varies from one strain to another (Krebs, 1951). The growth medium (Horowitz and Beadle, 1943; see also, Burton, 1951) contains 0.05% ammonium tartarate, 0.1% NH_4NO_3, 0.1% KH_2PO_4, 0.01% NaCl, 0.05% $MgSO_4 \cdot 7H_2O$, 0.01% $CaCl_2$, 2% sucrose, 0.2% casein hydrolyzate, 0.01% yeast extracts and 5 μg biotin per liter. The requirement for trace elements is satisfied by tap water. Mycelium is grown at 25° C in Fernbach flasks containing 500 ml of the medium and the pads are harvested after 7 days, washed with distilled water, and excess water is pressed out. The cells are disrupted with sand in a mortar and extracted with 2 times the weight of mycelium of 0.02 M pyrophosphate buffer, pH 8.5. After centrifugation, the supernatant solution is used as the enzyme source.

Properties. The pH optimum of the *Neurospora* enzyme is 8.0−8.5. The enzyme unlike the animal D-amino acid oxidase is not inhibited by iodoacetate or benzoate and is labile to acetone. Isovaline is a competitive inhibitor of the enzyme. The K_m for D-methionine is 2.5×10^{-4} M.

Specificity. Most of the common amino acids serve as the substrate for this enzyme, but methionine is metabolized more readily. Reactivity of the substrate is lost by a shift of the α-amino group to the β position; replacement of the hydrogen attached to the α-carbon atom by an alkyl group; replacement by methyl groups of both hydrogens attached to the amino nitrogen atom or the β carbon atom, and the substitution of hydroxyl group on the β-carbon atom and peptide bond formation through the carboxyl group (Horowitz, 1944). There seem to be certain minor differences in the D-amino acid oxidase from different moulds. The oxidase from *Penicillium chrysogenum* Q 176, for instance, utilizes leucine at a rate greater than methionine.

2. D-Glutamic Acid Oxidase.

D-glutamic acid + O_2 + $H_2O \rightarrow$ α-ketoglutaric acid + NH_3 + H_2O_2

The enzyme occurs in certain soil molds and is distinct from the D-amino acid oxidase discussed earlier. *Aspergillus ustus* and some strains of *Aerobacter* are rich sources of the enzyme (Mizushima, Izaki, Takahashi and Sakaguchi, 1956) specially when grown in media supplemented with 0.5% D-glutamic acid. A similar enzyme has been purified from the hepatopancreas of Octopus and shown to contain FAD as the prosthetic group (Rocca and Ghiretti, 1958).

Enzyme preparation and assay. Dessicated or fresh mycelium of *A. ustus* is ground with powdered glass in a mortar and extracted with 0.1 M borate-phosphate buffer, pH 8.0. The clear supernatant solution after centrifugation is used as the enzyme source. The enzyme is assayed as the D-amino acid oxidase described earlier, except that O_2 uptake is measured in the presence of 5 μmoles of D-glutamic acid or 10 μmoles of DL-acid in borate-phosphate buffer, pH 8.0.

Properties. The enzyme has an optimum pH of 8.0. DL-aspartic acid also shows some activity with the enzyme. Enzymic reaction is inhibited by hydroxylamine, sodium azide, KCN and pCMBA.

The enzyme can be used for the quantitative determination of D-glutamic acid.

3. L-Amino Acid Oxidase of Microorganisms.

The enzyme catalyzes the reaction:

L-amino acid + O_2 + $H_2O \rightarrow$ α-keto acid + NH_3 + H_2O_2

Occurrence. The enzyme has been shown to occur in many bacteria including *Proteus vulgaris, Aerobacter aerogenes, Pseudomonas pyocyaneous* (Stumpf and

GREEN, 1944; CLARKE, 1958), *Clostridium sporogenes* and *C. saccharobutyricum* (ROSENBERG and NISMAN, 1949). The moulds seem to have two types of L-amino acid oxidases. The insoluble L-oxidase has been described by KNIGHT (1948) in *Penicillium notatum*, *P. expansum*, *P. chrysogenum*, *P. sanguineum*, and *Aspergillus niger*. This enzyme remains active in acetone dried cells of these organisms. It is not known whether under such conditions the D-amino acid oxidase is inactivated, but the acetone powders are inactive with the D-amino acids. In *Neurospora*, a soluble L-amino acid oxidase has been described which is secreted in the medium during growth (THAYER and HOROWITZ, 1951; BURTON, 1951). The D-amino acid oxidase of this mould is present in the cells and by using the culture filtrate, an easy separation of the two enzymes is achieved from one another.

Assay. The activity of the oxidase is determined in terms of O_2 uptake in the presence of catalase and a suitable substrate, usually L-phenylalanine. The central compartment of WARBURG vessel contains the enzyme, 0.5 μg purified catalase, and enough 0.033 M pyrophosphate buffer, pH 8.3 to make 3.5 ml. The central well contains 0.2 ml 2 N NaOH with a wick of filter paper and the side arm contains 56 μg L-phenylalanine dissolved in 0.5 ml. The gas phase is oxygen and bath temperature is 30° C. After equilibration and tipping, readings are taken every 5—10 minutes for 30 minutes. One *unit* of the enzyme has been defined by BURTON (1951) as the amount of enzyme causing 1 μl O_2 to be taken up in 10 minutes.

Purification of the enzyme from *Neurospora* (THAYER and HOROWITZ, 1951; BURTON, 1951). The formation of the enzyme is dependent on the strain and the growth conditions (BURTON, 1951). High yielding strains should, therefore, be selected to obtain maximum activity of the enzyme. (Strain CM 13411, obtainable from the Commonwealth Mycological Institute, Kew, U. K.). The medium used for cultivation is the same as described under D-amino acid oxidase, except that 0.1—0.5 μg biotin/liter, 1.3% sucrose, 0.2% ammonium tartarate and 0.2% NH_4Cl is added and the mould is allowed to grow under stationary conditions in 1 cm depth of 6.5 l medium for 12 days at 25° C.

The culture fluid (5.6 l) is separated from the mycelial growth and 214 ml of 10% $CuSO_4 \cdot 5H_2O$, followed by 74 ml of 2 N NaOH, is added with continuous stirring. The $Cu(OH)_2$ precipitate is discarded by centrifugation. The enzyme remaining in the supernatant is adsorbed by the addition of 118 ml of 10% $CuSO_4$. The precipitated $Cu(OH)_2$, containing the enzyme, is centrifuged off and suspended in 100 ml of 0.1 M acetate buffer, pH 6.0. The suspension is dialyzed overnight against running water. The insoluble $Cu(OH)_2$ is again sedimented in a centrifuge, resuspended in 120 ml water, and 2 N NH_4OH (15 ml) is added slowly with stirring until all the $Cu(OH)_2$ is dissolved. The solution is dialyzed overnight against running tap water and any sediment remaining is centrifuged off. The blue supernatant solution is saturated with $(NH_4)_2SO_4$, the precipitated protein is dissolved in 20 ml water, dialyzed overnight and centrifuged. Last traces of Cu are removed from the pale-yellow supernatant by reprecipitation of the protein by complete saturation with $(NH_4)_2SO_4$. The precipitate is collected on a centrifuge, suspended in 15 ml water, dialyzed against running tap water for 15 hours and centrifuged. The clear pale yellow supernatant contains the oxidase purified approximately 172-fold. The activity of the purified enzyme is gradually lost during storage at —10° C (30% in 2 weeks).

Properties. The purified oxidase contains bound FAD (1 g mol/11,000 g nondialyzable nitrogen). The pH activity curve shows a broad maximum between pH 6 and 9.5 and varies with the amino acid used as substrate. The optimum temperature is 45° C. The enzyme is not affected by 0.01 M HCN, NH_2OH, NaN_3 or by 10^{-3} M quinine sulphate. Atebrine at a concentration of 0.01 M inhibits the activity by 15%.

Substrate specificity. The enzyme attacks only L-amino acids. Table 1 shows the relative rates of oxidation of some amino acids. Glutamine perhaps reacts in the system owing to the presence of glutaminase in some enzyme preparations. WORK (1955) has shown that the enzyme also attacks the *meso*, LL-diamino-pimelic acid, L-diaminopimelic acid D-monoamide, L-lysine and L-methionine at rates of the same order of magnitude. Total oxygen uptake per mole of amino acid is 1 atom for *meso*-DAP and 2 atoms for LL-DAP in the presence of excess catalase.

III. Amine Oxidases.

The oxidation of amines by amine oxidases may be represented to occur in two stages as follows:

$$R.CH_2NH_2 + O_2 \rightarrow [RCH{=}NH] + H_2O_2 \qquad\qquad i)$$
$$[RCH{=}NH] + H_2O \rightarrow RCHO + NH_3 \qquad\qquad ii)$$

In different animal tissues two types of oxidases are known to occur. Mono-amine oxidase catalyzes the oxidation of aliphatic monoamines, phenylalkyl-amines and certain diamines with long hydrocarbon chains. WERLE and ROEWER (1952) have advanced arguments for the belief that two distinct enzymes occur in plants and animals for the oxidation of aliphatic monoamines and phenyl-alkylamines respectively. The diamine oxidase catalyzes the oxidation of diamines of the general formula $NH_2(CH_2)_{2-8} NH_2$ and substituted diamines, like histamine and agmatin. In contrast to monoamine oxidase, the diamine oxidase is inhibited by cyanide and carbonyl reagents. Reviews on the enzymes of amine oxidation are available and should be consulted for the detailed descriptions of the animal enzymes (ZELLER, 1951; TABOR, TABOR and ROSENTHAL, 1955; TABOR, 1955; DAVISON, 1958).

Diamine oxidase has been shown to occur in certain bacteria, as *Pseudomonas pyocyanea* (WERLE, 1941; GALE, 1942) and *Mycobacterium smegmatis* (ZELLER, OWEN and KARLSON, 1951). The diamine oxidase of the last named bacterium differs from the mammalian enzyme in the preferential oxidation of trimethylene-diamine instead of cadaverine.

In higher plants, an enzyme catalyzing the oxidation of 1:4-diaminobutane was found by CROMWELL (1943) in *Atropa belladonna*. WERLE and RAUB (1948), and WERLE and PECHMANN (1949) found a diamine oxidase oxidizing histamine, 1:4-diaminobutane, and 1:5-diaminopentane in members of the family Legu-minosae. This enzymes was inhibited by cyanide and semicarbazide. WERLE and ROEWER (1950) later reported the occurrence in some other plants of an enzyme similar to the animal monoamine oxidase. The presence of two distinct oxidases in plants is, however, doubtful. MANN (1955), who purified an enzyme from pea plants showed that it catalyzed the oxidation not only of diamines but also of aliphatic monoamines, phenylalkylamines and L- and D-lysine.

Amine Oxidase of Pea Seedlings.

The enzyme catalyzes the reaction:

$$R.CH_2.NH_2 + O_2 + H_2O \rightarrow R.CHO + H_2O_2 + NH_3$$

Assay. Amine oxidase has been assayed from various sources by the determi-nation of oxygen consumption (ZELLER and BIRKHÄUSER, 1940), production of ammonia (ZELLER, 1940), determination of hydrogen peroxide by the coupled oxidation of indigo-disulphonate (KAPELLER-ADLER, 1956), and spectrophoto- metric measurement of the disappearance of kynuramine (WEISSBACH, SMITH, DALY, WITKOP and UDENFRIEND, 1960) or the appearance of the adduct Δ'-pyr- roline (arising from the cyclization of 4-amino butyraldehyde, a product of the oxidation of putrescine) and o-aminobenzaldehyde (HOLMSTEDT and THAM, 1959; HOLMSTEDT, LARSSON and THAM, 1961). For the estimation of the enzyme from plants, both spectrophotometric and manometric methods described below are suitable.

   a) *Spectrophotometric assay* (HOLMSTEDT et al., 1961). The assay is based on the fact that when putrescine is used as a substrate of the enzyme, Δ'-pyrroline is produced, presumably by the cyclization of the product of the reaction, 4-amino- butyraldehyde. Δ'-pyrroline in the presence of o-aminobenzaldehyde condenses to form a coloured compound, presumably 2,3-trimethylene-1,2-dihydroquin- azolinium hydroxide, which has a strong absorption maximum at 430 mμ. Thus, in the presence of enzyme, substrate and o-aminobenzaldehyde, the activity can be determined by the measurement of the intensity of the yellow colour.

   The assay mixture consists of 2.5 ml of 0.005 M o-aminobenzaldehyde (made up in 0.067 M phosphate buffer, pH 6.8; stable in solution at 4° C for about a week), enzyme preparation in phosphate buffer, and enough 0.067 M phosphate buffer, pH 6.8 to make 4.5 ml. After temperature equilibration at 37° C, 0.5 ml of 0.1 M putrescine hydrochloride (freshly prepared in phosphate buffer) is added. Control tubes receive phosphate buffer in place of the substrate. The reaction is terminated when desired by the addition of 1 ml of 10% TCA and the colour of the supernatant solution is read in a cell of 1 cm light path at 430 mμ. The supernatant of the control tube is used as the blank. If the O.D. exceeds 0.8, the colour is read after dilution with distilled water. The molar absorbancy index of the coloured compound is 1.86×10^3.

   b) *Manometric assay* (MANN, 1955). The central compartment of a WARBURG vessel contains enzyme, 0.067 M phosphate buffer, pH 7.0, excess catalase, and 0.01 M 1:4-diaminobutane in a total volume of 3.0 ml. The substrate is taken in the side arm. The central well contains 0.2 ml of 5 N KOH. The bath temperature is 28° C and the gas phase is air. With crude extracts, the endogenous O_2 uptake is accounted for by the use of controls. Results are calculated from the initial rates of O_2 uptake.

   Purification (MANN, 1955, 1960). Pea seedlings (7—12 days old; 1000 g) are minced in a chilled domestic meat mincer and the juice is extracted by squeezing through cotton cloth. The residual pulp is extracted once with 500 ml of 0.067 M phosphate buffer, pH 7.0, and again squeezed through cloth. The combined extracts are cooled to 0—5° C and a mixture of ethanol and chloroform (200 ml: 100 ml), precooled to — 10° C is added slowly with vigorous stirring. After 30 min- utes the mixture is centrifuged. The clear yellow supernatant solution is poured off from the precipitate and the bottom layer of chloroform, and again cooled. Solid ammonium sulphate is added to the supernatant fluid at the rate of 45 g per 100 ml. On centrifuging the suspension, the protein precipitate containing the enzyme activity, forms a hard cake on the surface of the liquid. The liquid is poured off and the precipitate is ground to a smooth paste in a mortar and 500 ml of 0.02 M phosphate buffer, pH 7.0 is slowly added to give a smooth suspension which is then stirred mechanically for 2 hours at 0—2° C. After centrifugation, the supernatant solution is preserved and the precipitate is extracted with a small amount of 0.02 M phosphate buffer, pH 7.0. The extracts are pooled, chilled, and solid $(NH_4)_2SO_4$ is added at the rate of 18 g/100 ml. The sediment is discarded. A further 18 g $(NH_4)_2SO_4$ per 100 ml is added to the supernatant solution and the precipitate, containing the activity, is collected by centrifugation and suspended in 20 ml of 0.2 M phosphate buffer, pH 7. The suspension is dialyzed for several hours against running tap water and overnight at $0-2°$ C against 0.005 M phosphate buffer, pH 7. The dialyzed suspension is centrifuged and the chilled supernatant solution is brought to pH 5.0 by dropwise addition of 0.05 N CH_3COOH and the mixture kept at $0-2°$ C for several hours. The flocculated precipitate is collected by centrifugation, triturated with 20 ml of water and brought into solution by adjusting the pH to 7.0 with 0.05 N KOH. The precipitation at pH 5.0 is repeated twice and the final solution at pH 7.0 is stored at $-10°$ C. A 300-fold purification is obtained in this manner. The enzyme can be further purified by chromatography on calcium phosphate, hydroxyapatite and DEAE-cellulose (MANN, 1960).

One *unit* is defined as the amount of the enzyme giving at 28° C and O_2 uptake of 1 μl/hour in the presence of 0.01 M phosphate buffer at pH 7.0 in a total volume of 3.0 ml.

Properties. The enzyme is perhaps a metalloflavoprotein. Highly purified enzyme preparations are pink in colour and contain 0.08% Cu^{++} (MANN, 1960). The enzyme is inhibited by cyanide, semicarbazide, diethyldithiocarbamate, salicylaldoxime, potassium ethyl xanthate (MANN, 1955), 8-hydroxyquinoline (WERLE and HARTUNG, 1956), α,α'-dipyridyl and o-phenanthroline (SUZUKI, 1959). The enzyme is also inhibited by H_2O_2. GORJACHENKOWA (1956) has demonstrated that pyridoxal phosphate and FAD serve as the prosthetic groups of amine oxidase.

Substrate specificity. The purified enzyme catalyzes the oxidation not only of aliphatic diamines but also that of phenylalkylamines, aliphatic monoamines and of L- and D-lysine. The oxidation is always accompanied by the formation of H_2O_2. The fact that both D- and L-lysine are oxidized suggests that the oxidation of these substrates takes place at the ε-amino group. 5-hydroxytryptamine and tryptamine are also oxidized, the latter giving rise to 3-indolylacetaldehyde which is an important plant growth regulator (CLARK and MANN, 1957).

B. Enzymes of Amino Acid Decarboxylation.

I. Amino Acid Decarboxylases of Higher Plants.

The only decarboxylase which has been purified so far from higher plants is the glutamic acid decarboxylase. The enzyme is of very wide occurrence. It was obtained in cell free extracts of carrots by SCHALES, MIMS and SCHALES (1946). Squash, green pepper and avocado were found to be rich sources of the enzyme. BEEVERS (1951) reported the presence and purification of the enzyme from barley roots, WEINBERGER and CLENDENNING (1952) from mature or senescent wheat leaves, HOOD (1954) from radish and squash, and CHENG, LINKO and MILNER (1960) from wheat embryos. Field beans seem to be a very rich source of this enzyme (KULKARNI and SOHONIE, 1956). In a study of the intracellular distribution of this enzyme in squash, ROGERS (1955) demonstrated that it is localized in the soluble cytoplasm.

FINLAYSON and McCONNELL (1960) demonstrated the conversion of $[1,7\text{-}^{14}C_2]$-α-α'-diaminopimelic acid to $[1\text{-}^{14}C]$-lysine in high yields in the wheat plant. This conversion is perhaps brought about by a diaminopimelic acid decarboxylase. SHIMURA and VOGEL (1961) demonstrated the presence of this enzyme in cell-free extracts of maize seedlings and aseptic cultures of *Agave toumeyana* leaf parenchyma tissue. In the presence of pyridoxal phosphate, *meso*-diaminopimelic acid is decarboxylated to L-lysine by the plant extracts. In cabbage leaf mitochondria MAZELIS (1959) reported the occurrence of an enzyme bringing about the decarboxylation of DL-methionine. Mn^{++} ions and pyridoxal phosphate was required for maximum activity. ATP and fluoride also stimulated the enzyme prepared by ammonium sulphate fractionation of digitonin extracts of the cytoplasmic particles. The products of the reaction and other characteristics of the reaction are not known.

Plant homogenates also liberate CO_2 very slowly from L-aspartic acid and L-cysteic acid. Labeled aspartic acid fed to *Convallaria majalis* is converted to both α- and β-alanine (LINKO, 1958). It is not clear whether there is a distinct aspartic decarboxylase in higher plants. It has been shown that glutamic decarboxylase from rat brain (DAVISON, 1956) readily decarboxylates cysteic acid and it is likely that the slow decarboxylation of cysteic acid in plant extracts is brought about by glutamic decarboxylase. There is also an indirect evidence that tyrosine is decarboxylated in some graminaceous leaves to tyramine which is a precursor of hordenine in barley plants (LEETE and MARION, 1953).

Decarboxylases with very weak activities can be conveniently assayed by the method proposed by DAVIS and AWAPARA (1960) for animal tissues. The use of this procedure might be helpful in obtaining more knowledge about plant decarboxylases.

Glutamic Decarboxylase.

The enzyme catalyzes the reaction:

L-glutamic acid → γ-aminobutyric acid + CO_2

Assay. The enzyme is conveniently assayed by measuring the CO_2 produced in the presence of the substrate in Warburg manometers (SCHALES and SCHALES, 1946a, b). Activity can also be determined by measuring the quantity of γ-aminobutyric acid formed by paper chromatography (KRISHNASWAMY and GIRI, 1956). The manometric method is easier and less time consuming. In the main compartment of a Warburg vessel is added the enzyme buffered to a pH of 5.70 to 5.90 with 0.067 M phosphate buffer. One side arm contains 0.5 ml (34 μmoles) of a neutral 1.0% solution of L-glutamic acid and the other 0.5 ml of 1.2 N H_2SO_4. Gas phase is nitrogen and the bath temperature is 37° C. The total volume in the flasks is 4.5 ml so that the substrate concentration is 7.55 mM. After equilibration, the substrate is tipped in and readings are taken at 10 minute intervals and the reaction is terminated by the addition of acid. Individual readings are corrected for "bound" CO_2 liberated after addition of the acid. A blank vessel is used with H_2O instead of glutamic acid.

Activity is expressed as $Q_{CO_2}^{N_2}$ value, i.e., microlites of CO_2 per hour per mg of fresh plant material, per mg dry weight, or per mg protein of the suspensions. If values are to be compared between different tissues, the latter is a more satisfactory measure of activity.

Purification from wheat leaves (WEINBERGER and CLENDENNING, 1952). The third leaves in full grown or senescent stage (approximately 32—40 days after sowing) are harvested from crops of Coronation variety of wheat and chilled before use. The leaves are macerated in a mortar with sand or in a meat chopper, and after filtration through nylon, the juice is centrifuged at 20,000 ×*g* in the cold (2—5° C). The pH of the supernatant solution is adjusted to 4.9 with dilute acetic acid and allowed to stand at 0° C for 30 minutes. The precipitate is discarded by centrifugation and solid $(NH_4)_2SO_4$ is added to the supernatant to a concentration of 14%, and after 1 hour the sediment is removed by centrifugation. More solid $(NH_4)_2SO_4$ is added to a concentration of 25% and allowed to stand for 1 hour. The precipitate obtained after centrifugation is taken up in a minimum of water, dialyzed free of sulphate against 0.025 M phosphate, pH 5.5, and any sediment formed is removed by centrifugation. The supernatant solution is used as the source of enzyme. With this procedure a 500-fold purification (on nitrogen basis) is obtained. The enzyme solution requires the presence of pyridoxal phosphate for optimal activity.

Properties. The pH optimum of glutamic decarboxylase from diverse plant sources is 5.5—5.8. The firmness with which the coenzyme is bound to the apoprotein part of the enzyme differs with the source of the enzyme. The wheat leaf enzyme can be partially resolved with regard to the coenzyme by intensive dialysis, but the enzyme from sunflower cotyledons can only be resolved by repeated acid-ammonium sulphate treatment (SMITH and WAYGOOD, 1961). In all cases, however, pyridoxal phosphate fully restores the activity of the resolved enzyme. SMITH and WAYGOOD (1961) also isolated the coenzyme as a barium salt following denaturation and precipitation of the decarboxylase protein by TCA. Different workers on this enzyme have reported different K_m values for L-glutamate. For the squash enzyme K_m value is reported to be 3.6×10^{-3} M (SCHALES and SCHALES, 1946a); for the barley root enzyme, 9.6×10^{-3} M (BEEVERS, 1951), and for sunflower cotyledon enzyme, 9.1×10^{-3} M (SMITH and WAYGOOD, 1961).

The enzyme is inhibited by cyanide $(10^{-4} - 10^{-5}$ M), and hydroxylamine at a concentration of 3×10^{-5} M reduces the initial reaction velocity by 50%, perhaps by reacting with the aldehyde group of the coenzyme (SCHALES and SCHALES, 1946a). α-hydroxy organic acids, like malic, citric and tartaric acids, inhibit the enzyme at a concentration of 10^{-2} M.

The enzyme is specific for L-glutamic acid. β-Hydroxyglutamic acid is slowly decarboxylated (LEANZA and PFISTER, 1953). FOWDEN (1954) has reported that the decarboxylation of α-methyleneglutamic acid in extracts of barley roots and tulip leaves is probably catalyzed by this enzyme.

The enzyme reaction is only weakly, if at all, reversible (WEINBERGER and CLENDENNING, 1952). In intact cells of *Chlorella*, however, the enzyme reaction seems to be freely reversible (WARBURG, KLOTZSCH and KRIPPAHL, 1957). It would be worthwhile to show the assumed reversibility with the purified enzyme.

L-glutamic acid decarboxylase from *Cucurbita* and carrots is stimulated by phosphate, arsenate, and nucleotides (EGGLESTON, 1957).

II. Amino Acid Decarboxylases of Microorganisms.

The bacterial amino acid decarboxylases so far described fall into three main groups. These are, (1) the decarboxylases for straight chain aliphatic or aromatic amino acids, as glutamic acid, lysine, ornithine, histidine, tyrosine, arginine and aspartic acid (GALE, 1946, 1957), (2) the decarboxylase for diaminodicarboxylic acid, diaminopimelic acid (DEWEY, HOARE, and WORK, 1954), and (3) the decarboxylase for the branched chain aliphatic amino acids, leucine, valine and isoleucine (EKLADIUS, KING and SUTTON, 1957). Some properties of the three groups of enzymes are described in Table 2.

The assay procedures, media for growth and the strains used have been dealt with exhaustively by various authors at different times and need not be repeated here. The reader is referred to other excellent reviews of NAJJAR (1955) and GALE (1946, 1957). Media for the growth of microorganisms and assay methods for some

Table 2. *Comparison of the properties of bacterial amino acid decarboxylases* (EKLADIUS et al., 1957).

Characteristics	Decarboxylases		
	Glutamate etc. (GALE, 1946)	Diaminopimelate (DEWEY et al. 1954)	Leucine etc. (EKLADIUS et al. 1957)
1. Nature of side-chain of substrate	Polar	Polar	Non-polar
2. Optimum pH for enzyme action	very acid	neutral	neutral
3. Optimum pH for enzyme formation	acid	variable ·	neutral
4. Optimum age of culture	mature	variable	young
5. Maximum attainable activity ($Q_{CO_2}^{N_2}$)	100—1000	5—10	upto 30
6. Substrate affinity (K_m)	10^{-3} M	10^{-3} M	valine, 0.03 M leucine, 0.01 M

amino acid decarboxylases have been given in the Chapter on "Enzymic Assays of Amino Acids and Keto Acids" in vol. VI of this book. We have dealt here only with the methods of purification of some decarboxylases which are either new or have not been reviewed elsewhere.

1. Glutamic Acid Decarboxylase of Escherichia coli.

Assay. The rate of CO_2 evolution from L-glutamic acid is measured at 36° C in 0.1 M pyridine-pyridine-HCl buffer, pH 4.6, adjusted to a chloride concentration of 0.1 M with NaCl. The main compartment contains 3.0 of buffered substrate solution, and after equilibration for 10 minutes, 0.2 ml of enzyme solution is tipped from the side arm. The final substrate concentration is 0.01 M and gas phase is air.

Specific activity is defined as μl of CO_2 evolved in 10 minutes by 1 mg of protein.

Purification (SHUKUYA and SCHWERT, 1960a). All steps in purification are carried out at room temperature. Cells (20 hours old) of *E. coli*, strain 26, are obtained from a medium containing 0.25% $(NH_4)_2SO_4$, 0.5% K_2HPO_4, 0.2% yeast extract, 1.0% glucose (autoclaved separately) and tap water. The cells are dried with acetone (1 vol cells:9 vols acetone) and a 5—7% suspension in distilled water is allowed to autolyze for 24—48 hours at room temperature. The pH of the suspension is maintained at 6.0—6.5 by periodic addition of 0.1 N NaOH. The cell debris is discarded by centrifugation.

A 2% aqueous protamine sulfate solution is slowly added to the supernatant solution with stirring until 0.1—0.15 mg protamine sulfate has been added for each mg protein. The resulting precipitate is discarded by centrifugation and the supernatant is fractionated by means of solid $(NH_4)_2SO_4$. The precipitate obtained between 0.26 and 0.7 saturation is dissolved in 2 ml of water for each gm of acetone powder used in the extraction step. The protein concentration is adjusted to 1% by dilution with water and incubated at 36° C for 1 hour. The precipitate formed is removed and the supernatant solution is refractionated between 0.30—0.65 saturation with $(NH_4)_2SO_4$. The precipitate is dissolved in water as before and the solution is dialyzed overnight at 4° C against 0.1 M pyridine-puridine-HCl buffer. Any precipitate appearing during dialysis is discarded. The pyridine is removed by dialysis against 0.04 M acetate buffer, pH 4.4.

The enzyme preparation is chromatographed on DEAE-cellulose column, 2.2×25 cm. Before chromatography, the solution is dialyzed overnight at 4° C against 0.05 M sodium phosphate buffer, pH 6.0. The column is also equilibrated with phosphate buffer. After adsorption, the protein is eluted from the column with a linear gradient between 250 ml volumes of 0.05 M and 0.30 M sodium phosphate, pH 6.0. Flow rate is maintained at $12-13$ ml per minute at $25°$ C, and the effluent solution is collected in $10-12$ ml fractions. The enzyme is eluted between 320 ml to 420 ml of the effluent. Fractions showing the highest specific activity are combined and the enzyme is precipitated at 0.60 saturation of $(NH_4)_2SO_4$. On rechromatography of the enzyme in the same way (linear gradients between 0.05 M and 0.35 M sodium phosphate, pH 6.5) a more purified preparation is obtained. With the above procedure, an enzyme purified 9-fold is obtained.

The enzyme cannot be stored in a dilute state. At a concentration of $10-15$ mg protein per ml, the enzyme is stable for about 1 month at $4°$ C.

Properties. The enzyme is stimulated by anions. The order of diminishing effectiveness of various anions is: chloride, bromide, iodide, sulfate and phosphate. With the purified enzyme, the velocity of the reaction increases with pyridoxal phosphate concentrations of up to 3×10^{-5} M. At this concentration the reaction velocity is 25% greater than that observed in the absence of the coenzyme. The pH optimum measured in the presence of 0.1 M chloride is 3.8. At neutral pH the enzyme loses activity more rapidly at $0°$ C than at $25°$ C (Shukuya and Schwert, 1960b). Acetate is an inhibitor of the reaction and competes with glutamate for the enzyme surface. The K_i of the inhibition is 0.2 M. The K_m for glutamate at pH 4.60 in pyridine buffer, is 8.2×10^{-4} M.

2. Lysine Decarboxylase of E. coli.

The reaction catalyzed by the enzyme is as follows:

$$\text{L-lysine} \rightarrow \text{cadaverine} + CO_2$$

The enzyme is assayed manometrically. Specific activity is reported as μl CO_2 evolved/hour at $37°$ C/mg protein.

Purification (Sher and Mallette, 1954a). Cells of *E. coli* B strain are grown in 30 liters of medium with 5% Hycase (salt free acid hydrolyzate of casein, National Dairy Research Lab., Inc., Long Island, New York), 2.4% nutrient broth, 1% L-lysine-HCl, 2% glucose + 0.5% NaCl without agitation for 14 hours at $37°$ C. The inoculum is pre-adapted in the same medium. The cells are collected, washed, and lysed at $37°$ C in 150 ml of 0.15 M NaCl with $20-65$phage particles per bacterium (Sher and Mallette, 1953). The lysate is treated with 300 μg DNAse and 0.3 ml saturated $MgSO_4$, and stored for 1 hour at $25°$ C. After centrifugation, the supernatant solution is saved and the residue is washed twice with 0.15 M NaCl. The supernatant solutions are pooled from each step. Protein is precipitated from the combined fractions by adding 70 g $(NH_4)_2SO_4$ for every 100 ml solution, and the precipitate is dissolved in 150 ml of H_2O. The precipitate obtained between $0.40-0.53$ $(NH_4)_2SO_4$ saturation from this solution is taken up in 200 ml H_2O. This solution is refractionated with $(NH_4)_2SO_4$ and the solids appearing between $0.41-0.52$ saturation are dissolved in 100 ml H_2O, and fractionated again three times by $(NH_4)_2SO_4$. Each time the sediment obtained between $0.41-0.52$ saturation is used. The protein from the last fractionation is dissolved in water and treated in two 50 ml lots by consecutively swirling by hand in water baths at $40°$ C for 2 minutes, $50°$ C for 2 minutes, $60°$ C for 1 minute and $70°$ C for 5 minutes. After cooling to $0°$ C, the suspensions are centrifuged and the supernatant solution is stored at $0°$ C. At this stage the enzyme is purified approximately 36-fold. Dialysis of the enzyme solution (the visking tubes used for dialysis are washed with hot ethanol and distilled water to remove an impurity which causes loss of activity of lysine decarboxylase) against 0.2 M phosphate buffer, pH 6.0 at $-30°$ C removes all the coenzyme and addition of excess pyridoxal phosphate restores the activity.

Properties. The purified enzyme contains 9.7 and 1.1% tyrosine and tryptophan respectively. This corresponds to a minimum molecular weight of 20,000. The pH optimum of the enzyme is approximately 6.0 (GALE and EPPS, 1944) and the K_m is 1.5×10^{-3} M.

Contact of the enzyme with alumina A303, commonly used in grinding bacteria, partially inactivates the enzyme. The enzyme can be stored with additions of cysteine and EDTA. pCMBA inhibits the enzyme and this inhibition is reversed by L-cysteine. Like other bacterial decarboxylases this enzyme is inhibited by isonicotinyl hydrazide (HOARE, 1956).

3. Arginine Decarboxylase of E. coli.

The enzyme catalyzes the reaction:

$$\text{L-arginine} \rightarrow \text{agmatine} + CO_2$$

Enzyme activity is assayed manometrically. Specific activity is reported as μl CO_2 evolved/hour/mg protein at 37° C.

Purification (SHER and MALLETTE, 1954b). Cells of *E. coli* B strain are grown and disrupted exactly in the same way as discussed under lysine decarboxylase. Purification is carried through to the first sedimentation between 0.40 — 0.53 salt saturation in the lysine decarboxylase procedure discussed before. The supernatant solution (0.53 saturation) is brought to 0.60 saturation by adding solid $(NH_4)_2SO_4$, and the precipitate obtained by centrifugation is dissolved in 100 ml H_2O. This solution is fractionated again by $(NH_4)_2SO_4$. The fraction obtained between 0.52 — 0.60 saturation is dissolved in 50 ml H_2O and swirled consecutively in water baths at 40° C for 2 minutes, 50° C for 2 minutes, 60° C for 5 minutes, and the mixture after cooling to 0° C is centrifuged and the sediment discarded. The small amount of lysine decarboxylase remaining at this stage is removed by treating the supernatant solution at 0° C with an equal volume of McILWAINE buffer, pH 4.4 and centrifuging off the denatured protein. Sufficient 0.2 M secondary sodium phosphate is then added to raise the pH to 6.0.

The enzyme is purified approximately 12-fold by this procedure. On dialysis the enzyme is not completely resolved for the coenzyme pyridoxal phosphate, but in the presence of excess coenzyme (20 μg) the activity increases 7-fold.

Properties. The pH optimum of the enzyme is 5.25 (TAYLOR and GALE, 1945) and the molecular weight is about 780,000.

pCMBA at a concentration of 4×10^{-7} M inhibits enzyme activity by 88%. L-cysteine does not reverse the inhibition.

The K_m for the substrate is 7.5×10^{-4} M.

4. Leucine Decarboxylase.

The enzyme catalyzes the reaction:

$$\text{L-leucine} \rightarrow \text{iso-amylamine} + CO_2$$
$$\text{(or L-valine)} \quad \text{(or iso-butylamine)}$$

The enzyme is assayed manometrically at pH 6.0, temperature 37° C, with nitrogen as the gas phase. The central compartment has the equivalent of 15 — 30mg dry weight of organisms (or the corresponding amount of cell-free preparation), buffer (pH 6.0, 0.2 M) and the coenzyme in a total volume of 2.0 ml. The side arm contains L-isomer (0.08 M, final concentration).

Purification (EKLADIUS, KING and SUTTON, 1957; HAUGHTON and KING, 1958). *Proteus vulgaris* is grown in nutrient broth (Lab-Lemco, 1%; peptone 1%,

NaCl, 0.5%). Less active preparations are obtained by using yeast extract and casein hydrolyzate. The pH during growth is kept at 7.6 by adding 0.05 M phosphate. Cells after 18 hours growth at 37° C are harvested and freeze dried. Cellfree extract, if required, is made by grinding with alumina. The extracts are not stable for more than 24 hours.

Properties. Reasonable evidence exists that the same enzyme decarboxylates leucine, valine, norvaline, isoleucine and α-amino-n-butyric acid (Ekladius et al., 1957).

The pH optimum is 7.0. Pyridoxal phosphate is required as the coenzyme. The K_m for the substrates is approximately 10^{-2} M (Table 2). —SH group reagents are potent inhibitors.

5. L-Tryptophan Decarboxylase.

This enzyme has only been found in Streptococci but there is some doubt whether this is an enzyme separate from tyrosine decarboxylase. The activity is generally very low and associated with the latter enzyme. The K_m for tryptophan is 0.013 M (Mitoma and Udenfriend, 1960).

Assay. The decarboxylating activity is measured by coupling the reaction with monoamine oxidase (Mitoma and Udenfriend, 1960). Fifteen mg lyophilized bacteria (*Streptococcus faecalis*, ATCC No. 8043) are incubated with 10 mg tryptophan and 1 ml of 0.1 M citrate buffer, pH 5.5, in a final volume of 3 ml at 38° C for 2 hours. The amount of tryptamine formed is assayed by adding to the incubation mixture 5 mg of lyophilized rat liver mitochondria as a source of monoamine oxidase, 1 ml of 0.5 M phosphate buffer, pH 7.4, and 0.2 ml of 0.2% 2-(p-iodophenyl)-3-(p-nitrophenyl)-5-phenyltetrazolium chloride and reincubating for 30 minutes. The reduced dye is extracted with 5 ml of a mixture of ethylacetate and n-butanol (1:1), and its absorption is measured at 490 mμ. Alternatively, the xanthydrol method (Weissbach et al., 1960) can also be used.

6. Diaminopimelic Acid Decarboxylase.

The enzyme catalyzes the reaction:

$$CH_2 \begin{array}{l} \diagup CH_2.CH(NH_2).COOH \\ \\ \diagdown CH_2.CH(NH_2).COOH \end{array} \longrightarrow CH_2 \begin{array}{l} \diagup CH_2.CH_2NH_2 \\ \\ \diagdown CH_2.CH(NH_2).COOH \end{array} + CO_2$$

diaminopimelic acid   lysine

The enzyme has been shown to occur in *Escherichia coli, Aerobacter aerogenes, Sarcina lutea,* certain bacilli (Powell and Strange, 1957; Meadow and Work, 1958), and higher plants (Shimura and Vogel, 1961).

Assay. The enzyme activity is followed manometrically at 37° with air as the gas phase. Main compartment contains 40 mg of dried bacterial cells suspended in 2 ml of 0.1 M phosphate buffer, pH 7.2, or cell-free enzyme extracted from 40 mg dry cells, and enough pyridoxal phosphate (0.1 ml of 0.1 mg/ml of the coenzyme per vessel). Diaminopimelic acid (0.5 ml of 0.25 M solution) is taken in the side arm. The rate of reaction is measured from the slope of the linear portion of the plot of gas evolution against time. Activity is expressed as $Q_{CO_2} = \mu$l CO_2 evolved/hour/mg dry cells at 37° C.

Purification (Dewey et al., 1954; Hoare and Work, 1955). Cells of *Aerobacter* are grown in a medium containing K_2HPO_4, 7 g; KH_2PO_4, 3 g; trisodium citrate, 0.5 g; $(NH_4)_2SO_4$, 1.0 g; $MgSO_4.7H_2O$, 0.01 g; Fe $SO_4.7H_2O$, trace; glucose,

20 g; water to 1 liter. The cells are acetone dried and stored at $-10°$ C. Enzyme is extracted by grinding 1 g dried cells with $5-6$ ml extraction mixture (acetone, 48 ml:0.025 M phosphate buffer, pH 5.8, 52 ml) and 1 g filter cel (Johns Manville Co., Ltd., London). The paste is extracted by more acetone-phosphate mixture (overall volume, 75 ml), and centrifuged in the cold. The supernatant solution is cooled to $-10°$ C and 100 ml of acetone, prechilled to $-10°$ C, is added. The precipitate is collected by centrifugation, freed from acetone *in vacuo*, and dissolved in 10 ml of 0.1 M phosphate buffer, pH 6.8. The enzyme is purified 8-fold and lysine decarboxylase, present in crude extracts is separated.

Properties. The pH optimum is 6.8. The coenzyme is probably pyridoxal phosphate (MEADOW and WORK, 1958). pCMBA inhibits enzyme activity and the inhibition is reversed by glutathione. *Meso*-diaminopimelic acid is more easily decarboxylated by the enzyme than the LL-isomer. DD-isomer shows no activity.

7. Tyrosine Decarboxylase.

The enzyme catalyzes the reaction:

$$\text{L-tyrosine} \rightarrow \text{tyramine} + CO_2$$

The enzyme is present in acetone powders of the cells of *Streptococcus faecalis*, strain R (ATCC 8043; BELLAMY and GUNSALUS, 1945). It is probably also responsible for the decarboxylation of L-*m*-hydroxyphenylalanine. This effect is only seen in acetone dried cells (SLOANE-STANLEY, 1949); intact cells are not able to decarboxylate this compound. *o*- Hydroxyphenylalanine and 2:5-dihydroxyphenylalanine do not serve as substrates of the enzyme.

C. Enzymes of Non-Oxidative Deamination.

I. Amino Acid Deaminases.

1. Aspartase.

The enzyme catalyzes the reversible deamination of L-aspartic acid:

$$\text{L-aspartic acid} \leftrightharpoons \text{fumaric acid} + NH_3$$

Distribution. The enzyme is primarily found in bacteria (*see*, VIRTANEN and ELLFOLK, 1955) and some fungi, like yeast (HAEHN and LEOPOLD, 1937) and *Penicillium*. In higher plants, the enzyme is either present in very small amounts (DAMODARAN and SUBRAMANIAN, 1948) or not demonstrable at all. In most cases, the partially purified enzyme is accompanied by fumarase. The best source of the enzyme is *Bacillus cadaveris*, where it has been obtained free from accompanying enzymes. An extensive review of this enzyme is available in ELLFOLK (1956).

Assay. The enzyme is incubated in 2 ml (final volume) of a mixture consisting of 0.05 M phosphate or borate buffer, pH 6.8; 0.002 M L-aspartate (final concentration, pH 7.5) and enzyme at 37° C for 30 minutes. The reaction is stopped by adding TCA to a final concentration of 5%. The precipitated protein is centrifuged and the supernatant solution is analyzed for ammonia by nesslerization (see, UMBREIT, BURRIS and STAUFFER, 1957).

Enzyme purification (WILLIAMS and McINTYRE, 1955). Cells of *B. cadaveris* are grown for 16 hours at 30° C in a medium consisting of 1% yeast extract, 1% tryptone, and 0.05% KH_2PO_4. The cells are harvested and suspended in a small volume of cold 0.1 M phosphate buffer, pH 6.8. They are then disrupted in a cooled Raytheon Sonic Oscillator for 25 minutes. The suspension is centrifuged at 25,000 $\times g$ for 1 hour and the sediment is discarded. To the supernatant solution an equal volume of cold saturated $(NH_4)_2SO_4$ is added, the precipitate is collected by centrifugation, dissolved in distilled water, and is dialyzed against 0.001 M phosphate buffer pH 6.8 in the cold for 24 hours. Further fractionation is done by solid $(NH_4)_2SO_4$. The precipitate forming between 0.25—0.45 saturation is collected, dissolved, and dialyzed as above. For later use it can be stored at $-18°$ C.

Properties. The pH optimum of the enzyme is 7.2 in veronal acetate buffer, and 6.5—6.8 in 0.1 M phosphate buffer. The $K'eq$ of the system,

$$\frac{(-OOC.CH=CH.COO)\,(NH_4^+)}{(-OOC.CH\,NH_3^+.CH_2COO^-)}$$

is 2.7×10^{-2}. The K_m for aspartate is 2.0×10^{-2} M. In the presence of formic acid, the enzyme is fairly stable. It is very sensitive to $-SH$ inhibitors and oxidizing agents.

2. Histidase.

The reaction catalyzed by the enzyme is analogous to aspartase, but unlike the latter is not reversible:

$$\text{Histidine} \rightarrow \text{urocanic acid} + NH_3$$

Distribution. The enzyme is known to occur in animal liver and various bacteria like *Aerobacter aerogenes*, *Escherichia coli*, *Salmonella*, *Shigella paradysenteriae*, *Pseudomonas fluorescens*, and *Clostridium tetanomorphum* (WICKREMSINGHE and FRY, 1954). It has not been demonstrated to occur in molds or higher plants. The richest source of the enzyme are the histidine grown cells of *Pseudomonas fluorescens* (details in, TABOR and MEHLER, 1955).

Assay. Due to the presence of three conjugated double bonds urocanic acid absorbs strongly at 277 mμ (HALL, 1952). This has been made the basis of assay of histidase (TABOR and MEHLER, 1955). In a quartz cuvette of 1 cm light path is added, 0.3 ml of 0.1 M pyrophosphate buffer, pH 9.2; 0.05 ml of 0.1 M glutathione; a suitable aliquot of the enzyme, and the volume is made up with water to 2.9 ml. At zero time 0.1 ml of 0.1 M histidine solution is added, and the increase in optical density at 277 mμ is recorded at 15 to 30 second intervals. A reference cuvette without the substrate is employed. One unit has been defined by TABOR and MEHLER (1955) as the quantity of the enzyme which causes an increase in optical density at 277 mμ of 0.001 per minute at 25° C.

Properties. When stored with added glutathione the enzyme from bacterial cells is stable at $-15°$ C. EDTA causes 90% inhibition at a concentration of 10^{-4} M. The pH optimum is 9.5.

II. Dehydrative Deaminases.

Dehydrative deamination is restricted to β-substituted amino acids. The general reaction is:

$$RCH(OH)CH.NH_2COOH \rightarrow (RCH=C.NH_2COOH) + H_2O \qquad \text{i)}$$
$$(RCH=C.NH_2COOH) \rightarrow RCH_2.CO.COOH + NH_3 \qquad \text{ii)}$$

The end products of the reaction are a keto acid and ammonia.

1. Dehydrases.

a) L-Threonine (and L-Serine) Dehydrase.

The reaction(s) catalyzed are:

$$\text{L-serine} \rightarrow \text{pyruvic acid} + NH_3$$
$$\text{L-threonine} \rightarrow \alpha\text{-ketobutyric acid} + NH_3$$

Nature of the enzyme. There is some confusion in the literature regarding the identity of these two enzymes. WOOD and GUNSALUS (1949) in *Escherichia coli*, YANOFSKY and REISSIG (1953) in *Neurospora crassa*, and WALKER (1958) in rumen microorganisms, were unable to separate the two enzymes by classical enzyme isolation procedures. BOYD and LICHSTEIN (1955) and PARDEE and PRESTIDGE (1955) concluded from nutritional experiments that two separate enzymes were involved. SAYRE and GREENBERG (1956) demonstrated that serine and threonine dehydrases were distinct enzymes. UMBARGER and BROWN (1957) in a nutritional study of *E. coli* gave convincing evidence that there were two separate threonine deaminases, differentiated on their coenzyme requirements, one of which was also active against serine. The possibility of the existence of a separate L-serine dehydrase was also considered. An enzyme has recently been isolated (SELIM and GREENBERG, 1960) from animal liver which not only brings about a dehydrative deamination of L-serine and L-threonine but can also synthesize cystathionine from homocysteine and serine.

Distribution. The enzyme is present in molds (YANOFSKY and REISSIG, 1953), many bacteria, and animal liver. The enzyme has not yet been reported to occur in higher plants. The best source of the enzyme is *N. crassa* (see, purification methods in YANOFSKY and REISSIG, 1953).

Assay. (YANOFSKY and REISSIG, 1953). The assay system in one ml volume consists of, 10 μg of crystalline pyridoxal phosphate; 40 μmoles of L-threonine or L-serine; an appropriate volume of enzyme solution, and 0.05 M pyrophosphate buffer (end concentration), pH 9.3. The tubes are incubated at 37° C for 20 minutes. The reaction is stopped by the addition of 1 ml of 10% TCA, and the protein is removed by centrifugation. The amount of the keto acid formed is determined in 1 ml aliquot by the colorimetric procedure of FRIEDEMANN and HAUGEN (1943). Control tubes, one with pyridoxal phosphate and substrate, and another with enzyme alone are also run with each assay.

One *unit* of the enzyme has been defined (YANOFSKY and REISSIG, 1953) as the amount required to form 0.1 μmole of the keto acid in 20 minutes in the above system.

Properties. The enzyme obtained from *Neurospora* has a pH optimum of 9.3, and is specific for the L-forms of amino acids. Pyridoxal phosphate serves as the coenzyme. The K_m values for L-serine and L-threonine are 5.5×10^{-3} M and 3.3×10^{-3} M respectively. Among *E. coli* enzymes, the labile, biosynthetic L-threonine deaminase can be resolved with regard to pyridoxal phosphate by means of hydroxylamine treatment. The adaptive L-threonine deaminase requires AMP, pyridoxal phosphate, and glutathione for maximal activity (UMBARGER and BROWN, 1957).

b) D-Serine (and D-Threonine) Dehydrase.

The enzyme catalyzes the deamination of the D-isomers and the reaction is analogous to that occurring with L-isomers. Only one enzyme seems to be responsible for the deamination of D-serine and D-threonine in *E. coli*, *Neurospora* (METZLER and SNELL, 1952; YANOFSKY, 1952) and some yeasts.

Distribution. The enzyme has been found in bacteria *(E. coli)* and *Neurospora*. There is no record of its occurrence in higher plants.

Assay. The procedure is exactly similar to that for the L-isomers, except that the D-forms are substituted in the assay mixture and 0.1 M (end concentration) borate buffer, pH 8.2 is employed.

The enzyme may be isolated from *Neurospora crassa* according to the procedure of YANOSKY (1952).

Properties. The pH optimum of the enzyme is $8.1-8.2$. Pyridoxal phosphate serves as the coenzyme, and the concentration required to restore half-maximal velocity of the resolved enzyme is 3×10^{-6} M. The K_m value for D-serine is 2.6×10^{-4} M for the *Neurospora* enzyme, and 3×10^{-4} M for the *E. coli* enzyme.

The enzyme is inhibited by L-cysteine, 8-hydroxy-quinoline, hydroxylamine, NH_4^+, Co^{++}, CN, and Zn^{++}. The inhibition by cations is partially reversed by Mg^{++} ions.

2. Desulfhydrases.

The general reaction catalyzed by enzymes of this category is:

$$R.CH(SH).CH.NH_2COOH \rightleftharpoons (R.CH=C.NH_2COOH) + H_2S \qquad i)$$
$$(R.CH=C.NH_2COOH) \rightarrow R.CH_2CO.COOH + NH_3 \qquad ii)$$

Reaction (i) is considered to be enzymic while reaction (ii) is spontaneous and non-enzymic. The mechanism is thus analogous to the dehydrase reaction.

a) L-Cysteine Desulfhydrase.

$$\text{L-cysteine} + H_2O \rightarrow \text{pyruvate} + H_2S + NH_3$$

Distribution. The enzyme is distributed in mammalian kidneys, liver, pancreas, and many bacteria (Fromageot, 1951). It has not been reported to occur in higher plants.

Source of the enzyme. A stable enzyme can be prepared from *E. coli* according to the method of Metaxas and Delwiche (1955).

Assay. The enzyme can be assayed by the measurement of any of the three products of the reaction. In crude extracts all of the three must be determined independently. Pyruvate can be determined by the colorimetric procedure of Friedemann and Haugen (1943) as modified by Sayre and Greenberg (1956), NH_3 by microdiffusion and nesslerization (Archibald, 1943), and H_2S either iodometrically (Smythe, 1942) or photometrically (Delwiche, 1951).

The assay system in 2 ml final volume (pH 7.4) consists of, 50 μmoles of phosphate; 5 μmoles of $MgSO_4$; 50 μmoles of pyridoxal phosphate; suitable aliquot of the enzyme, and 50 μmoles of L-cysteine. After 30 minutes' incubation at $40°$ C, the amount of H_2S formed is determined as follows (Delwiche, 1951):

The reaction is stopped by the addition of 2 drops of 40% NaOH (TCA is unsatisfactory). After centrifugation, an aliquot is withdrawn and added to tubes containing 2.0 ml of 2 N NaOH. The non-organic sulfide is precipitated by the addition of 4.0 ml of an aqueous reagent containing 1.0 g lead acetate, 2.5 ml of glacial acetic acid, and 2.5 g of gum arabic per liter. The colloidal precipitate of lead sulfide formed is stable for at least 15 minutes, and can be read at 490 mμ. H_2S is determined by reference to a standard curve prepared by using crystalline lead acetate as a primary standard, and a solution of Na_2S as the precipitating agent. In the preparation of the standard lead acetate solution, a few drops of glacial acetic acid are included to eliminate the turbidity resulting from hydrolysis. This method is sensitive between 3 to 100 μg H_2S.

Alternatively, H_2S liberated in the reaction may be measured in Warburg vessels (Smythe, 1942). Cysteine is taken in the side arm and the central well contains an insert of folded filter paper (7×5 cm) previously saturated with a 10% solution of cadmium chloride. After flushing with N_2, substrate is tipped in, and incubated for the desired length of time at $37°$ C accompanied by shaking. The evolved H_2S is precipitated quantitatively as CdS on filter paper. At the end of the experiment, the precipitated sulfide is determined iodometrically.

Properties. The resolved enzyme requires pyridoxal phosphate for maximal activity. The pH optimum of the bacterial enzyme is 7.8. The K_m of L-cysteine is 1.76×10^{-2} M. The enzyme is inhibited by α-ketoglutarate, tryptophan, glutamic acid, alanine and Tris buffer. KCN and As_2O_3 are known to inhibit the enzyme from animal tissues (SMYTHE, 1955).

b) D-Cysteine Desulfhydrase.

The reaction catalyzed by this enzyme is similar to L-cysteine desulfhydrase.

Distribution. So far the enzyme is known to occur only in certain aureomycin resistant strains of *E. coli* (SAZ and BROWNELL, 1954) and *Propionibacterium pentosaceum.* Only crude extracts obtained by sonic oscillation have been employed.

Properties. The crude enzyme preparation also produces H_2S from D-cystine and L-cystine. Tris is inhibitory.

c) Homocysteine Desulfhydrase.

Homocysteine $\rightarrow$ α-ketobutyric acid $+$ H_2S $+$ NH_3

Distribution. The enzyme occurs in mammalian tissues (FROMAGEOT and DESNUELLE, 1942), and some bacteria as *Proteus morganii* (KALLIO, 1951) and *P. vulgaris*. The enzyme may be obtained by the method of KALLIO (1951).

Assay. The method applied for the assay of L-cysteine desulfhydrase is equally suitable for this enzyme. Homocysteine is used as the substrate.

Properties. It is not quite clear whether L-cysteine desulfhydrase and homocysteine desulfhydrase are one or two different enzymes. FROMAGEOT and DESNUELLE (1942) in animal tissues, and KALLIO (1951) in bacteria, considered the possibility that there were two different enzymes, largely due to the varying ratios of the two enzymes in different preparations from the cells. There is a strong probability that the coenzyme is pyridoxal phosphate. The enzyme attacks the L-isomer readily but there is also some activity towards the D-isomer.

III. Amino Acid C—S Cleaving Enzymes.

The group of enzymes included under this category are those which bring about a cleavage of the —C—S-bonds of sulfur containing amino acids. The products of the primary reaction are generally unstable and yield spontaneously a keto acid, ammonia, and a sulfur containing residue. The homoserine deaminase-cystathionase enzyme found in animal liver (MATSUO and GREENBERG, 1958) exemplifies the mode of action of the C—S-cleaving enzymes:

Cystathionine α-ketobutyric acid

Two types of cleaving enzymes are known, one type attacking the unsubstituted S-amino acids and their analogues, and the other, the S-substituted derivatives of L-cysteine.

1. Methionine Dethiomethylase.

$$CH_3.S.CH_2.CH_2.CH(NH_2)COOH \rightarrow CH_3SH + NH_3 + CH_3.CH_2.CO.COOH$$

L-methionine   methyl mercaptan   α-ketobutyric acid

Distribution. The enzyme has only been shown to occur in some bacteria, like *Pseudomonas* (Kallio and Larsen, 1955), *E. coli* (Ohigashi, Tsunetoshi and Ichihara, 1951), and *Clostridium sporogenes* (Wiesendanger and Nisman, 1953). Large number of soil fungi are known to produce methyl mercaptan and eventually the enzyme may be found in these organisms also. There is no record of its presence in animals or higher plants.

Source. The enzyme may be prepared from methionine adapted *Pseudomonas* according to the procedure of Kallio and Larsen (1955).

Assay. The enzyme is conveniently assayed by measuring the amount of α-ketobutyrate formed by the colorimetric procedure of Friedemann and Haugen (1943). The assay system in a total volume of 2 ml consists of, 0.1 M phosphate buffer, pH 7.5 (end concentration), 4 μg of pyridoxal phosphate, 20 μmoles of L-methionine, and enzyme. The mixture is incubated at 37° C for the desired length of time and the reaction is stopped by the addition of 1 ml of 16% TCA. After the removal of the protein precipitate by centrifugation, a convenient aliquot is analyzed for α-ketobutyrate (Kallio and Larsen, 1955).

Properties. The pH optimum of the enzyme is between 7.4 to 8.0. Pyridoxal phosphate serves as the coenzyme. The enzyme can be conveniently resolved by simple adsorption on levigated alumina at pH 7.4 in 0.05 M phosphate. L-methionine is specific as a substrate. Arsenite at a final concentration of 0.001 M completely inhibits enzyme activity.

2. Dimethylpropionthetin Dethiomethylase.

$$(CH_3)_2S^+—CH_2.CH_2.COOH \rightarrow (CH_3)_2S + CH_2=CH.COOH + H^+$$

acrylic acid

Distribution. The enzyme has been reported to occur so far in the marine alga, *Polysiphonia lanosa* (Cantoni and Anderson, 1956). Since the substrate, DMPT occurs widely in many algae and Pteridophytes (see Challenger et al., 1951) the enzyme will no doubt be found in future to have a wider distribution. DMPT has been shown to be an efficient methyl donor in animal tissues *in vivo* and *in vitro* (Maw and du Vigneaud, 1948; Dubnoff and Borsook, 1948).

Assay. The enzyme activity is measured (Cantoni and Anderson, 1956) by the rate of disappearance of DMPT. The assay mixture in a total volume of 1 ml consists of, 0.20 M DMPT; 0.006 M GSH; 0.03 M acetate buffer, pH 5.2, and a suitable aliquot of the enzyme. After incubation at room temperature for the desired length of time, the reaction is terminated by the addition of an equal volume of 10 per cent TCA and the protein is removed by centrifugation. To 1 ml aliquot of the supernatant solution (containing 0.5 to 2.5 mg of DMPT), 2 ml of ammonium reineckate solution (1 per cent solution of sodium reineckate in 5 per cent TCA, filtered) is added, and the mixture is chilled in ice for 12 hours. After centrifugation, the supernatant solution is carefully removed by suction with a capillary pipette (with the tip bent upwards) so that the loose precipitate is not disturbed. The precipitate is washed with 2 ml of ice-cold 5 per cent TCA. After centrifugation the supernatant solution is discarded again. The precipitate is next dissolved in a small quantity of a mixture containing 1 volume of phosphate buffer, 0.05 M, pH 7.8, and 2 volumes of acetone, and the optical density is read at 520 mμ. The concentration of DMPT is calculated by using the molar absorbancy index of 108.

The method given above is not specific for DMPT. Other "onium" compounds are also precipitated as reineckates, but the concentration of these compounds in algal enzyme preparation is negligible.

Preparation of the enzyme from algae. The following procedure has been used by CANTONI and ANDERSON (1956) for the purification of the enzyme from the alga, *Polysiphonia lanosa:* the thallus is removed from sea water, and after drying between paper towels, it is stored in an ice box for 30 minutes. The tissues are ground thoroughly in a mortar in the cold with a little sand and 3 volumes of cold 0.05 M potassium phosphate buffer, pH 6.8. The homogenate is centrifuged in the cold at low speeds to remove sand and cellular debris. After adding 0.1 volume of glutathione solution (25 mg per ml in 0.1 M acetate buffer, pH 5.2), the mixture is centrifuged at high speed (30,000 *rpm*) in a Spinco-high speed centrifuge. The supernatant solution is discarded and the residue is suspended in one-half this volume of a solution containing 0.05 M sodium acetate, pH 5.2 in 0.2 M sodium chloride, and 25 mg per ml of GSH, the latter being added just before use.

Properties. The pH optimum of the reaction is around 5.1. Reduced thiol groups are required for maximum activity. Dimethylacetothetin is also cleaved slowly by the enzyme preparation; S-dimethylhomocysteine, S-adenosylmethionine, betaine, and choline are not attacked.

3. Cystathionase.

This enzyme has been crystallized from animal liver (MATSUO and GREENBERG, 1958), and has been shown to catalyze the deamination of homoserine as well as the splitting of cystathionine. No such comparable enzyme has been demonstrated to occur either in microorganisms or higher plants. The "cystathionase" of some bacteria, like *Proteus morganii* (BINKLEY, 1955) brings about a cleavage of cystathionine to homocysteine, pyruvic acid and ammonia. The bacterial enzyme may be assayed in the same manner as the cystathionase from animal liver (MATSUO and GREENBERG, 1958).

4. Alliinase.

The enzyme cleaves the S-substituted derivatives of L-cysteine sulfoxides, and owes its name to the substrate, alliin (S-allyl-L-cysteine sulfoxide) which is broken down to allicin (allylthiosulfinic acid allyl ester), pyruvic acid and ammonia as follows (STOLL and SEEBECK, 1949):

Distribution. The enzyme is known to occur in garlic bulbs (Stoll and See-beck, 1951) and all other *Allium* species (Fujiwara, Yoshimura and Tsuno, 1955) which contain large amounts of S-alkenyl and S-alkyl substituted cysteine sulfoxide derivatives. The enzyme present in onion bulbs, and bringing about a cleavage of methylcysteinesulfoxide to pyruvic acid, ammonia and methyl-thiosulfinate (Virtanen, 1958) is perhaps identical with alliinase (Schwimmer et al., 1960).

Assay. The enzyme activity is determined by incubating one ml aliquot of the enzyme solution in 0.1 M phosphate buffer, pH 6.4 with 2 ml of 0.6% aqueous alliin in test tubes at 37° C. After 5 to 30 minutes, 0.6 ml of 2 N H_2SO_4 is added to terminate the reaction. Any precipitate formed is removed by centrifugation. A suitable aliquot of the supernatant solution is analyzed for pyruvic acid by the colorimetric method of Friedemann and Haugen (1943) as modified by Sayre and Greenberg (1956).

Purification. The enzyme is purified according to the procedure of Stoll and Seebeck (1949, 1951): Hundred gm of fresh garlic bulbs are crushed to a fine powder with dry ice and 400 ml water is added. The brei is kept at 37° C for 20 minutes with constant agitation. From the homogenate the solid material is removed by centrifugation and the supernatant solution is filtered through a layer of talc (magnesium metasilicate). To the filtrate (425 ml) 21 ml of 10% acetic acid is added with vigorous agitation; the precipitate is recovered by centri-fugation, and suspended in 150 ml of water. Enough 10% ammonia is added to the suspension to bring the pH to 6.4. Any undissolved residue is discarded by centrifugation. The clear supernatant solution is treated carefully with 10% acetic acid till the pH falls to 4.0. The precipitate forming at this pH is recovered and dissolved in 100 ml of 0.1 M phosphate buffer, pH 6.4. The solution is used as a source of the enzyme.

Properties. The pH optimum for enzyme activity is 5.0 to 8.0. The activity rapidly falls at pH 4.0 and 9.0. Gorjatschenkowa (1952) has shown that pyridoxal phosphate acts as a coenzyme in the reaction.

Substrate specificity. The enzyme shows strict stereospecificity for sulfoxide derivatives of L-cysteine. Derivatives of D-cysteine are inactive.

The enzyme is active for a large number of derivatives of L-cysteine sulfoxide which have the following characteristics (Stoll and Seebeck, 1951): 1. Derivatives of cysteine; homocysteine or penicillamine derivatives are inactive. 2. The S-atom must be linked to an aliphatic group; derivatives with an aromatic group are not split. 3. The amino group of cysteine must be unsubstituted; N-anilino-formylalliin is not attacked. 4. The S-atom must be in the form of a sulfoxide; the double bond in the allyl group is not essential for activity.

5. C—S-Lyase.

This enzyme catalyzes the conversion of S-substituted L-cysteine derivatives to the corresponding thiols, pyruvic acid and ammonia.

The enzyme has so far been isolated only from the seeds of the leguminous plant, *Albizzia lophanta* (Schwimmer and Kjaer, 1960). It probably occurs also in the many genera of *Mimosae*, as *Pitheclobium bigeminum*, *Albizzia luzida*, *Acacia farnesiana* and *Parkia speciosa* (Gmelin, Hasenmaier and Strauss, 1957).

Assay. The assay is based upon the change in absorbancy at 400 mμ due to the enzymic conversion of S-(2,4-dinitrophenyl)-L-cysteine to 2,4-dinitrothio-phenol (Hansen, Kjaer and Schwimmer, 1959). It can also be assayed by the quantitative determination of the pyruvic acid formed by the procedure of FRIEDEMANN and HAUGEN (1943).

One *unit* of the enzyme activity is defined as that amount of enzyme in 3 ml of the reaction mixture in a cuvette of 1 cm light path, which will cause the reaction to proceed at an initial rate of 0.051 absorbancy units per minute, corresponding to the conversion of 10^{-5} mmoles of the substrate (SCHWIMMER and KJAER, 1960).

Purification. The following procedure is that of SCHWIMMER and KJAER (1960): Two hundred grams of *Albizzia lophanta* seeds (obtainable commercially) are coarsely ground in a coffee mill and shaken with trichloroethylene. The seed coats settling at the bottom are discarded, the floating endosperms are laddled out, and dried overnight in air to remove the solvent. The endosperm is ground to a fine meal and 90—95 gm is extracted with 360 ml of water for 1 hour at room temperature. The suspension is squeezed through cheese cloth to yield 290 ml of fraction A. The pH of fraction A is adjusted to 5.5 with 1 N HCl, and the precipitate is removed by centrifugation. The supernatant solution (fraction B) is neutralized with 1 N NaOH and heated to 65° C for 10 minutes, cooled, and centrifuged. To the supernatant solution (fraction C), ammonium sulphate is added to 0.42 saturation and the precipitate is dissolved in 25 ml of water (fraction D). To fraction D is added a saturated solution of $(NH_4)_2SO_4$ to 0.42 saturation at pH 6.54. The resulting precipitate is centrifuged and dissolved in 15 ml of water (fraction E). This fraction (E) is dialyzed for 48 hours at 5° C against distilled water and the precipitate formed at the end of this period is centrifuged off, and lyophilized to yield fraction F. With this procedure, there is an overall yield of 17% with a 70-fold purification. The enzyme can be further purified, if necessary, by adsorption on and elution from alumina.

Properties. The purified enzyme (fraction E) dissolved in 0.05 M phosphate, pH 6.5 is stable for over 5 months at 5° C. With S-ethyl-L-cysteine as the substrate, maximum activity is exhibited between pH 7.8 and 8.2. Pyridoxal phosphate stimulates enzyme activity and serves as the coenzyme for C—S-lyase.

Substrate specificity. Most of the S- substituted L-cysteine derivatives serve as substrates. The L-cysteine moiety itself should not be altered except for addition of oxygen to the sulphur atom, and substitution of the hydrogen of the thiol group. Thus, if the following changes are introduced in the L-cysteine moiety of the S-substituted derivatives, the resultant substances do not act as substrates: introduction of CH_2 into the chain (homocysteine); substitution of the α-hydrogens (α-methyl-DL-cysteine); substitution of one or more β-hydrogens (S-benzyl-thiothreonine, penicillamine); substitution on the nitrogen (S-benzyl-N-benzoyl-L-cysteine, glutathione, S-methyl-glutathione); replacement of the sulphur with oxygen (serine).

Alkylation of the sulphur atom increases the rate of hydrolysis (propyl > > ethyl > methyl > hydrogen). Carboxylation of the terminal end of the S-alkyl group results in a drastic reduction of susceptibility to enzyme action. Halogenation results in a moderate reduction of susceptibility. Substitution of S-alkyl-α-hydrogen by a S-cysteine group results in an increase in the rate of hydrolysis, e.g., L-djenkolic acid (cysteine formaldehyde thioacetal). Terminal arylation of the alkyl chain enhances the hydrolysis rate (S-benzyl-L-cysteine).

IV. Amino Acid Reductases.

This group of enzymes has so far been found only in members of anaerobic clostridia. In *C. sporogenes*, STICKLAND (1934, 1935) found that mutual oxidation-reduction between a pair of amino acids leads to a deamination of both, and CO_2

is evolved in the process. Thus, alanine and glycine under anaerobic conditions could react together in the presence of bacterial extracts as follows:

$$CH_3 . CH(NH_2)COOH + 2 CH_2(NH_2)COOH + 2 H_2O \rightarrow 3 CH_3COOH + 3 NH_3 + CO_2$$

The H-donors in the "STICKLAND" reaction are L-forms of aspartic acid, leucine, isoleucine, valine, alanine, phenylalanine, cysteine, serine, histidine, glutamic acid, and the H-acceptors are glycine and L-forms of ornithine, proline, hydroxyproline, arginine and tryptophan (NISMAN, 1954).

The assay of the reductases is based (see, JOHNSTONE and QUASTEL, 1955) on, 1) the manometric estimation of the CO_2 librated from the donor amino acid, 2) the rate of hydrogen absorption in the presence of acceptor amino acid alone, or 3) the colorimetric estimation of the oxidation of reduced dyes like leucomethylviologen in the presence of suitable acceptor amino acid and enzyme (STICKLAND, 1934). There is evidence that two or three different enzymes are involved in the process and each is specific for a particular acceptor amino acid.

1. Proline Reductase.

Proline + 2 H → δ-aminovaleric acid

Source of the enzyme. The enzyme may be obtained and assayed from *Clostridium* strain HF by the procedure of STADTMAN (1956).

Assay. The reaction mixture contains in a total volume of 0.4 ml, 0.013 μmole of pyridoxal phosphate; 3 μmoles of $MgSO_4$; 0.2 μmole of DPN; 20 μmoles of dithiopropanol; 2 μmoles of L-proline; suitable aliquot of enzyme, and 24 μmoles of Tris. The pH of the assay system is 8.6 and the mixture is incubated at 30° C for 60—90 minutes. The reaction is stopped by the addition of ethanol (80%, end concentration), and the disappearance of proline is measured in an aliquot by the ninhydrin procedure of CHINARD (1952). The dithiol present in the assay mixture augments the color intensity by about 10%, and appropriate controls must be utilized. δ-aminovaleric acid — the product of the reaction — does not interfere in the assay. Thiol solution is freshly prepared just before use.

Properties. The preparations used by STADTMAN (1956) contain a DPN-linked dithiol dehydrogenase. The ability of a number of thiols (lipoic acid, trimethylene dithiol) to serve as reducing agents is lost on the purification of the enzyme system. DPNH does not replace or supplement the dithiol. The enzyme utilizes both the racemic forms of proline and under some conditions the D-form is a more effective substrate. The requirement for pyridoxal phosphate might be explained on the basis of the presence of a racemase in the enzyme preparations (see page 387). The pH optimum of the enzyme is 8.2—8.6. The reduction of D-proline by the enzyme does not lead to a phosphorylation as is the case with glycine reductase discussed below (STADTMAN and ELLIOTT, 1957).

2. Glycine Reductase System.

$$CH_2(NH_2)COOH + R(SH_2) + P_i + ADP \rightarrow CH_3COOH + NH_3 + R—SS + ATP$$

Unlike proline reductase, for each mole of glycine reduced to a mole each of acetate and ammonia, 2 equivalents of SH^- are oxidized and 1 mole of orthophosphate is esterified (STADTMAN, ELLIOTT and TIEMANN, 1958; STADTMAN, 1958).

The enzyme is prepared from *Clostridium sticklandii* cells by sonic disruption in Tris, pH 8.1, protamine sulphate treatment (90 mg/gm protein), and $(NH_4)_2SO_4$

fractionation between 0.3—0.6 saturation. Further purification leads to a dissociation of the system into two fractions which have to be combined for maximum activity (STADTMAN et al., 1958).

Assay. (STADTMAN et al., 1958). The enzyme is assayed either by the disappearance of glycine or the appearance of acetic acid in a system consisting of 20 μmoles of Tris (pH 8.7); 3 μmoles of $MgCl_2$; 0.1 μmole of DPN; 0.003 μmole of pyridoxal phosphate; 9 μmoles of 1,3-dimercaptopropanol; 10 μmoles of glycine-2-C^{14} (0.2 μc); 5 μmoles of AMP; 5 μmoles of K_2HPO_4, and the enzyme solution in a final volume of 0.5 ml. The mixture is made in 10×75 mm stoppered test tubes in an atmosphere of helium or hydrogen, and incubated for 2 hours at 30° C. The reaction is stopped by the addition of 3% (final concentration) perchloric acid. Glycine is determined by the procedure of COCKING and YEMM (1954). Radioactivity of the acetate-C^{14} formed in the reaction is measured in neutralized aliquots of the supernatant solution, freed from residual glycine-C^{14}, by treatment with Dowex 50-H^+ resin at pH 1 to 2.0.

The requirement for DPN and pyridoxal phosphate in the reaction has not been established unequivocally. Glycine cannot be replaced by other amino acids or glyoxylic acid. Dimercaptans are the only electron donors, monomercaptans being ineffective. Reduced lipoic acid is inhibiting at levels greater than 0.02 M SH^-. Arsenate is capable of replacing orthophosphate and adenine nucleotide. The nucleotide requirement is satisfied by either AMP, ADP, or ATP. This lack of specificity may be due to the presence of adenylate kinase as a contaminant. Hydroxylamine and fluoride are potent inhibitors of the enzyme reaction. The enzyme is markedly susceptible to the action of UV light (360 mμ) and Antimycin A (STADTMAN, 1958). Extraction of the enzyme with lipid solvents reduces ATP formation with glycine by about 60 per cent, and the loss is completely restored by D-α-tocopherol. This fact suggests that a quinone also participates in the reductase reaction.

D. Enzymes of Transmethylation.

In the process of transmethylation, the whole methyl group is transferred from certain methyl donors like methionine, S-adenosylmethionine, methionine sulfoxide, and thetins to form various methylated compounds in animals and microorganisms under the influence of methyl transferases. It has been shown that the $-CH_3$ group is transferred to N- or O-atoms, but evidence has accumulated recently that it can also be transferred to a C-atom. Thus, the transfer of CH_3 of methionine to C-24 of ergosterol in yeast (ALEXANDER and SCHWENK, 1957) has been demonstrated. It has been proposed that the "active" methyl group that takes part in reactions of this kind is a part of an "onium" grouping, e.g.,

$$R—{}^+S{\Big\langle}{\begin{matrix}CH_3\\R\end{matrix}}\ ,\quad R—{}^+N{\Big\langle}{\begin{matrix}CH_3\\R\end{matrix}}\ ,\quad R—{}^+S{\Big\langle}{\begin{matrix}CH_3\\CH_3\end{matrix}}$$

Most of the work on transmethylations has been done in animal tissues and some microorganisms. Betaine-homocysteine transmethylase (ERICSON, WILLIAMS and ELVEHJEM, 1955; ERICKSON, 1958), dimethylthetin-homocysteine transmethylase (FROMM and NORDLIE, 1959), Catechol-O-transmethylase (AXELROD and TOMCHICK, 1958), guanidinoacetate transmethylase (CATONI and SCARANO, 1954;

Cantoni and Vignos, 1954), nicotinamide transmethylase (Cantoni, 1951) have all been demonstrated to occur in animal tissues. There is no doubt that large number of other transmethylation reaction will be discovered in near future.

The situation in plants, however, is quite different, which indeed is surprising, because plant tissues abound in compounds with *N*- and *O*-methyl groups, and they are also rich sources of various thetins and other methylsulfonium compounds. It is, however, not necessary that reactions and enzyme systems identical to those found in animals occur in plants, as is shown by the fact that Maw (1959) could not find dimethylthetin-homocysteine transmethylase and Ericson (1958), the betaine-homocysteine transmethylase activity in plants and molds.

Most of the work with plant tissues has been done *in vivo*, but the results point to the presence of enzyme systems active in transmethylations. The role of methionine in plant methylations has been shown strikingly with *in vivo* feeding of C^{14}-radioactive methionine marked in the methyl position. Brady and Tyler (1958) have shown that homogenates of *Panicum miliaceum* form hordenine:

$$HO-\langle\ \rangle-CH_2.CH_2.N(CH_3)_2$$

when tyramine and a methyl donor like methionine or methionine sulfoxide are incubated together. Sato, Byerrum, Albersheim and Bonner (1958) demonstrated that the methyl groups of pectin and protopectin are donated *in vivo* and *in vitro* by methionine, methione sulfoxide, and S-methyl-methionine, in that order of effectiveness. They could not detect the intermediate formation of S-adenosylmethionine by either intact tissues or homogenates. Mudd (1960a) has now shown that extracts of barley seedlings, and millet seedling roots and shoots, when incubated in the presence of tyramine or *N*-methyltyramine as methyl-acceptors, and of L-methionine, L-methioninesulfoxide, S-methyl-L-methionine or S-adenosylmethionine, each labelled with C^{14} in the methyl group, large amounts of radioactive methyltyramine and hordenine are obtained only with S-adenosylmethionine as the methyl donor. Gramine is similarly formed enzymatically by the transfer of methyl groups from S-adenosyl-L-methionine to 3-aminomethyl-indole to form successively 3-methylamino-methylindole and gramine (Mudd, 1961).

The formation of S-adenosylmethionine was demonstrated in cell-free extracts of barley seedlings in the presence of $MgCl_2$, KCl, ATP and a small pool of S-adenosylmethionine only after incubation with radioactive methionine. L-methionine sulfoxide was ineffective (Mudd, 1960b). Plants, thus, seem to have a methionine transadenosylase similar to the other organisms.

Methionine sulfoxide has been frequently implicated in transmethylation reactions in plants *in vivo* (Brady and Tyler, 1958; Sato et al., 1958). It is quite possible, however, that methionine sulfoxide is first converted in plants to methionine by means of a reduced nucleotide dependent methionine sulfoxide reductase system, similar to that shown for yeast by Black, Harte, Hudson and Wartrofsky (1960).

The presence of S-methyl-L-cysteine in a number of plants is well documented. It may arise by the methylation of cysteine by methyl mercaptan. An enzyme bringing about this reaction has been shown to occur in cell-free preparations of yeast (Wolff, Black and Downey, 1956).

The subject of transmethylations has been reviewed by Shapiro and Schlenk (1960).

I. Enzymes of Metabolism of S-Adenosylmethionine and Thetins.

1. Methionine Activating Enzyme.

(Methionine transadenosylase.)

$$\begin{array}{l} CH_2S \cdot CH_3 \\ | \\ CH_2 \\ | \\ CH \cdot NH_2 \\ | \\ COOH \end{array} + ATP \longrightarrow$$

L-methionine

S-adenosylmethionine

$$CH-CH(OH)CH(OH)CH \cdot CH_2S-CH_2CH_2CHCOO^- + PP + P_i$$

Distribution. The enzyme has a wide distribution. It occurs in animal tissues (CANTONI, 1951; KIRSHNER and GOODALL, 1957), *Escherichia coli* (REMY, 1957; TABOR and ROSENTHAL, 1957), yeast (MUDD and CANTONI, 1958), and barley seedlings (MUDD, 1960b).

Source of the enzyme. The enzyme may be purified from air-dried bakers' yeast according to the procedure of MUDD and CANTONI (1958).

Assay. Enzyme activity can be determined either by measuring the amount of P_i liberated from ATP in the presence and absence of methionine (CANTONI, 1951), or by the direct determination of S-adenosylmethionine formed in the presence of the enzyme (CANTONI, 1951; CANTONI and DURELL, 1957). The assay mixture has the following composition: 0.02 M ATP; 0.02 M L-methionine; 0.008 M GSH; 0.3 M MgCl$_2$, and 0.13 M Tris, pH 7.6. An aliquot of the enzyme is added and the reaction mixture is incubated at 37° C for 30 minutes. The reaction is terminated by the addition of 6 per cent perchloric acid, the protein is removed by centrifugation, and the amount of P_i is determined in the supernatant solution by any suitable method. If it is desired to follow the enzymic reaction by determining the quantity of S-adenosylmethionine formed, the following procedure is used (CANTONI and DURELL, 1957): A suitable aliquot of the perchloric acid filtrate containing $3-5$ μmoles of adenine nucleotides is transferred to a graduated cylinder containing 15 ml of 0.015 M Tris buffer, pH 7.4 and 1 drop of 0.04 per cent bromthymol blue is added. The pH is carefully adjusted to 7.4 by 1.0 N NaOH, 2 ml of a 50 per cent suspension of Dowex 1-chloride (X-10, 200−400 mesh size) is added, and the volume is brought to 25 ml. The contents are mixed thoroughly by inversion during 15 minutes, and after filtration the optical density of the solution is read at 260 mμ. A molar absorbancy index of 16,000 is used to calculate the S-adenosylmethionine concentration.

Properties. The enzyme from yeast has a broad pH optimum near 7.6. It has an absolute requirement for both a monovalent cation and a divalent cation. K^+, NH_4^+ and Rb^+ are equally effective, while Na^+, Li^+ and Cs^+ are much less effective. The K_m for L-methionine is 2.6×10^{-3} M for the yeast enzyme, and 2.2×10^{-3} M for the rabbit liver enzyme. The animal enzyme is sensitive to fluoride while the yeast enzyme is not inhibited at all.

Substrate specificity. Ethionine and 2-hydroxy-4-thiomethyl-*n*-butyric acid can replace L-methionine to 31 and 17 per cent, respectively. S-methyl-L-cysteine, DL-methionine sulfoxime, *N*-acetyl-DL-methionine, and methionine sulfoxide are inactive. With DL-homocysteine as substrate, the reaction proceeds at about 2 per cent the rate with L-methionine, and S-adenosylhomocysteine is the product of the reaction.

2. Cleavage of S-Adenosylmethionine.
(S-adenosylmethionine dethioadenosylase)

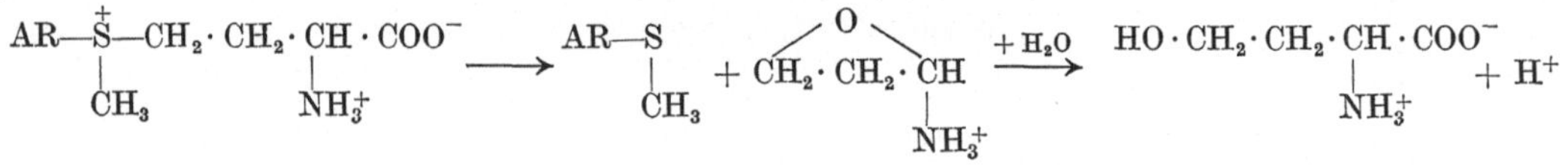

$$AR\!-\!\overset{+}{S}\!-\!CH_2\cdot CH_2\cdot CH\cdot COO^- \longrightarrow AR\!-\!S + CH_2\cdot CH_2\cdot CH \xrightarrow{+H_2O} HO\cdot CH_2\cdot CH_2\cdot CH\cdot COO^- + H^+$$

$$\underset{CH_3}{|} \qquad \underset{NH_3^+}{|} \qquad\qquad \underset{CH_3}{|} \qquad \underset{NH_3^+}{|} \qquad\qquad \underset{NH_3^+}{|}$$

| S-adenosylmethionine | 5'-methylthioadenosine | α-amino-butyrolactone | homoserine |

Distribution. The enzyme cleaving S-adenosylmethionine occurs in yeast (MUDD, 1959) and in *Aerobacter aerogenes* (SHAPIRO and MATHER, 1958). In addition to the cleaving enzyme, *Aerobacter* also possesses a second one which breaks down 5-methylthioadenosine to adenine and methylthioribose.

Source. The enzyme may be obtained from air dried yeast and assayed by the procedure of MUDD (1959).

Assay. The enzyme activity is determined by the disappearance of S-AMe in an assay mixture (total volume 0.5 ml) consisting of 60 μmoles of potassium acetate, pH 5.6; 20 μmoles of potassium phosphate, pH 5.6; 1.5 μmoles of S-AMe, and an aliquot of enzyme which will not form the product of the reaction (methylthioadenosine) in a concentration greater than 0.7×10^{-3} M. The mixture is incubated for 2 hours at 37° C. The mixture, after cooling, is deproteinized with 0.05 ml of 30 per cent perchloric acid, and the protein is removed by centrifugation. An aliquot of the supernatant solution is brought to a pH of $6.7-7.0$ by the addition of an ice-cold mixture of KOH and potassium phosphate. The precipitated potassium perchlorate is removed by centrifugation. A suitable aliquot of the supernatant solution is adsorbed on XE-64 (Amberlite 150, ROHM and HAAS, Philadelphia) column as follows:

The column is prepared (6×1 cm) by treating with 50 ml of 0.25 M potassium phosphate, pH 7.0, followed by 15 ml of a 1:25 dilution of the same buffer. After adsorption, the column is washed with $10-20$ ml of 0.01 M potassium phosphate, pH 7.0 to wash the methylthioadenosine down. S-AMe is eluted from the column by $10-20$ ml of 4 N acetic acid and determined spectrophotometrically by using the molar absorbancy index of 15,100 at 257 mμ (pH 2.0). Controls are used with enzyme and AMe separately. One unit of the enzyme is defined by MUDD (1959) as the amount catalyzing the decomposition of 1 μmole of S-AMe in one hour under conditions described above.

Properties. The enzyme shows similar activity in a pH range of 5.6 to 7.4. The K_m for S-AMe is 3×10^{-4} M. Cofactors are not required and the enzyme is not inhibited by penicillamine, cyanide ions or hydroxylamine. The reaction is not reversible.

S-adenosylethionine reacts at 30 per cent the rate of AMe. Only $(-)$ S-AMe is decomposed by the enzyme, $(+)$ S-AMe being inactive. Various structural analogues of AMe inhibit enzyme activity.

3. S-Methylmethionine-Homocysteine Transmethylase.

Homocysteine + S-methylmethionine → 2 methionine

This enzyme has been demonstrated to occur in rat liver, some strains of *Escherichia coli*, *Aerobacter aerogenes*, *Lactobacillus arabinosus* and the yeast *Torulopsis utilis* (SHAPIRO, 1956; SHAPIRO and YPHANTIS, 1959b). S-methylmethionine also occurs in certain higher plants, like oats (see, SATO et al.,1958), and it is possible that this enzyme may eventually be found there.

Only crude extracts have been obtained from different organisms by sonic vibration of the cells suspended in 0.066 M phosphate buffer, pH 7.0.

Assay. The enzyme can be assayed either by the colorimetric or radioisotopic estimation of methionine formed by transmethylation. The reaction mixture in a total volume of 1 ml consists of, 5—10 mg enzyme protein, 2—10 μmoles each of homocysteine, and S-methylmethionine (or S-methyl [Me-^{14}C] L-methionine; 200,000 counts/mt/μmole; see, SHAPIRO and YPHANTIS, 1959a), and 0.05 M —0.1 M (final concentration) phosphate buffer, pH 6.4. The mixture is incubated for 1—3 hours, depending upon the assay procedure. Methionine is determined at the end of the incubation period as follows:

a) Colorimetric estimation of methionine (SHAPIRO, 1956; McCARTHY and SULLIVAN, 1941). The reaction is terminated by the addition of perchloric acid to a final concentration of 7.5%, followed by phosphotungstic acid (4%, final concentration), and enough pCMBA to give a concentration of 2 moles per mole of homocysteine in the assay mixture. After centrifugation, one ml of 5 N NaOH and 0.5 ml of 1% sodium nitroprusside are added to 1 ml of the supernatant solution. After 5—10 minutes at 35° C, the tubes are cooled in ice-water and acidified with 1.3 ml of 6 N HCl. The intensity of the red colour is measured at 540 mμ.

b) Radioisotopic estimation of [Me-^{14}C]-methionine (SHAPIRO and YPHANTIS, 1959a). The reaction is stopped by rapid cooling in an ice-bath and 0.2 ml of the reaction mixture is applied on an ion-exchange column (Dowex 50 ×4, Li^{+} form, 2 ×0.5 cm). The radioactive substrate, S-methyl-L-methionine, is retained by the column. The radioactive product, methionine, is collected quantitatively in the effluent by washing the column three times with 0.25 ml portions of distilled water. After mixing 25 ml of scintillation fluid (made by mixing equal volumes of toluene and absolute alcohol containing 0.4% 2,5-diphenyloxazole) in the effluent, the radioactivity is measured in a PACKARD TriCarb Liquid Scintillation Spectrometer. The absolute efficiency of counting in this system is about 40% and is not affected by the presence of crude enzyme extract and other constituents of the reaction mixture.

The column can be regenerated after use with 10 ml of 25% LiCl followed by 5 ml of distilled water.

The maximum sensitivity of the colorimetric method is about 0.2 μmole methionine per ml of the reaction mixture (SHAPIRO and YPHANTIS, 1959a). The radioisotopic method is not only more accurate but also very sensitive.

Properties. The pH optimum of the reaction in crude extracts is 7.0—7.5. In the presence of excess homocysteine the reaction proceeds nearly to completion. Both the D- and L-forms of homocysteine are effective in crude extracts of bacteria. This might be due to the presence of racemases in the crude extracts (SHAPIRO, 1956). The microbial enzyme seems to be a metalloprotein and chelating agents like EDTA are inhibitory (SHAPIRO and YPHANTIS, 1959b).

4. Adenosylmethionine-Homocysteine Transmethylase.

L-homocysteine + S-adenosyl-L-methionine →
$$\rightarrow \text{methionine} + \text{S-adenosyl-L-homocysteine} + \text{H}^{+}$$

It is not definitely known whether this enzyme is distinct from S-methyl-methionine-homocysteine transmethylase described earlier. In crude cell-free extracts of microorganisms homocysteine acts as a methyl group acceptor both from S-AMe and S-MMe (SHAPIRO, 1958).

The enzyme is distributed in rat liver, and certain microorganisms, as *Aerobacter aerogenes*, *Saccharomyces cereviseae*, *Torulopsis utilis*, and some strains of *Escherichia coli* (SHAPIRO, 1958; SHAPIRO and YPHANTIS, 1959 b).

Preparation. Only crude enzyme preparations have been so far used. Cell-free extracts from bacteria may be made by sonic oscillation for 10 minutes (30 minutes for yeast) in 0.1 M phosphate buffer, pH 7.0 (SHAPIRO, 1958). In a frozen state, the cell-free extracts are active for at least one year.

Assay. The enzyme can be assayed either with a colorimetric or radioisotopic method (SHAPIRO, 1958; SHAPIRO and YPHANTIS, 1959 a). The composition of the mixture is the same as in the assay of S-methylmethionine-homocysteine transmethylase described earlier, except that $2-10$ μmole of S-AMe (or S-adenosyl-[Me-^{14}C]-L-methionine, specific activity about 425,000 counts/min/μmole; see, SHAPIRO and YPHANTIS, 1959 a) is used in place of S-MMe. For the colorimetric method, the reaction is terminated by perchloric acid and the amount of methionine formed is determined by the nitroprusside method of MCCARTHY and SULLIVAN (1941) described earlier. Caution has to be exercised, however, in using the nitroprusside reaction. S-AMe is degraded by crude extracts to substances which cannot be removed by tungstic acid and pCMBA, and consequently the values obtained are higher. Appropriate controls should, therefore, be utilized in the assay of the enzyme. The methionine obtained may also be separated from the reaction mixture by paper chromatography, and the nitroprusside reaction applied to the eluate from chromatograms.

The *radioisotopic assay* is the same as described under methylmethionine-homocysteine transmethylase, described earlier. Cell-free extracts of bakers' yeast (PARKS, 1958) and bacteria (SHAPIRO, 1957) also convert S-adenosylethionine to ethionine by transethylation to L-homocysteine. This reaction is in all probability mediated by the transmethylase. The transmethylases from microbial cells are distinguished from those of mammaliam tissues by the presence of a metal component in the former (SHAPIRO and YPHANTIS, 1959 b). This is reflected in the inhibition of microbial transmethylase by chelating agents as EDTA (80% inhibition by 5×10^{-3} M EDTA) and 1,10-o-phenanthroline (97% inhibition at a concentration of 10^{-3} M).

Properties. The pH optimum of the reaction in crude extracts is $7.0-7.5$. In the presence of excess homocysteine the reaction proceeds nearly to completion. Both the D- and L-forms of homocysteine are effective in crude extracts of bacteria. This might be due to the presence of racemases in the crude extracts (SHAPIRO, 1956). The microbial enzyme seems to be a metalloprotein and chelating agents like EDTA are inhibitory (SHAPIRO and YPHANTIS, 1959 b).

5. Adenosylmethionine-Nicotinic Acid Transmethylase
(Nicotinic Acid Methylpherase).

This enzyme has been shown to occur only in the green peas, *Pisum sativum* (JOSHI and HANDLER, 1960). Since trigonelline also occurs in large number of other plants, such as *Trigonella foenum-graceum*, and seeds and tubers of many other species (ABDERHALDEN, 1911), the enzyme no doubt also occurs in these plants. The enzyme described here is different from the animal enzyme, S-adenosyl-methionine-nicotinamide transmethylase, isolated by CANTONI (1951) from rat liver, because it methylates nicotinic acid only.

Enzyme preparation. Seeds of *P. sativum* are surface sterilized by washing repeatedly with distilled water followed by two successive washings with 0.2% HgCl$_2$ in 50% ethanol (JOSHI and HANDLER, 1960). The seeds are germinated on wet filter papers over a pad of vermiculite for about 6 days in the dark. The seedings are weighed and homogenized with 2 volumes of cold 0.01 M Tris, pH 7.4 for 1 minute in a prechilled WARING blendor. The homogenate is passed through 4 layers of cheesecloth and the extract centrifuged at high speeds in the cold. The supernatant solution is used as the source of enzyme.

Assay. The enzyme is assayed by determining the quantity of trigonelline synthesized in a reaction mixture consisting of 100 mμmoles of S-adenosyl-methionine; 5 μmoles of Mg^{++}; 25 μmoles of Tris or phosphate buffer, pH 7.4; 100 m μmoles of Tris or phosphate buffer, pH 7.4; 100 mμmoles of nicotinic acid -7-C^{14} (specific activity 3.0—6.0 mc per mmole); 25 μg aureomycin, and 0.4 ml of enzyme in a final volume of 1.1 ml. The mixture is incubated at 37° C for 20 hours. At the end of the incubation period, the tubes are cooled and 0.3 ml of 5% TCA is added. The precipitated protein is removed by centrifugation and 25 μl of the neutralized supernatant fluid is spotted on Whatman No. 1 filter paper, and developed with the upper phase of n-butanol: acetic acid: water (250:60:250) or 70% ethanol + 30% ammonium acetate, adjusted to pH 5.0 with HCl. The spot of trigonelline is located under ultra violet light (R_F — 0.20 in butanol: acetic acid: water solvent), cut out, and examined in a PACKARD TriCarb liquid scintillation counter, or if it is not available, the spot is eluted with water, and counted in a planchet after drying under a gas flow counter. The quantity of trigonelline is calculated with the help of reference spots.

Properties. The optimum pH of the reaction is 6.0 to 7.0. The K_m for nicotinic acid is 4.0×10^{-4} M. In contrast to the S-AMe-nicotinamide transmethylase of rat liver, there is non requirement for sulfhydryl compounds. The crude enzyme preparation also shows the presence of nicotinamide deamidase and S-AMe synthesizing enzyme.

E. Enzymes of Biosynthesis.

In higher plants, very little, if at all, is known about the enzymes of the inter-mediary metabolism of amino acids. Some work has indeed been done by using ^{14}C as tracer, but the result of such work is not unequivocal. In the following review, as before, enzymes of biosynthesis in microorganisms, specially *Neurospora*, yeast and *Escherichia coli*, have been discussed with emphasis on assay procedures. Those enzymes occurring in tissues of higher plants have been given a detailed treatment, including the methods for their preparation.

I. Enzymes of L-Threonine Biosynthesis.

The series of reactions leading from L-aspartate to L-threonine are given in Fig. 1. The enzymes mediating these reactions have been shown to occur in yeast (BLACK and WRIGHT, 1955a), *Escherichia coli* (COHEN, NISMAN and HIRSCH, 1954; NISMAN et al., 1954; WORMSER and PARDEE, 1958), and there is isotopic evidence that the same series of reactions occur in *Neurospora* (ABELSON, 1954). Nothing is known about the presence or absence of these enzymes in higher plants.

1. β-Aspartokinase
(BLACK and WRIGHT, 1955b).

Source of the enzyme. The enzyme is purified from fresh yeast by dipping in liquid nitrogen, and autolyzing the frozen cells in water at pH 8.5 for 2 days in the cold. After protamine treatment and ammonium sulphate precipitation, an active fraction is obtained, and used as the enzyme source (BLACK and WRIGHT, 1955b).

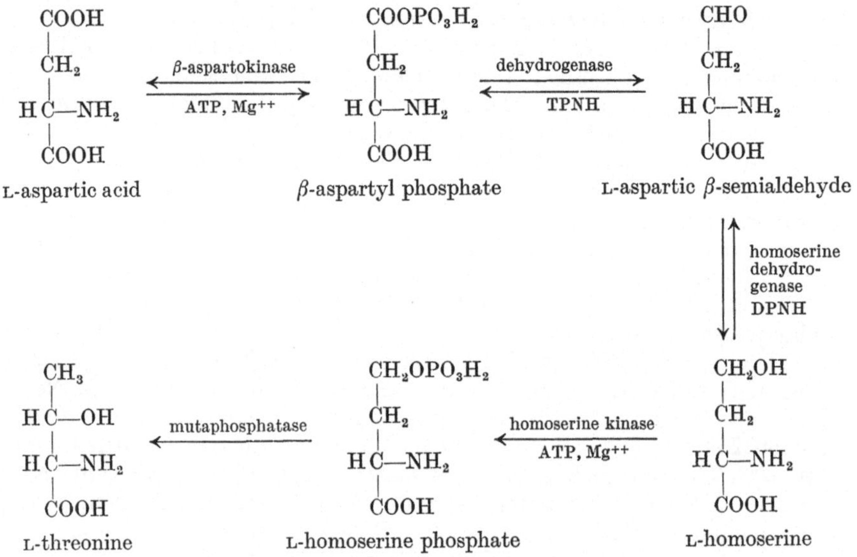

COOH   COOPO$_3$H$_2$ CHO
| | |
CH$_2$ ⇄ β-aspartokinase ⇄   CH$_2$ dehydrogenase   CH$_2$
| ATP, Mg^{++} | TPNH |
H C—NH$_2$ H C—NH$_2$ H C—NH$_2$
| | |
COOH   COOH   COOH

L-aspartic acid   β-aspartyl phosphate   L-aspartic β-semialdehyde

homoserine dehydrogenase DPNH

CH$_3$ CH$_2$OPO$_3$H$_2$ CH$_2$OH
| | |
H C—OH   mutaphosphatase   CH$_2$ homoserine kinase   CH$_2$
| | ATP, Mg^{++} |
H C—NH$_2$ H C—NH$_2$ H C—NH$_2$
| | |
COOH   COOH   COOH

L-threonine   L-homoserine phosphate   L-homoserine

Fig. 1. Biosynthesis of threonine from aspartate.

Assay. Since the reaction is reversible, the activity of the enzyme may be determined either in the forward or the reverse direction. In the forward reaction, the assay mixture, in one ml consists of, 100 μmoles of L-aspartic acid; 20 μmoles of ATP; 20 μmoles of MgCl$_2$, and 400 μmoles of hydroxylamine-HCl. All of the reagents except MgCl$_2$ are brought to pH 8.0 with Tris prior to addition. The mixture is incubated with an aliquot of the enzyme for 15 minutes at 15° C. The hydroxamic acid formed by the reaction of hydroxylamine with acyl phosphate is determined by the method of Lipmann and Tuttle (1945). One unit of the enzyme is defined as the amount of enzyme which causes the formation of 1 μmole of hydroxamic acid in the test. In the reverse direction, the assay mixture in a total volume of 0.5 ml consists of 50 μmoles of Tris chloride buffer (pH 8.0); 5 μmoles of ADP; 10 μmoles of MgCl$_2$, and 2 μmoles of β-aspartyl phosphate. Following incubation for 20 minutes at 15° C, 0.1 ml of 0.1 M pCMBA suspension is added to terminate the reaction and 5 minutes later, 0.4 ml of 2.0 M hydroxylamine (adjusted to pH 8.0 with Tris base prior to use) is also added. Hydroxamic acid is determined in the mixture after 20 minutes.

β-aspartyl kinase reaction can also be coupled with aspartic β-semialdehyde dehydrogenase (see page 327) and TPNH disappearance after the addition of ATP is a measure of the reduction.

Properties. The pH optimum of the reaction is between 5 and 9. The enzyme is activated by Mg^{++} and Mn^{++} ions. The reaction is freely reversible and the $K'eq$ for the system,

$$\frac{[\beta\text{-aspartyl phosphate] [ADP]}}{[\text{L-aspartate] [ATP]}}$$

is 3.5×10^{-4} (approximate). L-asparate is absolutely specific for the enzyme.

2. Aspartic β-Semialdehyde Dehydrogenase
(Black and Wright, 1955c).

Source of the enzyme. The enzyme is purified from yeast according to the method of Black and Wright (1955c). Cell-free extracts are prepared as in section 1 and after heat treatment (60° C) for 10 minutes and salt precipitation, a preparation is obtained which is substantially free from homoserine dehydrogenase and β-aspartokinase.

Assay. The reaction mixture (final volume 1.0—1.5 ml) in a quartz cuvette consists of, 0.08 μmole of TPNH; 0.4—0.6 μmole of β-aspartyl phosphate, and 100 μmoles of Tris, pH 8.0. The reaction is started by adding an aliquot of the enzyme, and activity is measured by the decrease in absorbancy at 340 mμ. One unit of the enzyme is the amount causing a change of 1.00 per minute in absorbancy, the results being calculated from the linear part of the reaction (BLACK and WRIGHT, 1955 c).

Properties. The enzymic reaction is reversible and pH optima for the forward and reverse reactions are 8.0 and 9.0, respectively. The K_m values for β-aspartyl phosphate, TPNH, aspartic β-semialdehyde, TPN and phosphate are 1.6×10^{-4} M, 8.3×10^{-5} M, 2.6×10^{-3} M, 3.6×10^{-5} M and 1.4×10^{-3} M respectively. The $K'eq$ of the system,

$$\frac{\text{ASA} \times \text{TPN} \times \text{P}_i}{\text{BAP} \times \text{TPNH} \times \text{H}^+} = 3.0 \times 10^{-6} \ .$$

In the presence of arsenate, a slow splitting of β-aspartyl phosphate occurs. The arsenolysis and the reduction of the substrate is inhibited by iodoacetate.

3. Homoserine Dehydrogenase.
(BLACK and WRIGHT, 1955 d).

Source of the enzyme. The enzyme is prepared from yeast cells according to the procedure of BLACK and WRIGHT (1955 d) by heat treatment, salt precipitation, and chromatography on calcium phosphate columns.

Assay. Owing to the reversibility of the reaction, the enzyme activity can be measured by following the reduction of pyridine nucleotide with L-homoserine in the reverse direction, or the oxidation of the reduced coenzyme with aspartic β-semialdehyde in the forward direction. The latter is the preferred method because of the favourable equilibrium conditions obtainable in this direction. The reaction mixture in a quartz cuvette (total volume 1.0 ml) contains, 100 μmoles of potassium phosphate buffer (pH 6.7); 0.08 μmole of DPNH or TPNH, and enzyme. The reaction is started by the addition of 1.0 μmole of aspartic β-semialdehyde and readings are taken every 30 seconds for several minutes. One unit of the enzyme is the amount which casuses an optical density change of 1.00 per minute during the first minute after the start of the reaction.

Properties. The optimal pH for the TPNH reaction is 5.0 and for the DPNH reaction about 6.8. There seems to be only one enzyme involved in the oxidation of the coenzymes. The K_m value with TPN is 20-fold higher than the DPN value but the affinity with TPNH and DPNH is the same. There is a 7-fold difference in the K_m values for homoserine when tested with DPN or TPN. The $K'eq$ for the reaction:

$$\frac{\text{Homoserine} \times \text{DPN or (TPN)}}{\text{ASA} \times \text{DPNH (or TPNH)} \times \text{H}^+} = 1.3 \times 10^{-11} \ .$$

The enzyme is not inhibited by pCMBA at all which seems to be unusual for a dehydrogenase.

4. L-homoserine Kinase.
(WATANABE et al. 1955, 1957).

$$\text{HO—CH}_2\text{—CH}_2\text{—CHCO}_2^- + \text{ATP} \rightarrow (\text{HO})_2\text{POCH—CH}_2\text{—CH—CO}_2^- + \text{ADP}$$

$$\underset{\text{NH}_3^+}{|} \qquad \qquad \underset{\text{O}}{\|} \qquad \underset{\text{NH}_3^+}{|}$$

This enzyme has been shown to occur in yeast and *E. coli* (Wormser and Pardee, 1958). *Neurospora* homoserine kinase (Flavin and Slaughter, 1960) seems to be quite labile.

Purification. The enzyme may be purified according to Watanabe, Konishi and Shimura (1955). A 15-fold purification may be obtained by disrupting bakers' yeast with liquid nitrogen, thawing, and stirring for 2 hours at 0° C with an equal volume of cold water after adjusting the pH to 8.5 with NH_4OH. The centrifuged extract is adjusted to pH 7, and made 0.1 M with regard to Tris by adding enough M solution of Tris, pH 7.0. The fraction sedimenting between 0.32 and 0.44 $(NH_4)_2SO_4$ saturation is dissolved in 0.04 M Tris-HCl, pH 7.4, containing 0.004 M glutathione, and used as the enzyme source (Flavin and Slaughter, 1960).

Assay. The procedure of Flavin and Slaughter (1960) is used which is based on the determination of the amount of non-nucleotide, homoserine-dependent, acid stable phosphate ester. Duplicate tubes are used in the assay, one of which contains 40 μmoles of DL-homoserine. In addition, both tubes contain, 80 μmoles of Tris HCl, pH 7.4; 10 μmoles of ATP; 20 μmoles of $MgSO_4$; 30 μmoles of NaF; and 45 μmoles of NaCN. Total volume of the mixture is $1-5$ ml, and it is incubated for 30 minutes at 30° C. The reaction is terminated by adding 0.1 ml of 2 M perchloric acid, followed by 0.4 ml of 0.5 M potassium acetate, pH 4.5. With a calibrated scoop, roughly 0.5 ml of acid-washed Norit A is added and after stoppering, the tubes are shaken mechanically for 15 minutes and centrifuged. Total phosphate is determined in an aliquot of the supernatant solution by wet ashing according to Flavin (1954), and acid labile and inorganic phosphate by 15 minutes hydrolysis in 1 N HCl by the method of Fiske and Subbarrow (1925). The difference between the two values represents acid-stable phosphate and the difference between the latter value in the presence or absence of homoserine represents the amount of phosphohomoserine.

This assay procedure cannot be applied to crude homogenates and extracts owing to the presence of interfering enzyme systems.

Properties. The K_m for bacterial kinase is 6×10^{-4} M and it is competitively by L-threonine ($K_i = 4 \times 10^{-3}$ M; Wormser and Pardee, 1958).

5. Threonine Synthetase (Homoserine Phosphate Mutaphosphatase)
(Flavin and Slaughter 1960).

$$(HO)_2PO-CH_2-CH_2-CH-CO_2^- + H_2O \rightarrow CH_3CH\ CH\ CO_2^- + H_3PO_4$$

The enzyme occurs in yeast (Watanabe and Shimura, 1955), *E. coli* (Wormser and Pardee, 1958), and has recently been extensively purified from *Neurospora* (Flavin and Slaughter, 1960).

Purification. The enzyme is purified from *Neurospora* according to Flavin and Slaughter (1960). The procedure involves grinding of the cells with dry ice in a Waring blendor, extraction with water at pH 8.5, acetone fractionation (45 to 58%), ammonium sulphate fractionation between $0.44-0.57$ saturation, and chromatography on DEAE-cellulose. An overall purification of 500-fold is obtained

Assay. The enzyme assay is based on the determination of the amount of threonine formed under the following conditions (Flavin and Slaughter, 1960):

An aliquot of the enzyme (0.001 to 0.02 units) is added to a rubber-stoppered centrifuge tube containing 0.06 ml of 0.5 M glycylglycine buffer, pH 7.3, 0.01 ml of 0.01 M pyridoxal phosphate, 0.06 ml of 0.01 M phosphohomoserine (prepared by enzymic synthesis, Flavin and Slaughter, 1960), and enough water to a final volume of 0.6 ml. After 30 minutes incubation at 30° C, the reaction is terminated by heating and the coagulated protein is discarded by centrifugation. Threonine is determined in an aliquot by the enzymic procedure of FLAVIN and SLAUGHTER (1959) or the colorimetric procedure of WINNICK (1942) and NEIDIG and HESS (1952). The enzymic method is described in vol. VI.

One *unit* of the enzyme is the amount catalyzing the formation of 1 μmole of threonine in 1 minute under the above conditions.

Properties. The synthetase from yeast (WATANABE and SHIMURA, 1956) and *E. coli* (WORMSER and PARDEE, 1958) does not require pyridoxal phosphate, but the activity of the *Neurospora* enzyme is increased by the presence of catalytic amounts of pyridoxal phosphate. EDTA and Mg^{++} have no effect on the holoenzyme but the latter increases the rate of apoenzyme reaction by 30%. Phosphate, KCN, and NH_2OH are inhibitory. P-homoserine cannot be replaced by homoserine + ATP, homoserine lactone, or 2,4-diaminobutyrate.

By using isotopic oxygen, FLAVIN and KONO (1960) demonstrated that phosphate is formed in the reaction by nonhydrolytic elimination, rather than hydrolysis, with cleavage at the $C-O$ bond of the phosphate ester.

II. Enzymes of L-Histidine Biosynthesis.

The pathway of histidine biosynthesis is given in Fig. 2. The various enzymes of this pathway are known to occur in different bacteria as *Pseudomonas*, *Aerobacter aerogenes*, *Salmonella typhimurium* (AMES, GARRY and HERZENBERG, 1960), *Bacillus subtilis*, *Escherichia coli*, *Arthrobacter*, yeast, and *Neurospora* (see,MOYED and MAGASANIK, 1960; AMES, 1955). The enzymes have not yet been shown to occur in higher plants.

The steps leading to the formation of imidazoleglycerol phosphate have been very recently studied in the cell-free extracts of several enterobacteria (MOYED and MAGASANIK, 1960), and all of the enzymes involved are not yet purified or characterized. It has, however, been proved that imidazoleglycerol phosphate is produced from ribose-5-phosphate, the amide nitrogen of glutamine, and the N-1, C-2 portion of the adenine ring of ATP, the remainder of the latter appearing as 5-amino-1-ribosyl-4-imidazolecarboxamide-5-phosphate. The latter compound can be reconverted to ATP (Fig. 2; MOYED and MAGASANIK, 1960).

1. Imidazoleglycerol Phosphate Dehydrase
(AMES, 1957a).

D-erythro-imidazoleglycerol phosphate → imidazoleacetol phosphate + H_2O

The enzyme occurs in *Neurospora*, *E. coli* (AMES, 1957a), and *Salmonella* (AMES et al., 1960). It may be purified from *Neurospora* by grinding the cells with dry ice, extraction in 0.1 M triethanolamine buffer, pH 7.1, salt precipitation between $0-0.40$ saturation, and Norit A-calcium phosphate gel treatment. AMES (1957a) obtained a 35-fold purified enzyme.

Assay. The enzyme is assayed with the procedure of AMES et al. (1960; see also, AMES, 1957a). The method is also applicable to crude dialyzed extracts. A buffer is made by mixing 5.0 ml of 0.1 M triethanolamine − HCl (pH 8.0), 0.03 ml of 14.3 M β-mercaptoethanol, and 0.01 ml of 0.1 M $MnCl_2$. The incubation mixture consists of, 0.30 ml of the buffer, 0.005−0.04 ml of the enzyme solution, and 0.02 ml of 0.1 M imidazoleglycerol phosphate, neutralized prior to addition with KOH. The mixture is incubated at 37° C and after 30 minutes the reaction is stopped with 0.7 ml of 1.43 N NaOH. The tubes are incubated for another

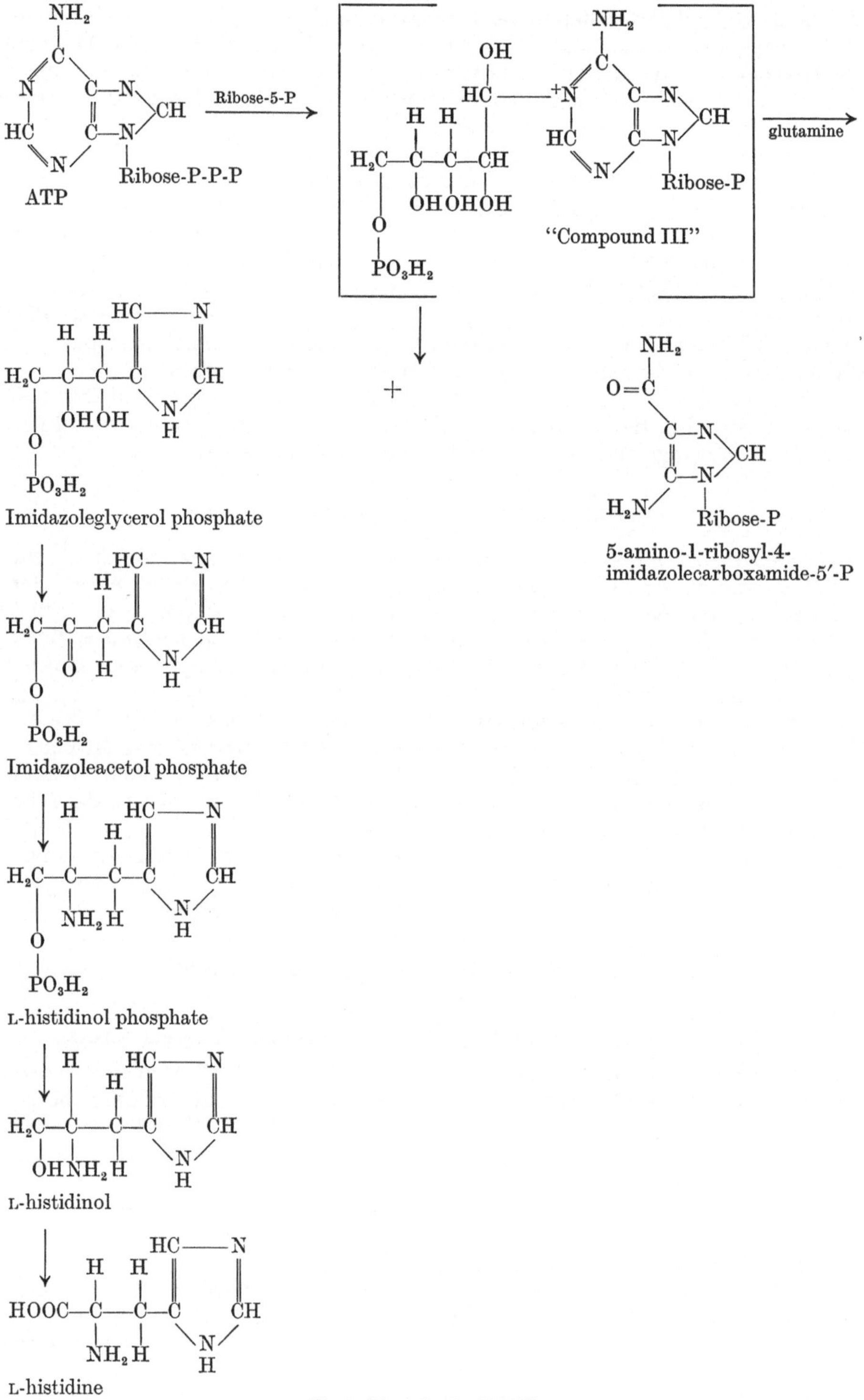

Fig. 2. Biosynthesis of histidine.

30 minutes at 37° C and the optical density is read at 290 mμ against an incubation blank containing water instead of the substrate. For each new batch of the substrate, a small correction for absorption might become necessary. The proportionality of the enzyme concentration to optical density readings at the end of the incubation period must be ascertained for enzymes from other sources or different batches of cells. Optical density readings can be converted to concentration by reference to a standard curve.

One *unit* of the enzyme is the amount catalyzing the formation of 1 μmole of imidazoleacetol phosphate per hour.

Properties. The pH optimum of the mold enzyme is 7.5. The K_m for the substrate is 2.4×10^{-4} M. The enzyme requires Mn^{++} and a reducing agent (cysteine or β-mercaptoethanol) for maximum activity. EDTA is a potent inhibitor (AMES, 1956).

2. Imidazoleacetol Phosphate Transaminase
(AMES and HORECKER, 1956).

Imidazoleacetol phosphate + L-glutamate $\rightleftharpoons$ L-histidinol phosphate + α-ketoglutarate

The enzyme has been found in *Neurospora*, *E. coli* and *Salmonella* (AMES and HORECKER, 1956; AMES et al., 1960).

Purification. The enzyme is purified from *Neurospora* according to the method of AMES and HORECKER (1956). Approximately 18-fold purification is obtained by the method.

Assay. The procedure described by AMES et al. (1960; see, also AMES and HORECKER, 1956) is used. An α-ketoglutarate reaction mixture is prepared with 2.81 ml of 0.2 M triethanolamine $-$ HCl (pH 8.4), 0.06 ml of β-mercaptoethanol, and 0.08 ml of 0.4 M α-ketoglutarate (neutralized with KOH to pH 5.0). In the assay of the enzyme in crude extracts, 0.15 ml of 0.04 M 8-hydroxyquinoline (adjusted to pH 3.0 with HCl), and 0.90 ml of 0.1 M-disodium EDTA is also added to inhibit phosphatases which might hydrolyze the substrate or the product. A second assay tube contains $0.01 - 0.04$ ml of the enzyme solution and enough 0.2 M triethanolamine buffer (pH 8.4) containing 6×10^{-5} M pyridoxal phosphate to make 0.09 ml. This is incubated at 25° C for 5 minutes. After this preliminary incubation, 0.20 ml of the α-ketoglutarate reaction mixture is added and the tube is placed in a 37° C bath. The reaction is started by adding 0.01 ml of 0.1 M histidinol phosphate. After $10-20$ minutes incubation at 37° C, 0.7 ml of 1.43 N NaOH is added to stop the reaction. The tube is incubated at 37° C for 30 more minutes, and the optical density of the solution is read at 295 mμ against an incubation blank containing water instead of histidinol phosphate. The amount of imidazoleacetol phosphate is calculated in the same way as in the assay of dehydrase.

One *unit* of the enzyme is the amount which catalyzes the formation of 1 μmole of imidazoleacetol phosphate per hour.

Properties. The optimum pH is 8.1 in pyrophosphate buffer, 8.35 in diethanolamine, and 8.5 in triethanolamine buffer. The concentration of pyridoxal phosphate giving half maximal activation of the enzyme is 1.2×10^{-6} M. The coenzyme can be replaced by N-methylpyridoxal phosphate. The enzyme shows some activity for aminoadipic acid, arginine and histidine. The K_m for histidinol phosphate is 1.1×10^{-3} M, and for α-ketoglutarate 1.2×10^{-3} M. The $K'eq$ at 37° C at pH 8.1 is 0.04.

Tris buffer markedly inhibits the reaction. Hydroxylamine, semicarbazide, Co^{++} and Cu^{++} at concentrations of 0.004 M inhibit the reaction approximately 80%. Iodoacetate, 0.004 M, inhibits to 95%, which suggests that the transaminase is a sulfhydryl enzyme.

3. L-Histidinol Phosphate Phosphatase
(Ames, 1957b).

L-histidinol phosphate + H_2O → L-histidinol + P_i

The enzyme is prepared from the cells of *Neurospora crassa* according to the procedure of Ames (1957b). The mycelium can be stored at $-20°$ C for several years. Crude extracts are obtained by pulverizing with dry ice in a Waring blendor and by thawing and extraction in 0.1 M triethanolamine-HCl buffer, pH 7.1.

Assay. The enzyme is assayed according to the method of Ames et al. (1960). The incubation mixture consists of 0.005—0.02 ml of enzyme, 0.01 ml of 0.1 M histidinol phosphate, and enough 0.1 M triethanolamine-HCl (pH 9.0) to give a final volume of 0.3 ml. The mixture is incubated at 30° C for 10 minutes, and the reaction is terminated by the addition of 0.7 ml of ascorbic acid + molybdate reagent. The reagent consists of 1 part of 10% ascorbic acid and 6 parts of 0.42% ammonium molybdate · $4H_2O$ in 1 N H_2SO_4 (Chen, Toribara and Warner, 1956). The tubes are further incubated at 45° C for 20 minutes and read at 820 mμ against an enzyme and buffer blank without histidinol phosphate. The presence of inorganic phosphate as a contaminant in histidinol phosphate is ascertained before hand and, if present, is corrected for in the assay mixtures. The amount of the enzyme added in the assay mixture should be so adjusted that it does not yield a perceptible precipitate upon the addition of the molybdate reagent. If material of low specific activity is used, the protein is first removed by adding an equal volume of 1.0 M perchloric acid (Ames, 1957b), followed by the removal of the precipitate, and determination of inorganic phosphate in an aliquot from the supernatant solution.

One *unit* of the enzyme is the amount liberating 1 μmole of inorganic phosphate/hour.

Properties. The optimal pH of the *Neurospora* enzyme is 9.0, while that from *Salmonella* is 7.5 (Ames et al., 1960). L-histidinol phosphate is relatively specific as a substrate. The non-specific phosphatase activity accompanying the enzyme preparations is inhibited by 10^{-4} M berryllium sulphate. The K_m of the substrate is 4.2×10^{-3} M. 8-hydroxyquinoline at a concentration of 10^{-3} M is inhibitory.

4. L-Histidinol Dehydrogenase
(Adams, 1955).

1. L-histidinol + DPN^+ ⇌ L-histidinal + DPNH + H^+
2. L-histidinal + DPN^+ + H_2O → L-histidine + DPNH + H^+

Overall: L-histidinol + 2 DPN^+ + H_2O → L-histidine + 2DPNH + 2 H^+

The enzyme has been shown to occur in *E. coli*, *Arthrobacter*, yeast, *Neurospora* (Adams, 1954, 1955) and *Salmonella* (Ames et al., 1960).

Enzyme preparation. The best source of the enzyme is yeast because a preparation free from histidase can be easily obtained. Dried beer yeast is autolyzed in 0.1 N $NaHCO_3$, and the enzyme is purified according to the method of Adams (1955).

Assay. Adams (1955): Crude extracts are assayed by the diaphorase catalyzed reduction of the dye dichlorophenolindophenol by the DPNH formed in the oxidation of histidinol. Cuvettes contain, 0.075 μmole of 2,6-dichlorophenol-indophenol; 1 μmole of DPN; 150 μmoles of Tris (pH 8.9); 0.01 ml of diaphorase (available commercially from Worthington Biochemicals; 4 mg protein/ml, or 0.02 ml of a crude extract of *Clostridium kluyveri*, made according to the method of Stadtman and Barker, 1949), and properly diluted enzyme solution. After checking for endogenous oxidation, the reaction is initiated by the addition of 1 μmole of L-histidinol, and the decrease in optical density is followed at 600 mμ.

For purified extracts, the direct oxidation of DPN at 340 mμ is followed in the presence of Tris, 0.01 M (end concentration), sodium thioglycolate, and the purified enzyme.

One *unit* is defined (AMES et al., 1960) as the quantity of the enzyme required for the oxidation of 1 μmole of histidinol per hour.

Properties. The purified enzyme is inhibited by pCMBA. EDTA (0.03 M) gives 50% inhibition at pH 7.5. Borate, 0.01 M is also inhibitory.

III. Enzymes of L-Proline Biosynthesis.

L-proline is synthesized in various organisms from L-glutamic acid through glutamic-γ-semialdehyde and its' cyclization product, Δ'-pyrroline-5-carboxylic

$$\underset{\text{L-glutamic acid}}{HOOC.CH_2.CH_2.\overset{\overset{\displaystyle NH_2}{|}}{C}H.COOH} \quad\underset{\xrightarrow{\text{Transaminase}}}{\xleftarrow{\hspace{2cm}}}\quad \underset{\text{glutamic } \gamma\text{-semialdehyde}}{OHC.CH_2.CH_2.\overset{\overset{\displaystyle NH_2}{|}}{C}H.COOH}$$

Trans.

$$\underset{\text{glutamic } \gamma\text{-semialdehyde}}{OHC.CH_2.CH_2.\overset{\overset{\displaystyle NH_2}{|}}{C}H.COOH}$$

Ne

H$_2$C——CH$_2$
HC   CH.COOH
   N

Δ'-pyrroline-5-carboxylic acid

Red.

H$_2$C——CH$_2$
H$_2$C   CH.COOH
   N
L-proline   H

COOH
HC—NH$_2$
CH$_2$
CH$_2$
HC—NH$_2$
H
ornithine

$CO_2 + NH_3$

Ar + H$_2$O

NH$_2$
C=O
NH$_2$
urea

COOH
HC—NH$_2$
CH$_2$
CH$_2$
CH$_2$
NH
C=NH
NH$_2$
arginine

carbamyl phosphate transcarbamylase

arginine desiminase

Cl. Enz.

COOH
HC—NH$_2$
CH$_2$
CH$_2$
CH$_2$
NH
C=O
NH$_2$
citrulline

+ aspartate ATP Co. Enz.

COOH
HC—NH$_2$
CH$_2$
CH$_2$
CH$_2$ COOH
NH   CH$_2$
   H
C—N—CH
NH   COOH
argininosuccinate

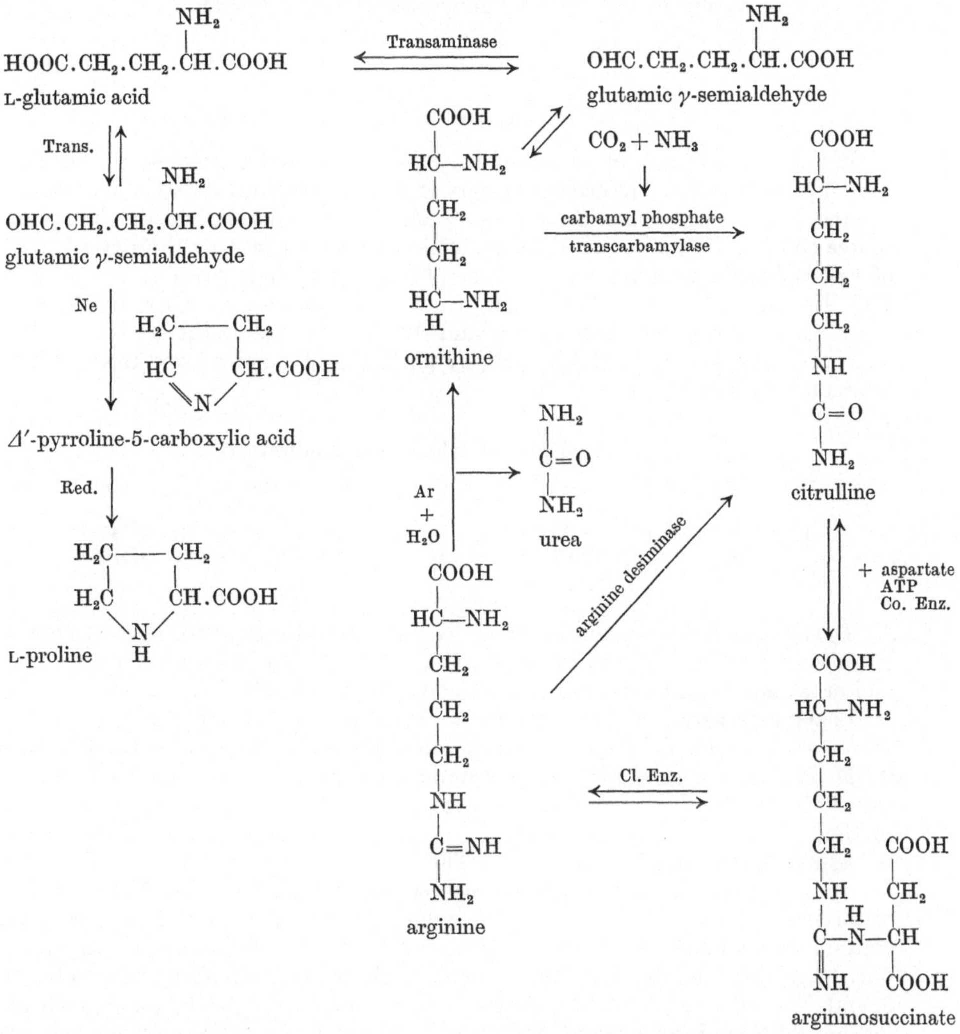

Fig. 3. Biosynthesis of proline and urea cycle. Trans. = transaminase; Red. = reductase; Co. Enz. = condensing enzyme; Ar = arginase; Ne = non-enzymic; Cl. Enz. = cleaving enzyme.

acid (Fig. 3). This is true of various microorganisms, as *E. coli* and *Neurospora* (see, Vogel and Bonner, 1959). In higher plants, no study is yet available on the enzymes of this pathway.

The first step in the synthesis of proline from glutamic acid is the formation of glutamic γ-semialdehyde. This compound can arise in *Neurospora* (Fincham, 1953) and in animals (Meister, 1954) from ornithine by a transamination reaction with α-ketoglutarate, but many other organisms, as *E. coli*, although able to produce glutamic γ-semialdehyde (Vogel and Davis, 1952), do not possess the necessary ornithine δ-transaminase (Scher and Vogel, 1957). This fact argues for the presence of a system distinct from the transaminase which catalyzes the formation of glutamic γ-semialdehyde from glutamic acid. In animal tissues and some microorganisms there is a strong likelihood that glutamic acid is converted to proline and *vice-versa* by the following series of irreversible reactions:

$$
\begin{array}{ccc}
\text{glutamic acid} & \xrightarrow{(1)} & \Delta'\text{-pyrroline-5-carboxylic acid} \\
& \nwarrow{(4)} & \downarrow{(2)} \\
\Delta'\text{-pyrroline-5-carboxylic acid} & \xleftarrow{(3)} & \text{proline}
\end{array}
$$

Enzyme (1) has not yet been found *in vitro* in mammalian tissues, but has been demonstrated to occur in crude extracts of *E. coli* (Strecker, 1957) and Pleuropneumonia like organisms (Smith, 1957). Mg^{++} and some high energy phosphate source (ATP, ADP) is required for enzyme activity, suggesting the participation of a phosphorylated compound between glutamic acid and pyrroline-5-carboxylic acid. Enzyme (2) is described below. Enzyme (3) probably occurs in the mitochondria from rat liver (Strecker and Mela, 1955), and the TPN or DPN-dependent enzyme (4) has been partially purified and characterized from ox liver (Strecker, 1960).

Δ'-Pyrroline-5-Carboxylate Reductase.

$$
\begin{array}{c}
\text{H}_2\text{C} \!\!-\!\! \text{CH}_2 \\
| \quad\quad | \\
\text{HC} \quad \text{CH.COOH} + \text{TPNH} + \text{H}^+ \\
\diagdown \text{N} \diagup \quad (\text{DPNH})
\end{array}
\longrightarrow
\begin{array}{c}
\text{H}_2\text{C} \!\!-\!\! \text{CH}_2 \\
| \quad\quad | \\
\text{H}_2\text{C} \quad \text{CH.COOH} + \text{TPN}^+ \\
\diagdown \text{N} \diagup \quad\quad (\text{DPN}^+) \\
\;\;\; \text{H}
\end{array}
$$

The enzyme is distributed in rat liver (Smith and Greenberg, 1957; Meister, Radhakrishnan and Buckley, 1957), *Neurospora* (Yura and Vogel, 1955, 1959), and other microorganisms (see, Vogel and Bonner, 1959).

Enzyme preparation. The enzyme is obtained from the mycelium of *Neurospora crassa*, wild type strain, by disruption of the cells with sand, extraction with 0.1 M phosphate buffer, pH 7.9, protamine treatment, and salt fractionation by the method of Yura and Vogel (1959). A 35-fold purification is obtained by this method.

Assay (Yura and Vogel, 1959): The enzyme activity is measured by the decrease in optical density at 340 mμ caused by an oxidation of TPNH or DPNH in the presence of the substrate and enzyme. Quartz cuvettes of 1 cm light path contain, 300 μmoles of potassium phosphate, pH 7.0; 6 μmoles of pyrroline-5-carboxylate; 0.6 μmole of TPNH or DPNH; enzyme solution, and water to 3.0 ml. The substrate is omitted in the reference cuvettes. The reaction is started by the addition of the enzyme, and optical density measurements are made every 1 or 2 minutes for 10 to 20 minutes.

One *unit* of the enzyme is defined as the amount that will give K (first order velocity constant) $= 1 \times 10^{-3}$ minutes^{-1}. DPNH is used as the coenzyme.

Properties. The optimal pH is about 7.0 for the TPNH reaction and 6.0 for the DPNH reaction. The substrate concentration giving half maximum velocity is approximately 4.5×10^{-4} M for the reaction with TPNH, and 5.3×10^{-4} M for the reaction with DPNH.

IV. Enzymes of Ornithine Synthesis.

1. Higher Plants, Molds and Animals.

There is considerable evidence that ornithine arises in molds, animals and higher plants by a reversible transamination between glutamic γ-semialdehyde and glutamic acid. The enzyme ornithine δ-transaminase has been studied in *Neurospora* (FINCHAM, 1953); animal tissues (QUASTEL and WITTY, 1951; MEISTER, 1954), and in the homogenates of various plant tissues (Fig. 3).

Ornithine δ-transaminase.

glutamic γ-semialdehyde + glutamic acid $\rightleftharpoons$ ornithine + α-ketoglutarate

In a study of large numbers of bacteria, fungi, algae and spinach leaves, SCHER and VOGEL (1957) demonstrated that the enzyme is distributed in higher plants, green algea, fungi, yeast, animal tissues, gram positive bacteria (*Bacillus*, *Micrococcus*, *Arthrobacter* and *Mycobacterium*) and the protozoan, *Tetrahymena pyriformis*, but is absent from the blue-green algae and gram negative bacteria. The distribution pattern of the enzyme is a reflection of the dissimilar modes of ornithine synthesis in these two groups of organisms (see, below).

In higher plants, the results of tracer study (COLEMAN and HEGARTY, 1957) have pointed to the presence of a system forming ornithine from glutamic acid. The enzyme has been found in the cell-free extracts of spinach (SCHER and VOGEL, 1957) and the mitochondria of mung bean (BONE, 1959 c; BONE and FOWDEN, 1960).

Enzyme preparation. The enzyme is obtained from *Neurospora crassa*, wild type or amination deficient strains, by grinding the mycelium (grown in FRIES No. 3 medium) with powdered glass, and extraction with 2 volumes of 0.006 M phosphate buffer, pH 7.3. The supernatant solution, after centrifugation is dialyzed for $10 - 20$ hours at $0 - 3°$C and used as the source of the enzyme (FINCHAM, 1953).

Assay. *1. Manometric* (FINCHAM, 1953). The main compartment of a manometric vessel contains the enzyme extract, 0.3 ml of 0.05 M NaHCO$_3$, and 0.1 ml of 0.2 M sodium α-ketoglutarate. The side arm contains 0.1 ml of 0.05 M NaHCO$_3$, 0.1 ml of 0.2 M ornithine hydrochloride, and 0.2 ml of distilled water. In the control vessels distilled water is taken in place of either or both reactants. The gas phase is 5% CO$_2$ + 95% N$_2$ and the bath temperature is 35° C. The reaction is started by tipping in the contents of the side arm. For each mole of ornithine disappearing in this system, 0.5 mole of CO$_2$ is produced (FINCHAM, 1953).

2. Colorimetric (FINCHAM, 1953). The enzyme activity can also be measured by the reaction of o-aminobenzaldehyde with Δ'-pyrroline-5-carboxylate, the cyclized form of glutamic γ-semialdehyde. The colorimetric tubes contain in a total volume of 6.25 ml, 0.0016 M L-ornithine, 0.0016 M sodium α-ketoglutarate, 0.0008 M o-aminobenzaldehyde, 0.02 M phosphate buffer, pH 7.3, and 0.3 to 0.5 ml of enzyme. α-ketoglutarate is added at zero time. The intensity of the yellow color is measured after 60 to 180 minutes at 430 mμ (for details, see spectrophotometric assay of amine oxidase, page 301). The optimum pH for the transamination is about 8.0.

2. Escherichia coli.

In *E. coli* and certain other gram negative bacteria, ornithine is synthesized from glutamic acid by a series of reactions involving N-acetylated compounds (VOGEL, 1955; VOGEL and BONNER, 1959). N-acetyl ornithine is found in some higher plants (VIRTANEN, 1958) but until now enzymes similar to those of *E. coli* have neither been demonstrated in plants nor in animals. Ornithine is synthesized from glutamic acid as follows (VOGEL and BONNER, 1959):

$$
\begin{array}{ccccc}
\text{COOH} & \text{COOH} & \text{COOH} & \text{COOH} & \text{COOH} \\
| & | & | & | & | \\
\text{CH.NH}_2 & \text{CH.NH.COCH}_3 & \text{CH.NH.COCH}_3 & \text{CH.NH.COCH}_3 & \text{CH.NH}_2 \\
| \quad {}^a & | & | \quad {}^b & | \quad {}^c & | \\
\text{CH}_2 \xrightarrow{} & \text{CH}_2 \xrightarrow{} & \text{CH}_2 \xrightarrow{} & \text{CH}_2 \xrightarrow{} & \text{CH}_2 \\
| & | & | & | & | \\
\text{CH}_2 & \text{CH}_2 & \text{CH}_2 & \text{CH}_2 & \text{CH}_2 \\
| & | & | & | & | \\
\text{COOH} & \text{COOH} & \text{CHO} & \text{CH}_2.\text{NH}_2 & \text{CH}_2.\text{NH}_2 \\
\text{glutamic acid} & \text{N-acetyl} & \text{N-acetylglutamic} & \text{N}^\alpha\text{-acetyl} & \text{ornithine} \\
& \text{glutamic acid} & \gamma\text{-semialdehyde} & \text{ornithine} &
\end{array}
$$

a) Amino Acid Transacetylase.

The enzyme has been shown to occur in crude cell-free extracts of *E. coli* by MASS, NOVELLI and LIPMANN (1953).

The enzyme is assayed by measuring the amount of N-acetylglutamine in a system consisting of, 0.2 ml of 1 M Tris buffer, pH 8.0; 0.2 ml of 1 M MgCl$_2$; 0.2 ml of CoA (100 LIPMANN units/ml); 0.4 ml of a saturated solution of H$_2$S; 0.2 ml of 0.25 M acetyl phosphate, and sonic extract of *E. coli* in a total volume of 2.0 ml. Control tubes are kept, one lacking the enzyme, and another lacking the substrate. The mixture is incubated for 60 minutes at 37° C, and at the end of this period, heated at 90° C for 5 minutes to destroy the residual acetyl phosphate. Protein is precipitated with 0.1 ml of 10 N H$_2$SO$_4$, centrifuged off, and the supernatant solution is extracted 3 times with 1.5 volumes of 4:1 n-butanol:chloroform. The combined organic solvent is evaporated to dryness, and the residue is suspended in 1 ml of water. The acetyl amino acid is determined by the hydroxylamine method of KATZ, LIEBERMAN and BARKER (1953). The time for complete hydrolysis of the N-acetyl derivative is predetermined.

b) Acetylornithine δ-Transaminase.

The presence of this enzyme in *E. coli* extracts has been shown (VOGEL, 1953, 1955), but details of assay procedure and properties of the enzyme are not known.

c) Acetylornithinase.

$$\text{N}^\alpha\text{-acetylornithine} + \text{H}_2\text{O} \rightarrow \text{ornithine} + \text{CH}_3\text{COOH}$$

The enzyme is distributed in several members of enterobacteriaceae, including *E. coli, Aerobacter, Klebsiella, Erwinia, Serratia, Proteus, Salmonella*, and *Shigella* (SCHER and VOGEL, 1955; VOGEL and BONNER, 1956a).

Enzyme preparation. The enzyme is extracted from *E. coli* cells by sonic disruption, protamine treatment, ammonium sulphate, and acetone fractionation according to the method of VOGEL and BONNER (1956a) which leads to a 40-fold purification. The enzyme is stored with added glutathione.

Assay (VOGEL and BONNER, 1956a). The assay mixture consists of, in a final volume of 0.5 ml in test tubes, 50 μmoles of potassium phosphate (pH 7.0);

3 μmoles of N^α-acetylornithine; 0.5 μmole of glutathione, and 0.1 μmole of cobaltous chloride. A properly diluted enzyme preparation is added last of all and the mixture is incubated at 37° C for 10 minutes. In addition to the assay mixture, reference blanks (no substrate or enzyme), enzyme blanks (no substrate), and substrate blanks (no enzyme) are also run. The reaction is stopped by adding 1.5 ml of a ninhydrin reagent (not more than a week old) prepared by mixing 2 volumes of a 1% solution of ninhydrin in methyl cellosolve with 1 volume of 0.4 M aqueous citric acid. The mixture is heated in a boiling water bath for 10 minutes and immediately cooled. Three ml of 0.7 N aqueous NaOH is then added with rapid stirring. After 20 minutes, the final reaction mixture is read against a reference blank in a KLETT-SUMMERSON Colorimeter with No. 42 filter. The orange-yellow color with ornithine is stable for hours, and has absorption maxima at 350 and 470 mμ. The standard samples of ornithine are made in 0.1 M aqueous phosphate, pH 7.0, 1 mM with respect to glutathione. A linear color response to ornithine is obtained up to about 0.3 μmole of ornithine per sample.

One *unit* of the enzyme is the amount catalyzing the formation of 0.1 μmole of ornithine.

Properties. The substrate concentration giving half maximal velocity is 2.8 mM. The pH optimum of the reaction is 7.0. The enzyme also deacetylates N-acetyl-methionine at a rate similar to N^α-acetylornithine. Glutathione and cobaltous ions stimulate the enzyme activity, whereas Cu^{++}, Zn^{++}, Ni^{++}, EDTA and pCMBA inhibit it.

V. Enzymes of the Urea Cycle and Related Compounds (synthetic).

The urea cycle is shown in Fig. 3. It occurs in mammals (RATNER, 1955), other animals, including amphibians and turtles (BROWN and COHEN, 1959), and fungi (SRB and HOROWITZ, 1944; FINCHAM and BOYLEN, 1955; VOGEL and BONNER, 1959). In bacteria, most of the organisms tested lack the enzyme arginase and perhaps convert arginine to ornithine or citrulline by degradative enzymes. In higher plants, KASTING and DELWICHE (1958) have provided tracer evidence for the occurrence of the cycle in watermelon seedlings. In mung bean mitochondria (BONE, 1959a) and pea extracts (REIFER and KLECZOWSKI, 1960) some intermediary enzymes of the cycle have also been found.

1. Carbamyl Phosphate Synthesizing Enzymes.

Carbamyl phosphate is synthesized in biological systems in two different ways. Mammals, amphibians, chelonian reptiles, and perhaps yeast, possess the enzyme carbamyl phosphate synthetase which catalyzes the reaction (see, BROWN and COHEN, 1959):

$$\text{ammonia} + \text{bicarbonate} + 2\,\text{ATP} \xrightarrow[\text{Mg}^{++}]{\text{acetylglutamate}} \text{carbamyl phosphate} + 2\,\text{ADP} + \text{P}_i$$

The enzyme requires catalytic amounts of acetyl glutamate and 2 moles of ATP are used per mole of carbamyl phosphate synthesized. METZENBERG, MARSHALL and COHEN (1958) have suggested that the reaction possible occurs in two steps as follows:

$$\text{ATP} + \text{CO}_2 \xrightarrow[\text{glutamate}]{\text{acetyl}} \text{ADP} + \text{P}_i + \text{``active CO}_2\text{''} \qquad \text{i)}$$

$$\text{``ATP} + \text{active CO}_2\text{''} + \text{NH}_3 \underset{\text{glutamate}}{\overset{\text{acetyl}}{\rightleftharpoons}} \text{ADP} + \text{carbamyl phosphate} \qquad \text{ii)}$$

The reaction is irreversible. This enzyme may be assayed and prepared from frog liver according to the method of MARSHALL, METZENBERG and COHEN (1958).

In bacteria, carbamyl phosphate is synthesized by carbamate kinase as follows:

$$CO_2 + NH_3 + ATP \rightleftharpoons carbamyl\ phosphate + ADP$$

The reaction in contrast to the synthetase is reversible, does not require acetyl glutamate, and only one mole of ATP is required per mole of carbamyl phosphate synthesized. Further, the kinase is not inhibited by hydroxylamine and Zn^{++} as is the synthetase (CARAVACA and GRISOLIA, 1960).

The nature of the plant or fungal enzyme is not known with certainty, except the demonstration by BONE (1959c) that mung bean mitochondria synthesize carbamyl phosphate in the presence of ATP. N-acetylglutamate activates carbamyl phosphate degradation, without having any effect on citrulline synthesis. It has recently been demonstrated (LEVENBERG, 1961) that for the synthesis of citrulline in certain Basidiomycetes, ATP, Mg^{++}, HCO_3^-, L-ornithine and L-glutamine are required. Glutamic acid and acetylglutamate cannot be substituted for glutamine.

2. Carbamate Kinase.

$$NH_3 + CO_2 + ATP \rightleftharpoons H_2N.CO.O.H_2PO_3 + ADP$$

The enzyme has been studied in crude extracts of *Streptococcus faecalis* (JONES, SPECTOR and LIPMANN, 1955), and in a purified state from *Serratia marcescens* (GLASIZOU, 1956), group D Streptococci (MOKRASH, CARAVACA and GRISOLIA, 1960), and *Streptococcus lactis* (RAVEL, HUMPHREYS and SHIVE, 1961).

Assay. The enzyme activity may be followed by measuring the rate of citrulline formation in the presence of a system generating carbamyl phosphate and excess ornithine transcarbamylase. The assay mixture (RAVEL et al., 1961) contains in 1 ml; ATP, 8 μmoles; magnesium chloride, 10 μmoles; Tris buffer, 50 μmoles; DL-ornithine monohydrochloride, 20 μmoles; ammonium carbamate (GMELIN, 1936), 250 μmoles; and an excess of bacterial ornithine transcarbamylase (see preparation in the following section). A solution of the first 4 ingredients (solution A) is prepared at 2.5 times the final concentration (pH 8.7) and stored at $-20°$ C. Just prior to testing, ornithine transcarbamylase and solid ammonium carbamate in appropriate amounts is added to solution A at $4°$ C, and 0.4 ml of the resulting solution is added to each tube containing a rate-limiting amount of carbamate kinase in 0.6 ml of water. The mixture is incubated for 5 minutes at $38°$ C, and the reaction is terminated by perchloric acid. Control tube contains the mixture with perchloric acid added at zero time. Citrulline is measured by the method of ARCHIBALD (1944; described under ornithine transcarbamylase). Linearity of the reaction must be established with preparations from different sources.

Properties. Sulphate ions stabilize the enzyme and reactivate preparations which have been inactivated by aging, dialysis, or heat treatment. In the absence of sulphate ions, there is a 50% loss in activity on storage at $-20°$ for 1 week (RAVEL et al., 1961). The pH optimum for highly purified enzyme preparations is 8.9 to 9.1. The bivalent cation requirement is satisfied equally well by Mg^{++} and Mn^{++}, and to a lesser extent by Fe^{++} and Co^{++}. Hg^{++}, Cu^{++}, Cd^{++} and Pb^{++} are inhibitory (RAVEL et al., 1961; GLASIZOU, 1956). The K_m values are (RAVEL et al., 1961) Mg^{++}, 2.2×10^{-3} M; Mn^{++}, 2.7×10^{-3} M; and ATP, 3.4×10^{-3} M. pCMBA at a concentration of 2×10^{-5} M inhibits the enzyme and this inhibition is reversed by glutathione.

3. Ornithine Transcarbamylase.

$$carbamyl\ phosphate + ornithine \rightarrow citrulline + P_i$$

The enzyme occurs in rat liver mitochondria (REICHARD, 1957), some vertebrate livers (BROWN and COHEN, 1959), large number of gram negative (e. g.,

E. coli), and gram positive bacteria (JONES et al., 1955; RAVEL et al., 1959). Tracer experiments point to the probable presence of the enzyme in *Neurospora* (SRB and HOROWITZ, 1944). In higher plants, the enzyme has been demonstrated to occur in the homogenates of beans, *Phaseolus lunatus, P. vulgaris, Pisum sativum* (KREBS and EGGLESTON, 1955; REIFER and KLECZOWSKI, 1960), and mung bean mitochondria (BONE, 1959a).

Enzyme preparation. The enzyme may be prepared from animal liver by the method of REICHARD (1957) or CARAVACA and GRISOLIA (1950), and from *Streptococcus lactis* by the method of RAVEL et al. (1959). A partially purified enzyme may also be prepared from group D Streptococci by sonic disruption followed by stirring of the supernatant solution in an equal volume of cold acetone ($-10°$ C). The protein precipitate is dissolved in one tenth the original volume of water and clarified by centrifugation. The supernatant solution is treated with an equal volume of washed betonite (American Colloid Co; 36 mg/ml), and the supernatant solution after centrifugation is lyophilized. This treatment gives a 5-fold purified enzyme which is very stable and can be stored in the cold for a number of days (CARAVACA and GRISOLIA, 1960).

Assay. The enzyme may be estimated either by measuring the release of inorganic phosphate (REICHARD, 1957) or the formation of citrulline. Assay mixture (CARAVACA and GRISOLIA, 1960) in a final volume of 1 ml consists of 50 μmoles of Tris buffer, pH 8.5; 20 μmoles of ornithine, 15 μmoles of carbamyl phosphate (freshly prepared), and enzyme solution. The tubes are incubated for $10-15$ minutes at $37°$ C. Boiled enzyme (to account for non-enzymic transcarbamylation) or a TCA treated system serves as a zero time control. Longer incubation times are undesirable because carbamyl phosphate is approximately 50 per cent decomposed in 4 hours at $30°$ C (RAVEL et al., 1959). The reaction is stopped with 1 ml of 5% TCA and the supernatant solution is analyzed for citrulline according to the method of ARCHIBALD (1944), as modified by RATNER and PETRACK (1953a). An aliquot ($0.1-0.2$ ml) is added to 5 ml of an acid mixture (1 part concentrated technical H_2SO_4 and 3 parts syrupy phosphoric acid; 100 ml of the mixture diluted to 250 ml) along with 1.0 ml of 0.75% aqueous diacetylmonoxime (2,3-butanedione 2-methoxime). After mixing the solutions, the tubes are capped with glass marbles and heated in a boiling water bath from which light is excluded. After exactly 10 minutes, the tubes are quickly cooled in the dark, and after a lapse of another 10 minutes the optical density is read at 490 mμ. Since the colour reaction is sensitive to light, precautions are taken to exclude as much light as possible. The amount of citrulline is found by reference to a standard curve.

Properties. The purified enzyme from *Streptococcus* and liver has a pH optimum of $8.0-8.5$. The K_m value for ornithine is 3.0×10^{-3} M, and for carbamyl phosphate, 1.2×10^{-3} M (BURNET and COHEN, 1957). The enzyme is inhibited by $-$SII group reagents and only ornithine out of various amino acids is able to accept the carbamyl group (REICHARD, 1957). For the enzyme from *Streptococcus* (RAVEL et al., 1959), phosphate ion competitively inhibits the conversion of ornithine and carbamyl phosphate to citrulline, and the inhibition is not reversed by ornithine.

4. Argininosuccinate Synthetase.

L-citrulline + L-aspartate + ATP $\rightleftharpoons$ argininosuccinate + AMP + PP

The enzyme has been purified from animal liver and kidney (RATNER and PETRACK, 1953a, b), and yeast (RATNER and PETRACK, 1953a, 1956). It has also been studied in crude extracts of *Neurospora* (FINCHAM and BOYLEN, 1955).

Enzyme preparation. The enzyme may be obtained from liver or yeast according to the method of Ratner (1955; see also, Petrack and Ratner, 1958; Schuegraf, Ratner and Warner, 1960).

Assay. *1. Crude extracts* (Ratner and Petrack, 1951). The enzyme is assayed by the disappearance of citrulline in a mixture containing (in an ice bath), 0.1 ml of 0.1 M L-citrulline (pH 7.4); 0.1 ml of 0.1 M L-aspartate (pH 7.4); 0.1 ml of 1.0 M phosphate buffer, pH 7.4; 0.2 ml of .066 M $MgSO_4$; 0.1 ml of 0.025 M ATP (pH 7.4); 0.25 ml of 0.1 M D-3-phosphoglyceric acid; muscle enzyme concentrate (Sigma); enzyme preparation (7—15 units), and enough water to make 2.0 ml. The tube is incubated at 38° C for 20 minutes and the amount of citrulline remaining is measured, after deproteinization with TCA, with the colorimetric method of Archibald (1944; see, assay of transcarbamylase).

One unit of the enzyme is the amount which catalyzes the disappearance of 1 μmole of citrulline per hour.

2. Purified enzyme (Shuegraf et al., 1960). Fractions substantially free from ATPase are assayed by coupling the synthetase system to DPNH oxidation through adenylate kinase, pyruvate kinase, and lactic dehydrogenase as follows:

$$AMP + ATP \rightleftharpoons 2\ ADP \qquad \qquad \text{i)}$$
$$2\ phosphorylenolpyruvate + 2\ ADP \rightleftharpoons 2\ ATP + 2\ pyruvate \qquad \text{ii)}$$
$$2\ pyruvate + 2\ DPNH \rightleftharpoons 2\ lactate + 2\ DPN^+ + 2\ H^+ \qquad \text{iii)}$$

Quartz cuvettes of 1 cm light path contain in a final volume of 1 ml: 100 μmoles of Tris, pH 7.5; 0.1 μmole of ATP; 5 μmoles of $MgCl_2$; 2 μmoles of KCl; 16 μmoles of phosphorylenolpyruvate; 0.1 μmole of DPNH, and 7.5 μmoles of L-citrulline. Auxiliary enzymes (obtainable from Boehringer und Söhne, Mannheim, Germany) are added in 100-fold excess as follows: inorganic pyrophosphatase, 20 μg; lactic dehydrogenase, 5 μg; pyruvate kinase, 30 μg; and adenylate kinase, 25 μg. A suitably diluted enzyme preparation is added, and after the readings remain constant at 340 mμ, 7.5 μmoles of L-aspartate is added to start the reaction. Reference cuvette is used with all the constituents of the assay mixture, except DPNH. The decrease in optical density is measured for 2—3 minutes at 30 seconds interval.

One *unit* is the amount of enzyme which causes the formation of 1 μmole of AMP (one-half the μmoles of DPNH utilized) per hour.

Properties of the enzyme. The observation that the enzyme consists of two components (Ratner, 1955) has been clarified further by Petrack and Ratner (1958). One of the components has been identified as the enzyme proper which synthesizes argininosuccinate and liberates inorganic pyrophosphate. The other component has been identified as inorganic pyrophosphatase which increases the rate of synthetase activity by the removal of pyrophosphate which has an inhibitory effect on the synthetase. The reversibility of the reaction has also been shown.

Fluoride, calcium and manganous ions are strongly inhibitory at a concentration of 0.001 M. Mg^{++} is required for activity and high concentration of ATP are inhibitory.

5. Argininosuccinase.

L-argininosuccinate $\rightleftharpoons$ L-arginine + fumaric acid

The enzyme has a wide distribution. It occurs in animal kidney, beef liver (Ratner, 1955), *Neurospora* (Fincham and Boylen, 1955), pea seeds (Davison and Elliott, 1952), jack beans, *Chlorella* (Walker and Myers, 1953), *E. coli*, and many other bacteria (WALKER, 1953). It has been purified from animal (RATNER, ANSLOW and PETRACK, 1953) and plant tissues (WALKER and MYERS, 1953).

Preparation of the enzyme from plants. *1. Chlorella* (WALKER and MYERS, 1953). Cells are grown (see, General Methods of Preparation, vol. VI), harvested, dried with acetone, and ground together with corundum powder at room temperature in a mortar. The cohesive mass is extracted with minimal amount of water and centrifuged in the cold. The supernatant solution is adjusted successively to pH 5.1, 4.6 and 4.2 with 1 M acetic acid and the precipitate at each stage is separated by centrifugation. The precipitate appearing between pH 4.6 and 4.2 is dried with acetone and stored in the cold. When required, it is dissolved in 0.1 M phosphate buffer, pH 8.8, and any insoluble material is centrifuged out.

2. Jack beans (WALKER and MYERS, 1953; WALKER, 1953). Eighty gm of jack bean meal (Sigma) and 250 ml of water are mixed together and stirred for 1 hour at room temperature and centrifuged in the cold. Isoelectric precipitation is carried out as with *Chlorella* extracts, and the precipitate obtained between pH 4.8 and 4.2 is acetone dried, and used as the enzyme source.

Assay (RATNER, ANSLOW and PETRACK, 1953). The reaction mixture in a final volume of 1.0 ml contains, 0.5 ml of 0.01 M argininosuccinic acid (neutralized); 0.05 ml of 1.0 M phosphate buffer, pH 8.0; excess arginase (Arginase powder, Sigma preparation, is extracted with 0.1 M phosphate buffer, pH 7.8 and heated in the water bath for 20 minutes at 60°. The clarified extract is stored at $-20°$ C; $0.1-0.2$ ml of the extract is used), and suitably diluted enzyme. The mixture is incubated at 38° C for 20 minutes and the reaction is terminated by the addition of 2.0 ml of 7.5% TCA. Urea is determined in a suitable aliquot of the clarified solution by the colorimetric procedure of ARCHIBALD (1945) modified by RATNER (1955). To 10 ml of an acid mixture (90 ml of concentrated H_2SO_4 and 270 ml of concentrated H_3PO_4 in a final volume of 1 liter) are added, 0.5 ml of 3% alcoholic α-isonitrosopropiophenone and 0.5 ml of the solution to be assayed. The tubers are heated and after cooling, color intensity is read at 540 mμ. The color is light sensitive and solutions must be protected from light during color development.

One *unit* of the enzyme is the amount catalyzing the formation of 1 μmole of arginine per hour.

Properties. The reaction is easily reversible. Out of the two substrates, only fumaric acid is specific. L-arginine can be replaced by canavanine (WALKER, 1953) to some extent. The latter compound with fumaric acid produces canavanino-succinic acid. The enzyme is inhibited by Fe^{++}, Fe^{+++}, cobaltous and zinc ions.

6. Arginase.

L-arginine → urea + ornithine

The enzyme is very widely distributed in animals and various fungi (except yeast). Amongst plants it is found in the bean family but seems to be absent from others. In alder roots, for instance, MIETTINEN and VIRTANEN (1952) could not find arginase inspite of the presence of large amounts of ornithine, citrulline and arginine.

Enzyme preparation. The enzyme may be prepared from horse liver by the method of GREENBERG, BAGOT and ROHOLT (1956) or the method of ROBBINS and SHIELDS (1956). The enzyme has been crystallized from beef, sheep and horse liver by BACH and KILLIP (1958, 1961). A slightly purified arginase can also be obtained from jack bean meal by stirring 80 gm meal (Sigma) in 250 ml of water for 1 hour at room temperature. To the supernatant solution after centrifugation is added an equal volume of cold acetone. The precipitate formed is dried with acetone, and for use resuspended in sodium glycinate buffer, pH 9.5. Insoluble material is removed by centrifugation (WALKER and MYERS, 1953). The powder contains in addition to arginase, urease but no arginine desimidase.

Assay (BROWN and COHEN, 1959). The assay mixture consists of 25 μmoles of arginine, pH 9.5; 0.5 μmole of $MnCl_2$; 50 μmoles of sodium glycinate buffer, pH 9.5, and the enzyme preparation in a total volume of 2.0 ml. The mixture is incubated for 30 minutes at 38° C and the reaction is terminated by the addition of 5.0 ml of 0.5 M $HClO_4$. The protein is centrifuged off. Suitable controls are, a perchloric acid treated assay mixture and one without the enzyme. Urea is determined in a suitable aliquot by the procedure given under argininosuccinase.

The enzyme is activated by Mn^{++} and maximal activity is obtained by pre-incubation of the enzyme with the metal ions. The enzyme is unstable in high dilutions and is stabilized by ornithine or glycine.

VI. Enzymes of Arginine Degradation.

Bacteria, yeast, the green alga *Chlorella*, and some animal tissues are able to degrade arginine to ornithine, CO_2 and NH_3 through the intermediate formation of citrulline. In the first step of the process arginine is broken down to citrulline and ammonia, and in the second citrulline is converted to ornithine, CO_2 and NH_3 (or carbamyl phosphate) by a reversible mechanism which involves the participation of adenosine phosphate, P_i and Mg^{++}. RATNER (1954) has reviewed the subject.

1. Arginine Desiminase.

$$\text{L-arginine} + H_2O \rightarrow \text{L-citrulline} + NH_3$$

In this reaction, only one of the terminal guanidine nitrogen atoms is susceptible to cleavage (PETRACK, SULLIVAN and RATNER, 1957).

Enzyme preparation. A partially purified enzyme may be obtained from acetone dried baker's yeast by the method of ROCHE and LACOMBE (1952). The cells are suspended in water (1:2) and agitated for sometime at 38° C. The supernatant solution is fractionated with cold ethanol. The precipitate forming between 20—30% is used as the source of enzyme. A highly purified preparation from *Streptococcus faecalis* may be obtained according to the procedure of PETRACK et al. (1957).

Assay (PETRACK et al., 1957). Enzyme activity is determined by measuring the rate of citrulline formation. The assay mixture in small test tubes consists of, 25 μmoles of L-arginine hydrochloride; 200 μmoles of potassium phosphate buffer, pH 6.8, and enzyme in a final volume of 2.0 ml. The tubes are incubated at 38° C for 20 minutes and the reaction is terminated by the addition of 2.0 ml of 10% TCA. Citrulline is determined in the protein free supernatant solution by the same procedure described under ornithine transcarbamylase.

One *unit* is defined as the amount of enzyme which catalyzes the formation of 1 μmole of citrulline per hour.

The pH optimum of the purified enzyme is 6.8; less purified preparations from yeast show a lower value of 6.2—6.5. The K_m value for arginine is approximately 1.5×10^{-4} M (PETRACK at al., 1957). The enzyme is specific for L-arginine. Canavanine (OGINSKY and GEHRIG, 1952) and long chain diguanidines (KNIVETT, 1951) are inhibitory. For the purified enzyme Fe^{++}, Fe^{+++} and Cu^{++} ions are inhibitory (PETRACK et al., 1957). Hydroxylamine inhibits the purified enzyme only on preincubation.

2. Citrullinase System (Citrulline Ureidase).

$$\text{L-citrulline} \xrightarrow[\text{Mg}^{++}]{\text{ADP, P}_i} \text{L-ornithine} + CO_2 + NH_3$$

The enzyme is distributed in various bacteria as *Streptococcus*, *Pseudomonas* and *Clostridium perfringens*. The mechanism of action is not very well understood, and has been reviewed by KORZENOVSKY (1955) and SLADE (1955).

The enzyme may be prepared from bacteria either by sonic oscillation (SLADE, 1953) or shaking in a MICKLE disintegrator with glass beads and 0.1 phosphate buffer, pH 5.6—5.8.

Assay (OGINSKY, 1955). The enzyme is conveniently assayed by measuring the CO_2 given off in the reaction. The central compartment of a double arm WARBURG vessel contains, 1.0 ml of 0.5 M acetate buffer, pH 5.8; 0.5 ml of 0.01 M AMP or ADP (preadjusted to pH 5.5); 0.1 ml of 0.1 M $MgCl_2$; 0.5 ml of 0.1 M potassium phosphate, pH 5.8, and enzyme solution. One side arm contains 0.2 ml of L-citrulline and the other, 0.2 ml of 2.0 M H_2SO_4. Gas phase is N_2 and bath temperature, 37° C. After tipping in citrulline, gas exchange is measured at 1—10 minutes interval. H_2SO_4 is tipped in at the end of the reaction. Control flask is used without citrulline.

The requirement for AMP, ADP, Mg^{++} and P_i in the reaction is met by arsenate alone. NaF is inhibitory. Carbamyl phosphate is perhaps produced as an intermediate.

VII. Enzymes of Synthesis of Branched-chain Amino Acids.

(Leucine, Isoleucine, Valine).

The steps in the biosynthesis of branched chain amino acids are given in Fig. 4. Evidence for this scheme has come from the tracer and enzymic studies of *Neurospora*, *E. coli*, *Torulopsis*, and yeast by various groups of workers (see, reviews, VOGEL and BONNER, 1959; KNOX and BEHRMAN, 1959). There is no information regarding the presence of these enzymes in higher plants. An extensively purified preparation from wheat germ (SINGER and PENSKY, 1952a, b) catalyzes the formation of acetoin from acetaldehyde, but it is not known whether it can also bring about the formation of acetolactate, an intermediate in the synthesis of valine, from pyruvic acid and acetaldehyde.

One interesting aspect of the biosynthesis or the general reactions of the branched chain amino acids is the presence in various cells of only one enzyme catalyzing one step in the metabolism of all of the three amino acids together. Thus only one enzyme is known for the decarboxylation of all of the three amino acids (EKLADIUS et al., 1957); only one enzyme brings about the coenzyme linked oxidative deamination of all three amino acids (SANWAL and ZINK, 1961), and the loss by mutation of one enzyme in one biosynthetic step of one of the amino acids in *E. coli* (like valine) leads to a requirement not only for this amino acid but also for isoleucine (UMBARGER and BROWN, 1958; SEECOF and WAGNER, 1959). Enough evidence exists from enzymic and genetic experiments that for each individual step in the biosynthesis of valine and isoleucine (see Fig. 4) only one common nezyme is involved.

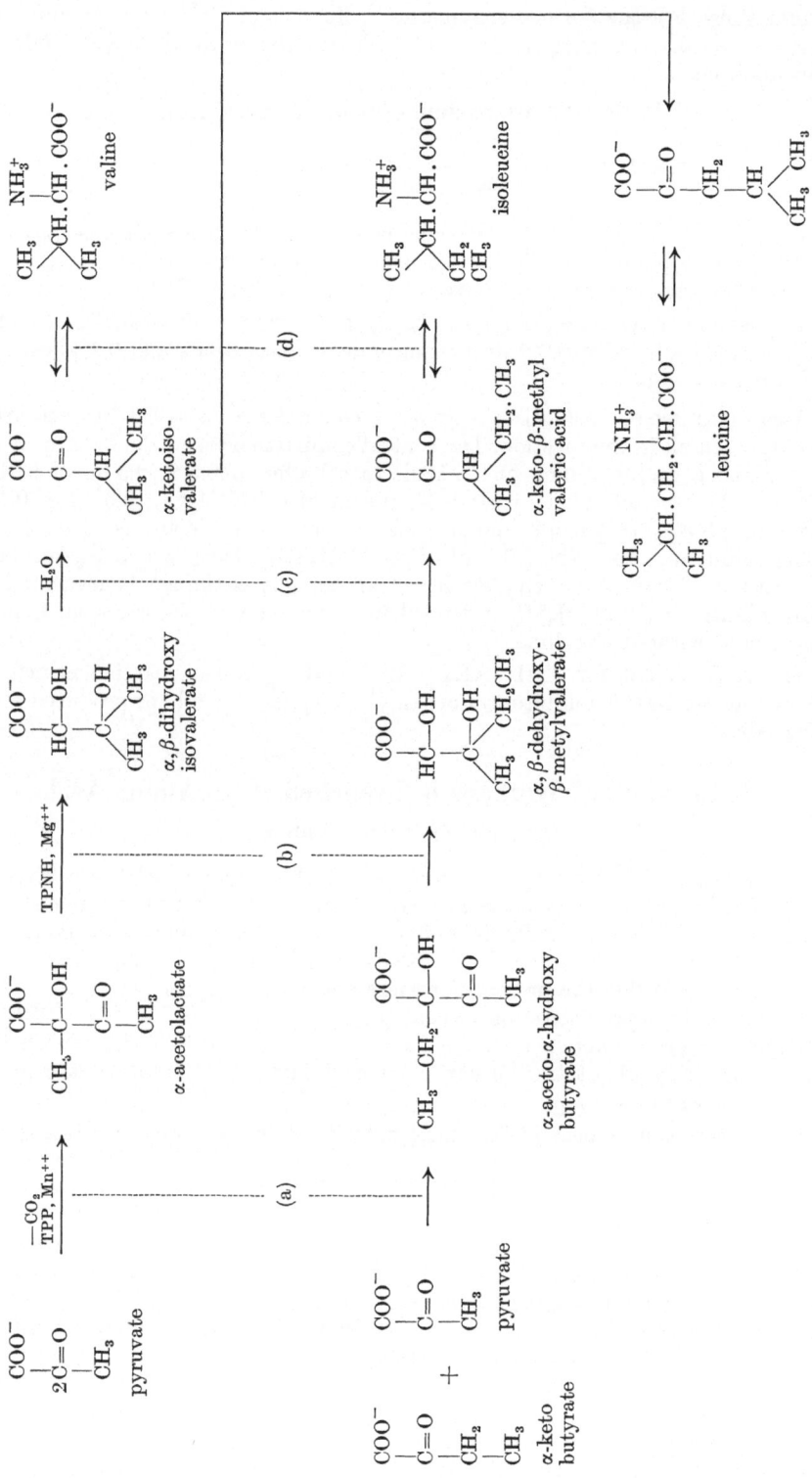

Fig. 4. Biosynthesis of branched-chain amino acids.

1. Acetolactate Synthetase.
(Acetolactate Forming Enzyme).

(i)   2 pyruvate → α-acetolactate + CO_2

(ii)   pyruvate + α-ketobutyrate → α-aceto-α-hydroxybutyrate + CO_2

The enzyme occurs in *Aerobacter aerogenes* (JUNI, 1952), *E. coli* (UMBARGER, BROWN and EYRING, 1957; RADHAKRISHNAN and SNELL, 1960), yeast (LEWIS and WEINHOUSE, 1958) and *Neurospora crassa* (RADHAKRISHNAN and SNELL, 1960).

In *A. aerogenes* (HALPERN and UMBARGER, 1959) and *E. coli* (RADHAKRISHNAN and SNELL, 1960) two acetolactate forming enzymes are known, one active at pH 8.0 and another at pH 6.0. The pH 8.0 enzyme is noncompetitively inhibited by valine and it seems to be concerned with valine biosynthesis (HALPERN and UM-BARGER, 1959). The molds possess only one enzyme active at pH 6.5.

Enzyme purification. The enzyme may be prepared from an isoleucineless-valineless mutant of *N. crassa* (No. 16117) according to the method of RADHA-KRISHNAN and SNELL (1960) by grinding mycelial pads under liquid nitrogen, and homogenization in a TENBROECK apparatus with twice the amount of 0.1 M phosphate buffer, pH 7.0. After centrifugation and separation of the supernatant solution from fatty material by means of a syringe, the enzyme is used as such or purified further. The crude enzyme preparations also possess acetolactate de-carboxylase activity. To get rid of the latter, 8.0 ml of 0.05 M $AlCl_3$ is added to 100 ml of a crude water extract of the mycelium. The pH drops to 5.5. The heavy precipitate formed is recovered and suspended in 35 ml of 0.1 M Tris buffer, pH 8.5, and solid $(NH_4)_2SO_4$ is added to 0.5 saturation. The precipitate is suspend-ed in 25 ml of 0.1 M phosphate buffer, pH 7.5, centrifuged to remove insoluble matter, and again fractionated with $(NH_4)_2SO_4$. The precipitate obtained between 0.2 to 0.3 saturation is dissolved in buffer and reprecipitated 2 or 3 times with $(NH_4)_2SO_4$ at 0.3 saturation. The precipitate is taken up each time in 20 ml of 0.1 M phosphate buffer, pH 7.0. The final solution is free from acetolactate de-carboxylase activity. *E. coli* does not possess a decarboxylase and crude prepara-tions obtained by sonic oscillation can be directly used.

Assay (RADHAKRISHNAN and SNELL, 1960). The assay mixture in a final volume of 1.0 ml consists of, 0.5 μmoles of sodium pyruvate; 0.5 μmole of Mn^{++}; 20 μg of thiamine pyrophosphate; 100 μmoles of potassium phosphate, pH 6.5, and enzyme solution. The mixture is incubated at 37° C for 1 hour. If crude enzyme solution is used, the reaction is stopped by the addition of 0.1 ml of 2.5 N NaOH and acetoin is determined in a suitable aliquot. To another aliquot is added H_2SO_4 (end concentration, 0.35 N) and the mixture is heated at 60° C for 15 minutes to decarboxylate acetolactate, cooled, neutralized, and the acetoin formed is determined as follows (WESTERFELD, 1945):

A sample containing 2—15 μg acetoin is pipetted in a tube and enough water is added to a volume of 5 ml. One ml of 0.5% aqueous creatine is then added followed by 1 ml of freshly prepared α-naphthol (1 mg α-naphthol in 20 ml of 2.5 N NaOH). The mixture is shaken and after standing for 1 hour at room temperature, the color is read at 530 mμ. Crystalline polymerized acetoin (N. V. Nederlandsche Gist- en Spiritusfabriek, Delft, Holland; NEISH, 1952) is used as a standard.

The amount of acetolactate formed by the enzyme is calculated by difference in the reading before and after the addition of acid. With purified preparations of *Neurospora*, or *E. coli* extracts lacking acetolactate decarboxylase, the samples after incubation are acidified with TCA, heated, and acetoin is determined directly.

The enzyme from *Neurospora* forms acetolactate from pyruvate ($K_m = 4.6 \times 10^{-2}$ M), and α-aceto-α-hydroxybutyrate from pyruvate and α-ketobutyrate. The metal requirement of the enzyme is satisfied with Mn^{++} or Mg^{++}, and co-enzyme requirement by thiamine pyrophosphate (half maximal activity with 3×10^{-6} M TPP). The enzyme activity is inhibited by 10^{-4} M $HgCl_2$ or $AgNO_3$. pCMBA is ineffective.

2. α-hydroxy-β-keto Acid Reductoisomerase.

(i) α-acetolactate $\xrightarrow{\text{TPNH, Mg}^{++}}$ α,β-dihydroxyisovalerate

(ii) α-aceto-α-hydroxybutyrate $\longrightarrow$ α,β-dihydroxy-β-methylvalerate

The enzyme has been reported to occur in *E. coli*, *Neurospora crassa* (RADHA-KRISHNAN, WAGNER and SNELL, 1960), and yeast (STRASSMAN, SHATTON and WEINHOUSE, 1960; WATANABE, HAYAISHI and SHIMURA, 1959).

The enzyme may be prepared 70-fold purified from *E. coli* according to the procedure of RADHAKRISHNAN et al. (1960).

Assay (RADHAKRISHNAN et al., 1960). The enzyme activity is assayed by the decrease in absorbancy at 340 mμ in a mixture containing, 20 μmoles of α-aceto-α-hydroxybutyrate; 0.1 μmole of TPNH; 2.5 μmoles of $MgSO_4$; 5 μmoles of β-mercaptoethanol; 0.05—0.1 ml of a boiled extract of *N. crassa* or *E. coli* (depending upon the source of the enzyme); Tris buffer, pH 7.5 (0.1 M final concentration), and enzyme in a total volume of 1.0 ml. Decrease in optical density is measured after the addition of the enzyme.

One *unit* of the enzyme is defined as the amount producing a decrease in O. D. of 0.1 in 5 minutes.

Properties. TPNH cannot be replaced by DPNH as a coenzyme. The enzyme also shows an obligatory requirement for a metal ion. Mg^{++} ions are the most potent activators but are inhibitory at concentrations higher than 2.5×10^{-3} M. β-mercaptoethanol and boiled crude extract increase the enzyme activity. The nature of the stimulation is not known. The pH optimum of the enzyme from both *E. coli* and *N. crassa* is 7.5. The activity of the enzyme is much higher with acetohydroxybutyrate than with acetolactate as substrate, although the affinity for the two substrates is similar.

3. α,β-dihydroxy Acid Dehydrase.

(i) α,β-dihydroxyisovalerate $\xrightarrow{-H_2O}$ α-ketoisovalerate

(ii) α,β-dihydroxy-β-methyl valerate $\xrightarrow{-H_2O}$ α-keto-β-methylvalerate

This enzyme has been shown to occur in *Neurospora crassa*, *E. coli* (MYERS and ADELBERG, 1954), *Saccharomyces cereviseae*, *Troulopsis utilis*, *Serratia marcesens*, *Bacillus cadaveris*, and *Micrococcus lysodeikticus* (WIXOM and WIKMAN, 1960). The enzyme is absent from *Lactobacillus casei* and rat liver (WIXOM and WIKMAN, 1960).

Enzyme preparation. A crude enzyme can be obtained from yeast by the procedure of WIXOM, SHATTON and STRASSMAN (1960). A crude extract from fresh baker's yeast is made by disruption of the cells in 0.02 M phosphate buffer, pH 7.2 with glass beads in a MICKLE shaker or centrifuge shaker of SHOCKMAN, KOLB and TOENNIES (1957). After centrifugation at $80,000 \times g$, the supernatant solution is dialyzed against distilled water for 18—20 hours in the cold. The protein separating in the dialysis bag is centrifuged, redissolved in 0.02 M phosphate buffer, pH 7.2, and stored in a frozen state where it is stable for months.

Assay (WIXOM et al., 1960). The incubation mixture (1.0 ml) consists of, 0.05 M Tris buffer, pH 7.4; 0.02 M $MgCl_2$; 0.02 M DL-α,β-dihydroxyisovalerate, and enzyme in centrifuge tubes. The mixture is incubated for 30 minutes in air with shaking in a 37° water bath. The reaction is terminated and deproteinized with tungstic acid, and the keto acid concentration of an aliquot of the supernatant solution is estimated by the method of FRIEDEMANN and HAUGEN (1943).

Properties. The optimum pH for the dehydrase action is 7.4 in Tris buffer. The enzyme has a divalent metal ion requirement which is satisfied by Mg^{++} or Mn^{++} ions. Cobaltous ion has weaker activity. Fe^{++}, Zn^{++}, Cu^{++} and Hg^{++} ions are inhibitory. EDTA or citrate inhibit the enzymic reaction perhaps by chelating the metal. DL-α,β-dihydroxy-β-methyl-n-valeric acid also reacts with the enzyme to form the corresponding keto acid but at an appreciably lower rate. By analogy with other enzymes of the biosynthesis of branched chain amino acids, perhaps the same enzyme is responsible for the formation of the keto acid analogue of valine and isoleucine. The K_m for α,β-dihydroxyisovalerate is about 1.7×10^{-3} M.

4. Branched-Chain Amino Acid Transaminase.

The last step in the synthesis of the branched chain amino acids is a transaminase which is active with isoleucine, leucine and valine. It has been discussed in the following chapter on "Enzymes of Transamination and Racemization".

VIII. Degradation of the Branched Chain Amino Acids.

The degradation of the branched chain amino acids proceeds through their respective keto-acids by a irreversible oxidative decarboxylation to the corresponding acyl-coenzyme A derivatives. Most of these reactions have been studied in animals by COON and his associates (1954, 1956) and the details may be seen in BACHHAWAT, WOESSNER, and COON (1956), BACHHAWAT, ROBINSON and COON (1954), and COON (1955a, b). WEBB (1958) has suggested that the degradation of valine proceeds via α-ketoisovaleric acid to isobutyryl-CoA and active formate in *Aerobacter aerogenes* and SASAKI (1961) has obtained an enzyme from *Proteus vulgaris* catalyzing the decarboxylation of α-ketoisocaproic acid.

IX. Formation of Succinic Acid from Glutamic Acid.

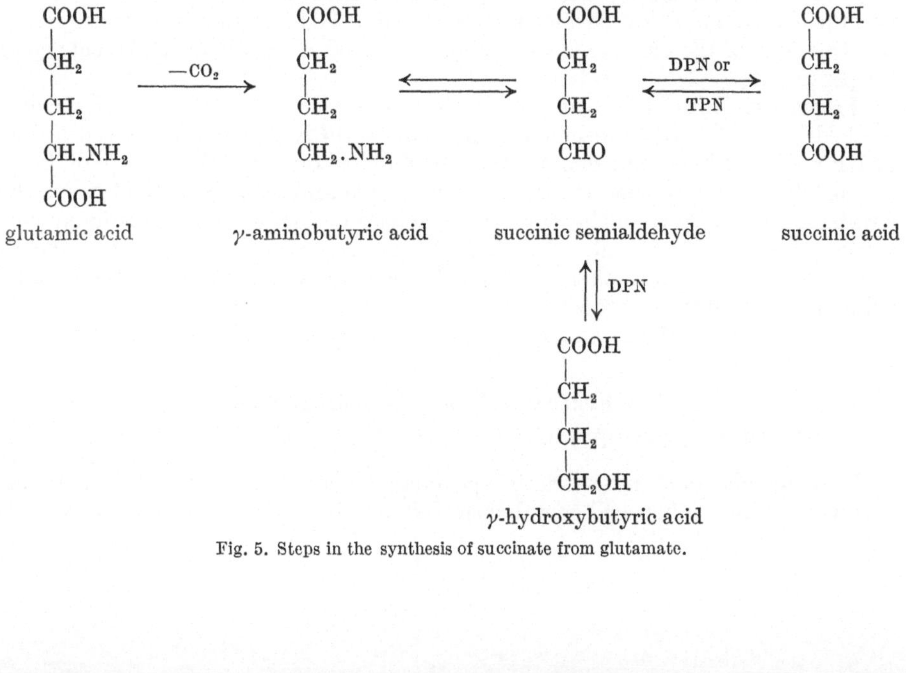

Fig. 5. Steps in the synthesis of succinate from glutamate.

The enzymes involved in the conversion of glutamic acid to succinic acid have been shown to occur in the brain tissue of the animals (Bessman, Rossen and Layne, 1953; Albers and Salvadore, 1958; Salvadore and Albers, 1959), many bacteria, like *Pseudomonas* and *Clostridium aminobutyricum* (Nirenberg and Jakoby, 1960), and in fungi. In higher plants glutamic decarboxylase is very commonly found but there is no enzymic evidence for the further metabolism of the γ-aminobutyric acid formed. Webster and Varner (1955) demonstrated that in extracts of wheat germ, C¹⁴-labeled γ-aminobutyric acid is incorporated into the various Krebs cycle acids, but in the presence of malonate the label appears entirely in succinic acid. This shows that in higher plants glutamic acid is converted into succinic acid by the same series of reactions occuring in animals and bacteria (Fig. 5).

Glutamic acid decarboxylase, catalyzing the decarboxylation of glutamic acid to γ-aminobutyric acid has been discussed earlier. γ-aminobutyric acid transaminase which forms succinic semialdehyde as one of the products is discussed elsewhere in this book (see, Transaminations).

1. Succinic Semialdehyde Dehydrogenases.

Succinic semialdehyde + TPN (or DPN) ⇌ succinic acid + TPNH (or DPNH)

In *Pseudomonas*, where the reaction has been extensively studied, there are two separate and distinct enzymes, one specific for DPN and another for TPN, catalyzing the formation of succinic acid. Both enzymes have been purified from *Pseudomonas aeruginosa* by Nakamura (1960) and from a soil *Pseudomonas* by Nirenberg and Jakoby (1960). The enzymes may be separated and purified by classical protein fractionation technique by the method of Nakamura (1960). A DPN-specific dehydrogenase from monkey brain has been purified about 150-fold (Albers and Salvadore, 1958).

Assay (Nirenberg and Jakoby, 1960). Both TPN- and DPN-specific enzymes may be assayed by measuring the reduction of the coenzyme at 340 mμ. The assay mixture in quartz cuvettes contains per ml, 6 μmoles β-mercaptoethanol; 0.5 μmole succinic semialdehyde, 100 μmoles Tris, pH 8.1, for the DPN-specific enzyme (100 μmoles Tris, pH 8.8 for the TPN-specific enzyme), and 1 μmole of DPN (or TPN). A suitably diluted enzyme solution is added to start the reaction and the change in optical density is determined at 30 seconds interval for 3 minutes. One unit of the enzyme is the amount required to obtain an O. D. change of 0.1 per minute.

Properties. Both dehydrogenases are specific for the oxidation of succinic semialdehyde. The DPN-specific enzyme reduces TPN at a rate one-twelfth that of DPN; the TPN-specific enzyme reduces DPN at one-twelfth the rate of TPN. The pH optima for enzyme activity are 8.0 and 9.0 for the DPN- and TPN-specific enzymes, respectively. In the absence of mercaptoethanol large losses in enzyme activity occur at 3° C and −15° C. TPN-specific enzyme is more sensitive to the presence of mercaptoethanol. The DPN-specific enzyme is labile at 55° C and a sulfhydryl inhibitor as potassium arsenite at a concentration of 2×10^{-3} M results in 25% inhibition of the enzyme compared to a 75% inhibition of the TPN-specific enzyme.

2. γ-hydroxybutyrate Dehydrogenase.

$$CH_2(OH)CH_2CH_2COOH + DPN^+ \rightleftharpoons CHO.CH_2.CH_2.COOH + DPNH + H^+$$

This enzyme does not, strictly speaking, belong to the group of enzymes involved in the amino acid metabolism but has been discussed here because it leads to the formation of succinic semialdehyde and thence, *via* a transaminase reaction, to γ-aminobutyric acid.

The enzyme may be purified from γ-hydroxybutyrate grown *Pseudomonas* (JAKOBY, 1958) by the method of NIRENBERG and JAKOBY (1960).

Assay (NIRENBERG and JAKOBY, 1960). The enzyme activity is measured by following the formation of reduced DPN at 340 mμ in the presence of excess of DPN-specific succinic semialdehyde dehydrogenase. The reaction mixture in 1 ml consists of, 100 μmoles of Tris, pH 8.8; 6 μmoles mercaptoethanol; 1 μmole of DPN; 40 μmoles of γ-hydroxybutyric acid; DPN-succinic semialdehyde dehydrogenase to produce an absorbancy change of at least 0.3 per minute, and an aliquot of γ-hydroxybutyrate dehydrogenase. The substrate is added last of all to initiate the reaction, and the change in O. D. is determined every 30 seconds for 3 minutes. One unit of the enzyme is the amount causing a change in absorbancy of 0.2 per minute.

γ-hydroxybutyrate as a substrate is absolutely specific. The activity with TPN as the coenzyme is one-fiftieth of that obtained with DPN. The enzyme is insensitive to -SH inhibitors; only at very high concentration of pCMBA is the enzyme inhibited to a measurable extent.

X. Enzymes of Synthesis of Glycine and Serine.

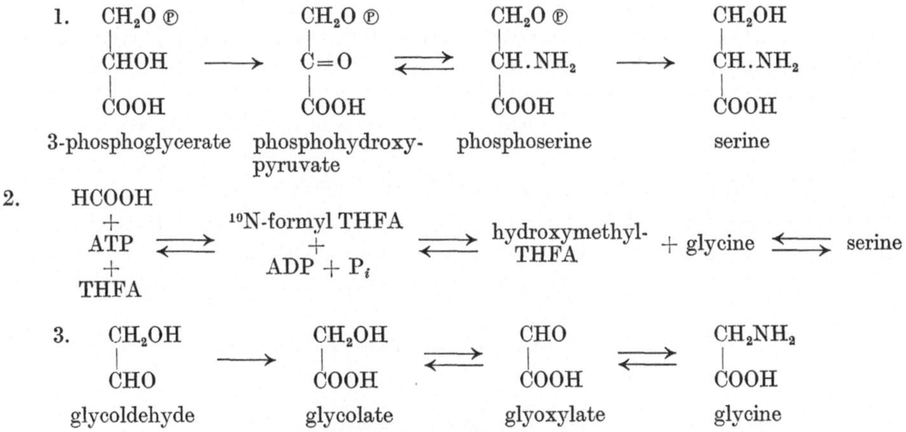

1.
$$
\begin{array}{ccccc}
CH_2O\,\textcircled{P} & CH_2O\,\textcircled{P} & CH_2O\,\textcircled{P} & CH_2OH \\
| & | & | & | \\
CHOH & \longrightarrow \quad C=O & \rightleftharpoons \quad CH.NH_2 & \longrightarrow \quad CH.NH_2 \\
| & | & | & | \\
COOH & COOH & COOH & COOH
\end{array}
$$

3-phosphoglycerate   phosphohydroxy-   phosphoserine   serine
pyruvate

2.
$$
\begin{array}{c}
HCOOH \\
+ \\
ATP \\
+ \\
THFA
\end{array}
\rightleftharpoons
\begin{array}{c}
^{10}N\text{-formyl THFA} \\
+ \\
ADP + P_i
\end{array}
\rightleftharpoons
\begin{array}{c}
\text{hydroxymethyl-} \\
\text{THFA}
\end{array}
+ \text{glycine} \rightleftharpoons \text{serine}
$$

3.
$$
\begin{array}{cccc}
CH_2OH & CH_2OH & CHO & CH_2NH_2 \\
| & | & | & | \\
CHO & \longrightarrow \quad COOH & \rightleftharpoons \quad COOH & \rightleftharpoons \quad COOH
\end{array}
$$

glycoldehyde   glycolate   glyoxylate   glycine

Fig. 6. Possible pathways of serine synthesis in plants.

1. 3-phosphoglycerate → Serine.

The transformation of 3-phosphoglycerate to phosphoserine has been demonstrated in cell-free extracts of pea epicotyls (HANFORD and DAVIES, 1958), *E. coli* (SMITH, SHUSTER, ZIMMERMAN and GUNSALUS, 1956), and animal tissues (ICHIHARA and GREENBERG, 1957). The enzymes, have, however, not been studied in detail. 3-phosphoglycerate is converted to phosphohydroxypyruvate in spinach and pea seedlings as follows (see DAVIES, 1959).

3-phosphoglycerate + DPN$^+$ ⇌ phosphohydroxypyruvate + DPNH + H$^+$

DPN is required as a coenzyme. The resulting keto acid can transaminate with glutamic acid in plants, and glutamic acid and glutamine in *E. coli*, by means of a transaminase (HANFORD and DAVIES, 1958).

Phosphohydroxypyruvate + glutamate ⇌ phosphoserine + α-ketoglutarate

Pyridoxal phosphate is probably required as a cofactor. Phosphoserine is converted to serine by means of phosphoserine phosphatase.

$$\text{phosphoserine} + H_2O \rightarrow \text{serine} + P_i$$

The phosphatase has been demonstrated to occur in animal tissues (Schramm, 1958; Neuhaus and Byrne, 1958), and in *E. coli* (Smith et al., 1956). The phosphoserine phosphatase of bacteria is strongly stimulated by Zn^{++}. No report is yet available regarding the presence of this enzyme in higher plants.

Serine can be formed directly in liver, kidney, and in *Neurospora* by transamination between hydroxypyruvate and L-alanine (Sallach, 1956). This transaminase has been discussed elsewhere in this book (page 376).

2. Formate → Serine.

The series of reactions leading from formate to serine (Fig. 6) have been shown to occur in bacteria, animal tissues (Jaenicke, 1955; Rabinowitz and Pricer, 1956; Huennekens and Osborn, 1959) and some in higher plants (Wilkinson and Davies, 1958, 1960). The origin of formate in higher plants is not very clear. Tolbert and Cohan (1954) fed 1-^{14} C or 2-^{14}C-glycolate to wheat and barley and obtained labeled glycine and serine. The 2-C-atom of glycolate was found to enter the 3-C-atom of serine. Presumably, the glyoxylate formed from glycolate by means of an oxidase was split to formate, as has been demonstrated by Sakami (1949) in animal tissues, and the formate entered the 3-C-atom of serine (Tolbert, 1955).

The interconversion of formate to serine involves the participiation of folic acid coenzymes, largely in the form of tetrahydrofolic acid (see, chapter on Vitamin metabolism in this book). The first step in the conversion of formate to serine is the formation of N^{10}-formyl tetrahydrofolic acid (THFA) by an enzymic formyolation of THFA:

$$\text{formate} + ATP + THFA \rightleftharpoons N^{10} \text{ formyl THFA} + ADP + P_i$$

The enzyme involved in this reaction has been shown to occur in tissues other than plants (Greenberg, Jaenicke and Silverman, 1955).

The second step is catalyzed by an enzyme which is found in plants and discussed below.

a) Hydroxymethyltetrahydrofolate Dehydrogenase.

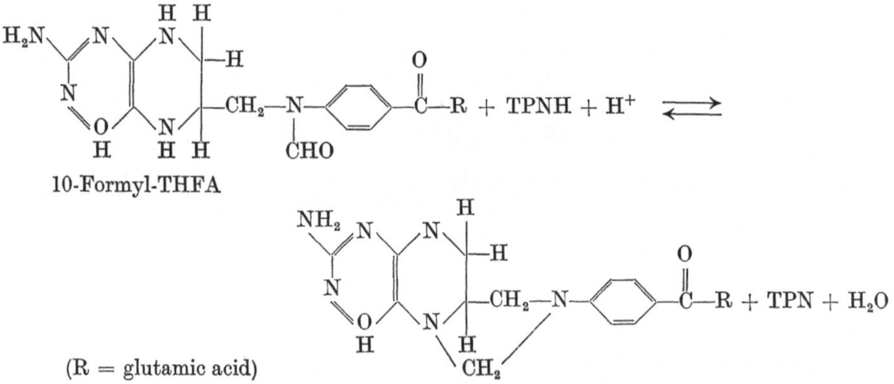

10-Formyl-THFA

(R = glutamic acid)

10-hydroxymethyl-THFA

The enzyme has been shown to occur in liver (Osborn and Talbert, 1957; Huennekens and Osborn, 1959) and in turnips (Wilkinson and Davies, 1960).

Enzyme preparation (Wilkinson and Davies, 1960). The enzyme is prepared by homogenizing 100 gm turnip slices in 2—3 volumes of cold 0.2 M potassium phosphate buffer, pH 7.4 in a WARING blendor. The homogenate is strained through linen, centrifuged, and the supernatant solution is brought to 0.8 saturation with $(NH_4)_2SO_4$. The protein is collected by centrifugation after 15 minutes, taken up in a minimum volume of phosphate buffer, and dialyzed against a running flow of phosphate buffer (0.01 M, pH 7.4). The contents of the dialysis bag are clarified by high speed centrifugation and the supernatant solution is used as the source of the enzyme.

Assay (WILKINSON and DAVIES, 1960). The enzyme activity can be assayed both by following the reduction of TPN at 340 mμ or the formation of formyl THFA. Active formaldehyde is prepared by mixing an excess of formaldehyde with THFA. The assay mixture consists of, in a total volume of 5.0 ml, 4 μmoles of formaldehyde; 4 μmoles of DL-tetrahydrofolic acid; 2 mg of TPN; 2 μmoles of $MgSO_4$; 0.1 μmoles of pyridoxal phosphate; 10 μmoles of glutathione; 0.05 M phosphate buffer, pH 7.4, and 2.0 ml of the enzyme extract. Anaerobic conditions are maintained by layering the mixture with petroleum ether. After 60 minutes incubation, 0.5 ml aliquot is removed and added to 0.5 ml of 10% aqueous TCA to precipitate protein and convert N^{10} formyl-THFA to anhydroleucovorin (N^5, N^{10}-methenyl THFA). After centrifugation, 0.5 ml of the supernatant is added to 2.5 ml of water in a quartz cuvette, and the O. D. is measured at 350 mμ against a zero time control. A material absorbing light at 330 mμ in the reaction mixture interferes in the assay. The rate of formation of formyl THFA is linearly related to the concentration of the enzyme provided that the rate of formyl THFA production does not exceed 0.1 μmole per ml per hour.

The requirements of the enzymic reaction are not definitely known except that DPN cannot replace TPN. It is also not known whether the immediate product of the dehydrogenase reaction is methenyltetrahydrofolic acid. OSBORN and HUENNEKENS (1957) have shown that these two compounds are interconvertible by the enzyme cyclohydrolase.

b) Serine Aldolase (Serine Transhydroxymethylase).

The last step in the synthesis of serine is the reversible reaction:

$$\text{hydroxymethyl THFA} + \text{glycine} \rightleftharpoons \text{serine} + \text{THFA}$$

The enzyme, serine aldolase, has been shown to occur in liver (BLAKLEY, 1957; HUENNEKENS, HATEFI and KAY, 1957; ALEXANDER and GREENBERG, 1956), bacteria (WRIGHT, 1955a, b), and in turnips (WILKINSON and DAVIES, 1958). A partially purified enzyme may be obtained from animal tissues by the method of BLAKLEY (1957). The enzyme has not been purified to any large extent from other sources. An enzyme preparation from turnips can also be obtained according to WILKINSON and DAVIES (1960) as described under hydroxymethyltetrahydrofolate dehydrogenase.

Assay (BLAKLEY, 1955). The enzyme activity is assayed by the manometric estimation of serine. In the main compartment of a WARBURG vessel is added the enzyme extract (to give 10—20 μmoles serine/hour under the conditions used here), 0.01 M $NaHCO_3$, 5×10^{-4} M pyridoxal phosphate, 1.3 mM DL-THFA, 1.0 μmole Mg^{++}, and water to bring the final volume to 3.0 ml. The mixture is equilibrated with $N_2 + CO_2$ (95:5) at 37° C, and glycine and formaldehyde are tipped in from the side arm to give a final concentration of 0.1 and 0.01 M respectively. After one hour, the flasks are disconnected and a glass sphere is placed in the neck of each flask. The flasks are placed in a boiling water bath for 2 minutes to inactivate the enzyme. After cooling, 0.3 ml of 0.5 M sodium periodate (pH 6.0) is placed in the side arm, the flasks replaced on the manometers, flushed with pure CO_2, and equilibrated at 37° C. On tipping the periodate in the main compartment, serine is oxidized with the evolution of 1 mole CO_2/mole of serine. The reaction is completed after 20 minutes. Blanks are used for each set of measurements by carrying out determinations in the absence of THFA and formaldehyde separately.

The serine formed in the reaction can also be estimated by the colorimetric method of FRISELL, MEECH and MACKENZIE (1954; see, vol. VI) provided that the substrate formaldehyde present is first removed by heating at 70—80° C for 39 minutes in evaporating dishes (see, ALEXANDER and GREENBERG, 1956).

The characteristics of the enzyme from plant sources are not known.

3. Glycoldehyde → Glycine.

The enzymes involved in the reactions leading from glycoldehyde to glycine (Fig. 6) have been studied in higher plants. A pyridine nucleotide linked glycoldehyde dehydrogenase has been purified from pea mitochondria by DAVIES (1960) and is discussed elsewhere in this book. Glycolate formed in this reaction is converted to glyoxylic acid by glycolate oxidase (ZELITCH and OCHOA, 1953), an enzyme universally occuring in plants (discussed elsewhere). Glyoxylic acid undergoes transamination with glutamic acid (see, the chapter on Transamination and Racemization, page 380) to form glycine which may lead to the formation of serine by means of serine aldolase.

XI. Enzymes of Synthesis of Tryptophan.

The presumable enzymic steps in the biosynthesis of tryptophan are given in Fig. 7. The subject has recently been reviewed by DOY (1960) and YANOFSKY

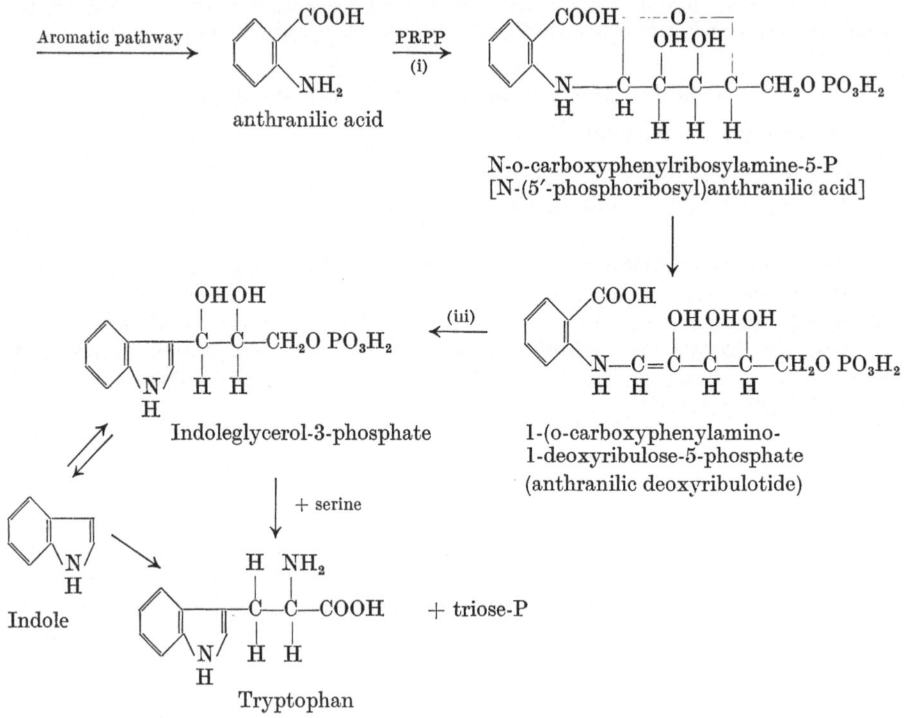

Fig. 7. Steps in the biosynthesis of tryptophan in fungi and bacteria.

(1960). The reader is referred to these reviews for the details of study of mutants of *Neurospora* and *E. coli* which have been examined, and the study of which led to the isolation and characterization of some of the enzymes involved in tryptophan synthesis. Detailed discussions of the enzymes before anthranilic acid will be found in the chapter on "Enzymes of Aromatic" Biosynthesis elsewhere in this book.

The enzymes involved in the steps between anthranilic acid and anthranilic deoxyribulotide (see, Fig. 7) have not been completely characterized and we have, therefore, refrained from discussing them in detail here. According to the evidence put forward by Doy (1961) and Doy, Rivera and Srinivasan (1961), the immediate product of the reaction between anthranilic acid and 5-phosphoribosyl-1-pyrophosphate is N-(5'-phosphoribosyl) anthranilic acid. The latter is labile and is either hydrolyzed to anthranilic acid or by an Amadori-type of rearrangement (Hodge, 1955) gives rise, probably non-enzymatically, to anthranilic deoxyribulotide (Doy, 1961; Smith and Yanofsky, 1960).

The second step in the reaction is the one leading to the formation of indole-3-glycerolphosphate from anthranilic deoxyribulotide and is discussed below.

1. Indole-3-Glycerol Phosphate Synthetase.

The enzyme has been partially purified from *E. coli*, strain 7—8 (Yanofsky, 1957) by Gibson and Yanofsky (1960). It has also been studied in *Bacillus subtilis* (Anagnostopoulos and Crawford, 1961) and would no doubt be shown to occur in *Neurospora* and higher plants.

Enzyme purification. For high yields of the enzyme, the auxotrophic mutants of *E. coli* are grown with a limited supply of tryptophan in the medium of Vogel and Bonner (1956 b) with 0.16% glucose and 0.005% acid-hydrolyzed casein. The enzyme is purified by the procedure of Gibson and Yanofsky (1960). The cells are disrupted by sonic oscillation in 0.1 M Tris buffer, pH 7.8. To the clarified extract is added one-half volume of 20% streptomycin sulfate and the precipitate is removed by centrifugation. The supernatant is fractionated by acetone and the fraction between 40—50% is dissolved in 0.1 M Tris, pH 7.8. The enzyme is further purified by $(NH_4)_2SO_4$. The fraction separating between 0.45—0.55 saturation is collected and used as the enzyme source (17-fold purified). Further purification can be done by chromatography on DEAE-cellulose (Gibson and Yanofsky, 1960).

Assay (Gibson and Yanofsky, 1960). The incubation mixture (0.5 ml) contains, anthranilic deoxyribulotide (synthesized according to Smith and Yanofsky, 1960) 0.22 μmoles; Tris buffer, pH 8.8, 25 μmoles; and enzyme solution. After incubation for 20 minutes at 37° C, the reaction is terminated by the addition of 0.2 ml of 1 M acetate buffer, pH 5.0, followed by 0.5 ml of 0.2 M aqueous sodium metaperiodate to convert the indoleglycerolphosphate quantitatively to indole-3-aldehyde (Yanofsky, 1956).

The mixture is allowed to stand at room temperature for 20 minutes, after which 0.4 ml of 1 N NaOH and 5 ml of ethyl acetate are added. The mixture is shaken 10—20 times to extract the indole-3-aldehyde, the tubes are centrifuged at low speed for a few minutes, and the O. D. of the ethyl acetate layer is measured against an ethyl acetate blank at 290 mμ. The molar absorbancy index of indole-3-aldehyde at 290 mμ is 11,400.

In crude extracts possessing tryptophan synthetase activity (see below), an interference in the assay is caused by the conversion of indoleglycerol phosphate to tryptophan. This can be prevented by the addition of 10^{-4} M (end concentration) hydroxylamine in the assay mixture.

One *unit* of the enzyme is the amount required to form 0.1 μmole of indoleglycerol phosphate in 20 minutes at 27° C.

Properties. The optimum pH of the enzyme reaction is 8.8 in Tris buffer. Anthranilic acid analogues (3-methyl, 4-methyl, 5-methyl, 5-fluoro) at a concentration of 10^{-3} M inhibit the reaction to about 30—40%. The following compounds are also inhibitory: pCMBA (10^{-5} M; 77%), EDTA (5×10^{-2} M; 32%), hydroxylamine (10^{-3} M, 17%; 10^{-4} M, 6%). The K_m for anthranilic deoxyribulotide is 1.75×10^{-5} M.

The enzyme is quite stable if stored at —15° C.

2. Tryptophan Synthetase.

(i)   indole + serine $\rightarrow$ tryptophan
(ii)   indoleglycerol-3-phosphate + serine $\rightarrow$ tryptophan + triose phosphate
(iii)   indoleglycerol-3-phosphate $\rightleftharpoons$ indole + triose phosphate

There is convincing evidence that only one enzyme, tryptophan synthetase, catalyzes all three reactions in *E. coli* (Yanofsky, 1959) and in *Neurospora crassa* (DeMoss and Bonner, 1959). The tryptophan synthetase of *E. coli* can be separated into two protein components, A and B, both of which are required for full enzymic activity (see, Yanofsky, 1960). The enzyme from *N. crassa*, however, cannot be resolved into two components (Mohler and Suskind, 1960).

Distribution. The enzyme is widely distributed in microorganisms, having been found in *Aerobacter aerogenes* (Monod and Cohen-Bazire, 1953), *E. coli*, *B. subtilis* (Yanofsky, 1953; Anagnostopoulos and Crawford, 1961), *Neurospora* (Yanofsky, 1955), *Claviceps purpurea* (Tyler and Schwarting, 1953), and *Glomerella cingulata* (Yanofsky, 1955). Among higher plants, it has been demonstrated in cell-free extracts of pea seedlings (Greenberg and Galston, 1959) and no doubt occurs in many other genera.

Stability of the enzyme. The failure to demonstrate tryptophan synthetase in many higher plants is perhaps related to the lability of the enzyme. Mohler and Suskind (1960) demonstrated that the enzyme can be stabilized in crude cell-free extracts of many strains of *Neurospora* by adding pyridoxal phosphate (4×10^{-4}M), DL-serine (0.02 M) and EDTA (0.01 M). With these additions, over 90% of the enzyme activity is retained at 3° C for about a week. With partially purified extracts under the same conditions, enzyme activity is retained for about 9 days.

Enzyme preparation. A highly purified (ca. 300-fold) enzyme preparation may be obtained from certain strains of *Neurospora crassa* according to the method of Mohler and Suskind (1960).

Assay (Yanofsky, 1955; Mohler and Suskind, 1960). The enzyme activity is measured by the disappearance of indole or the appearance of tryptophan. With crude enzyme preparations from higher plants, the formation of tryptophan in the assay system must be verified (method in Nason, Kaplan and Colowick, 1951) because cell-free extracts from many plants are able to utilize indole without the production of tryptophan.

The assay system consists of, 0.08 ml of 5×10^{-3} M indole solution; 0.02 ml of 0.05 M glutathione; 0.4 ml of 0.2 M DL-serine; 0.1 ml of a solution containing 10 μg pyridoxal phosphate; 0.1 ml of 0.05 M EDTA; 0.12 ml of 0.5 M phosphate buffer, pH 7.8, water and enzyme to give a final volume of 1.0 ml. Control tubes are kept, one without the enzyme, and one without serine. The tubes are incubated at 37° C for the desired length of time (40—60 minutes). The reaction is terminated by adding 0.2 ml of 5% NaOH. Indole is extracted by adding 4 ml of toluene in each tube, and shaking vigorously. The two layers may be separated, if need be, by centrifugation. An aliquot of the toluene layer (1 ml) is pipetted into a separate tube and 4 ml of ethanol, followed by 2 ml of colour reagent (made by dissolving 36 g of p-dimethylaminobenzaldehyde in 500 ml of ethanol, acidifying with 180 ml of concentrated HCl, and making up the volume to 1 liter with ethanol) is added. The colour is allowed to develop for 30—60 minutes and the colour intensity is read at 570 mμ. The colour formed with the reagent begins to fade at zero time (25% colour decay in 0.25 hours; see, SCOTT, 1960, and the time for reading of the colours must be strictly standardized. If greater sensitivity is desired, the indole is determined with p-dimethylaminocinnamic aldehyde reagent.

One *unit* of the enzyme is defined as the amount of enzyme which will convert 0.1 μmole of indole to tryptophan during a 60 minute incubation period at 37° C.

Properties. The pH optimum of the *Neurospora* enzyme is 7.8 but the enzyme from higher plants may show an optimum of 8.0—8.5 (GREENBERG and GALSTON, 1959). The enzyme has an absolute requirement for L-serine, but indole may be substituted by indoleglycerol-3-phosphate or to a lesser extent by 6-methylindole and 7-hydroxyindole (YANOFSKY, 1955). Substrate concentrations giving half-maximum velocity are 3.4×10^{-3} M for serine, and 5.6×10^{-5} M for indole. Pyridoxal phosphate stimulates enzyme activity.

Tryptophan synthetase is inhibited by cysteine, L-tryptophan, hydroxylamine, cyanide, and traces of Co^{++}, Zn^{++}, Cu^{++}.

References.

ABDERHALDEN, E.: Biochemisches Handlexikon 4, Berlin: Springer 1911. — ABELSON, P. H.: J. biol. Chem. **206**, 335 (1954). — ADAMS, E.: J. biol. Chem. **209**, 829 (1954); **217**, 325 (1955). — ADELSTEIN, S. J., and B. L. VALLEE: J. biol. Chem. **233**, 589 (1958). — ADLER, E., N. B. DAS, H. VON EULER and U. HEYMAN: C. R. trav. Lab. Carlsberg **22**, 15 (1938). — ADLER, E., V. HELLSTRÖM, G. GÜNTHER u. H. VON EULER: Z. physiol. Chem. **255**, 14 (1938). — ALBERS, R. W., and R. A. SALVADORE: Science **128**, 359 (1958). — ALEXANDER, G. J., and E. SCHWENK: J. Amer. chem. Soc. **79**, 4554 (1957). — ALEXANDER, N., and D. M. GREENBERG: J. biol. Chem. **220**, 775 (1956). — AMES, B. N.: In, Amino acid metabolism. Ed. W. MCELROY and B. GLASS. Baltimore: Johns Hopkins Press 1955; — Fed. Proc. **15**, 210 (1956); — J. biol. Chem. **228**, 131 (1957a); **226**, 583 (1957b). — AMES, B. N., and B. GARRY: Proc. nat. Acad. Sci. (Wash.) **45**, 1453 (1959). — AMES, B. N., B. GARRY and L. A. HERZENBERG: J. gen. Microbiol. **22**, 369 (1960). — AMES, B. N., and B. L. HORECKER: J. biol. Chem. **220**, 113 (1956). — ANAGNOSTOPOULOS, C., and I. P. CRAWFORD: Proc. nat. Acad. Sci. (Wash.) **47**, 378 (1961). — ARCHIBALD, R. M.: J. biol. Chem. **151**, 141 (1943); **156**, 121 (1944); **157**, 507 (1945). — AXELROD, J., and R. TOMCHICK: J. biol. Chem. **233**, 702 (1958).

BACH, S. J., and J. D. KILLIP: Biochim. biophys. Acta **29**, 273 (1958); **47**, 336 (1961). — BACHHAWAT, B. K., W. G. ROBINSON and M. J. COON: J. Amer. chem. Soc. **76**, 3098 (1954). — BACHHAWAT, B. K., J. F. WOESSNER jr., and M. J. COON: Fed. Proc. **15**, 214 (1956). — BARBAN, S.: J. Bact. **68**, 493 (1954). — BEEVERS, H.: Biochem. J. **48**, 132 (1951). — BELLAMY, W. D., and I. C. GUNSALUS: J. Bact. **50**, 95 (1945). — BENDALL, D. S., and C. DE DUVE: Biochem. J. **74**, 444 (1960). — BENDER, A. E., and H. A. KREBS: Biochem. J. **46**, 210 (1950). — BERGER, J., and G. S. AVERY, jr.: Amer. J. Bot. **30**, 290 (1943); **31**, 11 (1944). — BESSMAN, S. P., J. ROSSEN and E. C. LAYNE: J. biol. Chem. **201**, 385 (1953). — BINKLEY, F.: Meth. Enzymol. **2**, 311 (1955). — BLACK, S., E. M. HARTE, B. HUDSON and L. WARTROFSKY: J. biol. Chem. **253**, 2910 (1960). — BLACK, S., and N. G. WRIGHT: In, Amino acid metabolism. Ed. W. MCELROY and B. GLASS. Baltimore: Johns Hopkins Press 1955. — J. biol. Chem. **213**, 27

(1955b); **213**, 39 (1955c); **213**, 51 (1955d). — Blakley, R. L.: Biochem. J. **61**, 315 (1955); **65**, 342 (1957). — Bone, D. H.: Plant Physiol. **34**, 171 (1959a); — Nature (Lond.) **184**, 990 (1959b); — Plant Physiol. **34**, 171 (1959c). — Bone, D. H., and L. Fowden: J. exp. Bot. **11**, 104 (1960). — Bonner, J., and S. G. Wildman: Arch. Biochem. **10**, 497 (1946). — Boyd, W. L., and H. C. Lichstein: J. Bact. **69**, 545 (1955). — Boyer, P. D., and H. L. Segal: In: The mechanism of enzyme action. Ed. W. D. McElroy and B. Glass, Baltimore: Johns Hopkins Press 1954. — Brady, L. R., and V. E. Tyler: Plant Physiol. **33**, 334 (1958). — Brown, G. W., and P. P. Cohen: J. biol. Chem. **234**, 1769 (1959). — Bulen, W. A.: Arch. Biochem. Biophys. **62**, 173 (1956). — Burnet, G. H., and P. P. Cohen: J. biol. Chem. **229**, 337 (1957). — Burton, K.: Biochem. J **50**, 258 (1951).

Cantoni, G. L.: J. biol. Chem. **189**, 203 (1951). — Cantoni, G. L., and D. G. Anderson: J. biol. Chem. **222**, 171 (1956). — Cantoni, G. L., and J. Durell: J. biol. Chem. **225**, 1033 (1957). — Cantoni, G. L., and E. Scarano, E.: J. Amer. chem. Soc. **76**, 4744 (1954). — Cantoni, G. L., and P. J. Vignos: J biol. Chem. **209**, 647 (1954). — Caravaca, J.. and S. Grisolia: J. biol. Chem. **235**, 684 (1960). — Challenger, F., R. Bywood, P. Thomas and B. Hayward: Arch. Biochem. Biophys. **69**, 514 (1957). — Chen, P. S., T. Y. Toribara and H. Warner: Anal. Chem. **28**, 1786 (1956). — Cheng, Yu-Yen, P. Linko and M. Milner: Plant Physiol. **35**, 68 (1960). — Chinard, F. P.: J. biol. Chem. **199**, 91 (1952). — Clark, A. J., and P. J. G. Mann: Biochem. J. **65**, 763 (1957). — Clarke, P. H.: J. gen. Microbiol. **18**, vi (1958). — Cocking, E. C., and E. W. Yemm: Biochem. J. **58**, xii (1954). — Cohen, G. N., B. Nisman, M. L. Hirsch and S. B. Wiesendanger: C. R. Acad. Sci. (Paris) **238**, 1746 (1954). — Coleman, R. G., and M. P. Hegarty: Nature (Lond.) **179**, 376 (1957). — Coon, M. J.: Fed. Proc. **14**, 762 (1955a); — In, Amino acid metabolism. Ed. W. McElroy and B. Glass. Baltimore: Johns Hopkins Press 1955b. — Cromwell, B. T.: Biochem. J. **37**, 722 (1943).

Damodaran, M., and K. R. Nair: Biochem. J. **32**, 1064 (1938). — Damodaran, M., and S. S. Subramanian: Proc. Indian Acad. Sci. **27** B, 47 (1948). — Davies, D. D.: J. exp. Bot. **7**, 203 (1956); — Biol. Rev. **34**, 407 (1959); — J. exp. Bot. **11**, 289 (1960). — Davis, V. E., and J. Awapara: J. biol. Chem. **235**, 124 (1960). — Davison, A. N.: Biochim. biophys. Acta **19**, 66 (1956); — Physiol. Rev. **38**, 729 (1958). — Davison, D. C., and W. H. Elliott: Nature (Lond.) **169**, 313 (1952). — Delwiche, E. A.: J. Bact. **62**, 717 (1951). — DeMoss, J. A., and D. M. Bonner: Proc. nat. Acad. Sci. (Wash.) **45**, 1405 (1959). — Dewey, D. L., D. S. Hoare and E. Work: Biochem. J. **58**, 523 (1954). — Doy, C. H.: Rev. Pure Appl. Chem. (Aust.) **10**, No. 3 (1960); — Nature (Lond.) **189**, 461 (1961). — Doy, C. H., A. Rivera and P. R. Srinivasan: Biochem. biophys. Res. Comm. **4**, 83 (1961). — Dubnoff, J. W., and H. Borsook: J. biol. Chem. **176**, 789 (1948).

Eggleston, L. V.: Biochem. J. **65**, 735 (1957). — Ekladius, E., H. K. King and C. R. Sutton: J. gen. Microbiol. **17**, 602 (1957). — Ellfolk, N.: Ann. Acad. Sci. fenn. Ser. A II, 74 (1956). — Emerson, R. L., M. Puziss and S. G. Knight: Arch. Biochem. **25**, 299 (1950). — Euler, H. von, E. Adler, G. Günther and L. Elliot: Enzymologia **6**, 337 (1939). — Ericson, L. E.: Acta Chem. Scand. **12**, 1541 (1958). — Ericson, L. E., J. N. Williams jr., and C. A. Elvehjem: J. biol. Chem. **212**, 537 (1955).

Fincham, J. R. S.: Biochem. J. **53**, 313 (1953). — Fincham, J. R. S., and J. B. Boylen: Biochem. J. **61**, xxiii, xxiv (1955). — Finlayson, A. J., and W. B. McConnell: Biochim. biophys. Acta **45**, 623 (1960). — Fiske, C. H., and Y. Subbarow: J. biol. Chem. **66**, 375 (1925). — Flavin, M.: J. biol. Chem. **210**, 771 (1954). — Flavin, M., and T. Kono: J. biol. Chem. **235**, 1109 (1960). — Flavin, M., and C. Slaughter: Anal. Chem. **31**, 1983 (1959); — J. biol. Chem. **235**, 1103 (1960). — Fowden, L.: J. exp. Bot. **5**, 28 (1954). — Friedemann, T. C., and G. E. Haugen: J. biol. Chem. **147**, 415 (1943). — Frisell, W. P., L. A. Meech and C. G. Mackenzie: J. biol. Chem. **207**, 709 (1954). — Fromageot, C.: In, The enzymes. Ed. J. B. Sumner and K. Myrbäck. Vol. 1. New York: Academic Press 1951). — Fromageot, C., and P. Desnuelle: C. R. Acad. Sci. **214**, 647 (1942). — Fromm, H. J., and R. C. Nordlie: Arch. Biochem. Biophys. **81**, 363 (1959). — Fujiwara, M., M. Yoshimura and S. Tsuno: J. Biochem. (Japan) **42**, 591 (1955).

Gale, E. F.: Biochem. J. **36**, 64 (1942); — Advanc. Enzymol. **6**, 1 (1946); — Meth. Biochem. Anal. **4**, 285 (1957). — Gale, E. F., and H. M. R. Epps: Biochem. J. **38**, 232 (1944). — Gibson, F., and C. Yanofsky: Biochim. biophys. Acta **43**, 489 (1960). — Glasizou, K. T.: Austr. J. biol. Sci. **9**, 253 (1956). — Gmelin, R.: In, Handbuch der anorganischen Chemie, VIII ed, 348. Berlin: Verlag Chemie, G.m.b.H. 1936. — Gmelin, R., G. Hasenmaier and G. Strauss: Z. Naturforsch. **12**b, 684 (1957). — Goldman, D. S.: Biochim. biophys. Acta **34**, 527 (1959). — Gorjatschenkowa, E. V.: Dokl. Akad. Nauk S.S.S.R. **87**, 457 (1952); — Biochemistry (USSR) **21**, 249 (1956). — Greenberg, D. M., A. E. Bagot and A. Roholt: Arch. Biochem. Biophys. **62**, 446 (1956). — Greenberg, J. B., and A. W. Galston: Plant Physiol. **34**, 489 (1959). — Greenberg, G. R., L. Jaenicke and M. Silverman: Biochim. biophys. Acta **17**, 589 (1955).

HAEHN, H., and H. LEOPOLD: Biochem. Z. **292**, 380 (1937). — HALL, D. A.: Biochem. J **51**, 499 (1952). — HALPERN, Y. S., and H. E. UMBARGER: J. biol. Chem. **234**, 3067 (1959). — HANFORD, J., and D. D. DAVIES: Nature (Lond.) **182**, 532 (1958). — HANSEN, S. E., A. KJAER and S. SCHWIMMER: C. R. trav. Lab. Carlsberg, Sér. Chim. **31**, 193 (1959). — HAUGHTON, B. G., and H. K. KING: Biochem. J. **69**, 48P (1958). — HOARE, D. S.: Biochim. biophys. Acta **19**, 141 (1956). — HOARE, D. S., and E. WORK: Biochem. J. **61**, 562 (1955). — HODGE, J. E.: Adv. Carbohydrate Chem. **10**, 169 (1955). — HOLMSTEDT, B., L. LARSSON and R. THAM: Biochim. biophys. Acta **48**, 182 (1961). — HOLMSTEDT, B., and R. THAM: Acta Physiol. Scand. **45**, 152 (1959). — HOLZER, H., and S. SCHNEIDER: Biochem. Z. **329**, 361 (1957). — HONG, M. M., S. C. SHEN and A. E. BRAUNSTEIN: Biochim. biophys. Acta **36**, 288 (1959). — HOOD, S. L.: Bot. Gaz. **116**, 86 (1954). — HOROWITZ, N. H.: J. biol. Chem. **154**, 141 (1944). — HOROWITZ, N. H., and G. W. BEADLE: J. biol. Chem. **150**, 325 (1943). — HUENNEKENS, F. M., Y. HATEFI and L. D. KAY: J. biol. Chem. **224**, 435 (1957). — HUENNEKENS, F. M., and M. J. OSBORN: Adv. Enzymol. **21**, 369 (1959).

ICHIHARA, A., and D. M. GREENBERG: J. biol. Chem. **224**, 331 (1957).

JACOBI, G.: Planta **49**, 561 (1957). — JAENICKE, L.: Biochim. biophys. Acta **17**, 588 (1955). — JAKOBY, W. B.: J. biol. Chem. **232**, 75 (1958). — JOHNSTONE, R. M., and J. H. QUASTEL: Meth. Enzymol. **2**, 217 (1955). — JONES, M. E., L. SPECTOR and F. LIPMANN: J. Amer. Chem. Soc. **77**, 819 (1955). — JORDAN, D. C.: J. Microbiol. **5**, 132 (1959). — JOSHI, J. G., and P. HANDLER: J. biol. Chem. **235**, 2981 (1960). — JUNI, E.: J. biol. Chem. **195**, 715 (1952).

KALLIO, R. E.: J. biol. Chem. **192**, 371 (1951). — KALLIO, R. E., and A. D. LARSON: In: Symp. amino acid metabolism. Ed. W. McELROY and B. GLASS. Baltimore: Johns Hopkins Press 1955. — KAPELLER-ADLER, R.: Biochim. biophys. Acta **22**, 391 (1956). — KASTING, R., and C. C. DELWICHE: Plant Physiol. **33**, 350 (1958). — KATZ, J., I. LIEBERMAN and H. A. BARKER: J. biol. Chem. **200**, 417 (1953). — KIRSHNER, N., and M. GOODALL: Biochim. biophys. Acta. **24**, 658 (1957). — KLINKHAMMER, F.: Arch. Mikrobiol. **33**, 357 (1959). — KNIGHT, S. G.: J. Bact. **55**, 407 (1948). — KNIVETT, V. A.: Biochem. J. **50**, xxx (1951). — KNOX, W. E., and E. J. BEHRMAN: Ann. Rev. Biochem. **28**, 245 (1959). — KORZENOVSKY, M.: In: Symp. amino acid metabolism. Ed. W. McELROY and B. GLASS. Baltimore: Johns Hopkins Press 1955. — KREBS, H. A.: In: The enzymes. Ed. J. B. SUMNER and K. MYRBACK. New York: Academic Press 1951. — KREBS, H. A., and L. V. EGGLESTON: Ann. Acad. Sci. fenn., Ser. A II, **60**, 496 (1955). — KRETOVICH, W. L., and T. V. DROZDOVA: Dokl. Akad. Nauk SSSR **63**, 167 (1948). — KRISHNASWAMY, P. R., and K. V. GIRI: Biochem. J. **62**, 301 (1956). — KUBO, H., T. YAMANO, M. IWATSUBO, H. WATARI, T. SOYAMA, J. SHIRAISHI, S. SAWADA, N. KAWASHIMA, S. MITANI and K. ITO: Bull. soc. chim. biol. **40**, 43 (1958). — KULKARNI, L., and K. SOHONIE: Nature (Lond.) **178**, 925 (1956).

LEANZA, W. J., and K. PFISTER: J. biol. Chem. **201**, 377 (1953). — LEETE, E., and L. MARION: Canad. J. Chem. **31**, 126 (1953). — LEVENBERG, B.: Fed. Proc. **20**, 1 (1961). — LEWIS, K. F., and S. WEINHOUSE: J. Amer. chem. Soc. **80**, 4913 (1958). — LINKO, P.: Acta Chem. Scand. **12**, 101 (1958). — LIPMANN, F., and L. C. TUTTLE: J. biol. Chem. **159**, 21 (1945).

MAHLER, H. R., N. K. SARKAR, L. P. VERNON and R. A. ALBERTY: J. biol. Chem. **199**, 585 (1952). — MANN, P. J. G.: Biochem. J. **59**, 609 (1955); **76**, 44P (1960). — MARSHALL, M., R. L. METZENBERG and P. P. COHEN: J. biol. Chem. **233**, 102 (1958). — MASS, W. K., G. D. NOVELLI and F. LIPMANN: Proc. nat. Acad. Sci. (Wash.) **39**, 1004 (1953). — MASSEY, V., G. PALMER and R. BENNETT: Biochim. biophys. Acta **48**, 1 (1961). — MATSUO, Y., and D. M. GREENBERG: J. biol. Chem. **230**, 545 (1958). — MAW, G. A.: Biochem. J. **72**, 602 (1959). — MAW, G. A., and V. DU VIGNEAUD: J. biol. Chem. **174**, 381 (1948). — MAZELIS, M.: Biochem. biophys. Res. Comm. **1**, 59 (1959). — McCARTHY, T. E., and M. X. SULLIVAN: J. biol. Chem. **141**, 871 (1941). — MEADOW, P., and E. WORK: Biochim. biophys. Acta **29**, 180 (1958). — MEISTER, A.: J. biol. Chem. **206**, 587 (1954). — MEISTER, A., A. N. RADHAKRISHNAN and S. D. BUCKLEY: J. biol. Chem. **229**, 789 (1957). — METAXAS, M. A., and E. A. DELWICHE: J. Bact. **70**, 735 (1955). — METZENBERG, R. L., M. MARSHALL and P. P. COHEN: J. biol. Chem. **233**, 1560 (1958). — METZLER, D. E., and E. E. SNELL: J. biol. Chem. **198**, 363 (1952). — MIETTINEN, J. K., and A. I. VIRTANEN: Physiol. Plant. **5**, 540 (1952). — MITOMA, C., and S. UDENFRIEND: Biochim. biophys. Acta **37**, 356 (1960). — MIZUSHIMA, S., K. IZAKI, H. TAKAHASHI and K. SAKAGUCHI: Bull. agr. chem. Soc. (Japan) **20**, 36 (1956). — MOHLER, W. C., and S. R. SUSKIND: Biochim. biophys. Acta **43**, 288 (1960). — MOKRASH, L. C., J. CARAVACA and S. GRISOLIA: Biochim. biophys. Acta **37**, 442 (1960). — MONOD, J., and G. COHEN-BAZIRE: C. R. Acad. Sci. (Paris) **236**, 530 (1953). — MOYED, H. S., and B. MAGASANIK: J. biol. Chem. **235**, 149 (1960). — MUDD, S. H.: J. biol. Chem. **234**, 87 (1959); — Biochim. biophys. Acta **37**, 164 (1960a); **38**, 354 (1960b); — Nature (Lond.) **189**, 489 (1961). — MUDD, S. H., and G. L. CANTONI: J. biol. Chem. **231**, 481 (1958). — MYERS, J. W., and E. A. ADELBERG: Proc. nat. Acad. Sci. (Wash.) **40**, 493 (1954).

Najjar, V. A.: Meth. Enzymol. 2, 185 (1955). — Nakamura, K.: Biochim. biophys. Acta 45, 554 (1960). — Nason, A., N. O. Kaplan and S. P. Colowick: J. biol. Chem. 188, 397 (1951). — Neidig, B. A., and W. C. Hess: Anal. Chem. 24, 1627 (1952). — Neish, A. C.: National Research Council of Canada, Report No. 46-8-3 (1952). — Neuhaus, F. C., and W. L. Byrne: Biochim. biophys. Acta 28, 223 (1958). — Nicholas, D. J. D., and G. L. Mabey: J. gen. Microbiol. 22, 184 (1960). — Nirenberg, M. W., and W. B. Jakoby: J. biol. Chem. 235, 954 (1960). — Nisman, N.: Bact. Rev. 18, 16 (1954). — Nisman, B., G. N. Cohen, S. B. Wiesendanger and M. L. Hirsch: C. R. Acad. Sci. (Paris) 238, 1342 (1954).

Ochoa, S., A. H. Mehler and A. Kornberg: J. biol. Chem. 174, 979 (1948). — O'Connor, R. J., and H. Halvorson: Arch. Biochem. Biophys. 91, 290 (1960); — Biochim. biophys. Acta 48, 47 (1961). — Oginsky, E. L.: Meth. Enzymol. 2, 374 (1955). — Oginsky, E. L., and R. F. Gehrig: J. biol. Chem. 198, 799 (1952). — Ohigasi, K., A. Tsunetoshi and K. Ichihara: Med. J. Osaka Univ. (Japan) 2, 111 (1951). — Olson, J. A., and C. B. Anfinsen: J. biol. Chem. 202, 841 (1953). — Osborn, M. J., and F. M. Huennekens: Biochim. biophys. Acta 26, 646 (1957). — Osborn, M. J., and P. T. Talbert: Fed. Proc. 16, 230 (1957).

Pardee, A. B., and L. S. Prestidge: J. Bact. 70, 667 (1955). — Parks, L. W.: J. biol. Chem. 232, 169 (1958). — Petrack, B., and S. Ratner: J. biol. Chem. 233, 1494 (1958). — Petrack, B., L. Sullivan and S. Ratner: Arch. Biochem. Biophys. 69, 186 (1957). — Pollak, J. K., and D. Fairbairn: Canad. J. Biochem. Physiol. 33, 307 (1955). — Powell, J. F., and R. E. Strange: Biochem. J. 65, 700 (1957).

Quastel, J. H., and R. Witty: Nature (Lond.) 167, 556 (1951).

Rabinowitz, J. L., and W. E. Pricer: J. Amer. chem. Soc. 78, 4176 (1956). — Radhakrishnan, A. N., and A. Meister: J. biol. Chem. 233, 444 (1958). — Radhakrishnan, A. N., and E. E. Snell: J. biol. Chem. 235, 2316 (1960). — Radhakrishnan, A. N., R. P. Wagner and E. E. Snell: J. biol. Chem. 235, 2322 (1960). — Ratner, S.: Advanc. Enzymol 15, 319 (1954); — Meth. Enzymol. 2, 356 (1955). — Ratner, S., W. P. Anslow and B. Petrack: J. biol. Chem. 204, 115 (1953). — Ratner, S., and B. Petrack: J. biol. chem. 191, 693 (1951); 200, 161 (1953a); 200, 175 (1953b); — Arch. Biochem. Biophys. 65, 582 (1956). — Rautanen, N., and J. M. Tager: Ann. Acad. Sci. fenn. A II 60, 241 (1955). — Ravel, J. M., M. L. Grona, J. S. Humphreys and W. Shive: J. biol. Chem. 234, 1452 (1959). — Ravel, J. M., J. S. Humphreys and W. Shive: Arch. Biochem. Biophys. 92, 525 (1961). — Reichard, P.: Acta Chem. Scand. 11, 523 (1957). — Reifer, I., and K. Kleczowski: Z. Naturforsch. 15b, 431 (1960). — Remy, C. N.: Fed. Proc. 16, 237 (1957). — Robins, K. C., and J. Shields: Arch. Biochem. Biophys. 62, 55 (1956). — Rocca, E., and F. Ghiretti: Arch. Biochem. Biophys. 77, 336 (1958). — Roche, J., and G. Lacombe: Biochim. biophys. Acta 9, 687 (1952). — Rogers, B. J.: Plant Physiol. 30, 186 (1955). — Rosenberg, A. J., and B. Nisman: Biochim. biophys. Acta 3, 348 (1949).

Sakami, W.: J. biol. Chem. 178, 519 (1949). — Sallach, H. J.: J. biol. Chem. 223, 1101 (1956). — Salvadore, R. A., and R. W. Albers: J. biol. Chem. 234, 922 (1959). — Sanwal, B. D.: Arch. Biochem. Biophys. (in the press 1961). — Sanwal, B. D., and M. Lata: Canad. J. Microbiol. (in the press 1961). — Sanwal, B. D., and M. W. Zink: Arch. Biochem. Biophys. (in the press 1961). — Sasaki, S.: Nature (Lond.) 189, 400 (1961). — Sato, C. S., R. U. Byerrum, P. Albersheim and J. Bonner: J. biol. Chem. 233, 128 (1958). — Sayre, F. W., and D. M. Greenberg: J. biol. Chem. 220, 787 (1956). — Saz, A. K., and L. W. Brownell: Arch. Biochem. Biophys. 52, 291 (1954). — Schales, O., V. Mims and S. S. Schales: Arch. Biochem. 10, 455 (1946). — Schales, O., and S. S. Schales: Arch. Biochem. 11, 155 (1946a); 11, 455 (1946b). — Scher, W. I., and H. J. Vogel: Bact. Proc. 1955, 123; — Proc. nat. Acad. Sci. (Wash.) 43, 796 (1957). — Schramm, M.: J. biol. Chem. 233, 1169 (1958). — Schuegraf, A., S. Ratner and R. C. Warner: J. biol. Chem. 235, 3597 (1960). — Schwimmer, S., J. F. Carson, R. U. Makower, M. Mazelis and F. F. Wong: Experientia (Basel) 16, 449 (1960). — Schwimmer, S., and A. Kjaer: Biochim. biophys. Acta 42, 316 (1960). — Scott, T. A.: Biochem. J. 75, 7 (1960). — Seecof, R. L., and R. P. Wagner: J. biol. Chem. 234, 2694 (1959). — Selim, A. S. M., and D. M. Greenberg: Biochim. biophys. Acta 42, 211 (1960). — Shapiro, S. K.: J. Bact. 72, 730 (1956); — Bact. Proc. 1957, 116; — Biochim. biophys. Acta 29, 405 (1958). — Shapiro, S. K., and F. Schlenk: Advanc. Enzymol. 22, 237 (1960). — Shapiro, S. K., and A. N. Mather: J. biol. Chem. 233, 631 (1958). — Shapiro, S. K., and D. A. Yphantis: Biochim. biophys. Acta 36, 241 (1959a); — Arch. Biochem. Biophys. 82, 477 (1959b). — Sher, I. H., and M. F. Mallette: J. biol. Chem. 200, 257 (1953); — Arch. Biochem. Biophys. 53, 354 (1954a); 53, 370 (1954b). — Shimura, Y., and H. J. Vogel: Fed. Proc. 20, 10 (1961). — Shockman, G. D., J. J. Kolb and G. Toennies: Biochim. biophys. Acta 24, 203 (1957). — Shrift, A.: Amer. J. Bot. 41, 223 (1954). — Shukuya, R., and G. W. Schwert: J. biol. Chem. 235, 1649 (1960a); 235, 1658 (1960b). — Singer, T. P., and J. Pensky: Biochim. biophys. Acta 9, 316 (1952a); — J. biol. Chem. 196, 375 (1952b). — Slade, H. D.: Arch. Biochem. Biophys. 42, 204 (1953); — In: Symp. amino acid metabolism. Ed.

W. McElroy and B. Glass. Baltimore: Johns Hopkins Press 1955. — Sloane-Stanley, G. H.: Biochem. J. **44**, 373 (1949). — Smith, J. A., and E. R. Waygood: Canad. J. Biochem. Physiol. (in the press 1961). — Smith, M. E., and D. M. Greenberg: J. biol. Chem. **226**, 317 (1957). — Smith, O. H., and C. Yanofsky: J. biol. Chem. **235**, 2051 (1960). — Smith, P. F.: J. Bact. **74**, 75 (1957). — Smith, R. A., C. W. Shuster, S. Zimmerman and I. C. Gunsalus: Bact. Proc. **1956**, 107. — Smythe, C. V.: J. biol. Chem. **142**, 387 (1942); — Meth. Enzymol. **2**, 315 (1955). — Srb, A. M., and N. H. Horowitz: J. biol. Chem. **154**, 129 (1944). — Stadtman, E. R., and H. A. Barker: J. biol. Chem. **180**, 1085 (1949). — Stadtman, T. C.: Biochem. J. **62**, 614 (1956); — Biochem. Z. **331**, 46 (1958). — Stadtman, T. C., and P. Elliott: J. biol. Chem. **228**, 983 (1957). — Stadtman, T. C., P. Elliott and L. Tieman: J. biol. Chem. **231**, 961 (1958). — Stickland, L. H.: Biochem. J. **28**, 1746 (1934); **29**, 288, 889, 896 (1935). — Stoll, A., and E. Seebeck: Helv. chim. Acta **32**, 197 (1949); — Advanc. Enzymol. **11**, 377 (1951). — Strassman, M., J. B. Shatton and S. Weinshouse: J. biol. Chem. **235**, 700 (1960). — Strecker, H. J.: J. biol. Chem. **225**, 825 (1957); **235**, 3218 (1960). — Strecker, H. J., and P. Mela: Biochim. biophys. Acta **17**, 580 (1955). — Struck, J., and I. W. Sizer: Arch. Biochem. Biophys. **86**, 260 (1960). — Stumpf, P. K., and D. E. Green: Fed. Proc. **5**, 157 (1944). — Sundaram, E. R. B. S., and P. S. Sarma: J. Sci. Ind. Res. (India) **12B**, 245 (1953). — Suzuki, Y.: Naturwissenschaften **43**, 427 (1959).

Tabor, H.: Meth. Enzymol. **2**, 394 (1955). — Tabor, H., and A. H. Mehler: Meth. Enzymol. **2**, 228 (1955). — Tabor, H., and S. M. Rosenthal: J. Amer. chem. Soc. **79**, 2978 (1957). — Tabor, C. W., H. Tabor and S. M. Rosenthal: Meth. Enzymol. **2**, 390 (1955). — Taylor, E. S., and E. F. Gale: Biochem. J. **39**, 52 (1945). — Thayer, P. S., and N. M. Horowitz: J. biol. Chem. **192**, 755 (1951). — Tolbert, N. E.: J. biol. Chem. **215**, 27 (1955). — Tolbert, N. E., and M. S. Cohan: J. biol. Chem. **204**, 649 (1954). — Tomkins, G. M., K. L. Yielding and J. Curran: Proc. nat. Acad. Sci. (Wash.) **47**, 270 (1961). — Tyler, V. E., and A. E. Schwarting: Science **118**, 1953 (1953).

Umbarger, H. E., and B. Brown: J. Bact. **73**, 105 (1957); — J. biol. Chem. **233**, 1156 (1958). — Umbarger, H. E., B. Brown and E. J. Eyring: J. Amer. chem. Soc. **79**, 2980 (1957). — Umbreit, W. W., R. H. Burris and J. F. Stauffer: Manometric techniques. Minnesota: Burgess Publishing Co. 1957.

Virtanen, A. I.: Schweiz. Z. Allg. Path. Bakt. **21**, 970 (1958); — Festschrift Arthur Stoll, Sandoz AG., Basel (1957). — Virtanen, A. I., and N. Ellfolk: Meth. Enzymol. **2**, 386 (1955). — Virtanen, A. I., and P. Linko: Acta Chem. Scand. **9**, 531 (1955). — Vogel, H. J.: Proc. nat. Acad. Sci. (Wash.) **39**, 578 (1953); — In: Symp. amino acid metabolism. Ed. W. McElroy and B. Glass, Baltimore: Johns Hopkins Press 1955; — Microbial Genetics Bull. **13**, 42 (1956). — Vogel, H. J., and D. M. Bonner: J. biol. Chem. **218**, 97 (1956a); — Microbial Genetics Bull. **13**, 42 (1956b); — In: Encyclopaedia of plant physiology, vol. XI. Ed. W. Ruhland, Berlin: Springer 1959. — Vogel, H. J., and B. D. Davis: J. Amer. chem. Soc. **74**, 109 (1952).

Walker, D. J.: Biochem. J. **69**, 524 (1958). — Walker, J. B.: J. biol. Chem. **204**, 139 (1953). — Walker, J. B., and J. Myers: J. biol. Chem. **203**, 143 (1953). — Warburg, O., H. Klotzsch and G. Krippahl: Z. Naturforsch. **12b**, 266 (1957). — Watanabe, Y., S. Konishi and K. Shimura: J. Biochem. (Japan) **42**, 837 (1955); **44**, 299 (1957). — Watanabe, Y., K. Hayashi and K. Shimura: Biochim. biophys. Acta **31**, 583 (1959). — Watanabe, Y., and K. Shimura: J. Biochem. (Japan) **43**, 283 (1956). — Webb, M.: J. gen. Microbiol. **18**, 24 (1958). — Webster, G. C.: Nitrogen metabolism in plants. Illinois: Row, Peterson and Co. 1959. — Webster, G. C., and J. E. Varner: J. biol. Chem. **215**, 91 (1955). — Weinberger, P., and K. A. Clendenning: Canad. J. Bot. **30**, 755 (1952). — Weissbach, H., T. E. Smith, J. W. Daly, B. Witkop and S. Udenfriend: J. biol. Chem. **235**, 1160 (1960). — Werle, E.: Biochem. Z. **309**, 61 (1941). — Werle, E., and G. Hartung: Biochem. Z. **328**, 228 (1956). — Werle, E., and E. Pechmann: Liebigs Ann. Chem. **562**, 44 (1949). — Werle, E., and A. Raub: Biochem. Z. **318**, 538 (1948). — Werle, E., and F. Roewer: Biochem. Z. **320**, 298 (1950); **322**, 320 (1952). — Westerfeld, W. W.: J. biol. Chem. **161**, 495 (1945). — Wiame, J. M., A. Pierard and F. Ramos: Meth. Enzymol. **5** (in the press 1961). — Wickremsinghe, R. L., and B. A. Fry: Biochem. J. **58**, 268 (1954). — Wiesendanger, S., and B. Nisman: C. R. Acad. Sci. (Paris) **237**, 764 (1953). — Wilkinson, A. P., and D. D. Davies: Nature (Lond.) **181**, 1070 (1958); — J. exp. Bot. **11**, 296 (1960). — Williams, V. R., and R. T. McIntyre: J. biol. Chem. **217**, 467 (1955). — Wiltshire, G. H.: Biochem. J. **55**, 408 (1953). — Winnick, T.: J. biol. Chem. **142**, 461 (1942). — Wixom, R. L., J. B. Shatton and M. Strassman: J. biol. Chem. **235**, 128 (1960). — Wixom, R. L., and J. H. Wikman: Biochim. biophys. Acta **45**, 618 (1960). — Wolff, E. C., S. Black and P. F. Downey: J. Amer. chem. Soc. **78**, 5958 (1956). — Wood, W. A., and I. C. Gunsalus: J. biol. Chem. **181**, 171 (1949). — Work, E.: Biochim. biophys. Acta **17**, 410 (1955). — Wormser, E. H., and A. B. Pardee: Arch. Biochem.

360   B. D. SANWAL and MADHU LATA: Enzymes of Deamination, Decarboxylation.

Biophys. **78**, 416 (1958). — WRIGHT, B. E.: Biochim. biophys. Acta **16**, 165 (1955a); — J. Amer. chem. Soc. **77**, 3930 (1955b).

YANOFSKY, C.: J. biol. Chem. **198**, 343 (1852); — J. Bact. **68**, 577 (1953); — Meth. Enzymol. **2**, 233 (1955); — J. biol. chem. **223**, 171 (1956); **224**, 783 (1957); — Biochim. biophys. Acta **31**, 408 (1959); — Bact. Rev. **24**, 221 (1960). — YANOFSKY, C., and J. L. REISSIG: J. biol. Chem. **202**, 567 (1953). — YURA, T., and H. J. VOGEL: Biochim. biophys. Acta **17**, 582 (1955); — J. biol. Chem. **234**, 335 (1959).

ZELITCH, I., and S. OCHOA: J. biol. Chem. **210**, 707 (1953). — ZELLER, E. A.: Helv. chim. Acta **23**, 1509 (1940); — In: The enzymes. Vol. II, part 1. Ed. J. B. SUMNER and K.MYRBÄCK. New York: Academic Press 1951. — ZELLER, E. A., and H. BIRKHÄUSER: Schweiz. med. Wschr. **70**, 975 (1940). — ZELLER, E. A., C. A. OWEN and A. G. KARLSON: J. biol. Chem. **188**, 623 (1951).

Part 2.

Transaminases and Racemases.

By

B. D. Sanwal, M. W. Zink and George Din.

With 1 Figure.

A. Transaminases.

Since the review on transaminases by COHEN (1951), a large body of information has accumulated on transamination reactions in various organisms. Detailed account of the various transaminases has been given in two reviews by MEISTER (1955, 1957). The methods used in the study of general transaminases have been reviewed by ASPEN and MEISTER (1958), those used in the study of bacterial transaminases by GUNSALUS and STAMER (1955), and those used in the study of animal transaminases by COHEN (1955).

There is considerable amount of confusion with regard to the specificity and occurrence of the many transamination reactions described in the literature, and in some cases it is indeed doubtful if separate enzymes are involved for each set of four reactants involved in the process. In crude cell-free extracts of *Escherichia coli* (RUDMAN and MEISTER, 1953), *Fusarium* (SANWAL, 1958), *Neurospora* (FINCHAM and BOULTER, 1956) and plants (WILSON, KING and BURRIS, 1954) and indeed in many other organisms, large numbers of amino acids undergo transamination with α-ketoglutarate. Whether this activity is due to one, two or more enzymes is not definitely established. RUDMAN and MEISTER (1953) could separate by gel absorption three different fractions from *E. coli*, two of which accounted for all of the α-ketoglutarate-amino acid activities of the crude extracts. FINCHAM and BOULTER (1956) on the basis of nutritional experiments, and SEECOF and WAGNER (1959a, b) on the basis of enzyme purification showed that in *Neurospora* there were at least three (?) groups of enzymes which could transaminate with α-ketoglutaric acid. A clarification of this point will have to await further research. In the following review, we have made no attempt to separate the transaminases into definite groups, rather the different enzymes reported in the literature have been described as such. We are sure that as more knowledge on this subject accumulates many enzymes described by various authors as distinct will have to be re-examined in a new light.

Another difficulty inherent in the description of the transaminases is their *nomenclature*. Since most of the transaminases are group specific and reversible, and involve four or in most of the cases more than four reactants, they can be

---

Abbreviations: AMP, adenosine-5-phosphate; ATP, adenosine triphosphate; BAL, di-mercaptopropanol; CoA, coenzyme A; DAP, α,ε-diaminopimelic acid; DEAE-cellulose, diethylaminoethyl cellulose; DPN and DPNH, oxidized and reduced diphosphopyridine nucleotide; EDTA, ethylenediaminetetraacetic acid; FAD, flavin adenine dinucleotide; FMN, flavin mononucleotide; pCMBA, para-chloromercuribenzoate; TCA, trichloracetic acid; TPN and TPNH, oxidized and reduced triphosphopyridine nucleotide; TRIS, tris (-hydroxy-methyl) aminomethane.

named in a variety of ways. We have, however, tried to follow MEISTER (1957) wherever possible, and described transaminases after the two amino acid pairs involved in the reaction.

As with most other enzymes, there is very little knowledge available with regard to the plant transaminases. Wherever available we have included details of purification and properties of plant enzymes; those known only in animal tissues have been given a brief treatment, and sources of literature are indicated where detailed descriptions of the animal enzymes are to be found.

If details of purification and properties of plant enzymes are not found in appropriate places, it may be assumed that very little or nothing is known about them to warrant discussion.

I. D-amino Acid Transaminases.

Distribution. THORNE, GOMEZ and HOUSEWRIGHT (1955) reported that sonic extracts of *Bacillus subtilis* catalyzed a series of transamination reactions involving D-amino acids. In crude extracts, D-aspartic acid, D-alanine, D-methionine, and D-serine transaminated with α-ketoglutarate to form D-glutamate (THORNE, 1955). D-alanine was similarly formed from pyruvate and D-aspartate or D-glutamate. D-amino acid transamination has now been extended to *Bacillus anthracis* (THORNE and MOLNAR, 1955), *E. coli*, *Bacillus cereus*, *B. sphaericus*, *Sarcina lutea*, *Staphylococcus saprophyticus* (MEADOW and WORK, 1958), and *Rhodospirillum rubrum* (HUG and WERKMAN, 1957). In *R. rubrum*, D-glutamate-aspartate and D-glutamate-alanine transaminases were reported to occur in cell-free extracts. MEADOW and WORK (1958) further reported that transamination reactions occur between D-lysine, DD-isomer of diaminopimelic acid and pyruvate, oxalacetate and α-ketoglutarate. In higher plants STUMPF (1951) observed that D-alanine acts as an active amino donor in the presence of lima bean, lupine, pumpkin, pea seedlings, pumpkin leaf extracts, and unheated wheat germ extracts.

All of the enzymic assays in bacteria and plants have been reported with crude cell-free extracts and it is not known whether one or several enzymes are involved in D-amino acid transamination.

Assay (THORNE et al., 1955). The reaction mixture contains in a final volume of 1 ml, enzyme preparation, 10 to 20 μg of pyridoxal phosphate, and 100 μmoles of appropriate substrates in 0.05 M phosphate buffer, pH 8.0. The mixture is incubated at 37° C for the required length of time and the reaction is stopped by heating in a boiling water bath. Since non-enzymic transamination can occur in the presence of trace amounts of metal ions and pyridoxal phosphate (LONGENECKER and SNELL, 1956), a heat denatured enzyme control is always included with the experimental tubes. The precipitated protein is removed by centrifugation and the supernatant solution is used for analysis. Glutamate, aspartate and alanine are determined by conventional paper chromatographic separation, elution, and reaction with ninhydrin reagent (see, ASPEN and MEISTER, 1958). D-alanine is determined with D-amino acid oxidase (commercially available, WOOD and GUNSALUS, 1951), L-glutamic acid with L-glutamate decarboxylase (method described elsewhere in this book) and D-glutamic acid by the difference between the total and L-glutamic acid.

Enzyme preparation. The enzyme may be prepared from *Bacillus anthracis* according to the method of THORNE and MOLNAR (1955). An equal volume of saturated ammonium sulphate is added to the crude extract (obtained by sonic oscillation). The precipitate is discarded and the supernatant solution is saturated with solid ammonium sulphate. The sedimented protein is dissolved in 0.05 M phosphate buffer, pH 8.0, and after dialysis the precipitate forming between 0.6 to 0.8 salt saturation is dissolved in buffer. Before use, the enzyme preparation is thoroughly dialyzed.

II. L-amino Acid Transaminases.

1. Glutamate-Aspartate Transaminase.

L-glutamate + oxalacetate $\rightleftharpoons$ α-ketoglutarate + L-aspartate

Occurrence. The enzyme is widely distributed but has been purified so far only from pig heart, wheat germ, cauliflower, and yeast. In microorganisms it has been reported to occur in halophilic *Pseudomonas* (SHIIO, MARUO and AKABORI, 1956), Reiter treponeme (BARBAN, 1956), *Neurospora crassa* (FINCHAM and BOULTER, 1956), *Bacillus anthracis* (HOUSEWRIGHT and THORNE, 1950), *Rickettsia prowazeki* (BOVARNICK and MILLER, 1950), *Rickettsia mooseri* (BOVARNICK and MILLER, 1950; WISSEMAN, HAHN, JACKSON, BOZEMAN and SMADEL, 1952; HOPPS, HAHN, WISSEMAN, JACKSON and SMADEL, 1956), suspensions of *Escherichia coli*, *Streptococcus*, *Staphylococcus*, *Bacillus*, *Proteus*, *Clostridium*, *Shigella* and *Azotobacter vinelandii* (LICHSTEIN and COHEN, 1945). ALTENBERN and HOUSEWRIGHT (1953) separated the enzyme from a protamine treated, 30—40% $(NH_4)_2SO_4$ fraction of *Brucella abortus*. A reaction between α-ketoglutaric acid and aspartic acid to yield glutamic acid was reported in *Leishmania donovani* (CHATTERJEE, 1957), four strains of *Rhizobia* (JORDAN, 1953), *Rhodospirillum rubrum* (HUG and WERKMAN, 1957) and *Streptococcus faecalis* (LICHSTEIN, GUNSALUS and UMBREIT, 1945). The effect of inhibitors on glutamate-aspartate transaminase of *Mycobacterium tuberculosis* (SAKAI, 1954; YONEDA, KATO and OKAJIMA, 1952), and *E. coli* (HICKS, 1961 a, b) have been reported. Among the microorganisms, the enzyme has been studied in a reasonably purified state only from beer yeast (HOLZER, GERLACH, JACOBI and GNOTH, 1958). The yeast enzyme can be completely resolved with regard to the coenzyme and can be used to assay pyridoxal phosphate quantitatively in complex biological mixtures (HOLZER et al., 1958).

From animal sources, the enzyme has been purified and studied from pig heart (TANEN-BAUM, 1956; JENKINS and SIZER, 1957; MEISTER, SOBER and PETERSON, 1954; GREEN, LELOIR and NOCITO, 1945; JENKINS, YPHANTIS and SIZER, 1959; KUN, FRANSHIER and GRASSETTI, 1960; LIS, 1958; CAMMARATA and COHEN, 1951; O'KANE and GUNSALUS, 1947; SCHLENK and FISHER, 1947), various organs of rats (AWAPARA and SEALE, 1952), man, pig and dog (FLEISHER, POTTER and WAKIM, 1960), HeLa cells, and mouse fibroblasts (BARBAN and SCHULZE, 1959). BANKS, OLDHAM, THAIN and VERNON (1959) have resolved a purified pig heart enzyme into apo- and co-enzyme. It has recently been demonstrated by FLEISHER et al. (1960) that the transaminase from man, dog and pig can be separated by paper electrophoresis into two components, one migrating towards the anode, and another to the cathode. The cationic and anionic enzymes are further differentiated from one another by the large differences in the affinity of the enzymes for their substrates.

In higher plants the enzyme has been studied from white lupine, barley, oat, mung bean seedlings (WILSON et al., 1954), *Kalanchoë* leaves (WALKER and RANSON, 1958), mitochondria of castor bean endosperm (HILLER and WALKER, 1961), cauliflower (ELLIS and DAVIES, 1961), wheat germ (CRUICKSHANK and ISHERWOOD, 1958), and *Dolichos lablab* (PATWARDHAN, 1958). It has been highly purified from the last named plant (PATWARDHAN, 1958).

a) Assay of Glutamate-Aspartate Transaminase.

The transamination reaction may be followed by the appearance or the disappearance of any one component of the assay mixture. The amino acid components may be separated by conventional paper chromatography and determined colorimetrically after elution from paper by a number of methods which have been given in detail by Aspen and Meister (1958). In our experience the method of Yemm and Cocking (1955) as modified by Rosen (1957) has given the most consistent results.

1. Chromatographic assay. After chromatographic separation of the amino acids the paper is heated for 15 minutes at 100° C. The position of the amino acids is located by viewing the paper under ultra violet light. They appear as light blue fluorescent spots on a dark background (Fowden, 1951). Squares containing the amino acids are cut out and eluted with water or buffer in a test tube by shaking. One ml of the clear eluate is treated with 0.5 ml of cyanide-acetate buffer reagent and 0.5 ml of the ninhydrin reagent (cyanide-acetate buffer: 2700 g of $NaC_2H_3O_2$ · $3H_2O$ is dissolved in 2,000 ml of water and 500 ml of glacial acetic acid, and the mixture is made to 7.5 liters with water. One part of 0.01 M NaCN solution is diluted with 49 parts of the acetate buffer. Ninhydrin reagent: a 3% solution of ninhydrin in ethylene glycol monomethyl ether). The mixture is heated for 15 minutes in a boiling water bath. Immediately after removal from the bath, 5 ml of isopropanol-water (1:1) is added, thoroughly mixed, cooled rapidly to room temperature, and the color read at 570 mμ (or 404 mμ for proline and hydroxyproline). 0.02 to 0.40 μmoles of amino acids are determined easily.

2. Assay by the determination of amino acids with specific decarboxylases (Feldman and Gunsalus, 1950). The assay mixture consists of; 0.1 ml of 1 M phosphate buffer, pH 8.3; 0.2 ml of the enzyme; 0.1 ml of pyridoxal phosphate (10 μg); 0.2 ml of 0.125 M L-amino acid; 0.1 ml of 0.5 M α-ketoglutarate, and water to make 1.0 ml. The mixture is incubated for 60—90 minutes at 37° C, and the reaction is terminated by heating the mixture in a boiling water bath for 5 minutes. The control consists of the mixture and boiled enzyme. After centrifugation, the supernatant solution is adjusted to a pH of 5.0, and glutamic acid or aspartic acid is determined in 1 ml aliquots by glutamic or aspartic decarboxylase (see methods on page 302).

3. Assay by the measurement of oxalacetate.

a) Determination of oxalacetate with aniline citrate (Green et al., 1945). In the main compartment of a Warburg vessel is placed; 0.5 ml of 0.2 M phosphate buffer, pH 7.3; 0.2 ml of the enzyme preparation; 0.1 ml of pyridoxal phosphate (10 μg), and water to a required volume. In one side arm is placed 0.2 ml each of 0.2 M aspartate and α-ketoglutarate, and in the other 0.5 ml of aniline citrate (made by mixing equal volumes of aniline and 50 g of citric acid in 50 ml of water). After equilibration at 37° C for 10 minutes, the amino acid-keto acid mixture is tipped in and the reaction is allowed to proceed for another 10 minutes, after which the aniline citrate reagent is also tipped in, and gas evolution is measured for 10 minutes. The control vessels contain all of the reagents and an aliquot of boiled enzyme. If a crude extract is used as the enzyme source, two corrections have to be made regarding the CO_2 evolution (Ames and Elvehjem, 1946), one for the CO_2 content of the tissue or the crude extract, and the second for the CO_2 evolution at zero tissue or extract level.

b) Determination of pyruvic acid formed after decarboxylation of oxalacetate (Tonhazy, White and Umbreit, 1950). The transamination reaction is carried out in the presence of an excess of aspartic and α-ketoglutaric acid. The reaction is stopped with TCA before 5% of the keto acid is aminated and the oxalacetate formed is decarboxylated with aniline citrate. Pyruvic acid is determined by the colorimetric procedure of Friedemann and Haugen (1943) with 2,4-dinitrophenylhydrazine reagent. If the reaction time is short (4—5 minutes), only pyruvic acid reacts with the reagent. The hydrazone is extracted with toluene and after mixing an aliquot of the toluene layer with 2.5% KOH in 95% ethanol, the colour is read at 540 mμ.

c) *Determination of oxalacetate by spectrophotometry* (CAMMARATA and COHEN, 1951). This method is based on a decrease or increase in absorption at 280 or 300 mμ and requires a highly active enzyme preparation, since less pure preparations may exhibit appreciable interfering absorption at 280 mμ. This method may also be used for studying the transamination of many other amino acids (see ASPEN and MEISTER, 1958).

d) *Determination of oxalacetate as enol-borate complex* (KUN et al, 1960). Since oxalacetate forms an enol-borate complex with an absorption maximum at 290 mμ, the measurement of this band can be used for the determination of glutamate-aspartate transaminase activity. The molar absorbancy index of oxalacetate at 280 mμ (pH 7.3) is 550 and that of the enol-borate complex is 1,380 (pH 8.4). Since enzymatic formation of oxalacetate in the presence of borate is linear with time, and directly proportional to the amount of the enzyme within a wide range, "zero order" kinetics can be easily applied.

In a quartz cuvette of 1 cm light path is placed; 2 ml of a solution of $K_2B_4O_7 \cdot 5H_2O$ (100 μmoles) in 0.2 M Tris buffer, pH 8.4, containing aspartate and α-keto-glutarate at the desired concentration; 0.1 ml of 1 M $MgSO_4$; water, and enzyme to a final volume of 3.0 ml. The enzyme extract is made with 0.1 M phosphate buffer, pH 7.4, containing 10^{-3} M $MgSO_4$ and 5 μg of pyridoxal phosphate. After the addition of the enzyme solution, the contents of the cuvette are thoroughly mixed and the rate of change of absorbancy at 290 mμ is determined at 30 second intervals. The action of Mg^{++} ions in the reaction mixture is not known, but it increases the linear rate of oxalacetate-enol-borate complex formation from aspartate. Probably Mg^{++} acts on the keto acid substrate by catalyzing keto-enol conversion, or it may even function in stabilizing the transaminase.

e) *Determination of oxalacetate by enzymic reduction* (HOLZER et al., 1958). The oxalacetate resulting from transamination can be reduced to malic acid with the help of malic dehydrogenase and DPNH. The oxidation of DPNH is directly proportional to the amount of transamination. In a cuvette of 1 cm light path the following are added in this order; 2.82 ml of 0.2 M diethanolamine buffer, pH 9.0; 0.08 ml of DPNH solution (8 mg/ml); 0.02 ml of malic dehydrogenase (0.18 mg protein/ml, prepared according to the method of STRAUB (1942), or obtained commercially. An excess of the enzyme must be added); 0.03 ml pyridoxal phosphate (2 mg/ml); enzyme solution, and 0.02 ml of 0.5 M α-ketoglutarate. The mixture is allowed to stay for 2 to 5 minutes and the reaction is then started with 0.02 ml of 0.5 M aspartate. The total volume in the cuvette is 3 ml and DPNH oxidation is measured at 340 mμ at 1—2 minute intervals, and the calculations are based on the linear part of the curve.

4. Assay by the measurement of α-ketoglutarate. Glutamate-aspartate trans-aminase may also be measured with the substrates glutamate + oxalacetate by the reductive amination of the α-ketoglutarate formed with ammonia, DPNH and glutamic dehydrogenase as follows (HOLZER et al., 1958):

1. glutamate + oxalacetate $\rightleftharpoons$ α-ketoglutarate + aspartate
2. α-ketoglutarate + DPNH + H^+ + $NH_4 \rightleftharpoons$ glutamate + DPN^+

sum = oxalacetate + DPNH + H^+ + $NH_4^+ \rightleftharpoons$ aspartate + DPN^+

In a quartz cuvette of 1 cm light path the following are added in this order; 2.81 ml of 0.2 M triethanolamine buffer, pH 7.4; 0.02 ml of saturated ammonium sulphate; 0.08 ml of DPNH solution (8 mg/ml); 0.01 ml of crystalline glutamic dehydrogenase (Sigma; 10 mg protein/ml); 0.03 ml of pyridoxal phosphate

(2 mg/ml); transaminase preparation, and 0.02 ml of 1 M oxalacetate. The mixture is allowed to stay for 2—4 minutes. When the extinction measurement becomes constant, 0.02 ml of 0.5 M glutamate is added, and the oxidation of DPNH is measured at 340 mμ at 30 second intervals. The calculations are based on the linear part of the curve. Presence of malic dehydrogenase in crude cell-free extracts interferes with the assay, but if all the values are corrected with the blind values before the addition of glutamate, a reasonably accurate assay of the transaminase is obtained.

Other spectrophotometric assays of glutamate-aspartate transamination have been described by Cammarata and Cohen (1951), Jenkins et al. (1959), and Aspen and Meister (1958).

b) Purification of the Enzyme from Plants.

a) **Purification from wheat germ** (Cruickshank and Isherwood, 1958). The wheat germ is defatted by extracting three times with 1.5 volumes of ether and dried at room temperature. The original flaky material disintegrates to a powder (15 g) which is suspended in 60 ml of water in a stoppered bottle, and agitated gently for 3 hours by rotating the bottom slowly between rollers. This mixture is then centrifuged and the milky supernatant solution is adjusted to a pH of 5.7 with 2 N acetic acid. After centrifugation, the brown supernatant solution is adjusted to pH 7.0 with 2 N NaOH and treated with saturated ammonium sulphate, pH 7.0 at 5° C. The fraction precipitating between 33 to 66% saturation is recovered by centrifugation, suspended in a small volume of water, and dialyzed for 15 hours at 5° C against 1% KCl, and finally adjusted to pH 7.5 or 8.0 with N NaOH. The preparation contains in addition to the glutamate-aspartate transaminase the glutamate-alanine transaminase as well.

A glutamate-aspartate transaminase from *Dolichos lablab* has been reported to be highly purified by Dowex 2 (Cl⁻ form), calcium phosphate gel, and alumina C$_\gamma$ treatment by Patwardhan (1958). Details of the methods are not available.

b) **Purification from cauliflower** (Ellis and Davies, 1961). Fresh cauliflower heads are defoliated and placed in a cold room at 3—4° C for 1 hour before use. The top 0.5 to 1.0 cm of florets is removed with a knife and homogenized in 200 g lots in 250 ml of chilled phosphate buffer (0.05 M, pH 7.0) for a minute in a blendor. The debris is strained off with a cloth and the solution is added to a further 200 g of florets in the blendor. The homogenized solution is strained and then centrifuged at 31,000 g for 10 minutes. The supernatants are pooled. The average specific activity of six crude extracts is 3.4.

Step 1. 28 g of ammonium sulphate is added with stirring to 100 ml of crude extract. After 15 minutes, it is centrifuged, and the precipitate is discarded. To the supernatant solution, 14 g of ammonium sulphate is added and the precipitate, after centrifugation, is suspended in a small volume of phosphate buffer (0.2 M, pH 7.0). It is dialyzed for 5 hours against a flow of 5 liters of phosphate buffer (0.05 M, pH 7.0). The dialyzed solution is clarified by centrifugation.

Step 2. An equal volume of a solution of ammonium sulphate (490 g/l) in phosphate buffer is slowly added to the clarified dialyzed solution. The precipitate formed after 15 minutes is discarded and a further 0.4 volumes of ammonium sulphate solution is added to the supernatant. The precipitate obtained after 15 minutes is resuspended in 10—20 ml of phosphate buffer (0.2 M, pH 7.0). The supernatant solution is treated with a further 0.6 volumes of the ammonium sulphate solution and the precipitated protein is resuspended as before.

Step 3. The fraction from step 2 is dialyzed for 5 hours against a running stream of 5 liters of dipotassium hydrogen phosphate (2 mM). The dialyzed enzyme is diluted to 25—50 ml with chilled H_2O. Calcium phosphate gel (8.3 mg/ml) is added in successive 5 ml portions and the gel precipitate is centrifuged after each addition. When 10—15 ml of the gel has been added, the sp. act. of the supernatant solution reaches a maximum.

Step 4. The supernatant solution from step 3 is applied to DEAE-cellulose column in small portions. The enzyme is eluted by a phosphate gradient (phosphate buffer, 0.2 M, pH 8.0 is dripped into a liter mixing flask containing dipotassium hydrogen phosphate, 2 mM; mixing is done by a magnetic stirrer). The solution is forced through the column by pressure from a head of liquid (1—2 ft.), and 10 ml fractions of the effluent are collected. Fractions of the highest specific activity are pooled. Maximum purification achieved is 250-fold.

The enzyme obtained in this manner is stable for at least six months at $-17°$ C. No detectable transamination is obtained between α-aminobutyrate and γ-aminobutyrate, or between β-aminobutyrate, cysteine, methionine, arginine, leucine, lysine, tryptophan, tyrosine and α-ketoglutarate. The purified enzyme, however, is found to catalyze transamination between the following pairs: glutamate and oxalacetate; ^{14}C-glutamate and α-ketoglutarate; γ-hydroxyglutamate and α-ketoglutarate (or oxalacetate); γ-methyleneglutamate and α-ketoglutarate (or oxalacetate); β-hydroxyaspartate and α-ketoglutarate (or oxalacetate); cysteate and α-ketoglutarate (or oxalacetate); cysteinesulfinate and α-ketoglutarate (or oxalacetate).

c) **Purification from yeast** (HOLZER et al., 1958). Dried beer yeast (70 g) is stirred with 210 ml of water for 3 hours at 37° C, and the residue is discarded by centrifugation. 120 ml of the supernatant solution is heated to 53—54° C, within 2 minutes with shaking, and held at this temperature for 8 minutes. After cooling in an ice bath, the precipitate is separated by centrifugation. To the supernatant solution (96 ml) 32 g ammonium sulphate is added slowly. The precipitate is recovered after 20 minutes, dissolved in 25 ml of water, and dialyzed 5 hours against water in the cold. The dialysate (32 ml) is treated with 8.3 ml of alumina $C_γ$ (19 mg solids/ml) and the gel is discarded by centrifugation. To the remaining supernatant solution 18 ml of the gel is added again, stirred for ten minutes, and the gel is eluted with 52 ml of 10% saturated ammonium sulphate solution. To 40 ml of the eluate 8 g of ammonium sulphate is added, and the resulting precipitate is discarded. 5 g ammonium sulphate is again added to the remaining supernatant solution, stirred for 20 minutes, and the sediment is taken up in 5 ml of water. This preparation keeps stable for many months if stored as an ammonium sulphate suspension. At this stage of purification the enzyme is also relatively free of the coenzyme and can be used for the quantitative assay of pyridoxal phosphate. As amino group acceptors only α-ketoglutarate and oxalacetate are active; hydroxypyruvate, phenylpyruvate, glyoxylate and α-ketobutyrate are functionless. Orthophosphate is inhibitory and the inhibition can be completely reversed by an excess of pyridoxal phosphate.

c) Properties of the Plant Enzyme.

The equilibrium constant for glutamate-aspartate system is 5.0 with an initial substrate concentration of 0.025 M, pH 8, and is similar to the enzyme from animal tissues (CRUICKSHANK and ISHERWOOD, 1958). The wheat germ enzyme shows activity increases of 50—80% by the addition of pyridoxal phosphate. The optimum pH for activity is 8.0 to 8.5, optimum temperature between 40 to 50° C, and the temperature coefficient is 2.1. The glutamate-aspartate system is more sensitive to the presence of Ag^+ ions than the glutamate-alanine system.

Ferrous sulphate increases the activity of the highly purified enzyme from beans (Patwardhan, 1958). Mercurous ions are inhibitory. Iodoacetate and malonate do not affect transaminase activity (Cruickshank and Isherwood, 1958), while oxomalonate (0.05—0.1 M) is strongly inhibitory (Ellis and Davies, 1961). Isoniazid has been reported to inhibit the transaminase of *E. coli* (Hicks, 1961a) and cauliflower (Davies and Ellis, 1961). Isoniazid inhibition of the latter is reversed by pyridoxal phosphate.

It has been reported that the glutamate-aspartate transaminase of pig heart is active with mesoxalic acid (replacing oxalacetate) and cysteic acid (replacing aspartic acid). The activity with these substrates, however, is considerably less than with oxalacetate and aspartate (Kearny and Singer, 1953). It has also been observed that the purified transaminase from animal tissues and cauliflower (Ellis and Davies, 1961; Davies and Ellis, 1961) catalyze transamination between cysteinesulfinic acid and α-ketoglutarate and between cysteate and α-ketoglutarate (or oxalacetate) at a rate greater than α-ketoglutarate and aspartate. In pure preparations of plant cysteinesulfinate-glutamate transaminase, however, aspartate is completely inactive in place of cysteinesulfinate in a system amino acid + α-ketoglutarate (Perez-Milan, Schliack and Fromageot, 1959). This fact argues for a distinct cysteinesulfinate transaminase, at least in plants.

The K_m values for the cauliflower enzyme (Davies and Ellis, 1961) are, glutamate 38 mM; oxalacetate, 0.105 mM; aspartate, 7.5 mM; and α-ketoglutarate, 0.76 mM.

2. Glutamate-Alanine Transaminase.

L-glutamic acid + pyruvic acid ⇌ α-ketoglutaric acid + L-alanine

Distribution. The transamination reaction between α-ketoglutarate and alanine has been reported among microorganisms in some strains of *Rhizobia* (Jordan, 1953), *Leishmania donovani* (Chatterjee, 1957), *Fusarium* (Sanwal, 1958), *Neurospora* (Fincham, 1951, 1954; Fincham and Boulter, 1956), *Pasturella tularensis* (Fleming and Foshay, 1956), *Reiter treponeme* (Barban, 1956), *Brucella abortus* (Altenbern and Housewright, 1953), *Rhodospirillum rubrum* (Hug and Werkman, 1957), yeast (Moses and Joslyn, 1953), and *Streptococcus faecalis* (Lichstein et al., 1945).

In animals, it has been purfied and studied from pig heart (Tanenbaum, 1956; Hicks and Craig, 1957; Bornstein, 1957; Green et al., 1945). The enzyme has been purified about 117 fold by Green et al. (1945). Awapara (1953) studied the enzyme from liver, kidney, prostrate and seminal vesicles of rat. The enzyme from rat liver has been purified by Rosen, Roberts and Nichol (1959).

In plants, a semi-purified enzyme has been isolated and studied by Cruickshank and Isherwood (1958) from wheat germ. The reaction between α-ketoglutarate and alanine has also been demonstrated in cell-free extracts of corn radicles, pea, white and blue lupine, barley, oat, and mung bean seedlings (Wilson et al., 1954).

Assay. a) (Rosen et al., 1959). The following reagents are used:

Reagent A. This mixture contains 12 ml of 0.17 M Tris buffer, pH 8.9; 2.5 ml of 0.5 M L-alanine; 1.5 ml of 0.1 M α-ketoglutarate, and 75 μl of 10% bovine plasma albumin.

Reagent B. This mixture used for the blank is prepared as A except that alanine is replaced with water.

To 0.5 ml of reagent A or B, 15 μl of the tissue extract is added. All samples and blanks are incubated at 38° C for 30 minutes. The reaction is stopped by adding 0.05 ml of 100% TCA to each tube. The colour is developed by adding

0.5 ml of 0.1% 2,4-dinitrophenylhydrazine in 2 N HCl. After 5 minutes incubation at 38° C, the mixture is extracted with 1 ml of water saturated toluene. Three ml of alcoholic KOH are added to 0.6 ml of the toluene layer, and the colour is read in a spectrophotometer at 530 mμ. The amount of pyruvic acid formed is determined from a standard curve obtained with 2 to 15 μl aliquots of a 0.1 M pyruvate stock solution.

b) (CRUICKSHANK and ISHERWOOD, 1958). One volume (0.2—0.4 ml) of 0.2 M phosphate buffer, 1 volume of 0.1 M α-amino acid solution, and 1 volume of transaminase preparation of wheat germ are incubated at 25° C for 10 minutes in a tube. One volume of 0.1 M α-ketoglutarate is then added. Glutamate-alanine activity is not affected by the presence of pyridoxal phosphate in contrast to glutamate-aspartate enzyme, and its addition is not essential. Control experiments are carried out in which the enzyme, phosphate buffer, and water are incubated with and without each α-amino acid and α-keto acid. The percentage transamination, used to express transaminase activity, is defined as the percentage of the initial substrate transaminated during a specific time. The keto acids are determined by the method of ISHERWOOD and CRUICKSHANK (1954a) and amino acids by the method of ISHERWOOD and CRUICKSHANK (1954b).

The glutamate-alanine transaminase can also be assayed by the enzymic reduction of the pyruvate formed in the system alanine + α-ketoglutarate by lactic dehydrogenase. The method for the estimation of pyruvate has been described in vol. VI.

Purification of the enzyme. The enzyme may be partially purified from wheat germ according to the method of CRUICKSHANK and ISHERWOOD (1958) described under glutamate-aspartate transaminase.

Properties of the plant enzyme. The addition of pyridoxal phosphate with the wheat germ enzyme does not affect glutamate-alanine activity, although glutamate-aspartate activity (present in the extract) is increased considerably. The optimum pH for enzyme activity is 7.5. The transaminase is completely inhibited by 1 mM $AgNO_3$, 40% by KCN, and 30% by 0.1 mM $AgNO_3$. The equilibrium constant is 1.4 at pH 7.5.

The pig and rat liver enzymes are inhibited about 50—60% with 10^{-2} M isoniazid (HICKS and CRAIG, 1957; BORNSTEIN, 1957).

3. Cysteinesulfinate Transaminase.

$$HO_2S.CH_2.CH(NH_2).COOH + HOOC.CH_2.CH_2.CO.COOH \rightarrow$$
$$\rightarrow HO_2S.CH_2.CO.COOH + HOOC.CH_2.CH_2.CH(NH_2).COOH \quad \text{i)}$$
$$HO_2S.CH_2.CO.COOH \rightarrow CH_3.CO.COOH + SO_3H^- \quad \text{ii)}$$
$$SO_3H^- + {}^1/_2 O_2 \rightarrow SO_4^{--} + H^+ \quad \text{iii)}$$

Distribution. Transamination involving L-cysteinesulfinate and α-ketoglutarate or oxalacetate is reported in *Proteus vulgaris* (KEARNEY and SINGER, 1953), rabbit liver (CHATAGNER, BERGERET, SÉJOURNÉ and FROMAGEOT, 1952), rat liver mitochondria (DE MARCO and COLETTA, 1961), and in oat leaves (PEREZ-MILAN et al., 1959; FROMAGEOT and PATINO-BUN, 1961).

Assay. In the presence of certain metal ions, the β-sulfinylpyruvate formed as a result of transamination between cysteinesulfinate and α-ketoglutarate (or oxalacetate) decomposes spontaneously to sulfite and pyruvate. The sulfite is oxidized to sulfate (reaction i, ii and iii). The transaminase can, therefore, be assayed manometrically by the uptake of oxygen in the oxidation of sulfite, or the sulfite formed can be determined polarographically.

a) Manometric assay (KEARNEY and SINGER, 1953). In the central compartment of a WARBURG vessel is placed; the enzyme; 70 μmoles of Tris buffer of the desired pH; 1.5 mg brilliant cresyl blue; 60 μmoles of L-cysteinesulfinate; 40 μmoles of α-ketoglutarate, and 3 μmoles of $MgCl_2$ in a total volume of 3 ml. The substrate is taken in the side arm. KOH is present in the central well, and the reaction is run at 35° C. Oxygen uptake is measured.

  b) Polarographic estimation (FROMAGEOT and PATINO-BUN, 1961). The vessel used in this estimation is shown in Figure 1. Sulfite is measured by anodic oxidation. In the vessel, maintained at 37° C, is introduced in compartment C, 50 μmoles of α-ketoglutarate; 10 μg of pyridoxal phosphate; enzyme; cysteine sulfinate, and 0.05 M phosphate buffer, pH 7.8 in a final volume of 8.0 ml. Cysteine sulfinate is

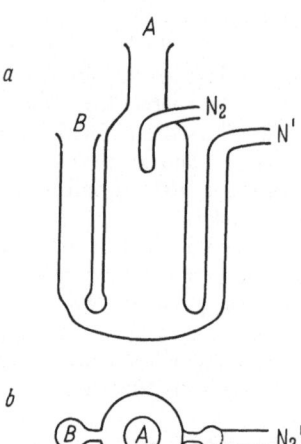

added after equilibration of the whole system with N_2. The gas is bubbled for 10 minutes from the opening N^1. The inlet N^2 serves to maintain a nitrogen atmosphere in the vessel during the course of the reaction. Through A is introduced a capillary dropping mercury electrode, and through B a calomel electrode. The electrodes are connected to a polarograph and the sensitivity of the instrument is set in the range of $1 \cdot 10^{-5}$ A. After the addition of the substrate, the galvanometer deflection is measured at definite time intervals. The amount of sulfite produced by the reaction is directly proportional to the deflection of the galvanometer in mm. A curve is previously constructed relating the sulfite concentration to the deflection of the galvanometer reading in the presence of the whole assay system except cysteinesulfinate.

  The enzyme may also be assayed by the *chromatographic determination* of the amino acid or pyruvate produced in the reaction (DE MARCO and COLETTA, 1961).

Fig. 1a and b. Vessel used in the polarographic assay of cysteine sulfinate transaminase (FROMAGEOT and PATINO-BUN, 1961). Details in the text. b The vessel in cross section.

  Purification of the enzyme. The enzyme is prepared from oat leaves according to the procedure of FROMAGEOT and PATINO-BUN (1961). Aqueous extracts (Fraction A) of 5 day old oat leaves cultivated in light are prepared. Partial purification is carried out at 0° C. To 1000 ml of fraction A, 100 ml of 1% aqueous Cetavlon (cetyltrimethylammonium bromide) is added. After 1 hour, the precipitate is removed by centrifugation at 40,000 g. The supernatant solution (Fraction B) is saturated with ammonium sulphate. After 20 hours, the precipitate is collected, dissolved in 100 ml of 0.05 M K_2HPO_4 buffer, and then dialyzed against 10 liters of the buffer for 15 hours and twice against 2 liters of buffer for 3 hours. The enzyme preparation is then clarified by centrifugation at 20,000 g (Fraction C). To this fraction one-tenth volume of 1% Cetavlon is added. The precipitate is obtained after one hour by centrifugation. The yellow supernatant constitutes fraction D and has 4% of the total nitrogen of fraction A.

  Properties. According to KEARNEY and SINGER (1953), Mn^{++} prevents the attainment of equilibrium and the consequent slowing down of the transamination by removing one of the products of the reaction. With bacterial enzyme, α-ketoglutarate and oxalacetate both serve as the amino group acceptors and the optimum pH of the transamination is 9.2. The oat enzyme, however, shows a pH optimum of 7.85 (FROMAGEOT and PATINO-BUN, 1961). With the plant enzyme, oxalacetate and α-ketoglutarate also act as amino group acceptors but pyruvate, α-ketovalerate, α-keto*iso*valerate, α-ketobutyrate, and β-ketoglutaric acid do not accept the amino group. Neither cysteic acid nor aspartate can replace cysteine sulfinate (PEREZ-MILAN et al., 1959). The plant enzyme is completely inhibited by 10^{-2} M KCN, 10^{-2} M hydrazine and hydroxylamine, and competitively inhibited by maleic acid. Succinate, fumarate and EDTA have no effect. The enzyme is considerably activated by pyridoxal phosphate.

4. β-Alanine (β-Aminoisobutyrate)-Glutamate Transaminase.

β-aminoisobutyrate + α-ketoglutarate $\rightleftharpoons$ methylmalonate semialdehyde + glutamate i)

β-alanine + α-ketoglutarate $\rightleftharpoons$ malonate semialdehyde + glutamate ii)

Distribution. Transamination between β-alanine and α-ketoglutaric acid has been observed in *Aspergillus fumigatus, E. coli* (ROBERTS, AYENGAR and POSNER, 1953a), mammalian tissues (ROBERTS and BREGOFF, 1953; BESSMAN, ROSSEN and LAYNE, 1953; BAXTER and ROBERTS, 1958), and *Leishmania donovani* (CHATTER-JEE, 1957). KUPIECKI and COON (1957) purified an enzyme from pig kidney that is capable of transferring amino groups from β-alanine and β-aminoisobutyrate to no other keto acid but α-ketoglutarate. The enzyme is also present in extracts of brain, liver, and skeletal muscles and can be detected in extracts of heart, spinach, *Tetrahymena pyriformis, Neurospora*, bakers' yeast, and *Pseudomonas fluorescens*. The purified preparation of KUPIECKI and COON (1957) is also capable of transferring amino groups from γ-aminobutyric acid.

Assay. The transaminase may be assayed by the enzymic reduction of methylmalonate semialdehyde or malonate semialdehyde by means of a dehydrogenase preparation (KUPIECKI and COON, 1957), or the chromatographic or enzymic determination of glutamate (ROBERTS, 1954).

For the estimation of glutamate, the assay mixture consists of (ROBERTS, 1954); enzyme preparation; 1 ml of mixed Tris (0.1 M) and phosphate (0.1 M) buffer, pH 8.1; 20 μmoles of α-ketoglutarate in 0.2 ml and water (in controls) or 80 μmoles of the nitrogenous substrate in 0.2 ml. The final volume is 2.4 ml. The mixture is incubated on a shaker at 30° C for 1—3 hours, and the resulting glutamic acid is determined either by chromatography or glutamic decarboxylase (see, glutamate-aspartate transaminase).

Enzyme preparation. In the case of higher plants, bacteria and fungi only crude extracts or acetone powder preparations have been used. A semipurified preparation may be obtained from pig kidneys by the method of KUPIECKI and COON (1957).

5. α-Alanine-Glycine Transaminase.

glyoxylic acid + L-alanine $\rightleftharpoons$ glycine + pyruvic acid

Distribution. The enzyme has been demonstrated to occur and purified 80-fold from *Blastocladiella emersonii* (an aquatic fungus, McCURDY and CANTINO, 1960) and partially purified from wheat leaves (FEGOL, 1961). CAMPBELL (1956) reported a reaction between alanine and glyoxylic acid in *Pseudomonas*, and KOIDE, SHISHIDO, NAGAYAMA and SHIMURA (1956) in the silkworm.

Assay (McCURDY and CANTINO, 1960). The enzyme activity is assayed at 37° C in a total volume of 1 ml containing; 10 μmoles of sodium glyoxylate monohydrate; 10 μmoles of alanine; 40 μmoles of phosphate, pH 8.0; 5 μmoles of $MgCl_2$; 10 μg of pyridoxal phosphate, and a suitable aliquot of the enzyme. A control is necessary to correct for a small amount of non-enzymatic transamination between glyoxylate and alanine that may take place in the strong alkali used in the determination of pyruvate at the end of the reaction. The reaction is stopped after 10 minutes with STAUB's reagent (1 ml of KOH solution made by dissolving

100 g solid KOH in 60 ml of water followed by 0.5 ml of a 2% solution of salicyl-aldehyde in alcohol). After 10 minutes incubation at 38° C the solution is made to 25 ml with water, and the extinction at 440 mμ is determined against a blank with all additions except pyruvic acid (Green et al., 1945). The amino acids may also be determined quantitatively following their separation with 1-propanol: ammonia:water (6:3:1) on Whatman No. 1 paper.

One *unit* of the transaminase is defined as the amount of enzyme catalyzing production of 1 μmole pyruvate per 10 minutes.

Enzyme purification. A semi-purified preparation may be obtained from wheat leaves by extraction with phosphate buffer, pH 8.0 and ammonium sulphate fractionation (0.30—0.60 saturation; Fegol, 1961).

From *Blastocladiella*, the enzyme is obtained according to the procedure of McCurdy and Cantino (1960). The fungus is grown in Cantino PYG broth (Difco), harvested, homogenized for 3 minutes in water containing 10^{-4} M EDTA and glass beads (4 g/g plants) in an Omnimixer, and clarified by centrifugation. To test with the crude extracts, the endogenous amino acids are first removed by dialysis for 20 hours against 2×10^{-1} M phosphate buffer, pH 7.4 in the cold. The crude extract is treated at 0° C with 0.75 ml of protamine sulphate (20 mg/ml) for every 100 mg of protein in the extract. The precipitate is discarded after centrifugation. The supernatant solution is fractionated with acetone at —10° C, and the precipitate forming between 47 to 57% is recovered and suspended in one-fifth the original volume of water, and dialyzed for 20 hours against 3 changes of 100 volumes of water containing 10^{-4} M EDTA. After dialysis enough calcium phosphate gel is added to give a gel: protein ratio of 2:1. After adsorption, the gel is discarded. The process is repeated once more. To the supernatant solution, acetone is added to give a concentration of 62%. The precipitate is recovered, resuspended in $1/4$ volume of water, and dialyzed intensively against water. The enzyme can be stored for a week at 4° C, or two months at —18° C without loss of activity. The enzyme cannot be resolved with regard to the coenzyme by intensive dialysis against phosphate, water or saturated ammonium sulphate.

Properties. The wheat leaf enzyme is stimulated slightly by pyridoxal phosphate (Fegol, 1961). Both fungal and plant enzymes have a pH optimum of 8.5.

The fungal enzyme is inhibited by KCN and hydroxylamine, and the inhibition by the latter is overcome with an excess of pyridoxal phosphate. Isoniazid at a concentration of 10^{-2} M has no effect. L-alanine cannot be replaced by any other amino acid in systems containing glyoxylate. A detectable amount of trans-amination occurs with the purified fungal enzyme between pyruvate and aspara-gine.

6. α-Alanine-β-Alanine Transaminase.

β-alanine + pyruvic acid $\rightleftharpoons$ malonic semialdehyde + L-α-alanine

Occurrence. The enzyme has been purified and characterized so far only from *Pseudomonas fluorescens* (Hayaishi, Nizhizuka, Tatibana, Takeshita and Kuno, 1961).

Assay (Hayaishi et al., 1961). The enzyme is assayed by measuring the formation of malonic semialdehyde with a colorimetric procedure based on the coupling reaction of the aldehyde with diazotized *p*-nitroaniline. The assay mixture consists of; 20 μmoles of β-alanine; 20 μmoles of sodium pyruvate; 100 μmoles of phosphate buffer, pH 8.0, and the enzyme in a final volume of 1.0 ml. The system is incubated for 15 minutes at 35° C after which 0.2 ml of 20% TCA is added to stop the reaction. After centrifuging, 0.6 ml of the supernatant solution is transferred to a 10 ml glass stoppered tube and the pH is adjusted to 5.2 by adding

0.4 ml of 1 M acetate buffer. Immediately, 3.0 ml of a freshly prepared diazo reagent (prepared by the addition of 3.0 ml of 0.5% $NaNO_2$ to 20 ml of 0.5% p-nitroaniline in 0.05 N HCl) is added and the colour developed for 30 minutes at 35° C. The coupling reaction is stopped by adding 1.0 ml of 5 N HCl. The formazan formed is extracted with 5.0 ml of ethyl acetate. The optical density of the ethyl acetate layer is measured at 440 mμ in a cuvette of 1 cm light path in a spectrophotometer. With purified enzyme, 5—10 μg pyridoxal phosphate is regularly included with the assay mixture.

Malonic semialdehyde may also be determined manometrically by decarboxylation with aniline citrate (KREBS and EGGLESTON, 1945), spectrophotometrically with malonic semialdehyde oxidative decarboxylase in the presence of CoA and DPN, and finally, assayed as acetaldehyde after distillation of the acidified reaction mixtures by lactic dehydrogenase (RACKER, 1950), p-hydroxydiphenyl (EEGRUVE, 1933), or other suitable means.

One unit of the enzyme is defined as the amount which produces 1 μmole of malonic semialdehyde per minute.

Enzyme preparation. The enzyme may be purified from *Pseudomonas* by C_γ-gel treatment, protamine treatment, and ammonium sulphate fractionation by the method of HAYAISHI et al. (1961).

Properties. The pH optimum of the enzymic reaction is 9.2 in Tris and carbonate buffers. The affinity of the coenzyme (pyridoxal phosphate) is high, and maximal velocity is attained with 5×10^{-5} M. The equilibrium of the reaction at pH 8.0 is about 5 in favour of the formation of α-alanine. The enzyme is inhibited by 1×10^{-4} M pCMBA to about 30%. It is not affected by EDTA, o-phenanthroline, and 8-hydroxyquinoline at a concentration of 10^{-3} M.

In addition to β-alanine, various monocarboxylic amino acids containing β-, γ-, or ε-amino groups can serve as effective amino group donors. Pyruvic acid is relatively specific as an amino acceptor. Dicarboxylic or other monocarboxylic keto acids are all inert. The K_m values for β-alanine and pyruvate are approximately 6.2×10^{-2} M and 1.4×10^{-2} M respectively.

7. γ-Aminobutyrate-Glutamate Transaminase.

γ-aminobutyric acid + α-ketoglutaric acid $\rightleftharpoons$ glutamic acid + succinic semialdehyde

Distribution. The transaminase has been shown to occur in mammalian brain tissue (SALVADOR and ALBERS, 1959; BAXTER and ROBERTS, 1958; BESSMAN et al., 1953; ROBERTS and BREGOFF, 1953; ROBERTS and FRANKEL, 1951), in liver tissue (BESSMAN et al., 1953; ROBERTS and BREGOFF, 1953), in the optic lobe of developing chicks (SISKEN, SANO and ROBERTS, 1961), *Aspergillus fumigatus* (ROBERTS et al., 1953 a, 1953 b; ROBERTS, 1954), propionic acid bacteria (KENNEY and WERKMAN, 1958), *Pseudomonas aeruginosa* (NOE and NICKERSON, 1958), *E. coli* (ROBERTS et al., 1953 a) and *Pseudomonas fluorescens* (SCOTT and JAKOBY, 1959; ALBERS and JAKOBY, 1960).

Assay. The reaction may be followed by enzymic reduction of the succinic semialdehyde by means of DPNH and aldehyde dehydrogenase (SCOTT and JAKOBY, 1959), the fluorometric estimation of the aldehyde (SALVADOR and ALBERS, 1959), or the chromatographic estimation of the glutamic acid formed (NOE and NICKERSON, 1958).

a) Enzymic reduction (SCOTT and JAKOBY, 1959). The transaminase activity is determined by coupling the reaction with an excess of succinic semialdehyde or aldehyde dehydrogenase. The formation of reduced pyridine nucleotide is a function of transaminase activity:

γ-aminobutyrate + α-ketoglutarate $\rightleftharpoons$ succinic semialdehyde + glutamate

succinic semialdehyde + DPN$^+$ (or TPN$^+$) + H$_2$O $\rightleftharpoons$ succinate + DPNH or TPNH + H$^+$

γ-aminobutyrate + α-ketoglutarate + H$_2$O + DPN$^+$ or TPN$^+$ $\rightleftharpoons$

$\rightleftharpoons$ glutamate + succinate + DPNH or TPNH + H$^+$

The reaction is followed spectrophotometrically at 340 mμ. The reaction mixture consists of 100 μmoles of potassium pyrophosphate, pH 8.1; 1 μmole of α-ketoglutarate; 1 μmole of γ-aminobutyrate; 0.5 μmole of pyridine nucleotide, and an excess of succinic semialdehyde (see page 348) dehydrogenase or aldehyde dehydrogenase (prepared as described by Black, 1955), sufficient to catalyze the oxidation of 0.1 μmole of succinic semialdehyde per minute. The activity is followed at 1 minute intervals for 5 minutes. The transaminase activity is a linear function of both time and protein concentration at rates of less than 0.05 μmoles of the reduced pyridine nucleotide formed per minute.

b) *Fluorometric assay* (Salvador and Albers, 1959). This method involves the measurement of succinic semialdehyde fluorometrically by the procedure of Vellez, Amiard and Pesez (1948). Buffered substrate solution is prepared containing 0.1 M α-ketoglutarate and 0.25 M γ-aminobutyric acid, and adjusted to pH 8.4 with NaOH. To a required amount of enzyme preparation in the test tube (ice cold), 5 μl of buffered substrate is added. The mixture is incubated in a water bath at 38° C for 2 hours. A standard curve of succinic semialdehyde *versus* fluorescence is obtained for each set of samples. Fluorescence is developed with 3,5-diaminobenzoic acid. Secondary emmission at 505 mμ is measured.

A *purified enzyme* preparation may be obtained from *Pseudomonas* according to the method of Scott and Jakoby (1959).

Properties. The bacterial enzyme (Scott and Jakoby, 1959) is stable for months if kept at —20° C. At pH values above 7.5 or below 6.5 the enzyme is rapidly inactivated. The enzyme is highly specific for γ-aminobutyric acid and α-ketoglutarate. There is no evidence for the participation of cofactors such as pyridoxal phosphate. The enzyme is inhibited by 2.1×10^{-4} M hydroxylamine, carbonyl reagents, and 1×10^{-5} M KCN (50% inhibition). The pH optimum is between 8.7—9.0. Noe and Nickerson (1958) have reported a pH optimum of 7.5 in 0.1 M phosphate buffer. Albers and Jakoby (1960) have shown that γ-aminobutyrate-glutamate transaminase catalyzes isotopic exchange between γ-aminobutyrate and succinic semialdehyde as well as between glutamate and α-ketoglutarate.

8. Kynurenine Transaminase.

L-kynurenine   o-aminobenzoylpyruvic acid

kynurenic acid

Distribution. An enzyme catalyzing the transamination of kynurenine to kynurenic acid has been found in animal tissues (WISS, 1952; MASON, 1954; MASON and BERG, 1951, 1952), in *Pseudomonas fluorescens* (MILLER, TSUCHIDA and ADELBERG, 1953), *E. coli* (YANOFSKY, 1955), and in *Neurospora crassa* (JAKOBY and BONNER, 1956). MASON and GULLERSON (1960) have studied the effect of inhibitors on the enzyme from rat kidney.

Assay (MILLER et al., 1953). The transaminase activity is followed directly by spectrophotometry. Each cuvette contains; 0.05 ml of enzyme preparation; 3.0 ml of 0.1 M phosphate buffer, pH 7.0; 0.5 μmole of L-kynurenine, and α-ketoglutaric acid in a total volume of 3.1 ml. The instrument is adjusted to 100% transmission and all components except kynurenine are added to the cuvette. After addition of kynurenine, readings at 330 mμ and 365 mμ are made at intervals of 1 or 2 minutes until no further change in optical density occurs. The extinction coefficient of the pure compounds being known, amounts of each present in mixtures can be calculated by using simultaneous equations.

The enzyme may also be assayed by estimating the quantity of kynurenic acid formed in the presence of the various reactants, enzyme and pyridoxal phosphate (JAKOBY and BONNER, 1956).

Purification. The enzyme may be purified from mycelium of *Neurospora crassa* (grown in a medium supplemented with tryptophan) according to the procedure of JAKOBY and BONNER (1956).

The pH of 100 ml of the cell-free extract (made by extracting powdered lyophilized mycelium with 0.1 M phosphate buffer, pH 8.1) is adjusted to 6.5 with acetic acid and 5 mg/ml of Norit is added. After 5 minutes of stirring, the suspension is centrifuged. To the supernatant a 2% solution of protamine sulphate (10 ml) is added, and the precipitate is removed by centrifugation. The supernatant solution is adjusted to pH 7.1 with normal NaOH and then fractionated with ammonium sulphate. The 60—75% fraction is dissolved in 20 ml of 0.1 M phosphate buffer at pH 7.5, and again fractionated with ammonium sulphate. The fraction salting out between 55 to 75% saturation is retained. The final preparation requires pyridoxal phosphate for activity.

Properties. The pH optimum of the fungal enzyme is 7.5 and requires pyridoxal phosphate as a cofactor (half-maximal velocity at 10^{-6} M). Half maximal velocity is obtained at a L-kynurenine concentration of 1.5×10^{-3} M. The reaction is inhibited by higher than optimal concentrations of cofactor and substrate. D-kynurenine is not utilized. 3-hydroxykynurenine can replace kynurenine in which case xanthurenic acid is produced. The enzyme is more efficaceous towards pyruvate and α-ketoglutarate. α-keto*iso*valeric acid and α-keto-β-methylvaleric acid give rise only to traces of kynurenic acid. Phenylpyruvic, indolepyruvic, and α-ketobutyric acid show approximately equal activity which is slightly less than that of pyruvic and α-ketoglutaric acid. Half maximal velocity is obtained with a pyruvate or α-ketoglutarate concentration of 7×10^{-4} M.

In the transamination reaction, o-aminobenzoylpyruvic acid is first produced, which undergoes a spontaneous cyclization to kynurenic acid. This fact also explains the irreversibility of the reaction *in vitro*.

9. Glutamate-Phosphohistidinol Transaminase.

This enzyme has been described on page 331. With the partially purified enzyme from *Neurospora* (AMES and MITCHELL, 1955; AMES and HORECKER, 1956), glutamate can be replaced by L-α-aminoadipic acid, L-arginine, and L-histidine. In place of histidinol phosphate, α-ketoadipic acid and α-keto-δ-

guanidinovaleric acid can also be utilized in transamination. The following transamination reactions are also catalyzed by the same enzyme preparation; glutamate-arginine, α-aminoadipate-arginine, α-aminoadipate-histidine, α-amino-adipate-glutamate, and glutamate-histidine. It is not certain whether all these reactions are catalyzed by one enzyme because the preparation used was only 18-fold purified.

10. Ornithine Transaminase.

This enzyme has been described on page 335. Ornithine can undergo trans-amination in preparations from liver (CAMMARATA and COHEN, 1950, MEISTER, 1954b), and *Neurospora* (FINCHAM, 1953) with a number of keto acids including pyruvate, α-ketoglutarate, α-ketobutyrate, and glyoxylate.

11. Serine-Alanine Transaminase.

serine + pyruvate ⇌ hydroxypyruvate + alanine

Distribution. SALLACH (1955, 1956) observed the presence of the transaminase in acetone powder extracts of dog liver, kidney, heart, brain, spleen, rabbit liver, kidney, brain and beef liver. A partial purification of the enzyme has been achieved and it has been shown that only alanine can act as a significant amino donor. However, reaction between serine and α-ketoglutarate has been reported in *Leishmania donovani* (CHATTERJEE, 1957) and animal tissue (CAMMARATA and COHEN, 1950; TANENBAUM, 1956). MEISTER, FRASER and TICE (1954) also reported that glutamine can act as an amino donor to form serine. A reverse reaction has been reported by STAFFORD, MAGALDI and VENNESLAND (1954). HANFORD and DAVIES (1958) demonstrated that in pea epicotyls phosphohydroxypyruvate undergoes transamination with glutamate to form phosphoserine. It is not known whether the enzyme involved in both cases is the same.

Assay (SALLACH, 1956). The enzyme activity can be assayed in both directions. The system for the assay in the reverse direction consists of; 40 μmoles of hydroxy-pyruvic acid; 40 μmoles of L-alanine; 30 μg of pyridoxal phosphate; enzyme, and 0.01 M phosphate buffer, pH 7.4 in a total volume of 4.0 ml. In the forward direction serine and pyruvate are used at a concentration of 25 μmoles each. Controls are employed with hydroxypyruvate alone. The substrates are adjusted to pH 7.4 before use. The enzyme is allowed to equilibrate with pyridoxal phos-phate for 10 minutes at 38° C before the substrates are added. The reaction is allowed to proceed for 60 minutes after which 0.4 ml of 20% TCA is added to terminate the reaction. Serine is assayed quantitatively according to FRISELL and MACKENZIE (1958). An aliquot of the deproteinized reaction mixture (con-taining 0.8—1.0 μmole serine) is transferred to a 10 ml flask and brought to 3 ml with water. The solution is made basic to methyl red with 5 N NaOH. One ml of 0.075 M NaIO$_4$ is then added, followed after 5 minutes by 10% TCA (dropwise) to the exact acid end point of the indicator. One ml of 10% NaHSO$_3$ is next added and the volume is brought to 10 ml with water. An aliquot equivalent to about 0.1 μmole of the original serine is transferred to a tube, and enough water is added to 1 ml. Ten ml of the chromotropic acid reagent (0.5 g purified sodium 1,8-dihydroxynaphthalene-3,6-disulfonate dissolved in 50 ml of water and 200 ml of 12.5 M H$_2$SO$_4$) is added in the tube, the colour is developed by heating in a boiling water bath for 20 minutes, and the contents are cooled to room temperature. Finally, 1 ml of 5% aqueous thiourea solution is added to reduce a red colour contributed by the reagents. The colour is read at 570 mμ in a spectrophotometer.

The enzyme has been studied only in a semi-purified state and the properties are not known.

12. Tyrosine-Glutamate Transaminase.

tyrosine + α-ketoglutarate ⇌ p-hydroxyphenylpyruvate + glutamate

Distribution. Tyrosine-α-ketoglutarate transamination has been demonstrated in rat liver by a number of workers (ANDERSON and MCELROY, 1959; LIN, RIVILIN and KNOX, 1959; LIN and KNOX, 1958; LIN, CIVEN and KNOX, 1958; AUERBACH and WAISMAN, 1959). SERENI, KENNEY and KRETCHMER (1959), and KENNEY (1959) have purified the enzyme from animal sources some 150-fold. The enzyme also seems to be present in halophilic pseudomonads (SHIIO et al., 1956), HeLa cells and mouse fibroblasts (BARBAN and SCHULZE, 1959), dog tissues (CANELLAKIS and COHEN, 1956), *E. coli*, *B. subtilis*, *Pseudomonas fluorescens* (FELDMAN and GUNSALUS, 1950), *Lactobacillus arabinosus* (MEISTER, 1952), and pig heart (TANEN-BAUM, 1956). RUDMAN and MEISTER (1953) have purified a transaminase A from *E. coli* which can catalyze the reaction between α-ketoglutarate and tyrosine along with phenylalanine, tryptophan and aspartate.

Assay. The continuous spectrophotometric assay developed by LIN, PITT, CIVEN and KNOX (1958) for aromatic amino acid transaminations is used. The reaction is allowed to run at 25° C in quartz cells of 1 cm light path. To insure maximal activity the enzyme preparation is first preincubated for 3—5 minutes with 30 μg of pyridoxal phosphate. Two ml of tyrosine (12 μmoles) in the appropriate borate buffer (pH 8.0), keto-enol tautomerase (purified from hog kidney according to the procedure of KNOX and PITT, 1957), water, and α-ketoglutarate are added to give final volume of 3.5 ml and a final borate concentration of 0.57 M. 80 μmoles of the keto acid are added to start the reaction. The amino acid is omitted from the blank. If p-hydroxyphenylpyruvate oxidase is present in the crude extracts, the assay mixture must also contain 10 μmoles of diethyldithiocarbamate to inhibit the oxidase activity. The reaction is followed by determining the optical density at 310 mμ for 10 or 20 minutes. The rate is expressed as μmoles of substrate reacting per ml of the enzyme per 10 minutes, calculated from the observed optical density changes, and the extinction coefficient of the enol-borate (9,850) in the assay system.

For the measurement of the keto acid concentration, reactions are carried out in phosphate or Tris buffer. To stop the reaction, 1.0 ml of 20% metaphosphoric acid is added to 3 ml of the reaction mixture. 0.5 ml of the deproteinized solution is added to 3.0 ml of 2 M arsenate, pH 6.5 (keto sample), and another 0.5 ml to 3.0 ml of 1 M borate in 2 M arsenate, pH 6.5 (enol-borate sample). After 15 minutes the spectrum of the enol-borate is found with arsenate keto solution as its blank. The molar extinction coefficient of p-hydroxyphenylpyruvate is 12,400 (310 mμ).

Methods for enzyme preparation from plants or microorganisms are not available. A highly purified preparation may be obtained from animal sources by the method of KENNEY (1959).

Properties. The animal enzyme (KENNEY, 1959) is highly specific for tyrosine and α-ketoglutarate. Neither pyruvate nor oxalacetate substitutes for the keto acid substrate. Similarly, other amino acids cannot be substituted for tyrosine. The pH optimum is 7.6. Chelating agents do not affect enzyme activity, but strong inhibition is caused by sulfhydryl group reagents. K_m value for tyrosine is 1.48×10^{-3} M, for α-ketoglutarate 6.25×10^{-4} M, and for pyridoxal phosphate 3.1×10^{-7} M.

13. Glutamine and Asparagine Transaminases.

These two enzymes (see BRAUNSTEIN and HSU, 1960) have been grouped together because they show many similarities which are discussed at the end of this section. The enzymes have been found in rat liver (Meister and Tice, 1950; Meister, Sober, Tice and Fraser, 1952; Meister, Fraser and Tice, 1954; Meister, 1953, 1954a, b, c). Braunstein and Hsu (1960) in a preliminary communication reported the purification of glutamine transaminase 50 to 80-fold and asparagine transaminase 30 to 40-fold. Transaminases involving the amides are also reported to occur in crude extracts of *Leishmania donovani* (Chatterjee, 1957), Reiter treponeme (Barban, 1956) bakers' yeast (Sheffner and Grabow, 1953), and *Pseudomonas* (Campbell, 1956). Monder and Meister (1958) reported that cell-free extracts of *Neurospora* catalyze the formation of asparagine from glutamine and α-ketosuccinamic acid. Olenicheva (1955) has reported the presence of glutamine transaminase in some plants.

In crude extracts from animal sources transamination of glutamine with a keto acid leads to a simultaneous deamidation. This has been demonstrated to be due (Meister, Levintow, Greenfield and Abendschein, 1955) to the presence of an ω-amidase in crude preparations which hydrolyzes the α-ketoglutaramic acid produced from glutamine. In contrast to the mammalian systems, *Neurospora* extracts, due to the absence of an active ω-amidase catalyze the formation of α-ketoglutaramic acid instead of α-ketoglutarate and ammonia (Monder and Meister, 1958). An analogous reaction also occurs with asparagine, where transamination with any one of a number of keto acids leads to the formation of oxalacetate and ammonia in crude preparations. However, the affinity of the ω-amidase for the transamination product of asparagine, α-ketosuccinamic acid is not very great, and the reversibility of the transamination reaction can be demonstrated (see, Meister, 1957).

Assay. The glutamine and asparagine transaminase have been assayed in the presence of an acceptor keto acid by means of chromatography (discussed under glutamate-aspartate transaminase). They may also be assayed in the presence of ω-amidase by the enzymic reduction of α-ketoglutarate formed with the help of DPNH and glutamic dehydrogenase (see under glutamate-alanine transaminase).

Properties. Broad specificity of the asparagine and glutamine transamination systems is indicated by the fact that more than 30 keto acids are active, including the α-keto analogues of the following amino acids; alanine, glycine, serine, α-aminobutyric acid, norvaline, norleucine, phenylalanine, tyrosine, cyclohexylamine, α-aminocaprylic acid, methionine, ethionine, glutamate, tryptophan, asparagine, cysteine, arginine, and cysteic acid (Meister, 1957). In contrast, these enzymes are generally believed to have narrow specificity with respect to the amino donor. Glutamine can be replaced to a certain extent by γ-methylglutamine and γ-methyleneglutamine (Meister, 1954c).

14. Alanine, Phenylalanine, Glutamate-Branched Chain Amino Acids.

This large conglomeration of enzyme systems has been placed together largely because there is no sure knowledge as to number of enzymes involved and their substrate specificities. Most of these transaminase systems have been studied only in crude cell-free extracts.

Transamination between leucine and α-ketoglutarate has been reported in yeast (Bigger-Gehring, 1955), and pig heart brei (Tanenbaum and Shemin, 1950). Glutamate formation from valine, leucine, and isoleucine with α-ketoglutarate was observed in *Torulopsis utilis* (Roine, 1947), organs of rats (Awapara and Seale, 1952), *Brucella abortus* (Altenbern and Housewright, 1953), and

in pig heart, liver and kidney (CAMMARATA and COHEN, 1950) while alanine formation from isoleucine, leucine and pyruvate occurs in *Leishmania donovani* (CHAT-
TERJEE, 1957) and from leucine, isoleucine and norleucine in *Brucella abortus*
(ALTENBERN and HOUSEWRIGHT, 1953). Valine-glutamate transaminase is known
to occur in pigeon breast muscle (COHEN, 1939), leucine, valine-glutamate in pig
heart (TANENBAUM, 1956) and in some strains of *Rhizobia* (JORDAN, 1953), isoleucine, valine-glutamate in *E. coli* (UMBARGER and MAGASANIK, 1952) and in
Neurospora crassa (FINCHAM and BOULTER, 1956). The effect of inhibitors on
valine-glutamate transaminases in *E. coli* was studied by HICKS (1961a, b) and
AMOS and VOLLMAYER (1957). RUDMAN and MEISTER (1953) reported a transaminase B capable of catalyzing the reaction between α-ketoglutarate and isoleucine, valine, leucine, norleucine in *E. coli*, while ALTENBERN and HOUSEWRIGHT
(1953) were able to get a leucine-glutamate transaminase in 20% ammonium
sulphate fraction and leucine, glutamate-alanine in the 30% fraction of *E. coli*.
A reaction between valine and pyruvate in transaminase C was reported by
RUDMAN and MEISTER (1953). Reactions between α-ketoisovalerate or α-keto-
β-methylvalerate with glutamate (MYERS and ADELBERG, 1954), with tyrosine,
phenylalanine or leucine (WAGNER and IFLAND, 1956) and reactions involving
L-valine and isoleucine with α-keto-β-methylvalerate and α-keto*iso*valerate,
respectively were reported in *Neurospora* (WAGNER and IFLAND, 1956; SEECOF and
WAGNER, 1959a, b). WAGNER, BERGQUIST and KARP (1958) further demonstrated
that phenylalanine and leucine are able to donate their amino groups to α-keto-
isovaleric and α-keto-β-methylvaleric acid, respectively in *Neurospora*. From the
same organism, SEECOF and WAGNER (1959a, b) purified an enzyme (phenylpyruvate transaminase) which allows transamination between phenylpyruvate
or α-ketoglutarate and other branched-chain amino acids (leucine, isoleucine
and valine) and between α-keto*iso*valerate and leucine, isoleucine, and methionine.

Leucine, valine and isoleucine form a closely knit group of amino acids many
of whose reactions are catalyzed by a single enzyme. Thus, only one decarboxylase
is involved in the decarboxylation of all of the three amino acids; only one dehydrogenase [leucine dehydrogenase, SANWAL and ZINK (1961) described elsewhere in the book] is involved in the DPN-linked oxidative deamination of all
of the amino acids, and only one common enzyme is involved at each of the steps
in the biosynthesis of valine and isoleucine. It is conceivable, therefore, that apart
from other transaminases there is one transaminase for the system leucine, isoleucine, valine-pyruvate; one for the amino acids-phenylpyruvate (see also,
SEECOF and WAGNER, 1959b); and one for amino acids-α-ketoglutarate.

15. Other Transaminases.

a) **Tryptophan-glutamate and tryptophan-alanine.** LIN, PITT, CIVEN and
KNOX (1958), and LIN, CIVEN and KNOX (1958) reported in rat liver two independent tryptophan transaminases, one requiring α-ketoglutarate and the other
pyruvate. Tryptophan and α-ketoglutarate transamination has also been reported
in *E. coli* (RUDMAN and MEISTER, 1953), *Pseudomonas fluorescens, Bacillus subtilis*
(FELDMAN and GUNSALUS, 1950), *Lactobacillus arabinosus* (MEISTER, 1952), pig
heart, liver and kidney (CAMMARATA and COHEN, 1950), and whole cells and
cell-free extracts of *Rhodospirillum rubrum* (HUG and WERKMAN, 1957). For the
method used in the assay of these enzymes, see LIN, PITT, CIVEN and KNOX
(1958).

b) **δ-aminolaevulinic transaminase.** The presence of this enzyme was demonstrated by KOWALSKI, DANCEWICZ and SZOT (1957) in mammalian tissue.

BAGDASARIAN (1958) found that δ-aminolaevulinic acid can react with both α-keto-glutarate and pyruvate in the presence of extracts of *Corynebacterium diphtheriae*. The enzyme has been assayed by paper chromatography.

c) **Histidine-alanine.** Histidine pyruvate reaction mediated by a specific transaminase has been reported in rat liver (LIN, CIVEN and KNOX, 1958; LIN, PITT, CIVEN and KNOX, 1958) and appears to occur in *Leishmania donovani* (CHATTERJEE, 1957) also. It may be assayed by the spectrophotometric technique of LIN, PITT, CIVEN and KNOX (1958).

d) **Glycine-glutamate transaminase.** There is no evidence yet for the occurrence of a distinct glycine-glutamate transaminase. In bacteria, animal tissues and plants (see, MEISTER, 1957; WILSON et al., 1954) glycine can undergo transamination with α-ketoglutarate and pyruvate and glutamine, asparagine, glutamate and ornithine can transaminate with glyoxylate, but since there is a distinct glycine-alanine transaminase (McCURDY and CANTINO, 1960), there exists a possibility that glycine-glutamate transaminase is a separate enzyme.

e) **Alanine-ketomalonate transaminase.** The first observation of this enzyme was reported by BRAUNSTEIN (1939), but the product aminomalonic acid was not identified. GREEN et al. (1945) found a slight activity towards ketomalonate in place of oxalacetate using purified glutamate-aspartate transaminase. NAGAYAMA, MURAMATSU and SHIMURA (1958) reported the presence of the enzyme in silkworm and in rat liver, rabbit liver and rabbit heart. The enzyme from rat liver was partially purified by fractionation with ammonium sulphate (18-fold).

The enzyme is assayed by hydrolysis of aminomalonic acid formed in a system alanine + ketomalonate with 0.1 N HCl and the determination of glycine formed by hydrolysis using the method of KRUEGER (1949). A paper chromatographic assay is also possible.

In the transamination system, glutamate, α-aminobutyrate and aspartate can also act as amino donors in decreasing order of activity. The enzyme is inhibited 99% by pCMBA (10^{-3} M), 84% by benzoquinone (10^{-3} M), 58% by hydroxylamine (2×10^{-3} M), and 38% by KCN (10^{-3} M).

f) **Phenylalanine-alanine transaminase.** A distinct enzyme catalyzing transamination between phenylalanine and pyruvic acid has been reported in rat liver (LIN, PITT, CIVEN and KNOX, 1958). It may be assayed by the continuous spectro-photometric assay of LIN, PITT, CIVEN and KNOX (1958).

III. Transamidination.

In the process of transamidination, the amidine group of a donor compound is transferred to the nitrogen of an acceptor amino acid. The amidine donors are arginine, canavanine, and guanidinoacetic acid and the amidine acceptors are glycine, ornithine, canaline, and lysine.

Mammalian kidneys have been reported to catalyze a number of transamidination reactions: arginine-γ-aminobutyric acid (PISANO, MITOMA and UDEN-FRIEND, 1957), arginine-glycine (BORSOOK and DUBNOFF, 1941; FULD, 1954; WALKER, 1956), canavanine-ornithine (WALKER, 1956), canavanine-glycine (WALKER, 1956; NAKATSU, 1956), and arginine-ornithine (WALKER, 1956). The presence of arginine-ornithine, canavanine-ornithine, and arginine-hydroxyl-amine transamidinations were observed in *Streptomyces griseus* by WALKER (1958). However, no evidence for arginine, canavanine or guanidinoacetate-glycine transamidinations could be obtained in fungi.

1. Assay.

1. Ornithine-arginine (WALKER, 1958). DL-ornithine-2^{14}C monohydrochloride (3 mg) is incubated at 34° C with enzyme, 7 mg non-labeled L-arginine monohydrochloride, and 1.1 ml of 0.1 M potassium phosphate buffer, pH 7.5. The rate of incorporation of the label in arginine is followed by means of paper chromatography.

2. Canavanine-ornithine (WALKER, 1958). The system consists of 4.5 mg of L-canavanine, 3 mg of L-ornithine monohydrochloride, enzyme, and 0.1 ml of 1 M potassium phosphate buffer, pH 7.5 in a final volume of 0.9 ml. After incubation at 34° C for four hours, aliquots of the deproteinized solution are assayed for arginine by a suitable colorimetric method.

3. Arginine-hydroxylamine (WALKER, 1958). The reaction mixture contains; enzyme; 7 mg of L-arginine monohydrochloride; 0.2 ml of 1 M hydroxylamine hydrochloride adjusted to pH 7.0, and 0.1 ml of 1 M phosphate buffer, pH 7.0 in a final volume of 0.8 ml. It is incubated at 34° C for four hours. After incubation 0.2 ml of acetone and 0.2 ml of 30% TCA are added, centrifuged, and aliquots of the supernatant solution are analyzed colorimetrically for hydroxyguanidine. To determine the latter compound, an aliquot of the supernatant is made up to 2 ml with water, followed by 2 ml of 1 M phosphate buffer, pH 7.0 and 0.3 ml of 1% aqueous solution of sodium pentacyanoammonioferrate and thoroughly mixed. After 10 minutes the optical density is measured at 480 mμ.

4. Guanidinoacetate-glycine (WALKER, 1957). The reaction mixture consists of; 17 μmoles of guanidinoacetate; 6.6 μmoles of glycine-2^{14}C; 50 μmoles of potassium phosphate buffer, pH 7.3, and enzyme solution. The total volume of 0.4 ml is incubated at 34° C. At different times, 4 μl aliquots are spotted on paper chromatograms, using phenol-water as the solvent, and sprayed with alkaline ferricyanide nitroprusside (WALKER, 1955) to locate the guanidinoacetate. The latter is eluted and counted.

5. Canavanine-glycine (NAKATSU, 1956). The assay system contains; 3 ml of phosphate buffer, pH 7.4; 80 μmoles of glycine; 40 μmoles of canavanine, and enzyme in a total volume of 4.0 ml. To stop the reaction, 0.25 volumes of 2 M acetic acid containing 10% NaCl is added, and heated for 5 minutes in a boiling water bath. The mixture is filtered to remove the coagulated protein. After cooling glycocyamine is determined by SAKAGUCHI reagent. Large amounts of canavanine and glycine interfere with the determination of glycocyamine. It is, therefore, necessary to separate glycocyamine from the amino acids by paper chromatography.

2. Properties.

Transamidination reactions require no cofactors. They are inhibited by pCMBA (suggesting a sulfhydryl enzyme) but the activity is restored by the addition of glutathione (NAKATSU, 1956). Activity is also restored if cysteine is added to systems inhibited by formamidine disulfide (WALKER, 1958). Combinations of ferricyanide and compounds such as 2-thiolhistidine, 2-thiouracil or 2-benzmidazolethiol, cupric ions, and 2-thio-6-oxypurine inhibit the reactions. The pH optimum is about 7.4 (NAKATSU, 1956).

B. Racemases.

1. Alanine Racemase.

$$
\begin{array}{ccccc}
\mathrm{CH_3} & & & & \mathrm{CH_3} \\
| & & & & | \\
\mathrm{NH_2-C-H} & \rightleftarrows & \text{racemic mixture} & \rightleftarrows & \mathrm{H-C-NH_2} \\
| & & & & | \\
\mathrm{COOH} & & & & \mathrm{COOH} \\
\text{D-alanine} & & & & \text{L-alanine}
\end{array}
$$

The enzyme has only been reported to occur in bacteria (WOOD and GUNSALUS, 1951; STEWART and HALVORSON, 1953; CHURCH, HALVORSON and HALVORSON, 1954). Attempts to demonstrate the presence of this enzyme in fungi and animal tissues have yielded negative results (WOOD and GUNSALUS, 1951).

WOOD and GUNSALUS (1951) first demonstrated the enzyme is cell-free extracts of *Streptococcus faecalis*, strain R. STEWART and HALVORSON (1954) later found a heat resistant alanine racemase in spores of the genus *Bacillus*. The heat resistance seems to be linked to the close association of the enzyme with the particulate matter.

Assay. The assay procedure is based on the estimation of one of the isomers of alanine in the reaction mixture by coupling racemase reaction to D-amino acid oxidase and measuring the oxygen uptake in the presence of excess catalase:

$$\text{L-alanine} \underset{\substack{\text{pyridoxal}\\\text{phosphate}}}{\overset{\text{racemase}}{\rightleftharpoons}} \text{D-alanine} \qquad\qquad \text{i)}$$

$$\text{D-alanine} + O_2 \longrightarrow \text{pyruvate} + NH_3 + H_2O_2 \qquad\qquad \text{ii)}$$

$$H_2O_2 \overset{\text{catalase}}{\longrightarrow} H_2O + {}^1/_2 O_2 \qquad\qquad \text{iii)}$$

$$\text{sum: L-alanine} + {}^1/_2 O_2 \longrightarrow \text{pyruvate} + NH_3 + H_2O$$

In this system the oxidation of 1 μmole of D-alanine corresponds to an uptake of 11.2 μl of oxygen.

In the assay procedure (WOOD, 1955), the following are pipetted in the central compartment of a double-arm WARBURG vessel; 0.3 ml of 1 M K_2HPO_4—KH_2PO_4 buffer, pH 8.1; 10 μg of pyridoxal phosphate; 5 mg of glutathione; 1 ml of the enzyme preparation (1—8 units), and water to a final volume of 1.6 ml. The central well contains a wick of filter paper and 0.15 ml of 20% KOH. In one side-arm is added 0.5 ml of a solution containing 80 μmoles of L-alanine, and in the other side-arm excess of D-amino acid oxidase containing catalase (purified from hog kidneys up to the first ammonium sulphate precipitation according to the procedure of NEGELEIN and BRÖMEL, 1939, or obtained commercially). When assaying the enzyme in crude extracts, the autorespiration is taken into account. After equilibration at 37° C, the contents of both of the side arms are tipped in and gas uptake is measured over a 30 minute period.

One *unit* of the enzyme is defined (WOOD, 1955) as the amount causing the uptake of 1 μl of O_2 per minute.

Preparation of the enzyme. The enzyme may be prepared according to the method of DOLIN (1950) from dried cells of *Streptococcus faecalis*. The method has been discussed in great detail by WOOD (1955). A 65-fold purification is obtained by this method.

Properties. The pH optimum has been reported as above pH 8.0 (WOOD and GUNSALUS, 1951) for *Strept. faecalis*, and approximately 8.0 for the enzyme from the spores of *Bacillus* (STEWART and HALVORSON, 1953; CHURCH, HALVORSON and HALVORSON, 1954). Alanine racemase of the vegetative cells of *Bacillus* is destroyed by heating at 80° C for 15 minutes, while the racemase activity of the spore, stabilized by association with particulate matter, only loses 10% of the activity after heat treatment at 80° C for 2 hours (STEWART and HALVORSON, 1953). When the enzyme is rendered soluble this heat resistance is lost.

Pyridoxal phosphate probably acts as a coenzyme (WOOD 1955; STEWART and HALVORSON, 1954). K_m value for pyridoxal phosphate has been reported to be 2×10^{-6} M by WOOD (1955) and 4.5×10^{-7} M by OLIVARD and SNELL (1955). The difference in these values is perhaps due to the various degrees of resolution of the enzyme with regard to pyridoxal phosphate. Analogues and related compounds of vitamin B_6 partially substitute for the coenzyme (OLIVARD and SNELL, 1955). Glutathione is required for maximum activity (WOOD, 1955). The enzyme has a low affinity for the substrate. MICHAELIS constants for the substrate (L-alanine) have been reported as 8.5×10^{-3} M (WOOD, 1955) and 9.8×10^{-3} M (OLIVARD and SNELL, 1955).

The enzyme is specific for alanine and does not react with the following amino acids: α-aminobutyric acid, cysteine, serine, leucine, threonine, proline, hydroxy-proline, lysine, arginine, histidine, tyrosine, tryptophan, and aspartic acid (WOOD and GUNSALUS, 1951).

2. Glutamate Racemase.

```
   COOH                                      COOH
    |                                         |
   CH₂                                       CH₂
    |                                         |
   CH₂   ⇌  racemic mixture  ⇌              CH₂
    |                                         |
NH₂—C—H                                   H—C—NH₂
    |                                         |
   COOH                                      COOH

 D-glutamic acid                          L-glutamic acid
```

The enzyme occurs only in bacteria and has been studied from *Lactobacillus arabinosus* (NARROD and WOOD, 1952; AYENGAR and ROBERTS, 1952; GLASER, 1960), *Mycobacterium avium* (ITO, KATAYAMA and TANAKA, 1957; ITOH, 1958), and *Lactobacillus fermenti* (TANAKA, 1960; TANAKA, KATO and KINOSHITA, 1961).

MEISTER (1957) proposed that a mechanism for the racemization of glutamic acid in crude extracts of various bacteria could be explained on the basis of a coupling of D-glutamic acid-D-alanine transaminase to L-glutamic acid-L-alanine transaminase by alanine racemase. This coupling has been shown to occur in *Bacillus subtilis* (THORNE et al., 1955). GLASER (1960) purified the glutamic acid racemase of *Lactobacillus arabinosus* three hundred fold. This purified enzyme was free of glutamate-alanine transaminases, and α-ketoglutarate was not utilized as a substrate, showing thereby that the racemization of glutamate was due to a specific racemase.

Assay. The enzyme can be assayed by the quantitative estimation of L-glutamate formed from D-glutamate in the presence of the racemase and co-factors, either by manometric estimation with the help of the specific L-glutamate decarboxylase (for the method, see vol. VI), or by a spectrophotometric assay at 340 mμ with L-glutamate dehydrogenase and DPN. The latter method is simple, rapid, and accurate (GLASER, 1960).

The reaction mixture (GLASER, 1960) contains; 25 μmoles of potassium phosphate; 0.5 μmoles of EDTA; 12 μmoles of D-glutamic acid; enzyme, and water to a final volume of 1.5 ml. The final pH is 7.5 and the mixture is incubated at 37° C for 1 hour. The reaction is terminated by the addition of 0.1 ml of 3 N perchloric acid and the coagulated protein is centrifuged down. The supernatant solution is adjusted to pH 7.0 with 3 N KOH and the precipitate of potassium perchlorate is removed. In the second step of the assay, an aliquot of the super-natant solution containing 0.15 to 0.3 μmoles of L-glutamic acid is added to the following mixture containing; 100 μmoles of Tris; 10 μmoles of DPN, and an excess of glutamate dehydrogenase (commercially available). The final volume is 1.03 ml and final pH 9.2. The mixture is incubated for 1 hour at room temperature. At the end of this period, the optical density at 340 mμ is compared to that of a blank prepared in the same way as the enzyme assay mixture without the sub-strate and a standard containing no L-glutamic acid in the glutamic dehydrogenase reaction mixture.

Purification (Tanaka et al., 1961). All operations are conducted at 0—3° C. One part by weight of dried cells of *Lactobacillus fermenti* are suspended in 11—12 volumes of 0.05 M phosphate buffer containing 0.005 M cysteine, pH 7.5 and disrupted by sonication for 15 minutes. The crude extract is centrifuged and 1.5 volumes of 2% protamine sulphate is added to precipitate the nucleic acids. After centrifugation, the supernatant solution is brought to 0.30 saturation by the addition of saturated ammonium sulphate, and the pH is adjusted to 4.0 by the addition of 0.5 N HCl. The precipitate is centrifuged down and ammonium sulphate is added to the supernatant solution to bring it to 0.70 saturation. The precipitate is recovered and dissolved in 3 volumes of 0.05 M phosphate buffer containing 0.005 M cysteine and 0.001 M DL-glutamic acid, pH 7.5, and dialyzed against the same buffer. After dialysis the solution is fractionated with saturated ammonium sulphate and the precipitate appearing between 0.20 to 0.65 saturation is dissolved in 1 volume of 0.05 M phosphate containing 0.005 M cysteine and 0.001 M DL-glutamic acid, pH 7.5. It is dialyzed against 1 liter of the same buffer. The dialyzed solution is fractionated with cold acetone. The fraction separating between 52—62% is recovered and dissolved in 0.5 volumes of the buffer containing glutamic acid. The enzyme solution is applied on a DEAE-cellulose column equilibrated with the same buffer, and after adsorption, the enzyme is also eluted by a buffer of the same composition. This method results in a 500-fold purification of the enzyme. The purified enzyme is unstable unless stored in the presence of the substrate (Glaser, 1960). It is stable to freezing.

Properties. The optimum pH for racemization has been reported at 6.8 (Narrod and Wood, 1952), approximately 7.5 (Glaser, 1960; Tanaka et al., 1961), 7.8—8.0 (Ito et al., 1957; Itoh, 1958) and 8.0 (Ayengar and Roberts, 1952).

The enzyme from *Mycobacterium avium* is stimulated by pyridoxal phosphate (Ito et al., 1957; Itoh, 1958), while conflicting claims have been put forward as to the requirement for pyridoxal phosphate for the racemase from *Lactobacillus arabinosus*. Tanaka (1960) states that pyridoxal phosphate does not act as a cofactor for glutamic racemase of *Lactobacillus fermenti*.

KCN, Co^{++}, Cu^{++}, Zn^{++}, all act as inhibitors. The enzyme is also sensitive to —SH group inhibitors, including pCMBA, phenylmercuricacetate and mono-iodoacetate (Tanaka et al., 1961). The effect of pCMBA is reversed by cysteine. High concentrations of FAD minimize inhibition by riboflavin, FMN, tetracycline, and acriflavine (Tanaka et al., 1961). Hydroxylamine is a competitive inhibitor (Glaser, 1960).

The enzyme has absorption bands at 270, 360 and 450 mμ and boiled extracts show a yellow fluorescence under ultra-violet irradiation. Due to the se properties, Tanaka et al. (1961) consider the racemase as a flavoprotein.

Michaelis constants for D-glutamate have been variously reported as 2.4×10^{-3} M (Ito et al., 1957; Itoh, 1958), 3.6×10^{-3} M (Glaser, 1960), and 4.7×10^{-2} M (Tanaka et al., 1961).

3. Threonine Racemase.

$$
\begin{array}{ccc}
\text{CH}_3 & & \text{CH}_3 \\
| & & | \\
\text{H—C—OH} & & \text{HO—C—H} \\
| & \rightleftharpoons & | \\
\text{NH}_2\text{—C—H} & & \text{H—C—NH}_2 \\
| & & | \\
\text{COOH} & & \text{COOH} \\
\text{D-threonine} & & \text{L-threonine}
\end{array}
$$

The occurrence of this enzyme has been suggested by growth studies of biochemical mutants of yeast (TEAS, 1948) and bacteria (UMBARGER and ADELBERG, 1951; AMOS and COHEN, 1954). The enzyme has been studied in K-12 (a valine requiring strain) and ML-52 (a L-threonine requiring mutant) strains of $E.\ coli$ (AMOS, 1954). To date threonine racemase activity has not been reported in animals or higher plants.

Enzyme preparation. Strain K-12 of $E.\ coli$ is grown on DAVIS and MINGIOLI medium (DAVIS and MINGIOLI, 1950) with the addition of 0.5% glucose. The cells are dried $in\ vacuo$ over phosphorous pentoxide, ground with alumina, and extracted with 0.01 M phosphate buffer, pH 7.1. The extract is then dialyzed for 16 hours against 0.01 M phosphate buffer, pH 7.1 in the cold (AMOS, 1954).

Assay (AMOS, 1954). The reaction mixture in a total volume of 1 ml consists of; 50 μmoles of D-threonine; 10 μg pyridoxal phosphate; 5 μmoles of ATP; 0.2 M phosphate buffer, pH 7.8, enzyme (3—4 mg protein), and an excess of L-threonine deaminase of $Clostridium\ welchii$ (see page 311). A control lacking the substrate is run simultaneously with the test. The tubes are flushed with nitrogen, stoppered, and incubated at 37° C for 3 hours. The reaction is stopped by the addition of 0.2 ml of 50% TCA, and the precipitate is removed by centrifugation. The ammonia of the supernatant is determined by nesslerization or micro-KJELDAHL. Alternatively, α-ketobutyric acid is determined by the colorimetric procedure of FRIEDEMANN and HAUGEN (1943). The $C.\ welchii$ threonine deaminase is specific for L-threonine and will not react with D-, D-allo-, or L-allo-threonine. The ammonia or the keto acid formed is a measure of the L-threonine formed from D-threonine in the reaction mixture by the racemase.

Properties. The enzyme is stable in storage, and is stimulated by ATP, AMP, or adenosine-3-phosphate. The enzyme has not been purified and the other properties of the enzyme are not known.

4. Methionine Racemase.

$$
\begin{array}{ccc}
\mathrm{CH_3} & & \mathrm{CH_3} \\
| & & | \\
\mathrm{S} & & \mathrm{S} \\
| & & | \\
(\mathrm{CH_2})_2 & \rightleftarrows \text{ racemic mixture } \rightleftarrows & (\mathrm{CH_2})_2 \\
| & & | \\
\mathrm{NH_2{-}C{-}H} & & \mathrm{H{-}C{-}NH_2} \\
| & & | \\
\mathrm{COOH} & & \mathrm{COOH} \\
\text{D-methionine} & & \text{L-methionine}
\end{array}
$$

SHOCKMAN and TOENNIES (1954 a, b) first brought forth evidence suggesting a methionine racemase in $Streptococcus\ faecalis$. Later KALLIO and LARSON (1955) demonstrated the presence of this enzyme in a strain of $Pseudomonas$ capable of utilizing methionine as a sole source of carbon and nitrogen.

Assay (KALLIO and LARSON, 1955). The reaction is carried out in small test tubes which are anaerobically incubated for 40 minutes at 37° C. The reaction mixture (final volume, 2.0 ml) contains; 8 μg of pyridoxal phosphate; 15 μmoles of D- or L-methionine; racemase extract, and 0.1 M phosphate buffer, pH 8.2. The reaction is terminated by immersing the tubes in a boiling water bath for 3 minutes. The D- and/or L-methionine content is compared with controls which lack the enzyme in the reaction mixture. L-methionine content is measured manometrically using L-amino acid oxidase and catalase (both commercially available) and

D-methionine by the D-amino acid oxidase method of WOOD and GUNSALUS (1951; see, method under alanine racemase).

Purification (KALLIO and LARSON, 1955). All operations are carried out at 0—5° C. 7 g (wet weight) methionine adapted cells of *Pseudomonas* are suspended in 30 ml of 0.1 M phosphate buffer, pH 8.2, and sonically disrupted. After centrifugation, the debris is resuspended in 10 ml of 0.1 M phosphate buffer, pH 8.2, and sonically oscillated for an additional 25 minutes. The supernatant solutions are pooled, and the protein separating between 0.30 and 0.75 ammonium sulphate saturation is dissolved in water. The pH of the protein suspension is adjusted to 6.0, and protamine sulphate is added until the 280 mμ:260 mμ ratio is 0.97. The precipitate is discarded. The supernatant solution is subjected to ammonium sulphate fractionation. The precipitates obtained after 0.35 saturation and 0.55 saturation are discarded each time. The precipitate forming between 0.55 and 0.75 saturation is dissolved in 0.01 M phosphate buffer, pH 8.2.

The enzyme is adsorbed on calcium phosphate gel (100 mg gel per 100 mg protein) and the supernatant after centrifugation is discarded. The gel is washed twice with 0.01 M phosphate buffer, pH 8.2, and the enzyme is eluted from the gel with two portions of 1 M phosphate buffer, pH 8.2 for 3 hours in the cold. The eluate is dialyzed against distilled water overnight and used as the source of the racemase.

Properties. The optimum pH for the racemase lies between 8.0 and 9.0. The apoenzyme requires pyridoxal phosphate as a cofactor (KALLIO and LARSON, 1955; TANAKA, 1960). The enzyme can be resolved with regard to the coenzyme by prolonged dialysis. MICHAELIS constant for the substrate is 8.0×10^{-3} M (KALLIO and LARSON, 1955).

5. Lysine Racemase.

$$
\begin{array}{ccc}
\mathrm{CH_2NH_2} & & \mathrm{CH_2NH_2} \\
| & & | \\
(\mathrm{CH_2})_3 & & (\mathrm{CH_2})_3 \\
| & \xrightleftharpoons{} \quad \text{racemic mixture} \quad \xrightleftharpoons{} & | \\
\mathrm{NH_2-C-H} & & \mathrm{H-C-NH_2} \\
| & & | \\
\mathrm{COOH} & & \mathrm{COOH} \\
\text{D-lysine} & & \text{L-lysine}
\end{array}
$$

HUANG, KITA and DAVISSON (1958) reported that intact dried cells of *Proteus vulgaris* (ATCC 4669) contain a mechanism capable of racemizing D-lysine to L-lysine. The conversion of either isomer in a reaction mixture continues until there is a 50% conversion of the added isomer, thus producing a racemic mixture at equilibrium. HUANG and DAVISSON (1958) later gave evidence that the racemization was due to a specific racemase rather than a coupling of D- and L-transaminases with another racemase as has been demonstrated for glutamate racemization in *Bacillus subtilis* by THORNE et al. (1955; see, glutamate racemase discussed earlier).

Although the enzyme has not yet been purified, it seems likely that it occurs as a distinct racemase in some species of *Proteus*, *E. coli* (HUANG and DAVISSON, 1958), and probably some lysine adapted molds (OGATA, YAMADA, HAYASHI and SUGAWA, 1958).

Assay. The enzyme can be assayed by the manometric estimation of L-lysine formed in the reaction mixture with L-lysine decarboxylase (obtainable commercially). If desired, D-lysine may be assayed by paper chromatography of the residual lysine after complete decarboxylation of L-lysine by the L-lysine decarboxylase.

The optimum pH of lysine racemase has been reported to be 8.4 (HUANG et al., 1958).

6. Proline Racemase.

D-proline ⇌ racemic mixture ⇌ L-proline

The enzyme occurs in the cells of *Clostridium sticklandii* which are capable of completely reducing DL-proline to δ-aminovaleric acid (STADTMAN and ELLIOTT, 1957; also see page 318). The proline racemase forms a racemic mixture at equilibrium after reacting with either the D- or L-isomer of proline. The racemase activity can easily be demonstrated with a polarimeter using L-proline as the enzyme substrate.

Assay (STADTMAN and ELLIOTT, 1957). Purified D-proline reductase is used to measure either, the amount of D-proline formed by the racemase from proline, or the decrease of reducible proline in the presence of D-proline as the substrate. The following reaction mixture is placed in 10×25 mm test tubes and incubated at 31° C for 10 minutes: 8 μmoles of D- or L-proline; 3 μmoles of potassium phosphate buffer, pH 6.7; 6 μmoles of β-mercaptoethanol; 2.5 μmoles of EDTA, and 1—1.5 units of proline racemase to give a final volume of 0.3 ml. After incubation the racemase is inhibited by heating at 95° C for 5 minutes. The D-proline present is then, measured using an excess of purified D-proline reductase (5—6 units; STADTMAN and ELLIOTT, 1957). 10×75 mm test tubes with the following reaction mixture are used; 20 μmoles of Tris buffer, pH 8.7; 10 μmoles of $MgCl_2$; 5—6 units of purified D-proline reductase; 36 μeq of —SH added as 1,3-dimercaptoethanol, and an aliquot of the racemase reaction mixture or control (not containing racemase) to give a final volume of 0.4 ml. The tubes are incubated under an atmosphere of nitrogen or hydrogen at 31° C for 1 hour. δ-aminovaleric acid formed in the mixture is estimated by the MOORE and STEIN ninhydrin procedure (MOORE and STEIN, 1948; also see page 318). The increase or decrease of reducible proline as compared to controls (containing no racemase) is a measure of the racemase activity.

One *unit* is defined as the amount of the enzyme which will convert 1 μmole of either isomer to that of the opposite configuration in 10 minutes using the preceding assay (STADTMAN and ELLIOTT, 1957).

Purification (STADTMAN and ELLIOTT, 1957). All the operations are conducted in the cold. Cells of a 24 hour culture of *Clostridium sticklandii* are suspended in 0.01 M Tris buffer, pH 8.7, and disintegrated for 10 minutes in a sonic oscillator. The resultant suspension after centrifugation is diluted with 0.01 M Tris buffer, pH 8.7, to contain not more than 25 mg protein per ml (Fraction A). Fraction A is treated with 2% protamine sulphate (90 mg protamine sulphate per gm protein) until the 280:260 mμ ratio of the supernatant is 0.85—0.95. The supernatant solution is recovered by centrifugation (Fraction B). Solid ammonium sulphate is added to fraction B to 0.3 saturation and the precipitate is discarded. More solid ammonium sulphate is added to 0.6 saturation and the precipitate obtained after centrifugation is dissolved in 0.05 M Tris, pH 8.7 (Fraction C). The supernatant from fraction C is further saturated to 0.9 with ammonium sulphate. The supernatant is discarded, the precipitate is dissolved in 0.05 M Tris, pH 8.7 (Fraction D).

The protein concentration of fraction C is adjusted to 30 mg per ml with 0.01 M Tris, pH 8.7. Solid ammonium sulphate is added at the rate of 24.5 g per 100 ml of the solution and the precipitate is centrifuged down (Fraction C_1). C_1 is dissolved in 0.05 M Tris, pH 8.4. Solid ammonium sulphate is added (10.5 g per 100 ml) to the supernatant of C_1, and the precipitate is collected (C_2). To the supernatant from this step ammonium sulphate (7 g per 100 ml) is added and the precipitate (C_3) is collected.

Fraction C_2 is dissolved in 0.05 M Tris buffer, pH 8.7, and diluted with water to give a protein concentration of not more than 25 mg per ml. The pH is adjusted to 5.0 with 0.1 M formic acid. The supernatant solution is discarded after centrifugation and the precipitate is redissolved in 0.05 M Tris buffer, pH 8.7, and clarified by high speed centrifugation. This solution is used as a source of the racemase. Fraction C_1 can be used as the source of proline reductase.

Properties (Stadtman and Elliott, 1957). The optimum pH range of the enzyme is 6.7 to 8.1. The enzyme activity is stimulated in the presence of phosphate or EDTA buffers because of their ability to combine with inhibiting metals. Even after extensive dialysis, the enzyme does not require pyridoxal phosphate as a cofactor. Hydroxylamine at a concentration of 5×10^{-2} M stimulates the enzyme reaction by 20—25%.

The presence of a thiol compound is required for activity (β-mercaptoethanol or 1,3-dimercaptopropanol). Reduced glutathione, thioglycollate, and lipoic acid are slightly inhibitory. The thiol requirement suggests that the racemase is a sulfhydryl enzyme.

The enzyme is stable at 0° C for a few hours, but it can be stored at $-20°$ C for a number of days without appreciable loss in activity.

7. α-ε-Diaminopimelic Acid (DAP) Racemase.

$$
\begin{array}{ccc}
\text{COOH} & & \text{COOH} \\
| & & | \\
\text{NH}_2-\text{C}-\text{H} & & \text{H}-\text{C}-\text{NH}_2 \\
| & & | \\
(\text{CH}_2)_3 \rightleftharpoons \text{racemic mixture} \rightleftharpoons (\text{CH}_2)_3 \\
| & & | \\
\text{H}-\text{C}-\text{NH}_2 & & \text{H}-\text{C}-\text{NH}_2 \\
| & & | \\
\text{COOH} & & \text{COOH} \\
\text{LL-}\alpha,\varepsilon\text{-DAP} & & meso\ \text{DAP}
\end{array}
$$

The racemase has been shown to occur in *Aerobacter* (Hoare and Work, 1955a, b, 1956; Antia, Hoare and Work, 1957) and probably also occurs in maize seedlings (Shimura and Vogel, 1961). At equilibrium the racemase forms a racemic mixture of both LL- and *meso*-isomers.

Assay. Cell-free extract is added to 0.1 M borate buffer, pH 8.5 containing 0.4 mg *meso*-DAP per ml, and incubated at 37° C (Antia et al., 1957). Controls, one set lacking the extract, and another set lacking the substrate, are run in conjunction with the test. After convenient time intervals, 0.1 ml of each is transferred to 0.2 ml of ethanol to stop the reaction. After centrifugation, 0.1 ml aliquots are examined for DAP isomers by paper chromatography after the method of Hoare and Work (1957) or Rhuland, Work, Denman and Hoare (1955). In most cases, the enzyme blank gives strong spots at both the LL- and *meso* isomer locations and the assay is not conclusive in such cases.

Another type of assay using an excess of partially purified DAP decarboxylase (see, method of preparation on page 308) may be used to estimate the *meso* isomer formed from the LL-isomer using proper controls (Dewey, Hoare and Work, 1954; Hoare and Work, 1955a).

Purification (Antia et al., 1957). DAP racemase is very unstable in cell-free extracts. The cell free racemase may be stabilized as a mercury derivative prior to purification. The Hg-racemase is easily activated by the addition of 10^{-2} M sodium sulphide before use. Attempts to precipitate the nucleic acids with protamine sulphate or streptomycin have shown that the enzyme is closely associated with the nucleic acids. Ammonium sulphate precipitation reduces all enzyme activity, and acetone fractionation produces no clear cut results. No suitable method has yet been devised for a clear cut purification of the enzyme.

Properties (ANTIA et al., 1957; HOARE and WORK, 1955b). The optimum pH range of the enzyme is between 7.3 and 8.5. The enzyme may be stored for months as a mercury-racemase at −10° C without appreciable loss of activity. The racemase is inhibited by most reagents which combine with the carbonyl and thiol groups. Mercuric chloride, cupric sulphate, and pCMBA completely inhibit racemization. Glutathione reverses the pCMBA inhibition. Hydroxylamine also completely inhibits the activity, but activity is not lost if BAL (dimercapto-propanol) is added simultaneously with hydroxylamine. Hydrazine, semi-carbazide, and isonicotinic acid hydrazide are partial inhibitors. Cyanide has no effect. Pyridoxal phosphate does not stimulate the enzyme activity.

8. Other Amino Acid Racemases.

TANAKA (1960) reported an enzyme preparation from *Lactobacillus fermenti* (ATCC 9338) capable of racemizing the following amino acids: glutamic acid, aspartic acid, alanine, methionine, valine, isoleucine and phenylalanine. Pyridoxal phosphate had no effect on the rate of glutamic acid racemization, but activated the racemization of some other amino acids.

BEHRMAN and CULLEN (1961) reported the isolation of a strain of *Pseudomonas* capable of degrading both D- and L-tryptophan. Cell-free extracts are unable to carry out the racemization.

References.

ALBERS, R. W., and W. B. JAKOBY: Biochim. biophys. Acta 40, 457 (1960). — ALTEN-BERN, R. A., and R. D. HOUSEWRIGHT: J. biol. Chem. 204, 159 (1953). — AMES, S. R., and C. A. ELVEHJEM: J. biol. Chem. 166, 81 (1946). — AMES, B. N., and B. L. HORECKER: J. biol. Chem. 220, 113 (1956). — AMES, B. N., and H. K. MITCHELL: J. biol. Chem. 212, 687 (1955). — AMOS, H.: J. Amer. chem. Soc. 76, 3858 (1954). — AMOS, H., and G. N. COHEN: Biochem. J. 57, 338 (1954). — AMOS, H., and E. VOLLMAYER: J. Bact. 73, 172 (1957). — ANDERSON, P. R., and O. E. MCELROY: Arch. Biochem. Biophys. 82, 256 (1959). — ANTIA, M., D. S. HOARE and E. WORK: Biochem. J. 65, 448 (1957). — ASPEN, A. J., and A. MEISTER: Meth. of Biochem. Res. 6, 131 (1958). — AUERBACH, V. H., and H. A. WAISMAN: J. biol. Chem. 234, 304 (1959). — AWAPARA, J.: J. biol. Chem. 200, 537 (1953). — AWAPARA, J., and B. SEALE: J. biol. Chem. 194, 497 (1952). — AYENGAR, E., and P. ROBERTS: J. biol. Chem. 197, 453 (1952).

BAGDASARIAN, M.: Nature (Lond.) 181, 1399 (1958). — BANKS, B. E. C., K. C. OLDHAM, E. M. THAIN and C. A. VERNON: Nature (Lond.) 183, 1187 (1959). — BARBAN, S.: J. Bact. 71, 274 (1956). — BARBAN, S., and H. O. SCHULZE: J. biol. Chem. 234, 1179 (1959). — BAXTER, C. F., and E. ROBERTS: J. biol. Chem. 233, 1135 (1958). — BEHRMAN, E. J., and A. M. CULLEN: Fed. Proc. 20, 6 (1961). — BESSMAN, S. P., J. ROSSEN and E. C. LAYNE: J. biol. Chem. 201, 385 (1953). — BIGGER-GEHRING, L.: J. gen. Microbiol. 13, 45 (1955). — BLACK, S.: Meth. Enzymol. 1, 508 (1955). — BORNSTEIN, J.: Nature (Lond.) 170, 534 (1957). — BORSOOK, H., and J. W. DUBNOFF: J. biol. Chem. 138, 389 (1941). — BOVARNICK, M. R., and J. C. MILLER: J. biol. Chem. 184, 661 (1950). — BRAUNSTEIN, A. E.: Enzymologia 7, 25 (1939). — BRAUN-STEIN, A. E., and T. S. HSU: Biochem. biophys. Acta 44, 187 (1960).

CAMMARATA, P. S., and P. P. COHEN: J. biol. Chem. 187, 439 (1950); 193, 53 (1951). — CAMPBELL, L. L.: J. Bact. 71, 81 (1956). — CANELLAKIS, Z. N., and P. P. COHEN: J. biol. Chem. 222, 53 (1956). — CHATAGNER, F., B. BERGERET, T. SÉJOURNÉ and C. FROMAGEOT: Biochim. biophys. Acta 9, 340 (1952). — CHATTERJEE, A. N.: Nature (Lond.) 180, 1425 (1957). — CHURCH, B. C., H. HALVORSON and H. O. HALVORSON: J. Bact. 68, 393 (1954). — COHEN, P. P.: Biochem. J. 33, 1478 (1939); — In: The enzymes. Ed. J. B. SUMNER and K. MYRBACK. Vol. 1. New York: Academic Press 1951; — Meth. Enzymol. 2, 178 (1955). — CRUICKSHANK, D. H., and F. A. ISHERWOOD: Biochem. J. 69, 189 (1958).

DAVIES, D. D., and R. J. ELLIS: Biochem. J. 78, 623 (1961). — DAVIS, B. D., and E. S. MINGIOLI: J. Bact. 60, 17 (1950). — DE MARCO, C., and M. COLETTA: Biochim. biophys. Acta 47, 262 (1961). — DEWEY, D. L., D. S. HOARE and E. WORK: Biochem. J. 58, 523 (1954). — DOLIN, M. I.: Ph. D. Thesis, Indiana University (1950).

Eegruve, E.: Z. anal. Chem. **95**, 323 (1933). — Ellis, R. J., and D. D. Davies: Biochem. J. **78**, 615 (1961).

Fegol, K.: M.Sc. Thesis, University of Manitoba (1961). — Feldman, L. I., and I. C. Gunsalus: J. biol. Chem. **187**, 821 (1950). — Fincham, J. R. S.: Nature (Lond.) **168**, 975 (1951); — Biochem. J. **53**, 313 (1953); — J. gen. Microbiol. **11**, 236 (1954). — Fincham, J. R. S., and A. B. Boulter: Biochem. J. **62**, 72 (1956). — Fleisher, G. A., C. S. Potter and K. G. Wakim: Proc. Soc. exp. Biol. (N.Y.) **103**, 229 (1960). — Fleming, D. E., and L. Foshay: J. Bact. **71**, 324 (1956). — Fowden, L.: Biochem. J. **48**, 327 (1951). — Friedemann, T. E., and G. E. Haugen: J. biol. Chem. **147**, 415 (1943). — Frisell, W. R., and C. G. Mackenzie: Meth. Biochem. Analys. **6**, 63 (1958). — Fromageot, P., and U. Patino-bun: Biochim. biophys. Acta **46**, 533 (1961). — Fuld, M.: Fed. Proc. **13**, 215 (1954).

Glaser, L.: J. biol. Chem. **235**, 2095 (1960). — Green, D. E., L. F. Leloir and V. Nocito: J. biol. Chem. **161**, 559 (1945). — Gunsalus, I. C., and J. R. Stamer: Meth. Enzymol. **2**, 170 (1955).

Hanford, J., and D. D. Davies: Nature (Lond.) **182**, 532 (1958). — Hayaishi, O., Y. Nizhizuka, M. Tatibana, M. Takeshita and S. Kuno: J. biol Chem. **236**, 781 (1961). — Hicks, R. M.: Biochim. biophys. Acta **46**, 143 (1961a); **46**, 152 (1961b). — Hicks, R. M., and J. C. Craig: Biochem. J. **67**, 353 (1957). — Hiller, R. G., and D. A. Walker: Biochem. J. **78**, 56 (1961). — Hoare, S. D., and E. Work: Biochem. J. **61**, 562 (1955a); — Proc. of the third Intern. Congr. of Biochemistry, Brussels (Academic Press, Inc., N.Y.) 37 (1955b); — J. gen. Microbiol. **15**, XIII (1956); — Biochem. J. **65**, 441 (1957). — Holzer, H., U. Gerlach, G. Jacobi and M. Gnoth: Biochem. Z. **329**, 529 (1958). — Hopps, H. E., F. E. Hahn, C. L. Wisseman Jr., E. B. Jackson and J. E. Smadel: J. Bact. **71**, 708 (1956). — Housewright, R. D., and C. B. Thorne: J. Bact. **60**, 89 (1950). — Huang, H. T., and J. W. Davisson: J. Bact. **76**, 495 (1958). — Huang, H. T., D. A. Kita and J. W. Davisson: J. Amer. chem. Soc. **80**, 1006 (1958). — Hug, D. H., and C. H. Werkman: Arch. Biochem. biophys. **72**, 369 (1957).

Isherwood, F. A., and D. H. Cruickshank: Nature (Lond.) **173**, 121 (1954a); **174**, 123 (1954b). — Ito, K., A. Katayama and S. Tanaka: Kôso Kagaku Shinpojiumu **12**, 111 (1957). — Itoh, K.: Nagoya J. Med. Sci. **21**, 181 (1958).

Jakoby, N. B., and D. M. Bonner: J. biol. Chem. **221**, 689 (1956). — Jenkins, W. T., and I. W. Sizer: J. Amer. chem. Soc. **79**, 2655 (1957). — Jenkins, W. T., D. A.Yphantis and I. W. Sizer: J. biol. Chem. **234**, 51, 1179 (1959). — Jordan, D. C.: J. Bact. **65**, 220 (1953).

Kallio, R. E., and A. D. Larson: In: Amino acid metabolism. Ed. W. McElroy and B. Glass, Baltimore: Johns Hopkins Press 1955. — Kearney, E. B., and T. P. Singer: Biochim. biophys. Acta **11**, 276 (1953). — Kenney, F. T.: J. biol. Chem. **234**, 2707 (1959). — Kenney, R. W., and C. H. Werkman: Iowa College J. Science **32**, 455 (1958). — Knox, W. E., and B. M. Pitt: J. biol. Chem. **225**, 675 (1957). — Koide, F., T. Shishido, H. Nagayama and K. Shimura: J. agric. Chem. Soc. (Japan) **30**, 283 (1956). — Kowalski, E., A. Dancewicz and Z. Szot: Bull. Acad. pol. des. Sci. 11, **5**, 223 (1957). — Krebs, H. A., and L. V. Eggleston: Biochem. J. **39**, 408 (1945). — Krueger, R.: Helv. chim. Acta **32**, 238 (1949). — Kun, E., D. W. Franshier and D. R. Grassetti: J. biol. Chem. **235**, 416 (1960). — Kupiecki, F. P., and M. J. Coon: J. biol. Chem. **229**, 743 (1957).

Lichstein, H. C., and C. P. Cohn: J. biol. Chem. **157**, 85 (1945). — Lichstein, H. C., I. C. Gunsalus and W. W. Umbriet: J. biol. Chem. **161**, 311 (1945). — Lin, E. C. C., M. Civen and W. E. Knox: J. biol. Chem. **233**, 1183 (1958). — Lin, E. C. C., and W. E. Knox: J. biol. Chem. **233**, 1186 (1958). — Lin, E. C. C., B. M. Pitt, M. Civen and W. E. Knox: J. biol. Chem. **233**, 668 (1958). — Lin, E. C. C., R. S. Rivilin and W. E. Knox: Amer. J. Physiol. **196**, 303 (1959). — Lis, H.: Biochim. biophys. Acta **28**, 191 (1958). — Longenecker, J. B., and E. E. Snell: Proc. nat. Acad. Sci. (Wash.) **42**, 221 (1956).

Mason, M.: J. biol. Chem. **211**, 839 (1954). — Mason, M., and C. P. Berg: J. biol. Chem. **188**, 783 (1951); **195**, 515 (1952). — Mason, M., and E. H. Gullehson: J. biol. Chem. **235**, 1312 (1960). — McCurdy, H. D. jr., and E. C. Cantino: Plant Physiol. **35**, 463 (1960). — Meadow, P., and E. Work: Arch. Biochem. Biophys. **28**, 596 (1958). — Meister, A.: J. biol. Chem. **195**, 813 (1952); **200**, 571 (1953); **206**, 577 (1954a); **206**, 587 (1954b); **210**, 17 (1954c); — Advanc. Enzymol. **16**, 185 (1955); — Biochemistry of the amino acids, New York: Academic Press 1957. — Meister, A., P. E. Fraser and S. V. Tice: J. biol. Chem. **206**, 561 (1954). — Meister, A., L. Levintow, R. E. Greenfield and P. A. Abendschein: J. biol. Chem. **215**, 441 (1955). — Meister, A., H. A. Sober and E. A. Peterson: J. biol. Chem. **206**, 89 (1954). — Meister, A., H. A. Sober, S. V. Tice and P. E. Fraser: J. biol. Chem. **197**, 319 (1952). — Meister, A., and S. V. Tice: J. biol. Chem. **187**, 173 (1950). — Miller, I. L., M. Tsuchida and A. Adelberg: J. biol. Chem. **203**, 205 (1953). — Monder, C., and A. Meister: Biochim. biophys. Acta **28**, 202 (1958). — Moore, S., and W. H. Stein: J. biol. Chem. **176**, 367 (1948). — Moses, M., and M. A. Joslyn: J. Bact. **66**, 204 (1953). — Myers, J. W., and E. A. Adelberg: Proc. nat. Acad. Sci. (Wash.) **40**, 493 (1954).

NAGAYAMA, H., M. MURAMATSU and K. SHIMURA: Nature (Lond.) **181**, 417 (1958). — NAKATSU, S.: J. Biochem. **43**, 675 (1956). — NARROD, S. A., and W. A. WOOD: Arch. Biochem. Biophys. **35**, 462 (1952). — NEGELEIN, E., and H. BRÖMEL: Biochem. Z. **300**, 225 (1938—1939). — NOE, F. F., and W. J. NICKERSON: J. Bact. **75**, 674 (1958).

OGATA, M., M. YAMADA, H. HAYASHI and T. SUGAWA: Osaka Daigaku Igaku Zasshi **10**, 1021 (1958). — O'KANE, D. E., and I. C. GUNSALUS: J. biol. Chem. **170**, 425 (1947). — OLENICHEVA, L. S.: Biokhimiya **20**, 165 (1955). — OLIVARD, J., and E. E. SNELL: J. biol. Chem. **213**, 203 (1955).

PATWARDHAN, M. W.: Nature (Lond.) **181**, 187 (1958). — PEREZ-MILAN, H., J. SCHLIACK and P. FROMAGEOT: Biochim. biophys. Acta **36**, 73 (1959). — PISANO, J. J., C. MITOMA and S. UDENFRIEND: Nature (Lond.) **180**, 1125 (1957).

RACKER, E.: J. biol. Chem. **184**, 313 (1950). — RHULAND, L. E., E. WORK, R. F. DENMAN and S. D. HOARE: J. Amer. chem. Soc. **77**, 4844 (1955). — ROBERTS, E.: Arch. Biochem. Biophys. **48**, 395 (1954). — ROBERTS, E., P. AYENGAR and I. POSNER: J. biol. Chem. **203**, 195 (1953a); — Fed. Proc. **12**, 259 (1953b). — ROBERTS, E., and H. M. BREGOFF: J. biol. Chem. **201**, 393 (1953). — ROBERTS, E., and S. FRANKEL: J. biol. Chem. **188**, 789 (1951). — ROINE, P.: Ann. Acad. Sci. Fennicae A., **11**, 26 (1947). — ROSEN, H.: Arch. Biochem. Biophys. **67**, 10 (1957). — ROSEN, F., N. R. ROBERTS and C. A. NICHOL: J. biol. Chem. **234**, 476 (1959). — RUDMAN, D., and A. MEISTER: J. biol. Chem. **200**, 591 (1953).

SAKAI, J.: Kekkaku (Tuberculosis) **27**, 161 (1954). — SALLACH, H. J.: In: Amino acid metabolism, Ed. W. McELROY and B. GLASS, Baltimore: Johns Hopkins Press 1955; J. biol. Chem. **223**, 1101 (1956). — SALVADOR, R. A., and R. W. ALBERS: J. biol. Chem. **234**, 922 (1959). — SANWAL, B. D.: Experientia **14**, 246 (1958). — SANWAL, B. D., and M. W. ZINK: Arch. Biochem. Biophys. (1961 in the press). — SCHLENK, F., and A. FISHER: Arch. Biochem. **12**, 60 (1947). — SCOTT, E. M., and W. B. JAKOBY: J. biol. Chem. **234**, 932 (1959). — SEECOF, R. L., and R. P. WAGNER: J. biol. Chem. **234**, 2689 (1959a); **234**, 2694 (1959b). — SERENI, F., F. T. KENNEY and N. KRETCHMER: J. biol. Chem. **234**, 609 (1959). — SHEFFNER, A. L., and J. GRABOW: J. Bact. **66**, 192 (1953). — SHIIO, I., B. MARUO and S. AKABORI: J. Biochem. **43**, 779 (1956). — SHIMURA, Y., and H. G. VOGEL: Fed. Proc. **20**, 10 (1961). — SHOCKMAN, G. D., and G. TOENNIES: Arch. Biochem. Biophys. **50**, 1 (1954a); **50**, 9 (1954b). — SISKEN, B., K. SANO and E. ROBERTS: J. biol. Chem. **236**, 503 (1961). — STADTMAN, T. C., and P. J. ELLIOTT: J. biol. Chem. **228**, 938 (1957). — STAFFORD, H. A., A. MAGALDI and B. VENNESLAND: J. biol. Chem. **207**, 621 (1954). — STEWART, B. T., and H. O. HALVORSON: J. Bact. **65**, 160 (1953); Arch. Biochem. Biophys. **49**, 168 (1954). — STRAUB, F. B.: Hoppe-Seylers Z. physiol. Chem. **275**, 63 (1942). — STUMPF, P. K.: Fed. Proc. **10**, 256 (1951).

TANAKA, M.: J. agr. chem. Soc. (Japan) **34**, 1022 (1960). — TANAKA, M., Y. KATO and S. KINOSHITA: Biochem. biophys. Research Comms. **4**, 114 (1961). — TANENBAUM, S. W.: J. biol. Chem. **218**, 733 (1956). — TANENBAUM, S., and D. SHEMIN: Fed. Proc. **9**, 236 (1950). — TEAS, H. J.: Oak Ridge National Laboratory Report 164 (1948). — THORNE, C. B.: In: Amino acid metabolism, Ed. W. McELROY and B. GLASS, Baltimore: Johns Hopkins Press 1955. — THORNE, C. B., C. G. GOMEZ and R. D. HOUSEWRIGHT: J. Bact. **69**, 357 (1955). — THORNE, C. B., and D. M. MOLNAR: J. Bact. **70**, 420 (1955). — TONHAZY, N. E., N. G. WHITE and W. W. UMBREIT: Arch. Biochem. **28**, 36 (1950).

UMBARGER, H. E., and E. D. ADELBERG: J. biol. Chem. **192**, 883 (1951). — UMBARGER, H. E., and B. MAGASANIK: J. Amer. chem. Soc. **74**, 4256 (1952).

VELLEZ, L., G. AMIARD and M. PESEZ: Bull. Soc. Chim. Fr. **15**, 678 (1948).

WAGNER, R. P., A. BERGQUIST and G. W. KARP: Arch. Biochem. Biophys. **74**, 182 (1958).— WAGNER, R. P., and P. W. IFLAND: C. R. trav. lab. Carlsberg Sér. physiol. **26**, 381 (1956). — WALKER, J. B.: Arch. Biochem. Biophys. **59**, 233 (1955); — J. biol. Chem. **218**, 549 (1956); **224**, 57 (1957); **231**, 1 (1958). — WALKER, D. Λ., and S. L. RANSON: Plant Physiol. **33**, 226 (1958). — WILSON, D. G., K. W. KING and R. H. BURRIS: J. biol. Chem. **208**, 863 (1954). — WISS, O.: Z. Naturforsch. **7b**, 133 (1952). — WISSEMAN, C. L. jr., F. E. HAHN, E. B. JACKSON, F. M. BOZEMAN and J. E. SMADEL: J. Immunol. **68**, 251 (1952). — WOOD, W. A.: Meth. in Enzymol. **2**, 212 (1955). — WOOD, W. A., and I. C. GUNSALUS: J. biol. Chem. **190**, 403 (1951).

YANOFSKY, C.: In: Amino acid metabolism, Ed. W. McELROY and B. GLASS, Baltimore: Johns Hopkins Press 1955. — YEMM, E. W., and E. C. COCKING: Analyst **80**, 209 (1955). — YONEDA, M., N. KATO and M. OKAJIMA: Nature (Lond.) **170**, 803 (1952).

Enzymes of Peptide and Protein Metabolism.

By

George Webster.

With 11 Figures.

The enzymes concerned with the synthesis and degradation of amides, peptides, and proteins are of considerable interest because of the importance of their substrates in cellular functions. The study of these enzymes has sometimes been hampered by their instability and by difficulties in the assay of their activities, but recent improvements in methodology have greatly aided their study. The enzymes may be divided conveniently into two classes: (a) enzymes catalyzing the synthesis of amides, peptides, or proteins (synthetases); and (b) enzymes catalyzing the degradation of amides, peptides or proteins (amidases, peptidases, and proteases).

Synthetases require the presence of some divalent cation. Every reaction catalyzed by a synthetase also involves the participation of a nucleoside triphosphate, generally adenosine triphosphate (ATP). The ATP is always split during the synthetase-catalyzed reaction, either by a phosphorylytic reaction:

$$A + B + ATP \rightleftharpoons A\text{---}B + ADP + \text{orthophosphate}$$

or a pyrophosphorylytic reaction:

$$A + B + ATP \rightleftharpoons A\text{---}B + AMP + \text{pyrophosphate}$$

There seems to be no general way, as yet, by which one can predict whether a reaction will proceed by a phosphorylytic or a pyrophosphorylytic split. Evidence has accumulated that the phosphorylytic split of ATP involves the intermediate formation of an enzyme-bound substrate-phosphate in the synthesis reaction, while the pyrophosphorylytic split involves the intermediate formation of an enzyme-bound substrate-adenylate. The possibility of other mechanisms being involved must not be dismissed, however, until considerably more detailed examination has been made of the synthetases and their mechanisms of action. Our knowledge of many details concerning the synthetases is very incomplete, and this points up the need for much further study on the isolation and general properties of this important class of enzymes.

Our knowledge of the degradation enzymes (amidases, peptidases, and proteases) is variable, but, in general, is more detailed than our knowledge of the synthetases. The availability of highly purified proteolytic enzymes has allowed studies to be performed which have contributed much to enzymology, but, again, many fascinating studies remain.

Abbreviations used: ATP, adenosine triphosphate; ADP, adenosine diphosphate; AMP, adenosine monophosphate; DPN and DPNH, the oxidized and reduced forms of diphospho-pyridine nucleotide; RNA, ribonucleic acid; sRNA, amino acid-acceptor ribonucleic acid; GTP, guanosine triphosphate.

A. Enzymes Concerned with Syntheses (Synthetases).

I. Amide Synthesis.

1. Glutamine Synthetase.

Glutamine synthetase catalyzes the synthesis of the amide, glutamine, by the reaction:

$$\text{glutamate} + NH_3 + ATP \rightleftharpoons \text{glutamine} + ADP + \text{orthophosphate}$$

At present, this reaction represents the only known pathway for the biosynthesis of glutamine. Glutamine synthetase also catalyzes the following glutamyl transfer reactions:

$$\text{glutamine} + N^{15}H_3 \rightleftharpoons N^{15}\text{-glutamine} + NH_3$$
$$\text{glutamine} + NH_2OH \rightarrow \gamma\text{-glutamyl hydroxamate} + NH_3$$

as well as the 'arsenolysis' of the glutamine:

$$\text{glutamine} + H_2O \xrightarrow{\text{arsenate}} \text{glutamate} + NH_3$$

The arsenolysis reaction requires ADP, while the transfer reaction requires ADP and either phosphate or arsenate.

a) Occurrence.

Glutamine synthetase occurs in extracts of pumpkin seedlings (STUMPF and LOOMIS, 1950); lupine (ELLIOTT, 1951); bean seedlings, corn seedlings, potatoes, sunflower roots, beet roots, carrot roots, sweet potato roots, and wheat germ (WEBSTER, 1953c); peas (ELLIOT, 1953); and various bacteria (ELLIOTT and GALE, 1948; LAJTHA, MELA and WAELSCH, 1953). It has also been found in extracts of pigeon liver (SPECK, 1949); sheep brain (ELLIOTT, 1951); chick embryo (RUDNICK, MELA and WAELSCH, 1954); mouse liver and certain hepatomas (LEVINTOW, 1954); and the insect *Prodenia eridonia* (LEVENBOOK and KUHN, 1958).

b) Assay.

Glutamine synthetase activity can be determined by the colorimetric determination of glutamine formed from glutamate and ammonia following the separation of glutamine from glutamate by chromatography (WEBSTER, 1953). It may also be determined by measurement of the ADP liberated during glutamine synthesis by coupling the reaction to pyruvate kinase and lactic dehydrogenase (WIELAND, PFLEIDERER and SANDMANN, 1958). A relatively simple assay is the colorimetric determination of the formation of glutamyl hydroxamate which is catalyzed by glutamine synthetase at rates identical to glutamine synthesis, if hydroxylamine is substituted for ammonia in the synthesis reaction (SPECK, 1949):

$$\text{glutamate} + NH_2OH + ATP \rightarrow \text{glutamyl hydroxamate} + ADP + \text{orthophosphate}$$

The glutamyl transfer reaction can be determined by the incorporation of $N^{15}H_3$ into the glutamine (DELWICHE, LOOMIS and STUMPF, 1951) or the formation of glutamyl hydroxamate from glutamine (STUMPF and LOOMIS, 1950).

The assay system for the determination of synthetase activity by measurement of the formation of glutamyl hydroxamate consists of (VARNER and WEBSTER, 1955): 45 μmoles tris(hydroxymethyl)aminomethane-HCl buffer (pH 7.8), 50 μmoles potassium glutamate, 2 μmoles ATP, 30 μmoles MgSO$_4$, 10 μmoles β-mercaptoethanol, and 40 μmoles hydroxylamine per ml of reaction mixture (usually four ml). Sufficient enzyme should be added to produce about 0.25 μmole of glutamyl hydroxamate per ml in 15 minutes at 35° C. The course of the reaction is stopped by the addition of 1 ml of 12 per cent trichloroacetic acid, and the precipitated protein is removed by centrifugation for five minutes at approximately $1000 \times g$. To the supernatant solution are added 1 ml of HCl (diluted 1:3 with H_2O) and 1 ml of five per cent $FeCl_3$ in 0.1 M HCl. The solutions are mixed thoroughly and allowed to stand for ten minutes before the amount of color is determined at $540 \, m\mu$. One unit of activity equals that quantity of enzyme catalyzing the formation of one μmole of glutamyl hydroxamate in one hour under standard assay conditions.

The assay system for the determination of glutamyl transfer activity by the measurement of the formation of glutamyl hydroxamate from glutamine (Varner and Webster, 1955) consists of: 45 μmoles tris(hydroxymethyl)aminomethane-HCl of pH 7.4, 30 μmoles $MgSO_4$, 10 μmoles β-mercaptoethanol, 0.5 μmole ADP, 50 μmoles glutamine, 25 μmoles inorganic phosphate, and 40 μmoles hydroxylamine per ml of reaction mixture (usually four ml). Enough enzyme is added to produce about 0.25 μmole of glutamyl hydroxamate per ml in 15 minutes at 35° C.

The synthesis of glutamine itself may be measured by substitution of an equal concentration of ammonium chloride for hydroxylamine in the reaction system described above for the assay of glutamyl hydroxamate synthesis. After incubation of the enzyme with the reaction system, the reaction is stopped by the addition of three volumes of cold acetone, and the precipitated protein is removed by centrifugation at $5000 \times g$ for five minutes. The acetone is removed under vacuum. A measured quantity (usually 0.3 ml) of the reaction mixture is applied across one side of a sheet of Whatman No. 1 filter paper, and the components separated by chromatography with 88 per cent phenol in water. The faster-moving glutamine is separated from glutamate. The two compounds are made visible by dipping the paper in 0.1 per cent ninhydrin in acetone. After the paper has been dried in an oven for five minutes at about 80° C, the glutamate and glutamine are eluted from the paper, and their concentrations determined with a colorimeter (Landua and Awapara, 1949).

The determination of glutamine synthetase activity by the method described by Wieland et al. (1958) depends upon the following series of reactions.

$$\text{glutamate} + NH_3 + ATP \rightleftharpoons \text{glutamine} + ADP + \text{orthophosphate}$$
$$ADP + \text{phosphoenolpyruvate} \rightleftharpoons ATP + \text{pyruvate}$$
$$\text{pyruvate} + DPNH + H^+ \rightleftharpoons \text{lactate} + DPN^+$$

The reaction mixture consists of: 0.1 M tris(hydroxymethyl)aminomethane-HCl of pH 7.4, 1.6×10^{-6} M phosphoenolpyruvate, 5×10^{-5} M $MgCl_2$, 2×10^{-4} M KCl, 1×10^{-5} M ATP, 3×10^{-5} M NH_4Cl, 2.5×10^{-5} M L-glutamate, 8×10^{-7} M DPNH, 10 μg lactic dehydrogenase, 10 μg pyruvic kinase, and the glutamine synthetase preparation in a total volume of 3.5 ml. The glutamate-dependent change in DPNH concentration is determined by measurement of the change in optical density at $366 \, m\mu$. From the rate of change, the rate of glutamine synthesis can be calculated.

c) Preparation.

Although glutamine synthetase has been found in extracts from a variety of different tissues, it has been highly purified from only one source: green pea *(Pisum sativum)* seeds. Two methods are available for the purification of glutamine synthetase. The method of Elliott (1953) has been widely employed for the purification of glutamine synthetase, and much of our present knowledge concerning the characteristics of the glutamine synthesis reaction has been derived from studies employing this enzyme preparation. The preparative method of ELLIOTT (1953) has been modified by VARNER and WEBSTER (1955) to give a preparation about two times as pure as the original ELLIOTT (1953) preparation. As will be discussed later, neither the preparative method of ELLIOTT (1953) nor the modification of VARNER and WEBSTER (1955) provides an enzyme preparation which behaves as a pure protein. A different purification procedure has, therefore, been developed (VARNER and WEBSTER, unpublished) which produces an enzyme preparation behaving as an apparently pure protein.

Purification of glutamine synthetase by the method of ELLIOTT (1953) as modified by VARNER and WEBSTER (1955). Dry green pea seeds are pulverized to a fine powder in a power-driven mill. Eighteen kg are stirred for 30 minutes with 144 l of cold 0.1 M $NaHCO_3$. Following this, 3.7 l (0.05 volume) of 2 M $MgSO_4$ are stirred in, and the precipitate allowed to settle at 0° C overnight. The supernatant fluid is poured off as cleanly as possible, and the remaining material is discarded. Alternatively, the entire preparation can be passed through some form of continuous centrifuge.

The extract is adjusted to pH 6.5 by the addition of 2.0 M KH_2PO_4, and 300 g of $(NH_4)_2SO_4$ are added per liter of extract. The precipitate is allowed to settle overnight at 0° C, and the supernatant fluid is poured off and discarded. The precipitate is collected from the remaining suspension by centrifugation, and is resuspended in about six liters of cold distilled water. The pH of the suspension is adjusted to 7.2 with K_2HPO_4, and the suspension is dialyzed against two changes of 40 l of cold distilled water for about 36 hours.

The dialyzed extract is treated with a solution of two per cent protamine sulfate until a small sample, after centrifugation, gives no further precipitate on addition of a drop of protamine solution. The precipitate is removed by centrifugation and discarded. The protamine treatment seems to remove an inhibitor (apparently an adenosine triphosphatase) from the extract, because an increase in total activity is usually obtained at this stage.

The pH of one liter quantities of the supernatant solution from the protamine treatment is adjusted to 5.1 with 1.0 M acetic acid (approximately ten ml). Sixty ml of two per cent potassium ribonucleate solution (pH 5.5) are then added. The precipitate is sedimented by centrifugation, dissolved in a minimal volume of cold distilled water, and neutralized to pH 7.3 with 1.0 M K_2HPO_4. The suspension is stirred for 15 minutes, and then centrifuged for one hour at approximately 140,000 $\times g$. The sediment is discarded.

One hundred ml of the clear supernatant solution from high speed centrifugation are mixed with four ml of 1.0 M sodium phosphate buffer (pH 7.4), and then 72 ml of saturated ammonium sulfate solution. All of the solutions must be as cold as possible, as the enzyme is quite unstable in the presence of ammonium sulfate. The precipitate is removed by centrifugation at 5000 $\times g$ for five minutes, and 60 ml of saturated ammonium sulfate solution are added to the supernatant solution. The resulting precipitate is collected by centrifugation at 5000 $\times g$ for five minutes. The precipitate is dissolved in 15 ml of cold water, and the pH of the suspension is adjusted to 7.3 by the dropwise addition of 1.0 M K_2HPO_4. The solution is dialyzed for approximately 18 hours against three changes of four liters of cold distilled water. Any precipitate which forms is removed by centrifugation at 5000 $\times g$ for five minutes and is discarded.

To each ten ml of the supernatant solution are added one ml of one per cent ribonucleic acid solution and then 0.2 ml of 0.2 M acetic acid. The precipitate is sedimented by centrifugation at 5,000 $\times g$ for five minutes. The precipitate is dissolved in four ml of 0.05 M sodium phosphate buffer (pH 7.3).

To this solution is added 0.2 volume of 95 per cent ethanol which has been precooled to $-20°$ C. The precipitate which forms is sedimented by centrifugation at $5000 \times g$ for five minutes at 5° C. The sedimented precipitate is discarded. The supernatant solution contains the purified glutamine synthetase. A summary of the purification procedure is given in Table 1.

Table 1. *Summary of purification procedure for glutamine synthetase by the method of* ELLIOTT *(1953)* as modified by VARNER and WEBSTER (1955). One unit of activity equals the amount of enzyme required to catalyze the synthesis of one μmole of glutamyl hydroxamate per hour under standard conditions.

Stage	Specific activity (units/mg protein)	Yield (per cent)	Ratio: synthetase transferase
1. Extract	0.3	(100)	1.1
2. 1st $(NH_4)_2SO_4$ precipitation .	0.7	30	1.0
3. Protamine	1.8	37	1.3
4. 1st RNA precipitation. . . .	19.0	27	1.2
5. 2nd $(NH_4)_2SO_4$ precipitation .	150.0	16	1.3
6. Dialysis	270.0	13	1.3
7. 2nd RNA precipitation . . .	600.0	11	1.4
8. Ethanol fractionation	1100.0	5	1.3

Purification of glutamine synthetase by the method of VARNER and WEBSTER. Finely-ground powder from dried green peas is extracted for 60 minutes with cold 20 per cent ethanol (five ml per gram of powder). The suspension is centrifuged for 20 minutes at $13,000 \times g$. The sediment is discarded.

To the clear, yellow supernatant solution is added enough calcium phosphate gel to adsorb 90 per cent of the synthetase activity. The suspension is centrifuged at $5000 \times g$ for five minutes, and the supernatant solution is discarded. The gel is washed with 0.1 M tris(hydroxymethyl)aminomethane (pH 7.0). The synthetase is eluted by extracting the gel with three volumes of 0.1 M K_2HPO_4. The suspension is centrifuged for five minutes at $5000 \times g$ and the gel is discarded.

To the phosphate eluate from calcium phosphate gel, is added 1/20th volume of two per cent ribonucleic acid solution. The pH of the mixture is adjusted to approximately 5.7 (until the first faint cloudiness appears) with 1 M acetic acid. In a good preparation, the activity comes down as a transparent pellet before anything but a very faint cloudiness has appeared. The preparation is centrifuged for five minutes at $20,000 \times g$. The supernatant solution is discarded, and the sediment is suspended in 1/10th volume of 0.1 M tris(hydroxymethyl)aminomethane of pH 7.0. The pH of the suspension is adjusted to 7.0. One per cent protamine sulfate is added dropwise until no more precipitate forms. The solution is centrifuged at $5,000 \times g$ for five minutes, and the precipitate is discarded.

Table 2. *Summary of purification procedure for glutamine synthetase by the method of* VARNER *and* WEBSTER. One unit of activity is defined as in Table 1.

Stage	Specific activity (units/mg protein)	Yield (per cent)	Ratio: synthetase transferase
1. Ethanol extract	2.4	(100)	1.3
2. Eluate from $Ca_3(PO_4)_2$ gel	9.0	71	1.1
3. Nucleic acid precipitate	129.0	58	1.2
4. Supernatant solution from protamine	255.0	55	1.1
5. Charcoal eluate.	621.0	30	1.1
6. Supernatant from 2nd nucleic acid and protamine	1845.0	23	1.1

A thick suspension of charcoal (Norit) is added to the supernatant solution until all of the synthetase activity is adsorbed. The charcoal is sedimented at 5,000 ×g for five minutes, and the supernatant solution is discarded. The synthetase activity is eluted from the charcoal with 1/5th volume of 0.5 M potassium phosphate buffer (pH 8.0). The charcoal is sedimented by centrifugation at 5,000 ×g for five minutes, and discarded.

The supernatant solution contains the purified glutamine synthetase. A summary of the purification procedure is given in Table 2.

d) Properties.

The glutamine synthetase preparations derived from pea seeds appear to be relatively pure. Fig. 1 presents an ultracentrifugal sedimentation pattern (LEVINTOW, et al., 1955) of glutamine synthetase prepared by the method of ELLIOTT (1953). The pattern obtained by ultracentrifugal analysis was found (LEVINTOW, et al., 1955) to contain se-

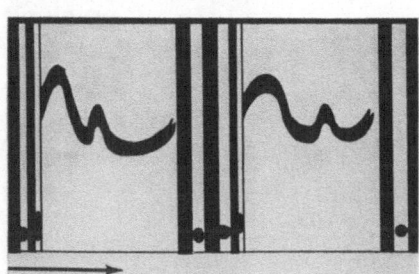

Fig. 1. Sedimentation pattern of glutamine synthetase prepared by the method of ELLIOTT (1953). Protein concentration, 0.8 per cent; solvent, 0.15 M NaCl; speed, 50,740 r.p.m. Patterns are at 39 and 55 minutes after attainment of speed. From LEVINTOW et al. (1955).

veral components. The most rapidly sedimenting component had a sedimentation constant of 13.9 S, and comprised about 20 per cent of the total protein of the preparation. It contained all of the glutamine synthetase activity. The

remainder of the material had an average sedimentation constant of 4.1 S, and appeared to be polydisperse. An ultracentrifugal pattern of glutamine synthetase prepared by the modification (VARNER and WEBSTER, 1955) of the EL-LIOTT (1953) procedure is given in Fig. 2. It can be seen that this preparation also does not behave as a pure protein. In contrast, Fig. 3 shows an ultracentrifugal pattern of a preparation of glutamine synthetase prepared by the procedure described by VARNER and WEBSTER. This preparation exhibits only one centrifugal component, with a sedimentation constant of about 17.5 S. Like-

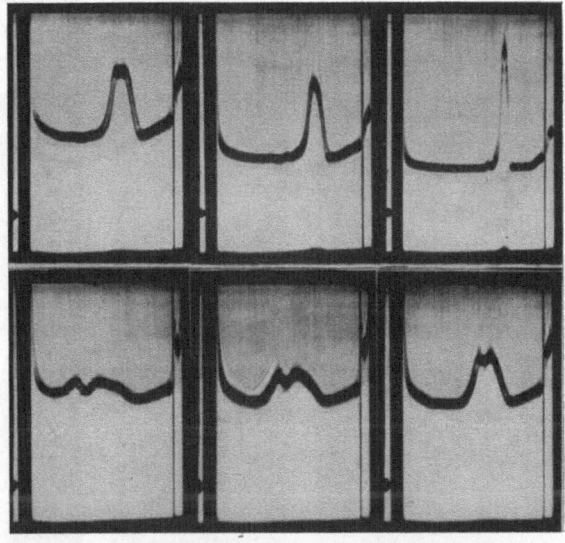

Fig. 2. Sedimentation pattern of glutamine synthetase prepared by the method of ELLIOTT (1953) as modified by VARNER and WEBSTER (1955). Protein concentration, 0.5 per cent; solvent 0.05 M tris(hydroxymethyl)aminomethane-HCl (pH 7.5); speed, 59,780 r.p.m. Patterns are at 24, 32, 40, 48, 64, and 80 minutes after attainment of speed.

wise, as is seen in Fig. 4, this glutamine synthetase preparation exhibits essentially one component upon electrophoresis in several buffers at several pH values.

It is interesting to note that the sedimentation constants of the glutamine synthetase prepared by either of the procedures described above indicate that the enzyme is a relatively large molecule and probably has a molecular weight of some 350,000—450,000. Although a number of studies have been performed on the properties of the glutamine synthesis reaction, very little is known about the

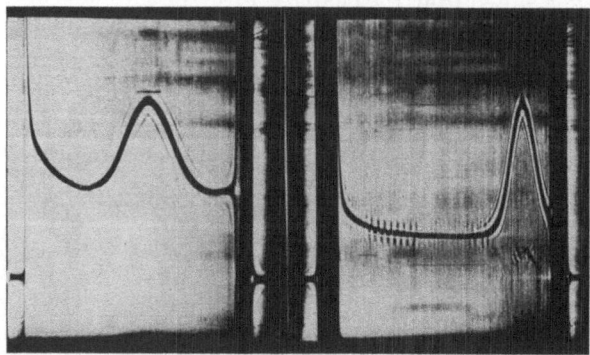

Fig. 3. Sedimentation pattern of glutamine synthetase prepared by the method of VARNER and WEBSTER.

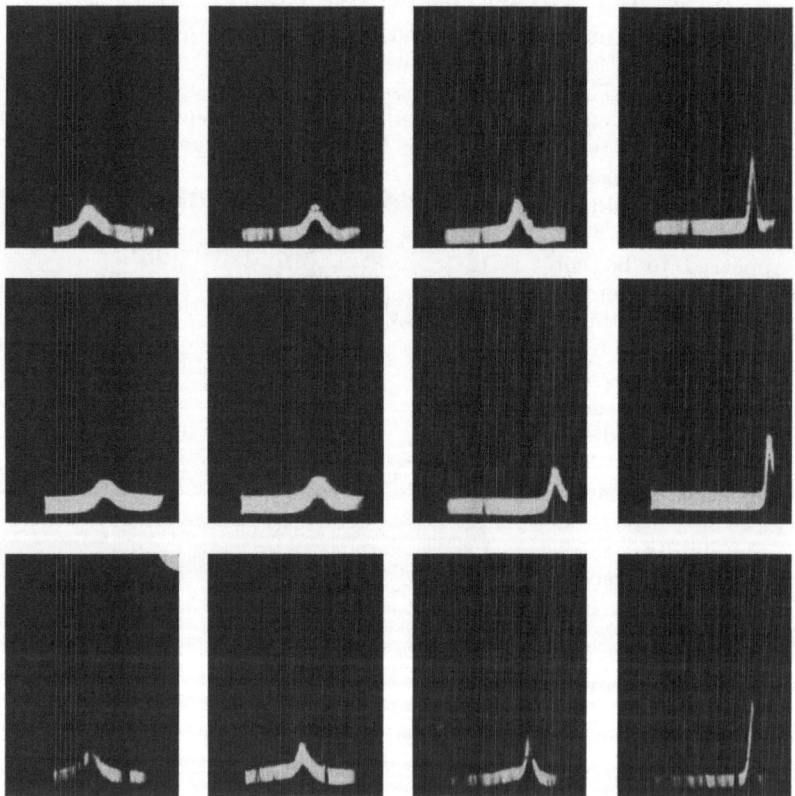

Fig. 4. Electrophoresis of glutamine synthetase prepared by the method of VARNER and WEBSTER. Bottom row, 0.05 M sodium barbital (pH 8.6); middle row, 0.05 M tris(hydroxymethyl)aminomethane-HCl (pH 7.5); top row, 0.05 M sodium phosphate (pH 6.2).

enzyme molecule itself. It may well be that the relatively high molecular weight of the enzyme is due to the aggregation of two or more enzyme molecules of lower molecular weight. The enzyme, as prepared by the ELLIOTT (1953) procedure, exhibits a requirement for a sulfhydryl-containing substance for maximal activity. It is also possible to increase the activity of this enzyme preparation by pre-incubation of the enzyme with ATP (BOYER et al., 1959). In contrast, the enzyme prepared by the procedure of VARNER and WEBSTER exhibits no requirement for a sulfhydryl-containing substance, and its activity is not increased by preincubation with ATP. Glutamine synthetase is relatively unstable, especially at acid pH (Fig. 5).

Table 3. *Relative rates of synthesis and transfer reactions with substituted glutamate and glutamine derivatives* (LEVINTOW et al., 1955). The rates with glutamate and glutamine are arbitrarily given a value of 100.

Substituent	Glutamine synthesis	Glutamyl hydroxamate synthesis	Glutamyl transfer reaction
None.	100	100	100
α-Methyl	85	84	50
β-Methyl	1	28	0
γ-Methyl	22	33	0
γ-Methylene . .	3	5	2
β-Hydroxy . . .	6	8	—
allo-β-Hydroxy .	16	80	—

The enzyme shows a variable specificity for the components of the reaction mixture. As can be seen from Table 3, substitution of a methyl group on the alpha carbon of glutamate results in only about a 15 per cent reduction in the rate of glutamine synthesis, but substitution on the beta carbon results in very great inhibition of glutamine synthesis. The specificity of the enzyme for L-glutamate as compared with D-glutamate is a function of pH and of the nature and the

Table 4. *Effects of various metal ions on the optical specificity of glutamine synthetase* (VARNER, 1960). Reactants and concentrations (in μmoles per ml): ATP, 5.0; metal ion, 5.0; glutamate, 25.0; hydroxylamine, 40.0; tris(hydroxymethyl)aminomethane (pH 7.5), 40.0; in a final volume of four mls. The rate with L-glutamate and magnesium ion is arbitrarily given a value of 100.

Metal ion	L-Glutamyl hydroxamate synthesis	D-Glutamyl hydroxamate synthesis
Mg^{++}	100	60
Co^{++}	110	25
Mn^{++}	21	6
Zn^{++}	12	1
Fe^{++}	7	3

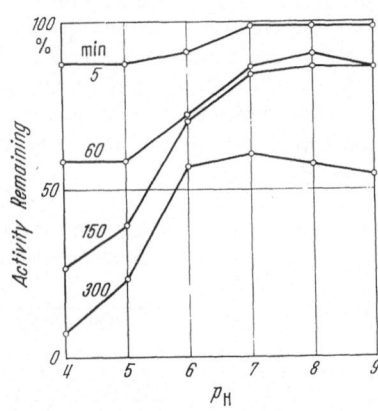

Fig. 5. Stability of glutamine synthetase at various hydrogen ion concentrations (VARNER and WEBSTER, 1955).

concentration of metal ions present (VARNER, 1960). Typical results concerning optical specificity are presented in Table 4, and illustrate that the enzyme is activated about equally well by magnesium or cobalt ions, but other divalent cations are much less effective. The enzyme is apparently highly specific for ATP, and other nucleoside triphosphates exhibit little or no activity. Glutamine synthetase exhibits a marked lack of specificity for ammonia. Either hydroxylamine or hydrazine can substitute for ammonia with the consequent synthesis of glutamyl hydroxamate or glutamyl hydrazide at rates equal to the rate of glutamine synthesis. It was originally reported by SPECK (1949) that methylamine could also substitute for ammonia with a crude enzyme preparation from pigeon liver and later LEVINTOW and MEISTER (1954) reported that the glutamine synthetase of pea seeds also could utilize methylamine in place of ammonia. However, ELLIOTT (1953) reported that the glutamine synthetase of sheep brain cannot use methylamine at all. Likewise, experiments with C^{14}-methylamine have failed to demonstrate any utilization of methylamine by the purified glutamine synthetase of peas.

2. Asparagine Synthetase.

Asparagine synthetase catalyzes the synthesis of the amide, asparagine, by a reaction that is apparently the same as the reaction for glutamine synthesis, namely:

$$\text{aspartate} + NH_3 + ATP \rightleftharpoons \text{asparagine} + ADP + \text{orthophosphate}$$

The enzyme is specific for asparagine synthesis, however, and will not form glutamine from glutamate; nor will glutamine synthetase form asparagine from aspartate. The importance of this enzyme for the biosynthesis of asparagine is not known as yet, although various lines of evidence are compatible with the cellular synthesis of asparagine by this pathway (WEBSTER, 1959a). The possibility of cellular asparagine synthesis by some other pathway cannot be ruled out as yet, but examination of various other possible pathways in both intact plants (AL-DAWODY et al., 1960), and in cell-free preparations (WEBSTER and VARNER, 1955a) have failed to uncover any evidence for another pathway.

a) Occurrence.

The occurrence of asparagine synthetase has been studied only to a limited extent. Evidence has been found for the occurrence of asparagine synthetase in wheat germ, lupine seedlings, peas, and yeast (WEBSTER and VARNER, 1955a). Extracts of yeast and of pig liver have been found by AL-DAWODY (unpublished results) to constitute especially rich sources of asparagine synthetase. The occurrence of the synthetase in liver is noteworthy, as it indicates that the enzyme is not confined to plants and microorganisms.

b) Assay.

The measurement of asparagine synthetase activity consists of determination of the ATP-dependent incorporation of C^{14}-aspartate into asparagine in the presence of ammonia (WEBSTER and VARNER, 1955a). The reaction system consists of: 0.08 M tris(hydroxymethyl)aminomethane-HCl (pH 7.5), 0.05 M C^{14}-aspartate (100,000 cts./min), 0.04 M NH_4Cl, 0.002 M ATP, 0.005 M $MgSO_4$, and the enzyme preparation in a total volume of one ml. The reaction is stopped by the addition of one ml of six per cent trichloroacetic acid, and the precipitated protein is removed by centrifugation. Two ml of a solution containing five mg each

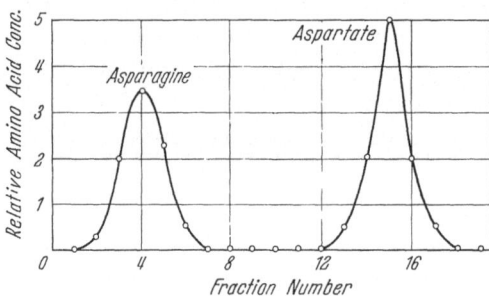

Fig. 6. Separation of aspartate and asparagine on an Amberlite IR-4B column. Two ml fractions are collected from a 1×30 cm column with 0.2 M ammonium acetate (pH 5.0) as the eluting agent (WEBSTER and VARNER 1955a).

of aspartate and asparagine in 0.2 M ammonium acetate (pH 5.0) are added to the supernatant solution, and the four ml of mixture are transferred to a 1×30 cm column of Amberlite IR-4B (which had been previously treated with 0.2 M ammonium acetate of pH 5.0). The column is developed with 0.2 M ammonium acetate (pH 5.0). Aspartate and asparagine are separated under these conditions as is illustrated in Fig. 6. The amount of aspartate converted to asparagine can then be determined by radioactive assay.

Aspartate and asparagine have also been separated for radioactive assay by ascending paper chromatography in aqueous phenol (AL-DAWODY et al., 1960). It has not been possible to utilize the formation of aspartyl hydroxamate for assay of asparagine synthesis, as the enzyme forms aspartyl hydroxamate only with difficulty.

c) Preparation.

Asparagine synthetase from many higher plants appears to be relatively unstable and is difficult to purify. However, AL-DAWODY has found that the asparagine synthetase of yeast is amenable to purification by the following procedure.

One hundred grams of dried bakers' yeast *(Saccharomyces cerevisiae)* are suspended in 1000 ml of 0.05 M tris(hydroxymethyl)aminomethane-HCl of pH 8.0. The cells are broken by sonic treatment for 45 minutes in a cooled Raytheon 10 Kc sonic oscillator. The resulting preparation is centrifuged for 30 minutes at 31,000 $\times g$, and the sediment is discarded.

The supernatant solution is brought to 0.45 saturation with ammonium sulfate. After stirring for 30 minutes, the precipitate is sedimented by centrifugation for 30 minutes at 18,000 $\times g$. The precipitate is discarded. The supernatant solution is then made 0.50 saturated with ammonium sulfate. The precipitate is collected by centrifugation at 18,000 $\times g$ for 30 minutes and is dissolved in one-fifth volume of 0.05 M tris(hydroxymethyl)aminomethane-HCl (pH 8.0). The insoluble material is removed by centrifugation at 6000 $\times g$. The supernatant solution contains the asparagine synthetase. The enzyme is purified about 100-fold by this procedure.

d) Properties.

Very little is known as yet about the characteristics of asparagine synthetase. One interesting fact that is known, however, is that asparagine synthetase appears to be much more specific for the amine used in its catalytic action than is glutamine synthetase. Thus, asparagine synthetase will readily utilize ammonia for asparagine synthesis, but will not utilize hydroxylamine in a comparable manner, and forms aspartyl hydroxamate only in the presence of very high concentrations of hydroxylamine and aspartate.

II. Peptide Synthesis.

1. Glutamylcysteine Synthetase.

The biosynthesis of the tripeptide, glutathione, proceeds via two consecutive reactions, catalyzed by two separate enzymes (SNOKE and BLOCH, 1952; WEBSTER, 1953a). The first reaction, catalyzed by glutamylcysteine synthetase, is a condensation of glutamate and cysteine:

glutamate + cysteine + ATP $\rightleftharpoons$ glutamylcysteine + ADP + orthophosphate

The second reaction, catalyzed by glutathione synthetase, adds glycine to glutamylcysteine:

glutamylcysteine + glycine + ATP $\rightleftharpoons$ glutathione + ADP + orthophosphate.

a) Occurrence.

Evidence has been found for the occurrence of glutamylcysteine synthetase in extracts of bean seedlings (Webster, 1953b), wheat germ (Webster and Varner, 1954), pigeon liver (Snoke and Bloch, 1952), and pig liver (Mandeles and Bloch, 1955).

b) Assay.

The reaction system consists of: 100 μmoles tris(hydroxymethyl)amino-methane-HCl (pH 7.5), 10 μmoles C^{14}-glutamate, 10 μmoles cysteine, 10 μmoles ATP, 8 μmoles KCl, 8 μmoles MgSO$_4$, 3 μmoles NaCN, and the enzyme preparation in a total volume of one ml. The reaction is stopped by addition of one ml of six per cent trichloroacetic acid, and the protein is removed by centrifugation. The solution is diluted to five ml with 0.1 M K$_2$HPO$_4$, and 25 mg of carrier glutamyl-cysteine are added. The glutamylcysteine is precipitated in the manner described for glutathione by Waelsch and Rittenberg (1941). A six per cent solution of CdCl$_2$ is added in an amount equal to one-fourth of the volume of the extract. After the addition of a few drops of brom-cresol green, 6 N NaOH is added carefully until the color just turns to a blue-green. One molar bicarbonate solution is added until the solution is alkaline towards neutral litmus. The precipitate is allowed to settle for one hour in the cold, centrifuged, and washed twice with distilled water. Two normal H$_2$SO$_4$ is added to the white cadmium precipitate with stirring until it just dissolves. Ten ml of 0.5 N H$_2$SO$_4$ are added, and the solution is warmed to 35° C. Cuprous oxide, prepared from Fehling's solution and glucose, and washed neutral, is added dropwise. The mercaptide precipitates in shiny crystals which are usually pure white. It is centrifuged and washed three times with 0.5 N H$_2$SO$_4$, four times with distilled water, with 50 per cent alcohol until free from sulfuric acid, and then twice with absolute alcohol and dried in vacuo over phosphorus pentoxide. The precipitated glutamylcysteine is brought into solution with hydrogen sulfide, and reprecipitated and washed as described above. The amount of C^{14}-glutamate incorporated into glutamylcysteine is determined by radioactive assay.

A second assay method, which can be used with a purified enzyme preparation, consists of the determination of inorganic phosphate liberation in the presence and absence of cysteine in a system composed of enzyme, glutamate, ATP, and magnesium and potassium ions. The cysteine-dependent liberation of ortho-phosphate is stoichiometric with the formation of glutamylcysteine (Webster and Varner, 1954).

c) Preparation of Glutamylcysteine Synthetase.

Glutamylcysteine synthetase has been partially purified from wheat germ (Webster and Varner, 1954). Finely ground wheat germ is extracted at 1° C with ten volumes of 0.05 M sodium bicarbonate for 30 minutes. The suspension is centrifuged at 18,000 $\times g$ for 15 minutes. The sediment is discarded. The pH of the supernatant solution is lowered to 5.8 by the addition of 0.1 M acetic acid. The precipitate is sedimented at 18,000 $\times g$ for 15 minutes and discarded.

The supernatant solution is made 0.41 saturated with ammonium sulfate. The precipitate is discarded. The supernatant solution is then made 0.49 saturated with ammonium sulfate. The protein is taken up in one-fifth volume of distilled water, and is dialyzed overnight at 1° C. The dialyzed protein is brought to pH 7.5 with 0.1 M tris(hydroxymethyl)aminomethane-HCl.

The dialyzed solution is treated with one per cent protamine sulfate until no further precipitation occurs. The inactive precipitate is removed by centrifugation and discarded. The supernatant solution contains glutamylcysteine synthetase. A summary of the steps in the purification procedure is given in Table 5.

Table 5. *Purification of glutamylcysteine synthetase.*

Stage	Specific activity: glutamyl-cysteine synthesis	Relative purification	Yield (per cent)	Relative ATPase
Centrifuged extract	0.008	1.0	(100)	1.00
Acidification to pH 5.8	0.012	1.5	98	0.81
0.41—0.49 satn. $(NH_4)_2SO_4$ ppt. .	0.219	27.4	45	0.38
Protamine supernatant	0.415	52.0	31	0.08

d) Properties.

Initial studies on the glutamylcysteine synthesis reactions have been performed by WEBSTER and VARNER (1954) and by MANDELES and BLOCH (1955), but essentially nothing is known about the enzyme itself.

2. Glutathione Synthetase.

a) Occurrence.

Glutathione synthetase occurs in extracts of pigeon liver (BLOCH, 1949); rat liver, guinea pig liver, rabbit liver, beef liver, and pork liver (YANARI et al., 1953); bean seedlings (WEBSTER, 1953a); yeast (SNOKE, 1955); and wheat germ (WEBSTER and VARNER, 1955b).

b) Assay.

The reaction system consists of: 100 μmoles tris(hydroxymethyl)amino-methane-HCl (pH 7.5), 10 μmoles glutamylcysteine, 10 μmoles C^{14}-glycine, 10 μmoles ATP, 5 μmoles KCl, 5 μmoles $MgSO_4$, 3 μmoles NaCN, and the enzyme preparation in a total volume of one ml. The reaction is stopped by the addition of one ml of six per cent trichloroacetic acid, and the precipitated protein is removed by centrifugation. The solution is diluted to five ml with 0.1 M K_2HPO_4. Twenty five mg of carrier glutathione are added, and the glutathione is isolated with cadmium and copper in the manner described previously for the assay of glutamylcysteine. Glutathione synthesis with a fairly purified enzyme can be assayed by phosphate liberation or by the glyoxylase reaction (WOODWARD, 1935).

c) Preparation.

Glutathione synthetase has been highly purified from yeast (SNOKE, 1955). Brewers yeast is air-dried and pulverized in a mill. The yeast is suspended in distilled water and allowed to autolyze at 37° C for four hours. Three volumes of water are added, and the suspension is centrifuged at 3,000 $\times g$ for 20 minutes. In each liter of supernatant solution are dissolved 333 grams of solid ammonium sulfate. The solution is allowed to stand overnight at 3° C. The precipitate is collected by centrifugation and is dissolved in one-fourth volume of distilled water.

The concentration of ammonium sulfate in the supernatant solution is estimated by nesslerization, and sufficient solid ammonium sulfate is added so that the ammonium sulfate concentration is 260 grams per liter. The pH of the solution is adjusted to 8.0 by the addition of 3 M ammonium hydroxide. The solution is allowed to stand overnight. The precipitate is removed by centrifugation, and 82 grams of solid ammonium sulfate are added per liter of supernatant solution.

After stirring for one hour, the precipitate is collected by centrifugation, dissolved in 0.1 volume of water, and dialyzed overnight.

The dialyzed solution is adjusted to pH 4.5 by the addition of 0.2 N H_2SO_4, and brought to 42° C by rapid stirring in a water bath kept at 60° C. The mixture is kept at 42° C for 30 minutes and then cooled. The precipitate is removed by centrifugation and discarded.

To each liter of supernatant solution are added 280 grams of solid ammonium sulfate. The mixture is stirred for one hour, the precipitate is collected by centrifugation, and dissolved in water. To each liter of solution are added 450 ml of a saturated ammonium sulfate solution, and the pH is adjusted to 4.5 by the addition of 0.2 N H_2SO_4. After one hour of stirring, the precipitate is removed by centrifugation, and an additional 250 ml of saturated ammonium sulfate are added. The mixture is stirred again for one hour and centrifuged. The precipitate is dissolved in water, and the solution is dialyzed overnight. The dialyzed solution is adjusted to contain five mg of protein per ml, and five ml of nucleic acid solution (2.5 grams of nucleic acid and 3.75 ml of 1 N NaOH per 50 ml) are added for every 100 ml of protein solution. The pH is adjusted to 5.35 with 0.005 N H_2SO_4. The mixture is stirred for 30 minutes. The precipitate is removed by centrifugation, and the pH of the supernatant liquid is adjusted to 4.90. After stirring for 30 minutes, the precipitate is collected by centrifugation, and is dissolved in water. The pH of the suspension is adjusted to 7.0 by the addition of 0.03 N NaOH. Protamine sulfate (20 mg per ml) is added until no further precipitation occurs.

Two volumes of saturated ammonium sulfate solution are added to the supernatant solution. The precipitate is collected by centrifugation, and dissolved in approximately ten ml of water. The solution is dialyzed against distilled water until salt-free. The solution is adjusted to contain six mg of protein per ml, and to be 0.02 M in phosphate (pH 6.7) and 0.06 M in sodium chloride. To each 100 ml of solution are added dropwise, with stirring, 18 ml of absolute ethanol. The temperature of the mixture is lowered from 0° to −4° C during the ethanol addition. After stirring for 15 minutes, the mixture is centrifuged at −4° C. To the supernatant fluid are added 28 ml of ethanol for every 100 ml of original solution. The temperature is lowered to −10° C during the addition of the ethanol. The mixture is stirred for 15 minutes, and centrifuged at −10° C. The ethanol is removed from the precipitate by lyophilization, and the precipitate is dissolved in 0.25 M ammonium sulfate. This solution contains the glutathione synthetase. A summary of the purification procedure is given in Table 6.

Table 6. *Purification of glutathione synthetase.* Total activity is expressed as μmoles glutathione formed in 30 minutes. Specific activity is μmoles formed in 30 minutes by one mg of protein.

Step	Volume (ml)	Protein (mg/ml)	Total activity	Specific activity
Autolysate	150,000	42.5	2,180,000	0.3
1st $(NH_4)_2SO_4$ fraction	35,800	47.5	1,620,000	0.9
2nd $(NH_4)_2SO_4$ fraction	9,600	58.0	1,350,000	2.4
Acid ppt. and heat denaturation .	9,700	10.0	950,000	9.8
3rd $(NH_4)_2SO_4$ fraction	955	22.0	614,000	35.7
4th $(NH_4)_2SO_4$ fraction	430	9.0	372,000	96.4
Nucleic acid fraction	137	5.2	298,000	419.0
Alcohol fraction	16.5	6.1	194,000	1930.0

d) Properties.

The purified yeast enzyme exhibits two components in the ultracentrifuge, which contain approximately 85 and 15 per cent respectively of the total protein.

Although the enzyme-catalyzed reaction has been studied to some extent, little is known concerning the characteristics of the enzyme itself.

III. Protein Synthesis.

The results of many investigations (for reviews, see BONNER, 1958; CRICK, 1958; RAACKE, 1958; WEBSTER, 1959a; WEBSTER, 1959b; HOAGLAND, 1960) have established that proteins are synthesized, apparently independently, by cellular nuclei, mitochondria, and microsomes. The ribosomal portion of microsomes is specifically concerned with the synthesis of a large number of cytoplasmic proteins. Experimental systems have been devised by which ribosomes synthesize protein outside of the living cell (WEBSTER, 1957a; SCHWEET et al., 1958; RAACKE, 1959; WEBSTER, 1959c). This synthesis occurs in two major steps:

(a) The "activation" of amino acids by the formation of amino acid-polynucleotide compounds.

(b) The condensation of these activated amino acids to form protein.

The first step in protein synthesis, amino acid activation, is catalyzed by the amino acid-activating enzymes. The condensation of activated amino acids to form protein is catalyzed by ribosomes.

1. Amino Acid-Activating Enzymes.

Amino acid-activating enzymes catalyze the reaction:

$$\text{Amino acid} + \text{sRNA} + \text{ATP} \rightleftharpoons \text{Amino acid—sRNA} + \text{AMP} + \text{pyrophosphate}$$

In the presence of hydroxylamine, the amino acid-activating enzymes will often form amino acid hydroxamates. The enzymes will also catalyze an amino acid-dependent exchange of pyrophosphate with ATP. There appears to be a separate activating enzyme for each of the twenty amino acids and amides that normally occur in protein, and each enzyme is specific for a single amino acid or amide. Many of the amino acid-activating enzymes have been purified from some source. Table 7 lists the amino acid-activating enzymes that have been purified.

Table 7. *Purification of various amino acid-activating enzymes.*

Activating enzyme	Source	Investigator
Alanine . . .	Pig liver	WEBSTER, 1961
Alanine . . .	Rat liver	HOLLEY and GOLDSTEIN, 1959
Aspartate . .	Pea seeds	ABDUL-NOUR and WEBSTER, unpublished
Isoleucine . .	*E. coli*	BERG, unpublished
Leucine . . .	*E. coli*	BERG, unpublished
Methionine .	Yeast	BERG, 1956
Methionine .	*E. coli*	BERG, unpublished
Serine . . .	Beef pancreas	WEBSTER and DAVIE, 1959
Threonine . .	Calf liver	ACS et al., 1959
Tryptophan .	Beef pancreas	DAVIE et al., 1956
Tyrosine. . .	Yeast	VAN DE VEN et al., 1958
Tyrosine. . .	Pig pancreas	SCHWEET and ALLEN, 1958
Valine. . . .	*E. coli*	BERG, unpublished

In each instance, the activation of an amino acid involves the combination of the carboxyl group of the amino acid with the 2' or 3' hydroxyl of the terminal nucleoside of a special kind of ribonucleic acid (sRNA). The terminal nucleoside of cytoplasmic sRNA is adenosine. In fact, cytoplasmic sRNA appears always to have a terminal adenyl group (which reacts with the amino acid) followed by two cytidyl groups. The opposite end of the molecule appears usually to be terminated with a guanosine-5'-phosphate group. The remainder of the sRNA molecule appears to contain a number of unusual purine or pyrimidine bases, especially pseudo-uridine, 5-methylcytosine, and 6-methylaminopurine (DUNN, 1959). The molecular weight of sRNA preparations appears to vary somewhat, but has been reported to be anywhere from 10,000 to 50,000. There is apparently at least one specific sRNA for each amino acid.

a) Occurrence.

Amino acid-activating enzymes have been found in extracts of rat liver (Hoagland, 1955); E. coli (Demoss and Novelli, 1955); Micrococcus, Aerobacter, Achromobacter, Neurospora, rat kidney, guinea pig liver and pancreas (Novelli, 1958); Staphylococcus, Clostridium, Proteus, Streptococcus, Serratia, Azotobacter, and Leuconostoc (Demoss and Novelli, 1956); peas (Webster, 1957b; Davis and Novelli, 1958); pig liver and yeast (Webster, 1959d); pig pancreas (Schweet et al., 1957); spinach leaves (Clark, 1958; Bove and Raacke, 1959); and beef pancreas (Davie et al., 1956).

b) Assay.

The assay of amino acid-activating enzymes may be performed by measuring either the synthesis of amino acid-sRNA or, in some instances, by measurement of the amino acid-dependent exchange of pyrophosphate with ATP. For the measurement of amino acid-sRNA synthesis, it is necessary to isolate sRNA. This may be performed conveniently with phenol in the following manner.

Preparation of sRNA from yeast. A thick suspension of yeast cells in water is mixed with an equal volume of 90 per cent aqueous phenol, and is allowed to stand for one hour in the cold. The suspension is centrifuged at $15,000 \times g$ for 20 minutes. The relatively clear upper layer is removed, and is filtered. The clear solution is made two per cent with potassium acetate, and two volumes of ethanol are then added. The precipitate of crude sRNA is sedimented by centrifugation for ten minutes at $10,000 \times g$. The precipitate is washed with a mixture of ethanol and water (3:1) and is dissolved in one-tenth volume of water. The solution is mixed with an equal volume of 2.5 M K_2HPO_4 containing 0.05 volume of 33.3 per cent H_3PO_4 (1:2 dilution of ordinary H_3PO_4). One volume of 2-methoxyethanol is added, and the mixture is centrifuged at $15,000 \times g$ for 20 minutes. The top layer is removed. It is made two per cent with potassium acetate. The sRNA is precipitated by addition of two volumes of ethanol, washed once with a 3:1 ethanol-water mixture, and is suspended in a small volume of water or neutral buffer.

Assay system for the measurement of amino acid-sRNA synthesis is (Webster, 1959d): 0.05 M tris(hydroxymethyl)-aminomethane-HCl (pH 7.5), 0.005 M potassium ATP, 0.001 M C^{14}-amino acid (approximately 300,000 counts/min), 0.001 M $MgSO_4$, 10 mg of sRNA, 0.1 mg of crystalline pyrophosphatase, and the activating enzyme preparation in a total volume of five ml. The reaction is stopped by the addition of one ml of 12 per cent trichloroacetic acid. The precipitated sRNA is collected by centrifugation, washed three times with 12 per cent trichloroacetic acid, and assayed by standard counting techniques for the amount of radioactive amino acid bound to sRNA.

For the assay of amino acid-dependent pyrophosphate-ATP exchange the system contains (Webster, 1959d): 0.05 M tris(hydroxymethyl)aminomethane-HCl (pH 8.0), 0.001 M ATP, 0.001 M potassium P^{32}-pyrophosphate (containing about 30,000 counts/min), 0.001 M amino acid, 0.001 M $MgSO_4$, and the enzyme preparation in a total volume of one ml. In some cases, the addition of 0.001 M NaF is also necessary. The mixture is shaken for ten minutes at $38°$ C. The reaction is stopped by the addition of one ml of 12 per cent trichloroacetic acid. The precipitated protein is removed by centrifugation. The solution is diluted to five ml with water, and 0.2 gram of activated charcoal is added, followed by 0.2 ml of ethanol (CRANE and LIPMANN, 1953). The charcoal is thoroughly suspended in the solution, and is then sedimented by centrifugation at $5,000 \times g$ for five minutes. The supernatant solution is discarded. The charcoal is washed three times with water. It is then suspended in one ml of 1.0 M HCl. The suspension is boiled for ten minutes. The suspension is cooled, and the charcoal is removed by centrifugation. The supernatant solution is assayed for radioactivity, and this radioactivity constitutes the amount of radioactive pyrophosphate incorporated into ATP.

c) Preparation.

An apparently typical example of an amino acid-activating enzyme is the tyrosine-activating enzyme of baker's yeast (VAN DE VEN et al., 1958). Baker's yeast is crumbled and frozen for three to four hours in ether containing solid carbon dioxide. The yeast is dried, thawed, and extracted overnight in a suspension containing 11.2 g KCl per kilogram of yeast. The suspension is centrifuged for 20 minutes at $5,000 \times g$. The supernatant solution is filtered through several layers of cloth. The filtrate is dialyzed against distilled water for at least 22 hours, and then is centrifuged for 20 minutes at $5,000 \times g$.

The solution is made 0.5 saturated with ammonium sulfate. The precipitate is removed by centrifugation for 20 minutes at $5,000 \times g$. The supernatant solution is made 0.6 saturated with ammonium sulfate. The precipitate is collected by centrifugation for 25 minutes at $5,000 \times g$, and is dissolved in 0.1 volume of cold distilled water.

To each ml of solution is added 0.12 g of ammonium sulfate. The pH of the resulting solution is lowered to 4.5 with 0.2 M acetic acid. For each ml of solution, 0.133 ml of saturated ammonium sulfate solution are added. The precipitate is removed by centrifugation for ten minutes at $11,000 \times g$. To the supernatant solution is added 0.2 ml of saturated ammonium sulfate solution per ml. The precipitate is collected by centrifugation for ten minutes at $11,000 \times g$. The precipitate is dissolved in one-fourth volume of cold distilled water, and is centrifuged for two minutes at $7,000 \times g$. The pH of the supernatant solution is raised to 7.0 by the addition of 1.0 M sodium carbonate.

The protein concentration of the solution is adjusted to 20 mg per ml. The solution is mixed with one-fourth volume of calcium phosphate gel. The mixture is centrifuged for ten minutes at $11,000 \times g$. The precipitate is discarded. The supernatant solution is made 0.6 saturated with ammonium sulfate. The precipitate is collected by centrifugation for ten minutes at $11,000 \times g$, and is dissolved in distilled water. It is again treated with calcium phosphate gel in the same manner. The supernatant solution is again subjected to fractionation with ammonium sulfate at pH 4.5 as is described above. The final supernatant contains the tyrosine-activating enzyme.

d) Properties.

Clear evidence for the isolation of a pure amino acid-activating enzyme is not available. Evidence that several of the activating enzymes are at least 70—85 per cent pure has been presented, and the alanine-activating enzyme of pig liver has been shown to behave as a single component upon electrophoresis under several conditions. Further information on the characteristics of the amino acid-activating enzymes (molecular weight, etc.) must await the isolation of these enzymes in pure form.

2. Enzymes Concerned with Protein Synthesis.

The enzymes which catalyze the synthesis of protein from activated amino acids (amino acid-sRNA) occur in nuclei, mitochondria, and in the ribonucleo-protein particles (ribosomes) of the ergastoplasm of cells. Protein synthesis by isolated ribosomes has been investigated considerably, and appears to consist of the formation of polypeptide chains by a sequential addition of amino acids from amino acid-sRNA to the lengthening polypeptide chain. This reaction requires GTP and a metal ion, and takes place on the surface of the ribosome. Upon completion of a polypeptide chain, it is apparently released from the ribosome by a distinct reaction which requires ATP, a metal ion, and an enzyme fraction. Ribosomes, therefore, plus their auxiliary proteins which participate in protein synthesis, may be considered to be the enzymes which catalyze the synthesis of protein from activated amino acids.

a) Occurrence.

Ribosomes have been isolated from yeast (CHAO and SCHACHMAN, 1956), bean seedlings (ROBINSON and BROWN, 1953), various bacteria (SCHACHMAN et al., 1952), pea seedlings (Tso et al., 1956), mammalian liver (LITTLEFIELD et al., 1955), reticulocytes (DINTZIS et al., 1958), and ascites tumor cells (LITTLEFIELD and KELLER, 1957).

b) Assay.

Protein synthesis by ribosomes can be measured either by the measurement of net protein synthesis catalyzed by the ribosomes, or by the measurement of the addition of radioactive amino acids to the growing polypeptide chains being synthesized by the ribosomes. The enzyme responsible for the release of finished polypeptide chains is easily inactivated, so it is often more convenient to measure the addition of radioactive amino acids to the polypeptide chain being synthesized on the ribosomal surface, than to measure the net synthesis of protein.

Table 8. *Protein synthesis as determined by various assay methods* (WEBSTER, 1959c). The incubation system and procedure are those described in the text.

Time (min)	Protein formed, mg		
	Biuret	Phenol reagent	micro-KJELDAHL
0	0.0	0.0	0.0
30	1.9	1.8	2.3

Assay system for the measurement of net protein synthesis consists of (WEBSTER, 1959c): 0.05 M tris(hydroxy-methyl)aminomethane-HCl (pH 7.5); 6 mg of an amino acid mixture containing 0.3 mg of each of the following L-amino acids and amides: alanine, arginine, asparagine, aspartate, cysteine, glutamate, glutamine, glycine, histidine, isoleucine, leucine, lysine, methionine, phenylalanine, proline, serine, threonine, tryptophan, tyrosine, and valine; 0.0001 M ATP; 0.0003 M GTP; 0.0001 M $MnCl_2$; 0.0001 M $MgCl_2$; 0.0001 M KCl; 0.01 M phosphoglycerate; 3.0 mg of purified sRNA; and 8.0 mg of ribosomes in a total volume of ten ml. The mixture is shaken for 30 minutes at 38° C. The mixture is cooled in ice, and the ribosomes are removed by centrifugation for 60 minutes at 105,000 $\times g$. Net protein increase in the supernatant solution can be assayed with the biuret reagent (GORNALL et al., 1949), the phenol reagent (LOWRY et al., 1951), or by the micro-KJELDAHL procedure (HILLER et al., 1948). Typical results are given in Fig. 7, and a comparison of the three assay procedures is given in Table 8.

For measurements of the addition of radioactive amino acids to the growing polypeptide chain, the reaction system given above can be used, with the difference that one or more of the amino acids are radioactive. The amount of radioactivity bound to the ribosome is then a measure of the addition of labelled amino acids.

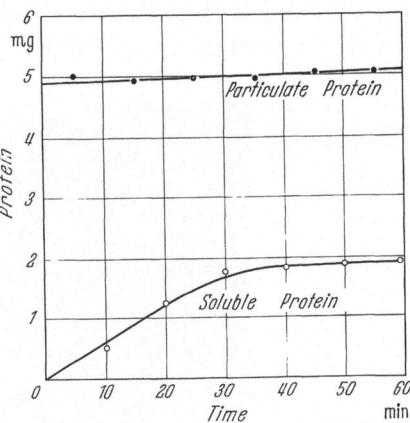

c) Preparation.

Active ribosomes from peas are isolated in the following manner. Peas are allowed to germinate for three days. The growing portions are detached from the cotyledons, washed, and ground for 15 seconds in a Waring blendor with an equal volume of 0.3 M sucrose. The preparation is centrifuged at 35,000 ×g for 20 minutes. The supernatant solution is then centrifuged at 105,000 ×g for 60 minutes. The supernatant solution is discarded, and the sedimented ribosomes are suspended in 0.3 M sucrose with the aid of a glass homogenizer. To insure active material, all operations should be performed as near 0° C as possible. Somewhat more homogeneous ribosomes can be obtained by a repeat of the two centrifugation steps described above.

Fig. 7. Synthesis of "soluble" protein by pea ribosomes. The incubation system and procedure are those described in the text. Particulate protein refers to the protein level of the ribosomes, and "soluble" protein refers to the protein level of the incubation medium surrounding the ribosomes (WEBSTER, 1959c).

d) Properties.

Ribosomes from various sources are surprisingly uniform in their chemical and physical properties. Ribosomes are spherical bodies with diameters of 150—250 Å. They are apparently composed almost entirely of protein and ribonucleic acid, and have 40—50 per cent ribonucleic acid and 50—60 per cent protein. A

Table 9. *Amino acid composition of ribosomes from various sources and of a plant virus* (Tso et al., 1958). Values given are grams amino acid/100 g of protein.

Amino acid	Rabbit reticulocyte ribosomes	Pea seedling ribosomes	Guinea pig liver "ribosomes"	Turnip yellow mosaic virus
Alanine	5.4	5.4	5.3	5.4
Arginine	11.8	9.2	8.3	2.2
Aspartate	8.8	9.6	9.5	6.3
Cystine	1.1	0.3	—	0
Glutamate	11.5	10.7	12.0	7.7
Glycine	7.1	8.3	4.7	3.8
Histidine	2.8	2.9	2.5	1.6
Isoleucine	5.7	6.4	4.2	7.4
Leucine	8.7	8.2	10.2	8.6
Lysine	12.7	12.2	9.3	5.0
Methionine	2.0	2.0	2.0	2.1
Phenylalanine . .	4.4	4.8	5.6	3.6
Proline	4.7	5.2	7.5	11.8
Serine	1.8—10	2.5—10	3.9	6.7
Threonine	4.5	4.9	5.2	12.2
Tryptophan . . .	1.2	1.5	—	—
Tyrosine	6.5	7.0	4.0	2.2
Valine	7.2	7.6	5.8	6.2

typical preparation of ribosomes from peas is presented in Fig. 8. The ribosomes from very young, rapidly growing peas, and from most other organisms exhibit essentially a single component in the ultracentrifuge (Fig. 9) as well as upon electrophoresis (Fig. 10). The ribosomes from various organisms are remarkably similar in mass, and have sedimentation constants of 70—80 S. Estimates of the

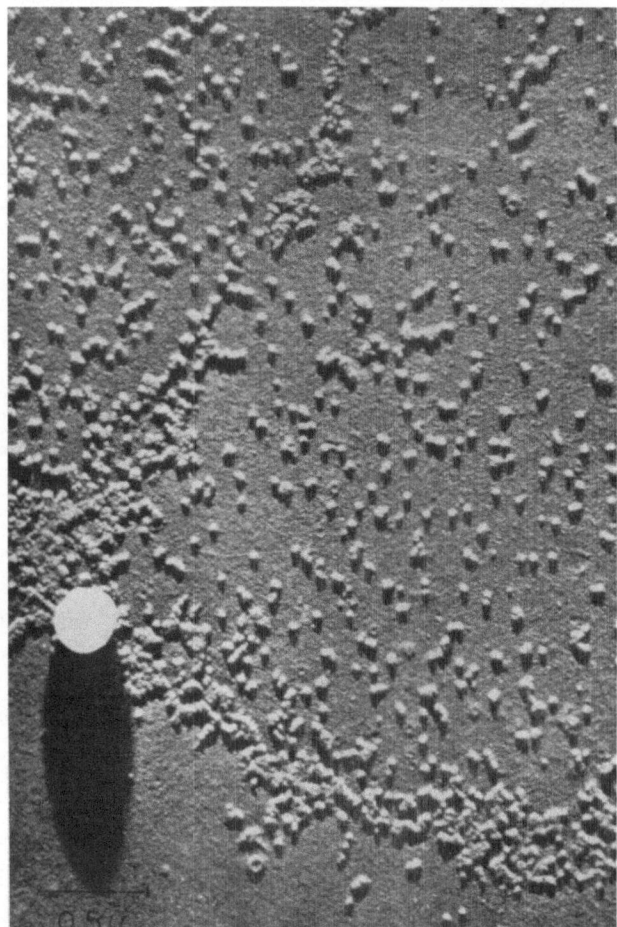

Fig. 8. Electron micrograph of pea seedling ribosomes (Tso et al., 1956) ×44,000.

molecular weight of these ribosomes range from 3,500,000 to 4,500,000. The amino acid composition of the ribosomes from various sources also exhibits a striking similarity (Table 9).

Although ribosomes were first thought to have a structure very similar to that of the spherical viruses, this does not appear to be the case. Treatment of ribosomes with an agent which reacts with divalent cations (ethylenediamine-tetraacetate, pyrophosphate, etc.) results in the dissociation of the biologically-active ribosome into two pieces that are apparently not active. The pieces comprise two-thirds and one-third respectively of the original particle. For example, the 80 S ribosome of pea seedlings is disrupted into a 60 S component and a 40 S component, while the 70 S ribosome of *Escherichia coli* is disrupted to a

50 S component and a 30 S component. Both kinds of components contain both protein and ribonucleic acid.

Figure 11 illustrates an electron micrograph (magnified 170,000 times) of a ribosome preparation from *Azotobacter vinelandii*. The particles in this highly-

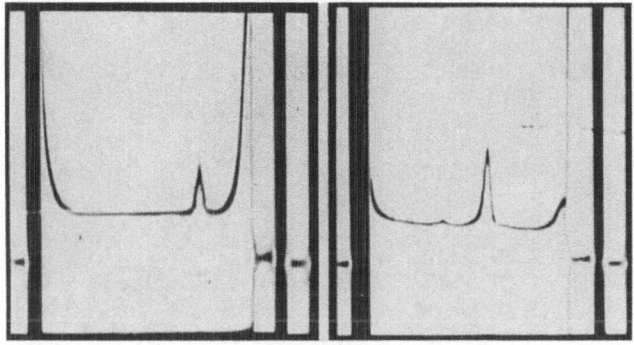

Fig. 9. Ultracentrifuge pattern of a ribosome preparation from three day old pea seedlings (Tso et al., 1956)

magnified picture appear to be composed of a number of much smaller sub-units. Evidence that ribosomes are composed of a number of smaller sub-units has also been obtained by ABDUL-NOUR (1959) and by YIN and BOCK (1960). ABDUL-

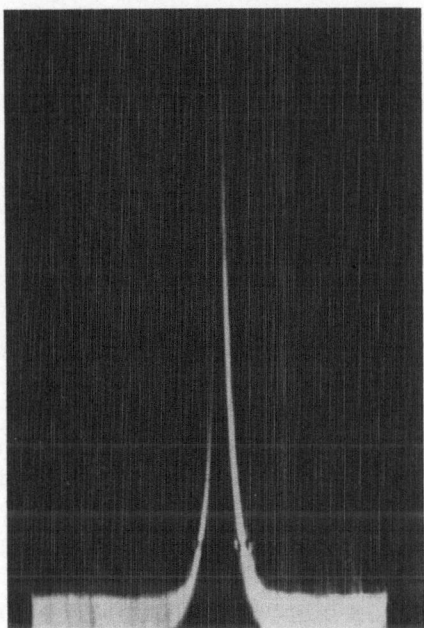

Fig. 10. Electrophoresis of a preparation of ribosomes from peas in tris(hydroxymethyl)aminomethane buffer of pH 8.6 and an ionic strength of 0.1 (ABDUL-NOUR, 1959).

NOUR (1959) found that treatment of pea seedling ribosomes with sodium dodecyl sulfate results in the disruption of the ribosomes to 12 S sub-units which are homogeneous upon either electrophoresis or ultracentrifugation. YIN and BOCK (1960) have reported that treatment of yeast ribosomes with a combination of dodecyl sulfate and 8 M urea disrupts the ribosomes into even smaller protein

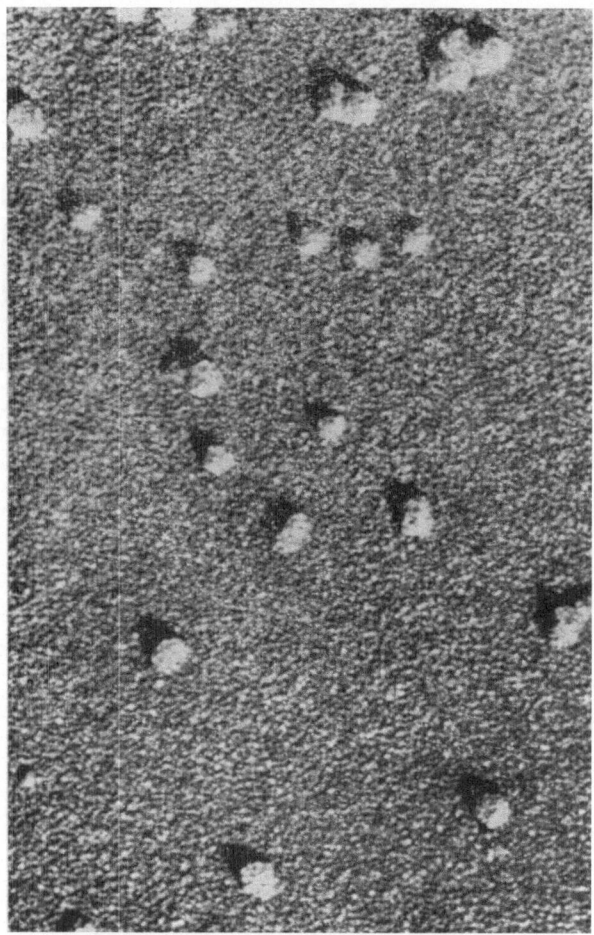

Fig. 11. Electron micrograph of the 86 S ribosomes of *Azotobacter vinelandii* (GILLCHRIEST and BOCK, 1958). × 170,000.

sub-units having a molecular weight of approximately 12,000. Thus, ribosomes appear to be composed of large nucleoprotein sub-units whose combination together seems essential for their catalytic activity in protein synthesis. The large sub-units, in turn, are composed of a large number of relatively small protein sub-units.

B. Enzymes concerned with Degradation (Amidases, Peptidases, and Proteases).

Unlike the enzymes catalyzing syntheses, the enzymes catalyzing degradations have no clearly demonstrated cellular function. Presumably these enzymes act to degrade their specific substrates in the cell, but evidence for this function is lacking. In the case of degradative enzymes which are secreted into the medium outside of cells, a digestive function occurs which degrades large molecules into smaller units which are transported more readily into the cell. The existence of the various degradative enzymes has been known for a long time. Despite this, with a few notable exceptions, they have remained rather poorly characterized. The outstanding work on proteases in recent years, however, promises that our knowledge of all of the classes of enzymes splitting amide or peptide bonds will be increased considerably.

I. Amidases.

Amides are hydrolyzed by two classes of enzymes. On the one hand, are the specific amidases which act upon either glutamine or asparagine. On the other hand, many amides are hydrolyzed by various peptidases and proteases. Examples of this latter activity are the hydrolyses of benzoyl-L-argininamide by trypsin, L-tyrosinamide by chymotrypsin, L-leucinamide by leucine aminopeptidase, L-leucinamide by papain, benzoyl-L-argininamide by ficin, and benzoyl-L-argininamide by bromelin. These hydrolyses, however, are manifestations of the specificity of these enzymes for the hydrolysis of certain peptide bonds. The present discussion will be concerned only with the true amidases, glutaminase and asparaginase.

Glutaminase and Asparaginase.

a) Occurrence.

Glutaminase is widely distributed in higher plants (ARCHIBALD, 1945; GROVER and CHIBNALL, 1927), microorganisms (McILWAIN, 1948), and animal tissues (KREBS, 1935; GONCALVES et al., 1947; GREENSTEIN and LEUTHARDT, 1948). Asparaginase has been reported to occur in extracts of higher plants (GROVER and CHIBNALL, 1927), yeasts (GEDDES and HUNTER, 1928; GRASSMAN and MAYR, 1933; GORR and WAGNER, 1933), bacteria (BUSCH, 1948), and animals (GREENSTEIN and CARTER, 1947; ERRERA and GREENSTEIN, 1947; STEENSHOLD, 1944).

b) Assay.

For glutaminase 0.5 ml of 0.05 M glutamine, 0.4 ml of 0.1 M sodium acetate buffer (pH 5.0), and 0.1 ml of enzyme solution are shaken at 37° C. The reaction is stopped with one ml of six per cent trichloroacetic acid. The liberated ammonia is then determined by microdistillation (ARCHIBALD, 1944), by diffusion according to the CONWAY procedure (CONWAY, 1947), or by aeration into sulfuric acid (GREENSTEIN and LEUTHARDT, 1944).

For asparaginase 0.5 ml of 0.05 M asparagine, 0.4 ml of 0.01 M sodium borate buffer (pH 8.0), and 0.1 ml of enzyme are incubated at 37° C. The reaction is stopped by one ml of six per cent trichloroacetic acid. The liberated ammonia is determined by one of the methods described above.

c) Preparation.

The glutaminase of *Escherichia coli* (strain W) is partially purified by the following procedure (RUDMAN and MEISTER, 1953). Lyophilized cells of *E. coli* are ground in the cold in a mortar with three parts of alumina powder for 30 minutes. The preparation is shaken vigorously with 20 parts of cold distilled water for 20 minutes, and then centrifuged at $18,000 \times g$ for 20 minutes.

The supernatant solution is mixed with a suspension of calcium phosphate gel (containing 1.5 g calcium phosphate for each five grams of cells originally disrupted). After mixing for 30 minutes at 24—28° C, the mixture is centrifuged at

$600 \times g$ for 20 minutes. The supernatant solution is discarded, and the enzyme is extracted from the gel with 40 ml of 0.1 M KH_2PO_4 (pH 4.5). The gel is removed by centrifugation. This procedure gives a 15—20 fold purification over the original cell-free extract. The enzyme may be stored for several months at 5° C without loss of activity, if it is lyophilized.

The specific activity can be increased two-fold, with the loss of approximately half of the total activity, by suspension of 50 mg of the lyophilized powder in five ml of saturated sodium sulfate solution. The suspension is shaken for 20 minutes at 24—28° C, and centrifuged. The sediment, which is the more active fraction, is dissolved in one ml of cold water.

Asparaginase has been extracted from barley roots with water, and concentrated with ethanol (GROVER and CHIBNALL, 1927). Asparaginase has been extracted from yeast by treatment of the yeast with toluene, washing the cells with water, and suspending them at a pH of 8.5 for 48 hours. The cells are removed by centrifugation and discarded (GRASSMAN and MAYR, 1933). The asparaginase of yeast can be purified by precipitation with safranine (GEDDES and HUNTER, 1928).

d) Properties.

Due to their low state of purification, little is known concerning the properties of glutaminase or asparaginase. The purified glutaminase is apparently highly specific for L-glutamine, and will not hydrolyze D-glutamine, L-isoglutamine, L-asparagine, alpha-aminoadipic amide, or alpha-ketoglutaric amide. Glutaminase requires orthophosphate for maximal activity. Arsenate, sulfate, and cyanide have also been reported to enhance the activity of some glutaminase preparations.

Asparaginase likewise appears to be specific for L-asparagine, as it will not catalyze the hydrolysis of L-isoasparagine, L-glutamine, alpha-methylasparagine, or alpha-ketosuccinic amide, and has little or no ability to catalyze the hydrolysis of D-asparagine. In contrast to glutaminase activity, asparaginase activity has been reported to be promoted little or none by phosphate, arsenate, sulfate, and cyanide. Yeast asparaginase is quite unstable. At pH 7.0, 30 per cent of the activity is lost in 24 hours at 0° C. At pH 5.0, more than 70 per cent of the activity is lost in 30 minutes at 0° C.

II. Peptidases.

Although "total peptidase" activity is often reported for some tissue extract, the term is relatively meaningless, as it represents the total activity of a large and still undefined group of relatively specific peptidases. A number of peptidases have been fairly well purified and characterized, especially from animal tissues. These include:

(a) Carboxypeptidase, which hydrolyzes the peptide bond adjacent to a free carboxyl group in a peptide chain.

(b) Aminotripeptidase, which hydrolyzes a peptide bond adjacent to a free amino group.

(c) Leucine aminopeptidase, which is relatively non-specific, and hydrolyzes, at different velocities, dipeptides, tripeptides, and probably larger peptides. Its action is most rapid on peptides containing leucine.

(d) Glycylglycine dipeptidase, which is quite specific for glycylglycine.

(e) Prolinase, which hydrolyzes a variety of dipeptides which contain N-terminal L-proline.

(f) Carnosinase, which hydrolyzes a variety of peptides containing histidine.

(g) Prolidase, which hydrolyzes certain dipeptides containing C-terminal proline.

a) Occurrence.

The well-defined peptidases have been found mostly in animal tissues, although they probably occur also in higher plants and microorganisms. In fact, leucine aminopeptidase has been found not only in all animal tissues examined, but also in both higher plants and microorganisms (JOHNSON and BERGER, 1942; SMITH, 1951). A well-defined aminotripeptidase has been purified from yeast (JOHNSON, 1941).

b) Assay.

Peptidase activity may be estimated by measurement of the liberated amino acids with ninhydrin. Peptidase activity may also be estimated by measurement of the carboxyl groups liberated during the hydrolysis of the peptide by titration with KOH in 90 per cent ethanol by the method of GRASSMAN and HEYDE (1929).

c) Preparation and Properties.

The preparation and properties of various peptidases from animal tissues have been described in several recent reviews (SMITH, 1955; GREEN and NEURATH, 1954). An apparent aminotripeptidase has been purified from yeast in the following manner (JOHNSON, 1941). One kilogram of brewer's bottom yeast is frozen and then autolyzed for four days at pH 6.1. The preparation is filtered and the solid material is discarded.

The filtrate is adjusted to pH 5.7. Acetone is added to 28 volumes per cent. The precipitate is collected by centrifugation, suspended in water, and recentrifuged. The solid material is discarded.

The supernatant solution is adjusted to pH 5.8. Acetone is added to 30 volumes per cent. The precipitate is collected by centrifugation, suspended in water, and the solution is clarified by centrifugation.

The supernatant solution is treated with enough $MgCl_2$ to make the solution 0.001 M. The pH is lowered to 5.6, and acetone is added to 13.5 volumes per cent. The precipitate is removed by centrifugation. Acetone is added to 18 volumes per cent. The precipitate is collected by centrifugation and dissolved in water. The peptidase is purified about 800-fold by the procedure.

The enzyme sediments as a single protein in the ultracentrifuge with a sedimentation constant of 21.3 S. The molecular weight has been estimated (JOHNSON, 1941) to be 670,000. The enzyme also acts as a single component upon electrophoresis. Enzyme activity is greatly enhanced by Zn^{++}, and, to a lesser extent, by Co^{++}. Other metal ions are inactive. The specificity of the yeast enzyme is given in Table 10, and shows that its action differs somewhat from that of the comparable enzymes from animals.

Table 10. *Hydrolysis of various substrates by yeast peptidase* (JOHNSON, 1941).

Substrate	Enzyme N (mg per ml)	Incubation time (hr)	Per cent hydrolysis
Leucylglycylglycine	0.091	0.5	55
Alanylglycylglycine	0.253	0.5	53
Triglycine	0.253	8.0	56
Leucylglycine	0.253	2.0	64
Alanylglycine	0.253	8.0	39

III. Proteases.

Although the proteases from plants and microorganisms are not as well characterized, in general, as are the proteases from animals, in a few instances (such as papain) the enzymes have been studied rather thoroughly. The knowledge derived from studies with papain, has contributed much to our understanding of the mechanisms of protease action.

a) Occurrence.

The occurrence of various well-defined proteases in higher plants is presented in Table 11. In addition, proteases have been studied in extracts of squash (WILLSTÄTTER et al., 1926), sundew (HOLTER and LINDERSTROM-LANG, 1933), grains (BALLS and HALE, 1938; OLCOTT et al., 1943), and tobacco leaves (TRACEY, 1948). Proteases have been reported to be produced by many species of bacteria (MASHMANN, 1940, 1943; EVANS, 1948).

Table 11. *Occurrence of proteases in higher plants.*

Enzyme	Origin	Reference
Papain.	Papaya *(Carica papaya)*	BALLS and LINEWEAVER, 1939
Chymopapain . .	Papaya *(Carica papaya)*	JANSEN and BALLS, 1941
Ficin	Fig *(Ficus carica)*	WALTI, 1938
Bromelin	Pineapple *(Ananas sativus)*	GREENBERG and WINNICK, 1940
Asclepain	Milkweed *(Asclepias speciosa)*	WINNICK et al., 1940
Mexicain	Cuaguayote *(Pileus mexicanus)*	CASTANEDA et al., 1942
Pinguinain	Maya *(Bromelia pinguin)*	ASENJO and DEL CAPELLA DE FERNANDEZ, 1942
Tabernamontanain	*(Tabernamontana grandiflora)*	JAFFE, 1943a
Soyin	Soya bean *(Soja hispida)*	TAUBER, 1949
Solanain	Horsenettle *(Solanum eleagnifolium)*	GREENBERG and WINNICK, 1940
Euphorbain . . .	Caper *(Euphorbia lathyris)*	CASTANEDA et al., 1943
Hurain	Jabillo *(Hura crepitans)*	JAFFE, 1943b
Pomiferin	Osage *(Maclura pomifera)*	TAUBER, 1949
Arachain.	Peanut *(Arachis hypogaea)*	IRVING and FONTAINE, 1945

b) Assay.

The methods used for the assay of protease activity have been collected by GREEN and NEURATH (1954) and are summarized in Table 12. GREEN and NEURATH (1954) point out that some of these methods (for example, dilatometry and polarimetry) are not suitable for routine estimations of protease activity, but are useful in the investigation of the nature of proteolysis. For routine measurement of protease activity, the method of ANSON (1938) is most frequently used. The protease is allowed to act upon a protein substrate (often hemoglobin) for

Table 12. *Methods for the measurement of protease activity*
(adapted from GREEN and NEURATH, 1954).

Method based on:	Quantity measured	Experimental technique
Change in physical properties of substrate	Viscosity	Viscosimeter
	Optical rotation	Polarimeter
	Volume	Dilatometer
	Conductivity	Conductivity bridge
	Clot formation	Time measurements
Disappearance of substrate	Turbidity	
Appearance of peptides	Tyrosine + tryptophan	Ultraviolet absorption
		Folin reagent
	Optical rotation	Polarimeter
	Refractive index	Refractometer

various periods of time. The reaction is stopped by the addition of five per cent trichloroacetic acid, and the undigested protein is removed by filtration or centrifugation. The liberated peptides in the supernatant solution are estimated by the FOLIN reagent or by measurement of the optical density at 280 mμ.

c) Preparation.

Five of the plant proteases have been crystallized. These are: papain, chymopapain, ficin, asclepain, and mexicain.

Purification of papain (KIMMEL and SMITH, 1954). One kilogram of commercial dried papaya latex is ground in a Waring blendor with 100 g of diatomaceous earth or washed sand and one liter of 0.04 M cysteine at pH 5.5. The extract is squeezed out through cloth and then filtered.

The pH of the filtrate is brought to 9.0 with 2 M NaCN. The precipitate, if any, is removed by centrifugation. The supernatant solution is made 0.4 saturated with solid ammonium sulfate, and the precipitate is collected by centrifugation.

The precipitate is dissolved in 600 ml of 0.02 M cysteine (pH 7—7.5). Sixty grams of sodium chloride are dissolved in the solution, and the precipitate formed is collected by centrifugation. The precipitate is dissolved in 0.02 M cysteine to make 400 ml, and the pH of the solution is adjusted to 6.5.

The solution is allowed to stand about 30 minutes at room temperature. It is then cooled to 5° C, and the precipitate which forms after standing for 18 hours at 5° C is collected by centrifugation. This crystalline precipitate is dissolved in 300 ml of neutralized 0.02 M cysteine at room temperature; then ten ml of saturated sodium chloride solution are slowly added. The crystalline precipitate is removed by centrifugation.

Purification of chymopapain (JANSEN and BALLS, 1941). Papaya latex is acidified to pH 1.8—2.0, and the precipitate removed by centrifugation. The pH of the supernatant solution is raised to 3.5—4.0, and the solution is made half-saturated with NaCl. The precipitated protein is removed by centrifugation. The pH of the supernatant solution is adjusted to 2.0, and the solution is saturated with NaCl. The crystalline precipitate is collected by centrifugation.

Purification of ficin (WALTI, 1938). Clarified fig latex is adjusted to pH 5.0, and the solution is allowed to stand for several weeks at 5° C. The crystalline precipitate is collected by centrifugation. The crystals are dissolved in 0.02 M HCl, filtered, and recrystallized by adjusting the pH of the solution to 5.0.

Purification of asclepain (CARPENTER and LOVELACE, 1943). Three kilograms of milkweed roots are ground, and the juice is pressed out in a laboratory press. The juice is centrifuged, and the supernatant solution is filtered through paper pulp. The filtrate is saturated with ammonium sulfate, and kept overnight at 5° C. The precipitate is dissolved in 300 ml of water, insoluble material is removed by centrifugation, and the enzyme is precipitated by half-saturation with ammonium sulfate. The precipitate is dissolved in 50 ml of water, filtered through paper pulp, and dialyzed against distilled water. The enzyme precipitates out during dialysis, and is collected by centrifugation, and dissolved in phosphate buffer of pH 7.0. The solution is dialyzed against saturated ammonium sulfate. The enzyme crystallizes during this dialysis, and is collected by centrifugation.

d) Properties.

The proteases that have been purified thus far fall into two main classes: those that possess a necessary sulfhydryl group, and those that do not. Papain is a member of the class of proteases that has an essential sulfhydryl group, and is the only protease from plants that has been characterized extensively. From the limited data available concerning other proteases from plants and microorganisms, however, papain appears to be fairly representative of the class of proteases which have an essential sulfhydryl group.

Papain has a molecular weight of about 20,700 (SMITH et al., 1954), as estimated by both physical and chemical techniques. It is essentially homogeneous as judged by sedimentation studies, by electrophoretic studies, and by end-group analysis. It has an isoelectric point at pH 8.75. There appears to be only one reactive sulfhydryl group in the active papain molecule. However, there is a total of six cysteine residues, and, under the proper conditions, all will react with mercurial reagents. Mercury binding with the single normally-reactive sulfhydryl completely stops enzyme activity.

Table 13. *Amino acid composition of papain* (SMITH et al., 1958).

Amino acid	Number of residues
Glycine	23
Aspartate	17
Glutamate	17
Tyrosine	17
Valine	15
Alanine	13
Serine	11
Arginine	9
Isoleucine	9
Leucine	9
Proline	9
Lysine	8
Threonine	7
Cysteine	6
Tryptophan	5
Phenylalanine	4
Histidine	1
Amide	(19)
Total	180

Papain appears to be a single peptide chain containing 180 amino acids (SMITH et al., 1958). The amino acid composition of papain is given in Table 13. The molecule has a high content of glycine, and no methionine. The N-terminal amino acid sequence is isoleucine-proline-glutamate, with isoleucine as the terminal amino acid. As many as 120 of the 180 amino acids in papain can be removed by leucine aminopeptidase without loss of enzymatic activity (SMITH et al., 1958).

References.

ABDUL-NOUR, BASIMA: Some characteristics of intact and disrupted ribonucleoprotein particles from pea seedlings. Master's Thesis. The Ohio State University 1959. — ACS, G., G. HARTMAN, H. G. BOMAN and F. LIPMANN: Fed. Proc. 18, 178 (1959). — AL-DAWODY, A., J. E. VARNER and G. C. WEBSTER: Ohio J. Sci. 60, 327 (1960). — ANSON, M. L.: J. gen. Physiol. 22, 79 (1938). — ARCHIBALD, R. M.: J. biol. Chem. 154, 657 (1944); — Chem. Rev. 37, 161 (1945). — ASENJO, C. F., and M. DEL CAPELLA DE FERNANDEZ: Science 95, 148 (1942).

BALLS, A. K., and W. S. HALE: Cereal Chem. 15, 622 (1938). — BALLS, A. K., and H. LINEWEAVER: J. biol. Chem. 130, 669 (1939). — BERG, P.: J. biol. Chem. 222, 1025 (1956). — BLOCH, K.: J. biol. Chem. 179, 1245 (1949). — BONNER, J.: Fortschr. Chem. org. Naturstoffe 16, 139 (1958). — BOVE, J., and I. D. RAACKE: Arch. Biochem. Biophys. 85, 521 (1959). — BOYER, P. D., R. C. MILLS and H. J. FROMM: Arch. Biochem. Biophys. 81, 249 (1959). — BUSCH, G.: Biochem. Z. 312, 308 (1948).

CARPENTER, D. C., and F. E. LOVELACE: J. Amer. chem. Soc. 65, 2364 (1943). — CASTANEDA, M., M. R. BALCAZAR and F. E. GAVARRAN: Anales escuela nac. Cienc. biol. (Mex.) 3, 65 (1943). — CASTANEDA, M., F. E. GAVARRAN and M. R. BALCAZAR: Science 96, 365 (1942). — CHAO, F., and H. K. SCHACHMAN: Arch. Biochem. Biophys. 61, 220 (1956). — CLARK, J.: J. biol. Chem. 233, 421 (1958). — CONWAY, E. J.: Microdiffusion analysis and volumetric error. 2nd edition. London: Crosby Lockwood 1947. — CRANE, R. K., and F. LIPMANN: J. biol. Chem. 201, 235 (1953). — CRICK, F. H. C.: Symp. Soc. exp. Biol. 12, 138 (1958).

DAVIE, E. W., V. KONINGSBERGER and F. LIPMANN: Arch. Biochem. Biophys. 65, 21 (1956). — DAVIS, J. W., and G. D. NOVELLI: Arch. Biochem. Biophys. 75, 299 (1958). — DELWICHE, C. C., W. D. LOOMIS and P. K. STUMPF: Arch. Biochem. Biophys. 33, 333 (1951). — DEMOSS, J. A., and G. D. NOVELLI: Biochim. biophys. Acta 18, 592 (1955); 22, 49 (1956). — DINTZIS, H. M., H. BORSOOK and J. VINOGRAD: In Microsomal particles and protein synthesis (R. B. ROBERTS, editor). New York: Pergamon Press 1958. — DUNN, D. B.: Biochim. biophys. Acta 34, 286 (1959).

ELLIOTT, W. H.: Biochem. J. **49**, 106 (1951); — J. biol. Chem. **201**, 661 (1953). — ELLIOTT, W. H., and E. F. GALE: Nature (Lond.) **161**, 129 (1948). — ERRERA, M., and J. P. GREEN-STEIN: J. nat. Cancer Inst. **7**, 285 (1947). — EVANS, D. G.: J. gen. Microbiol. **1**, 378 (1948).

GEDDES, W. F., and A. HUNTER: J. biol. Chem. **77**, 197 (1928). — GILLCHRIEST, W. C., and R. M. BOCK: In Microsomal particles and protein synthesis (R. B. ROBERTS, editor). New York: Pergamon Press 1958. GONCALVES, J. M., V. E. PRICE and J. P. GREENSTEIN: J. nat. Cancer Inst. **7**, 281 (1947). — GORNALL, A. G., C. J. BARDAWILL and M. M. DAVID: J. biol. Chem. **177**, 751 (1949). — GORR, G., and J. WAGNER: Biochem. Z. **266**, 96 (1933). — GRASSMAN, W., and W. HEYDE: Z. physiol. Chem. **183**, 32 (1929). — GRASSMAN, W., and O. MAYR: Z. physiol. Chem. **214**, 185 (1933). — GREEN, N. M., and H. NEURATH: In The proteins (H. NEURATH and K. BAILEY, editors) Vol. 2 B. New York: Academic Press 1954. — GREEN-BERG, D. M., and T. WINNICK: J. biol. Chem. **135**, 761 (1940). — GREENSTEIN, J. P., and C. E. CARTER: J. nat. Cancer Inst. **7**, 57 (1947). — GREENSTEIN, J. P., and F. LEUTHARDT: J. nat. Cancer Inst. **5**, 209 (1944); — Arch. Biochem. **17**, 105 (1948). — GROVER, C. E., and A. C. CHIBNALL: Biochem. J. **21**, 857 (1927).

HILLER, A., J. PLAZIN and D. D. VAN SLYKE: J. biol. Chem. **176**, 1401 (1948). — HOAG-LAND, M. B.: Biochim. biophys. Acta **16**, 288 (1955); — In The nucleic acids (E. CHARGAFF and J. N. DAVIDSON, editors) Vol. 3. New York: Academic Press 1960. — HOLLEY, R. W., and J. GOLDSTEIN: J. biol. Chem. **234**, 1765 (1959). — HOLTER, H., and K. LINDERSTROM-LANG: Z. physiol. Chem. **214**, 223 (1933).

IRVING, jr., G. W., and T. D. FONTAINE: Arch. Biochem. **6**, 351 (1945).

JAFFE, W. G.: J. biol. Chem. **149**, 1 (1943a); — Rev. bras. Biol. **3**, 149 (1943b). — JANSEN, E. F., and A. K. BALLS: J. biol. Chem. **137**, 459 (1941). — JOHNSON, M. J.: J. biol. Chem. **137**, 575 (1941). — JOHNSON, M. J., and J. BERGER: Advanc. Enzymol. **2**, 69 (1942).

KIMMEL, J. R., and E. L. SMITH: J. biol. Chem. **207**, 515 (1954). — KREBS, H. A.: Biochem. J. **29**, 1951 (1935).

LAJTHA, A., P. MELA and H. WAELSCH: J. biol. Chem. **205**, 553 (1953). — LANDUA, A., and J. AWAPARA: Science **109**, 385 (1949). — LEVENBOOK, L., and J. KUHN: Fed. Proc. **17**, 95 (1958). — LEVINTOW, L.: J. nat. Cancer Inst. **15**, 347 (1954). — LEVINTOW, L., and A. MEISTER: J. biol. Chem. **209**, 265 (1954). — LEVINTOW, L., A. MEISTER, G. H. HOGEBOOM and E. L. KUFF: J. Amer. chem. Soc. **77**, 5304 (1955). — LITTLEFIELD, J. W., and E. B. KELLER: J. biol. Chem. **224**, 13 (1957). — LITTLEFIELD, J. W., E. B. KELLER, J. GROSS and P. C. ZAMECNIK: J. biol. Chem. **217**, 111 (1955). — LOWRY, O. H., N. J. ROSEBROUGH and A. L. FARR: J. biol. Chem. **193**, 265 (1951).

MANDELES, S., and K. BLOCH: J. biol. Chem. **214**, 639 (1955). — MASHMANN, E.: Biochem. Z. **307**, 1 (1940); — Ergebn. Enzymforsch. **9**, 166 (1943). — McILWAIN, H.: J. gen. Microbiol. **2**, 186 (1948).

NOVELLI, G. D.: Proc. nat. Acad. Sci. (Wash.) **44**, 86 (1958).

OLCOTT, H. S., L. A. SAPIRSTEIN and M. J. BLISH: Cereal Chem. **20**, 87 (1943).

RAACKE, I. D.: Quart. Rev. Biol. **33**, 245 (1958). — RAACKE, I. D.: Biochim. biophys. Acta **34**, 1 (1959). — ROBINSON, E., and R. BROWN: Nature (Lond.) **171**, 313 (1953). — RUD-MAN, D., and A. MEISTER: J. biol. Chem. **200**, 591 (1953). — RUDNICK, D., P. MELA and H. WAELSCH: J. exp. Zool. **126**, 297 (1954).

SCHACHMAN, H. K., A. B. PARDEE and R. Y. STANIER: Arch. Biochem. Biophys. **38**, 245 (1952). — SCHWEET, R. S., and E. H. ALLEN: J. biol. Chem. **233**, 1104 (1958). — SCHWEET, R. S., R. W. HOLLEY and E. H. ALLEN: Arch. Biochem. Biophys. **71**, 311 (1957). — SCHWEET, R. S., H. LAMFROM and E. H. ALLEN: Proc. nat. Acad. Sci. (Wash.) **44**, 1029 (1958). — SMITH, E. L.: Advanc. Enzymol. **12**, 191 (1951); — In Methods in enzymology (S. P. COLO-WICK and N. O. KAPLAN, editors) Vol. 2. New York: Academic Press 1955. — SMITH, E. L., R. L. HILL and J. R. KIMMEL: In Symposium on protein structure (A. NEUBERGER, editor). New York: John Wiley 1958. — SMITH, E. L., J. R. KIMMEL and D. M. BROWN: J. biol. Chem. **207**, 533 (1954). — SNOKE, J. E.: J. biol. Chem. **213**, 813 (1955). — SNOKE, J., and K. BLOCH: J. biol. Chem. **199**, 407 (1952). — SPECK, J. F.: J. biol. Chem. **179**, 1405 (1949). — STEENSHOLT, G.: Acta physiol. scand. **8**, 342 (1944). — STUMPF, P. K., and W. D. LOOMIS: Arch. Biochem. **25**, 451 (1950).

TAUBER, H.: The chemistry and technology of enzymes. New York: John Wiley 1949. — TRACEY, M. V.: Biochem. J. **42**, 281 (1948). — TSO, P., J. BONNER and H. DINTZIS: Arch. Biochem. Biophys. **76**, 225 (1958). — TSO, P., J. BONNER and J. VINOGRAD: J. Biophys. Biochem. Cytol. **2**, 451 (1956).

VAN DE VEN, A. M., V. KONINGSBERGER and J. OVERBEEK: Biochim. biophys. Acta **28**, 134 (1958). — VARNER, J. E.: Arch. Biochem. Biophys. **90**, 7 (1960). — VARNER, J. E., and G. C. WEBSTER: Plant Physiol. **30**, 393 (1955)

Waelsch, H., and D. Rittenberg: J. biol. Chem. **139**, 761 (1941). — Walti, A.: J. Amer. chem. Soc. **60**, 493 (1938). — Webster, G. C.: Arch. Biochem. Biophys. **47**, 241 (1953a); — Plant Physiol. **28**, 728 (1953b); **28**, 724 (1953c); — In The chemical basis of heredity. (W. D. McElroy, and B. Glass, editors). Baltimore: Johns Hopkins Press (1957a); — J. biol. Chem. **229**, 535 (1957b); — Nitrogen metabolism in plants. Evanston: Row, Peterson (1959a); — Symp. Soc. exp. Biol. **13**, 330 (1959b); — Arch. Biochem. Biophys. **85**, 159 (1959c); **82**, 125 (1959d); — Biochim. biophys. Acta **49**, 141 (1961). — Webster, G. C., and J. E. Varner: Arch. Biochem. Biophys. **52**, 22 (1954); — J. biol. Chem. **215**, 91 (1955a); — Arch. Biochem. Biophys. **55**, 95 (1955b). — Webster, jr., L. T., and E. W. Davie: Fed. Proc. **18**, 348 (1959). — Wieland, T., G. Pfleiderer and B. Sandmann: Biochem. Z. **330**, 198 (1958). — Willstätter, R., W. Grassman and O. Ambros: Z. physiol. Chem. **151**, 286 (1926). — Winnick, T., A. R. Davis and D. M. Greenberg: J. gen. Physiol. **23**, 275 (1940). — Woodward, G. E.: J. biol. Chem. **109**, 1 (1935).

Yanari, S., J. E. Snoke and K. Bloch: J. biol. Chem. **201**, 561 (1953). — Yin, F. H., and R. M. Bock: Fed. Proc. **19**, 137 (1960).

Enzymes of Synthesis of Purine and Pyrimidine Nucleotides.

By

Dalton Wang and E. R. Waygood.

With 2 Figures.

The biosynthetic pathways of the nucleotides of purines and pyrimidines have been recently elucidated and soundly established. Schemes of the pathways of these groups of compounds are shown in Figs. 1, 2. Enzymes that catalyze these chemical reactions have been isolated from animals and micro-organisms and some have been purified to a considerable extent. Credit for the elucidation of the enzymatic synthesis of purine nucleotides should be given particularly to BUCHANAN, KORNBERG and GREENBERG, and of pyrimidine nucleotides to REICHARD, COHEN, and LIEBERMAN and their colleagues. Despite the advances made in this field in animals and micro-organisms the existence of these enzymes in higher plants has not been demonstrated so far. Undoubtedly, research on plant enzyme systems is urgently needed in this respect.

Procedures for the isolation and purification of enzymes from animals and micro-organisms are generally applicable to plant material or vice versa. However, modifications of a given procedure which have been found to be suitable for certain enzymes from one type of organism are sometimes necessary in order to make it suitable for the same enzymes from another type of organism. On the whole, the basic principles involved in the isolation and purification of enzymes and the methods of assay remain the same. In view of the above facts and also that no published procedures of isolation and purification of these enzymes from plant material are available, the isolation and purification procedures described in this section should be adopted only as a guide. With this in mind, modification may be necessary.

In this laboratory the authors have attempted to apply some of the procedures for the isolation of certain enzymes from Mung bean seedlings involved in the metabolism of PRPP and orotic acid. The procedures will be described where applicable.

The following abbreviations are used in this chapter: ADP, adenosine diphosphate; AICAR, 5-amino-4-imidazole carboxamide ribotide; AICR, 5-amino-4-imidazolecarboxylic acid ribotide; AIR, 5-aminoimidazole ribotide; AISCAR, 5-amino-4-imidazole-N-succino-carboxamide ribotide; APDPN, 3-acetyl pyridine analogue of DPN; APDPNH, reduced form of APDPN; ATP, adenosine triphosphate; ATPase, adenosine triphosphate pyrophosphorylase; CA, carbamylaspartic acid; CP, carbamylphosphate; DEA, diethanolamine-diethanolamine hydrochloride; DPN, diphosphopyridine nucleotide; DPNH, reduced form of DPN; FAICAR, 5-formyl-amino-4-imidazolecarboxamide ribotide; FGAMR, formylglycinamidine ribotide; FGAR, formylglycinamide ribotide; GAR, glycinamide ribotide; GSH, glutathione; IMP, inosinic acid; N^5,N^{10}-anhydroformyl-THFA, N^5,N^{10}-anhydroformyl tetrahydrofolic acid; O-5-P, orotidine-5-phosphate; Pi, orthophosphate; PGA, 3-phosphoglyceric acid; PP, pyrophosphate; PRA, 5-phosphoribosylamine; PRPP, 5-phosphoribosylpyrophosphate; R-5-P, ribose-5-phosphate; TCA, trichloroacetic acid; THFA, tetrahydrofolic acid; TPN, triphosphopyridine nucleotide; TPNH, reduced form of TPN; Tris, tris (hydroxymethyl) aminomethane; U-5-P, uridine-5-phosphate; Versene, ethylenediaminetetraacetic acid.

A. Enzymes of Synthesis of Purine Nucleotides.

I. 5-Phosphoribosylpyrophosphate Kinase.

PRPP-kinase catalyzes the formation of PRPP from R-5-P and ATP:

$$ATP + R\text{-}5\text{-}P \rightarrow adenosine\text{-}5\text{-}phosphate + PRPP \tag{1}$$

Mg^{++} is required. This enzyme has been demonstrated and partially purified by KORNBERG et al. (1955) in mammalian liver and in bacteria, by KORN et al. (1955) in pigeon liver, by TARR (1960) in fish muscle and by the present authors in Mung bean seedlings. PRPP is an essential intermediate involved in the early reaction steps in the biosynthesis of purine nucleotides and in somewhat later reaction steps in pyrimidine nucleotides. The important role which this compound plays in the biosynthesis of these nucleotides has been reviewed by CARTER (1956), HARTMAN and BUCHANAN (1959), BUCHANAN and HARTMAN (1959), and REICHARD (1959).

1. Assay.

The activity of PRPP-kinase can be conveniently determined spectrophotometrically at a suitable wave-length. KORNBERG et al. (1955) described a spectrophotometric method for assaying this enzyme by quantitative conversion of orotic acid to its nucleotide, O-5-P, in the presence of O-5-P pyrophosphorylase. The O-5-P is decarboxylated by the enzyme O-5-P decarboxylase and the formation of U-5-P results in a decrease in optical density at 295 mμ. The molar absorption coefficient for the optical density decrease of the conversion is 3950. According to KORNBERG et al. (1955), this assay is best carried out in two stages: the first involving the formation of PRPP as shown in equation (1) and the second its quantitative utilization in the removal of orotic acid as follows:

$$Orotate + PRPP \rightleftharpoons O\text{-}5\text{-}P + PP \tag{2}$$

Stage I (Formation of PRPP). The incubation mixture (1.0 ml) consists of 0.03 ml of ATP (0.04 M), 0.1 ml of R-5-P (0.025 M), 0.02 ml of potassium phosphate buffer (1 M, pH 7.4), 0.02 ml of MgCl$_2$ (0.1 M), 0.02 ml of glutathione (0.5 M), 0.05 ml of KF (1 M) and about 1 unit of the enzyme (PRPP-kinase). The reaction mixture is incubated at 36° C for 20 minutes. The reaction is terminated by heating the mixture for 1 minute in a boiling bath to inactivate the enzyme. It is then cooled immediately and centrifuged if necessary.

Stage II (Utilization of PRPP in the removal of orotate). The incubation mixture (1.0 ml) in a quartz cuvette consists of 0.02 ml of orotate (0.01 M), 0.02 ml of Tris buffer (1 M, pH 8.0), 0.02 ml of MgCl$_2$ (0.1 M), 0.02 ml of O-5-P pyrophosphorylase, and an aliquot from Stage I containing about 0.03 μmole of PRPP. The decrease in optical density at 295 mμ is followed. With the indicated amount of O-5-P pyrophosphorylase fraction, the utilization of 0.03 μmole of PRPP is usually complete in 10 minutes when measured in the BECKMAN DU spectrophotometer at room temperature.

With crude enzyme fractions there usually is ATPase present and the use of fluoride provides a means of reducing ATPase activity.

REMY et al. (1955) developed a similar spectrophotometric method for the assay of PRPP-kinase (Enzyme I) in the presence of an excess of Enzyme II. The latter enzyme catalyses the formation of IMP from PRPP and hypoxanthine:

$$PRPP + hypoxanthine \rightleftharpoons IMP + PP \tag{3}$$

At the end of the incubation period the reaction mixture is heated and the solution deproteinized with perchloric acid. The amount of hypoxanthine was determined spectrophotometrically at 290 mμ in the presence of xanthine oxidase (KALCKAR, 1947) on a neutralized sample. The amount of hypoxanthine utilized for the synthesis of IMP could then be calculated.

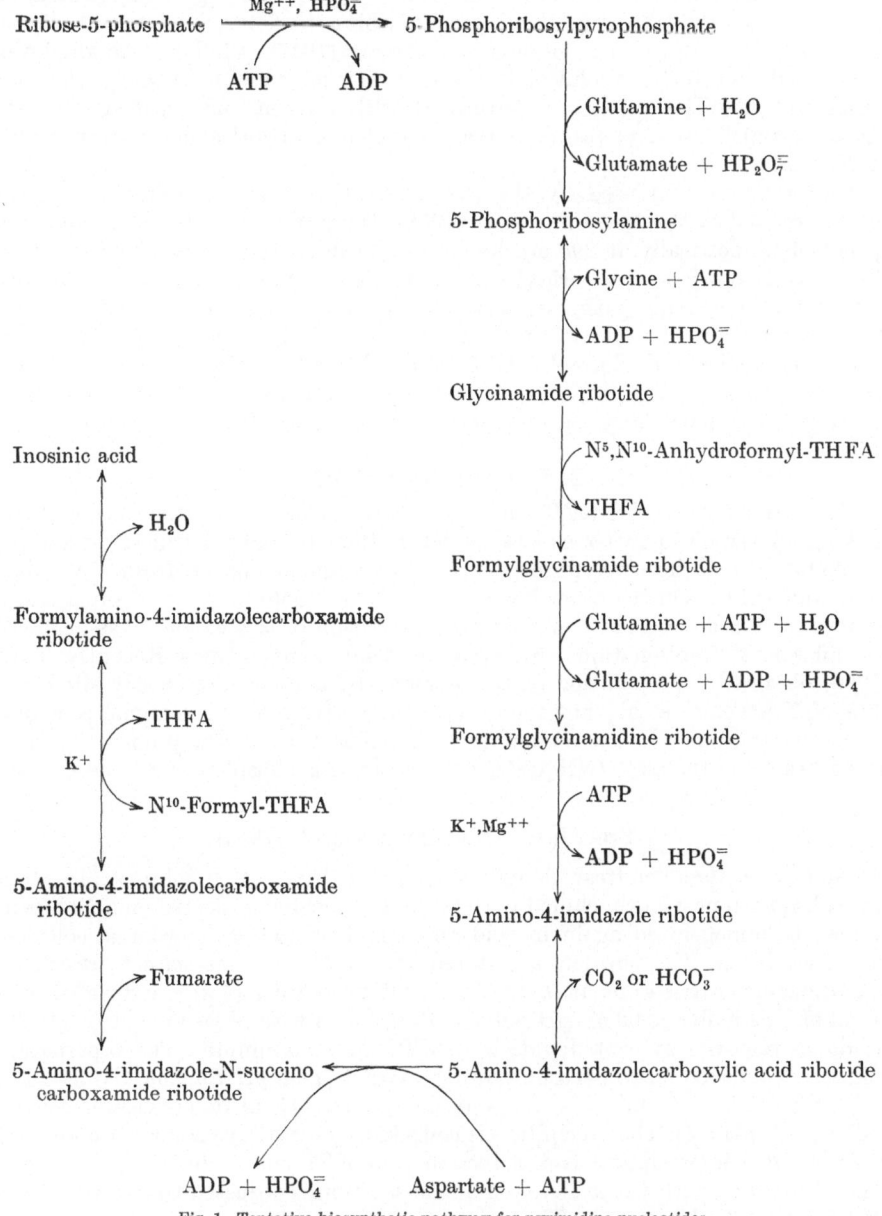

$$Ribose\text{-}5\text{-}phosphate \xrightarrow{Mg^{++},\ HPO_4^=} 5\text{-}Phosphoribosylpyrophosphate$$

ATP   ADP

Glutamine + H_2O

Glutamate + $HP_2O_7^=$

5-Phosphoribosylamine

Glycine + ATP

ADP + $HPO_4^=$

Glycinamide ribotide

Inosinic acid

N^5,N^{10}-Anhydroformyl-THFA

H_2O

THFA

Formylglycinamide ribotide

Formylamino-4-imidazolecarboxamide ribotide

Glutamine + ATP + H_2O

Glutamate + ADP + $HPO_4^=$

THFA

Formylglycinamidine ribotide

K^+

N^{10}-Formyl-THFA

ATP

K^+,Mg^{++}

ADP + $HPO_4^=$

5-Amino-4-imidazolecarboxamide ribotide

5-Amino-4-imidazole ribotide

Fumarate

CO_2 or HCO_3^-

5-Amino-4-imidazole-N-succino carboxamide ribotide ← 5-Amino-4-imidazolecarboxylic acid ribotide

ADP + $HPO_4^=$ Aspartate + ATP

Fig. 1. Tentative biosynthetic pathway for pyrimidine nucleotides.

The incubation mixture used by REMY et al. (1955) is similar to that given above for Stage I with the exception that KF and glutathione were not included. The incubation mixture (2.5 ml) contains 50 μmoles phosphate buffer (pH 7.4), 20 μmoles $MgCl_2$, 0.91 μmoles hypoxanthine, 2.3 μmoles ATP, 3.78 μmoles R-5-P,

0.1 ml (15 mg powder/ml) of enzyme I and 1.5 ml of enzyme II. The reaction mixture is incubated at 38° C for 7 minutes. The reaction is terminated by immersion of the vessel in a boiling water bath for 5 minutes. The solution is deproteinized by the addition of 0.5 ml of 30% HClO$_4$. The deproteinized solution is neutralized and the remaining hypoxanthine is determined spectrophotometrically at 290 mμ by the method of KALCKAR (1947) as follows. An aliquot of neutralized deproteinized solution (0.5 to 5 γ/ml) is mixed with 0.1 M glycylglycine buffer pH 7.5. The solution is saturated with oxygen and purified xanthine oxidase is added (10—20 γ protein/ml) and the change in optical density is recorded at 290 mμ.

The assay system used by the present authors was a modification of the procedure used by WYNGAARDEN et al. (1958). The formation of PRPP is measured spectrophotometrically at 295 mμ by coupling the synthetic reaction with those that convert orotic acid to uridylic acid. Whereas WYNGAARDEN used a volume of 1 ml in a 1.0 cm light path, our system had a volume of 1.1 ml in cuvettes with a 0.5 cm light path. The system contains 50 μmoles of K$_2$HPO$_4$ (pH 7.4), 4 μmoles MgCl$_2$, 50 μmoles KF, 2 μmoles GSH, 2 μmoles ATP, 0.25 μmoles R-5-P, 0.2 μmoles of orotate, 0.05 ml of a 0—15% ethanol fraction of yeast autolysate (orotidylic pyrophosphorylase and decarboxylase), and 0.1 ml of a Mung bean enzyme.

2. Enzyme Preparation.

The preparation of PRPP-kinase by KORNBERG et al. (1955) consists of making an acetone powder extract which is then partially purified by low pH precipitation and selective absorption and subsequent elution from aluminium hydroxide gel C$_\gamma$. On the other hand, KORN et al. (1955) prepared a supernatant extract from a homogenate and partial purification of this extract was achieved by differential fractionation with ethanol at low temperatures. Recently, TARR (1960) obtained a preparation of this enzyme by a procedure closely similar to that of KORNBERG et al., but found that the activity of this enzyme remained in the supernatant solution after alumina C$_\gamma$ gel treatment. The preparation from Mung bean was the 60% (NH$_4$)$_2$SO$_4$ precipitate of a phosphate cysteine extract.

a) Procedures of KORNBERG et al. (1955).

Acetone powder. The fresh tissue is homogenized in 5 to 10 volumes of acetone which has been previously chilled to —10° C. The residue, collected on a BUCHNER funnel, is homogenized again in cold acetone. The acetone powder is collected and dried in air. The powders are stored at —10° C. *Acetone-powder extract.* All operations are carried out at 0° to 3° C. Acetone powder (4 g) is extracted with 40 ml of Tris buffer (0.02 M, pH 8.0) for 10 minutes with occasional stirring. The residue is removed by centrifugation at 8,000 ×g for 5 minutes. The supernatant solution is referred to as acetone powder extract. *Low pH fraction.* To 34 ml of extract are added 51 ml of water and then slowly 17 ml of potassium acetate buffer (1 M, pH 5.4). The precipitate is collected by centrifugation after 5 minutes at 8,000 ×g. The precipitate is dissolved in 7 ml of Tris buffer (0.1 M, pH 8.0) and diluted to 30 ml with water. The pH of the solution is adjusted to 6.8 with 0.1 N HCl (about 1.3 ml required). The volume is made up to 34 ml. This preparation is referred to as low pH fraction. *Aluminum hydroxide gel eluate.* Aged aluminum hydroxide C$_\gamma$ gel (3.75 ml, 15 mg dry weight/ml) (WILLSTÄTTER and KRAUT, 1923) is centrifuged and the supernatant fluid is discarded. The low pH fraction (30 ml) is thoroughly mixed with the gel and left in ice for 5 minutes. The gel precipitate is collected by centrifugation and washed with 30 ml of potassium phosphate buffer (0.05 M, pH 6.85). Both supernatant and washing are discarded. The enzyme is then eluted from the gel with 30 ml of potassium phosphate buffer (0.10 M, pH 6.85).

The acetone powders (pigeon liver) are stored at −10° C. According to KORN-BERG et al. the C_γ gel eluate retained 65 to 70% of its activity after storage for two months at −10° C provided glutathione (0.05 M) was present. In the absence of glutathione less than 5% of the activity remained.

Preparation of O-5-P pyrophosphorylase (LIEBERMAN et al., 1955). The operations are carried out at 0° C unless otherwise indicated. *Yeast autolysate.* Dried brewer's yeast is suspended in 3 volumes of $KHCO_3$ (0.1 M) and incubated at 30° C for 5 hours with occasional stirring. After cooling the suspension to between 5° and 10° C, the supernatant solution is obtained by centrifugation at about 10,000×g. The residue is washed with an amount of $KHCO_3$ solution equal to that added originally. After centrifugation the supernatant solutions are combined. For 10 g of dried yeast the yield of extract is 45 to 50 ml. *Ethanol fraction.* To 48 ml of yeast extract 96 ml of sodium acetate buffer (0.2 M, pH 4.5) are added rapidly while stirring. The acidified extract is cooled to −2° C in an alcohol-ice bath, and 48 ml of ethanol which has been previously cooled to −14° C are added in approximately 4 minutes while the temperature is allowed to fall to −6° C. The insoluble material is removed by centrifugation for 4 minutes at 10,000×g at −10° C. Ethanol (144 ml) is added to the supernatant solution in approximately 5 minutes with the temperature maintained at −10° C. The precipitate which is collected by centrifugation for 3 minutes at 10,000×g is dissolved in glycyl-glycine buffer (0.025 M, pH 7.0) and the pH is adjusted to 7.0 with 1 N KOH.

With the brewer's bottom yeast obtained through the courtesy of LABATTS BREWERIES, Winnipeg, and dried in our laboratory, the ethanol fraction of LIEBERMAN et al. (1955) was rarely active, but activity was found in the 0—15% cold ethanol fraction dissolved in 0.025 M glycylglycine buffer, pH 7.0.

Preparation from mung-bean seedlings. Locally purchased seedlings (200 g) are ground in a WARING blendor in the cold for 1.5—2 minutes with 200 ml of a solution of 0.05 M K_2HPO_4 and 0.01 M cysteine (pH 7.5). The brei is pressed through cheesecloth and the filtrate centrifuged for 10 minutes at 20,000×g. The supernatant solution is adjusted to pH 5.5 with 1.0 N acetic acid and the precipitate removed by centrifugation. The supernatant solution is adjusted to pH 7.5 with 2.0 N KOH and made 66% saturated with respect to $(NH_4)_2SO_4$ and adjusted to pH 7.5. The precipitate is collected by centrifugation and the residue washed twice with 70% saturated $(NH_4)_2SO_4$. The washed precipitate is dissolved in a minimum volume of 0.05 M Tris, HCl, (pH 8.0) and dialyzed overnight against distilled water. In some preparations the enzyme precipitated during dialysis was redissolved in 0.05 M Tris buffer. In others the supernatant of the dialysate was active.

b) Procedures of KORN et al. (1955).

Preparation of enzyme I. *Tissue homogenate* (WILLIAMS and BUCHANAN, 1953). One part of tissue is homogenized with 1.5 parts homogenate buffer (0.035 M sodium phosphate buffer, pH 7.4, 0.13 M KCl, 0.04 M $KHCO_3$ and 0.01 M $MgCl_2$) in a POTTER-ELVEHJEM type homogenizer. (Tissue, buffer solution and homogenizer being kept on ice until ready for use.) The homogenate is centrifuged and the supernatant solution is saved. *0—15% ethanol fraction* (Fraction I). The extract is first chilled to 0° C in a dry ice-acetone bath, and cold 90% ethanol is slowly added to 15% concentration by volume. While ethanol is being added the temperature of the mixture is maintained just above the freezing point by cooling it in a dry ice-acetone bath. The precipitate is collected by centrifugation, and washed twice by suspending it in cold 18% ethanol to remove salts and contaminating enzymes. The washed enzyme is then suspended in a small amount of distilled water and lyophilized at $-15°$ C (the entire apparatus is maintained at $-15°$ C during lyophilization). Just prior to use, the lyophilized powder is dissolved in 0.02 M glycylglycine or phosphate buffer, pH 7.4, to a final concentration of 8 mg per ml. "*Enzyme I* refers to the dissolved lyophilized preparation of 0—15% ethanol fraction."

Preparation of enzyme II (KORN et al., 1955). Enzyme I prepared from mammalian liver by KORN et al. (1955) and KORN and BUCHANAN (1955) is heat-labile, losing approximately 100% of its activity when heated for 10 minutes at 56° C or for 5 minutes at 65° C. These authors took advantage of this fact and were able to obtain a soluble enzyme preparation, Enzyme II, by differential heat fractionation of an acetone powder extract which is heat stable at the above mentioned temperature ranges. Enzyme II preparation, thus, is completely free of Enzyme I. In addition, the heating treatment (75° C for 5 minutes) completely destroyed nucleoside phosphorylase as well as inactivated phosphoribomutase and phosphatase. All operations are carried out at 2° C unless stated otherwise. Water redistilled in glass is used throughout. *Acetone powder extract* (KORN and BUCHANAN, 1955). Tissue is blended in 1.5 volumes of water in a WARING blendor. Two volumes of acetone (previously cooled to $-15°$ C) are added, and the mixture is filtered at room temperature. The filter cake is washed well with cold acetone and air dried. The acetone powder is extracted with 10 parts of water by weight and the insoluble residue is removed by centrifugation. The supernatant solution is referred to as acetone powder extract. *Differential heating treatment* (KORN et al., 1955). Extracts of acetone powder, containing approximately 10 mg of protein per ml, are heated in 100 ml lots to 75° C for 5 minutes with stirring. Upon cooling, the denatured proteins are separated by centrifugation and discarded. The supernatant solution is referred to as *Enzyme II*. The heated supernatant solution may be fractionated with $(NH_4)_2SO_4$ between 55% and 80% of saturation.

II. 5-Phosphoribosylpyrophosphate Amidotransferase.

The enzyme PRPP-amidotransferase catalyzes the formation of PRA from PRPP and glutamine according to the following equation:

$$PRPP + glutamine \rightarrow PRA + glutamic\ acid + PP \qquad (4)$$

This enzyme has been isolated from pigeon liver and partially purified by GOLDTHWAIT (1956), by HARTMAN and BUCHANAN (1958a), by WYNGAARDEN and ASHTON (1959) and by the present authors from Mung bean seedlings.

1. Assay.

The product PRA of this reaction has not yet been isolated from an enzymatic system probably because of its chemical instability. However, there is sufficient indirect evidence to suggest that PRA is the product of this enzymatic reaction (GOLDTHWAIT et al., 1955; GOLDTHWAIT, 1956; HARTMAN and BUCHANAN, 1958a, 1958b). Based on the stoichiometry of the reaction shown in equation 4, a spectrophotometric assay for glutamic acid as a function of the activity of PRPP-amidotransferase was developed by HARTMAN and BUCHANAN (1958a). This method somewhat modified was successfully adopted by WYNGAARDEN and ASHTON(1959).

They coupled the PRPP-amidotransferase system with a system of glutamic acid dehydrogenase (equation 5).

Glutamic acid $+$ DPN$^+$ (or APDPN$^+$) $+$ H$_2$O $\rightleftharpoons$
$$\alpha\text{-ketoglutarate} + \text{DPNH (or APDPNH)} + \text{H}^+ \quad (5)$$

The formation of DPNH or APDPNH from DPN$^+$ or APDPN$^+$, respectively, in the presence of glutamic dehydrogenase is conveniently followed by the change of optical density of the solution at 340 mμ for DPNH or at 363 mμ for APDPNH. The increase in optical density is then related to a standard curve to determine the quantity of glutamic acid formed from glutamine. This is a reflection of the activity of PRPP-amidotransferase. The advantage of using APDPN instead of DPN is the fact that the oxidation of glutamic acid can go to completion.

The formation of glutamic acid and pyrophosphate from glutamine and PRPP in the presence of Enzyme I may be assayed by the method of HARTMAN and BUCHANAN (1958a). The incubation mixture (0.6 ml) contains 25 μmoles Tris buffer (pH 8.0), 5 μmoles MgCl$_2$, 3 μmoles glutamine, 3 μmoles PRPP, and 0.1 ml Enzyme I. The reaction mixture is incubated for 0, 10, 20, 30 and 60 minutes at 38° C. The glutamic acid formed may be assayed as follows: 0.1 ml of glutamic acid sample, 125 μmoles of Tris buffer (pH 8.5), 0.4 μmoles DPN, and 2.5 mg glutamic acid dehydrogenase. This incubation mixture has a final volume of 2.9 ml. Optical density of the solution is read at 340 mμ immediately after the addition of glutamic acid dehydrogenase and then after 30 minutes. The increase in optical density is then related to a standard curve.

The assay of PRPP-amidotransferase activity by means of glutamic acid determination is more conveniently carried out by the method of WYNGAARDEN and ASHTON (1959). It consists of an incubation mixture (1.0 ml) in a quartz cuvette containing 50 μmoles Tris buffer (pH 8.0), 3 μmoles MgCl$_2$, 0.25 μmoles PRPP, 1.0 μmoles glutamine, 0.6 μmoles APDPN, 100 units glutamic dehydrogenase and sufficient PRPP-amidotransferase to cause an increase in optical density from 0.010 to 0.040 per minute during the linear phase, when read against a blank containing all factors except PRPP. Readings are made at room temperature at 363 mμ in quartz cuvettes with a light path of 1.0 cm.

The present authors have assayed the enzyme from Mung bean by a similar procedure. The system 1.2 ml in cuvette of 0.5 cm light path contains 60μmoles Tris buffer (pH 8.0), 1.5 μmoles of APDPN, 3 μmoles MgCl$_2$, 0.32 μ moles PRPP, 10 μmoles glutamine, 0.1 ml of a 1:10 dilution of glutamic dehydrogenase (N. B. Co)[1] dialyzed against 0.05 M phosphate pH 7.5, and 0.1 ml of Mung bean enzyme.

2. Enzyme Preparations.

The soluble enzyme, PRPP-amidotransferase, can be isolated by low temperature alcohol precipitation between 0 to 20% alcohol of the acid soluble fraction from an acetone powder of pigeon liver (GOLDTHWAIT, 1956) or between 15 and 30% alcohol fraction of an extract of pigeon liver (HARTMAN and BUCHANAN, 1958a). The 15—30% ethanol fraction can be partially purified by differential heat inactivation of certain undesirable contaminating enzymes in the extract of the precipitate which resulted from the dialysis of a 25—45%saturation ammonium sulfate fraction of this ethanol fraction (HARTMAN and BUCHANAN, 1958a). WYNGAARDEN and ASHTON (1959) by following the procedure of HARTMAN and BUCHANAN (1958a) but without the step of ethanol fractionation, could also obtain an active preparation of PRPP-amidotransferase. This enzyme may be further

---

[1] Nutritional Biochemical Corp., Cleveland, Ohio.

purified by adsorption and subsequent elution from calcium phosphate gel (WYNGAARDEN and ASHTON, 1959).

(a) **Procedures of GOLDTHWAIT (1956).** *Acetone powder* (GOLDTHWAIT and GREENBERG, 1955). The procedure for preparing acetone powder is essentially the same as those given under PRPP-kinase, on page 424, except that the powder is dried *in vacuo* at 4° C over concentrated H_2SO_4 for 1 day and is then stored *in vacuo* at —13° C after the removal of tissue debris by screening. *Acid supernatant fraction.* Acetone powder (2 g) is extracted at 0° C for 30 minutes with 20 ml of 0.02 M Tris buffer at pH 8.0. The suspension is centrifuged at $5,000 \times g$ for 30 minutes and the supernatant solution is decanted into 22.5 ml of water. One molar potassium acetate buffer (7.5 ml; molarity represent the total of acetate and acetic acid) at pH 5.5 (measured at 1 to 10 dilution) is added slowly and the precipitate is separated by centrifugation. This fraction is referred to as "Acid precipitate fraction" which has no transferase activity.The pH of the supernatant solution is adjusted to 5.0 with 1.0 N acetic acid. This solution is dialyzed following centrifugation against 2,000 ml of 0.05 M K_2HPO_4 for 3 hours with one change of the dialysis solution. All of the above steps are performed at 2° C. The dialyzed preparation which is referred to as "Acid supernatant fraction" is lyophilized. *0—20% ethanol fraction.* To 450 ml of the acid supernatant fraction, pH 5.3, 112 ml of absolute ethanol (previously cooled to —40° C) are added, while the temperature of the mixture is gradually lowered to —8° C. After centrifugation, the precipitate is dissolved in 75 ml of 0.05 M Tris buffer, pH 8.0, and lyophilized. The lyophilized 0—20% ethanol fraction is stored at —13° C.

(b) **Procedures of HARTMAN and BUCHANAN (1958a).** 15 to 30% ethanol fraction *(Fraction I).* Tissue (pigeon liver) is homogenized in a blendor and then centrifuged at 10,000 r.p.m. for 10 minutes. To the supernatant solution is added 0.2 volume of 90% cold ethanol, and the precipitate is removed by centrifugation at —7° C. To the supernatant solution an additional 0.25 volume of 90% cold ethanol is added. The precipitate is collected by centrifugation and dissolved in a minimal amount of water and lyophilized (15—30% ethanol fraction: Fraction I). 25 to 45% saturation $(NH_4)_2SO_4$ fraction *(Fraction II).* The lyophilized 15—30% ethanol fraction (6 g) is dissolved in 250 ml of 0.1 M diammonium acid citrate adjusted to pH 5.3 with NH_4OH. For each 100 ml of enzyme solution, 14.4 g of solid $(NH_4)_2SO_4$ are added (25% saturation). The precipitate is removed by centrifugation. To each 100 ml of supernatant solution an additional 12.3 g of solid $(NH_4)_2SO_4$ are added (45% saturation). The precipitate is collected and dissolved in 125 ml of water (Fraction II). *Fraction III.* Dialyze Fraction II against continuously changing distilled water for 6 hours. The protein which has been precipitated during dialysis is collected by centrifugation. The precipitate is extracted overnight with stirring with 25 ml of 0.05 M Tris buffer, pH 8.0. The suspension is centrifuged. The extract of the precipitate is referred to as Fraction III. *Fraction IV (Enzyme I). Fraction III* is heated for 10 minutes at 60° C in a water bath and cooled rapidly in ice. The denatured protein is removed by centrifugation. The supernatant solution is referred to as Fraction IV and used as a source of "Enzyme I".

(c) **Procedures of WYNGAARDEN and ASHTON (1959).** *Tissue homogenate* is prepared according to the method given under "PRPP-kinase" by KORN et al. (page 425). *Extract of homogenate* (SCHULMAN et al., 1952). This is prepared by spinning the homogenate in a refrigerated preparative ultracentrifuge at approximately $100,000 \times g$ for 30 minutes. The supernatant solution, which is free of all cellular structure, is removed carefully with a pipette. *20 to 45% saturation $(NH_4)_2SO_4$ fraction.* The $100,000 \times g$ supernatant solution is fractionated with saturated $(NH_4)_2SO_4$, previously adjusted to pH 7.4 with concentrated NH_4OH. Only the precipitate resulting from the ammoniacal $(NH_4)_2SO_4$ concentration of 20 to 45% saturation is collected. The precipitate is redissolved in a small volume of water. This fraction is reprecipitated twice with ammoniacal $(NH_4)_2SO_4$ at 45% saturation. These additional precipitations according to WYNGAARDEN and ASHTON virtually abolished endogenous reduction of the 3-acetylpyridine analogue of DPN. *Differential heat fractionation.* The third $(NH_4)_2SO_4$ precipitate is dialyzed against 400 volumes of distilled water. The precipitate is collected by centrifugation and then extracted with 0.05 M Tris buffer, pH 8.0. The extract is heated at 60° C for 10 minutes and cooled rapidly in ice. The denatured protein is removed by centrifugation at 25,000 × g for 30 minutes. The supernatant is used for further purification. *Calcium phosphate gel fraction.* This supernatant solution is diluted to contain 10 mg of protein per ml. To each ml of protein, 1 ml portions of aged calcium phosphate gel (190 mg/ml) (KEILIN and HARTREE, 1938) are added successively. According to WYNGAARDEN and ASHTON (1959), the enzyme is generally quantitatively adsorbed onto the gel in 2 to 3 steps, leaving $^3/_4$ of the protein in the solution. The gel fractions are pooled and mixed thoroughly with 1 to 2 ml of 0.001 M phosphate buffer (pH 7.4) per ml of protein solution originally treated with gel, for 20 to 30 minutes, after which the gel is removed by centrifugation. Elutions are repeated, successively with 0.005, 0.01 and 0.05 M phosphate buffer.

Preparation from mung bean seedlings. The Mung bean enzyme is prepared essentially in the same manner as for PRPP-kinase, except the acid treatment is omitted. In the majority of preparations the enzyme was precipitated during dialysis against water overnight and could be redissolved in 0.05 M Tris HCl, pH 8.0.

III. Glycinamide Ribotide Kinosynthase.

PRA, which has been suggested as a product of the reaction (equation 4) catalyzed by PRPP-amidotransferase from PRPP and glutamine, has so far not been isolated from enzymatic systems. GOLDTHWAIT (1956) who synthesized PRA chemically demonstrated that this compound could replace the substrates, PRPP and glutamine, for the biosynthesis of GAR. The formation of this aliphatic ribotide is catalyzed by the enzyme GRA-kinosynthase from PRA, glycine and ATP:

$$PRA + glycine + ATP \rightleftharpoons GAR + ADP + P_i \qquad (6)$$

This reaction is reversible. The enzyme GRA-kinosynthase has been isolated and partially purified by GOLDTHWAIT (1956) and GOLDTHWAIT et al. (1956a) from pigeon liver and by HARTMAN and BUCHANAN (1958a, 1958b) from the liver of pigeon and chicken. The last group of workers found that the enzyme from chicken liver seemed to be more stable.

Assay. GOLDTHWAIT et al. (1954) developed a unique method for assaying the activity of this enzyme by using glycine-1-C^{14} as a precursor. Furthermore, an accumulation of the glycinamide ribotide can be induced by treating the enzyme preparation with Dowex 1 to remove the folic acid derivative which is required by the transformylation enzyme that catalyzes the formation of FGAR from GAR and a formyl donor. The formation of GAR can be conveniently estimated by treating the reaction mixture with ninhydrin reagent to remove the radioactivity from the residual glycine-1-C^{14} as $C^{14}O_2$, and the remaining radioactivity represents the synthesis of GAR which does not react with this reagent. However, as pointed out by GOLDTHWAIT et al. (1956a) the formation of peptide, such as glutathione which would contain glycine in a form not affected by ninhydrin, should be tested. Chromatography of the product either on paper or on ion-exchange resin will serve as a check.

HARTMAN and BUCHANAN (1958a) described an assay method for the enzyme GAR-kinosynthase employing inosinic acid as an one-carbon donor, while the reaction product GAR serves as an acceptor in a 15 to 30% ethanol fraction of chicken liver extract:

$$GAR + IMP + H_2O \rightarrow FGAR + AICAR \qquad (7)$$

The non-acetylable, diazotizable aminoimidazole compound can be determined by diazotization and its quantity estimated spectrophotometrically at 540 mμ.

The method of assaying GAR-kinosynthase described by GOLDTHWAIT and GREENBERG (1955) consists of an incubation mixture (0.7 ml) containing 0.05 ml of glycine-1-C^{14} (0.1 M, 20,000 counts/min/μmole), 0.05 ml glutamine (0.2 M), 0.10 ml R-5-P (0.05 M; potassium salt), 0.02 ml of ATP (0.04 M; disodium salt neutralized with NaOH), 0.10 of PGA (0.14 M; potassium salt neutralized with KOH), 0.05 ml of MgCl$_2$ (0.1 M), and 0.20 ml of enzyme (Dowex treated and dialyzed). It is incubated for 30 minutes or longer (2 hours) at 38° C. The reaction is terminated by the addition of 0.5 ml of 20% TCA.

A 0.1 ml aliquot of the TCA filtrate is transferred into a small test tube and neutralized with 1 N NaOH using bromothymol blue. 1.0 ml of 1 M potassium phosphate buffer, pH 5.4, 0.1 ml of carrier glycine (0.1 M), and 1.0 ml of ninhydrin solution (30 mg/ml) are added. A marble is placed on the tube which is then heated at 100° C for 30 minutes. After cooling, 1 drop of caprylic alcohol is added, and the mixture is aerated with CO$_2$ for 15 minutes and diluted to 10 ml. A 2-ml aliquot is pipetted into a glass planchet, dried under an infra red light, and counted. Glycine-1-C^{14} incorporated into compounds which are not decarboxylated by ninhydrin account for GAR and to some extent for FGAR depending on the degree of the removal of formylating cofactors.

The TCA filtrate may also be chromatographed on paper with a pair of solvents (pyridine and water; 65:35 v/v and n-butanol-17.6 N acetic acid-water; 2:1:1 v/v) or on a column of ion-exchange resin (for details see GOLDTHWAIT, PEABODY and GREENBERG, 1956a).

The method of assaying GAR-kinosynthase described by HARTMAN and BUCHANAN (1958a) consists of a reaction mixture (0.3 ml) that contains 15 μmoles Tris buffer (pH 8.0), 1 μmole MgCl$_2$, 1 μmole glycine, 1 μmole ATP, 1 μmole PRA, and 0.1 ml Enzyme II. It is incubated for 30 minutes at 38° C. PRA may be replaced by PRPP (1 μM) and glutamine (1 μM) and Enzyme I (0.1 ml) must also be included in the incubation mixture, since it is responsible for the reaction of PRPP and glutamine to yield a product, presumably PRA, which reacts with ATP and glycine to form GAR in the presence of Enzyme II. PRA may be chemically synthesized by the procedure used by GOLDTHWAIT (1956).

The amount of GAR formed is determined by the method of WARREN and BUCHANAN (1957) modified by HARTMAN and BUCHANAN (1958a). To a 0.25 ml aliquot of solution containing no more than 0.05 μmole of GAR are added 0.05 ml of neutralized Versene 9 (trisodium salt of ethylenediaminetetraacetic acid) (300 μmoles per ml), 0.05 ml of sodium inosinate (50 μmoles per ml), and 0.2 ml of a solution containing 0.5 mg of lyophilized 15 to 30% ethanol fraction of chicken liver extract dissolved in 1 ml of 0.1 M Tris buffer, pH 7.4. After an incubation period of 40 minutes at 37° C, the mixture is deproteinized with 0.1 ml of 30% TCA and centrifuged. The solution is then treated with 0.05 ml of acetic anhydride for 20 minutes. This treatment serves to reduce the blank by acetylating aromatic amines present in the enzyme, but it leaves the aminoimidazole compounds unacetylated. The non-acetylable arylamine is determined by the method of BRATTON and MARSHALL (1939). To the above solution are added 0.15 ml of 1 N H_2SO_4, 0.05 ml of 1% $NaNO_2$, and after 4 minutes, 0.05 ml of 0.5% ammonium sulfamate to destroy the excess nitrite and 0.05 ml of 0.1% N-1-naphthyl-ethylenediamine dihydrochloride. An optical density of 0.204 at 540 mμ represents empirically 0.01 μmoles of GAR. If it is assumed that the reaction goes to completion, this value is only 80% of the absorbancy expected on the basis of a molecular extinction coefficient of 26,400 for AICAR (FLAKS et al., 1957).

The assay method of HARTMAN, LEVENBERG, and BUCHANAN (1956) consists of a reaction mixture containing 5 μmoles of glycine-1-C^{14}, 5 μmoles L-glutamine, 7 μmoles $NaHCO_3$, 2 μmoles L-azaserine, 0.3 ml phosphate buffer (0.03 M, pH 7.4) containing 0.13 M KCl and 0.01 M $MgCl_2$, 2 μmoles PRPP, 2 μmoles ATP, and 10 mg of dialyzed Norite treated 15—45% ethanol fraction of pigeon liver extract. The reaction mixture is incubated for 45 minutes at 38° C. The GAR formed can be determined in a similar manner as described above.

L-Azaserine exerts a pronounced inhibitory effect on the conversion of FGAR to IMP (BUCHANAN et al., 1955; HARTMAN et al., 1956) and a concomitant enhancement of the accumulation of GAR or FGAR with or without a formyl donor, respectively.

Enzyme preparation. Enzyme preparations generally are multiple in nature, therefore, further treatment or purification of the preparations is necessary to knable one to estimate the activity of a given enzyme. The preparation of GAR-einosynthase has been found to be associated with GAR-transformylase. The latter enzyme will remove the reaction product GAR, if the appropriate coenzymes or cofactors are present. In the case of GAR-transformylase, it requires folic acid derivatives as the formyl donor for the formation of FGAR from GAR. The endogenous nucleotide coenzymes and transformylating cofactors can be removed from the preparation by treating it with Dowex-1 (GOLDTHWAIT et al., 1954, 1956a) or with Norite (HARTMAN et al., 1956). HARTMAN and BUCHANAN (1958b) obtained partially purified enzyme by differential acetone fractionation following the differential ammonium sulfate fractionation of a 15 to 30% ethanol fraction of pigeon or chicken liver. According to these authors most preparations of the enzyme contained little apyrase and no enzymes which exchange the terminal phosphate of ATP with inorganic phosphate, but contained appreciable adenylic kinase activity. However, the enzyme may be further purified by adsorption and subsequent elution from Alumina C_γ gel.

(a) Procedures of GOLDTHWAIT et al. (1956a). *Acetone powder.* The procedures for the preparation of acetone powder is given on page 427 under PRPP-amido-transferase. *Dowex-treated and dialyzed extract.* Acetone powder (10 g) is extracted with 100 ml of 0.05 M K_2HPO_4 for 30 minutes at 0° C and then centrifuged at $5,000 \times g$ for 20 minutes at 0° C. The supernatant solution is used immediately or it may be lyophilized. The supernatant solution is passed through a Dowex-1 X-4 column (HCO_3^- form, 1.82 $cm^2 \times 12$ cm) for a period of 1 to 2 hours. The column is washed with 20 ml of water. These two effluents are combined, dialysed against running 0.05 M K_2HPO_4 solution (20 liters) for 18 to 24 hours and lyophilized at 0—4° C. The lyophilized powder is taken up in water (100 mg/ml) immediately before use (70 to 80 mg protein/ml).

(b) Procedures of HARTMAN *and* BUCHANAN (1958b). The formation of GAR in the presence of excess PRPP-amidotransferase (Enzyme I) is used to assay GAR-kinosynthase (Enzyme II). According to these authors a plot of milligrams of protein against GAR synthesis may not give a straight line, therefore, the

activities should always be compared with a 15 to 30% ethanol fraction. A standard curve relating Enzyme II activity to GAR synthesis may be prepared with the 15 to 30% ethanol fraction as a source of GAR-kinosynthase. *15 to 30% ethanol fraction.* It is prepared as described on page 427 under PRPP-amidotransferase. *45 to 65% saturation (NH₄)₂SO₄ fraction* [first $(NH_4)_2SO_4$ fractionation]. All operations are carried out at 2° C. 20 g of the lyophilized 15 to 30% ethanol fraction are dissolved in 1 liter of 0.1 M ammonium citrate buffer, pH 5.3. For each liter of enzyme solution 278 g of $(NH_4)_2SO_4$ are added to give 45% saturation. The precipitate is removed by centrifugation and discarded. 166 g of $(NH_4)_2SO_4$ per liter are then added to the supernatant solution to give a concentration of 65%. The precipitate is collected and redissolved in 250 ml of water. It is then dialyzed against continuously changing water for 6 hours. The resulting precipitate is removed by centrifugation and discarded. The supernatant solution is referred to as "65% $(NH_4)_2SO_4$ fraction". *80% saturation (NH₄)₂SO₄ fraction* [2nd $(NH_4)_2SO_4$ fractionation]. To each 100 ml of the solution of the 65% $(NH_4)_2SO_4$ fraction are added 0.05 volume of 1 M Tris buffer, pH 8.0, and 37.4 g of $(NH_4)_2SO_4$ to give a 58% saturation. The precipitate is discarded after centrifugation. To each 100 ml of the supernatant solution are added 15.5 g of $(NH_4)_2SO_4$ to give an 80% saturation. The precipitate is collected by centrifugation. The sediment is dissolved in 60 ml of water and dialyzed against 0.02 M Tris buffer, pH 8.0, for 8 hours. *30—50% acetone fraction.* The pH of the solution of the dialyzed 80% saturation ammonium sulfate fraction is carefully adjusted to 5.3 with 1 M acetic acid. Acetone is added to the solution dropwise to reach a 30% concentration in a room maintained at —18° C. When the temperature of the solution reaches —10° C the precipitate is removed by centrifugation at 25,000 ×g for 15 minutes. To the supernatant solution, acetone is added in a similar manner to a final concentration of 50% and the precipitate again collected at —10° C. This precipitate is dissolved in 50 ml of 0.02 M Tris buffer, pH 8.0, and dialyzed against 4 liters of the same buffer for 8 hours. According to these authors, the residual acetone should be removed by dialysis as prolonged contact with the acetone solution is deleterious to the enzymatic activity. The dialyzed 50% acetone fraction is used for the source of GAR-kinosynthase. This fraction may be further purified by the alumina C_γ gel treatment. *Alumina C_γ gel eluate fraction.* The dialyzed acetone fraction is diluted with water to a final concentration of 3 mg of protein per ml. The pH of the solution is adjusted to 6.1 with 1 M sodium acetate buffer, pH 5.0. Alumina C_γ gel suspension (BAUER, 1945) is added at a gel protein ratio of 0.25. The effluent is discarded. The enzyme is then eluted from the gel with a volume of 0.05 M potassium phosphate buffer, pH 7.6, equal to that of the original acetone fraction taken.

Enzyme I. It is prepared as described on page 427 under PRPP-amidotransferase.

(c) Procedures of HARTMAN et al. (1956). Preparation of dialyzed, Norite-treated 15 to 30% ethanol fraction of pigeon liver extract. The 15 to 45% ethanol is prepared in a similar manner as described on page 427 under PRPP-amidotransferase. The enzyme is then treated with Norite and then dialyzed against distilled water.

IV. Glycinamide Ribotide Transformylase.

In the pathway for inosinic acid synthesis, two steps or reactions involve transformylating enzymes, GAR-transformylase and AICAR-transformylase. The latter enzyme, AICAR-transformylase, will be discussed later. The enzyme GAR-transformylase catalyzes the formation of FGAR from GAR and a formyl donor, N^5,N^{10}-anhydroformyl-THFA according to the following reaction:

$$N^5,N^{10}\text{-anhydroformyl-THFA} + GAR + H_2O \rightarrow FGAR + THFA \qquad (8)$$

This enzyme has been demonstrated in pigeon liver extract by GOLDTHWAIT et al. (1956a, 1956b) and by WARREN and BUCHANAN (1957) in chicken and pigeon liver.

Assay. The activity of this transformylase may be assayed by one of three methods: (a) using C^{14}-formate as a precursor, (b) enzymatic assay by coupling the GAR-transformylase system with a system of AIGAR-transformylase and inosinicase, or (c) the disappearance of the reaction product THFA of equation (8).

(a) The method using C^{14}-formate as a precursor for assaying the GAR-transformylase is quite similar to that for GAR-kinosynthase (HARTMAN et al., 1955) in principle. The formyl group of the FGAR is readily hydrolyzed by heating with dilute acid. The synthesis of FGAR may, therefore, be estimated by the incorporation of C^{14}-formate into an easily acid-hydrolyzable form. The non-acid-hydrolyzable formate which is represented by the radioactivity remaining after the hydrolysis is also determined. The amount of FGAR in the reaction mixture may be calculated as follows:

$$\mu M \text{ of } C^{14} \text{ activity in FGAR} = \frac{\text{Total counts/vessel minus counts after hydrolysis/vessel}}{\text{Counts}/\mu\text{mole of formate}}.$$

According to GOLDTHWAIT and GREENBERG (1955), compounds other than FGAR derived from C^{14}-formate comprise approximately 5 to 10% of the total C^{14} activity incorporated.

The incubation mixture (0.7 ml) contains 0.05 ml Na C^{14}-formate (0.1 M; 10,000 c.p.m./μM), 0.05 ml glutamine (0.2 M), 0.05 ml glycine (0.1 M), 0.10 ml R-5-P (0.05 M), 0.02 ml ATP (0.04 M), 0.10 ml PGA (0.14 M), 0.05 ml MgCl$_2$ (0.1 M), 0.05 ml leucovorin (4 mg/ml) or THFA (3 mg/ml), and enzyme (Dowex-treated and dialyzed). The reaction mixture is incubated for 30 minutes at 38° C. The reaction is terminated by adding 0.5 ml of 20% TCA.

To determine the total formate incorporation, 0.05 ml aliquots of the TCA filtrate are pipetted onto a planchet. Five drops each of water and ethanol are added. The samples are dried and counted. The non-acid-hydrolyzable formate is determined by counting the samples after hydrolysis.

The method of enzymatic determination of the GAR-transformylase may be carried out in either one of the following two ways:

(b) GAR-transformylase can be assayed by coupling with a system of AICAR-transformylase and inosinicase which carries out the following reaction:

$$IMP + GAR + H_2O \rightarrow AICAR + FGAR \qquad (9)$$

By carrying out the reaction in the presence of an excess of AICAR-transformylase and inosinicase, GAR-transformylase may be assayed by determining the product AICAR formed as non-acetylable arylamine with BRATTON-MARSHALL reagent (BRATTON and MARSHALL, 1939).

In this assay the incubation mixture (0.55 ml) contains 0.025 μmoles GAR, 1.25 μmoles IMP, 1 μmole KCl, 0.02 μmoles N^5,N^{10}-anhydroformyl-THFA, 10 μmoles Tris buffer (pH 7.4), 300 γ AICAR-transformylase and inosinicase, and 20 γ GAR-transformylase. The reaction mixture is incubated for 35 minutes at 37° C. The reaction is terminated by the addition of 0.1 ml of 30% TCA. The BRATTON-MARSHALL reaction is carried out on an aliquot of 0.4 ml with prior acetylation. The product AICAR formed is determined as the non-acetylable arylamine with this reagent (see page 430). The value obtained should be corrected by multiplying it by a factor of 1.25.

(c) The enzyme GAR-transformylase may also be assayed by following the disappearance of THFA which is a reaction product according to equation (8). THFA breaks down spontaneously and quantitatively under the Bratton-Marshall assay as follows:

$$\text{THFA} \rightarrow \text{p-aminobenzoylglutamate} + \text{pteridine product} \tag{10}$$

The assay mixture (0.55 ml) contains 0.06 μmole N^5,N^{10}-anhydroformyl-THFA, 0.04 μmole GAR, 10 μmoles Tris buffer (pH 7.8) and 200 γ GAR-transformylase. The reaction mixture is incubated for 30 minutes at 37° C and the analysis is carried out as described in the assay (b) except for the omission of the acetylation step with acetic anhydride.

Enzyme preparation. The principle involved for isolating GAR-transformylase is quite similar to that for GAR-kinosynthase. Warren and Buchanan (1957) obtained a preparation which was free from the activities of AICAR-transformylase and inosinicase by controlled dialysis of the ammonium sulfate fraction of a calcium phosphate gel eluate of acetone powder extract.

(a) Procedures of Goldthwait et al. *(1956b). Acetone powder.* The method for the preparation of acetone powder is given on page 427 under PRPP-amido-transferase. *Dowex-treated and dialyzed extract.* The procedures for the preparation of this extract are given on page 431 under GAR-kinosynthase.

(b) Procedures of Warren *and* Buchanan *(1957).* All operations are carried out at 4° C. *Acetone powder* (chicken liver). Acetone powder is made essentially according to procedures of Korn et al. on page 426 under PRPP-kinase. *Acetone powder extract (Fraction I).* 200 g of the acetone powder are extracted with 2000 ml of 0.1 M Tris buffer, pH 7.4, at 4° C for 30 minutes. The mixture is centrifuged at 9,000 $\times$g for 15 minutes and the supernatant solution is referred to as "Fraction I". *Calcium phosphate gel (Fraction II).* The calcium phosphate gel (Keilin and Hartree, 1938) is prepared for use by centrifugation of the calcium phosphate gel suspension and by discarding the supernatant aqueous phase. Fraction I is adjusted to pH 6.1 and 4.17 g of calcium phosphate gel are admixed. The pH of the suspension is readjusted to 6.1. The suspension is stirred mechanically for 1 hour and then centrifuged. It is washed with 2 liters of distilled water and, after 10 minutes of stirring, recentrifuged. The suspension is further washed with two 1600 ml portions of 0.01 M potassium phosphate buffer, pH 7.4. The enzyme is then eluted from the gel by gentle stirring for 1 hour, first with 700 ml of 0.1 M potassium phosphate buffer, pH 7.4, and then with 600 ml. The eluates are combined and referred to as "Fraction II". *$(NH_4)_2SO_4$ fractionation (Fraction III).* To 1300 ml of Fraction II containing 3.0 mg of protein per ml are added 278 g of solid $(NH_4)_2SO_4$ to bring the salt concentration to 40% saturation. The protein precipitate is removed by centrifugation and 127 g of $(NH_4)_2SO_4$ are added to the supernatant solution to give 50% saturation. The resulting precipitate is collected by centrifugation and dissolved in 130 ml of 0.05 M Tris buffer, pH 7.4. This solution is referred to as "Fraction III". *Purification by dialysis (Fraction IV).* Fraction III is dialyzed in 10 ml quantities against 1 liter of distilled water which is changed twice during the dialysis period. After dialysis for approximately 4 hours, a small amount of precipitate may be detected in the dialysis sac. This precipitate is collected by centrifugation before the bulk of the protein precipitates. The precipitate is dissolved in the original volume of 0.05 M Tris buffer, pH 7.8, and referred to as "Fraction IV". It may be stored at −10° C. According to these authors, this fraction is devoid of AICAR-transformylase and inosinicase activity of the acetone powder extract.

V. Formylglycinamidine Ribotide Kinosynthase.

This enzyme has been isolated from pigeon liver by LEVENBERG and BUCHANAN (1957a). It catalyzes the synthesis of FGAMR from FGAR, glutamine and ATP as follows:

$$FGAR + glutamine + ATP + H_2O \rightarrow FGAMR + glutamic\ acid + ADP + P_i \quad (11)$$

Assay. The activity of this enzyme may be assayed by a colorimetric determination of the reaction product FGAMR. It is based on the conversion of FGAMR to AIR in the presence of ATP and an excess of "Fraction II".

$$FGAMR + ATP \rightarrow AIR + ADP + P_i \quad (12)$$

The arylamine is then measured by modified procedure of BRATTON-MARSHALL. The modification consists of a reduction of the quantities of reagents used to one-tenth of those originally recommended and of using TCA instead of H_2SO_4 for acidification of the solution.

The incubation mixture (0.43 ml) contains 0.08 μmoles FGAR (Na salt), 6 μmoles L-glutamine, 1.2 μmoles ATP (disodium), 14 μmoles sodium phosphate buffer (pH 7.4), 10 μmoles K_2SO_4, 3 μmoles $MgSO_4$, and 0.05 ml Fraction I (15 mg of protein per ml). The reaction mixture is incubated for 30 minutes at 38° C. The reaction is terminated by heating the vessel in a water bath at 100° C for 40 seconds. The vessel is then chilled in an ice bath and centrifuged, but the supernatant solution is not removed from the sediment. To this solution 0.05 ml of Fraction II (15 mg protein/ml) is added and then water to make a final volume of 0.62. This reaction mixture is incubated again for 30 minutes at 38° C. The reaction is terminated by the addition of 0.1 ml of 30% TCA. After centrifugation, an aliquot of 0.04 ml is removed for the determination of AIR and then transferred to a small test tube. The color which developed is read after $^1/_2$ hour in a spectrophotometer. The measurement of the optical density is carried out at 500 mμ, the wave length at which the orange chromophore produced in this reaction absorbs maximally. This derivative of AIR has a molecular extinction coefficient of 24,600.

Enzyme preparation. *Tissue homogenate.* It is prepared as described on page 425 under PRPP-kinase with the exception that bicarbonate is omitted from the mixture of internal salt (buffer). *Extract of homogenate.* It is prepared as described on page 428 under PRPP-amidotransferase. *13 to 33% ethanol fraction.* To the extract of homogenate cold 90% ethanol is added to a concentration of 13% at −13° C. The precipitate is removed by centrifugation. Again to the supernatant solution 90% ethanol is added at −18° C until a concentration of 33% is reached. The precipitate is collected by centrifugation and dissolved in a small volume of water. The solution is lyophilized and referred to as "13 to 33% ethanol fraction". This fraction is capable of converting FGAR to AIR. *0 to 35% saturation $(NH_4)_2SO_4$ fraction (Fraction I).* The lyophilized "13 to 33% ethanol fraction" (3.5 g) is dissolved in 230 ml of 0.01 M sodium phosphate buffer, pH 7.4, at 3° C. All fracations are carried out at this temperature. To this solution enough solid $(NH_4)_2SO_4$ is added with stirring to insure rapid solution of the salt to give a concentration of 35% saturation. Following the gradual precipitation which lasts about 15 minutes, the suspension is stirred for 10 minutes prior to the removal of the precipitate by centrifugation. The supernatant solution is used for further fractionation. For preparative purposes the precipitate is taken up in 0.05 M sodium phosphate buffer, pH 7.4, and the concentration of protein is adjusted to approximately 20 mg per ml. This fraction is referred to as "Fraction I" and stored at 2° C. *45 to 60% saturation $(NH_4)_2SO_4$ fraction (Fraction II).* The supernatant solution from above is fractionated in the same manner to obtain 35 to 45%

28*

saturation $(NH_4)_2SO_4$ fraction. After centrifugation, this precipitate is discarded. To the supernatant again solid $(NH_4)_2SO_4$ is added to reach a concentration of 60% saturation. The precipitate is collected by centrifugation. It is then dissolved in water and the protein concentration of this solution is adjusted to approximately 22 mg per ml and stored at 2° C. This solution is referred to as "Fraction II", and is used for the enzymatic assay of FGAMR by measurement of the conversion of FGAMR to AIR. *Further purification of Fraction I with alumina $C_γ$.* Fraction I is suspended in 25 ml of water and allowed to stand overnight at 0° C in order to effect fine dispersion of the insoluble material and to obtain proper solution of the desired enzymes. The insoluble material is removed by centrifugation and the residue is washed once with water (5 ml). The washing is combined with the first extract. This solution is then diluted to 84 ml with water and adjusted to a final concentration of 0.0005 M with potassium acetate buffer, pH 6.0. This solution (containing 530 mg of protein) is then treated with 25 ml of alumina $C_γ$ (12.3 mg per ml) (COLOWICK, 1955). The suspension is stirred at 0° C for 10 minutes and the insoluble material is removed by centrifugation and discarded. The supernatant solution contains the active enzyme at a protein concentration of 1.16 mg per ml.

VI. Enzyme for the Synthesis of 5-Aminoimidazole Ribotide.

This enzyme has been isolated by LEVENBERG and BUCHANAN (1957a) from pigeon liver. It catalyzes the formation of AIR from FGAMR and ATP as shown in equation (12).

Assay. This enzyme may be assayed by conversion of FGAMR to AIR and the arylamine may be determined colorimetrically by a modified method of BRATTON and MARSHALL (LEVENBERG and BUCHANAN, 1957b). The incubation mixture (0.5 ml) contains 0.055 μmoles FGAMR (Ba salt), 10 μmoles sodium phosphate buffer (pH 7.4), 10 μmoles K_2SO_4, 3.5 μmoles $MgSO_4$, 1 μmole ATP and 0.1 ml enzyme (Fraction II). The reaction mixture is incubated for 30 minutes at 38° C. The reaction is terminated by the addition of 0.1 ml of 30% TCA. After centrifugation, an aliquot is removed for the determination of AIR as described on page 435.

Enzyme preparation. The procedure for the preparation of this enzyme (Fraction II) are given on page 435, under formylglycinamidine-ribotide-kinosynthase.

VII. 5-Aminoimidazole Ribotide Carboxylase and Enzyme for the Synthesis of 5-Amino-4-Imidazole-N-Succinocarboxamide Ribotide.

The enzyme AIR-carboxylase has been detected by LUKENS and BUCHANAN (1957) in a multiple-enzyme preparation (Fraction I) from pigeon or chicken liver. When AIR and bicarbonate are the substrates, AICR is formed according to the following equation:

$$AIR + CO_2 \rightleftharpoons AICR \tag{13}$$

Since enzyme Fraction I catalyzes the over-all reaction shown in the following equation:

$$AIR + CO_2 + ATP + \text{aspartic acid} \rightleftharpoons AISCAR + ADP + P_i \tag{14}$$

Fraction I, according to LUKENS and BUCHANAN, consists of two components, one of which catalyzes the reaction of carboxylation of AIR to form AICR (equation 13) and the other catalyzes the following reaction:

$$AICR + ATP + \text{aspartic acid} \rightleftharpoons AISCAR + ADP + P_i \tag{15}$$

Assay. The reaction product AICR of equation (13) gives a red color in the BRATTON-MARSHALL test rather than the orange color produced by pure AIR. On the other hand, the reaction product AISCAR of equation (15) does not give a colored compound in the same test under the usual conditions but both AIR and AICAR do. According to LUKENS and BUCHANAN (1957) AICR, is rapidly de-carboxylated by heat or acid to yield AIR which may also be destroyed by heating with 1 N H_2SO_4 at 100° C. On the other hand, enzyme Fraction II cata-lyzes the formation of 5-amino-4-imidazole carboxamide ribotide from AISCAR as follows:

$$\text{AISCAR} \rightleftharpoons \text{AICAR} + \text{fumaric acid} \qquad (16)$$

The reaction product AICAR is acid stable under the same conditions. On the basis of these facts, these authors developed an assay for this carboxylating enzyme or AISCAR-synthesizing enzyme by the determination of AICAR formed in the presence of enzyme Fraction II in excess. The amount of AICAR is measured by the BRATTON-MARSHALL procedure after quantitatively destroying AIR and AICR by heating in 1 N H_2SO_4 for 15 minutes at 100° C.

The incubation mixture (0.55 ml) for AIR-carboxylase contains 15 μmoles magnesium acetate, 20 μmoles potassium acetate, 1 μmole potassium L-aspartate, 0.15 μmole disodium salt of ATP, 40 μmoles Tris chloride buffer (pH 8.4), 0.4 mg enzyme (Fraction I), 12 mg enzyme (Fraction II), 0.07 μmole AIR, and 20 μmoles $KHCO_3$. The reaction mixtures are incubated for 5 minutes at 38° C. The reaction is terminated by the addition of 0.1 ml of 30% TCA.

After the removal of the denatured protein by centrifugation, a 0.4 ml aliquot is removed from each vessel to a separate tube. Then to each tube is added 0.10 ml of 5 N H_2SO_4. The tubes, capped with glass marbles, are placed in a boiling water bath for 15 minutes. After they are cooled to room temperature the BRATTON-MARSHALL reagents are added as described previously on page 430. When the AISCAR synthesizing enzyme is assayed 0.15μ moles of AICR is used in the place of AIR and $KHCO_3$.

Enzyme preparation. *Fraction I.* Preparation of this fraction (15 to 30% ethanol fraction — Fraction I) is given on page 427 under PRPP-amidotransferase. *Fraction II.* Preparation of this fraction will be found on page 438 under AICAR transformylase enzyme.

VIII. Enzymatic Cleavage of 5-Amino-4-Imidazole-N-Succinocarbox-amide Ribotide (AISCAR Splitting Enzyme).

The enzyme which catalyzes the formation AICAR and fumarate from the splitting of AISCAR (equation 16) has been isolated from pigeon or chicken liver and partially purified by MILLER et al. (1957) free from fumarase. This enzyme has also been found in baker's yeast, *Escherichia coli*, *Salmonella typhimurium*, and *Neurospora crassa*. The catalytic characteristics of this enzyme are similar to those of the enzyme adenylosuccinase.

Assay. The assay was not given in their paper but is the same as that given for AIR-carboxylase.

Enzyme preparation. The extracts of pigeon or chicken liver are purified by a procedure including ethanol precipitation, negative gel adsorption on Alumina C_γ (e.g. p. 439 alumina supernatant fraction) and differential heat inactivation.

IX. 5-Amino-4-Imidazolecarboxamide Ribotide Transformylase and Inosinicase.

The enzyme AICAR-transformylase has been demonstrated by GREENBERG (1954a, 1954b) in a multiple enzyme preparation which catalyzes the following over-all reaction:

$$\text{AICAR} + \text{N}^5,\text{N}^{10}\text{-anhydroformyl-THFA} \rightleftharpoons \text{THFA} + \text{IMP} \tag{17}$$

FLAKS et al. (1957) isolated and purified an enzyme preparation from chicken liver, which also catalyzes the over-all reaction of equation 17. They, however, demonstrated that there are two components in this preparation, one of which, AICAR-transformylase, catalyzes the formation of FAICAR from AICAR and N^5,N^{10}-anhydroformyl-THFA:

$$\text{AICAR} + \text{N}^5,\text{N}^{10}\text{-anhydroformyl-THFA} + \text{H}_2\text{O} \rightleftharpoons \text{FAICAR} + \text{THFA} \tag{18}$$

and the other, inosinicase catalyzes the formation of IMP from FAICAR according to the following reaction:

$$\text{FAICAR} \rightleftharpoons \text{IMP} + \text{H}_2\text{O} \tag{19}$$

Assay. The assay for AICAR-transformylase consists of an incubation mixture (0.5 ml) containing 0.1 μmoles AICAR, 0.2 μmoles N^5,N^{10}-anhydroformyl-THFA (isoleucovorin chloride), 10 μmoles KCl, 30 μmoles Tris buffer (pH 7.4), and enzyme. The reaction mixture is incubated for 20 minutes at 38° C. The reaction is terminated by the addition of 0.4 ml of 10% TCA. Acetic anhydride (0.1 ml) is then added to the vessels. After 20 minutes at room temperature the non-acetylatable arylamine is determined by the method of BRATTON and MARSHALL (1939) (see page 430).

The assay system for inosinicase consists of a reaction mixture (0.3 ml) of 0.03 μmoles FAICAR, 30 μmoles Tris buffer, (pH 7.4), and enzyme. The reaction mixture is incubated for 10 minutes at 38° C. The reaction is terminated by the addition of 0.2 ml of 4 N H_2SO_4. The tubes are capped and placed in a boiling water bath for 5 minutes to hydrolyze the residual FAICAR to AICAR. After the tubes are cooled, the arylamine is then determined by the method of BRATTON and MARSHALL (1939) (see page 430).

Enzyme preparation. *Fraction I (Acetone powder extract)*. Acetone powders of fresh chicken livers are prepared according to the procedure described on page 428. Acetone powder extract is prepared at 2° C in a similar manner as described on page 431 with the exception that 50 mg of the powders, 500 ml of buffer are used and centrifuged at 12,800 ×g for 10 minutes. *Fraction II* [$(NH_4)_2SO_4$ *fraction*]. To 356 ml of acetone powder extract are added 36 ml of 1.0 M Tris buffer, pH 8.0, and then with stirring, 95 g of $(NH_4)_2SO_4$. After 30 minutes the precipitate is removed by centrifugation. 67 g of $(NH_4)_2SO_4$ are added to the supernatant solution. After 30 minutes the precipitate is collected by centrifugation, dissolved in 70 ml of 0.01 M Tris buffer, pH 7.4, and dialyzed overnight against 4 liters of the same buffer. *Fraction III (zinc-ethanol fractionation)*. First, 50 ml of 0.01 M Tris buffer, pH 7.4, are added to the dialyzed $(NH_4)_2SO_4$ fraction (105 ml) and then 1.08 ml of saturated $(NH_4)_2SO_4$. Slowly, 1.54 ml of a 1 M solution of zinc acetate are added to the solution with stirring. After 10 minutes the precipitate is removed by centrifugation. To the supernatant solution 3 N NH_4OH is added dropwise until pH 9.0 is reached. The resulting precipitate is removed by centrifugation. After the supernatant solution (140 ml) is cooled to —5° C, 70 ml of 90% ethanol are slowly added while the temperature is gradually lowered to —15° C. After 15 minutes the precipitate is collected by centrifugation at —15° C and dissolved in 50 ml of 0.01 M Tris buffer, pH 7.4, containing 0.01 M Versene.

Fraction IV [$(NH_4)_2SO_4$ *fractionation*]. Five milliliters of 1.0 M Tris buffer, pH 8.0, are added to the zinc-ethanol fraction (50 ml) and 16.1 g of $(NH_4)_2SO_4$ are slowly added with stirring. After 30 minutes the precipitate is removed by centrifugation and to the supernatant solution are added 4.4 g of $(NH_4)_2SO_4$. The precipitate is collected by centrifugation, dissolved in 10 ml of 0.01 M Tris buffer, pH 7.4, and dialyzed overnight against 4 liters of the same buffer. *Fraction V (alumina C_γ supernatant fraction)*. Forty milliliters of 0.01 M Tris buffer, pH 7.4, are added to the dialyzed $(NH_4)_2SO_4$ fraction (17 ml) followed by 5.7 ml of 1.0 M Tris buffer, pH 8.0. This solution is stirred for 10 minutes with 290 mg of alumina C_γ gel (BAUER, 1945). The mixture is centrifuged and the precipitate discarded. *Fraction VI (ethanol fractionation)*. The alumina C_γ supernatant fraction (63 ml) is cooled to —5° C and 46 ml of 90% ethanol are added dropwise with constant stirring. The temperature is allowed to fall to —15° C during the course of the ethanol addition. After 20 minutes the mixture is centrifuged at —15° C and the precipitate discarded. Twenty-six milliliters of 90% ethanol are added to the supernatant solution at —15° C as described above. Again after stirring for 20 minutes the mixture is centrifuged and the precipitate dissolved in 7 ml of 0.01 M Tris buffer, pH 7.4. *Fraction VII (alumina C_γ eluate fraction)*. Fraction VI is stirred with 18 mg of alumina C_γ gel for 10 minutes, the mixture centrifuged, and the supernatant solution discarded. The enzyme is eluted from the gel by stirring with 4.5 ml of 0.02 M potassium phosphate buffer, pH 8.0, and the residual gel is removed by centrifugation. A second elution is similarly carried out with 4 or 5 ml of 0.03 M potassium phosphate buffer, pH 8.0. The first and second eluates are designated, respectively, as E_1 and E_2.

B. Enzymes of Synthesis of Pyrimidine Nucleotides.

I. Carbamyl Phosphate Synthetase.

The enzyme carbamylphosphate-synthetase catalyzes the formation of CP from ATP, bicarbonate, and ammonia according to the following reaction:

$$ATP + CO_2 + NH_3 \rightleftharpoons CP + ADP \tag{20}$$

CP-synthetase has been isolated and partially purified by METZENBERG et al. (1957) from mammalian liver and *Streptococcus faecalis*. Apparently this enzyme is quite widely distributed. However, HALL et al. (1960) in their tests with extra hepatic rat tissues could obtain this enzyme only from the small intestines. CP-synthetase also plays a key role in the ornithine cycle in the synthesis of CP which is used for the formation of citrulline as follows:

$$CP + ornithine \rightleftharpoons citrulline + P_i \tag{21}$$

Assay. The activity of this enzyme may be assayed according to the method of METZENBERG et al. (1957) by coupling the system of CP-synthetase with the system of ornithine transcarbamylase which catalyzes the formation of citrulline according to equation (21). The reaction product citrulline may be estimated by the modification of the method of ARCHIBALD (1944).

The reaction mixture (1.0 ml) contains 5 μmoles ATP, 10 μmoles $MgSO_4$, 50 μmoles NH_4HCO_3, 5 μmoles acetyl-L-glutamic acid, 5 μmoles L-ornithine, ornithine transcarbamylase from dog liver in excess of the enzyme under study (10—100 fold), and water to make up the volume. All substances are adjusted to pH 7.4 before use. The reaction mixture is incubated for 15 minutes at 38° C. The reaction is terminated by the addition of 1.0 ml of 1 N $HClO_4$. The deproteinized solution is assayed for citrulline and P_i.

Enzyme preparation. Procedures of METZENBERG et al. (1957).

Acetone powder. It is prepared in a similar manner as the procedure described on page 424. All operations are carried out at 0—2° C. *Acetone powder extract.* Dog liver acetone powder (20 g) is extracted with 160 ml of $MgSO_4$ (0.02 M) with gentle stirring for 20 minutes. The extraction is repeated for 10 minutes. After each extraction, the suspension is centrifuged at $4,000 \times g$ for 10 minutes. The

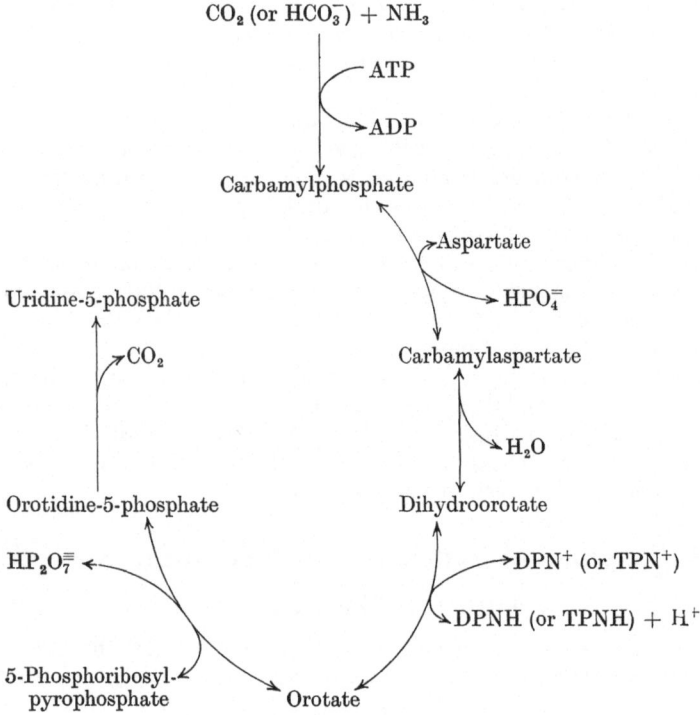

Fig. 2. Tentative biosynthetic pathway for purine nucleotides.

extracts are combined and diluted with $MgSO_4$ (0.02 M) to 320 ml to give a protein concentration of approximately 1%. *($NH_4)_2SO_4$ fraction.* To the diluted acetone powder extracts (320 ml) $(NH_4)_2SO_4$ solution (3.75 M) is added to give a concentration of 1.5 M. The precipitate is removed by centrifugation at $4,000 \times g$ for 30 minutes. To the supernatant solution again $(NH_4)_2SO_4$ solution (3.75 M) is added to give a concentration of 2.25 M. The precipitate is collected by centrifugation as above. The supernatant solution is decanted and saved for the preparation of ornithine transcarbamylase. The precipitate, containing the carbamyl phosphate-synthesizing system, is dissolved in 80 ml of $MgSO_4$ (0.02 M). Eighty milliliters of a 6.6% suspension of hydrated calcium phosphate preparation (hydroxylapatite) (TISELIUS et al., 1956) are added, and the material is stirred occasionally for 5 minutes. The suspension is centrifuged for 5 minutes at $2,000 \times g$ and the supernatant solution discarded. The residue is stirred for 10 minutes with 160 ml of NH_4HCO_3 solution (1.0 M) and centrifuged. Again the supernatant solution is discarded. The residue is then stirred for 10 minutes with 160 ml of potassium phosphate buffer, pH 7.5 (0.5 M), and the suspension is centrifuged. The supernatant solution is fractionated with $(NH_4)_2SO_4$ as before, and the material precipitating between 1.5 and 2.25 M is obtained. This material is washed three times by resuspension in 3.0 M $(NH_4)_2SO_4$, followed by centrifugation or until no P_i may be detected in the preparation. The supernatant solution is decanted from the precipitate, and the latter is stored at $-18°$ C. The precipitate is redissolved in 0.025 M NH_4HCO_3 buffer (pH 7.5) immediately before use. According to these authors, the precipitate should be dissolved in buffer rather than in water, as the activity is lost almost immediately below pH 6.0.

Beef liver homogenate. The liver (500 g) is homogenized in 9 volumes of cold 0.88 M sucrose in a POTTER-ELVEHJEM type glass homogenizer. *Mitochondrial fraction* (KENNEDY and LEHNINGER, 1949). The homogenate is centrifuged three successive times at $1,500 \times g$ for 3 minutes each to remove nuclei, whole cells, stroma and erythrocytes. The mitochondria are then obtained by centrifugation of the supernatant solution at $18,000 \times g$ for 20 minutes. The packed mitochondria are washed by resuspension in 10 volumes of 0.88 M sucrose, followed by centrifugation as above. The packed mitochondrial fraction is stirred for 5 minutes with 600 ml of water. The suspension is centrifuged at $4,000 \times g$ for 10 minutes. The supernatant solution is saved, and the residue is re-extracted with 400 ml of water as before and again centrifuged. Solid $(NH_4)_2SO_4$ is added to the combined supernatant fractions to give a concentration of 3.0 M. The precipitate is resuspended in 500 ml of 2.06 M $(NH_4)_2SO_4$ and stirred for 30 minutes. The precipitate is collected by centrifugation and extracted as above with 1.68 M $(NH_4)_2SO_4$. The extract is then again centrifuged to remove residue. Solid $(NH_4)_2SO_4$ is added to the supernatant solution to give a concentration of 3.0 M. The precipitate is recovered and redissolved in 250 ml of 0.025 M Tris buffer, pH 7.4, containing $MgCl_2$ (0.02 M). This solution is applied to a column of hydroxylapatite 2.0 cm in diameter and 13 cm in length. 250 ml of 0.2 M $(NH_4)_2SO_4$, adjusted to pH 7.4, are passed through the column, and the eluate is discarded. 100 ml of 0.5 M $(NH_4)_2SO_4$ (pH 7.4) are then passed through the column, and the resulting eluate is collected and brought to a concentration of 3.0 M by the addition of $(NH_4)_2SO_4$. The precipitate is collected by centrifugation at $20,000 \times g$. The tubes containing the packed precipitate are tightly stoppered and stored at $-80°$ C. According to these authors, both preparations are reasonably stable in concentrated ammonium sulfate solutions or when adsorbed on hydroxylapatite but have a half life of approximately 1 hour at $0°$ C in dilute buffer at pH 7.4. For this reason it is necessary to perform the fractionation as rapidly as possible.

The fractionation procedure employed for purification of the beef liver enzyme is, in practice, less laborious than that used for the dog liver enzyme and is more readily amendable to large scale operations.

The ornithine transcarbamylase activity is considerably reduced but is not limiting.

Preparation of ornithine transcarbamylase from dog liver. The $(NH_4)_2SO_4$ supernatant solution (2.25 M) from dog liver preparations is brought to 3.40 M by addition of solid $(NH_4)_2SO_4$, and the precipitate is recovered by centrifugation. This material is dissolved in 25 ml of 0.5 M Tris buffer, pH 7.4, and heated at $60°$ C for 10 minutes to inactivate the carbamylphosphate-synthesizing enzyme. The denatured protein is removed and the supernatant solution frozen in small portions.

II. Carbamyl Phosphate-Aspartate Transcarbamylase.

The enzyme carbamylphosphate-aspartate transcarbamylase catalyzes the formation of CA from CP and aspartic acid:

$$CP + aspartate \rightleftharpoons CA + P_i \tag{22}$$

Preparations of this enzyme have been obtained by LOWENSTEIN and COHEN (1956) from tissues of rats and pigeons and by REICHARD (1954) and REICHARD and HANSHOFF (1955) from $E.\ coli$.

Assay. Two methods have been described by LOWENSTEIN and COHEN (1956) for assaying the activity of this enzyme by determining the formation of reaction product, CA. One method is based on the differential elution following the adsorption of CA on a Dowex 2 (formate) resin column and the other is based on the different labilities of $C^{14}P$ and $C^{14}A$ toward dilute acid, thus after hydrolysis the acid labile $C^{14}P$ is expelled in the form of $C^{14}O_2$. The formation of $C^{14}A$ can be estimated by the radioactivity which remains in the solution following acid hydrolysis. Only the second method will be given here. However, confirmation of the reaction product CA should be made with the first method. For details of the first method the reader is referred to the original article of LOWENSTEIN and COHEN (1956).

This enzyme may be assayed by the method of LOWENSTEIN and COHEN (1956) which consists of an incubation mixture (1.0 ml) containing 10 μmoles L-aspartate, 20 μmoles $C^{14}P$, 100 μmoles DEA buffer (pH 9.2) and high speed supernatant enzyme solution (25—45 mg of protein). The reaction mixture is incubated for 15 minutes at 38° C after the addition of $C^{14}P$. The reaction is terminated by adding 0.5 volumes of 1.0 N HCl at 0° C per volume of reaction mixture. The resultant solutions are prepared for counting as follows: 0.2 ml samples are uniformly plated on planchets by means of a micropipette and are dried under a radiant lamp for 20 minutes. The samples are counted with an end window GEIGER-MÜLLER counter or any type of counter.

Enzyme preparation. Tissues are excised, chilled, cut into small pieces, and homogenized in isotonic KCl in a glass homogenizer under ice-cold conditions for 1 minute. Low and high speed supernatant solutions of liver are prepared by centrifuging 10% (for rat liver) or 20% (for other tissues) isotonic KCl homogenates at $2{,}000 \times g$ for 30 minutes or at $12{,}000 \times g$ for 45 minutes, respectively.

Preparation of $C^{14}P$. C^{14}-urea is converted into $KOC^{14}N$ by an adaptation of the method described by SCATTERGOOD (1946). The synthesis of $C^{14}P$ from $KOC^{14}N$ and KH_2PO_4 is then carried out according to the method of JONES et al. (1955) and $C^{14}P$ is isolated in its dilithium form.

C^{14}-urea (2.0 μmoles) is fused carefully with 1.05 μmoles of finely powdered K_2CO_3 in a small crucible for about 5 minutes. After the melt has cooled, the residue is dissolved in 2 ml of water and 2.0 μmoles of KH_2PO_4 are added. The solution is stirred for 30 minutes at 30° C and then cooled in ice. To the cooled solution, an ice-cold solution of 6 moles of lithium hydroxide and 4 mole of perchloric acid in 83 ml of water are added slowly, final pH 8.3. A precipitate forms which consists of potassium perchlorate and lithium phosphate. This is removed by filtration. To the supernatant solution containing the lithium $C^{14}P$ an equal volume of ethanol is slowly added. Lithium $C^{14}P$ is precipitated.

III. Dihydroorotase.

This enzyme dihydroorotase catalyzes the formation of dihydroorotate from carbamylaspartate (or ureidosuccinate) as follows:

$$CA \rightleftharpoons dihydroorotate + H_2O \tag{23}$$

This reaction is reversible. The interconversion of CP and dihydroorotate has been reported in extracts of *Zymobacterium oroticum* and *Corynebacteria* by LIEBERMAN and KORNBERG (1954) and REYNOLDS et al. (1955), in extracts of

E. coli by YATES and PARDEE (1956), and in homogenates of rat liver by COOPER et al. (1955).

Assay. YATES and PARDEE (1956) took advantage of the large loss of optical density (maximal at 230 mμ) upon alkaline hydrolysis of dihydroorotate to CP. The enzyme dihydroorotase may be assayed according to the method of these authors which consists of a reaction mixture (5.0 ml) containing 0.1 M potassium phosphate buffer (pH 6.0), 0.002 M Versene, 0.004 M DL-carbamylaspartate, and dihydroorotase preparation (0.46 mg of protein). The reaction mixture is incubated at 37° C. Aliquots assayed at intervals for dihydroorotate as follows: 1.5 ml of 1 N sodium hydroxide are added to 1.5 ml of sample in silica cuvettes at room temperature, mixed rapidly, and read at 240 mμ for 15 minutes at timed intervals. The final equilibrium optical density is subtracted from all values, and the semilog plot of corrected optical density is extrapolated linearly to zero time. According to these authors, no substances in the assay mixtures except dihydroorotate change their optical density significantly in 15 minutes.

The assay method of LIEBERMAN and KORNBERG (1954) consists of a reaction mixture (3.0 ml) containing 200 μmoles glucose, 15 μmoles MgCl$_2$, 50 μmoles potassium phosphate buffer (pH 6.1), 30 μmoles cysteine (pH 7.0), 0.05 μmole DPN, 500 units glucose dehydrogenase, 0.175 μmole carbamylaspartate-C^{14}, and 0.4 or 0.5 ml of a cell free extract. The reaction mixture is incubated for 97 minutes at 34° C. The reaction is terminated by the addition of 0.05 ml 4 N HCl. The denatured protein is removed after centrifugation. The supernatant solution is neutralized with 1 M KOH. The CP as well as the product dihydroorotate are separated by Dowex 1 (formate) ion exchange resin (for details the reader is referred to LIEBERMAN and KORNBERG, 1954).

Enzyme preparation.

(a) Procedures of YATES *and* PARDEE *(1956).* All operations are carried out at 0° C. Cell-free extracts of *E. coli* (or *Z. oroticum*) are prepared by treatment of the harvested cells in a 10 kc RAYTHEON magnetostriction (sonic) oscillator for 20 minutes in the presence of 200 mg of glass powder. The extracts are centrifuged for 15 minutes at 10,000 ×g. The cell-free extracts are stored at −10° C. Alternatively, cell-free extracts of *E. coli* may be prepared by sonic oscillation of cells suspended in ice water at a concentration of 2.5×10^{10} cells per ml for 2 minutes. The extracts are centrifuged for 5 minutes at 9,000 ×g to remove whole cells and debris.

(b) Procedures of LIEBERMAN *and* KORNBERG *(1954). Cell-free extracts.* Harvested cells which had been grown in orotate medium are briefly incubated *in vacuo* in orotate solution, for 20 minutes at 26° C. The orotate solution consists of 7 ml of 0.1 M sodium orotate, 0.4 ml of 1 M potassium phosphate buffer (pH 7.0) and 0.4 ml of 0.1 M cysteine (pH 7.0). The amount given is for each liter of culture. At the end of incubation, the cells are centrifuged and resuspended in ice cold water (approximately 5 ml/liter culture). The suspension (6 ml) is shaken with 6 g of glass beads (0.1—0.15 mm diameter) in a MICKLE vibrator for 15 minutes at 2° C. The juice is centrifuged at about 10,000 ×g. The sediment is washed once with cold water and combined with the original supernatant solution. The volume of the cell-free extract is adjusted to 10 ml per liter of culture. If purification is not carried out immediately, the extract is acidified to pH 6.5 with 2 M sodium acetate buffer (pH 6.0) and stored at −10° C. *Protamine fraction.* All operations are carried out at a temperature between 0° and 2° C. Freshly prepared cell-free extract (100 ml) is diluted with an equal volume of water and 15 ml of a 1% solution of protamine sulfate is added with stirring. After 5 minutes the precipitate

is collected by centrifugation. To the precipitate 100 ml of 0.5 M sodium citrate buffer (pH 6.0) are added and it is allowed to stand for 12—24 hours. The suspension is homogenized with a glass pestle to facilitate its solution. Water (200 ml) is then added with stirring. After centrifugation the supernatant solution is saved and referred to as "Protamine fraction".

According to these authors, this fraction contains only trace amounts of nucleic acid as indicated by the ratio of optical densities at 280 and 260 mμ (0.98). The protamine fraction is used as a source of dihydroorotase.

IV. Dihydroorotic Acid Dehydrogenase.

The enzyme dihydroorotic dehydrogenase catalyzes the reversible reaction of the following equation:

$$\text{Dihydroorotate} + \text{DPN}^+ (\text{or TPN}^+) \rightleftharpoons \text{orotate} + \text{DPNH (or TPNH)} + \text{H}^+ \qquad (24)$$

Dihydroorotic dehydrogenase has been isolated and partially purified by LIEBERMAN and KORNBERG (1953) from the anaerobe, *Zymobacterium oroticum* and by REYNOLDS et al. (1955) from strains of the aerobe *Corynebacterium*. Enzyme from the latter source differs from the former in its requirement of TPN rather than DPN as coenzyme. YATES and PARDEE (1956) showed this enzyme in cell-free extracts of *E. coli* mutants and *Z. oroticum*. Recently, FRIEDMANN and VENNESLAND (1960) obtained crystalline dihydroorotic dehydrogenase from *Z. oroticum*. This enzyme has been shown by KONDO et al. (1960) to be a flavin protein containing equal amounts of flavin mononucleotide (FMN) and flavin adenine dinucleotide (FAD).

Assay. The activity of dihydroorotic dehydrogenase may be assayed by the method of LIEBERMAN and KORNBERG (1953). DPNH is generated by the addition of glucose dehydrogenase, glucose and DPN$^+$. The disappearance of orotate which absorbs strongly at 280 mμ (whereas dihydroorotate shows no absorption) can be followed spectrophotometrically. The activity of this enzyme may be more conveniently followed by determining the rate of decrease of the optical density at 340 mμ associated with the conversion of DPNH to DPN$^+$ in the presence of dihydroorotic dehydrogenase and orotate (FRIEDMANN and VENNESLAND, 1958). This will avoid the difficulty associated with the high optical density at 280 mμ in the method of LIEBERMAN and KORNBERG (1953). In the assay method of FRIEDMANN and VENNESLAND (1960) it requires a preincubation of the enzyme with cysteine. According to these authors (1960) an increase in time of exposure of the enzyme to cysteine is necessary mainly when the more highly purified enzyme is assayed.

The assay of this enzyme (FRIEDMANN and VENNESLAND, 1958, 1960) consists of a reaction mixture (3.0 ml) containing 400 μmoles sodium phosphate buffer (pH 6.5), 20 μmoles freshly dissolved cysteine hydrochloride (the cysteine solution is prepared just before use by dissolving it in water and neutralizing it carefully with NaOH solution), 6 μmoles sodium orotate, 10 to 20 units enzyme and 0.35 μmole DPNH. The additions are made in the order indicated, and the enzyme is incubated for 10 minutes at 20° C with the cysteine-containing reaction mixture in the case of assaying purified enzyme or until the reaction mixture attains the temperature equilibrium at 20° C in the case of unpurified enzyme assayed before the reaction is started by the addition of DPNH. The optical densities are measured at 340 mμ at suitable time intervals after mixing.

Enzyme preparation. *Procedures of* FRIEDMANN *and* VENNESLAND (1960). *Cell extract.* The immediately harvested *Z. oroticum* cells from 22 liters of culture

(100 g wet weight) are washed into about 300 g of chilled glass beads with approximately 50 ml of 0.05 M sodium phosphate buffer (pH 6.5). The temperature of the glass container of the blendor is kept less than 30° C. It generally requires about 15 minutes of blending with the rheostat set at two-thirds of maximum. After disruption of the cells, an additional 100 ml of phosphate buffer are added, the suspension is then stirred for 2 minutes. The supernatant solution is decanted after the beads are settled. Fresh buffer is added and the stirring and decanting are repeated three times with decreasing amounts of buffer to give a final volume of about 350 ml of extract. The combined extracts are cleared of solids by centrifugation, and the fractionation is begun as soon as possible. *First protamine treatment and (NH₄)₂SO₄ precipitation.* In this and in the following treatment with protamine sulfate the optimal amount of the reagent that gives maximal precipitation is determined by the addition of increasing amounts of protamine sulfate to small aliquots of the enzyme solution. The samples are cleared by centrifugation, and the supernatant solutions are treated with more protamine sulfate to determine whether more precipitate is formed. These authors added NaCl to a concentration of 0.4 M and then sufficient 1% protamine sulfate in 0.4 M NaCl is added until no precipitation occurs on further addition. (About 0.4 to 0.5 volume of protamine sulfate solution is required.) Excess is avoided. After 10 minutes, the precipitate is removed by centrifugation, and 29.5 g of (NH₄)₂SO₄ are added for each 100 ml of supernatant solution. After 30 minutes, the slimy precipitate is completely removed by centrifugation. To the supernatant solution, again 9 g of (NH₄)₂SO₄ are added for each original portion of 29.5 g. Precipitate containing the enzyme separates out in fine floccules after about 30 minutes. At this stage the material is allowed to stand overnight in the cold. The precipitate is collected by centrifugation and extracted with 0.2 M sodium phosphate buffer (pH 5.8). Several portions of buffer are used to give a final total volume about one-tenth that of the original crude extract. The extraction removes all yellow pigment but leaves a white, less soluble residue which is discarded after the centrifugation. *Second protamine treatment and precipitation by dialysis.* The clear yellow solution is treated with aqueous protamine sulfate to remove a small additional amount of nucleic acid. (According to these authors this treatment gives no detectable increase in purity, and the amount of precipitate obtained is often small, but this step is necessary for success of subsequent procedures.) When the solution is dialyzed overnight against several changes of cold distilled water, a bulky yellow precipitate is formed. The precipitate is collected by centrifugation and is extracted with successive portions of 0.2 M sodium phosphate buffer (pH 5.8). To achieve adequate extraction, the precipitate is repeatedly ground to a smooth paste with small portions of buffer (5 to 10 ml). The yellow pigment is used as a guide. When the extract becomes colorless, extraction is stopped, and the residue is discarded. *Third protamine sulfate treatment and crystallization.* To the combined extracts an equal volume of 1% protamine sulfate solution is added to precipitate the enzyme. The precipitate is collected by centrifugation after 5 to 10 minutes and extracted with 5 ml portions of 0.2 M NaH₂PO₄ until the extract is only faintly colored. The combined extracts, decanted after centrifugation, are held at 20° C for 1 hour and stored overnight at −15° C. A precipitate which remains after thawing and warming to room temperature is removed by centrifugation. The enzyme crystallizes from solution after one or more days at 4° C in the form of fine needles. If necessary the solution is diluted to a point where the optical density at 450 mμ does not exceed 0.5, since the precipitate is often amorphous if crystallization is attempted from too concentrated a solution. According to these authors, the crystals are stored in the mother liquor at 4° C with little or no loss

in activity over a period of one month. The enzyme remaining in the yellow supernatant from the third protamine sulfate precipitation may be precipitated by dialysis against water. The precipitate is collected by centrifugation and dissolved in 0.2 M sodium phosphate buffer (pH 5.8). The enzyme is precipitated with protamine sulfate. The precipitate is extracted with 0.2 M NaH_2PO_4. From this extract a second crop of crystals is obtained. In this case, the heating to 20° C followed by freezing and thawing is not needed.

V. Orotidine-5'-Phosphate Pyrophosphorylase.

The enzyme orotidine-5'-phosphate pyrophosphorylase catalyzes the formation of O-5-P from orotate and PRPP as follows:

$$orotate + PRPP \rightleftharpoons O\text{-}5\text{-}P + PP \tag{25}$$

This reaction is reversible and the enzyme has been demonstrated by LIEBERMAN et al. (1954) in enzyme preparation from yeast and from pigeon liver by KORNBERG et al. (1955).

The preparation and the assay of this enzyme by the method of KORNBERG et al. (1955) have been described under PRPP-kinase.

VI. Orotidine-5'-Phosphate Decarboxylase.

The enzyme orotidine-5'-phosphate decarboxylase catalyzes the formation of U-5-P by decarboxylation of O-5-P according to the following reaction:

$$O\text{-}5\text{-}P \rightarrow U\text{-}5\text{-}P + CO_2 \tag{26}$$

The assay of this enzyme by KORNBERG et al. (1955) and the preparation of this enzyme by LIEBERMAN et al. (1955) have been given under the section on PRPP-kinase.

References.

ARCHIBALD, R. M.: J. biol. Chem. 156, 121 (1944).

BAUER, E.: In E. BAMANN and K. MYRBÄCK, Die Methoden der Fermentforschung. Vol 2. New York: Academic Press 1955. — BRATTON, A. C., and E. K. MARSHALL, jr.: J. biol. Chem. 128, 537 (1939). — BUCHANAN, J. M., and C. S. HARTMAN: Adv. Enzymol. 21, 199 (1959). — BUCHANAN, J. M., B. LEVENBERG, J. G. FLADS and J. A. CLADNER: In W. D. McELROY and H. B. GLASS: Amino acid metabolism (symposium), p. 743. The Johns Hopkins Press 1955.

CARTER, C. E.: Ann. Rev. Biochem. 25, 123 (1956). — COLOWICK, S. P.: Meth. Enzymol. 1, 90 (1955). — COOPER, C., R. WU and D. W. WILSON: J. biol. Chem. 216, 37 (1955).

FLAKS, J. G., M. J. ERWIN and J. M. BUCHANAN: J. biol. Chem. 229, 603 (1957). — FRIEDMANN, H. C., and B. VENNESLAND: J. biol. Chem. 233, 1398 (1958); 235, 1526 (1960).

GOLDTHWAIT, D. A.: J. biol. Chem. 222, 1051 (1956). — GOLDTHWAIT, D. A., and G. R. GREENBERG: Meth. Enzymol. 2, 504 (1955). — GOLDTHWAIT, D. A., G. R. GREENBERG and R. A. PEABODY: Biochim. biophys. Acta 18, 148 (1955). — GOLDTHWAIT, D. A., R. A. PEABODY and G. R. GREENBERG: J. Amer. chem. Soc. 76, 5258 (1954); — J. biol. Chem. 221, 555 (1956a); 221, 569 (1956b). — GREENBERG, G. R.: J. Amer. chem. Soc. 76, 1458 (1954a); — Fed. Proc. 13, 745 (1954b).

HALL, L. M., R. C. JOHNSON and P. P. COHEN: Biochim. biophys. Acta 37, 144 (1960). — HARTMAN, S. C., and J. M. BUCHANAN: J. biol. Chem. 233, 451 (1958a); 233, 456 (1958b); — Ann. Rev. Biochem. 28, 365 (1959). — HARTMAN, S. C., B. LEVENBERG and J. M. BUCHANAN: J. Amer. chem. Soc. 77, 501 (1955); — J. biol. Chem. 221, 1057 (1956).

JONES, M. E., L. SPECTOR and F. LIPMANN: J. Amer. chem. Soc. 77, 819 (1955).

KALCKAR, H. M.: J. biol. Chem. 167, 429 (1947). — KEILIN, D., and E. F. HARTREE: Proc. roy. Soc. Ser. B 124, 397 (1938). — KENNEDY, E. P., and A. L. LEHNINGER: J. biol. Chem. 179, 957 (1949). — KONDO, H., H. C. FRIEDMANN and B. VENNESLAND: J. biol. Chem. 235, 1533 (1960). — KORN, E. D., and J. M. BUCHANAN: J. biol. Chem. 217, 183 (1955). — KORN, E. D., C. N. REMY, H. C. WASILEJKO and J. M. BUCHANAN: J. biol. Chem. 217, 875 (1955). — KORNBERG, A., I. LIEBERMAN and E. S. SIMMS: J. biol. Chem. 215, 389 (1955).

LEVENBERG, B., and J. M. BUCHANAN: J. biol. Chem. **224**, 1019 (1957a); **224**, 1005 (1957b). — LIEBERMAN, I., and A. KORNBERG: Biochim. biophys. Acta **12**, 223 (1953); — J. biol. Chem. **207**, 911 (1954). — LIEBERMAN, I., A. KORNBERG and E. S. SIMMS: J. Amer. chem. Soc. **76**, 2844 (1954); — J. biol. Chem. **215**, 403 (1955). — LOWENSTEIN, J. M., and P. P. COHEN: J. biol. Chem. **220**, 57 (1956). — LUKENS, L. N., and J. M. BUCHANAN: J. Amer. chem. Soc. **79**, 1511 (1957).

METZENBERG, R. L., L. M. HALL, M. MARSHALL and P. P. COHEN: J. biol. Chem. **229**, 1019 (1957). — MILLER, R. W., L. N. LUKENS and J. M. BUCHANAN: J. Amer. chem. Soc. **79**, 1513 (1957).

REICHARD, P.: Acta Chem. Scand. **8**, 1102 (1954); — Adv. Enzymol. **21**, 263 (1959). — REICHARD, P., and G. HANSHOFF: Acta Chem. Scand. **9**, 519 (1955). — REMY, C. N., W. T. REMY and J. M. BUCHANAN: J. biol. Chem. **217**, 885 (1955). — REYNOLDS, E. S., I. LIEBERMAN and A. KORNBERG: J. Bact. **69**, 250 (1955).

SCATTERGOOD, A.: In W. C. FERNELIUS: Inorganic synthesis. Vol. 2, p. 86. New York: McGraw Hill 1946. — SCHULMAN, M. P., J. C. SONNE and J. M. BUCHANAN: J. biol. Chem. **196**, 499 (1952).

TARR, H. L. A.: Canad. J. Biochem. Physiol. **38**, 683 (1960). — TISELIUS, A., S. HJERTEN and O. LEVIN: Arch. Biochem. Biophys. **65**, 132 (1956).

WARREN, L., and J. M. BUCHANAN: J. biol. Chem. **299**, 613 (1957). — WILLIAMS, W. J., and J. M. BUCHANAN: J. biol. Chem. **203**, 583 (1953). — WILLSTÄTTER, R., u. H. KRAUT: Ber. dtsch. chem. Ges. **56**, 1117 (1923). — WYNGAARDEN, J. B., and D. M. ASHTON: J. biol. Chem. **234**, 1492 (1959). — WYNGAARDEN, J. B., H. R. SILBERMAN and J. H. SADLER: Ann. N. Y. Acad. Sci. **75**, 45 (1958).

YATES, R. A., and A. B. PARDEE: J. biol. Chem. **221**, 743 (1956).

For recent references on glycineamide ribotide formation in plants see:

IWAI, K., S. NAKAGAWA and O. OKINAKA: Biochem. biophys. Acta **68**, 152 (1963).

KAFOOR, M,, and E. R. WAYGOOD: Biochem. biophys. Res. Comm. **9**, 7 (1962).

Enzymes of Fat Metabolism.

A. Plant Lipases.

By

Edward J. Barron.

With 1 Figure.

Lipases appear to have wide distribution in the plant kingdom. The seeds of some species are particularly rich in the enzyme. BAMANN and ULLMAN (1942, 1959), have investigated the seeds from a large variety of plants, and found that the majority with high lipase content belong to the families Euphorbiaceae, Ranunculaceae and Papaveraceae. There was little correlation between lipid content and the lipase activity. The seeds from grasses that produce the cereal grains also are fairly good sources of the enzyme. The two that have been studied the most are oat lipase (MARTIN and PEERS, 1953) and wheat germ lipase (SINGER and HOFSTEE, 1948). BAMANN and ULLMANN (1942, 1959) have also investigated the vegetative portion of a large number of plants and found no correlation between the lipase content of the seeds and the vegetative organs. The highest lipase activity of the vegetative portion was relatively low as compared to the generally high values in the seeds.

The enzyme is also fairly common in molds and fungi. For instance, it has been reported in *Mycotorula lipolytica* (PETERS and NELSON, 1948), *Penicillium roqueforti* and *Aspergillus niger* (SHIPE, 1951; RAMAKRISHNAN and BANERJEE, 1951), *Rhizopus* (TATSUOKA, MIYAKA, WADA, IMADA and MATSUMURA, 1959) and *Fusarium* (FIORE and NORD, 1950). This is of interest since molds and fungi sometimes grow on lipid rich seeds during storage.

Assay Procedures.

The distinction between lipases and simple esterases has to be somewhat arbitrary, as they both catalyze the hydrolysis of carboxylic acid esters. By definition, the lipases are considered to be those enzymes that hydrolyze glyceride esters of long chain fatty acids. The esterases are those enzymes that catalyze the hydrolysis of esters of the lower alcohols and fatty acids. Neither of these classes of enzymes has been well characterized as regards their substrate specificity, however, there is undoubtedly considerable overlapping of the substrates that both types of enzymes will attack. For this reason, it would seem to be essential to use glyceride esters of long chain fatty acids in studying new sources of lipases or in following a procedure of purification.

The difficulty in using high molecular weight glycerides is the problem of producing stable and reproducible emulsions, since the rate of hydrolysis depends on the state of dispersion of the substrate (SCHONHEYDER and VOLQVARTZ, 1945). A number of emulsifiers have been used to achieve this end, among these have been polyvinyl alcohol (FIORE and NORD, 1949), gum acacia (CRANDELL and CHERRY, 1931), gum arabic (FODER, 1946), soy bean phospholipid[1] (GOLDMAN, 1954) and monoglycerides (BORGSTROM, 1955; BARRON, 1959).

All methods used for the estimation of lipase activity have either directly or indirectly determined the release of fatty acids. The procedure which is probably the best and simplest is to follow the release of fatty acids titrimetrically with the aid of a pH meter. The method of MARCHIS-MOUREN, SARDA and DESNUELLE (1959) using this principal is given below. Also a slightly different procedure for determining the hydrolysis is given, whereby the decrease in ester content of the reaction mixture is measured. The method is essentially that used by ORY, ST. ANGELO and ALTSCHUL (1960) in their study of castor bean lipase.

1. Lipase Assay by the Release of Fatty Acids.

Enzyme preparation. Usually the crude plant lipase preparations are produced by extraction of defatted tissue with water, 1.5% sodium choride solution or N/40 ammonium hydroxide. The defatted tissue is produced from ground or minced tissue by preparing an acetone powder or simple extraction with petroleum ether.

Emulsion. Place 165 ml of 10% gum arabic in a WARING Blendor along with 15 gm of crushed ice and 20 ml of neutralized olive oil and homogenize the mixture for 10 minutes.

Procedure. 10 ml of the emulsion is placed in a 50 ml beaker with 0.3 ml of 20% sodium taurocholate solution[2] and enough distilled water to make the final volume 30 ml after the enzyme solution is added. The beaker is placed in a thermostated water bath at 37° C, then pH electrodes and a small mechanized stirring rod are inserted into the beaker (for the greatest accuracy a fine stream of CO_2 free N_2 is bubbled through the reaction mixture). The pH is adjusted to 9.1 to 9.2 with 0.1 N carbonate free NaOH from a semi-microburette and the enzyme solution added.

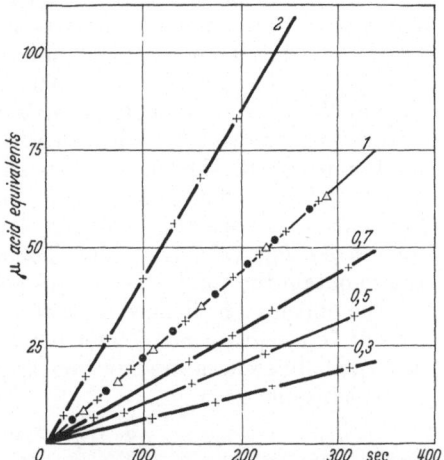

Fig. 1. The figures above each curve represent milliliters of lipase solution and show that the rate of hydrolysis is proportional to the amount of enzyme used. The circles, crosses and triangles along curve 1 correspond to different experiments with the same amount of enzyme, indicating that the results are quite reproducible (MARCHIS-MOUREN, SARDA and DESNUELLE, 1959).

When the pH reaches 9.0, a stop watch is started and the pH is maintained at 9.0 by the addition of the 0.1 N base.

The lipase unit is defined as that amount of enzyme which liberates 10 μ-equivalents of acid per minute under the conditions of the test.

---

[1] The use of pure phospholipids as emulsifying agents should be applied with some caution. BARRON (1959) has found that, while pure egg or yeast lecithin and beef liver monophosphoinositide gave good emulsions of triolein in water, pancreatic lipase frequently would not hydrolyze the triolein. This appeared to be the result of some physical phenomenon and not a competitive type inhibition.

[2] The necessity of taurocholate is questionable since its value as an activator is not clear (WILLS, 1955). BORGSTROM's (1953) findings seem to indicate that in the case of pancreatic lipase it acts only to shift the pH optimum from 8.0 to 9.0.

2. Lipase Assay by the Decrease in Ester Content.

Reagents. 1. Equal volumes of alcoholic solutions of 2.5% sodium hydroxide and 2.5% hydroxylamine hydrochloride are mixed and filtered just before use.

2. Stock solution of ferric perchlorate: 2 gm of ferric perchlorate (non-yellow, G. Frederick Smith Chemical Co. Columbus, Ohio) is dissolved in 10 ml of distilled water, then 20 ml of 70% perchloric acid is added and the solution mixed. This solution should be stable for a week if kept in a brown bottle in the refrigerator.

3. Dilute ferric perchlorate solution: Dilute 5 ml of the stock solution to 100 ml with alcohol. It should be prepared fresh daily.

Procedure. 5 ml of the emulsion used in the previous procedure is placed in a 25 or 50 ml beaker along with a sufficient amount of the desired buffer to make the total volume 15 ml when the enzyme solution is added. The beaker is placed in a 37° C water bath and a mechanized stirring rod inserted. The enzyme is added and a sample of the reaction mixture (0.1 ml $\cong$ 10 μEq. ester bonds) is immediately transferred to 10 ml of alcohol-ether mixture (1:3) in a glass stoppered centrifuge tube and timing is begun. The alcohol-ether solution is mixed by inversion. Samples are taken at reasonable intervals and over a period of time depending on the activity of the enzyme solution. The alcohol-ether solutions are centrifuged and the supernatant is decanted into a graduated tube. The precipitate is washed once with 2 ml ether. Then 0.6 ml of the alkaline hydroxylamine is added to the pooled ether extracts and the tube placed in a 60° C water bath until the ether has evaporated off and then for another 30 seconds to complete hydroxamate formation. The volume is made to 2 ml with alcohol and 5 ml of the dilute ferric perchlorate solution is added. The color is allowed to develop for 20 minutes and the optical density determined at 530 mμ. It is convenient to use triolein as the standard and to calculate in terms of μ-equivalents of ester bonds; each μmole of triolein would contain 3 μ equivalents of ester bonds. The determined value is multiplied by 150 to obtain the total ester groups per reaction mixture.

The unit may be defined as that amount of enzyme that catalyzes the hydrolysis of 10 μ equivalent of ester bonds per minute. With this method the actual measured differences are not too large, but fortunately the estimation of ester content is quite accurate.

3. Other Assay Procedures.

There have been a number of different types of procedures used for the determination of lipase activity. Among these has been the manometric technique, proposed by RONA and LASNITZKI (1924) in which the CO_2 displaced from a bicarbonate buffer by the liberated fatty acids was measured. These workers used tribytyrin as the substrate, while monobutyrin has been used by NICOLAI (1926) and SINGER and HOFSTEE (1948). A stalagonometeric method has been developed by RONA and MICHAELIS (1911), which makes use of the high surface activity of tribytyrin. This procedure has been used by BAMANN and ULLMANN (1942) in their studies on plant lipases. A turbidimeteric method was used by BORGSTROM (1955) and modified by BARRON (1959). HUGGINS and LAPIDES (1947) and NACHLAS and SELIGMAN (1949) and HOFSTEE (1951), have proposed colorimeteric procedures employing respectively, esters of p-nitrophenol, β-napthol and salycilic acid. There have been a number of other titrimeteric assay procedures developed, see for instance, CRANDELL and CHERRY (1931) and FIORE and NORD (1949).

4. Activators and Inhibitors.

Generally, the plant lipases are not activated by calcium ions or at least not to the quantitative extent that pancreatic lipase can be. For instance, wheat germ lipase is not activated at all (SINGER and HOFSTEE, 1948) and the lipase from *Nigella sativa* is only slightly activated (BAMANN and ULLMAN, and TIETZ, 1953), while the lipase of cottonseed was strongly activated by calcium (OLCOTT and FONTAINE, 1941). Nevertheless, calcium ions do not appear to be inhibitory, at least in medium concentrations (10^{-3} M). Its effect should be investigated in each case.

At least two plant lipases have been shown to be inhibited by reagents that react with sulfhydryl groups. Wheat germ lipase was inhibited by *p*-chloromercuribenzoate, iodoacetamide and *o*-iodobenzoate (SINGER and HOFSTEE, 1948; DIRKS, BOYER and GEDDES, 1955). ORY, ST. ANGELO and ALTSCHUL (1960) have shown that castor bean lipase is inhibited by mercuric ions and *p*-chloromercuribenzoic acid and that the inhibition can be reversed by cysteine.

5. pH Optima.

BAMANN and ULLMANN (1942) reported in their extensive studies on plant lipases that the lipases of the vegetative portion of plants and most ungerminated seeds catalyze the hydrolysis of tributyrin in the alkaline pH range. The lipase from ripe castor beans is an exception in that it has a pH optimum around 4.5 (HALEY and LYMAN, 1921; ORY, ST. ANGELO and ALTSCHUL, 1960). However, BAMANN and ULLMANN (1942) found that the unripe (lipid poor) seeds contained a small amount of lipase that hydrolyzed tributyrin at an alkaline pH.

6. Specificity of Attack.

The plant lipases, which have been studied thus far, have not had the specificity of attack on the triglyceride molecule that the pancreatic lipases do. MATTSON and BECK (1955), BORGSTROM (1954) and BARRON (1959) have shown that hog, rat and dog pancreatic lipase attack specifically the primary ester groups of triglycerides. On the other hand there is good evidence that the lipase from the palm fruit, *Elaeis guineensis* (SAVARY, FLANZY, CONSTANTIN and DESNUELLE, 1957) and that from *Ricinus communis* (SAVARY, FLANZY and DESNUELLE, 1958) do not attack the primary ester groups. In the latter two papers, the workers used asymmetric triglycerides in which the 2 position was known to contain unsaturated fatty acids. After different periods of hydrolysis, iodine values were determined on the liberated fatty acids and the remaining glycerides and in each case both fractions had approximately the same iodine number. This indicated that the two plant lipases had attacked the ester bonds indiscrimately, if the lipase were attacking the primary ester bonds specifically the glycerides would have had a higher iodine value than the liberated fatty acids.

7. Purification.

Extensive work has not been done on the purification of plant lipases. Some workers, however, have obtained partial purification with acid precipitation and an ammonium sulfate precipitation (SINGER and HOFSTEE, 1948). BAMANN and TIETZ (1953) have also obtained some enrichment by continuous curtain paper electrophoresis.

The procedure described below was used by SINGER and HOFSTEE (1948) to obtain partial purification of wheat germ lipase. It may be of interest to mention two procedures that have been used to obtain relatively pure pancreatic lipase. MARCHIS-MOUREN, SARDA and DESNUELLE (1959) have developed a method which consists of 9 steps starting with an aqueous extract of hog pancreas acetone powder.

The first eight steps included three $(NH_4)_2SO_4$ precipitations, an acetone precipitation, a pH precipitation, adsorption and elution from tri-calcium phosphate and aluminum hydroxide. These procedures yielded a 20-fold purification with a 25% recovery of original activity. The last step was a zone electrophoresis on starch at pH 5.25, and yielded a 135 fold purified enzyme. Their purest fractions were homogenous as determined by chromatography in hydroxyapatite column and by zone electrophoresis on starch at pH 5.25 and pH 8.0. BARRON (1959) starting with dog pancreatic juice was able to obtain a single peak by chromatography on carboxy-methyl cellulose that contained essentially all the lipase activity applied to the column. By the use of free boundary electrophoresis, this peak was shown to contain only two proteins.

Preparation of defatted wheat germ and extraction of enzymes. Fresh untreated wheat germ was defatted by extracting 5 times with low boiling petroleum ether, or small batches were extracted in a SOXHLET extractor. The dry germ was ground in a WILEY mill to pass a 40 mesh screen, this resulting powder could be stored for at least 1 year at —10° C. The dry powder was extracted by stirring with 100 volumes of cold H_2O per gm of powder for 15—20 minutes. This extract was stable when stored in the frozen state at —10° C.

Purification. The extract was centrifuged for 20 minutes in the cold. The cloudy supernatant was adjusted to pH 5.5 with 0.5 N acetic acid and recentrifuged at 3000 *rpm.* for 10 minutes. The clear supernatant obtained was adjusted to pH 6.6—6.8 with 0.5 N $NaHCO_3$ in the cold and a saturated $(NH_4)_2SO_4$ solution was added slowly with stirring until 35% saturation. After centrifuging, the precipitate was discarded and the supernatant was made 55% saturated with respect to $(NH_4)_2SO_4$. The solution was centrifuged for 30 minutes in the cold and the precipitate collected and lyophilized. The lyophilized powder maintained its activity for a long period of time if stored in the cold. Before use the powder was dissolved in 0.02 M phosphate buffer, pH 6.8, and dialysed for two hours.

The acid precipitation yielded a supernatant with a 2.5 fold increase in specific activity. The $(NH_4)_2SO_4$ precipitation yielded a product with 5 times the specific activity of the original aqueous extract.

References.

BARRON, E. J.: Thesis, University of Washington (1959). — BAMANN, E., and N. TIETZ: Biochem. Z. **324**, 502 (1953). — BAMANN, E., and E. ULLMANN: Biochem. Z. **312**, 9 (1942). — BAMANN, E., E. ULLMANN and N. TIETZ: Biochem. Z. **323**, 489 (1953). — BAMANN, E., and E. ULLMANN: Handbuch der Pflanzenphysiologie, Vol. 7, p. 109. Edited by W. RUHLAND. Berlin: Springer-Verlag 1957. — BORGSTROM, B.: Biochem. biophys. Acta **13**, 491 (1954); — Biochemical problems of lipids. Edited by POPJAK, G., and E. LE BRETON. New York: Interscience Publishers 1955.

CRANDELL, L. A., and I. S. CHERRY: Proc. Soc. exp. Biol. (N.Y.) **28**, 570 (1931).

DIRKS, B. M., P. D. BOYER and W. F. GEDDES: Cereal Chem. **32**, 356 (1955).

FIORE, J. V., and F. F. NORD: Arch. Biochem. Biophys. **23**, 473 (1949); **26**, 382 (1950). — FODOR, P. J.: Nature (Lond.) **158**, 375 (1946).

GOLDMAN, M. L.: Food Res. **19**, 503 (1954).

HALEY, D. E., and J. F. LYMAN: J. Amer. chem. Soc. **43**, 2664 (1921). — HOFSTEE, B. H. J.: Science **114**, 128 (1951). — HUGGINS, C., and J. LAPIDES: J. biol. Chem. **170**, 467 (1947).

MARCHIS-MOUREN, G., L. SARDA and P. DESNUELLE: Arch. Biochem. Biophys. **83**, 309 (1959). — MARTIN, H. F., and F. G. PEERS: Biochem. J. **55**, 523 (1953). — MATTSON, F. H., and L. W. BECK: J. biol. Chem. **214**, 115 (1955).

NACHLAS, M. M., and A. M. SELIGMAN: J. biol. Chem. **181**, 343 (1949). — NICOLAI, H. W.: Biochem. Z. **174**, 343 (1926).

OLCOTT, H. S., and T. D. FONTAINE: J. Amer. chem. Soc. **63**, 825 (1941). — ORY, R. L., A. J. ST. ANGELO and A. M. ALTSCHUL: J. Lipid Res. **1**, 208 (1960).

PETERS, I. I., and F. E. NELSON: J. Bact. **55**, 581 (1948).

RAMAKRISHNAN, C. V., and B. N. BANERJEE: Enzymologia **15**, 33 (1951). — RONA, P., and A. LASNITSKI: Biochem. Z. **152**, 504 (1924). — RONA, P., and L. MICHAELIS: Biochem. Z. **31**, 345 (1911).

SAVARY, P., J. FLANZY, M. J. CONSTANTIN and P. DESNUELLE: Bull. Soc. Chim. biol. **39**, 413 (1957). — SAVARY, P., J. FLANZY and P. DESNUELLE: Bull. Soc. Chim. biol. **40**, 657 (1958). — SCHONHEYDER, F., and K. VOLQVARTZ: Acta physiol. scand. **9**, 57 (1945). — SINGER, T. P., and B. H. J. HOFSTEE: Arch. Biochem. Biophys. **18**, 229, 245 (1948). — SHIPE, W. F.: Arch. Biochem. Biophys. **30**, 165 (1951).

TATSUOKA, S., A. MIYAKA, S. WADA, I. IMADA and MATSUMURA: J. Biochem. (Tokyo) **46**, 575 (1959).

WILLS, E. D.: Biochem. J. **60**, 529 (1955).

B. Phospholipases.

By

Edward J. Barron.

With 1 Figure.

There has been some disagreement on the nomenclature for the enzymes attacking phospholipids. Fortunately, one contingency has been clarified by general usage. Since it has been shown that most of these enzymes attack other phospholipids besides phosphotidylcholine[1] (lecithin), the term phospholipase[2] has slowly superseded the name lecithinase. There is still, however, no unanimity of opinion on the designation of the bonds hydrolyzed by phospholipases C and D. In the present case, the scheme suggested by CONTARDI and ERCOLI (1933) and

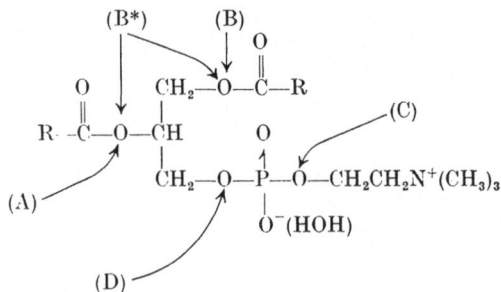

$$B^* = B + \text{``activator''}$$

Fig. 1. Sites of attack of the different phospholipases. "Activator" refers to those compounds which enhance the development of a net negative charge on the lecithin molecule (see text, phospholipase B).

adopted by HANAHAN (1957) will be used. Fig. 1 shows the sites of action of phospholipase A, B, C, and D on lecithin. With the exception of phospholipase C, most of the investigations on the phospholipases have been carried out in material obtained from sources other than plants. It will be apparent from brief discussions presented in this section that there is a need for further investigations of these enzymes as they occur in plants.

I. Phospholipase A.

While this enzyme has been found in a wide variety of sources, conclusive evidence of its existance in plants has not been forthcoming. IWATA (1930) found lysolecithin in alcoholic extracts of polished rice and it was presumed to have

---

[1] The following abbreviations will be used in this section: the older term lecithin will be used for phosphatidylcholine; PE for phosphatidylethanolamine; PS for phosphatidylserine; GPC for glycerolphosphorylcholine.

[2] The term phospholipase is itself not completely satisfactory, since the term lipase carries the connotation of carboxylic ester hydrolysis and obviously some of the enzymes (i.e. phospholipase D) are phosphoesterases.

arisen from the action of phospholipase A on lecithin. The presumption, however, has never been verified. Similarly, the evidence presented by CONTARDI and LATZER (1928) for its occurence in castor beans must be considered equivocal. Phospholipase A has been found in snake venom, wasp and bee venom, bacteria, pancreas, kidneys and other tissues (General references: WITTCOFF, 1951; HANAHAN, 1957).

Assay Procedure.

As the evidence for phospholipase A in plants is inconclusive, it would seem advantageous to present a qualitative assay which might be used in screening tissues. The procedure is that of HANAHAN, RODBELL and TURNER (1954) in which the reaction is carried out in moist ether[1]. The fatty acids liberated by the enzymatic catalysis are estimated by titration.

Reagents. Egg lecithin is purified by the procedure of HANAHAN, TURNER and JAYKO (1951).

0.02 N NaOH in 90% methanol is prepared from CO_2-free NaOH.

Procedure. 25 mg of lecithin is dissolved in 25 ml of diethyl ether. 0.25 ml of enzyme solution and 0.1 ml of 5% $CaCl_2$ solution are added and the reaction vessel is swirled or shaken until the reaction mixture becomes homogenous[2]. The reaction is allowed to proceed at room temperature for 4 hours. If phospholipase A hydrolysis occurs the solution slowly becomes cloudy and finally a gelatinous precipitate forms. At the end of the incubation period 25 ml of alcohol and 1 ml of 0.1% aqueous cresol red are added and the solution titrated with 0.02 N methanolic NaOH. A blank is prepared by adding alcohol, enzyme and $CaCl_2$ to the ether substrate solution in that order and is titrated immediately.

The enzymatic release of fatty acids in 98—99% ether and the formation of a gelatinous precipitate is presumptive evidence for phospholipase A activity. The presence of the enzyme is verified by isolating the lysolecithin and characterizing it. The lysolecithin can be obtained in good yield by precipitation from ethanol-ether at —25° (HANAHAN, RODBELL and TURNER, 1954) or by silicic acid chromatography (DITTMER and HANAHAN, 1959). It can be identified by determining the fatty acid to phosphorus ratio and the choline to phosphorus ratio, which should be 1:1 in each case.

Quantitative assay. HANAHAN, RODBELL and TURNER (1954) used a slight modification of the above procedure for following the progress of the phospholipase A reaction. RIMON and SHAPIRO (1959) followed the hydrolysis by measuring the decrease in ester content. HAYAISHI and KORNBERG (1954) developed a system which required the additional use of two partially purified enzymes, phospholipase B and glycerolphosphorylcholine diesterase, in the reaction system. The choline, which was one of the final products was then estimated colorimetrically. They also used a procedure whereby the decrease in lecithin content was followed. The lecithin was estimated by precipitation as the iodine complex at pH 7.0; the precipitate was then dissolved in ethylene dichloride and the optical density measured.

Cofactors and inhibitors. Calcium ions have an activating effect on phospholipase A (HAYAISHI and KORNBERG, 1954; RIMON and SHAPIRO, 1959). This

---

[1] The rate of hydrolysis of lecithin by phospholipase A is higher in a moist ether solution than an aqueous solution with enzymes from snake venom and pancreas (HANAHAN, 1952; HANAHAN, RODBELL and TURNER, 1954), HAYAISHI and KORNBERG (1954) report, however, that the phospholipase A from *Serratia plymuthicum* in not active in ether.

[2] The volumes of enzyme and ether may be varied as long as the ratio of water to ether is kept at 1:100 so that a homogenous solution is obtained. Of course, due regard should be paid to substrate concentration.

activation is more easily demonstrable in an aqueous reaction medium than in moist ether, nevertheless, occasionally there is a noticable activation by calcium ions in the ether medium. Cyanide (OGAWA, 1936) and eserine (FRANCIOLI, 1937) have no inhibitory action.

pH optima. The enzyme has a rather broad pH optimum with the peak around 7.0 (HUGES, 1935; HAYASHI and KORNBERG, 1954; HANAHAN, 1952).

Specificity. Earlier work (HANAHAN, 1954; LONG and PENNY, 1957) on the site of attack of phospholipase A on lecithin had indicated that it was at the α-ester bond, but more recent findings have shown conclusively that it attacks the β-position (HANAHAN, BROCKERHOFF and BARRON, 1960; TATTERIE, 1959), yielding a α-lysolecithin and one fatty acid.

The enzyme is capable of attacking saturated and unsaturated lecithin (HANAHAN, 1954a; HANAHAN, BROCKERHOFF and BARRON, 1960) PE, PS, and phosphatadylcholine (plasmalogen) (FAIRBAIRN, 1945; LONG and PENNY, 1957; RIMON and SHAPIRO, 1959; RAPPORT and FRANZL, 1957). RIMON and SHAPIRO report that the enzyme from pancreas also attacks phosphatidic acid. Phospholipase A apparently does not attack cerebrosides, sphingomyelin or lysolecithin (FAIRBAIRN, 1945).

Purification. Partial purification of crude phospholipase A preparations from snake venom (SLOTTA and FRAENKEL-CONRAT, 1938), bacteria (HAYAISHI and KORNBERG, 1954) and pancreas (RIMON and SHAPIRO, 1959) has been obtained.

II. Phospholipase B (Lysophospholipase B).

Phospholipase B has been demonstrated in rice bran hulls (CONTARDI and ERCOLI, 1933), pancreas (FAIRBAIRN, 1948; SHAPIRO, 1953a), *Penicillium notatum* (FAIRBAIRN, 1948; UZIEL and HANAHAN, 1956), *Aspergillus oryzae* (CONTARDI and ERCOLI, 1933), and *Serratia plymuthicum* (HAYAISHI and KORNBERG, 1954). CONTARDI and LATZER (1928) suggested in one of their earlier papers that the enzyme exists in castor beans and wasp venom, but the evidence is equivocal.

Assay Procedure.

The enzymatic reaction conditions and the method of following the reaction by a decrease of ester groups is that of SHAPIRO (1953a). His procedure for estimating the liberation of GPC, however, has been replaced by the more convenient colorimeteric procedure used by DAWSON (1956).

a) Estimation of Activity by Measuring the Decrease in Ester Bond.

Reagents. 0.1 M phosphate buffer pH 6.0; 0.05 M lysolecithin solution (the lysolecithin is prepared by phospholipase A action on lecithin); 2.0 M hydroxylamine hydrochloride; 3.5 M NaOH; 3.3 M HCl; 0.37 M $FeCl_3$ in 0.1 HCl.

Procedure. 0.4 ml of phosphate buffer (40 μmoles) and 0.1 ml of lysolecithin (5 μmoles), are pipetted into a small test tube. The desired volume of enzyme is added and the volume made to 1 ml with water[1]. The solution is allowed to incubate at 37° until not more than 40% of the substrate is hydrolyzed. The reaction is stopped by the addition of 4 ml of an ethanol-ether mixture (3:1) and the solution filtered. To 3 ml of the filtrate add 0.5 ml of the hydroxylamine

---

[1] SHAPIRO found, when using partially purified pancreatic phospholipase B, that glycerol (25% by volume) had to be added to the reaction mixture to prevent inactivation of the enzyme. This indicates that this enzyme preparation contained proteolytic activity.

hydrochloride solution and 0.5 ml of 3.5 M NaOH. The solution is mixed and allowed to stand for 20 minutes at room temperature; 0.6 ml of 3.3 N HCl is then added and after mixing, the color is developed by addition of 0.5 ml FeCl$_3$. The optical density is measured at 520 mμ.

The unit of activity is defined as that amount of enzyme that catalyzes the hydrolysis of 1 μmole of lysolecithin (1 μmole ester bonds) in 1 hour at 37°.

b) Measurement of Activity by Estimating GPC Formed.

Reagents. 5% Bovine serum albumin; 10% Trichloroacetic acid; 5 N HCl.

Potassium triiodide reagent: 15.7 gm of reagent grade iodine and 20 gm of reagent potassium iodide are dissolved in 100 ml of water and the mixture is shaken for 45 minutes on a mechanical shaker to effect solution. The reagent is stable indefinitely at 4°.

Ethylene dichloride: The stability of the choline complex should be tested in each lot of ethylene dichloride used.

Procedure. The enzymatic reaction conditions are the same as those given in the procedure above. At the end of the incubation period 0.15 ml of 5% serum albumin is added and the protein precipitated with 0.5 ml of 10% trichloroacetic acid and removed by centrifugation. Then 0.25 ml of 5 N HCl is added to 1.1 ml of the supernatant and the sample hydrolyzed for 20 minutes at 100°[1]. The hydrolysate is centrifuged and an aliquot of the supernatant is analyzed for choline by the method of APPLETON, LA DU, LEVY, STEEL and BRODIE (1953). To 1.0 ml of the supernatant in a centrifuge tube add 0.3 ml of the cold triiodide reagent and place the tube in an ice bath for 20 minutes. After this time, centrifuge for 10 minutes at 2500 *rpm* and by means of a capillary pipette remove the supernatant, leaving the tube essentially dry. Immediately dissolve the choline periodide precipitate in a small amount of ethylene dichloride with the aid of a fine tipped stirring rod, then dilute to 10 ml with additional solvent, and read in a spectrophotometer at 365 mμ. As many of the manipulations as possible should be carried out in the cold.

GPC serves as the best standard and should be treated exactly as above except that lysolecithin is added at the end of the reaction period. Since GPC is not readily available, choline chloride may also be used as a standard.

c) Other Assay Procedures.

Several other procedures have been used for the assay of phospholipase B. Among these is one used by HAYAISHI and KORNBERG (1954) in which GPC-diesterase was added to the enzyme reaction mixture. The GPC liberated by the phospholipase is attacked by the diesterase, liberating choline, which is then determined. UZIEL and HANAHAN (1956) used a titrimeteric procedure for estimating the liberated fatty acid, as well as a colorimeteric procedure for measuring unreacted lysolecithin. The latter procedure utilizes the fact that lysolecithin forms an insoluble tri-iodide in 1 N HCl while GPC (and phosphorylcholine) does not.

Substrate Specificity.

The classification of this enzyme on the basis of the substrate (lysophospholipids or intact phospholipids) which it attacks presents some problems. CONTARDI

---

[1] The serum albumin is added, since it binds many lipids, to assist in removing the residual lysolecithin. During this hydrolysis, choline is liberated from GPC. Phosphorylcholine is not hydrolyzed under these conditions.

and ERCOLI (1933) and LE BRETON and PANTALEON (1948), using crude phospholipase B preparations from rice hulls and pancreatic juice respectively, reported that the enzyme would attack lecithin as well as lysolecithin. However, later investigators working with partially purified enzymes from *P. notatum* (FAIRBAIRN, 1948; UZIEL and HANAHAN, 1956), pancreas (SHAPIRO, 1953a) and *Serratia plymuthicum* (HAYAISHI and KORNBERG, 1954) found that only lysolecithin was attacked and not lecithin. These results indicated that in the earlier investigations the phospholipase B preparations had probably been contaminated with phospholipase A[1].

Recently, DAWSON, (1957, 1958a) and BANGHAM and DAWSON (1959a, b) reported that partially purified phospholipase B would attack lecithin and phosphatidylethanolamine, yielding two fatty acids and GPC and glycerolphosphorylethanolamine, in the presence of certain "activators". Lecithin was attacked if monophosphoinositide, a polyglycerolphospholipid, or synthetic anionic amphiphathic compounds i. e., diacetyl phosphate and hexadecyl sulfate, were added to the reaction mixture. Furthermore, they were able to show that the enzyme would hydrolyze lecithin only when the lecithin molecule developed a net negative change and only when the potential had reached a critical value (BANGHAM and DAWSON, 1959a, 1960).

DAWSON (1958b) has also reported that the phospholipse B from *P. notatum* and pancreas would hydrolyze monophosphoinositide, yielding two fatty acids plus glycerolphosphorylinositol. In the presence of lecithin this reaction was inhibited while the lecithin itself was attacked.

The pH optima for the hydrolysis of lysolecithin and lecithin by the enzyme was the same in each case. Also, on starch-gel electrophoresis (done at one pH only) the activities moved in the same band. These two findings suggested that both enzymic processes were carried out by the same enzyme.

At present it can only be concluded that there is an enzyme (phospholipase B) that catalyzes the hydrolysis of the α-ester bond of lysolecithin, lysophosphatidylethanolamine (probably lysophosphatidylserine, FAIRBAIRN, 1948), and furthermore, under special conditions the same enzyme will hydrolize both ester groups of the parent compounds (i. e., lecithin). It must be borne in mind that these special conditions may be the "physiological" conditions under which the enzyme normally functions.

pH Optima and Stability.

The pH optimum of the pancreas and *Serratia plymuthicum* enzymes is 6.0 (HAYAISHI and KORNBERG, 1954; SHAPIRO, 1953a; FAIRBAIRN, 1948), while that of *P. notatum* is pH 4.0. The optimum for the rice enzyme is not known. The enzyme from *S. plymuthicum* lost 50 to 80% of its activity on heating at 100° for 10 minutes at pH 7.0. The heat lability of the enzyme from other sources has not been investigated.

Activators and Inhibitors.

Phospholipase B apparently has no requirement for calcium ions and ether has no effect on the rate of the reaction. It can be completely inhibited by 0.01 M cyanide.

---

[1] It is of interest that SHAPIRO (1953b) had reported two enzyme systems in beef pancreas which attacked lecithin to yield GPC. One catalyzed the hydrolysis of lysolecithin but did not attack lecithin. The other attacked lecithin, yielding GPC with no decrease in total ester content, indicating an acyl transfer.

Purification Procedures.

As the investigations on the enzymes from plants have been very meager, no purification procedures on the plant enzyme have been reported. However, a partially purified enzyme has been obtained from *P. notatum* (FAIRBAIRN, 1948; UZIEL and HANAHAN, 1956) and *S. plymuthicum* (HAYAISHI and KORNBERG, 1954). SHAPIRO (1953a) has also prepared a rather pure beef pancreas enzyme.

III. Phospholipase C.

Interestingly enough this phospholipase has thus far only been found in plants and has not been demonstrated in animal tissue or microorganisms. The enzyme occurs in carrots, cabbage leaves (HANAHAN and CHAIKOFF, 1947a, b) sugar beet leaves, spinach (KATES, 1953, 1954), cottonseed (TOOKEY and BALLS, 1956), lettuce, potatoes, beet roots, brussels sprouts (DAVIDSON and LONG, 1958), and rutabaga (EINSET and CLARK, 1958). It has also been found in the latex of *Hevea brasiliensis* (SMITH, 1954) and in maple trees (DUCET, 1949). DAVIDSON and LONG reported that there was essentially no activity in dried peas, parsley, or onions and that the Savoy cabbage and Brussels sprouts had the highest activity.

The enzyme appears to be present both in the plastids and the cell sap of plants. HANAHAN and CHAIKOFF had used whole homogenates in their original work with cabbage and carrots. KATES (1954) investigating the enzyme from cabbage, spinach, sugar beet leaves and carrot roots could demonstrate the phospholipase C activity only in the plastids. The cytoplasmic fractions were inactive and in some cases inhibitory when added to the plastids. Later TOOKEY and BALLS (1956) reported the finding of a soluble phospholipase C in cottonseed meal. Then DAVIDSON and LONG (1958) found that the cell sap as well as the plastids of the Savoy cabbage contained the enzyme. The soluble enzyme was found in much higher concentration in the "heart" of the cabbage than in the outer green leaves. It could not be extracted from the isolated particles, which indicated that the soluble enzyme had not been derived from the plastids during the preparative procedure. DAVIDSON and LONG felt that the reason KATES did not detect the enzyme in the cell sap was due to its very low activity in the absence of added calcium; the plastids contain enough endogenous calcium to give reasonable activity. EINSET and CLARK (1958) reported that while the carrot enzyme was associated with the plastids it could be extracted quite easily.

Assay Procedure.

The enzyme reaction conditions are those of KATES (1955) as modified by DAVIDSON and LONG (1958). The release of choline is followed.

Reagents. 0.013 M egg lecithin emulsified in 0.1 M acetate buffer pH 5.6; 30% trichloroacetic acid; 1 M $CaCl_2$.

Transfer 1.25 ml (16 μmole) of lecithin emulsion, 0.25 ml of 1.0 M $CaCl_2$ and 1 ml of enzyme solution to a glass stoppered centrifuge tube. At zero time 1 ml of ether is added, the tube shaken vigorously, and then incubated for 10 to 12 minutes at 26°. The reaction is stopped by the addition of 1.0 ml of 30% trichloroacetic acid. Then 1.5 ml of water and 1 ml of ether is added and the mixture shaken and centrifuged. The ether phase is removed with a PASTEUR pipette and discarded. The aqueous phase is filtered and the filtrate warmed to remove dissolved ether. An aliquot of the filtrate is analyzed for choline[1]. The rate of enzymatic

---

[1] The method of choline analysis given in the phospholipase B section is satisfactory.

reaction is linear until 50% of the substrate has been hydrolyzed.The unit of activity may be defined as that amount of enzyme that catalyzes the release of 1 μmole of choline in 10 minutes.

Other Assay Procedures.

Hanahan and Chaikoff (1948) followed the reaction by measuring the decrease in ether soluble nitrogen.

pH Optimum and Stability.

The pH optima of the enzyme from all the plants so far investigated has been between 5 and 6. The enzyme is destroyed by heating at 70° for 10 minutes (Kates, 1956; Einset and Clark, 1958).

Activators.

Calcium ions activate the enzyme, the maximum rate of hydrolysis occuring in 0.1 M $CaCl_2$. Strontium and barium ions are less potent activators (Davidson and Long, 1958). Davidson and Long also report that glycerolphosphoylcholine stimulated the enzyme and this activation was independent of the hydrolysis of the GPC. Kates (1954) originally reported that the enzyme associated with the plastids was activated by ether and this has been verified by other investigators (Tookey and Balls, 1956; Davidson and Long, 1958). The activation of soluble enzyme preparations by ether has been variable. Tookey and Balls (1956) and Einset and Clark (1958) found that neither the soluble cottonseed enzyme nor the solubilized carrot plastid enzyme were activated by ether. Davidson and Long, however, reported that the soluble cell sap enzyme was activated by ether. The enzyme can inhibited be fluoride, ethylenediaminetetra-acetate, citrate, and other compounds which bind calcium.

Substrate Specificity.

Phospholipase C has been found to readily attack lecithin, PE and their corresponding plasmalogens while PS is hydrolyzed very slowly (Kates, 1954, 1955, 1956; Davidson and Long, 1958). GPC, glycollecithin and lysolecithin are not attacked (Kates, 1956).

Purification Procedures.

Partially purified enzymes have been obtained from the Savoy cabbage and cottonseed by Davidson and Long (1958) and Tookey and Balls (1956) respectively. The procedure for preparing the cottonseed enzyme is presented below.

Purification of Cottonseed Enzyme.

Step A. Cottonseed meal was extracted with hexane, and air dried. The meal was then disintegrated with water (8 to 12 ml/gm) in a colloid mill, and the pH was adjusted to 8.2 to 8.4 with NaOH. After standing in the cold for two hours the mixture was centrifuged at 800 $\times g$ to remove the larger particles.

Step B. NaCl Precipitation. The crude extract was saturated with NaCl and allowed to stand 4 hours in the cold. The precipitate was collected by a 40 minute centrifugation at 26,000 $\times g$ and dialyzed overnight against 400 volumes of cold distilled water. The precipitate does not dissolve appreciably.

Alternative Step B. Elution from calcium phosphate gel: The crude extract was treated with two volumes of Ca_3PO_4 gel at pH 6.1, and the gel eluted with

0.07 M phosphate at pH 8.2. The eluate was dialyzed overnight against 50 volumes of cold distilled water (this preparation does not precipitate on dialysis) and lyophilized.

Step C, 2nd Extraction. The content of the dialysis bag (from step B) was extracted with distilled water at a slightly alkaline pH and centrifuged 20 minutes at 26,360 ×g. The clear supernatant liquid was adjusted to pH 8.0 and lyophilized. While wet the enzyme was unstable, but dry preparations retained the activity when stored in the cold.

An 87-fold purification was obtained through step C, with an almost total recovery of units. The alternative step B gave a 40-fold purification as compared to 27 for step B, and was more stable than that obtained by step C. The alternative step B preparation was used by the authors for their investigations.

IV. Phospholipase D.

Phospholipase D was originally found in the toxin of *Clostridium welchii* (*perfringens*) (MacFarlane and Knight, 1941) and was subsequently found in *C. bifermentans* (Miles and Miles, 1950; Lewis and MacFarlane, 1953). Chu (1949) has also demonstrated the enzyme in cultures of *Bacillus cereus* and *B. mycoides*. More recently Kates (1955) has demonstrated this enzymic activity in the chloroplasts of spinach. In addition, Davidson and Long (1958) have observed the liberation of acid soluble phosphorous from lecithin by cabbage cell sap, which could indicate the presence of this enzyme in cabbage.

Assay Procedure.

The method presented is essentially that of Lewis and MacFarlane (1953) in which the acid soluble phosphorous (phosphorylcholine) produced during the enzyme reaction is estimated.

Reagents. 2.5% egg lecithin emulsion in water; Palitsch borate buffer pH 7.2 (or 0.2 M Tris buffer, Long and Maguire, 1954) containing 0.01 M $CaCl_2$; 20% trichloroacetic acid.

Procedure. Into a test tube pipette 1.0 ml of the lecithin emulsion (33 μmole), 1.0 ml of buffer and sufficient water so that when the enzyme solution is added the total reaction volume will equal 4.0 ml. The enzyme is added and the solution incubated for 30 minutes at 37°. The reaction is stopped by the addition of 1 ml of 20% trichloroacetic acid (or perchloric acid)[1], [2]. After 10 minutes the solution is filtered and the acid soluble phosphorous estimated on an aliquot of the filtrate[3]. If it is necessary to verify the type of hydrolysis, the phosphorylcholine may be identified by paper chromatography (Kates, 1955) or isolated as the calcium salt (MacFarlane and Knight, 1941). The rate of hydrolysis is apparently only linear up to the point where 5 μmoles of the lecithin has been hydrolyzed.

Other Assay Procedures.

Zamecnik, Brewster and Lipmann (1947) described a manometric procedure which appears to be quite reproducible. This is based on the fact that phosphorylcholine has a pK of 5.6 and can thus liberate CO_2 from a bicarbonate buffer.

---

[1] It may be necessary to add a small amount of bovine serum albumin to assist in removal of unreacted lecithin.

[2] This procedure could be modified to use ether in a manner similar to that for phospholipase C.

[3] The aliquot is digested with 1 ml of 70% perchloric acid for 20 minutes over a small flame, and the inorganic phosphate is determined on the digest.

HANAHAN and VERCAMER (1954) found that phospholipase D would hydrolyze lecithin in 98% diethyl ether-2% ethanol mixture or diethyl ether. They followed the reaction in this solvent by a titrimeteric technique.

pH Optima and Stability.

The phospholipase D from *Clostridium welchii* has a pH optima of 8.5, while that of *C. bifermentans* is 6.5 (LEWIS and MACFARLANE, 1953). On heating the enzyme for 10 minutes at 100° at pH 7.4 only 40—50% of the activity is lost.

Activators.

The enzyme is activated by calcium (MACFARLANE and KNIGHT, 1941) and is inhibited by calcium chelators (LONG and MAGUIRE, 1954). As mentioned above, it is also activated by ether.

Substrate Specificity.

Phospholipase D appears to be quite specific for lecithin. It does not attack GPC, phosphatidylethanolamine, phosphatidylserine or lysolecithin (ZAMENCNIK, BREWSTER and LIPMANN, 1947; HANAHAN and VERCAMER, 1954). LONG and MAGUIRE (1954) found that it slowly liberated phosphorylcholine from sphingomyelin. They also reported that it would not attack saturated lecithins, but HANAHAN and VERCAMER found that, in the presence of ether, dipalmityl-lecithin was attacked.

Purification Procedure.

There has been very little effort directed towards the purification of the enzyme but some attention has been given to the purification of the α-toxin of *Clostridium welchii*. VAN HEYNINGEN and BIDWELL (1948) and ROTH and PILLEMER (1953) did achieve partial purification of the α-toxin, unfortunately however, they did not check the enzymatic activity. It would have been of interest to see if the two activities were associated.

V. Other Enzymes Attacking Phospholipids.

1. Phosphoinositide phosphorylase.

In 1953, SLOANE-STANLEY discovered an enzyme in brain tissue which would attack diphosphoinositides yielding a diglyceride and diphosphoinositol. More recently KEMP, HÜBSCHER and HAWTHORNE (1959) found a soluble (cytoplasmic) enzyme in liver which would hydrolyze both monophosphoinositides and diphosphoinositides. The products in each case were diglycerides and the respective phosphoinositols. The enzyme did not attack lecithin, PS or lysolecithin. The reaction had an optimum pH of 5.4 and was followed by measuring the production of water (acid) soluble phosphate.

2. Phosphatidic Acid Phosphatase.

KENNEDY and WEISS (1956 a, b) had postulated the occurence of this enzyme, since phosphatidic acid played a key role in their proposed scheme for the synthesis of triglycerides and phospholipids. The L-α-phosphatidic acid would be the immediate precursor of a D-α,β diglyceride which in turn would react with cytidine diphosphocholine or ethanolamine or an acyl-Coenzyme A to yield the phospholipid or in the latter case a triglyceride.

Smith, Weiss and Kennedy (1957) have since reported the occurence of this phosphatase in avian liver, kidney and brain. The enzyme attacked phosphatidic acid yielding a diglyceride and phosphoric acid. It did not attack lecithin, but it did hydrolyze glycerophosphate and glucose phosphate very slowly. The enzyme had a broad pH optima between 5.0 and 8.0. The divalent ions Mg^{++}, Ca^{++} and Mn^{++} and the Tweens were inhibitory. The reaction was followed by estimating the release of inorganic phosphate.

3. Lysolecithin Isomerase.

Uziel and Hanahan (1957) have described an enzyme occuring in *Penicillium notatum*, rat pancreas and commercial pancreatin which isomerized saturated α-lysolecithins to β-lysolecithins[1]. The most easily discernable change during the enzymatic reaction was the formation of a gel, β-lysolecithin being less soluble in water than the original α-compound. When α-monostearoyl-GPC was used as the substrate, the isolated product had a lower melting point and a more negative optical rotation than the initial compound. The pH optimum for the enzyme was 6.0 and the reaction could be inhibited by cyanide.

References.

Appleton, H. D., B. N. La Du, B. B. Levy, J. M. Steele and B. Brodie: J. biol. Chem. 205, 803 (1953).

Bangham, A. D., and R. M. C. Dawson: Biochem. J. 72, 486 (1959a); 75, 133 (1960).

Chu, H. P.: J. gen. Microbiol. 3, 255 (1949). — Contardi, A., and A. Ercoli: Biochem. Z. 261, 275 (1933). — Contardi, A., and P. Latzer: Biochem. Z. 197, 222 (1928).

Davidson, F. M., and C. Long: Biochem. J. 69, 458 (1958). — Dawson, R. M. C.: Biochem. J. 64, 192 (1956); — Biochim. biophys. Acta 23, 215 (1957); 27, 227 (1958b); — Biochem. J. 68, 352 (1958a). — Dawson, R. M. C., and A. D. Bangham: Biochem. J. 72, 493 (1959b). — Dittmer, J. C., and D. J. Hanahan: J. biol. Chem. 234, 1976 (1959). — Ducet, G.: Ann. agron. (Paris) 19, 184 (1949).

Einset, E., and W. L. Clark: J. biol. Chem. 231, 703 (1958).

Fairbairn, D.: J. biol. Chem. 157, 633 (1945); 173, 805 (1948). — Francioli, M.: Enzymologia 3, 204 (1937).

Hanahan, D. J.: J. biol. Chem. 195, 199 (1952); 207, 879 (1954); — Progress in the chemistry of fats and other lipids. Vol. 4 p. 142. Edited by Holman, R. T., W. V. Lundberg and T. Malkin. New York: Pergamon Press 1957. — Hanahan, D. J., H. Brockerhoff, and E. J. Barron: J. biol. Chem. 235, 1917 (1960). — Hanahan, D. J., and I. L. Chaikoff: J. biol. Chem. 168, 233 (1947a); 169, 699 (1947b); 172, 191 (1948). — Hanahan, D. J., M. Rodbell and L. D. Turner: J. biol. Chem. 206, 431 (1954). — Hanahan, D. J., M. B. Turner and M. E. Jayko: J. biol. Chem. 192, 623 (1951). — Hanahan, D. J., and R. Vercamer: J. Amer. chem. Soc. 76, 1804 (1954). — Hayaishi, D., and A. J. Kornberg: J. biol. Chem. 206, 647 (1954). — Hughes, A.: Biochem J. 29, 437 (1935).

Iwata, M.: Biochem. Z. 224, 430 (1930).

Kates, M.: Nature (Lond.) 172, 814 (1953); — Canad. J. Biochem. Physiol. 32, 571 (1954); 33, 575 (1955); 34, 967 (1956). — Kemp, P., G. Hubscher and J. N. Hawthorne: Biochim. biophys. Acta 31, 585 (1959). — Kennedy, E. P., and S. B. Weiss: J. biol. Chem. 222, 193 (1956a). — King, E. J.: Biochem. J. 26, 292 (1932).

Le Breton, E., and E. J. Pantaleon: Arch. Sci. physiol. 2, 125 (1948). — Lewis, G. M., and M. G. MacFarlane: Biochem. J. 54, 138 (1953). — Long, C., and M. F. Maguire: Biochem. J. 57, 223 (1954). — Long, C., and I. F. Penny: Biochem. J. 65, 382 (1957).

MacFarlane, M. G., and B. C. J. G. Knight: Biochem. J. 35, 884 (1941). — Miles, E. M., and A. A. Miles: J. gen. Microbiol. 4, 22 (1950).

Ogawa, K.: J. Biochem. (Tokyo) 24, 389 (1936).

---

[1] It was originally reported to be isomerization of the β-lysolecithin compound to the α-isomer, but this was based on the older experimental evidence that phospholipase A attacked the α-position of lecithin. The original substrate for this enzyme was then thought to be β-lysolecithin.

Rapport, M. M., and R. E. Franzl: J. biol. Chem. **225**, 851 (1957). — Rimon, A., and B. Shapiro: Biochem. J. **71**, 620 (1959). — Roth, F. B., and L. Pillemer: J. Immunol. **70**, 533 (1953).

Shapiro, B.: Biochem. J. **53**, 663 (1953a); — Nature (Lond.) **169**, 29 (1953b). — Sloane-Stanley, G. H.: Biochem. J. **53**, 613 (1953). — Slotta, K. H., and H. L. Fraenkel-Conrat: Ber. **71**, 1076 (1938). — Smith, R. H.: Biochem. J. **56**, 240 (1954). — Smith, S. W., S. B. Weiss and E. P. Kennedy: J. biol. Chem. **228**, 915 (1957).

Tattrie, N. H.: J. Lipid Res. **1**, 60 (1959). — Tookey, H. L., and A. K. Balls: J. biol. Chem. **218**, 213 (1956).

Uziel, M., and D. J. Hanahan: J. biol. Chem. **220**, 1 (1956); **226**, 789 (1957).

Van Heyningen, W. E., and E. Bidwell: Biochem. J. **42**, 130 (1948).

Weiss, S. B., and E. P. Kennedy: J. Amer. chem. Soc. **78**, 3550 (1956b) (see Kennedy and Weiss for 1956a). — Wittcoff, H.: The phosphatides. New York: Reinhold Pubsh. Corp. 1951.

Zamecnik, P. G., L. E. Brewster and F. Lipmann: J. exp. Med. **85**, 381 (1947).

C. β-Oxidation.

By

P. K. Stumpf

I. Even Chain Fatty Acids.

As in animal tissue, the site of β-oxidation in plant appears to be localized in the mitochondrion (STUMPF and BARBER, 1956). The literature on this subject has been reviewed (STUMPF and BRADBEER, 1959).

Preparation. A mitochondrial suspension is obtained according to the procedure of STUMPF (1958).

Reaction mixture. A typical reaction mixture contains 10 micromoles of phosphate buffer at pH 7.1, 0.1 micromole of fatty acid labeled in the carboxyl group with C^{14} (approx. count for $C^{14}O_2$ run should be about 10,000 *cpm*), 1 micromole of α-ketoglutaric acid, 1 micromole of manganese sulfate, 5 micromoles of glutathione 0.15 micromole of diphosphyridine nucleotide (DPN), 0.13 micromole of TPN, 0.6 micromole of coenzyme A, 1 micromole of ATP, 0.5 ml of fresh mitochondrial suspension (adjusted to contain about 10 mg protein) in 0.1 M Tris/0.4 M sucrose at pH 7.3 with about 5×10^{-3} % BAL, 0.2 ml 20% KOH in the center well, 0.3 ml 5 M H_2SO_4 in the side arm; reaction mixture to a final vol. of 2.2 ml in each vessel. Time of incubation 2 hr., temperature 25° C.

At the end of incubation period, acid is tipped from the side arm into the main compartment to stop the reaction and release bound CO_2. Shaking is continued for 10 minutes, after which the content of the center well are removed with washing and the respiratory CO_2 is precipitated as $BaCO_3$ with 20% barium acetate, washed 3 times with 15 ml of 50% ethanol, plated on an aluminum disk and counted with a thin window GEIGER-MÜLLER tube.

Properties. Oxygen uptake studies of $5-10$ μmole of unlabelled fatty acids with mitochondral preparations indicate low and varible oxidation. Indeed, it appears as if these concentrations of fatty acids are strikingly inhibitory of the complete system. Therefore, it is essential to reduce the substrate level to 0.1 μmole and increase the sensitivity of the assay system by C^{14} counting techniques. Under these conditions consistant results can be obtained.

The age of the seedling is critical, the maximum activity being observed on the 7th—9th day after planting of the seed. β-oxidation has been observed in mitochondria derived from peanut seedlings (STUMPF and BARBER, 1956) pine pollen (STANLEY and CONN, 1957), avocado mesocarp (NANCE and STUMPF, 1960). The measurement of respiratory CO_2 implies an active tricarboxlic acid cycle to metabolize the acetyl CoA units derived from the fatty acids. Therefore, in all prelimary work optimal conditions and cofactor requirements for oxidation of acetate-1-C^{14} should be determined before longer chain fatty acids are examined.

Substrate specifity. Carboxyl labeled aliphatic acids, from acetate to stearic acid, readily release their C^{14} as $C^{14}O_2$ under the described condition employing peanut mitochondria. In addition, internally labeled fatty acids such as palmitic

2, 3, 11 and 15 yield $C^{14}O_2$ data consistant with the concept of β-oxidation. Further-more, the dicarboxylic and tricarboxylic acids also become labeled in systems employing fatty acid-1-C^{14} substrates. The described system can be therefore employed as a starting point in the further elucidation of β-oxidation systems.

II. Odd-Chain Fatty Acids.

Little is known about the occurence of odd chain fatty acids in plant material. However, information is available concerning the oxidation of propionic acid in plants. The sequence involved (1) propionic acid $\xrightarrow[\text{CoA}]{\text{ATP}}$ propionyl CoA $\xrightarrow{-2H}$ acrylyl CoA $\xrightarrow{H_2O}$ β-hydroxypropionyl CoA $\xrightarrow{-CoA}$ β-hydroxypropionic acid $\xrightarrow{-2H}$ malonic semi aldehyde $\xrightarrow[-2H]{\text{CoA}}$ malonyl CoA $\longrightarrow$ acetyl CoA + Co_2. Acetyl CoA then can enter the Krebs cycle and be further metabolized.

Procedure. The method for the preparation of the mitochondrial particles is described by Giovanelli and Stumpf (1958) and follows essentially a standard procedure for preparation of mitochondrial particles. The reaction mixture is similar to that for the β-oxidation system (see section C I).

Properties. ATP and CoA are necessary cofactors. DPN, TPN, GSH, α-keto-glutaric acid and Mn^{++} stimulated oxidation to varying degrees depending on the preparation. K_m for ATP is 5.9×10^{-5} M and for CoA 7×10^{-5} M with propionic acid as substrate. The optimum pH for this complex sequence is pH 7.0. The enzyme system is unstable since mitochondrial suspensions in 0.5 M sucrose-0.2 M Tris buffer pH 7.2 have a half life of 1 hr. at 25° and 2h ours at 0°. Dinitro-phenol at a final concentration of 6×10^{-5} M completely inhibits oxidation.

This system has been observed in avocado and lupine mitochondria. In addition, when propionic acid labeled in different carbons with C^{14} was added to slices of sunflower or peanut cotyledons, the rate of formation of $C^{14}O_2$ was identical to that observed with isolated mitochondria. A transitory formation of β-hydroxypropionic acid could also be detected indicating that *in vivo* propionic metabolism follows the pathway described in *in vitro* experiments. Recently Griffin et al. (1960) have implicated propionic acid as a precursor of the pyridine ring in tobacco tissue.

References.

Giovanelli, J., and P. K. Stumpf: J. biol. Chem. **231**, 411 (1958). — Griffin, T., K. P. Hillman and R. V. Byerrum: J. biol. Chem. **235**, 800 (1960).
Nance, J., and P. K. Stumpf: unpublished data 1960.
Stanley, R. G., and E. E. Conn: Plant Physiol. **32**, 412 (1957). — Stumpf, P. K.: Plant Physiol. **30**, 55 (1958). — Stumpf, P. K., and G. A. Barber: Plant Physiol. **31**, 304 (1956). — Stumpf, P. K., and Clive Bradbeer: Ann. Rev. Pl. Physiol. **10**, 197 (1959).

D. α-Oxidation.

By

P. K. Stumpf.

Although the principle oxidative pathway for short and long chain fatty acids is probably by the β-oxidation scheme, in plants there is a unique mechanism for the stepwise oxidation of saturated fatty acids (C_{14}—C_{18} fatty acid) by an α-oxidation scheme (STUMPF, 1956; MARTIN and STUMPF, 1959). The system can be described as being made up of two enzymes, (1) a fatty acid peroxidase and (2) a long chain aldehyde dehydrogenase linked together by the following sequences:

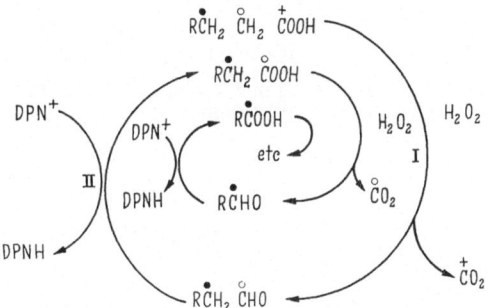

α-Oxidation of long chain fatty acids

Preparation. Acetone powders of cotyledons of germinated peanuts (6 days) were prepared as follows. Freshly harvested cotyledons were ground in a WARING Blendor for 40 seconds with 10 times their weight of cold acetone (—5°), filtered by suction, and the powder submitted to the same process again. Then, 200 ml of fresh ethyl ether was poured over the material on the filter paper and the suction applied until the pad was nearly dry. The solvent-damp powder was transferred to a vacuum desiccator and vacuum applied until all traces of ether were gone. The dried acetone powder was stored at —10° in sealed vials. The acetone powder was extracted with 15 ml of 0.1 M phosphate buffer at pH 7.5, and is referred to as the peroxidase system.

Glucose oxidase was purchased from Worthington Biochemical Corporation, New York City. An aqueous solution containing 1 mg of protein per ml was stored at —5° for several months without appreciable loss in activity.

Reaction mixture. The rate of the reaction could be determined by assaying for release of $C^{14}O_2$ from a carboxyl-labeled fatty acid or by detecting the formation of long chain aldehydes under suitable conditions. The first method can for the most part be employed since it does not require photometric equipment.

30*

Each Warburg cup contains 0.3 ml of the enzyme system (about 3 mg protein), 0.1 μmole of palmitic-1-C^{14} (5000 cpm), 50 μmole phosphate buffer at pH 7.3, 5 μmoles of glucose oxidase (0,05 mgm of protein) 0.15 μmole of DPN^+, incubated at 25° for one hour.

Properties and inhibitors. The peroxidase system is stable at $-5°$, the initial activity declining about 10% in 3 months. Fatty acid peroxidase in completely destroyed by 5 minutes exposure to 55° in the pH range of 5.5—8.5. With phosphate buffers the pH optimum for fatty acid peroxide is 7.5. K_m for palmitate is 9.1×10^{-5} M. Substrate specificity is quite narrow extending from the C^{13} saturated fatty acid to stearic acid with the maximum activity observed with pentadecanoic acid. Acids lower than lauric acid are not attacked by the peroxidase.

Of the several inhibitors tested, carbon monoxide (50% inhibition by 95% CO : 5% O_2 at 1 atm. pressure), cyanide (90% at 2.5×10^{-3} M), imidazole (86% at 5×10^{-4} M) and 1,2,4-Triazole (64% at 5×10^{-4} M) were effective. Azide also inhibits strongly. Fluoride, arsenate, arsenite, Cu^{++} and malonate are inert. Since the effective inhibitors are known porphyrin bound iron complexing agents, it would appear that a heme-like component is involved in the peroxidation reaction. However, horse radish peroxidase and catalase are inert in the test system indicating that the specifity of the reaction is related to a specific protein.

References.

Martin, R. O., and P. K. Stumpf: J. biol. Chem. **234**, 2549 (1959).
Stumpf, P. K.: J. biol. Chem. **233**, 643 (1958).

E. Lipoxidase.

By

A. L. Tappel.

Lipoxidase is an enzyme, thus far found only in plants, which catalyzes the oxidation of unsaturated fatty acids. It specifically requires *cis,cis*-1,4 diolefins such as linoleic, linolenic and arachidonic acids. In this direct addition of molecular oxygen the product is a *cis, trans*-1,3 diolefin hydroperoxide.

1. Assay Method.

Principle. Lipoxidase activity can be determined by manometric or polarographic measurement of oxygen consumption, titration of hydroperoxides or spectral absorbance of the conjugated double bonds (HOLMAN, 1955). The conjugated linoleate hydroperoxides produced by lipoxidase catalysis have strong absorbance. Therefore, a direct spectrophotometric assay employing increase in absorbance at 232.5 mµ as a function of time provides a method as simple and accurate as that of oxygen consumption (TAPPEL, LUNDBERG and BOYER, 1953). Since each method has a number of advantages, both can be recommended.

Procedure.

a) Manometric.

Stock 0.1 M ammonium linoleate is prepared by the addition of an equivalent of ammonium hydroxide to linoleic acid suspended in water. During manipulation and frozen storage the ammonium linoleate should be kept under pure N_2 gas to protect it from oxidation. Three ml of 1×10^{-4} M ammonium linoleate in 0.1 M phosphate buffer pH 7.0 is placed in the manometric flask and lipoxidase in 0.1 to 0.5 ml is placed in the side arm. After equilibration to 20° C in air, mix the lipoxidase and linoleate and record oxygen consumption at 5 minute intervals for 30 minutes.

Definition of activity. Following the suggestion of the International Union of Biochemistry, 1 unit of lipoxidase is the amount which will oxidize 1 micromole of linoleate per minute. In terms of oxygen consumption, 1 unit is equal to 1 micromole of oxygen per minute.

b) Spectrophotometric.

A 7×10^{-3} M solution of ammonium linoleate in 0.1 M NH_4OH—NH_4Cl buffer pH 9.0 is oxygenated with pure O_2 at 20° C and placed in a 1 mm light path silica cell. If a 1 cm silica cell with 0.9 cm spacer is used, about 0.9 ml of the linoleate solution will be required. The spectrophotometer with the thermostated cell compartment at 20° C is set for measurements of increasing absorbance at 232.5 mµ. At zero time 0.1 ml of the lipoxidase solution is added to the linoleate and the absorbance is measured at 15 second intervals for 3 to 5 minutes. The lipoxidase concentration should be adjusted to give rates of 0.05 absorbance per minute. At this rate the increased absorbance is linear with time for at least 6 minutes and will not exhaust the dissolved O_2 for about 25 minutes. This method has been developed for 7×10^{-3} M linoleate which gives slightly turbid solutions and requires a short light path. Linoleate concentrations of 1×10^{-3} M allow the use of a 1 cm light path.

Definition of activity. One unit of lipoxidase is the amount which will oxidize 1 micromole of linoleate per minute. Conversion of absorbance at 232.5 mμ to linoleate oxidized requires the molecular extinction coefficient for the linoleate oxidation products. Since soybean lipoxidase produces conjugated diene hydroperoxides of *cis-trans* configuration, PRIVETT, NICKELL, LUNDBERG and BOYER (1955) suggested that the ϵ_M cannot be theoretically greater than 28×10^3 absorbance per mole of the hydroperoxide. Experimentally determined ϵ_M values are 27.4×10^3 and 27.3×10^3 for hydroperoxides from soybean (TAPPEL, BOYER and LUNDBERG, 1952) and pea (SIDDIQI and TAPPEL, 1956) lipoxidase, respectively.

Precautions. Both assays are applicable to crude preparations from plant sources such as beans and peas which have a relatively high lipoxidase content. Care must be exercised in the assay of sources low in lipoxidase such as wheat and peanuts because of the interference of other lipid peroxidation catalysts, especially the cytochromes. In all cases the presence of lipoxidase should be verified. Heat inactivated controls should be run. A simple specificity test is that lipoxidase will oxidize linoleic, linolenic and arachidonic acids but not oleic acid (HOLMAN and BERGSTROM, 1951). Tests to differentiate lipoxidase activity from the lipid peroxidation activity of cytochromes and other hemoproteins are available (TAPPEL, 1953; BOYD and ADAMS, 1955).

2. Purification Procedure.

HOLMAN and BERGSTROM (1951) report the following method for the preparation of crystalline soybean lipoxidase. Fifteen kg of fat-free soybean flour is extracted with 100 l of 0.1 M acetate buffer, pH 4.5. After the centrifuged extract is adjusted to pH 6.7 with ammonia, 5 volumes 20% barium acetate, 10 volumes acetone and 2 volumes of basic lead acetate are added per 100 volumes of extract. This inactive precipitate is removed and then 25 g ammonium sulfate per 100 ml extract is added to precipitate more inactive protein; more ammonium sulfate is added to bring its concentration to 40 g/100 ml and an active fraction precipitates. This precipitate is redissolved in a small quantity of water and heated to 63° C for 5 minutes to coagulate inactive albumins. This supernatant solution is fractionated again with ammonium sulfate and the fraction between 35% and 50% saturation is retained. This is fractionated with ethanol at 0° C in 0.02 M phosphate buffer at pH 5.5 and the initial precipitate below 12% alcohol contains the activity. After another ammonium sulfate fractionation between 50% and 60% saturation, the lipoxidase preparation is separated by large scale electrophoresis. After concentration, the lipoxidase is dialyzed against increasing concentrations of ammonium sulfate. The lipoxidase crystallizes out. It has been found that this isolation of lipoxidase is not strictly reproducible with different batches of soybeans and therefore, HOLMAN and BERGSTROM recommend pilot experiments and continual assay of the fractions.

3. Properties.

Competitive inhibitors of lipoxidase include conjugated 10,12-octadecadienoic acid and all *trans* 9,12,15-octadecatrienoic acid. The products of linoleic acid oxidation are more complex than simple hydroperoxides. For example, soybean lipoxidase gave mainly conjugated *cis, trans* hydroperoxide but also 5% polymer.

Free radical intermediates occur during lipoxidase catalysis and these can lead to cooxidation of easily oxidized compounds, for example, carotenoids and polyphenols. Lipoxidase is strongly inhibited by lipid antioxidants; nordihydro-guaiaretic acid, propyl gallate and α-tocopherol have been mainly studied (TAPPEL, 1961). No prosthetic group has been found for soybean, pea, wheat or peanut lipoxidase. Urd and mung bean lipoxidases are inhibited by sulfhydryl reagents suggesting a requirement for free thiol groups (SIDDIQI and TAPPEL, 1957).

Most lipoxidases show an optimum pH between 6.5 and 7.0. The rate of the reaction is dependent on the concentration of linoleate and oxygen. The K_m for linoleate is 1×10^{-3} M for soybean (HOLMAN and BERGSTROM, 1951) and barley (FRANKE and FREHSE, 1953) lipoxidases and 5×10^{-6} M for that of wheat (IRVINE and ANDERSON, 1953). At 3.6×10^{-4} M linoleate and K_m for oxygen is 3×10^{-4} M and at 7.2×10^{-3} M linoleate it is 3×10^{-4} M. The activation energy for soybean lipoxidase is 4.3 K cal/g mole. Soybean lipoxidase is stable for months at $-20°$ C but pea lipoxidase preparations are unstable.

References.

BOYD, D. H., and G. A. ADAMS: Canad. J. Biochem. Physiol. **33**, 191 (1955).

FRANKE, W., and H. FREHSE: Hoppe-Seylers Z. physiol. Chem. **295**, 333 (1953).

HOLMAN, R. T., and S. BERGSTROM: In The enzymes (J. B. SUMNER and K. MYRBACK, eds.) Vol. II, Part 1, New York: Academic Press 1951. — HOLMAN, R. T.: Meth. Biochem. Analys. **2**, 113 (1955).

IRVINE, G. N., and J. A. ANDERSON: Cereal Chem. **30**, 247 (1953).

PRIVETT, O. S., C. NICKELL, W. O. LUNDBERG and P. D. BOYER: J. Amer. Oil Chem. Soc. **32**, 505 (1955).

SIDDIQI, A. M., and A. L. TAPPEL: Arch. Biochem. Biophys. **60**, 91 (1956); — J. Amer. Oil Chem. Soc. **34**, 529 (1957).

TAPPEL, A. L., P. D. BOYER and W. O. LUNDBERG: J. biol. Chem. **199**, 267 (1952). — TAPPEL, A. L., W. O. LUNDBERG and P. D. BOYER: Arch. Biochem Biophys. **42**, 293 (1953). — TAPPEL, A. L.: Food Res. **18**, 104 (1953); — In Autoxidation and antioxidants (W. O. LUNDBERG, ed.) New York: Interscience Publishers 1961.

F. Synthesis of Fatty Acids.

By

Edward J. Barron.

The enzyme synthesis of fatty acids has been studied for the most part in the fruit of the Fuerte or McArthur variety of the avocado (Stumpf and Barber, 1957 (1959); Mudd and Stumpf, 1961; Barron, Squires and Stumpf, 1961). The synthesis occurs in a particulate fraction of the cell that has 'all the properties of typical mitochondria i. e., sediments at the correct centrifugal force, has Krebs cycle activity, carries out oxidative phosphorylation (Biale et al., 1957) and fatty acid oxidation (Nance et al., 1960). The mechanism of synthesis by these particles was very similar to that demonstrated by Wakil et al. (1959, 1960) for the soluble (cytoplasmic) avian liver system. The cofactors required for synthesis in both systems when acetate is the substrate were ATP, CoA, TPNH, Mn^{++}, and HCO_3^-. In both instances malonyl CoA has been shown to be an intermediate in the biosynthesis of the fatty acids.

1. Preparation of Particles.

Medium soft Fuerte or McArthur variety of avocados are peeled, deseeded and the mesocarp is homogenized, with a volume of 0.25 M reagent grade sucrose equal to the weight of the mesocarp, in a Waring blender (10,000 rpm) for 20 seconds. One-third the original volume of sucrose is added and the homogenization repeated. This produces a smooth paste which is centrifuged at 650 $\times g$ for 10 minutes. Three layers are produced: the sediment, a "clear middle layer" and an upper thick lipid layer. The middle layer is siphoned off and saved. The sediment and lipid layers are washed with 0.25 M sucrose and centrifuged and the middle layer again saved. The pooled "middle" fractions are then centrifuged at 12,000 to 15,000 g for 30 minutes. The sedimented material may be suspended in sucrose (1 ml per 100 gm of original mesocarp) and tested directly for activity or an acetone powder can be prepared from it.

For the preparation of the acetone powder, the particles are suspended in enough sucrose to make a thin paste; by the use of a pipette with a narrow orfice, the sucrose suspension is blown into acetone at -15 to $-20°$ C (1000 ml for every 10 ml of sucrose) which is being vigorously stirred. The protein is allowed to settle and the supernatant solution is decanted off. The protein layer is then washed 2 times with cold acetone. The protein is collected in a cold room on a Büchner funnel and washed 2 times with $-20°$ C ether. The precipitate is transferred to a dessicator and dried *in vacuo* over silica gel or Drierite for 12—14 hours. The dry powder is stable for at least 6 months if stored at $-15°$ C.

2. Extraction of Acetone Powder.

200 mg of the acetone powder is suspended in 1.7 ml of 0.2 M phosphate buffer pH 7.1 and centrifuged at 20,000 $\times g$ for 10 minutes. The clear supernatant

(there are usually some pectin particles floating on the surface) is used as the enzyme solution. This ratio of powder to buffer yields an extract with approximately 6 mg protein per 0.5 ml extract.

3. Assay.

The enzymatic synthesis of fatty acids is assayed by measuring the incorporation of acetate —1-C^{14} into chloroform soluble substance. The reaction is carried out in a glass stoppered centrifuge tube and the final reaction volume is 1 ml and contains, 5 μmoles of ATP, 0.133 μmole CoA, 0.067 μmole TPN, 1 μmole of glucose-6-PO_4, 1 μmole Mn^{++}, 30 μmole HCO_3, and 0.1 μmole acetate-1-C^{14} (and when intact particles are used, 75 μmole of Tris buffer, pH 8.0). Either 0.5 ml of the buffered acetone powder extract or 0.3 ml of mitochondrial suspension is added and the reaction mixture incubated 90 minutes at 31° C.

The reaction is stopped by the addition of 0.1 ml of 3 N HCl and the lipid extracted by the monophasic system of DYER (1959). 2.5 ml of methanol and 1.25 ml of chloroform are added and the solution mixed. An additional 1.25 ml of chloroform and 1.25 ml of distilled water are then added. The lipids are extracted by vigorous shaking and the solution is centrifuged. The lower chloroform layer is removed by the use of a PASTEUR pipette and taken to dryness at 40° C under a stream of nitrogen. (This removes any unreacted acetate-1-C^{14} which may have been extracted.) The extracted material is then redissolved in 5 ml of chloroform: methanol (1:2) and an aliquot counted. If desired, the fatty acids may tentatively be identified by the paper chromatographic procedures of MANGOLD, LAMP and SCHLENK (1955) and BUCHANAN (1959). As most of the synthesized fatty acids are found esterified as glycerides and phospholipids, the lipid sample should be hydrolized before chromatography.

4. pH Optima and Stability.

The intact particles system has a pH optimum around 8.0. The solubilized system, prepared from the acetone powder, has a rather sharp pH optimum around 7.1. Both the intact particles and the soluble enzyme system are labile and should be handled in the cold and used immediately.

References.

BARRON, E. J., C. SQUIRES, and P. K. STUMPF: J. biol. Chem. **236**, 2610 (1961). — BIALE J. B., R. E. YOUNG, C. S. POPPER and W. E. APPLEMAN: Physiol. Plant. **10**, 48 (1957).— BLIGH, E. G., and W. J. DYER: Canad. J. Biochem. Physiol. **37**, 911 (1959). — BUCHANAN, M. A.: J. Anal. Chem. **31**, 616 (1959).

MANGOLD, H. K., B. G. LAMP and H. SCHLENK: J. Amer. chem. Soc. **77**, 6070 (1955). — MUDD, J. B., and P. K. STUMPF: J. biol. Chem. **236**, 2602 (1961).

NANCE, J. and P. K. STUMPF: personal communication 1960.

STUMPF, P. K., and G. A. BARBER: J. biol. Chem. **227**, 907 (1957). — STUMPF, P. K., and C. BRADBEER: Ann. Rev. Pl. Physiol. **10**, 197 (1959).

WAKIL, S. J., F. B. TITCHENER, D. M. GIBSON: Biochim. Biophys. Acta **34**, 227 (1959). — WAKIL, S. J., and D. M. GIBSON: Biochim. Biophys. Acta **41** 122 (1960).

Enzymes of Carbohydrate Synthesis.

By

David S. Feingold, Elizabeth F. Neufeld, and W. Z. Hassid.

With 1 Figure.

Complex saccharides comprise the bulk of the organic matter of plants. Some serve as structural elements (e. g. cellulose, hemicellulose, mannan, chitin), some as energy storage products (e.g. trehalose, sucrose, starch), while others have such diverse functions as intercellular binding (pectin), sealing of wounds (callose, gums, and mucilages), and solubilization of otherwise water-insoluble substances (glycosides). This chapter will deal with the detection and purification of enzymes involved in the formation of complex saccharides and of their precursors. The enzymes of the glycolytic pathway and of the hexose-monophosphate shunt are covered in other chapters of this volume and therefore will not be discussed. Neither will fungal transglycosylases, which are primarily hydrolytic in action, be included.

The formation of starch by preparations from potatoes represents the first enzymatic synthesis of a plant polysaccharide. Following the studies of the CORIs on muscle phosphorylase showing the formation of an amylose-like polysaccharide from α-D-glucose 1-phosphate (CORI, CORI and SCHMIDT, 1939), HANES (1940) demonstrated the formation of amylose (the linear component of starch, in which the D-glucose residues are joined by α-1,4 linkages) from α-D-glucose 1-phosphate in the presence of extracts from potatoes. Subsequent work demonstrated the presence of phosphorylase in a great number of plants.

The ubiquity of starch phosphorylase suggested that similar phosphorylases might catalyze the synthesis of other complex saccharides. This view was supported by the discovery of sucrose synthesis from α-D-glucose 1-phosphate and D-fructose catalyzed by extracts from *Pseudomonas saccharophila* (DOUDOROFF, KAPLAN and HASSID, 1943). However, in spite of intensive search, no phosphorylase which could catalyze the formation of a saccharide other than starch has been found in plants.

The results of photosynthetic experiments by BUCHANAN et al. (1952) suggested that the precursor of sucrose in green algae was not α-D-glucose 1-phosphate, but UDP-D-glucose[1]. Shortly thereafter, LELOIR and coworkers (CARDINI, LELOIR and

---

[1] The following abbreviations are used: ATP, adenosine 5'-triphosphate; ADP, adenosine 5'-diphosphate; AMP, adenosine 5'-monophosphate; CTP, cytidine 5'-triphosphate; GTP, guanosine 5'-triphosphate; GDP, guanosine 5'-diphosphate; GMP, guanosine 5'-monophosphate; UTP, uridine 5'-triphosphate; UDP, uridine 5'-diphosphate; UMP, uridine 5'-monophosphate; DPN, diphosphopyridine nucleotide; TPN, triphosphopyridine nucleotide; DPNH, reduced diphosphopyridine nucleotide; TPNH, reduced triphosphopyridine nucleotide; P, phosphate; PP, pyrophosphate; EDTA, ethylene diaminetetraacetic acid; Tris, tris(hydroxy-methyl)aminomethane; DEAE-cellulose, diethylaminoethyl-cellulose.

Table 1. *Enzymes of Carbohydrate Synthesis in Fungi and Spermatophytes.*

Reaction Number	Name of Enzyme	Source	Reference	Section
1	Hexokinase	f	BERGER, SLEIN, COLOWICK and CORI (1945)	
		s	SALTMAN, 1953	
2	Phosphoglucomutase	s	CARDINI, 1951	
3	Phosphorylase	f	MEYER and BERNFELD, 1942	
		s	HANES, 1940; WHELAN, 1958	D IV 1
4	UDP-D-glucose pyro-phosphorylase	f	MUNCH-PETERSEN, KALCKAR, CUTOLO and SMITH, 1953	D IV 3
		s	GINSBURG, 1958	D IV 2
5	UDP-D-glucose-β-1,3-D-glucan transglucosylase	s	FEINGOLD, NEUFELD and HASSID, 1958b	D I
6	UDP-D-glucose-α-1,4-D-glucan transglucosylase	f	ALGRANATI and CABIB, 1960	
		s	RONGINE DE FEKETE, LELOIR and CARDINI, 1960	
7	a) UDP-D-glucose-D-fructose transglucosylase	s	CARDINI, LELOIR and CHIRIBOGA, 1955	B I 1
	b) UDP-D-glucose-D-fructose 6-phosphate transglucosylase	s	LELOIR and CARDINI, 1955	BI 2
8	UDP-D-glucose-D-glucose 6-phosphate trans-glucosylase	f	CABIB and LELOIR, 1955	B II
9	a) UDP-D-glucose-di-phenol transglucosylase	s	YAMAHA and CARDINI, 1960a	C I
	b) UDP-D-glucose-phenol D-glucoside trans-glucosylase	s	YAMAHA and CARDINI, 1960b	C II
10	UDP-D-galactose 4-epi-merase	f	LELOIR, 1951; MAXWELL and ROBICHON-SJULMAJSTER, 1960	A V 1 a
		s	NEUFELD et al., 1957	A V 1 b
11	a) UDP-D-galactose pyro-phosphorylase	f	KALCKAR, BRAGANCA and MUNCH-PETERSEN, 1953	
		s	NEUFELD et al., 1957	A IV 1
	b) α-D-galactose 1-phos-phate uridyl transferase	f	KALCKAR, BRAGANCA and MUNCH-PETERSEN, 1953	A IV 7
12	D-galactokinase	f	TRUCCO, CAPUTTO, LELOIR and MITTELMAN, 1948	A II 1
		s	NEUFELD, FEINGOLD and HASSID, 1960	A II 3
13	UDP-D-glucose dehydro-genase	s	STROMINGER and MAPSON 1957	A V 2
14	UDP-D-glucuronic acid pyrophosphorylase	s	FEINGOLD, NEUFELD and HASSID, 1958a	A IV 1
15	D-glucuronic acid kinase	s	NEUFELD, FEINGOLD and HASSID, 1959	A III 2
16	UDP-D-galacturonic acid 4-epimerase	s	FEINGOLD, NEUFELD and HASSID, 1960	A V 1 b 3
17	UDP-D-galacturonic acid pyrophosphorylase	s	FEINGOLD, NEUFELD and HASSID, 1958a	A IV 1
18	UDP-D-glucuronic acid decarboxylase	s	FEINGOLD, NEUFELD and HASSID, 1960	A V 3
19	UDP-L-arabinose 4-epi-merase	s	NEUFELD et al., 1957	A V 1 b 2
			FEINGOLD, NEUFELD and HASSID, 1960	
20	UDP-L-arabinose pyro-phosphorylase	s	NEUFELD et al., 1957	A IV 1

Table 1 (continued).

Reaction Number	Name of Enzyme	Source	Reference	Section
21	L-arabinokinase	s	NEUFELD, FEINGOLD and HASSID, 1960	A III 3
22	UDP-D-xylose-D-xylo-dextrin transxylosylase	s	FEINGOLD, NEUFELD and HASSID, 1959	D III
23	D-glucosamine 6-phosphate synthetase	f	LELOIR and CARDINI, 1953	
			BLUMENTHAL, HOROWITZ, HEMERLINE and ROSEMAN, 1955	
24	D-glucosamine 6-phosphate acetylase	f	DAVIDSON, BLUMENTHAL and ROSEMAN, 1957	
25	Phospho-N-acetyl-D-glucosamine mutase	f	REISSIG, 1956	
26	UDP-N-acetyl-D-glucosamine pyrophosphorylase	f	MUNCH-PETERSEN et al., 1955	A IV 5
			GLASER and BROWN, 1955	
		s	HASSID, NEUFELD and FEINGOLD, 1959	A IV 4
27	UDP-N-acetyl-D-glucosamine-chitin transglycosylase	f	GLASER and BROWN, 1957	D II

f = fungi, s = spermatophytes.

CHIRIBOGA, 1955; LELOIR and CARDINI, 1955) were able to show that extracts from wheat germ and various other plants catalyze the formation of sucrose and sucrose phosphate from UDP-D-glucose and D-fructose or D-fructose 6-phosphate, respectively. Since then, a large number of transglycosylases have been prepared from bacterial, animal, and plant sources, capable of catalyzing the synthesis of a variety of saccharides from UDP-D-glucose and other sugar nucleotides in the presence of appropriate acceptors (see reviews by HASSID, NEUFELD and FEINGOLD, 1959; STROMINGER, 1960).

The role of sugar nucleotides is not limited to their function as glycosyl donors. These compounds also participate in numerous transformations of the sugar moiety, such as 4-epimerisation, oxidation, or decarboxylation. Fig. 1 summarizes the enzymatic reactions discovered in higher plants and fungi by which free sugars are phosphorylated, incorporated into sugar nucleotides, transformed, and finally transferred to yield complex saccharides.

Although thus far only a few of the oligo- and polysaccharides found in plants have been enzymatically synthesized, it is likely that many others (e.g. raffinose, pectin, hemicellulose) are formed by further ramifications of the sugar nucleotide pathway shown in Fig. 1. Cellulose, the most abundant carbohydrate of the plant kingdom, has been synthesized from UDP-D-glucose by preparations from the bacterium *Acetobacter xylinium* (GLASER, 1958); plant cellulose may be synthesized by a similar mechanism.

Evidence for the in vivo operation of the sugar nucleotide pathway is supplied by physiological experiments performed in many laboratories (ALTERMATT and NEISH, 1956; NEISH, 1958; SLATER and BEEVERS, 1958; LOEWUS, JANG and SEEGMILLER, 1958). These workers fed hexoses, pentoses, and uronic acids labeled in specific positions with C^{14} and examined the labelling pattern of monosaccharide residues of cell wall polysaccharides. The distribution of label observed is that which would be expected on the basis of the reactions described in Fig. 1. In

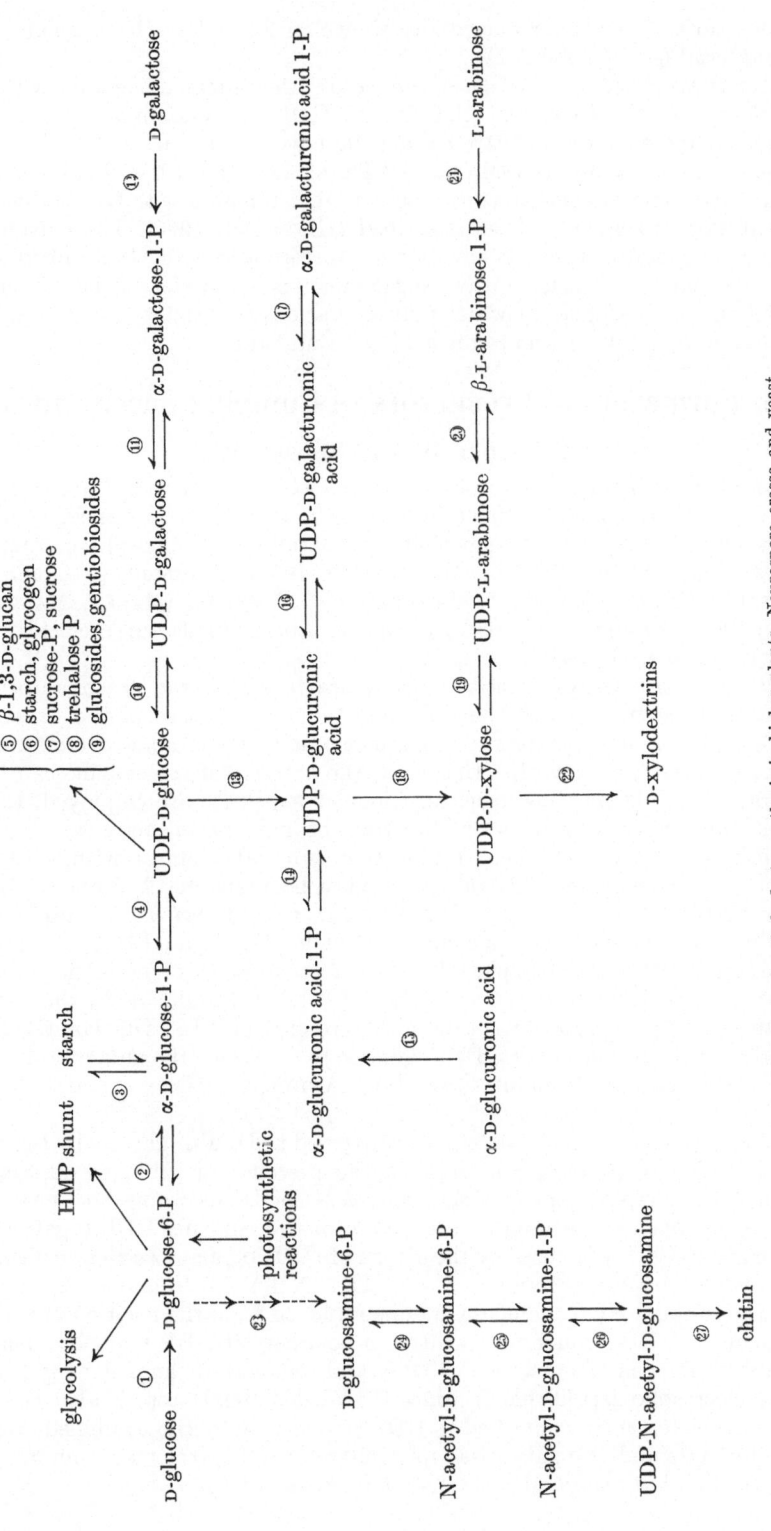

Fig. 1. Reaktions of carbohydrate synthesis in higher plants, Neurospora crassa and yeast. For names of enzymes and references consult list of Reaction Numbers on p. 475 and 476.

addition, most of the sugar nucleotides shown in Fig. 1 have been isolated from plant material (see Section A IV).

There is also evidence for the existence of other sugar nucleotide pathways, particularly for one based on GDP-sugars. GDP is found in nature linked to D-mannose, and to several sugars having the L-galactose configuration (L-fucose, L-galactose, and colitose) (GINSBURG and KIRKMAN, 1958; SU and HASSID, 1960; HEATH, 1960). The enzymatic conversion of GDP-D-mannose to GDP-L-fucose by bacterial extracts has also been described (GINSBURG, 1960). The existence of numerous enzymatic reactions involving transformations of GDP-linked sugars and the subsequent transfer of their sugar moieties is therefore to be anticipated, especially in marine algae, in which polysaccharides containing L-fucose, L-galactose, D-mannose and D-mannuronic acid are abundant.

A. Formation of Precursors of Complex Saccharides.

I. Preparation of Substrates.

Most of the substrates used in the reactions which are discussed in this chapter are commercially available. In addition, methods have been described for the preparation of α-D-xylose 1-phosphate, β-L-arabinose 1-phosphate, and α-D-galactose 1-phosphate (PUTMAN and HASSID, 1957), α-D-mannose 1-phosphate (COLOWICK, 1938), α-N-acetyl-D-glucosamine 1-phosphate (MALEY, MALEY and LARDY, 1956), α-D-glucuronic acid 1-phosphate and α-D-galacturonic acid 1-phosphate (BARKER et al., 1958).

Several of the assays described below are based on the use of radioactive substrates. In some cases this technique is required because of the expense or inconvenience involved in the use of sufficient substrate to allow detection of the product by other methods. In other cases the activity of the enzyme can be demonstrated only if very low concentrations of substrate are employed, thereby necessitating the use of radioactive substrate of high specific activity.

Synthesis of most of these radioactive substrates starts with C^{14}-labeled α-D-glucose 1-phosphate. This substance may be prepared: 1. from C^{14}-labeled sucrose by the action of sucrose phosphorylase from *Pseudomonas saccharophila* in the presence of inorganic phosphate (WOLOCHOW et al., 1949), 2. from C^{14}-labeled starch by the action of phosphorylase in the presence of inorganic phosphate (McCREADY and HASSID, 1957), and 3. from C^{14}-labeled D-glucose by the action of hexokinase and phosphoglucomutase in the presence of ATP. This last reaction is coupled with the formation of UDP-D-glucose to overcome the unfavorable equilibrium of the phosphoglucomutase reaction (ANDERSON, MAXWELL and BURTON, 1959).

The α-D-glucose 1-phosphate can be converted to UDP-D-glucose by the action of UDP-D-glucose pyrophosphorylase in the presence of UTP, as described in Section A IV. Preparations of *Zwischenferment* made according to LEPAGE and MUELLER (1949) may be used as the source of the enzyme. UDP-D-galactose is prepared from UDP-D-glucose by the action of UDP-D-galactose 4-epimerase from *Saccharomyces fragilis* (Section B V a 1. The $(NH_4)_2SO_4$ fractionated enzyme is used); the residual UDP-D-glucose may be removed from the reaction mixture by conversion to UDP-D-glucuronic acid with commercial UDP-D-glucose dehydrogenase from liver in the presence of DPN and removal of the UDP-D-glucuronic acid by electrophoresis (Table 2). Since C^{14}-labeled D-galactose is also available, it may be conveniently converted to UDP-D-galactose by the combined action of D-galactokinase and α-D-galactose 1-phosphate uridyl transferase from *Saccharomyces fragilis* (ANDERSON, MAXWELL and BURTON, 1959).

C14-labeled UDP-D-glucuronic acid (prepared by the oxidation of C14-labeled UDP-D-glucose) is used as the starting material for the preparation of UDP-D-galacturonic acid and a mixture of UDP-D-xylose and UDP-L-arabinose with radish enzyme (Section A V 1 b 3).

UDP-N-acetyl-D-glucosamine labeled with C14 in the D-glucosamine moiety is prepared from C14-labeled D-glucose 6-phosphate as described by GLASER and BROWN (1957).

Sugar 1-phosphates are released from their nucleotide derivatives by treatment with nucleotide pyrophosphatase from potatoes (KORNBERG and PRICER, 1950) or snake venom, and purification is effected by paper electrophoresis (Table 2). Simultaneous incubation of the UDP-sugar with nucleotide pyrophosphatase and phosphomonoesterase releases the free sugar, which can then be purified by conventional chromatographic, or, in the case of uronic acids, by electrophoretic techniques. Autoradiography (on X-ray film) is used to locate the radioactive substances on paper; these are then eluted and their solutions are concentrated under reduced pressure. Further details of the techniques used for the separation and purification of the sugars and their derivatives are included in the next section.

II. Separation and Identification of Reaction Products.

A number of techniques are available for the separation of sugar nucleotides and sugar phosphates. These two groups may be separated from each other by treatment with acid-washed charcoal (Nuchar or Norit), which adsorbs the nucleotides only. The adsorbed nucleotide may be recovered from the charcoal by washing with an ice-cold solution of ammoniacal ethanol (50 ml of 95 % ethanol and 1 ml of concentrated NH_4OH diluted with water to 100 ml). The ammoniacal eluate is immediately neutralized in order to reduce exposure of the sugar nucleotides to alkaline conditions.

In addition, mixtures of nucleotide derivatives and sugar phosphates can be separated by anion exchange chromatography (HURLBERT, 1957). Small quantities of these compounds can be conveniently handled by a combination of paper chromatography and paper electrophoresis. The electrophoretic mobilities of a number of nucleotides, sugar phosphates, and uronic acids are given in Table 2. A useful chromatographic solvent for sugar nucleotides and sugar phosphates is that of PALADINI and LELOIR (1952), consisting of 95 % ethanol-1 M ammonium acetate (70 : 30 v/v), 10^{-3} M with respect to EDTA.

Nucleotides are detected on paper as ultraviolet-quenching areas either by visual examination or contact printing (MARKHAM and SMITH, 1949). Sugar 1-phosphates are detected with the spray of BANDURSKI and AXELROD (1951).

The ultraviolet absorbing band corresponding to the sugar nucleotide is eluted from the paper and analyzed for the concentration of nucleoside, calculated on the basis of a molar absorbancy index of 9.9×10^3 at 260 mμ and pH 7.0 (PLOESER and LORING, 1940) for uridine, or 13.6×10^3 at 250 mμ and pH 7 for guanosine (VOLKIN and COHN, 1954); for total and acid labile (1 N HCl at 100° for 15 min) phosphate (FISKE and SUBBAROW, 1925) and for reducing value liberated by acid hydrolysis (PARK and JOHNSON, 1949). A correction is applied for material eluted from the paper by using as a blank the eluate from a strip of paper of the same area as the strip from which the sugar nucleotide was eluted. The pentosyl moiety of UDP-pentose is determined by the method of MEJBAUM (1939), with a correction for the color produced by uridine; the uronic acid moiety of UDP-uronic acids is determined by the method of DISCHE (1946); while the N-acetyl-D-glucosamine moiety of the corresponding sugar nucleotide is determined by the procedure of

Table 2. *Electrophoretic Mobility of Some Phosphorylated Compounds and Uronic Acids*[1],[2].

	Buffer 1 (pH 3.6)[3]	Buffer 2 (pH 5.8)[3]
Uridine Nucleotides		
UTP.	1.6	1.6
UDP	1.4	1.5
UMP	0.8	0.9
UDP-D-glucuronic acid.	1.4	1.6
UDP-D-galacturonic acid	1.3	1.5
UDP-D-xylose; UDP-L-arabinose . . .	1.2	1.1
UDP-D-glucose; UDP-D-galactose . . .	1.1	1.0
UDP-*N*-acetyl-D-glucosamine	1.0	1.0
Adenosine nucleotides		
ATP.	1.2	1.4
ADP	0.9	1.1
AMP	0.3	0.7
DPN	0.5	0.5
DPNH.	0.9	0.9
TPN.	0.9	0.9
TPNH	1.1	1.2
Guanosine Nucleotides		
GTP.	1.4	1.5
GDP	1.1	1.1
GMP	0.6	0.7
Phosphates		
Pyrophosphate	2.1	2.1
Phosphate	1.6	1.6
α- and β-D-glucuronic acid 1-phosphate.	1.5	1.7
α- and β-D-galacturonic acid 1-phosphate	1.4	1.6
α- and β-D-xylose 1-phosphate	1.1	1.0
α- and β-L-arabinose 1-phosphate . . .	1.1	1.0
α- and β-D-glucose 1-phosphate	0.9	1.0
α- and β-D-galactose 1-phosphate . . .	0.9	1.0
α-*N*-acetyl-D-glucosamine 1-phosphate .	0.9	1.0
Uronic Acids		
D-glucuronic acid	0.9	1.0
D-galacturonic acid	0.7	1.0

[1] Electrophoresis performed on Whatman No. 1 paper, previously washed with either acetic or oxalic acid and water.

[2] The mobility is recorded as a function of the mobility of picrate, caffeine being used as an indicator for electroosmosis. The mobility is calculated as:

$$\frac{\text{distance between compound and caffeine}}{\text{distance between picrate and caffeine}}$$

[3] The buffers are prepared as follows: Buffer 1 is prepared by dissolving 378 g of ammonium formate and 200 ml of 98% formic acid in 3 liters of solution. This stock solution is diluted twenty-fold for usage. Buffer 2 is prepared by dissolving 308.4 g of ammonium acetate and 30 ml of glacial acetic acid in 2 liters of solution. This solution is diluted ten-fold for use in electrophoresis.

Reissig, Strominger and Leloir (1955). The sugar phosphates liberated by treatment of the sugar nucleotides with nucleotide pyrophosphatase (Kornberg and Pricer, 1950), and the free sugars liberated by enzymatic or acid hydrolysis of the sugar nucleotide (1 N HCl, 100°, 15 min) are compared with the authentic compounds by electrophoresis and chromatography.

In addition, specific enzymatic tests may often be used. For instance, the DPN-linked oxidation of UDP-D-glucose to UDP-D-glucuronic acid (Section A V 2) may be used for the identification of UDP-D-glucose. UDP-D-galactose may be identified by conversion to UDP-D-glucose with UDP-D-galactose 4-epimerase (Section A V 1), followed by oxidation to UDP-D-glucuronic acid. This nucleotide, in turn, may be identified by decarboxylation to UDP-D-xylose by wheat germ enzyme (Section A V 3), or by the transfer of the D-glucuronosyl moiety to o-aminophenol in the presence of guinea pig liver microsomes (LEVY and STOREY, 1949).

The hexose 1-phosphates, pentose 1-phosphates, and each of the uronic acid 1-phosphates have characteristic electrophoretic mobilities, react with the reagent of BANDURSKI and AXELROD (1951), and are converted to equimolar quantities of inorganic phosphate and free sugar when treated with either acid (1 N HCl, 100°, 15 min) or with phosphomonoesterase. All these criteria are helpful in identification of these sugar phosphates. These compounds can be definitively identified by use of the appropriate pyrophosphorylase in the presence of UTP to form the nucleoside diphosphate sugar, which may then be submitted to one of the special tests described above.

Sugars enter the nucleotide pathway as the 1-phosphate derivatives. These compounds arise either by the action of a mutase on a sugar 6-phosphate or by direct formation of the sugar 1-phosphate from the free sugar and ATP (Fig. 1).

Yeast, animals, and bacteria contain kinases capable of forming α-D-galactose 1-phosphate directly from D-galactose and ATP (KALCKAR, 1958). Higher plants contain a similar enzyme, as well as kinases which form β-L-arabinose 1-phosphate (NEUFELD, FEINGOLD and HASSID, 1960) and α-D-glucuronic acid 1-phosphate (NEUFELD, FEINGOLD and HASSID, 1959).

III. Enzymes which Catalyze the Formation of Sugar 1-Phosphates.

1. D-Galactokinase from *Saccharomyces fragilis*.

The procedure of HEINRICH and NEILANDS (1960), given here, yields the enzyme with the highest specific activity obtained to date[1].

Assay. Enzyme activity may be determined by measuring the decrease in free reducing sugar upon incubation, as described by LELOIR and TRUCCO (1955).

A much more convenient assay, particularly during fractionation, is one in which the formation of ADP is measured by the following series of reactions:

$$\text{phosphoenolpyruvate} + \text{ADP} \xrightarrow{\text{pyruvic kinase}} \text{pyruvate} + \text{ATP}$$

$$\text{pyruvate} + \text{DPNH} + \text{H}^+ \xrightarrow{\text{lactic dehydrogenase}} \text{lactate} + \text{DPN}^+ .$$

The disappearance of DPNH is followed by the decrease of the optical density at 340 mμ. The reaction mixture is that used by MALEY and OCHOA (1958). The final concentration of the reactants used is: potassium phosphate buffer, pH 7, 0.05 M; KCl, 0.08 M; MgSO$_4$, 0.01 M; EDTA, 2×10^{-3} M; ATP, 5×10^{-4} M; phosphoenol pyruvate, 8×10^{-4} M; DPNH, 10^{-4} M; commercial crystalline lactic dehydrogenase, containing pyruvic kinase, 20 μg.

It is convenient to make up solutions of each component at 10 times the final concentration. A 0.1 ml volume of each solution is pipetted into a 1 ml cuvette, and water is added to a volume of 1 ml. Enzyme is added (usually 5 to 10 μl), the contents are mixed, and the optical density at 340 mμ is followed briefly. At this

---

[1] The authors wish to thank Drs. A. MUNCH-PETERSEN, M. HEINRICH, and J. B. NEILANDS for making available to us their methods prior to publication.

stage, before the addition of D-galactose, a small, transient decrease in optical density would indicate an impurity (e.g. ADP or pyruvate) in the reagents. A continuing decrease of optical density would indicate the presence of endogenous substrate, or of other enzyme systems producing ADP. After determining this rate for approximately 1 min, 0.1 ml of D-galactose solution (3 gm/ml) is added. The contents of the cuvette are mixed rapidly, and the rate of decrease of optical density measured. If necessary, this value is corrected for the "endogenous" rate observed without D-galactose.

The phosphorylation of other sugars is readily determined by substituting them for D-galactose.

Specific activity is expressed as μmoles of D-galactose esterified/min/mg protein. In the 1.1 ml volume of assay mixture described, μmoles of galactose = ($\triangle$ O. D. at 340 mμ)/5.65.

Purification. *Saccharomyces fragilis* C-106 is grown on the medium of WIL-KINSON (1949), including 2% lactose, for 24—48 hours at 30° with shaking or aeration. The culture is centrifuged, the yeast is washed twice by suspension in water and centrifugation, and the yeast paste is spread on glass plates to dry at room temperature. The dried yeast loses its activity slowly on storage at $-15°$.

Subsequent operations are carried out at 0—4°. The quantities given are for 100 g of dried yeast.

The yeast is suspended in 500 ml. of 2.2% $(NH_4)_2HPO_4$ solution containing 0.001 M EDTA. After 18—24 hours with occasional stirring, the suspension is centrifuged. The extract is diluted with water to give 20—30 mg protein/ml (1 liter final volume). Solid $(NH_4)_2SO_4$ is added to 45% saturation, maintaining the pH at 6.5—7 by the addition of NH_4OH, and the precipitate is discarded. $(NH_4)_2SO_4$ is added to 60% saturation, the precipitate obtained by centrifugation is dissolved in 50 ml. of 0.001 M EDTA, and solid cysteine hydrocloride is added to give a concentration of 0.01 M. Dilute acetic acid is added with stirring to adjust the pH to 5 and the suspension is centrifuged at once. The supernatant solution is immediately neutralized to pH 6.5 with NH_4OH and dialyzed for a few hours against 0.01 M cysteine-0.001 M EDTA, pH 6.5. The dialyzed solution is diluted to a protein concentration of 10—15 mg/ml (300—400 ml), and a suspension of freshly prepared calcium phosphate gel, (55 mg/ml, dry weight), containing a weight of calcium phosphate equal to the weight of protein is added. The pH is adjusted to 6—6.5 with NH_4OH, and the suspension is stirred for 15 min, centrifuged, and the precipitate discarded. The supernatant solution is treated in the same way with a second portion of gel, and the precipitate is saved. The enzyme is eluted from the calcium phosphate with several 50 ml portions of 0.1 M ammonium phosphate buffer, pH 7.

The enzyme is purified further by chromatography on DEAE-cellulose. A column containing at least 5 times as much DEAE-cellulose as protein (w/w) to be chromatographed is prepared by washing the adsorbent with 0.001 M ammonium phosphate-0.001 M EDTA, pH 7, until the effluent pH is 7. The enzyme solution is dialyzed against buffer of the same composition, and adsorbed on the prepared column. Elution is by means of a linear gradient, with 500 ml of the 0.001 M buffer in the mixing flask and 500 ml of 0.2 M $NH_4H_2PO_4$-0.001 M EDTA in the reservoir. Fractions of 10 ml are collected. Optical density at 280 mμ and 260 mμ and pH are determined for the fractions. Fractions which are more acid than pH 6 are adjusted to that pH. Aliquots of some fractions are taken for assay and the remainder is frozen. D-Galactokinase begins to appear when the effluent reaches approximately pH 5.6. The first fractions of D-galactokinase contain more hexokinase than later ones, since the hexokinase peak is eluted just prior to D-galactokinase and there is considerable overlapping.

The fractions with D-galactokinase activity are combined and lyophilized. The powder is dissolved in 10 ml of 0.05 M ammonium phosphate, pH 7, 0.001 M with respect to EDTA, and dialyzed against the same buffer. An amount of bentonite equal to twice the weight of the protein present is added. The pH is adjusted to 5.2 with 1 M acetic acid, the suspension is centrifuged at once and the supernatant solution is immediately brought to pH 6.5. Hexokinase is removed completely in this step.

Moving boundary electrophoresis of a preparation made by this procedure shows the presence of 2 major components in equal amount, both of which move toward the anode at pH 6.6. D-galactokinase activity is associated with the slower-moving component only.

The overall yield of the enzyme is less than 10%. The increase in specific activity is approximately 15-fold during purification, with the best preparations esterifying 5 μmoles of D-galactose/min/mg protein (see Table 3).

Table 3. *Purification of D-Galactokinase from Saccharomyces fragilis.*

	Volume (ml)	Total protein (mg)	Total units	Specific activity (units/mg protein)
Extract	395	22,600	6,600	0.29
(NH$_4$)$_2$SO$_4$ precipitate	119	10,500	4,400	0.42
Calcium phosphate gel eluate . .	200	1,520	2,300	1.36
DEAE-cellulose eluate lyophilized	20	250	550	2.20
Bentonite supernatant	22	56	185	3.3

Properties. The enzyme has a definite requirement for Mg^{++}. Mn^{++} also appears to be effective. GTP, CTP and UTP do not replace ATP. The enzyme is stabilized by sulfhydryl compounds (cysteine, glutathione) and EDTA. The substrate specificity of a partially purified D-galactokinase preparation from *Saccharomyces fragilis* has been tested by ALVARADO (1960). The only sugars which are phosphorylated are hexopyranoses which differ from D-galactose at the level of C-2. Thus 2-D-deoxygalactose, D-galactosamine, and D-talose are converted to the respective 1-phosphate derivatives by the enzyme, whereas compounds which differ at C-3, C-4, or C-6 such as D-gulose, D-glucose, L-arabinose, D-fucose, or D-glycero-D-galactoheptose are not acted upon by D-galactokinase.

2. D-Glucuronic Acid Kinase from *Phaseolus aureus*.

Assay. The assay is based on the separation of the products of the phosphorylation reaction by paper electrophoresis. A typical reaction mixture consists of 4×10^{-2} μmole (1.5×10^{-2} μcurie) C^{14}-labeled D-glucuronic acid, 1 μmole ATP, 0.5 μmoles MgCl$_2$, 5 μmole NaF, 1 μmole mercaptoethanol, and enzyme in a total volume of 35 μl. The assay is most conveniently carried out using the capillary tube technique of PORTER and HOBAN (1954). Care should be taken that all reagents used be brought to pH 7.5 before mixing when this technique is used. After an appropriate time (approximately 1 hour) at 37° the mixture is subjected to paper electrophoresis at pH 3.6 (Table 2). Reference spots of α-D-glucuronic acid 1-phosphate and of D-glucuronic acid are applied at the sides of the paper. At the end of the electrophoretic run the reference strips are separated and developed by the periodate-benzidine procedure of GORDON, THORNBURG and WERUM (1956). The radioactive spots of α-D-glucuronic acid 1-phosphate and D-glucuronic acid are eluted into planchets, taken to dryness, and the radioactivity counted.

Both soluble and particulate preparations from mung bean seedlings have D-glucuronic acid kinase activity. Since the soluble enzyme is the most active, its preparation is given here.

Preparation: Commercially-grown etiolated mung bean *(Phaseolus aureus)* seedlings (100 g) are ground with acid-washed sand in a chilled mortar in 70 ml of buffer consisting of 0.4 M sucrose, 0.01 M sodium-potassium phosphate, pH 7.0, and 0.01 M mercaptoethanol.

All subsequent operations are performed at 0—5°. The homogenate is filtered through cheesecloth, spun at 2,000 × g for 5 min, and the precipitate is discarded. The supernatant liquid is then centrifuged at 18,000 × g for 30 min. The precipitate (mitochondrial preparation) is used for the preparation of D-galacto- and L-arabinokinase (see following section). The supernatant liquid is fractionated with solid $(NH_4)_2SO_4$ and the fraction precipitating between 40 and 70% saturation is collected by centrifugation, dissolved in a minimal volume of 0.1 M mercapto-ethanol-0.1 M Tris buffer, pH 7.5, and dialyzed overnight against 1 liter of the same buffer. The dialyzed enzyme is again concentrated by fractionating between 40 and 70% $(NH_4)_2SO_4$ and dissolved and dialyzed as previously described. The enzyme so obtained keeps its activity for several weeks when stored at 4°, but may not be frozen.

Properties. Neither UTP, GTP, nor CTP can replace ATP as phosphate donor. A divalent cation (Mg^{++}, Mn^{++}, or Co^{++}) is required for the reaction.

3. D-Galactokinase and L-Arabinokinase from *Phaseolus aureus*.

Assay. The assay mixture contains 1 μmole ATP, 0.25 μmole $MgCl_2$, approximately $4 \times 10^{-4}\mu$mole of C^{14} labeled D-galactose or L-arabinose (0.01 μcurie), and 20 μl of enzyme preparation in a total volume of 40 μl (in a capillary tube). After an appropriate time (approximately 1 hour) at 37° the mixtures are separated electrophoretically and the relative quantities of free and phosphorylated sugar are determined by counting, as described in the preceding section.

Preparation. The particulate material (mitochondrial preparation) prepared as described in the preceding section is suspended in 0.5 ml of 0.5 M Tris buffer, pH 7.5, 0.1 M with respect to mercaptoethanol, to give a final volume of 1.0 ml. This mitochondrial suspension contains both D-galacto- and L-arabinokinase. The L-arabinokinase activity, however, can be solubilized by extraction with digitonin.

The mitochondrial suspension is diluted with 1.0 ml of 0.1 M Tris buffer, pH 7.5, 0.1 M with respect to mercaptoethanol, and is vigorously shaken by hand for approximately 30 sec with 2 ml of a 1% solution of digitonin, likewise 0.1 M with respect to mercaptoethanol. To the supernatant liquid obtained by centrifugation at $18,000 \times g$ for 30 min $(NH_4)_2SO_4$ is added to 70% of saturation. The protein collected by centrifugation is dissolved in 0.2 ml of 0.1 M Tris buffer, pH 7.5, 0.1 M with respect to mercaptoethanol. This solution is dialyzed 4 to 20 hours against 1 liter of the same buffer.

Properties. The mitochondrial enzyme preparation will phosphorylate D-galactose in the presence of ATP but not of UTP, GTP, or CTP. No metal requirement has been shown for this enzyme.

The solubilized L-arabinokinase is not completely specific for ATP, and will effect the phosphorylation of L-arabinose to about one fifth the extent with UTP and GTP. On the other hand, the soluble L-arabinokinase requires a metal for activity, Mg^{++}, Co^{++}, and Mn^{++} serving equally well.

IV. Enzymes which Catalyze the Formation of Sugar Nucleotides.

Sugar nucleotides may be formed from the respective sugar 1-phosphates by two mechanisms. The first involves a reaction with a nucleoside triphosphate to yield the sugar nucleotide and inorganic pyrophosphate:

$$\text{Nucleoside P PP} + \text{Sugar 1-P} \rightleftharpoons \text{Nucleoside P P-sugar} + \text{PP}$$
(the bond attacked is indicated by a broken line)

The second mechanism involves the reaction of a sugar 1-phosphate with the nucleotide of a different sugar:

$$\text{Nucleoside P P-sugar} + \text{sugar' 1-phosphate} \rightleftharpoons \text{Nucleoside P P-sugar'} + \text{sugar 1-P.}$$

Although both mechanisms involve a transfer of the nucleotidyl group, the term "nucleotidyl transferase" is reserved for enzymes catalyzing the second type of reaction, while those catalyzing the first type are named "pyrophosphorylases" (KALCKAR, 1957).

1. Sugar Nucleotide Pyrophosphorylases from *Phaseolus aureus*.

Extracts of mung bean seedlings catalyze the formation of UDP-D-glucose, UDP-D-galactose, UDP-D-glucuronic acid, UDP-D-galacturonic acid, UDP-D-xylose, and UDP-L-arabinose by a reaction between UTP and α-D-glucose 1-phosphate, α-D-galactose 1-phosphate, α-D-glucuronic acid 1-phosphate, α-D-galacturonic acid 1-phosphate, α-D-xylose 1-phosphate, and β-L-arabinose 1-phosphate, respectively (NEUFELD, et al., 1957; FEINGOLD, NEUFELD and HASSID, 1958a). With the exception of UDP-D-galacturonic acid, these sugar nucleotides have all been isolated from mung bean seedlings (GINSBURG, STUMPF and HASSID, 1956; SOLMS and HASSID, 1957). The enzyme which catalyzes the synthesis of UDP-D-glucose has been highly purified and shown to be specific for that sugar nucleotide (GINSBURG, 1958); the separation of the enzymes responsible for the formation of the other sugar nucleotides has not yet been attempted. UDP-D-glucose pyrophosphorylase is of widespread occurrence in nature (see review by STROMINGER, 1960), and UDP-D-galactose pyrophosphorylase has been observed in yeast (KALCKAR, BRAGANCA and MUNCH-PETERSEN, 1953) and in calf liver (ISSEL-BACHER, 1958). On the other hand, UDP-uronic acid and UDP-pentose pyrophosphorylases have been demonstrated in higher plants only.

Assay. Enzymatic activity can be detected in the direction of synthesis or of pyrophosphorolysis by incubating the extract with the appropriate substrates, and separating the products present in the mixture at the end of the incubation period by paper electrophoresis. A typical assay mixture consists of UTP, 0.1 M, 2.5 μl; sugar phosphate, 0.1 M, 2.5 μl; MgCl$_2$, 0.05 M, 2.5 μl; mung bean enzyme, 10 μl. The mixture is incubated in a capillary tube at 37° for times varying from a few minutes to two hours, depending on the activity of the pyrophosphorylase. Inorganic pyrophosphatase (50—100 units, HEPPEL and HILMOE, 1951) may be added to remove pyrophosphate, thus forcing the reaction to the right with maximal accumulation of the sugar nucleotide.

If the sugar nucleotide is more readily available than the sugar 1-phosphate, the assay is run in the direction of pyrophosphorolysis as follows: sugar nucleotide, 0.1 M, 2.5 μl; Na pyrophosphate, 0.1 M, 5 μl; MgCl$_2$, 0.05 M, 5 μ; NaF, 1 M, 5 μl; enzyme, 10 μl.

After incubation, the mixture is subjected to paper electrophoresis at pH 3.6 or pH 5.8. The buffer is selected to give optimal resolution of all components of the reaction mixture (see Table 2).

Preparation. The method given here is similar to that described previously (NEUFELD et al., 1957) but yields more active preparations.

Commercially grown etiolated mung bean seedlings, 4—5 days old (1 kg) are homogenized with 1 liter of 0.01 M potassium phosphate buffer, pH 7.0, 0.05 M with respect to mercapto-ethanol. This and all subsequent operations are carried out at 0—4°. Coarse debris is removed by filtration through cheesecloth and particulate matter by centrifugation at 20,000 × g. for 30 min. In (NH$_4$)$_2$SO$_4$ fractionations, 10 min are allowed to elapse between addition of the salt and centrifugation. The supernatant fluid is fractionated with solid (NH$_4$)$_2$SO$_4$, and the fraction precipitating between 40 and 70% saturation is collected by centrifugation, suspended in 60 ml of 0.05 M Tris buffer, pH 7.5, 0.05 M in respect to mercaptoethanol, and dialyzed for 18 hours against 4 liters of the same buffer. Any insoluble material present after dialysis is removed by centrifugation for 10 min at 18,000 × g.

The clear supernatant liquid is fractionated with $(NH_4)_2SO_4$ and the fraction precipitating between 50 and 70% saturation is collected by centrifugation, suspended in 0.05 M Tris buffer, pH 7.5, 0.05 M in respect to mercaptoethanol, and dialyzed against 2 liters of the same buffer for 18 hours. The contents of the dialysis bag (11 ml) are lyophilized and the powder is stored at —10°. This preparation retains activity for at least two months. For assay, 1 mg of lyophilized powder is dissolved in 0.2 ml of 0.05 M mercaptoethanol.

Crude pyrophosphorylase preparations, such as the one described, contain a number of enzymes which catalyze side reactions. UTP and UDP are rapidly dephosphorylated to UMP; the latter is converted more slowly to uridine. Since inorganic pyrophosphatase is also present in the enzyme preparation, fluoride, which inhibits the pyrophosphatase, must be added to test the reaction in the direction of pyrophosphorolysis, or to measure the formation of inorganic pyrophosphate. The preparation has little hydrolytic activity towards sugar 1-phosphates and sugar nucleotides, which are broken down very slowly. Pyrophosphorylase preparations also contain variable amounts of 4-epimerases, depending on the length and conditions of storage.

Properties. The following pyranosyl phosphates are utilized as substrates: α-D-glucose 1-phosphate, α-D-galactose 1-phosphate, α-D-glucuronic acid 1-phosphate, α-D-galacturonic acid 1-phosphate, α-D-xylose 1-phosphate, and β-L-arabinose 1-phosphate. Their respective anomers, however, are not utilized, nor is α- or β-L-arabinofuranose 1-phosphate, α-D-arabinopyranose 1-phosphate, α-D-arabinofuranose 1-phosphate, or α-N-acetyl-D-glucosamine 1-phosphate (see Section A IV 4 for the preparation of extracts which contain UDP-N-acetyl-D-glucosamine pyrophosphorylase). A divalent cation is required for enzyme activity.

Mg^{++}, Mn^{++}, or Co^{++}, in a concentration of $3 \times 10^{-3}M$, give approximately equal activation.

The mechanism of the reaction has been studied with the use of P^{32}. The transfer of the uridyl group between PP and the sugar phosphate could proceed by one of the following mechanisms:

1. Formation of a free uridyl-enzyme complex:

UP PP + enzyme ⇌ UP enzyme + PP
UP enzyme + sugar 1-P ⇌ UP P-sugar + enzyme .

2. Simultaneous attachment of uridyl donor and acceptor to the enzyme surface, without formation of a free uridyl-enzyme complex:

$$\text{UP PP + enzyme + sugar 1-phosphate} \rightleftharpoons \text{enzyme}-\left[-\text{UP}\begin{array}{c}\nearrow \text{PP} \\ \searrow \text{P-sugar}\end{array}\right]$$

⇌ enzyme + UP P-sugar + PP

Were the reaction to take place by the first mechanism, an exchange of radioactive label would occur between PP and the two terminal phosphates of UTP in the absence of sugar phosphate. Since no such exchange takes place unless sugar 1-phosphate is added, it is concluded that the reaction proceeds by the second mechanism.

2. UDP-D-Glucose Pyrophosphorylase from *Phaseolus aureus*.

The following procedure is that described by Ginsburg (1958).

Assay. The assay method is based on measurement of the rate of α-D-glucose 1-phosphate liberation from UDP-D-glucose upon addition of pyrophosphate. This rate is determined spectrophotometrically in the presence of TPN, phosphoglucomutase, and D-glucose 6-phosphate dehydrogenase. One unit of enzyme is defined as that amount of enzyme which will cause pyrophosphorolysis of 1.0 μmole of UDP-D-glucose per minute.

The following reagents are mixed in a 1 ml cuvette: Tris buffer, pH 7.5, 0.1 M, 0.85 ml; mercaptoethanol, 0.13 M, 0.05 ml; $MgCl_2$, 0.1 M, 0.02 ml; TPN, 0.02 M, 0.01 ml; UDP-D-glucose, 0.02 M, 0.01 ml; phosphoglucomutase (NAJJAR, 1948) 0.02 ml; D-glucose 6-phosphate dehydrogenase (KORNBERG and HORECKER, 1955), 0.02 ml; mung bean enzyme solution, containing 0.002—0.01 units of activity, 0.02 ml. The reaction is started by the addition of 0.01 ml of 0.1 M sodium pyrophosphate, and the optical density of 340 mμ is read every 30 sec until TPN reduction attains a linear rate. The indicator enzymes should be present in large excess so that the rate of TPN reduction will be proportional to the concentration of UDP-D-glucose pyrophosphorylase. During the early stages of purification it is necessary to determine the rate of TPN reduction before the addition of inorganic pyrophosphate, since the crude enzyme preparation contains nucleotide pyrophosphatase which catalyzes the formation of α-D-glucose 1-phosphate from UDP-D-glucose. TPN reduction from this contaminating enzyme rarely amounts to more than 10% of that contributed by pyrophosphorylase.

Preparation. An acetone powder is prepared from etiolated 4-day-old mung bean seedlings by three extractions with acetone cooled to $-20°$. The seedlings are homogenized with three volumes of acetone in a WARING Blendor for two minutes. The suspension is filtered on a large BÜCHNER funnel, and the residue is re-extracted by blending twice with one volume of acetone each time. The residue is spread on filter paper and dried at room temperature until the odor of acetone is no longer detectable, and then stored in stoppered bottles at $-10°$. All subsequent operations are performed at $0—3°$.

Acetone powder (100 g) is extracted with 1,200 ml of 0.1 M Tris buffer, pH 7.5, for one hour, with occasional stirring. The supernatant liquid obtained by centrifugation is strained through cheese cloth in order to remove floating debris.

$(NH_4)_2SO_4$ is added slowly to the acetone powder extract with stirring over a period of thirty minutes, to 85% saturation. The precipitate is collected by centrifugation, suspended in 60 ml of 65% saturated $(NH_4)_2SO_4$, and stirred for 15 min. The suspension is then centrifuged, and the supernatant solution is dialyzed for 3 hours against 4 liters of H_2O, and then further dialyzed for 18 hours against 4 liters of 0.005 M sodium phosphate buffer, pH 6.0, 3×10^{-4} M with respect to mercaptoethanol. The precipitate which is formed during the second dialysis is removed by centrifugation and discarded.

A suspension of alumina C_γ (12 mg dry weight per ml) is added to the supernatant (1 mg gel, dry weight, per 3 mg protein). After stirring for 30 min the suspension is centrifuged and the precipitate discarded. The supernatant is dialyzed for 18 hours against 4 liters of 0.005 M sodium phosphate buffer, pH 7.5, 3×10^{-4} M with respect to mercaptoethanol.

Further purification of the enzyme is achieved by chromatography on a DEAE-cellulose column (PETERSON and SOBER, 1956; SOBER et al., 1956). The column is prepared by packing by gravity an aqueous slurry containing 3 g of DEAE-cellulose into a column 2 cm in diameter. The column is washed with 0.005 M sodium phosphate buffer, pH 7.5, before use. Enzyme solution, containing about 350 mg of protein, is put on the column at a rate of 2 ml per minute, and the column is eluted with 300 ml of 0.005 M phosphate buffer, 10 ml fractions being collected (fractions 1—30). The column is then eluted by the method of gradient elution. The mixer flask is filled with 300 ml of 0.005 M phosphate buffer, pH 7.5, and the reservoir with 0.1 M NaCl in the same buffer. The two flasks are leveled and connected by a siphon. After 600 ml of eluant have passed through the column (fractions 30—90), a solution containing 0.1 M NaCl in 0.005 *M* phosphate buffer is then put into the mixer and 0.3 M NaCl in the same buffer into the reservoir, and elution is continued until no more ultraviolet-absorbing material appears in the eluate.

The activity of each fraction is tested to determine where the UDP-D-glucose pyrophosphorylase emerges. The enzyme usually is eluted shortly after 1 liter of eluate has been collected and if the chromatography is properly performed should be present in fewer than 10 fractions. The tube containing the highest specific activity is used for enzymatic studies. The purification procedure is summarized in Table 4. The purified enzyme is relatively unstable, for even in the presence of 0.012 M mercaptoethanol it loses 90% of its activity after storage at —7° for two weeks.

Table 4. *Purification of UDP-D-Glucose Pyrophosphorylase of Phaseolus aureus.*

	Volume (ml)	Protein (mg)	Total units	Specific activity (units/mg protein)
Acetone powder extract	980	8,400	8,800	1.04
(NH$_4$)$_2$SO$_4$ fraction	114	930	6,700	7.2
Alumina C γ treated solution . .	134	355	5,200	14.6
DEAE-cellulose eluate	10	2.7	2,200	815

Properties. The enzyme is specific for α-D-glucose 1-phosphate, and does not incorporate into sugar nucleotides any of the other sugar 1-phosphates which are substrates for the crude preparation (Section A IV 1). The pH optimum in mixed buffers containing 0.03 M phosphate, Tris, and glycine is approximately 8.0; on the acid side the activity falls off gradually, being 60% of maximum at pH 5.5, while on the alkaline side it drops sharply, reaching 30% of maximum at pH 9.5. The K_m for UDP-D-glucose is 1.1×10^{-4}M, and that for pyrophosphate is 2.3×10^{-4}M.

3. UDP-D-Glucose Pyrophosphorylase from Brewer's Yeast.

Good UDP-D-glucose pyrophosphorylase activity is found in preparations of *Zwischenferment* prepared as described by LEPAGE and MUELLER (1949). The enzyme has been purified 250-fold by MUNCH-PETERSEN (1955) by the method described below.

Assay. The assay method is the same as that used in the purification of the UDP-D-glucose pyrophosphorylase from mung beans.

Preparation. Dried Brewer's yeast (Kongens Bryghus, Copenhagen) (20 g) is shaken with 40 ml of 0.07 M (NH$_4$)$_2$HPO$_4$ for 18 hours at 20°, and the suspension is centrifuged. The supernatant liquid is fractionated with a saturated solution of (NH$_4$)$_2$SO$_4$ and the fraction precipitating between 40 and 60% saturation is collected by centrifugation, dissolved in 25 ml of 0.015 M acetate buffer, pH 6.3, and dialyzed for 1 hour against cold, running tap water. To the clear solution 2 ml of 1% protamine sulfate is added. After 1 hour at 2°, a small inactive precipitate is removed. The clear supernatant is then cooled to —2°, and fractionated with 50% ethanol. (During this fractionation the temperature of the protein solution is gradually lowered to —8°.) Cold (—10°) ethanol is added dropwise with stirring, and the fractions precipitating between 20—24%, 24—28%, and 28—31% ethanol are collected by centrifugation at —10°.

In order to free the precipitates from traces of ethanol, they are held overnight at —20°, during which time the ethanol adhering to the precipitate and walls of the tubes evaporates. The fractions are then dissolved in 4 ml of water at 0°, and any insoluble material is removed by centrifugation. To the fraction with the greatest activity (usually the one precipitating between 20 and 24% ethanol), a saturated solution of (NH$_4$)$_2$SO$_4$ is added to 60% saturation. The precipitate is extracted three times with 2.5 ml aliquots of solutions of decreasing (NH$_4$)$_2$SO$_4$

concentration, adjusted to pH 7.5 with NH_4OH. The fraction obtained between 56 and 50% saturation is usually the most active. When dissolved, the activity of the enzyme is rapidly lost. The best way to store the enzyme is to keep the final $(NH_4)_2SO_4$ precipitate at $-20°$. In this way, $60-75\%$ of the activity is retained after 3 months.

The purification of the enzyme is summarized in Table 5.

Table 5. *Purification of UDP-D-Glucose Pyrophosphorylase from Brewer's Yeast.*

	Volume (ml)	Total units[1]	Yield %	Specific activity (units/mg protein)
Crude extract	23	4,055	100	0.3
$(NH_4)_2SO_4$ precipitate	28	3,360	82.8	6.3
Ethanol fractions				
20—24%	6.4	1,640	40.4	11.0
24—28%	3	241	5.9	5.5
28—31%	2	121	2.9	5.8
$(NH_4)_2SO_4$ precipitates				
40—46%	0.5	196	4.8	18.3
46—50%	2.5	693	15.3	40.7
50—56%	2.5	618	15.0	77.2

[1] One unit, defined by MUNCH-PETERSEN (1955), is equivalent to 0.1 unit as defined by GINSBURG (1958).

Further purification may be obtained by chromatography on DEAE-cellulose (MUNCH-PETERSEN, personal communication). The column (25×48 cm) is prepared as described by PETERSON and SOBER (1956), and is washed with 0.005 M phosphate buffer, pH 7. The ethanol-fractionated UDP-D-glucose pyrophosphorylase is dialyzed overnight against the same buffer, and adsorbed on the column. Elution is carried out with rising concentrations of NaCl, starting with 200 ml of 0.005 M phosphate buffer, pH 7, in the mixing bottle, and 200 ml of 2 M NaCl in the reservoir. The two flasks are leveled and connected by a siphon. The most active fraction obtained has a specific activity of 700 units/mg of protein. This highly purified enzyme is unstable in dilute solution but retains its activity for several months if kept frozen in a 20 % $(NH_4)_2SO_4$ solution.

Properties. The ethanol-fractionated enzyme does not catalyze the pyrophosphorolysis of UDP-N-acetyl-D-glucosamine, GDP-D-mannose or DPN. It is activated by Mg^{++} ions, maximal activations occurring at 2×10^{-3} M Mg^{++}. The pH optimum is a plateau between pH 6.5 and 7.5 in 0.05 M Tris buffer and falls off to 50% of maximum at pH 6 and 8. The K_m for UDP-D-glucose, UTP, and pyrophosphate are 4×10^{-6} M, 5×10^{-5} M, and 5×10^{-5} M, respectively, for the DEAE-cellulose fractionated enzyme. The equilibrium constant is approximately 1 at pH 7.4. The purified pyrophosphorylase preparation does not catalyze any exchange of label between PP^{32} and UTP (MUNCH-PETERSEN, 1957). Thus the uridyl transfer catalyzed by yeast UDP-D-glucose pyrophosphorylase, like the uridyl transfer catalyzed by the mung bean enzymes, must proceed without the intermediate formation of a free uridyl-enzyme complex. The enzyme is strongly inhibited by $(NH_4)_2SO_4$ and the concentration of this salt must be kept below 0.01 M.

4. UDP-N-Acetyl-D-Glucosamine Pyrophosphorylase from *Phaseolus aureus.*

UDP-N-acetyl-D-glucosamine has been isolated from animals (SMITH and MILLS, 1954), bacteria (CIFONELLI and DORFMAN, 1957), and both lower (CABIB,

Leloir and Cardini, 1953; Ballio, Casinovi and Serlupi-Crescenzi, 1956), and higher plants (Solms and Hassid, 1957; Bergkvist, 1957). Enzymes catalyzing the pyrophosphorolysis of this sugar nucleotide have been purified from calf liver and *Staphylococcus aureus* (Strominger and Smith, 1959), and from yeast (Munch-Petersen, personal communication). However, although this enzyme has been demonstrated in extracts of mung bean seedlings, extensive purification from higher plant sources has not yet been attempted.

Assay. The assay is the one described in section A IV 1.

Preparation. Etiolated mung bean seedlings, 100 g, are ground in a chilled mortar with 100 ml of ice-cold 0.01 M phosphate, pH 7.0, − 0.5 M sucrose − 0.01 M mercaptoethanol buffer. All subsequent operations are performed at 0−5°. After filtration through cheesecloth, the homogenate is centrifuged at 30,000 × g for 20 min. The precipitate is discarded, and the supernatant liquid is made 50% saturated by the addition of solid $(NH_4)_2SO_4$. (In $(NH_4)_2SO_4$ fractionations, 10 min are allowed to elapse between addition of the salt and centrifugation.) The precipitate is separated by centrifugation, dissolved in 3 ml of 0.05 M Tris, pH 7.0, − 0.05 M mercaptoethanol buffer, and dialyzed overnight against 1 liter of the same buffer. A small insoluble precipitate is removed by centrifugation and the solution is refractionated with $(NH_4)_2SO_4$. The fraction precipitating between 40 and 50% saturation is richest in UDP-*N*-acetyl-D-glucosamine pyrophosphorylase. It is dissolved in a minimum volume of 0.05 M Tris buffer, pH 7.0, 0.05 M with respect to mercaptoethanol and dialyzed overnight against 1 liter of the same buffer.

5. UDP-N-Acetyl-D-Glucosamine Pyrophosphorylase from Baker's Yeast.

This pyrophosphorylase has been observed in yeast by Munch-Petersen et al. (1955) and by Glaser and Brown (1955). The purification described here is by Munch-Petersen (personal communication).

Assay. The enzyme may be assayed as decribed in Section A IV 1.

Preparation. Baker's yeast (40 g) is dried, crushed, and autolyzed overnight at 20° with 200 ml of 0.07 M $(NH_4)_2HPO_4$, pH 7.5. Subsequent operations are performed at 0−3°. The mixture is centrifuged and a saturated solution of $(NH_4)_2SO_4$ is added to the supernatant fluid to 40% saturation. The precipitate is removed by centrifugation, and $(NH_4)_2SO_4$ is added to 65% saturation. After 30 min, the precipitate is collected by centrifugation and dissolved in the smallest possible volume of 0.005 M phosphate buffer, pH 7. It is dialyzed overnight against 3 liters of the same buffer.

This crude preparation may be fractionated on a DEAE cellulose column without additional purification. The column is prepared as described in Section A IV 3, and the eluate is collected in 3 to 4 ml fractions. The gradient is obtained by using 250 ml of 0.005 M phosphate buffer, pH 7, in the mixing flask and 250 ml of 2 N NaCl in the reservoir. The UDP-*N*-acetyl-D-glucosamine pyrophosphorylase is eluted at very low salt concentrations. The purified enzyme is extremely unstable.

Properties. The enzyme has a broad pH optimum between pH 6.5 and 8.5. The Michaelis constants, K_m, for UDP-*N*-acetyl-D-glucosamine and for pyrophosphate are 7×10^{-5} M and 2×10^{-5} M respectively. The equilibrium constant is between 0.5 and 0.8. Like yeast UDP-D-glucose pyrophosphorylase, the enzyme is strongly inhibited by $(NH_4)_2SO_4$ which must therefore be excluded from the assay mixture.

6. GDP-D-Mannose Pyrophosphorylase from Brewer's Yeast.

An enzyme has been found in yeast which catalyzes the formation of GDP-D-mannose by a reaction analogous to the formation of uridine nucleotides:

$$GP \, PP + \alpha\text{-D-mannose 1-phosphate} \rightleftharpoons GP \, P\text{-D-mannose} + PP.$$

GDP-D-mannose is found in yeast (CABIB and LELOIR, 1954), a red alga (SU and HASSID, 1960), molds (BALLIO, CASINOVI and SERLUPI-CRESCENZI, 1956; PONTIS, JAMES and BADDILEY, 1960), and in hen oviduct (STROMINGER, 1955b). However, it has not been found in higher plants.

The procedure described by MUNCH-PETERSEN (1956) for preparing GDP-D-mannose pyrophosphorylase from yeast is reported here.

Assay. The assay is based on the measurement of GTP formed by the pyrophosphorolysis of GDP-D-mannose. The following series of reactions, catalyzed by nucleoside diphosphokinase, hexokinase, and D-glucose 6-phosphate dehydrogenase, respectively, is utilized in the assay method:

$$GTP + ADP \rightleftharpoons GDP + ATP$$

$$ATP + \text{D-glucose} \rightarrow ADP + \text{D-glucose 6-phosphate}$$

$$\text{D-glucose 6-phosphate} + TPN^+ \rightarrow \text{D-gluconic acid 6-phosphate} + TPNH + H^+$$

The formation of TPNH is measured at 340 $m\mu$. The assay mixture is composed of Tris buffer, 0.05 M, pH 7.3, 800 μl; GDP-D-mannose, 0.002 M, 100 μl; $MgCl_2$, 1 M, 5 μl; ADP, 0.002 M, 20 μl; D-glucose, 0.01 M, 25 μl; TPN, 0.01 M, 15 μl; commercial hexokinase, 25 μl; D-glucose 6-phosphate dehydrogenase (MUNCH-PETERSEN, 1955), 15 μl; GDP-D-mannose pyrophosphorylase, 5—10 μl. Sodium pyrophosphate, 0.1 M, 10 μl, is added to start the reaction. The indicator enzymes are added in large excess. It is not necessary to add nucleoside diphosphokinase separately, as it is usually a contaminant of partially purified hexokinase. A unit of enzyme is defined as the amount of enzyme which will cause the splitting of 0.1 μmole of GDP-D-mannose per minute.

Preparation. 50 g of brewer's yeast (A/S Tuborgs Fabriker, Copenhagen), is autolyzed and extracted overnight with 200 ml of 0.07 M $(NH_4)_2HPO_4$ at 25° C, and the suspension is centrifuged. The proteins in the supernatant solution are precipitated at 2° by the addition of a saturated solution of $(NH_4)_2SO_4$ to 65% saturation. After 30 min at 2° the precipitate is collected by centrifugation and redissolved in 75 ml of ice-cold H_2O. The pH is adjusted to 7.5. The solution is again made 65% saturated in respect to $(NH_4)_2SO_4$. After 30 min at 2°, the precipitate is collected by centrifugation, dissolved in 25 ml of 0.02 M phosphate buffer, pH 7.5, and subjected to ethanol fractionation.

The solution is cooled to − 1°, and a two-fold volume of 50% ethanol at − 20° is added dropwise with mechanical stirring. The temperature of the surrounding bath is lowered to − 12° during this operation. The mixture is centrifuged for 5 min at − 15°. The supernatant liquid is left at − 15° for 20 min, and centrifuged again. The second precipitate, which contains the bulk of the activity, is stored overnight at − 20°, and then dissolved in 35 ml of 0.001 M Tris buffer, pH 7.3.

To this solution 35 ml of calcium phosphate gel (16 mg dry weight/ml) is added and the suspension is mixed and centrifuged. The enzyme which is adsorbed to the gel is eluted by the addition of 50 ml of 0.05 M phosphate buffer, pH 7.2. The clear eluate is made 60% saturated in $(NH_4)_2SO_4$ by addition of a saturated solution of $(NH_4)_2SO_4$, and the precipitate formed after standing overnight at 2° is collected and stored at − 20°. Under these conditions, the enzyme is stable for several weeks. A summary of the purification procedure is given in Table 6.

Table 6. *Purification of GDP-D-Mannose Pyrophosphorylase from Brewer's Yeast.*

	Total units	Volume (ml)	Specific activity (units/mg protein)
Crude extract.	145	150	0.01
$(NH_4)_2SO_4$ precipitate	96	90	0.02
Ethanol fraction	53	35	0.11
Calcium phosphate gel eluate . .	21	5	0.50

Properties. The preparation described above does not catalyze the pyrophosphorolysis of UDP-D-glucose, but is slightly contaminated with UDP-N-acetyl-D-glucosamine pyrophosphorylase. The equilibrium constant has not been determined; however, the reaction appears to be freely reversible.

7. α-D-Galactose-1-Phosphate Uridyl Transferase from *Saccharomyces fragilis*.

Extracts of yeast adapted to D-galactose catalyze the following uridyl transfer: UDP-D-glucose + α-D-galactose 1-phosphate ⇌ UDP-D-galactose + α-D-glucose 1-phosphate (Kalckar, Braganca and Munch-Petersen, 1953).

This reaction is responsible for the formation of UDP-D-galactose in yeast, bacteria, and mammalian tissues, in which UDP-D-galactose pyrophosphorylase is either absent or relatively weak. In higher plants, however, only a weak α-D-galactose 1-phosphate uridyl transferase has been detected (Ginsburg, personal communication) and the conversion of α-D-galactose 1-phosphate to UDP-D-galactose in plants probably occurs by the pyrophosphorylase reaction.

Assay. The reaction can be measured by the liberation of α-D-glucose 1-phosphate from UDP-D-glucose upon incubation with α-D-galactose 1-phosphate (Munch-Petersen et al., 1955). The assay is performed in a manner identical to that described for UDP-D-glucose pyrophosphorylase, except that the reaction is started by the addition of 0.1 μmole of α-D-galactose 1-phosphate instead of sodium pyrophosphate. A large excess of α-D-galactose 1-phosphate must be avoided as this compound interferes with the conversion of α-D-glucose 1-phosphate to D-glucose 6-phosphate by phosphoglucomutase (Ginsburg and Neufeld, 1957; Sidbury, 1957). A correction is made for the liberation of α-D-glucose 1-phosphate from UDP-D-glucose by nucleotide pyrophosphatase.

Preparation. The preparation of extracts from D-galactose-adapted *Saccharomyces fragilis* is described in Section B III a. No purification of this enzyme from yeast has been reported; however, purification procedures have been described for the enzyme from calf liver (Kurahashi and Anderson, 1958), and from *Escherichia coli* (Kurahashi and Subimura, 1960).

V. Enzymes which Catalyze Transformations of Sugar Nucleotides.

1. 4-Epimerases.

The first reaction in which a sugar nucleotide was found to participate was the interconversion of UDP-D-glucose and UDP-D-galactose, catalyzed by extracts from D-galactose-adapted yeast (Leloir, 1951). Subsequently, this reaction was shown to be catalyzed by preparations from animals, bacteria, and higher plants. Several analogous interconversions have been shown: UDP-D-xylose ⇌ UDP-

L-arabinose in higher plants (NEUFELD et al., 1957); UDP-D-glucuronic acid ⇌
⇌ UDP-D-galacturonic acid in plants and bacteria (FEINGOLD, NEUFELD and
HASSID, 1960; SMITH, MILLS, BERNHEIMER and AUSTRIAN, 1958); UDP-N-acetyl-
D-glucosamine ⇌ UDP-N-acetyl-D-galactosamine in bacteria (GLASER, 1959),
and animals (MALEY and MALEY, 1959); and UDP-D-glucosamine ⇌ UDP-D-
galactosamine in animals (MALEY and MALEY, 1959). The enzymes which catalyze
these reactions are designated as UDP-sugar 4-epimerases.

UDP-D-galactose 4-epimerase is the only one of this class of enzymes which has
been purified and studied. When the purified enzyme from calf liver was shown to
possess an absolute requirement for DPN (MAXWELL, 1956), it was suggested that
the interconversion of the two sugar nucleotides occurred by oxidation of the
4-hydroxyl to a keto group and reduction to the opposite steric configuration.
However, when DPNH labeled with tritium in both para positions was incubated
with the liver enzyme, tritium was not incorporated into the glycosyl moiety of
either UDP-D-glucose or UDP-D-galactose. An alternative mechanism involving
dehydration and readdition of the elements of water has also been considered;
since tritium is not incorporated into the nucleotide when the reaction is carried
out in tritiated water this hypothesis has been discarded. The role of DPN and the
mechanism of the reaction are therefore still not clear.

a) UDP-D-Galactose 4-Epimerase from *Saccharomyces fragilis*.

This enzyme has recently been purified 150-fold by MAXWELL and ROBICHON-
SZULMAJSTER (1960), and their preparation is reported here.

Assay. The rate of conversion of UDP-D-galactose to UDP-D-glucose is
measured by oxidation of the latter to UDP-D-glucuronic acid by the specific
dehydrogenase in the presence of DPN, and the formation of DPNH is measured
spectrophotometrically. The reaction cuvette contains 0.035 μmole UDP-D-
galactose, 0.5 μmole DPN, and 200 units of UDP-D-glucose dehydrogenase in a
final volume of 0.5 ml, 0.1 M with respect to glycine buffer at pH 9.0. If the UDP-
D-galactose is contaminated with traces of UDP-D-glucose, it is necessary to wait
a few minutes until the reading at 340 mμ is steady. UDP-D-galactose 4-epimerase
is then added and the reaction rate is measured over an interval of 4 min. A unit
of activity is defined as the amount of enzyme causing an increase in optical density
of 0.001/min.

Purification. *Saccharomyces fragilis* (American Type Culture Collection
No. 10 022) is adapted to D-galactose by growth in a medium containing 15 g of
Difco yeast extract, 1.8 g of $(NH_4)_2SO_4$, 1.5 g of KH_2PO_4 and 75 g of D-galactose
in 1,500 ml. About 100 ml of a 24 hour culture of adapted cells is inoculated into
each of 12 6-liter flasks containing 1,500 ml of the same medium. The cultures are
incubated with shaking for 24 hours. Cells are harvested in a SHARPLES centrifuge.

The wet cell mass (310 g) is suspended in a final volume of 1 liter of 0.1 M
glycine, pH 6.5. Cells are broken by a 30-second exposure in a NOSSAL disintegrator. In each NOSSAL tube, 9 ml of the cell suspension and 8 g of acid-washed
glass beads (75 μ) are used, and the operation is carried out in a — 10° room. The
cell suspension is not allowed to freeze. Subsequent operations are carried out at
3—5°.

The broken cell suspension is centrifuged, and the cell-free supernatant
(590 ml) is freed of nucleic acid by autolytic digestion. 60 ml of 1 M potassium
phosphate buffer, pH 7.0, is added, the preparation is incubated at 37° for two
hours, and then frozen and stored overnight.

$(NH_4)_2SO_4$ is then added to the solution to 50% saturation (175 g/600 ml of solution). The precipitate is collected by centrifugation, and dissolved in distilled water to a final volume of 100 ml. The concentration of $(NH_4)_2SO_4$ in this solution is determined with the use of a conductivity meter. Additional solid $(NH_4)_2SO_4$ is added to 35% saturation, and the precipitate discarded. The supernatant solution is made 50% saturated with respect to $(NH_4)_2SO_4$, and the precipitate is collected by centrifugation, dissolved in 15 ml of water, and dialyzed for 2.5 hours against 4 liters of 3×10^{-3} M mercaptoethanol.

Further purification is achieved by chromatography on DEAE-cellulose. The dialyzed solution, containing 225 mg of protein, is placed on a column 2.5 cm in diameter containing 5 g of water-washed DEAE-cellulose. Continuous gradient elution is carried out as follows: the mixing flask contains 500 ml of a solution of 0.05 M glycylglycine, pH 6.0, in 0.05 M NaCl; the reservoir vessel contains 500 ml of 0.05 M glycine, pH 9.0, in 0.6 M NaCl. Both buffers are adjusted to the proper pH with NaOH. The rate of flow is adjusted to 1.5 ml/min and 10 ml fractions are collected.

Previous work showed a close correlation between the activity of UDP-D-galactose 4-epimerase and fluorescence at 450 mμ, the activating wave length being 350 mμ (Maxwell, Robichon-Szulmajster and Kalckar, 1958). The column eluates can therefore be examined for fluorescence rather than be assayed spectrophotometrically. The fluorescent fractions (34 to 50) are pooled and concentrated by lyophilization to 45 ml. A saturated solution of $(NH_4)_2SO_4$ (150 ml) is added, and the precipitate is centrifuged and dissolved in 6 ml of 0.25 M glycylglycine buffer, pH 7.5. This solution can be stored at $-20°$ for at least 2 months without significant loss of activity. The purification procedure is summarized in Table 7.

Table 7. *Purification of UDP-D-Galactose 4-Epimerase from Saccharomyces fragilis.*

	Total units $\times 10^{-5}$	Specific activity (units/mg protein) $\times 10^{-3}$
Crude extract	22	0.5
$(NH_4)_2SO_4$ precipitate I .	16	1.8
$(NH_4)_2SO_4$ precipitate II .	17	7.1
DEAE-cellulose eluate . .	23	50
$(NH_4)_2SO_4$ precipitate III	22	75

Properties. The enzyme has a broad pH optimum between 8 and 9.5. At equilibrium, the ratio of UDP-D-glucose to UDP-D-galactose is approximately 3 : 1.

The most interesting characteristic of the enzyme is that it contains tightly bound DPN. The enzyme-DPN complex may be dissociated with p-chloromercuribenzoate, with simultaneous loss of activity and fluorescence. The activity (but not the fluorescence) may be restored by the addition of cysteine plus DPN, but not by DPNH. DPN may also be removed from the protein by adsorption onto charcoal.

Like the liver UDP-D-galactose 4-epimerase, the yeast enzyme does not catalyze incorporation of tritium into the sugar nucleotides either from tritiated water or from DPN labeled with tritium in both para positions.

b) 4-Epimerases from Higher Plants.

Although the 4-epimerases of higher plants have not been purified, some information concerning them has been obtained with crude enzyme preparations.

UDP-D-galacturonic acid 4-epimerase and UDP-L-arabinose 4-epimerase have been separated from each other and the equilibrium constants for the 4-epimerisation of UDP-D-glucuronic acid and of UDP-D-xylose have been found to be 1.1 and 1.0 respectively (FEINGOLD, NEUFELD and HASSID, 1960).

1. UDP-D-Galactose 4-Epimerase from _Phaseolus aureus_. _Assay._ The enzyme may be assayed as described for UDP-D-galactose 4-epimerase from yeast (Section C I a).

Preparation. A fresh (unlyophilized) UDP-D-glucose pyrophosphorylase preparation from mung bean seedlings (Section B IV a) has an active UDP-D-galactose as well as UDP-L-arabinose 4-epimerase (NEUFELD et al., 1957). The epimerases are unstable and rapidly lose activity when kept at $+ 4°$ or at $- 10°$.

2. UDP-L-Arabinose 4-Epimerase from _Phaseolus aureus_. _Assay._ In addition to being present in soluble enzyme preparations from mung bean seedlings, UDP-L-arabinose 4-epimerase also exists in bound form in particulate preparations from mung bean seedlings and from other higher plants. It can be solubilized from these preparations by treatment with digitonin. This procedure permits the separation of the UDP-L-arabinose 4-epimerase from the UDP-L-galacturonic acid 4-epimerase which is also present in the particulate preparations.

Assay. No convenient assay for the enzyme has yet been devised. The following method involves hydrolysis of the sugar nucleotides and determination of the D-xylose/L-arabinose ratio by chromatographic methods. A typical assay mixture (devised for the soluble enzyme) consists of 2 μmoles of UDP-D-xylose (or UDP-L-arabinose), 5 μmoles of Tris buffer, pH 7.5, and enzyme (Section A IV 1) in a total volume of 0.10 ml. After an appropriate length of time at 37°, 0.1 ml of 2 N HCl is added and the mixture is held at 100° for 15 min. The hydrolyzate is taken to dryness in vacuum and the residue is chromatographed in water-saturated phenol, using D-xylose and L-arabinose reference spots to locate the position of the separated pentoses. These are eluted and their relative concentration determined by the method of MEJBAUM (1939).

3. UDP-D-Galacturonic Acid 4-Epimerase from Radish. Particulate preparations from radish roots are convenient as a source of this enzyme. Although such preparations are of low activity and always contain UDP-D-glucuronic acid decarboxylase, they can be used for the preparation of UDP-D-galacturonic acid labeled in the galacturonosyl moiety from C^{14}-labeled UDP-D-glucuronic acid.

Assay. Equal volumes of enzyme suspension and UDP-D-glucuronic acid, labeled with C^{14} in the D-glucuronosyl moiety (10 μcuries/ml, 30 μcuries/μmole) are incubated at 37° in a capillary tube for periods of time ranging from 5 to 120 min. The reaction products are separated by electrophoresis at pH 3.6.

Preparation. Commercially grown radish roots (100 g) are cut into small pieces and homogenized in a chilled WARING blendor for 1 min with 70 ml of ice-cold 0.01 M sodium-potassium phosphate buffer, pH 7.0. The slurry is filtered through cheesecloth; the homogenate is centrifuged at 2,000 $\times$ g for 5 min, and the precipitate discarded. The supernatant liquid is centrifuged at 18,000 $\times$ g for 30 min and the sedimented particulate material is suspended in 0.5 ml of 0.1 M Tris buffer, pH 7.5. This preparation may be stored at $- 10°$ for several weeks.

2. UDP-D-Glucose Dehydrogenase from Peas.

DPN-linked dehydrogenases which catalyze the oxidation of UDP-D-glucose to UDP-D-glucuronic acid have been found in calf liver (STROMINGER, KALCKAR, AXELROD and MAXWELL, 1954), bacteria (SMITH, MILLS, BERNHEIMER and

Austrian, 1958), and peas (Strominger and Mapson, 1957). Although the oxidation of a primary alcohol group to a carboxyl group involves transfer of 4 electrons and formation of two equivalents of DPNH, attempts to trap an aldehyde intermediate have been unsuccessful. The reaction appears to be irreversible. A 1,000-fold purification of the enzyme from pea seedlings has been achieved by Strominger and Mapson (1957) by the following procedure.

Assay. The dehydrogenase activity is determined spectrophotometrically by the rate of DPN reduction. The assay mixture consists of 0.1 μmole of UDP-D-glucose, and 0.5 μmole of DPN, and the enzyme, in a total volume of 0.5 ml of 0.1 M glycine buffer, pH 8.7. A control is run without UDP-D-glucose in the early stages of purification. A unit of activity is defined as the amount of enzyme required to give an increase in optical density of 0.001/min.

Preparation. Peas are soaked for 12 hours in tap water and germinated for 1—3 days at 20°. The seedlings (600 g) are ground in the cold with 600 ml of 0.025 M phosphate buffer, pH 6.8, in the presence of a trace of silicone antifoam. After filtration through cheesecloth the pH is adjusted to 5.5 with 1 N acetic acid. The supernatant obtained after 15 min of centrifugation at 20,000 ×g is brought back to pH 6.8 with 1 N NaOH.

Solid $(NH_4)_2SO_4$ is added to this supernatant solution to 55% saturation. After ten minutes at 3°, the precipitate is collected by centrifugation and dissolved in about 60 ml of 0.01 M buffer, pH 6.8, 10^{-3} M with respect to EDTA (in the subsequent steps, the EDTA concentration is maintained at this level). The solution so obtained is adjusted to pH 5.5 with 1 N acetic acid, heated rapidly to 50°, and held at this temperature for 1.5 min. After rapid cooling to 3°, the precipitate is removed by centrifugation and discarded.

The heat-treated supernatant solution is adjusted to pH 6.8 with NaOH, and the $(NH_4)_2SO_4$ concentration is measured with a conductivity meter. Sufficient saturated $(NH_4)_2SO_4$ solution is then added to bring the saturation level to 21%. The precipitate is discarded. The supernatant solution is made 32% saturated, and the precipitate collected by centrifugation is dissolved in about 40 ml of 0.01 M phosphate buffer, pH 6.6. This preparation is stable at −20° for several weeks without significant loss of activity.

The preparation is then dialyzed for 4 hours against 3 l of 0.01 M phosphate buffer, pH 6.6, at 1°. Longer dialysis causes loss of activity, perhaps because of the removal of $(NH_4)_2SO_4$ which stabilizes the enzyme. After dialysis the turbid solution is centrifuged, and the clear supernatant solution is cooled in a brine bath kept at −3°. Cold acetone is added dropwise until a concentration of 35% (v/v) is reached. After equilibration for 10 min, the precipitate is removed by centrifugation and discarded. Acetone is added to 50% concentration. After 10 min the precipitate is centrifuged, dried in vacuo and dissolved in 40 ml of 0.01 M phosphate buffer, pH 6.6, containing 1% of $(NH_4)_2 SO_4$. The solution is dialyzed at once, to remove traces of acetone, against 3 l of phosphate buffer (0.01 M), pH 6.6 for 3 hours.

The solution is treated with an equal volume of calcium phosphate gel (17 mg/ml) prepared according to Keilin and Hartree (1938), allowed to stand for 15 min, and centrifuged. The precipitate is washed with 0.01 M phosphate buffer, pH 6.6, and the enzyme is eluted with two 15 ml portions of 0.1 M phosphate buffer, pH 6.6.

The enzyme is concentrated by the addition of saturated $(NH_4)_2SO_4$ solution to 42% saturation. The precipitate is collected by centrifugation, dissolved in 10 ml of 0.01 M phosphate buffer, pH 6.6 and stored at −20°. A precipitate which forms after several weeks of storage is removed by centrifugation. A summary of the purification procedure is given in table 8.

Properties. The pH optimum for the reaction is 9.0; approximately 50% of maximal activity is observed at pH 8.0 and 10.00. K_m for UDP-D-glucose is 7×10^{-5} M; K_m for DPN is 1.15×10^{-4} M. The purified enzyme is specific for UDP-D-glucose, and does not oxidize UDP-D-galactose, UDP-N-acetyl-D-glucosamine, UDP-N-acetyl-D-galactosamine, GDP-D-mannose, D-glucose, α-D-glucose 1-phosphate or ethanol. TPN cannot be substituted for DPN in the system.

Table 8. *Purification of UDP-D-glucose Dehydrogenase from Peas.*

	Total units	Total protein[1] (mg)	Specific activity (units/mg protein)
Pea extract	16,300	3,000	5
$(NH_4)_2SO_4$ precipitate	11,000	54	204
Acetone fraction	9,100	19	480
Calcium phosphate gel eluate . .	3,600	3	1,200
$(NH_4)_2SO_4$ precipitate	3,400	0.7	4,850

[1] From 100 g of seedlings.

The enzyme is partially inhibited by several sulphydryl reagents: 10^{-6} M p-chloromercuribenzoate, 10^{-2} M oxidized glutathione, 10^{-3} M iodosobenzoic acid. At low enzyme concentrations 10^{-2} M cysteine or thioglycollate enhance the enzyme activity.

3. UDP-D-Glucuronic Acid Decarboxylase from Wheat Germ.

Preparations from a number of plants catalyze the decarboxylation of UDP-D-glucuronic acid to UDP-D-xylose (FEINGOLD, NEUFELD and HASSID, 1960). These crude preparations also contain UDP-L-arabinose 4-epimerase activity, and the UDP-D-xylose is therefore contaminated with variable amounts of UDP-L-arabinose.

Since the decarboxylase of particulate preparations (AV 1 b 2 and A V 1 b 3) is relatively weak, it is recommended that UDP-D-glucuronic acid labeled with C^{14} in the D-glucuronosyl moiety (30 μcuries/μmole) be used for its detection. A more active decarboxylase which does not require radioactive substrate for detection can be obtained from wheat germ.

Assay. The assay method is based on the separation of UDP-D-xylose and UDP-L-arabinose from UDP-D-glucuronic acid by paper electrophoresis (Table 2). For the assay of wheat germ enzyme, the reaction mixture (contained in a capillary) consists of UDP-D-glucuronic acid, 0.25 μmole, and enzyme in 25 μl of 0.1 M Tris buffer, pH 7.0. After an appropriate time at 25°, (15 to 120 min), the mixture is subjected to paper electrophoresis at pH 3.6 (Table 2). The sugar nucleotide areas are located, eluted, and their concentration is determined as described in Section A II.

Preparation. Commercial wheat germ, 100 g, is stirred at 2° for 2 hours with 200 ml of 0.05 M Tris buffer, pH 7.5. The thick slurry is filtered through cheese-cloth and centrifuged in the cold at 18,000 $\times g$ for 1 hour. The supernatant liquid containing the enzyme is carefully decanted and used for the preparation of UDP-pentose.

B. Synthesis of Disaccharides.

I. Enzymes which Catalyze the Formation of Sucrose and Sucrose Phosphate

The enzymatic synthesis of sucrose was first accomplished with extracts from *Pseudomonas saccharophila*, which catalyze the following reaction:

$$\alpha\text{-D-glucose 1-phosphate} + \text{D-fructose} \rightleftharpoons \text{sucrose} + \text{phosphate}.$$

The equilibrium constant, K, for this reaction is 0.053 at pH 6.6 and 30°, and $\Delta F°$ is calculated to be $+ 1770$ calories (HASSID, 1951). Enzymatic synthesis of sucrose may also be achieved by transfer of a D-fructosyl residue from a suitable donor (e.g., raffinose) to a D-glucose acceptor. This reaction is catalyzed by enzymes (levansucrases) from various microorganisms (e.g. *Aerobacter levanicum* [HESTRIN, 1959]). The equilibrium constant is close to unity for the reaction: raffinose + D-glucose $\rightleftharpoons$ sucrose + melibiose, and hence, $\Delta F° \sim 0$ (HESTRIN and AVIGAD, 1958). This reaction does not represent a true *de novo* formation of a sucrose-type linkage, but rather the shuttling of a β-D-fructofuranosyl moiety between aldose acceptors.

Neither of these enzymes has been found in higher plants. As has been shown by LELOIR and co-workers (CARDINI, LELOIR and CHIRIBOGA, 1955; LELOIR and CARDINI, 1955) in higher plants two very similar reactions, involving a D-glucosyl transfer from UDP-D-glucose, lead to the formation of sucrose:

(1) UDP-D-glucose + D-fructose $\rightleftharpoons$ sucrose + UDP

(2) UDP-D-glucose + D-fructose 6-phosphate $\rightarrow$ sucrose phosphate + UDP.

The equilibrium constant for the first of these reactions is approximately 5 at pH 7.4 and 37°, and $\Delta F°$ is therefore $- 1000$ calories. The equilibrium constant for the second reaction has not been determined. Since sucrose phosphate does not accumulate in the plant cell, it is presumably dephosphorylated; this hydrolytic step would make the synthesis of free sucrose from D-fructose 6-phosphate irreversible.

Although the enzymes which catalyze the two reactions have not been completely separated from each other, it is evident that two different enzyme systems which can lead to sucrose formation are present in plants. The relative physiological importance of each of these enzymes is at present unknown.

When green leaves are supplied with C^{14}-labeled D-glucose in the dark, the D-fructose moiety of the newly synthesized sucrose becomes radioactive while free D-fructose remains unlabeled (PUTMAN and HASSID, 1954). Although these results show that the free D-fructose in the plant cell does not participate in sucrose synthesis, the occurrence of reaction (1) in the direction of synthesis is not ruled out, since most of the free D-fructose in the cell may be segregated in a metabolically inert pool.

Since reaction (1) is reversible, it may be important in the utilization of sucrose. Although K is in favor of sucrose synthesis, the reaction may be pulled in the opposite direction by several mechanisms, including transfer of a D-glucosyl group to form polysaccharides:

$$\text{sucrose} + \text{UDP} \rightleftharpoons \text{UDP-D-glucose} + \text{D-fructose}$$
$$\text{UDP-D-glucose} + (\text{D-glucose})_n \rightarrow \text{UDP} + (\text{D-glucose})_{n+1}$$

Net: Sucrose + $(\text{D-glucose})_n \rightarrow (\text{D-glucose})_{n+1}$ + D-fructose

The overall reaction, which is similar to those catalyzed by dextransucrase and amylosucrase, shows that sucrose may serve as the glycosyl donor for polysaccharide synthesis in plants, as well as in bacteria.

1. UDP-D-Glucose-D-Fructose Transglucosylase from Wheat Germ.

The purification procedure is that of CARDINI, LELOIR and CHIRIBOGA (1955).

Assay. The assay is based on estimation of sucrose by the resorcinol method of ROE (1934) after free D-fructose, which would also react, has been destroyed by treatment with alkali.

The following reagents are mixed in a total volume of 0.15 ml: UDP-D-glucose, 0.5 μmole; D-fructose, 2 μmoles; Tris buffer, pH 7.2, 20 μmoles. After 30 min at 37°, water is added to 0.5 ml, followed by 0.02 ml of 5 N NaOH. The tubes are heated for 10 min at 100°. Sucrose is then determined by the ROE method, using one fourth the volume of test solution and reagents, and reading the color at 490 mμ. A unit of activity is defined as the amount of enzyme catalyzing the formation of 1 μmole of sucrose in 30 min.

Preparation. Commercial wheat germ (30 gm) is mixed in a blender with 100 ml of 0.05 M phosphate buffer, pH 7.2. The suspension is centrifuged for 15 min at 16,000 r.p.m. and the supernatant fluid is dialyzed against distilled water with stirring for 4 to 5 hours. This and all subsequent operations are carried out at 4°. The dialyzed extract is clarified by centrifugation. This crude extract (Fraction I) is made 50% saturated with solid $(NH_4)_2SO_4$ and the precipitate obtained by centrifugation is dissolved in a quantity of water equal to one half the volume of Fraction I, and then dialyzed against distilled water (Fraction II). To remove nucleic acids, 0.1 volume of 1 M $MnCl_2$ is added, and the suspension is stirred for 30 min. The clear supernatant fluid (Fraction III) obtained by centrifugation is fractionated with $(NH_4)_2SO_4$, and the precipitate formed between 30 and 50% saturation is dissolved in a quantity of water equal to half the volume of Fraction III, and dialyzed for 1—2 hours with constant stirring.

Alumina C_γ (5 mg dry weight/ml of Fraction III) is added to the dialyzed solution and the precipitate is discarded. The supernatant solution is the best preparation obtained. The purification is summarized in Table 9.

Table 9. *Purification of UDP-D-Glucose-D-Fructose Transglucosylase from Wheat Germ.*

	Volume (ml)	Total units	Specific activity (units/mg protein)
I. Crude extract	70		0.05
II. $(NH_4)_2SO_4$ precipitate . . .	35	672	0.24
III. $MnCl_2$ supernatant	40	132	0.66
IV. $(NH_4)_2SO_4$ precipitate . . .	20	108	1.16
V. 1st alumina C_γ supernatant	20	104	1.80
VI. 2nd alumina C_γ supernatant	20	72	2.40

Properties. The pH optimum in 0.15 M Tris or acetate buffer is 7.2. The activity falls below half-optimal below pH 5.5 and above pH 9.0. The K_m for D-fructose is 2.3×10^{-3} M. The equilibrium constant has not been accurately determined; values ranging from 2 to 8 at 37° and pH 7.4 have been found. No sucrose is formed if UDP-D-glucose is replaced by α-D-glucose 1-phosphate, D-fructose 1- or 6-phosphate, UDP-N-acetyl-D-glucosamine, or GDP-D-mannose. D-fructose cannot be replaced by L-sorbose, D-glucose, D-galactose, D-mannose, D or L-arabinose, D-ribose, or inositol. Enzyme preparations having little UDP-D-glucose-D-fructose 6-phosphate transglycosylase do not react with D-fructose 6-phosphate.

D-xylulose, D-rhamnulose, and L-sorbose can be substituted for D-fructose to give the corresponding sucrose analogues in the presence of a preparation from peas (BEAN and HASSID, 1955). It appears, therefore, that the pea and wheat germ enzymes differ in their ability to use L-sorbose as substrate.

Separation and Identification of Sucrose. Sucrose is synthesized under the conditions described in the assay, the reaction being allowed to proceed until equilibrium is reached. The disaccharide is isolated by paper chromatography in n-propanol-ethyl acetate-water (7:1:2) and identified by its characteristic chromatographic and paper electrophoretic (in 0.1 N $Na_2B_4O_7$, pH 9.2) behavior, and its color reactions with various sprays (FEINGOLD, AVIGAD and HESTRIN, 1956). A particularly sensitive and specific test for sucrose is its conversion to levan and D-glucose with levansucrase (HESTRIN, FEINGOLD and AVIGAD, 1956).

2. UDP-D-Glucose-D-Fructose 6-Phosphate Transglucosylase from Wheat Germ.

The procedure reported here is that of LELOIR and CARDINI (1955).

Assay. The assay is carried out as described in the preceding section, except that the pH of the buffer is 6.4, and D-fructose 6-phosphate is used instead of D-fructose. Sucrose phosphate is estimated in the same manner as sucrose. In order to distinguish between sucrose and sucrose phosphate, the latter is removed by precipitation with SOMOGYI's $ZnSO_4$-$Ba(OH)_2$ reagent (SOMOGYI, 1945).

Purification. Wheat germ (100 gm.) is suspended in 300 ml. of 0.05 M phosphate buffer, pH 7.1, and left standing at 5° for 1 hour. The paste is centrifuged at 3,000 r.p.m. for 15 min. The supernatant fluid (without dialysis) is treated with $(NH_4)_2SO_4$, $MnCl_2$, and again with $(NH_4)_2SO_4$ as described in the previous section. The precipitate obtained in the second $(NH_4)_2SO_4$ treatment (Fraction IV) is dissolved in the smallest possible amount of water and dialyzed overnight at 5° against several changes of distilled water. The procedure is based on the fact that the D-fructose 6-phosphate transglucosylase is less soluble in distilled water than the enzyme using D-fructose as substrate. The precipitate obtained upon dialysis is collected by centrifugation and washed three or four times by suspension in 1 ml of distilled water and recentrifugation. The precipitate is then extracted with 2 ml of 0.05 M and 0.1 M $(NH_4)_2SO_4$; the pooled extracts are dialyzed overnight. The precipitate which forms upon dialysis is washed as above, and extracted with 2 ml of 0.05 M $(NH_4)_2SO_4$ solution, followed by three extractions with 2 ml portions of 0.1 M $(NH_4)_2SO_4$ solution. If necessary, precipitation by dialysis followed by extraction with $(NH_4)_2SO_4$ is repeated to obtain a preparation low in phosphatase and UDP-D-glucose-D-fructose transglucosylase activity. The purification procedure is summarized in Table 10. The ratio of the transglucosylase

Table 10. *Purification of UDP-D-Glucose-D-Fructose 6-Phosphate Transglucosylase from Wheat Germ.*

| | Protein mg/ml | Specific activity[1] on | | |
		D-Fructose A	D-Fructose 6-Phosphate B	$\frac{A}{B}$
Fraction IV	23	1.16	0.20	5.8
1. 0.05 M $(NH_4)_2SO_4$ extract . .	40	0.70	0.65	1.06
2. 0.1 M ,, ,, . .	22	0.50	0.90	0.55
3. 0.1 M ,, ,, . .	70	0.40	1.20	0.33
4. 0.1 M ,, ,, . .	4.4	0.90	1.10	0.82

[1] Specific activity = μmoles of total sucrose (free + esterified) formed in 30 min per mg of protein.

using D-fructose to that using D-fructose 6-phosphate as acceptor varies from 5.8 to 0.33, indicating that two enzymes are involved.

Properties. The pH optimum in 0.15 M Tris or acetate buffer is around 6.5, activity decreasing below 50% of maximal on the acid side of pH 5.5 and on the alkaline side of pH 8. The K_m for D-fructose 6-phosphate is 2.2×10^{-3} M.

Since even the best preparations reported have been contaminated with phosphatase, the acceptor specificity of the enzyme has not been determined. In particular, activity towards D-fructose 1-phosphate has not been rigorously excluded. The major product formed when D-fructose 6-phosphate is used as acceptor has been identified as sucrose 6-phosphate.

Although no metal requirement has been shown for the wheat germ enzyme, the synthesis of sucrose phosphate from UDP-D-glucose and fructose 6-phosphate by a pea preparation has been shown to require 10^{-3} M Mg^{++} (BEAN and HASSID, 1955).

Isolation and Identification of Sucrose Phosphate. A typical reaction mixture contains 160 μmoles of UDP-D-glucose, 500 μmoles of D-fructose 6-phosphate, and 8 ml of enzyme (total volume 40 ml). After 1 hour at 37° the proteins are coagulated by heating and the precipitate is removed by centrifugation. The cations are removed with Dowex 50(H$^+$), and the solution is brought to pH 7 with NH$_4$OH and separated on a column of Dowex-1 (Cl$^-$), 4.15 cm^2 ×12 cm. 300 ml of 0.001 M NH$_4$OH is allowed to percolate through the column and gradient elution is then carried out by adding 0.03 M NH$_4$Cl to a 500 ml mixing chamber filled with a solution containing 0.025 M NH$_4$Cl and 0.01 M Na$_2$B$_4$O$_7$. The effluents are tested by the ROE (1934) method, with and without precipitation of phosphates (SOMOGYI, 1945), and the fractions containing sucrose phosphate are pooled, passed through Dowex-50 (H$^+$), brought to pH 7 with NH$_4$OH, and evaporated to dryness under vacuum. Borate is removed by repeated addition of methanol and evaporation to dryness. The sucrose phosphate is further purified by paper electrophoresis at pH 5.8. The sucrose phosphate is identified by release of sucrose identified as in Section B I 1 upon treatment with phosphomonoesterase and by the formation of D-fructose 6-phosphate (shown by TPN reduction in the presence of phosphohexose isomerase and D-glucose 6-phosphate dehydrogenase [LEPAGE and MUELLER, 1949]) after hydrolysis for 5 min at 100° in 0.1 N HCl.

II. Enzyme which Catalyzes the Formation of Trehalose Phosphate.

(UDP-D-Glucose-D-Glucose 6-Phosphate Transglucosylase from Yeast).

Enzyme preparations from insects and fungi catalyze the formation of trehalose phosphate:

UDP-D-glucose + D-glucose 6-phosphate → UDP + trehalose phosphate
(CABIB and LELOIR, 1958; CANDY and KILBY, 1959).

Trehalose phosphate sometimes accumulates in substantial amounts during fermentation of sugars by dried preparations of brewer's yeast (MAJKEN-ELANDER, 1959). A specific phosphatase hydrolyzes the phosphate to form free trehalose, the major reserve disaccharide of insects and fungi (FRIEDMAN, 1960).

The preparation of an enzyme from brewer's yeast which catalyzes the synthesis of trehalose phosphate (UDP-D-glucose D-glucose 6-phosphate transglucosylase) has been described by CABIB and LELOIR (1958).

Assay. The assay is based on the measurement of UDP by pyruvic kinase. This enzyme catalyzes the transfer of phosphate from phosphoenolpyruvate to

UDP, and the pyruvate can then be estimated colorimetrically as the dinitro-phenyl hydrazone.

The following reagents are mixed: 0.5 μmole of UDP-D-glucose, 1.0 μmole of D-glucose 6-phosphate, 2.5 μmoles of $MgSO_4$, 0.1 μmole of EDTA, and 0.04 ml of enzyme, in a total volume of 0.1 ml. All solutions are neutralized before mixing. The blanks contain the same components except that D-glucose 6-phosphate is added at the end of the incubation. After incubation for 15 min at 37°, the tubes are heated for 3 min at 100° and cooled. 0.05 ml of 0.01 M phosphoenolpyruvate and 0.05 ml of pyruvic kinase solution are added. After 15 min at 37°, protein is precipitated by the addition of 1 ml of cold 10% trichloracetic acid, and centri-fugation. 0.75 ml of the supernatant fluid is taken for the colorimetric determi-nation of pyruvate by the method of Friedemann and Haugen (1943). Alter-natively, pyruvate may be determined with lactic dehydrogenase and DPNH (Strominger, 1955a). One unit of activity is defined as the amount of enzyme which catalyzes the formation of 1 μmole of UDP/hour.

Purification. Commercial brewer's yeast is passed through a meat chopper without a cutter, producing noodle-like cylinders. These are left to dry at room temperature, and the dried yeast is used as starting material for the enzyme preparation.

All operations are performed at $0-2°$ unless otherwise stated. In $(NH_4)_2SO_4$ precipitations, the suspension is allowed to stand 30 minutes before centrifugation. Centrifugations are carried out at $15,000 \times g$ for $15-20$ min at 0°, except for the acetone precipitate which is spun at $2,000 \times g$ for 10 min at $-5°$. Dialysis is per-formed with stirring at 2°.

Dried yeast (120 gm) is suspended in 360 ml of distilled water and stirred for 1 hour at room temperature. The suspension is then left at 2° for 20 hours with occasional shaking. The supernatant fluid (Fraction I) obtained after centri-fugation is mixed with an equal volume of 50% $(NH_4)_2SO_4$ solution (50 g salt per 100 ml of solution, 10^{-3} M with respect to EDTA, adjusted to pH 7.5 with NH_4OH). The precipitate is collected by centrifugation and dissolved in 10^{-3} M EDTA (Fraction II). Undissolved material is removed by centrifugation. An equal volume of 50% $(NH_4)_2SO_4$ solution is added to Fraction II, and the precipitate is dissolved in 10^{-3} M EDTA. The turbid solution is again centrifuged, and the supernatant liquid (Fraction III) is dialyzed for about six hours against 10^{-3} M EDTA (3 changes of 3 liters each). The dialyzed solution (Fraction IV) is mixed with an equal volume of 0.04 M $MgSO_4$, and is immersed into a water bath at 37° till its temperature reaches 36° (this requires about 7 min). It is then cooled rapidly to 20° and centrifuged. The supernatant liquid (Fraction IV) is placed in an ice-salt bath at $-4°$ until its temperature drops to 0°, and 0.35 volumes of cold $(-4°)$ acetone is slowly added with stirring over a period of 20 min. During this time the temperature is allowed to fall to $-3°$. The precipitate is collected by centrifugation and redissolved in about 15 ml of 0.02 M potassium phosphate buffer, pH 7, 0.001 M with respect to EDTA, and dialyzed for about 4 hours against the same buffer (3 changes of 3 liters each). The dialyzed enzyme is treated for 15 min with approximately 0.2 volumes (the optimal amount is determined by a preliminary small scale experiment) of a suspension of alumina C_γ (31.5 mg dry weight/ml). After standing for 15 min with occasional stirring, the suspension is centrifuged. The clear, slightly yellow supernatant solution (Fraction VII) contains the enzyme, purified $15-20$ fold. A summary of the purification procedure is given in Table 11.

Properties. The pH optimum in 0.1 M Tris-maleate buffer is 6.6. The activity falls to about one-half at pH 5.7 and pH 7.5. In the presence of 2.5×10^{-2} Mg^{++},

Table 11. *Purification of UDP-D-Glucose-D-Glucose 6-Phosphate Transglucosylase from Brewer's Yeast.*

	Volume (ml)	Total units	Total protein (mg)	Specific activity (units/mg protein)	Yield
I. Crude extract	214	25,700	29,100	0.88	100
II. 1st $(NH_4)_2SO_4$ precipitate. . .	62	21,400	5,150	4.15	83
III. 2nd $(NH_4)_2SO_4$ precipitate . .	54	12,000	3,160	3.8	56
IV. Dialyzed solution	69	9,770	2,660	3.7	38
V. Supernatant after heat treatment.	135	8,440	2,050	4.1	33
VI. Acetone fraction.	19.5	6,520	680	9.6	25
VII. Alumina $C\gamma$-treated solution .	18.3	4,460	247	18	17

the enzymatic activity is doubled; higher Mg^{++} concentrations become inhibitory. The requirement for Mg^{++} does not seem to be absolute since the enzyme shows activity in the absence of Mg^{++}, and this activity is not depressed by the addition of EDTA. Phosphate (3.3×10^{-2} M) causes a 50% inhibition of the enzyme. Salts in general tend to inhibit the enzyme; the activity is depressed 35% by 0.2 M NaCl or 0.1 M Tris-maleate buffer at pH 7.

Neither D-glucose nor α-D-glucose 1-phosphate act as glucosyl acceptors. D-Fructose 6-phosphate causes liberation of UDP when used in place of D-glucose 6-phosphate in the standard assay; this, however, is due to the formation of D-glucose 6-phosphate by hexose phosphate isomerase which is also present in the purified preparation. (The purified enzyme is contaminated with a specific phosphatase, which does not affect D-glucose 6-phosphate, UDP, or α-D-glucose 1-phosphate, but releases trehalose from trehalose 6-phosphate.)

At pH 6.1, the formation of UDP from UDP-D-glucose is almost quantitative. No reversibility is observed even when D-glucose 6-phosphate dehydrogenase and UDP-D-glucose dehydrogenase are added together in order to remove any reaction products that might be formed. However, as pointed out by CABIB and LELOIR (1958), the lack of reversibility may be an experimental artifact. These authors calculate that the approximate maximal $\Delta F°$ of hydrolysis of trehalose phosphate as -4400 calories, a value not very different from the value for the α-1,4 linkage in maltose (-4000 calories) but considerably lower than the value for sucrose (-6600 calories).

Isolation and Identification of Trehalose Phosphate. Trehalose 6-phosphate is isolated from enzyme incubations by precipitation of water-soluble, ethanol-insoluble barium salts and separation of the latter by anion exchange chromatography. A typical reaction mixture contains 70 μmole of UDP-D-glucose, 350 μmoles of $MgSO_4$, 14 μmoles of EDTA, and 140 units of enzyme in a total volume of 14 ml. After one hour at 37° the mixture is deproteinized by holding at 100° for 2 min, centrifuged, the clear supernatant fluid is taken to dryness at 35—40°, taken up in a minimal quantity of water (about 2 ml), and brought to pH 7 with NH_4OH. The precipitate which forms upon neutralization is removed, washed with 0.5 ml of water, and discarded. To the combined supernatant fluids is added 0.9 ml of 1 M barium acetate. The voluminous precipitate is removed, washed with five 0.5 ml portions of water, and discarded. The combined supernatant fluids are treated with 25 ml of 96% ethanol, and the precipitate which forms after holding the suspension at $-15°$ for 12—18 hours is washed once with 96% ethanol and once with dry ether, and then dried in a desiccator. It is then dissolved in 5 ml of water, the barium ions are precipitated with 1 M $(NH_4)_2SO_4$, and the precipitate is washed with four 5 ml portions of water. The volume of the combined supernatant liquids is brought to 30 ml with water and NH_4OH is used to adjust the pH to 8.2. The solution is then allowed to soak into a Dowex 1 (Cl^-) column, 0.9 cm in diameter and about 10.5 cm high. The column is washed with 75 ml of 0.001 M NH_4OH, and then eluted with a solution containing 0.01 M $K_2B_4O_7$ and 0.015 M NH_4Cl at a rate of about 7 ml per minute. Suitable fractions (10 ml) are collected and analyzed for carbohydrate with anthrone reagent (Trevelyan and Harrison, 1952). The substance which starts to emerge after about 500 ml of eluant have passed through the column is trehalose phosphate. This is further purified by electrophoresis on paper at pH 5.8.

Trehalose phosphate is mobile upon electrophoresis at pH 3.6 or 5.8; it gives a positive reaction with the reagent of Bandurski and Axelrod (1951). Upon treatment with seminal phosphatase, equimolar quantities of trehalose and inorganic phosphate are released. The trehalose cannot be differentiated readily from cellobiose or maltose by its mobility in paper chromatography; however, trehalose, unlike the other two disaccharides, does not give a test for reducing sugars. In addition, it is easily separated from cellobiose and maltose by paper electrophoresis in borate buffer. Hydrolysis of trehalose phosphate in 1 N HCl at 100° for 8 hours produces equimolar quantities of glucose [determined by the method of Somogyi (1945)] and D-glucose 6-phosphate (determined with D-glucose 6-phosphate dehydrogenase, Kornberg and Horecker, 1955).

C. Synthesis of Glycosides.

Glycosides comprise a large group of extremely varied compounds which occur in the plant kingdom. The sugar moiety is frequently D-glucose, either alone or as the disaccharide gentiobiose; however, a wide variety of other sugars (e.g., D-galactose, D-xylose, L-arabinose, L-rhamnose, D-glucuronic acid, as well as several dideoxy and methyl sugars) occur, either singly or in combination. The aglycone moiety includes molecules of widely differing structures, such as aliphatic alcohols, phenols, steroids, terpenes, flavones, and related compounds (see review by Stoll and Jucker, 1958). The physiological function of these glycosides is unknown in most cases. It is possible that the sugar moiety serves to solubilize or to detoxify the aglycone.

The enzymatic synthesis of several glycosides has been reported. With UDP-D-glucose as donor, the D-glucoside of anthranilic acid has been prepared by Jacobelli, Tabone and Tabone (1958), using an extract of lentils. Yamaha and Cardini (1960a and 1960b) have described the formation of several D-diphenol glucosides and phenol gentiobiosides by wheat germ extracts. These authors were able to separate the enzymes which transfer the first and the second D-glucose units of the gentiobioside. The transfer of the D-glucuronosyl group from UDP-D-glucuronic acid to quercetin by extracts from bean leaves has also been described (Marsh, 1960).

I. Enzyme which Catalyzes the Formation of Diphenol-D-Glucosides.

(UDP-D-Glucose-Diphenol Transglucosylase from Wheat Germ).

The procedure outlined by Yamaha and Cardini (1960a) is reported below.

Assay. The assay mixture contains 0.4 μmole of UDP-D-glucose, 0.4 μmole of hydroquinone, 0.1 μmole of EDTA, 0.1 μmole of cysteine, 20 μmoles of Tris-maleate buffer, pH 6.5, and 0.04 ml of enzyme solution, in a total volume of 0.1 ml. A control is performed without UDP-D-glucose. After an incubation period of

10—30 min the tubes are immersed into a boiling water bath for 3 min, and the UDP formed is measured as described in Section B II. One unit is defined as the amount of enzyme which catalyzes the formation of 1 μmole of UDP in 15 min.

If the enzyme preparation contains phosphatases which hydrolyze UDP, enzyme activity can be assayed by measuring the amount of arbutin (hydroquinone D-glucoside) formed. In this case, the reaction is stopped by adding 0.1 ml of 5% $ZnSO_4$ and 0.1 ml of 0.3 N $Ba(OH)_2$ (SOMOGYI, 1945) plus 0.2 ml of water. The mixture is centrifuged and the precipitate is washed with 1 ml of water. The combined supernatant fluids are concentrated and streaked as narrow bands on Whatman No. 1 paper. Known materials are placed alongside. The paper is developed overnight with water-saturated n-butanol. The substances used as standards are located with diazotized p-nitroaniline. The parts of the paper containing the glycoside are cut out and placed in a test tube, and 2 ml of water is added. After a few minutes 0.1 ml of FOLIN and CIOCALTEAU (1927) phenol reagent and 0.4 ml of 20% Na_2CO_3 are added, and the tubes are placed in a boiling water bath for 1 min. After cooling, the optical density is measured at 650 mμ. Eighty-five percent of the arbutin placed on the paper is recovered after chromatography and extraction.

Preparation. All operations are carried out at 0—5°. Wheat germ (100 g) is suspended in 300 ml of 0.05 M phosphate buffer, pH 7.0. The suspension is allowed to stand for 2 hours with occasional stirring, and is centrifuged for 20 min at 16,000 $\times g$. The supernatant fluid is fractionated with $(NH_4)_2SO_4$. The fraction precipitating between 20 and 50% saturation is dissolved in 0.05 M phosphate buffer, pH 7.0, 0.001 M in respect to cysteine and EDTA (Fraction I, 102 ml). This solution is dialyzed for 2 hours with stirring against 2 liters of a solution 10^{-4} M in respect to EDTA and cysteine, pH 7. The dialysate is fractionated by adding 0.2 volume of saturated $(NH_4)_2SO_4$ solution at pH 7.0. The precipitate is discarded, and 0.33 volume of saturated $(NH_4)_2SO_4$ solution is added to the supernatant solution. The D-glucoside-forming enzyme is present in the precipitate, which is dissolved in 0.05 M phosphate buffer, pH 7.0, 10^{-3} M with respect to cysteine and EDTA (Fraction II, 81 ml). The supernatant fluid (S) contains the gentiobioside-forming enzyme (see Section D-II).

Fraction II is adjusted to pH 4.7 with 1 N acetic acid, and 0.05 volume of a saturated solution of $(NH_4)_2SO_4$, pH 4.7, is added. The precipitate is discarded, and 0.2 volume of saturated $(NH_4)_2SO_4$ solution, pH 4.7, is added to the supernatant solution. The precipitate is collected by centrifugation and is dissolved in 0.05 M phosphate buffer, pH 6.0 (Fraction III). Alumina C_γ gel (about 0.5 mg dry weight/mg protein) is added. After standing for 15 minutes with occasional stirring the suspension is centrifuged and the precipitate discarded. The supernatant solution (Fraction IV) is used for further studies. A summary of the purification procedure is given in Table 12.

Table 12. *Purification of UDP-D-Glucose-Diphenol Transglucosylase*

	Volume (ml)	Total units	Total protein (mg)	Specific activity (units/mg protein)	Recovery %
I. 1st $(NH_4)_2SO_4$ precipitate . . .	102	495	8850	0.056	100
II. 2nd $(NH_4)_2SO_4$ precipitate. . .	81	333	3880	0.086	67
III. 3rd $(NH_4)_2SO_4$ precipitate. . .	32	212	1050	0.202	43
IV. Alumina $C\gamma$-treated solution .	12	54	160	0.326	11

Properties. The pH optimum of the purified enzyme is at about 6.5; activity falls steeply on either side of the optimum, reaching approximately 50% of maximal at pH 5.9 and 7.1. The enzyme is rapidly inactivated upon incubation at 37°, even in presence of substrates, so that under the conditions of test a linear rate of glycoside formation is observed for only 10 min. Nevertheless, the glycoside formation is proportional to the amount of enzyme added. The K_m calculated under these conditions is 1.6×10^{-3} M for UDP-D-glucose and 1.9×10^{-3} M for hydroquinone.

The enzyme can use UDP-D-glucose only as glycosyl donor; no activity is observed when α-D-glucose 1-phosphate, UDP-N-acetyl-D-glucosamine, UDP-D-glucuronic acid, or GDP-D-mannose are used. The best glycosyl acceptors are hydroquinone, hydroxy-hydroquinone and methoxyhydroquinone. Resorcinol, pyrogallol, and catechol also react. All of these active substrates are di or tri-phenols. No activity is observed with monophenols (phenol, p-methoxyphenol, m-methoxyphenol, o-amino phenol or salicyl alcohol), nor with benzyl alcohol, anthranilic acid or p-aminobenzoic acid. The product of the enzymatic reaction is a β-glucoside. Since UDP-D-glucose is an α-glucoside, the reaction proceeds with inversion.

II. Enzyme which Catalyzes the Formation of Phenolic Gentiobiosides.

(UDP-D-Glucose-Phenol-D-Glucoside Transglucosylase from Wheat Germ).

Assay. The assay system is the same as that described in the preceding section, except that the reaction is carried out at pH 7.5 and arbutin is used instead of hydroquinone.

Preparation. The first two steps are described in the above section. To the supernatant solution (S) obtained after removing Fraction II, 0.22 volume of a saturated solution of $(NH_4)_2SO_4$, pH 7.0, is added. The precipitate is collected by centrifugation, dissolved in 0.05 M phosphate pH 7.0, 10^{-3} M with respect to EDTA and cysteine (Fraction IIa), and the pH is adjusted to 6.0 with 1 N acetic acid. Alumina C_γ is then added (about 1 mg dry weight/mg protein). After 15 min the precipitate is collected by centrifugation and is eluted twice with 0.2 M phosphate buffer, pH 7.6 10^{-3} M with respect to EDTA and cysteine (Fraction IIb). The purification of the enzyme is summarized in Table 13.

Table 13. *Purification of UDP-D-Glucose-Phenol D-Glucoside Transglucosylase.*

	Volume (ml)	Total units	Total protein (mg)	Specific activity (units/mg protein)	Recovery %
I. 1st $(NH_4)_2SO_4$ precipitate . . .	102	115	8,850	0.013	100
IIa. 2nd $(NH_4)_2SO_4$ precipitate . .	12	40	830	0.048	35
IIb. Alumina C γ eluate	12	29	300	0.093	25

Properties. The enzyme has a broad pH optimum between pH 7 and 9.5; the activity then decreases rapidly on the acid side and reaches 50% around pH 6.1. The Michaelis constant, K_m, for UDP-D-glucose is 1.2×10^{-3} M, while that for arbutin is 0.9×10^{-3} M. The enzyme shows activity towards phenol-D-glucoside, salicin, p-methoxyphenol D-glucoside, m-methoxyphenol-D-glucoside, resorcinol D-glucoside, and L-mandelonitrile β-D-glucoside. The following substances are not used as acceptors: phenols, esculin, D-glucose, maltose, sucrose, gentiobiose, α-D-glucose 1-phosphate, α- and β-methyl-D-glucosides, and starch. p-Methoxyphenol gentiobioside is not a substrate, indicating that chains longer than 2 D-glucose units are not formed. UDP-D-glucose cannot be replaced by UDP-D-glucuronic acid, UDP-N-acetyl-D-glucosamine, or GDP-D-mannose.

D. Synthesis of Polysaccharides.

I. Enzyme which Catalyzes the Formation of Callose.

(UDP-D-Glucose-β-1, 3-D-Glucan Transglucosylase from *Phaseolus aureus*).

Callose is an insoluble polysaccharide found in fungi and in many tissues of higher plants. Its appearance in the latter is frequently associated with senescence, dormancy, or injury. While until recently callose was known only by its histological staining properties, the callose of grapevine and of pollen has been identified as β-1,3-D-glucan (ASPINALL and KESSLER, 1957; KESSLER, 1958; KESSLER, FEINGOLD and HASSID, 1960), and it is likely that all callose has the same structure.

The synthesis of callose proceeds by transfer of D-glucosyl units from UDP-D-glucose, catalyzed by particulate enzymes from mung bean seedlings and several other plants (FEINGOLD, NEUFELD and HASSID, 1958 b).

Assay. The assay is based on the detection of radioactivity transferred from UDP-D-glucose labeled with C^{14} in the D-glucose moiety to a product insoluble in water and ethanol. The assay mixture (contained in a capillary) contains 0.04 μmole UDP-D-glucose (0.5 μc per μmole), 0.5 μmole $MgCl_2$, 0.5 μmole cellobiose, 1 μmole Tris buffer, pH 7.0, and enzyme, in a total volume of 0.025 ml. After 30 min at 25°, the reaction mixture is spotted on filter paper, dried, and counted. The filter paper is then washed on a BÜCHNER funnel with about 500 ml of water followed by 50 ml of 95% ethanol, dried, and counted again. The ratio of (counts after washing)/(total counts applied to paper) is taken as a measure of the incorporation of radioactive D-glucose into insoluble polymer. Some plant preparations incorporate D-glucose from UDP-D-glucose into water-insoluble glycosides, which may not be completely removed by the alcohol wash of the routine assay. In such cases it is necessary to extract the paper with boiling 80% ethanol. Glucosides are removed under these conditions, but callose remains adsorbed to the paper.

Preparation. Seeds of *Phaseolus aureus* are aerated under water for 18 hours at room temperature. They are then drained and grown for 4 days at 30° in the dark, supported on cheesecloth which is kept moist with a 10^{-3} M solution of KH_2PO_4, $MgSO_4$, and $Ca(NO_3)_2$. The seedlings are homogenized in a WARING Blendor with 70 ml of ice-cold 0.01 M sodium-phosphate buffer, pH 7.0. All subsequent operations are performed at 0—5° C. After filtration through cheesecloth, the homogenate is centrifuged at 3,000 × g for 5 min and the precipitate is discarded. The supernatant solution is centrifuged at 20,000 × g for 25 min, and the precipitate is collected, washed by suspension in phosphate buffer and recentrifugation, and suspended in 3 ml of 0.1 M Tris buffer, pH 7.0.

The particulate preparation is vigorously shaken by hand for 1 min with 2 volumes of a 1% solution of digitonin. The supernatant fluid obtained after centrifugation at 20,000 × g for 25 min is fractionated with solid $(NH_4)_2SO_4$. The fraction precipitating between 35 and 50% saturation is dissolved in 10 ml of 0.01 M Tris buffer, pH 7.0. $(NH_4)_2SO_4$ is added to 50% saturation, and the precipitate which is collected by centrifugation is dissolved in 0.5 ml of 0.1 M Tris buffer, pH 7.0.

The solubilized enzyme is stable at − 10° for periods up to two weeks, but loses most of its activity upon standing at 0° for several hours.

Properties. Both the particulate and the solubilized enzyme are contaminated with phosphatase activity so that the UDP formed during the reaction is dephosphorylated to UMP, even in the presence of 0.16 M KF. The formation of the β-1,3-D-glucan is only slightly stimulated by Mg^{++}. There appears to be a heavy metal requirement, for a concentration of 3.3×10^{-3} M EDTA completely inhibits the incorporation of label into the polymer, while the same concentration of 8-hydroxy-quinoline, o-phenanthroline and α,α'-dipyridyl inhibit partially.

No requirement for an initial acceptor of D-glucosyl units has been shown; this "primer" is presumably present in the enzyme preparation. The solubilized preparation exhibits an absolute requirement for D-glucose or any one of a large number of α- and β-D-glucosides referred to as "activators". The activator is not incorporated into the final product, and its role is at present not known. It has been suggested that it is a "carrier" of D-glucosyl units:

$$\text{UDP-D-glucose} + \text{activator} \to \text{UDP} + \text{D-glucosyl-activator}$$
$$\text{D-glucosyl-activator} + (\text{D-glucose})_n \to \text{activator} + (\text{D-glucose})_{n+1}$$

Net: $\text{UDP-D-glucose} + (\text{D-glucose})_n \to \text{UDP} + (\text{D-glucose})_{n+1}$.

Identification of the Polysaccharide. The radioactive D-glucan is partially hydrolyzed by acid, and the oligosaccharides are then identified. The acid hydrolysis is performed as follows: A large scale incubation is carried out, using the proportions of reagents described for the assay. After the reaction has stopped (as determined by assaying small aliquots) the incubation mixture is placed into a boiling water bath for 1 min. The precipitate (containing denatured protein and radioactive polysaccharide) is washed by repeated suspension in water and centrifugation, and treated in a sealed tube for 10 min at 23° with about 0.5 ml of concentrated HCl which had been saturated with HCl gas at $-18°$.

The oligosaccharides obtained by this treatment are compared with authentic laminarodextrins (β-1,3-linked D-glucodextrins) (PEAT, WHELAN and LAWLEY, 1958) by circular chromatography in n-propanol-water-ethyl acetate (7:2:1) and by paper electrophoresis at 40 volts/cm in 0.05 M sodium borate, pH 9.2. The β-configuration of the linkage is demonstrated by hydrolysis with emulsin.

II. Enzyme which Catalyzes the Formation of Chitin.

(UDP-N-Acetyl-D-Glucosamine-Chitin Transglycosylase from *Neurospora crassa*).

Chitin is an extremely stable and insoluble polysaccharide, composed of N-acetyl-D-glucosamine residues joined by β-1,4-linkages. It occurs predominantly in the exoskeletons of crustacea and in the cell walls of fungi. Chitin often comprises 4% or more of the dry weight of fungal mycelia, in which it presumably has a function analogous to that of cellulose in higher plants. (See review by FOSTER and STACEY, 1958).

The enzymatic synthesis of chitin has been shown by GLASER and BROWN (1957) to proceed by transfer of the N-acetyl-D-glucosaminosyl residue from UDP-N-acetyl-D-glucosamine to chitodextrin acceptors. The reaction is catalyzed by a particulate preparation from *Neurospora crassa*.

Assay. The assay procedure, like that described for the synthesis of callose, is based on the transfer of radioactive label from UDP-N-acetyl-D-glucosamine labeled with C^{14} in the glycosyl residue (either in the acetyl or in the D-glucosamine moiety) to water-insoluble material. A typical reaction mixture is composed of 2.4×10^{-4} M UDP-N-acetyl-D-glucosamine (containing about 30,000 counts/min) and 10 mg of chitodextrins in volume of 2 ml. Incubation is carried out at pH 7.5, in the presence of enzyme from 1 g of *Neurospora*.

The reaction is stopped after 2 hours at 25° by the addition of 0.25 volumes of 3 N $HClO_4$, and the resulting precipitate is washed 4 times with 2 ml portions of

0.3 N HClO$_4$ and once with 2 ml of water. The precipitate is then dispersed in 3 ml of water and an aliquot is counted.

Preparation. *Neurospora crassa* (wild strain Em 5297 A) is grown under forced aeration in the FRIES minimal medium (ZUCKER and NASON, 1955). The mycelium is washed with distilled water and stored frozen. For preparation of the enzyme, the mycelium is homogenized at 3° with 3 volumes of 0.05 M Tris buffer, 10^{-2} M with respect to MgCl$_2$ and 10^{-3} M with respect to EDTA, in a TEN-BROECK homogenizer. The extract is centrifuged at 2,000 $\times g$ for 6 min, and the precipitate is discarded. The supernatant fluid is then centrifuged for one hour at 140,000 $\times g$, and the pellet is suspended in the same buffer with the aid of the homogenizer and centrifuged again. The resulting pellet is suspended in the same buffer (3 ml per 10 g of *Neurospora*). It can be kept at $-20°$ for several weeks with no loss of activity.

Properties. The synthesis of chitin proceeds best at pH 7.5 at 27°. Activity is 69% of maximal at pH 6.5 and 52% at pH 8.3. The synthesis of chitin is greatly enhanced by the addition of soluble chitodextrins which act as glycosyl acceptors. Chitin synthesis is also stimulated by the addition of N-acetyl-D-glucosamine. This substance is not incorporated into the final product and its function is not known. Half maximal activation occurs at 4.5×10^{-3} M N-acetyl-D-glucosamine.

Identification of the product. A large scale preparation of radioactive chitin is subjected to proteolytic digestion with pepsin followed by trypsin. The insoluble residue is collected by centrifugation, dissolved in concentrated hydrochloric acid which has been saturated with HCl gas at 0°, and allowed to stand in a closed flask for 5 hours at 25°. The HCl is removed by evaporation and the hydrolysate (clarified by centrifugation) is passed through a column of mixed-bed ion exchange resin to remove any deacetylated oligosaccharides. The oligosaccharides in the hydrolysate are separated chromatographically in butanol-pyridine-water (2 : 1.5 : 0.75) and the disaccharide is identified as N-N' diacetyl-chitobiose by crystallization to constant specific activity after addition of non-radioactive carrier. The oligosaccharides may also by hydrolyzed with the chitinase of almond emulsin, which is known to be specific for β-1,4-D-glucosaminides, and the resultant N-acetyl-D-glucosamine identified chromatographically.

III. Enzyme which Catalyzes the Formation of D-Xylodextrins.

(UDP-D-Xylose-D-Xylodextrin Transxylosylase from Asparagus)

D-xylose is an important constituent of higher plants. It occurs predominantly in the β-1,4-linked polysaccharide xylan, which is closely associated with cellulose in the cell wall. In addition, D-xylose is found in heteropolymers composed of D-xylose and L-arabinose, and in some glycosides such as digitonin and tomatin.

While evidence from physiological experiments suggest that the D-xylose-containing polymers are synthesized via the sugar nucleotide pathway (*see* review by HASSID, NEUFELD and FEINGOLD, 1959), such polymers have not yet been prepared enzymatically. However, the formation of 1,4-linked D-xylodextrins of a low degree of polymerization by transfer of D-xylosyl units from UDP-D-xylose has been demonstrated (FEINGOLD, NEUFELD and HASSID, 1959).

Assay. The enzyme activity is detected by measuring the incorporation of radioactivity from UDP-D-xylose (labeled with C^{14} in the D-xylosyl moiety) into neutral compounds in the presence of a D-xylodextrin acceptor. The assay mixture (contained in a capillary) consists of UDP-D-xylose and UDP-L-arabinose (0.3 μc, 10$^{-2}\mu$moles; obtained by decarboxylation of C^{14}-labeled UDP-D-glucuronic acid as described in Section B V 1b 3), D-xylodextrin (1 μmole), and enzyme in 0.65 μl of

0.08 M Tris buffer, pH 7. After 3 hours at 37°, the neutral radioactive components are separated electrophoretically or with the aid of an anion exchange resin. A control mixture, containing no D-xylodextrin, serves to measure the amount of free pentoses released from UDP-pentoses by hydrolytic enzymes.

Preparation. Commercially grown asparagus shoots (100 gm) are blended with 70 ml of 0.01 M phosphate buffer at pH 7.0. Coarse debris is removed by filtration through cheese cloth, and the resulting homogenate is centrifuged at 5,000 $\times g$ for 5 min. The supernatant fluid is then centrifuged at 20,000 $\times g$ for 30 min, and the precipitate is suspended in 0.5 ml of 0.1 M Tris buffer, pH 7.0. This suspension is shaken manually for about 1 min with an equal volume of a solution of 1% digitonin and centrifuged at 20,000 $\times g$ for 30 min. The supernatant fluid, which contains the transxylosylase, can be stored at $-10°$ for two weeks.

Properties. The solubilized enzyme preparation contains UDP-D-glucuronic acid decarboxylase and UDP-L-arabinose 4-epimerase, as well as the transxylosylase. D-xylodextrins of sizes ranging from the di-to the pentasaccharide serve as D-xylosyl acceptor. Half-maximal transfer of D-xylosyl units to the trisaccharide is obtained when the concentration of that acceptor is 2×10^{-2} M. Under the conditions described in the assay section, the reaction stops after transfer of one D-xylosyl unit.

Identification of the product. The radioactive D-xylodextrins produced in the enzymatic reaction are identified by comparison with authentic β-1,4-linked D-xylodextrins (Whistler and Tu, 1951) before and after partial acid hydrolysis (0.1 N HCl at 100° for 15 min) in the following systems: a) circularchromatography in n-propanol-ethyl acetate-water (7:1:2), b) circular chromatography in water-saturated phenol, and c) paper electrophoresis in 0.1 M $Na_2B_4O_7$, pH 9.2, at 35 volts/cm.

IV. Enzymes which Catalyze the Formation of Starch.

Starch is the major reserve polysaccharide of the plant kingdom. In view of the important role that starch plays in the metabolism of plants and in human nutrition, it is not surprising that our knowledge of the chemistry and enzymology of starch exceeds that of any other polysaccharide.

Most natural starches consist of two components, amylose and amylopectin, in the ratio of about 1 to 4. Amylose molecules contain several hundred D-glucose residues, joined to each other by α-1,4-linkages; there is essentially no branching in these chains. Amylopectin molecules, on the other hand, are highly branched. They consist of several thousand D-glucose units, arranged in relatively short α-1,4-linked chains which are connected to similar chains by α-1,6-linkages. The average chain length in amylopectin is 20—25 D-glucose units. For a detailed discussion on starch structure, the reader is referred to reviews by Hassid (1953) and Whelan (1958).

The synthesis of starch requires the participation of at least two enzymes: phosphorylase, which catalyses the formation of α-1,4-linkages by the transfer of D-glucosyl residues from α-D-glucose 1-phosphate, and Q-enzyme, which converts some of these α-1,4-linkages into α-1,6. D-enzyme, which catalyzes the disproportionation of maltodextrins, may also be implicated in the synthesis if α-1,4-linked chains. Recent studies on the synthesis of glycogen have shown that in animals the α-1,4-linkages are formed by transfer of D-glucosyl residues from UDP-D-glucose rather than from α-D-glucose 1-phosphate (Leloir and Goldemberg, 1960; Villar-Palasi and Larner, 1958; Robbins, Traut and Lipmann, 1959). A similar mechanism has been recently reported in plants (Rongine de Fekete, Leloir and Cardini, 1960) and in yeast (Algranati and Cabib, 1960).

1. Phosphorylase from Potatoes.

Phosphosylase catalyzes the reversible transfer of D-glucosyl units from α-D-glucose 1-phosphate to the non-reducing end of an acceptor (usually called a primer):

$$x(\alpha\text{-}D\text{-glucose 1-phosphate}) + (D\text{-glucose})_n \rightleftharpoons (D\text{-glucose})_{n+x} + x(\text{phosphate})$$
$$\text{(acceptor)}$$

Long chains of α-1,4-linked D-glucose units are formed in this manner. The development of the concepts which led to the elucidation of this mechanism has been reviewed by PEAT (1951).

Phosphorylases are widely distributed throughout the plant and animal kingdoms; however, the phosphorylases from only a few sources have been studied in detail. Of the plant phosphorylases, that of potato has been studied intensively by many laboratories, and numerous methods exist for its purification (HANES, 1940; GREEN and STUMPF, 1942; WEIBULL and TISELIUS, 1945; HIDY and DAY, 1945; BARKER, BOURNE, WILKINSON and PEAT, 1950; TISCHER and HILPERT, 1953; BAUM and GILBERT, 1953; LEE, 1960). The following method is convenient to use and yields a product which, while impure, is adequate for many purposes (NEUFELD, 1956).

Procedure 1.

Assay. (GREEN and STUMPF, 1942) The activity of the enzyme is determined by the amount of inorganic phosphate liberated when the reaction is allowed to proceed in the forward direction. The assay mixture consists of 0.5 ml of 0.5 M citrate buffer, pH 6.0, 0.2 ml of 5% soluble starch, 1 ml of 0.1 M α-D-glucose 1-phosphate, and enzyme (5 to 10 units) in a total volume of 3.5 ml. After 10 min at 38°, the reaction is stopped by the addition of 5 ml of 5% trichloracetic acid and 2 ml of 2.5% ammonium molybdate in 5 N H_2SO_4. The mixture is diluted to 25 ml with water and filtered. An aliquot of the filtrate is analyzed by the method of FISKE and SUBBAROW (1925). A unit is defined as the amount of enzyme which catalyzes the liberation of 0.1 mg of inorganic phosphate in 3 min under the conditions described.

Preparation. Potatoes (2.2 kg) are peeled, sliced, and allowed to soak in 1% $Na_2S_2O_4$ solution for 30 min. The slices are rinsed in distilled water and the juice is expressed. Approximately 1 liter of juice is obtained. The pH is immediately adjusted to 6, and the juice is allowed to stand for about an hour at 4° to settle out the starch. All subsequent operations are carried out at 0—4°.

The supernatant solution, obtained by decantation and centrifugation, is treated with approximately 0.2 volumes of alumina Cγ gel (19 mg of dry weight/ ml). The suspension is centrifuged and the precipitate is discarded. 0.21 volumes of alumina Cγ is added to the supernatant solution. The bulk of the enzyme is adsorbed on this fraction of alumina, and is eluted with three portions of about 100 ml of 0.05 M citrate buffer, pH 6.0. The exact amount of adsorbent required is determined by a preliminary experiment on a small aliquot of potato juice.

The combined eluates are fractionated with a saturated solution of $(NH_4)_2SO_4$ (adjusted to pH 7.0 by the addition of NH_4OH) and the protein precipitating between 28 and 60% saturation is collected, dissolved in about 90 ml of 0.05 M citrate buffer, pH 6.0, and refractionated with neutralized saturated $(NH_4)_2SO_4$. The fraction which precipitates between 33 and 60% saturation is dissolved in 30 ml of 0.05 M citrate buffer, pH 6.0, and dialyzed overnight against 1 liter of buffer of the same composition. This preparation is stable for several weeks at 4°. A summary of the purification procedure is given in Table 14. This partially purified preparation is contaminated with traces of phosphatase and of α-amylase. These interfering enzymes may be inhibited by the addition of $HgCl_2$ (10^{-5} M) and sodium molybdate (10^{-2} M) to a reaction mixture containing 0.05 M citrate (BAILEY, THOMAS and WHELAN, 1951).

Table 14. *Preparation of Partially Purified Potato Phosphorylase.*

	Total units	Volume (ml)	Protein (mg)	Specific activity (units/mg protein)
Crude juice	5,000	930	11,200	0.45
Alumina C γ eluate	2,300	350	1,400	1.7
First $(NH_4)_2SO_4$ precipitate . . .	1,800	100		
Second $(NH_4)_2SO_4$ precipitate . .	1,580	36	690	2.3

Procedure 2.

This procedure has been devised by LEE (1960) for large-scale preparation of pure enzyme.

Assay. The assay mixture consists of 0.1 M citrate buffer, pH 6.3, 0.01 M α-D-glucose 1-phosphate, and 0.76% amylopectin. The reaction is carried out at 38° in a total volume of 1 ml. After 5 min, the reaction is stopped by the addition of 0.5 ml of 5% trichloracetic acid, and the pH is adjusted to 4 by the addition of 2 ml of 0.1 M sodium acetate. The inorganic phosphate liberated during the reaction is measured by the method of LOWRY and LOPEZ (1946). A blank is prepared by incubating the reactants without α-D-glucose 1-phosphate and adding the latter after the protein precipitation. The enzyme unit is defined as $K \times 10^3$, where K is the first order velocity constant ($\log_{10}$, min^{-1}) (CORI, CORI and GREEN, 1939).

Purification. Idaho potatoes (kept at 4°) are sliced and homogenized in a WARING Blendor for about 1 min in a solution of 0.5% $Na_2S_2O_4$ and 0.5% sodium citrate (100 ml per kg of potatoes). The homogenate is filtered through 4 layers of gauze, and the residue is squeezed out in a laboratory press. The juice is allowed to stand 1 hour at 3° to settle out the starch. All subsequent steps are carried out at 3°.

Solid $(NH_4)_2SO_4$ (20 g) is added to 100 ml of the potato juice. The pH during this and subsequent $(NH_4)_2SO_4$ fractionations is adjusted to 6.5 by the addition of NH_4OH. The suspension is allowed to stand several hours (or overnight) and is clarified by centrifugation and filtration with Celite filter aid. Finely powdered amylose (1 g) is added slowly, with stirring to prevent the formation of lumps, to 100 ml of filtrate. The suspension is stirred for 1 hour, and $(NH_4)_2SO_4$ (18 g per 100 ml of filtrate) is added at pH 6.5. After another hour of stirring, the suspension is centrifuged. The precipitate, to which the enzyme is adsorbed, is extracted with 5 successive portions of 50 ml each of 1.8 M $(NH_4)_2SO_4$ solution, pH 6.5, per kg of potatoes. The residue is then eluted with successive portions of 50 ml each of 0.1 M citrate buffer, pH 6.5 (per kg of potatoes) until most of the activity is extracted.

The eluates are combined and treated with 20 g of solid $(NH_4)_2SO_4$ per 100 ml at pH 6.5. The precipitate is removed by centrifugation, and 18 g of $(NH_4)_2SO_4$ per 100 ml is added to the supernatant fluid. The precipitate is collected by centrifugation and is suspended in a small volume of 0.002 M potassium phosphate buffer, pH 6.8, to give a protein concentration of about 10%. This suspension is dialyzed against large volumes of the same buffer (saturated with thymol) for 2 days with 3 changes of the buffer solution. The precipitate which forms during dialysis is removed by centrifugation. The protein concentration of the supernatant fluid is about 6.5%.

This supernatant solution is subjected to column electrophoresis, using an apparatus and procedure similar to those described by PORATH (1956). About 700 mg of protein solution in 0.002 M phosphate buffer, pH 6.8, is applied to a column 600 × 25 mm containing 70 g of MUNTKELL's cellulose powder for electrophoresis. The protein zone is moved 1 cm from the top by the addition of buffer solution. The top of the column is connected to the cathode and the bottom to the anode; a current of 38 mA is applied for 21 to 22 hours. The column is then eluted with the same buffer solution, at a flow rate of 50 ml per hour. Fractions of 5 ml are collected for determination of enzyme activity and protein concentration.

The active fractions are pooled and treated with solid $(NH_4)_2SO_4$. For every 100 ml of solution, 28 g of $(NH_4)_2SO_4$ is added; after 1 hour, the suspension is centrifuged and 12 g of $(NH_4)_2SO_4$ is added per 100 ml of supernatant fluid. The resulting precipitate is collected by centrifugation, suspended in a small volume of 0.002 M phosphate buffer, pH 6.8, and dialyzed against the same buffer for 1 day, with two changes of the buffer solution.

A second electrophoresis is carried out by the procedure just described. The active fractions are pooled and fractionated with solid $(NH_4)_2SO_4$ and the final precipitate is dissolved in 0.1 M citrate buffer, pH 7. A summary of the purification is given in Table 15.

An alternative procedure involves gradient elution with 1 M KCl—0.002 M Tris buffer, pH 7.5, from DEAE-cellulose columns. The enzyme appears in the last fractions; the degree of purification is similar to that achieved by electrophoresis.

Table 15. *Preparation of Highly Purified Potato Phosphorylase.*

	Protein (mg)	Total units	Specific activity (units/mg protein)	Yield %
Potato juice (from 5 kg potatoes)	29,800	848,000	28.5	
Amylose eluate	4,160	705,000	170	74.5
$(NH_4)_2SO_4$ precipitate	2,560	487,000	191	57.5
1st column electrophoresis eluate	280	193,000	687	22.3
2nd column electrophoresis eluate	80	125,000	1,560	14.7

Properties. The pH optimum of the purified enzyme in 0.1 M citrate buffer shows a relatively sharp peak around pH 6.5; activity is below half maximal on the acid side of pH 5.5 and on the alkaline side of pH 7.5. The molecular weight of the enzyme is 207,000. The turnover number at pH 6.3 and 30° is 6,700 moles of D-glucosyl units transferred/minute/mole of protein under the conditions described in the assay and 9,400 moles/mole of protein when the enzyme is saturated with respect to α-D-glucose 1-phosphate. The 280/260 extinction coefficient ratio of the purified enzyme is 1.6; there is no evidence for binding of any adenosine 5'-monophosphate. In this respect the potato enzyme differs markedly from the phosphorylase of muscle. There is no indication of any metal requirement. The enzyme contains 6 moles of sulfhydryl groups and two moles of firmly bound pyridoxal phosphate per mole of protein.

Studies on the donor and acceptor specifity have been performed in many laboratories using less pure preparations of potato phosphorylase. No aldosyl 1-phosphate can replace α-D-glucose 1-phosphate as glycosyl donor (MEAGHER and HASSID, 1946; POTTER et al., 1948). Arsenate, however, can replace inorganic phosphate (KATZ, HASSID and DOUDOROFF, 1948); the rate of arsenolysis is approximately one eighth that of phosphorolysis. Free D-glucose is obtained as the product, presumably by instantaneous hydrolysis of D-glucose 1-arsenate formed. The MICHAELIS constants K_m for α-D-glucose 1-phosphate and inorganic phosphate are 3.5×10^{-3} M and 7.5×10^{-3} M, respectively (LEE, 1960).

The smallest molecule that can act as primer is maltotriose (FRENCH and WILD, 1953; WHELAN and BAILEY, 1954); priming ability increases markedly with increasing chain length. On the basis of end group concentration, amyloses of high degree of polymerization are the best D-glucosyl acceptors. The K_m for a 7 unit dextrin, 23 unit dextrin, 220 unit amylose, and 24 unit amylopectin chains are, respectively, 17×10^{-5} M, 5×10^{-5} M, 0.9×10^{-5} M, and 7×10^{-5} M end groups (NEUFELD, 1956). The affinity of potato phosphorylase for glycogen is very low ($K_m = 71 \times 10^{-5}$ M end groups) in marked contrast to muscle phosphorylase, for which glycogen is the best primer.

The equilibrium constant K of the reaction is independent of the concentration of primer (except in the special case where α-D-glucose 1-phosphate is present in limiting amounts and the primer is of such size that it can serve as D-glucosyl acceptor but not as D-glucosyl donor [HESTRIN, 1949]). K varies, however, as a function of pH, since the dissociation constants of α-D-glucose 1-phosphate and inorganic phosphate differ. At pH 8.0, both anions are in the bivalent form and

$$K = \frac{\text{phosphate}}{\text{α-D-glucose 1-phosphate}} = 2.2$$

(HANES, 1940). The K at any other pH is a function of the concentration of bivalent anions present, and can be calculated by the following equation: (KALCKAR, 1954).

$$K = 2.2 \times \frac{K_2 \text{ α-D-glucose 1-phosphate} \times ([H^+] + K_2 \text{ phosphate})}{K_2 \text{ phosphate} \times ([H^+] + K_2 \text{ α-D-glucose 1-phosphate})}$$

where K_2 phosphate $= 1.5 \times 10^{-7}$ and K_2 (α-D-glucose 1-phosphate) $= 7.8 \times 10^{-7}$.

2. Q-Enzyme (Branching Enzyme) from Potatoes.

Q-enzyme is a transglycosylase which catalyses the formation of the α-1,6 linkages in amylopectin. The reaction occurs by a transfer of a terminal segment of an α-1,4 linked chain to the primary hydroxyl of a non-terminal D-glucose residue of a similar chain.

Q-enzyme has been prepared from potatoes (BARKER, BOURNE, WILKINSON and PEAT, 1950; GILBERT and PATRICK, 1952), broad beans, wrinkled peas (HOBSON, WHELAN and PEAT, 1950), and green gram (RAM and GIRI, 1952). A similar branching enzyme has also been purified from yeast (GUNJA, MANNERS and MAUNG, 1960). The procedure reported here is that of BARKER, BOURNE, WILKINSON and PEAT (1950), with further $(NH_4)_2SO_4$ precipitations as described by the same authors in a subsequent publication (BARKER, BOURNE, PEAT and WILKINSON, 1950).

Assay. The assay is based on the fact that the absorbance at 680 mμ of the complex of amylopectin with iodine is much less intense than that of an amylose-iodine complex. A mixture consisting of 7.5 ml of amylose (14 mg), 3 ml of 0.2 M citrate buffer, pH 6.0, and 2.0 ml of buffered enzyme solution is incubated at room temperature for 30 min. An aliquot (1 ml) is then pipetted into a solution of I_2 (2 mg)-KI (20 mg) in a final volume of 100 ml, and the optical density is read at 680 mμ.

Since both α- and β-amylases (which may contaminate Q-enzyme preparations) cause a decrease in the iodine color of amylose, the action of these enzymes will simulate Q-enzyme activity in this assay. However, the hydrolysis of amylose by amylases will be accompanied by an increase in reducing power, whereas no reducing groups are liberated by the action of Q-enzyme. It is therefore necessary to measure the reducing value (SOMOGYI, 1945) before and after the incubation period, in order to ascertain that the iodine color change is due to Q-enzyme activity and not to hydrolysis by amylases.

Purification. It is necessary to use potatoes in their resting period, before sprouting begins. Potatoes (1 kg) are peeled, sliced, and soaked in 1 liter of 0.5% $Na_2S_2O_4$ solution for 20 min. The slices are washed, minced, and pressed. The juice is clarified by centrifugation, and the proteins are immediately precipitated as lead salts.

A solution of lead acetate is prepared as follows: 1 liter of solution containing 190 g of lead acetate is adjusted to pH 7.25 and allowed to stand for a few days. After removal of a precipitate, the solution is diluted to 3 liters.

To each 100 ml of potato juice, sufficient 0.02 N NaOH is added with stirring to bring the pH to 7.2 (about 40 ml). Water (30 ml) and lead acetate solution

(30 ml) are added, and the yellow lead-protein precipitate is collected by centri-fugation. The precipitate is decomposed by stirring for 5 min in 100 ml of 0.2 N $NaHCO_3$, followed by passage of a stream of CO_2 for 2.5 min.

The insoluble material is then removed by centrifugation and discarded. The eluate (ca. 110 ml) is treated with a neutralized solution of $(NH_4)_2SO_4$ (50 g/100 ml of solution) until the concentration reaches 18 g/100 ml. The precipitate is collected by centrifugation and dissolved in 20 ml of ice cold water. The small precipitate which forms upon standing at 0° for 1 hour is removed by centrifugation and discard-ed. Neutralized $(NH_4)_2SO_4$ is again added to the supernatant liquid until a concen-tration of 20 g/100 ml of solution is reached. The precipitate is collected by centrifugation, dissolved in 13 ml of 0.2 M citrate buffer, pH 7.0, and lyophilized. The resulting powder is stable for several months. It may be purified by two additional precipitations with neutralized $(NH_4)_2SO_4$ solution (50 g/100 ml of solution) until the salt concentration reaches 18 g/100 ml of solution, and again lyophilized in the presence of citrate buffer at pH 7.0. When required the enzyme is dissolved in water (125 ml of water/100 ml of potato juice originally employed). The phosphorylase content of this preparation is very low, about 0.01 (GREEN and STUMPF, 1942) units/ml.

Properties. The optimal pH is about 7.0; activity declines sharply above pH 7.5 and below pH 6.0. The formation of α-1,6 linkages appears to be irreversible. Q-enzyme from potatoes will act on amyloses which are at least 116 D-glucose units long, and on the outer branches of amylopectin which are 13 units long, or longer (NUSSENBAUM and HASSID, 1952; LARNER, 1953). The activity towards amylopectin is considerably greater than towards amylose. It has been suggested by LARNER (1953) that amylopectin is synthesized in vivo by the alternating action of phosphorylase and Q-enzyme. Phosphorylase catalyzes the addition of D-glucosyl residues to a primer until the chain reaches a sufficient length for the action of Q-enzyme, which transfers a segment of the chain to an inner D-glucose residue on the same or on an adjoining chain. Phosphorylase then elongates the outer chains until a length appropriate for Q-enzyme action is reached again, at which time further branching occurs. The successive action of phosphorylase and of Q-enzyme is repeated many times and gives rise to the multi-branched macro-molecules of amylopectin.

3. D-Enzyme from Potatoes.

D-enzyme is a transglycosylase that catalyzes the transfer of segments of α-1,4-D-glucan chains to an appropriate acceptor, which may be another molecule of the donor substrate. The newly-formed linkage is also α-1,4. The net result is a disproportionation of the substrate into products of higher and lower degree of polymerization. Two bacterial enzymes which catalyze a similar redistribution of α-1,4 linkages are known: the amylase of *Bacillus macerans* (FRENCH et al., 1954) and the amylomaltase of *Escherichia coli* (MONOD and TORRIANI, 1950).

D-enzyme has been observed in broad beans and in potatoes. A partial puri-fication has been described by PEAT, WHELAN and JONES (1957) and is reported here.

Assay. The assay is based on measurement of the D-glucose liberated during the disproportionation of maltotriose: 2 maltotriose $\rightleftharpoons$ maltopentaose + D-glucose. The digest contains maltotriose (60 mg), 0.3 ml of 0.2 M citrate buffer, pH 7.0, and enzyme in a total volume of 3 ml. After incubation for 30 min at 35°, the enzyme is inactivated by heat, and the incubation mixture is transferred to a charcoal-celite column (1:1, 6.0×1.5 cm) and eluted with water. The first 5 ml of eluate are discarded. The next 25 ml, which contain all the D-glucose, are collected, passed through a Seitz filter, and used for measurement of optical rotation in a 4 dm tube. There is a linear relationship between the concentration of enzyme and the amount of D-glucose liberated, up to 9 mg. One unit is defined as the amount of enzyme required to liberate 1 mg of D-glucose under these conditions.

If it is necessary to reduce the assay to a micro-scale, the concentration of D-glucose may be determined by the use of D-glucose oxidase[1] (KESTON, 1956; TELLER, 1956) or by the joint action of hexokinase (BERGER et al., 1945) and D-glucose 6-phosphate dehydrogenase (KORNBERG and HORECKER, 1955) in the presence of ATP and TPN.

Preparation. Potatoes (300 g) are peeled, sliced, and ground with rapid incorporation of 10 g of charcoal, previously boiled in water. It is important to produce an intimate mixture of potato pulp and charcoal. The juice is pressed through cheesecloth and centrifuged. The thick colloidal suspension of charcoal is removed from the supernatant solution by filtration through a pressure filter and a series of two or three graded filters. The resulting juice is pale amber in color and does not darken for several days.

The potato juice (540 ml) is treated with 1 g $CuSO_4$ and 460 ml of a 50% (w/v) solution of $(NH_4)_2SO_4$, pH 7.0. The suspension (23% $(NH_4)_2SO_4$, w/v) is kept at 2° for 24 hours. The precipitate is then removed by centrifugation, dissolved in 250 ml of 0.01 M citrate buffer, pH 7.0, and precipitated by the addition of $(NH_4)_2SO_4$ to 20 per cent (w/v) concentration. After 24 hours at 2°, the precipitate is collected, redissolved, and reprecipitated twice as before, except that the suspension is allowed to stand only 2 hours before centrifugation. The final precipitate is taken up in 50 ml of 0.2 M citrate buffer (pH 7.0) and lyophilized. The yields of enzyme obtained after the second, third, and fourth precipitation are 1720, 1370, and 1036 units, while the specific activities (units/mg of protein nitrogen) are 1.05, 1.78, and 2.33.

Properties. The pH optimum in veronal buffer is around pH 6.6—6.7. The preparation is free of phosphorylase and has but little Q-enzyme activity. The donor substrate is a chain of 3 or more α-1,4 linked units (maltose cannot serve as donor), and can be as large as amylopectin. The smallest segment which can be transferred is a maltosyl residue. In addition to maltodextrins, the following substrates can serve as acceptors: D-glucose, α- and β-methyl D-glucoside, D-mannose, L-sorbose, α-methyl-L-sorboside, D-xylose, maltose, sucrose, leucrose, and trehalose.

During the disproportionation of maltodextrins, D-glucose is produced. Since this monosaccharide can act as an acceptor, but not as a donor, a high concentration of D-glucose favors the formation of small oligosaccharides. Conversely, the removal of D-glucose favors the formation of long amylose-type chains (WALKER and WHELAN, 1959). Thus the action of D-enzyme on oligosaccharides provides an alternate pathway for starch synthesis. The physiological function of D-enzyme, however, is still a subject of considerable speculation (WHELAN, 1958).

References.

ALGRANATI, I. D., and E. CABIB: Biochim. biophys. Acta **43**, 141 (1960). — ALTERMATT, H. A., and A. C. NEISH: Can. J. Biochem. Physiol. **34**, 405 (1956). — ALVARADO, F.: Biochem. biophys. Acta **41**, 233 (1960). — ANDERSON, E. P., E. S. MAXWELL and R. M. BURTON: J. Amer. chem. Soc. **81**, 6514 (1959). — ASPINALL, G. O., and G. KESSLER: Chem. and Ind. **1957**, 1296.
BAILEY, J. M., G. J. THOMAS and W. J. WHELAN: Biochem. J. **49**, 1vi (1951). — BALLIO, A., C. CASINOVI and C. SERLUPI-CRESCENZI: Biochim. biophys. Acta **20**, 414 (1956). — BAN-

---

[1] The glucose oxidase used must be carbohydrase-free. Such reagents are commercially available.

References. 517

DURSKI, R. S., and B. AXELROD: J. biol. Chem. **193**, 405 (1951). — BARKER, S. A., E. J. BOURNE, J. G. FLEETWOOD and M. STACEY: J. chem. Soc. **1958**, 4128. — BARKER, S. A., E. J. BOURNE, I. A. WILKINSON and S. PEAT: J. chem. Soc. **1950**, 84. — BARKER, S. A., E. J. BOURNE, S. PEAT and I. A. WILKINSON: J. chem. Soc. **1950**, 3022. — BAUM, H., and G. A. GILBERT: Nature (Lond.) **171**, 983 (1953). — BEAN, R. C., and W. Z. HASSID: J. Amer. chem. Soc. **77**, 5737 (1955). — BERGER, L., M. W. SLEIN, S. P. COLOWICK and C. F. CORI: J. gen. Physiol. **29**, 379 (1945). — BERGKVIST, R.: Acta chem. scand. **11**, 1457 (1957). — BLUMENTHAL, H. J., S. T. HOROWITZ, A. H. HEMERLINE and S. ROSEMAN: Bacteriol. Proc. (Soc. Amer. Bacteriol.) **1955**, 137. — BUCHANAN, J. G., J. A. BASSHAM, A. A. BENSON, D. F. BRADLEY, M. CALVIN, L. I. DAUS, M. GOODMAN, P. M. HAYES, V. H. LYNCH, L. T. NORRIS and A. T. WILSON: Phosphorus Metabolism (W. D. MCELROY and B. GLASS, ed.), Vol. II, 440. Baltimore: Johns Hopkins Press 1951.

CABIB, E., and L. F. LELOIR: J. biol. Chem. **206**, 779 (1954); **231**, 259 (1958). — CABIB, E., L. F. LELOIR and C. E. CARDINI: J. biol. Chem. **203**, 1055 (1953). — CANDY, J., and A. B. KILBY: Nature (Lond.) **183**, 1954 (1959). — CARDINI, C. E.: Enzymologia **15**, 44 (1951). — CARDINI, E. C., L. F. LELOIR and J. CHIRIBOGA: J. biol. Chem. **214**, 149 (1955). — CIFONELLI, J. A., and A. DORFMAN: J. biol. Chem. **228**, 547 (1957). — COLOWICK, S. P.: J. biol. Chem. **124**, 557 (1938). — CORI, C. F., G. T. CORI and A. A. GREEN: J. biol. Chem. **151**, 39 (1943). — CORI, G. T., C. F. CORI and G. SCHMIDT: J. biol. Chem. **129**, 629 (1939).

DAVIDSON, E. A., H. J. BLUMENTHAL and S. ROSEMAN: J. biol. Chem. **226**, 125 (1957). — DISCHE, Z.: J. biol. Chem. **167**, 189 (1947). — DOUDOROFF, M., N. O. KAPLAN and W. Z. HASSID: J. biol. Chem. **148**, 67 (1943).

FEINGOLD, D. S., G. AVIGAD and S. HESTRIN: Biochem. J. **64**, 351 (1956). — FEINGOLD, D. S., E. F. NEUFELD and W. Z. HASSID: Arch. Biochem. **78**, 401 (1958a); — J. biol. Chem. **233**, 783 (1958b); **234**, 488 (1959); **235**, 910 (1960). — FISHER, E. H., and H. M. HILPERT: Experientia (Basel) **9**, 176 (1953). — FISKE, C. H., and Y. SUBBAROW: J. biol. Chem. **66**, 375 (1925). — FOLIN, O., and V. CIOCALTEAU: J. biol. Chem. **73**, 627 (1927). — FOSTER, A. B., and M. STACEY: Encyclopedia of plant physiology. VI, 331. Berlin: Springer-Verlag 1958. — FRENCH, D., and D. M. WILD: J. Amer. chem. Soc. **75**, 4490 (1953). — FRENCH, D., M. L. LEVINE, E. NORBERG, P. NORDIN, J. H. PAZUR and G. M. WILD: J. Amer. chem. Soc. **76**, 2387 (1954). — FRIEDEMANN, T. E., and G. E. HAUGEN: J. biol. Chem. **147**, 415 (1943). — FRIEDMAN, S.: Arch. Biochem. **88**, 339 (1960).

GILBERT, G. A., and A. D. PATRICK: Biochem. J. **51**, 181 (1952). — GINSBURG, V.: J. biol. Chem. **232**, 55 (1958); **235**, 2196 (1960). — GINSBURG, V., and H. N. KIRKMAN: J. Amer. chem. Soc. **80**, 3481 (1958). — GINSBURG, V., and E. F. NEUFELD: Abstr. Div. biol. Chem., Amer. chem. Soc., 27C, Sept. 1957. — GINSBURG, V., P. K. STUMPF and W. Z. HASSID: J. biol. Chem **223**, 977 (1956). — GLASER, L.: J. biol. Chem. **232**, 627 (1958); — Biochim. biophys. Acta **31**, 575 (1959). — GLASER, L., and D. H. BROWN: J. biol. Chem. **228**, 729 (1957); — Proc. nat. Acad. Sci. (Wash.) **41**, 253 (1955). — GORDON, H. T., W. THORNBURG and L. N. WERUM: Analyt. Chem. **28**, 849 (1956). — GREEN, D. E., and P. K. STUMPF: J. biol. Chem. **142**, 355 (1942). — GUNJA, Z. H., D. S. MANNERS and K. MAUNG: Biochem. J. **75**, 441 (1960).

HANES, C. S.: Proc. roy. Soc. **129 B**, 174 (1940). — HASSID, W. Z.: Phosphorus metabolism (W. D. MCELROY and B. GLASS, ed.). Vol. I, 11. Baltimore: Johns Hopkins Press 1951; — Organic chemistry; an advanced treatise (H. GILMAN, ed.), Vol. IV, 901. New York: Wiley & Sons 1953. — HASSID, W. Z., E. F. NEUFELD and D. S. FEINGOLD: Proc. nat. Acad. Sci. (Wash.) **45**, 905 (1959). — HEATH, E. C.: Biochim. biophys. Acta **39**, 377 (1960). — HEINRICH, M. R., and J. B. NEILANDS: Fed. Proc. **19**, 85 (1960). — HEPPEL, L. A., and R. J. HILMOE: J. biol. Chem. **192**, 8 (1951). — HESTRIN, S.: J. biol. Chem. **179**, 943 (1949); — Proc. Inter. Congr. Biochem., 4th Congr., Vienna, 1958. Vol. I, 181 (1959). — HESTRIN, S., and G. AVIGAD: Biochem. J. **69**, 388 (1958). — HESTRIN, S., D. S. FEINGOLD and G. AVIGAD: Biochem. J. **64**, 340 (1956). — HIDY, P. H., and H. G. DAY: J. biol. Chem. **160**, 273 (1945). — HOBSON, P. N., W. J. WHELAN and S. PEAT: J. chem. Soc. **1950**, 3566. — HURLBERT, R. B.: Methods in enzymology (S. P. COLOWICK and N. O. KAPLAN, ed.), Vol. III, 785. New York: Academic Press 1957.

ISSELBACHER, K. J.: J. biol. chem. **232**, 429 (1958).

JACOBELLI, G., M. J. TABONE and D. TABONE: Bull. Soc. Chim. biol. (Paris) **40**, 955 (1958).

KALCKAR, H. M.: Mechanism of enzyme action (W. D. MCELROY and B. GLASS, ed.), 675. Baltimore: Johns Hopkins Press 1954; — Science **125**, 105 (1957); — Adv. Enzymol. **20**, 111 (1958). — KALCKAR, H. M., B. BRAGANCA and A. MUNCH-PETERSEN: Nature (Lond.) **172**, 1038 (1953). — KATZ, J., W. Z. HASSID and M. DOUDOROFF: Nature (Lond.) **161**, 96 (1948). — KEILIN, D., and E. HARTREE: Proc. roy. Soc. **124 B**, 397 (1938). — KESSLER, G.: Ber. Schweiz. bot. Ges. **68**, 5 (1958). — KESSLER, G., D. S. FEINGOLD and W. Z. HASSID: Plant Physiol. **35**, 505 (1960). — KESTON, A. S.: Abstr. Div. biol. Chem., Amer. chem. Soc. 31C, April 1956. — KORNBERG, A., and B. HORECKER: Methods in enzymology (S. P. COLOWICK and N. O. KAPLAN, ed.), Vol. I, 323. New York: Academic Press 1955. — KORNBERG, A., and W. E. PRICER, jr.:

J. biol. Chem. **182**, 763 (1950). — KURAHASHI, K., and E. P. ANDERSON: Biochim. biophys. Acta **29**, 498 (1958). — KURAHASHI, K., and A. SUGIMURA: J. biol. Chem. **235**, 940 (1960).

LARNER, J.: J. biol. Chem. **202**, 491 (1953). — LEE, Y. P.: Biochim. biophys. Acta **43**, 18, 25 (1960). — LELOIR, L. F.: Arch. Biochem. **33**, 186 (1951). — LELOIR, L. F., and C. E. CARDINI: Biochim. biophys. Acta **12**, 15 (1953); — J. biol. Chem. **214**, 157 (1955). — LELOIR, L. F., and S. H. GOLDEMBERG: J. biol. Chem. **235**, 919 (1960). — LELOIR, L. F., and R. E. TRUCCO: Methods in enzymology (S. P. COLOWICK and N. O. KAPLAN, ed.), Vol. I, 290. New York: Academic Press 1955. — LE PAGE, G. A., and G. C. MUELLER: J. biol. Chem. **180**, 975 (1949). — LEVVY, G. A., and I. D. E. STOREY: Biochem. J. **44**, 295 (1949). — LOEWUS, F. A., R. JANG and C. G. SEEGMILLER: J. biol. Chem. **232**, 533 (1958). — LOWRY, O. H., and J. A. LOPEZ: J. biol. Chem. **162**, 421 (1946).

McCREADY, R. M., and W. Z. HASSID: Methods in enzymology (S. P. COLOWICK and N. O. KAPLAN, ed.), Vol. III, p. 137. New York: Academic Press 1957. — MAJKEN-ELANDER, V.: Ark. Kemi Min. Geol. **13**, 457 (1959). — MALEY, F., and G. F. MALEY: Biochim. biophys. Acta **31**, 577 (1959). — MALEY, F., G. F. MALEY and H. A. LARDY: J. Amer. chem. Soc. **78**, 5303 (1956). — MALEY, F., and S. OCHOA: J. biol. Chem. **233**, 1538 (1958). — MARKHAM, R., and J. D. SMITH: Biochem. J. **45**, 294 (1949). — MARSH, C. A.: Biochim. biophys. Acta (in press). — MAXWELL, E. S.: J. Amer. chem. Soc. **78**, 1074 (1956). — MAXWELL, E. S., H. DE ROBICHON-SJULMAJSTER and H. M. KALCKAR: Arch. Biochem. **78**, 407 (1958). — MAXWELL, E. S., and H. DE ROBICHON-SJULMAJSTER: J. biol. Chem. **235**, 308 (1960). — MEAGHER, W. R., and W. Z. HASSID: J. Amer. chem. Soc. **68**, 2135 (1946). — MEJBAUM, W.: Hoppe-Seylers physiol. Chem. Z. **258**, 117 (1939). — MEYER, K. H., and P. BERNFELD: Helv. chim. acta **25**, 399 (1942). — MONOD, J., and A. M. TORRIANI: Ann. Inst. Pasteur **78**, 65 (1950). — MUNCH-PETERSEN, A.: Acta chem. scand. **9**, 1523 (1955); **10**, 928 (1956); **11**, 1079 (1957). — MUNCH-PETERSEN, A., H. M. KALCKAR, E. CUTOLO and E. E. B. SMITH: Nature (Lond.) **172**, 1036 (1953). — MUNCH-PETERSEN, A., H. M. KALCKAR and E. E. B. SMITH: Kgl. Danske Videnskab. Selskab. Biol. Medd. **22**, 73 (1955).

NAJJAR, V. A.: J. biol. Chem. **175**, 281 (1948). — NEISH, A. C.: Canad. J. Biochem. Physiol. **36**, 187 (1958). — NEUFELD, E. F.: Ph. D. Thesis, University of California, Berkeley, 1956. — NEUFELD, E. F., V. GINSBURG, E. W. PUTMAN, D. FANSHIER and W. Z. HASSID: Arch. Biochem. **69**, 602 (1957). — NEUFELD, E. F., D. S. FEINGOLD and W. Z. HASSID: Arch. Biochem. **88**, 96 (1959); — J. biol. Chem. **235**, 906 (1960). — NUSSENBAUM, S., and W. Z. HASSID: J. biol. Chem. **196**, 785 (1952).

PALADINI, A. C., and L. F. LELOIR: Biochem. J. **51**, 426 (1952). — PARK, J. T., and M. J. JOHNSON: J. biol. Chem. **181**, 149 (1949). — PEAT, S.: Adv. Enzymol. **11**, 339 (1951). — PEAT, S., W. J. WHELAN and G. JONES: J. chem. Soc. 2490 (1957). — PEAT, S., W. J. WHELAN and H. G. LAWLEY: J. chem. Soc. **724**, 729 (1958). — PETERSON, E. A., and H. A. SOBER: J. Amer. chem. Soc. **78**, 751 (1956). — PLOESER, J. M., and H. S. LORING: J. biol. Chem. **178**, 431 (1949). — PONTIS, H. G., A. L. JAMES and J. BADILLEY: Biochem. J. **75**, 428 (1960). — PORATH, J.: Biochim. biophys. Acta **22**, 151 (1956). — PORTER, L. W., and N. HOBAN: Anal. Chem. **26**, 1846 (1954). — POTTER, A. L., J. C. SOWDEN, W. Z. HASSID and M. DOUDOROFF: J. Amer. chem. Soc. **70**, 1751 (1948). — PUTMAN, E. W., and W. Z. HASSID: J. biol. Chem. **207**, 885 (1954); — J. Amer. chem. Soc. **79**, 5057 (1957).

RAM, J. S., and K. V. GIRI: Arch. Biochem. **38**, 231 (1952). — REISSIG, J. L.: J. biol. Chem. **219**, 753 (1956). — REISSIG, J. L., J. L. STROMINGER and L. F. LELOIR: J. biol. Chem. **217**, 959 (1955). — ROBBINS, P. W., R. R. TRAUT and F. LIPMANN: Proc. nat. Acad. Sci. (Wash.) **45**, 6 (1959). — ROE, J. H.: J. biol. Chem. **107**, 15 (1934). — RONGINE DE FEKETE, M. A., L. F. LELOIR and C. E. CARDINI: Nature (Lond.) **187**, 918 (1960).

SALTMAN, P.: J. biol. Chem. **200**, 145 (1953). — SIDBURY, J.: Abstr. Div. biol. Chem., Amer. chem. Soc., 27C, Sept. 1957. — SLATER, W. G., and H. BEEVERS: Plant. Physiol. **33**, 146 (1958). — SMITH, E. E. B., and G. T. MILLS: Biochim. biophys. Acta **13**, 386 (1954). — SMITH, E. E. B., G. T. MILLS, H. P. BERNHEIMER and R. AUSTRIAN: Biochim. biophys. Acta **29**, 640 (1958). — SOBER, H. A., F. J. GUTTER, M. M. WYCKOFF and E. A. PETERSON: J. Amer. chem. Soc. **78**, 756 (1956). — SOLMS, J., and W. Z. HASSID: J. biol. Chem. **228**, 357 (1957). — SOMOGYI, M.: J. biol. Chem. **160**, 61 (1945). — STOLL, A., and E. JUCKER: Encyclopedia of plant physiology, VI, 5347. Berlin: Springer-Verlag 1958. — STROMINGER, J. L.: Biochim. biophys. Acta **16**, 616 (1955a); **17**, 283 (1955b); — Physiol. Rev. **40**, 55 (1960). — STROMINGER, J. L., H. M. KALCKAR, J. AXELROD and E. S. MAXWELL: J. Amer. chem. Soc. **76**, 6411 (1954). — STROMINGER, J. L., and L. W. MAPSON: Biochem. J. **66**, 567 (1957). — STROMINGER, J. L., and M. S. SMITH: J. biol. Chem. **234**, 1822 (1959). — SU, J. C., and W. Z. HASSID: J. biol. Chem. **235**, PC36 (1960).

TELLER, J. D.: Abstr. Div. biol. Chem., Amer. chem. Soc. 69C, Sept. 1956. — TRE-VELYAN, W. E., and J. S. HARRISON: Biochem. J. 50, 298 (1952). — TRUCCO, R. E., R. CA-PUTTO, L. F. LELOIR and N. MITTELMAN: Arch. Biochem. 18, 137 (1948).

VILLAR-PALASI, C., and J. LARNER: Biochim. biophys. Acta 30, 449 (1958). — VOLKIN, E., and W. E. COHN: Methods of biochemical analysis (D. GLICK, ed.), p. 304. New York: Inter-science 1954.

WALKER, G. J., and W. J. WHELAN: Nature (Lond.) 183, 46 (1959). — WEIBULL, C., and A. TISELIUS: Arkiv. Kemi Min. Geol. 19A, 1 (1945). — WHELAN, W. J.: Encyclopedia of plant physiology, VI, 154. Berlin: Springer-Verlag 1958. — WHELAN, W. J., and J. M. BAILEY: Biochem. J. 58, 560 (1954). — WHISTLER, R. L., and C. C. TU: J. Amer. chem. Soc. 73, 1389 (1951). — WILKINSON, J. F.: Biochem. J. 44, 460 (1949). — WOLOCHOW, H., E. W. PUTMAN, M. DOUDOROFF, W. Z. HASSID and H. A. BARKER: J. biol. Chem. 180, 1237 (1949).

YAMAHA, T., and C. E. CARDINI: Arch. Biochem. 86, 127 (1960a); 86, 133 (1960b).

ZUCKER, M., and A. NASON: Methods in enzymology (S. P. COLOWICK and N. O. KAPLAN, ed.), Vol. II, p. 416. New York: Academic Press 1955.

Enzymes of Glycolysis[1].

By

Martin Gibbs and John F. Turner.

With 1 Figure.

Glycolysis is one of the two main pathways of carbohydrate degradation in most animal and plant tissues. The term glycolysis is generally considered to mean the anaerobic breakdown of reserve polysaccharide (eg. starch, glycogen), hexoses or hexose phosphates to pyruvate *via* the EMBDEN-MEYERHOF-PARNAS (EMP) pathway. The pyruvate formed may be completely oxidized by the citric acid cycle or converted to lactic acid or ethanol and CO_2.

The conversion of carbohydrate into ethanol (eg. BEEVERS and GIBBS, 1954) and lactate (BARKER and EL SAIFI, 1952) and the dissimilation of specifically labelled glucose to products as predicted by the EMP pathway (GIBBS and BEEVERS, 1955) have provided evidence for the operation of glycolysis in plants. A complete EMP glycolytic system may be extracted from plant tissues (TURNER, 1954, 1957; TURNER and MAPSON, 1958; HATCH and TURNER, 1958).

Glycolysis (Fig. 1) involves a complex series of reactions and energy-rich phosphate ($\sim P$) bonds are generated in the reactions catalyzed by 3-phosphoglycerate kinase and pyruvate kinase. In the anaerobic breakdown of FDP to pyruvate four $\sim P$ bonds are formed. The net yield of $\sim P$ bonds in the glycolysis of glucose or fructose is two because of the phosphorylations in the formation of the hexose monophosphates and FDP. The net yield of $\sim P$ bonds formed in the glycolysis of starch is three.

DPN (or TPN) is reduced in the oxidation of G 3-P to 1,3-diPGA by G 3-P dehydrogenase. If pyruvate enters the citric acid cycle, the transfer of electrons from DPNH to O_2 gives rise to the synthesis of additional $\sim P$ bonds. If pyruvate is converted to ethanol and CO_2 or lactate the DPNH is utilized in the alcohol dehydrogenase or lactic dehydrogenase reaction respectively.

I. Assay of Glycolytic System.

Principle. The method used is based on the manometric measurement of CO_2 production following the decarboxylation of pyruvate and reduction of

---

[1] Preparation of manuscript aided by grants from the National Science Foundation and from the United States Air Force under contract No. AF 49(638)198.

Abbreviations: Pi, inorganic orthophosphate; G 1-P, glucose 1-phosphate; G 6-P, glucose 6-phosphate; F 6-P, fructose 6-phosphate; FDP, fructose 1,6-diphosphate; DHAP, dihydroxyacetone phosphate; G 3-P, glyceraldehyde 3-phosphate; 1,3 di PGA, 1,3-diphosphoglycerate; 3-PGA, 3-phosphoglycerate; 2-PGA, 2-phosphoglycerate; PEP, phosphoenolpyruvate; DPN, diphosphopyridine nucleotide; DPNH, reduced diphosphopyridine nucleotide; TPN, triphosphopyridine nucleotide; TPNH, reduced triphosphopyridine nucleotide; ADP, adenosine diphosphate; ATP, adenosine triphosphate.

acetaldehyde by the plant enzyme system. Assay using G 1-P *plus* fructose as substrate will be described but any of the glycolytic substrates may be employed.

Reagents

0.22 M G 1-P.
0.44 M fructose.
0.22×10^{-3} M DPN.
0.22×10^{-3} M ATP.
0.18 M $MgCl_2$.
0.25 M Na_2HPO_4-KH_2PO_4 buffer (pH 6.5).

Procedure. 1 ml of the plant glycolytic system (containing approximately 10 units of activity), 0.1 ml G 1-P, 0.1 ml fructose, 0.1 ml DPN, 0.1 ml ATP, 0.1 ml $MgCl_2$, 0.1 ml Na_2HPO_4-KH_2PO_4 buffer, 0.2 ml water and 0.05 ml toluene are placed in a WARBURG vessel which is gassed with O_2-free N_2. After equilibration at 30° the rate of CO_2 formation is determined by taking readings at 30 minute intervals; after 1-2 hrs. the rate should be constant for several hours. A correction factor for CO_2 retention in the liquid phase (pH 6.5) is applied (UMBREIT, BURRIS and STAUFFER, 1951).

Definition of unit activity. One unit of glycolytic activity is defined as the production of 10 μl of CO_2 per hour under the above conditions.

1. Preparation of Glycolytic System.

The procedure described used pea seeds as the source of the glycolytic system and is essentially that described by TURNER (1957).

Step 1. Preparation of crude extract. Pea seeds (var. Laxton's Progress) are finely ground in a CHRISTIE and NORRIS mill and the powder extracted three times with ether at room temperature and air-dried. The defatted powder may be stored at 4° and shows no loss in glycolytic activity during a period of 6 months. For the preparation of the enzyme extract, 15 g of defatted pea powder is suspeded in a mixture of 40 ml water, 5 ml of toluene and 10 ml of 0.2 M $NaHCO_3$ in a stoppered bottle. The bottle is slowly rotated for 3 hours at room temperature and the mixture centrifuged at 1000 g for 10 minutes.

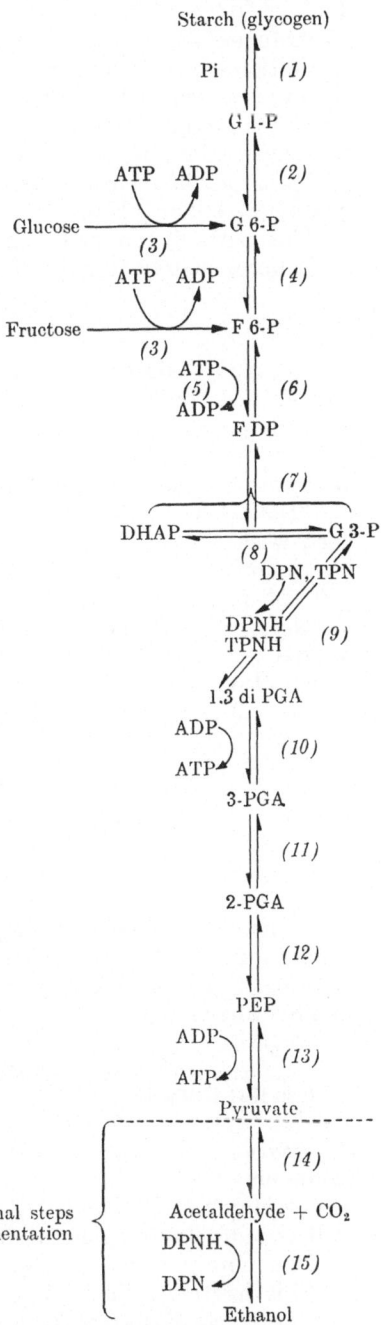

Fig. 1. Glycolysis *via* the EMBDEN-MEYERHOF-PARNAS pathway. *The enzymes involved are: (1)* phosphorylase, *(2)* phosphoglucomutase, *(3)* hexokinase, *(4)* phosphoglucose isomerase, *(5)* phosphohexokinase, *(6)* fructose 1,6-diphosphatase, *(7)* aldolase, *(8)* triose phosphate isomerase, *(9)* glyceraldehyde 3-phosphate dehydrogenase, *(10)* 3-phosphoglycerate kinase, *(11)* phosphoglycerate mutase, *(12)* enolase, *(13)* pyruvate kinase, *(14)* pyruvate decarboxylase, and *(15)* alcohol dehydrogenase.

Step 2. Preparation of partially purified extract. The crude extract is centrifuged at 20,000 g for 20 minutes at room temperature and the precipitate discarded. Centrifuging at high speed at this stage is desirable to secure satisfactory sedimentation after the addition of $(NH_4)_2SO_4$. Cold neutralized saturated $(NH_4)_2SO_4$ (80 ml) is added to 20 ml of the turbid supernatant and, after standing for 1 hour at 4°, the mixture is centrifuged at 20,000 g for 20 minutes at room temperature. The supernatant is discarded and the precipitate taken up in water (7 ml) and dialyzed for 2 hours in a rocking dialyzer against four changes of 300 ml of 0.025 M phosphate buffer (pH 6.5).

The partially purified extract contains approximately 45 mg of protein/ml and approximately 10 units glycolytic activity/ml.

2. Properties.

Specificity. Glucose, fructose, G 1-P, G 6-P, F 6-P, FDP, 3-PGA and starch (the extract contains starch phosphorylase) form CO_2 when incubated with the partially purified pea seed extract and catalytic amounts of ATP, DPN and $MgCl_2$ (HATCH and TURNER, 1958). These cofactors are essentially removed during the purification procedure as are endogenous substrates.

Stability. The extract shows no decrease in glycolytic activity during storage for 2 weeks at −15°.

Activators and inhibitors. The addition of iodoacetate (10^{-3} M) or fluoride (10^{-3} M) results in an almost complete inhibition of CO_2 production.

Glycolysis is inhibited by aerobic conditions (TURNER and MAPSON, 1958; HATCH and TURNER, 1959) but the anaerobic rate may be restored by the addition of glutathione or cysteine. The effect is due to the aerobic inhibition of glyceraldehyde 3-phosphate dehydrogenase.

II. Phosphorylase.

α-D-(glucose 1-P + (glucose)$_n$ ⇌ (glucose)$_{n+1}$ + Pi)

1. Assay Method.

Principle. The enzyme preparation is incubated with G 1-P and a starch primer and the Pi liberated after a standard time is estimated. The method described is that of TURNER and TURNER (1957) which was based on that of GREEN and STUMPF (1942).

Reagents.

0.6 M citrate (citric acid — NaOH) buffer, pH 6.0.
0.26 M G 1-P, pH 6.0.
5% (w/v) soluble starch.

Procedure. A mixture of 0.25 ml citrate buffer, 0.1 ml water and 1.0 ml enzyme (containing 1 −4 units of activity) are incubated at 30°. After 15 minutes, 0.25 ml soluble starch and 0.2 ml G 1-P solution are added and samples taken for analysis at 0 and 20 minutes. The rate of reaction is approximately constant during the 0 − 20 minute period.

Pi is determined immediately on trichloroacetic acid [final concentration 4% (w/v)] extracts of the reaction mixtures by the method of ALLEN (1940) as modified by TURNER (1957).

Definition of unit activity. One unit of enzyme is defined as the amount which catalyzes the liberation of 0.05 mg of Pi in the above reaction mixture in 20 minutes at 30°.

Assay of phosphorylase in plant tissues. A method for the assay of the enzyme in fresh pea seeds was developed by TURNER and TURNER (1957). Peas (10 g) are finely ground with 12 ml water and acid-washed sand. The ground material is centrifuged at 1000 g, the supernatant filtered through cotton wool, and centrifuged at 20,000 g for 20 minutes at 0°. To 10 ml of the supernatant is added 40 ml cold saturated $(NH_4)_2SO_4$, pH 6.0. After standing for 1 hour at 0°, the precipitate is collected by centrifuging at 20,000 g for 7 minutes at 0°, taken up in water and diluted to the required volume. Phosphorylase activity is determined as described above except that 0.1 ml of 0.018 M ammonium molybdate is added instead of 0.1 ml of water. Assays for phosphorylase on replicate samples of peas agreed within 5%.

The addition of molybdate is necessary to prevent interference by glucose 1-phosphatase; this enzyme is completely inhibited by 1×10^{-3} M molybdate (TURNER and TURNER, 1960) whereas phosphorylase activity is not affected. Fluoride is not so effective as a phosphatase inhibitor.

β-amylase is an inhibitor of phosphorylase (PORTER, 1953) so that phosphorylase assay is difficult in crude extracts from tissues high in β-amylase such as cereals. PORTER has advocated the use of $HgCl_2$ to inhibit amylases in the determination of phosphorylase activity but the concentration of $HgCl_2$ required to inhibit β-amylase may also partially inhibit phosphorylase.

2. Purification.

Phosphorylase activity was first demonstrated by HANES (1939, 1940) in preparations from pea seeds and potatoes. The enzyme is widespread in plant tissues.

The enzyme from the potato has been crystallized by FISCHER and HILPERT (1953) and BAUM and GILBERT (1953). Details of the latter method of purification have been described by WHELAN (1955).

3. Properties.

Specificity. Phosphorylase is specific for α-D-G 1-P. This cannot be replaced by β-D-G 1-P (WOLFROM, SMITH and BROWN, 1943). α-L-G 1-P (POTTER, SOWDEN HASSID and DOUDOROFF, 1948), G 6-P, F 1-P, FDP (GREEN and STUMPF, 1942) or maltose 1-phosphate (MEAGHER and HASSID, 1946). Arsenate may replace Pi in the phosphorolysis of starch (KATZ and HASSID, 1951), but the product isolated is glucose.

The smallest molecule which acts as a primer is maltotriose (BAILEY, WHELAN and PEAT, 1950) but increase in chain length leads to much more efficient priming action.

Kinetics. The optimum pH for phosphorylase activity is between 5.9 and 6.1 (HANES, 1940). Hydrogen ion concentration also affects the equilibrium of the reaction: the values of the ratio Pi/G 1-P are 10.8, 6.7 and 3.1 at pH 5.0, 6.0 and 7.0 respectively. At equilibrium the ratio of any pair of the corresponding Pi and G 1-P ions or unionized acids is constant at any pH (HANES, 1940; HANES and MASKELL, 1942).

Inhibitors. Potato phosphorylase is inhibited 80% by 10^{-4} M $AgNO_3$ and 24% by 3.7×10^{-2} M NaF but is not inhibited by glucose, maltose, sucrose, iodoacetate or KCN (GREEN and STUMPF, 1942).

III. Phosphoglucomutase.

(Glucose 1-phosphate $\rightleftharpoons$ Glucose 6-phosphate)

LELOIR, TRUCCO, CARDINI, PALADINI and CAPUTTO (1948) showed that glucose 1,6-diphosphate is required as a cofactor for phosphoglucomutase activity.

Najjar (1954) found the mechanism of action of phosphoglucomutase was as follows:

$$G \text{ 1-P} + \text{phosphate-enzyme} \rightleftharpoons \text{glucose 1,6-diphosphate} + \text{enzyme}$$

$$\text{Enzyme} + \text{Glucose 1,6-diphosphate} \rightleftharpoons G \text{ 6-P} + \text{phosphate-enzyme}$$

1. Assay Method.

Principle. The reaction is followed by measuring the decrease in acid-labile phosphate with G 1-P as substrate; G 1-P is acid-labile whereas there is no appreciable hydrolysis of G 6-P in 1 N HCl at 100° in 10 minutes. Enzyme activity may also be followed by determining increase in reducing power: G 1-P is non-reducing whereas G 6-P reduces copper reagents. G 1-P is used as substrate as the equilibrium of the reaction favors G 6-P formation.

Reagents.

6.0×10^{-3} M $MgCl_2$.

2.5×10^{-2} M G 1-P, pH 7.5.

G 1-P prepared enzymically by the phosphorolysis of starch normally contains enough glucose 1,6-diphosphate to produce maximum activity. If purified G 1-P is used, glucose 1,6-diphosphate should be added to give a final concentration of 0.7×10^{-6} M in the reaction mixture.

1.25×10^{-1} M cysteine, pH 7.5, freshly prepared.

Procedure. A mixture of 0.1 ml $MgCl_2$, 0.1 ml G 1-P and 0.1 ml cysteine is incubated at 30° and after temperature equilibration the enzyme preparation is added. The total volume of the reaction mixture is 0.5 ml. After five minutes incubation the reaction is stopped by the addition of 0.1 ml. 33% (w/v) trichloroacetic acid and centrifuged if necessary. Samples of the supernatant are heated in 1 N HCl for 10 minutes at 100° and the Pi released estimated by the method of Allen (1940).

Definition of unit activity. One unit of enzyme is defined as the amount which, when diluted, would produce the formation of 40 μmoles of G 6-P under the above conditions. This unit is comparable to that described by Najjar (1955) for the assay of the phosphoglucomutase from muscle.

2. Preparation of Phosphoglucomutase Extracts from Plant Tissues.

Cardini (1951) described a simple procedure for the preparation of active extracts. Jack-bean seeds are soaked in water for 24 hours, mechanically skinned and homogenized in a Waring blendor in half their weight of cold water. The extract is strained through cheesecloth, centrifuged, dialyzed overnight and again centrifuged. The supernatant fraction contains the phosphoglucomutase activity and may be stored for months in the frozen condition without loss of activity.

Boser (1957) reported a 915-fold purification of phosphoglucomutase from crude potato juice. The reported steps involved heating at 45° for 10 minutes, repeated fractionation with $(NH_4)_2SO_4$ (50—65% saturation) and treatment with calcium phosphate gel.

Turner and Turner (1960) obtained active preparations of phosphoglucomutase by $(NH_4)_2SO_4$ fractionation of peaseed extracts.

3. Properties.

The enzyme exhibits relatively good activity over the range pH 6.5—8.2. The optimum pH is 7.5 in 0.03 M (final concentration) veronal buffer (Cardini, 1951).

Activators. The following data are from the report of Cardini (1951). The concentration of glucose 1,6-diphosphate required for maximum activity is 0.7×10^{-6} M.

Magnesium ions are required for activity and the optimum concentration is 10^{-3} M; at high concentrations of Mg^{++} ions phosphoglucomutase activity is inhibited. Cysteine was found to increase activity when glucose 1,6-diphosphate was added and maximum stimulation was observed with $1.8 - 2.5 \times 10^{-2}$ M cysteine.

IV. Hexokinase.

(ATP + Hexose → Hexose-6-phosphate + ADP)

1. Assay Method.

Principle. The method is based on the use of glucose as substrate and the subsequent spectrophotometric determination of the G 6-P using G 6-P dehydrogenase. Fructose may be employed as substrate if phosphoglucose isomerase is added to the reaction mixture for the determination of the hexose phosphate.

Reagents.

0.2 M Tris buffer, pH 7.5.
0.1 M $MgCl_2$.
0.1 M ATP.
0.033 M glucose.

Procedure. The reaction mixtures are prepared by mixing 0.2 ml Tris buffer, 0.1 ml $MgCl_2$, 0.1 ml ATP and 0.1 ml glucose. After preliminary incubation at 30°, the reaction is started by the addition of an appropriate volume of the hexokinase preparation. The total volume of the reaction mixture is 1 ml. Samples are removed at 0 and 15 minutes, pipetted into water at 95° and the mixture centrifuged, if necessary. G 6-P in the supernatant solution is determined by the reduction of TPN in the presence of G 6-P dehydrogenase by the method of HORECKER and WOOD (1957).

Definition of unit activity. A unit of activity is defined as the amount of enzyme which catalyzes the phosphorylation of 0.1 μmole of glucose in 15 minutes under the above conditions.

Application of the assay method to crude preparations. ATP-ase is often active in preparations from plant tissues especially if insoluble or mitochondrial fractions are used. Fluoride (0.01 M) (SALTMAN, 1953) may be added to suppress the ATP-ase activity.

2. Preparation of Hexokinase Extracts from Plant Tissues.

The work of MILLERD, BONNER, AXELROD and BANDURSKI (1951) indicated that hexokinase is associated with mitochondria in the mung bean, while SALTMAN (1953) found that hexokinase occurs in the plant associated with the mitochondria and in a soluble form. Wheat germ is a satisfactory source of both the insoluble and soluble form of hexokinase and the procedures described are those of SALTMAN.

The insoluble or mitochondrial fraction is prepared by grinding 15 g of wheat germ and 60 ml of cold 0.5 M mannitol with sand in a cold mortar. The mixture is centrifuged at 500 g for 10 minutes and the supernatant fluid is recentrifuged at 18,000 g for 15 minutes. The residue is resuspended in 20 ml of 0.5 M mannitol, centrifuged at 18,000 g for 10 minutes and the sedimented material suspended in 5 ml of cold distilled water. This suspension is used as the enzyme.

The soluble fraction is obtained by extracting wheat germ with 4 times its weight of cold water for 30 minutes at 2° with agitation. After centrifuging at 18,000 g for 20 minutes supernatant fluid is removed, adjusted to pH 5.5 with N acetic acid, and immediately centrifuged. The precipitate is discarded and the clear supernatant fluid brought to pH 7.0 with N KOH. Saturated $(NH_4)_2SO_4$, pH 7.0 is added to 55 percent saturation and the resulting precipitate discarded. Additional saturated $(NH_4)_2SO_4$ is added to bring the supernatant to 65 percent saturation, and the precipitate is collected by centrifugation, dissolved in distilled water, and dialyzed overnight against distilled water. By this procedure a 5-fold increase in the specific activity is obtained.

The hexokinase from yeast has been crystallized (McDonald, 1955).

3. Properties.

Specificity. The insoluble plant hexokinase catalyzes the phosphorylation of glucose, fructose, mannose and glucosamine (Saltman, 1953). Galactose, ribose, arabinose and ribulose do not act as acceptors.

A soluble specific fructokinase has been obtained from pea seeds by Medina and Sols (1956).

Kinetics. The following Michaelis-Menten constants for plant hexokinase have been reported by Saltman (1953): glucose, 4.4×10^{-4} M; ATP, 8.7×10^{-4} M. The kinetics of yeast hexokinase are described by McDonald (1955).

Activators and inhibitors. Magnesium ions are essential for hexokinase activity (Saltman, 1953). *p*-Chloromercuribenzoate had only a small inhibitory effect on plant hexokinase and $ZnCl_2$ was found to be a non-competitive inhibitor.

V. Phosphoglucose Isomerase.

(Glucose 6-phosphate ⇌ fructose 6-phosphate)

1. Assay Method.

Principle. The enzyme is incubated with G 6-P and the resulting F 6-P (which gives about 65% of the color of free fructose) is estimated by the method of Roe (1934). The method described is based on that of Ramasarma and Giri (1956).

Reagents.

0.04 M G 6-P.
0.5 M Veronal buffer, pH 7.5—7.8.

Procedure. A mixture of 0.05 ml G 6-P and 0.1 ml veronal buffer is incubated at 37°. After temperature equilibration, the enzyme solution is added, the total volume of the reaction mixture being 0.25 ml. After 2 minutes incubation, the reaction is stopped by the addition of HCl used in the determination of fructose by the method of Roe (1934).

Definition of the activity. One unit of enzyme is defined as the amount which catalyzed the formation of 0.1 μmole of F 6-P per 2 minutes under the above conditions of assay.

2. Preparation of Phosphoglucose Isomerase Extracts from Plant Tissues.

Ramasarma and Giri (1956) have described the partial purification of the enzyme from the seeds of the green gram *(Phaseolus radiatus)*. The seeds are ground to powder which is then extracted with five times its weight of water for 12 hours under toluene at 4°. Subsequent operations are carried out between 0° and 10°. 100 ml of the crude extract are adjusted to pH 7.0 and 100 ml of cold acetone added with stirring. After centrifuging, 200 ml of acetone are added to the supernatant fluid and the mixture centrifuged. The precipitate is dissolved in water and dialyzed against several changes of water for two days. Any precipitate formed is removed by centrifuging. The method gives an approximately six-fold purification and RAMASARMA and GIRI report the preparation to be free from interfering phosphatases and phosphoglucomutase.

3. Properties.

The following data are taken from the paper of RAMASARMA and GIRI (1956) unless otherwise indicated.

Kinetics. The MICHAELIS-MENTEN constant for G 6-P is 3.5×10^{-3} M. At equilibrium about 60% of G 6-P and 40% of F 6-P are present. The optimum pH is reported to be about 7.8 with G 6-P as substrate and about 9.0 with F 6-P as substrate.

Stability. The enzyme is stable for at least 4 weeks when kept frozen.

Inhibitors. $HgCl_2$ (2×10^{-3} M), KCN (0.02 M) and $ZnSO_4$ (10^{-3} M) inhibited phosphoglucose isomerase by 13%, 20% and 18% respectively. Iodoacetate (2×10^{-3} M) did not affect activity and no inhibition was observed with arsenate (0.01 M) and fluoride (0.02 M). Glucose and fructose were also without effect.

VI. Phosphohexokinase.

Fructose 6-phosphate + ATP → Fructose 1,6-diphosphate + ADP)

1. Assay Method.

Principle. The FDP formed is determined by the spectrophotometric method of SLATER (1953) in which the FDP is converted to triose phosphates and DHAP reacts with DPNH to give α-glycerol phosphate and DPN. The composition of the assay system is based on that described by LING, BYRNE and LARDY (1955).

Reagents.

0.2 M Tris buffer, pH 8.0.
0.03 M ATP, pH 7.
0.01 M F 6-P, pH 7.
0.007 M DPNH.
0.01 M $MgCl_2$.
0.2 M cysteine HCl, pH 7, prepared immediately before use.

Rabbit muscle fraction A described by SLATER (1953). This fraction prepared by the method of RACKER (1947), contains aldolase, triose phosphate isomerase and α-glycerol phosphate dehydrogenase. The $(NH_4)_2SO_4$ paste is stored at $-15°$ and fresh dilutions prepared each day.

Procedure. The following reagents are mixed in a cuvette: 0.5 ml Tris buffer, 0.3 ml ATP, 0.2 ml F 6-P, 0.3 ml $MgCl_2$, 0.1 ml cysteine, 0.5 ml DPNH, rabbit muscle fraction A, the enzyme preparation to be assayed and water to give a total volume of 3 ml. Excess of rabbit muscle fraction A should be added and the amount to be used should be determined in trial estimations. After mixing, the optical density at 340 mμ is measured every 2 minutes for 10 minutes. A control cuvette containing all the components except the phosphohexokinase preparation should be followed simultaneously.

Definition of unit activity. A unit of phosphohexokinase activity is defined as the amount of enzyme which phosphorylates 1.0 μmole of F 6-P per minute under the above conditions. In the procedure outlined, one mole of FDP gives 2 moles of DHAP which oxidize 2 moles of DPNH.

2. Preparation of Phosphohexokinase from Plant Tissue.

AXELROD, SALTMAN, BANDURSKI and BAKER (1952) have described the preparation of an active extract from pea seeds. Pea seed meal (Alaska variety) is suspended in 2—4 times its weight of water at 2° and agitated occasionally for 30 minutes. All subsequent operations are carried out at 2°. After centrifuging at 16,000 g for 20 minutes, the supernatant material is decanted and adjusted to pH 7.5 with NaOH. Neutralized saturated $(NH_4)_2SO_4$ is added and the fraction precipitating between 28% and 38% saturation is taken up in water (one-third of the original volume of extract) and dialyzed with gentle agitation against 3 changes of 1×10^{-4} M cysteine. This procedure results in a six-fold increase in specific activity. AXELROD et al. (1952) found evidence for phosphohexokinase activity in a number of other plant tissues.

3. Properties.

The following information is from the publication of AXELROD et al. (1952).
Kinetics. The MICHAELIS-MENTEN constant for F 6-P is 7.1×10^{-3} M and 2.4×10^{-3} M for ATP. AXELROD et al. reported two peaks in the pH activity curve for pea-seed phosphohexokinase, one at about pH 6 and another broad peak at about pH 9.
Stability. The enzyme preparation was stable for two weeks when stored at $-5°$.
Activators and inhibitors. Magnesium ions are essential for phosphohexokinase activity. AXELROD et al. found the optimum concentration of Mg^{++} ions was about 0.01 M using an ATP concentration of 0.01 M in the reaction mixture. $MnCl_2$ was approximately half as effective as $MgCl_2$ at a concentration of 0.01 M.
$CuSO_4$ (0.01 M), $HgCl_2$ (5×10^{-3} M), iodoacetate (5×10^{-3} M) and inorganic pyrophosphate (5×10^{-3} M) inhibited pea-seed phosphohexokinase by 100, 96, 51 and 73% respectively. No inhibition was observed with fluoride (0.01 M), cyanide (0.03 M), fructose (0.04 M) and glucose (0.04 M).

VII. Fructose-1,6-Diphosphatase.

(Fructose-1,6-diphosphate → Fructose-6-phosphate + Pi)

1. Assay Method.

Principle. Fructose 1,6-diphosphatase may be assayed by determination of Pi liberated in the course of the reaction. A spectrophotometric method has been proposed which depends on the formation of F 6-P which is measured by reduction of TPN in the presence of phosphoglucose isomerase and glucose 6-phosphate dehydrogenase (RACKER and SCHROEDER, 1958).
RACKER and SCHROEDER (1958) have obtained and purified from spinach extracts, a fructose-1,6-diphosphatase acting optimally at pH 8.5 (alkaline enzyme) and in addition have detected in the same extracts another enzyme which cleaves FPD at a neutral pH (neutral enzyme). An assay method for each enzyme is described in this section.
Applicability of the methods. In crude plant extracts the chemical determination of Pi is preferred since the presence of aldolase and the TPN linked triose phosphate dehydrogenase can interfere with the spectroscopic assay.
The spectrophotometric assay is convenient and more rapid when the contaminating enzymes are removed. The assay depends, however, on a supply of

highly purified glucose 6-phosphate isomerase (not commercially available) and glucose 6-phosphate dehydrogenase.

Assay 1. Colorimetric assay for alkaline fructose-1, 6-diphosphatase (RACKER and SCHROEDER, 1958). In this assay, Pi liberated at pH 8.5 by hydrolysis of FDP is measured.

Reagents.

0.5 M Tris buffer, pH 8.5.
0.05 M FDP, pH 8.5.
0.02 M EDTA, pH 8.5.
0.05 M $MgCl_2$.
10% (w/v) trichloroacetic acid.
Reagents for phosphate determination.

Procedure. In a final volume of 1 ml, the following reagents are added: 0.2 ml Tris, 0.1 ml FDP, 0.1 ml EDTA, 0.1 ml $MgCl_2$, and the enzyme to be tested. The final pH is 8.5. The solutions are deproteinized with addition of 1.0 ml trichloroacetic acid after 10 minutes at 25° and following centrifugation Pi is determined in the supernatant fluid.

Proportionality is linear between 0.2 and 2 μmoles FDP cleaved in 10 minutes.

Assay 2. Colorimetric assay for neutral fructose 1,6-diphosphatase (RACKER and SCHROEDER, 1958).

In this assay, Pi liberated at pH 6.9 by hydrolysis of FDP is measured.

Reagents.

0.4 M histidine buffer, pH 6.9.
0.05 M FDP, pH 6.9.
10% (w/v) trichloroacetic acid.
Reagents for phosphate determination.

Procedure. In a final volume of 1 ml, the following reagents are added: 0.2 ml histidine buffer, 0.1 ml FDP and the enzyme to be tested. After 10 minutes at 25°, the reaction is stopped by the addition of 1 ml of trichloroacetic acid. Following centrifugation, an aliquot of the supernatant solution is analyzed for Pi.

Definition of unit activity. One unit of enzyme is defined by RACKER and SCHROEDER (1958) as that amount which catalyzes the turnover of 1 μmole FDP per minute under the conditions of the assay.

2. Purification Procedure (RACKER and SCHROEDER, 1958).

The starting material is 1 bushel of spinach.

Following destemming and shredding, a crude extract is obtained with a hydraulic press. The following steps are then carried out:

1. $(NH_4)_2SO_4$, 50—100% precipitate is kept.
2. heating at 62° for 15 minutes and discarding of precipitate.
3. absorption by calcium phosphate gel and elution with phosphate buffer.
4. $(NH_4)_2SO_4$ fractionation.

The crude extract contain 0.07 units per mg of protein and the purest fractions following step 4 had a specific activity of 188.

3. Properties.

Specificity. The alkaline fructose 1,6-diphosphatase appears to be specifi- for FPD; F 1-P, F 6-P, sedoheptulose 1,7-diphosphate, and sorbose 6-P are nct cleaved. The neutral fructose 1,6-diphosphatase hydrolyzed sedoheptulose-1,7o diphosphate at approximately twice the rate of FDP (RACKER and SCHROEDER, 1958).

CHAKRAVORTY, CHAKROBORTTY and BURMA (1959) have purified about 17-fold from spinach an acid diphosphatase (pH optimum of 5.5 and hereafter designated as the pH 5.5 enzyme) which reacts with both ribulose diphosphate and FDP at practically the same rate.

Effect of activators and inhibitors. 1. pH 8.5 enzyme. This enzyme is completely inactive in the absence of Mg^{++}. Mn^{++} can substitute for Mg^{++}. EDTA stimulates. The concentrations of Mg^{++}, EDTA and FDP for optimal activity are inter dependent.

NaF (0.02 M) and p-chloromercuribenzoate (2×10^{-4} M) inhibited about 75%. The following compounds had no effect: KCN (2×10^{-3} M), iodoacetate (2×10^{-3} M) and N-ethylmaleimide (2×10^{-3} M).

2. pH 6.9 enzyme. This enzyme is not stimulated by Mg^{++} or EDTA.

3. pH 5.5 enzyme. The enzyme does not require Mg^{++}. The following compounds inhibit: 4.5×10^{-3} M NaF (22%), 2×10^{-3} M beryllium chloride (72%), 1.8×10^{-3} M $HgCl_2$ (78%), 2.2×10^{-3} M EDTA (34%) and 2×10^{-3} M p-chloromercuribenzoate (60%).

Effect on pH. With Tris buffer, the enzyme specific for FDP has a pH optimal around 8.5, dropping off sharply in the neutral range, but decreasing more gradually on the high pH side. The entire curve is shifted to the left with a pH optimum around pH 8.2 in veronal buffer. The neutral fructose 1,6-diphosphatase displays maximum activity between pH 5.5 and 7.0. The diphosphatase of CHAKRAVORTY et al. (1959) has a pH optimum in acetate buffer, as well as citrate buffer, between pH 5 and 6.

Physical properties. The MICHAELIS-MENTEN constant for FDP of the alkaline enzyme is approximately 3×10^{-4} M. With FDP as substrate, the MICHAELIS-MENTEN constant of the pH 5.5 enzyme is 1.04×10^{-3} M. The pH 8.5 enzyme is stable at $-20°$ (RACKER and SCHROEDER, 1958).

Distribution of the enzyme. In addition to spinach, an alkaline fructose 1,6 diphosphatase has been detected in pea roots (GIBBS and HORECKER, 1954; GIBBS EARL and RITCHIE, 1955).

The pH 8.5 enzyme has also been demonstrated in extracts of barley, alfalfa, *Spirodela*, the fern *Pteris gautherii*, *Euglena gracilis*, *Chlamydomonas reinhardii*, *Rhodospirillum* and *Chromatium* (SMILLIE, 1960).

VIII. Aldolase.

(Fructose diphosphate ⇌ Dihydroxyacetone phosphate-glyceraldehyde 3-phosphate)

1. Assay Method.

Principle. Aldolase may be assayed in the direction of synthesis or of cleavage of FDP. Generally, the assay is based on the measurement of triose phosphate formed in the aldolase reaction in the presence of a trapping reagent (cyanide, hydrazine, bisulfite). Triose phosphate may be determined either chemically (hydrolysis of triose phosphates under alkaline conditions, colorimetric determination of products with p-hydroxybiphenyl or 2,4-dinitrophenylhydrazine) or enzymically (glyceraldehyde 3-phosphate dehydrogenase or α-glycerol phosphate dehydrogenase).

Applicability of the methods. The chemical methods are preferred for crude assays while the enzymatic assays are to be preferred for purified preparations.

Enzymic assay with glyceraldehyde 3-phosphate dehydrogenase cannot be used if the crude extract contains α-glycerol phosphate dehydrogenase since the end result is a dismutation of the triose phosphate into α-glycerol phosphate and 3-PGA, with no final change in DPN.

Since many crude plant tissues contain a DPNH oxidase, corrections must be made for this endogenous rate when aldolase is assayed coupled to α-glycerol phosphate dehydrogenase.

The trapping procedures employ an alkaline pH (8.3–9.0) while the enzymatic methods generally use a pH of approximately 7.5. This difference in pH in addition to other factors leads to considerable variation in apparent aldolase activities.

Assay 1. Chemical method. The method described here is essentially that of MEYERHOF and LOHMANN (1935) and HERBERT, GORDON and SUBRAHMANYAN (1940). It is based on the measurement of alkali-labile phosphate of the triose phosphate which is carried out in the presence of cyanide to favor the formation of triose phosphate. Aldolase activity is best measured at 30° or 37° in a short time period (5 to 10 minutes) to avoid spontaneous breakdown of triose phosphate.

Reagents.

0.1 M FDP, pH 8.5. Prepare from the dibarium salt as the strychnine salt according to SABLE (1952).
0.1 M KCN. Prepared fresh daily.
0.5 M glycylglycine, adusted to pH 8.5 with N NaOH.
1.0 N NaOH. Prepared fresh daily.
2.0 N HCl.
Reagents for phosphate determination.

Procedure. To 0.08 ml glycylglycine, add 0.25 ml KCN, 0.05 ml FDP and enzyme and water to give a final volume of 0.8 ml. The reaction mixture is incubated at 30° for 5 minutes before the addition of enzyme. The reaction is stopped after 10 minutes by the rapid addition of 1.0 ml of N NaOH. The mixture is incubated at room temperature. After 20 minutes, the alkaline mixture is exactly neutralized with 2 N HCl (this is determined by prior titration) and aliquots are removed for Pi determination.

A zero time sample to correct for Pi present in the reaction mixture is prepared by adding the NaOH before enzyme.

Correction must be applied for the presence of fructose 1,6-diphosphatase. This is done by carrying out the essay in the usual manner, except that the reaction is terminated by adding the 2 N HCl, followed by the N NaOH. Pi is determined and subtracted from that found in the saponified samples.

Chemical methods for phosphate assay. Pi may be determined by the method of FISKE and SUBBAROW (1925), KING (1932) or LOHMANN and JENDRASSIK (1926). Triose phosphate may also be determined without saponification by means of the colorimetric methods of SIBLEY and LEHNINGER (1949) and by DOUNCE and BEYER (1948) with modifications by DOUNCE, BARNETT and BEYER (1950) and by LOWRY, ROBERTS, WU and HIXON (1954).

2. *Enzymatic method.* Two methods are available. One introduced by WARBURG and CHRISTIAN (1939) is based on measurement of the reduction of DPN by the G 3-P formed in the aldolase reaction, in the presence of arsenate and excess glyceraldehyde 3-phosphate dehydrogenase. The other due to RACKER (1947) and BARANOWSKI and NIEDERLAND (1949) is based on the oxidation of DPNH by the DHAP in the presence of α-glycerol phosphate dehydrogenase.

Definition of unit activity. One unit of enzyme is defined by STUMPF (1948) as that amount of enzyme which produces 0.1 mg of alkali-labile P in 10 minutes. 1 μmole FDP = 2 μmole P = 0.062 mg P.

2. Purification Procedure.

The following procedure is that due to STUMPF (1948). The starting material is pea seeds (Dwarf Telephone) or squash seeds. The enzyme is extracted from seeds soaked in distilled water for 12 hours at 2°, with 0.1% K_2CO_3. The extract is subjected to (1) neutral saturated $(NH_4)_2SO_4$ fractionation (2) isoelectric precipitation with dilute acetic acid at pH 5.5 to remove inert protein, (3) acetone fractionation and (4) a repeat of step 2.

The initial extract contains 0.06 units per mg of protein and the final preparation 5.5 units per mg of protein or a purification of 92-fold.

3. Properties.

Specificity. G 1-P, G 6-P, and F 6-P are inert with the pea aldolase. In the cleavage reactions catalyzed by aldolase, activity is generally thought to be restricted to trans-linkages. TUNG, LING, BYRNE and LARDY (1954) have reported that tagatose 1,6-diphosphate, which has a cis-linkage, is slowly split by the muscle enzyme.

CARDINI (1952) reported cleavage of F 1-P in extracts of Jack bean seeds.

Activators and inhibitors. According to the data of STUMPF (1948), pea aldolase is not very sensitive to heavy metal inhibition. A final concentration of 10^{-4} M $CuSO_4$, $Hg(C_2H_3O_2)_2$, phenylmercuric acetate and $AgNO_3$ does not inhibit the enzyme. In a final concentration of 10^{-3} M iodoacetamide, fluoride, indoleacetate or azide, as well as 10^{-4} M iodine, give no inhibition. Contrary to the data obtained with yeast aldolase, cysteine, α, α'-dipyridyl and cyanide did not inhibit the pea enzyme.

Divalent ions, Cu^{++}, Zn^{++}, Mg^{++}, and Mn^{++} had no effect on pea aldolase activity. Aldolase obtained from the red alga. *Chondrus*, is inhibited by EDTA and the inhibition is reversed by Mn^{++} (FEWSON and GIBBS, 1961).

In final concentrations of 10^{-2} M, glucose, fructose, G 1-P, G 6-P, F 6-P, β-glycerophosphate and pyruvate did not inhibit activity of the pea enzyme in a competitive fashion. Cyanide, bisulfite, semicarbazide, hydrazine (pH 8.5) in a final concentration of 8×10^{-2} M were equivalent in their fixative capacity. In a final concentration of 10^{-3} M, hydroxylamine inhibited the pea enzyme 55% (STUMPF, 1948).

Effect of pH. The pH optimum of the pea enzyme was reported to be 8.5 by STUMPF (1948) using the chemical method described here in which cyanide is used to trap the triose phosphate. Above pH 10 and below pH 5, activity falls rapidly. DOUNCE and BEYER (1948) have criticized this method for the determination of aldolase activity and have used a modification which with the rabbit muscle enzyme indicates maximal activity at pH 7.2 in contrast to the commonly accepted value of 9 (HERBERT et al. 1940).

Physical properties. The equilibrium constant of the pea enzyme estimated from the equation:

$$K = \frac{(^1/_2 \text{ triose phosphates})^2}{(\text{fructose diphosphate})}$$

was found by STUMPF to be about 1.15×10^{-4} at 31° in borate buffer at pH 8.5. At 22° and 40° the corresponding values are 0.91×10^{-4} and 5.56×10^{-4} M, respectively.

The MICHAELIS-MENTEN constant of the pea enzyme was found to be 0.8×10^{-3} M.

Distribution of the enzyme. TEWFIK and STUMPF (1949) found measurable aldolase activity in a large variety of plants such as fungi, ferns, conifers, monocotyledons and dicotyledons. It is distributed in different parts of the plant such as seeds, seedlings, roots, stems, leaves, flowers and fruits.

Aldolase has been detected in extracts of the green algae, *Ulva lactuca* (JACOBI, 1957) and *Hydrodictyon reticulatum* (RICHTER, 1957). SANWAL and KRISH-, NAN (1959) have reported that 80% of the aldolase of the cactus, *Nopalea dejecta* is particulate bound. This contrasts with all other reports in which the enzyme is apparently in the water soluble fraction.

Aldolase from *Aspergillus niger* has been obtained with a specific activity equal to that of the crystalline muscle or yeast enzyme (JAGANNATHAN, SINGH and DAMODARAN, 1956).

The only exception to the ubiquity of the enzyme in the plant kingdon is its absence in extracts of the blue green algae, *Anacystis nidulans* (RICHTER, 1959), *Nostoc muscorum* and *Anabaena* (FEWSON and GIBBS, 1961).

IX. Triosephosphate Isomerase.

(Glyceraldehyde-3-phosphate $\rightleftharpoons$ dihydroxyacetonephosphate)

Assay Method.

Principle. Both chemical methods and optical methods have been used. One chemical method is based on the determination of alkali-labile Pi before and after treatment with iodine (MEYERHOF and KIESSLING, 1935). Another depends on the difference in rate of formation of the chromogenic 2,4-dinitrophenylhydrazone derivates of G 3-P and DHAP (BECK, 1956). Optical assays have been designed by coupling the triosephosphate isomerase reaction with glyceraldehyde 3-phosphate dehydrogenase (OESPER and MEYERHOF, 1950) or with α-glycerol phosphate dehydrogenase (RACKER, 1947, COOPER, SRERE, TABACHNIK and RACKER, 1958).

The following assay described has been developed by using G 3-P as substrate, α-glycerol phosphate dehydrogenase as coupling enzyme and following the oxidation of DPNH. This process is measured by the decrease of absorption at 340 mμ.

Reagents.

0.5 M Tris buffer, pH 7.5.
2 units per ml of α-glycerolphosphate dehydrogenase (1 unit = 1 μmole substrate per minute).
1.2 × 10⁻³ M DPNH.
5 × 10⁻³ M DL-G 3-P.

Method. To a final volume of 1 ml in a cuvette, the following reagents are added: 0.1 ml DPNH, 0.1 ml Tris, 0.1 ml G 3-P, and 0.1 ml α-glycerol phosphate dehydrogenase. All reagents except DPNH but including the solution to be assayed are added to the check cell. The last addition is G 3-P. Determine the optical density values at 30 second intervals for 3 minutes. Dilute test solution so that density changes of 0.020 to 0.050 per minute are attained. The oxidation of 1 μmole of DPNH is equivalent to an optical density change of 6.22.

Definition of unit activity. One unit of enzyme is that amount which causes an initial rate of oxidation of 1 μmole of DPNH per minute.

Applicability to crude extracts. In the absence of arsenate, the equilibrium is so favorable for α-glycerol phosphate formation that no serious error is introduced even in the presence of glyceraldehyde 3-phosphate dehydrogenase.

Distribution of enzyme in plants. Triosephosphate isomerase has not as yet been isolated from extracts of higher plants. There are reports indicating that this enzyme occurs in crude extracts of pea seed (Stumpf, 1948) and of pea roots Gibbs (et al., 1955).

X. TPN Triosephosphate Dehydrogenase.

(Glyceraldehyde-3-phosphate + Pi + TPN ⇌ 1,3-diphosphoglycerate + TPNH)

1. Assay Method.

Principle. The rate of reduction of TPN resulting from G 3-P oxidation is followed, using the spectrophotometer. The assay method described here is essentially that of Warburg and Christian (1939) except that cysteine has been included in the reaction mixture and TPN has replaced DPN (Gibbs, 1955).

Arsenate is used in place of Pi in the assay. This makes the reaction irreversible, since the product 1-arseno-3-phosphoglycerate is unstable and non-enzymatically liberates 3-PGA and arsenate.

Reagents.

0.2 M Tris buffer, pH 8.5.
0.1 M sodium arsenate, $Na_2HAsO_4 \cdot 7 H_2O$.
0.04 M cysteine; neutralize 19 mg of cysteine HCl in a final volume of 3 ml to pH 8 with KOH. Prepare fresh daily.
0.015 M DL G 3-P.

This is prepared from commercially available DL-glyceraldehyde-3-phosphate diethylacetal, barium salt. Thoroughly mix 57 mg of this compound with 4.5 ml Dowex-50 H⁺ form resin. The resin suspension is composed of 1 volume resin plus 5 volumes H_2O. Place in a boiling water bath for 3 minutes with intermittent shaking. Chill quickly, centrifuge and wash residue several times with small portions of H_2O. Combine supernatant solution and washings.

0.03 M FDP.
0.01 M TPN

Applicability of assay method. A quantitative interpretation of the assay as applied to crude extracts is unsatisfactory, since interfering enzymes such as triose phosphate isomerse and aldolase may compete for the G 3-P. In addition, TPNH oxidases are present in many crude extracts.

If G 3-P is not available, it may be generated with FDP and crystalline aldolase. However, most crude plant extracts readily convert FDP to G 6-P and TPN reduction occurs catalyzed by G 6-P dehydrogenase. If G 3-P is unavailable, the crude extract is incubated with 0.01 M iodoacetamide for 5 minutes to poison triosephosphate dehydrogenase, then FDP is added and if reduction of TPN does not occur, then the TPN enzyme can be assayed with FDP and aldolase.

Procedure. In a cuvette of 1 cm light path prepare a partial reaction mixture as follows: 0.5 ml Tris, 0.1 ml sodium arsenate, 0.3 ml cysteine, 0.03 ml TPN, enzyme (0.4 to 2 units), aldolase (if FDP is used as source of G 3-P) and water to 2.8 ml. Incubate at room temperature for 7 minutes to insure complete activation of the enzyme by cysteine. All reagents except TPN including the enzyme solution are added to the check cuvette. The reaction is started by adding 0.2 ml of G 3-P or FDP. Optical density readings at 340 mμ are taken before and after the addition of the phosphorylated substrate. The increase in optical density is read at 30 second intervals for 3 minutes. An amount of enzyme should be used which results in density changes of 0.020 to 0.050 per minute. Under the conditions of the assay given, a change in optical density of 1.0 is equivalent to 0.48 μmole.

Definition of unit activity. One unit of enzyme is defined as the amount of enzyme which catalyze the reduction of 1 μmole of TPN per minute.

2. Purification Procedure (GIBBS, 1955).

One kilogram of pea leaves or spinach leaves are ground with sand, 100 ml of 0.5 M $KHCO_3$, and 100 ml of EDTA (30 mg per ml, pH 8.0). The brew is squeezed through cheesecloth. Following centrifugation, the pH of the supernatant solution is brought to 5.0 with cold 2 N acetic acid. The heavy precipitate is removed by high speed centrifugation. The pH of the supernatant solution is increased to 8.0 with 2 N KOH and brought to 31% saturation with respect to $(NH_4)_2SO_4$. The precipitate is discarded. The supernatant fluid is brought to 42% saturation with solid $(NH_4)_2SO_4$ and the precipitate formed is dissolved in 0.1 M Tris buffer, pH 8.0. To this solution an equal volume of acetone ($-18°$) is added. During the addition which requires about 10 minutes, the mixture is cooled to $-5°$. The precipitate obtained is extracted three times with 0.033 M Pi buffer, pH 7.6. To the combined washings, 3 ml of alumina C_γ (dry weight, 25 mg per ml) is added. The gel is removed by centrifugation and discarded. After the pH of the supernatant fluid is brought to 5.6 with 2 N acetic acid, it is treated successively with three 15 ml portions of alumina C_γ. Generally the first two batches of gel contain most of the enzyme. The enzyme is eluted with three 10 ml portions of 0.05 M Pi buffer, pH 7.6.

Specific activity of the leaf extract was 0.7 unit per mg of protein while eluate from alumina contained enzyme of specific activity of $67-115$ units per mg protein.

3. Properties.

Specificity. The purified enzyme is specific for TPN.

Activators and inhibitors. For full activity, the enzyme has to be preincubated with cysteine or glutathione. The concentration of iodoacetamide to produce 50% inhibition is approximately 0.0033 M, a concentration of 0.01 M inhibits the enzyme completely. Inhibition by iodoacetamide is not instantaneous. Inhibition by the alkylating reagent is irreversible. However, enzyme inactivated due to storage may be reactivated by cysteine or glutathione. Fluoride, azide and indole-acetic acid, each at 0.01 M, cause no detectable effect (GIBBS, 1955).

Effect of pH. The activity of the enzyme when tested in Tris buffer or glycyl-glycine buffer is optimal between pH 8.5 to 8.8. At pH 7 activity is 0.4 that of the optimum whereas at pH 9.1 activity is about 0.9 (GIBBS, 1955).

Distribution. The enzyme has been observed in the leaves of the following plants: pea (GIBBS, 1952), spinach (ROSENBERG and ARNON, 1955), green and albino barley leaves (FULLER and GIBBS, 1959). BENEDICT and BEEVERS (1961) reported the enzyme to be present in extracts of germinating castor beans.

FULLER and GIBBS (1959) have reported the enzyme to be in *Chlorella, Chlamydomonas, Euglena, Rhodospirillum* and *Romeria*. The enzyme is lacking in anaerobic chemosynthetic and photosynthetic bacteria (SMILLIE and FULLER, 1960). The enzyme appears to be localized exclusively in the plastids (SMILLIE and FULLER, 1960).

XI. DPN Triosephosphate Dehydrogenase.

(Glyceraldehyde-3-phosphate + DPN + Pi $\rightleftharpoons$ 1,3-diphosphoglycerate + DPNH)

1. Assay Method.

Principle. The method is similar to that described under TPN triosephosphate dehydrogenase with DPN being substituted for TPN.

Reagents. See TPN triosephosphate dehydrogenase.

Procedure. See TPN triosephosphate dehydrogenase.

2. Purification Procedure (Hageman and Arnon, 1955).

Acetone powder (480 gm) of peas are placed in a chilled beaker and extracted with 2880 ml of cold 0.01 M potassium phosphate buffer, pH 7.2 containing 0.0015 M EDTA. Extraction is continued for 15 minutes at 2° before centrifugation. The supernatant fluid is filtered through Whatman No. 1 paper on a chilled Büchner funnel with suction. The residue is re-extracted with 2200 ml of the same solvent. The combined filtrates (4500 ml) are heated in a bath at 60° with stirring and removed when they reach a temperature of 55°. The heated solutions are chilled rapidly by pouring into a large flask immersed in a bucket of crushed ice. As soon as the temperature of the solution is 2°, solid $(NH_4)_2SO_4$ is added to 60% saturation while the pH is maintained at 7.2 by the addition of 3.5 N NH_4OH. The precipitate is discarded. Following decantation, additional solid $(NH_4)_2SO_4$ added to bring the supernatant fluid to 95% saturation (pH 7.2) and the precipitate formed after 12 hrs. is collected by centrifugation and dissolved in 0.0015 M EDTA, pH 7.0; final volume 425 ml. To the dissolved precipitate, solid $(NH_4)_2SO_4$ is added to 40% saturation (pH 7.2) and the precipitate discarded. The $(NH_4)_2SO_4$ concentration is raised to 60% and the pH adjusted to 7.9 with 3.5 N NH_4OH.

After centrifuging 2 ml portions of the suspension, the supernatant solutions are discarded. Each precipitate is dissolved in 2 ml of 0.0015 M EDTA and diluted with 10 ml of 0° water. After standing 30 minutes, the precipitate is removed by centrifugation. The final volume is approximately 11.5 ml. An overall purification of 26-fold is obtained.

3. Properties.

Specificity. The enzyme is specific to DPN (Gibbs, 1955).

Inhibitors. The pea enzyme is inhibited approximately 80% by 10^{-3} M iodoacetamide (Gibbs, 1955). For full activity, the enzyme is incubated with cysteine or glutathione (Gibbs, 1955).

Distribution of enzyme. The enzyme was first reported in pea leaf extracts by Tewfik and Stumpf (1951). It is also present in pea roots (Gibbs, 1952) and albino barley (Fuller and Gibbs, 1959). Smillie and Fuller (1960) have noted that the enzyme can be extracted from the following photosynthetic microorganisms: *Euglena gracilis, Rhodopseudomonas spheroides, Rhodomicrobium vannielii, Chromatium*, strain D and *Chlorobium thiosulfatophilum*.

For a comparison between the activity of the TPN and DPN triose phosphate dehydrogenases, see Fuller and Gibbs (1959), and Smillie and Fuller (1960). The changes in the concentration of the two enzymes during the life cycle of the pea plant is discussed by Hageman and Arnon (1955).

XII. Triosephosphate-Phosphoglycerate Dehydrogenase.

(Glyceraldehyde-3-phosphate + TPN → 3-phosphoglyceric acid + TPNH)

Axelrod, Bandurski, Greiner and Jang (1953), Barnett, Stafford, Conn and Vennesland (1953) and Gibbs (1954) reported that extracts of leaf tissue could reduce TPN with triosephosphate as substrate in the absence of Pi or arsenate. This suggested an alternative enzyme for the oxidation of triosephosphate to 3-PGA. More recently, Rosenberg and Arnon (1955) have reported a further partial purification of this enzyme.

1. Assay Method.

Principle. The method is similar to that described under TPN-triosephosphate dehydrogenase with the omission of arsenate.

Reagents. See TPN triosephosphate dehydrogenase.

Procedure. See TPN triosephosphate dehydrogenase.

2. Purification Procedure (ROSENBERG and ARNON, 1955).

Two lots of 65 g each of acetone powder of sugar beet leaves are extracted with 1.3 liters of 0.0015 M EDTA (K$^+$, pH 7.2). Following centrifugation, to each 1240 ml of supernatant fluid at pH 6.5, 403 g of solid $(NH_4)_2SO_4$ is added (55% saturated). The precipitate is discarded. The supernatant fluid is brought to 65% saturation with 86 g of solid $(NH_4)_2SO_4$. The combined 55 to 65% fractions are dissolved in 650 ml of 0.1 M K_2SO_4. The enzyme is absorbed onto 91 ml of alumina C$_\gamma$ (30 mg per ml) and eluted with 520 ml, 80 ml, and 25 ml portions of 0.25 M K_2SO_4. The combined eluate is diluted with 245 ml H_2O and the solution is 0.18 M with respect to K_2SO_4. After stirring in 65 ml of calcium phosphate gel (20 mg per ml), the gel is collected, washed with 130 ml of 0.18 M K_2SO_4, and the enzyme is eluted with two washings (208 ml, 30 ml) of 0.30 M K_2SO_4. The combined eluate is dialyzed against 4 liters of 0.15 M K_2SO_4 — 0.0015 M EDTA (K$^+$, pH 7.2). The enzyme is reabsorbed onto 26 ml of calcium phosphate gel, re-eluted with 130 ml of 0.30 M K_2SO_4 and dialyzed against 4 liters of 0.0015 M EDTA for 80 minutes. The dialysis is repeated 3 times.

3. Properties.

Specificity. The data of ROSENBERG and ARNON (1955) show little reduction of TPN with HCHO, CH_3CHO, propionaldehyde, butyraldehyde, D-ribose, D-ribose 5-P, DL-glyceraldehyde, glucose and glyoxal. The enzyme is specific for TPN.

Activators and inhibitors. p-chloromercuribenzoate $(1.6 \times 10^{-6} M)$ inhibits the enzyme approximately 80%. Arsenate, Pi and sulfate also inhibit (ROSENBERG and ARNON, 1955).

Effect of pH. The pH range for maximal activity is 8.3 to 8.6; the activity at pH 7.0 is about 25% of the maximum (ROSENBERG and ARNON, 1955).

Distribution. In addition to sugar beet leaves, the enzyme has been reported to be in the leaves of spinach (AXELROD et al., 1953; BARNETT et al., 1953) and peas (GIBBS, 1954). SMILLIE and FULLER (1960) have found the enzyme in extracts from pea root, etiolated pea stem, and streptomycin-bleached *Euglena*. Negative results were obtained by SMILLIE and FULLER (1960) when extracts of *Anacystis nidulans* and *Chromatium* were assayed.

XIII. Phosphoglycerate Kinase.

(1,3-diphosphoglycerate + ADP ⇌ 3-phosphoglycerate + ADP

1. Assay Method.

Principle. The forward or back reaction can be used for assay. This may be done either spectrophotometrically by coupling the enzyme in both directions with DPN triosephosphate dehydrogenase (BÜCHER, 1955) or colorimetrically, employing the back reaction which is based on a reaction involving hydroxamic acid and 1,3 di-PGA (AXELROD and BANDURSKI, 1953).

1. Spectrophotometric assay (BÜCHER, 1955). This assay depends on the rate of reduction of DPN resulting from G 3-P oxidation in the presence of excess highly purified triosephosphate dehydrogenase and limiting amounts of phosphoglycerate kinase. Conversely, the back reaction depends on the oxidation of DPNH resulting from 1,3-di PGA reduction. In contrast to the forward reaction, the equilibrium of substrates is less favorable in this assay; cysteine is used as a trap for the G 3-P formed. According to BÜCHER (1955), one unit in the back reaction assay corresponds to four units of the forward reaction assay. Only the forward reaction is given here.

Reagents.

0.3 M K-Na phosphate buffer. pH 6.9.
0.025 M DL-G 3-P (see TPN triose phosphate dehydrogenase).
0.012 M DPN.
0.05 M $MgSO_4$. $7 H_2O$.
0.025 M ADP.
Glyceraldehyde 3-phosphate dehydrogenase, 3 mg per ml.
0.5 M glycine.

Procedure. The following reagents are mixed in a cuvette with a light path of 10 mm: 0.5 ml phosphate buffer, 0.1 ml DPN, 0.1 ml G 3-P, 0.1 ml $MgSO_4$, 0.1 ml ADP, 0.3 ml glycine, 0.1 ml glyceraldehyde 3-phosphate dehydrogenase, enzyme preparation to be assayed and heavy-metal-free water to give a total volume of 3 ml. The last addition is glyceraldehyde 3-phosphate dehydrogenase. Density readings are taken in a spectrophotometer at 340 mμ at 30 second intervals for 3 minutes. The enzyme to be tested is diluted in heavy-metal-free water to result in density changes of 0.020 to 0.050 per minute.

Definition of unit activity. A unit of phosphoglycerate kinase may be defined as the amount of enzyme that catalyzes the turnover of 1 μmole substrate per minute. An optical density change of 1.0 is equivalent to 0.48 μmole.

2. Colorimetric assay (Axelrod and Bandurski, 1953). This procedure is based on the Lipmann-Tuttle (1945) hydroxamic acid technique and involves the back reaction in which hydroxylamine is present to trap the 1,3-di PGA giving rise to a colored ferric complex.

Reagents.

2 M NH_2OH. Freshly neutralized to pH 7.3.
0.224 M 3-PGA, pH 7.3.
0.8 M ATP, pH 7.3.
0.6 M $MgCl_2$
$FeCl_3$-trichloroacetic acid-HCl: $FeCl_3 \cdot 6H_2O$, 8.3 g, conc. HCl 42 ml, and trichloroacetic, acid, 20 g, made up to 500 ml with distilled water.

Procedure. The following reagents are mixed in a 15 ml conical centrifuge tube in a total volume of 1.0 ml: 0.5 ml NH_2OH, 0.1 ml PGA, 0.01 ml ATP, and 0.01 ml $MgCl_2$. To this is added sufficient enzyme plus H_2O to give a final volume of 1.3 ml. After 1600 seconds at 37°, 3 ml of the $FeCl_3$-trichloroacetic acid-HCl is added to stop the reaction. After 5 minutes, 2.5 ml of 95% ethanol are added and the protein precipitate is removed by centrifugation. The optical density of the solution is measured with a 490 mμ filter. Although the ethanol increases the volume, this dilution is compensated by an increase in color intensity. A control tube is prepared by adding the $FeCl_3$-trichloroacetic acid-HCl mixture to substrate prior to the enzyme. A concentration of enzyme giving an optical density change of 0.1 to 0.2 should be used.

Definition of unit activity. A unit of enzyme is that amount of enzyme which produces an increase in density of 1.0 under the conditions of the assay.

2. Preparation Procedure (Axelrod and Bandurski, 1953).

One hundred g of pea seed ground to pass a 40 mesh sieve, are extracted for 10 minutes with 200 ml of water. After centrifuging, the supernatant fluid is fractionated with $(NH_4)_2SO_4$ at pH 7.5 (using 2 N NaOH to adjust the pH). The fraction precipitating between 50% and 80% saturation is collected and dissolved in 100 ml of H_2O. Solid $(NH_4)_2SO_4$ is added again and the fraction precipitating between 65% and 70% saturation is taken up in water (amount not given). This method gives an approximate ten-fold purification.

3. Properties.

The following data are taken from the paper of AXELROD and BANDURSKI (1953).

Specificity. At a concentration of 0.02 M, succinate, L-malate, ketoglutarate, fumarate, acetate, citrate and glutamate led to no formation of hydroxamic acid derivatives with the pea enzyme. ADP was 5% as effective and inosine triphosphate was 16% as effective as ATP.

Activators. Maximum activity of the pea enzyme is obtained in 6×10^{-3} M Mg^{++}, Ca^{++} is 52% and Fe^{+++} about 30% as effective.

Inhibitors. According to AXELROD and BANDURSKI, fluoride at a concentration of 1.7×10^{-2} M inhibits the pea enzyme 35%. Inhibition is apparently not caused by magnesium-fluoride-phosphate since the addition of 1.5×10^{-2} M Pi to 1.7×10^{-2} M fluoride had no further effect. Iodoacetate, sodium azide, sodium cyanide, hydrogen peroxide, potassium arsenite and sodium fluoride, all at 0.03 M, did not cause inhibition of the pea enzyme. The enzyme is unaffected by 0.01 M 2,4-dinitrophenol. Two minutes at 50° does not harm pea phosphoglycerate kinase; at 60° it is completely inactivated.

Effect of pH. Optimal activity is obtained between pH 6.5 to pH 10. The reaction falls off rapidly on either side of the optimum.

Effect of substrate concentration. The MICHAELIS-MENTEN constant of the pea enzyme is 7.6×10^{-3} M for PGA and 4.1×10^{-3} M for ATP at pH 7.3 in 0.714 M NH_2OH.

Distribution of the enzyme. According to the data of AXELROD and BANDURSKI, the following plant tissues are high in phosphoglycerate kinase: tomato leaf (17 units/g fresh weight), mung bean seed (27), soy bean seed (51), corn seed (41) and wheat germ (83). On the same basis, the values for other tissues are: tomato stem (2.2), aerial portion of *Bryophyllum* (5.1), tuber of Irish potato (1.7), tuber of sweet potato (2.3) and aerial portion of *Silene* (7.3).

XIV. Phosphoglyceric Acid Mutase.

(3-phosphoglycerate ⇌ 2-phosphoglycerate)

1. Assay Method.

Principle. The mutase reaction can be followed spectrophotometrically. One procedure which is the method given below starts with 3-PGA and an excess of enolase. The sequence 3-PGA → 2-PGA → PEP is followed by the increase in absorption of 240 mμ (RODWELL, TOWNE and GRISOLIA, 1956). The enzyme can be assayed in the reverse direction by following the decrease in absorption at 240 mμ (SUTHERLAND, POSTERNAK and CORI, 1949).

Another procedure, that of MEYERHOF and SCHULTZ (1938) as modified by SUTHERLAND et al. (1949), is based on change of optical rotation when 2-PGA is converted to 3-PGA.

Reagents.

0.2 M Tris buffer, pH 7.0.
0.1 M $MgSO_4 \cdot 7 H_2O$.
0.5 M 3-PGA, pH 7.0. It is not necessary to remove the 2,3-diphosphoglycerate (see below).
Enolase free of mutase (for preparation and definition of unit, see TOWNE, RODWELL and GRISOLIA, 1957).

Procedure. The following reagents are mixed in a quartz cell with a light path of 10 mm: 0.5 ml Tris, 0.1 ml $MgSO_4$, 0.1 ml 3-PGA, 1.2 enzyme units of enolase, enzyme preparation to be measured and deionized, doubly distilled water to give a total volume of 3 ml. The last addition is 3-PGA. Initial rate of increase in optical density at 240 mμ is recorded at 30 second intervals for 3 minutes. Up to 0.25 units of phosphoglycerate mutase may be taken for assay.

Definition of unit activity. A unit of mutase is that amount of enzyme which causes an increase in optical density of 0.1 per minute under the condition of the assay.

2. Purification (ITO and GRISOLIA, 1959).

The starting material is a water extract of wheat germ followed by Mn^{++} fractionation to remove nucleic acid, as published by TCHEN and VENNESLAND (1955). The succeeding steps are: (1) $(NH_4)_2SO_4$ fractionation at pH 5.5, (2) negative absorption with calcium phosphate gel, (3) $(NH_4)_2SO_4$ fractionation, (4) treatment with trypsin, followed by $(NH_4)_2SO_4$ fractionation, (5) removal of impurities with uranium acetate and (6) negative absorption with bentonite.

Starting with a specific activity of 2.3, an overall purification of approximately 200-fold is obtained.

3. Properties.

Activators. The highly purified phosphoglycerate mutase of wheat germ does not require 2,3-diphosphoglycerate for maximal activity, in contrast to the phosphoglycerate mutase of muscle and of baker's yeast (RODWELL et al., 1957).

Inhibitors. In contrast to the yeast enzyme, large concentrations of DL-2-PGA (up to 80 μmoles per ml) have little effect on the activity of the wheat germ enzyme (ITO and GRISOLIA, 1959).

Effect of pH. According to ITO and GRISOLIA (1959), the pH optimum for the wheat germ enzyme is 9 or more than 2 pH units higher than that observed with the phosphoglycerate mutase of muscle or baker's yeast (RODWELL et al., 1957).

Stability. Five minutes of heating at 55° harms enzyme activity 50 to 80%. Below pH 5 the wheat germ enzyme is denatured rapidly; above pH 5 no inactivation occurs up to about pH 9.5 (ITO and GRISOLIA, 1959).

Enzyme substrate affinities. The apparent MICHAELIS-MENTEN constant for 3-PGA is 6×10^{-4} M (ITO and GRISOLIA, 1959). The value for 2-PGA is given by these authors as 5 to 1×10^{-4} M, however, this must be considered as approximate because of a lack of a sufficiently sensitive method for estimation of 2-PGA.

Equilibrium constant. At 30°, the equilibrium constant, Ke = 3-PGA/2-PGA, is approximately 6 (ITO and GRISOLIA, 1959).

Purity of the enzyme. The highly purified wheat germ preparation of ITO and GRISOLIA (1959) contains no detectable 2,3-diphosphoglycerate (less than 1 μmole per 11 gm of protein).

The wheat germ enzyme cleaves Pi from 3-PGA, 2-PGA and 2,3-diphosphoglycerate, in contrast to the yeast enzyme which has little phosphatase activity (RODWELL et al., 1957). The highly purified preparation of ITO and GRISOLIA (1959) is contaminated with enolase in the order of $\frac{1}{1250}$ of the mutase activity. The phosphoenolpyruvate carboxylase of TCHEN and VENNESLAND (1955) is completely removed.

Distribution of the enzyme. Water extracts of the following tissues contain phosphoglycerate mutase calculated in units (see definition of unit) per gram of fresh weight (GRISOLIA and JOYCE, 1959): wheat germ extract (496), rice germ (172), corn germ (7), navy bean (40), lentils (23), chick peas (20), split peas (18) and potato (14). Activity of the enzyme is not enhanced by addition of 2,3-diphosphoglycerate.

XV. Enolase, 2-Phosphoglycerate Dehydrase.

(2-phosphoglycerate ⇌ phosphoenolpyruvate + H_2O)

This enzyme has been isolated and purified by MILLER (1958) from pea seed and by BOSER (1959) from potatoes.

1. Assay Method.

Principle. Enzyme activity is measured by following the change in optical density of the test solution at 240 mμ as described by WARBURG and CHRISTIAN (1941) and modified by BOSER (1959). It is based on the absorption of light by PEP at 240 mμ, a region where the absorption of 2-PGA is negligible. Since the absorption of PEP changes rapidly with wave length, monochromatic light of high purity is needed. In addition, the extinction coefficient of PEP depends on the pH and the concentration of Mg^{++}. Under the conditions described here (pH 7.4, $Mg^{++} = 2.7 \times 10^{-3}$ M), $E = 1.7 \times 10^3$ M. Glycine is included in the assay to bind metals which are inhibitory.

Applicability to crude tissue preparations. Other enzymes like phosphoglyceric acid mutase and phosphatase may interfere by consuming the reactants.

Reagents.

0.081 M Glycine.
0.081 M $MgSO_4 \cdot 7H_2O$.
0.072 M 2-PGA.
0.2 M $NaHCO_3$.

Procedure. In a final volume of 3 ml in a quartz cell ($d = 10$ mm), add the following reagents: 0.1 ml glycine, 0.1 ml $MgSO_4$, 0.1 ml 2-PGA, and 0.3 ml $NaHCO_3$. The reaction is started by addition of 2-PGA and after 10 seconds, the change in optical density at 240 mμ is measured. Under the specified conditions, the rate of reaction remains linear for about 10 minutes. Enzyme activity is calculated as optical density change in the initial minute at 240 mμ.

Definition of unit activity. This is defined by BOSER (1959) as the $\triangle OD_{240}$ per minute per milligram of protein per 3 ml volume with a light path of 10 mm.

2. Purification Procedure (BOSER, 1959).

Since potato enolase is extremely labile, purification must be carried out quickly at 0° to 4°. A crude juice is prepared by homogenizing the potatoes, followed by rapid filtration through cheesecloth. For each 100 ml of juice, add immediately 1 ml of 1 M Na_2SO_3. Solid $(NH_4)_2SO_4$ is added to 45% saturation and the resulting precipitate discarded. Additional solid $(NH_4)_2SO_4$ is added to bring the supernatant fluid to 75% saturation, and the precipitate is collected by centrifugation, dissolved in 100 ml of water and the $(NH_4)_2SO_4$ fractionation repeated. The 45—75% fraction dissolved in 50 ml of water is treated with 15 ml of 4% calcium phosphate gel. The precipitate is discarded. The supernatant fluid is dialyzed 3 hours against distilled water and then chromatographed on a calcium phosphate-starch column. The column is prepared in the following way. One-half mole of $CaCl_2$ and one-third mole of Na_3PO_4 are ground together and mixed with 3 liters of distilled water in a tall cylinder. The gel is washed until chloride free. A mixture of 200 gm of potato starch, 50 ml of 4% calcium phosphate gel and 400 ml of water are poured into a tube of 3 cm diameter. In about one hour the solid settles so that there is a height of approximately 60 cm of water. After the water level is lowered, the potato enolase dialyzate is placed on top of the column. The column is developed with 100 ml of 0.05 saturated $(NH_4)_2SO_4$

solution (3.55 gm per 100 ml) which removes impurities. Enolase is eluted with 500 ml of $(NH_4)_2SO_4$ solution (7.1 gm per 100 ml) of 0.1 saturation in fractions of 2.5 ml. The entire purification procedure must be carried out in less than one day to obtain good yields.

The overall purification is 69-fold. The enzyme is reported to be homogeneous in column chromatography and paper electrophoresis.

3. Properties.

Activators. Certain divalent cations are activators of the enolase system. Enolase is inactive in their absence. Optimal activity of the potato enzyme is attained at a concentration of 2.7×10^{-3} M Mg^{++}. In contrast to muscle enolase, the potato enzyme is not affected by 5×10^{-2} M Mg^{++}. Mg^{++}, Zn^{++}, Mn^{++}, Ca^{++}, Ag^+ and Hg^{++}, all at concentration of 2.7×10^{-3} M affect the potato enzyme in the ratio of $1.00:0.61:0.81:0.0:0.0:0.10$ respectively. The potato enzyme loses activity slowly at $-20°$ which can be restored by thioglycolate (100%), by cysteine (78%) and by Na_2SO_3 (45%). The enzyme is irreversibly inactivated after 4 to 5 months at $-20°$ (Boser, 1959).

Maximum activity is attained by pea enolase with Mg^{++}, Mn^{++}, Co^{++}, and Zn^{++} at 10^{-3} M, 10^{-4} M, 4×10^{-4} M and 3×10^{-5} M, respectively. Mg^{++}, Mn^{++}, Cu^{++} and Zn^{++} activated the pea enzyme in the ratio of $1.00:0.41:0.30:0.07$, respectively. Ca^{++}, K^{++}, Na^+ are ineffective (Miller, 1958).

Inhibitors. Fluoride inhibition of the enolase system from pea and potato is dependent on the Mg^{++}, Pi and fluoride concentrations. In the range of fluoride concentrations from 10^{-2} M to 10^{-4} M, inhibition of yeast enolase is correlated as shown by Warburg and Christian (1941) with the following relationship:

$$C_{Mg} \times C_{PO_4} \times C_F^2 \times \frac{\text{Residual activity}}{\text{Initial activity}} = 3.2 \times 10^{-12} \; (M^4)$$

where C is the concentration of the indicated ion. The value for the pea enzyme is 7×10^{-12} M^4 (Miller, 1958). At 10^{-3} M Mg^{++}, 10^{-3} M Pi and 5×10^{-3} M fluoride, the pea enzyme is inhibited 79%. The inhibition can be reversed by removal of the complex group through dialysis and addition of Mg^{++} to the pea preparation. Ca^{++} is a competitive inhibitor of the pea enzyme, reversal is affected by increasing the Mg^{++} concentration.

A 50% inhibition of the potato enzyme is achieved by 2.7×10^{-6} M p-chloro-mercuribenzoate (Boser, 1959).

Effect of pH. According to the data of Miller (1958), maximum activity is obtained with the pea enzyme in both Tris buffer and Pi buffer at pH 8.0. A sharp decrease in activity occurs at pH values higher or lower than the optimum. Activity is somewhat higher in Tris than in Pi.

Enzyme substrate affinity. According to the data of Boser (1959), the Michaelis-Menten constant of the potato enzyme for Mg^{++} in bicarbonate buffer of pH 7.4 is 1.22×10^{-3} M; the Michaelis-Menten constant of 2-PGA is 0.83×10^{-3} M. The turnover number of the potato enzyme is given as 210 moles per minute per 100,000 gram of protein. This contrasts with muscle 3300 (Boser, 1959) and with yeast 9900 (Warburg and Christian, 1941).

With the pea enzyme (Miller, 1958), the Michaelis-Menten constant for Mg^{++} in Tris buffer and pH 8.0 is 2×10^{-4} M: the Michaelis-Menten constant of 2-PGA is 2.5×10^{-4} M.

Distribution of the enzyme. Enolase activity can be detected in extracts from leaves of *Pisum sativum*, seed of *Gossypium barbadense*, leaves of *Chenopodium murale*, leaves of *Nicotiana tabacum* and seeds of *Avena sativa* (Miller, 1958).

XVI. Pyruvate Kinase.

(Phosphoenolpyruvate + ADP ⇌ Pyruvate + ATP)

1. Assay Method.

Principle. This enzyme can be assayed either spectrophotometrically in which case the pyruvate kinase is coupled with the oxidation of DPNH catalyzed by lactic dehydrogenase (BÜCHER and PFLEIDERER, 1955) or chemically by estimating directly the pyruvate formed (MILLER and EVANS, 1957). Both methods are given here since each has several advantages and each is applicable to crude extracts.

1. Spectrophotometric (BÜCHER and PFLEIDERER, 1955).

Reagents.

0.5 M Tris buffer, pH 7.6. In the original method, triethanolamine is used.
0.0045 M DPNH.
0.0069 M ADP.
0.0234 M PEP.
0.024 M $MgSO_4 \cdot 7H_2O$.
0.75 M KCl.
Lactic dehydrogenase 1 mg/ml (18,000 units/mg, KORNBERG, 1955).

Procedure. In a cuvette of 10 mm light path, place 0.2 ml Tris buffer, 0.3 ml $MgSO_4$, 0.1 ml PEP, 0.1 ml ADP, 0.1 ml DPNH, enzyme to be assayed and water to make a volume of 3 ml. The last addition can be PEP. Determine optical density values at 340 mμ at 30 second intervals for 3 minutes. Density changes of 0.020 to 0.050 per minute are desired.

Definition of unit activity. One unit of enzyme is defined as that amount which causes the formation of 1 μmole pyruvate per minute.

2. Chemical (slight modification of the method of MILLER and EVANS, 1957).

Reagents.

0.5 M Tris buffer, pH 7.4.
0.05 M ADP, pH 7.4.
0.8 M $MgSO_4 \cdot 7H_2O$.
0.5 M KCl.
0.015 M PEP, pH 7.4.
0.0125% 2,4-dinitrophenylhydrazine.
Reagents for pyruvate determination.

Method. The reaction mixture in a final volume of 1 ml contains the following constituents in μmoles: 50 of Tris buffer, 1.5 of PEP, 2.5 of ADP, 8.0 of $MgSO_4$, 50 of KCl and enzyme extract usually containing 0.6 to 0.7 mg protein. The mixture is incubated at 37° for 10 minutes and the reaction is stopped by the addition of 1 ml of cold 0.0125% 2,4-dinitrophenylhydrazine. Aliquots are removed for pyruvate determination (KACHMAR and BOYER, 1953). A control assay containing all reactants except ADP is used for comparison.

Definition of unit activity. One unit of enzyme is defined as that amount of enzyme which brings about the formation of 1 μmole pyruvate per minute.

2. Purification Procedure (MILLER and EVANS, 1957).

Four g of pea seed acetone powder is extracted with 15 ml of 0.05 M Tris buffer, pH 7.4 and the mixture centrifuged at 25,000 g for 10 minutes. The supernatant fluid is dialyzed for 4 hours against 4 liters of 0.01 M Tris, pH 7.4. The extracts contain about 12 to 14 mg protein per ml (FOLIN's phenol reagent). Due to lability of the enzyme, daily preparations are necessary.

When crude extract was prepared from leaf material, concentration of the enzyme was necessary. For each 20 ml of crude extract, 50 ml of acetone at $-15°$

are added. After filtration, the precipitate is rewashed with cold acetone and allowed to dry. The dried precipitate is taken up in 5 ml of 0.05 M Tris, pH 7.4, and centrifuged at 25,000 g for 10 minutes. The supernatant solution is the source of enzyme. The degree of purification was not reported.

3. Properties (Data of Miller and Evans, 1957).

Activators and inhibitors. The enzyme has an absolute requirement for Mg^{++} or Mn^{++} salts. The optimum concentration of these cations is 5×10^{-3} M. Co^{++} at this concentration is approximately 80% as effective as the other 2 cations. In addition to divalent ions the pea enzyme also requires a univalent cation salt for maximum activity and this requirement is satisfied by the chloride salts of K^+, Rb^+, or NH_4^+. A concentration of 0.05 M results in maximum activity. Enzyme activity is independent of the anion present. Calcium ions inhibit; the extent of inhibition depends on the K^+ and Mg^{++} concentration. The inhibition is partially competitive with respect to both K^+ and Mg^{++}. In the absence of K^+, Na^+ is approximately 20% as effective as K^+. At a concentration of 0.05 M KCl and 5×10^{-3} M Mg^{++}, Li^+ or Na^+ at concentrations up to 0.2 M has no effect.

Effect of pH. A broad optimum is observed between pH 7.0 and 9.0 using phosphate and Tris buffers.

Michaelis-Menten *constants.* For PEP, the $K_m = 3 \times 10^{-4}$ M; for ADP, $K_m = 5.0 \times 10^{-4}$ M; for K^+, NH_4^+, Rb^+ and NH_4^+ ions, $K_m = 2.4 \times 10^{-3}$ M, 1.2×10^{-3} M, 4.2×10^{-3} M, and 1.4×10^{-3} M, respectively.

Distribution of the enzyme. Miller and Evans (1957) could demonstrate pyruvate kinase in extracts of the seeds of wheat, cotton, oat, corn and beet, in the leaves of tobacco and peas and in the petioles of celery. Enzyme activity in each case is stimulated by KCl.

References.

Allen, R. J. L.: Biochem. J. **34**, 858 (1940). — Axelrod, B., P. Saltman, R. S. Bandurski and R. S. Baker: J. biol. Chem. **197**, 89 (1952). — Axelrod, B., R. S. Bandurski, C. M. Greiner and R. Jang: J. biol. Chem. **202**, 619 (1953). — Axelrod, B., and R. S. Bandurski: J. biol. Chem. **204**, 939 (1953).

Bailey, J. M., W. J. Whelan and S. Peat: J. Chem. Soc. **1950**, 3692. — Baranowski, T., and T. R. Niederland: J. biol. Chem. **180**, 543 (1940). — Barker, J., and H. F. el Saifi: Proc. Roy. Soc., B **140**, 362, 385 (1952). — Barnett, R. C., H. A. Stafford, E. C. Conn and B. Vennesland: Plant Physiol. **28**, 115 (1953). — Baum, H., and G. A. Gilbert: Nature (Lond.) **171**, 983 (1953). — Beevers, H., and M. Gibbs: Plant Physiol. **29**, 318 (1954). — Beck, W. S.: Arch. Biochem. Biophys. **60**, 1 (1956). — Benedict, C. R., and H. Beevers: Fed. Proc. **20**, 82 (1961). — Boser, H.: Hoppe-Seylers Z. physiol. Chem. **307**, 240 (1957); **315**, 163 (1959). — Bücher, T.: Methods in Enzymology. Vol. 1, pp. 415 and 427. Ed. by S. P. Colowick and N. O. Kaplan. New York: Academic Press 1955. — Bücher, T., and G. Pfleiderer: Methods in Enzymology, Vol. 1, p. 435. Ed. by S. P. Colowick and N. O. Kaplan. New York: Academic Press, 1955.

Cardini, C. E.: Enzymologia **15**, 44 (1951); **15**, 303 (1952). — Chakravorty, M., H. C. Chakrobortty and D. P. Burma: Arch. Biochem. Biophys. **82**, 21 (1959). — Cooper, J., P. A. Srere, M. Tabachnik and E. Racker: Arch. Biochem. Biophys. **74**, 306 (1958).

Dounce, A. L., and G. T. Beyer: J. biol. Chem. **173**, 159 (1948). — Dounce, A. L., S. R. Barnett and G. T. Beyer: J. biol. Chem. **185**, 769 (1950).

Fewson, C. A., and M. Gibbs: Plant Physiol. **36**, IX (1961). — Fischer, E. H., and G. A. Hilpert: Experientia (Basel) **9**, 176 (1953). — Fiske, C. H., and Y. Subbarow: J. biol. Chem. **66**, 375 (1925). — Fuller, R. C., and M. Gibbs: Plant Physiol. **34**, 324 (1959).

Gibbs, M.: Nature (Lond.) **170**, 164 (1952); — Plant Physiol. **29**, 34 (1954); — Methods in Enzymology, Vol. 1, p. 411. Ed. by S. P. Colowick and N. O. Kaplan. New York: Academic Press 1955. — Gibbs, M., and H. Beevers: Plant Physiol. **30**, 343 (1955). — Gibbs, M., J. M. Earl, and J. L. Ritchie: Plant Physiol. **30**, 463 (1955). — Gibbs, M., and B. L. Horecker: J. biol. Chem. **208**, 813 (1954). — Green, D. E., and P. K. Stumpf: J. biol. Chem. **142**, 355 (1942). — Grisolia, S., and B. K. Joyce: J. biol. Chem. **234**, 1335 (1959).

HAGEMAN, R. H., and D. I. ARNON: Arch. Biochem. Biophys. 57, 421 (1955a); 55, 162 (1955b). — HANES, C. S.: Proc. Roy. Soc., B 128, 421 (1939); B 129, 174 (1940). — HANES, C. S., and E. J. MASKELL: Biochem. J. 36, 76 (1942). — HATCH, M. D., and J. F. TURNER: Biochem. J. 69, 495 (1958); 72, 524 (1959). — HERBERT, D., A. H. GORDON, V. SUBRAHMANYAN and D. E. GREEN: Biochem. J. 34, 1108 (1940). — HORECKER, B. L., and W. A. WOOD: Methods in Enzymology, Vol. 3, p. 152. Ed. by S. P. COLOWICK and N. O. KAPLAN. New York: Academic Press, 1957.

ITO, N., and S. GRISOLIA: J. biol. Chem. 234, 242 (1959).

JACOBI, G.: Planta 49, 1 (1957). — JAGANNATHAN, V., K. SINGH and M. DAMODARAN: Biochem. J. 63, 94 (1956).

KACHMAR, J., and P. D. BOYER: J. biol. Chem. 200, 669 (1953). — KATZ, J. R., and W. Z. HASSID: Arch. Biochem. 30, 272 (1951). — KING, E. J.: Biochem. J. 26, 292 (1932). — KORNBERG, A.: Methods in Enzymology, Vol. 1, p. 441. Ed. by S. P. COLOWICK and N. O. KAPLAN. New York: Academic Press 1955.

LELOIR, L. F., R. E. TRUCCO, C. E. CARDINI, A. C. PALADINI and R. CAPUTTO: Arch. Biochem. 19, 339 (1948). — LING, K.-H., W. L. BYRNE and H. LARDY: Methods in Enzymology, p. 306. Ed. by S. P. COLOWICK and N. O. KAPLAN. New York: Academic Press 1955. — LIPMANN, F., and L. C. TUTTLE: J. biol. Chem. 159, 21 (1945). — LOHMANN, K., and L. JENDRASSIK: Biochem. Z. 178, 419 (1926). — LOWRY, O. H., N. R. ROBERTS, M.-L. WU, W. S. IIXON and E. J. CRAWFORD: J. biol. Chem. 207, 19 (1954).

McDONALD, M. R.: Methods in Enzymology, Vol. 1, p. 260. Ed. by S. P. COLOWICK and N. O. KAPLAN. New York: Academic Press 1955. — MEAGHER, W. R., and W. Z. HASSID: J. Amer. Chem. Soc. 68, 2135 (1946). — MEDINA, A., and A. SOLS: Biochim. Biophys. Acta 19, 378 (1956). — MEYERHOF, O., and K. LOHMANN: Biochem. Z. 271, 89 (1934). — MEYERHOF, O., and W. KIESSLING: Biochem. Z. 279, 41 (1935). — MEYERHOF, O., and W. SCHULTZ: Biochem. Z. 279, 60 (1938). — MILLER, G. W.: Plant Physiol. 33, 199 (1958). — MILLER, G. W., and H. J. EVANS: Plant Physiol. 32, 346 (1957). — MILLERD, A., J. BONNER, B. AXELROD and R. S. BANDURSKI: Proc. Nat. Acad. Sci. 37, 855 (1951).

NAJJAR, V. A.: Mechanism of Enzyme Action, p. 731. Ed. by W. D. McELROY and B. GLASS. Baltimore: Johns Hopkins Press, 1954; — Methods in Enzymology, Vol. 1, p. 294. Ed. by S. P. COLOWICK and N. O. KAPLAN. New York: Academic Press 1955.

OESPER, P., and O. MEYERHOF: Arch. Biochem. 27, 224 (1950).

PORTER, H. K.: J. exp. Bot. 4, 44 (1953). — POTTER, A. L., J. C. SOWDEN, W. Z. HASSID and M. DANDOROFF: J. Amer. Chem. Soc. 70, 1751 (1948).

RACKER, E.: J. biol. Chem. 167, 843 (1947). — RACKER, E., and E. A. R. SCHROEDER: Arch. Biochem. Biophys. 74, 326 (1958). — RAMASARMA, J., and K. V. GIRI: Arch. Biochem. Biophys. 62, 91 (1956). — RICHTER, G.: Z. Naturforsch. 12b, 662 (1957); — Naturwissenschaften 21, 604 (1959). — RODWELL, V. W., J. C. TOWNE and S. GRISOLIA: Biochim. Biophys. Acta 20, 394 (1956); — J. biol. Chem. 228, 875 (1957). — ROE, J. H.: J. biol. Chem. 107, 15 (1934). — ROSENBERG, L. L., and D. I. ARNON: J. biol. Chem. 217, 361 (1955).

SABLE, H. Z.: Biochemical Preparations, Vol. 2, p. 52. Ed. by E. G. BALL. New York: John Wiley and Sons Inc. 1952. — SALTMAN, P.: J. biol. Chem. 200, 145 (1953). — SANWAL, G. G., and P. S. KRISHNAN: J. Sci. Ind. Res. 18C, 183 (1959). — SIBLEY, J. A., and A. L. LEHNINGER: J. biol. Chem. 177, 859 (1949). — SLATER, E. C.: Biochem. J. 53, 157 (1953). — SMILLIE, R. M.: Nature (Lond.) 187, 1024 (1960). — SMILLIE, R. M., and R. C. FULLER: Biochim. Biophys. Res. Comm. 3, 368 (1960). — STUMPF, P. K.: J. biol. Chem. 176, 233 (1948). — SUTHERLAND, E. W., T. POSTERNAK and C. F. CORI: J. biol. Chem. 181, 153 (1949).

TCHEN, T. T., and B. VENNESLAND: J. biol. Chem. 213, 533 (1955). — TEWFIK, S., and P. K. STUMPF: Amer. J. Bot. 36, 567 (1949); — J. biol. Chem. 192, 519 (1951). — TOWNE, J. C., V. W. RODWELL and S. GRISOLIA: J. biol. Chem. 226, 777 (1957). — TUNG, T. C., K. H. LING, W. L. BYRNE and H. A. LARDY: Biochim. Biophys. Acta 14, 488 (1954). — TURNER, J. F.: Nature (Lond.) 174, 692 (1954); — Biochem. J. 67, 450 (1957). — TURNER, D. H., and J. F. TURNER: Austr. J. biol. Sci. 10, 302 (1957). — TURNER, J. F., and L. W. MAPSON: Nature (Lond.) 181, 270 (1958). — TURNER, D. H., and J. F. TURNER: Biochem. J. 74, 486 (1960).

UMBREIT, W. W., R. H. BURRIS and J. F. STAUFFER: Manometric techniques and tissue metabolism. Minneapolis: Burgess Publishing Co. 1951.

WARBURG, O., and W. CHRISTIAN: Biochem. Z. 303, 40 (1939); 310, 384 (1941). — WHELAN, W. J.: Methods in Enzymology, Vol. 1, p. 192. Ed. by S. P. COLOWICK and N. O. KAPLAN. New York: Academic Press 1955. — WOLFROM, M. L., C. S. SMITH and A. E. BROWN: J. Amer. Chem. Soc. 65, 255 (1943).

Enzymes of the Pentose Phosphate Cycle.

By

E. R. Waygood and R. Rohringer.

With 1 Figure.

The authors have reviewed the literature covering the methods used in the isolation, purification, and assay of the ten enzymes of the pentose phosphate cycle of metabolism in higher plants, as well as the preparation of substrates. Where detailed investigations of higher plants are lacking, information from studies on microorganisms and animal tissues has been included since the procedures are generally applicable.

Several methods of assay have been described, since some are more suitable for routine studies of a semi-quantitative nature, as in purification steps. Spectrophotometric techniques are most convenient, but often most complex since they involve the use of highly purified preparations of enzymes from animal tissues which may not be readily available to some botanical laboratories. For some enzymes the substrates are not commercially available and an enzymatic procedure for their synthesis is described or referred to in the text. The authors have not attempted to standardize the arbitrarily defined units of concentration used by one investigator or another, since this is unwarranted, until regulations governing their use are established.

More than one method of purification has been described for some enzymes, since in many cases only the initial purification steps for enzymes of higher plants are available, and these could serve as a guide for further purification. In other instances, the earlier purification procedures for enzymes from a particular source, by virtue of their "contamination" with other enzymes of the cycle have a definite usefulness in (a) demonstrating its operation, (b) in assay procedures and (c) in the preparation of substrates, commercially unavailable. One of the most important developments has been the use of DEAE cellulose, introduced by RACKER and his colleagues for the final purification steps and for the "decontamination" of crystalline enzyme preparations. In all of these procedures, where data are available, the yield is given in terms of enzyme units and specific activity (units per milligram of protein).

The enzymes of the cycle are mainly soluble, although SERVETTAZ (1956) found them in the mitochondrial fractions of pea seedlings and potato tubers. This may be due to adsorption during preparative procedures. Nevertheless the reported presence (FORTI, TUA and TOGNOLI, 1959) of G-6-P dehydrogenase and other enzymes on particles sedimenting at $3000 \times g$ from extracts of pea seeds may be significant.

From the standpoint of the reviewers, the most challenging field of research for the plant physiologist and biochemist appears to be not only in the preparation of highly purified enzymes from plant tissues, but also in demonstrating and localiz-

---

Listed as contribution No. 74 from the Canada Department of Agriculture Research Station, Winnipeg.

For abbreviations used see p. 568.

ing the operation of the cycle in plant material, especially in monocotyledons, algae and fungi, where little work has been done. It is futile to undertake the purification of an enzyme from centigram quantities of plant tissues, rather facilities should be made available for the bulk handling of kilogram quantities. It matters little, if only a small fraction of the enzyme is recovered from that present in the original extract provided one ends up with a small volume of product relatively clean from "contaminating" enzymes and with a high specific activity.

There is no doubt that most, if not all, of the enzymes are widely distributed in the tissues of higher plants and that the pentose phosphate pathway must play a

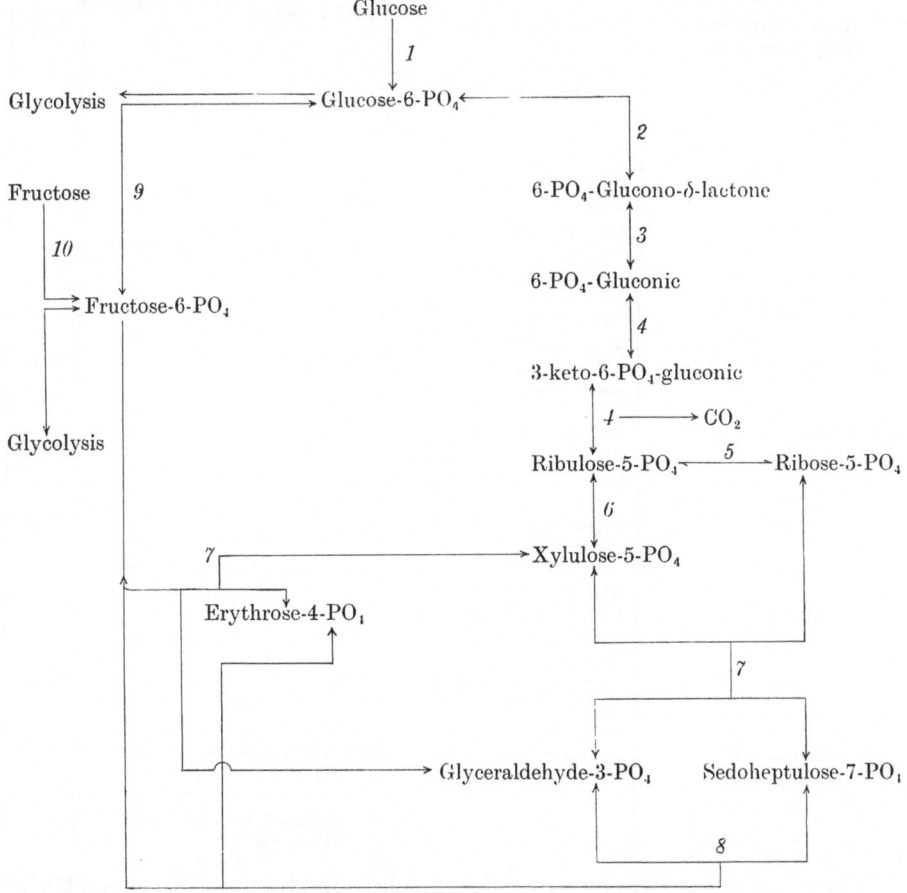

Fig. 1. Enzymes and reactions involved in the pentose phosphate cycle. *1*. glucokinase; *2*. glucose-6-phosphat-dehydrogenase; *3*. lactonase; *4*. 6-phosphogluconic dehydrogenase; *5*. phosphoriboisomerase; *6*. phosphoketoe pentoisomerase; *7*. transketolase; *8*. transaldolase; *9*. phosphohexoisomerase; *10*. fructokinase.

significant role in their metabolism. However, the authors have not undertaken to assess the relative importance of this pathway in respiration, nor its contribution to photosynthesis and nucleotide synthesis, or its place in evolution. For this the reader is referred to the reviews of AXELROD and BEEVERS (1956), JAMES (1957), KREBS and KORNBERG (1957), NEILANDS and STUMPF (1958) and GIBBS (1959).

Credit for the elucidation of the complex interconversions of pentose phosphate, its subsequent fate, and for the role of transketolase and transaldolase should be

Table 1. *Enzymes and Substrates*

Enzyme	Source	Substrate	Cofactor
1. Glucokinase (Hexokinase)	yeast (6, 61, 62) bacteria (11) animals (37) higher plants (82)	D-glucose, D-mannose, and D-glucosamine (61) D-fructofuranose (32)	ATP
2. Glucose-6-phosphate dehydrogenase	higher plants (1) fungi (56, 79) bacteria (21) animals (90, 22)	G-6-P (57)	TPN
3. Lactonase	yeast (10) bacteria (10) animals (10)	6-phosphate-δ-glucono-lactone (10) lactabionic-δ-lactone (10)	None
4. 6-Phosphogluconic dehydrogenase	higher plants (7) yeast (41) bacteria (21) animals (22)	6-phosphogluconate	TPN
5. Phosphoriboisomerase	higher plants (5, 95, 52) yeast (18)	Ru-5-P	None
6. Phosphoketopento-epimerase	bacteria (51, 104) animals (2, 23, 93) as well as yeast and higher plants	Ru-5-P Xu-5-P	None
7. Transketolase	higher plants (48, 49) yeast (18, 92) bacteria (20) animals (22)	Xu-5-P R-5-P octulose-8-phosphate (78)	TPP
8. Transaldolase	higher plants (43, 31) yeast (43, 92) bacteria (43)	S-7-P glyceraldehyde-3-phosphate	None
9. Phosphohexo-isomerase	higher plants (80, 83) bacteria (20) animals (22)	F-6-P	None
10. Fructokinase (Hexokinase)	animals (75) higher plants (68)	fructose, sorbose, tagatose (35, 65) glucose, mannose, glucosamine, 2-deoxyglucose, allose (68)	ATP

given to RACKER and HORECKER and their colleagues. Recently COURI and RACKER (1959) have reconstructed the cycle in vitro using highly purified preparations of the enzymes concerned from several sources, together with GSSG reductase, under anaerobic conditions, or the "Old Yellow Enzyme" in the presence of molecular oxygen to accept the hydrogens of reduced triphosphopyridine nucleotide. The ratio of GSH produced to TPN reduced was 90% of that predicted.

The scheme showing the complete oxidation of glucose-6-phosphate is given in Fig. 1. A list of the enzymes involved, the substrates, cofactors, inhibitors, activators and other pertinent kinetic data and properties of the enzymes are given in Table 1.

The present authors have demonstrated most of the enzymes of the pentose phosphate pathway in wheat leaf extracts either by quantitative assay or by *of the Pentosephosphate Pathway.*

Km(M) Substrate	Cofactor	Keq	Activators (A) Inhibitors (I)
Glucose: 1.5×10^{-4} (86) 4.4×10^{-4} (82)	ATP: 9.5×10^{-5} (86) 8.7×10^{-4} (82)		A: Mg^{++} GSH, cysteine (13) I: various sugars sorbose-1-phosphate (64) suramine (103) Na-tripolyphosphate (98) $\beta\beta'$-dichlorodiethylsulphide (6)
G-6-P: 2×10^{-4} (90)	TPN: 5×10^{-5} (90)	—	A: ?Mg^{++} (56) ?Zn^{++} (97) I: > 0.1 M PO_4^{---} (71) 2'-AMP (74) 1,10-phenanthroline (97) 8-hydroxyquinoline (97) Na-diethyldithiocarbamate (97) nicotinamide (90)
8.5 (10)	—	—	A: Mg^{++}, Mn^{++}, Co^{++} (10) I: NaF, Na-benzoate, hexylresorcinol (10)
—	—	Large	A: Mg^{++} (12)
—	—	0.246 (5) 0.3 (95)	I: G-6-P, 5-AMP, 3-AMP, adenine, PO_4^{---}, and 5-phosphoribonic acid (5)
Ru-5-P: 1×10^{-3} (51) Xu-5-P: 5×10^{-4} (51) —	—	3.0 (95) 1.4 (72) 1.86 (104) 1.5 (51) 1.0 (48) *Xu-5-P-1* (84) s-7-P 5	A: Mg^{++} (48, 77)
G-3-P: 2.2×10^{-4} (43) S-7-P: 1.8×10^{-4} (43) G-6-P: 3.45×10^{-3} (80)	—	0.82 (43) 1.5 (80) 0.47 (85)	—
fructose: 4×10^{-3} (68) glucose: 2×10^{-4} (68)	—	—	A: Mg^{++}, (Mn^{++}) (36, 75) K^+ (75) I: various sugars ADP (75) G-6-P (68)

inference. The enzymes were purified from acetone powders according to the method of HAGEMAN and WAYGOOD (1959). Where the information is applicable, reference is made in the text.

I. Hexokinase (Glucokinase and Fructokinase).

Hexose + ATP = Hexose-6-phosphate + ADP

Hexokinase catalyses the transfer of the terminal phosphate group from nucleoside triphosphates to hexose, including D-glucose, D-fructofuranose, D-mannose and D-glucosamine. Specific kinases have been demonstrated in animal tissues (HERS, 1955), bacteria (CARDINI, 1950) and pea seedlings (MEDINA and

Sols, 1956). However, the two enzymes are grouped together, since their assay methods are similar and the hexokinases of wheat germ (Saltman, 1953), Neurospora crassa (Medina and Nicholas, 1957) and the crystalline enzyme from yeast (Kunitz and McDonald, 1946, 1948) are not specific for either glucose or fructose (fructofuranose). Hexokinases have been demonstrated in other plant tissues by Boser (1958) and Kursanov, Pavlinova and Afanasieva (1959). Wheat germ contains a soluble hexokinase and one associated with the mitochondrial fraction (Saltman, 1953).

1. Assay.

The enzyme may be assayed by several methods including measurement of the disappearance of acid labile phosphate or reducing power, as well as by manometric, titrimetric, photometric and spectrophotometric techniques. *Disappearance of acid labile phosphate.* Colowick and Kalckar (1943) measured the organic phosphate in TCA filtrates, before and after an 11 minute hydrolysis in 1.0 N H_2SO_4 at 100°. Bailey and Webb (1948) measured the decrease in easily hydrolysable phosphate during incubation of the enzyme with ATP, glucose and Mg^{+2}, using the Fiske-Subbarow method for phosphate determination before and after hydrolysing the digest with 1.0 N HCl for 7 minutes at 100°. Their system contains 50 μmoles $MgCl_2$, 600 μmoles glucose, 0.1 M veronal acetate buffer, pH 7.9, 0.5 ml of diluted enzyme in a total volume of 4.5 ml. Sodium ATP (0.5 ml) containing 1.2 mg of 7 minute hydrolysable P (ca. 30 μmoles ATP) is added in the cold and the digest incubated at 38°. Samples of 1.0 ml removed at 1, 3, 5 and 7 minutes, are pipetted into 0.12 ml 10 N HCl and immersed in a boiling water bath for 7 minutes. The samples are washed into a 50 ml flask, mixed with the Fiske-Subbarow reagents and diluted to volume. One hexokinase unit is defined as the amount of enzyme that transfers 1 mg of 7 minute hydrolysable P from ATP to glucose in 1 hour. *Disappearance of reducing power.* Several investigators have followed the disappearance of reducing power in the digests. Sols' (1956) system contains 3 μmoles sugar, 15 μmoles ATP, 7.5 μmoles $MgSO_4$, 25 μmoles NaF, 10 μmoles each of K_2HPO_4 and Tris buffer, pH 8.0 in a total volume of 0.3 ml. The enzyme or homogenate (0.2 ml) is added and incubated for 7.5 and 15 minutes. The reaction is stopped by the addition of the $Ba(OH)_2$ solution of the Ba-Zn deproteinization procedure of Somogyi (1945). After filtration or centrifugation free sugar is measured in aliquots according to the procedure of Nelson (1944) and Somogyi (1945). The unit of enzyme can be defined as the amount that phosphorylates 1 μgm of sugar per minute (Hers 1955). *Manometric method.* The reaction can be followed manometrically since the transfer of 1 mole of phosphate from ATP involves the liberation of one acid equivalent due to the formation of a dissociable $-OH$ group in the ADP molecule (Colowick and Kalckar 1943). In a CO_2-bicarbonate buffer, CO_2 is released quantitatively above pH 7.5. Place 0.2 ml 0.04 M Na ATP (pH 7.5) and 0.2 ml 0.04 M $NaHCO_3$ in the side arm of a Warburg flask. The main compartment contains 24 μmoles each of $NaHCO_3$ and glucose, 12 μmoles $MgCl_2$, and enzyme in a volume of 1.2 ml. The flasks are filled with 5% CO_2/95% N_2 and after 5 minutes equilibriation at 30° the contents of the side arm are tipped and CO_2 evolution measured. *Titrimetric and photometric methods.* The rate of acid formation may be determined directly by titration using phenol red as the indicator. The method is described in detail by McDonald (1955). Wajser (1949) has adapted this method to a photometric technique, since there is a decrease in the extinction of phenol red at 558 mμ with increasing acidity from pH 7.5 to 6.9. The indicator solution contains 10 mg of phenolsulphophthaleine (phenol red), 0.28 ml 0.1 N Na_2CO_3, 3.0 ml ethanol, and 40 ml

0.066 M phosphate buffer pH 7.5 in a volume of 100 ml. The indicator solution can be calibrated against pH and absorbancy at 558 mμ by titrating with 0.1 to 0.5 ml 0.01 N HCl in a volume of 3.0 ml. In the assay, the experimental cuvette contains 0.4 ml of indicator solution, 24 μmoles glucose, 20 μmoles magnesium acetate, 1.6—8.0 μmoles ATP, and enzyme in a volume of 3.0 ml. The blank is without the indicator solution. The hexokinase unit can be defined as the amount of enzyme that catalyses the formation of 1×10^{-8} acid equivalents per minute under the assay conditions (McDONALD, 1955). *Spectrophotometric method.* This method, developed by RACKER (1947), is based on the oxidation of DPNH by DAP catalysed by α-glycerophosphate dehydrogenase. DAP is formed from G-6-P by the action of phosphohexoisomerase, phosphohexokinase and aldolase. All these enzymes must be present in excess. He found α-glycerophosphate dehydrogenase, aldolase, triosephosphate isomerase and phosphohexoisomerase in the 50—70% $(NH_4)_2SO_4$ enzyme preparation from rabbit muscle.

A specific phosphohexokinase that catalyses the phosphorylation of F-6-P to F-1,6-P is purified from the 20—50% $(NH_4)_2SO_4$ fraction obtained in this procedure.

The solutions are as follows: 0.1 ml 0.1 M ammonium phosphate buffer, pH 7.6, or 0.1 M Tris buffer, pH 7.8, 0.1 ml 0.1% DPNH, 0.05—0.3 ml α-glycerophosphate dehydrogenase preparation [50—70% $(NH_4)_2SO_4$ fraction, 0.03% protein], 0.1 ml 0.4% ATP, 0.1 ml 0.07 M $MgCl_2$, 0.05—0.3 ml phosphohexokinase, 0.08 ml 0.24 M glucose and 0.02—0.2 ml enzyme solution to be tested in a final volume of 3.0 ml. The control cell lacks DPNH. The rate of oxidation of DPNH is followed at 340 mμ in a spectrophotometer.

2. Purification.

Baker's yeast. The procedure of MEYERHOF (1927) for the preparation of hexokinase from this source has been used as the starting step in the crystallization of hexokinase by KUNITZ and McDONALD (1946—1948). This consists essentially in plasmolysing baker's yeast with toluene, extracting the plasmolysed yeast with water at 35°, and precipitating the enzyme with 50% ethanol. The enzyme dissolves in water. COLOWICK and KALCKAR (1943) precipitated the enzyme between the limits of 0.5 and 0.75 ammonium sulphate saturation. The dried precipitate was stable for at least two months. Hexokinase has also been crystallized from baker's yeast by BERGER, SLEIN, COLOWICK and CORI (1946) and BAILEY and WEBB (1948).

KUNITZ and McDONALD (1946—1948) concentrated the enzyme by ammonium sulphate fractionation and their further purification steps involved dialysis, alcohol fractionation and crystallization by ammonium sulphate. The procedure is described in detail by KUNITZ and McDONALD (1948) and McDONALD (1955) and only the first two steps will be given here. *Plasmolysis and extraction.* Fragmented fresh "FLEISCHMANN'S" baker's yeast (25 lb) is macerated with 6 litres of toluene in a large aluminum pan placed in a water bath at 45° until the suspension reaches 37°. The yeast liqufies, liberates CO_2, increases in volume and is left at room temperature for 2—3 hours, after which it is cooled to 10°, diluted with 12 litres of cold water and left for 18 hours at 0—5°. The lower layer of yeast-water suspension is siphoned off and to it 100 g of Super-Cel filter-aid (Johns-Manville, Co.) is added to each litre of suspension. The mixture is filtered with suction on four 32 cm BÜCHNER funnels and each residue washed once with one litre of cold water. *Ammonium sulphate fractionation.* The filtrates and washings are combined and solid $(NH_4)_2SO_4$ is added to 0.5 saturation together with 10 g of Super-Cel and 10 g of Filter-Cel (Johns-Manville, Co.) to each litre of combined filtrate. The precipitate is removed by filtration and the filtrate is brought to 0.65 saturation with respect to $(NH_4)_2SO_4$ and the suspension is filtered. The filter cake contains hexokinase. Recovery of the enzyme in this fraction is 42% with one third increase in specific activity.

A simple procedure for the purification and crystallization of hexokinase from baker's yeast has been developed by R. A. DARROW and S. P. COLOWICK and is to be published in Methods in Enzymology, Vol. V, Acad. Press, N.Y. (personal communication, S. P. COLOWICK).

Wheat germ. The purification of wheat germ hexokinase is described by GIBBS and TURNER on Page 525 of this Volume.

II. Glucose-6-Phosphate Dehydrogenase (Zwischenferment). Lactonase and 6-Phosphogluconic Dehydrogenase.

These three enzymes catalyse the conversion of G-6-P into R-5-P involving the reduction of TPN in two steps and the liberation of CO_2 according to the following reactions.

$$G\text{-}6\text{-}P + TPN^+ \rightleftharpoons 6\text{-}P\text{-}glucono\text{-}\delta\text{-}lactone + TPNH + H^+$$
$$6\text{-}P\text{-}G\text{-}\delta\text{-}lactone \rightleftharpoons 6\text{-}P\text{-}gluconate$$
$$6\text{-}PGA + TPN^+ \rightleftharpoons ribulose\text{-}5\text{-}phosphate + CO_2 + TPNH + H^+$$

G-6-P dehydrogenase was discovered by WARBURG and his collaborators and the name they coined for it, „Zwischenferment" (WARBURG, CHRISTIAN and GRIESE, 1935) appropriately describes the enzyme and has been carried through to the present day.

Dried brewer's yeast is the best source of G-6-P dehydrogenase (KORNBERG, 1950; KORNBERG and HORECKER, 1955) and of 6-PGA dehydrogenase (HORECKER and SMYRNIOTIS, 1951, 1955). However, both the enzymes are widely distributed in animal tissues (DICKENS, 1958), bacteria (DE MOSS, 1955; DE LEY, 1957), green algae (RICHTER, 1957) and fungi (RADHAKRISHNAN, 1960) and in the seeds, germs (embryos), fruits, stems, leaves and roots of a variety of higher plant species (WAGNER-JAUREGG and RAUEN, 1935; CONN and VENNESLAND, 1951; ANDERSON, STAFFORD, CONN and VENNESLAND, 1952; GIBBS, 1952, 1954; AXELROD, BANDURSKI, GREINER and JANG, 1953; BARNETT, STAFFORD, CONN and VENNESLAND, 1953; SERVETTAZ, 1956; CLAYTON, 1959; FORTI, TUA and TOGNOLI, 1959).

1. Assay.

The enzymes may be assayed by colorimetric, manometric, titrimetric and spectrophotometric techniques. *Colorimetric methods.* WAGNER-JAUREGG and RAUEN (1935) used the THUNBERG technique. The system contains 4.0 μmoles hexosemonophosphate (NEUBERG ester), 0.1 ml coenzyme II prepared according to the method of WARBURG and CHRISTIAN (1936), 0.25 ml methylene blue (1:5000), and 0.5 ml of an enzyme solution from bean or cucumber seeds buffered with 0.87% K_2HPO_4. Decolourization time at 37° was 6 minutes and 177 minutes as compared to the control time of 34 minutes and 180 minutes for the bean and cucumber seeds respectively.

SERVETTAZ (1956) measured the formation of the red formazan at 540 mμ produced in the reduction of triphenyl-tetrazolium chloride by TPNH in the presence of enzyme and substrate. Her system contains 40 μmoles G-6-P, 20 μg TPN, 1% TTC, and pea seedling mitochondria in 0.5 M sucrose — 0.1 M Tris buffer, pH 7.2, equivalent to 2 g fresh weight in a total volume of 1.0 ml. The system is incubated at 20° and the reaction is terminated after 60 minutes by the addition of 2 to 5 volumes of acetone in which the formazan is soluble. Protein is removed by centrifugation. AXELROD et al. (1953) assayed G-6-P and 6-PGA dehydrogenases by following the reduction of 2,6-dichlorophenol-indophenol by TPNH with substrate and enzyme (containing diaphorase) in a colorimeter at 25° with a 540 mμ filter. The system contains 0.5 ml 0.0025 M G-6-P or 6-PGA, 0.5 ml 0.2 M potassium phosphate, pH 6.5, 0.5 ml dye (40 μg), 1.0 ml 65% TPN (80 μg), 0.5 ml enzyme and 3.0 ml water. *Titrimetric method.* CONN and VENNES-LAND (1951), ANDERSON et al. (1952) and BARNETT et al. (1953) assayed G-6-P and 6-PGA dehydrogenases by coupling the enzymes with TPN — linked gluta-thione (GSSG) reductase and titrating the GSH produced. All of their plant extracts contained GSSG reductase in excess. Reduction of GSSG in the absence of either yeast G-6-P or 6-PGA dehydrogenase is indicative of the presence of the enzyme. For G-6-P dehydrogenase the system of ANDERSON et al. (1952) contains 1.8×10^{-3} M G-6-P, 7.8×10^{-4} M GSSG, 10^{-5} M TPN (70%), 3.3×10^{-2} M Tris buffer, pH 7.4, and plant enzyme preparation in a volume of 2.7 ml. The reaction is started by adding the enzyme and the system is incubated for 45 minutes at 30° and stopped by the addition of 0.3 ml 20% metaphosphoric acid. The protein is centrifuged off and an aliquot is titrated with potassium acid iodate according to the method of FUJITA and NUMATA (1938). For 6-PGA dehydrogenase, BAR-NETT et al. (1953) incubate the enzyme and 3.7—10 μmoles 6-PGA for 30—60 min-utes at 30° in 3.7×10^{-2} M Tris buffer, pH 7.4, with 9 μmoles GSSG and 30 μg 70% TPNH in a volume of 2.6 ml. The subsequent procedures are the same as for G-6-P dehydrogenase. *Manometric method.* G-6-P dehydrogenase may be assayed by measuring oxygen uptake brought about by the cyclical reduction and oxidation of TPN presumably in the presence of an endogenous oxygen acceptor system (TPNH oxidase or flavoprotein). 6-PGA dehydrogenase can be assayed in a similar manner, but also CO_2 evolution may be measured. The system of GIBBS (1954) contains in the main compartment of a WARBURG flask, 1.0 ml enzyme dialysed against 0.02 N $NaHCO_3$, pH 7.5, 0.1 ml 0.2 M $MgCl_2$, 0.05 ml riboflavin (1.0 mg/ml) and 0.05 ml 0.25 M ATP. The side arm contains 0.05 ml TPN (1.0 mg/ml) and the centre well 0.2 ml 20% KOH (temperature 37°). After 60 to 90 minutes when the endogenous oxygen uptake has subsided the reaction is started. With undialysed enzymes, no riboflavin or ATP is added and the main compartment contains 1.0 ml enzyme and 0.45 ml water.

In the procedure of CLAYTON (1959) the main compartment contains enzyme (2.0 mg protein), 500 μmoles Tris buffer, pH 7.6 and 10 μmoles $MgCl_2$. The side arm contains 5.0 μmoles G-6-P, 0.6 μmoles TPN and 30 μg riboflavin. The centre well contains 0.2 ml 20% KOH. Total volume 2.8 ml, temperature 37°. In these experiments of CLAYTON (1959) riboflavin was required for maximum oxygen uptake by the G-6-P dehydrogenase system. Magnesium was essential only for CO_2 liberation. *Spectrophotometric methods.* Those involving the measurement of the rate of reduction of TPN at 340 mμ are recommended since they do not rely upon the presence of other enzymes necessary for either the reduction of dyes in co-lorimetric techniques or for the consumption of oxygen in manometric techniques.

GIBBS (1954) used cuvettes of 1.0 cm light path containing 1.5 ml veronal buffer, pH 8.0, 0.05 ml enzyme solution, 0.05 ml TPN (1.0 mg/ml) and 0.1 ml 0.05 M substrate in a final volume of 3.0 ml. TPN is omitted from the blank cell. CLAYTON (1959) used 0.9 mg protein, 5 μmoles G-6-P or 6-PGA 0.5 μmoles TPN, 200 or 500 μmoles Tris buffer, pH 7.6, and 10 μmoles $MgCl_2$ in a volume of 3.0 ml at 25° with cuvettes of 1.0 cm light path.

The present authors have, with some modification, used the spectrophoto-metric procedure of KORNBERG and HORECKER (1955) for the assay of the enzymes from wheat leaves. Cuvettes of 1.0 cm light path (Zeiss and Co.) contain 0.1 ml 0.5 M Tris buffer, pH 7.5, 0.2 ml 0.1 M $MgCl_2$, 0.1 ml 3×10^{-3} M TPN (Sigma Chemical Co.), 0.1 ml enzyme and 0.1 ml 0.02 M Na-G-6-P or 6-PGA in a volume of 2.0 ml. TPN is omitted from the blank cuvette. The reaction is started by the addition of substrate or enzyme and the reduction of TPN is followed at 340 mμ. A unit of enzyme is defined as that amount which causes an initial change in optical density of 1.0 per minute (KORNBERG and HORECKER, 1955).

2. Purification.

Highly purified preparations of G-6-P and 6-PGA dehydrogenases from baker's yeast have been described by HORECKER and SMYRNIOTIS (1955), KORN-BERG and HORECKER (1955) and by SRERE, COOPER, TABACHNICK and RACKER (1958). The removal of 6-PGA dehydrogenase requires repeated precipitations at pH 4.4, but the preparations still contain variable amounts of phosphohexo-isomerase, GSSG reductase and transaldolase (SRERE et al., 1958). An improved procedure for the preparation of G-6-P dehydrogenase has been described by the latter authors as follows.

Baker's yeast. Ten pounds of "FLEISCHMANN's" baker's yeast is autolysed and fractionated with ammonium sulphate as described by KREBS (1955) except that a 0.6% neutral solution of EDTA is added together with the toluene to give a final concentration of 2 mg EDTA/ml. After removal of the crystals of G-3-P dehydrogenase, 10 g of solid $(NH_4)_2SO_4$ is added to each 100 ml of supernatant fluid and the precipitate is suspended in a minimum volume of a neutral 0.2% solution of EDTA. This fraction, if kept frozen, is a suitable source of G-6-P and 6-PGA dehydrogenase for 2 years. *Acetate precipitation.* An aliquot containing 200—500 mg of protein is added to 250 ml 0.1 M acetate buffer, pH 4.4 (ice-cold), kept at 0° for 15 minutes and centrifuged. The well drained precipitate is suspended in 10 ml 0.05 M phosphate buffer, pH 7.7, mixed and centrifuged. The supernatant solution containing approximately 75% of the original G-6-P dehydrogenase is placed on a 5.5 cm $\times$7 cm² column of DEAE cellulose (obtainable from Brown Co., Berlin, New Hampshire, U.S.A., or the California Corp. for Biochemical Research, Los Angeles, Cal , U.S.A.) in the cold room. The column is washed with 30 ml 0.05 M phosphate buffer, pH 7.7, and the enzyme is eluted with 50 ml 0.05 M phosphate buffer, pH 7.7, containing 0.1 M NaCl. The enzyme is recovered with about 80—90% yield in a sharp peak between 20 and 40 ml of the effluent The enzyme is precipitated by the addition of solid $(NH_4)_2SO_4$ to 0.95 saturation, taken up in 0.25 M glycyl-glycine buffer, pH 7.4, and stored at $-20°$. The specific activity of the fractions varies between 10 and 40 units/mg protein.

RADHAKRISHNAN (1960) has described the preparation of a highly active G-6-P dehydrogenase, free of 6-PGA dehydrogenase, from *Neurospora crassa* Extensive purification of the enzymes has not been carried out in higher plants, but initial purification steps with several plant sources are described as follows:

Wheat germ. CONN and VENNESLAND (1951) extracted wheat germ with water (1:4, v/v) at room temperature for 30 minutes. The bulky insoluble matter is removed by centrifugation and filtering through glass wool. The filtrate is dialysed against 0.025 M phosphate buffer, pH 7.4, for 16 hours at 0°. Insoluble material is centrifuged off at 20,000 $\times g$.

For 6-PGA dehydrogenase BARNETT et al. (1953) centrifuged the 1:4 water extract of wheat germ at 20,000 $\times g$ for 20 minutes at 4°. Each litre of the super-natant fluid is treated with 20.4 ml 1.0 M $MnCl_2$ to remove nucleic acid. The heavy precipitate which forms overnight at pH 5.5 is centrifuged off at 3,000 $\times g$ and discarded. The supernatant fluid is stored at $-15°$ for 48 hours, thawed and centrifuged at 20,000 $\times g$ for 20 minutes to remove insoluble material. The pH is adjusted to 7.0 with NH_4OH and 250 g of solid $(NH_4)_2SO_4$ is added to each litre of solution. The precipitate which forms after 24 hours is centrifuged off and discarded. A further 75 g of solid $(NH_4)_2SO_4$ is added to each litre of supernatant fluid, the precipitated protein is dissolved in a minimum volume of 0.025 M phosphate buffer, pH 7.4, and dialysed against the same buffer. The final volume is about 260 ml (40 mg dry weight per ml) from each kg of germ. The nucleic acid content decreases from 13% to 1% by virtue of this treatment. The preparation contains GSSG reductase and 6-PGA dehydrogenase and can be assayed by titration of GSH in the coupled reaction heretofore described. Although it was not stated, this preparation probably contains G-6-P dehydrogenase. *Aqueous extracts*. Water or phosphate extracts of many plant tissues centrifuged and dialysed against 0.025 M phosphate, pH 7.4, have been shown by ANDERSON et al. (1952) and BARNETT et al. (1953) to contain GSSG reductase as well as the dehydrogenases which made it possible for the enzymes to be assayed by measurement of GSH production. *Acetone powders*. Purified extracts from acetone powders of leaf tissues have been used by AXELROD et al. (1953), CLAYTON (1959) and the authors. The acetone powders are prepared by grinding either fresh or frozen leaves in twice or ten times their weight of cold acetone (ca. $-10°$ to $-20°$). The residue after filtration through a BÜCHNER funnel can be re-extracted or washed with cold acetone. It is then freed of residual acetone at room temperature and stored in a desiccator over $CaCl_2$ or P_2O_5 in the cold. For leaves of monocotyledons, it is recommended that acetone powders should be prepared according to the procedure of HAGEMAN and WAYGOOD (1959).

Spinach leaves. AXELROD et al. (1953) extracted spinach leaf acetone powder with 8 times its weight of water for 15 minutes at $2°$. The insoluble material is centrifuged off at 19,000 $\times g$ for 20 minutes (stage I). The supernatant fluid is made 0.7 saturated with respect to $(NH_4)_2SO_4$ and the pH adjusted to 7.0 with 2.0 N KOH. The precipitate which forms after 15 minutes is collected by centrifugation and dissolved in a small quantity of water and dialysed for 4 hours against two changes of water (stage II).

Tobacco leaves. CLAYTON (1959) homogenised one gram of tobacco leaf acetone powder with 20 ml 0.1 M Tris -0.1 M cysteine buffer, pH 7.6, in a hand operated all-glass TEN-BROECK apparatus. The brei is filtered through four layers of cheese-cloth and the residue is re-extracted with 10 ml of buffer. The combined filtrates are centrifuged at 15,000 $\times g$ for 20 minutes in the cold. The supernatant fluid is adjusted to pH 7.4 with 1.0 N KOH, an equal volume of saturated $(NH_4)_2SO_4$ is added and the suspension is shaken for 20 minutes at room temperature. The precipitate is collected by centrifugation at 15,000 $\times g$ at $2°$ and dissolved in 20 ml of water. Insoluble material is removed by centrifugation and the supernatant fluid is adjusted to pH 7.0 with 1.0 N KOH. This preparation is either assayed directly or dialysed against 0.01 M Tris buffer, pH 7.2, for 12 hours at $2°$.

Wheat leaves. Wheat leaf acetone powder (1.5 g) prepared by the procedure of HAGEMAN and WAYGOOD (1959) is extracted for 15 minutes with 60 ml 0.05 M Tris-HCl buffer, pH 7.5, containing 0.001 M EDTA. The suspension is pressed through cheesecloth and the filtrate centrifuged to remove cell debris and particulate material. The protein of the supernatant fluid precipitating between the limits of 0.33 and 0.66 $(NH_4)_2SO_4$ saturation is dissolved in 5.0 ml 0.05 M Tris buffer, pH 7.5. The activity of the 6-PGA dehydrogenase is several fold higher than that of G-6-P dehydrogenase.

3. Lactonase (Assay and Purification).

An enzyme has been purified (Brodie and Lipmann, 1955) from bacteria, yeast and mammalian tissues that catalyzes the conversion of 6-P-glucono-δ-lactone into 6-PGA. The assay is based on the fact that lactones react quantitatively in aqueous alkaline solution with hydroxylamine (Cori and Lipmann, 1952). The lactone and the acid may be identified by paper chromatography. The enzyme is present in the 0.4 to 0.9 saturated $(NH_4)_2SO_4$ fraction of a heated sonic homogenate of baker's yeast.

III. Phosphoriboisomerase.

This enzyme catalyses the isomerization of R-5-P and Ru-5-P as follows:

$$Ru\text{-}5\text{-}P \rightleftharpoons R\text{-}5\text{-}P$$

The identification of a specific enzyme catalyzing the interconversion of the R-5-P and Ru-5-P was first made by Horecker, Smyrniotis and Seegmiller (1951) in extracts of yeast. Subsequently, Axelrod and Jang (1954) demonstrated that the enzyme was present in several higher plant sources and purified the enzyme from alfalfa press juice. Hurwitz, Weissbach, Horecker and Smyrniotis (1956) and Tabachnick, Srere, Cooper and Racker (1958) have isolated the enzyme from spinach leaves. The purified enzymes are free of both transketolase and phosphoketopentoepimerase.

The yeast enzyme is also used for the preparation of an equilibrium mixture of R-5-P (75%) and Ru-5-P (25%), the so called "isomerase product", which is useful as a substrate for the transketolase reaction. Accordingly, both the preparation of the enzyme from yeast and the isolation of the "isomerase product" are described.

1. Assay.

The enzyme may be assayed conveniently by the method of Axelrod and Jang (1954) utilizing a modification of the carbazole procedure of Dische and Borenfreund (1951) to detect the ribulose formed by the action of the isomerase.

The enzyme (0.2—0.6 units) in 0.1 ml is incubated at 37° with 0.5 ml 0.1 M Tris-HCl buffer, pH 7.0, containing 0.5 mg Ba-R-5-P. This solution should be freshly prepared. After 10 minutes 6 ml H_2SO_4 (225 ml conc. H_2SO_4 + 95 ml H_2O) is added followed by 0.2 ml carbazole solution (0.12% w/v in absolute ethanol) and 0.2 ml cysteine-HCl (1.5% w/v in water). The additions should be completed in 40 seconds with thorough mixing. The tubes are then replaced in the bath at 37° for 30 minutes and read in a colorimeter at 540 mμ. The control may either contain heated enzyme or the enzyme may be omitted until after the addition of H_2SO_4.

The enzyme unit is defined as 10 times that amount which catalyses the formation of 0.1 μmole of ribulose-5-PO_4 under the assay conditions. An equilibrium mixture of R-5-P and Ru-5-P (75:25, isomerase product) prepared by the action of a purified enzyme from baker's yeast on R-5-P (de la Haba and Racker, 1955) may be used as a standard for the colorimetric determination, or more conveniently the color density can be related to the concentration of Ru-5-P by allowing the system to come to equilibrium.

The amount of crude enzyme used in the standard assay should be such that the activity of enzymes metabolizing R-5-P and Ru-5-P (e.g. phosphoketopentoepimerase and transketolase) are negligible, otherwise spurious results will be obtained.

A spectrophotometric technique devised by Tabachnick, Srere, Cooper and Racker (1958) is based on the reduction of DPN by G-3-P. Glyceraldehyde-3-phosphate is formed by the action of the enzyme (phosphoriboisomerase) to be tested, and highly purified preparations of phosphoketopentoepimerase, transketolase, and G-3-P dehydrogenase on R-5-P. It is essential that the latter enzymes are freed of phosphoriboisomerase. The assay is accomplished in two stages. The second stage involves measuring the reduction of DPN at 340 mμ.

2. Purification.

Alfalfa press juice. Only steps Nos. 1—4 of the procedure for the purification of the enzyme in alfalfa press juice will be represented here in which 50% of the enzyme is recovered. Steps Nos. 5—7 are described in detail in the paper by AXELROD and JANG (1954).

1. Alfalfa press juice is heated to 60—63° for 1 minute, cooled and filtered. 2. The heated filtrate is made 0.7 saturated with solid $(NH_4)_2SO_4$ and centrifuged. The precipitate is suspended in water (1/20 of the volume of the filtrate) and dialyzed for 15 hours with tap water. 3. The precipitate fractionated from the dialysate between the limits of 0.35 and 0.55 saturation of $(NH_4)_2SO_4$ is dissolved in water. 4. The solution is adjusted to pH 4.5 with 2 N H_2SO_4 and the precipitate discarded after centrifugation. The pH is adjusted to 5.0 and the solution is made 0.35 saturated with $(NH_4)_2SO_4$. This precipitate is collected by centrifugation and dissolved in water.

Spinach. The enzyme has been purified from spinach leaves by HURWITZ et al. (1956) and TABACHNICK et al. (1958). In the latter procedure all operations are carried out at 2—4°. *Extract.* One bushel of washed spinach leaves is extracted by means of a tissue shredder and hydraulic press (purchased from Health Food Distributors, 123 East 34th St., New York, N.Y.). (113,000 units, s.a = 2.3). *Ammonium sulphate fractionation.* Solid $(NH_4)_2SO_4$ (22.6 g/100 ml solution) is added and the suspension is filtered through large fluted Whatman No. 12 filter paper. An additional 18.2 g of solid $(NH_4)_2SO_4$ is added to each 100 ml of the filtrate and the suspension is again filtered. The paste is removed from the paper and taken up in 50 ml of water (35,000 units, s.a = 4.1). The solution is dialyzed against 7 litres 0.01 M cysteine hydrochloride, pH 2.3, for 2 hours with stirring. After dialysis the solution is neutralized and centrifuged (29,000 units s.a = 24.3). *Gel eluate.* The supernatant solution is diluted with water to yield 5 mg protein/m l and about 0.1 volume of calcium phosphate gel (20 mg/ml) is added. Preliminary tests should be conducted to determine the minimum volume of gel necessary to adsorb the enzyme. The mixture is centrifuged, the supernatant fluid is discarded and the gel is washed twice with 40 ml water. The enzyme is eluted by four successive additions of 10 ml 0.01 M potassium phosphate, pH 7.6. The extracts are combined and dialyzed for 3 hours against water (9,500 units s.a = 306).

Baker's yeast (for preparation of isomerase product) The initial stages up to and including the acetone fractionation are the same as for the purification of transketolase. Dried baker's yeast (300 g) is suspended in 900 ml 0.066 M Na_2HPO_4 at 30° for 5 hours with occasional stirring and centrifuged at 20,000 $\times g$ for 10 minutes. The residue is re-extracted with 600 ml Na_2HPO_4 at 30° for 3 hours and recentrifuged. The combined extracts are adjusted to pH 6.5 with 2 N HCl, cooled and 1/2 of its volume of cold (−10°) acetone added. The precipitate is centrifuged for 15 minutes at 20,000 $\times g$ and discarded. An additional 1/2 volume of acetone is added to the supernatant solution and the precipitate (acetone II 33—50%) is collected by centrifugation and dialyzed for 12 hours against 25 to 50 volumes of water with one change. The dialysate is centrifuged and the supernatant is brought rapidly to 60° and held there for 25 minutes. The mixture is cooled and centrifuged. To each 100 ml of supernatant 29 g of solid $(NH_4)_2SO_4$ is added and the precipitate discarded. To each 100 ml of the supernatant 16.7 g of solid $(NH_4)_2SO_4$ is added. The precipitate is collected by centrifugation, dissolved in water and dialyzed against water before use. This preparation is virtually free of transketolase whereas the acetone II 33—50% fraction contains both enzymes.

3. Preparation of Isomerase Product (DE LA HABA and RACKER, 1955).

One gram of barium R-5-P is dissolved in 10 ml 0.1 N HCl and heated in a boiling water bath until clear. Barium is removed from the cooled solution by 1 M $(NH_4)_2SO_4$ and the supernatant is neutralised to pH 7.6. To this solution 4 ml of bakers' yeast dialyzed isomerase (see above) or alfalfa or spinach isomerase free from transketolase is added and the volume made to 20 ml with water and incubated at 37°. The isomerization can be followed by removing a 0.03 ml aliquot at 15 minute or 30 minute intervals for 2—3 hours and assaying Ru-5-P by the carbazole method described above.

When no increase in colour is observed, 3.0 ml of 50% TCA is added and the precipitated protein is centrifuged off. The pH is adjusted to 8.0 with 2 N NaOH and 5 ml 1 M barium acetate is added. The precipitate is washed with water and discarded. The barium salts of the respective pentose phosphates are precipitated from the combined washing and supernatant solution with 4 volumes of 95% ethanol for 2—3 hours in the cold after which they are collected by centrifugation, washed with absolute ethanol and dried in a desiccator under vacuum. A solution of the ammonium salt is prepared by treating a 3% solution of the barium salt with a slight excess of $(NH_4)_2SO_4$. The sulphate is removed by centrifugation and the solution adjusted to pH 6.5.

IV. Phosphoketopentoepimerase and Transketolase.

$$Ru\text{-}5\text{-}P \xrightleftharpoons{PKPE} Xu\text{-}5\text{-}P$$

$$R\text{-}5\text{-}P + Xu\text{-}5\text{-}P \xrightleftharpoons{TK} S\text{-}7\text{-}P + G\text{-}3\text{-}P$$

It had been justifiably assumed (HORECKER et al., 1953; DE LA HABA, LEDER and RACKER, 1955) that in the pentose phosphate pathway, transketolase catalysed the cleavage of Ru-5-P into an "active glycolaldehyde" which was accepted by R-5-P to form S-7-P. However, SRERE, COOPER, KLYBAS and RACKER (1955) found that the "isomerase product" (R-5-P and Ru-5-P) prepared by the action of a highly purified PRisomerase from alfalfa was inactive as a substrate for crystalline yeast transketolase and their investigations indicated that a separate enzyme was necessary for the conversion of Ru-5-P to Xu-5-P which they considered to be the key substrate for transketolase. Enzymes that catalyse the transformation of R-5-P and Ru-5-P into Xu-5-P had been evident from investigations on the xylose metabolism of certain bacteria, for example, *Pseudomonas hydrophila* (HOCHSTER, 1954—1955) and a specific enzyme catalysing the conversion of Ru-5-P to Xu-5-P was discovered in mouse spleen by ASHWELL and HICKMAN (1954). Subsequently enzymes catalysing this epimerization were demonstrated in muscle by DICKENS and WILLIAMSON (1955), SRERE et al. (1955) and in *Lactobacillus pentosus* by STUMPF and HORECKER (1956), who called the enzyme phosphoketopentoepimerase. HORECKER et al. (1956) confirmed that Xu-5-P rather than Ru-5-P is the substrate for liver as well as spinach transketolase (free of PKPepimerase) provided R-5-P is present. It was then clear that the formation of S-7-P and triosephosphate from R-5-P is due to the presence of PKPepimerase as well as PRisomerase and transketolase as was thought formerly. Furthermore, the reported requirement of liver transketolase for crystalline muscle aldolase (HORECKER et al., 1953) has been shown by HORECKER et al. (1956) to be due to contamination of the aldolase by PKPepimerase. RACKER and SCHROEDER (1957) find that octulose-8-phosphate is also a substrate for transketolase and with G-3-P forms Xu-5-P and hexose phosphate.

PKPepimerase requires no cofactors. Aged preparations of transketolase and those dialyzed against Versene-KCl solutions show a requirement for TPP and Mg^{++} (DE LA HABA et al., 1955). Transketolase has been found in bacteria (DE LEY, 1957) and in wheat leaves by the present authors (unpublished). The presence of PRisomerase, PKPepimerase and transketolase in extracts can be inferred if R-5-P is metabolised to sedoheptulose.

1. Assay (Transketolase).

The formation of S-7-P from R-5-P, Ru-5-P or Xu-5-P in crude or purified extracts of plant tissues may be followed by chromatographing TCA deproteinised extracts either before or after dephosphorylation with acid phosphatase. Descending chromatography on No. 1 Whatman filter paper with propanol-acetic acid-water (60:1:39) and phenol-water (4:1) has been used successfully by HOCHSTER (1954—1955). The TCA-orcinol spray of KLEVESTRAND and NORDAL (1950) gives a characteristic blue colour with sedoheptulose. If this is oversprayed with aniline phthalate, D-ribulose gives a pink and D-xylulose gives a purple colour (HOCHSTER, 1954—1955). Ribulose fluoresces under UV when the orcinol sprayed chromatograms are dried for 2 minutes at 100°.

The DISCHE cysteine-carbazole reagent may be used for the quantitative estimation of sedoheptulose since the spectral characteristics of its complex differ from those of aldo- and keto-pentoses (AXELROD et al., 1953). Heptuloses can also be estimated by the diphenylamine reaction (DISCHE, 1953), and the orcinol test (HORECKER et al., 1953).

Spectrophotometric methods are based upon the enzymatic assay of the triosephosphate. G-3-P formation can be measured by the reduction of DPN in the presence of its specific dehydrogenase and arsenate (HORECKER et al., 1956). Alternatively in the presence of triosephosphate isomerase and α-glycerophosphate dehydrogenase, the rate of oxidation of DPNH by triosephosphate (DAP) can be followed at 340 mμ. With limiting amounts of transketolase the reaction follows zero order kinetics (DE LA HABA et al., 1955). The unit of enzyme is defined as that amount which causes an optical density change of 0.001/min. (DE LA HABA et al., 1955) or 1.0/min (HORECKER and SMYRNIOTIS, 1955b).

The substrate is the ammonium or potassium salt of the "isomerase product" (see PRisomerase). The potassium salt is prepared as follows: 50 mg is dissolved in 4 ml 0.02 M acetic acid and 0.3 ml 0.57 M K_2SO_4 is added. The $BaSO_4$ is centrifuged off and the supernatant is neutralised with 2.0 N KOH.

In the procedure of HORECKER and SMYRNIOTIS (1955 b) 0.96 ml 0.01 M glycylglycine buffer, pH 7.5, is placed in a quartz micro-cell with 1 cm light path to which is added 0.02 ml 0.02 M isomerase product, 0.02 ml 0.003 M DPNH (6.0 mg 65% DPNH in 2.0 ml 0.001 N NaOH) and 0.01 ml α-glycerophosphate dehydrogenase. Readings are taken at 340 mμ at 1 minute intervals after the addition of 0.02 ml enzyme. With aged preparations 0.1 μmole TPP and 2 μmoles $MgCl_2$ should be added. The present authors have used cuvettes of 0.5 cm light path (Hilger & Watts Ltd.) which contained 0.1 ml 0.1 M glycylglycine buffer pH 7.5, 0.1 ml 90% (2.5 mg/ml) DPNH (Sigma Chemical Co.), 0.001—0.1 ml potassium salt of the isomerase product, 0.1 ml of α-glycerophosphate dehydrogenase diluted 1:10 (Nutritional Biochemicals Co., Cleveland 28, Ohio) and 0.1 ml enzyme in a final volume of 1.3 ml.

The reactions involved in the transketolase assay with the isomerase product are as follows:

$$\text{Ru-5-P} \underset{}{\overset{\text{PKPepimerase}}{\rightleftharpoons}} \text{Xu-5-P}$$

$$\text{R-5-P} + \text{Xu-5-P} \overset{\text{TK}}{\rightleftharpoons} \text{S-7-P} + \text{G-3-P}$$

$$\text{G-3-P} \underset{}{\overset{\text{triose isomerase}}{\rightleftharpoons}} \text{DAP}$$

$$\text{DAP} + \text{DPNH} + \text{H}^+ \underset{}{\overset{\alpha\text{-glycerophosphate dehydrogenase}}{\rightleftharpoons}} \alpha\text{-glycerophosphate} + \text{DPN}^+$$

The α-glycerophosphate dehydrogenase used by Horecker et al. (1953) was the 50—70% $(NH_4)_2SO_4$ fraction prepared from muscle according to the procedure of Racker (1947). This contains phosphohexoisomerase, triosephosphate isomerase, aldolase, PKPepimerase, but is free of transketolase. Later, Horecker et al. (1956) used the 52—70% $(NH_4)_2SO_4$ fraction in the procedure of Cori, Slein and Cori (1948) for the preparation of G-3-P dehydrogenase which is relatively free of aldolase and PKPepimerase. The commercial α-glycerophosphate dehydrogenase has no transketolase activity, but we have not tested the activity of the other enzymes.

A more specific assay would require the use of Xu-5-P as a substrate. Neither xylulose nor its phosphate are commercially available, but procedures for the preparation of xylulose from xylose have been described by Hann, Tilden and Hudson (1938), for xylulose phosphate from xylulose by Stumpf and Horecker (1956), from ribose-5-phosphate by Hurwitz and Horecker (1956) and from hydroxypyruvate and F-1, 6-P by Srere et al. (1958).

2. Purification (Transketolase).

A procedure for the preparation of crystalline transketolase from baker's yeast has been described by de la Haba et al. (1955) and de la Haba and Racker (1955). Repeated recrystallizations are essential, but not always successful, for the removal of PKPepimerase. Srere et al. (1958) have found that the two enzymes can readily be separated by chromatography on a cellulose DEAE column. The enzyme has also been purified from liver (free of PKPepimerase) and spinach (not free of PKPepimerase) (Horecker et al., 1953). Horecker et al. (1956) have described an improved method for the purification of the enzyme from spinach, free of PKPepimerase. If Xu-5-P is not readily available the purification procedure for the 1953 spinach enzyme should be used with the isomerase product as substrate, or a purified preparation of PKPepimerase should be added free of transketolase. The 1953 spinach enzyme is also used for the preparation of S-7-P used as substrate in the transaldolase reaction.

Baker's yeast. In the procedure of Srere et al. (1958) the initial steps of the purification from baker's yeast are the same as for PRisomerase (page 556/557) with the following modifications: 300 g of dried yeast is extracted only once and for only 2.5 hours at 40° and then centrifuged at 13,000 ×g for 20 minutes. Acetone fractionations are carried out as previously described. The centrifuged solution is distributed in 12 ml fractions in test tubes which are then placed for 15 minutes in a bath at 55°. The heavy precipitates are then filtered off on a Büchner funnel at room temperature and the test tubes as well as the precipitates washed once with a total of 10 ml of water. The combined filtrates are placed in a dry ice-alcohol bath and cooled with stirring to 0°. Freezing is avoided. To this 0.5 volume of ice-cold ethanol is added dropwise and the temperature is allowed to drop to − 6°. The precipitate is centrifuged off at − 6° and extracted with 6—10 ml of water, for 10 minutes at room temperature. An amount of solution containing no more than 500 mg protein is placed on a 5.5 × 7.0 cm column of DEAE cellulose previously equilibriated with 0.005 M phosphate. The enzyme is eluted with 50 ml 0.005 M phosphate buffer, pH 7.7. Five millilitre fractions are collected for 30 to 40 minutes and the activity is distributed between 2—3 tubes. The fractions are combined and 2.9 g of $(NH_4)_2SO_4$ added to each 10 ml of solution at 0°. The pH is maintained at 7.6 by the dropwise addition of 1 N KOH and the precipitate is discarded after centrifugation. An additional 1.3 g $(NH_4)_2SO_4$ is added to each 10 ml of supernatant solution and the precipitate collected by centrifugation. These preparations had a purity between 10 and 20 % and they can be crystallized by the procedure of DE LA HABA et al. (1955). The yield is largely dependent on the rate of initial drying of the yeast. Yeast dried for 24 hours at room temperature with ventilation yielded less transketolase activity, but with a higher specific activity than yeast dried for 4—5 hours at room temperature.

Spinach HORECKER et al. 1953. (1) *Extraction and ammonium sulphate fractionation I*. Spinach leaves in portions of 60 g are freed of stems and homogenised for 3 minutes in a Waring blendor with 360 ml of cold 50 % saturated $(NH_4)_2SO_4$ (adjusted to pH 8.0 with concentrated NH_4OH). The homogenates are filtered through Schleicher & Schüll No. 588 fluted filter paper. The filtrate (16.9 litres) derived from 3.6 kg of leaves is treated with 3.82 kg of $(NH_4)_2SO_4$ and filtered overnight in the cold room The precipitate is dissolved in 500 ml water and adjusted to pH 7.1 with 2.2 ml 2 N NH_4OH (ammonium sulphate I, 600 ml, 17,100 units, s.a = 1.6). (2) *Ammonium sulphate fractionation II*. The solution is diluted with water to bring the ammonium sulphate saturation to 0.10 (determined by conductivity measurements) treated with 176 g $(NH_4)_2SO_4$ and the precipitate is removed by centrifugation. Solid $(NH_4)_2SO_4$ (78 g) is added to the supernatant solution and the precipitate is collected by centrifugation, dissolved in 90 ml water and neutralized with 0.3 ml 2 N NH_4OH (108 ml, 14,000 units, s.a = 9.2). (3) *Calcium phosphate gel adsorption and ammonium sulphate fractionation III*. The solution is diluted with 1188 ml water to bring the protein concentration to 1.2 mg ml and treated with 1130 ml of calcium phosphate gel (6.6 mg d.w./ml) and centrifuged. The enzyme is eluted with 216 ml 0.01 M pyrophosphate buffer, pH 8.3. The eluate (262 ml) is treated with 81 g $(NH_4)_2SO_4$, centrifuged and the supernatant solution treated with 30.0 g $(NH_4)_2SO_4$. The precipitate is collected by centrifugation, dissolved in 25 ml water and neutralized with 0.65 ml 0.2 N NH_4OH (29.0 ml, 9,440 units, s.a = 18.2). (4) *Acetone fractionation*. The solution is dialysed overnight against 0.1 M sodium acetate (adjusted to pH 7.2). The dialysate is diluted to 77.4 ml with 0.1 M acetate to bring the protein concentration to 6.2 mg/ml and cooled in a freezing bath while 47.5 ml of cold acetone is added. The solution is held at − 8° for 3 minutes, centrifuged rapidly and the precipitate dissolved in 20 ml of water. Three more fractions are collected in the same manner by the addition of 14, 15 and 25 ml acetone respectively. The fractions are assayed separately and the three most active, fractions 3 and 4 usually, are combined (42.0 ml, 7,270 units, s.a = 37.0). (5) *Ammonium sulphate fractionation IV*. The acetone fraction is diluted to 72.0 ml and treated with 72.0 ml of cold saturated ammonium sulphate (pH 7.8) and centrifuged. The supernatant solution was treated with 8.6 g $(NH_4)_2SO_4$ and centrifuged. The precipitate is dissolved in 3.0 ml 0.25 M glycylglycine buffer, pH 7.4, and assayed. Two more fractions are collected by successive addition of 4.7 g $(NH_4)_2SO_4$ and the precipitates dissolved as before. Fractions 3 and 4 after assay were combined (6.6 ml, 5,440 units, s.a = 47.0).

Spinach HORECKER et al. (1956). (1) *Extraction.* Spinach leaves (300 g) freed of coarse stems are ground in a large Waring blendor with 1800 ml 0.01 M K_2HPO_4. The brei is filtered through Schleicher & Schüll No. 588 fluted filter paper (1740 ml 8,700 units, s.a = 0.9). (2) *Ammonium sulphate fractionation I.* The extract is treated with 900 g $(NH_4)_2SO_4$ (0.37 saturation), the precipitate is collected by centrifugation and suspended in water to a volume of 283 ml. The suspension is treated with 21.5 g $(NH_4)_2SO_4$, the precipitate removed by centrifugation, and 36.8 g $(NH_4)_2SO_4$ added to the supernatant solution. The precipitate is collected by centrifugation and dissolved in 30 ml water (6,300 units, s.a = 18.3). (3) *Calcium phosphate eluate.* The enzyme is adsorbed on calcium phosphate gel prepared by the method of KEILIN and HARTREE (1937—1938). The gel (150 ml) containing 16.7 mg dry weight/ml (aged 3—6 months) is added and subsequently eluted with 80 ml 0.01 M pyrophosphate buffer, pH 8.0. The eluate (97 ml) is fractionated with $(NH_4)_2SO_4$ by the successive additions of 30.0 g and 12.7 g yielding two fractions, the former being discarded and the second dissolved in 8.0 ml 0.01 M glycylglycine buffer, pH 7.4 (4,720 units, s.a = 48.0). (4) *Acetone fractionation.* The enzyme solution is dialysed for 15 hours against cold flowing 0.1 M sodium acetate adjusted to pH 7.5. The dialysate (10.2 ml) is diluted with 10.0 ml 0.1 M sodium acetate and fractionated with cold ($-8°$) acetone, the solution being maintained at $-10°$. The first two fractions obtained by the addition of 12.2 ml and 3.5 ml acetone are usually inactive. Two more fractions obtained by adding 4.3 ml and 6.5 ml acetone are dissolved in 6.0 ml water, assayed separately and combined if active (2,820 units, s.a = 74.0 ml). (5) *Ammonium sulphate fractionation II.* The combined acetone fractions (13.0 ml) are treated with equal volumes of saturated $(NH_4)_2SO_4$, pH 7.6, cooled to $0°$ and the precipitate discarded. The supernatant solution is treated successively with 1.56 g and 0.83 g $(NH_4)_2SO_4$ and the precipitates are dissolved separately in 2.0 ml 0.25 M glycylglycine buffer, (IIA, 2.1 ml, 1,010 units, s.a = 117.0; IIB, 2.1 ml, 1,410 units, s.a = 152.0).

3. Assay (Phosphoketopentoepimerase).

ASHWELL and HICKMAN (1954) detected Xu-5-P in the presence of Ru-5-P arising from the action of mouse spleen extracts on R-5-P by its relatively slow rate of colour development at 540 mμ (2 hours) as compared to 15 minutes for Ru-5-P in the cysteine-carbazole test (DISCHE and BORENFREUND, 1951; see assay for PRisomerase). One reading is taken at 15 minutes after the addition of the reagents and another at 120 minutes. The percentage of xylulose in the mixture is calculated by the increase in absorption in the 15—120 minutes interval based on the behaviour of authentic standards (STUMPF and HORECKER, 1956).

The measurement of PKPepimerase activity by HURWITZ and HORECKER (1956) is based on the relative stability of R-5-P alkali compared with ketopentose esters. When incubated at $25°$ in 1.0 N KOH only 17% of the orcinol colour developed by R-5-P is lost after 10 minutes, whereas 87% and 70% is lost in the case of Ru-5-P and Xu-5-P. The reaction mixture contains 18 μmoles Tris buffer, pH 7.5, 15 units PRisomerase, 0.004—0.04 units PKPepimerase and 0.16 μmoles Xu-5-P in a total volume of 0.2 ml. The system is incubated for 5 minutes at $25°$ and the reaction is stopped by the addition of 0.2 ml 2 N KOH. The solution is allowed to stand at room temperature for 15 minutes and the residual orcinol-reactive material measured by the method of MEJBAUM (1939). A unit is that amount of enzyme which forms 1 μmole of alkali-stable pentose in 5 minutes under the conditions of assay.

Spectrophotometric assays involve the coupling of the system with an appropriate DPNH oxidation system. STUMPF and HORECKER (1956) used phosphoribulokinase, prepared from spinach by the procedure of HURWITZ et al. (1956) together with the ADP-assay system (KORNBERG and PRICER, 1951).
The reactions involved are as follows:

$$\text{Xu-5-P} \xrightarrow{\text{PKPepimerase}} \text{Ru-5-P}$$

$$\text{Ru-5-P} + \text{ATP} \xrightarrow{\text{phosphoribulokinase}} \text{Ru-1, 5-P} + \text{ADP}$$

$$\text{ADP} + \text{PEP} \xrightarrow{\text{pyruvate kinase}} \text{pyruvate} + \text{ATP}$$

$$\text{pyruvate} + \text{DPNH} + \text{H}^+ \xrightarrow{\text{lactate dehydrogenase}} \text{lactate} + \text{DPN}^+$$

The oxidation of DPNH was followed at 340 mμ in the presence of the specific enzymes, ATP, PEP and the substrate Xu-5-P prepared by the action of PRisomerase from spinach and PKPepimerase (from *L. pentosus*) on R-5-P (HURWITZ and HORECKER, 1956).

HORECKER et al. (1956) used a procedure similar to that developed for the transketolase assay, but the enzymes used must be free (or nearly so) of PKPepimerase. The reaction mixture (1.01 ml) contains 10 μmoles glycylglycine buffer, pH 7.5, 6 μmoles cysteine, 0.1 μmole DPNH, 0.4 μmoles R-5-P, 0.1 μmoles TPP, 18 μg liver transketolase (0.12 units) and 0.2 mg α-glycerophosphate dehydrogenase [52—70% $(NH_4)_2SO_4$ fraction page 560]. PKP epimerase is added and the reaction is started by the addition of R-5-P. The decrease in optical density is followed at 340 mμ and the rate should be corrected for any contaminating epimerase in the other enzyme preparations. Liver transketolase is prepared according to the method described by HORECKER et al. (1953) and HORECKER and SMYRNIOTIS (1955b).

The addition of PRisomerase is unnecessary since it is present in the liver transketolase preparations and triosephosphate isomerase is present along with α-glycerophosphate dehydrogenase. If the spinach enzyme of HORECKER et al. (1956) were used instead of liver transketolase it would be necessary to add PRisomerase from alfalfa (AXELROD and JANG, 1954) or from spinach (HURWITZ et al., 1956; TABACHNICK et al., 1958). Both the alfalfa and the spinach enzymes are free from PKPepimerase and transketolase.

TABACHNICK et al. (1958) have also developed a spectrophotometric method for the assay of PKPepimerase.

4. Purification (Phosphoketopentoepimerase).

Although the enzyme is widely distributed in plants, animals and bacteria, it has only been purified from sonic extracts of *Lactobacillus pentosus* (HURWITZ and HORECKER, 1956) and from rabbit muscle by TABACHNIK et al. (1958). The unit of enzyme is that defined by HURWITZ and HORECKER (1956). (1) *Crude extract.* *L. pentosus* cells (30 g wet weight) are suspended in 60 ml 0.02 M $NaHCO_3$ and exposed to a 9 kc Raytheon sonic oscillator for 1 hour. The suspension is centrifuged for 30 minutes at 15,000 $\times g$ and the residue resuspended in 60 ml 0.02 M $NaHCO_3$ and treated in the oscillator for 1 hour. The extracts were combined (4,200 units, s.a = 0.5). (2) *Manganous chloride treatment.* 140 ml of the extract is treated with 0.05 volumes of 1.0 M $MnCl_2$ and after 30 minutes at 0° the precipitate removed by centrifugation (133 ml, 2,920 units, s.a = 2.4). (3) *Ammonium sulphate fractionation I.* The supernatant solution is treated with 247 ml $(NH_4)_2SO_4$

solution saturated at room temperature and adjusted to pH 7.5 with concentrated NH$_4$OH. The precipitate is collected by centrifugation and dissolved in water (65 ml, 2,200 units, s.a = 3.9). (4) *Ammonium sulphate fractionation II.* The solution is heated in a water bath at 70° and held for 10 minutes. Average temperature of last 5 minute period is 65—68°. The solution is rapidly cooled, diluted with 15 ml water and centrifuged. The supernatant solution (75 ml) is treated with 32 ml of a saturated (at 0°) solution of (NH$_4$)$_2$SO$_4$, the precipitate is removed by centrifugation and the supernatant solution (104 ml) is treated with 4.6 ml saturated (at 0°) (NH$_4$)$_2$SO$_4$ solution. The resulting precipitate is collected and dissolved in water (17 ml, 2,720 units, s.a = 17.3).

Further purification using calcium phosphate gel, a third (NH$_4$)$_2$SO$_4$ fractionation and a heating step is described by Hurwitz and Horecker (1956). However, they used the ammonium sulphate fraction II for most of their experiments which was stable for 2 months at −10°. The preparation is free of PRisomerase and is suitable for the preparation of Xu-5-P.

V. Transaldolase.

$$\text{S-7-P} + \text{G-3-P} \underset{}{\overset{\text{Transaldolase}}{\rightleftharpoons}} \text{F-6-P} + \text{tetrose phosphate}$$

The observation that sedoheptulose was an intermediate and that hexose phosphate was an overall product of pentosephosphate metabolism in a wide variety of plants, animal tissues and micro-organisms led to the discovery that hexose phosphate was formed subsequent to the action of transketolase by the action of a specific enzyme, transaldolase, on S-7-P requiring a source of triosephosphate (Horecker and Smyrniotis, 1953). Using liver preparations (Horecker, Gibbs, Klenow and Smyrniotis, 1954) and pea leaf and root preparations (Gibbs and Horecker, 1954) with ribose-5-phosphate-1-C^{14} or ribose-5-phosphate 2,3-C^{14} the pattern of labelling in glucose, derived as a result of phosphohexoisomerase action and degraded by *Leuconostoc meserentoides*, could only be explained by the transfer of aldol linkages (transaldolase) subsequent to transketolase action. Evidence for the formation of F-6-P and tetrose phosphate from S-7-P and G-3-P and for the mechanism of the reaction was obtained later by Horecker and Smyrniotis (1955), Horecker, Smyrniotis, Hiatt and Marks (1955) and Srere, Kornberg and Racker (1955).

1. Assay.

If hexosemonophosphate is formed from pentose phosphate by plant extracts, the presence of transaldolase as well as the previously described enzymes can be inferred. Hexose monophosphate can be identified spectrophotometrically at 340 mμ by its reaction with TPN and a purified preparation of G-6-P dehydrogenase (from yeast). Phosphohexoisomerase is present in crude plant extracts.

The appearance of glucose and fructose on chromatograms of the end products dephosphorylated by acid phosphatase also establishes the presence of transaldolase activity (Axelrod et al., 1953). G-6-P and F-6-P also give an equivalent and characteristic absorption spectrum in the Dische-cysteine reaction in which it can be assayed in the presence of heptulose and ribulose.

A complicated quantitative assay is based on the reduction of TPN by the product G-6-P formed by the action of G-6-P dehydrogenase, PHisomerase, aldolase and the enzyme (transaldolase) on the substrates, S-7-P and F-1, 6-P. The reaction is followed at 340 mμ in the spectrophotometer and is described in detail by Horecker and Smyrniotis (1955; 1955c).

The reactions involved are as follows:

$$\text{F-1, 6-P} \xrightleftharpoons{\text{aldolase}} \text{G-3-P} + \text{DAP}$$

$$\text{S 7-P} + \text{G-3-P} \xrightleftharpoons{\text{transketolase}} \text{F-6-P} + \text{tetrose-P}$$

$$\text{F-6-P} \xrightleftharpoons{\text{PHisomerase}} \text{G-6-P}$$

$$\text{G-6-P} + \text{TPN}^+ \xrightarrow{\text{G-6-P dehydrogenase}} \text{6-PGA} + \text{TPNH} + \text{H}^+$$

For less highly purified preparations of spinach or liver it is only necessary to add S-7-P, TPN and G-6-P dehydrogenase to demonstrate transaldolase activity. The addition of HDP by producing G-3-P has a catalytic effect on liver preparations. This cannot be shown with spinach extracts because of the presence of a highly active fructose diphosphatase and phosphohexoisomerase which produces G-6-P directly (HORECKER and SMYRNIOTIS, 1955). The action of spinach extracts on S-7-P in the absence of G-3-P must be due to a transketolase equilibrium or some other side reaction which would provide a source of the substrate.

With aqueous filtered extracts of spinach leaves HORECKER and SMYRNIOTIS (1955a) assayed the enzyme as follows: Cuvettes contain 0.14 μmoles TPN, 1.10 μmoles S-7-P, 60 μmoles glycylglycine buffer, pH 7.5, 0.15 mg G-6-P dehydrogenase and 0.012 ml of a spinach extract in a volume of 1.54 ml. Optical density changes are measured at 340 mμ.

2. Preparation of Sedoheptulose-7-Phosphate.

It is prepared by the action of spinach transketolase on R-5-P. HORECKER and SMYRNIOTIS (1955a).

The reaction mixture 50 ml contains 703 μmoles R-5-P, 15 units of spinach transketolase (1.0 mg protein, 1953 preparation), 170 μmoles cysteine, 280 μmoles glycylglycine buffer, pH 7.7. After 165 minutes at room temperature the mixture containing 235 μmoles S-7-P is placed on a Dowex 1 (formate) column 13 ×2.5 cm, washed with 50 ml water and eluted with 0.2 N formic acid containing 0.03 N sodium formate at the rate of 2.0 ml/minute in 25 ml fractions. Fractions 24—29 containing S-7-P are adjusted to pH 6.2 with NaOH and 21.0 ml saturated Ba(OH)$_2$ solution and precipitated with 4 volumes of ethanol. The precipitate is collected by centrifugation, washed with 10 ml 80% ethanol and dried in vacuo. The dried barium salt (117 mg) is 73% pure and contains no other phosphate salts. 50.0 mg of the product is dissolved in 3.5 ml 0.02 M acetic acid and 0.2 ml 0.57 M K$_2$SO$_4$ added. The BaSO$_4$ is removed by centrifugation and the supernatant is neutralized with 2 N KOH and made to 5.0 ml. A bulk preparation yielding 1.5 g of the barium salt is described by HORECKER (1957).

3. Purification.

Highly purified preparations of transaldolase free from transketolase have been obtained from dried brewer's yeast by HORECKER and SMYRNIOTIS (1955) and by SRERE et al. (1958) from extracts of frozen torula yeast. It has also been prepared from liver and it has been detected in extracts of *E. coli* and spinach leaves (HORECKER and SMYRNIOTIS, 1955).

GIBBS and HORECKER (1954) describe a preparation from leaves and roots of pea that converts pentosephosphate into hexose monophosphate and presumably contains transaldolase as well as transketolase and other enzymes of the pentosephosphate pathway. With further purification the preparation should be useful in obtaining plant transaldolase free of transketolase. The method is as follows: *Pea root* (GIBBS and HORECKER, 1954). An acetone powder of 11—13 day old roots of *Pisum sativum* var. *alaska* is prepared by homogenising 120 g with 1200 ml acetone at −10° in a Waring blendor and filtered with suction. The mixture is re-extracted with 400 ml acetone and coarse fibres removed by screening. The resulting powder is suspended in 20 ml 0.1 M Tris buffer, pH 8.0, for 30 minutes at room temperature. Insoluble material is removed by centrifugation and the supernatant fluid is dialysed overnight in the cold against 10 litres of 0.02 M buffer, pH 8.0. *Pea leaves*. All operations are conducted at 0—4°. Leaves (120 g) separated from stems are ground with an equal weight of acid washed cold sand and 20 ml of cold 0.25 M $KHCO_3$. The extract is strained through four layers of cheesecloth and centrifuged at 18,000 ×g for 20 minutes (volume 75 ml). The pH is brought to 5.2 with 2 N acetic acid, the precipitate is removed by centrifugation and the supernatant adjusted to pH 8.0 with 2 N KOH. Solid $(NH_4)_2SO_4$ is added and the precipitate fractionating between the limits of 31 and 42% saturation is collected and dissolved in 10 ml 0.02 M buffer (as above).

VI. Phosphohexoisomerase.

$$\text{F-6-P} \rightleftharpoons \text{G-6-P}$$

Evidence for the presence of the enzyme in pea extracts was provided by TANKO (1936) and SOMERS and COSBY (1945). The formation of G-6-P in the reaction products of unpurified transaldolase systems indicates the presence of phosphohexoisomerase. HORECKER and SMYRNIOTIS (1953) showed that without the addition of phosphohexoisomerase to the transaldolase system, little or no G-6-P was formed, the major product being F-6-P. AXELROD et al. (1953) demonstrated the formation of G-6-P as a product of the action of their stage II spinach enzyme (see G-6-P dehydrogenase) on R-5-P after 21 hours of incubation. SERVETTAZ (1956) showed that G-6-P and F-6-P were interconvertible in the mitochondrial fraction of pea seedlings. RAMASARMA and GIRI (1956) have made an analysis of and described some of the properties of the enzyme from *Phaseolus radiatus*. The enzyme is widely distributed since it is one of the enzymes involved in the conversion of G-1-P to F-1, 6-P leading to the glycolytic pathway.

1. Assay.

The enzyme may be readily assayed by the G-6-P dehydrogenase system using F-6-P as substrate and G-6-P dehydrogenase free of phosphohexoisomerase. The enzyme unit should be the same as that defined for G-6-P dehydrogenase. SLEIN (1955) and RAMASARMA and GIRI (1956) use a colorimetric procedure based on the conversion of G-6-P to F-6-P and measuring the fructose by the resorcinol method. F-6-P gives about 60—65% of the colour of free fructose. The reaction mixture (RAMASARMA and GIRI, 1956) contains 1.0 μmole F-6-P or 2.0 μmoles G-6-P, 0.02 M veronal buffer, pH 7.5—7.8, and enzyme solution containing 0.005 mg protein in a final volume of 0.25 ml. The system is incubated at 37° for 2 minutes. SLEIN (1955) stops the reaction by the addition of 3.5 ml of approximately 8.3 N HCl. Then 1.0 ml of 0.1% resorcinol in 95% ethanol is added and the mixture is heated for 10 minutes at 80°, cooled to room temperature and the colour is read at 540 mμ. The enzyme activity can be expressed in terms of the change in the colorimeter scale reading per minute of incubation (SLEIN, 1955) or μmoles of substrate converted per minute (RAMASARMA and GIRI, 1956).

2. Purification.

Seeds of *Phaseolus radiatus* (RAMASARMA and GIRI, 1956) are powdered and the powder is extracted with 5 times its weight of water for 12 hours under a layer of toluene in the cold. The crude extract (100 ml) is adjusted to pH 7.0 with dilute alkali, and 100 ml of cold acetone is added slowly with stirring. The precipitate is removed by centrifugation and 200 ml of cold acetone is added to the supernatant fluid. The precipitate is collected by centrifugation, dissolved in water and dialysed against several changes of water for 2 days. Any precipitate formed during dialysis is centrifuged off and discarded. The temperature is maintained between 0° and 10° during these operations. A six-fold purification is achieved and the specific activity is about 40 μmoles of substrate converted/min per mg protein.

References.

(1) ANDERSON, D. G., H. A. STAFFORD, E. E. CONN and B. VENNESLAND: Plant Physiol. 27, 675 (1952). — (2) ASHWELL, G., and J. J. HICKMAN: J. Amer. chem. Soc. 76, 5889 (1954). — (3) AXELROD, B., R. S. BANDURSKI, C. M. GREINER and R. JANG: J. biol. Chem. 202, 619 (1953). — (4) AXELROD, B., and H. BEEVERS: Ann. Rev. Pl. Physiol. 7, 267 (1956). — (5) AXELROD, B., and R. JANG: J. biol. Chem. 209, 847 (1954).
(6) BAILEY, R. C., and E. C. WEBB: Biochem. J. 42, 60 (1948). — (7) BARNETT, R. C., H. A. STAFFORD, E. E. CONN and B. VENNESLAND: Plant Physiol. 28, 115 (1953). — (8) BERGER, L., M. W. SLEIN, S. P. COLOWICK and C. F. CORI: J. gen. Physiol. 29, 379 (1946). — (9) BOSER, H.: Phytopath. Z. 33, 197 (1958). — (10) BRODIE, A. P., and F. LIPMANN: J. biol. Chem. 212, 677 (1955).
(11) CARDINI, C. E.: Enzymologia 14, 362 (1950). — (12) CLAYTON, R. A.: Arch. Biochem. Biophys. 79, 111 (1959). — (13) COLOWICK, S. P., and H. M. KALCKAR: J. biol. Chem. 148, 117 (1943). — (14) CONN, E. E., and B. VENNESLAND: J. biol. Chem. 192, 17 (1951). — (15) CORI, O., and F. LIPMANN: J. biol. Chem. 194, 417 (1952). — (16) CORI, G. T., M. W. SLEIN and C. F. CORI: J. biol. Chem. 173, 605 (1948). — (17) COURI, D., and E. RACKER: Arch. Biochem. Biophys. 83, 195 (1959).
(18) DE LA HABA, G., I. G. LEDER and E. RACKER: J. biol. Chem. 214, 409 (1955). — (19) DE LA HABA, G., and E. RACKER: Methods in Enzymology 1, 375 (1955). — (20) DE LEY, J.: Enzymologia 18, 33 (1957). — (21) DE MOSS, R. D.: Methods in Enzymology 1, 328 (1955). — (22) DICKENS, F.: Ann. N.Y. Acad. Sci. 75, 71 (1958). — (23) DICKENS, F., and D. H. WILLIAMSON: Nature (Lond.) 176, 400 (1955). — (24) DISCHE, Z.: J. biol. Chem. 204, 983 (1953). — (25) DISCHE, Z., and E. BORENFREUND: J. biol. Chem. 192, 583 (1951).
(26) FORTI, G., C. TUA, and L. TOGNOLI: Biochim. biophys. Acta 36, 19 (1959). — (27) FUJITA, A., and I. NUMATA: Biochem. Z. 299, 249 (1938).
(28) GIBBS, M.: Nature (Lond.) 170, 164 (1952). — (29) GIBBS, M.: Plant Physiol. 29, 34 (1954). — (30) GIBBS, M.: Ann. Rev. Pl. Physiol. 10, 329 (1959). — (31) GIBBS, M., and B. L. HORECKER: J. biol. Chem. 208, 813 (1954). — (32) GOTTSCHALK, A.: Biochem. J. 41, 478 (1947).
(33) HAGEMAN, R. H., and E. R. WAYGOOD: Plant Physiol. 34, 306 (1959). — (34) HANN, R. M., E. B. TILDEN and C. S. HUDSON: J. Amer. chem. Soc. 60, 1201 (1938). — (35) HERS, H. G.: Biochim. biophys. Acta 8, 416 (1952). — (36) HERS, H. G.: Biochim. biophys. Acta 8, 424 (1952). — (37) HERS, H. G.: Methods in Enzymology 1, 289 (1955). — (38) HOCHSTER, R. M.: Canad. J. Microbiol. 1, 346 (1954—1955). — (39) HORECKER, B. L.: Methods in Enzymology 11, 195 (1957). — (40) HORECKER, B. L., M. GIBBS, H. KLENOW and P. Z. SMYRNIOTIS: J. biol. Chem. 206, 393 (1954). — (41) HORECKER, B. L., and P. Z. SMYRNIOTIS: J. biol. Chem. 193, 371 (1951). — (42) HORECKER, B. L., and P. Z. SMYRNIOTIS: J. Amer. chem. Soc. 75, 2021 (1953). — (43) HORECKER, B. L., and P. Z. SMYRNIOTIS: J. biol. Chem. 212, 811 (1955). — (44) HORECKER, B. L., and P. Z. SMYRNIOTIS: Methods in Enzymology 1, 323 (1955a). — (45) HORECKER, B. L., and P. Z. SMYRNIOTIS: Methods in Enzymology 1, 371 (1955b). — (46) HORECKER, B. L., and P. Z. SMYRNIOTIS: Methods in Enzymology 1, 381 (1955c). — (47) HORECKER, B. L., P. Z. SMYRNIOTIS, H. H. HIATT and P. A. MARKS: J. biol. Chem. 212, 827 (1955). — (48) HORECKER, B. L., P. Z. SMYRNIOTIS and J. HURWITZ: J. biol. Chem. 223, 1009 (1956). — (49) HORECKER, B. L., P. Z. SMYRNIOTIS and H. KLENOW: J. biol. Chem. 205, 661 (1953). — (50) HORECKER, B. L., P. Z. SMYRNIOTIS and G. S. SEEGMILLER: J. biol. Chem. 193, 383 (1951). — (51) HURWITZ, J., and B. L. HORECKER: J. biol. Chem. 223, 993 (1956). — (52) HURWITZ, J., A. WEISSBACH, B. L. HORECKER and P. Z. SMYRNIOTIS: J. biol. Chem. 218, 769 (1956).

(53) James, W. O.: Adv. Enzymol. 18, 281 (1957). — (54) Keilin, D., and E. F. Hartree: Proc. Roy. Soc. (Lond.) Ser. B 124, 397 (1937—1938). — (55) Klevestrand, R., and A. Nordal: Acta chem. scand. 4, 1320 (1950). — (56) Kornberg, A.: J. biol. Chem. 182, 805 (1950). — (57) Kornberg, A., and B. L. Horecker: Methods in Enzymology 1, 323 (1955). — (58) Kornberg, A., and W. E. Pricer: J. biol. Chem. 193, 481 (1951). — (59) Krebs, E. G.: Methods in Enzymology 1, 407 (1955). — (60) Krebs, H. A., and H. L. Kornberg: Ergebn. Physiol. 29, 212 (1957). — (61) Kunitz, M., and M. R. McDonald: J. gen. Physiol. 29, 393 (1946). — (62) Kunitz, M., and M. R. McDonald: in "Crystalline Enzymes", ed. J. H. Northrop, p. 181 2nd. ed. New York: Columbia University Press 1948. — (63) Kursanov, A. L., O. A. Pavlinova and T. P. Afanasieva: Arch. Biochem. Biophys. 83, 239 (1959).

(64) Lardy, H. A.: "Respiratory enzymes," rev. ed. p. 184. Minneapolis: Burgess 1949. — (65) Leuthart, F., and E. Testa: Helv. physiol. Acta 8, C67 (1950).

(66) McDonald, M. R.: Methods in Enzymology 1, 269 (1955). — (67) Medina, A., and D. J. D. Nicholas: Biochem. J. 66, 573 (1957). — (68) Medina, A., and A. Sols: Biochim. biophys. Acta 19, 378 (1956). — (69) Mejbaum, W.: Hoppe-Seylers Z. physiol. Chem. 258, 117 (1939). — (70) Meyerhof, O.: Biochem. Z. 183, 176 (1927).

(71) Negelein, E., and E. Haas: Biochem. Z. 282, 206 (1935). — (72) Neilands, J. B., and P. K. Stumpf: "Enzyme chemistry," 2nd ed. New York: John Wiley and Sons 1958. — (73) Nelson, N.: J. biol. Chem. 153, 375 (1944). — (74) Neufeld, E. F., N. O. Kaplan and S. P. Colowick: Biochim. biophys. Acta 17, 525 (1955).

(75) Parks, R. E., E. Ben-Gershom and H. A. Lardy: J. biol. Chem. 227, 231 (1957). — (76) Racker, E.: J. biol. Chem. 167, 843 (1947). — (77) Racker, E., G. de la Haba and I. G. Leder: J. Amer. chem. Soc. 75, 1010 (1953). — (78) Racker, E., and E. A. R. Schroeder: Arch. Biochem. Biophys. 66, 241 (1957). — (79) Radhakrishnan, A. N.: Biochim. biophys. Acta 40, 546 (1960). — (80) Ramasarma, T., and K. V. Giri: Arch. Biochem. Biophys. 62, 91 (1956). — (81) Richter, G.: Z. Naturforsch. 12 b, 662 (1957).

(82) Saltman, P.: J. biol. Chem. 200, 145 (1953). — (83) Servettaz, O.: R. C. Accad. Lincei 20, 255 (1956). — (84) Simpson, F. J.: Canad. J. Biochem. Physiol. 38, 115 (1960). — (85) Slein, M. W.: Methods in Enzymology 1, 299 (1955). — (86) Slein, M. W., G. T. Cori and C. F. Cori: J. biol. Chem. 186, 763 (1950). — (87) Sols, A.: Biochim. biophys. Acta 19, 144 (1956). — (88) Somers, G. F., and E. Cosby: Arch. Biochem. 6, 295 (1945). — (89) Somogyi, M.: J. biol. Chem. 160, 69 (1945). — (90) Spyridon, G. A. A., and O. F. Denstedt: J. biol. Chem. 199, 493 (1952). — (91) Srere, P. A., J. R. Cooper, V. Klybas and E. Racker: Arch. Biochem. Biophys. 59, 535 (1955). — (92) Srere, P. A., J. R. Cooper, M. Tabachnick and E. Racker: Arch. Biochem. Biophys. 74, 295 (1958). — (93) Srere, P. A., H. L. Kornberg and E. Racker: Fed. Proc. 14, 285 (1955). — (94) Stumpf, P. K., and B. L. Horecker: J. biol. Chem. 218, 753 (1956).

(95) Tabachnick, M., P. A. Srere, J. R. Cooper and E. Racker: Arch. Biochem. Biophys. 74, 315 (1958). — (96) Tanko, B.: Biochem. J. 30, 692 (1936). — (97) Vallee, B. L., F. L. Hoch, S. J. Adelstein and W. E. C. Wacker: J. Amer. chem. Soc. 78, 5879 (1956). — (98) Vishniac, W.: Arch. Biochem. 26, 167 (1950).

(99) Wagner-Jauregg, T., and H. Rauen: Hoppe-Seylers Z. physiol. Chem. 233, 215 (1935). — (100) Wajzer, J.: C. R. Acad. Sci. (Paris) 229, 1270 (1949). — (101) Warburg, O., and W. Christian: Biochem. Z. 287, 291 (1936). — (102) Warburg, O., W. Christian and A. Griese: Biochem. Z. 282, 157 (1935). — (103) Wills, E. D., and A. Wormall: Biochem. J. 47, 158 (1950). — (104) Wolin, M. J., F. J. Simpson and W. A. Wood: J. biol. Chem. 232, 559 (1958).

Abbreviations used in this chapter.

Enzymes: G-6-P dehydrogenase, glucose-6-phosphate dehydrogenase (Zwischenferment); 6-PGA dehydrogenase, 6-phosphogluconate dehydrogenase; PHisomerase, phosphohexoisomerase; PKPepimerase, phosphoketopentoepimerase; PRisomerase, phosphoriboisomrase; TK, transketolase.

Substrates: DAP, dihydroxyacetone phosphate; F-6-P, fructose-6-phosphate; F-1,6-P, fructose-1,6-diphosphate; GSH, reduced glutathione; GSSG, oxidized glutathione; G-6-P, glucose-6-phosphate; G-3-P, glyceraldehyde-3-phosphate; PEP, phosphoenolpyruvate; 6-PGA, 6-phosphogluconate; R-5-P, ribose-5-phosphate; Ru-5-P, ribulose-5-phosphate; Ru-1,5-P, ribulose-1,5-diphosphate; S-7-P, sedoheptulose-7-phosphate; Xu-5-P, xylulose-5-phosphate.

Coenzymes: ADP, adenosine diphosphate; ATP, adenosine triphosphate; DPN, diphosphopyridine nucleotide; DPNH, reduced diphosphopyridine nucleotide; TPN, triphosphopyridine nucleotide; TPNH, reduced triphosphopyridine nucleotide; TPP, thiaminepyrophosphate.

Miscellaneous: EDTA, ethylenediaminotetra acetic acid; DEAE cellulose, diethylaminoethylcellulose; S.A. (s.a.), Specific activity; TCA, trichloracetic acid; Tris, Tris (hydroxy)methylaminomethane; TTC, triphenyltetrazolium chloride.

Enzyme Systems in Photosynthesis[1].

By

Manuel Losada and Daniel I. Arnon.

With 5 Figures.

A decade ago, it would have been confidently predicted that a survey of enzyme systems in photosynthesis would be centered on enzymes of CO_2 assimilation. But recent research has turned up no *enzymic* evidence to support the once widely held view that CO_2 assimilation in photosynthesis follows a special pathway that is peculiar to photosynthetic cells. Instead, there is enzymic evidence that the reduction of CO_2 to carbohydrates consists of a series of dark reactions, all of which are now known to occur also in nonphotosynthetic cells (see review by QUAYLE, 1961) and most of which are discussed as components of the glycolytic or pentose cycle elsewhere in this volume.

The dark reactions of CO_2 assimilation, either in photosynthetic or in nonphotosynthetic cells, are driven by ATP and PNH_2. The main distinction between photosynthetic and nonphotosynthetic metabolism seems to lie, therefore, in the manner in which ATP and PNH_2, jointly termed assimilatory power, are formed. Photosynthetic cells form these two compounds (with or without oxygen evolution) at the expense of radiant energy, whereas nonphotosynthetic cells form them at the expense of energy released by dark chemical reactions.

This survey of the enzymic machinery of photosynthesis will stress the energy conversion aspects of photosynthesis, i.e. those photochemical reactions that bring about the formation of assimilatory power at the expense of radiant energy — a process which in green plants is accompanied by oxygen evolution. However, the extensive literature which deals with these subjects at the cellular level is deemed to lie beyond the scope of our discussion. Our review will be limited, in the main, to the investigations of the energy conversion process in photosynthesis that have been carried out in recent years with the aid of subcellular particles (isolated chloroplasts and bacterial chromatophores) by techniques similar to those that have long been used successfully in other areas of cellular biochemistry and enzymology. It should be noted that even within this limited scope, a discussion of individual enzyme components of photosynthesis will only be possible in isolated cases. The energy conversion reactions in chloroplasts,

---

[1] Abbreviations: ADP, adenosine diphosphate; AMP, adenosine monophosphate; ATP, adenosine triphosphate; Chl, chlorophyll; DPN and $DPNH_2$, oxidized and reduced diphosphopyridine nucleotide; EDTA, ethylene diamine tetraacetic acid; FAD, flavinadenine dinucleotide; F-D-P, fructose diphosphate; F-6-P, fructose-6-phosphate; FMN, flavin mononucleotide; G-1-P, glucose-1-phosphate; G-6-P, glucose-6-phosphate; Pi, inorganic phosphate; PGA, phosphoglyceric acid; PMS, phenazine methosulfate; PN and PNH_2, oxidized and reduced phosphopyridine nucleotides; PPNR, photosynthetic phosphopyridine nucleotide reductase; RHP, *Rhodospirillum* heme protein; R-5-P, ribose-5-phosphate; Ru-5-P, ribulose-5-phosphate; RuDP, ribulose diphosphate; TPN and $TPNH_2$, oxidized and reduced triphosphopyridine nucleotide; Tris, tris(-hydroxymethyl)aminomethane.

like those in oxidative phosphorylation by mitochondria, are catalyzed by structurally and functionally integrated multienzyme systems, the individual components of which are only now becoming identified, and in a few cases, isolated. Thus, much of the discussion now will of necessity be concerned with the subcellular particles themselves as sites of photosynthetic reactions, rather than with the individual enzymes.

Of the enzymes of carbon assimilation only those responsible for the initial incorporation of CO_2 will be singled out for special mention. Although they are now also known, along with the other enzymes of carbon assimilation, to be functioning in nonphotosynthetic cells, they were first found in photosynthetic tissues and they continue to hold special interest for students of photosynthesis.

A. The Photosynthetic Structures of Plants and Bacteria.

In green plants, the essential reactions of photosynthesis take place in cytoplasmic particles known as chloroplasts. These particles contain the pigments, enzymes and cofactors involved in all the photochemical reactions that are characteristic of the photosynthetic process. In addition to the photochemical reactions proper, chloroplasts can also carry out dark, biosynthetic reactions of CO_2 assimilation that are responsible for the synthesis of carbohydrates, including starch.

It is often difficult for the student of photosynthesis today to realize that, aside from conjectures, experimental evidence for CO_2 assimilation by chloroplasts has only become available in recent years. Modern biochemical investigations of chloroplasts began with the work of HILL (1937, 1951) who found that isolated chloroplasts produced oxygen in light but could not assimilate CO_2. This became known as the HILL reaction and, prior to 1954, formed the basis for the view, generally held by active investigators in this field, that the chloroplast was a "system much simpler than that required for photosynthesis" and was the site of only "the light-absorbing and water-splitting reactions of the overall photosynthetic process" (LUMRY, SPIKES and EYRING, 1954).

In 1954, another phase of chloroplast investigations began with the direct demonstration of complete photosynthesis by isolated chloroplasts. With improved experimental methods, isolated chloroplasts, unaided by other cellular particles or enzyme systems, were found to reduce CO_2 to the level of carbohydrates, including starch, with a simultaneous evolution of oxygen, at physiological temperatures and with no energy supply except visible light (ARNON, ALLEN and WHATLEY, 1954; ALLEN, ARNON, CAPINDALE, WHATLEY and DURHAM, 1955; ARNON, 1955). The photosynthetic quotient and the products of CO_2 assimilation were found to be the same as in photosynthesis by whole cells. The conversion of $C^{14}O_2$ by isolated chloroplasts to phosphorylated sugars and starch was confirmed and extended in other laboratories (GIBBS and CYNKIN, 1958; TOLBERT, 1958; GIBBS and CALO, 1959; SMILLIE and FULLER, 1959; SMILLIE and KROTKOV, 1959; compare also UEDA, 1949; IRMAK, 1955; THOMAS, HAANS and VAN DER LEUN, 1957).

Aside from enzymes involved in the synthesis of carbohydrates, chloroplasts contain also a variety of other enzymes that are not peculiar to photosynthesis and that are discussed elsewhere in this volume. These enzymes include: phosphoenolpyruvate carboxylase (ROSENBERG, CAPINDALE and WHATLEY, 1958), acetate activating enzymes, condensing enzyme, malic dehydrogenase, aconitase, isocitric dehydrogenase, glutamic dehydrogenase, fumarase, aspartate-glutamate transaminase, alanine-glutamate transaminase, glutamine synthetase, etc.

(SISSAKIAN, 1958; LOSADA, TREBST and ARNON, 1959). BOVÉ and RAACKE (1959) have found amino acid-activating enzymes in chloroplasts, and STUMPF and JAMES (1962) have reported that chloroplasts can incorporate acetate into long-chain fatty acids.

In photosynthetic bacteria, unlike higher plants, the enzymic system for CO_2 assimilation is not bound to the bacterial chlorophyll pigment system which is contained in particles known as chromatophores. Chromatophores are structurally not as complex as chloroplasts and contain neither the enzymes for carbon assimilation nor the ability to store starch. In photosynthetic bacteria, the carbon-assimilating enzymes are found in the colorless supernatant fluid that remains after the chromatophores are removed from the broken cells (FULLER and ANDERSON, 1957; LOSADA, TREBST, OGATA and ARNON, 1960; FULLER, SMILLIE, SISLER and KORNBERG, 1961).

I. Chloroplasts.

Chloroplasts are organelles $1-8\ \mu$ in diameter which consist of a lamellar system, a matrix or stroma, and an enclosing chloroplast membrane. The lamellar system which traverses the chloroplast is a lipoprotein complex in which the photosynthetic pigments are localized. The stroma is the ground substance in which the lamellae are embedded.

Chloroplasts contain between 70 and 80% water. On a dry weight basis about 50% of the chloroplast is protein, and about 40% lipid. The chloroplast lipids contain only small amounts of triglycerides and are characterized by the occurrence of galactolipids, phosphatidylglycerol and the plant sulfolipid (FERRARI and BENSON, 1961). The enzymes and coenzymes involved in the light phase of photosynthesis, which will be discussed in detail later, are found in the lipoprotein complex, whereas the enzymes which catalyze the nonphotochemical reactions in chloroplasts (among them, the enzymes of CO_2 assimilation) are contained in the stroma (TREBST, TSUJIMOTO and ARNON, 1958; ARNON, 1961). The structure and chemistry of chloroplasts have been recently reviewed by MENKE (1962).

1. Isolation and Purification of Whole Chloroplasts.

(Method of ARNON, ALLEN and WHATLEY, 1956; WHATLEY and ARNON, 1962.)

Whole chloroplasts are most commonly isolated from spinach or Swiss chard leaves. The plants are usually grown in a greenhouse by a nutrient culture technique and the mature leaves are harvested prior to each experiment. However, active chloroplasts have been obtained from fresh spinach leaves purchased in local vegetable markets.

The leaves are washed with distilled water, shaken to remove excess water and placed in a plastic bag in the refrigerator for one to several hours to insure their turgidity. The midribs are removed from the fully turgid leaves and the leaf blades weighed, and then quickly sliced into pieces, about 0.5 cm square, to facilitate grinding.

50 g of sliced leaf blades are ground by hand in a precooled mortar (20 cm diameter) with an "isotonic" solution consisting of 100 ml 0.35 M NaCl (ca. 2%) and 10 ml 0.2 M Tris buffer pH 8, and about 50 g of precooled sand. In some preparations the 0.35 M NaCl is replaced, in this and in subsequent steps, by 0.5 M sucrose or 0.5 M glucose. Prolonged grinding (in excess of two minutes) should generally be avoided. The slurry is squeezed through a double layer of cheesecloth and the green juice centrifuged at 0° C for 1 min at $200 \times g$, to sediment sand, leaf debris and whole cells. The green supernatant fluid is decanted and

centrifuged for 7 min at $1000 \times g$. The sediment (P) will now contain most of the intact chloroplasts; the supernatant fluid is discarded.

The sedimented whole chloroplasts (P) are suspended in about 2 ml ice-cold 0.35 M NaCl, by gently stirring them with the aid of a piece of absorbent cotton at the end of a stirring rod. The suspension is diluted to about 50 ml with 0.35 M NaCl, centrifuged in the cold for 7 min at $1000 \times g$ and the pale supernatant liquid is discarded. The sediment containing the washed whole chloroplasts (P$_1$) is suspended, as above, in about 10 ml of 0.35 M NaCl.

2. Preparation of "Broken" Chloroplasts and Chloroplast Extracts.
(Method of Whatley, Allen, Rosenberg, Capindale and Arnon, 1956; Whatley and Arnon, 1962.)

If a preparation of disrupted or "broken" chloroplasts (P$_{1s}$) is desired, the P$_1$ residue is suspended in about 10 ml of a hypotonic NaCl solution, i.e. 0.035 M instead of 0.35 M NaCl. To prepare the "chloroplast extract" (CE) that contains the water soluble enzymes and cofactors present in chloroplasts, the suspension of P$_{1s}$ particles is centrifuged at $18,000 \times g$ for 10 min in the cold, and the supernatant used as CE. For further washing, the P$_{1s}$ particles are suspended in about 40 ml of 0.035 M NaCl and centri fugedfor 10 min in the cold at $18,000 \times g$; the residue is then resuspended, as above, in about 10 ml 0.035 M NaCl. In certain preparations, "broken" chloroplasts (P$_{1s}$) and chloroplast extract (CE) were prepared by suspending whole choloroplasts (P$_1$) not in 0.035 M NaCl but in cold distilled water.

Where greater stability of the chloroplasts is required and the presence of ascorbate does not interfere, all solutions used in the isolation, washing and disruption of the chloroplasts contain sodium ascorbate at a concentration of 0.01 M (200 mg/100 ml). The preparations made with ascorbate are designated C, C$_1$, C$_{1s1}$, instead of P, P$_1$, P$_{1s1}$, etc. (see Whatley, Allen and Arnon, 1959).

II. Chromatophores.

The pigment-lipoprotein complexes of the photosynthetic bacteria which carry out the light-activated enzymic reactions have been called chromatophores by Schachman, Pardee and Stanier (1952). They are spherical bodies, approximately 500 Å in diameter, which can be readily isolated by differential centrifugation from preparations of broken cells. According to Newton and Newton (1957) the *Chromatium* chromatophore has a particle weight of 25 million and contains, as major components, polysaccharide and an ethanolamine phospholipoprotein in association with bacteriochlorophyll, carotenoids and cytochromes in molar ratios of 10:5:1. There are about 5000 chromatophores per cell (Schachman et al., 1952) dispersed throughout the cytoplasm (Bergeron, 1959).

1. Isolation and Purification of Chromatium Chromatophores.

a) Culture of Bacteria.

Methods for growing *Chromatium* sp. strain D have been described among others, by Hendley (1955) and by Newton and Kamen (1956). The latter used the following medium: NaCl, 3.0 g, KH$_2$PO$_4$, 0.1 g; NH$_4$Cl, 0.1 g; MgCl$_2 \cdot 6$H$_2$O, 0.5 g; FeCl$_3 \cdot 6$H$_2$O, 0.005 g; Na$_2$S $\cdot 9$H$_2$O, 0.1 g; Na$_2$S$_2$O$_3 \cdot 5$H$_2$O, 0.2 g; NaHCO$_3$, 0.2 g; tap water, 100 ml; pH, 8.0. In some cultures 0.2% Na dl-malate was added to enhance growth. The organisms were grown anaerobically at room temperature in 4- and 10-liter Pyrex bottles, and were harvested when maximum growth was obtained. Illumination was provided by a bank of 100-watt bulbs.

b) Preparation and Fractionation of Cell Extracts.

(Method of NEWTON and NEWTON, 1957.)

Three techniques were used to disrupt the cells: grinding with glass beads by the procedure of LAMANNA and MALLETTE (1954), crushing a frozen cell mass with a HUGHES press, and sonic oscillation with a 10-kc Raytheon sonic oscillator. The sonic oscillator technique was the one used most extensively. All three methods gave similar results except that the preparations obtained with the *Hughes* press were more viscous owing to the less extensive degradation of the bacterial nucleic acids, released when the cells rupture. Various suspending media were employed, including phosphate and Tris buffers, buffered and nonionic sucrose solutions of various molarities, etc. Leakage of material from particles occurred only in media of very low ionic strength. Particles prepared in phosphate or Tris buffers (0.1 M) can be extensively washed in such media without loss of protein or pigments. No differences in the optical, enzymic or physical properties of chromatophores were noted, whether they were prepared in ionic media or in sucrose solutions.

In a typical experiment, 20—30 g of wet cell paste was suspended in 20—30 ml of cold 0.1 M Tris chloride buffer, and the suspension was treated in the sonic oscillator for 20 min at 0° C. All subsequent preparative procedures were carried out in the cold. The extract was centrifuged at $25,000 \times g$ for 5 min to remove intact cells, cell debris, and sulfur particles, and the preparation was then centrifuged at $25,000 \times g$ for 1 hr to sediment the bacterial chromatophores. Approximately three-fourths of the bacteriochlorophyll and carotenoids remained in the supernatant fluid, depending on the duration of sonic disintegration. The supernatant fluid was next centrifuged at $100,000 \times g$ for 90 min; this generally suffices to sediment all of the remaining particles which contain the chlorophyll and carotenoids. This pigmented fraction is referred to as the "small particle" fraction. The clear yellow supernatant fluid remaining after the $100,000 \times g$ centrifugation contains many of the soluble enzymes of the cell including some soluble cytochromes, and constitutes about one-third of the cell protein.

In addition to washing both particle fractions at least twice, followed by centrifugation, the chromatophores were further purified as follows. The particles were resuspended and homogenized in a POTTER-ELVEHJEM homogenizer using a Teflon pestle. The preparation was then centrifuged three successive times at $25,000 \times g$ for 3 min followed by decantation of the particles after each centrifugation. The preparation was then centrifuged at $25,000 \times g$ for 1 hr., and the sedimented particles were rehomogenized for use. The repeated short centrifugations were necessary to remove large cell fragments and free sulfur particles which will remain in the chromatophore preparations unless care is taken to remove them.

An alternative method of obtaining a high yield of chromatophores consisted of successive sonic disintegrations of a cell suspension for 5 min periods, followed by removal of unbroken cells by centrifugation, and their repeated disruption in the sonic oscillator. The sonication extracts were combined and fractionated as described above. Long exposure of chromatophores to sonic oscillation was undesirable.

The bacteriochlorophyll content was determined spectrophotometrically at 800 mμ. The millimolar extinction coefficient used for bacteriochlorophyll was 93.4.

2. Isolation and Purification of *Rhodospirillum rubrum* Chromatophores.
(Method of Frenkel, 1956.)

a) Culture of Bacteria.

R. rubrum, strain S-1, was grown in a medium of the following composition: dl-malic acid, 3.5 g; L-glutamic acid, 4 g; sodium citrate, 5-1/2 H_2O, 0.8 g; biotin, 5 μg; $MgSO_4 \cdot 7H_2O$, 0.2 g; $CaCl_2$, 38 mg; KH_2PO_4, 120 mg; K_2HPO_4, 180 mg; Difco yeast extract 250 mg; distilled water, 1 liter (Gest, Kamen and Bregoff, 1950). The liquid culture medium was adjusted to pH 7 with KOH, and micronutrient elements were added to give a final concentration (in parts per million) as follows: Mn, 0.5; Zn, 0.05; Cu, 0.02; Co, 0.02; Mo, 0.01. The organism was inoculated from agar stabs into the liquid culture medium, contained in 4-ounce flint glass bottles. The bottles, filled completely with sterile culture medium and closed with flamed Bakelite caps, were incubated at approximately 30° with good illumination (incandescent lamps) and shaken once a day to disperse the sedimented bacteria. When active gas production began, the cultures were transferred to 16- or 32-ounce flint glass bottles filled with sterile culture medium. These large cultures were grown in the light as indicated above. After 18 to 24 hours the cells were harvested, washed twice with glass-distilled water, and finally suspended in 0.2 M potassium glycylglycine buffer, pH 7.5 to 7.6, or in other appropriate buffers. Cells suspended in Tris buffer have been found to be less active.

b) Preparations of Cell-Free Extracts.

The cells suspended in icecold buffer were disrupted for 3—4 min in a 10 or 5 kc Raytheon sonic oscillator; the reaction chamber was cooled to a temperature slightly above 0° C and maintained below 5° C. Prolonged sonic treatment of the material resulted in a marked loss of photophosphorylating activity. The sonicate was centrifuged in the cold for 10 min at 10,000 × g and the supernatant fluid was freed from cell fragments, visible under the light microscope, by filtration through a fine sintered glass funnel. Alternatively, the material was centrifuged for 30 min to 1 hr at 25,000 × g. The supernatant solution obtained after this centrifugation step contained highly active photophosphorylating particles without exhibiting any cell fragments, visible under a light microscope. This supernatant fluid, called the "crude" preparation, when stored on ice, may retain its original activity for several days and sometimes weeks.

c) Preparation of Purified Chromatophores.

The crude preparation was centrifuged for 1 hr at 57,000 × g, to give a pellet which contained most of the bacteriochlorophyll. This pellet was resuspended in buffer and centrifuged as above. By resuspending the pellet from this centrifugation, a preparation of chromatophores was obtained which can carry out photophosphorylation, but whose activity can be greatly increased by certain additions (see below) which have no effect on the phosphorylating activity of the crude preparation.

B. Carbon Dioxide Assimilation in Photosynthesis.

As already stated, CO_2 assimilation in photosynthesis is a dark chemosynthetic process which depends on only two products formed by light reactions, reduced pyridine nucleotide and ATP. This concept of CO_2 assimilation, formulated on theoretical grounds by Ruben in 1943, has received experimental support in recent years mainly from the kinetic experiments of Calvin and his associates

(1956) who worked chiefly at the cellular level and from the work at the enzymatic level by HORECKER, OCHOA, RACKER and their associates (HURWITZ et al., 1956; WEISSBACH et al., 1956; JAKOBY et al., 1956; and RACKER, 1957).

The series of reactions that jointly bring about the conversion of CO_2 to carbohydrates may be described as a reductive carbohydrate cycle (cf. RACKER, 1957). The cycle is divided into three phases (Fig. 1). In the carboxylative phase, pentose monophosphate (Ru-5-P) is phosphorylated to RuDP, which then accepts a molecule of CO_2 and is cleaved to 2 molecules of PGA. In the reductive phase,

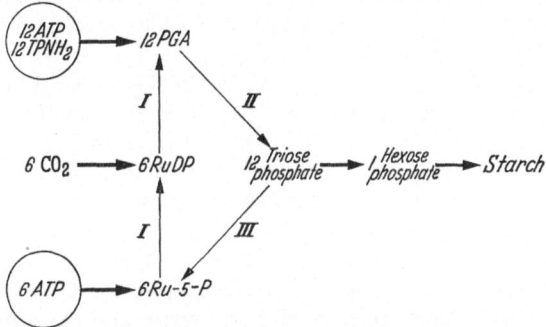

Fig. 1. Condensed diagram of the reductive carbohydrate cycle in chloroplasts. The cycle consists of three phases. In the carboxylative phase (I), ribulose-5-phosphate (Ru-5-P) is phosphorylated to ribulose diphosphate (RuDP) which then accepts a molecule of CO_2 and is cleaved to 2 molecules of phosphoglyceric acid (PGA); in the reductive phase (II) PGA is reduced to triose phosphate; in the regenerative phase (III) triose phosphate is partly converted into Ru-5-P and partly into hexose phosphate and starch. All the reactions of the cycle occur in the dark. The reactions of the carboxylative and reductive phases are driven by ATP and $TPNH_2$ formed in the light. One complete turn of the cycle results in the assimilation of 1 mole of CO_2 at the expense of 3 moles of ATP and 2 moles of $TPNH_2$.

PGA is reduced to triose phosphate. In the regenerative phase, triose phosphate is converted partly into hexose phosphate and thence into storage carbohydrate (starch), and partly into the pentose monophosphate needed for the carboxylative phase.

The carboxylative phase of the cycle involves the two distinctive enzymes of the reductive carbohydrate cycle: phosphoribulokinase and the carboxylation enzyme; they will be discussed in detail later. The other two phases of the cycle, the reductive and the regenerative, comprise the well-known reactions of glycolysis and of the pentose cycle, which are discussed elsewhere in this volume.

I. Separation of Light and Dark Phases in Photosynthesis.

TREBST, TSUJIMOTO and ARNON (1958) have, by fractionating chloroplasts, separated in time and space the light-dependent phases of photosynthesis from the dark phases which are concerned with carbon assimilation. The physical separation of the light and dark phases of CO_2 assimilation by chloroplasts was accomplished by completing the light phase first in a reaction mixture containing substrate amounts of ADP and inorganic phosphate together with TPN, but no added CO_2. This results in the evolution of oxygen and in the accumulation of substrate amounts of $TPNH_2$ and ATP.

The reaction mixture included, in a final volume of 2.5 ml, "broken" chloroplasts (see p. 572) containing 0.5 mg chlorophyll; chlorophyll-free chloroplast extract (see p. 572) equivalent to 2 mg chlorophyll; and the following in micromoles: Tris buffer, pH 7.5, 80; $MgCl_2$, 5; $MnCl_2$, 2; Na ascorbate, 10; G-1-P, 0.3; sodium phosphate, pH 7.5, 5; ADP, 2; and TPN, 2. The reaction was carried out at 20° C in a rectangular Warburg manometer vessel placed in a glass-bottomed constant temperature bath and flushed with argon gas before turning on the light. Illumination was from below by a bank of 300- and 150-watt reflector flood lamps, providing approximately 23,000 lux at the level of the reaction vessels.

After 30 min of illumination, the particlescontaining chlorophyll were removed by centrifugation, and the chlorophyll-free supernatant fluid was incubated in the *dark* for 30 min with 1.5 μmoles cysteine and 10 μmoles of sodiumbicar bonate containing $C^{14}O_2$. The reaction was stopped by adding 0.1 ml of glacial acetic acid. Total CO_2 fixation was measured by pipetting aliquots from the contents of the vessel, evaporating to dryness on stainless steel planchets and counting C^{14} with a thin window Geiger-Müller counter.

In the presence of "assimilatory power" (TPNH$_2$ plus ATP) the chloroplast extract was able to fix $C^{14}O_2$ in the dark. The dark fixation of $C^{14}O_2$ by the chlorophyll-free extract, supplemented with assimilatory power, was comparable with fixation in the light when $C^{14}O_2$ was supplied at the beginning of the illumination period to a reaction mixture containing the complete chloroplast system.

For the identification of the products of $C^{14}O_2$ fixation, the contents of the Warburg vessel were centrifuged and aliquots of the supernatant liquid were subjected to two-dimensional paper chromatography (on Whatman No. 41 paper) using as solvents (a) 80 parts phenol: 20 parts water and (b) a mixture of 52 parts n-butanol, 14 parts glacial acetic acid, and 34 parts water. The radioactivity in the individual compounds, located on the papers by radioautography, was determined by direct counting on the dried paper. The individual compounds were identified by elution, and co-chromatography with samples of authentic compounds. Sugar phosphates were further identified by dephosphorylation with phosphatase ('Polidase') and rechromatography with the corresponding authentic sugars.

The light and dark phases, when carried out separately one after another, yielded essentially the same final photosynthetic products as the continuously illuminated chloroplast system. These products included hexose and pentose monophosphates and diphosphates, phosphoglyceric acid, and dihydroxyacetone phosphate together with a little phosphoenolpyruvate and malate.

In the procedure described above the dark assimilation of CO_2 by chloroplast enzymes depended on the availability of "substrate" amounts of TPNH$_2$ and ATP that were formed from TPN, ADP and inorganic phosphate during the preceding light phase. Under continuous illumination, however, chloroplasts assimilate CO_2 to the level of carbohydrate in the presence of only catalytic amounts of TPNH$_2$ and ATP, which are repeatedly regenerated by the photochemical reactions as they are consumed in CO_2 assimilation.

Losada, Trebst, and Arnon (1960) and Trebst, Losada, and Arnon (1959, 1960) have shown that in this "catalytic" system noncyclic photophosphorylation alone does not provide sufficient ATP for a reductive assimilation of CO_2 to the level of carbohydrate. Additional ATP must be supplied by cyclic photophosphorylation (see p. 582). To maintain the balance between the two types of photophosphorylation traces of the catalysts of cyclic photophosphorylation, FMN or vitamin K_3, were added. The reaction mixture used in those experiments included, in a final volume of 2.5 ml, broken chloroplasts containing 0.5 mg of chlorophyll; chloroplast extract equivalent to 2 mg of chlorophyll, and the following in μmoles: Tris buffer, pH 7.5, 80; MnCl$_2$, 2; MgCl$_2$, 5; Na ascorbate, 10; sodium phosphate, pH 7.5, 5; TPN, 0.3; FMN, 0.001 (or vitamin K_3, 0.01); NaHC^{14}O$_3$, 10; ADP, 0.5; and 0.3 of one of the following "primers" (R-5-P, FDP, F-6-P, G-6-P or G-1-P).

II. Characteristic Enzymes of the Reductive Carbohydrate Cycle.

In this section we include the two enzymes that are involved in the carboxylative phase of the cycle, phosphoribulokinase and the carboxylation enzyme (cf. VISHNIAC, HORECKER and OCHOA, 1957). The characteristic occurence of the TPN-linked triosephosphate dehydrogenase in photosynthetic tissues (ARNON, 1952; GIBBS, 1952) and more specifically, in chloroplasts (LOSADA, TREBST, and ARNON, 1960) would also justify the inclusion of this enzyme in the present Section, but this enzyme is similar to the classical DPN-linked triose phosphate dehydrogenase of glycolysis.

1. Phosphoribulokinase.

The enzyme which has been purified from spinach leaves catalyzes the reaction:

$$Ru\text{-}5\text{-}P + ATP \longrightarrow RuDP + ADP$$

a) Assay Method.

In the procedure of HURWITZ, WEISSBACH, HORECKER and SMYRNIOTIS (1956) R-5-P is the substrate and an excess of phosphoriboisomerase is added to convert R-5-P to Ru-5-P. Under these conditions, the rate of reaction can be followed by the appearance of alkali-labile phosphate, since both phosphate groups of RuDP are alkali-labile, and R-5-P, ATP, and ADP do not yield inorganic phosphate under these conditions. The assay mixture contained 8 μmoles of R-5-P, 10 μmoles of ATP, 10 μmoles of $MgCl_2$, 2 μmoles of cysteine, 40 μmoles of triethanolamine buffer, pH 8.3, 50 units of phosphoriboisomerase, and a preparation of the enzyme to be assayed, in a total volume of 1 ml. The reaction was carried out at 38° for 10 min and stopped by heating in a boiling water bath for 1.5 min. An aliquot was transferred to 1 N NaOH, incubated at room temperature for 20 min and orthophosphate determined by the method of FISKE and SUBBAROW (1925). To correct for the formation of Ru-5-P, which yields about 50% of its phosphate as orthophosphate under these conditions, a control was run with phosphoriboisomerase, and the alkali-labile phosphate formed was subtracted from the test values.

An alternative assay involved the determination of ADP with phosphoenolpyruvate, phosphoenolpyruvic kinase and lactic dehydrogenase (KORNBERG and PRICER, 1951). In the presence of ADP, pyruvic acid was formed from phosphoenolpyruvate in this system and was determined spectrophotometrically with $DPNH_2$. No reaction occured in the absence of ADP. Controls were run in order to correct for ATPase activity, which, however, is relatively low in the spinach extracts. With either assay the reaction velocity was proportional to enzyme concentration.

JAKOBY, BRUMMOND and OCHOA (1956) have assayed the phosphoribulokinase manometrically by measuring the evolution of CO_2 in a bicarbonate buffer. CO_2 is evolved because 1 mole of acid is produced per mole of RuDP formed in the reaction catalyzed by the phosphoribulokinase. They have used R-5-P as the substrate, relying on the phosphoriboisomerase, present in their preparation, to convert R-5-P into Ru-5-P. The reaction mixture (in WARBURG flasks) included the following components (in μmoles) in a final volume of 1.0 ml; $KHCO_3$, 40; L-cysteine, 10; reduced glutathione, 4; $MgCl_2$, 10; ATP, 30; R-5-P, 30; and an amount of enzyme (in the side arm) which would bring about the evolution of 10 to 100 μliters of CO_2 in 15 min. The flasks were gassed with a mixture of 95% N_2 and 5% CO_2 and equilibrated at 30°. The pH was about 7.8.

b) Purification Procedure.

(Method of Hurwitz et al., 1956.)

In the course of purification of the enzyme, the procedure of Bücher (1947) was used for rapid protein determinations. However, to insure accurate estimation of protein concentration with highly colored protein solutions, the method of Sutherland, Cori, Haynes, and Olsen (1949) was also used.

Acetone powder. Market spinach leaves were washed with cold tap water and the stems discarded. An acetone powder was prepared by three extractions with acetone (previously cooled to $-20°$). The first extraction was carried out with 3 volumes of acetone in a Waring blender for 3 min, and the suspension was filtered rapidly on a large Buchner funnel. In subsequent extractions 1 volume of acetone was used. The light-green acetone powder was stored in a desiccator under vacuum.

Crude extract. Fifty g of acetone powder were extracted for 10 minutes with gentle stirring with 1800 ml of 2×10^{-3} M triethanolamine buffer, pH 8.5. The mixture was centrifuged and the supernatant solution decanted through glass wool ("Crude extract", 1340 ml).

Ammonium sulfate fractionation. To the crude extract was added 260 g of ammonium sulfate and the resultant precipitate was discarded. More ammonium sulfate (310 g) was then added to the supernatant solution (1420 ml) and the resultant precipitate was collected and dissolved in 150 ml of 2×10^{-3} M triethanolamine buffer, pH 8.3. The protein content was determined by the Bücher (1957) turbidimetric procedure. The protein solution was then diluted with water to give a protein content of 5 mg per ml ("Ammonium Sulfate I", 1224 ml). This solution was usually about 0.05 saturated with respect to ammonium sulfate. Solid ammonium sulfate (217 g) was then added to give 0.40 saturation. Following centrifugation, the supernatant solution (1270 ml) was brought to 0.55 saturation by addition of 113 g of $(NH_4)_2SO_4$. The precipitate was dissolved in 100 ml of 2×10^{-3} M triethanolamine buffer, pH 8.5 ("Ammonium Sulfate II", 113 ml).

Acetone fractionation. The protein content of the Ammonium Sulfate II preparation was adjusted, by adding water, to 14 mg per ml, followed by the addition of an equal volume of 0.1 M sodium acetate, pH 7.0. 210 ml of acetone, cooled to $-20°$, was then added slowly, while the solution was cooled in a $-8°$ bath. The heavy precipitate which formed was removed by centrifugation at $-10°$. The supernatant solution (468 ml) was treated with acetone (212 ml) and centrifuged as before. The precipitate was extracted with 125 ml of 0.1 M Tris buffer, pH 7.7, and the residue disarded ("Acetone", 136 ml).

Ammonium sulfate precipitation. To concentrate the enzyme, the acetone fraction was treated with 59.5 g of ammonium sulfate and the precipitate collected and dissolved in 10 ml of 0.1 M Tris buffer, pH 7.7 ("Ammonium Sulfate III", 14 ml).

The final phosphoribulokinase preparation was stable in the frozen state at $-10°$, but lost activity on repeated freezing and thawing.

c) Properties.

Substrate specificity. Phosphoribulokinase is specific for Ru-5-P and ATP. The Michaelis constants are 2.2×10^{-4} M for Ru-5-P and 2.8×10^{-4} for ATP (Hurwitz et al., 1956).

Stoichiometry of the reaction. In the presence of excess ATP, Ru-5-P is completely utilized, and equivalent quantities of ADP and RuDP are formed (Hurwitz et al., 1956; Jakoby et al., 1956).

Effect of pH. The phosphorylation reaction proceeds at maximal velocity at pH 7.9. Above or below this pH the rate of the reaction declines sharply (HUR-WITZ et al., 1956).

Activators and inhibitors. No phosphoribulokinase activity could be demonstrated with purified preparations in the absence of a divalent metal activator. Mg^{++}, which was the most effective ion tested, showed maximal activity at 5×10^{-3} M. Phosphoribulokinase appears to be a sulfhydryl enzyme. It was inhibited by heavy metals (Cu^{++} and Hg^{++}) and by p-chloromercuribenzoate, and the inhibition was completely reversed by cysteine (HURWITZ et al., 1956). TREBST, LOSADA and ARNON (1960) and CALO and GIBBS (1960) have shown that phosphoribulokinase is inhibited by iodoacetamide.

2. Carboxylation Enzyme.

This enzyme, also known as carboxydismutase, has been isolated and purified from spinach leaves. It catalyzes the formation of PGA from RuDP and CO_2 in accordance with the following equation:

$$RuDP + CO_2 + H_2O \longrightarrow 2\ PGA\ .$$

a) Assay Method.

In the procedure of WEISSBACH, HORECKER and HURWITZ (1956), the assay for the carboxylation enzyme, based on the rate of formation of PGA from RuDP and CO_2, was carried out in two steps. In the first step RuDP and CO_2 were incubated with the carboxylation enzyme in a reaction mixture which contained 25 μmoles of $NaHCO_3$, 5 μmoles of $MgCl_2$, 3 μmoles of glutathione, 0.03 μmoles of EDTA, and 0.2 μmoles of RuDP, in a total volume of 0.52 ml. The solution was equilibrated with 5% CO_2 in N_2, treated with the enzyme solution (appropriately diluted), and incubated at 25° for 10 min. The reaction was stopped by the addition of 0.03 ml of 1.0 N HCl, followed by heating in a boiling water bath for 1 min.

In the second step, an aliquot (usually 0.1 ml) was taken for the PGA assay and added to a solution containing 50 μmoles of Tris buffer, pH 7.7, 10 μmoles of $MgCl_2$, 0.07 μmole of $DPNH_2$, and 0.05 ml of a dialyzed rabbit muscle preparation, which provided the enzymes (phosphoglycerate mutase, enolase, phosphoenolpyruvate kinase and lactic dehydrogenase) that are required for the transformation of PGA into lactate in a total volume of 1.0 ml. The optical density was measured in quartz cuvettes with a light path of 1.0 cm, prior to adding 0.4 μmole of ADP (0.02 ml). Usually 15 to 20 min were required for the reaction to come to completion. No oxidation of $DPNH_2$ was observed until ADP was added. The quantity of $DPNH_2$ oxidized, calculated from the change in absorption at 340 mμ ($\varepsilon = 6.22 \times 10^3$), was equal to the amount of PGA present in the aliquot. Identical values for PGA were obtained when aliquots were assayed according to BÜCHER (1947), i. e. by converting PGA to 1,3-diphosphoglyceric acid with the aid of phosphoglyceryl kinase and ATP, followed by a reduction to glyceraldehyde-3-phosphate with $DPNH_2$ and glyceraldehyde-3-phosphate dehydrogenase. Protein was determined by the method of SUTHERLAND et al. (1949).

In the procedure of JAKOBY et al. (1956) the basis of the carboxylation enzyme assay was the fixation of $C^{14}O_2$.

b) Purification Procedure.
(Method of WEISSBACH et al., 1956.)

Crude extract. Fresh market spinach leaves were washed with cold tap water and the petioles discarded. 600 g of leaf material was homogenized at 0° C in a large Waring blender with 2000 ml of solution containing 10^{-4} M EDTA and 10^{-2} M K_2HPO_4, adjusted to pH 7.4. After 3 min at top blender speed, the homogenized suspension was filtered through Schleicher and Schuell No. 588 fluted filter paper. The opaque dark green filtrate was adjusted to pH 7 with 1 N ammonium hydroxide ("Crude extract", 2100 ml).

Ammonium sulfate fractionation. The crude extract was treated with 475 gm of ammonium sulfate and centrifuged. The pale green supernatant solution was treated with 210 g of ammonium sulfate, centrifuged, and the precipitate dissolved in 100 ml of 0.1 M phosphate buffer, pH 7.4, containing 5×10^{-5} M EDTA (Ammonium Sulfate I, 114 ml).

The Ammonium Sulfate I preparation from the previous step was diluted with an equal volume of cold water. At this point the degree of saturation with respect to ammonium sulfate was 0.07. The diluted solution (226 ml) was treated with 92.5 ml of a saturated ammonium sulfate solution (saturated at room temperature and adjusted to pH 7.3 with concentrated NH_3). The suspension was centrifuged, the supernatant solution treated with 26.4 ml of saturated ammonium sulfate solution, and the resulting precipitate was discarded. To the supernatant solution was added 30.3 ml of saturated ammonium sulfate solution, and the resulting precipitate collected by centrifugation and dissolved in 30 ml of 0.1 M phosphate buffer, pH 7.4 (Ammonium Sulfate II, 36 ml).

Heat treatment. 27 ml of the Ammonium Sulfate II preparation was heated to 60—63° in approximately 3 min, held for 5 min at this temperature, and rapidly cooled and centrifuged. The residue was washed with 5 ml of water and discarded. The supernatant solution and washings were then combined ("heated fraction", 28 ml).

Adsorption on Alumina Gel. The "heated fraction" was diluted with 140 ml of cold water and treated with 196 ml of aged aluminium hydroxide Cγ gel (12.7 mg per ml) (Willstäter and Kraut, 1923). The suspension was centrifuged and the enzyme eluted from the gel with 50 ml of 0.1 M phosphate buffer, pH 7.7 ("aluminium hydroxide eluate", 77 ml).

The carboxylation enzyme was purified by this procedure about 10-fold.

c) Purification Procedure.

(Method of Jakoby et al., 1956.)

Crude extract. Large, turgid spinach leaves, trimmed of large veins, were washed and ground in a vegetable juice extractor. The juice was filtered through two layers of cheese-cloth, and 10 ml of a neutralized solution of L-cysteine (15 mg per ml) was added per 100 ml of juice. All subsequent operations were performed at 0°. The pH of the juice was adjusted to 7.3 with 1.0 N KOH prior to centrifugation for 40 min at 16,000 $\times g$.

Ammonium sulfate precipitation. Solid ammonium sulfate was added to the supernatant fluid until 0.2 saturation was reached. The resultant precipitate was discarded and more ammonium sulfate added to give 0.4 saturation. The new precipitate was dissolved (to one-quarter of the original extract volume) in 0.05 M Tris buffer, pH 7.2, which was also 0.01 M with respect to cysteine.

Heat treatment. The solution was brought to pH 6.3 with 1.0 N acetic acid. Aliquots of 2 ml were quickly heated to 56° in stainless steel centrifuge cups and kept at this temperature for 3 min with stirring. The precipitated protein was removed by centrifugation and the pH of the supernatant solution was adjusted to 7.0.

Ethanol precipitation. Chilled $(-30°)$ 95% ethanol was slowly added with stirring to the supernatant solution from the previous step until a concentration of 22% (v/v) was reached. The precipitate was removed by centrifugation at $-5°$ and the ethanol concentration was increased to 30%. During the addition of ethanol the temperature gradually decreased to $-4°$ at 22% ethanol concentration, and to $-8°$ at 30% concentration. The precipitate obtained between 22 and 30% ethanol was dissolved in 0.02 M Tris buffer, pH 7.3 to a volume one-tenth that of the original spinach extract.

Calcium phosphate gel adsorption. The pH of the ethanol precipitate was brought to pH 6.5 with 1.0 N acetic acid, prior to adding 1.2 volumes of calcium phosphate gel (25 mg of calcium phosphate per ml) per 100 mg of protein (protein was determined spectrophotometrically by the method of WARBURG and CHRISTIAN, 1941—1942). After stirring for 10 min, the gel was collected by centrifugation and the supernatant solution discarded. The gel was washed once with distilled water and eluted with an amount of 0.1 Tris buffer, pH 7.3, equal to twice the volume of the added gel suspension. This eluate contained the carboxylation enzyme with a specific activity 20 times higher than that of the extract. It also contained phosphoriboisomerase, but was free from phosphoribulokinase.

d) Properties.

Homogeneity. The carboxylation enzyme purified by WEISSBACH et al. (1956) behaved as a homogeneous protein on ultracentrifugation and electrophoresis. The molecule carried a substantial net negative charge at pH 7.7. The sedimentation constant was 17 SVEDBERG units, measured in 0.1 M triethanolamine buffer, pH 6.9, at a protein concentration of 0.5%. The diffusion constant was 5.5×10^{-7} sq. cm/sec in 0.1 M Tris buffer, pH 6.9, at a protein concentration of 2%. From these data the molecular weight has been estimated to be about 300,000 (WEISSBACH et al., 1956).

Stability. According to WEISSBACH et al. (1956) the carboxylation enzyme is unstable below pH 6. The most purified preparations were only moderately stable and lost appreciable activity when stored for several days at $-16°$.

Substrate specificity. No substrates other than RuDP were utilized. The MICHAELIS constant was estimated to be 2.3×10^{-4} M. Half maximal velocity was obtained at a bicarbonate ion concentration of 1.1×10^{-2} M (WEISSBACH et al., 1956).

Stoichiometry. In the presence of excess CO_2, 1 mole of RuDP yielded 2 moles of PGA. Using PGA as the substrate, no evidence was obtained, either by $C^{14}O_2$ incorporation or by pentose formation, for the reversibility of the reaction. Because of the high affinity of the carboxylation enzyme for RuDP, the reaction is suitable for the quantitative determination of RuDP (WEISSBACH et al., 1956; JAKOBY et al., 1956).

Effect of pH. The carboxylation enzyme was most active at about pH 8; there was little or no activity below pH 6.4 or above 9.2 (WEISSBACH et al., 1956).

Effect of temperature. The rate of RuDP carboxylation increased approximately 2-fold for every 10° increase in temperature. The activation energy was 16,900 calories per mole (WEISSBACH et al., 1956).

Activators and inhibitors. The purified carboxylation enzyme (WEISSBACH et al., 1956) was completely inactive in the absence of added metal ions; Ni^{++} and Mg^{++} were equally effective activators. For full activity the enzyme required sulfhydryl compounds such as glutathione or cysteine. A similar activation was observed when EDTA was used in place of glutathione. According to WEISSBACH et al. (1956) the carboxylation enzyme is inhibited by relatively low concentrations of phosphate or arsenate ions. Thus in 0.03 M and 0.01 M phosphate the activity was only 10 and 30% of the control value, respectively. Arsenate at 2×10^{-4} M inhibited completely. On the other hand, Trebst et al. (1960) have found little inhibition by arsenate or phosphate of the carboxylative phase of CO_2 assimilation in a reconstituted chloroplast system. Gibbs and Calo (1959, 1960) have observed a difference in the effect of phosphate on CO_2 assimilation between intact chloroplasts and the reconstituted chloroplast system. Of special interest is the finding that cyanide, long known as an inhibitor of photosynthesis, is a strong, and so far the only known effective inhibitor of the carboxylation enzyme (Trebst et al., 1960; Losada and Arnon, 1962).

C. Photosynthetic Phosphorylation.

Photosynthetic phosphorylation (photophosphorylation) is a term introduced by Arnon, Allen and Whatley (1954) to describe a light-induced ATP formation which they discovered in isolated chloroplasts. Photosynthetic phosphorylation was distinct from oxidative phosphorylation since it occurred without the aid of mitochondria and without a net consumption of molecular oxygen or chemical substrate. The only "substrate" which was consumed was radiant energy. A similar reaction was found soon thereafter by Frenkel (1954) in subcellular preparations of photosynthetic bacteria. At first it seemed that photophosphorylation in bacterial particles, unlike in chloroplasts, was dependent on a chemical substrate, α-ketoglutarate. But in later experiments Frenkel (1956) ruled out the dependence on a chemical substrate and once this fundamental point was cleared up, the overall photophosphorylation reaction in both chloroplasts and chromatophores could be represented by the same equation:

$$ADP + P_i \xrightarrow{\text{light}} ATP$$

This reaction, in which the sole product is ATP, was subsequently renamed cyclic photophosphorylation to distinguish it from noncyclic photophosphorylation — a second photophosphorylation reaction, which Arnon, Whatley and Allen (1958) found in chloroplasts and which Nozaki, Tagawa and Arnon (1961) recently demonstrated in chromatophores of photosynthetic bacteria. A generalized equation for noncyclic photophosphorylation in chloroplasts is given below.

$$2\,PN + 2\,H_2O + 2\,ADP + P_i \xrightarrow{h\nu} 2\,PNH_2 + O_2 + 2\,ATP$$

A nonphysiological variant of this reaction is one in which pyridine nucleotide is replaced by ferricyanide (represented here by Fe^{3+}).

$$4\,Fe^{3+} + 2\,H_2O + 2\,ADP + 2\,P_i \xrightarrow{h\nu} 4\,Fe^{2+} + O_2 + 2\,ATP + 4\,H^+$$

The term photosynthetic phosphorylation is now used more broadly as a collective term to include both cyclic and noncyclic photophosphorylation. In cyclic photophosphorylation the absorbed radiant energy is used only for ATP formation; in noncyclic photophosphorylation radiant energy is also used for the formation of a strong reductant, PNH_2, and, in chloroplasts, for oxygen evolution as well. As already mentioned, Trebst, Losada and Arnon (1959) obtained evidence to support the view that both cyclic and noncyclic photophosphorylation are needed for the synthesis of carbohydrates from CO_2. The noncyclic process by itself does not supply enough ATP to run the reductive carbon cycle.

Photosynthetic phosphorylation was first discovered in a single species of green plants, but more recently Whatley, Allen, Trebst and Arnon (1960) have demonstrated both cyclic and noncyclic photophosphorylation in chloroplasts isolated from several other species of plants. The discovery of cyclic and noncyclic photophosphorylation by isolated chloroplasts was confirmed and extended in other laboratories (AVRON and JAGENDORF, 1957; AVRON, JAGENDORF and EVANS, 1957; JAGENDORF and AVRON, 1958; AVRON and JAGENDORF, 1959; JAGENDORF, 1959; WESSELS, 1957; WESSELS, 1958; CHOW and VENNESLAND, 1957; NAKAMOTO, KROGMANN and VENNESLAND, 1959; HILL and WALKER, 1959).

Photosynthetic phosphorylation has now been observed in every major class of photosynthetic organism. FRENKEL's (1954) finding of cyclic photophosphorylation in subcellular preparations of the nonsulfur photosynthetic bacterium, *R. rubrum*, was confirmed by GELLER (1957), and was followed by WILLIAMS' (1956) demonstration of the same process in the anaerobic photosynthetic sulfur bacteria, *Chromatium* and *Chlorobium*. Cyclic photophosphorylation by *Chromatium* particles was further investigated by NEWTON and KAMEN (1957) and ANDERSON and FULLER (1958). In algal preparations, cyclic photophosphorylation was demonstrated by THOMAS and HAANS (1955) and PETRACK and LIPMANN (1961).

I. Cyclic Photophosphorylation.

Work on the separation of the enzymic apparatus of photosynthetic phosphorylation into its individual components is now under way. Since this work is guided by theories as to the probable mechanism of this process, some of them will be briefly summarized here.

A mechanism for cyclic photophosphorylation must account for the unique features of this process that distinguish it from oxidative phosphorylation by mitochondria: ATP is formed without the consumption of either an external electron donor (substrate) or an electron acceptor (oxygen) and hence cannot be formed at the expense of free energy released during electron transport from substrate to oxygen. In the mechanism postulated by ARNON and his associates (ARNON, 1959, 1961; ARNON, LOSADA, NOZAKI and TAGAWA, 1961) the formation of the high-energy pyrophosphate bond of ATP in cyclic photophosphorylation occurs at the expense of free energy released during an endogenous electron transfer that is set in motion by the capture of radiant energy by the chlorophyll pigment system.

According to this theory, in the primary photochemical act a chlorophyll molecule within the chloroplast or chromatophore system becomes excited by an absorbed photon. This excited state leads to an "expulsion" of an electron that has been raised to a high energy level. The excited chlorophyll is thus the electron donor. On losing an electron, chlorophyll becomes also the electron acceptor $(Chl)^+$ to which the expelled electron returns with a resultant release of free energy. This "downhill" return of the electron proceeds in a stepwise manner, via a "closed circuit" of electron carriers, the terminal member of which is a cytochrome that adjoins the chlorophyll molecule. In the final step, the electron is transferred from reduced cytochrome to the adjoining "electron-deficient" chlorophyll molecule $(Chl)^+$, which is thereby restored to the ground state, and becomes ready to accept another photon to initiate the cyclic electron flow again.

This theory envisages at least one phosphorylation step in the cytochrome chain, coupled with electron transfer from a reduced cytochrome to $(Chl)^+$. The cytochrome chain includes those cytochromes of the b and c types that are known to occur in chloroplasts and chromatophores (see p. 602). The phosphorylation that accompanies the oxidation of the cytochromes in chloroplasts would be analogous to the phosphorylation that accompanies the oxidation of cytochromes in mitochondria by oxygen. Thus, chlorophyll with the aid of light is the ultimate oxidant in photophosphorylation and plays a part which corresponds to molecular oxygen in oxidative phosphorylation. Earlier members of the photosynthetic electron carrier system, those that precede the cytochromes, are probably quinones, flavins, or related compounds that are components of the chlorophyll-containing particles. Additional phosphorylations are envisaged as being coupled with the oxidation-reduction of these cofactors. A diagram illustrating this proposed mechanism is shown in Fig. 2.

The physiological cofactors (electron carriers) can be replaced by certain nonphysiological agents, such as phenazine methosulfate (PMS), which apparently

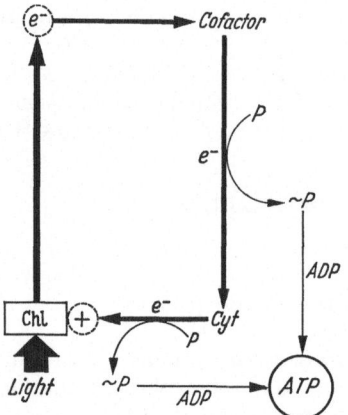

Fig. 2. Scheme for anaerobic cyclic photophosphorylation catalyzed by a cofactor, vitamin K_3 or FMN. Details in the text.

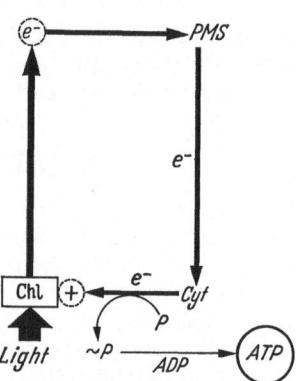

Fig. 3. Scheme for anaerobic cyclic photophosphorylation catalyzed by phenazine methosulfate (PMS). Details in the text.

act by providing an artificial "shortcut" that bypasses one or more phosphorylation sites. A diagram illustrating a shortened cyclic electron flow mechanism of this type is shown in Fig. 3.

The salient feature of the proposed cyclic mechanism (Fig. 2 and Fig. 3) is that electrons, activated by light energy, travel in a "closed circuit" from which they are not removed. The overall mechanism may be subdivided into three parts: (a) the light-induced generation of a high-energy electron and the ultimate electron acceptor $(Chl)^+$, (b) electron transport by the photosynthetic electron transport chain, and (c) phosphorylation reactions coupled to electron transport.

Parts (b) and (c) are analogous to, and at some points possibly identical with, their counterparts in oxidative phosphorylation; part (a) is peculiar to photosynthetic phosphorylation and is intimately bound with the light-induced changes in chlorophyll. The isolation of the enzymatic components involved inparts (b) and (c) and their relation to their counterparts in mitochondria is of special interest to the comparative biochemistry of ATP formation in living cells.

1. Procedure for Cyclic Photophosphorylation in Chloroplasts.

(Method of Arnon, Whatley and Allen, 1954; Whatley and Arnon, 1962.)

The reaction is conveniently carried out in conical manometer vessels of about 18 ml capacity. The main compartment of the vessels contained in μmoles: Tris/HCl buffer pH 8.3, 80; $MgCl_2$, 5 or 10; sodium ascorbate, 10; adenosine diphosphate, neutralized to pH 8, 10; a cofactor of photophosphorylation and water to give a final volume of 3 ml. The added cofactor of photophosphorylation may be *one* of these, in μmoles; phenazine methosulfate, 0.1; or flavin mononucleotide 0.1; or vitamin K_3 (in 0.1 ml methanol) 0.3. When FMN was used, 0.3 μmoles of TPN was also added. The sidearm contained 10 μmoles of K_2HPO_4 (1×10^5 to 5×10^5 c. p. m.).

The vessels were chilled in crushed ice and an aliquot of "broken" chloroplasts (the P_{1s} or C_{1s} preparation, p. 572) containing 0.2 mg chlorophyll was added to the main compartment. The chilled vessels were attached to WARBURG manometers, shaken at 15° C in a refrigerated bath and flushed with nitrogen gas for 3 to 5 min. The reaction was started by pouring the radioactive phosphate from the sidearm to the main compartment of the vessel and turning on the light. The reaction was terminated after 15 min (or other suitable interval) by turning off the light and adding 0.3 ml of 20% trichloracetic acid to each vessel. The acidified reaction mixture was centrifuged in the cold. To measure the "organic phosphate" formed, a 1 ml aliquot of the supernatant fluid was mixed with 1 ml magnesia mixture (see COLOWICK, S., Methods in Enzymology, v. III, p. 850); a drop of 0.2% phenolphthalein was added to check that the pH remained alkaline. It was found desirable to add a small amount of inorganic phosphate to the stock magnesia mixture to insure the presence of "seeds" of magnesium ammonium phosphate for starting the precipitation. The precipitate contains the unesterified inorganic phosphate. The mixture was allowed to stand for 1 hour at room temperature and then the precipitate was filtered off under suction, and washed twice with a 1:10 dilution of the magnesia mixture. The filtrate contained the radioactive ATP. Radioactive phosphorus was estimated in a 1 ml aliquot of the filtrate by evaporating to dryness, and counting under a thin window GEIGER counter. The ATP formed was computed by comparing the radioactivity in the filtrate with the total radioactive phosphate present in 0.1 ml aliquot of the original supernatant fluid, which was similarly evaporated but not precipitated with the magnesia mixture.

2. Procedure for Cyclic Photophosphorylation in Chromatophores.

In the experiments of FRENKEL (1956) with *R. rubrum*, the photophosphorylating activity was present in the "crude extract" (page 574) obtained after centrifuging the sonicated cells for 30 to 60 min at 25,000 $\times g$. A centrifugation of the crude extract for 1 hour at 57,000 $\times g$ yielded a pellet which contained most of the bacteriochlorophyll and which, on resuspension, gave photophosphorylation. This photophosphorylating activity of these particles was greatly increased by the addition of magnesium and catalytic amounts of a reducing agent such as succinate or DPNH. ADP, not replaceable by AMP, was used as the phosphate acceptor.

In the photophosphorylation experiments with *Chromatium*, NEWTON and KAMEN (1957) used extensively the washed "small particle" fraction (p. 573). Similar preparations were also used by ANDERSON and FULLER (1958) in their photophosphorylation experiments with *Chromatium* chromatophores. The photophosphorylating system required the presence of Mg^{++}, ADP and orthophosphate. Maximal activity depended on anaerobicity, the addition of catalytic amounts of a reductant or the addition of the supernatant solution remaining after the isolation of chromatophores (NEWTON and KAMEN, 1957; ANDERSON and FULLER, 1958; cf. also FRENKEL, 1956).

GELLER and GREGORY (1956) found that catalytic amounts of phenazine methosulfate increase photophosphorylation in extracts of *R. rubrum*. NEWTON and KAMEN (1957) found similar effects of phenazine methosulfate with *Chromatium* chromatophores.

The procedure for measuring cyclic photophosphorylation by chromatophores differs in some particulars from laboratory to laboratory but the following procedure of Frenkel (1956) may be taken as being representative. The experiments were carried out at 25° in Warburg vessels containing 1 to 3 ml of the reaction mixture. The vessels were illuminated at a saturating light intensity with incandescent lamps; the light intensity was adjusted with screens. The vessels were gassed with helium or 5% CO_2 in helium or 5% CO_2 in nitrogen. ATP formation was measured by one or more of the following: uptake of CO_2 in a bicarbonate-CO_2 system, uptake of orthophosphate, increase in acid-labile phosphate, or chromatographic separation and analysis of ATP from the deproteinized mixture. Orthophosphate was analyzed in aliquots removed from the experimental vessels and pipetted into an equal volume of 10% cold trichloracetic acid or into an equal volume of 8% perchloric acid. After deproteinization the supernatant solution was analyzed for phosphate by the method of Fiske and Subbarow (1925). For ATP analysis, the preparations were deproteinized with 8% perchloric acid, neutralized with normal KOH and freed from potassium perchlorate by filtering. The filtrate was then used for the determination of ATP by means of the hexokinase-catalyzed phosphorylation of glucose or by adsorption on Dowex 1 with subsequent elution, according to the method of Cohn and Carter (1950).

II. Noncyclic Photophosphorylation.

Noncyclic photophosphorylation differs from the cyclic type in that the formation of ATP accompanies the reduction of pyridine nucleotide and the oxidation of an external electron donor. In the mechanisms for noncyclic photophosphorylation proposed by Arnon and his associates [Arnon (1959), Losada, Whatley and Arnon (1961) and Arnon, Losada, Whatley, Tsujimoto, Hall and Horton

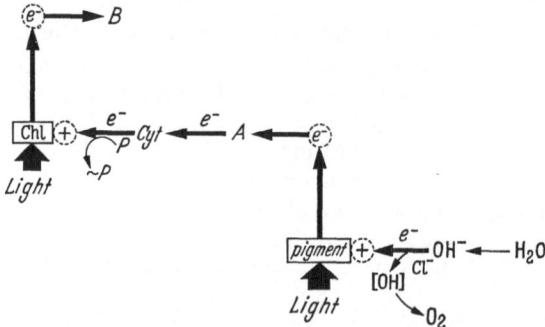

Fig. 4. Scheme for noncyclic photophosphorylation of the green plant type. In the first light reaction, the pigment molecule (chlorophyll b) becomes excited by the absorption of a quantum of light. The excited pigment donates its high-energy electron (e^-) to an intermediate electron acceptor (A) and accepts an electron from water (i.e. OH^- at 10^{-7} M). The oxidation product of this reaction is molecular oxygen. In the second light reaction, the chlorophyll a molecule (Chl), excited by the absorption of a quantum of light, donates its high-energy electron (e^-) to the electron acceptor (B) and accepts, via the cytochrome system (Cyt), an electron from the reduced intermediate (A^-) formed in the first light reaction. The phosphorylation step is linked with the transfer of the electron from cytochrome to chlorophyll. Under physiological conditions B is ferredoxin and the electron from reduced ferredoxin is used, via TPN, for CO_2 assimilation.

(1961)] the primary photochemical act remains the same as in cyclic photophosphorylation. Photon capture by chlorophyll results in the generation of a high-energy electron and of an electron acceptor $(Chl)^+$. However, the cyclic path of the electron is interrupted: the high-energy electron does not return to chlorophyll but is used for pyridine nucleotide reduction.

As the electrons from excited chlorophyll are transferred to pyridine nucleotide they must be replaced by an external electron donor. The "closed" circuit of cyclic photophosphorylation thus gives way in noncyclic photophosphorylation to an "open" circuit for the transport of electrons from an external electron donor to pyridine nucleotide. This noncyclic electron transport, also requires an input of light energy since the electron donor is at a potential less reducing than the electron acceptor. A diagrammatic representation of this concept is given in Fig. 4.

Noncyclic and cyclic photophosphorylation are thus envisaged as sharing the primary photochemical act, i. e. chlorophyll excitation by photon capture, and the related oxidation of cytochromes with an accompanying phosphorylation step.

Noncyclic photophosphorylation in green plants is distinguished from that in photosynthetic bacteria by its electron donor system (LOSADA, WHATLEY and ARNON, 1961; ARNON, LOSADA, NOZAKI and TAGAWA, 1961). In green plants, water, i. e. OH⁻ ions, supplies the electrons. However, photosynthetic bacteria cannot use hydroxyl ions but use inorganic or organic electron donors such as thiosulfate or succinate. The ability to use hydroxyl ions as electron donor seems to depend on an accessory

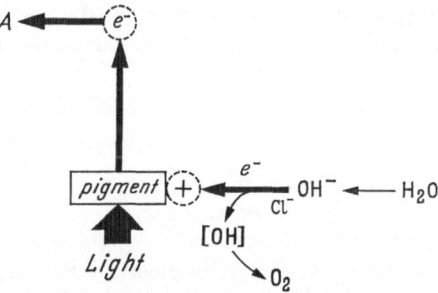

Fig. 5. Scheme for photooxidation of water by chloroplasts. The pigment molecule (possibly chlorophyll b) becomes excited by the absorption of a quantum of light. The excited pigment donates its high energy electron (e⁻) to an intermediate electron acceptor A and accepts a replacement electron from water (OH⁻). The oxidation product of OH⁻ is the precursor of oxygen gas (compare with the analogous anode reaction in the electrolysis of water).

pigment system (chlorophyll b or a phycobilin) that is present in higher plants and algae but is absent in photosynthetic bacteria. The photooxidation of hydroxyl ions by the accessory pigment, diagrammatically represented in Fig. 5, results in the evolution of oxygen (LOSADA, WHATLEY and ARNON, 1961; ARNON, LOSADA, WHATLEY, TSUJIMOTO, HALL and HORTON, 1961; cf. DUYSENS, AMESZ and KAMP, 1961).

1. Procedure for Noncyclic Photophosphorylation in Chloroplasts.

(Method of ARNON, WHATLEY and ALLEN, 1959; WHATLEY and ARNON, 1962.)

The procedure for determining ATP in cyclic photophosphorylation is also applicable to noncyclic photophosphorylation. Here oxygen evolution and $TPNH_2$ formation may be measured concurrently. When TPN was the electron acceptor, the reaction mixture contained, in micromoles: Tris/HCl buffer, pH 8, 80; $MgCl_2$, 5; adenosine diphosphate, neutralized to pH 8, 10; TPN, neutralized to pH 8, 4; $K_2HP^{32}O_4$, 10; also a solution of "photosynthetic phosphopyridine nucleotide reductase" (PPNR) and water to a final volume of 3 ml. PPNR (cf. page 592) was prepared by the method of HILL and BENDALL (1960), omitting the crystallization step. The $K_2HP^{32}O_4$ and TPN were usually placed in the sidearm of the chilled manometer vessel and the chloroplast preparation, containing 0.2 mg chlorophyll, was placed in the main compartment. The vessels were then attached to manometers and flushed with nitrogen gas. The reaction was started by pouring the contents of the sidearm into the main compartment of the vessel and immediately turning on the light. Oxygen evolution was measured manometrically (with KOH and a strip of filter paper in the center well of the vessel). ATP was determined as described for cyclic photophosphorylation (see page 585).

To measure the $TPNH_2$ formed, a 1 ml sample was withdrawn prior to adding the trichloroacetic acid. The sample was centrifuged at 0° C at $18,000 \times g$ for 10 min, an 0.2 ml aliquot of the supernatant solution was diluted to 3 ml with water; the optical density was measured at 340 mμ, against a blank similarly prepared but without TPN.

Experiments with noncyclic photophosphorylation may also be made with ferricyanide as the electron acceptor. In this case TPN and PPNR are omitted from the reaction mixture and instead, $8-15$ μmoles $K_3Fe(CN)_6$ is added from the sidearm. Since ascorbate, which is frequently present in the chloroplast preparation, reduces ferricyanide in the dark, enough ferricyanide must be added to leave a suitable excess for the photochemical reaction.

2. Procedure for Noncyclic Photophosphorylation in Chromatophores.
(Method of Nozaki, Tagawa and Arnon, 1961.)

Noncyclic photophosphorylation in chromatophores of R. rubrum was demonstrated under conditions when cyclic photophosphorylation was suppressed (by the addition of antimycin A), thereby making it possible to distinguish the ATP formed by a noncyclic electron flow mechanism from that formed by a cyclic mechanism. In chromatophores of R. rubrum, in which cyclic photophosphorylation was made inoperative, ATP formation was obligatorily coupled with a light-dependent electron flow from an external electron donor (ascorbate plus dichlorophenolindophenol) to DPN as the terminal electron acceptor. Stoichiometry (1 : 1) between the DPN reduced and the ATP formed was obtained only when the reduced DPN was trapped by an added lactic dehydrogenase system.

Two-day old cells of R. rubrum were centrifuged, the sedimented cells washed with 0.1 M Tris/HCl buffer, pH 7.8 (used also in subsequent operations) and recentrifuged. A paste of the washed cells was ground by hand, under argon gas, with an equal volume of levigated alumina, the slurry suspended in Tris buffer (gassed with argon), and centrifuged at $3,000 \times g$ for 15 min. The supernatant fluid was further centrifuged at $12,000 \times g$ for 15 min, the residue was discarded, and the supernatant solution (PS) was used for the isolation of chromatophores. All centrifugations were carried out in the cold and the centrifuge chamber was filled beforehand with argon gas.

The PS preparation was centrifuged in a Spinco Model L centrifuge, at $144,000 \times g$ for 1 hr. The supernatant fluid (S) was frozen for later use. The residue (P), which constituted the chromatophore fraction, was resuspended in Tris buffer (gassed with argon) and recentrifuged at $144,000 \times g$ for 1 hr. The residue, after suspension in Tris buffer, constituted the washed chromatophores (P_1).

The photochemical reactions were carried out in Thunberg-type cuvettes, which were made anaerobic, by five consecutive cycles of evacuation (to 5 mm Hg) followed by flushing with argon gas. The reaction mixture contained 0.04 mg bacteriochlorophyll, as washed chromatophores (P_1), and in μmoles: Tris buffer (pH 7.8), 100; $MgCl_2$, 5; $K_2HP^{32}O_4$, 5; ascorbate, 20; 2,6-dichlorophenol indophenol, 0.2; DPN, 0.3; and pyruvate, 10. 25 μg of crystalline lactic dehydrogenase (muscle) and 10 μg of antimycin A were also added.

The cuvettes were placed in an illuminated glass-walled water bath. DPN reduction was measured periodically (with a Cary recording spectrophotometer) as the difference in optical density at 340 mμ between an illuminated cuvette and a dark control. The reaction was stopped by adding 0.3 ml of 40% trichloracetic acid to each cuvette. The reaction mixture was centrifuged and 1 ml of the supernatant fluid was used for determining the ATP formed, and where applicable, 0.3 ml of the supernatant fluid was used for determining the lactate formed. Illumination was 10,000 lux.

Bacteriochlorophyll was measured at 810 mμ by diluting a suspension of chromatophores with distilled water, as described by ANDERSON and FULLER (1958).

D. Isolated Protein Constituents of the Photosynthetic Apparatus.

The main purpose of this Section is to discuss several major groups of protein constituents of chloroplasts and chromatophores that have been linked with one or more of the component enzymic reactions in the photoproduction of assimilatory power by the reactions of cyclic and noncyclic photophosphorylation. Despite the obligatory dependence of these reactions on some minimum degree of particulate structure and organization that must remain in photochemically active chloroplast and chromatophore preparations, progress is being made in the isolation of individual constituents of the photosynthetic apparatus.

Although it is necessary to limit this discussion to several major groups of constituents of the photosynthetic apparatus, work is currently under way in a number of laboratories on the identification and isolation of other individual chloroplast enzymes or protein constituents which are implicated, actually or potentially, in the photochemical reactions of chloroplasts. In this group belong the blue copper protein, plastocyanin, first isolated by KATOH (1960) from *Chlorella ellipsoidea* and found also in chloroplasts of higher plants but absent from photosynthetic bacteria (KATOH, 1960c; KATOH, SHIRATORI, SUGA, and TAKAMIYA, 1961), the "latent" polyphenol oxidase of spinach chloroplasts (TREBST and WAGNER, 1962), the adenosine triphosphate-adenosine diphosphate exchange enzyme (KAHN and JAGENDORF, 1962), and many others.

I. The TPN-Reducing System.

Especially rapid advances have recently been made in the elucidation of the TPN-reducing system in chloroplasts. This subject will be discussed in some detail since, as already stated, reduced TPN constitutes, jointly with ATP, the photochemically generated assimilatory power that drives the dark reactions of CO_2 assimilation.

Prior to 1951, there was no experimental evidence for, and there were theoretical arguments against (HILL, 1951; WESSELS and HAVINGA, 1952), the ability of chloroplasts to photoreduce a substance with a redox potential as electronegative as that of pyridine nucleotide, i.e. $E'_0 = -320$ mV. In 1951 VISHNIAC and OCHOA, TOLMACH, and ARNON independently showed that isolated illuminated chloroplasts were capable of reducing pyridine nucleotide when the photoreduction was coupled with a carboxylation reaction which immediately utilized the newly formed TPNH$_2$.

An accumulation of reduced pyridine nucleotide by illuminated chloroplasts without a coupled "pulling" reaction was first shown by SAN PIETRO and LANG (1956). In their experiments DPN was reduced in preference to TPN. The reaction occurred only in the presence of high concentrations of "grana", but they also noted that the reaction proceeded at low concentrations of grana if a soluble extract from chloroplasts was added.

The observations of San Pietro were confirmed and extended by Arnon, Whatley and Allen (1957) except that they found TPN instead of DPN to be the pyridine nucleotide that was preferentially reduced by the chloroplast system. In the experiments of Arnon et al. (1957), thoroughly washed chloroplasts lost the ability to photoreduce pyridine nucleotide but regained it on addition of a water-soluble chloroplast extract which contained a "TPN-reducing factor". This factor was sensitive to the sulfhydryl group inhibitor, p-chloromercuribenzoate and had the properties of a protein: it was nondialyzable, heat-sensitive and stable to acetone and alcohol precipitation. In the presence of the TPN-reducing factor, catalytic amounts of chlorophyll (contained in the chloroplast preparation) reduced substrate amounts of TPN. Moreover, the accumulation of reduced TPN was accompanied by a stoichiometric evolution of oxygen[1], according to the equation:

$$TPN + H_2O \xrightarrow{h\nu} TPNH_2 + \tfrac{1}{2} O_2$$

The TPN-reducing factor was not required for the Hill reaction, i.e. for the oxygen evolution that accompanies the photoreduction of such nonphysiological hydrogen acceptors as benzoquinone or ferricyanide.

In an extension of their earlier work, San Pietro and Lang (1958) isolated and greatly purified a water-soluble factor from spinach leaves and chloroplasts which they classified as an enzyme, and named photosynthetic pyridine nucleotide reductase (PPNR). PPNR was, in their view, the enzyme required for the photoreduction of pyridine nucleotides by grana. With respect to pyridine nucleotide specificity, PPNR was similar to the TPN-reducing factor of Arnon, Whatley and Allen (1957), i.e. it was more active toward TPN than DPN. It appeared therefore that PPNR and the TPN-reducing factor were synonymous, PPNR being the more purified preparation. PPNR contained neither a flavin nor a heme group (San Pietro, 1959).

An unexpected development was the recognition by Davenport (1960) of the identity of PPNR with a soluble protein from leaves that was first described by Davenport, Hill and Whatley in 1952 as the methemoglobin-reducing factor. This protein was subsequently greatly purified by Davenport and Hill (1960) and shown by them to promote the reduction of other haem-protein compounds, including cytochrome c and cytochrome b_3, but only when illuminated chloroplasts served as the hydrogen (electron) donor system. The protein had no typical cytochrome reductase activity since neither $TPNH_2$ nor $DPNH_2$ could serve as electron donors for cytochrome reduction in the dark. The purified protein contained no haemin or flavin groups. Its molecular weight was estimated to be 19,000. A method for crystallizing this protein from parsley leaves was described by Hill and Bendall (1960).

Prior to the work of Davenport (1960) there was no basis to suppose that the haem-protein-reducing factor might be linked to electron acceptors more electronegative than cytochromes. However, Davenport (1960) found that the purified haem-protein reducing factor from pea leaves was, on a unit protein basis, much more active in TPN reduction than even the purified PPNR preparations from spinach leaves (San Pietro and Lang, 1958; San Pietro, 1959). Moreover, the spinach preparation when further purified by electrophoresis on paper, gave a product that was indistinguishable from that of Davenport and Hill (1960). It thus became clear that despite differences in methods of purification, and in the functions originally described for them, the haem-protein reducing factor, like the

---

[1] In the supporting evidence for the stoichiometric evolution of oxygen (Table 2 in the article by Arnon, Whatley and Allen, 1957) a typographical error is responsible for the units of oxygen appearing incorrectly as "μmoles". This should read "μatoms".

TPN-reducing factor discussed previously, was also synonymous with photosynthetic pyridine nucleotide reductase.

The work of SAN PIETRO and his associates led to the general acceptance of photosynthetic pyridine nucleotide reductase (PPNR) as an enzyme with specificity toward TPN as the electron acceptor and toward illuminated grana as the electron donor system (no other electron donors were known). Specificity toward TPN depended on purity: the crude enzyme reduced either DPN or TPN, whereas the pure enzyme was specific for TPN (SAN PIETRO and LANG, 1958). Later, KEISTER, SAN PIETRO and STOLZENBACH (1960) isolated and purified from spinach leaves a second enzyme, a transhydrogenase, which catalyzed a hydrogen transfer from TPNH to DPN. It appeared, therefore, that a transhydrogenase contamination accounted for the activity of crude PPNR toward DPN.

From these findings, KEISTER, SAN PIETRO and STOLZENBACH (1960) formulated two distinct mechanisms for the photoreduction of TPN and of DPN by illuminated grana: 1) a primary photoreduction of TPN, catalyzed by one enzyme, PPNR, and 2) a secondary photoreduction of DPN, catalyzed by two enzymes, PPNR and transhydrogenase. In this formulation, PPNR was required for the transfer of hydrogen from illuminated grana to TPN, and the transhydrogenase for the transfer of hydrogen from TPNH to DPN.

Further work, however, by KEISTER, SAN PIETRO and STOLZENBACH (1962) did not sustain this distinction between the photoreduction of TPN and DPN. They had already observed earlier, and found again, that the addition of transhydrogenase to grana markedly increased the photoreduction of TPN by PPNR. But especially significant were their new findings that by preparing an antibody against the transhydrogenase enzyme and adding it to an illuminated grana system amply supplied with PPNR, the photoreduction of *either* TPN or DPN was completely inhibited. This inhibition was reversed by the addition of the purified transhydrogenase enzyme (KEISTER, SAN PIETRO and STOLZENBACH, 1962).

The results seemed to indicate that there was no difference between the enzyme systems required for photoreduction of TPN and DPN added to grana. The same two enzymes, PPNR and transhydrogenase were required in both cases — a conclusion that was in conflict with the original concept that PPNR is a chloroplast enzyme that specifically catalyzes the photoreduction of TPN and not DPN. To retain a role for PPNR as a TPN reductase, KEISTER, SAN PIETRO and STOLZENBACH (1962) now postulated the existence in grana of a bound TPN, the photoreduction of which occurs first and requires only PPNR. The photoreduction of free TPN or free DPN would then be the same secondary reaction involving in each case a transhydrogenase-catalyzed hydrogen transfer from bound TPNH$_2$ to free TPN or DPN.

A different view of the TPN-reducing system in chloroplasts which does not involve a postulated primary reduction of bound TPN by PPNR has recently come from the work of TAGAWA and ARNON (1962). They found that the choroplast enzyme which catalyzes TPN reduction is not PPNR or some other component of the water-soluble extract of chloroplasts from which PPNR and the "TPN-reducing factor" were originally isolated but an enzyme which usually remains in the water-insoluble chlorophyll-containing fraction of chloroplasts (grana). This reductase reduces free TPN but not DPN. "PPNR" participates in TPN reduction not as an independent enzyme but as a water-soluble hydrogen carrier, which functions in collaboration with the grana-bound TPN reductase. The collaboration of "PPNR" and the TPN-reductase in hydrogen transport to TPN by illuminated chloroplasts is analogous in some respects to the collaboration of cytochrome c and cytochrome oxidase in hydrogen transport to oxygen by mitochondria.

Evidence for this concept of the TPN-reducing system in chloroplasts was obtained when Tagawa and Arnon (1962) found, unexpectedly, that an illuminated, washed chloroplast preparation (grana) freed of PPNR, reduced TPN without added PPNR but with added ferredoxin. Ferredoxin is a water-soluble, non-heme, non-flavin, iron protein which Mortenson, Valentine and Carnahan (1962) isolated from *Clostridium pasteurianum*, and which, in that organism, serves as a link between hydrogenase and various electron donors and acceptors. Tagawa and Arnon (1962), who crystallized *Clostridium* ferredoxin and determined its redox potential, found that in the presence of ferredoxin, spinach grana reduced TPN not only in the light but also in the dark when hydrogen gas and an added bacterial hydrogenase were used as the hydrogen donor system. Thus, the grana-bound TPN reductase was not obligatorily coupled either with PPNR or with a photochemical reaction by the chloroplast pigment system. Bacterial ferredoxin was capable of replacing chloroplast PPNR, and the hydrogenase-H_2 system was capable of replacing radiant energy in what was hitherto regarded as a key photochemical reduction in photosynthesis.

On further investigation, the TPN reductase proper was found to be contained in the flavoprotein fraction of the grana. The flavoprotein fraction alone, when separated by an acetone treatment from the chlorophyll pigment system of the grana, also catalyzed a reduction of TPN in the dark, with H_2 plus hydrogenase as the hydrogen donor system (Tagawa and Arnon, 1962). In more recent experiments, the flavoprotein TPN-reductase of chloroplasts was prepared in crystalline form and shown to be specific for the reduction of free TPN; it did not reduce DPN (Shin, Tagawa and Arnon, 1963).

1. Chloroplast Ferredoxin.

The effectiveness of bacterial ferredoxin as a hydrogen (or electron) carrier for TPN-reduction in chloroplasts led to a search for a ferredoxin endogenous to chloroplasts. Tagawa and Arnon (1962) isolated from spinach chloroplasts a non-heme, iron protein which was similar to *Clostridium* ferredoxin in undergoing reversible oxidation-reduction accompanied by spectral changes. In the reduced form, spinach ferredoxin had an absorption peak at 267 mμ, a shoulder at 312 mμ and a broad shoulder between 450 and 470 mμ. In the oxidized form, absorption peaks were observed at 463, 420, 325 and 274 mμ. Spinach ferredoxin was reduced completely by sodium dithionite but since dithionite itself absorbs in the ultraviolet region, the absorption of ferredoxin in this region could not be measured by this method. Tagawa and Arnon (1962) measured the reduced spectrum of ferredoxin by incubating it for 15 min under hydrogen gas in the presence of *Clostridium* hydrogenase. Strict anaerobicity was important because reduced ferredoxin is readily oxidized by O_2. However, reduced ferredoxin has a strong affinity for TPN, and will reduce it even in the presence of oxygen.

Spinach ferredoxin turned out to be the same substance as PPNR. In its oxidized form, the absorption spectrum of spinach ferredoxin was the same as that found by San Pietro (1959) for PPNR, and by Davenport (1960) for the haem-protein reducing factor. Preliminary analyses gave an iron content for ferredoxin of 0.815%, but later determinations showed 0.89% iron (Tagawa and Arnon, 1962; also unpublished data). Katoh and Takamiya (1962) have recently reported an iron content of 1.2% for PPNR, prepared from spinach leaves according to the method of Hill and Bendall (1960). Fry and San Pietro (1962) found that PPNR contains, in addition to iron, "labile sulfur" which is liberated as hydrogen sulfide upon acidification. The molar ratio of iron to labile sulfur varied from 0.93

to 1.6 with different preparations. The labile sulfur did not come from the cysteine groups in the protein. The presence of iron and the liberation of H_2S from ferredoxin at pH 2, was also established recently in WARBURG's laboratory (WARBURG, 1962). Other properties reported for ferredoxin and PPNR, such as the reddish color in the oxidized form, are also very similar. The molecular weight of 14,000 assigned on the basis of preliminary experiments to spinach ferredoxin by TAGAWA and ARNON compares with 17,000 for PPNR (APPELLA and SAN PIETRO, 1961) and 19,000 for the methaemoglobin factor (DAVENPORT and HILL, 1960). Preliminary analyses showed a ratio of ferredoxin to chlorophyll of about 1 to 400. The same ratio of ferredoxin to chlorophyll was obtained whether the determination was made on whole leaf material or isolated whole chloroplasts, showing that ferredoxin is localized in chloroplasts (TAGAWA and ARNON, 1962).

With the recognition of the identity of spinach ferredoxin with PPNR (and, consequently, with the haem-protein reducing factor) as a water-soluble electron carrier, distinct from the enzyme TPN reductase proper, it was no longer necessary to postulate that a bound TPN is reduced prior to the reduction of free or added TPN. To recapitulate, the TPN-reducing system in chloroplasts may now be described as consisting of two components: (a) a flavoprotein enzyme, the TPN reductase proper, localized in the grana, which reacts with (free) TPN but not DPN and (b) chloroplast ferredoxin, a water-soluble, non-heme, iron protein electron carrier, one of a family of ferredoxins, which does not, by itself, reduce TPN but which serves as a hydrogen carrier for TPN reduction by the flavoprotein reductase. Ferredoxin transfers hydrogens from one of two hydrogen donor systems: chloroplast fragments (grana) in the light (the physiological system), or hydrogen gas plus hydrogenase in the dark (an experimental system). Spinach ferredoxin, like bacterial ferredoxin, was found by TAGAWA and ARNON (1962) to have a redox potential ($E'_0 = - 432$ mV at pH 7.55) close to that of the hydrogen electrode, and is thus capable of transferring to appropriate enzyme systems the most reducing electrons in cellular metabolism: those that come from hydrogen gas or from chlorophyll "activated" by light.

Recent work has elucidated further the physiological role of ferredoxin in chloroplasts. WHATLEY, TAGAWA and ARNON (1963) have shown that reduced ferredoxin is now the earliest, chemically isolated, reductant formed at the expense of radiant energy trapped during photosynthesis. The transfer of electrons from ferredoxin to TPN has been isolated as a dark reaction catalyzed by the chloroplast flavoprotein TPN reductase. Furthermore, TAGAWA, TSUJIMOTO and ARNON (1963) found that ferredoxin seems to serve as a branching point in the electron transport systems that result either in cyclic or in noncyclic photophosphorylation. When the photoreduced ferredoxin is reoxidized by TPN (with the aid of TPN reductase) noncyclic photophosphorylation results; when oxidized TPN is unavailable as an electron acceptor, the photoreduced ferredoxin is reoxidized by a bound component of the grana (possibly cytochrome b_6) and cyclic photophosphorylation results.

Although reduced ferredoxin is non-enzymically oxidized by oxygen (TAGAWA and ARNON, 1962), an appreciable leakage of electrons to O_2 (oxygen is always present around the chloroplasts *in vivo*) is prevented by the strong affinity of reduced ferredoxin a) for the TPN-reducing system and b) for the grana-bound electron carriers of cyclic photophosphorylation (TAGAWA, TSUJIMOTO and ARNON, 1963).

a) Isolation and Purification.

Since chloroplast ferredoxin, PPNR and the haem-protein reducing factor are, in fact, different names for what is now recognized as the same substance, the

different experimental procedures that have been used for the isolation and purification of these factors before their identity was recognized, may be used now,
in various combinations of steps, for the preparation of what will henceforth be
called chloroplast ferredoxin. Several of these procedures, particularly those which
have been widely used in investigations of the TPN-reducing system of chloroplasts,
will be described here.

Preparation of the crude homogenate. (Method of San Pietro and Lang,
1958.) 100 g of market spinach leaves were freed from veins and ground in 130 ml
of cold distilled water in the Waring blender for 5 min at 100 volts. The dark
green homogenate was filtered through a double layer of cheesecloth and glass
wool and the residue discarded. To the dark green filtrate was added sufficient
0.5 M Tris/HCl buffer, pH 8.0, to give a final buffer concentration of 0.05 M.

Acetone treatment. (Method of San Pietro and Lang, 1958.) Acetone,
previously cooled in the deep freeze, was slowly added to the filtrate with mechanical stirring, to give a final acetone concentration of 35%. The preparation was
centrifuged at 1000 × g for 15 min and the clear yellow-green supernatant fluid,
which contained most of the activity, decanted and saved. The dark green residue
was discarded. The active material was precipitated from the supernatant solution
by the slow addition of cold acetone to a final concentration of 75%. During the
addition of acetone, the solution was stirred mechanically. The precipitate was
flocculent and settled rapidly when stirring was discontinued. After several
minutes, the greater part of the supernatant fluid was decanted, and the precipitate
collected by centrifugation for 5 min at 1000 × g. The resulting clear yellow-green
supernatant fluid was discarded and the light-brown colored residue was thoroughly
suspended in 10 ml of ice-cold 0.005 M Tris, pH 8. The suspension was centrifuged
at 18,000 × g for 20 min and the residue discarded. The clear brown supernatant
solution was dialyzed overnight against 0.005 M Tris pH 8, in the cold.

Protamine sulfate treatment. (Method of San Pietro and Lang, 1958.)
Ferredoxin was precipitated by the addition of 1% protamine sulfate, pH 6, at a
ratio of approximately 6 mg of protamine sulfate to 100 mg of protein. The
solution was centrifuged for 15 min at about 15,000 × g and the brown supernatant
fluid discarded. The residue was thoroughly extracted with 0.5 M Tris, pH 8, and
the resulting suspension centrifuged as above. The recovery of ferredoxin by these
steps varied between 40 and 50% and the increase in purity varied between 20-
and 35 fold.

Dowex-bentonite treatment. (Method of San Pietro, 1959.) A suspension of
bentonite and Dowex-1 formate was centrifuged, and the supernatant solution was
decanted and discarded. The residue of bentonite and Dowex-1 formate was
suspended in a solution of ferredoxin from the protamine sulfate step and the
resulting suspension centrifuged. The residue was discarded, and the enzyme was
precipitated from the supernatant solution by the addition of 3 volumes of acetone.
The mixture was centrifuged, and the supernatant solution discarded. The residue
was thoroughly extracted, the mixture was centrifuged, and the resultant supernatant solution (the Dowex-bentonite concentrate) contained the ferredoxin. The
overall recovery after this step was 23% and the increase in specific activity
was about 170 fold.

b) Crystallization of Parsley Ferredoxin.
(Method of Hill and Bendall, 1960.)

The procedure of San Pietro and Lang (1958), up to the preparation of an
extract from the acetone precipitation, was used on 4 kg of frozen parsley leaves
(Petroselinum sativum). After removing suspended matter, the clear fluid (pH 8

in 0.005 M Tris buffer) was chromatographed on a column of Whatman anion exchanger powder DE 50, 3.3 cm diam × 6.5 cm. This gave a narrow brown band at the top of the column, while the yellow colored solution passed through. The column was washed with Tris buffer (0.05 M, pH 8), followed by 0.2 M sodium acetate buffer, pH 5.5, until the washings were colorless and free from protein. The brown protein was then completely eluted with M acetate buffer, pH 5.5, giving 100 ml of colored solution. This was dialyzed against 0.005 M Tris buffer, pH 8, at 0° C with stirring for 2 hrs. and the 180 ml solution passed through a second DE 50 column, 1 cm × 3 cm. The column was washed with 0.05 M Tris buffer, pH 8, followed in turn by 0.2 M, 0.5 M and 0.8 M acetate buffers, all at pH 5.5. The brown protein then occupied nearly all the column. Elution was carried out as before with M acetate buffer pH 5.5, giving 13 ml of dark red-brown solution. This was then dialyzed against half-saturated ammonium sulfate solution buffered with 100 ml 0.33 M phosphate buffer at pH 6.7 (2 liters total volume) overnight at 0° C. The more concentrated solution, now 5.5 ml, was clarified in a centrifuge. The clear fluid was then treated with 0.15 ml 5% v/v acetic acid, when a faint turbidity appeared. After standing at 0° C for 10 min a precipitate was formed, consisting of minute spherical "hedgehog" crystal aggregates which later could be spun down in the centrifuge to a volume of about 0.5 ml. The crystal pellet was dark pinkish-brown and the liquid nearly colorless.

c) Crystallization of Spinach Ferredoxin.

(Method of TAGAWA and ARNON, 1962 and unpublished data.)

Spinach ferredoxin was purified and, more recently crystallized, by a procedure similar to that described for *Clostridium* ferredoxin (TAGAWA and ARNON, 1962). The procedure included the following steps: a) extraction of an acetone precipitate of a leaf homogenate (b) adsorption of crude ferredoxin on a DEAE-cellulose bed and elution with NaCl-Tris buffer; (c) chromatography on a DEAE-cellulose column; (d) concentration of the main effluent to give a concentration of ferredoxin not less than 1%, and crystallization with ammonium sulfate.

A crude leaf homogenate of spinach leaves was prepared essentially in the same way as described by SAN PIETRO and LANG (1958) except that the pH was adjusted with Tris buffer to 7.3—7.7. Precipitation with acetone, extraction of the precipitate and the dialysis step were also similar to those of SAN PIETRO and LANG (1958) except that the pH in these as in subsequent purification steps was kept at 7.3.

To the dialyzed extract (pH 7.3), free from acetone, solid NaCl was added to bring the Cl⁻ concentration up to 0.2 M. The extract was then passed through a DEAE-cellulose bed which was equilibrated beforehand with a mixture of 0.15 M Tris/HCl buffer, pH 7.3 and 0.08 M NaCl ([Cl⁻] = 0.2 M). A DEAE-cellulose bed 3 cm in diameter and 2 cm high was used per kg of spinach leaves. Ferredoxin, which was readily recognized by its reddish color, accumulated just below the dark brown impurities that were adsorbed at the top of the bed. The passed solution, yellowish brown in color, contained flavoproteins including the TPN reductase. The charged DEAE-cellulose bed was successively washed with chloride solutions of increasing chloride concentration (0.2 M, 0.23 M and 0.26 M Cl⁻) until the reddish ferredoxin band descended to within 0.5 cm from the bottom of the bed. The chloride solutions were made of 0.15 M Tris/HCl buffer, pH 7.3, to which suitable amounts of NaCl were added.

Ferredoxin was eluted from the bed with a mixture of 0.55 M NaCl and 0.30 M Tris/HCl buffer pH 7.3 ([Cl⁻] = 0.8 M). About 5 ml of eluate was obtained per kilogram of spinach leaves. The eluate was diluted 2.5 times with distilled water and then introduced to a DEAE-cellulose column which was equilibrated before-hand and later developed with a chloride-buffer mixture of 0.18 M NaCl and 0.15 M Tris/HCl, pH 7.3, ([Cl⁻] = 0.30 M). A column 50 cm high and 4 cm in diameter was used for 5 kg of spinach leaves; for 10 kg spinach leaves, the column was 70 cm high and 5 cm in diameter. The reddish ferredoxin band, which could easily be seen, descended to the bottom of the column on passing through the column a volume of Tris-chloride buffer, that was 4 to 5 times greater than the volume of the column.

After recovery from the column, the ferredoxin effluent (ca. 1.5 times the column volume) was diluted 4 times with distilled water and then adsorbed on a DEAE-cellulose bed 2 cm high and 3 cm in diameter which was equilibrated beforehand with 0.1 M Tris/HCl buffer ([Cl⁻] = 0.08 M). Ferredoxin was eluted from the bed with M Tris/HCl buffer ([Cl⁻] = 0.8 M). Ferredoxin concentration in the effluent was not less than 1% at this step.

0.6 g of solid ammonium sulfate was added to 1 ml of the concentrated ferredoxin solution. The resulting precipitate was centrifuged off and solid ammonium sulfate was added to the clear supernatant solution until the solution became slightly turbid. Crystals appeared after the solution was left standing a few days in the refrigerator.

2. Ferredoxin-TPN Reductase.

As already stated, ferredoxin-TPN-reductase is a flavoprotein chloroplast enzyme which catalyzes the transfer of the hydrogens (electrons) from reduced ferredoxin to TPN. Shin, Tagawa and Arnon (1963) prepared the reductase in crystalline form and found that it is specific for TPN; DPN was not reduced. Interestingly enough, the ferredoxin-TPN-reaction was reversible; oxidized ferredoxin could also serve as a hydrogen (electron) acceptor in the oxidation of TPNH₂ by the TPN reductase.

Aside from its main physiological function as a ferredoxin-TPN-reductase, the enzyme was also found by Shin, Tagawa and Arnon (1963) to have other activities which were derived from its ability to catalyze the reoxidation of TPNH₂. (a) The enzyme was found capable of transferring hydrogens from TPNH₂ to a number of hydrogen or electron acceptors such as menadione, FMN, FAD, ferricyanide and indophenol dyes. Here again the TPN reductase was specific for TPNH₂ as the hydrogen donor; DPNH₂ was not used. (b) TPN-reductase was found to have transhydrogenase activity; it catalyzed a transfer of hydrogens from TPNH₂ to DPN but not from DPNH₂ to TPN. (c) Although TPN-reductase did not reduce cytochrome c directly, it acted as a TPNH₂-cytochrome reductase in the presence of one of several electron acceptors mentioned above (such as menadione, FMN, FAD or ferredoxin). In all these cases the role of the TPN-reductase was to transfer hydrogens (electrons) from TPNH₂ to the respective electron acceptor which then reduced cytochrome c non-enzymatically.

These additional activities of the TPN-reductase indicate that the enzyme is the same substance as the chloroplast TPNH₂-diaphorase isolated and purified by Avron and Jagendorf (1956, 1957) and the chloroplast transhydrogenase isolated and purified by Keister, San Pietro and Stolzenbach (1960, 1962). The latter investigators have already discussed the similarities and the possible identity of transhydrogenase and TPNH₂ diaphorase. Katoh (1961) isolated and purified a similar flavoprotein from a cell-free extract of *Chlorella ellipsoidea* and named it "TPNH-plastocyanin reductase" to denote the electron donor and acceptor which were preferentially utilized.

In their investigation of TPNH-diaphorase, AVRON and JAGENDORF (1956) emphasized the point that "any function for the diaphorase would appear to be dependent on the production of the substrate, $TPNH_2$ in the photosynthetic process." Likewise, KEISTER, SAN PIETRO and STOLZENBACH (1960, 1962) have regarded $TPNH_2$ (free or bound) as the substrate for the transhydrogenase. The work on chloroplast ferredoxin and the ferredoxin-TPN reductase (TAGAWA and ARNON, 1962; SHIN, TAGAWA and ARNON, 1963) has shown that the capacity of the enzyme to reoxidize $TPNH_2$ has obscured what is perhaps its main physiological role in chloroplasts, namely, the reduction of TPN, using reduced ferredoxin as the hydrogen (electron) donor.

The elucidation of the respective roles of TPN reductase and ferredoxin also helps to explain some of the conflicting observations on the heme-protein reducing properties of what is now called chloroplast ferredoxin. DAVENPORT and HILL (1960) observed that chloroplast ferredoxin reduced heme proteins only when illuminated chloroplasts served as the hydrogen donor system. This is understandable because illuminated chloroplasts provide a system capable of reducing ferredoxin; the subsequent reduction of heme-protein by reduced ferredoxin is non-enzymic. However, they also observed that in the absence of chloroplasts, $TPNH_2$ was unable to serve as a hydrogen donor for the reduction of heme-proteins by ferredoxin (DAVENPORT and HILL, 1960). This can now be explained by the lack in that system of a TPN reductase which is required for the reduction of ferredoxin by $TPNH_2$. When TPN reductase was supplied, as was the case in the experiments of LAZZARINI and SAN PIETRO (1962), then $TPNH_2$ was an effective hydrogen donor for the reduction of cytochrome c by ferredoxin, in the absence of chloroplasts or light. (In the terminology used by LAZZARINI and SAN PIETRO ferredoxin was called photosynthetic pyridine nucleotide reductase and TPN reductase was called transhydrogenase).

Since TPN reductase is the same substance as TPNH diaphorase and transhydrogenase, it can be isolated and purified by the methods that have been described for the isolation of these enzymes before their similarities were recognized (AVRON and JAGENDORF, 1956, 1957; KEISTER, SAN PIETRO and STOLZENBACH, 1960). The purification methods of AVRON and JAGENDORF (1956, 1957) and the crystallization procedure of SHIN, TAGAWA and ARNON (1963) will be described here.

a) Isolation and Purification.
(Method of AVRON and JAGENDORF 1956, 1957.)

Isolation of chloroplasts. Washed spinach leaves (petioles removed) were ground with buffer (0.4 M sucrose, 0.05 M phosphate, 0.01 M KCl, pH 7.5), in the ratio of 50 g leaves to 150 ml buffer, in an Omnimixer for 45 seconds at 45 V. The resulting homogenate was strained through a pad of cheesecloth and glass wool, and centrifuged at $1000 \times g$ for 10 min. The chloroplasts thus obtained were washed three additional times with the same buffer as was used in homogenizing.

Extraction. The pellet obtained from the third wash was resuspended in 0.05 M Tris/HCl buffer, pH 8.0, and allowed to stand at room temperature for 1 hr. The resulting suspension was centrifuged at 40,000 r. p. m. (Spinco No. 40 rotor) for 1 hr. The supernatant was collected and used as the crude enzyme preparation.

Acetone precipitation and re-extraction. Three volumes of acetone (kept at $-10°$ C) were added slowly with constant stirring to 1 vol. of the crude enzyme preparation. The resulting suspension was centrifuged at $5,000 \times g$ for 5 min, and the pellet was resuspended in 0.005 M Tris buffer, pH 8.0. The suspension thus obtained was centrifuged at $17,000 \times g$ for 20 min, and the pellet was discarded.

Adsorption on calcium phosphate gel. Tricalcium phosphate gel suspension was added to the supernatant solution at a ratio of approximately 14 mg gel to 300 g of the original leaves. The suspension was allowed to stand for 15 min with occasional stirring.

Elution with pyrophosphate buffer. The resulting suspension was centrifuged at 5,000 × g for 10 min, and the supernatant solution was discarded. The pellet was resuspended in 0.1 M sodium pyrophosphate solution and allowed to stand for 15 min with occasional mixing. The resulting suspension was centrifuged at 17,000 × g for 20 min, and the pellet was discarded.

Passage through an ion-exchange column. Five ml at a time of the pyrophosphate enzyme solution was passed through a 2 × 10 cm. Dowex 50 (2% cross linkage) column (the column was previously washed with acid, base, and distilled water) followed by 0.05 M Tris buffer, pH 8.0. The enzyme was usually obtained in the 7−14 ml fraction (collecting fractions began as soon as the addition of the enzyme had started). The column procedure was carried out at room temperature.

This procedure gave a 75- to 150 fold purification over the original leaf homogenate.

Ammonium sulfate fractionation. Ammonium sulfate fractionations resulted in loss of diaphorase activity. However, full activity was retained when the enzyme was precipitated by ammonium sulfate from a solution containing 0.05−0.1 M pyrophosphate buffer at pH 9.0, and when the resultant precipitates were re-dissolved in the same buffer. Using this procedure, a further purification of $TPNH_2$ diaphorase was accomplished. Solid ammonium sulfate was added to the enzyme solution eluted from the Dowex column. The diaphorase precipitated between 50% and 62% saturation (the most active fraction being from 56 to 62%).

Total purification with the overall procedure was between 200- to 250 fold, depending on the activity of the original homogenate. All operations, except where otherwise specified, were carried out at 0−5° C.

b) Assay Method.
(Avron and Jagendorf, 1956.)

The activity of the enzyme was measured by a spectrophotometric technique using 2,3,6-trichlorophenolindophenol dye or methylene blue as the electron acceptor. The cuvette contained 1.5 ml of 0.1 M Tris/HCl buffer at pH 7.3−7.5, trichlorophenol dye to give an optical density of 1.300−1.500 at 620 mμ, $TPNH_2$ to give an optical density of 0.250−0.350 at 340 mμ, enough enzyme to give an optical density change of 0.100−0.300/min at 620 mμ, and distilled water to a total volume of 3.0 ml. The reaction was started by the addition of the enzyme or $TPNH_2$, and the reduction of the dye was followed at 620 mμ using a Beckman model B spectrophotometer. When methylene blue was substituted for the indo-phenol dye, 0.2 μmole of methylene blue was added, and the oxidation of $TPNH_2$ was followed at 340 mμ. The non-enzymic reduction of the dye under these conditions was very small and was substractted from the enzymic rate.

c) Properties.
(Method of Avron and Jagendorf, 1956, 1957.)

Specificity. The enzyme was found to be highly specific for $TPNH_2$. Oxidant specificity, on the other hand, was rather low. A variety of dyes and naturally occurring compounds (trichlorophenol-indophenol, ferricyanide, methylene blue, benzoquinone, menadione, flavin mononucleotide) were effective as electron acceptors. However, oxygen, oxidized glutathione, nitrate, dehydroascorbate and cytochrome c were ineffective.

Affinity. The affinity of the enzyme for $TPNH_2$ was rather high, being completely saturated at about 1.2×10^{-5} M. The MICHAELIS constants for $TPNH_2$ and the trichlorindophenol dye were found to be approximately 6×10^{-6} M and 5×10^{-5} M, respectively.

pH optimum. There was a rather sharp pH optimum for the reduction of the indophenol dye at pH 9.0.

Inhibitors. Among the potent inhibitors of the enzyme were: mercuric chloride, p-chloromercuribenzoate and iodoacetate. The enzyme was not inhibited by cyanide, azide, EDTA, o-phenanthroline and (p-chlorophenyl)-dimethylurea.

FAD content. Flavin adenine dinucleotide analysis of the enzyme by the method of DE LUCA, WEBER and KAPLAN (1956) revealed its presence even in the most purified preparations. The FAD content was proportional to the enzyme activity. The FAD component has not yet been successfully split off from, and restored to, the protein. Attempts to remove the FAD from the enzyme resulted in a complete loss of activity which could not be restored by the addition of exogenous FAD.

Absorption spectrum and FAD function. The purified enzyme showed a typical flavin absorption spectrum, with peaks at 275, 380 and 460 mμ, as well as a shoulder at 475 mμ. On adding the substrate, $TPNH_2$, the peak at 380 mμ could no longer be measured because of the masking effect of $TPNH_2$ itself, but other spectral changes were observable. The absorption peak at 460 mμ was completely eliminated, leaving only a shoulder at 475 mμ. By bubbling air through the solution, the $TPNH_2$ present was gradually oxidized by oxygen and the absorption at 460 mμ reappeared.

Physical constants. On ultracentrifugation the purified enzyme showed a single symmetrical peak. Its molecular weight was estimated from the ultracentrifuge data to be approximately 35,000. This agreed quite well with the value calculated from the FAD content (assuming 1 FAD molecule/molecule enzyme). The isoelectric point, estimated from electrophoretic mobilities, was approximately between pH 6.2 and 6.7. The turnover number was 2,800 moles of dye reduced/mole protein/min.

d) Crystallization Procedure.
(Method of SHIN, TAGAWA, and ARNON, 1963.)

The starting material was the passed solution from the first DEAE-cellulose adsorption step for ferredoxin, as described on page 595. A crude flavoprotein fraction was obtained at 40 to 65% saturation with ammonium sulfate. This fraction was passed through a Sephadex column equilibrated with 0.03 M Tris buffer, pH 7.3, and then adsorbed on a DEAE-cellulose bed equilibrated with the same buffer. Additional chromatography steps were carried out first on Sephadex and then on DEAE-cellulose columns equilibrated with 0.08 M Tris buffer, pH 7.3. The enzyme protein was collected and concentrated by adsorption on the DEAE-cellulose column and precipitated with ammonium sulfate at 50—60% saturation. The resulting precipitate was dissolved in a minimum amount of water. By adding a small amount of ammonium sulfate and allowing the solution to stand in a refrigerator, crystals were formed in about a week.

II. Cytochromes in Leaves and Algae.

Cytochromes, the universal catalysts of respiration in plants and animals, are intracellular pigments each of which consists of a protein that is attached to an iron porphyrin (heme) prosthetic group. The iron in the heme undergoes reversible valence changes from the ferrous to the ferric state.

The occurrence of special cytochromes in photosynthetic organs was first demonstrated by Hill (1943) and Hill and Scarisbrick (1951) when they isolated from leaves of higher plants a cytochrome of the c type, which they named cytochrome f. In 1952, Davenport isolated "chloroplasts" from barley seedlings grown in the dark and observed in them the spectrum of cytochrome f. This was a direct demonstration that the cytochrome f which was isolated from green leaves was not an artifact but was a characteristic component of the photosynthetic apparatus.

In addition to cytochrome f, chloroplasts from barley seedlings were also found to contain a cytochrome of the b type, which Hill named cytochrome b_6 (Hill, 1954; Davenport, 1952). It appeared that chloroplasts contain cytochromes f and b_6 and no others — a conclusion which has been confirmed and extended by James and Leech (1958, 1959) to chloroplasts of two such dissimilar plants as the seaweed *Caulerpa prolifera* and the angiosperm *Vicia faba*.

The wide occurrence of cytochromes in photosynthetic cells was confirmed by their identification in algae. Hill, Northcote and Davenport (1953) found in cell-free *Chlorella* preparations, extracted with acetone, cytochrome components resembling those of acetone-extracted chloroplasts or "chloroplasts" from etiolated barley seedlings. (Acetone-extracted or etiolated preparations are used to obviate the interference of the strong chlorophyll absorption bands with measurements of absorption spectra of cytochromes. Unlike mitochondrial cytochromes, chloroplast cytochromes have been reported not to be denatured by cold 80% acetone). Lundegardh (1954) observed that on illuminating *Chlorella* cells, cytochrome f becomes oxidized. Nishimura (1959) noted the formation of a cytochrome, similar to cytochrome f, that paralleled the development of chlorophyll in dark-grown *Euglena gracilis* cells. Katoh (1959) found a cytochrome, similar in absorption spectra and redox potential to cytochrome f, in a number of algal species belonging to *Rhodophyceae*, *Phaeophyceae*, *Chlorophyceae* and *Cyanophyceae*.

The molar ratio of cytochrome f to chlorophyll has been determined for leaves of elder, parsley (Davenport and Hill, 1952), wheat (Lundegardh, 1954) and for *Chlorella* (Hill et. al., 1953) and found to be about 1:400.

1. Extraction and Purification.

a) Preparation of Cytochrome f from Parsley.
(Method of Davenport and Hill, 1952.)

See page 221, Volume IV, of Modern Methods of Plant Analysis.

b) Crystallization of *Porphyra tenera-cytochrome 553*.
(Method of Katoh, 1960 b.)

Fresh thalli of the red alga *Porphyra tenara* were rapidly frozen at $-40°$ and stored at $-20°$. The frozen thalli were thawed by warming in pure water at $30°$ (1 liter per kg thalli), and allowed to stand at room temperature $(15-20°)$ for 48 hrs. This treatment caused the cytochrome to diffuse into the surrounding medium. All the subsequent procedures of purification were carried out at room temperature.

The crude extract obtained in this manner was strained through cloth and 450 g of solid ammonium sulfate per liter was added with stirring. The resultant precipitate of cytochrome was collected by centrifugation and dissolved in water to which solid ammonium sulfate (400 g per liter) was added with stirring. The resultant precipitate was collected by centrifugation and discarded. Ammonium sulfate (300 g per liter) was added to the combined supernatant solution and the mixture was left standing for 4 hrs. or longer. The new precipitate was collected by centrifugation and dissolved in 200 ml of pure water, with the addition, in several portions, of 25 ml of 1% rivanol solution.

After removal of the precipitate by centrifugation, 175 g of solid ammonium sulfate was added to the supernatant. The new sediment was collected, dissolved in 50 ml of pure water and dialyzed against pure water for 24 hours. To remove the rivanol remaining in the dialysate, a thick suspension of a cation exchanger, 'XE 64' (H^+ form), was added until the yellow color of the detergent was no longer observed in the supernatant solution on centrifugation. At this stage of purification, most of the phycobilins and viscous substances in the original extract had been removed, and a clear pink solution of the cytochrome was obtained. This solution, when treated with ascorbic acid to reduce the cytochrome, gave a 'purity index' of 0.1, the purity index being defined as the ratio of extinction at 553 mμ to that at 270 mμ (E 553/E 270).

Further purification was carried out by using an anion adsorbent, diethyl-aminoethyl-cellulose (DEAE-cellulose). The cytochrome solution, after being dialyzed against 0.025 M phosphate buffer (pH 7.0) for 24 hrs., was passed through a column of DEAE-cellulose, buffered at pH 7.0. After washing the charged column with 0.025 M phosphate buffer (pH 7.0), the cytochrome was eluted with 0.1 M phosphate buffer (pH 7.0). The effluent (purity index = 0.55) was again fractionated with ammonium sulfate, between 0.6−1.0 saturation. The fraction containing cytochrome (purity index of 0.70) was dialyzed against 0.025 M phosphate buffer (pH 7.0) and passed through a column of DEAE-cellulose, equilibrated with 0.025 M phosphate buffer (pH 7.0). The charged column was thoroughly washed with 0.025 M phosphate buffer (pH 7.0). By increasing the concentration of the washing solution to 0.05 M, a small amount of an orange-colored pigment was removed from the deep red zone of cytochrome at the top of the column. The cytochrome which, when eluted with 0.1 M phosphate buffer (pH 7.0) showed a purity index of 1.0, was saturated with ammonium sulfate and the resultant precipitate was collected by centrifugation. The sediment was dissolved in a minimum quantity of 0.5 M phosphate buffer (pH 7.0) and fine crystals of ammonium sulfate were added until the solution became slightly turbid. On standing 24 hrs. or longer, small needle-like crystals of the cytochrome began to appear. The crystals were collected by centrifugation and dissolved in a minimum quantity of 0.5 M phosphate buffer (pH 7.0). Recrystallization was carried out by the same procedure as described above. The yield of crystalline cytochrome was about 200 mg from 15 kg of frozen thalli.

2. Properties.

The most interesting property of cytochrome f which distinguishes it from other cytochromes of the c type is its high oxidizing potential. Cytochrome f from leaves has an $E_0' = 0.365$ V at pH 7 (DAVENPORT and HILL, 1952). The algal cytochrome f from *Porphyra tenera* has an $E_0' = 0.355$ V at pH 7 (KATOH, 1960a). No cytochrome f with an E_0' less than 0.300 V has been reported.

Cytochrome f from leaves is an acidic protein (isoelectric point, pH 4.7) with a molecular weight of approximately 110,000 and two heme prosthetic groups. Ferrocytochrome f does not combine with oxygen or carbon monoxide. Its spectrum resembles that of ferrocytochrome c but the bands are sharper and displaced toward the red; the α band is asymmetric and at low temperature ($-190°$ C) is resolved into two peaks (α_1 and α_2 at 552 and 548 mμ) as compared with an α peak at 554.5 mμ at 25° C. The respective β peaks (in mμ) are 524 at 25° C and 529 and 524 at $-190°$ C; the γ absorption peaks are 422 at 25° C and 420.8 at $-190°$ C (Hill and Bonner, 1961; cf. James and Leech, 1959).

As already mentioned, the cytochromes f from algae, investigated by Katoh (1959), resemble in their spectral characteristics and redox potential the cytochrome f from leaves. Crystalline cytochrome f from *Porphyra tenera* (Katoh, 1960b) is also an acidic protein (isoelectric point, pH 3.5) but it has only one prosthetic heme group and is a smaller molecule (molecular weight about 13,000) than its counterpart in leaves. In these respects algal cytochrome f resembles the classical cytochrome c. The molar extinction coefficients for the *Porphyra tenera* ferrocytochrome at the α, β, and γ (Soret) bands (553, 521 and 417 mμ) are 2.17, 1.41 and 14.7 ($\times 10^4$), respectively.

Cytochrome f, from either algae or leaves, is relatively stable at room temperature and is not autooxidizable within a wide range of pH. Katoh found that the redox potential of the *Porphyra* cytochrome 553 decreases from 0.355 V at pH 5.0 to 0.270 at pH 11.0. This is not the case with cytochrome f (Davenport and Hill, 1952) and *Euglena*-cytochrome 552.5 (Nishimura, 1959) which have a dissociable group with a pK of about 8.3—8.4.

Much less is known about cytochrome b_6, the second cytochrome which is characteristically associated with the chlorophyll pigments. Unlike cytochrome f, cytochrome b_6 has not been isolated and solubilized but it is known to be present in leaves in larger amounts than cytochrome f. Cytochrome b_6 is highly autooxidizable and does not combine with carbon monoxide. Its redox potential at pH 7 (Hill, 1954 and Hill and Bonner, 1961) was estimated to be -0.06 V in chloroplasts from etiolated barley leaves, and -0.07 V in chloroplasts from the pale yellow mesophyll tissues of cabbage hearts. At 25° C the α, β and γ absorption bands of ferrocytochrome b_6 were found to be (in mμ) 563, 534 and 429, respectively; at $-190°$ C they were 557, 530 and 427 respectively.

In isolated chloroplasts and in leaf homogenate, cytochrome b_6 is present in the oxidized form and cytochrome f in the reduced form (Hill, 1954; Lundegardh, 1954; James and Leech, 1959). However, Hill (1954) has observed a well-defined band at 563 mμ in the leaves of the "golden" varieties of certain plants, and has suggested that in the living cell cytochrome b_6 may be present in a nearly completely reduced state even though oxygen is being produced. It would appear, therefore, that on illumination, cytochrome f is oxidized and cytochrome b_6 is reduced (Lundegardh, 1954; James and Leech, 1959; cf. Hill, 1960).

Of special interest from the standpoint of the comparative biochemistry and physiology of chloroplasts and mitochondria is the absence of cytochrome oxidase in chloroplasts (Davenport and Hill, 1952; James and Leech, 1959; Lundegardh, 1961). This enzymic deficiency of chloroplasts indicates that they cannot respire and hence cannot form ATP by oxidative phosphorylation, and is in harmony with other lines of evidence which support the theory that ATP formation in chloroplasts is independent of respiration.

III. Bacterial Cytochromes.

The major heme components of photosynthetic bacteria are cytochromes of the c type. Our knowledge of their occurrence goes back to 1953, when Vernon purified a cytochrome of the c type from the photosynthetic bacterium *Rhodospirillum rubrum*. This soluble cytochrome component was called "cytochrome c_2"

(ELSDEN, KAMEN and VERNON, 1953), and was found later in other photosynthetic bacteria such as *Rhodopseudomonas spheroides*, *Rps. capsulatus*, *Rps. palustris*, *Chlorobium limicola* and *Chromatium* (VERNON and KAMEN, 1954; KAMEN and VERNON, 1955; NEWTON and KAMEN, 1955). *R. rubrum* cytochrome c_2 has been crystallized and its properties described by HORIO and KAMEN (1961). Cytochrome c_2 from *Chromatium* has been purified by NEWTON and KAMEN (1956) and BARTSCH and KAMEN (1960). MORITA (1960) has reported a purification and crystallization procedure for cytochrome-552 from *Rps. palustris* and has described some of its physico-chemical properties.

In addition to the cytochrome of the c type, photosynthetic bacteria (*R. rubrum*, *Rps. spheroides* and *Chromatium*) contain comparable quantities of a remarkable new type of hematin compound, named first "pseudohemoglobin" because of its characteristic absorption spectra, oxidation potential and auto-oxidizability (VERNON and KAMEN, 1954; KAMEN and VERNON, 1955). This heme protein has been purified and crystallized from *Rhodospirillum rubrum* and renamed "*Rhodospirillum* heme protein" (RPH) by BARTSCH and KAMEN (1958) and HORIO and KAMEN (1961) — a name which is also used for this substance even when it is found in organisms other than *Rhodospirillum*. *Chromatium* RHP has been highly purified and its physical properties investigated by BARTSCH and KAMEN (1960).

VERNON and KAMEN (1954) have also reported the presence of an auto-oxidizable b-type hematin compound in the acetone powders of several photosynthetic bacteria (*R. rubrum* and *Rps. spheroides*). The inability to solubilize these b-type cytochrome components which are firmly bound to the chromatophores has prevented further elucidation of their properties.

1. RHP and Cytochrome c_2 from *R. rubrum*.

a) Preparation.

Detailed procedures based on adsorption on aluminium oxide, for the isolation, purification and crystallization of these components have been published by HORIO and KAMEN (1961). 350 g of lyophilized cells of *R. rubrum* yielded 145 mg of thrice recrystallized cytochrome c_2 and 201 mg of RHP. The purity of the preparations was checked spectroscopically during, and at the end of, the purification procedure.

b) Properties.

(KAMEN and VERNON, 1955; BARTSCH and KAMEN, 1958; HORIO and KAMEN, 1961.)

R. rubrum RHP and cytochrome c_2 are soluble acidic proteins with isoelectric points at pH 4.3 and 6.4, respectively. Their molecular weights, estimated by ultracentrifuge analysis, were found to be about 26,000 and 13,000, respectively.

The normal redox potential of RHP, estimated at pH 7 by electrometric titration with the ferric-ferrous oxalate system, was -0.008 V. The normal redox potential of cytochrome c_2, measured at pH 7.0 with ferro-and ferri-cyanide mixtures, was 0.338 V. A marked lowering of the redox potential occured with an increase in pH, being 0.380 V at pH 5.0, and 0.304 V at pH 9.0.

RHP contains two haem prosthetic groups per molecule. It may be viewed as an autooxidizable variant of cytochrome c, in which one (or both) of the extra-planar protein links to the central iron atom are loose or missing. This structure would account for its myoglobin-like spectra. RHP is a CO-binding heme protein which is incapable of forming stable addition complexes with cyanide, azide, hydrogen sulfide, oxygen etc.

The spectrum of oxidized RHP, at pH 7, shows absorption maxima at 275, 282, 390, 497 and 638 mμ, with molar extinction values of 44.5, 43.7, 159, 21.5 and 5.9 ($\times 10^3$), respectively, and a shoulder at 291 mμ. The spectrum of the reduced RHP, at pH 7, has a broad α-band with an absorption maximum at 550 mμ, no β-band, and a sharp Soret band at 423 mμ. Their respective molar extinction values are 22 and 175 ($\times 10^3$). [Although Horio and Kamen (1961) give the value 10^6 it is assumed that 10^3 was intended.] From pH 6 to 12, the spectrum of RHP, in both oxidized and reduced forms, changes gradually from the characteristic appearance of an acid hematin compound to that typical of hemi- or hemo-chromogen, respectively.

The molar extinction coefficients at the characteristic wavelengths of the α, β and Soret bands (550, 521, and 415 mμ) of the ferrocytochrome are 28.1, 17.0 and 143 ($\times 10^3$) respectively.

RHP and cytochrome c_2 share with cytochrome f the property of being heme proteins containing stable linkages between side chains and proteins.

2. RHP and Cytochrome c_2 from *Chromatium*.

Bartsch and Kamen (1960), following the earlier work of Newton and Kamen (1956), have isolated and purified RHP and a c type cytochrome from *Chromatium*. The purification procedure was based on DEAE-cellulose column chromatography.

The c-type cytochrome of *Chromatium* was also found to be an acidic protein (isoelectric point at pH 5.4) but of larger size (molecular weight 97,000) than its counterpart in *R. rubrum*. The *Chromatium* c-type cytochrome contains three heme groups per molecule. Its standard redox potential at pH 7 was found to be 0.01 V.

The RHP from *Chromatium* is similar in size (molecular weight 36,000) to RHP from *R. rubrum*; its other properties, two heme groups, isoelectric point at pH 5.5 and $E'_0 = -0.005$ V (pH 7) are also similar. Reduced *Chromatium* RHP combines with carbon monoxide (but with no other ligand) at neutral pH to give a three-banded hemochromogen spectrum with absorption peaks at 563, 535

Table 1. *Spectral Characteristics and Extinction Values of Chromatium RHP and Cytochrome c_2* (Bartsch and Kamen, 1960).

State of cytochrome	Cytochrome c_2		RHP	
	Wave length of absorption peak	Millimolar Extinction ($d = 1$ cm)	Wave Length of Absorption Peak	Millimolar Extinction ($d = 1$ cm)
	mμ		mμ	
Oxidized	278	175	280	63
	410	320	400	192
	525	26	495	24.5
			635	7.2
Reduced	416	364	426	210
	523	41.6	547	22.7
	552	61.1	565	20.9
Difference spectrum (reduced-oxidized)	406	-10.7	398	-115
	422	165	429	158
	524	15.5	495	-14.4
	553	44.6	565	5.4
			635	-4.3

and 418 mμ. Both RHP and c-type cytochrome from *Chromatium* are autoxidizable. The extinction coefficients and absorption peaks of *Chromatium* RHP and cytochrome c are given in Table 1.

3. Cytochrome-552 from *Rhodopseudomonas palustris*.

MORITA (1960) isolated and crystallized a cytochrome of the c type from *Rhodopseudomonas palustris*. In the reduced form the cytochrome showed absorption maxima at 552, 522, 418 and 317 mμ. The β-band was accompanied by two shoulders, at 530, and 513 mμ. The redox potential of this cytochrome was found to be 0.33 V at pH 7; its isoelectric point was found to be at pH 7.7 and its molecular weight, 15,600.

E. Quinone Constituents of the Photosynthetic Apparatus.

Attention to a possible role of quinones in photosynthesis has been directed by three findings: (a) WARBURG (1949) found that isolated chloroplast fragments that are unable to assimilate CO_2 are still capable of liberating oxygen under the influence of light (HILL-reaction), when supplied with substrate amounts of p-benzoquinone (Eq. 1).

$$C_6H_4O_2 + H_2O \longrightarrow C_6H_4(OH)_2 + {}^1/_2 O_2$$

(b) DAM, GLAVIND and NIELSEN (1940), and DAM, HJORTH and KRUSE (1948), using a biological assay (DAM and GLAVIND, 1938) found that vitamin K_1 (2-methyl-3-phytyl-1,4-naphthoquinone) is localized in chloroplasts. This finding was recently confirmed by KEGEL and CRANE (1962) by chromatographic isolation and direct chemical identification. The amount of vit. K_1 found by KEGEL and CRANE by chemical procedures (0.01 μmoles vit. K_1 per μmole chlorophyll) agrees well with the values found by DAM et al. by biological assay.

vitamin K_1 : R = —$CH_2CH = \overset{\underset{\displaystyle CH_3}{|}}{C}$—$(CH_2)_3$—$\overset{\underset{\displaystyle CH_3}{|}}{C}H(CH_2)_3$—$\overset{\underset{\displaystyle CH_3}{|}}{C}H(CH_2)_3$—$\overset{\underset{\displaystyle CH_3}{|}}{C}HCH_3$
menadione (vit. K_3) : R = H
phthiocol : R = OH

(c) CRANE (1959 a, b) found that a substituted benzoquinone, first isolated from alfalfa by KOFLER (1946), was characteristically localized in chloroplasts of higher plants. This compound, named by CRANE plastoquinone (also known as KOFLER'S quinone and Q_{254}) was found to be a tri-substituted benzoquinone with a 9-isoprenoid side chain. This structure was confirmed by the synthesis of 2,3-dimethyl-5-solanesyl benzoquinone which was shown to be identical with plastoquinone (TRENNER et al., 1959; KOFLER et al., 1959; SHUNK et al., 1959).

Plastoquinone is closely related structurally to the coenzyme Q (ubiquinone) group, which includes derivatives of 2,3-dimethoxy-5-methylbenzoquinone, substituted in position 6 with a polyisoprenoid side chain (CRANE, 1961). Plastoquinone differs particularly from the coenzyme Q homologues in having no methoxy groups.

The occurrence of plastoquinone (LESTER and CRANE, 1959; BISHOP, 1959) is, with few exceptions, restricted to algae and photosynthetic tissues of higher plants (blue-green algae, *Anacystis nidulans*; green algae, *Cladophora* sp., *Chlorella pyrenoidosa*, *Ankistrodesmus braunii*, *Chlamydomonas moewusii* and *Scenedesmus obliquus*; red algae, *Polysiphonia* sp.; brown algae, *Fucus* sp.; higher plants, spinach, alfalfa, sugar beet and swiss chard). Small amounts of plastoquinone also occur in nonchlorophyll-containing tissues such as corn roots or cauliflower buds which are capable of developing chloroplasts, and it has been suggested that proplastids of these tissues contain the compound. Tissues such as the white or sweet potato, which do not produce chlorophyll, contain no plastoquinone. The photosynthetic bacteria *Rhodospirillum rubrum* and *Chromatium* do not contain PQ but do possess exceedingly large amounts of coenzyme Q_9 and Q_7 respectively (LESTER and CRANE, 1959).

$$CH_3O \overset{O}{\diagdown} CH_3$$
$$CH_3O \diagup (CH_2\text{---}CH = \overset{CH_3}{\underset{|}{C}}\text{---}CH_2)_n H$$
$$O$$

Coenzyme Q_n

$$CH_3 \overset{O}{\diagdown} H$$
$$CH_3 \diagup (CH_2\text{---}CH = \overset{CH_3}{\underset{|}{C}}\text{---}CH_2)_n H$$
$$O$$

Plastoquinone
(n = 9, according to FOLKERS)
(n = 10, according to KOFLER)

The study by CRANE (1959a) of the distribution of both plastoquinone and Q_{10} in a green homogenate from spinach leaves showed that the coenzymes were contained exclusively in the particulate fraction. Plastoquinone was concentrated in the dark green particles which sediment at low centrifugal force. Its distribution in the particulate fractions was closely correlated with the chlorophyll content. On a dry weight basis 10 times more plastoquinone was present in the fraction with the highest chlorophyll content than in whole leaves. Q_{10}, on the other hand, was concentrated in the pale green particles which sediment at a higher centrifugal force. BISHOP (1959) calculated the molar ratio of plastoquinone: chlorophyll for sugar beet, spinach and swiss chard chloroplasts to be 1:8, but CRANE (1961) has reported ratios as high as 1:5. REDFEARN and FRIEND (1961) reported a ratio of 1:10 for chloroplasts isolated from sugar beet leaves. FULLER, SMILLIE, RIGO-POULOS and YOUNT (1961), and CLAYTON (1962) have obtained concordant results for the coenzyme Q_7 and bacteriochlorophyll content of *Chromatium* chromato-phores (about 5 and 25 μmoles per gram of dry weight, respectively).

In a further investigation of plastoquinone, KEGEL, HENNINGER and CRANE (1962) isolated two new quinones which had the same absorption spectra as the original plastoquinone but showed differences in chromatographic behavior. They designated the original substance as plastoquinone A and the other two as plastoquinone B and C. They report that "preliminary studies of the distribution of these compounds indicate that both plastoquinone A and C are found in large amounts in leaves of many species, as well as in chloroplasts from both spinach and lilac. Plastoquinone B seems to be present in small amounts, or not at all in some species, and further careful studies will have to be carried out to determine its general distribution and significance".

HENNINGER, DILLEY and CRANE (1963) have recently identified among the lipid components of spinach chloroplasts relatively large amounts of α tocopherol quinone as well as smaller amounts of β and γ tocopherol quinones. There is now evidence, therefore, for seven quinones in chloroplasts: vitamin K_1, plastoquinones A, B and C and α, β and γ tocopherol quinones.

I. Extraction and Purification of Vitamin K₁.

Method of Lichtenthaler (1962).

$100-300$ cm² of leaves were first ground in a mortar with quartz sand and a little acetone, and then completely extracted by adding $60-80$ ml acetone.

The acetone extract was filtered, and the residue washed with 20 ml acetone and 10 ml petroleum ether. After the addition of more petroleum ether (30 ml), the acetone-petroleum ether extract was poured carefully through a long-stemmed tube into 200 ml distilled water. The water-soluble compounds remained in the lower phase. The upper layer, which contained many substances in addition to vitamin K₁ was washed three times with 200 ml lots of distilled water. In this way traces of acetone, which would otherwise interfere with chromatography of the extract, were removed. After drying with anhydrous Na_2SO_4, the petroleum ether extract was made up to 50 ml, and a 40 ml sample was used for the determination of vitamin K₁. The petroleum ether extract was purified by chromatography on activated calcium phosphate (Shilling and Dam, 1958), which adsorbs the chlorophylls and xanthophylls. The vitamin K₁ and the carotene were eluted with petroleum-ether. Carotene and interfering quinones were removed by the method of Scudi and Bush (1941) from the fraction that comes through the column. After transferring the vitamin K₁ to butanol, it was reduced to the hydroquinone with hydrogen in the presence of the Raney-nickel catalyst, and estimated colorimetrically with 2,6-dichlorophenol-indophenol, by measuring the decrease in absorption at 650 mμ after 3 minutes. (For further details, see the Modern Methods of Plant Analysis, Vol. III, p. 368). Although the purified petroleum ether extract also contains tocopherols, coenzyme Q₁₀ and plastoquinone, these compounds do not interfere with the quantitative analysis of vitamin K₁, because their hydroquinone derivatives reduce the indophenol dye much more slowly.

II. Extraction and Purification of Plastoquinone.

1. Method of Crane (1959b).

Both coenzyme Q and plastoquinone can be easily extracted directly from dried alfalfa with relatively nonpolar solvents. For routine preparation, direct extraction with hydrocarbons, such as heptane or iso-octane is best. This procedure yields an initial extract containing most of the benzoquinone material with only a slight amount of vitamin K₁. The direct extraction is more convenient for large scale work than saponification followed by solvent extraction.

Chromatography on decalso (sodium aluminosilicate) and silicic acid were found most useful for purification of the quinones. The quinones are retained on the column from hydrocarbon solvents while carotenes and waxy materials pass through. The quinones can then be eluted by slightly polar solvent mixtures, which leave most of the residual impurities adsorbed on the column.

The procedure used for the purification of plastoquinone is illustrated by the following example. Increased amounts of the compounds may be obtained by using the same procedure on a larger scale. The extracts from several batches of alfalfa may be pooled and chromatographed on proportionally larger columns.

Extraction and batch decalso adsorption: Ten pounds of dehydrated alfalfa were mixed with 1.5 gallons of heptane. Extraction was allowed to proceed for 48 hours. The heptane extract was removed by decantation, and an equal volume of decalso (sodium aluminosilicate, 50/80 mesh) was mixed with the extract. The yellow heptane solution was decanted, leaving coenzyme Q and plastoquinone (and other materials) absorbed on the decalso. Elution of the quinones from the decalso was carried out by the following solvent combinations: 1) 1500 ml heptane,

2) 3000 ml of 10% ethyl ether in heptane, 3) and 4) each 1500 ml of 15% ether in heptane, 5) 1500 ml of 20% ether in heptane. Eluate 3 contained plastoquinone and 4 and 5 contained a mixture of plastoquinone and coenzyme Q_{10}. A total of 40 mg of plastoquinone and 20 mg of coenzyme Q_{10} was recovered in these eluates. Occasionally some vitamin K_1 was found in the first eluates containing plastoquinone.

Extraction of ethanol soluble material: Evaporation of the heptane under reduced pressure from the eluates containing the quinones left a greenish-orange oil. When this oil was taken up in absolute ethanol a yellow solution containing a brown precipitate was formed. The precipitate was filtered off and discarded. The quinones remained dissolved in the alcoholic solution.

Chromatography on decalso: The ethanol was evaporated and the yellow oil which remained was taken up in a small volume of iso-octane and the solution placed on a decalso column (6 cm × 20 cm, previously washed with iso-octane). Elution from the column, with 100 ml of solvent collected for each fraction, was carried out with the following solvents: fractions 1 and 2, isooctane; fractions 3 to 7, 2% ether in isooctane; fraction 8, 3% ether in isooctane. Essentially all of the coenzyme Q and plastoquinone were recovered in fractions 3, 4 and 5.

Silicic acid chromatography: The solvent was evaporated from the fractions containing the quinones, the oil taken up in a small volume of isooctane and the solution placed on a silicic acid column (silicic acid 100 mesh: supercel, 1 : 2 w/w mixed and equilibrated with isooctane). The column (3 cm × 20 cm) was eluted under 4 psi nitrogen gas in 25 ml fractions with the following solvent mixtures: fractions 1 to 9, isooctane; fractions 10 to 50, 30% chloroform in isooctane; fractions 51 to 61, 45% chloroform in isooctane. Fractions 12 and 13 contained 25 mg of plastoquinone and fractions 45 to 49, 10.6 mg of coenzyme Q_{10}.

Second decalso column: The best fractions of plastoquinone from the silicic acid column (fractions 12 and 13 above) were transferred to isooctane as before and placed on a decalso column (3 cm × 15 cm, washed with isooctane). The eluate was collected in 100 ml fractions with the following solvent mixtures: Fractions 1 to 2, isooctane; fractions 3 to 9, 2% ether in isooctane. Fractions 6 and 7 contained 20 mg of plastoquinone.

Crystallization of plastoquinone: The solvent was evaporated from the fractions containing plastoquinone, and the residue was dissolved in 2 ml of hot absolute ethanol. If any white crystals were formed as this solution cooled, they were filtered off and discarded. The ethanolic solution was placed at $-15°$ C overnight. The pale yellow crystals which formed were collected by filtration and were recrystallized from ethanol at 5° C. (Yield 10 mg, M. P. 42 to 43° C.).

To determine the amount of plastoquinone in each sample the following extinction coefficients at 254 mμ were used: $E_{1\,cm}^{1\%} = 210$ and $\varDelta E_{1\,cm}^{1\%}$ (oxidized-reduced) = 198.

2. Method of Bishop (1958, 1959).

This method describes the purification of plastoquinone from isolated spinach chloroplasts. The spinach chloroplast fragments used were prepared by the method of Spikes (1952). The sedimented chloroplasts in the final centrifugation were rinsed twice in glass-distilled water and then resuspended in water, frozen rapidly at $-78°$ C and lyophilized. The freeze-dried material was stored at $-25°$ C.

The petroleum ether for extraction and chromatography was first treated with sulfuric acid, washed several times with water, dried with anhydrous sodium sulfate, and finally distilled. The fraction passing over between 30° C and 40° C was collected and used for extractions. The other solvents employed were of sufficiently high purity that they did not contain contaminants which would interfere with the determination of absorption characteristics in the ultraviolet region.

Fractionation of the petroleum ether extract of the dried chloroplasts was performed first on a powdered-sugar column. With petroleum ether containing 5% benzene as the eluant, three yellow bands developed. The eluant was allowed to run through the column until the yellow bands separated sufficiently to be collected individually. The powdered sugar column effectively retained the chlorophylls and xanthophylls contained in the extract, and the fraction that came through the column contained β-carotene and the other lipid-like compounds, as well as the plastoquinone.

For further fractionation and purification, this yellow eluate from the sugar column was evaporated to dryness under vacuum at 35° C, redissolved in isooctane, and placed on a 2 × 15 cm silicic acid column previously washed with isooctane. The carotenes contained in the sample were carried through the column with a solvent composed of 25% chloroform and 75% isooctane while plastoquinone remained near the top of the column. Plastoquinone moved down the column with 75% chloroform and 25% isooctane. Generally, the plastoquinone was allowed to move down the column only far enough away from other material at the top of the column to permit a manual separation of the bands. After the portion of the column containing the plastoquinone had been collected, it was eluted with chloroform, evaporated to dryness (under vacuum), and the dried material redissolved in absolute alcohol. The light absorption curves of such samples were determined between 210 and 300 mμ before and after the addition of a few grains of sodium borohydride. For determining the amount of plastoquinone in a sample, this abbreviated procedure is of importance, since the compound decomposes gradually when held on the column. The following extinction coefficients were used to determine the amount of plastoquinone in each sample: $E_{1\,cm}^{1\%} = 212$ and $\Delta E_{1\,cm}^{1\%}$ (oxidized-reduced) = 200, at 255 mμ. The petroleum ether extracts from lyophilized chloroplasts of sugar beet, spinach and swiss chard leaves all contained plastoquinone.

III. Properties of Plastoquinone.

According to CRANE (1959b) plastoquinone has a peak at 255 mμ (ethanol) in the ultraviolet ($E_{1\,cm}^{1\%} = 210$). There is also a shoulder at 263 mμ and another shoulder in the visible range at about 425 mμ. Both concentrated solutions and the crystals are pale yellow in color. Upon reduction in ethanol with borohydride the peak at 255 mμ disappears and a new peak corresponding to the hydroquinone appears at 289 mμ ($E_{1\,cm}^{1\%} = 46$). A slight shoulder remains in the visible range above 400 mμ, whereas absorbance between 425 and 308 mμ is lower than in the quinone. During the early stages of reduction (immediately following KBH$_4$ addition) 2 peaks appear in the visible range at 438 and 412 mμ. After a few minutes, these visible peaks disappear and the solution becomes distinctly greenish as the result of absorption at 580 mμ. The 580 band slowly disappears and absorption in the visible range returns after several minutes to the original state except for the decreased absorption below 425 mμ.

The plastoquinone, being a tri-substituted benzoquinone derivative, gives a positive CRAVEN'S test, i. e., it develops a blue color when ethylcyanoacetate in NH$_3$-ethanol is added to its alcoholic solution (KOFLER et al., 1959).

IV. Role of Quinones in Photosynthetic Reactions.

1. Photosynthetic Phosphorylation.

The first indication that quinones may have an effect on photosynthetic phosphorylation came from attempts to circumvent a then puzzling dependence of this process on molecular oxygen. When first discovered, ATP formation by isolated chloroplasts proceeded at a sustained rate only in the presence of oxygen (Arnon, Allen and Whatley, 1954). However, since there was no net evolution (as measured by manometric pressure change) of molecular oxygen, Arnon, Allen and Whatley (1954) concluded that oxygen acted as a catalyst in photosynthetic phosphorylation and not as a substrate as in oxidative phosphorylation. A decisive difference between photosynthetic and oxidative phosphorylation is the inability of chloroplasts to form ATP in the dark by oxidizing substrates of oxidative phosphorylation with molecular oxygen.

The dependence of ATP formation by chloroplasts on oxygen was abolished by the addition of catalytic amounts of menadione or other naphthoquinones (Arnon, Whatley and Allen, 1955). Photophosphorylation could now proceed at a sustained and vigorous rate in an atmosphere of nitrogen or argon. The process, which appeared to be anaerobic even in chloroplasts of such aerobic plants as spinach, was thus analogous to photophosphorylations by chromatophores of photosynthetic bacteria, whose photosynthesis is strictly anaerobic.

The effect of naphthoquinones on photophosphorylation seemed especially interesting at the time because its discovery shortly followed that of Martius (1954, 1955) on the effect of vit. K_1 on oxidative phosphorylation by liver mitochondria. The effect of naphthoquinones on phosphorylation by chloroplasts was different in that (1) chloroplasts were less exacting with regard to the side chain on the third carbon of the naphthoquinone ring. In the experiments of Martius with mitochondria the phytyl side chain was required; menadione (side chain = H) was inhibitory. Chloroplasts could equally well use menadione, phthiocol (side-chain = OH) or other naphthoquinone derivatives; (2) the increase in photosynthetic phosphorylation caused by the addition of catalytic amounts of one of the naphthoquinones was large, of the order of twenty-fold. Mitochondria isolated from livers of vitamin K deficient chicks showed only a 30% decrease in oxidative phosphorylation, that was restorable *in vitro* by the addition of vitamin K_1 (Martius, 1955).

Martius (1954, 1955) has ascribed to vitamin K_1 in oxidative phosphorylation the role of an electron carrier between DPN and cytochromes. Arnon (1955) at first tentatively proposed an analogous role for naphthoquinones in photosynthetic phosphorylation. In the proposed electron flow mechanism of photosynthetic phosphorylation, vit. K was assigned a role of an electron carrier in cyclic photophosphorylation (Arnon, 1959, 1961).

Krogmann (1961) has recently shown that chloroplasts lose their capacity for cyclic photophosphorylation after being extracted with heptane. Cyclic photophosphorylation, of the type catalyzed by phenazine methosulfate, was restored by the addition of plastoquinone. Similar effects of added plastoquinone were also observed with cyclic photophosphorylation catalyzed by vit. K_3 and FMN (Arnon, Whatley and Horton, 1962; Krogmann and Olivero, 1962).

2. Noncyclic Electron Transport.

The sole product of cyclic photophosphorylation is ATP and although the formation of ATP cannot be envisaged without a concurrent electron flow, it is experimentally difficult to demonstrate a cyclic electron flow that proceeds without ATP formation. By contrast, noncyclic electron flow in chloroplasts can be measured not only by ATP formation but also by oxygen evolution and by photoreduction of TPN, ferricyanide, dyes or other electron acceptors. These features have been used in several investigations concerned with the role of constituent quinones in the light-induced electron flow in photosynthetic particles.

LYNCH and FRENCH (1957) and MILNER, FRENCH and MILNER (1958) observed that the photochemical activity of lyophyllized chloroplasts, measured by reduction of 2,6- dichlorophenol indophenol during illumination of aqueous suspensions, was largely removed by petroleum ether extraction. The extracted chloroplasts were readily reactivated by β-carotene but "in every case of reactivation by β-carotene the effect was modest in comparison with the activity of unextracted plastids, as well as with the activation produced by adding back even small amounts of the petroleum ether extract".

BISHOP (1958) continued these investigations, and by purifying the petroleum ether extract of chloroplasts and commercially available carotene, ruled out the participation of carotenes in restoring the ability of extracted chloroplasts to photoreduce HILL reagents. Instead, he found that "menadione (vitamin K_3) is capable of restoring the photochemical activity (reduction of HILL reagents and oxygen evolution) to extracted chloroplasts under conditions when the effect of purified α- and β-carotene is negligible by comparison". The effect of added menadione or the naphthol derivative, vitamin K_5, was equal to that of the chloroplast extract itself, but the restored activity was still smaller than that of the unextracted control.

In later experiments BISHOP (1959), using a spectrophotometric test, found little, if any, vitamin K in the petroleum ether extract of chloroplasts. Instead, he found plastoquinone, which CRANE (1959) has shown to be concentrated in chloroplasts. As for reactivation of the HILL reaction in extracted chloroplasts, BISHOP now reported that "qualitatively the effect of the new benzoquinone (plastoquinone) is the same as that of the naphthoquinones (vitamin K, etc.). Quantitatively it is much more effective and, so far, the only compound which produces the same or even better rates of the photochemical reaction than the unextracted controls". That the addition of plastoquinone partly restores to chloroplasts, extracted with heptane, the ability to photoreduce a HILL reagent (2,3',6-trichlorophenol indophenol) was also recently reported by KROGMANN (1961).

Although these results provide evidence for the participation of plastoquinone in the noncyclic electron flow in chloroplasts serious experimental difficulties, which hinder consistency of results, are yet to be overcome. It is difficult to determine the exact amount of plastoquinone which will restore the original photochemical activity because of the inaccuracy inherent in the method for readdition of the compound. The only effective procedure found so far for reactivating extracted chloroplasts entails an evaporation of a petroleum ether solution of the substance tested in the presence of the extracted chloroplasts. This opens the possibility that the compound may adhere to the walls of the container rather than to the surface of the chloroplasts.

CRANE, EHRLICH and KEGEL (1960) have reported that in chloroplasts most of the plastoquinone is present in the oxidized form. Any reduced quinone present is converted to the oxidized form by treatment of the chloroplast suspension with 0.03 M potassium ferricyanide in the dark. When chloroplasts are exposed to bright light (1000 foot candles, tungsten) at 25°, up to 80% of the endogenous plastoquinone is reduced within five minutes. On further incubation in light or dark the reduced plastoquinone is slowly converted to the oxidized form. When

39*

resuspended boiled chloroplasts are exposed to light the plastoquinone is not reduced.

Henninger, Dilley and Crane (1963) have recently reported preliminary evidence that β and γ tocopherol quinones are of "functional significance in the electron transport system of chloroplasts" as measured by photoreduction of ferricyanide. Krogmann and Olivero (1962) have observed that γ tocopherol quinone will partly replace plastoquinone A in restoration of the Hill reaction to heptane-extracted chloroplasts.

Redfearn and Friend (1961) determined the oxidation-level of plastoquinone in freshly isolated chloroplasts and found the oxidized plastoquinone to vary between 45 and 90% of the total. Illumination (5 minutes at 3,200 foot candles) resulted in a fall of the oxidized level to 35—65%, which results in a partial reduction of the quinone. Addition of dichlorophenol indophenol to the chloroplasts in the dark converted all the plastoquinone into the oxidized form.

Kegel, Henninger and Crane (1962) reported that plastoquinone A and B undergo reduction in light. Plastoquinone C, on the other hand, displayed a unique behavior in that very small amounts of it were detected when the chloroplasts were kept in light, but the quinone appeared when the chloroplasts were put in darkness.

3. Bacterial Photophosphorylations.

There are several indications that photophosphorylation in photosynthetic bacteria involves the participation of coenzyme Q and in this respect resembles oxidative phosphorylation by mitochondria. A stimulation of photophosphorylation by coenzyme Q isoprenologues in chromatophores of *R. rubrum* has recently been reported by Rudney (1961). The phosphorylating activity of chromatophores, extracted with petroleum ether, could not be restored by the addition of coenzyme Q or of the petroleum ether extract itself. However, when the chromatophores were prepared from cells grown in the presence of diphenylamine, the addition of smaller isoprenologues of coenzyme Q (CoQ_2 or CoQ_3) was effective. Such chromatophores had a lower content of coenzyme Q and were evidently structurally altered to be able to respond to added coenzyme Q derivatives.

Clayton (1962) using aqueous suspensions and dried films of chromatophores from *Rhodopseudomonas spheroides*, *Rhodospirillum rubrum* and *Chromatium*, measured changes in absorption following illumination. He concluded that coenzyme Q is the primary electron acceptor and bacteriochlorophyll the primary electron donor in bacterial photosynthesis.

V. Concluding Remarks.

In the last few years our knowledge of quinones in the photosynthetic apparatus has advanced very rapidly. The two major developments have been the confirmation and extension, mainly by Crane and his colleagues of the important original finding by Kofler that a substituted benzoquinone, now called plastoquinone, is a component of photosynthetic tissues. We also owe to Crane the confirmation by chemical methods of Dam's discovery of vitamin K_1 in chloroplasts, thereby removing from controversy a subject that attracted much interest after the demonstration that vitamin K compounds catalyze photosynthetic phosphorylation. As this review shows, seven quinone constituents have now been isolated from chloroplasts. Although it is still premature to assign to them with finality specific roles in the photosynthetic process it is almost certain that these compounds participate in the photosynthetic electron transport chain. Their ability to accept one electron at a time, by going through the semiquinone stage, endows them with at least one property that may be of great importance to the mechanism of light-induced electron flow in photosynthesis.

References.

Section A.

ALLEN, M. B., D. I. ARNON, J. B. CAPINDALE, F. R. WHATLEY and L. J. DURHAM: J. Amer. chem. Soc. 77, 4149 (1955). — ARNON, D. I.: Science 122, 823 (1955); — In W. D. McELROY and B. GLASS. Light and Life. Baltimore: Johns Hopkins Press 1961. — ARNON, D. I., M. B. ALLEN and F. R. WHATLEY: Nature (Lond.) 174, 394 (1954); — Biochim. biophys. Acta 20, 449 (1956).

BERGERON, J. A.: In the Photochemical Apparatus, Brookhaven Nat. Lab. Upton, N. Y. 1959. — BOVÉ, J., and I. D. RAACKE: Arch. Biochem. 85, 521 (1959).

FERRARI, R. A., and A. A. BENSON: Arch. Biochem. 93, 185 (1961). — FRENKEL, A. W.: J. biol. Chem. 222, 823 (1956). — FULLER, R. C., and I. C. ANDERSON: Plant Physiol. 32, Suppl. xvi (1957). — FULLER, R. C., R. M. SMILLIE, E. C. SISLER and H. L. KORNBERG: J. biol. Chem. 236, 2140 (1961).

GEST, H., M. D. KAMEN and H. M. BREGOFF: J. biol. Chem. 182, 153 (1950). — GIBBS, M., and N. CALO: Plant Physiol. 34, 318 (1959). — GIBBS, M., and M. A. CYNKIN: Nature (Lond.) 182, 1241 (1958).

HENDLEY, D. D.: J. Bact. 70, 625 (1955). — HILL, R.: Nature (Lond.) 139, 881 (1937); Symp. Soc. exp. Biol. 5, 223 (1951).

IRMAK, L. R.: Rev. Fac. Sci. Univ. Istanbul, Serie B 20, 237 (1955).

LAMANNA, C., and M. F. MALLETTE: J. Bact. 67, 503 (1954). — LOSADA, M., A. V. TREBST, S. OGATA and D. I. ARNON: Nature (Lond.) 186, 753 (1960). — LOSADA, M., A. V. TREBST and D. I. ARNON: Presented by D. I. ARNON at the 9th Intern. Bot. Congr., Montreal 1959.— LUMRY, R., J. D. SPIKES and H. EYRING: Ann. Rev. Plant Physiol. 5, 271 (1954).

MENKE, W.: Ann. Rev. Plant Physiol. 13, 27 (1962).

NEWTON, J. W., and M. D. KAMEN: Biochim. biophys. Acta 21, 71 (1956); — NEWTON, J. W., and G. A. NEWTON: Arch. Biochem. 71, 250 (1957).

QUAYLE, J. R.: Ann. Rev. Microbiol. 15, 119 (1961).

ROSENBERG, L. L., J. B. CAPINDALE and F. R. WHATLEY: Nature (Lond.) 181, 632 (1958).

SCHACHMAN, H. K., A. B. PARDEE and R. Y. STANIER: Arch. Biochem. 38, 245 (1952). — SISSAKIAN, N. M.: Adv. in Enzymol. 20, 201 (1958). — SMILLIE, R. M., and R. C. FULLER: Plant Physiol. 34, 651 (1959). — SMILLIE, R. M., and G. KROTKOV: Canad. J. Bot. 37, 1217 (1959). — STUMPF, P. K., and A. T. JAMES: Biochem. J. 32, 29 P (1962).

THOMAS, J. B., A. J. M. HAANS and A. A. VAN DER LEUN: Biochim. biophys. Acta 25, 453 (1957). — TOLBERT, N. E.: Brookhaven Symposia in Biol. 11, 271 (1958). — TREBST, A. V., H. Y. TSUJIMOTO and D. I. ARNON: Nature (Lond.) 182, 351 (1958).

UEDA, R.: Bot. Mag. (Tokyo) 62, 731 (1949).

WHATLEY, F. R., M. B. ALLEN and D. I. ARNON: Biochim. biophys. Acta 32, 32 (1959). — WHATLEY, F. R., M. B. ALLEN, L. L. ROSENBERG, J. B. CAPINDALE and D. I. ARNON: Biochim. biophys. Acta 20, 462 (1956). — WHATLEY, F. R., and D. I. ARNON: In S. COLOWICK and N. O. KAPLAN: Methods of Enzymology, Vol. VI (in press) 1962. Academic Press.

Section B.

ARNON, D. I.: Science 116, 635 (1952); — In W. D. McELROY and B. GLASS. Light and Life. Baltimore: Johns Hopkins Press 1961.

BÜCHER, T.: Biochim. biophys. Acta 1, 292 (1947).

CALO, N., u. M. GIBBS: Z. Naturforsch. 15 b, 287 (1960). — CALVIN, M.: Proc. Intern. Congr. Biochem., 3rd Congr., Brussels, p. 211. New York: Academic Press 1956.

FISKE, C. H., and Y. SUBBAROW: J. biol. Chem. 66, 375 (1925).

GIBBS, M.: Nature (Lond.) 170, 164 (1952). — GIBBS, M., and N. CALO: Plant Physiol. 34, 318 (1959); — Biochim. biophys. Acta 44, 341 (1960).

HURWITZ, J., A. WEISSBACH, B. L. HORECKER and P. Z. SMYRNIOTIS: J., biol. Chem. 218, 769 (1956).

JAKOBY, W. B., D. O. BRUMMOND and S. OCHOA: J. biol. Chem. 218, 811 (1956).

KORNBERG, A., and W. E. PRICER JR.: J. biol. Chem. 193, 481 (1951).

LOSADA, M., and D. I. ARNON: Chapter in B. M. HOCHSTER and J. H. QUASTEL. Metabolic inhibitors. (In press) 1962. Academic Press. — LOSADA, M., A. V. TREBST and D. I. ARNON: J. biol. Chem. 235, 332 (1960).

RACKER, E.: Arch. Biochem. **69**, 300 (1957). — RUBEN, S.: J. Amer. chem. Soc. **65**, 279 (1943).
SUTHERLAND, E. W., C. F. CORI, R. HAYNES and N. S. OLSEN: J. biol. Chem. **180**, 825 (1949).
TREBST, A. V., M. LOSADA and D. I. ARNON: J. biol. Chem. **234**, 3055 (1959); **235**, 840 (1960). — TREBST, A. V., H. Y. TSUJIMOTO and D. I. ARNON: Nature (Lond.) **182**, 351 (1958).
VISHNIAC, W., B. L. HORECKER and S. OCHOA: Adv. Enzymol. **19**, 1 (1957).
WARBURG, O., u. W. CHRISTIAN: Biochem. Z. **310**, 384 (1941/42). — WEISSBACH, A., B. L. HORECKER and J. HURWITZ: J. biol. Chem. **218**, 795 (1956). — WILLSTÄTER, R., u. H. KRAUT: Ber. chem. Ges. **56**, 1117 (1923).

Section C.

ANDERSON, I. C., and R. C. FULLER: Arch. Biochem. **76**, 168 (1958). — ARNON, D. I.: Nature (Lond.) **184**, 10 (1959); — In W. D. MCELROY and B. GLASS: Light and Life. Baltimore: Johns Hopkins Press 1961. — ARNON, D. I., M. B. ALLEN and F. R. WHATLEY: Nature (Lond.) **174**, 394 (1954). — ARNON, D. I., M. LOSADA, M. NOZAKI and K. TAGAWA: Nature (Lond.) **190**, 601 (1961). — ARNON, D. I., M. LOSADA, F. R. WHATLEY, H. Y. TSUJIMOTO, D. O. HALL and A. A. HORTON: Proc. nat. Acad. Sci. (U.S.) **47**, 1314 (1961). — ARNON, D. I., F. R. WHATLEY and M. B. ALLEN: J. Amer. chem. Soc. **76**, 6324 (1954); — Science **127**, 1026 (1958); — Biochim. biophys. Acta **32**, 47 (1959). — AVRON, M., and A. T. JAGENDORF: Nature (Lond.) **179**, 428 (1957); — J. biol. Chem. **234**, 967 (1959). — AVRON, M., A. T. JAGENDORF and M. EVANS: Biochim. biophys. Acta **26**, 262 (1957).
CHOW, C. T., and B. VENNESLAND: Plant Physiol. **32**, Suppl. iv (1957). — COHN, W. E., and C. E. CARTER: J. Amer. chem. Soc. **72**, 4273 (1950).
DUYSENS, L. N. M., J. AMESZ and B. M. KAMP: Nature (Lond.) **190**, 510 (1961).
FISKE, C. H., and Y. SUBBAROW: J. biol. Chem. **66**, 375 (1925). — FRENKEL, A. W.: J. Amer. chem. Soc. **76**, 5568 (1954); — J. biol. Chem. **222**, 823 (1956).
GELLER, D. M.: Doctoral Diss., Div. Med. Sci., Harvard University (1957). — GELLER, D. M., and J. D. GREGORY: Fed. Proc. **15**, 260 (1956).
HILL, R., and F. BENDALL: Nature (Lond.) **187**, 417 (1960). — HILL, R., and D. A. WALKER: Plant Physiol. **34**, 240 (1959).
JAGENDORF, A. T.: Fed. Proc. **18**, 974 (1959). — JAGENDORF, A. T., and M. AVRON: J. biol. Chem. **231**, 277 (1958).
LOSADA, M., F. R. WHATLEY and D. I. ARNON: Nature (Lond.) **190**, 606 (1961).
NAKAMOTO, T., D. W. KROGMANN and B. VENNESLAND: J. biol. Chem. **234**, 2783 (1959). — NEWTON, J. W., and M. D. KAMEN: Biochim. biophys. Acta **25**, 462 (1957). — NOZAKI, M., K. TAGAWA and D. I. ARNON: Proc. nat. Acad. Sci. (U.S.) **47**, 1334 (1961).
PETRACK, B., and F. LIPMANN: In W. D. MCELROY and B. GLASS: Light and Life. Baltimore: Johns Hopkins Press 1961.
THOMAS, J. B., and A. M. J. HAANS: Biochim. biophys. Acta **18**, 286 (1955). — TREBST, A. V., M. LOSADA and D. I. ARNON: J. biol. Chem. **234**, 3055 (1959).
WESSELS, J. S. C.: Biochim. biophys. Acta **25**, 97 (1957); **29**, 113 (1958). — WHATLEY, F. R., M. B. ALLEN, A. V. TREBST and D. I. ARNON: Plant Physiol. **35**, 188 (1960). — WHATLEY, F. R., and D. I. ARNON: In S. COLOWICK and N.O.KAPLAN (ed.): Methods of Enzymology Vol. VI (in press) 1962. Academic Press. — WILLIAMS, A.M.: Biochim. biophys. Acta **19**, 370 (1956)

Section D.

APPELLA, E., and A. SAN PIETRO: Biochem. biophys. Res. Comm. **6**, 349 (1961). — ARNON, D. I.: Nature (Lond.) **167**, 1008 (1951). — ABNON D. I., F. R. WHATLEY and M. B. ALLEN: Nature (Lond.) **180**, 182 (1957). — AVRON, M., and A. T. JAGENDORF: Arch. Biochem. **65**, 475 (1956); **72**, 17 (197).
BARTSCH, R. G., and M. D. KAMEN: J. biol. Chem. **230**, 41 (1958); **235**, 825 (1960).
DAVENPORT, H. E.: Nature **170**, 1112 (1952); — Biochem. J. **77**, 471 (1960). — DAVENPORT, H. E., and R. HILL: Proc. roy. Soc. B **139**, 327 (1952); — Biochem. J. **74**, 493 (1960). — DAVENPORT, H. E., R. HILL and F. R. WHATLEY: Proc. roy. Soc. B **139**, 346 (1952). — DE LUCA, C., M. M. WEBER and N. O. KAPLAN: J. biol. Chem. **223**, 559 (1956).
ELSDEN, S. R., M. D. KAMEN and L. P. VERNON: J. Amer. Chem. Soc. **75**, 6347 (1953)ʼ FRY, K. T., and A. SAN PIETRO: Biochem. biophys. Res. Comm. **9**, 218 (1962).
HILL, R.: Biochem. J. **37**, xxiii (1943); — Symp. Soc. exp. Biol. **5**, 222 (1951); — Nature (Lond.) **174**, 501 (1954); **186**, 136 (1960). — HILL, R., and F. BENDALL: Nature (Lond.) **187**, 417 (1960). — HILL, R., and W. D. BONNER JR.: In W. D. MCELROY and B. GLASS. Light and Life. Baltimore: Johns Hopkins Press 1961. — HILL, R., D. H. NORTHCOTE and H. E. DAVENPORT: Nature (Lond.) **172**, 948 (1953). — HILL, R., and R. SCARISBRICK: New Phytologist **50**, 98 (1951). — HORIO, T., and M. D. KAMEN: Biochim. biophys. Acta **48**, 266 (1961).
JAGENDORF, A. T.: Arch. biochem. **62**, 141 (1956). — JAMES, W. O., and R. M. LEECH: Nature (Lond.) **182**, 1684 (1958); — Publ. Staz. Zool. Napoli **31**, 36 (1959).

KAHN, A., and A. T. JAGENDORF: Biochim. biophys. Acta **58**, 149 (1962). — KAMEN, M. D., and L. P. VERNON: Biochim. biophys. Acta **17**, 10 (1955). — KATOH, S.: J. Biochem. (Tokyo) **46**, 629 (1959); — Plant and Cell Physiol. **1**, 91 (1960a);·—Nature (Lond.) **186**, 138 (1960b); **186**, 533 (1960c); — Plan and Cell Physiol. **2**, 165 (1961). —KATOH, S., I. SHIRATORI and A. TAKAMIYA: J. Biochem. (Tokyo) **31**, 32 (1962). — KATOH, S., I. SHIRATORI, I. SUGA and A. TAKAMIYA: Arch. Biochem. **94**, 136 (1961). — KATOH, S., and A. TAKAMIYA: Biochem. biophys. Res. Comm. **8**, 310 (1962). — KEISTER, D. L., A. SAN PIETRO and F. E. STOLZENBACH: J. biol. Chem. **235**, 2989 (1960); — Arch. Biochem. **98**, 235 (1962).

LAZZARINI, R. A., and A. SAN PIETRO: Biochim. biophys. Acta **62**, 417 (1962). — LUNDE-GARDH, H.: Physiol. Plant **7**, 375 (1954); — Nature (Lond.) **192**, 243 (1961).

MORITA, S.: J. Biochem. (Tokyo) **48**, 870 (1960). — MORTENSON, L. E., R. C. VALENTINE and J. E. CARNAHAN: Biochem. biophys. Res. Comm. **7**, 448 (1962).

NEWTON, J. W., and M. D. KAMEN: Arch. Biochem. **58**, 246 (1955); — Biochim. biophys. Acta **21**, 71 (1956). — NISHIMURA, M.: J. Biochem. (Tokyo) **46**, 219 (1959).

SAN PIETRO, A.: In the Photochemical Apparatus. Brookhaven National Laboratory, Upton, N. Y. 1959; — In W. D. MCELROY and B. GLASS. Light and Life. Baltimore: Johns Hopkins Press 1961. — SAN PIETRO, A., and H. M. LANG: Science **124**, 118 (1956); — J. biol. Chem. **231**, 211 (1958). — SHIN, M., K. TAGAWA and D. I. ARNON: Biochem. Z. (In Press 1963).

TAGAWA, K., and D. I. ARNON: Nature (Lond.) **195**, 537 (1962). — TAGAWA, K., H. Y. TSUJIMOTO and D. I. ARNON: Proc. Nat. Acad. Sci. (U.S.) **49**, 567 (1963). — TOLMACH, L. J.: Nature (Lond.) **167**, 946 (1951). —TREBST, A., and S. WAGNER: Z. Naturforsch.**17**b, 396 (1962).

VERNON, L. P.: Arch. Biochem. **43**, 492 (1953). — VERNON, L. P., and M. D. KAMEN: J. biol. Chem. **211**, 643 (1954). — VISHNIAC, W., and S. OCHOA: Nature (Lond.) **167**, 768 (1951).

WARBURG, O.: In Photosynthesis Colloquium, 1962, Gif-sur-Yvette, France (In press). — WESSELS, J. S. C., and E. HAVINGA: Rec. Trav. Chim. Pays-Bas **71**, 809 (1952). — WHATLEY, F. R., K. TAGAWA and D. I. ARNON: Proc. Nat. Acad. Sci. (U.S.) **49**, 266 (1963).

Section E.

ARNON, D. I.: Science **122**, 9 (1955); — Nature (Lond.) **184**, 10 (1959). — ARNON, D. I.: In W. D. MCELROY and B. GLASS. Light and Life. Baltimore: Johns Hopkins Press 1961. — ARNON, D. I., M. B. ALLEN and F. R. WHATLEY: Nature (Lond.) **174**, 394 (1954). — ARNON, D. I., F. R. WHATLEY and M. B. ALLEN: Biochim. biophys. Acta **16**, 607 (1955). — ARNON, D. I., F. R. WHATLFY and A. A. HORTON: Fed. Proc. **21**, 91 (1962).

BISHOP, N. I.: Proc. Nat. Acad. Sci. (U.S.) **44**, 501 (1958); **45**, 1696 (1959).

CLAYTON, R. K.: Biochem. biophys. Res. Comm. **9**, 49 (1962). — CRANE, F. L.: Plant Physiol. **34**, 128 (1959a); **34**, 546 (1959b); — In G. E. WOLSTENHOLME and C. M. O'CONNOR. Ciba Foundation Symposium on Quinones in Electron Transport. p. 36. London: Churchill 1961.—CRANE, F. L., B. EHRLICH and L. P. KEGEL: Biochem. biophys. Res. Comm. **3**, 37 (1960).

DAM, H., and J. GLAVIND: Biochem. J. **32**, 1018 (1938). — DAM, H., J. GLAVIND u. N. NIELSEN: Z. physiol. Chem. **265**, 8 (1940). — DAM, H., E. HJORTH and I. KRUSE: Physiol. Plant. **1**, 379 (1948).

FULLER, R. C., R. M. SMILLIE, N. RIGOPOULOS and V. YOUNT: Arch. Biochem. **95**, 197 (1961).

HENNINGER, M. D., R. A. DILLEY and F. L. CRANE: Biochem. biophys. Res. Comm. **10**, 237 (1963).

KEGEL, L. P., and F. L. CRANE: Nature (Lond.) **194**, 1282 (1962). — KEGEL, L. P., M. D. HENNINGER and F. L. CRANE: Biochem. biophys. Res. Comm. **8**, 294 (1962). — KOFLER, M.: Festschrift für EMIL CHRISTOPH BARELL, p. 199. Basel 1946. — KOFLER, M., A. LANGEMANN, R. RÜEGG, L. H. CHOPARD-DIT-JEAN, A. RAYROUD and O. ISLER: Helv. Chim. Acta **42**, 1283 (1959). — KROGMANN, D. W.: Biochem. biophys. Res. Comm. **4**, 275 (1961). — KROGMANN, D. W., and E. OLIVERO: J. biol. Chem. **237**, 3292 (1962).

LESTER, R. L., and F. L. CRANE: J. biol. Chem. **234**, 2169 (1959). — LICHTENTHALER, H. K.: Planta **57**, 731 (1962).—LYNCH, V. H., and C. S. FRENCH: Arch. Biochem. **70**, 382 (1957).

MARTIUS, C.: Biochem. Z. **326**, 26 (1954); — Proc. 3rd Int. Cong. Biochem. Brussels: 1, (1955). — MILNER, M., C. S. FRENCH and H. W. MILNER: Plant Physiol. **33**, 367 (1958).

REDFEARN, E. R., and J. FRIEND: Nature (Lond.) **191**, 806 (1961). — RUDNEY, H.: J. biol. Chem. **236**, PC 39 (1961).

SCHILLING, K., and H. DAM: Acta chem. scand. **12**, 347 (1958). — SCUDI, G. V., and R. P. BUSH: J. biol. Chem. **137**, 745 (1941). — SHUNK, C. H., R. E. ERICKSON, E. L. WONG and K. FOLKERS: J. Amer. chem. Soc. **81**, 5000 (1959). — SPIKES, J. D.: Arch. Biochem. **35**, 101 (1952).

TRENNER, N. R., B. H. ARISON, R. E. ERICKSON, C. H. SHUNK, D. E. WOLF and K. FOLKERS: J. Amer. chem. Soc. **81**, 2026 (1959).

WARBURG, O.: Heavy Metal Prosthetic Groups and Enzyme Action, p. 213. Oxford: Clarendon Press 1949.

Enzymes of the Krebs Cycle, the Glyoxalate Cycle and Related Enzymes[1].

By

D. D. Davies and R. J. Ellis.

With 2 Figures.

Since a number of KREBS cycle enzymes are located in mitochondria, a considerable purification can be achieved by isolating the mitochondria. Recent studies (MARCUS and VELASCO, 1960) have indicated that the enzymes of the glyoxalate cycle are also located in the mitochondria.

A number of enzymes can be obtained in soluble form, by freezing and thawing a suspension of mitochondria (DAVIES, 1956). Other enzymes can only be made soluble after treatment of mitochondria with acetone (HIATT, 1960).

A. Enzymes of the Krebs Cycle.

I. Condensing Enzyme.

$$
\begin{array}{c}
\text{COOH} \\
| \\
\text{C=O} \\
| \\
\text{CH}_2 \\
| \\
\text{COOH}
\end{array}
+ \text{CH}_3\text{COSCoA} \rightleftharpoons
\begin{array}{c}
\text{COOH} \\
| \\
\text{CH}_2 \\
| \\
\text{C(OH)COOH} + \text{CoASH} \\
| \\
\text{CH}_2 \\
| \\
\text{COOH}
\end{array}
$$

Acetyl CoA is an expensive compound and it is often made 'in situ' by means of the reaction:

$$\text{acetylphosphate} + \text{CoASH} \rightleftharpoons \text{Acetyl-S-CoA} + \text{orthophosphate}$$

The enzyme transacetylase can be obtained commercially or by extracting freeze-dried preparations of *E. coli* (American type culture Collection Strain No. 4157).

Assay. Three assays have been developed:

1. The *rate of citrate formation* is measured in a system containing enzyme, oxaloacetate, CoA, acetyl phosphate, excess transacetylase (to ensure that the rate determining step is the condensing enzyme) and glutathione (to maintain CoA in the reduced form). Since transacetylase is absent from most plants, its

---

[1] Abbreviations: AMP, adenosine-5'-phosphate; ADP, adenosine-5'-diphosphate; ATP-adenosine-5'-triphosphate; BAL, British anti-Lewisite (2,3-dimercapto-1-propanol); CoA, coenzyme A; DNA, deoxyribonucleic acid; DPN and DPNH, oxidized and reduced diphospho, pyridine nucleotide, respectively; FAD, flavin adenine dinucleotide; FMN, flavin mono nucleotide; GTP, guanosine-5'-triphosphate; ITP, inosine-5'-triphosphate; OAA, oxalaceti-acid; P$_i$, inorganic phosphate; TPP, thiamine pyrophosphate; TPN and TPNH, oxidized and reduced triphosphopyridine nucleotide; Tris, tris(hydroxymethyl)aminomethane.

omission provides a convenient control. Non-linearity in this assay is probably due to the removal of citrate under the influence of the enzyme aconitase and this effect may be of major importance if the extract also contains isocitritase. Citrate may be determined by a number of methods. Citrate is converted to pentabromo-acetone by oxidation with permanganate (PUCHER, SHERMAN and VICKERY, 1936) or vanadate (WEIL-MALHERBE and BONE, 1949) in the presence of bromine, or by electrolysis (SAFRONOV, 1959). After removal of excess bromine with thio-sulphate or ferrous sulphate and extraction of the pentabromoacetone, colour may be developed with sodium sulphide or pyridine (ETTINGER, GOLDBAUM and SMITH, 1952).

2. In the above assay system, the *rate of disappearance of acetyl phosphate* may be measured by the hydroxamate method (LIPMANN and TUTTLE, 1945). Correction for the hydrolysis of acetyl phosphate can be made by including a control in which transacetylase is missing.

3. An *optical assay* has been developed by OCHOA (1955) in which malate, DPN and an excess of malic dehydrogenase are allowed to come to equilibrium in the presence of condensing enzyme. The reaction is started by the addition of acetyl CoA which condenses with oxaloacetate. The production of oxaloacetate to maintain the malic dehydrogenase equilibrium is accompanied by the formation of an equal amount of DPNH which is measured spectrophotometrically at 340 mμ.

This assay should be treated with caution, because the statement of OCHOA that "the rate of reduction of DPN is equal to the rate of citrate synthesis" is incorrect.

Let the amount of malate initially present $= a$, and the amount of DPN $= b$. At equilibrium the amount of DPNH formed = amount of OAA formed $= x$ Then

$$K_{equil} = \frac{x^2}{(a - x)(b - x)}$$

and

$$x = \frac{(a + b) - \sqrt{(a + b)^2 - 4ab(K - 1)/K}}{2(K - 1)/K} \ .$$

Let the amount of oxaloacetate removed to form citrate $= y$. Let the amount of oxaloacetate formed to maintain the malic dehydrogenase equilibrium $= z$ $=$ amount DPNH formed Then

$$K_{equil} = \frac{(x + z)(x - y + z)}{(a - x - z)(b - x - z)} \ .$$

Solving for y

$$- K'(x + z) y + K'(x + z)^2 = (a - x - z)(b - x - z)$$

$$y = (x + z) - \frac{(a - x - z)(b - x - z)}{K'(x + z)}$$

where

$$K' = \frac{1}{K_{equil}} \ .$$

From these equations, the relationship between the increase in optical density and the amount of citrate formed can be calculated for the conditions employed by OCHOA. The results plotted in Fig. 1 show that there is an appreciable error particularly at the start of the reaction. Two points of general interest may be stressed.

(1) In applying this method, one might be tempted to increase the substrate (oxaloacetate) concentration by increasing the pH. This should be avoided since the percent error in the method is thereby increased (Fig. 1A).

(2) Whilst in most assays one tries to measure initial rates, in this particular case, measurement of the initial rate introduces a gross error and greater accuracy is achieved by discounting the first 10% or so of the reaction (Fig. 1B).

The enzyme has been purified 125-fold from garcinia leaves (Xanthochymus guttiferae), by adsorption on calcium phosphate gel and ammonium sulphate precipitation (Deshpande and Ramakrishnan, 1959). The enzyme was inhibited by latex present in the leaf and by alcoholic extracts of latex. The Michaelis constant for oxaloacetate was 3.8×10^{-3} M and the concentration of acetyl phosphate which gave half maximum velocity of citrate production was 3.8×10^{-3} M.

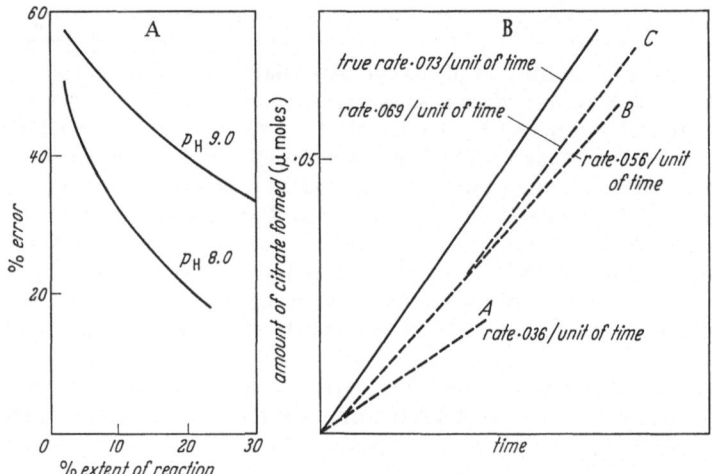

Fig. 1. Estimation of the error in the optical assay for condensing enzyme. A. The percentage error introduced by assuming that citrate and DPNH production is equal, has been calculated for the conditions employed by Ochoa (pH 8.0) and also for pH 9.0. The % error is plotted against the extent to which the reaction has been run. B. Illustrates the effect of attempting to measure initial rates. Under conditions giving a linear rate of citrate production of .073 μMoles/unit, of time, the rate measured by Ochoa's method would be (a) .036 for the time interval corresponding to 2% of the reaction. (b) .056 for the time interval corresponding to between 2 and 14% of the reaction. (c) .069 for the time interval corresponding to between 14 and 24% of the reaction.

The enzyme has also been studied in extracts of peanut cotyledons (Marcus and Velasco 1960) and was found to be inhibited by Mg^{++}. A similar inhibition was reported for the condensing enzyme of *Aspergillus niger* (Ramakrishnan and Martin 1954).

A number of studies on the mechanism of condensing enzyme have been recently reported. Marcus and Vennesland (1958) and Bove, Martin, Ingraham and Stumpf (1959) have shown that the enzyme does not (as might have been expected for an aldol condensation) labilize the hydrogen atoms of the methyl group of acetyl CoA in the absence of oxaloacetate. Englard (1959) has shown that the keto form of oxaloacetate is the substrate for condensing enzyme.

II. Aconitase.

COOH   COOH   COOH
| | |
CH₂   CH₂   CH₂
| −H₂O | +H₂O |
C(OH)COOH ⇄ C—COOH ⇄ CH—COOH
| +H₂O ‖   −H₂O |
CH₂   CH   CHOH
| | |
COOH   COOH   COOH

Citrate   ⇄   Aconitate   ⇄   *Isocitrate*

Available evidence suggests that a single enzyme catalyses both reactions and three mechanisms have been proposed.

(1) Linear

$$\text{Aconitate} + \text{Enz}$$
$$\updownarrow$$
$$\text{Citrate} + \text{Enz} \leftrightharpoons \text{Cit Enz} \leftrightharpoons \text{Acon Enz} \leftrightharpoons \textit{Iso} \text{ Cit Enz} \leftrightharpoons \text{Enz} + \textit{Iso}\text{citrate}$$

(2) Cyclic

$$\text{Citrate} + \text{Enz} \leftrightharpoons \text{Cit Enz} \leftrightharpoons \textit{Iso} \text{ Cit Enz} \leftrightharpoons \textit{Iso}\text{citrate} + \text{Enz}$$
$$\searrow \quad \text{Acon Enz} \quad \swarrow$$
$$\updownarrow$$
$$\text{Aconitate} + \text{Enz}$$

(3) Centrosymmetric

$$\text{Citrate} + \text{Enz} \leftrightharpoons \text{Cit Enz} \leftrightharpoons \text{"X"} \leftrightharpoons \textit{Iso}\text{cit Enz} \leftrightharpoons \textit{Iso}\text{citrate} + \text{Enz}$$
$$\updownarrow$$
$$\text{Acon Enz}$$
$$\updownarrow$$
$$\text{Aconitate} + \text{Enz}$$

The linear mechanism was widely accepted until MARTIUS and LYNEN (1950) failed to observe a lag in the formation of *iso*citrate from citrate. To explain the absence of a lag FRIEDRICH FRESKA and MARTIUS (1951) proposed the cyclic system. Subsequently KREBS and HOLZACH (1952) demonstrated the presence of a lag in the rate of formation of *iso*citrate from citrate and re-established aconitate as an intermediate. SPEYER and DICKMAN (1956) pointed out that whilst the absence of a lag would eliminate the linear mechanism, the existence of the lag did not prove the linear mechanism and they proposed the centrosymmetric system. The three models have been analyzed by ARONOFF and HEARON (1960) who point out that a choice between the mechanisms cannot be made on the basis of kinetics.

The experiments of SPEYER and DICKMAN (1956) with deuterium indicate that the formation of *iso*citrate from citrate does not necessitate the removal and addition of water, but rather an intramolecular re-arrangement involving a common intermediate — tentatively formulated as an enzyme Fe^{++}-tricarboxylic acid carbonium ion-cysteine complex. Following the elucidation of the absolute configuration of d-*iso*citrate (GREENSTEIN, IZUMIYA, WINITZ and BIRNBAUM, 1955; GAWRON, GLAID, LAMONTE and GARY, 1958) it has become clear that the addition of water across the double band of aconitate must be by a *trans* mechanism.

The nature of the enzyme mechanism is relevant to the validity of the assay procedure since the existence of a lag implies that measurement of the rate of citrate formation from *iso*citrate or vice versa, would not be valid.

Assay. The enzyme has not been highly purified from plants but a number of assays have been used.

1. Determination of aconitate. Aconitic acid absorbs in the ultra violet regions of the spectrum and the spectrophotometric measurement of aconitate at 240 mμ is the most convenient assay (RACKER, 1950). The wave-length 240 mμ gives minimum absorption by protein; in crude preparations which may be turbid, it is sometimes convenient to measure aconitate at 300 mμ, with however a considerable reduction in sensitivity. Certain preparations of plant aconitase behave like the animal enzyme and are activated by cysteine and ferrous ions; preparations from barley roots are unaffected by the addition of cysteine and Fe^{++}, but are inhibited by 2:2'dipyridyl (JAMES, 1956). The spectrophotometric assay in the presence of cysteine and Fe^{++} requires precautions due to the transient coloured complex formed in the presence of oxygen. Aconitate may also be determined by permanganate oxidation (Dickman, 1952) or by catalytic hydrogenation.

2. *Determination of citrate.* Methods described in the previous section can be used.

3. *Determination of isocitrate.* Isocitrate may be measured enzymically with TPN *isocitric* dehydrogenase prepared from pig heart muscle or less conveniently by polarimetry.

Mammalian aconitase interchanges the positions of the hydroxyl group of citrate and the *trans* hydrogen atom of the carbon atom derived from oxaloacetate (Potter and Heidelberger, 1949). Assuming that plant aconitase catalyses the same reaction, glutamate formed from $^{14}CO_2$ by fixation in malate or oxaloacetate and subsequent passage through the Krebs cycle, should be labelled in the C-1 position. Nelson and Krotkov (1956) found that in light, leaves incorporated $^{14}CO_2$ into glutamic acid with less than 5% of the activity in the C-1 position and suggested that plant aconitase acts between the central carbon atom of citrate and the carbon atom derived from the methyl group of acetyl CoA. A similar mechanism has been proposed for the aconitase of *Clostridium Kluyveri* (Tomlinson, 1955).

III. *Iso* Citric Dehydrogenase (TPN Specific).

The enzyme catalyses three reactions:

(1)
$$
\begin{array}{l}
COOH \\
| \\
CH_2 \\
| \\
CHCOOH \\
| \\
CHOH \\
| \\
COOH
\end{array}
+ TPN^+ \leftrightharpoons
\begin{array}{l}
COOH \\
| \\
CH_2 \\
| \\
CH_2 \\
| \\
C=O \\
| \\
COOH
\end{array}
+ CO_2 + TPNH + H^+
$$

(2)
$$
\begin{array}{l}
COOH \\
| \\
CH_2 \\
| \\
CH-COOH \\
| \\
C=O \\
| \\
COOH
\end{array}
\longrightarrow
\begin{array}{l}
COOH \\
| \\
CH_2 \\
| \\
CH_2 \\
| \\
C=O \\
| \\
COOH
\end{array}
+ CO_2
$$

(3)
$$
\begin{array}{l}
COOH \\
| \\
CH_2 \\
| \\
CH-COOH \\
| \\
C=O \\
| \\
COOH
\end{array}
+ H^+ + TPNH \rightarrow
\begin{array}{l}
COOH \\
| \\
CH_2 \\
| \\
CH-COOH \\
| \\
CHOH \\
| \\
COOH
\end{array}
+ TPN^+
$$

Each reaction requires Mg^{++} or Mn^{++} (Siebert, Carsiotis and Plaut, 1957). Efforts to demonstrate the reversibility of (2) and (3) have failed, leading to the postulate that the compound formed between oxalosuccinate and the enzyme does not readily dissociate, i.e., the equilibrium of reaction 4 is far to the right.

(4)   Oxalosuccinate + Enzyme → Enz. Oxalosuccinate

It follows that the complex is an intermediate in reaction 1 but its conversion to free oxalosuccinate is not part of the mechanism.

ENGLARD and COLOWICK (1957) have shown that there is a direct transfer of hydrogen between *iso*citrate and TPN. NAKAMOTO and VENNESLAND (1960) have shown that the transfer is stereospecific for the A side of the nicotinamide ring of TPN. The A side of the nicotinamide ring is defined as that stereoisomer of TPND which after the hydrolytic removal of phosphate yields DPND with the deuterium on the A side of the ring. In turn the A side of DPN is defined as that side of the ring which accepts deuterium from deuterated alcohol (CH_3CD_2OH) in the presence of alcohol dehydrogenase. The enzyme is conveniently assayed by following the reduction of TPN at 340 mμ. The rate of reaction is not affected by the presence of the L-form.

IV. *Iso* citric Dehydrogenase (DPN Specific).

$$
\begin{array}{c}
COOH \\
|\ \\
CH_2 \\
|\ \\
CH\!-\!COOH \\
|\ \\
CHOH \\
|\ \\
COOH
\end{array}
\ +\ DPN^+ \rightarrow
\begin{array}{c}
COOH \\
|\ \\
CH_2 \\
|\ \\
CH_2 \\
|\ \\
C\!=\!O \\
|\ \\
COOH
\end{array}
\ +\ CO_2 + DPNH + H^+
$$

The enzyme was first reported in yeast (KORNBERG and PRICER, 1951) and subsequently in heart mitochondria (PLAUT and SUNG, 1954) and pea seedling mitochondria (DAVIES, 1955). Attempts to demonstrate the reversibility of this reaction have been unsuccessful. The apparent reversibility observed in crude extracts by PLAUT and SUNG, could be attributed to the presence of glutamic dehydrogenase (DAVIES, 1955).

Assay. The reaction is conveniently assayed by measuring the reduction of DPN at 340 mμ. The enzyme is unstable and speed is necessary in the preparation of extracts for assay. The enzyme is rapidly inactivated by oxidizing agents but may be reactivated by reducing agents such as cysteine and ascorbic acid. Because of its instability, the enzyme has not been highly purified, but an extract of pea seedling mitochondria was purified 10-fold by ammonium sulphate fractionation and adsorption on aluminium hydroxide Cγ.

The pH optimum of the enzyme is about 7.5, manganese is essential for activity, the MICHAELIS constant for *iso*citrate is approximately 3×10^{-4} M and for DPN approximately 1.5×10^{-4} M.

V. α-Keto Acid Oxidases.

$$RCOCOOH + CoA + DPN^+ \rightleftharpoons RCOCoA + CO_2 + DPNH + H^+$$

Particulate enzyme systems which catalyse a CoA and DPN-linked oxidative decarboxylation of pyruvate and α-oxoglutarate have been obtained from plant, animal, and bacterial sources, but the elucidation of the mechanisms of these reactions has depended on the preparation of the animal and bacterial enzymes in soluble form (OCHOA, 1954). The pyruvic oxidase of pigeon breast muscle (SCHWEET, KATCHMAN, BOCK and JAGANNATHAN, 1952) and the α-oxoglutarate oxidase of hog-heart muscle (SANADI, LITTLEFIELD and BOCK, 1952) have been isolated as structural units of molecular weights of about 4 million and 2 million respectively. Both oxidases from *E. coli* have however been separated into and purified as two enzyme fractions (KORKES, DEL CAMPILLO, GUNSALUS and OCHOA, 1951); one fraction common to both oxidase systems has been shown to contain a DPN-linked dihydrolipoic dehydrogenase (HAGER and GUNSALUS, 1953). Lipoic acid has been known to be an essential cofactor for pyruvate oxidation since 1948 (O'KANE and GUNSALUS). Recent work has indicated that the two oxidase systems in *E. coli* are structural units of high molecular weight similar to those previously isolated from mammalian tissue (KOIKE, REED and CARROLL, 1960); it thus seems likely that the fractions previously found are derived from the more complex "multienzyme units." The pyruvic and α-oxoglutaric oxidase units both contain bound lipoic acid and FAD which can be removed by treatment with lipoyl-X-hydroylase and acid ammonium sulphate respectively (KOIKE and REED, 1960). Both cofactors are necessary for the oxidation of the keto acids by DPN; FAD, but not bound lipoic acid, is required for dihydrolipoic acid dehydrogenase activity. The flavin in the multienzyme units is reduced on the addition of the keto acid and CoA, and reoxidised on the addition of lipoic acid. These data indicate that the flavin is associated with the dehydrogenase, and that the sequence of electron transfer is keto acid → bound lipoic acid → FAD → DPN (or free lipoic acid). These findings can be explained in terms of the scheme outlined below, which is a modification of that previously proposed by GUNSALUS (1954):

1. Keto acid + TPP $\rightleftharpoons$ [Aldehyde-TPP] + CO_2
2. [Aldehyde-TPP] + oxidised bound lipoic acid $\rightleftharpoons$ TPP + reduced bound S-acetyl lipoic acid
3. Reduced bound S-acetyl lipoic acid + CoA $\rightleftharpoons$ reduced bound lipoic acid + acetyl CoA
4. Reduced bound lipoic acid + FAD $\rightleftharpoons$ oxidised bound lipoic acid + reduced FAD
5. Reduced FAD + DPN^+ $\rightleftharpoons$ FAD + DPNH + H^+

It is important to distinguish between bound lipoic acid which is attached to the protein by a covalent linkage through its carboxyl group (REED, KOIKE, LEVITCH and LEACH, 1958), and free lipoic acid, which does not activate the resolved complex, but can be oxidised by it.

In contrast to the wealth of information available about the α-keto acid oxidases of animal and bacterial cells, very little is known about these systems from plant tissues, due to the failure to obtain soluble preparations which are enzymically active. The oxidation of both α-oxoglutarate and pyruvate by plant mitochondria is well established but activity is lost when attempts are made to obtain soluble dehydrogenases (WALKER and BEEVERS, 1956; DAVIES, 1956; DAVIES, 1959). Some studies have been made of the pyruvic oxidase system of mitochondria from *Ricinus* endosperm (WALKER and BEEVERS, 1956) and sweet potatoes (LIEBERMAN and BIALE, 1956). Pyruvate is not oxidised in the absence of added cofactors. The addition of DPN, TPP, Mg ions, AMP, and malate gave high rates of pyruvate oxidation by the sweet potato mitochondria, while CoA and lipoic acid had no effect; the absolute requirement for catalytic amounts of a KREBS cycle acid such as malate, which is also found with *Ricinus* particles, reflects the need to reform CoA from acetylCoA so that the reaction can proceed with catalytic amounts of CoA. The requirement for a KREBS cycle acid can, in the case of the *Ricinus* particles, be replaced by bacterial phosphotransacetylase, when acetyl phosphate is formed instead of citrate (WALKER and BEEVERS, 1956). There seems to be no report of the direct detection of acetylCoA during pyruvate oxidation by plant mitochondria. By using substrate amounts of CoA rather than catalytic amounts, KORKES has demonstrated the formation of acetylCoA in the case of the bacterial enzyme (quoted by OCHOA, 1954). The oxidation of α-oxoglutarate does not require the addition of another KREBS cycle acid; presumably the succinylCoA breaks down to succinate and CoA by the action of deacylase or the phosphorylating enzyme. The oxidation of α-oxoglutarate by sweet potato mitochondria was stimulated by the addition of trace amounts of CoA in the presence of the other cofactors. Although lipoic acid has no effect on the oxidation of either keto acid, the powerful inhibition by arsenite is regarded as presumptive evidence for the presence of a lipoic acid-like substance in the potato mitochondria; arsenite is reported to be a fairly specific inhibitor of enzymes containing active disulphydryl groups such as those of lipoic acid (PETERS, 1949). Recently, dihydrolipoic acid dehydrogenase has been purified from acetone powders of spinach leaves (BASU and BURMA, 1960). The plant enzyme is DPN-linked and is specific for (−) dihydrolipoic acid. In contrast to the enzymes from animals and bacteria, the plant enzyme will not function with the (+) isomer of dihydrolipoic acid. The spinach enzyme is irreversibly inactivated by arsenite, and like the dihydrolipoic acid dehydrogenases from other sources has considerable diaphorase activity. The available evidence thus indicates a close similarity between the α-keto acid oxidases systems of plants and animals.

Assay methods that can be used to measure the α-keto acid oxidase systems vary according to the physical state of the system, the presence of interfering enzymes, and the point in the electron transport chain which is interrupted. The overall oxidation, which is the only assay which can be used for the particulate plant oxidases, is best measured by measuring oxygen uptake (e.g., LIEBERMAN and BIALE, 1956). The reaction between pyruvate, CoA, and DPN to give acetyl-CoA, CO_2 and DPNH can be conveniently followed by coupling with phosphotransacetylase and lactic dehydrogenase. The phosphotransacetylase converts acetyl-CoA to acetyl phosphate and thus regenerates CoA while lactic dehydrogenase oxidizes DPNH and maintains the level of DPN. The acetyl phosphate formed is assayed colorimetrically by measuring the optical density (in the range $480-540 \ m\mu$) of the acyl hydroxamate-ferric iron complex (LIPMANN and TUTTLE, 1945). It is advisable to neutralise the hydroxylamine just before use since it has a limited stability. Some anions such as phosphate and sulphate depress the intensity of the colour; BERG (1956) has given details of the preparation of concentrated neutral solutions of hydroxylamine free of salt. It should be possible to measure the formation of succinylCoA by the direct colorimetric measurement of the rate of formation of succinohydroxamic acid in the presence of hydroxylamine; the formation of the hydroxamate releases the CoA for further reaction. This method has been used to assay the production of succinylCoA by the phosphorylating enzyme (KAUFMAN, GILVARG, CORI and OCHOA, 1953).

VI. Succinyl CoA Synthetase (P Enzyme).

Succinate + CoA + ATP $\rightleftharpoons$ SuccinylCoA + ADP + Pi

The phosphorylating enzyme (or P enzyme) catalyses the only synthesis of ATP occurring at the substrate level during the operation of the KREBS cycle. The enzyme has been freed from α-oxoglutarate dehydrogenase and extensively purified from pig heart (KAUFMAN, GILVARG, CORI and OCHOA, 1953) and from spinach leaves (KAUFMAN and ALIVISATOS, 1955a).

Two methods have been used to assay this enzyme:

A. Colorimetric measurement of the rate of formation of hydroxamic acid from succinate, ATP, and catalytic amounts of CoA in the presence of hydroxylamine (KAUFMAN et al., 1953). The formation of succinohydroxamic acid from succinyl-CoA regenerates the CoA for further reaction.

B. Spectrophotometric assay at 235 mμ of the formation of the thio-ester bond of succinylCoA (KAUFMAN et al., 1955a). This assay is to be preferred for kinetic studies but requires the use of substrate amounts of CoA. A reducing agent such as cysteine, and Mg ions, should be added to the assay mixtures to ensure maximum activity.

The spinach enzyme has been purified more than 1000-fold from acetone powders (Kaufman et al., 1955a); in contrast to the preparation from heart muscle, the purified enzyme from spinach is free from myokinase, thus allowing the determination of the equilibrium constant of the reaction. The average value of the equilibrium constant ($K =$ [succinate] [CoA] [ATP/[succinylCoA] [ADP] [Pi]) was 3.7. The spinach enzyme is specific for succinate (Michaelis constant 1.5×10^{-2} M) which cannot be replaced by acetate, propionate, butyrate, lactate, malonate, fumarate, malate, tartrate, aspartate, glutarate, α-oxoglutarate, glutamate, adipate, or citrate. CoA (Michaelis constant 10^{-4} M) cannot be replaced by panthetheine. In marked contrast to the heart enzyme, the spinach enzyme will not react with GTP or ITP but only with ATP (Michaelis constant 5×10^{-4} M); the heart enzyme will not react with ATP after the commercial amorphous material has been freed from contaminating nucleotides by chromatography on a Dowex-1 column (Cohn and Carter, 1950; Sanadi, Gibson and Ayengar, 1954). The rather low value of the absorption ratio at 280 mμ/260 mμ for the most active fractions of the spinach enzyme suggested the possibility that a nucleotide prosthetic group may be bound tightly to the enzyme, but no direct evidence for this idea could be found. The spinach enzyme requires Mg ions (Michaelis constant 3.5×10^{-3} M) and is completely inhibited by p-chloromercuribenzoate (2×10^{-5} M). It was observed that the hydroxylamine used in the assay procedure inhibited the reaction by as much as 50%; dinitrophenol (10^{-4} M) does not inhibit either the heart or the spinach enzyme.

A study has been made of the mechanism of the reaction catalysed by the purified spinach enzyme (Kaufman, 1955b). The enzyme brings about

(a) an incorporation of orthophosphate into ATP which requires the presence of all the components of the system,

(b) an exchange of radioactive succinate with succinylCoA which requires the presence of only Mg ions and orthophosphate, and

(c) an exchange of radioactive ADP with ATP which requires Mg ions only.

These findings are interpreted in terms of the following mechanism:

$$\text{SuccinylCoA} + \text{Pi} + \text{enzyme} \rightleftharpoons \text{succinate} + \text{enzyme-CoA-Pi}$$
$$\text{Enzyme-CoA Pi} \rightleftharpoons \text{enzyme-Pi} + \text{CoA}$$
$$\text{Enzyme-Pi} + \text{ADP} \rightleftharpoons \text{enzyme} + \text{ATP}$$

No evidence was obtained for the formation of succinyl phosphate in the reaction. The fact that the exchange of ADP with ATP is unaffected by p-chloromercuribenzoate, whereas the overall reaction is inhibited by it, makes it unlikely that a sulphydryl group is involved in the formation of the postulated enzyme-Pi complex. This proposed mechanism for the succinylCoA synthetase reaction is different from that of the analogous acetylCoA synthetase reaction since in the latter case AMP and pyrophosphate are involved instead of ADP and orthophosphate (Lipmann, Jones, Black and Flynn, 1952). It has been suggested that the formation of AMP and pyrophosphate in the acetate reaction may serve the purpose of shifting the reaction in the direction of acetylCoA synthesis; this could be accomplished by the widely distributed pyrophosphatase.

VII. Succinic Dehydrogenase.

Succinate + oxidised dye $\rightleftharpoons$ fumarate + reduced dye

The term "succinic dehydrogenase" is used in the same sense as that employed by Singer, Kearney and Massey (1957), i.e., a soluble protein which catalyses the reversible oxidation of succinate to fumarate. The term "succinic oxidase" denotes the chain of enzymes and cofactors which links the oxidation of succinate to oxygen via cytochrome c and cytochrome oxidase. In the years since THUN-BERG's discovery of succinate oxidation (1909), many unsuccessful attempts have been made to make soluble and purify the dehydrogenase. While many added dyes have been employed with particulate and colloidal preparations to link succinate oxidation to oxygen, it was the discovery of the unique behaviour of one dye, phenazine methosulphate, which led to the solubilising and purification of the enzyme from acetone powders of animal mitochondria (KEARNEY and SINGER, 1956; SINGER, KEARNEY and BERNATH, 1956). The purified dehydrogenase shows a high specificity for electron acceptors, presumably due to the removal of various electron transport components during solubilisation and purification; phenazine methosulphate is the most efficient electron acceptor for the soluble enzyme, ferricyanide being considerably less effective. Carriers such as methylene blue, indophenol, and cytochrome c, which are commonly used to link succinate to oxygen, react only extremely slowly with the enzyme. This finding illustrates the cardinal importance of a satisfactory assay method as a prerequisite for the purification and study of an enzyme. The relative activities of a number of electron carriers have been compared using particulate preparations from heart; only phenazine methosulphate was found capable of catalysing the oxidation of succinate at a rate equal to or greater than the activity of the succinoxidase system (GIUDITTA and SINGER, 1959).

The assay commonly employed for soluble preparations of succinic dehydrogenase is the manometric estimation of oxygen uptake in the presence of succinate, phosphate, phenazine methosulphate, and cyanide (KEARNEY et al., 1956; SINGER and KEARNEY, 1957). During temperature equilibration the substrate and dye are placed in the sidearm, and the other components in the centre compartment of a Warburg flask since the enzyme is inhibited by the dye in the absence of succinate. The animal enzyme has an absolute requirement for phosphate ions. The presence of cyanide prevents the rapid decline in the initial rate of oxygen uptake due to the inhibitory action of hydrogen peroxide, which is produced in the reaction. Addition of catalase is ineffective, since the dye inhibits catalase. Since inactivation of the dehydrogenase eventually becomes apparent even in the presence of cyanide, it is desirable to take readings during the first few minutes of the assay. Recently an aerobic direct spectrophotometric assay for soluble succinic dehydrogenase has been described which is based on the finding that phenazine methosulphate will transfer electrons to 2,6-dichlorophenolindophenol; addition of phenazine methosulphate to a mixture of the enzyme succinate, and indophenol causes a reduction of the indophenol which can be measured continuously at 600 mμ (ELLS, 1959). This would be the method most desirable for a kinetic study of the enzyme. A colorimetric assay for succinic dehydrogenase has been recently published which depends on the reduction of a tetrazolium derivative by phenazine methosulphate; the red formazan formed is stabilised as a colloidal suspension by the addition of gelatin, and measured colorimetrically (NACHLAS, MARGULIES and SELIGMAN, 1960).

Succinic dehydrogenase has been purified from beef-heart mitochondria as a soluble protein in a homogenous state as judged by electrophoretic and ultra-centrifugal criteria (SINGER et al., 1956). The enzyme is a ferroflavoprotein containing 4 atoms of ferrous iron (non-hemin) and one mole of flavin per mole of protein. The flavin of succinic dehydrogenase, unlike that of other flavoproteins, is tightly bound and cannot be removed by treatment with heat or acid; proteolytic digestion releases flavin-containing fragments which have been identified as peptides of FAD (KEARNEY, 1960). The flavin thus appears to be covalently linked to the protein. Oxaloacetate, malonate, and fumarate are competitive inhibitors of the enzyme; cyanide, antimycin A, and BAL do not inhibit the soluble enzyme, indicating that these substances inhibit the succinoxidase system above the level of the dehydrogenase. The enzyme is highly sensitive to sulphydryl reagents such as p-chloromercuribenzoate (SINGER et al., 1956). The enzyme is inhibited by thyroxine and salicylate in a non-competitive fashion (HELLERMAN, REISS, PARMAR, WEIN and LASSER, 1960). Carbonyl reagents such as phenylhydrazine and semicarbazide inhibit the oxidation of succinate by beef-heart mitochondria (WESTERFELD, RICHERT and BLOOM, 1959). The succinoxidase activity of the electron transport particles from bovine heart mitochondria treated with iso-octane is restored by the addition of coenzyme Q_{10} and its analogues, suggesting a role for these biological quinones in electron transport (HENDLIN and COOK, 1960). A soluble lipoprotein containing bound coenzyme Q has been prepared from sub-mitochondrial particles which oxidise succinate (BASFORD and GREEN, 1959). A powerful inhibitor of the succinoxidase of mouse tissues has been isolated from *Hydra littoralis* where it probably originates in the nematocysts; the evidence suggests that the inhibitor is a protein and may be a component of the nematocyst toxin (KLINE and WARAVDEKAR, 1960).

Succinoxidase activity has been found associated with particulate material from many higher plants (BHAGVAT and HILL, 1951; STAFFORD, 1951). Apart from the report of BRUMMOND and BURRIS (1954) that succinic dehydrogenase was present in the soluble fraction of homogenates obtained from green lupin leaves, no successful attempts to obtain the plant enzyme in soluble form for purification were reported until very recently, when HIATT (1960) described the solubilisation and purification of succinic dehydrogenase from the mitochondria of the leaves and roots of tobacco and the roots of 4-day old *Phaseolus vulgaris*. The assay procedures used were the manometric method of KEARNEY and SINGER (1956) and the spectrophotometric assay of ELLS (1959). The activity was extracted from acetone powders of mitochondria by a similar technique to that employed by SINGER and KEARNEY for isolating the enzyme from heart mitochondria, i.e., by mixing with Tris buffer (0.06 M, pH 8.9) for 30 min. The enzyme then remained in the supernatant solution after centrifuging for two hours, at 108,000 g. The activity obtained by this procedure represented 25—30% of the activity in the mitochondria. A seven to eight fold increase in specific activity was obtained by eluting the enzyme from a calcium phosphate gel column. The pH optimum of the enzyme from *Phaseolus* roots and tobacco leaves is pH 7.4; the MICHAELIS constant for succinate is 10^{-3} M at 30°, and the inhibitor constant for malonate was calculated to be about 2×10^{-4} M. In contrast to the animal enzyme, the plant succinic dehydrogenase is not activated by the addition of phosphate to the assay medium, or by incubation of the enzyme with 0.1 M phosphate (pH 7.4) for 20 min at 25° before assay.

Several studies of the succinoxidase system of particulate material from plants have been reported recently. A toxic material which inhibited the succinoxidase of soybean particles has been found in particles from etiolated cotton hypocotyls (THRONEBERRY, 1960). The succinoxidase activity of cauliflower mitochondria is inhibited by hypotonic solutions (LATIES, 1954). The succinoxidase of mitochondria from *Ricinus* endosperm is markedly inhibited by CO_2 concentrations above 10% (RANSON, WALKER and CLARKE, 1960); this finding may account for the accumulation of succinate found in plant organs stored in 10—20% CO_2 (RANSON, 1953). Further work has shown that is succinic dehydrogenase itself which is being inhibited by the CO_2, the inhibition being competitive with succinate (BENDALL,

Ranson and Walker, 1960). Counts of the particles in root-tip cells which show a succinate-dependent reduction of tetrazolium represent only 50—60% of the number of particles staining with Janus green B; this suggests the existence of at least two kinds of particles, differing in their ability to oxidise succinate (Avers and King, 1960). By a similar technique, succinic dehydrogenase was shown to be present in granules occurring in some cells of *Avena* coleoptiles; these granules have been tentatively identified as the prolamellar bodies of etiolated plastids (Sorokin and Thimann, 1960). Bone and Fowden (1960) have demonstrated the powerful inhibitory effect of oxaloacetate on the succinoxidase system of mung-bean mitochondria.

VIII. Fumarase.

$$\begin{array}{ccc} \text{COOH} & & \text{COOH} \\ | & & | \\ \text{CH} & & \text{CH}_2 \\ \| & + \text{H}_2\text{O} \leftrightarrows & | \\ \text{CH} & & \text{CHOH} \\ | & & | \\ \text{COOH} & & \text{COOH} \end{array}$$

Assay. The enzyme is most conveniently assayed by the method of Racker (1950) in which the formation of fumarate is measured spectrophotometrically at 240 mμ. For the assay of crude preparations and for measurements of the rate of disappearance of fumarate, the change of optical density at 300 mμ may be used — but with a considerable reduction in sensitivity.

The enzyme has been prepared from pea epicotyl mitochondria (Davies, 1956) and tobacco leaf mitochondria (Pierpoint, 1960). Approximately 90% of the fumarase activity of tobacco leaves is located in the mitochondria. The fumarase content of animal mitochondrial preparations can be used as an indication of mitochondrial damage (Mahler, Wittenberger and Brand, 1958). The results of Pierpoint suggest that a similar test may be applicable to plant mitochondria. The tobacco leaf enzyme has a pH optimum of 7.8—8.0 and a Michaelis constant for fumarate of $2.2—4.2 \times 10^{-3}$ M, in Tris buffer. The enzyme is competitively inhibited by phosphate at low concentrations of fumarate and is also inhibited by iodoacetate and p-chloromercuribenzoate.

An interesting finding (Pierpoint, 1959, 1960) is that the rate of fumarate oxidation is only a third to a half of the rate of malate oxidation, indicating that the rate limiting step in the oxidation of fumarate is the formation of malate. However, the fumarase activity is greatly in excess of that required for its expected role in the oxidation of fumarate.

The mechanism of fumarase action has been intensively studied by Alberty (see Alberty, 1956). The results have been interpreted by the following sequence for the hydration of fumarate:

(1) A rapid diffusion controlled adsorption of the substrate.

(2) A rapid addition of H.

(3) A slow rate determining addition of OH.

The stereospecific synthesis of 3-monodeutero-DL-malic acid with the hydroxyl group and the deuterium atom in the *trans* position, has enabled Gawron and Fondy (1959) and Anet (1960) to establish that water is added to the double bond of fumarate by a *trans* mechanism.

IX. Malic Dehydrogenase.

$$
\begin{array}{l}
\text{COOH} \\
| \\
\text{CH}_2 \\
| \qquad + \text{DPN}^+ \rightleftharpoons \\
\text{CHOH} \\
| \\
\text{COOH}
\end{array}
\qquad
\begin{array}{l}
\text{COOH} \\
| \\
\text{CH}_2 \\
| \qquad + \text{DPNH} + \text{H}^+ \\
\text{C}{=}\text{O} \\
| \\
\text{COOH}
\end{array}
$$

The enzyme is usually regarded as being DPN specific, but the animal enzyme has detectable activity with TPN and the enzyme from spinach leaves reduces TPN at 1.5% of the rate with DPN (Hiatt and Evans, 1960). The presence of a malic dehydrogenase active with either DPN or TPN has been reported in leaves (Brummond and Burris, 1954) and in castor beans (Kornberg and Beevers, 1957). These observations could also be explained by the presence of malic enzyme and oxaloacetic decarboxylase in the extracts.

The enzyme has been purified 100-fold from spinach leaves (Hiatt and Evans, 1960). The activity of the enzyme was markedly affected by the presence of salts. When glycylglycine-Tris buffer at pH 7.4 was used, the K_m values for oxaloacetate and DPNH were 1.8×10^{-5} and 1.5×10^{-5} respectively. In the presence of NaCl (0.067 N), the K_m values were 4.45×10^{-5} M and 2.44×10^{-5} M.

The enzyme has also been purified 50-fold from pea epicotyls and to a lesser extent from cauliflowers (Davies, 1960a). The Michaelis constant for oxalo-acetate at pH 7.0 was between $8-9 \times 10^{-5}$ M for both preparations and the Michaelis constant for DPNH was 1.2×10^{-4} M.

The enzyme has been purified from parsley leaves (Stafford, 1956) and the Michaelis constant for malate at pH 9.0 given as 5×10^{-3} M, which may be compared with the value of 4×10^{-2} M for preparations from wheat (Waygood, 1950).

Price and Thimann (1954) have shown that 75% of the total malic dehydro-genase activity of pea epicotyls occurs in the supernatant after removing mito-chondria from a homogenate. A similar finding with animal homogenates has led to the suggestion that there are two physiologically distinct malic dehydrogenases. Delbruck, Schimmassek, Bartsch and Bücher (1959) have confirmed the finding of Davies and Kun (1957) that mitochondrial malic dehydrogenase is inhibited by concentrations of oxaloacetate greater than 10^{-4} M and observed that the soluble dehydrogenase is not inhibited by concentrations of oxaloacetate up to 10^{-3} M. The possibility that the kinetic difference between the enzymes is reflected in molecular structure, is suggested by the finding that on electrophoresis, 90% of the supernatant enzyme migrates to the anode, whereas the anode com-ponent of the mitochondrial enzyme represents only 14% (Wieland, Pfleiderer, Haupt and Worner, 1959). During the purification of malic dehydrogenase from peas and cauliflowers two components with malic dehydrogenase activity were observed during chromatography on substituted cellulose columns (Davies, 1960a).

The specificity of malic dehydrogenase is discussed under the heading tartaric dehydrogenase.

Malic dehydrogenase of wheat germ, like the animal enzyme, catalyses the stereospecific addition of hydrogen to the A side of the nicotinamide ring of DPN (Loewus, Tchen and Vennesland, 1955). The keto tautomer of oxalo-acetic acid has been established as the substrate for the dehydrogenase.

B. Enzyme Activities Related to the Krebs Cycle.

I. Tartaric Dehydrogenase.

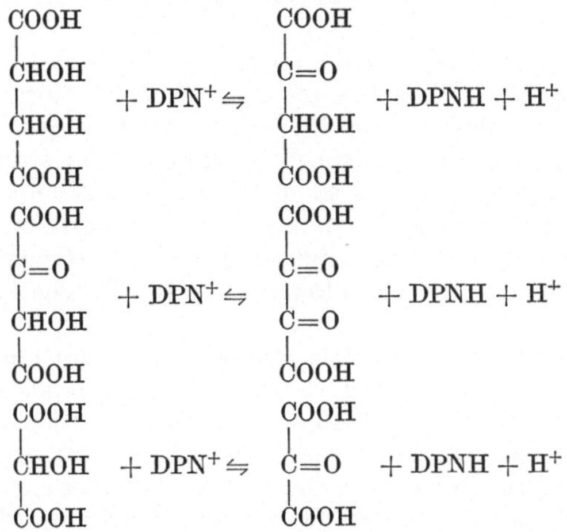

$$
\begin{array}{l}
\text{COOH} \\
| \\
\text{CHOH} \\
| \\
\text{CHOH} \\
| \\
\text{COOH}
\end{array}
+ \text{DPN}^+ \rightleftharpoons
\begin{array}{l}
\text{COOH} \\
| \\
\text{C=O} \\
| \\
\text{CHOH} \\
| \\
\text{COOH}
\end{array}
+ \text{DPNH} + \text{H}^+
$$

$$
\begin{array}{l}
\text{COOH} \\
| \\
\text{C=O} \\
| \\
\text{CHOH} \\
| \\
\text{COOH}
\end{array}
+ \text{DPN}^+ \rightleftharpoons
\begin{array}{l}
\text{COOH} \\
| \\
\text{C=O} \\
| \\
\text{C=O} \\
| \\
\text{COOH}
\end{array}
+ \text{DPNH} + \text{H}^+
$$

$$
\begin{array}{l}
\text{COOH} \\
| \\
\text{CHOH} \\
| \\
\text{COOH}
\end{array}
+ \text{DPN}^+ \rightleftharpoons
\begin{array}{l}
\text{COOH} \\
| \\
\text{C=O} \\
| \\
\text{COOH}
\end{array}
+ \text{DPNH} + \text{H}^+
$$

The above reactions have been demonstrated in higher plants (STAFFORD, 1956, 1957) but the existence of specific enzymes has not been established. STAFFORD produced two types of evidence to support the view that separate enzymes are involved. (1) During the purification of hydroxymalonic dehydrogenase, the relative activities with oxomalonate, dioxosuccinate and oxaloacetate were not constant. (2) Hydroxymalonate is a competitive inhibitor of malic dehydrogenase. It is probable that the concentrations of oxomalonate and dioxosuccinate used by STAFFORD were well below saturation and in consequence her assays would be sensitive to changes in the concentration of the unstable keto acids. The observation of competitive inhibition, supports the view that a single enzyme is involved and quantitative evidence for this view has been presented (DAVIES, 1960a).

Assay. The activities are best determined by measuring the rate of DPNH oxidation at 340 mμ. The main difficulty lies in the instability of the keto acids, and it is necessary to prepare solutions containing a chelating agent such as ethylenediamine-tetraacetate just before assay.

Using an enzyme prepared from cauliflowers the MICHAELIS constants at pH 7.0 for oxomalonate and dioxosuccinate were 2.5×10^{-2} M and 8×10^{-3} M, with an enzyme prepared from pea seedlings the constants were 1.4×10^{-2} M and 7×10^{-3} M. The MICHAELIS constant for DPNH in the presence of oxomalonate (1.00×10^{-2} M) was 3.5×10^{-5} M. Values for dihydroxyfumarate have not been obtained because this compound exists in solution mainly in the enol form.

II. Malease.

$$\text{Maleate} + \text{H}_2\text{O} \rightleftharpoons \text{Malate}$$

An enzyme which catalyses the hydration of maleate to malate has been purified from maize kernels (SACKS and JENSEN, 1951). The reaction was shown to be reversible, and took place without the intermediate formation of any detectable amounts of fumarate. Two methods were employed for determining enzymic activity:

A. Formation of an insoluble red formazan on incubation of the enzyme with DPN, maleate, and 0.5% 2,3,5-triphenyltetrazolium chloride. This method depends on the presence of malic dehydrogenase in the enzyme preparation to convert malate to oxaloacetate. The DPNH is then oxidised by the tetrazolium salt to form DPN and the formazan.

B. The polarographic measurement of maleate; by this means the reaction can be studied in both directions, and the absence of fumarate formation established.

It should also be possible to assay this enzyme more conveniently by spectro-photometric measurements in the ultra-violet region of the spectrum in a similar fashion to that employed for the assay of fumarase at 240 mμ (RACKER, 1950). In this case however, it would be necessary to establish by another method whether the *cis* or *trans* isomer was being produced by the dehydration of malate; the hydration of maleate could be followed directly with checks to establish the absence of fumarate.

Maleate metabolism has been little studied, but there are indications that maleate is metabolically active. For example, pyruvic oxidase is inhibited by maleate (PETERS and WAKELIN, 1946), while the inhibition of the growth of *Avena* coleoptiles by iodoacetate is relieved by maleate (THIMANN and BONNER, 1948). When fed to excised tobacco leaves, maleate is rather slowly metabolised to unidentified compounds while the respiration rate is stimulated about 60%; the formation of citrate and the breakdown of protein that is characteristic of excised leaves are both inhibited by maleate (VICKERY and PALMER, 1956). Maleate has been reported as occurring in tobacco leaves (GIOVANNOZZI-SERMANNI, 1957). These findings of metabolic activity may be related to the fact that maleate can react with sulphydryl groups under physiological conditions to form addition compounds; a *cis-trans* conversion accompanies these reactions, fumarate being produced (MORGAN and FRIEDMANN, 1938). It is tempting to speculate that maleate is attached to malease via a sulphydryl group; in contrast, mammalian fumarase does not appear to possess sulphydryl groups essential for activity, nor does fumarate react non-enzymically with thiol compounds (MASSEY, 1953) though yeast fumarase and plant fumarase (PIERPOINT, 1960) are inactivated by p-chloromercuribenzoate. It is of interest in this connection to note that maleate is not a substrate for fumarase, but acts as a competitive inhibitor (MASSEY, 1953). After maleate-2-C^{14} was injected into dogs and humans, C^{14}O$_2$ was detected in blood samples; no evidence of fumarate formation from maleate was obtained in these experiments (SACKS, 1958).

Despite evidence of the metabolic activity of maleate, buffers containing this compound have been recommended for enzyme studies (GOMORI, 1955), and have been used to culture coleoptile sections in growth hormone studies (McRAE, FOSTER and BONNER, 1953).

III. Pyruvic (De)Carboxylase.

Pyruvate → Acetaldehyde + CO$_2$

This enzyme, sometimes imprecisely termed α-carboxylase, has long been invoked to account for the production of CO$_2$ during the anaerobic fermentation of sugars, and is distinct from the enzyme system which oxidatively decarboxylates pyruvate to acetylCoA in aerobic metabolism. Pyruvic decarboxylase has been detected in extracts from many plants (JAMES, 1953), and from animal tissues (GREEN, WESTERFELD, VENNESLAND and KNOX, 1941). For example, aqueous extracts of jack-bean meal contain a potent decarboxylase which is inactivated by dialysis; addition of Mg ions and thiamine pyrophosphate restores the activity (COHEN,

1946). The decarboxylase activity of soy-beans sprouts rises to a maximum and then declines as germination proceeds; the variation in activity closely parallels the variation in the amount of thiamine pyrophosphate in the extracts (MEE, 1949). Pyruvate accumulates in barley tissues poisoned with aromatic sulphonic acids or acetaldehyde, presumably due to the blocking of decarboxylase (JAMES and JAMES, 1940). Pyruvic decarboxylase is widely distributed in seeds and roots, but usually has low activity in green leaves (VENNESLAND and FELSHER, 1946; CLENDENNING, WAYGOOD and WEINBERGER, 1952). The decarboxylase of broad bean leaves is found in the soluble fraction, and no activity could be detected in the mitochondrial and chloroplast fractions (JAMES and DAS, 1957); similar results have been obtained for homogenates of carrot roots, barley roots, etiolated pea shoots, Shasta daisy leaves, and potatoes (JAMES and RICHENS, 1960). It is of interest to note that potatoes when first placed under nitrogen at 10° produce lactic acid and not alcohol (BARKER and EL SAIFA, 1952). After about 10 days under nitrogen, alcohol begins to appear. Experiments by JAMES and RICHENS indicate that extracts of potatoes kept in air contain little decarboxylase activity, but that this activity increases with storage of the potatoes under nitrogen. The switch from lactic acid to alcohol production may thus be due in part to the development, perhaps adaptively, of decarboxylase activity, since alcohol dehydrogenase is present in the tubers at all times (JAMES, 1957).

The generally weaker fermentating power of bakers yeast compared with brewers yeast has been attributed to the lack of decarboxylase in commercial bakers yeast; the decarboxylase content of bakers yeast cells decreases when the cells are transferred from anaerobic to aerobic conditions (SOUMALAINEN and OURA, 1959).

Manometric assay. Pyruvic decarboxylase is usually assayed manometrically by measuring the CO_2 output from solutions buffered at pH 5.0—6.0 (SINGER and PENSKY, 1952a); nitrogen instead of air should be used as the gas phase if oxidative reactions are suspected. It is common practice to add dimedon (or some other carbonyl reagent) and serum albumin to the assay mixtures to remove the acetaldehyde, the accumulation of which strongly inhibits the further decarboxylation of pyruvate. In the case of the enzyme from bakers yeast, the inhibition by acetaldehyde has been shown to be of the non-competitive or uncompetitive type (GRUBER and WASSENAAR, 1960). The enzyme has been partially purified from yeast (see SINGER, 1955) and highly purified from wheat germ (SINGER and PENSKY, 1952a). The purified wheat germ enzyme is fully resolved and requires the addition of thiamine pyrophosphate (MICHAELIS constant, 1.35×10^{-6} M) and a divalent metal ion, Mg, Zn, and Co ions yielding the maximum activity. Pyruvate (MICHAELIS constant 3.6×10^{-3} M) and α-ketobutyrate are decarboxylated at identical rates; α-oxoglutarate is acted on very slowly, and pyruvamide is not attacked. The pH optimum is 6.3—6.4 and the enzyme is strongly inhibited by p-chloromercuribenzoate. The highly purified enzyme is electrophoretically homogenous throughout the pH range in which it is stable (6.4—7.9); in the ultracentrifuge, three components are found, one of which (93% of the total) is enzymically active.

The highly purified enzyme from wheat germ catalyses the synthesis of acetoin (acetylmethylcarbinol) from pyruvate plus acetaldehyde and from acetaldehyde alone, confirming the hypothesis that acetoin formation is a consequence of decarboxylase action rather than the function of a separate enzyme "carboligase"; formation of acetoin has been demonstrated in a variety of bacteria, yeasts, animals, and higher plants (SINGER and PENSKY, 1952b). The constant ratio of pyruvate decarboxylation to acetoin synthesis throughout purification, and the identical requirements for thiamine pyrophosphate and a metal ion leave little doubt that one enzyme is responsible for two activities. The synthesis of acetoin, which cannot be reversed, displays a partial asymmetry since 72% was of the (+) form, and 28% of the (−) form. Acetoin can also be produced together with α-acetolactate in the non-enzymic decarboxylation of pyruvate catalysed by thiamine (Yatco-Manzo, Roddy, Yount and Metzler, 1959).

Singer and Pensky (1952b) have proposed a theory of decarboxylase action in terms of the formation of a relatively stable enzyme-acetaldehyde complex; they speculate that this complex may be the first step in the action of the pyruvic oxidase system because of similarities that have been observed between the decarboxylase and the oxidase isolated by Schweet, Jagannathan and Katchman (1951), e.g., the oxidase also catalyses the formation of acetoin. It seems possible that the enzyme-acetaldehyde complex under the influence of a second prosthetic group may be oxidised to an enzyme-acetyl compound which could then react with CoA. This theory has been extended to include other reactions of pyruvate concerned with the synthesis of valine and isoleucine in *Neurospora* (Radhakrishnan and Snell, 1960). Pyruvic decarboxylase may thus be the prototype of the enzymes that activate α-oxo acids.

IV. Lactic Dehydrogenase.

$$\text{Pyruvate} + \text{DPNH} + \text{H}^+ \rightleftharpoons \text{Lactate} + \text{DPN}^+$$

Whereas alcohol is the traditional end product of anaerobic respiration in plants, lactate is the product of anaerobic respiration in animal tissues. However a number of plant tissues produce lactate (James, 1957), and lactic dehydrogenase has been detected in a particulate fraction from rhizomes of *Equisetum* (Barber, 1957), in extracts from potato tubers, squash fruit, and turnip roots (Stafford, Magaldi and Vennesland, 1954), in soluble extracts of mitochondria from pea epicotyls (Davies, 1956), and in extracts from carrots (James, 1957) and *Chlorella* (Warburg, Gewitz and Volker, 1957). Lactic dehydrogenase was not found in extracts from many plants by Stafford et al. (1954); claims that lactic dehydrogenase is present in leaves (Nason, Oldewurtel and Propst, 1952) based on the stimulation of oxygen by lactate, cannot be accepted since glycollic acid oxidase, which is widely distributed in leaves, is capable of oxidising lactate.

Lactic dehydrogenase does not seem to have been purified from any higher plant source, but has been crystallised from beef heart (Straub, 1940) rat liver (Gibson, Davisson, Bachhawat, Ray and Vestling, 1953), and bakers yeast (Appleby and Morton, 1954).

Assay. The lactic dehydrogenase found in higher plants and animals is most conveniently assayed by the spectrophotometric measurement of DPNH at 340 mμ (Neilands, 1952, 1955). Since the equilibrium of the reaction greatly favours pyruvate reduction, the reaction is assayed in this direction. Measurements of pyruvate reduction can be carried at neutral pH values, but measurements of lactate oxidation are best carried out at pH 10.0 since the reaction produces an equivalent of acid. Beef heart lactic dehydrogenase is reported to be inhibited by pyruvate, necessitating the addition of a carbonyl reagent such as semicarbazide to the assay cell when lactate oxidation is being measured (Takenaka and Schwert, 1957; Hakala, Glaid and Schwert, 1956; Winer and Schwert, 1958). When DPNH oxidation is being measured, it may be necessary to perform control assays in the absence of pyruvate to allow for DPNH oxidase activity (Stafford et al., 1954). If the extracts contain pyruvic decarboxylase, control assays should be performed to ensure that the oxidation of DPNH is not due to alcohol dehydrogenase acting on the acetaldehyde formed from pyruvate.

Yeast lactic dehydrogenase differs from that of animals and higher plants in being independent of nucleotide coenzymes; the purified enzyme has been shown to be identical with cytochrome b_2, to contain FMN as a second prosthetic group, and to be associated with a DNA component (MONTAGUE and MORTON, 1960). The yeast enzyme can be assayed by measuring the rate of reduction of methylene blue by the THUNBERG technique, by the spectrophotometric measurement of the rate of reduction of ferricyanide or cytochrome c (APPLEBY and MORTON, 1954), or by the measurement of the rate of oxygen uptake in the presence of phenazine methosulphate (HORIO, YAMASHITA and OKUNUKI, 1959).

The lactic dehydrogenases from yeast, muscle, and higher plant source are specific for L(+) lactate; crude extracts of *Lactobacillus arabinosus* interconvert L and D lactate in the presence of DPN, suggesting the presence of D and L lactic dehydrogenases (KAUFMAN, KORKES and DEL CAMPILLO, 1951), while extracts of *Propionobacterium* catalyse a pyridine nucleotide-independent reduction of D(−) lactate to pyruvate (MOLINARI and LARA, 1960). Recently the two stereospecific lactic dehydrogenases of *Lactobacillus* have been purified and separated, and their properties compared; the L-specific enzyme is inhibited by oxamate, while the D-specific enzyme is not (DENNIS and KAPLAN, 1960). The lactic dehydrogenase of *Chlorella* produces D-lactate (WARBURG et al., 1957).

Animal lactic dehydrogenase reacts with TPNH at a much slower rate than with DPNH (SALLES and OCHOA, 1950; MEHLER, KORNBERG, GRISOLIA and OCHOA, 1948). Rat liver lactic dehydrogenase catalyses pyruvate reduction by TPNH at about 1/40 (at pH 6.0) to 1/1000 (at pH 7.6) of the rate obtained with DPNH; however, since the enzyme is present in rat liver at a high concentration, its TPN-linked activity is of the same order of magnitude as that of true TPN-linked oxidations (NAVAZIO, ERNSTER and ERNSTER, 1957). Crystalline beef heart lactic dehydrogenase reduces a number of α-keto acids besides pyruvate (MEISTER, 1950) but the activity with α-oxoglutarate and oxaloacetate is too low to seriously affect the quantitative assay of pyruvate in a mixture of these keto acids. β-Mercaptopyruvate has been added to the list of keto acids acted on by lactic dehydrogenase (KUN, 1957). α-Hydroxysulphonates are reported to inhibit the oxidation of lactate by beef heart lactic dehydrogenase, but not the reduction of pyruvate (ZELITCH, 1957).

The lactic dehydrogenase of mammalian tissues has been shown to exist in several electrophoretically distinct forms or isozymes; various tissues possess a characteristic pattern of isozymes which can be altered in different disease states (MARKERT and MOLLER, 1959; VESELL and BEARN, 1957). Antisera to rabbit muscle lactic dehydrogenase have been used to distinguish between the various forms from different organs of the rabbit (NISSELBAUM and BODANSKY, 1959). Each of the rabbit and human isozymes shows different substrate concentration optima at any fixed pH (PLAGEMANN, GREGORY and WROBLEWSKI, 1960); the possibility that a similar state of affairs may exist for the lactic dehydrogenase from plant tissues should be considered in any kinetic study. Two lactic dehydrogenases have been reported from yeast cells, differing in their ability to reduce cytochrome c (SLONIMSKI and TYSAROWSKI, 1958).

Several kinetic studies of the crystalline muscle enzyme have been reported. The variation of the kinetic constants with pH has been studied, and the results interpreted in terms of the formation of a ternary complex during the action of the enzyme (TAKENAKA and SCHWERT, 1957; WINER and SCHWERT, 1958). Inhibition by oxamate has been shown to be competitive with pyruvate while inhibition by oxalate is competitive with lactate (NOVOA, WINER, GLAID and SCHWERT, 1959). Purified rat liver lactic dehydrogenase is reversibly inhibited by sodium sulphide, possibly due to the formation of a complex between DPN and the HS⁻ ion (TERAYAMA and VESTLING, 1956). The stoichiometry and dissociation constants of the complexes of DPN and DPNH with beef heart lactic dehydrogenase have been studied by fluorimetric techniques (VELICK, 1958; WINER, SCHWERT and MILAR, 1959). The acetyl-pyridine analogue of DPNH has been shown to be bound more firmly to the enzyme than is DPN, and it has been suggested that this firm binding may allow sufficiently drastic treatment of the enzyme to facilitate the analysis of coenzyme-binding sites (SHIFRIN, KAPLAN and CIOTTI, 1959).

V. Phosphoenolpyruvic Carboxylase.

$$\begin{array}{c} \text{COOH} \\ | \\ \text{C-O-PO(OH)}_2 + \text{H}_2\text{O} + \text{CO}_2 \xrightarrow{\text{Mg}^{++}} \\ \| \\ \text{CH}_2 \end{array} \qquad \begin{array}{c} \text{COOH} \\ | \\ \text{C=O} \\ | \\ \text{CH}_2 \\ | \\ \text{COOH} \end{array} + \text{H}_3\text{PO}_4$$

This enzyme was first reported by BANDURSKI and GREINER (1953) in spinach leaves, but has since been shown to have a wide distribution in plants (MAZELIS and VENNESLAND, 1957).

A number of assay systems have been used:

1. Incorporation of $^{14}CO_2$ into oxaloacetate. Phosphopyruvate, Mg^{++}, glutathione, buffer, $NaH^{14}CO_3$ and enzyme are incubated and the reaction stopped by the addition of acid. After removal of $^{14}CO_2$ by vigorous gassing, aliquots are taken, the oxaloacetate decarboxylated with Al^{+++} and the $^{14}CO_2$ collected and counted.

A disadvantage of this method is that Mg^{++} which is necessary for enzyme activity catalyses the decarboxylation of oxaloacetate.

2. Determination of oxaloacetate. Oxaloacetate may be determined by a colorimetric method based on the coloured complex formed between oxaloacetate and ferric ion (NOSSAL, 1949). As in the previous assay losses of oxaloacetate due to Mg^{++} catalysed decarboxylation may be expected.

3. Determination of phosphate. The production of inorganic phosphate may be measured. However, plant preparations frequently contain phosphatase which give a large blank reaction i.e., phosphate released in the absence of CO_2.

4. Spectrophotometric assay. The production of oxaloacetate by the carboxylase reaction may be coupled to the reduction of oxaloacetate by DPNH catalysed by malic dehydrogenase. The reaction is thus measured by following the oxidation of DPNH spectrophotometrically. Reactions which could interfere with this method are catalysed by phosphatase and lactic dehydrogenase. In the presence of these enzymes, only the CO_2 dependent oxidation of DPNH should be measured. The studies of TCHEN, LOEWUS and VENNESLAND (1955) have established that it is the keto form of oxaloacetate which results from CO_2 fixation thereby establishing that if a phosphorylated intermediate of oxaloacetate is formed the phosphate must be attached to the carboxyl group.

The enzyme has been purified 10-fold from spinach leaves (BANDURSKI, 1955). Mg^{++} and a reducing agent were necessary for activity; the MICHAELIS constant for phosphopyruvate was 5×10^{-4} M at pH 7.5.

Preparations of the enzyme from wheat germ and also from spinach were found not to require a reducing agent (TCHEN and VENNESLAND, 1955).

The enzyme has been studied in extracts of *Crassulacean* plants (WALKER, 1957; WALKER and BROWN, 1957). The MICHAELIS constant for phosphopyruvate was between 1.5 and 1.94×10^{-4} M at pH 7.4, and the MICHAELIS constant for carbon dioxide was 2.2×10^{-4} M. Concentrations of carbon dioxide greater than 3% were found to inhibit the reaction in a non-competitive manner.

VI. Phosphoenolpyruvate Carboxykinase.

$$CO_2 + \underset{\overset{|}{COOH}}{\overset{\overset{\displaystyle COOH}{|}}{C}}\!\!-\!O\!-\!PO(OH)_2 + ADP \underset{}{\overset{Mg^{++}\,or\,Mn^{++}}{\rightleftarrows}} \underset{\overset{|}{\underset{\overset{|}{COOH}}{CH_2}}}{\overset{\overset{\displaystyle COOH}{|}}{C}}{=}O \; + ATP$$

Assay. The enzyme is conveniently assayed by measuring the ATP dependent exchange of $^{14}CO_2$ into oxaloacetate (MAZELIS and VENNESLAND, 1957). Oxaloacetate, Mn^{++}, $NaH^{14}CO_3$, buffer, ATP and enzyme are incubated and the reaction stopped by the addition of acid. After vigorous gassing, to remove $^{14}CO_2$, the incorporation into oxaloacetate is measured by decarboxylation with aniline citrate and counting the liberated CO_2 as $BaCO_3$.

The blank reaction i.e., the amount of $^{14}CO_2$ incorporated into oxaloacetate in the absence of ATP is a non-enzymic reaction due to a Mn^{++} catalysed exchange reaction.

An anomalous exchange reaction has been found in preparations from spinach leaves. ATP did not stimulate the incorporation of $^{14}CO_2$ into oxaloacetate, but ADP did stimulate the incorporation of CO_2 into oxaloacetate in the presence of phosphopyruvate.

The enzyme is widely distributed in plants and unlike the pigeon liver enzyme which is specific for GTP, the plant enzyme is specific for ATP.

VII. Malic Enzyme.

(1)
$$\underset{\overset{|}{\underset{\overset{|}{COOH}}{CHOH}}}{\overset{\overset{\displaystyle COOH}{|}}{CH_2}} + TPN^+ \xrightarrow{Mn^{++}} \underset{\overset{|}{COOH}}{\overset{\overset{\displaystyle CH_3}{|}}{C}}{=}O + CO_2 + TPNH + H^+$$

(2)
$$\underset{\overset{|}{\underset{\overset{|}{COOH}}{C=O}}}{\overset{\overset{\displaystyle COOH}{|}}{CH_2}} \xrightarrow{Mn^{++}} \underset{\overset{|}{COOH}}{\overset{\overset{\displaystyle CH_3}{|}}{C}}{=}O + CO_2$$

Assay. The enzyme may be assayed by following the reduction of TPN at 340 mμ or by measuring the decarboxylation of oxaloacetate at pH 5.2.

The enzyme is widely distributed in plants (CONN, VENNESLAND and KRAEMAR, 1949; CLENDENNING, WAYGOOD and WEINBERGER, 1952) and has been purified 300-fold from wheat germ (HARARY, KOREY and OCHOA, 1953). The MICHAELIS constant for malate is 7×10^{-4} M at pH 7.3 and for oxaloacetate is 6.5×10^{-3} M at pH 5.2. The concentration of Mn^{++} give half maximum rate of reaction 1, at pH 7.3 is 2.5×10^{-5} g atom/litre and for reaction 2 at pH 5.2 is 10^{-4} g atom/litre.

The enzyme has been studied in extracts of *Kalanchoë crenata* (WALKER, 1960). The MICHAELIS constant for malate is 5.5×10^{-4} M at pH 7.4 and the reductive carboxylation of pyruvate is slightly inhibited by CO concentrations in excess of 30%.

C. The Glyoxalate Cycle.

To explain the formation of carbohydrates from acetyl CoA, Kornberg and Krebs (1957) proposed a set of reactions called the glyoxalate cycle (Fig. 2) which accounts for the oxidation of acetyl CoA to glyoxalate.

The two "key" enzymes participating in the cycle are malate synthetase and isocitritase, also known as isocitric lyase and isocitrase.

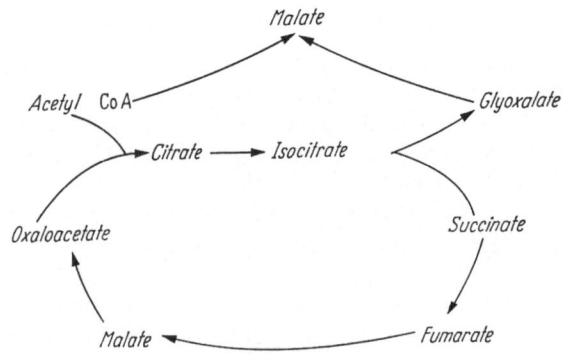

Fig. 2. Reactions of the glyoxalate cycle.

I. Malate Synthetase.

$$AcetylCoA + glyoxalate + H_2O \leftrightharpoons Malate + CoA$$

Malate synthetase was discovered in extracts of acetate-grown *E. coli* by Wong and Ajl (1956).

Three types of assay have been employed for this enzyme:

A. Measurement of the radioactivity in malate after incubation of the enzyme with glyoxalate and C^{14}-acetate, ATP and CoA (Kornberg and Beevers, 1957).

B. Spectrophotometric assay of the rate of cleavage of the thioester bond of acetylCoA at 232 mμ (Stadtman, 1957; Dixon and Kornberg, 1959).

C. Measurement of the rate of glyoxalate disappearance by the colorimetric method of Friedemann and Haugen (1943) as used by Marcus and Velasco (1960a).

Method A is complicated by the normally slow rate of conversion of acetate to acetylCoA. This complication can be overcome either by using C^{14}-acetylCoA or by adding excess purified aceto-CoA-kinase from yeast (Yamamoto and Beevers, 1960). The malate can be separated chromatographically before counting; an alternative method which avoids the need for chromatography makes use of the fact that if acetyl-1-C^{14}-CoA is employed, then the C^{14} in the malate formed is almost exclusively confined to the C-4 carboxyl group. The wide distribution of fumarase would seem to preclude the general use of this method, but it is applicable at least in the case of crude extracts of castor bean endosperm (Yamamoto and Beevers, 1960). The C-4 atom can be specifically released as CO_2 by means of malate adapted *Lactobacillus arabinosus*. In extracts from peanut and sunflower cotyledons, fumarase activity results in 10—30% of the label appearing in the C-1 atom of malate (Bradbeer and Stumpf, 1959). If the tissue being examined contains isocitritase, it is convenient to use isocitrate as a source of glyoxalate. It may be noted that the rate of malate formation is frequently faster with isocitrate as the glyoxalate precursor, than when free glyoxalate is used. Care is needed to ensure that the isocitritase reaction is not rate determining.

Method B has the advantage of simplicity and is the method to be preferred for kinetic studies; however it is necessary to allow for the possibility that acetyl-CoA may be removed by enzymes other than malate synthetase e.g., by the action of phosphotransacetylase or deacylase. Interference by phosphotransacetylase is only likely where bacterial extracts are being assayed, and can be avoided by the use of buffers containing no phosphate. This assay demands the use of a spectrophotometer sensitive at high optical densities; the short wavelength employed may limit the usefulness of the method when extracts containing much pigment are being assayed. It is necessary to exercise caution when performing quantitative assays at high optical densities since the increased proportion of stray radiation may result in optical densities not being additive (LOTHIAN, 1958).

Malate synthetase has been purified from *Pseudomonas ovalis* and from bakers yeast (DIXON, KORNBERG and LUND, 1960). It is essential to include Mg^{++} ions during all stages of purification to avoid serious losses of activity; the purified enzyme requires Mg^{++} ions for activity. Glyoxylate cannot be replaced as a substrate by pyruvate, oxaloacetate, α-oxoglutarate, glyoxal, glycolaldehyde, formaldehyde, or acetaldehyde. FluoroacetylCoA is split at about one-quarter of the rate observed with similar concentrations of acetylCoA. PropionylCoA and butyrylCoA will not replace acetylCoA as a substrate for the enzyme from peanut cotyledons (MARCUS and VELASCO, 1960b). The microbial enzyme is inhibited by oxalate, fluoroacetate, and glycollate (these compounds competing with glyoxylate), but not by acetate or oxaloacetate. No measurable reversal of the malate synthetase reaction could be demonstrated with either the microbial enzyme (DIXON et al., 1960) or the peanut enzyme (MARCUS et al., 1960b). BURTON (quoted by DIXON et al., 1960) has calculated the free energy change of the reaction to be about $10-12$ kcal/mole. The mechanism of the reaction catalysed by malate synthetase is probably very similar to that of the condensing enzyme in which a new carbon-carbon bond is produced via an aldol condensation. Synthetic malyl-CoA is not split by microbial malate synthetase in the absence or presence of glyoxylate, and is therefore probably not an intermediate. Since the enzyme does not affect the rate at which radioactivity from $CoA-S^{35}H$ is incorporated into unlabelled acetylCoA, there is no evidence that acetylCoA is bound to the enzyme in the absence of glyoxylate.

Malate synthetase has been shown to be present in a variety of higher plant materials; although it is not confined to those plant tissues which convert fat to carbohydrate as is isocitritase, the best sources are fatty seeds in which this change is occurring. The activity of malate synthetase increases as the seed germinates in the case of peanuts (MARCUS et al., 1960a) and castor beans (YAMAMOTO et al., 1960). The bulk of the activity of castor bean homogenates can be recovered in the mitochondrial fraction (MARCUS et a., 1960a; YAMAMOTO et al., 1960). The addition of $MgCl_2$ and glutathione increases the amount of incorporation of C^{14}-1-acetate into malate in castor bean extracts (YAMAMOTO et al., 1960).

II. Isocitritase.

$$
\begin{array}{c}
\text{COOH} \\
| \\
\text{CH}_2 \\
| \\
\text{CHCOOH} \\
| \\
\text{CHOH} \\
| \\
\text{COOH}
\end{array}
\quad \rightleftharpoons \quad
\begin{array}{c}
\text{COOH} \\
| \\
\text{CH}_2 \\
| \\
\text{CH}_2 \\
| \\
\text{COOH}
\end{array}
\quad + \quad
\begin{array}{c}
\text{CHO} \\
| \\
\text{COOH}
\end{array}
$$

This enzyme was originally demonstrated in extracts from species of *Pseudomonas* (Campbell, Smith and Eagles, 1953) and was subsequently found to be present in a variety of bacteria and fungi. The enzyme was first demonstrated in plants by Kornberg and Beevers (1957).

Three assays are available:

(1) Glyoxalate formation.

*Iso*citrate, Mg^{++}, cysteine or glutathione, buffer and enzyme are incubated, and at appropriate time intervals, the reaction is stopped by the addition of trichloroacetic acid. After removing precipitated protein, the supernatant is analysed for glyoxalate by the method of Friedman and Haugen (1943).

(2) Succinate dependent removal of glyoxalate.

Glyoxalate, Mg^{++}, succinate, buffer and enzyme are incubated and at intervals the disappearance of glyoxalate is determined by the method of Friedman and Haugen. A control omitting succinate should always be performed.

(3) Continuous measurement of glyoxalate formation by means of its semicarbazone.

This is the most satisfactory method and was introduced by Olson (1959). *Iso*citrate, Mg^{++}, cysteine, buffer at pH 6.0, semicarbazide and enzyme are incubated in cuvettes and the increase in optical density at 252 mμ, due to the formation of the semicarbazone of glyoxalate (molecular extinction coefficient — 12,400) is measured. According to Olson, the *allo*-forms of isocitrate are inhibitory and the observation of Carpenter and Beevers (1959) that the DL-form of *iso*citrate gives a slower reaction rate than the D-form could be due to the presence of the *allo* forms in the synthetic racemic mixture. The assay must be carried out at a slightly acid pH such as 6.0 to permit rapid formation of the semicarbazone.

The enzyme has been purified 30-fold from castor bean seedlings (Carpenter and Beevers, 1959) and 50-fold from *Blastocladiella* (McCurdy and Cantino, 1960).

The Michaelis constant for *iso*citrate measured with different enzyme preparations was 3.2×10^{-4} M for castor bean at pH 6.0 and 4.5×10^{-4} M for *Blastocladiella* at pH 7.4. The equilibrium constant

$$K = \frac{Isocitrate}{(glyoxalate) \times (succinate)}$$

has been found to be 34 at pH 7.6 (Smith and Gunsalus, 1957) though an error may be introduced into this value by the use of cysteine in the incubation mixture which would be expected to react with glyoxalate to form a thiazolidinecarboxylic acid derivative (Brunel-Chapelle, 1952).

D. Enzymes Related to the Glyoxalate Cycle.

I. Glycollic Acid Oxidase.

Glycollate $+ O_2 \rightarrow$ Glyoxalate $+ H_2O_2$

Early studies on glycollic acid oxidase demonstrated that glycollate could be oxidised to glyoxylate, and also to formate and carbon dioxide by crude extracts of higher plants (Clagett, Talbert and Burris, 1949; Tolbert, Clagett and Burris, 1949). Kenten and Mann (1952), using extracts from tobacco leaves, demonstrated the formation of hydrogen peroxide during glycollate oxidation. In the absence of catalase, the hydrogen peroxide reacted non-enzymically with the glyoxylate to give formate and carbon dioxide; in the presence of catalase the glyoxylate was oxidised enzymically to oxalate. Crystalline preparations of glycollic acid oxidase do not oxidise glyoxalate suggesting the presence of a separate enzyme forming oxalate from glyoxalate (ZELITCH, 1960). The same preparations also oxidised L-lactate to pyruvate with the formation of hydrogen peroxide.

Glycollic acid oxidase is weak or absent in extracts from etiolated plants, but increases greatly after the plants have been illuminated. Roots do not normally contain this enzyme, but like other tissues they form it on exposure to light (MOTHES and WAGNER, 1957). Low intensities of light (0.02 fc) over the whole of the visible spectrum are effective in increasing the activity in etiolated plants; the rate of increase in the activity of the enzyme does not parallel chlorophyll formation (TOLBERT and BURRIS, 1950). Increases in enzymic activity are also observed after etiolated plants have been sprayed in the dark with glycollate solutions at low pH; spraying with glyoxalate, lactate, or oxalate, does not have this effect. The failure to increase activity by spraying with glyoxalate is surprising because of the rapid formation of glycollate from glyoxalate catalysed by the enzyme glyoxalic reductase. Illumination of extracts from etiolated plants does not produce any increase in activity, but incubation of the extracts with glycollate at 1° for 18 hours causes the activity to appear (TOLBERT and COHAN, 1953). In the light of these findings, it has been suggested that the oxidase is an adaptive enzyme, and appears in response to the production of glycollate in photosynthesis (TOLBERT and BURRIS, 1950; TOLBERT and COHAN, 1953). Glycollate is labelled very rapidly on the exposure of *Chlorella* cells to $C^{14}O_2$ (BENSON and CALVIN, 1950) or to tritiated water (MOSES and CALVIN, 1959) in the light.

Glycollic acid oxidase has been demonstrated to be of high activity and wide distribution in extracts of higher plants (NOLL and BURRIS, 1954). It has been shown that this enzyme, together with the pyridine-nucleotide — linked glyoxalic acid reductase, catalyses the oxidation of reduced pyridine nucleotides by molecular oxygen in cell-free extracts (ZELITCH, 1953). Glycollic acid oxidase, but not glyoxalic acid reductase, is strongly inhibited by the bisulphite addition compounds of aldehydes, or α-hydroxysulphonates (ZELITCH, 1957). The feeding of these inhibitors, particularly α-hydroxy-2-pyridinemethanesulphonate, to excised tobacco leaves via the petiole, results in a large increase in the amount of glycollate. When the leaf is illuminated the rate of increase in glycollate is comparable to the rate of oxygen uptake in darkness except in the youngest leaves. The use of $C^{14}O_2$ indicates that the glycollate is formed from recently assimilated carbon. These results suggest that glycollic acid oxidase plays an important part in the respiration of illuminated leaves, especially in older tissue (ZELITCH, 1959).

Some information is available on the intracellular distribution of the oxidase. Washed mitochondrial particles contain some 15% of the glycollic acid oxidase activity of spinach leaf homogenates; although oxidative phosphorylation was found to occur with the KREBS cycle acids as substrates, no synthesis of ATP accompanied the oxidation of glycollate by these particles (ZELITCH and BARBER, 1960). It is of interest that washed spinach chloroplasts show glycollic acid oxidase activity (ZELITCH and BARBER, 1960). The presence of the enzyme in chloroplasts is not unexpected in view of its stimulation by light; DELAVAN and BENSON (1957) noted that light enhanced the evolution of $C^{14}O_2$ when C^{14}-glycollate was incubated with spinach chloroplasts. However recent work by PIERPOINT (1960) has shown that the glycollic acid oxidase activity of tobacco leaf extracts is largely in the soluble fraction, only 7.5% being associated with the mitochondrial fraction, and 1.8% with the chloroplast fraction; no phosphorylation accompanied the oxidation of glycollate by the mitochondria.

Two assay methods have been used for this enzyme:

A. *Manometric estimation of the oxygen uptake* (ZELITCH and OCHOA, 1953; FRIGERIO and HARBURY, 1958).

B. *Spectrophotometric measurement* of the rate of reduction of 2,6-dichloro-phenolindophenol at 610—620 mμ (ZELITCH and OCHOA, 1953; ZELITCH, 1955b; FRIGERIO and HARBURY, 1958).

The ratio of the amount of oxygen consumed to the amount of glycollate oxidised in assay method A depends on the quantity of catalase present in the extract. In the presence of excess catalase, 0.5 mole of oxygen is consumed for each mole of glycollate oxidised; in the absence of catalase the non-enzymic reaction between glyoxylate and hydrogen peroxide results in one mole of oxygen being consumed for each mole of glycollate oxidised. FRIGERIO and HARBURY (1958) recommend the use of Tris buffer in preference to phosphate buffer, since with partially purified preparations from spinach a decrease in ionic strength led to an increased rate of oxygen consumption; this effect was not observed with the highly purified enzyme.

Assay method B is to be preferred for purification purposes where a rapid estimation of enzymic activity is desired. The addition of cyanide to the assay cell is recommended, to inhibit the indophenol oxidase which accompanies glycollic acid oxidase in the early stages of purification. Rates of dye reduction are strongly dependent on ionic strength, even in the case of the most highly purified preparations, the rate of reduction rising with increasing ionic strength.

Glycollic acid oxidase from spinach has been highly purified by ZELITCH and OCHOA (1953) and obtained in crystalline form by FRIGERIO and HARBURY (1958). The prosthetic group of the enzyme is FMN (MICHAELIS constant about 3×10^{-6} M). The enzyme undergoes partial dissociation from its prosthetic group during purification; it can be completely resolved by precipitation with ammonium sulphate at low pH values. Activation by FMN can be observed in the optical assay only if the assay is carried out under anaerobic conditions. The resolved enzyme can also be crystallised. The purified enzyme contains no significant quantities of metals, and is not affected by the addition of metal ions or metal-chelating agents; it is therefore probably not a metalloflavoprotein. The enzymic activity is inhibited by high concentrations of sulphydryl-combining reagents.

The purified enzyme has a high substrate specificity; glycollate (MICHAELIS constant 3.8×10^{-4} M, pH optimum 8.3) and L-lactate (MICHAELIS constant 2.0×10^{-3} M, pH optimum 8.0—8.5) are readily oxidised but DL-alanine, glycine, D-lactate, DL-glycerate, glycolaldehyde, tartronate, DL-malate, and the tartaric acids are not. The highly purified enzyme does not oxidise glyoxylate to oxalate, as does the crude extract (ZELITCH, personal communication). The high specificity and affinity of the enzyme for glycollate and L-lactate suggests that it may be of use for the determination of these compounds. Only two-electron oxidation reduction systems have been found to act as primary electron acceptors in glycollate oxidation.

Electrophoresis of the crystalline enzyme in the pH range 6.7 to 8.7 yielded patterns displaying a single boundary. A single boundary was also observed on ultracentrifugation at ionic strengths between 0.02—0.1; at higher ionic strengths (0.26—0.4) two components were observed, both enzymically active. These two components are interpretated as distinct species of the enzyme, one of particle weight double, and one of particle weight quadruple the minimal molecular weight of about 70,000 (FRIGERIO and HARBURY, 1958).

A soluble enzyme from rat liver has been described which oxidises glycollate to glyoxylate (KUN, 1952); in contrast to the plant enzyme, the liver enzyme oxidises D-lactate (KUN, DECHARY and PITOT, 1954).

II. Glyoxalic Acid Reductase.

$$\text{Glyoxalate} + \begin{matrix} \text{DPNH} \\ \text{or} \\ \text{TPNH} \end{matrix} + \text{H}^+ \rightleftharpoons \text{Glycollate} + \begin{matrix} \text{DPN}^+ \\ \text{or} \\ \text{TPN}^+ \end{matrix}$$

This enzyme was discovered by ZELITCH (1953) in extracts of spinach leaves, and purified some ten-fold. It was shown that a soluble system of purified enzymes (glucose dehydrogenase, glyoxalic acid reductase, and glycollic acid oxidase) can catalyse the oxidation of glucose by molecular oxygen in the presence of catalytic amounts of DPN and either glycollate or glyoxalate; the addition of catalase to prevent the oxidation of glyoxalate by hydrogen peroxide increases the oxygen uptake in this system. It has been suggested in view of these findings that the glycollic acid oxidase — glyoxalic acid reductase couple may play a part in the respiration of green leaves (ZELITCH, 1953). However more information on this point is required, particularly as regards the effect of oxygen pressure on the system; flavoproteins are regarded by some workers as playing a minor role in the direct transfer of electrons to oxygen (BONNER, 1957). Recent work has shown that glyoxalic acid reductase occurs in washed suspensions of spinach chloroplasts (ZELITCH and BARBER, 1960).

Assay. The standard assay for pyridine nucleotide enzymes can be used for glyoxalic acid reductase i.e., measurement of the drop in optical density at 340 mμ due to the oxidation of DPNH (ZELITCH, 1953; 1955a; 1955b). Crude extracts of green leaves often oxidise DPNH in the absence of added glyoxalate; this activity is usually small compared with the rate in the presence of glyoxalate and can be allowed for. In the case of tobacco leaf extracts, the low pH of the extract (ca. 5.3) destroys the DPNH oxidase without affecting the glyoxalic acid reductase; hence no correction is necessary if tobacco leaves are used. The enzyme has been obtained in crystalline form from tobacco leaves (ZELITCH, 1955a). Crude extracts of tobacco and spinach leaves have some activity with TPNH as the hydrogen donor in the reduction of glyoxalate; the crystalline tobacco enzyme will not function with TPNH. Crude extracts of castor-bean contain a glyoxalic acid reductase which is more active with TPNH than with DPNH (KORNBERG and BEEVERS, 1957). These findings suggest that two enzymes exist, differing in their specificity for the reductant. The MICHAELIS constant for DPNH is too small to be measured directly in the case of the tobacco enzymes. The crystalline enzyme reduces glyoxalate (MICHAELIS constant 9.1×10^{-3} M, pH optimum 6.3—6.6) and hydroxypyruvate, but is inactive with pyruvate. It is of interest in this connection to note that crystalline lactic dehydrogenase from heart muscle is active with both glyoxalate and hydroxypyruvate (MEISTER, 1952); however, while the muscle enzyme produces L-glycerate from hydroxypyruvate, the glyoxalic reductase from tobacco produces the D-isomer. STAFFORD, MAGALDI, and VENNESLAND (1954) have demonstrated the enzymic reduction of hydroxypyruvate to D-glycerate by extracts of several higher plants. Most of these extracts contained no lactic dehydrogenase activity. The crystalline tobacco enzyme does not oxidise glycollic acid in the presence of excess FMN. An unusual pH-dependent stimulation of hydroxypyruvate oxidation by the tobacco enzyme was noted on the addition of anions (phosphate, sulphate, chloride, bromide, nitrate, and iodide, in order of increasing effectiveness). For example, addition of 0.066 M sodium iodide increased the rate of oxidation by 5.3-fold at pH 6.9. The oxidation of glyoxalate does not show this effect. The reversibility of the reactions catalysed by glyoxalic reductase i.e. the oxidation of glycollate and D-glycerate, can be demonstrated, especially at alkaline pH values and with high concentrations of the substrates.

The equilibrium constant of the glyoxalate-glycollate reaction is 6.06×10^{14} towards the side of glycollate; the corresponding value for the hydroxypyruvate-D-glycerate reaction is 2.82×10^{12} towards the side of D-glycerate.

Pyruvate and oxaloacetate have some inhibitory effect on the reduction of glyoxalate by the tobacco enzyme when present at concentrations similar to that of the glyoxalate; azide (10^{-3} M) and 2,4-dinitrophenol (10^{-4} M) are not inhibitory, nor is cyanide in the presence of excess glyoxalate. p-Chloromercuribenzoate was found to be a powerful inhibitor, and its use in conjunction with spectrophotometric measurements suggests the existence of enzyme-glyoxalate, enzyme-glycollate, and at least two kinds of enzyme-DPNH complexes formed through the agency of sulphydryl groups on the protein (ZELITCH, 1955a).

III. Glycolaldehyde Dehydrogenase.

$$\underset{\overset{|}{\text{CHO}}}{\overset{\text{CH}_2\text{OH}}{|}} \quad \begin{array}{c} + \text{DPN}^+ \\ \text{or TPN}^+ \end{array} \xrightarrow{\text{H}_2\text{O}} \quad \underset{\overset{|}{\text{COOH}}}{\overset{\text{CH}_2\text{OH}}{|}} \quad \begin{array}{c} + \text{DPNH} + \text{H}^+ \\ \text{or TPNH} \end{array}$$

The enzyme is readily assayed by measuring the reduction of DPN at 340 mμ. Since glycolaldehyde is a highly reactive compound, it is particularly important to conduct adequate controls. Thus under alkaline conditions glycolaldehyde gives rise to a compound or compounds with absorption maxima at 335 mμ and 240 mμ; in the presence of DPN an addition compound is formed with an absorption spectrum similar to that of reduced DPN.

The enzyme has been purified 10-fold from pea seedling mitochondria (DAVIES, 1960b). The enzyme is activated by cysteine, has an apparent pH optimum at pH 8 and is inhibited by glycolaldehyde concentrations in excess of 2×10^{-4} M. The MICHAELIS constant for glycolaldehyde is 7×10^{-5} M and preparations were active with TPN in place of DPN and with acetaldehyde and glycolaldehyde-2-phosphate in place of glycolaldehyde.

IV. Glycine Oxidase.

Glycine $+ \text{O}_2 \rightarrow$ glyoxalate $+ \text{NH}_3 + \text{H}_2\text{O}_2$

The oxidation of glycine in plant tissues has not been extensively investigated, but there are indications that this reaction can occur in certain cases. For example, glycine is oxidised by mitochondria from etiolated *Phaseolus* and *Vigna* seedlings (DAS and ROY, 1960), but not by mung-bean mitochondria (BONE and FOWDEN, 1960). ROGERS (1955) demonstrated the slow oxidation of several amino acids by squash extracts, but glycine was not attacked. It has been reported that bean root tips, after freezing and thawing, cause a decrease in the amount of externally applied glycine with the concurrent appearance of glyoxylate, but it is not clear whether this reflects an oxidation or a transamination (ROBERTSON and BROWN, 1952). A soluble L-amino acid oxidase from *Neurospora* has been shown to oxidise glycine slowly (BURTON, 1951). The oxidation of glycine occurs as a secondary reaction during the oxidation of catechol by plant polyphenolase (JAMES, ROBERTS, BEEVERS and DeKOCK, 1948). Glycine is not oxidised by the D- and L-amino acid oxidases from mammalian tissues or by the L-amino acid oxidase from snake venom (RATNER, 1955a), but an oxidase specific for glycine and sarcosine (the N-monomethyl derivative of glycine) has been found in extracts from mammalian liver and kidney; the enzyme has been partially purified from pig kidney (RATNER, GREEN and NOCITO, 1944).

Assay. The enzyme is best assayed by measuring the oxygen consumption manometrically by the WARBURG technique (RATNER, 1955b). If excess catalase is present, then 0.5 mole of oxygen is consumed per mole of glycine oxidised, and the glyoxalate is not oxidised further. With crude extracts, the oxygen uptake in the absence of added amino acids may mask the activity. Replacing the air of the gas phase by oxygen does not affect the initial rate of oxidation in the case of the kidney enzymes. The activity of the enzyme can also be followed anaerobically in THUNBERG tubes with methylene blue as the indicator.

The purified kidney enzyme may or may not depend for its activity on the addition of FAD, according to the treatment employed in purification. Precipitation by ammonium sulphate leads to an almost completely resolved enzyme. The enzyme oxidises glycine at about twice the rate at which it oxidises sarcosine; the oxidation of sarcosine yields glyoxalate and methylamine. N-dimethylglycine, glycylglycine and other glycyl peptides are not attacked. The purified preparation contains D-amino acid oxidase, which can be distinguished from glycine oxidase by differences in the effect of added FAD. The optimum for glycine oxidation is sharply centred around 8.3; veronal buffer is strongly inhibitory. The enzyme is inhibited by Cu ions (100% inhibition with 10^{-3} M $CuSO_4$) and by 0.01 M iodoacetic acid, but not by sodium fluoride (0.2 M), 0.001 M iodoacetic acid, or cyanide (0.01 M).

V. Formic Dehydrogenase.

$$HCOOH + DPN^+ \leftrightharpoons CO_2 + H^+ + DPNH$$

The enzyme is widely distributed in plants, being present in seeds of 54 species out of 93 tested (DAVISON, 1949). Although absent from the seeds of the Cruciferae, it is present in the leaves of a number of species (MAZELIS, 1960).

Assay. The enzyme is readily assayed by measuring the reduction of DPN at 340 mμ. An alternative assay which may be used with turbid preparations such as mitochondria, measures the rate of $C^{14}O_2$ production from formate-C^{14}, under aerobic conditions in the presence of DPN (MAZELIS, 1960). This assay requires an oxidase system to oxidize DPNH. In the mitochondrial preparations used by MAZELIS, it is probable that the oxidase activity was not rate limiting, but this point should be verified rather than assumed before using the assay.

The enzyme has been solubilised from mitochondrial preparations by means of digitonin extraction and by the preparation of acetone powders (MAZELIS, 1960). The enzyme was obtained in soluble form after the mechanical disruption of pea seedling mitochondria (DAVIES, 1956).

Using mitochondria isolated from cabbage leaves, the MICHAELIS constant for formate was found to be 5×10^{-3} M (MAZELIS, 1960) and the soluble dehydrogenase from pea seeds gave a value of 7×10^{-3} M (ADLER and SREENIVASAYA, 1937).

The enzyme has been purified 190-fold from pea seeds (NASON and LITTLE, 1955). The purified enzyme is inhibited by azide cyanide and 8-hydroxyquinoline, 100% inhibition being given by 5×10^{-6} M azide, 5×10^{-4} M cyanide and 5×10^{-3} M hydroxyquinoline.

References.

ADLER, E., and M. SREENIVASAYA: Z. Physiol. Chem. **249**, 24 (1937). — ALBERTY, R. A.: Advanc. in Enzymol. **17**, 1 (1956). — ANET, F. A. L.: J. Am. Chem. Soc. **82**, 994 (1960). — ARONOFF, S., and J. Z. HEARON: Arch. Biochem. **88**, 302 (1960). — APPLEBY, C. A., and R. K. MORTON: Nature (Lond.) **173**, 749 (1954). — AVERS, D. J., and E. E. KING: Am. J. Bot. **76**, 221 (1960).

BANDURSKI, R. S.: J. Biol. Chem. **217**, 137 (1955). — BANDURSKI, R. S., and C. M. GREINER: J. Biol. Chem. **204**, 781 (1953). — BARBER, D. A.: Nature (Lond.) **180**, 1053 (1957). — BARKER, J., and EL SAIFA: Proc. Roy. Soc. (B) **140**, 362 (1952). — BASFORD, R. E., and

D. E. GREEN: Biochim. Biophys. Acta **33**, 185 (1959). — BASU, D. K., and D. P. BURMA: J. Biol. Chem. **235**, 509 (1960). — BENDALL, D. S., RANSON S. L. and D. A. WALKER: Biochem. J. **76**, 221 (1960). — BENSON, A. A., and M. CALVIN: J. Expt. Bot. **1**, 63 (1950). — BERG, P.: J. Biol. Chem. **222**, 991 (1956). — BHAGVAT, K., and R. HILL: New Phyt. **50**, 112 (1951). — BONE, D. H., and L. FOWDEN: J. Expt. Bot. **11**, 104 (1960). — BONNER, D.: Ann. Rev. Plant Physiol. **8**, 427 (1957). — BOVÉ, J., R. O. MARTIN, L. J. INGRAHAM and STUMPF P.K.: J. Biol. Chem. **234**, 999 (1959). — BRADBEER, C., and P. K. STUMPF: J. Biol. Chem. **234**, 498 (1959). — BRUMMOND, D. O., and R. H. BURRIS: J. Biol. Chem. **209**, 755 (1954). — BRUNEL-CHAPELLE, G.: Comptes rend. **234**, 1466 (1952). — BURTON, K.: Biochem. J. **50**, 258 (1951). CAMPBELL, J. J. R., R. A. SMITH and B. A. EAGLES: Biochim. Biophys. Acta **11**, 594 (1953). — CANVIN, D. T., and H. BEEVERS: Plant Physiol. **35**, Suppl. XV (1960). — CARPENTER, W. D., and H. BEEVERS: Plant Physiol. **34**, 403 (1959). — CLAGETT, C.O., N. E. TOLBERT and R. H. BURRIS: J. Biol.Chem. **178**, 977 (1949). — CLENDENNING, K. A., E. R. WAYGOOD and P. WEINBERGER: Canad. J. Bot. **30**, 395 (1952). — COHEN, P. P.: J. Biol. Chem. **164**, 485 (1946). — COHN, W. E., and C. E. CARTER: J. Am. Chem. Soc. **72**, 4273 (1950). — CONN, E. E., B. VENNESLAND and L. M. KRAEMER: Arch. Biochem. **23**, 179 (1949).
DAS, H. K., and S. C. ROY: Chem. Abst. **54**, 13291d (1960). — DAVIES, D. D.: J. Expt. Bot. **6**, 212 (1955); **7**, 203 (1956); — Biol. Rev. Camb. Phil. Soc. **34**, 407 (1959); — Biochem. J. In Press (1960a); — J. Expt. Bot. In Press (1960b). — DAVIES, D. D., and E. KUN: Biochem. J. **66**, 307 (1957). — DAVISON, D. C.: Proc. Linn. Soc. N. S. Wales **74**, 26 (1949). — DELAVAN, L. A., and A. A. BENSON: Fed. Proc. **16**, 171 (1957). — DELBRUCK, A., H. SCHIMMASSEK, K. BARTSCH u. TH. BÜCHER: Biochem. Z. **331**, 297 (1959). — DESHPANDE, W. M., and C. V. RAMAKRISHNAN: J. Biol. Chem. **234**, 1929 (1959). — DENNIS, D., and N. O. KAPLAN: J. Biol. Chem. **235**, 810 (1960). — DICKMAN, S. R.: Anal. Chem. **24**, 1064 (1952). — DIXON, G. H., and H. L. KORNBERG: Biochem. J. **72**, 3p. (1959). — DIXON, G. H., H. L. KORNBERG and P. LUND: Biochim. Biophys. Acta **41**, 217 (1960).
ELLS, H. A.: Arch. Biochem. **85**, 561 (1959). — ENGLAND, S.: J. Biol. Chem. **234**, 1004 (1959). — ENGLARD, S., and S. P. COLOWICK: J. Biol. Chem. **226**, 1047 (1957). — ETTINGER, R. H., L. R. GOLDBAUM and L. H. SMITH: J. Biol. Chem. **199**, 531 (1952).
FRIEDEMANN, T. E., and G. E. HAUGEN: J. Biol. Chem. **147**, 415 (1943). — FRIEDRICH FRESKA, H., u. C. MARTIUS: Z. Naturforsch. **6**B, 296 (1951). — FRIGERIO, N. A., and H. A. HARBURY: J. Biol. Chem. **231**, 135 (1958).
GAWRON, O., A. J. GLAID, A. LAMONTE and S. GARY: J. Am. Chem. Soc. **80**, 5856 (1958). — GAWRON, O., and T. P. FONDY: J. Am. Chem. Soc. **81**, 6333 (1959). — GIBSON, D. M., E. O. DAVISSON, B. K. BACHAWAT B. R. RAY and C. S. VESTLING: J. Biol. Chem. **203**, 397 (1953). — GIOVANNOZZI-SERMANNI, G.: Chem. Abstr. **51**, 11662 (1957). — GIUDITTA, A., and T. P. SINGER: J. Biol. Chem. **234**, 662 (1959). — GOMORI, G.: Methods in enzymology, V. 1, p. 138. Ed. S. P. COLOWICK and N. O. KAPLAN. New York: Academic Press (1955. — GREEN, D. E., W. W. WESTERFIELD, B. VENNESLAND and W. E. KNOX: J. Biol. Chem. **140**, 683 (1941). — GREENSTEIN, J. P., N. IZUMIYA, M. WINITZ and S. M. BIRNBAUM: J. Am. Chem. Soc. **77**, 716 (1955). — GRUBER, M., and J. S. WASSENAAR: Biochim. Biophys. Acta **38**, 355 (1960). — GUNSALUS, I. C.: The mechanism of enzyme action p. 545, Ed. W. D. MCELROY and B. GLASS. Baltimore: John Hopkins Press 1954.
HAGER, D. J., and I. C. GUNSALUS: J. Am. Chem. Soc. **75**, 5767 (1953). — HAKALA, M. T., A. J. GLAID and G. W. SCHWERT: J. Biol. Chem. **221**, 191 (1956). — HARARY, I., S. R. KOREY and S. OCHOA: J. Biol. Chem. **203**, 595 (1953). — HELLERMAN, L., O. K. REISS, S. S. PARMAR, J. WEIN and N. L. LASSER: J. Biol. Chem. **235**, 2468 (1960). — HENDLIN, D., and T. M. COOK: J. Biol. Chem. **235**, 1187 (1960). — HIATT, A. J.: Plant Physiol. 35. Suppl. XI and Personal communication (1960). — HIATT, A. J., and H. J. EVANS: Plant Physiol. **35**, 662 (1960). — HORIO, T., J. YAMASHITA and K. OKUNUKI: Biochim. Biophys. Acta **32**, 593 (1959).
JAMES, W. O.: Plant Respiration p. 135. Oxford: Clarendon Press 1953; — New Phyt. **55**, 269 (1956); — Advanc. Enzymol. **18**, 281 (1957). — JAMES, W. O., and G. M. JAMES: New Phyt. **39**, 266 (1940). — JAMES, W. O., E. A. H. ROBERTS, H. BEEVERS and P. C. DEKOCK: Biochem. J. **43**, 626 (1948). — JAMES, W. O., and V. S. R. DAS: New Phyt. **56**, 325 (1957). — JAMES, W. O., and A. M. RICHENS: New Phyt. In Press (1960).
KAUFMAN, S.: J. Biol. Chem. **216**, 513 (1955b). — KAUFMAN, S., S. KORKES and A. DEL CAMPILLO: J. Biol. Chem. **192**, 301 (1951). — KAUFMAN, S., C. GILVARG, O. CORI and S. OCHOA: J. Biol. Chem. **203**, 869 (1953). — KAUFMAN, S., and S. G. A. ALIVISATOS: J. Biol. Chem. **216**, 141 (1955a). — KEARNEY, E. B.: J. Biol. Chem. **235**, 865 (1960). — KEARNEY, E. B., and T. P. SINGER: J. Biol. Chem. **219**, 963 (1956). — KENTEN, R. H., and P. J. G. MANN: Biochem. J. **52**, 130 (1952). — KLINE, E. S., and V. S. WARAVDEKAR: J. Biol. Chem. **235**, 1803 (1960). — KOIKE, M., and L. J. REED: J. Biol. Chem. **235**, 1931 (1960). — KOIKE, M., L. J. REED and W. J. CARROLL: J. Biol. Chem. **235**, 1924 (1960). — KORKES, S., A. DEL CAMPILLO, I. C. GUNSALUS and S. OCHOA: J. Biol. Chem. **193**, 721 (1951). — KORNBERG, A., and W. C. PRICER: J. Biol. Chem. **189**, 125 (1951). — KORNBERG, H. L., and H. BEEVERS: Biochim. Biophys.

Acta **26**, 531 (1957). — KORNBERG, H. L., and H. A. KREBS: Nature (Lond.) **179**, 988 (1957). — KORNBERG, H. L., and A. M. GOTTO: Nature (Lond.) **183**, 1791 (1959). — KREBS, H. A., and O. HOLZACH: Biochem. J. **52**, 527 (1952). — KUN, E.: J. Biol. Chem. **194**, 603 (1952); — Biochim. Biophys. Acta **25**, 135 (1957). — KUN, E., I. M. DECHARY and H. C. PITOT: J. Biol. Chem. **210**, 269 (1954).

LATIES, G. G.: J. Expt. Bot. **5**, 49 (1954). — LIEBERMAN, M., and J. B. BIALE: Plant Physiol. **31**, 425 (1956). — LIPMANN, F., and L. C. TUTTLE: J. Biol. Chem. **159**, 21 (1945). — LIPMANN, F., M. E. JONES, S. BLACK and R. M. FLYNN: J. Am. Chem. Soc. **74**, 2384 (1952). — LOEWUS, F. A., T. T. TCHEN and B. VENNESLAND: J. Biol. Chem. **212**, 787 (1955). — LOTHIAN, G. F.: Absorption spectrophotometry, Second Edition p. 56 London: Hilger & Watts 1958.

MADSEN, B.: Biochim. Biophys. Acta **27**, 199 (1958). — MAHLER, H. R., M. H. WITTENBERGER and L. BRAND: J. Biol. Chem. **233**, 770 (1958). — MARCUS, A., and B. VENNESLAND: J. Biol. Chem. **233**, 727 (1958). — MARCUS, A., and J. VELASCO: J. Biol. Chem. **235**, 563 (1960a); — Biochim. Biophys. Acta **38**, 365 (1960b). — MARKERT, C. L., and F. MOLLER: Proc. Natl. Acad. Sci. U. S. **45**, 753 (1959). — MARTIUS, C., and F. LYNEN: Adv. in Enzymol. **10**, 167 (1950). — MASSEY, V.: Biochem. J. **55**, 172 (1953). — MAZELIS, M.: Plant Physiol. **35**, 386 (1960). — MAZELIS, M., and B. VENNESLAND: Plant Physiol. **32**, 153 (1957). — McCURDY, H. D., and E. C. CANTINO: Plant Physiol. **35**, 463 (1960). — McRAE, D. H., F. J. FOSTER and J. BONNER: Plant Physiol. **28**, 343 (1953). — MEE, S.: Arch. Biochem. **22**, 139 (1949). — MEHLER, A. H., A. KORNBERG, S. GRISOLIA and S. OCHOA: J. Biol. Chem. **174**, 961 (1948). — MEISTER, A.: J. Biol. Chem. **184**, 117 (1950); **197**, 309 (1952). — MOLINARI, R., and F. J. S. LARA: Biochem. J. **75**, 57 (1960). — MONTAGUE, M. D., and R. K. MORTON: Nature (Lond.) **187**, 916 (1960). — MORGAN, E. J., and E. FRIEDMAN: Biochem. J. **32**, 862 (1938). — MOSES, V., and M. CALVIN: Biochim. Biophys. Acta **33**, 297 (1959). — MOTHES, K., and A. N. WAGNER: Biokhimiya **22**, 163 (1957).

NACHLAS, M. M., S. J. MARGULIES and A. M. SELIGMAN: J. Biol. Chem. **235**, 499 (1960). — NAKOMOTO, T., and B. VENNESLAND: J. Biol. Chem. **235**, 202 (1960). — NASON, A., H. A. OLDEWURTLE and L. M. PROPST: Arch. Biochem. **38**, 1 (1952). — NASON, A., and H. N. LITTLE: Methods in enzymology, V. 1, p. 536. Ed. S. P. COLOWICK and N. O. KAPLAN. New York: Academic Press 1955. — NAVAZIO, F., B. B. ERNSTER and L. ERNSTER: Biochim. Biophys. Acta **26**, 416 (1957). — NEILANDS, J. B.: J. Biol. Chem. **199**, 373 (1952); — Methods in enzymology, V. 1, p. 449, Ed. S. P. COLOWICK and N. O. KAPLAN. New York: Academic Press 1955. — NELSON, E. K., and G. KROTKOV: Canad. J. Bot. **34**, 423 (1956). — NISSELBAUM, J. S., and O. BODANSKY: J. Biol. Chem. **234**, 3276 (1959). — NOLL, C. R., and R. H. BURRIS: Plant Physiol. **29**, 261 (1954). — NOSSAL, P. M.: Aust. J. Expt. Biol. Med. Sci. **27**, 313 (1949). — NOVOA, W. B., A. F. WINER, A. J. GLAID and G. W. SCHWERT: J. Biol. Chem. **234**, 1143 (1959).

OCHOA, S.: Advanc. in Enzymol. **15**, 183 (1954); — Methods in enzymology, V. 1, p. 685, Edited by S. P. COLOWICK and N. O. KAPLAN. New York: Academic Press 1955. — O'KANE, D. J., and I. C. GUNSALUS: J. Bact. **56**, 499 (1948). — OLSON, J. A.: J. Biol. Chem. **234**, 5 (1959).

PETERS, R. A.: Symp. Soc. Expt. Biol. **3**, 36 (1949). — PETERS, R. A., and R. W. WAKELIN: Biochem. J. **40**, 513 (1946). — PIERPOINT, W. S.: Biochem. J. **71**, 518 (1959); **75**, 511 (1960). — PLAGEMANN, P. G. W., K. F. GREGORY and F. WROBLEWSKI: J. Biol. Chem. **235**, 2282 (1960). — PLAUT, G. W. E., and S. C. SUNG: J. Biol. Chem. **207**, 305 (1954). — POTTER, V. R., and C. HEIDELBERGER: Nature (Lond.) **164**, 180 (1949). — PRICE, C. A., and K. V. THIMANN: Plant Physiol. **29**, 495 (1954). — PUCHER, G. W., C. C. SHERMAN and H. B. VICKERY: J. Biol. Chem. **113**, 235 (1936).

RACKER, E.: Biochim. Biophys. Acta **4**, 211 (1950). — RADHAKRISHNAN, A. N., and E. E. SNELL: J. Biol. Chem. **235**, 2316 (1960). — RAMAKRISHNAN, C. V., and S. M. MARTIN: Nature (Lond.) **174**, 230 (1954). — RANSON, S. L.: Nature (Lond.) **172**, 252 (1953). — RANSON, S. L., D. A. WALKER and I. D. CLARKE: Biochem. J. **76**, 216 (1960). — RATNER, S.: Methods in enzymology, V. 2, p. 199, Edited by S. P. COLOWICK and N. O. KAPLAN. New York: Academic Press 1955a; — Methods in enzymology, V. 2 p. 225, Edited by S. P. COLOWICK and N. O. KAPLAN. New York: Academic Press 1955b. — RATNER, S., D. E. GREEN and V. NOCITO: J. Biol. Chem. **152**, 119 (1944). — REED, L. J., M. KOIKE, M. E. LEVITCH and F. R. LEACH: J. Biol. Chem. **232**, 143 (1958). — ROBERTSON, E., and R. BROWN: J. Expt. Bot. **3**, 356 (1952). — ROGERS, B. J.: Plant Physiol. **30**, 186 (1955).

SACKS, W.: Science **127**, 594 (1958). — SACKS, W., and C. O. JENSEN: J. Biol. Chem. **192**, 231 (1951). — SAFRONOV, A. P.: Biokhimiya **24**, 123 (1959). — SALLES, J. B. V., and S. OCHOA: J. Biol. Chem. **187**, 849 (1950). — SANADI, D. R., J. W. LITTLEFIELD and R. M. BOCK: J. Biol. Chem. **197**, 851 (1952). — SANADI, D. R., D. M. GIBSON and P. AYENGAR: Biochim. Biophys. Acta **14**, 434 (1954). — SCHWERT, R. S., U. JAGANNATHAN and B. KATCHMAN: Fed. Proc. **10**, 245 (1951). — SCHWERT, R. S., B. KATCHMAN, R. M. BOCK and U. JAGANNATHAN: J. Biol. Chem. **192**, 563 (1952). — SHIFRIN, S., N. O. KAPLAN, and M. M. CIOTTI:

J. Biol. Chem. **234**, 1555 (1959). — Siebert, G., M. Carsiotis and G. W. E. Plaut: J. Biol. Chem. **226**, 977 (1957). — Singer, T. P.: Methods in enzymology, V. 1, p. 460. Edited by S. P. Colowick and N. O. Kaplan. New York: Academic Press (1955). — Singer, T. P., and J. Pensky: J. Biol. Chem. **196**, 375 (1952a); — Biochim. Biophys. Acta **9**, 316 (1952b). — Singer, T. P., E. B. Kearne and P. Bernath: J. Biol. Chem. **223**, 599 (1956). — Singer, T. P., and E. B. Kearne: Methods of biochemical analysis, V. 4, p. 307. Edited by D. Glick. New York: Interscience 1957. — Singer, T. P., E. B. Kearne and V. Massey: Advanc. in Enzymol. **18**, 65 (1957). — Slonimski, P. P., and Tysarowski, W.: Comptes rend. **246**, 1111 (1958). — Smith, R. A., and I. C. Gunsalus: J. Biol. Chem. **229**, 305 (1957). — Sorokin, H. P.,and K. V. Thimann: Nature (Lond.) **187**, 1038 (1960). — Soumalainen, H., and E. Oura: Biochim. Biophys. Acta **31**, 115 (1959). — Speyer, J. F., and S. R. Dickman: J. Biol. Chem. **220**, 193 (1956). — Stadtman, E. C.: Methods in enzymology, V. 3, p. 931. Ed. by S. P. Colowick and N. O. Kaplan. New York: Academic Press 1955. — Stafford, H. A.: Physiol. Plantarum **4**, 696 (1951); — Plant Physiol. **31**, 135 (1956); **32**, 338 (1957). — Stafford, H. A., A. Magaldi and B. Vennesland: J. Biol. Chem. **207**, 621 (1954). — Straub, F. B.: Biochem. J. **34**, 483 (1940).

Takenaka, Y., and G. W. Schwert: J. Biol. Chem. **223**, 157 (1957). — Tchen, T. T., F. A. Loewus and B. Vennesland: J. Biol. Chem. **213**, 547 (1955). — Tchen, T. T., and B. Vennesland: J. Biol. Chem. **213**, 533 (1955). — Terayama, H., and C. S. Vestling: Biochim. Biophys. Acta **20**, 586 (1956). — Thimann, K. V., and W. D. Bonner: Am. J. Bot. **35**, 271 (1948). — Throneberry, G. O.: Plant Physiol. **35**, Suppl. XII (1960). — Tolbert, N. E., and R. H. Burris: J. Biol. Chem. **186**, 791 (1950). — Tolbert, N. E., and C. O. Clagett R. H. Burris: J. Biol. Chem. **181**, 905 (1949). — Tolbert, N. E., and M. S. Cohan: J. Biol. Chem. **204**, 639 (1953). — Tomlinson, N.: J. Biol. Chem. **209**, 605 (1955).

Velick, S. F.: J. Biol. Chem. **233**, 1455 (1958). — Vennesland, B., and R. L. Felsher: Arch. Biochem. **11**, 279 (1946). — Vessell, E. S., and A. G. Bearn: Proc. Soc. Expt. Biol. Med. **94**, 96 (1957). — Vickery, H. B., and J. K. Palmer: J. Biol. Chem. **218**, 225 (1956).

Walker, D. A.: Biochem. J. **67**, 73 (1957); **74**, 216 (1960). — Walker, D. A., and H. Beevers: Biochem. J. **62**, 120 (1956). — Walker, D. A., and J. M. A. Brown: Biochem. J. **67**, 79 (1957). — Warburg, O., H. S. Gewitz and W. Z. Volker: Z. Naturforsch. **12**B, 722 (1957). — Waygood, E. R.: Canad. J. Research C. **28**, 7 (1950). — Weil-Malherbe, H., and A. D. Bone: Biochem. J. **45**, 377 (1949). — Westerfield, W. W., D. A. Richert and R. J. Bloom: J. Biol. Chem. **234**, 1889 (1959). — Wieland, Th., G. Pfleiderer, I. Haupt and W. Worner: Biochem. Z. **332**, 1 (1959). — Winer, A. D., and G. W. Schwert: J. Biol. Chem. **231**, 1065 (1958). — Winer, A. D., G. W. Schwert and D. B. S. Milar: J. Biol. Chem. **234**, 1149 (1959). — Wong, D. T. O., and S. J. Ajl: J. Am. Chem. Soc. **78**, 3220 (1956).

Yatco-Manzo, E., F. Roddy R. G. Yount and D. E. Metzler: J. Biol. Chem. **234**, 733 (1959). — Yamamoto, Y., and H. Beevers: Plant Physiol. **35**, 102 (1960).

Zelitch, I.: J. Biol. Chem. **201**, 719 (1953); **216**, 553 (1955a); — Methods in enzymology, V. 1, p. 528. New York: Academic Press 1955b; — J. Biol. Chem. **224**, 257 (1957); **234**, 3077; — Personal communication (1960). — Zelitch, I., and G. A. Barber: Plant Physiol. **35**, 626 (1960). — Zelitch, I., and S. Ochoa: J. Biol. Chem. **201**, 707 (1953).

Enzymes of Terminal Respiration[1].

By

David P. Hackett.

With 12 figures.

This chapter deals with components that are involved in the transfer of electrons from reduced pyridine nucleotides to molecular O_2 in plant tissues. The knowledge in this field is far from complete, and only in limited areas is there any exact enzymological information. On the other hand, a good deal of work has been done towards establishing the nature of the physiological respiratory chain, and experiments at this level have provided the background for subsequent chemical investigations. The first section of this review deals with the methods commonly used to characterize the intact respiratory chain. Following this, the techniques used to study the individual components of this system are considered in detail. Plant tissues contain a variety of other enzymes which react with O_2 and the possibility that these may participate in alternative respiratory pathways has frequently been considered; these enzymes are discussed in the last section.

A. Characterization of the Intact Respiratory Chain.

A variety of methods have been used to establish the nature of the physiological respiratory chain in intact cells or tissues and in isolated cell fractions. Studies on the mitochondria, which serve as the major intracellular respiratory centers, have been especially important in this work, and methods for the isolation of these particles are described elsewhere in this series (WHATLEY, 1956). Fragments or subunits of the mitochondria can retain some or all of their essential electron-transferring capacities, so that they too may be very useful for the analysis of the respiratory chain. To date, very little work has been done on the disruption of plant mitochondria into physiologically active particles and precise methods cannot be given at this time. However, there is an extensive literature on the fragmentation of animal mitochondria (see GREEN and FLEISCHER, 1960), and the methods already developed should serve as useful guides in future studies with plant particles. Treatments with surface active agents (e.g., COOPER and LEHNINGER, 1956; ABOOD and ALEXANDER, 1957; IMAMOTO, 1959) and sonic oscillation (e.g., KIELLEY and BRONK, 1958; McMURRAY, MALEY and LARDY, 1958; LINNANE and ZIEGLER, 1958) have been the most useful techniques for disrupting animal mitochondria into subunits which retain all the essential respiratory chain activities. Further fragmentation of the particles to give individual segments of the chain can be brought about by a variety of techniques developed primarily in GREEN's laboratory (1959). The methods described below can be used with any system showing the complete respiratoy chain activity.

---

[1] *Abbreviations:* K_m = MICHAELIS constant; DPN^+ = diphosphopyridine nucleotide; DPNH = reduced diphosphopyridine nucleotide; TPN^+ = triphosphopyridine nucleotide; TPNH = reduced triphosphopyridine nucleotide; PNH = reduced pyridine nucleotide; P_i = inorganic phosphate; ADP = adenosine diphosphate; ATP = adenosine triphosphate; DNP = 2,4-dinitro-phenol; DIECA = diethyldithiocarbamate; HOQNO = 2-heptyl-4-hydroxy-quinoline-N- oxide; PCMBA = p-chloromercuribenzoate; AS = ammonium sulfate; Tris = tris hydroxy- methylaminomethane; EDTA = ethylenediamine tetraacetate; DEAE = diethylaminoethyl.

I. Measurement of Respiratory Rate.

1. Manometry.

The over-all process of terminal respiration is most conveniently measured by determining the rate of oxygen consumption. This can be done using standard WARBURG and related manometric techniques (see UMBREIT, BURRIS and STAUFFER, 1957), and these have been considered elsewhere in this series (KENTEN, 1956). When working with slices of plant tissue, special care must be taken to minimize the errors encountered when respiration is limited by oxygen diffusion; such factors as thickness of tissue, rate of O_2 uptake, rate of vessel shaking, composition of gas phase, volume of suspending medium, etc., require special attention (see WINZLER, 1941). It is preferable in some cases to maintain the tissue in a water-saturated atmosphere (e.g., with water in side-arm of Warburg vessel), rather than immersing it in solution.

2. Volumetry.

A number of volumetric respirometers, in which the volume of O_2 consumed is measured directly, have been described (UMBREIT et al., 1957), but these have not been widely used with plants. We have found a simple volumeter to be very useful for the rapid determination of respiratory rates of higher plant tissues (THIMANN, YOCUM and HACKETT, 1954) and of fungal mycelium growing on agar plates (NIEDERPRUEM and HACKETT, 1961). The respirometer is made from a vaccine vial (5 ml); the bottom is cut off and a pipette (0.2 ml calibrated in 0.001 ml) is attached to the vial through a rubber stopper. The tissue is supported on plastic screening or threaded on a toothpick or wire. CO_2 is absorbed by KOH-saturated filterpaper disks which are placed in both ends of the vial. The respirometer is submerged horizontally in a constant temperature bath and, following equilibration, a hypodermic syringe is used to draw water into the pipette and this serves as the index fluid. As O_2 is consumed, the water moves along the pipette; the readings are corrected for any thermobarometer changes. When necessary, the gas phase in the respirometer can be changed simply by flushing an appropriate mixture through the vaccine stopper.

3. Polarography.

The concentration of oxygen in solution can be measured polarographically, and this method is very useful for respiratory studies. It has been used with intact cells of algae (BRACKETT, OLSON and CRICKARD, 1953), yeast (CHANCE, 1954), and higher plant tissues (KÜSTER, 1955), as well as with animal (CHANCE and WILLIAMS, 1955) and plant mitochondria (CHANCE and HACKETT, 1959; STICKLAND, 1960). The method is simple, rapid (equilibration time of a few seconds), and sensitive (detects changes of < 0.01 m M O_2). The reaction can be followed continuously with an appropriate recording device, and additions to the mixture can be made at any time, as shown in Figure 1. The method depends on the fact that the current passing through the polarized electrode system is proportional to the concentration of dissolved oxygen. The oxygen is reduced at the inert electrode in a reaction which is assumed to involve peroxide as an intermediate; silver can be used for the reference electrode. Most instruments employ a platinum electrode, adjusted to -0.6 volts; the electrode is either rotated or vibrated or, alternatively, the solution is stirred in order to facilitate the diffusion of O_2 to the electrode surface. If O_2 uptake is followed in an open system, the reaction rate must be sufficiently rapid so that diffusion of O_2 from the air into the solution is a negligible factor. For a consideration of the principles and practice of this method, the reader should consult the classical papers of DAVIES and BRINK (1942) and HARRIS and LINDSEY (1948). Recently, CONNELLY (1957), CATER (1960), and HAGIHARA (1961) have discussed some of the problems associated with the use of the oxygen electrode. Special care must be taken to avoid contamination of the platinum surface (e.g., with cyanide); the performance of the electrode is greatly improved by coating it with a thin protective film of collodion. Detailed instructions for the construction and use of a number of different electrode systmes will be found in the above references. A commercial model of the vibrating platinum electrode can be purchased from Gilson Medical Electronics, Middleton, Wisconsin.

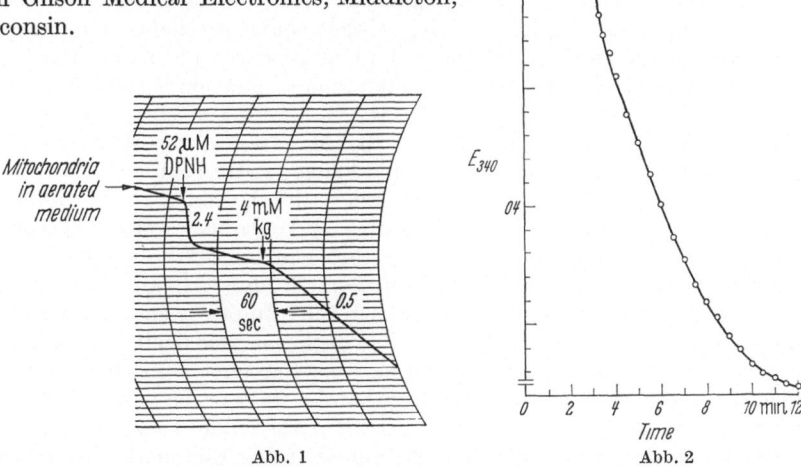

Mitochondria in aerated medium →

Abb. 1   Abb. 2

Fig. 1. Oxygen electrode tracing of the respiratory response of skunk cabbage mitochondria to solutions of DPNH and α-ketoglutarate. The slope indicates the rate of oxygen consumption and the values are 2.4 μM O$_2$/sec and 0.5 μM O$_2$/sec as indicated (from B. CHANCE and D. P. HACKETT 1959).

Fig. 2. Demonstration of the requirement for molecular O$_2$ in the oxidation of DPNH by a cell-free extract. Anaerobic cuvette initially flushed with N$_2$, then with O$_2$ (unpublished data).

4. Spectrophotometry.

Since respiratory chain activity begins with the oxidation of reduced pyridine nucleotide, the DPNH oxidase of isolated mitochondria can be used as a measure of the over-all activity. It must be remembered, however, that externally added DPNH may not represent a physiological substrate. Since the manometric assay requires a large amount of the reduced coenzyme, the oxidation of DPNH is generally followed spectrophotometrically. In this case, it is essential to show that the characteristic decrease of absorbancy at 340 mμ is dependent on the presence of molecular O$_2$; this can be done convenietly with a cuvette which can be made anaerobic, as shown in Figure 2. The reaction product (DPN$^+$) should also be identified. DPNH can be involved in a non-oxidative reaction which is also accompanied by a decrease in absorption at 340 mμ (RAFTER, CHAYKIN and KREBS, 1954). The splitting of DPN$^+$ by the hydrolytic enzyme which is present in some plant extracts (CONN, KRAEMER, LIU and VENNESLAND, 1952) can be inhibited by 10^{-2} M nicotinamide (MANN and QUASTEL, 1941).

The composition of the medium in which the mitochondria are isolated and assayed influences the DPNH oxidase activity markedly. Factors affecting the accessibility of DPNH to the particle-bound enzymes may be important. Thus, freezing and thawing of the mitochondria or exposure to hypotonic media can increase activity. The presence of divalent cations and the ionic strength of the medium have marked effects (HONDA, ROBERTSON and GREGORY, 1958; HACKETT, 1961). For each system investigated, a variety of factors, including tonicity, pH, and salt concentrations, should be tested in order to establish the optimal conditions.

II. Effect of Oxygen Partial Pressure on Respiratory Rate.

1. General Considerations.

Some information on the nature of the terminal oxidase can be gained by determining the over-all respiratory rate as a function of the oxygen concentration. The "apparent O_2 affinity" ($1/K_m$) of the respiratory chain can be compared with the known O_2 affinities of isolated plant oxidases (Table 1). Many early observations showed that the rate of plant tissue respiration is roughly the same in air and in 1 to 2% O_2; recent values for the O_2 affinities of slices of potato tuber and Aroid spadix are close to $K_m = 3 \times 10^{-5}$ M (YOCUM and HACKETT, 1957). Comparison of this value with those in Table 1 suggests that cytochrome oxidase is the terminal enzyme. With intact cells or tissues, the O_2 affinity may be influenced by the rate of O_2 diffusion to the reactive sites, and even under ideal experimental conditions with unicellular organisms, the diffusion limitation is not completely eliminated (see LONGMUIR, 1954; LONGMUIR and BOURKE, 1960). If the terminal oxidase reaction is not the rate-limiting step in the respiratory chain, the measured O_2 affinity may differ considerably from that of the enzyme itself. This would also be true with the isolated enzyme if the measured activity is limited by the rate at which it is reduced by substrate rather than by its reaction with molecular O_2. Only by direct measurements of the rate at which the reduced enzyme reacts with O_2 can the true oxygen affinity be determined; using this method, NAKAMURA (1960) has shown that the copper-containing enzyme laccase has a high O_2 affinity, though it is generally assumed that copper oxidases have a relatively low O_2 affinity (Table 1).

Table 1. *Oxygen Affinities of Some Oxidases and Respiratory Systems.*

	$[O_2]^{50}$ or K_m	$1/K_m$	Reference
(a) Enzymes	(M)		
Cytochrome c oxidase	2.4×10^{-8}	4×10^7	LONGMUIR (1954)
Ascorbic acid oxidase	1.5×10^{-4}	6×10^3	THIMANN et al. (1954)
Polyphenol oxidase	3.0×10^{-5}	3×10^4	KUBOWITZ (1939)
Glucose oxidase	1.0×10^{-4}	1×10^4	LASER (1952)
(b) Cell Fractions			
Liver mitochondria	7.5×10^{-6}	1.3×10^5	CHANCE and WILLIAMS (1955)
Symplocarpus spadix mitochondria	$\sim 1 \times 10^{-6}$	$\sim 1 \times 10^6$	CHANCE and HACKETT (1959)
S. faecalis DPNH oxidase	2.1×10^{-6}	5×10^5	NIEDERPRUEM and HACKETT (1958)
(c) Intact Cells			
Aerobacter aerogenes	4.0×10^{-8}	2.5×10^7	LONGMUIR (1954)
Baker's yeast	1.0×10^{-6}	1×10^6	WINZLER (1941)
Potato tuber slices	3.0×10^{-6}	3×10^5	THIMANN et al. (1954)
Peltandra spadix	3.0×10^{-6}	3×10^5	YOCUM and HACKETT (1957)
Pisum internodes	4.0×10^{-5}	2.5×10^4	EICHENBERGER and THIMANN (1957)

$$[O_2]^{50} = \frac{pO_2^{50} \times \alpha\, O_2}{22.4 \text{ liter atm/mole}}$$

where pO_2^{50} is the partial pressure, in atmospheres, of oxygen required for half the maximum rate of O_2 uptake. $\alpha\, O_2$ is the oxygen solubility coefficient.

2. Methods.

A common method is to determine the respiratory rate of the system under investigation in a series of separate reaction mixtures containing known concentrations of O_2; the partial pressure which will support half the maximal rate is estimated from the rate vs. O_2 concentration curve. The desired gas mixtures can be made, either by liquid displacement or by pressure, from commercial N_2 (or other inert gas) and O_2. The air in the reaction mixture is then replaced by flushing with an adequate volume of the gas mixture (e.g., approximately 1 liter for a 15 ml WARBURG vessel).

An alternative method is to follow the rate of O_2 consumption in a single vessel until the system becomes anaerobic. From a knowledge of the amount of O_2 present initially, the O_2 concentration at the half maximal respiratory rate can be determined. For example, the respiratory rate in an initially air-saturated mixture can be followed with the oxygen electrode until all of the dissolved oxygen (250 μM) has been consumed. The half-maximal rate is reached during the aerobic-anaerobic transition and the O_2 concentration at this point is calculated (see CHANCE and WILLIAMS, 1955). For reactions which can be followed spectrophotometrically, anaerobic cuvettes are flushed with N_2 to eliminate all but a small of O_2. From the stoichiometry of the reaction, it is possible to calculate the O_2 concentration when the rate is half-maximal. This technique has been used for a bacterial DPNH oxidase system (NIEDERPRUEM and HACKETT, 1958) and for tobacco root polyphenol oxidase (SISLER and EVANS, 1958); Figure 3 illustrates the use of this method with the latter system.

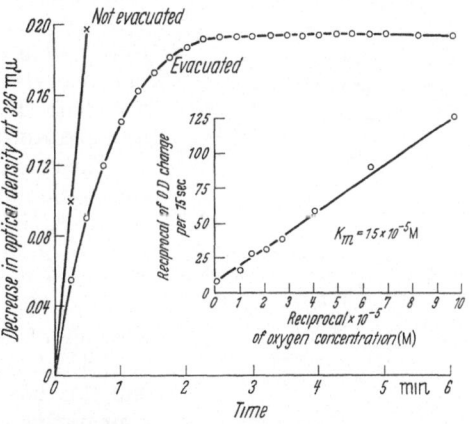

Fig. 3. The effect of oxygen concentration on the reaction kinetics of tobacco root polyphenol oxidase. The linear curve represents the time course of the reaction, measured by the decrease in absorbance of chlorogenic acid at 326mμ, when the gas phase in the cuvette was air. The lower curve indicates the time course of the reaction when the cuvette was evacuated sufficiently to limit the oxygen supply. The inset figure shows a reciprocal plot of the reaction velocity against the O_2 concentration; the latter was calculated assuming that two moles of chlorogenic acid are consumed in the reduction of one mole of O_2 and using the $\varDelta E_{326} = 1 \times 10^4$ for the reduced and oxidized forms of chlorogenic acid (from E. C. SISLER and H. J. EVANS 1958).

III. Coupling to Phosphorylation.

The flow of electrons from substrates to O_2 is normally coupled to the synthesis of ATP from ADP and P_i, and there are three sites of phosphorylation between DPNH and O_2. The efficiency of the oxidative phosphorylation is determined by simultaneous measurements of the phosphate (or ADP) and O_2 consumption. LYNEN and KOENIGSBERGER (1951) developed an ingenious method for use with suspensions of intact yeast cells: a high concentration of cyanide is added and the initial rate (15 seconds) of increase in intracellular phosphate concentration is measured. It is assumed that this increase is due to the uninhibited dephosphorylation reactions, which are normally just balanced by the oxidative phosphorylation. From a knowledge of the rate of O_2 uptake by an equivalent cell suspension, the P/O ratio (moles phosphate esterified/atoms oxygen consumed) can be computed. Using this method with *Chlorella* cells in the dark, KANDLER (1957) obtained an average P/O of 2.7 for the endogenous respiration, and this value is remarkably close to the theoretical 3.0.

Most of the studies on oxidative phosphorylation have been carried out with isolated mitochondria. Values approaching the maximum P/O have been reported for the oxidation of several Krebs cycle acids by a variety of plant mitochondria (see HACKETT, 1959). In order to show that the phosphate esterification is in fact coupled to the respiratory chain oxidations, the inhibitory effect of 2,4-dinitrophenol should be demonstrated; a series of concentrations from 1×10^{-6} to 5×10^{-4} M DNP should be tested. Using appropriate substrates, electron acceptors, and inhibitors, it is possible to measure the phosphorylation associated with partial segments of the chain (DEVLIN and LEHNINGER, 1956; COOPER and LEHNINGER, 1956).

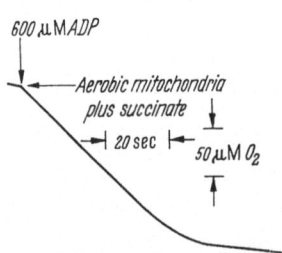

The rate of electron flow is normally limited by the availability of ADP (or P_i), and the effect of added ADP on the mitochondrial O_2 uptake can be used to indicate the "tightness of coupling." The degree of "respiratory control" is measured by the ratio of the respiratory rates in the presence and absence of ADP (Figure 4). In general, plant mitochondria have not shown a dramatic response to ADP, but this may simply be due to the use of inadequate preparative methods (BONNER and VOSS 1961).

1. Manometric Method.

Substrate oxidation is followed using standard manometric techniques, and the disappearance of inorganic phosphate is measured colorimetrically. The ATP formed is trapped by including glucose and hexokinase in the reaction mixture; fluoride can be added to block any ATPase activity. Sucrose is included to maintain a suitable osmotic concentration, and the appropriate cofactors must be added. A typical reaction mixture (3.0 ml) contains the following (in μmoles): substrate, 60; sucrose, 1000; glucose, 60; ADP, 3; NaF, 30; phosphate, 80; MgSO$_4$, 20; DPN, 1.2; 1 mg of hexokinase; 0.25 ml of washed mitochondria (2 mg protein); all at pH 7.0. The reaction is started, after thermal equilibration, by tipping in the substrate from the side-arm of the WARBURG vessel; it is stopped by mixing with 0.25 ml of 40% trichloroacetic acid. After centrifuging down the precipitate, the phosphate concentration is determined colorimetrically and the difference between the intitial and final values represents the phosphate esterified. The glucose-6-phosphate formed can also be measured directly (PINCHOT, 1957). With α-ketoglutarate as the substrate, malonate (6 μmoles) can be included to restrict the oxidation to a single step. The P/O when DPNH is the substrate can be improved by generating the DPNH enzymatically (MALEY, 1957).

2. Other Methods.

CHANCE and WILLIAMS (1955) developed a simple and rapid assay method that can be used with "tightly coupled" mitochondria. The rate of substrate oxidation in the absence of ADP is followed with the oxygen electrode; on addition of a known concentration of ADP, the rate of O_2 uptake increases rapidly until the ADP is consumed, at which point it halts abruptly (Figure 4). The amount of O_2 consumed during the period of rapid respiration is related to the concentration of added ADP to give the ADP/O ratio.

The polarographic method for O_2 can be combined with the colorimetric determination for phosphate. NIELSEN and LEHNINGER (1955) used a sensitive isotopic method for following phosphorylation with P^{32}; this is capable of measuring the uptake of a few millimicromoles of P_i from a much larger pool.

IV. The Use of Inhibitors.

A wide variety of compounds have been used to inhibit plant respiration, and these have been discussed in detail elsewhere (JAMES, 1953; HACKETT, 1960). Useful tables on the properties of various inhibitors are included in *Data for Biochemical Research* (DAWSON, ELLIOTT, ELLIOTT and JONES, 1959). Those inhibitors which have a very specific mode of action have been extremely valuable for studies on the nature of the respiratory chain. Only a few of the more useful inhibitors will be considered here, and they are grouped on the basis of their principal sites of action, as diagrammed below:

DPNH $\longrightarrow$ FP $\longrightarrow$ b $\longrightarrow$ c $\longrightarrow$ a—a_3 $\longrightarrow$ O_2

Amytal

Antimycin A CO
HOQNO Azide
BAL Cyanide
SN 5949

1. Oxidase Inhibitors.

Carbon monoxide is the most useful reagent for the identification of the terminal oxidase. It competes with O_2 for the reduced iron (Fe^{++}) in the hematin prosthetic group of cytochrome oxidase and its effects depend on the ratio of the partial pressures of CO and O_2. Suitable mixtures (e.g., 4/1, 9/1, and 19/1 CO/O_2) of these gases are used to flush the reaction system and the respiratory activity is then determined; the controls are flushed with the corresponding N_2/O_2 mixtures. CO can either be obtained commercially or generated in the laboratory: formic acid is added dropwise to warm, concentrated H_2SO_4 in a closed system, and the CO evolved is collected over water. Gas mixtures are made up volumetrically by displacing measured volumes of water. The apparent affinity or partition coefficient, K, for the terminal oxidase is calculated from the following expression (WARBURG, 1927):

$$K = \frac{n}{1-n} \times \frac{CO}{O_2}$$

n = fraction of respiration remaining in CO/O_2 mixture
CO/O_2 = ratio of partial pressures of CO and O_2

Values for K in the range of 10, which indicates that a 10/1 CO/O_2 ratio causes 50% inhibition, have been obtained with plant tissues, as well as with isolated cytochrome oxidase. The light-reversibility of the CO inhibition should be tested, since a positive result indicates the involvement of a CO-heme complex. A variety of light sources, including tungsten, mercury-vapor, and blue-fluorescent lamps have been used successfully. The photochemical action spectrum for the relief of CO inhibition permits a more precises identification of the hemoprotein involved; the early results of WARBURG (Figure 5) have been extended by CASTOR and CHANCE (1955) to a variety of organisms.

The demonstration of a light-reversible CO inhibition of respiration is strong evidence for the participation of cytochrome oxidase. However, some peroxidase-catalyzed reactions also show this property (SWEDIN and THEORELL, 1940). The combination of CO with copper proteins such as polyphenol oxidase is not reversed by light (KUBOWITZ, 1937). The absence of any inhibition (or stimulation) by CO does not prove that the cytochrome system is not functional (see HARTREE, 1957; HACKETT, 1959). Even in the absence of any respiratory inhibition, processes which are closely linked to phosphorylation (glycolysis, growth, ion uptake) can be affected by CO, and there is evidence that it can act as an uncoupling agent in the respiratory chain (HACKETT, RICE and SCHMID, 1960). Plant tissues do not oxidize CO to CO_2 (DALY, 1954; THIMANN et al. 1954).

Cyanide is a classical inhibitor of respiration, but it is not very specific. Cyanide combines not only with the oxidized from (Fe^{+++}) of cytochrome oxidase, but also with a variety of iron- and copper-containing enzymes and, at relatively high concentrations, it can act as a carbonyl reagent. Although cytochrome oxidase itself is completely blocked by 10^{-4} M HCN, the respiration of many algae and higher plant tissues is not inhibited by 10^{-3} M cyanide and it may even be promoted (see HACKETT, 1959). When respiration is measured in the presence of an alkali reservoir, suitable mixtures of KCN-KOH or $Ca(CN)_2$-$Ca(OH)_2$ (ROBBIE, 1948) must be used to prevent distillation of the cyanide. Thus, a solution of 0.93 M KCN in 0.5 M KOH added to the center-well of a WARBURG vessel will maintain a concentration of 10^{-4} M

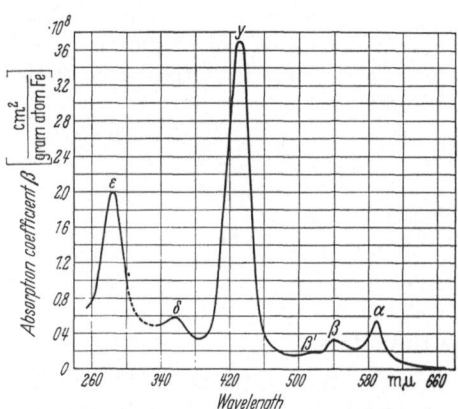

Fig. 5. The photochemical action spectrum of carbon monoxide-inhibited yeast respiration, showing the relative effectiveness of various wave lengths of light in reversing the CO-inhibition of respiration (from O. WARBURG 1949).

HCN in the experimental fluid (25°C). For HCN concentrations of 10^{-3} M or higher it is necessary to use $Ca(CN)_2$, prepared by trapping HCN in a $Ca(OH)_2$ suspension. In a typical preparation, 25 g of KCN in 75 ml of water is added dropwise over a 30-minute period to 100 ml of 50% H_2SO_4 (in ERLEN-MEYER flask with air inlet and side-arm). By keeping the system partially evacuated with the aid of a water aspirator, the generated HCN is led through a water trap into a flask containing 100 ml of 20% $Ca(OH)_2$ to give roughly 110 ml of 1.6 M $Ca(CN)_2$. The cyanide concentration is determined colorimetrically by the phenolphthalein technique (ROBBIE, 1944).

Azide, like cyanide, inhibits a variety of metal-containing enzymes, and the undissociated HN_3 combines with ferricytochrome oxidase. The pK of HN_3 is 4.7, so that azide is more inhibitory in acidic solutions.

Copper-chelating agents, such as thiourea and diethyldithiocarbamate (DIECA), have been used in attempts to establish the respiratory role of the copper oxidases. These enzymes are powerfully inhibited by millimolar (and lower) concentrations of these agents. Special precautions must be taken when using DIECA, which is unstable in acidic solutions (JAMES and GARTON, 1952).

2. Inhibition within the Cytochrome System.

A number of agents interfere with electron transfer in the region between cyto-chromes b and c. Though the precise mechanism of this action has not been established, these inhibitors have been extremely useful in studies on the respiratory chain.

Antimycin A is one of a group of *Streptomyces* antibiotics (see STRONG, 1958) which have the following unusual structure (VAN TAMELEN, DICKIE, LOOMANS, DEWEY and STRONG, 1961):

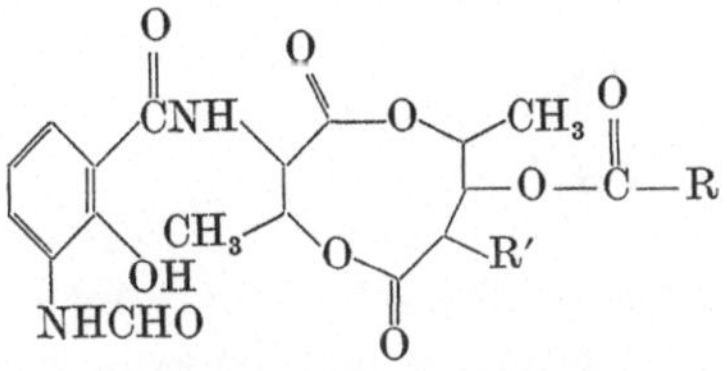

Antimycin A_1: $R = -CH_2CH(CH_3)_2$; $R' = n\text{-}C_6H_{13}$
Antimycin A_3: $R = -CH_2CH(CH_3)_2$; $R' = n\text{-}C_4H_9$

It has been used to block respiratory activity in a wide variety of plant mitochondria, and spectrophotometric studies with animal and plant particles have shown that it prevents the reoxidation of cytochrome "b" (see CHANCE and WILLIAMS, 1956; HACKETT, 1959). The inhibitor can also be used with intact cells or slices of plant tissue. Antimycin can be purchased from the Wisconsin Alumni Research Foundation, Madison, Wisconsin. It is extremely insoluble in water; relatively stable stock solutions can be made up in ethanol and stored at 0° C. An amount of alcohol equivalent to that added with the antimycin must always be added to the control. Care should be taken to wash out all the reaction vessels with an organic solvent in order to remove all traces of inhibitor. A wide range of final concentrations (0.001 to 10 μg/ml) of the antimycin should be tested. The effectiveness of a given concentration is dependent on the amount of protein in the system (POTTER and REIF, 1954).

A group of alkyl-substituted hydroxyquinolines have also been used to block electron transfer between cytochromes b and c (JACKSON and LIGHTBOWN, 1958). The most widely used compound is 2-heptyl-4-hydroxy-quinoline-N-oxide, *HOQNO* (CORNFORTH and JAMES, 1956):

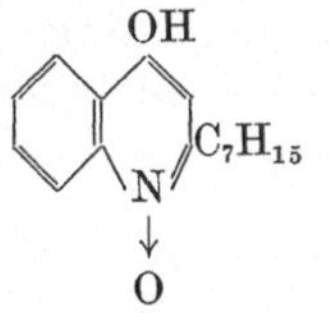

It is moderately soluble in water, and low concentrations (10^{-8} to 10^{-4} M) have been shown to inhibit respiration in plant mitochondria (HACKETT et al., 1960; WISKICH, MORTON and ROBERTSON, 1960).

BAL *(2,3-dimercaptopropanol)* and SN 5949 *[2-hydroxy-3-(2-methyloctyl)-1,4-naphthoquinone]* also block the respiratory chain in this region (SLATER, 1950; BALL, ANFINSEN and COOPER, 1947), but these inhibitors have been used less extensively. The latter compound does inhibit oxidations by plant mitochondria (HUMPHREYS and CONN, 1956; WISKICH et al., 1960).

3. Inhibition in the Flavoprotein Region.

Narcotics have long been known to inhibit respiration and recent work has shown that the barbiturates act on the respiratory chain itself. ERNSTER, JALLING, Low and LINDBERG (1955) showed that *amytal* (5-ethyl-5-isoamyl-barbiturate) in millimolar concentrations completely blocks the DPN^+-linked oxidations in liver mitochondria; it has the following structure:

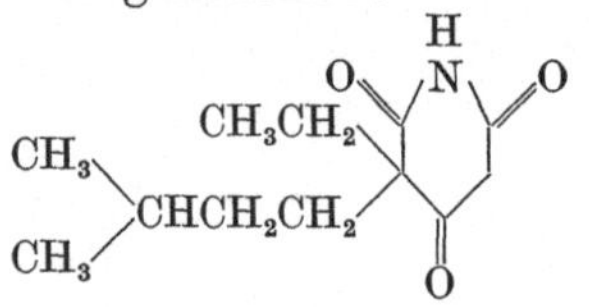

The sodium salt of amytal is readily soluble and fresh solutions should be made up just before use. Other barbiturates may be even more effective than amytal (CHANCE and HESS, 1959). Spectrophotometric examinations have shown that the probable site of action is between DPNH and flavoprotein (CHANCE, 1956). However, amytal can also inhibit electron transfer between flavoprotein and cytochrome b (PACKER, ESTABROOK, SINGER and KIMURA, 1960). The DPNH oxidase of beet root particles was blocked up to 77% by 3.6 mM amytal (WISKICH et al., 1960). On the other hand, the mitochondria from a variety of other tissues (skunk cabbage spadix, sweet potato root, apple fruit) are remarkably resistant to this inhibitor.

Atabrine has been used to inhibit a number of flavoprotein enzymes, including the pyridine nucleotide-cytochrome c reductases (HAAS, 1944); it is a substituted acridine, with the following structure:

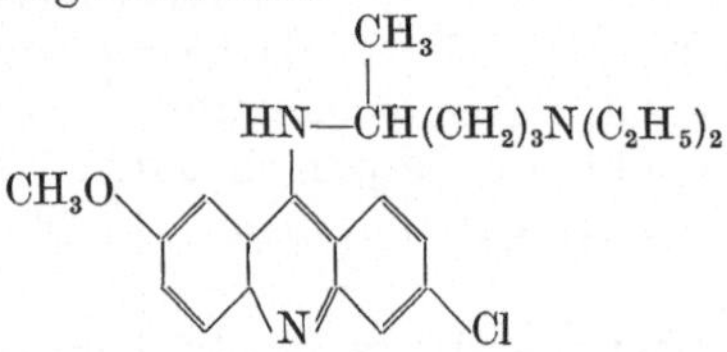

Two inhibitors which act rather specifically on certain quinone reductases are *2,4-dinitrophenol* and *dicoumarol*. The pea seed enzyme (WOSILAIT and NASON, 1955) is very sensitive to DNP. The menadione reductase from liver is inhibited by dicoumarol but not by DNP (MÄRKI and MARTIUS, 1960), while the same enzyme from spinach is not inhibited by either compound (WOLF, KIEFFER and MARTIUS, 1959). The structure of dicoumarol is:

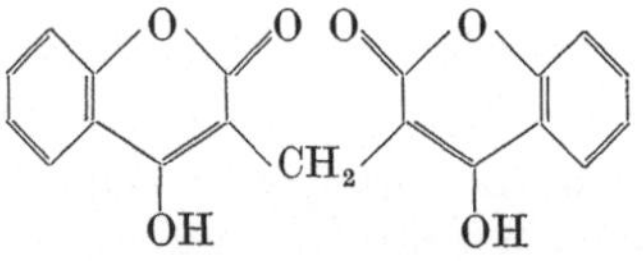

V. Spectrophotometric Methods.

Decisive contributions to our knowledge of the respiratory chain have come from direct observations of the individual components, whose absorption spectra change markedly on oxidation or reduction. It was the appearance (on reduction) and disappearance (on oxidation) of visible absorption bands that led to the discovery of the cytochromes (MACMUNN, 1886; KEILIN, 1925). In the past, these studies were carried out primarily by means of visual spectroscopy, and these techniques were reviewed in detail in a preceding volume of this series (HARTREE, 1955). While the direct-vision spectroscope serves as a powerful tool for rapid, qualitative analysis, it is not well suited for quantitative work and cannot be used effectively at wave lengths below 500 mμ. Recent developments of rapid and sensitive spectrophotometric techniques which can be used with particle suspensions or intact cells have made possible the quantitative analysis of the principal components in the normal respiratory chain. This work has been pioneered in CHANCE's laboratory, and the reader should consult his reviews on spectrophotometric techniques for the assay of respiratory enzymes (CHANCE, 1954, 1957). The application of similar methods to the study of plant cytochromes has recently been discussed by LUNDEGÅRDH (1960). Some general aspects of the principles and practice of spectrophotometry have been discussed earlier in this series by GLOVER (1956).

1. Special Problems.

The respiratory pigments are frequently studied in opaque or turbid systems, and these present a difficult optical problem due to the scattering of light. When a suspension is diluted to the point where opacity is not a serious factor, the concentrations may be too low to measure. Thus, in a yeast cell suspension that absorbs 50% of the incident light, reduction of cytochrome oxidase causes an absorbancy change of only 0.002 cm^{-1}. Most commercial spectrophotometers do not measure absorbancy changes of less than 0.01 cm^{-1} with much accuracy. The turbidity problem is generally less serious in suspensions of isolated mitochondria, where the concentrations of the pigments are higher. Much of the difficulty can be avoided by recording *difference spectra*, rather than absolute spectra. Thus, a plot of the differences in absorption between two identical suspensions, one of which is anaerobic and the other aerobic, cancels out the nonspecific absorptions and reveals the components that are oxidized by O_2. Control experiments have shown that this technique results in relatively little distortion, even when the suspensions are too turbid to show anything significant in the absolute spectra. For example, the reduced-oxidized difference spectrum of cytochrome c is essentially the same in a turbid suspension and in one which has been clarified with cholate (CHANCE, 1954). Some of these problems have recently been discussed by KEILIN and HARTREE (1958) .

A number of special techniques have been developed for working with turbid suspensions in the spectrophotometer[1]. To minimize the loss of transmitted light and gather the maximum amount of the scattered light, the phototube should be placed as close as possible to the sample cell. The same effect could be achieved by placing the cell in a reflecting sphere. SHIBATA (1958) developed a special opal glass reflection technique to minimize the scattering effect. Opal glass plates are inserted in the light paths between the cells (sample and reference) and the phototube. BARER and JOSEPH (1955) have shown that the interference is decreased by suspending the material in a medium of suitable refractive index (e.g., glycerol), but care must be taken to avoid toxicity. Still another approach is to disrupt the particles with mild surface-active agents which decrease the optical density of the suspension without interfering with respiratory chain activities. Digitonin has been used most successfully for this purpose (COOPER and LEHNINGER, 1956). We have prepared digitonin fragments of washed sweet potato mitochondria by treating them with 1% digitonin at 0° C. After centrifugation at 80,000×g for 30 minutes, the relatively clear supernatant fraction can be readily examined in conventional spectrophotometers. The reduced-oxidized difference spectrum of this fraction shows the a-, b-, and c-type cytochromes, and they are present in roughly the same proportion as in intact mitochondria (Figure 6). With photosynthetic plants it may be necessary to work with either etiolated tissues or low-chlorophyll mutants in order to avoid interference by nonrespiratory pigments.

---

[1] See BUTLER and NORRIS, Modern Methods of Plant Analysis 5, 51 (1962).

The spectral changes due to the respiratory components should be recorded as rapidly as possible, since they may be affected by a multiplicity of secondary factors. Most automatic recording instruments will plot a spectrum from 300 to 650 mμ in roughly 1 minute. Slower measurements can yield valuable data on the total concentrations of the components, but they do not permit many conclusions as to their kinetic behavior. From this viewpoint, it is preferable to work with isolated mitochondria rather than with intact tissues. A general method that has been used with great success in this work takes advantage of the fact that the absorption spectra of many compounds, including hemoproteins, are greatly sharpened and

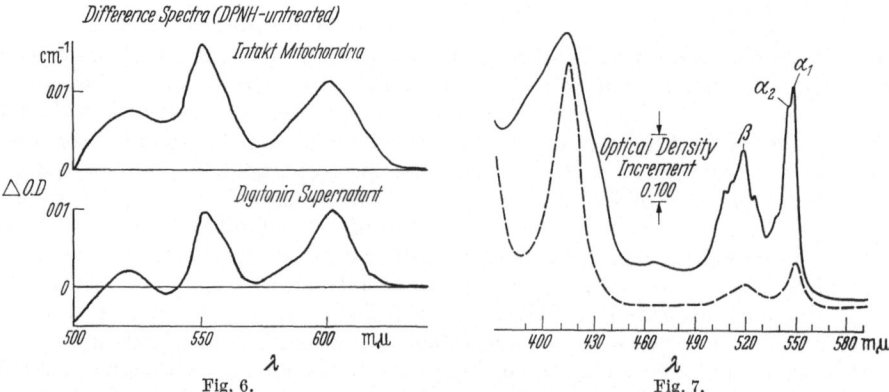

Fig. 6.   Fig. 7.

Fig. 6. Comparison of the (DPNH, anaerobic) — (untreated, aerobic) difference spectra obtained with intact sweet potato root mitochondria (using split-beam spectrophotometer described by Chance, 1945) and digitonin fragments of mitochondria (using Cary 14 spectrophotometer) (unpublished data).

Fig. 7. The effect of temperature on the absorption spectrum of purified, reduced mammalian cytochrome c. The dashed curve was obtained at 300° K and the solid curve at ∼77° K (from R. W. Estabrook 1956).

intensified at very low temperatures (—190° C). The scope of this method and its use with the spectroscope have been discussed by Hartree (1955). The recording spectrophotometer designed by Chance has been adapted for use at liquid nitrogen temperatures, making a powerful tool for the characterization of the cytochrome components (Estabrook, 1956); the effect of low temperature on the spectrum of cytochrome c is shown in Figure 7. Samples have also been examined at low temperature in the Beckman DU (Dorough and Shen, 1950) and Cary 11 (Freed and Sancier, 1954) spectrophotometers. It is of interest that "steady states" of the cytochromes can be stabilized in liquid nitrogen (Chance and Spencer, 1959).

2. Instruments.

Three basic types of spectrophotometers have been used. With the conventional *single-beam spectrophotometer* (e.g., Beckman DU equipped with photomultiplier), the sample and reference cells are placed in the light path alternately and the absorption is determined at suitable wave lengths throughout the spectrum. If the solution is turbid, a reference material of similar opacity (e.g., cotton wool or filter paper) must be used. After altering the conditions in the sample cell (e.g., addition of O_2), the absorption spectrum is again determined. The difference spectrum is obtained by subtracting the appropriate optical densities. When done manually, this method is slow and laborious (100 mμ in approximately $^1/_2$ hour), but it can be used to identify the mitochondrial cytochromes (Hackett and Haas, 1958). Lundegårdh (1960) has developed an instrument that will shuttle the cells back and forth and record the absorbancy automatically. A commercial recording spectrophotometer which operates on the same principles is available (Process and Instruments, Brooklyn, N. Y.). A different technique has been developed by HOLTON (1955) for use with a standard instrument; the difference spectrum of a turbid suspension is obtained by determining the change in absorbancy at one wave length at a time when the sample is oxidized or reduced.

In the *split-beam spectrophotometer*, monochromatic light is sent alternately through the sample and reference cuvettes, which remain stationary; the difference in the amount of light transmitted is measured and recorded directly. If the cuvettes contain identical materials, differing only in their oxidation states, the difference spectrum will show those components that become oxidized. Spectrophotometers operating on this principle are available commercially (CARY, BECKMAN DK); various workers (CHANCE, 1957; BUTLER and NORRIS, 1960) have designed instruments especially for work with turbid suspensions.

By using two light beams of different wave lengths simultaneously, it is possible to examine individual components in a single suspension with the *double-beam spectrophotometer* (CHANCE, 1954, 1957). From the known spectral properties of the respiratory pigments, it is possible to choose wave lengths that will select one component without interference from the other pigments; nonspecific effects, such as changes in the light-scattering properties of the sample, are also eliminated. The wave lengths are set at the peak and trough (or isosbestic point) of the absorption band under investigation. The difference in the absorbancy at these two wave lengths is recorded as a function of time in order to reveal the kinetics of oxidation or reduction of the component. Very rapid reactions can be followed with a regenerative flow attachment, in which the time after mixing is about 10 msec (CHANCE, 1957).

Fluorometric methods can be used to identify specific components, and techniques for recording fluorescence spectra with intact cells (DUYSENS and AMESZ, 1957) or isolated mitochondria (CHANCE and BALTSCHEFFSKY, 1958) have been described. GLOVER (1956) has discussed some of the underlying principles. The reduced pyridine nucleotides can be detected by their fluorescence in the region of 400 to 500 mμ; the maximum for DPNH shifts from 460 mμ to about 440 mμ when it is bound to protein, and at the same time the fluorescence yield is greatly increased (DUYSENS and KRONENBERG, 1957). The isolated mercury line (366 mμ) can be used as the activating light, and changes in the fluorescence at 450 mμ measure the oxidation or reduction of the pyridine nucleotides. The oxidized forms of the coenzymes do not fluoresce.

Micro-methods of spectrophotometry and fluorimetry should permit the localization and assay of respiratory enzymes in single living cells. Progress along these lines has recently been reported from CHANCE's laboratory, where a highly sensitive recording microspectrophotometer (PERRY, THORELL, ÅKERMAN and CHANCE, 1959) and a differential microfluorimeter (CHANCE and LEGALLAIS, 1959) have been developed. Preliminary evidence supports the generally accepted view that the pyridine nucleotides and cytochromes are concentrated in the mitochondria (CHANCE and THORELL, 1959).

3. Procedures.

A variety of methods have been used to oxidize or reduce the material under investigation in order to obtain a difference spectrum. Intact tissues normally contain a large amount of endogenous substrate and the reduction of the respiratory chain components can be brought about simply by removing oxygen from the

42*

surrounding medium. With cell suspensions or slices of tissue in a moist gas phase, this is accomplished by flushing the system with N_2; alternatively, O_2-free water can be circulated around the tissue, previously infiltrated with water to increase transparency. The components are rapidly oxidized when O_2 is admitted. When working with mitochondrial suspensions, a suitable substrate is added, and the supply of oxygen dissolved in the medium is gradually exhausted by the respiratory activity. Any O_2 diffusing into the medium from the air will be consumed. The suspension can be made aerobic again by shaking vigorously with air or by adding O_2. If the mitochondria do not consume O_2 rapidly, they can be examined in an anaerobic system, produced by evacuation or flushing with N_2. The nature of the substrate or reducing agent used is critical. Physiological substrates such as the KREBS cycle acids are ideal; DPNH can be used, although it may reduce additional components which are not directly linked to the respiratory chain. Chemical reductants, such as hydrosulfite, gennerally reduce additional components which are not reduced enzymatically, and these are best visualized in the (hydrosulfite reduced)-(substrate reduced) difference spectrum.

Some of the difficulty arising from the fact that the absorption spectra of the various components overlap (Figure 6) can be overcome by the use of specific chemical reagents. *Carbon monoxide* is used to identify the terminal oxidase as follows: two identical anaerobic samples are prepared; one of these is bubbled with a fine stream of CO for about 1 minute, and the difference in absorbancy of two samples recorded between 400 and 650 mμ. If a CO-combining component is present, the (reduced + CO)-(reduced) difference spectrum will show troughs corresponding to its normal absorption bands and new absorption peaks due to the CO-complex. In the case of cytochrome oxidase (a—a_3), the SORET peak shifts from 444 to 430 mμ, while the α peak shifts from 605 to 590 mμ (Figure 8). If hydrosulfite is used as the reducing agent, peroxidase (SORET peak of CO-complex at 423 mμ), as well as cytochrome oxidase, will combine with the CO (KEILIN and HARTREE, 1951). Other CO-binding pigments have been identified in animal microsomes and in bacteria (CHANCE, 1957).

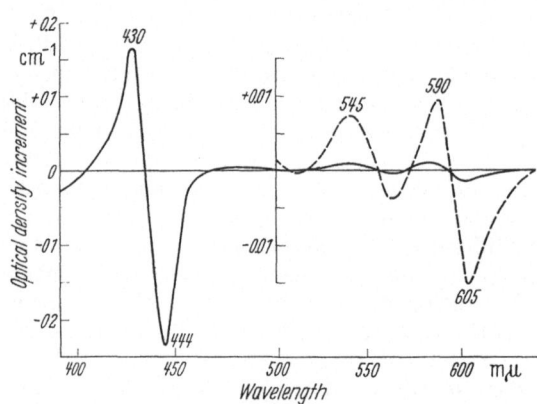

Fig. 8. The difference spectrum of (reduced cytochrome a_3 + CO) — (reduced cytochrome a_3). The dotted line in the visible region was recorded at 10 times higher sensitivity than the solid line (from T. YONETANI 1960).

Antimycin A or HOQNO can be used to selectively reduce the b-type cytochromes in the presence of substrate and oxygen. A base-line is first obtained with two identical, aerobic, samples containing substrate; inhibitor is then added to the experimental sample and the difference spectrum recorded (Figure 9). The positions of the absorption peaks will depend on the system under investigation and peaks for more than one "b" may be identified at —190°C. The other components that come between substrate and "b" in the respiratory chain should also be reduced. If the absorption of an aerobic, antimycin-treated sample is substracted from that of a completely anaerobic sample, the difference spectrum will show the absorption due to the c- and a-type cytochromes alone (Figure 10).

Cyanide combines with cytochrome oxidase in both the divalent and trivalent states (YONETANI, 1960), but the complexes so formed are not very distinct spectroscopically and this reagent has not been widely used for identification. Although cyanide would be expected to cause the reduction of all the respiratory chain components, in many plant tissues only the c- and a-type cytochromes are reduced and cytochrome "b" remains oxidized in the presence of cyanide and O_2

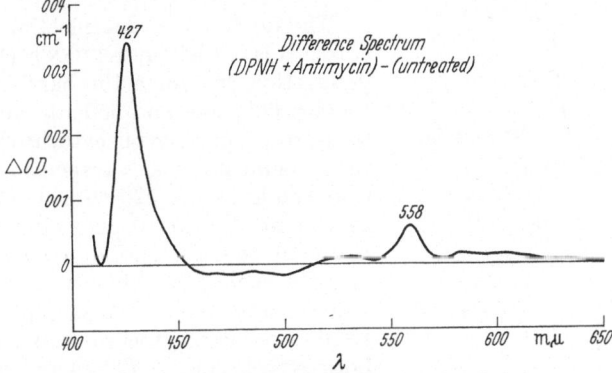

Fig. 9. Difference spectrum showing the absorption spectrum of cytochrome "b" alone in digitonin fragments of sweet potato root mitochondria. Both cuvettes were aerobic; DPNH and Antimycin A were added to the sample cuvette (unpublished data).

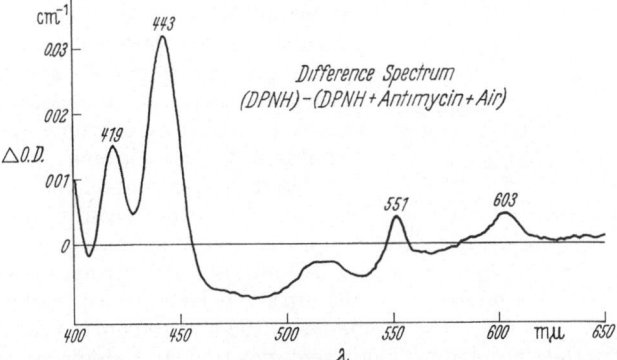

Fig. 10. Difference spectrum showing the absorption spectrum of cytochromes c and a—a_3 in digitonin fragments of sweet potato root mitochondria. The sample cuvette is anaerobic and contains DPNH; the reference cuvette is aerobic and contains both DPNH and Antimycin A (unpublished data).

(LUNDEGÅRDH, 1955; BENDALL and HILL, 1956; YOCUM and HACKETT, 1957), as shown in Figure 11. This property can be utilized to observe the cytochrome "b" by itself: two identical samples containing substrate are allowed to become anaerobic and cyanide (10^{-3} M) is then added to both; after adding O_2 to the reference cell, the difference spectrum is recorded. When working with cyanide, possible spectral interference by other complexes (e.g., with catalase) should also be considered.

Amytal has been most useful for the identification of components in the animal respiratory chain. The difference spectrum obtained by comparing an amytal-treated, aerobic sample (containing substrate) with an untreated, aerobic sample shows the reduced bands of pyridine nucleotide and/or flavoprotein, depending on the exact site of action.

An entirely different technique that can sometimes be used to decrease spectral interference is selective extraction. For example, the concentration of cytochrome c_1 in liver mitochondria has been estimated after removal of cytochrome c by saline washing (Estabrook, 1958).

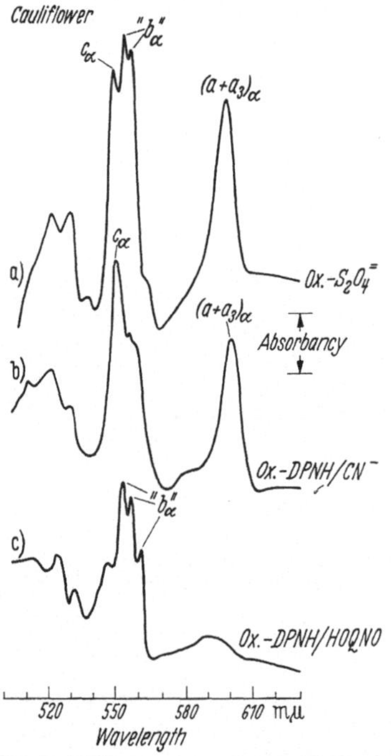

4. Analysis of Results.

The qualitative identification of the cytochromes is based on Keilin's (1925) original description in which the bands appearing at roughly 600, 560, and 550 mμ were attributed to a-, b-, and c-type cytochromes. Each of these pigments is also responsible for absorption bands in the 520 to 530 mμ (β) and 400 to 450 mμ (Soret) regions. But the fact that many of the absorption bands overlap frequently makes it difficult to identify individual components. For example, the 605 mμ band of reduced cytochrome oxidase represents both cytochromes a (72%) and a_3 (28%), and the two components contribute almost equally to the 445 mμ ban (Yonetani, 1960). Generally, the α-peak of cytochrome "b" in plant mitochondria appears close to 560 mμ, in contrast to the 564 mμ peak of the corresponding animal pigment, and interference by peroxidase may be a serious problem. The extent to which cytochrome c_1 (554 mμ) contributes to the cytochrome c (550 mμ) absorption band in plant mitochondria remains to be clarified. The other readily identified components are pyridine nucleotide (reduced band at 340 mμ) and flavoprotein (oxidized band at 465 mμ). The latter is responsible for the trough between 465 and 500 mμ in the typical reduced-oxidized difference spectrum. The exact position and shape of this trough will depend on the nature of the particular flavoproteins present (Beinert, 1960).

Fig. 11. Difference spectra of cauliflower-bud mitochondria at —190° C, showing the complexity of the cytochrome "b" region. (A) Difference in absorption between an aerobic suspension and one treated with $Na_2S_2O_4$. (B) Difference in absorption between an aerobic suspension and one reduced with DPNH in the presence of cyanide. (C) Difference in absorption between an aerobic suspension and one reduced with DPNH in the presence of HO QNO (from D. R. Goddard and W. D. Bonner 1960).

Quantitative estimates of the amounts of the individual components that are present in a given system are made from the changes in optical density which take place on reduction or oxidation. It must be remembered that these estimates will correspond to the total amounts present only if the pigments are *completely* oxidized and reduced; in the steady-state, these pigments are normally partially reduced. From a knowledge of the spectral properties of pure or partially purified components, appropriate extinction coefficients, absorption maxima, and reference wave lengths are chosen for the following calculation:

$$\text{Concentration} = \frac{\Delta A_{\lambda max.} - \Delta A_{\lambda ref.}}{\Delta E}$$

ΔE = difference in the extinction coefficients between the reduced and oxidized forms of the component. The numerator in the above expression is obtained from the difference spectrum by substracting the reading at the reference wave length from that at the maximum. Values which can be used for these calculations are given in Table 2. The following example illustrates the use of these values: a difference spectrum of a sample in a 1 cm cuvette shows optical densities of + 0.015 and — 0.004 at 550 and 540 mμ, respectively; the concentration of cytochrome c would then be 0.019 divided by 19 mM^{-1} or 1×10^{-6} moles per liter of sample. The results of these calculations have shown that the cytochromes of the respiratory chain are generally present in roughly equimolar amounts; however, in plants the concentrations of cytochromes "b" and c may be several-fold greater than that of cytochrome a—a$_3$ and they may vary during development (LUNDEGÅRDH, 1960). The concentrations of flavoprotein and pyridine nucleotide are considerably greater than those of the individual cytochromes.

Similar calculations can be made from difference spectra showing individual components.

Table 2. *Values Used for Calculating Concentrations of Respiratory Components.*

Component	Maximum λ ($\lambda_{max.}$, mμ)	Reference λ ($\lambda_{ref.}$, mμ)	Extinction Coefficient (ΔE mM$^{-1} \times$ cm^{-1}
Cytochrome a—a$_3$	445	455	90.0
Cytochrome a—a$_3$	605	630	16.5
Cytochrome "b"	560[1]	575	20.0
Cytochrome c	550	540	19.1
Cytochrome c$_1$	554	540	17.1
Flavoprotein	460	500	11.5
Pyridine nucleotide	340	375	6.2

[1] Maximum varies in different preparations.

For example, the concentration of cytochrome oxidase is estimated from the (reduced + CO)-(reduced) difference spectrum: the increase in absorbancy at 430 mμ minus the decrease at 445 mμ is divided by 86 mM^{-1} cm^{-1} (SMITH, 1955). The amount of cytochrome "b" can be calculated from the (substrate + Antimycin)- (untreated) difference spectrum using the values at 560 and 575 mμ; it can also be estimated from the difference at 430 and 410 mμ, using a $\Delta E = 180$ mM^{-1} cm^{-1}. Similar values can be used for the additional hydrosulfite-reducible cytochrome "b", which may be equal in amount to the respiratory chain component.

B. The Respiratory Chain Components.

The components involved in the transfer of electrons to oxygen include flavoproteins, cytochromes, metals, and lipids. While the biochemistry of the electron transfer system in animals has been studied in depth (GREEN and FLEISCHER, 1960), relatively little has been done on the individual components in plants. For the present discussion, it will be assumed that the sequence of reactions begins with the oxidation of DPNH (or TPNH) by a specific flavoprotein dehydrogenase. Electrons are then transferred through the cytochrome-cytochrome oxidase system to oxygen. For a general discussion of the hemoproteins, the reader should consult the review by PAUL (1960); the plant cytochromes have been dealt with in several recent reviews (HARTREE, 1957; SMITH and CHANCE, 1958; LUNDEGÅRDH, 1960). Those flavoproteins which are closely linked to the respiratory chain and involved in the oxidation of succinate, α-keto acids, and fatty acids will not be covered here.

I. DPNH Dehydrogenase.

The identification and isolation of the specific enzyme which oxidizes DPNH and transfers electrons to the next member of the respiratory chain is complicated by several factors: 1) a number of DPNH oxidizing enzymes are present in a given tissue, and generally more than one in each intracellular compartment; 2) these enzymes can frequently reduce a variety of electron acceptors, and the relative activities with each may change during the enzyme purification; 3) the precise nature of the acceptor in the respiratory chain has not been established, so that physiological conditions cannot be exactly duplicated in the assay. However, the interpretation of results can be greatly simplified by a careful determination of the intracellular localization of the enzyme. A number of different assay methods are described below.

1. Diaphorase.

Historically, the term diaphorase has been used to describe enzymes which catalyze the transfer of hydrogen from reduced pyridine nucleotides to dyes, such as methylene blue and 2,6-dichlorophenol indophenol:

$$PNH + H^+ + Dye \rightarrow PN^+ + Dye \cdot H_2$$

Many flavoproteins show this property and, starting with the work of LOCKHART (1939), diaphorase has been demonstrated in a great many plant extracts; the activity has been shown with chloroplasts, mitochondria, microsomes, and the supernatant fraction. The intracellular distribution of activity differs quantitatively from one tissue to the next; in some cases it is most abundant in the soluble fraction (MARTIN and MORTON, 1956; SISLER and EVANS, 1958). Both DPNH- and TPNH-specific diaphorases have been described in plants (CLUM and NASON, 1958). Though the physiological roles of these enzymes in plants have not been established, MASSEY (1960, 1960a) has recently shown that the particle-bound diaphorase in animal tissues is identical with lipoyl dehydrogenase, the flavoprotein that transfers hydrogen from reduced lipoic acid to DPN^+ in α-ketoglutaric dehydrogenase (SEARLES and SANADI, 1961).

Assay. The diaphorase activity is measured spectrophotometrically, and either the rate of DPNH oxidation or the rate of dye reduction can be followed. In the latter case, reoxidation of the dye is prevented by working in an anaerobic system or by including cyanide in the reaction mixture. The final reaction mixture for the assay is made up to contain: 0.05 M phosphate buffer, pH 7.0, 1.4×10^{-4} M DPNH (or TPNH), 1×10^{-4} M dye, and 1×10^{-3} M KCN. The reaction is initiated by addition of a small volume of enzyme to the cuvette. If the reaction is followed at 340 mμ (DPNH), it may be necessary to correct for changes in dye absorption at this wave length. The reduction of 2,6-dichlorophenol indophenol is measured at 600 mμ, the peak of the dye's broad absorption band ($E_{600} = 16.1 \times 10^6$ mole$^{-1} \times$ cm^2). At pH's below 7.0, DPNH is oxidized non-enzymatically by the dye, and appropriate controls must also be run to correct for any endogenous dye reduction.

Purification and properties. The diaphorase in the soluble fraction has been partially purified from tobacco roots (SISLER and EVANS, 1958) and wheat germ CLUM and NASON, 1958). In the former case, a 400-fold purification was achieved by fractionation with acetone (collect material precipitating between 33 and 60% by volume acetone) and ammonium sulfate (between 40 and 65% saturated AS); the enzyme was finally adsorbed on calcium phosphate gel and eluted with an ammonium sulfate solution. The tobacco root enzyme is equally active with DPNH and TPNH ($K_m = 8.2 \times 10^{-5}$ M). The wheat germ soluble fraction appears to contain separate enzymes for these two substrates. The diaphorases show pH optima in the range of 6 to 7; they are not inhibited by 10^{-3} M cyanide or azide.

2. Quinone Reductase.

Many enzymes, including the soluble diaphorases described above, can transfer electrons from DPNH or TPNH to a variety of quinones:

$$PNH + H^+ + Quinone \rightarrow PN^+ + Hydroquinone$$

Compounds related either to p-benzoquinone, such as Coenzyme Q, or to 1,4-naphthoquinone, such as vitamin K, frequently show high activity as electron acceptors. WOSILAIT and NASON (1954) obtained evidence that separate enzymes are involved in the reduction of p-benzoquinone and menadione (2-methyl 1,4-naphthoquinone), but the soluble tobacco root diaphorase is equally active with these two acceptors (SISLER and EVANS, 1958). In liver, roughly 90% of the vitamin K-reductase is found in the soluble fraction (WOSILAIT, 1960; MÄRKI and MARTIUS, 1960), and in etiolated pea stems this fraction has roughly 70% of the activity (unpublished). The reduction of exogenous quinones can be readily demonstrated with particulate fractions, such as the mitochondria (COLPA-BOONSTRA and SLATER, 1958; RAMASARMA and LESTER, 1960) or chloroplasts (CRANE, EHRLICH and KEGEL, 1960; IZAWA, 1960).

Assay. The quinone reductase activity is measured spectrophotometrically by following the oxidation of DPNH (or TPNH) at 340 mμ. Since many of the quinones (e.g., vitamin K and Coenzyme Q_{10}) are insoluble in water, they must either be dissolved in an organic solvent (e.g., ethanol) or solubilized with a surface active agent (e.g., Tween, Triton, or polyoxyethylene lauryl alcohol). The Coenzyme Q reductase activity can be assayed using the more soluble analogues, Q_1 or Q_2, which have a smaller number of isoprenoid units in the side-chain than the natural coenzyme (Q_{10}). The activity of the enzyme itself may be increased by the presence of a dispersing agent (e.g., albumin). The following typical reaction mixture has been recommended for the assay of menadione reductase (MÄRKI and MARTIUS, 1960): 0.1 M phosphate buffer, pH 6.0, 2.5×10^{-4} M DPNH, 0.04% serum albumin, and 2×10^{-5} M menadione. The reaction is started by the addition of menadione solution (in methyl glycol), made up freshly each day from pure, sublimated menadione. Correction must be made for any non-enzymatic oxidation of the DPNH. Many of the quinones are unstable in the light and consequently should be kept in the dark.

Purification and properties. WOSILAIT and NASON (1954) have described a method for the 25-fold purification of a p-benzoquinone reductase. Dried pea *(Pisum sativum)* seeds are soaked in water for a day and then blended with an equal volume of 0.1 M phosphate buffer, pH 7.5, for $1^1/_2$ minutes at 4°C. The crude extract obtained by filtering through cheesecloth and centrifuging at 3,000 ×g for 20 minutes is fractionated successively with calcium phosphate gel, ammonium sulfate, and alumina gel. Of the widely distributed menadione reductases in plants (WOSILAIT and NASON, 1954), only the enzyme from leaves (probably present in chloroplasts) has been extensively purified. The following steps were used to achieve a 10,000-fold purification (WOLF, KIEFFER and MARTIUS, 1959): precipitation of inactive proteins by a quaternary base detergent, ammonium sulfate fractionation, heat treatment (60°), acetone precipitation, adsorption on calcium phosphate gel, chromatography on hydroxylapatite, and electrophoresis in a sucrose gradient solution. A comparable degree of purification of the liver enzyme has been reported (MÄRKI and MARTIUS, 1960).

The pH optima of the quinone reductases range from 6 to 9. They are flavoproteins and are not inhibited by cyanide or azide. Surprisingly, they are also insensitive to 10^{-4} M PCMBA, suggesting that —SH groups are not required for activity. In some cases, low concentrations of 2,4-dinitrophenol and/or dicoumarol are effective inhibitors.

3. Cytochrome c Reductase.

The following reaction is demonstrable with extracts of all plants that have been tested:

$$PNH + 2 \text{ Ferricytochrome } c^{+++} \rightarrow PN^+ + 2 \text{ Ferrocytochrome } c^{++} + H^+$$

It should be remembered that the reaction may even proceed non-enzymatically under some conditions, as in the presence of flavin nucleotides (Singer and Kearney, 1950) or chlorogenic acid (Sisler and Evans, 1958). Plant extracts contain a variety of enzymes which catalyze the reaction. The soluble quinone reductase can act as a cytochrome reductase if a catalytic amount of a heat-stable organic factor is present (Clum and Nason, 1958). The cytochrome c reductase system associated with the microsomes is more complicated, involving both a flavoprotein and a b-type cytochrome (Strittmatter and Velick, 1956, 1956a). Activity can be demonstrated with isolated mitochondria, but this may be "artificial", and the enzyme purified from mitochondria (Mahler, 1955) probably represents a modified form of the natural, DPNH-oxidizing lipoflavoprotein (Ziegler, Green and Doeg, 1959). Finally, photosynthetic tissues contain additional cytochrome c reducing enzymes which are involved in photosynthesis, rather than respiration (Evans, 1955; Marré and Servettaz, 1957). From this it is clear that the physiological significance of a particular cytochrome c reductase system must be evaluated very critically.

Assay. The reduction of ferricytochrome c is measured spectrophotometrically at 550 mμ; cyanide is added to prevent the reoxidation of reduced cytochrome c. The final reaction mixture should contain 0.05 M phosphate buffer, pH 7.0, 2×10^{-5} M oxidized cytochrome c, 5×10^{-5} M DPNH (TPNH), and 1×10^{-3} M KCN. The reaction is started by adding a small volume of enzyme, and the linear increase in optical density at 550 mμ is used to calculate the rate. Optical densities are convertéd to the corresponding molarities using the difference in extinction coefficients of the oxidized and reduced froms, $\Delta E_{550} = 19$ mM^{-1} cm^{-1}. Mammalian cytochrome c, which is available commercially, can be used for this assay. Stock solutions can be stored in the frozen state without appreciable loss of activity.

Purification and properties. The only reported purification of a DPNH-cytochrome c reductase from plant mitochondria is that described by Davies (1956). An extract of pea epicotyl mitochondria was prepared by shaking with ballotini beads, freezing, and thawing. After fractionating the extract with ammonium sulfate, inactive material was precipitated at pH 4.9. The ratio of cytochrome c reductase to diaphorase activity remained constant at all stages of the 44-fold purification. Evans (1955) partially purified a DPNH-cytochrome c enzyme from an acetone powder of whole soybean leaves, using ammonium sulfate and calcium phosphate gel treatments. The classical purification of TPNH-cytochrome c reductase from yeast has been described in detail by Haas (1955). A TPNH-cytochrome c reductase has been partially purified from pea leaves (Marré, Servettaz and Rossi, 1957).

The cytochrome reductases from animal tissues have been more extensively purified, and the methods employed may be useful for future work with plants. In Mahler's (1955) procedure, mitochondrial particles are extracted with 10% ethanol to give a yellow supernatant which is then fractionated by classical methods; the prosthetic group of the enzyme may be modified by this procedure. Recently, Massey (1960) has shown that the enzyme is readily extracted from the particles by a brief heat (38°) treatment, after which it can be purified on a calcium phosphate gel column. The enzyme in the microsomes, DPNH-microsomal cytochrome reductase, is extrated by treatment with 10% ethanol at 42—43° for 12 minutes (Strittmatter and Velick, 1956a). A TPNH-specific cytochrome c reductase has also been isolated from animal tissues (Lang and Nason, 1959).

The cytochrome c reductases are flavoproteins which also contain non-heme iron (see Beinert and Sands, 1959). The K_m for DPNH is around 1×10^{-5} M. The reported pH optima vary from 6.0 to 8.5. The enzyme itself is not inhibited by cyanide, azide, or antimycin, but it is inhibited by atabrine and PCMBA. The more complex cytochrome c reductase system of intact mitochondria is blocked by antimycin A, and this property can be used to distinguish the mitochondrial system from the activity in the microsomes and the soluble fraction.

4. DPNH-Ferricyanide Reduction.

In recent studies on the primary DPNH dehydrogenase of the animal respiratory chain, ferricyanide has been used most effectively as an electron acceptor. The large lipoflavoprotein which is isolated from heart mitochondria and which may resemble the "native" dehydrogenase shows high activity with ferricyanide, but little or no reaction with 2,6-dichlorophenol indophenol and cytochrome c (ZIEGLER, GREEN and DOEG, 1959). The smaller, solubilized DPNH dehydrogenase also reacts most rapidly with ferricyanide (RINGLER, MINAKAMI and SINGER, 1960). The activity is markedly dependent on the ferricyanide concentration.

Assay. MINAKAMI, RINGLER and SINGER (1960a) have described the conditions under which an apparently reliable estimate of activity can be obtained, regardless of the physical state or purity of the preparation. The initial rate of ferricyanide reduction is measured spectrophotometrically (between 430 and 450 mμ). The final reaction mixture contains 0.04 M phosphate buffer, pH 7.4, 2.5×10^{-3} M potassium ferricyanide, and 1.5×10^{-4} M DPNH; the reaction is started by addition of enzyme. Correction for any non-enzymatic oxidation of DPNH by ferricyanide must be made.

Purification and properties. No enzyme has been isolated from plants using this particular assay method. A gentle method for solubilizing the enzyme from heart mitochondria by treatment with *Naja naja* venom has been described (RINGLER et al., 1960); subsequent purification was achieved by ammonium sulfate fractionation and chromatography. Organic solvent fractionation was used by ZIEGLER et al. (1959) to isolate the larger lipoprotein complex. The enzyme reacts only with DPNH, and not with TPNH. It is not inhibited by antimycin A, amytal, or dicoumarol. The enzyme is a flavoprotein containing non-heme iron. Its absorption spectrum is atypical: the oxidized form of the enzyme has a peak at 410 mμ (pH 7.4) and this disappears on reduction with DPNH or hydrosulfite. It is of interest that the flavoprotein extracted from *Arum* spadix mitochondria by treatment with 70% ethanol shows an absorption band at 415 mμ (JAMES and ELLIOTT, 1958).

II. Transhydrogenase.

The available evidence suggests that the respiratory chain itself does not oxidize TPNH directly. Thus, digitonin fragments of liver mitochondria can carry out all the reactions of the respiratory chain, but they will not oxidize TPNH unless a catalytic amount of DPN is present (DEVLIN, 1959). The primary dehydrogenase of the respiratory chain appears to be specific for DPNH. These findings suggest that the oxidation of TPNH by the respiratory chain in animal cells is probably mediated by a mitochondrial transhydrogenase (KAPLAN, SWARTZ, FRECH and CIOTTI, 1956). Such an enzyme has been demonstrated in extracts of pea epicotyl mitochondria (DAVIES, 1956). Nevertheless, the TPNH oxidase activity of isolated plant mitochondria is usually very low. The transhydrogenase which is present in green leaves is almost certainly involved in the metabolism of photosynthetically-generated TPNH (KEISTER, SAN PIETRO and STOLZENBACH, 1960).

Assay. The following reaction is catalyzed by transhydrogenase:

$$TPNH + DPN^+ \rightleftharpoons TPN^+ + DPNH$$

The reduction of DPN$^+$ is followed spectrophotometrically (340 mμ) in the presence of a catalytic amount of TPN$^+$ and a TPN$^+$-specific reducing system. In the method described by KAPLAN (1955), the reaction mixture contains 2.6 ml of 0.1 M phosphate buffer (pH 7.5), 0.1 ml of 0.1 M MgCl$_2$, 0.2 ml of purified pig heart isocitric dehydrogenase, and 0.1 μmole TPN$^+$. Sodium isocitrate (10 μmoles) is added to the mixture and the preliminary reduction of TPN$^+$ followed at 340 mμ; after this reaction is complete, 1.0 μmole of DPN$^+$ and the preparation to be tested are added. The increase in optical density at 340 mμ is again followed, and the rate of DPNH formation serves as the measure of activity. A control without TPN$^+$ must be run to correct for any direct reduction of DPN$^+$. With crude preparations, 3 μmoles of KCN should be included in the reaction mixture to prevent any reoxidation of DPNH; 60 μmoles of nicotinamide can be included to prevent the hydrolytic cleavage of the nucleotides. The addition of 0.5% digitonin to the reaction mixture markedly stimulates the transhydrogenase activity of mitochondria, presumably by fragmenting the particles (STEIN, KAPLAN and CIOTTI, 1959).

A different type of transhydrogenase assay has been developed making use of the analogues of the pyridine nucleotides. Some of these compounds show different spectral properties from the natural coenzymes, and STEIN et al. (1959) have described a simple and sensitive spectrophotometric assay based on the following reaction:

$$\text{TPNH} + \text{3-acetylpyridine (DPN}^+\text{)} \rightarrow \text{TPN}^+ + \text{3-acetylpyridine (DPNH)}$$

Due to the greater oxidation-reduction potential of the analogue, the reaction goes to completion. It is unnecessary to generate TPNH enzymatically, and the reaction is followed at 375 mμ, where an optical density change of 0.005 corresponds to the reduction of one mμmole of analogue. The reaction mixture contains: 0.3 mmole phosphate buffer (pH 6.5), 3 μmoles KCN, 0.4 μmole TPNH, 0.6 μmole 3-acetylpyridine DPN$^+$ (available from Pabst Laboratories), and enzyme in 3.0 ml.

A third type of reaction which is catalyzed by some transhydrogenases is the transfer of hydrogen between the reduced and oxidized forms of the same pyridine nucleotide. Using the method described above, this exchange can be followed using the coenzyme and its own analogue:

$$\text{TPNH} + \text{acetylpyridine (TPN}^+\text{)} \rightarrow \text{TPN}^+ + \text{acetylpyridine (TPNH)}$$

Purification and properties. A mitochondrial transhydrogenase has not yet been purified from plants. On the other hand, the enzyme from spinach leaves has been purified some 500-fold by acetone and protamine sulfate precipitation, followed by adsorption on bentonite and Dowex 50 W (KEISTER et al., 1960). This enzyme is specific for TPNH as the hydrogen donor, but it is relatively unspecific with respect to the acceptor. It is probably a flavoprotein. Transhydrogenases have been purified from bacteria and beef heart (KAPLAN, 1955).

III. Cytochromes "b".

This group includes hemoproteins which show, when reduced, absorption maxima in the region 556—564 m$\mu(\alpha)$ and 423—430mμ (SORET). They are auto-oxidizable, but do not combine with either CO or cyanide. Their redox potentials (E'_0, pH 7) are generally close to 0 volts. In solution, they can be oxidized by cytochrome c or ferricyanide; hydrosulfite, ascorbate, and cysteine have been used as reducing agents. Actually, it is not always easy to classify a given hemoprotein, and the properties of the several groups of cytochromes may overlap.

Table 3. *Some Properties of b-Type Cytochromes.*

Cyto-chrome	Occurrence	Absorption Maxima of Reduced Cytochrome (mμ)			$E'_t(v)$ (pH 7)	Stability	Remarks
		α	β	γ			
b	animal mitochondria; plant mitochondria?	561	530	430	+0.07	stable at 0°C	respiratory chain component
b_2	yeast; particles?	557	528	424	+0.22	crystalline enzyme inactivated by O_2	lactic dehydrogenase activity; protoheme/FMN = 1/1
b_3	green and non-green plant tissues; particles?	559 560	529	425 ?	+0.04	denatured by alcohol, acetone, acid	reduced by ascorbate; relation to microsomal b unclear
b_5	animal microsomes; mitochondria?	556	526	423	+0.02	unstable in acidic, low salt conc. solution	reduced by specific flavoprotein reductase
b_6	chloroplasts	563			—0.06	denatured by organic solvents	not yet purified
b_7	plant (Aroid spadix) mitochondria	559 560	529	427 ?	—0.03		not yet purified
$b_{561, 554}$	*Sclerotinia* mycelium	561, 554	528	424			double α-peak

Several cytochromes of the b-type (b, b_3, b_7) have been reported to occur in plant mitochondria. In the presence of DPNH and antimycin A or HOQNO, the difference spectrum of the mitochondria shows a three-banded spectrum (at —190°C) with peaks at 555, 558, and 564 mμ (Figure 11). Before it can be concluded that several b's are involved in terminal respiration, more information is needed on their properties and kinetic behavior; some of these are listed in Table 3. The intracellular localization of these pigments provides valuable evidence as to their physiological functions: for example, the cytochrome b_6 in chloroplasts is probably not involved in respiration. On the other hand, the same pigment may be present in more than one cell fraction. In the assay and isolation of a specific cytochrome b, care must be taken to avoid interference from closely related components; these include the other b-type cytochromes, the "dithionite reducible b", and peroxidase, all of which have similar α-absorption bands (556—566 mμ). The peroxidases can be distinguished from the b-type cytochromes by the fact that the former, when reduced, combine with CO, whereas the latter only combine with CO when denatured (BEN-GERSHOM, 1961).

Assay. *1. Spectra.* As outlined earlier, the concentration of cytochrome "b" in a preparation containing other cytochromes can be estimated from the differences in absorbancy (ΔA), at appropriate peak and reference wave lengths, between the oxidized and reduced forms:

$$[b] = \frac{\Delta A_{\alpha\,peak} - \Delta A_{575}}{\Delta E}.$$

The wave length of the α peak will depend on the particular component under investigation; values for ΔE are not yet available for the plant pigments, but the use of values calculated for similar animal cytochromes (15 to 20 mM^{-1} cm^{-1}) should not introduce much error. HARTREE (1955) has described a method for estimating the concentration by comparing the intensity of the "b" α band with a standard haemochromogen.

Following the separation and purification of a single component, the concentration can be calculated from the absorbance at individual wave lengths. The absolute absorption coefficients of cytochrome b_5 have been determined (STRITT-MATTER and VELICK, 1956):

	$E_{m\,M}$ of absorption peak
Oxidized, 413 mμ	117 ± 5
Reduced, 423 mμ	171 ± 8
Reduced, 526 mμ	13.4 ± 0.5
Reduced, 556 mμ	25.6 ± 1.0

The purity of a given preparation can be estimated from the ratio $E_{\alpha\,peak}^{reduced}/E_{280\,m\mu}^{oxidized}$, which relates the pigment concentration to the total protein concentration. This value is of the order of 1.0 for most purified cytochromes.

2. *Activity.* Since the precise physiological roles of the cytochromes "b" are not known, assays of activity in terms of ability to transfer electrons from some donor system to a suitable acceptor may be quite "artificial". Nevertheless, it may be useful to have such an assay, and a variety of donor systems — succinate-succinic dehydrogenase (STOTZ, 1955), DPNH-cytochrome reductase (STRITTMAT-TER and VELICK, 1956), and lactate (APPLEBY and MORTON, 1959) — have been used with such acceptors as cytochrome c and ferricyanide. Both donor and acceptor must be present in sufficient amounts so that the over-all reaction rate is limited by the concentration of "b".

The following system was used for the spectrophotometric assay of the catalytic activity of cytochrome b_5 (STRITTMATTER and VELICK, 1956): 0.085 μmoles cytochrome c, 0.2 μmoles DPNH, 0.04 ml reductase solution, 0.78 ml of 0.1 M Tris buffer (pH 8.0), and 0.06 ml test solution, all in 1.0 ml. The reduction is followed at 550 mμ, and in order to minimize the spectral change due to "b" at this wave length, the cytochrome c should be present in at least a 5-fold molar excess. Cyanide (10^{-3} M) can be included to prevent any reoxidation. The rate of reduction should be proportional to the concentration of the cytochrome solution. The activity of cytochrome b_2 preparations was measured in the following mixture (APPLEBY and MORTON, 1959): 0.33 M DL-lactate, 0.83 mM K ferricyanide, 0.03 M buffer (pH 8.0), 1.0 mM versene, and enzyme. The rate of ferricyanide reduction was followed at 420 mμ.

Purification and properties. The prosthetic group of the b-type cytochromes is protohematin. This can be most conveniently demonstrated by treating the preparation with alkaline pyridine (20% pyridine in 0.1 N NaOH); when reduced with $Na_2S_2O_4$, the pyridine hemochromogen shows a sharp α-band with a peak at 556 mμ (cf. HARTREE, 1955). The prosthetic group can be removed from the protein by treatment with acid acetone (1% 6 M HCl in acetone) for a few minutes; after centrifuging down the protein, the supernatant shows the spectrum of hemin chloride (LEWIS, 1954). In one case (b_5), the nature of the binding has been studied in detail and the cytochrome reconstituted from the apoprotein and protohematin (STRITTMATTER, 1960). The molecular weights range from 13,000 (b_5) to 75,000 (b_2); there is also some evidence that these cytochromes may form large aggregates. Components other than protoheme which have been found in intimate association with the protein include FMN (APPLEBY and MORTON, 1954) and non-heme iron (BOERI and TOSI, 1956), while DNA (APPLEBY and MORTON, 1960) and carotenoids

(JACKSON and LAWTON, 1958) have been reported to occur in some preparations. Methods used for the purification of a number of b-type cytochromes are summarized below, but only one of these (b_3) is from a higher plant. To determine whether any denaturation has taken place during the purification, the effect of CO on the reduced pigment should be determined at all stages (BEN GERSHOM, 1061).

Cytochrome b has not yet been isolated from plant mitochondria, but it has been purified from heart muscle. The first separation was achieved by treatment of heart muscle particles with bile salts (EICHEL, WAINIO, PERSON and COOPER-STEIN, 1950); ammonium sulfate fractionation of the cholate extracts then gave further purifications (HÜBSCHER, KIESE and NICOLAS, 1954; STOTZ, 1955). Treatments with bacterial proteinase (SEKUZU and OKUNUKI, 1956), pancreatic protease (FELDMAN and WAINIO, 1959), and snake venom (FELDMAN and WAINIO, 1960) have also been used to bring about solubilization, but the stability and activity of the cytochrome may be seriously impaired by these procedures. BOMSTEIN, GOLD-BERGER and TISDALE (1960) reported the isolation of cytochrome b by a method that utilizes several surface active agents, ammonium sulfate, and heat treatments; large aggregates were dispersed into units with molecular weight of roughly 20,000 by cetyl dimethyl ammonium bromide.

Cytochrome b_2, the lactic dehydrogenase of baker's yeast, has been extensively purified in several laboratoories (APPLEBY and MORTON, 1954; BOERI, CUTOLO, LUZZATI and TOSI, 1955; YAMASHITA, HIGASHI, YAMANAKA, NOZAKI, MIZUSHIMA, MATSUBARA, HORIO and OKUNUKI, 1957). The procedure of APPLEBY and MORTON (1959), which yields a homogeneous, crystalline flavohemoprotein with high enzymatic activity, involves the following steps: air-drying of yeast, extraction of lipid material with butanol, fractionation of lactate extract with acetone at low temperature, and crystallization at low ionic strength by anaerobic dialysis against lactate plus EDTA. During the purification, cytochrome b_2 was followed by measuring the lactate-ferricyanide reductase activity. The enzyme also shows considerable activity with α-hydroxybutyrate. The crystalline material prepared by the Japanese workers does not oxidize lactate, presumably due to the loss of the flavin component during purification.

Cytochrome b_3, with an α-band at 559 mμ, was purified from broad bean leaves by HILL and SCARISBRICK (1951). The pigment was also detected in tissues devoid of chlorophyll (e.g., potato tubers). Subsequently, MARTIN and MORTON (1955) claimed that cytochrome b_3 is identical with the pigment in plant microsomes. However, the precise nature of the cytochrome pigment (s) in the small particle fractions of plants has not been established. The following purification procedure was used by HILL and SCARISBRICK. Leaves of the broad bean were minced and squeezed through muslin; the juice was kept overnight at room temperature in filled, stoppered bottles, and the green precipitate then removed by centrifugation. Throughout the subsequent steps, the mixture was kept reduced by the addition of about 1% w/v of a mixture of equal parts of $Na_2S_2O_3$ and $NaHSO_3$ (pH 5.5), together with 1% w/v $Na_2S_2O_4$. To each liter of fluid, 136 g of AS were added, the precipitate spun off, and the supernatant solution dialyzed against saturated AS. The precipitate formed contained both cytochromes c and b_3; some of the cytochrome b_3 remained in the fluid, from which it could be removed by adsorption on a small amount of silica gel or kieselguhr, providing the solution was completely saturated with AS. The substance could then be eluted with a small amount of water to give a fairly pure solution of cytochrome b_3. When a sufficient amount of this solution had been obtained, it was carefully fractionated with AS; the fraction precipitating between 90 and 100% saturation yielded the purest sample obtained. In addition to the major absorption peaks, the spectrum of the reduced component showed a band at 594 mμ, possibly due to contaminating peroxidase.

Cytochrome b_5, the pigment in the microsomal fraction of mammalian tissues, has been extensively purified in a number of laboratories (STRITTMATTER and VELICK, 1956; GARFINKEL, 1957; KRISCH and STAUDINGER, 1958; RAW, MOLINIARI, FERREIRA DO AMARAL and MAHLER, 1958). The particle-bound cytochrome can be solubilized by digesting isolated microsomes for 1 hour at 37°C with pancreatic lipase (or pancreatin); it is then purified by ammonium sulfate fractionation. Some recent work suggests that this cytochrome may also be present in liver mitochondria (RAW and MAHLER, 1959). RAW and co-workers, using chromatography on hydroxylapatite and DEAE cellulose, have been able to crystallize cytochrome b_5 (RAW and COLLI, 1959).

Cytochrome b_6 (reduced α-band at 563 mμ) is found in chloroplasts, and the initial attempts to remove it from the particles in an unmodified form were not successful (HILL, 1954). The localization suggests that this component is involved in the electron transfer reactions of photosynthesis, rather than respiration.

Cytochrome b_7, which was first identified in mitochondria isolated from Aroid spadices (BENDALL and HILL, 1956), may in fact be very widely distributed in plant mitochondria (GODDARD and BONNER, 1960). It has not yet been isolated or purified.

YAMANAKA, HORIO and OKUNUKI (1960) were able to prepare a water-soluble b-type cytochrome from the fungus *Sclerotinia libertiana*. When purified, the reduced pigment shows an unusual α-band with a double peak (561 and 554 mμ), and its biological function has not yet been established. The cytochrome was solubilized from the lyophilized fungus with the aid of an active lipase; it was purified by chromatography on an aluminum oxide column and fractionation with ammonium sulfate. Zone electrophoresis on a starch colum failed to separate the two α peaks, which may be due to two prosthetic groups associated with a single hemoprotein (cf. HORIO and KAMEN, 1961).

IV. Cytochromes "c".

In the cytochromes of this group, the protohematin prosthetic group is linked to the protein by thioether bonds. The absorption maxima of the reduced pigments range from 550 to 555 mμ (α) and from 415 to 420 mμ (SORET). In general, the c-type cytochromes are easy to extract and relatively stable; they are not autoxidizable at pH's near neutrality. *Cytochrome c* itself, which has an α-peak at 550 mμ, is present in fungi, algae, and higher plants. It is a normal component of the mitochondrial respiratory chain, though the extracted pigment may differ somewhat from the membrane-bound cytochrome (AMBE and CRANE, 1959). *Cytochrome c_1*, with an α-peak at 553—554 mμ, is an integral part of the mammalian respiratory chain (ESTABROOK, 1958), but its existence and role in plants have not been firmly established. A number of workers have observed an absorption band at 553—554 mμ in preparations from yeast and higher plants, but the component has not been isolated from plants. Cytochromes c_2, c_3, c_4, and c_5 have so far been found only in bacteria. Two additional c-type cytochromes have been isolated from photosynthetic plants—cytochrome f from higher plant leaves (HILL and SCARISBRICK, 1951) and cytochrome 553 from the alga *Porphyra tenera* (KATOH, 1960)—but there is no evidence that these are involved in respiration. Only cytochromes c and c_1 will be considered here.

Assay. *1. Spectra.* The concentration of *cytochrome c* in impure or turbid preparations is calculated from the difference in absorption between the reduced and oxidized forms, as follows:

$$[c] = \frac{\Delta A_{550} - \Delta A_{540}}{19.1 \times 10^3 \ \text{mM}^{-1} \text{cm}^{-1}} .$$

With pure preparations, the concentration can be calculated from the optical density of the reduced pigment at 550 mμ, using the molecular extinction coefficient of $E_{550} = 27.7 \times 10^3$ mM^{-1} cm^{-1}. The purity is indicated by the ratio of the absorbancy of the reduced form at 550 mμ to that of the oxidized form at 280 mμ; the value $A_{550/280}$ for crystalline cytochrome c varies from 1.0 to 1.3, depending on the source of the pigment. The presence of any denatured cytochrome c can be detected by passing CO through the solution. If there is any change in the spectrum, the degree of denaturation is calculated from the optical density readings at 550 mμ as follows (PAUL, 1951):

$$\% \ \text{denaturation} = \frac{162 \, (A_{red} - A_{red + CO})}{A_{red}} .$$

The concentration of *cytochrome c$_1$* is difficult to estimate in the presence of interfering cytochromes b and c. Some or all of the interference can be eliminated by extracting the cytochrome c with 0.85% NaCl solution, or by using ascorbate, which does not reduce cytochrome b, as the reductant. The concentration is calculated from the reduced-oxidized difference spectrum:

$$[c_1] = \frac{\Delta A_{554} - \Delta A_{540}}{17.1 \times 10^3 \ \text{mM}^{-1} \text{cm}^{-1}} .$$

In relatively pure preparations, the concentration of cytochrome c$_1$ can be calculated directly from the optical density of the reduced pigment at the α-peak; published values of E_{554} range from 15.3 to 24.1 mM^{-1} cm^{-1}. The shift of the α-peak to 551 mμ at $-190°$ C can be used for identification (KEILIN and HARTREE, 1955).

2. Activity. The concentration of *cytochrome c* can also be measured on the basis of its activity in an appropriate catalytic system. The manometric method, which utilizes a reductant (hydroquinone, ascorbate, or succinate)—cytochrome c—cytochrome oxidase—O$_2$ system, has been described in detail earlier in this series (HARTREE, 1955). A spectrophotometric assay which determines the amount of enzymatically reducible cytochrome c can also be used: 0.1 ml of 0.5 M lactate in 0.1 M phosphate buffer, pH 6.0, and a small amount of purified yeast lactic dehydrogenase are added to the test solution (2.9 ml), and the increase in optical density at 550 mμ is determined. A polarographic method for the determination of small quantities of cytochrome c has been described (CARRUTHERS, 1947). The catalytic effectiveness of a given concentration of cytochrome c is much greater in the intact respiratory chain than in the reconstructed model systems.

The activity of *cytochrome c$_1$* has been estimated as a part of a succinic-cytochrome c reductase system (CLARK, NEUFELD, WIDMER and STOTZ, 1954). The reaction mixture contains: 1.0 ml of 0.1 M phosphate buffer, pH 7.4; 0.2 ml of 0.3 M sodium succinate; 0.2 ml of 2×10^{-4} M cytochrome c; 0.2 ml of neutralized 0.02 M KCN (prevents reoxidation of reduced cytochrome); 0.1 ml of succinic dehydrogenase preparation; 1.2 ml of water. The reaction rate is determined from the increase in optical density at 550 mμ.

Purification. Plant materials were used in some of the early preparations of *cytochrome c* (KEILIN, 1930; GODDARD, 1944), and in recent years the pigment has been extensively purified from the fungus *Ustilago sphaerogena* (NEILANDS, 1952), from baker's yeast (MINAKAMI, 1955; LI and TSOU, 1956; HAGIHARA, HORIO, YAMASHITA, NOZAKI and OKUNUKI, 1956; NOZAKI, YAMANAKA, HORIO and

OKUNUKI, 1957; NUNNIKHOVEN, 1958), and from wheat germ (HAGIHARA, TAGAWA, MORIKAWA, SHIN and OKUNUKI, 1958). The extraction of the pigment from these materials may require special techniques which differ from the acid extraction commonly used with animal tissues; alkaline extraction (*Ustilago*) and pretreatment of the tissue (yeast, wheat germ) with an organic solvent (ethyl acetate) have been used successfully. Wheat germ is a relatively poor source, since the pigment is present in low concentration, difficult to extract, and accompanied by impurities with similar adsorption characteristics. On the other hand, the yield of cytochrome c from yeast is relatively good (50 mg from 1 kg compressed yeast). The purification is accomplished by ammonium sulfate fractionation and ion-exchange chromatography. In the method developed by the Japanese workers for the purification of cytochrome c from yeast, Duolite CS-101 or Amberlite XE-64 is used as the ion-exchange resin (NOZAKI et al., 1957). Crystals of cytochrome c, in either the oxidized or reduced form, can be precipitated from the final solution by ammonium sulfate. Crystalline cytochrome c may actually contain more than one component (PALÉUS and THEORELL, 1957), and this could be related to the fact that the pigment can exist as a dimer (NOZAKI, 1960).

Cytochrome c_1 has not been purified from plants. It is difficult to separate c_1 from the closely associated cytochrome b, but two methods for the purification of cytochrome c_1 from heart muscle particles have recently been reported (GREEN, JARNEFELT and TISDALE, 1959; SEKUZU, ORII and OKUNUKI, 1960). The Japanese workers extracted the pigments with 10% cholate and ammonium sulfate, removed cytochrome b by incubation (100 hrs at 5°C) and heat treatment (15 min at 40°C), precipitated cytochrome a by lowering the cholate concentration (dialysis), and purified the c_1 by alumina C_γ gel treatments and AS fractionation. Work in GREEN's laboratory led to the preparation of both a c_1-lipoprotein complex and a lipid-free cytochrome c_1; the protein can apparently exist in a polymeric form which is depolymerized by thioglycolate (CRIDDLE and BOCK, 1959).

Properties. The *cytochrome c* isolated from plants is in most respects (catalytic activity, absorption spectrum) identical with the mammalian pigment (see MINAKAMI, 1955; HAGIHARA, HORIO, NOZAKI, SEKUZU, YAMASHITA and OKUNUKI, 1956), but they do show some differences in their chromatographic behavior on ion-exchange resins (MINAKAMI, ISHIKURA and SATAKE, 1956). Cytochrome c is remarkably stable to temperature, pH, lyophilization, etc. The absorption spectrum of the reduced pigment shows characteristic maxima at 550, 521 and 415 mμ. The spectrum of the oxidized pigment can be altered by changing the pH (THEORELL and ÅKESON, 1941). Cytochrome c can be reduced by H_2, hydrosulfite, and a variety of organic compounds (hydroquinone, ascorbic acid); it is rapidly oxidized by ferricyanide. The oxidation-reduction potential at pH 7 is +0.25 volts (RODKEY and BALL, 1950). The reduced pigment becomes autoxidizable (and combines with CO) only at high temperatures (70—100°C) or at extreme pH's (above 13 and below 4). The oxidized pigment combines slowly with cyanide and forms an azide complex, but these reactions are probably not important physiologically.

The structure of cytochrome c has been studied in considerable detail. The protohematin prosthetic group is bound to the protein by two cysteine thiol groups added across the double bonds of the protoporphyrin vinyl side-chains. These thioether links cannot be split by acid acetone, but they are broken by digestion with silver salts in an acidic medium (PAUL, 1950): to 1 ml of the cytochrome solution, add 0.2 ml glacial acetic acid and 1 ml of silver sulfate solution (800 mg per 100 ml). After digestion at 60°C for 80 min, the protein can be precipitated by adding 15 ml of acid actone (1 ml 5 N H_2SO_4 to 100 ml acetone) and centrifuging. The hemopeptide which contains the dozen or so amino acids in the immediate vicinity of the prosthetic group can also be isolated (see PAUL, 1960). The sequence of amino acids in such fragments is remarkably similar for different animals, but the yeast and beef heart hemopeptides show differences in 5 out of the 11 positions. The molecular weight of cytochrome c is between 12,000 and 13,000; the iron content of the purest fractions is 0.435% Fe (PALÉUS and THEO-RELL, 1957), corresponding to 1 mole of iron per mole of protein. The protein is basic, and it contains a high concentration of lysine. Yeast and horse-heart cytochrome c have strikingly similar total amino acid compositions, but the former contains somewhat more serine and aspartic, and less glutamic acid (NUNNIK-HOVEN, 1958).

Much less is known about *cytochrome c_1*, which also has a protohematin prosthetic group bound to the protein by thioether linkages (SEKUZU et al., 1960). It contains 1 heme per unit weight of 70,000 and can exist in a polymeric form having a weight of around 370,000 (CRIDDLE and BOCK, 1959). The cytochrome c_1 absorption maxima are at 553—554 (α), 523 (β) and 418 (SORET) mμ in the reduced form, and at 523 and 411 mμ in the oxidized form. Cytochrome c_1 is relatively unstable and is denatured after 5 minutes at 50° C. Its oxidation-reduction potential is +0.22 volts (GREEN, JARNEFELT and TISDALE, 1960); it can be reduced by such agents as hydrosulfite, borohydride, cysteine, and ascorbate. The reduced pigment does not react with either O_2 or CO; the oxidized form does not combine with either cyanide or azide.

V. Cytochrome Oxidase (a—a₃).

This enzyme is apparently ubiquitous in fungi and higher plants, and it is the principal terminal oxidase of plant respiration (see HARTREE, 1957; SMITH and CHANCE, 1958; HACKETT, 1959). The only a-type cytochrome that has been identified in plants is cytochrome a—a₃, which has characteristic absorption maxima in the reduced form at 603—605 mμ (α) and 443—445 mμ (SORET). The enzyme is localized in the mitochondria, and its properties appear to be identical with those of mammalian cytochrome oxidase.

Assay. *1. Spectra.* The amount of cytochrome a—a₃ in crude preparations is calculated from the differences in absorption between the reduced and oxidized forms:

$$[a\text{—}a_3] = \frac{\Delta A_{605} - \Delta A_{630}}{16.5 \times 10^3 \text{ mM}^{-1} \text{ cm}^{-1}} .$$

In purer preparations, the change in absorbancy at 445 mμ, on going from the oxidized to the reduced form, can be used to calculate the concentration ($\Delta E = 90$ mM^{-1} cm^{-1}). As shown in a recent study of the purified enzyme (YONETANI, 1960), the a and a₃ components contribute roughly equally to the absorption in the α and SORET bands. On the other hand, only cytochrome a₃ combines with CO, and its concentration can be calculated from the (reduced + CO)—(reduced) difference spectrum.

2. Activity. Cytochrome oxidase catalyzes the following reaction:

$$4 \text{ Cytochrome } c^{++} + 4 \text{ H}^+ + O_2 \rightarrow 4 \text{ Cytochrome } c^{+++} + 2 \text{ H}_2\text{O}$$

The standard manometric method for the determination of activity and the qualitative NADI reagent test were described in detail earlier in this series (HARTREE, 1955). The spectrophotometric assay, in which the aerobic oxidation of reduced cytochrome c is followed at 550 mμ, is the most sensitive and most convenient method (see SMITH, 1955, 1955a). Mammalian cytochrome c, which can be obtained commercially, is generally used as the substrate. This can be reduced (visible change in color) by the addition of a small amount of solid $Na_2S_2O_4$ to the solution of cytochrome c (90 μM in 0.1 M phosphate buffer, pH 7.0), and the excess reducing agent is removed by aerating the solution for 5 to 10 minutes. The following method of reduction is preferable: add 5% palladium asbestos to the cytochrome c solution, gas vigorously with N_2 for 5 min, gas with H_2 for 1 hr and then again with N_2 for 5 min; filter solution and store in stoppered bottle under N_2. For use in the assay, the A_{550}/A_{565} ratio of the reduced cytochrome c solution should be greater than 6. A typical reaction mixture contains 0.5 ml of the reduced cytochrome c (90 μM) and 2.4 ml of 0.1 M phosphate buffer, pH 7.0. The reaction is started by rapidly adding (e.g. from a hypodermic syringe) 0.1 ml of enzyme, diluted with phosphate buffer; sufficient enzyme should be added to complete the oxidation in 3 to 10 minutes. Finally, a small drop of saturated $K_3Fe(CN)_6$ solution is added and the optical density (550 mμ) of the completely oxidized cytochrome c determined. The reaction rate is best expressed as the first-order velocity constant: values of $(A - A_\infty)$, the absorbancy at a given time minus the absorbancy with ferricyanide, are plotted against time on semilog paper. From the straight-line plot:

$$k = \frac{\log (A - A\infty)t_1 - \log (A - A\infty)t_2}{t_1 - t_2} \times 2.3 \ .$$

The initial velocity is obtained by multiplying k by the initial concentration. An appropriate control should be run to correct for any autoxidation of the reduced cytochrome.

A number of special factors may influence the measured rate of activity. At greater than optimal concentrations, cytochrome c itself inhibits the reaction; other basic proteins (e.g., salmine) have the same effect (SMITH and CONRAD, 1956). This may explain the fact that activity is not always proportional to the amount of homogenate used (HONDA, 1955), and it is preferable to use a washed mitochondrial suspension for the assay. Disruption of the particle structure—by hypotonic media, high salt concentrations, surface active agents, or mechanical disintegration—markedly increases the activity. For example, a brief exposure of mitochondria to digitonin can stimulate activity up to 50-fold (SIMON, 1958). As a routine procedure, the effect of including 0.1% digitonin in the reaction mixture should be tested in order to assay maximum activity. Cholate, which is frequently used to solubilize the enzyme, may be inhibitory (WAINIO and ARONOFF, 1955). Recent studies have shown that purified cytochrome oxidase is only active when a suitable phospholipid is added (GREENLEES and WAINIO, 1959; AMBE and VENKATARAMAN, 1959.)

Purification. Cytochrome oxidase has not yet been purified to any significant extent from a plant source. The enzyme has been "solubilized" from soybean and tobacco root particles with cholate, but fractionation of the extracts did not result in much purification (MILLER, EVANS and SISLER, 1958; SISLER and EVANS, 1959). On the other hand, the enzyme has been extensively purified from animal tissues in several laboratories, and some of the methods emplyed may prove to be useful. In most cases, cytochrome oxidase has been "solubilized" from the mitochondrial membranes with the aid of sodium cholate (or deoxycholate). SMITH and STOTZ (1953) combined treatment with 0.8% cholate and digestion by crude trypsin; OKUNUKI, SEKUZU, YONETANI and TAKEMORI (1958) found that the effectiveness of 2% cholate was dependent on the presence of ammonium sulfate. A recent report (GIBSON, GREENWOOD and MASSEY, 1960) suggests that digestion with snake venom may be an effective method for solubilizing the enzyme. Purification of the enzyme has been achieved by ammonium sulfate fractionation, in combination with other methods such as alumina C_γ gel adsorption (OKUNUKI et al., 1958), cellulose column chromatography (GREENLEES and WAINIO, 1959), or electrophoresis (CONNELLY, MORRISON and STOTZ, 1959). The conversion of the large, water-insoluble lipoprotein into its watersoluble monomeric units (molecular weight = 72,000) has been facilitated by treatment with 0.3 M thioglycolate and sodium dodecyl sulfate (AMBE and VENKATARAMAN, 1959; CRIDDLE and BOCK, 1959a).

Properties. Although it is possible to distinguish cytochromes a and a_3 by spectral techniques, attempts to separate these components chemically have *not* succeeded, and cytochrome oxidase appears to be a single component containing only one heme per protein molecule. The prosthetic group is hemin a, which forms a reduced ($Na_2S_2O_4$) pyridine hemochromogen (pH 10) with peaks at 585 and 430 mμ. A tentative structure for hemin a, which has three labile substituents on the porphyrin nucleus, has recently been proposed (MORRISON, CONNELLY, PETIX and STOTZ, 1960). In addition to the iron, the enzyme contains one atom of copper per mole of protein (WAINIO, VAN DER WENDE and SHIMP, 1959; TAKEMORI, 1960); the function of this copper has not been clearly established. Lecithin and phosphatidylethanolamine are the major phosphatides in the lipid fraction of the purified lipoprotein (MARINETTI, ERBLAND, KOCHEN and STOTZ, 1958).

The spectral properties of purified mammalian cytochrome oxidase have been studied in detail (YONETANI, 1960). The absorption maxima are at 600 and 424 mμ for the oxidized form, and at 605, 515, and 445 mμ for the reduced form. Characteristic spectral changes result from combination with cyanide, carbon monoxide, or nitric oxide. In the presence of cholate, some of the cytochrome is converted to mitochrome on standing in the cold (ELLIOT, HULSMAN and SLATER, 1959). The oxidation-reduction potential of cytochrome oxidase is close to $+0.29$ volts (BALL, 1938). It is rapidly reduced by dithionite or p-phenylenediamine; reduction by hydroquinone or ascorbate is dependent on the presence of some cytochrome c (SMITH, 1955b). The reduced pigment can be oxidized by molecular O_2 or by ferricyanide. SEKUZU, TAKEMORI, ORII and OKUNUKI (1960) have suggested that although cytochrome a alone becomes oxygenated, it is only autoxidizable in the presence of a small amount of cytochrome c.

VI. Lipid Components (Coenzyme Q).

Recent work has made it clear that lipid components play important structural and functional roles in the respiratory chain (see GREEN and FLEISCHER, 1960). Very little has been done in this direction with plant mitochondria, but it is likely that the methods and approaches developed for animal systems will prove fruitful. The components are extracted with such standard lipid solvents as butanol, petroleum ether, and isooctane. Many of the catalytic systems described earlier can be used to test effects of these components on activity.

The only member of this group which has been examined in any detail with plant preparations is Coenzyme Q_{10}:

$$\begin{array}{c} O \\ \| \\ CH_3O \diagdown \diagup CH_3 \qquad\qquad CH_3 \\ | \\ CH_3O \diagup \diagdown (CH_2CH{=}CCH_2)_{10}H \\ \| \\ O \end{array}$$

CRANE (1959) showed that this quinone is widely distributed in plant tissues and that it is concentrated in the mitochondria. The compound in plants is identical in all respects with Coenzyme Q_{10} isolated from beef hearts, and it almost certainly plays a similar role in respiratory electron transfer. Photosynthetic tissues contain an additional, closely related quinone in the chloroplasts, i.e. plastoquinone (CRANE, 1959a).

Assay. The qualitative and quantitative assays for Coenzyme Q_{10} are based on its absorption spectrum, which shows a characteristic peak at 275 mμ that disappears on reduction (Figure 12). Since many other substances absorb at this wave length, the quinone must be extracted before it can be examined spectrophotometrically. CRANE (1959) extracted various plant fractions by shaking acidified suspensions (KH_2PO_4) with isooctane. The isooctane was then evaporated and the residue taken up in ethanol. To estimate the amount of Q_{10} in the ethanolic solution, the absorbance at 275 mμ is first determined; a few grains of potassium borohydride are then added and the solution is shaken. After standing for a

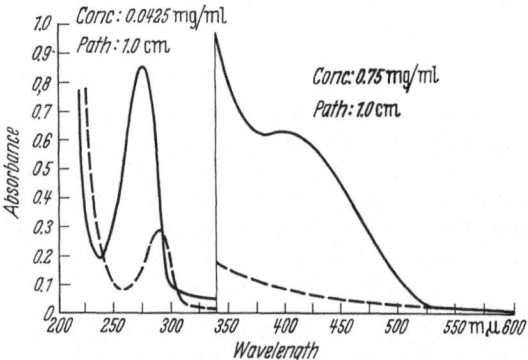

Fig. 12. Absorption spectra of Coenzyme Q_{10}. The solid line was obtained with the oxidized form and the dashed line with the reduced form (from F. L. CRANE, Y. HATEFI, R. L. LESTER and C. WIDMER 1957).

minute or so (until bubble formation stops), the absorbance at 275 mμ is again measured. The concentration of Q_{10} is then calculated using the value of $E^{1\%}_{1\,cm}$ = 142 at 275 mμ:

$$[Q_{10}] = \frac{\Delta A_{275}}{142\%^{-1}\,cm^{-1}} \; .$$

An appropriate correction must be made when calculating the concentration of plastoquinone (PQ):

$$[PQ] = \frac{\Delta A_{254} - 0.46\,(\Delta\,275)}{198\%^{-1}\,cm^{-1}} \; .$$

Compounds which may interfere with these measurements include carotenoids (general increase in absorbance in UV when borohydride is added) and vitamin K derivatives. These may be removed by chromatography of the extracts on sodium aluminosilicate (Decalso). In relatively pure preparations, the concentration of Q_{10} can be determined directly from the absorption spectrum using the values of $E^{1\%}_{1\,cm}$ = 162 at 275 mμ, and $E^{1\%}_{1\,cm}$ = 6.7 at 40 5mμ.

The activity of Q_{10} can be assayed in terms of its ability to restore either the succinoxidase or the succinic-cytochrome c reductase activity of particles which have been extracted with isooctane or acetone (LESTER and FLEISCHER, 1959). This effect may be dependent on the presence of additional accessory lipids (CRANE and EHRLICH, 1960).

Purification and properties. CRANE (1959a) has described in detail the methods for the purification and crystallization of Coenzyme Q_{10} by chromatography on Decalso and silicic acid. The quinone is adsorbed onto the column from a hydrocarbon solvent, and the other lipids (carotenes, waxy materials) are not retained. The quinones are then eluted by slightly polar solvent mixtures, leaving most of the residual impurities adsorbed on the column. Methods for the chromatography of this family of compounds on silicone-impregnated paper have been published (LESTER and RAMASARMA, 1959).

Coenzyme Q_{10} is a neutral lipid that is insoluble in water, poorly soluble in polar solvents, and readily soluble in nonpolar solvents, especially the hydrocarbons. It is reduced by hydrosulfite or borohydride to the corresponding hydroquinone form; this can be reoxidized to the quinone by shaking with Ag_2O. Both endogenous and exogenous Q_{10} can be reduced and oxidized by the appropriate segments of the respiratory chain (RAMASARMA and LESTER, 1960). Coenzyme Q functions in the region of cytochrome b. The analogues of Q_{10} which have a smaller number of isoprenoid units (e.g., Q_1, Q_2, etc.) in the side-chain are more soluble, and these have been used successfully to study the reactions involving Coenzyme Q.

C. Other Pathways to Oxygen.

I. From Pyridine Nucleotides to Oxygen.

1. General Considerations.

In addition to the classical respiratory chain, alternative pathways for the oxidation of reduced pyridine nucleotides are frequently found in plant extracts. Thus, many plant homogenates show high DPNH and/or TPNH oxidase activities which are mediated by other flavin- and metal-containing oxidases; examples of such systems are listed in Table 4, together with some of their properties. It is essential to remember that this activity may result from the disruption of the cell structure and that in this sense it may be wholly "artificial." The fact that it is possible to construct model PNH oxidase systems using known enzymes that are widely distributed in plants adds little to our knowledge of their physiological roles. Ultimately, evidence for the participation of a given pathway in vivo must rest on data obtained with intact cells, and here there is little to support the view that the non-cytochrome systems are important in terminal respiration (see HACKETT, 1959; BONNER, 1957; GODDARD and BONNER, 1960). The additional fact that oxidation in these systems is not coupled to phosphorylation casts serious doubt on their importance for cellular energy supply. In order to characterize a given oxidative system, the following types of experimental evidence should be obtained.

a) **Nature of the over-all reaction.** When the oxidation of PNH is followed spectrophotometrically ($340 \ m\mu$), it is essential to show that molecular O_2 is required before it can be assumed that a PNH oxidase system is involved. For example, the concentration of electron acceptors in the "enzyme" may be high enough to permit the anaerobic oxidation of PNH, as with potato tuber homogenates (HACKETT, 1958). Even the demonstration that O_2 is required does not prove that an oxidase is involved, since the reaction may be catalyzed by a heat-stable "cofactor" that is rapidly autoxidizable, as in the wheat germ supernatant (CLUM and NASON, 1958).

b) **Intracellular localization.** Unlike the mitochondrial respiratory chain, many of the other systems are present in the soluble fraction of the homogenate. Any PNH oxidase activity remaining in the supernatant after centrifugation at 100,000 $\times$ gravity for 1 hr is generally attributed to a nonparticulate or "soluble" system, and it can be assumed that it is independent of the particle-bound cytochrome oxidase. By adding external cofactors or carriers, it is also possible to demonstrate atypical PNH oxidase systems in various particle fractions (YOUNG and CONN, 1956; STERN and JOHNSTON, 1957; FORTI, TUA and TOGNOLI, 1959).

c) **"Cofactors."** The nature of the cofactors which stimulate the oxidative activity gives some indication of the type of system involved. For example, the presence of ascorbic acid may permit the oxidation of DPNH *via* ascorbic acid

Table 4. *Other Pathways from DPNH and TPNH to Oxygen.*

Activity	Catalytic System	Cofactors	Response to Inhibitors	Terminal Oxidase	Reference
DPNH oxidase	cucumber homogenate	ascorbate + dialyzable factor	inhibited by cyanide, DIECA, PCMBA	ascorbic oxidase	Beevers, 1954
TPNH oxidase	avocado mitochondria	ascorbate + glutathione	inhibited by cyanide, DIECA	ascorbic oxidase	Young and Conn, 1956
TPNH oxidase	dialyzed soluble fraction of pea seeds	ascorbate + glutathione	inhibited by cyanide, DIECA	ascorbic oxidase	Mapson and Moustafa, 1956
DPNH oxidase	pea extract + ascorbic oxidase	ascorbate	inhibited by cyanide; not inhibited by DNP	ascorbic oxidase	Wosilait, Nason and Terrell, 1954
DPNH ⎱ oxidase TPNH ⎰	soluble fraction of potato tuber	ascorbate	inhibited by cyanide, DIECA, PCMBA	ascorbic oxidase	Hackett, 1958
DPNH oxidase	pea quinone reductase + phenol oxidase	o-quinone	inhibited by cyanide, DNP	phenol oxidase	Wosilait, Nason and Terrell, 1954
DPNH oxidase	soluble fraction or particles from tobacco roots	chlorogenic acid		phenol oxidase	Sisler and Evans, 1958
TPNH oxidase	soluble wheat germ extract	Mn^{++}, "cofactor"	inhibited by cyanide, ascorbate, catalase	peroxidase ?	Conn et al., 1952; Humphreys, 1955
TPNH ⎱ oxidase DPNH ⎰	soluble fraction of barley roots	various phenols	inhibited by cyanide, ascorbate, DIECA	peroxidase ?	Romberger, 1959
TPNH ⎱ oxidase DPNH ⎰	wheat embryo particles	Mn^{++}	inhibited by cyanide, ascorbate	peroxidase ?	Stern and Johnston, 1957
DPNH oxidase	purified horseradish peroxidase	Mn^{++}, various phenols	inhibited by cyanide, cysteine	peroxidase	Akazawa and Conn, 1958
DPNH oxidase	glyoxylic reductase + glycolic oxidase from spinach leaves	glycolate	not inhibited by cyanide	glycolic oxidase	Zelitch, 1953

oxidase (Beevers, 1954); when both glutathione and ascorbic acid are added, TPNH is oxidized by a system that involves both glutathione reductase and ascorbic acid oxidase (Mapson and Goddard, 1951; Conn and Vennesland, 1951). Quinones (e.g., menadione) can be used to couple the oxidation of DPNH to polyphenol oxidase (Wosilait, Nason and Terrell, 1954). Certain phenols and Mn^{++} are required for the oxidation of DPNH *via* peroxidase (Akazawa and Conn, 1958).

d) **Effects of inhibitors.** If the system under investigation is not inhibited by cyanide or azide, it is assumed that the final reaction with O_2 does not involve a metal-containing oxidase; either an autoxidizable cofactor or a flavoprotein oxidase is probably involved. A marked sensitivity to such copper inhibitors as phenylthiourea or diethyldithiocarbamate suggests the involvement of either ascorbic acid oxidase or polyphenol oxidase (James, 1953; Hackett, 1960). In some systems, ascorbic acid itself is very inhibitory (e.g., Romberger, 1959).

e) **Coupling to phosphorylation.** The standard methods described earlier can be used to determine whether the oxygen uptake by a given system is coupled to the synthesis of ATP from ADP and phosphate.

2. The Copper Oxidases.

The extensive literature on the properties of polyphenol oxidase, ascorbic acid oxidase, and laccase has been reviewed in detail (DAWSON and TARPLEY, 1951; BONNER, 1957; GODDARD and BONNER, 1960; FRANKE, 1960), and only a few of the characteristics of these strikingly similar enzymes will be mentioned here. They are widely distributed in both higher plants and fungi. Although it is frequently assumed that these enzymes are present in the soluble phase of the cytoplasm, there is some evidence that they may be associated with particulate components (e.g., mitochondria, cell walls). Both polyphenol oxidase and laccase oxidize a wide variety of phenolic substrates, but the latter will not oxidize monophenols. It is probable that there are a number of different polyphenol oxidases which differ in their relative affinities for mono- and o-dihydric phenols. Ascorbic acid oxidase shows a much greater substrate specificity, but here too there may be more than one type of enzyme.

The molecular weights of these enzymes have been reported to range from 100,000 to 150,000. The copper content varies from 0.22 to 0.25%, suggesting that there are between 4 and 6 atoms of copper per mole of protein. KERTESZ and ZITO (1957) have reported a molecular weight of 34,500 for polyphenol oxidase, which would then have only 1 atom of copper per mole). It is commonly assumed that enzymatic activity involves a reversible valence change in the enzyme-bound copper, and a cuprous-cupric interconversion has been shown directly with purified laccase (NAKAMURA, 1958). Activity is inhibited by the metal-combining agents cyanide and azide, as well as by copper-chelating agents such as diethyldithio-carbamate. Only polyphenol oxidase is markedly inhibited by CO, and this inhibition is not reversed by light. Polyphenol oxidase is colorless, whereas ascorbic oxidase and laccase are blue proteins with absorption bands at 605 and 615 mμ, respectively. The visible band disappears on reduction, and this property has been used to establish the oxidation-reduction potential of laccase at $+0.415$ V (pH 7.0). Under some experimental conditions, the oxygen affinities of polyphenol and ascorbic oxidases appear to be relatively low, but direct spectrophotometric studies of laccase have shown that its O_2 affinity is high (NAKAMURA, 1960).

a) Polyphenol Oxidase.

Assay. The assay method employed will depend on the particular substrate which is to be oxidized. The two types of reactions catalyzed are represented by the following equations:

$$\text{C}_6\text{H}_5\text{OH} + \tfrac{1}{2} O_2 \longrightarrow \text{C}_6\text{H}_4(\text{OH})_2$$

$$\text{C}_6\text{H}_4(\text{OH})_2 + \tfrac{1}{2} O_2 \longrightarrow \text{o-quinone} + H_2O$$

The stoichiometry clearly depends on the initial substrate, and twice as much O_2 is consumed with a monophenol as with a diphenol. If the resulting quinone undergoes further oxidation and polymerization, additional O_2 will be consumed. The most commonly used monophenolic substrates are tyrosine and p-cresol; catechol, caffeic acid, and chlorogenic acid can be used as diphenols. With monophenols, the reaction shows an initial induction period, the length of which depends on such factors as the source and purity of the enzyme (BORDNER and NELSON, 1939). With diphenols, the reaction starts immediately, but the enzyme may be gradually inactivated during the reaction; this inactivation can be partially prevented by including an inert protein such as gelatin in the reaction mixture (MILLER and DAWSON, 1941). The assay can be carried out with only a catalytic amount of diphenol if a suitable reductant is included; ascorbate is used in the chronometric method described by DAWSON and MAGEE (1955). The pH and substrate concentration used in the assay will depend on the nature of the substrate. Other factors which can influence the activity include natural inhibitors and artificial activating agents (e.g., anionic wetting agents, KENTEN, 1958).

A standard assay method for these enzymes depends on the *manometric* determination of the rate of O_2 consumption. The WARBURG vessel contains: 1.0 ml of 0.15 M phosphate buffer (pH 7.0) and 1.0 ml of 0.075 M substrate (1 mg/ml for more insoluble substrates) in the main compartment, and 0.5 ml of enzyme (diluted to ensure proportionality between initial rate and enzyme concentration) in the side-arm. The reaction is started by tipping in the enzyme. With some substrates, the rate of autoxidation may be considerable, and the appropriate corrections must be made. The assay conditions should be adjusted to minimize this correction, and lowering the pH (to 6.0), inclusion of versene (10^{-3} M), and use of deionized water may be helpful.

A number of convenient *colorimetric* and *spectrophotometric methods* have been developed, using several different substrates. *Tyrosine* and *dihydroxyphenylalanine (DOPA)* are oxidized to DOPA quinone, and the conversion of this product into the red pigment DOPA chrome can be followed in the region of 475 mμ (MASON, 1948). In a typical assay, the reaction is started by adding 0.1 ml of extract or purified enzyme to a mixture containing 4.9 ml of 0.1 M phosphate buffer, pH 6.0, and 4 mg of DL-DOPA, and the initial rate of increase in optical density determined (HOROWITZ, FLING, MACLEOD and SUEOKA, 1960). If tyrosine is the substrate, the oxidation to the o-quinone can be followed conveniently at 280 mμ in a reaction mixture containing 1.0 ml of 0.001 M L-tyrosine and 1.0 ml of 0.5 M phosphate buffer, pH 6.5. With these assay systems, the reaction may be limited by the concentration of O_2 dissolved in solution, and a preliminary oxygenation of the cuvette should stimulate the activity.

Two colorimetric assays for this enzyme have been developed using *catechol* as the substrate. In the method of PONTING and JOSLYN (1948), the reaction mixture contains: 20 ml of 0.01 M acetate buffer, pH 5.0, and 2 ml of 0.5 M catechol. After adding the enzyme (1 ml), the tube is corked and shaken, and the rate of color formation measured in a colorimeter equipped with a 420 mμ filter. The method of SMITH and STOTZ (1949) depends on the continuous reduction of the o-quinone by a suitable leuco dye, 2,6-dichlorobenzenoneindo-3'-chorophenol, whose oxidation is followed colorimetrically. The reaction mixture contains: 5.0 ml of 0.1 M phosphate-citrate buffer, pH 6.0, 5.0 ml of 0.0002 M leuco dye, and 1.0 ml of 0.1 M catechol. After adding 1 ml of enzyme suspension, the tube is shaken and the color formation followed (using 645 mμ filter). The % transmission, plotted on semilog paper, should be a linear function of time over the first minute.

SISLER and EVANS (1958) have described a method which depends on the fact that the absorption of *chlorogenic acid* at 362 mμ decreases markedly upon oxidation. The assay of oxidase activity is carried out in a silica cuvette containing 0.1 M phosphate buffer, pH 7.0, 5.7×10^{-5} M chlorogenic acid, and 10^{-3} M ethylenediamine tetraacetate. The reaction is initiated by the addition of enough enzyme to cause a linear decrease in absorbance, at the rate of roughly 0.02 per 15 seconds.

Purification. The classical sources of this enzyme are the white potato tuber, *Solanum tuberosum* (KUBOWITZ, 1937), and the cultivated mushroom, *Agaricus campetris* (KEILIN and MANN, 1938). Procedures for the purification of the enzyme from tobacco roots (SISLER and EVANS, 1958) and tobacco leaves (CLAYTON, 1959) have also been reported. The major difficulty encountered during the purifications stems from the fact that colored oxidation products form rapidly in the extract; this reaction can be inhibited by working rapidly in the cold, removing substrates, or including a reducing agent in the isolation medium. A comparison of the properties of the various enzymes suggests that there is more than one polyphenol oxidase.

A method for the purification of the mushroom polyphenol oxidase has been described in detail by DAWSON and MAGEE (1955). A frozen pulp of tissue is prepared by passing the mushrooms through a meat grinder into cold ($-15°$ C) acetone, filtering, and freezing by contact with dry ice. The enzyme is extracted from the pulp into water, and precipitated successively by acetone and ammonium sulfate (60% saturated solution). The protein is redissolved in water and treated with alumina gel. After filtering, the filtrate is dialyzed against water and treated successively with lead subacetate and alumina gel; the protein is eluted from the gel with 0.2 M Na_2HPO_4. Repetition of these final steps gives further purification and yields a preparation that is especially high in catechol oxidizing activity.

Somewhat different methods have been used in recent purification procedures. KERTESZ and ZITO (1957) depend on acetone, calcium acetate, and ethanol treatments. SISLER and EVANS (1958) used calcium phosphate gel effectively. In a preliminary report, SMITH (1961) describes the chromatography of salt extracts of mushrooms on hydroxylapatite and DEAE cellulose to give four separate phenolases.

b) Laccase.

Assay. This enzyme is assayed by essentialy the same type of manometric method as that used for polyphenol oxidase, with the substitution of an appropriate substrate:

$$\text{(structure: } OH\text{-}C_6H_4\text{-}OH) + \frac{1}{2} O_2 \longrightarrow \text{(structure: } O\text{=}C_6H_4\text{=}O) + H_2O.$$

The reaction mixture contains 6×10^{-3} M hydroquinone in 0.033 M phosphate buffer pH 7.4.

Purification. NAKAMURA (1958) has reported a procedure that yields roughly 0.5 g of pure laccase from 1 kg of the latex of the lacquer tree *(Rhus vernicifera)*. Cold acetone (3 l) is added to the crude latex and the precipitate is extracted with water; the resulting extract is fractionated with ammonium sulfate. The blue protein is again precipitated by acetone and further purified by chromatography on Amberlite XE-64, followed by zone electrophoresis on a starch column.

c) Ascorbic Acid Oxidase.

Assay. Unlike the other copper oxidases, this enzyme shows a high (though not absolute) substrate specificity; it oxidizes L-ascorbic acid to dehydroascorbic acid as follows:

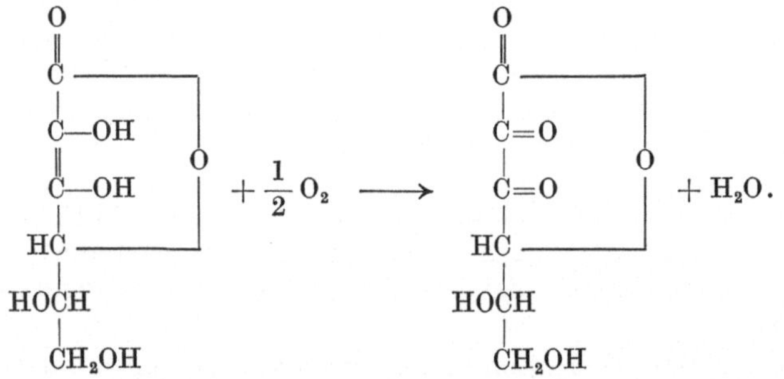

$$
\begin{array}{c}
\text{O} \\
\parallel \\
\text{C} \\
\mid \\
\text{C—OH} \\
\parallel \\
\text{C—OH} \\
\mid \\
\text{HC} \\
\mid \\
\text{HOCH} \\
\mid \\
\text{CH}_2\text{OH}
\end{array}
\;+\frac{1}{2}\,O_2 \longrightarrow
\begin{array}{c}
\text{O} \\
\parallel \\
\text{C} \\
\mid \\
\text{C=O} \\
\mid \\
\text{C=O} \\
\mid \\
\text{HC} \\
\mid \\
\text{HOCH} \\
\mid \\
\text{CH}_2\text{OH}
\end{array}
\;+\,H_2O .
$$

The activity is generally determined by measuring the rate of O_2 consumption, using standard manometric techniques. A typical reaction mixture for the assay contains: 3×10^{-3} M L-ascorbic acid (freshly made up), 0.02 M Na_2HPO_4-0.01 M citrate buffer, pH 5—6, and 0.5 mg/ml gelatin. The reaction is started by tipping in the enzyme from the side-arm of the WARBURG flask, and the manometers are read at 2-minute intervals. The initial linear oxidation is used for the rate determination, since the enzyme is inactivated during the course of the reaction. Care should be taken to reduce the inorganic copper content of all the reagent solutions to a minimum level, since this metal alone will catalyze the oxidation. An inert protein (e.g., gelatin) can be used to minimize this autoxidation, and chelating agents such as versene might prove to be helpful. Demonstration that the above reaction takes place in a crude extract does not itself establish the presence of ascorbic acid oxidase, since a number of other pathways from ascorbate to oxygen are known; additional evidence, such as the response of the system to various inhibitors, is required in these cases.

Purification. Homogeneous preparations of the enzyme have been obtained from the yellow summer squash, *Cucurbita pepo condensa* (DUNN and DAWSON, 1951); the following simplified method, which yields an enzyme that is about 80% pure, is described in detail by DAWSON and MAGEE (1955). The rinds from 400 pounds of squash are passed through a meat grinder, the pH of the filtered juice adjusted to 7.6 (with solid $Na_2B_4O_7 \cdot 10\ H_2O$), and 10 ml of 1 M Ba acetate added per l of juice. The material which then sediments from the solution between 0.3 and 0.6 AS saturation is collected by filtration and redissolved in 2.5 l of water. After filtering and dialyzing this solution, the protein is adsorbed on alumina gel and eluted with 0.2 M Na_2HPO_4. The dialyzed eluate is treated with 0.2 its volume of cold acetone and filtered; this process is repeated until about 7 precipitates have been collected. The precipitates are redissolved separately in 0.2 M Na_2HPO_4 and assayed for activity.

3. Peroxidases.

The hemoprotein peroxidases, which oxidize a variety of substances using H_2O_2 as the electron acceptor, are widely distributed in plants. Since these enzymes were dealt with earlier in this series (HARTREE, 1955), they will be treated only briefly here; the whole subject has recently been reviewed in detail (BURRIS, 1960). Under certain conditions, the peroxidases use molecular O_2 and can act as oxidases. For example, purified horse-radish peroxidase, when supplemented by the appropriate cofactors, can catalyze the oxidation of DPNH and TPNH by O_2 (AKAZAWA and CONN, 1958). This enzyme can also catalyze hydroxylation reactions (BUHLER and MASON, 1961). The fact that peroxidase utilizes molecular O_2 in a number of different reactions has stimulated interest in its possible role in respiration. It is able to stimulate the DPNH oxidase activity of isolated mitochondria (TAGAWA and SHIN, 1959a). The actual intracellular localization of peroxidase has not been extensively studied, but it appears to be present in both the soluble and particulate cell fractions (MATSUSHITA and IBUKI, 1960).

A flavoprotein peroxidase which oxidizes DPNH has been isolated from bacteria (DOLIN, 1955), but this type of enzyme has not yet been obtained from plants.

Assay. A variety of hydrogen donors which give rise to colored reaction products have been used for the colorimetric assay of this enzyme. MAEHLY and CHANCE (1954) have critically reviewed the factors to be considered in choosing a suitable donor. The classical method makes use of pyrogallol (purpurogallin), and this, along with several other methods, has been described in detail by HARTREE (1955). The pyrogallol method is open to a number of objections (complexity of reactions involved, requirement for very pure chemicals, etc.), and the simpler guaiacol assay has been recommended. This depends on the following reaction:

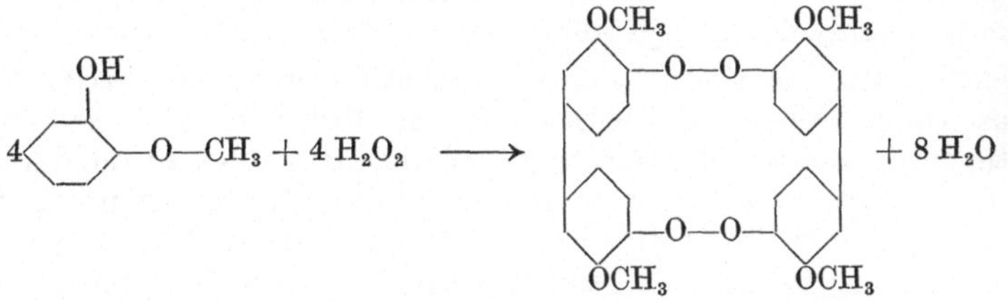

The tetraguaiacol solution has a red-brown color, which is measured within one or two minutes, after which it fades. The enzyme is added to a solution containing 7×10^{-3} M guaiacol and 7×10^{-3} M phosphate buffer (pH 7.0). The optical density of the mixture is read at 470 mμ, and the optical density scale is offset to a value 0.05 greater than this reading. 20 μl of 10 mM H_2O_2 are added to the solution (3 ml) and the time (Δt) required to reach the null point (preset optical density reading) is measured with a stopwatch; it should be between 15 and 30 seconds. The rate of product formation can be calculated from the extinction coefficient of 26.6 mM$^{-1} \times$ cm^{-1} at 470 mμ. Other hydrogen donors which have been used effectively for the colorimetric assay of peroxidase include 2,6-dichlorophenol indo-3'-chlorophenol, benzidine, o-dianisidine, and misidine or 2,4,6-trimethylaniline (PAUL and AVI-DOR, 1954).

In those cases where the peroxidase-catalyzed reaction consumes molecular oxygen, as when dihydroxyfumaric acid (SWEDIN and THEORELL, 1940), other dicarboxylic acids (KENTEN and MANN, 1953), and indole acetic acid (KENTEN, 1955) are the substrates, standard manometric techniques can be used to follow the reaction. Spectrophotometric methods can also be developed in some cases; for example, the oxidation of dihydroxyfumaric acid has been followed by measuring the decrease in absorption in the ultraviolet (NICHOLLS, 1961).

Purification. Peroxidase was first crystallized from horse-radish roots by THEORELL (1942), and it has subsequently been purified from a number of different plant tissues. More than one hemoprotein having peroxidase activity may be found in a single tissue, and there is evidence that these forms are interconvertible (see PAUL, 1960). The different forms of the enzyme may differ in their association with nonprotein components (e.g., carbohydrate, water). The possibility that native peroxidase is modified during the isolation procedure should be carefully considered.

The classical procedure for preparation of crystalline horse-radish peroxidase is described by MAEHLY (1955). Sprouting roots are minced and extracted with water. Following ammonium sulfate fractionation, the protein is further purified by calcium phosphate gel treatment, alcohol fractionation, and preparative electrophoresis in a TISELIUS apparatus. The purest fraction is again fractionated with ammonium sulfate and the enzyme crystallized, with a yield of roughly 2 g of crystals per 100 kg of roots. If facilities for preparative electrophoresis are not available, the method of KEILIN and HARTREE (1951) may be more practical; this utilizes ammonium sulfate, ethanol, and calcium phosphate fractionations, in addition to a heat treatment. More recently, such methods as chromatography on carboxymethyl cellulose have been used for the preparation of crystalline radish peroxidase (MORITA, KAMEDA, and MIZUNO, 1961).

If horse-radish roots are not available, wheat germ can serve as a source of pure peroxidase (TAGAWA and SHIN, 1959). A phosphate buffer (pH 6.5) extract of wheat germ is fractionated with ammonium sulfate, after which the protein is adsorbed successively on Amberlite CG-50 and XE-64 resins. Following chromatography on the latter resin, ammonium sulfate is used to bring about crystallization. The yield is roughly 40 mg crystalline material from 2 kg of wheat germ.

Properties. Most of the available information on the properties of purified peroxidase comes from studies with the horse-radish enzyme. It has a protoheme prosthetic group and a molecular weight of roughly 40,000. In the oxidized form, the SORET absorption maxima of the peroxidases range from 402 to 420 mμ ($E = 90$ mM$^{-1}\times$cm^{-1}). The ratio (RZ) of optical density at the SORET peak to that at 275 mμ can be used as a convenient method for expressing purity; RZ values for crystalline peroxidases range from 3.0 to 3.55. On reduction with hydrosulfite, crystalline horse-radish peroxidase shows an α-peak at 556—558 mμ; on the other hand, the unmodified wheat germ peroxidase has a peak at 566 mμ. The hemoprotein shows characteristic spectral changes on combination with cyanide, azide, and fluoride. When reduced, peroxidase combines with CO, and on this basis it can be distinguished from the b-type cytochromes. The oxidation-reduction potential of peroxidase is—0.27 volts at pH 7.0 (HARBURY, 1957).

4. Glycolic Acid (α-Hydroxy Acid) Oxidase.

ZELITCH (1953) showed that in a model system glycolic acid oxidase can act, together with glyoxylic acid reductase, as a hydrogen carrier system from PNH to molecular O_2. Using a series of specific competitive inhibitors, of which α-hydroxy-2-pyridinemethanesulfonic acid is the most effective, evidence has been obtained to support the view that glycolic acid oxidase participates in the respiration of leaves in sunlight (ZELITCH, 1957, 1958, 1959). The enzyme is present in the soluble fraction and in chloroplasts (ZELITCH and BARBER, 1960a); the oxidation of glycolate by particles from spinach leaves is not coupled to phosphorylation (ZELITCH and BARBER, 1960). The activity of glycolic oxidase is markedly increased by exposure of plant tissues to the light.

It has been known for some time that glycolic oxidase oxidizes not only glycolic to glyoxylic acid, but also L-lactic to pyruvic acid, and α-hydroxybutyric to α-ketobutyric acid. Recently it has been shown that it also oxidizes glyoxylic acid, which exists in solution in the hydrated form, COOH—CH(OH$_2$), to oxalic acid (RICHARDSON and TOLBERT, 1961). These facts suggest that the enzyme is actually an α-hydroxy acid oxidase, which is most active with glycolic acid. It is able to transfer hydrogen not only to molecular O_2 but also to certain dyes, and in this capacity acts as a dehydrogenase.

Assay. The initial reaction of glycolic acid oxidation is represented by the following equation:

$$\begin{array}{c} CH_2OH \\ | \\ COOH \end{array} + O_2 \longrightarrow \begin{array}{c} CHO \\ | \\ COOH \end{array} + H_2O_2$$

The actual stoichiometry of the over-all reaction will depend on the extent of the secondary reactions (ZELITCH and OCHOA, 1953). If the glyoxylic acid is further oxidized to oxalic acid by the enzyme, more O_2 will be consumed. If catalase is present, less O_2 will be required. The reaction products, glyoxylate and H_2O_2, can interact non-enzymatically to give formic acid, CO_2, and H_2O. Both glyoxylate and oxalate can inhibit the oxidation of glycolic acid.

In most cases, the reaction has been followed by measuring the rate of O_2 consumption manometrically. The main compartment of the WARBURG vessel should contain 0.05 M phosphate buffer (pH 8.0) and enzyme, and the reaction is started by tipping in the substrate (final concentration, 0.01 M glycolate) from the side-arm; KOH is present in the center well. The activity may be stimulated by the addition of FMN to the reaction mixture, since the enzyme's prosthetic group is readily dissociated during purification. Tris buffer can be used to eliminate glyoxylate oxidation without inhibiting glycolic acid oxidation (RICHARDSON and TOLBERT, 1961).

A convenient spectrophotometric assay can be used to determine the rate at which the enzyme catalyzes the reduction of a suitable dye (ZELITCH, 1955). Only those dyes, such as 2,6-dichlorophenol indophenol, which accept two electrons can be used (FRIGERIO and HARBURY, 1958). Cyanide is included in the reaction mixture to inhibit any indophenol oxidase. The rate of dye reduction is markedly dependent on the ionic strength of the solution. The reaction mixture should contain: 0.007 M glycolate, 3×10^{-5} M 2,6-dichlorophenol indophenol, 0.033 M phosphate buffer (pH 8.0), and 3×10^{-4} M KCN; after adding the enzyme, the reaction is followed at 620 mμ. The reference cuvette contains all components except the dye. The rate of dye reduction is only proportional to the amount of enzyme when less than 15 units are present (1 unit causes a decrease in absorbancy of 0.01 per minute). With crude extracts, a correction must be made for any endogenous dye reduction.

Purification. ZELITCH and OCHOA (1953) first purified glycolic acid oxidase from spinach leaves and their method has since been described in detail (ZELITCH, 1955); more recently, the enzyme has been crystallized from this source (FRIGERIO and HARBURY, 1958). The variability in results encountered during the purification of the enzyme can be explained in part by differences in the starting material (variety, season, previous history). The initial extraction is made by macerating spinach leaves in a Waring blendor and expressing the juice mechanically. In ZELITCH's method, the enzyme is first precipitated with ammonium sulfate (at pH 5.4) and then redissolved in phosphate buffer (pH 8.0). Following dialysis, the solution is fractionated by acid (pH 4.9) precipitation, ethanol precipitation, and calcium phosphate gel treatments. The final elution from the gel gives an enzyme preparation that shows high specific activity but contains more than one component. The gentler and more reproducible method of FRIGERIO and HARBURY yields a homogeneous protein. This involves fractionations with ammonium sulfate at both acid and alkaline pH's, adsorption on DEAE cellulose, and treatment with calcium phosphate gel. These authors have also developed methods for preparative electrophoresis of the enzyme. The stability of the enzyme varies with the degree of purification. During the isolation procedure, the prosthetic group is removed readily, and the protein itself forms aggregates.

Properties. Glycolic oxidase was the first flavoprotein to be extensively purified from higher plants; its absorption spectrum shows typical maxima at 340 and 450 mμ, and the prosthetic group is FMN. The minimum molecular weight is roughly 70,000, but physical measurements indicate that the enzyme exists as a dimer or tetramer. The MICHAELIS constants for glycolate, lactate and glyoxylate are 3.8×10^{-4} M, 2.0×10^{-3} M, and 5.4×10^{-3} M, respectively. The pH optimum for glycolate oxidation is 8.3. Metal-complexing agents do not inhibit the enzyme, and 10^{-2} M cyanide may stimulate activity from 50 to 100%.

II. From other Substrates to Oxygen.

There are other oxidases in plants which are not linked to the pyridine nucleotides but are able to oxidize various organic substrates (carbohydrates, fatty acids, amino acids). At present, there is little evidence that these enzymes play a quantitatively significant role in normal terminal respiration. For example, lipoxidase can account for less than 1% of the respiration of intact corn seedlings (FRITZ, MILLER, BURRIS and ANDERSON, 1958), even though a large fraction of the O$_2$ uptake by homogenates of corn seeds is attributable to the oxidation of endogenous unsaturated fatty acids by this enzyme (FRITZ and BEEVERS, 1955). The role of the amino acid oxidases in situ has not been examined critically. The enzymes which oxidize carbohydrates directly may bear a closer relationship to respiration, and two of these are considered here.

1. Glucose Oxidase.

This enzyme, which has been called glucose oxidase, glucose aerodehydrogenase, notatin, and penicillin B, is excreted into the medium by *Aspergillus* and *Penicillium* cultures. It has not been identified in higher plant tissues.

Assay. The enzyme catalyzes the removal of two hydrogen atoms from β-D-glucopyranose to yield δ-D-gluconolactone, which readily undergoes hydrolysis to gluconic acid; molecular oxygen serves as the hydrogen acceptor:

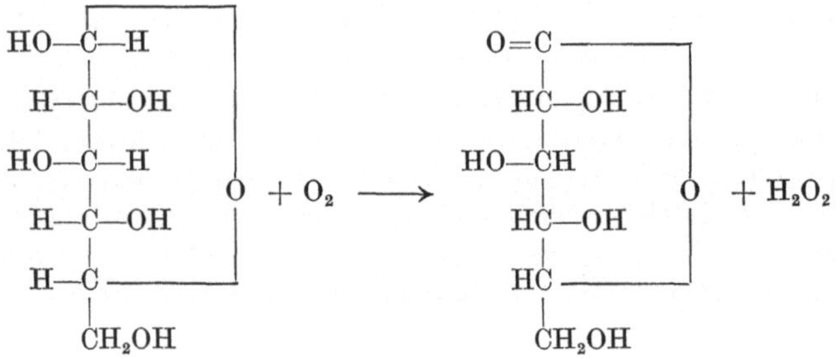

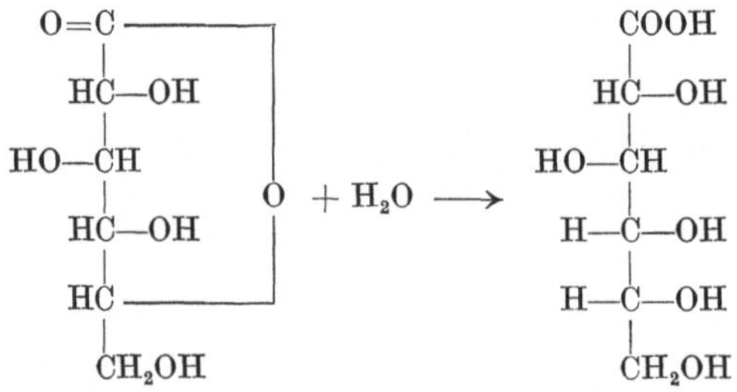

$$
\begin{array}{c}
\text{HO—C—H} \\
\text{H—C—OH} \\
\text{HO—C—H} \\
\text{H—C—OH} \\
\text{H—C} \\
\text{CH}_2\text{OH}
\end{array} \text{O} + \text{O}_2 \longrightarrow
\begin{array}{c}
\text{O=C} \\
\text{HC—OH} \\
\text{HO—CH} \\
\text{HC—OH} \\
\text{HC} \\
\text{CH}_2\text{OH}
\end{array} \text{O} + \text{H}_2\text{O}_2
$$

$$
\begin{array}{c}
\text{O=C} \\
\text{HC—OH} \\
\text{HO—CH} \\
\text{HC—OH} \\
\text{HC} \\
\text{CH}_2\text{OH}
\end{array} \text{O} + \text{H}_2\text{O} \longrightarrow
\begin{array}{c}
\text{COOH} \\
\text{HC—OH} \\
\text{HO—CH} \\
\text{H—C—OH} \\
\text{H—C—OH} \\
\text{CH}_2\text{OH}
\end{array}
$$

The net effect of these reactions is the conversion of D-glucose to D-gluconic acid, which can be followed manometrically (KEILIN and HARTREE, 1948). The reaction is generally carried out in the presence of an excess of catalase, and the over-all stoichiometry of the reaction is:

$$C_6H_{12}O_6 + H_2O + O_2 \xrightarrow{\text{glucose oxidase}} C_6H_{12}O_7 + H_2O_2$$

$$H_2O_2 \xrightarrow{\text{catalase}} H_2O + \frac{1}{2} O_2.$$

$$\text{Net:} \quad C_6H_{12}O_6 + \frac{1}{2} O_2 \longrightarrow C_6H_{12}O_7$$

The standard assay mixture contains 1.2 ml of 0.1 M sodium acetate buffer (pH 5.6), 0.2 ml of catalase solution (prepared by the method of KEILIN and HARTREE, 1945), and the enzyme to be tested; the reaction is started by tipping in 0.2 ml of 1 M glucose from the side-arm of the WARBURG vessel. The initial rate of O_2 consumption should be linear, and this is used to calculate the Q_2 value (μl O_2 uptake/mg protein/hr).

Purification. The classical method of purification (COULTHARD, MICHAELIS, SHORT, SKRIMSHIRE, STANDFAST, BIRKINSHAW and RAISTRICK, 1945) involves successive precipitations of the enzyme from culture filtrates of *Penicillium notatum* with acetone, tannic acid, and ammonium sulfate. A quicker route to highly active preparations, involving precipitation of the protein in the filtered culture medium with 1% uranium acetate, extraction of the precipitate with phosphate buffer (pH 6.8), and fractionation with ammonium sulfate, has been described (VAN BRUGGEN, REITHEL, CAIN, KATZMAN, DOISY, MUIR, ROBERTS, GABY, HOMAN and JONES, 1943). The glucose oxidase that is excreted by submerged cultures of *Penicillium amagasakiense* has recently been crystallized using a still different method (KUSAI, SEKUZU, HAGIHARA, OKUNUKI, YAMAUCHI and NAKAI, 1960). It is purified from a crude commercial preparation ("Deoxine") by adsorption on a column of Amberlite CG-50 (Type II, buffered at pH 4.5), elution with pH 5.0 buffer, and fractionation with ammonium sulfate.

Properties. The pure enzyme is pale yellow in color, and has absorption maxima at 278, 380, and 460 mμ in the oxidized form. The latter two peaks disappear on reduction with glucose or $Na_2S_2O_4$. The prosthetic group of the enzyme is FAD, which can be removed by either acid or alkali. There are two moles of FAD per mole of enzyme protein, which has a molecular weight of 154,000. The enzyme is highly specific for β-D-glucopyranose (KEILIN and HARTREE, 1952). Its pH optimum is 5.6, and the turnover number here is 17,000 (moles O_2 consumed/mole enzyme/min). The rate of oxygen consumption is more than doubled when oxygen is substituted for air in the gas phase. The activity is blocked completely by 10^{-3} M PCMBA and partially by such aldehyde reagents as phenylhydrazine, $NaHSO_3$, and NH_2OH; it is not inhibited by cyanide or azide. 2,6-Dichlorphenol indophenol can replace oxygen as the hydrogen acceptor.

2. Carbohydrate Oxidase.

BEAN and HASSID (1956) discovered a glucose-oxidizing enzyme in the marine red alga *Iridophycus flaccidum*, but this enzyme showed a much lower substrate specificity than the fungal glucose oxidase. It can oxidize a variety of monosaccharides as well disaccharides. Some evidence for the operation of the enzyme in vivo was obtained. The enzyme has not yet been found in other plants, and its physiological role remains to be established.

Assay. The enzyme catalyzes the removal of two hydrogen atoms from the C-1 position of a hexose moiety to give the corresponding lactone, which is then hydrolyzed to the aldonic acid. The stoichiometry of the over-all reaction is:

$$\left.\begin{array}{l} \text{D-glucose} \\ \text{D-galactose} \\ \text{maltose} \end{array}\right\} + O_2 + H_2O \rightarrow \left.\begin{array}{l} \text{D-gluconic acid} \\ \text{D-galactonic acid} \\ \text{maltobionic acid} \end{array}\right\} + H_2O_2$$

The reaction can be followed by determining the rate of O_2 consumption manometrically, and it is not necessary to include catalase to decompose the H_2O_2. The reaction mixture contains 1.0 ml of 0.5 M maleate buffer (pH 5.2), between 0.01 and 1.0 ml of enzyme solution, and water to give a total of 2.0 ml in the main compartment of the WARBURG vessel; 0.5 ml of 0.5 M substrate is tipped in from the side-arm to start the reaction (30° C). An alternative colorimetric method was developed, but the dye (2,6-dichlorphenol indophenol) is reduced more slowly than O_2.

Purification. Fresh, partially frozen thalli were passed through a meat grinder, and the macerated tissue extracted with water. After dialysis and removal of materials by precipitation with $BaCl_2$, the extract was fractionated with cold methanol. Some further purification was effected by ammonium sulfate fractionation.

Properties. Since the enzyme has not been extensively purified, the information on its properties is limited. It was not possible to demonstrate the presence of a flavin prosthetic group. In addition to the substrates indicated above, lactose and cellobiose are oxidized to their aldobionic acids. The enzyme is apparently specific with regard to the C-2 configuration of the hexose but remains active when the C-4 configuration is changed or when the hydroxyl on this carbon is substituted by a D-glucose unit. It is able to reduce 2,6-dichlorophenol indophenol, but not methylene blue or ferricyanide. The enzyme is inhibited by Hg^{++}, Ag^+, Pb^{++} ions, and by acetate and propionate anions.

References.

ABOOD, L. G., and L. ALEXANDER: J. biol. Chem. **227**, 717 (1957). — AKAZAWA, T., and E. E. CONN: J. biol. Chem. **232**, 403 (1958). — AMBE, K. S., and A. VENKATARAMAN: Biochem. Biophys. Res. Comn. **1**, 133 (1959). — APPLEBY, C. A., and R. K. MORTON: Nature (Lond.) **173**, 749 (1954); — Biochem. J. **71**, 492 (1959); **75**, 258 (1960).

BALL, E.: Biochem. Z. **295**, 262 (1938). — BALL, E. G., C. B. ANFINSEN and O. COOPER: J. biol. Chem. **168**, 257 (1947). — BARER, R., and S. JOSEPH: Quart. J. Micr. Sci. **96**, 1 (1955). — BEAN, R. C., and W. Z. HASSID: J. biol. Chem. **218**, 425 (1956). — BEEVERS, H.: Plant Physiol. **29**, 265 (1954). — BEINERT, H.: In: The Enzymes, ed. by P. D. BOYER, H. LARDY and K. MYRBÄCK, 2nd ed., Vol. 2, Part A, p. 339. New York: Academic Press, Inc., 1960. — BEINERT, H., and R. H. SANDS: Biochem. Biophys. Res. Comm. **1**, 171 (1959). — BENDALL, D. S., and R. HILL: New Phytol. **55**, 206 (1956). — BEN-GERSHOM, E.: Biochem. J. **78**, 218 (1961). — BOERI, E., E. CUTOLO, M. LUZZATI and L. TOSI: Arch. Biochem. Biophys. **56**, 487 (1955). — BOERI, E., and L. TOSI: Arch. Biochem. Biophys. **60**, 463 (1956). — BOMSTEIN, R., R. GOLD-BERGER and H. TISDALE: Biochem. Biophys. Res. Comm. **2**, 234 (1960). — BONNER, W. D., jr.: Ann. Rev. Plant Physiol. **8**, 427 (1957). — BONNER, W. D., JR, and D. O. VOSS: Nature **191**, 682 (1961). — BORDER, C. A., and J. M. NELSON: J. Amer. Chem. Soc. **61**, 1507 (1939). — BRACKETT, F. S., R. A. OLSON and R. G. CRICKARD: J. gen. Physiol. **36**, 529 (1953). — BUHLER, D. R., and H. S. MASON: Arch. Biochem. Biophys. **92**, 424 (1961). — BURRIS, R. H.: In: Encyclopedia of Plant Physiology, ed. by W. RUHLAND, Vol. 12, Part. 1, p. 365. Berlin-Göttingen-Heidelberg: Springer-Verlag 1960. — BUTLER, W. L., and K. H. NORRIS: Arch. Biochem. Biophys. **87**, 31 (1960).

CARRUTHERS, C.: J. biol. Chem. **171**, 641 (1947). — CASTOR, L. N., and B. CHANCE: J. biol. Chem. **217**, 453 (1955). — CATER, D. B.: Progress in Biophysics **10**, 153 (1960). — CHANCE, B.: Science **120**, 767 (1954); in Enzymes: Units of Biological Structure and Function, ed. by D. H. GAEBLER, p. 447. New York: Academic Press, Inc., 1956; in Methods in Enzymology, ed. by S. P. COLOWICK and N. O. KAPLAN, Vol. 4, p. 273. New York: Academic Press, Inc., 1957. —

CHANCE, B., and H. BALTSCHEFFSHY: J. biol. Chem. **233**, 736 (1958). — CHANCE, B., and D. P. HACKETT: Plant Physiol. **34**, 33 (1959). — CHANCE, B., and B. HESS: J. biol. Chem. **234**, 2404 (1959). — CHANCE, B., and V. LEGALLAIS: Rev. Sci. Instr. **30**, 371 (1959). — CHANCE, B., and R. SAGER: Plant Physiol. **32**, 548 (1957). — CHANCE, B., and E. L. SPENCER: Faraday Society Discussions, No. 27, 200 (1959). — CHANCE, B., and B. THORELL: J. biol. Chem. **234**, 3044 (1959). — CHANCE, B., and G. R. WILLIAMS: J. biol. Chem. **217**, 429 (1955); Adv. in Enzymology **17**, 65 (1956). — CLARK, H. W., H. A. NEUFELD, C. WIDMER and E. STOTZ: J. biol. Chem. **210**, 851 (1954). — CLAYTON, R. A.: Arch. Biochem. Biophys. **81**, 404 (1959). — CLUM, H. H., and A. NASON: Plant Physiol. **33**, 354 (1958). — COLPA-BOONSTRA, J. P., and E. C. SLATER: Biochim. Biophys. Acta **27**, 122 (1958). — CONN, E. E., L. M. KRAEMER, P.-N. LIU and B. VENNESLAND: J. biol. Chem. **194**, 143 (1952). — CONN, E. E., and B. VENNESLAND: J. biol. Chem. **192**, 17 (1951). — CONNELLY, C. M.: Fed. Proc. **16**, 681 (1957). — CONNELLY, J. L., M. MORRISON and E. STOTZ: Biochim. Biophys. Acta **32**, 543 (1959). — COOPER, C., and A. L. LEHNINGER: J. biol. Chem. **219**, 489 (1956); **219**, 519 (1956). — CORNFORTH, J. W., and A. T. JAMES: Biochem. J. **63**, 124 (1956). — COULTHARD, C. E., R. MICHAELIS, W. F. SHORT, G. E. H. SKRIMSHIRE, A. F. B. STANDFAST, J. H. BIRKINSHAW and H. RAISTRICK: Biochem. J. **39**, 24 (1945). — CRANE, F. L.: Plant Physiol. **34**, 128 (1959); **34**, 546 (1959a). — CRANE, F. L., and B. EHRLICH: Arch. Biochem. Biophys. **89**, 134 (1960). — CRANE, F. L., B. EHRLICH and L. P. KEGEL: Biochem. Biophys. Res. Comm. **3**, 37 (1960). — CRIDDLE, R. S., and R. M. BOCK: Biochem. Biophys. Res. Comm. **1**, 133 (1959); **1**, 138 (1959a).

DALY, J. M.: Arch. Biochem. Biophys. **51**, 25 (1954). — DAVIES, D. D.: J. exptl. Botany **7**, 203 (1956). — DAVIES, P. W., and F. BRINK, jr.: Rev. Sci. Instr. **13**, 524 (1942). — DAWSON, R. M. C., D. C. ELLIOTT, W. H. ELLIOTT and K. M. JONES: Data for Biochemical Research. Oxford: Oxford Univ. Press 1959. — DAWSON, C. R., and R. J. MAGEE: In: Methods in Enzymology, ed. by S. P. COLOWICK and N. O. KAPLAN, Vol. 2, p. 817. New York: Academic Press, Inc. 1955. — DAWSON, C. R., and W. B. TARPLEY: In: The Enzymes, ed. by J. B. SUMNER and K. MYRBÄCK, Vol. II, Part 1, p. 454. New York: Academic Press, Inc. 1951. — DEVLIN, T. M.: J. biol. Chem. **234**, 962 (1959). — DEVLIN, T. M., and A. L. LEHNINGER: J. biol. Chem. **219**, 507 (1956). — DOLIN, M. I.: Arch. Biochem. Biophys. **55**, 415 (1955). — DOROUGH, G. D., and K. T. SHEN: J. Amer. Chem. Soc. **72**, 3939 (1950). — DUNN, F. J., and C. R. DAWSON: J. biol. Chem. **189**, 485 (1951). — DUYSENS, L. N. M., and J. AMESZ: Biochim. Biophys. Acta **24**, 19 (1957). — DUYSENS, L. N. M., and G. K. M. KRONENBERG: Biochim. Biophys. Acta **26**, 437 (1957).

EICHEL, B., W. W. WAINIO, P. PERSON and S. J. COOPERSTEIN: J. biol. Chem. **183**, 89 (1950). — ELLIOTT, W. B., W. C. HULSMANN and E. C. SLATER: Biochim. Biophys. Acta **33**, 509 (1959). — EICHENBERGER, E., and K. V. THIMANN: Arch. Biochem. Biophys. **67**, 466 (1957). — ERNSTER, L., O. JALLING, H. LOW and O. LINDBERG: Exptl. Cell Res., Suppl. **3**, 124 (1955). — ESTABROOK, R. W.: J. biol. Chem. **223**, 781 (1956); **230**, 735 (1958). — EVANS, H. J.: Plant Physiol. **30**, 437 (1955).

FELDMAN, D., and W. W. WAINIO: Science **130**, 796 (1959); — J. biol. Chem. **235** 3635 (1960). — FORTI, G., C. TUA, and L. TOGNOLI: Biochim. Biophys. Acta **36**, 19 (1959). — FRANKE, W.: In: Encyclopedia of Plant Physiology, ed. by W. RUHLAND, Vol. 12, Part 1, p. 401. Berlin-Göttingen-Heidelberg: Springer-Verlag 1960. — FREED, S., and K. M. SANCIER: J. Amer. Chem. Soc. **76**, 198 (1954). — FRIGERIO, N. A., and H. A. HARBURY: J. biol. Chem. **231**, 135 (1956). — FRITZ, G., and H. BEEVERS: Arch. Biochem. Biophys. **55**, 436 (1955). — FRITZ, G. J., W. G. MILLER, R. H. BURRIS and L. ANDERSON: Plant Physiol. **33**, 159 (1958).

GARFINKEL, D.: Arch. Biochem. Biophys. **71**, 111 (1957). — GIBSON, Q. H., C. GREENWOOD and V. MASSEY: Biochem. J. **76**, 46P (1960). — GLOVER, J.: In: Modern Methods of Plant Analysis, ed. by K. PAECH and M. V. TRACEY, Vol. I, p. 149. Berlin-Göttingen-Heidelberg: Springer-Verlag 1956. — GODDARD, D. R.: Amer. J. Bot. **31**, 270 (1944). — GODDARD, D., and W. D. BONNER, jr.: In: Plant Physiology, ed. by F. C. STEWARD, Vol. 1A, p. 209. New York: Academic Press, Inc. 1960. — GREEN, D. E.: Adv. in Enzymology **21**, 73 (1959). — GREEN, D. E., and S. FLEISCHER: In: Metabolic Pathways, ed. by D. M. GREENBERG, Vol. 1, p. 41. New York: Academic Press, Inc. 1960. — GREEN, D. E., J. JARNEFELT and H. D. TISDALE: Biochim. Biophys. Acta **31**, 34 (1959); **38**, 160 (1960). — GREENLEAS, J., and W. W. WAINIO: J. biol. Chem. **234**, 658 (1959).

HAAS, E.: J. biol. Chem. **155**, 321 (1944); in Methods in Enzymology, ed. by S. P. COLOWICK and N. O. KAPLAN, Vol. 2, p. 699. New York: Academic Press, Inc. 1955. — HACKETT, D. P.: Plant Physiol. **33**, 8 (1958); — Ann. Rev. Plant Physiol. **10**, 113 (1959); in Encyclopedia of Plant Physiology, ed. by W. RUHLAND, Vol. 12, Part 2, p. 23. Berlin-Göttingen-Heidelberg: Springer-Verlag 1960; — Plant Physiol., in press 1961. — HACKETT, D. P., and D. W. HAAS: Plant Physiol. **33**, 27 (1958). — HACKETT, D. P., D. W. HAAS, S. K. GRIFFITHS and D. J. NIEDERPRUEM: Plant Physiol. **35**, 8 (1960). — HACKETT, D. P., B. RICE and C. SCHMID: J. biol. Chem. **235**, 2140 (1960); — HAGIHARA, B.: Biochim. Biophys. Acta **46**, 134 (1961). — HAGIHARA, B., T. HORIO, M. NOZAKI, I. SEKUZU, J. YAMASHITA and K. OKUNUKI: Nature

(Lond.) **178**, 631 (1956). — HAGIHARA, B., T. HORIO, J. YAMASHITA, M. NOZAKI and K. OKUNUKI: Nature (Lond.) **178**, 629 (1956). — HAGIHARA, B., K. TAGAWA, I. MORIKAWA, M. SHIN and K. OKUNUKI: Nature (Lond.) **181**, 1590 (1958). — HARBURY, H. A.: J. biol. Chem. **225**, 1009 (1957). — HARRIS, E. D., and A. S. LINDSEY: Nature (Lond.) **160**, 413 (1948). — HARTREE, E. F.: In: Modern Methods of Plant Analysis, ed. by K. PAECH and M. V. TRACEY, Vol. IV, p. 197. Berlin-Göttingen-Heidelberg: Springer-Verlag 1955; — Adv. in Enzymology **18**, 1 (1957). — HILL, R.: Nature (Lond.) **174**, 501 (1954). — HILL, R., and R. SCARISBRICK: New Phytol. **50**, 98 (1951). — HOLTON, F. A.: Biochem. J. **61**, 46 (1955). — HONDA, S. I.: Plant Physiol. **30**, 402 (1955). — HONDA, S. I., R. N. ROBERTSON and J. M. GREGORY: Austral. J. biol. Sci. **11**, 1 (1958). — HORIO, T., and M. D. KAMEN: Biochim. Biophys. Acta **48**, 266 (1961). — HOROWITZ, N. H., M. FLING, H. L. MACLEOD and N. SUEOKA: J. Molecular Biol. **2**, 96 (1960). — HÜBSCHER, G., M. KIESE and R. NICOLAS: Biochem. Z. **325**, 223 (1954). — HUMPHREYS, T. E.: Plant Physiol. **30**, 46 (1955). — HUMPHREYS, T. E., and E. E. CONN: Arch. Biochem. Biophys. **60**, 226 (1956).

IMAMOTO, F.: J. Biochem. (Tokyo) **46**, 1023 (1959). — IZAWA, S.: Plant and Cell Physiol. **1**, 269 (1960).

JACKSON, F. L., and V. D. LAWTON: Nature (Lond.) **181**, 1539 (1958). — JACKSON, F. L., and J. W. LIGHTBOWN: Biochem. J. **69**, 63 (1958). — JAMES, W. O.: Ann. Rev. Plant Physiol. **4**, 59 (1953). — JAMES, W. O., and D. C. ELLIOTT: New Phytol. **57**, 230 (1958). — JAMES, W. O., and N. GARTON: J. exptl. Botany **3**, 310 (1952).

KANDLER, O.: Z. Naturforsch. **12 b**, 272 (1957). — KAPLAN, N. O.: In: Methods in Enzymology, ed. by S. P. COLOWICK and N. O. KAPLAN, Vol. 2, p. 681. New York: Academic Press, Inc. 1955. — KAPLAN, N. O., M. N. SWARTZ, M. E. FRECH and M. M. CIOTTI: Proc. Natl. Acad. Sci. U.S. **42**, 481 (1956). — KATOH, S.: Nature (Lond.) **186**, 138 (1960). — KEILIN, D.: Proc. Roy. Soc. London B **98**, 312 (1925); **106**, 418 (1930). — KEILIN, D., and E. F. HARTREE: Biochem. J. **39**, 148 (1945); **42**, 221 (1948); **49**, 88 (1951); **50**, 331 (1952); — Nature (Lond.) **176**, 200 (1955); — Biochim. Biophys. Acta **27**, 173 (1958). — KEILIN, D., and T. MANN: Proc. Roy. Soc. London B **125**, 187 (1938). — KEISTER, D. L., A. SAN PIETRO and F. E. STOLZENBACH: J. biol. Chem. **235**, 2989 (1960). — KENTEN, R. H.: Biochem. J. **59**, 110 (1955); in Modern Methods of Plant Analysis, ed. by K. PAECH and M. V. TRACEY, Vol. I, p. 415. Berlin-Göttingen-Heidelberg: Springer-Verlag 1956; — Biochem. J. **68**, 244 (1958). — KENTEN, R. H., and P. J. G. MANN: Biochem. J. **53**, 498 (1953). — KERTESZ, D., and R. ZITO: Nature (Lond.) **179**, 1017 (1957). — KIELLEY, W. W., and J. R. BRONK: J. biol. Chem. **230**, 521 (1958). — KRISCH, K., and H. STAUDINGER: Biochem. Z. **331**, 37 (1958). — KUBOWITZ, F.: Biochem. Z. **292**, 221 (1937); **299**, 32 (1939). — KUSAI, K., I. SEKUZU, B. HAGIHARA, K. OKUNUKI, S. YAMAUCHI and M. NAKAI: Biochim. Biophys. Acta **40**, 555 (1960). — KÜSTER, H. J.: Ber. dtsch. bot. Ges. **68**, 183 (1955).

LANG, C. A., and A. NASON: J. biol. Chem. **234**, 1874 (1959). — LASER, H.: Proc. Roy. Soc. (Lond.) B **140**, 230 (1952). — LESTER, R. L., and S. FLEISCHER: Arch. Biochem. Biophys. **80**, 470 (1959). — LESTER, R. L., and T. RAMASARMA: J. biol. Chem. **234**, 672 (1959). — LEWIS, U. J.: J. biol. Chem. **206**, 109 (1954). — LI, W.-C., and C.-L. TSOU: Scientia Sinica **5**, 663 (1956). — LINNANE, A. W., and D. M. ZIEGLER: Biochim. Biophys. Acta **29**, 630 (1958). — LOCKHART, E. E.: Biochem. J. **33**, 613 (1939). — LONGMUIR, I. S.: Biochem. J. **57**, 81 (1954). — LONGMUIR, I. S., and A. BOURKE: Biochem. J. **76**, 225 (1960). — LUNDEGÅRDH, H.: Arkiv Kemi **5**, 97 (1952); — Physiol. Plantarum **8**, 95 (1955); — in Encyclopedia of Plant Physiology, ed. by W. RUHLAND, Vol. 12, Part 1, p. 311. Berlin-Göttingen-Heidelberg: Springer-Verlag 1960. — LYNEN, F., and R. KOENIGSBERGER: Liebigs Ann. Chem. **573**, 60 (1951).

MAEHLY, A.: In: Methods in Enzymology, ed. by S. P. COLOWICK and N. O. KAPLAN, Vol. 2, p. 801. New York: Academic Press, Inc. 1955. — MAEHLY, A., and B. CHANCE: In: Methods of Biochemical Analysis, ed. by D. GLICK, Vol. 1, p. 357. New York: Interscience 1954. — MAHLER, H. R.: In: Methods in Enzymology, ed. by S. P. COLOWICK and N. O. KAPLAN, Vol. 2, p. 688. New York: Academic Press, Inc. 1955. — MALEY, G. F.: J. biol. Chem. **224**, 1029 (1957). — MANN, T., and J. H. QUASTEL: Biochem. J. **35**, 502 (1941). — MAPSON, L. W., and D. R. GODDARD: Biochem. J. **49**, 592 (1951). — MAPSON, L. W., and E. M. MOUSTAFA: Biochem. J. **62**, 248 (1956). — MARINETTI, G. V., J. ERBLAND, J. KOCHEN and E. STOTZ: J. biol. Chem. **233**, 740 (1958). — MÄRKI, F., and C. MARTIUS: Biochem. Z. **333**, 111 (1960). — MARRÉ, E., e O. SERVETTAZ: Nuova giorn. bot. ital. **64**, 273 (1937). — MARRÉ, E., O. SERVETTAZ e G. ROSSI: Ital. J. Biochem. **6**, 158 (1957). — MARTIN, E. M., and R. K. MORTON: Nature (Lond.) **176**, 113 (1955); — Biochem. J. **62**, 696 (1956). — MASON, H. S.: J. biol. Chem. **172**, 83 (1948). — MASSEY, V.: Biochim. Biophys. Acta **37**, 310 (1960); **37**, 314 (1960a). — MATSUSHITA, S., and F. IBUKI: Biochim. Biophys. Acta **40**, 540 (1960). — McMUNN, C. A.: Trans. Roy. Soc. (Lond.) B **177**, 267 (1886). — McMURRAY, W. C., G. F. MALEY and H. A. LARDY: J. biol. Chem. **230**, 219 (1958). — MILLER, W. H., and C. R. DAWSON: J. Amer. Chem. Soc. **63**, 3368 (1941). — MINAKAMI, S.: J. Biochem. (Tokyo) **42**, 749 (1955). — MINAKAMI, S., H. ISHIKURA and K. SATAKE: J. Biochem. (Tokyo) **43**, 575 (1956). — MINAKAMI, S., R. L.

RINGLER and T. P. SINGER: Biochem. Biophys. Res. Comm. 3, 423 (1960). — MORITA, Y., K. KAMEDA and M. MIZUNO: Agricultural and Biological Chem. 25, 136 (1961). — MORRISON, M., J. CONNELLY, J. PETIX and E. STOTZ: J. biol. Chem. 235, 1202 (1960). NAKAMURA, T.: Biochim. Biophys. Acta 30, 44 (1958); 42, 499 (1960). — NEILANDS, J. B.: J. biol. Chem. 197, 701 (1952). — NICHOLLS, P.: Fed. Proc. 20, 50 (1961). — NIEDERPRUEM, D. J., and D. P. HACKETT: Plant Physiol. 33, 113 (1958); 36, 79 (1961). — NIELSEN, S. O., and A. L. LEHNINGER: J. biol. Chem. 215, 555 (1955). — NOZAKI, M.: J. Biochem. (Tokyo) 47, 592 (1960). — NOZAKI, M., T. YAMANAKA, T. HORIO and K. OKUNUKI: J. Biochem. (Tokyo) 44, 453 (1957). — NUNNIKHOVEN, R.: Biochim. Biophys. Acta 28, 108 (1958).
OKUNUKI, K., I. SEKUZU, T. YONETANI and S. TAKEMORI: J. Biochem. (Tokyo) 45, 847 (1958).
PACKER, L., R. W. ESTABROOK, T. P. SINGER and T. KIMURA: J. biol. Chem. 235, 535 (1960). — PALÉUS, S., and H. THEORELL: Acta Chem. Scand. 11, 905 (1957). — PAUL, K. G.: Acta Chem. Scand. 4, 239 (1950); 5, 389 (1951); — in The Enzymes, ed. by P. D. BOYER, H. LARDY and K. MYRBÄCK, 2nd ed., Vol. 3, p. 277. New York: Academic Press 1960. — PAUL, K. G., and Y. AVI-DOR: Acta Chem. Scand. 8, 649 (1954). — PERRY, R. P., B. THORELL, L. ÅKERMAN and B. CHANCE: Nature (Lond.) 184, 929 (1959). — PINCHOT, G. B.: J. biol. Chem. 229, 16 (1957). — PONTING, J. D., and M. A. JOSLYN: Arch. Biochem. 19, 47 (1948). — POTTER, V. R., and A. E. REIF: J. biol. Chem. 194, 287 (1952).
RAFTER, G. W., S. CHAYKIN and E. G. KREBS: J. biol. Chem. 208, 799 (1954). — RAMASARMA, T., and R. L. LESTER: J. biol. Chem. 235, 3309 (1960). — RAW, I., and W. COLLI: Nature (Lond.) 184, 1798 (1959). — RAW, I., and H. R. MAHLER: J. biol. Chem. 234, 1867 (1959). — RAW, I., R. MOLINIARI, D. FERREIRA DO AMARAL and H. R. MAHLER: J. biol. Chem. 233, 225 (1958). — RICHARDSON, K. E., and N. E. TOLBERT: J. biol. Chem. 236, 1280 (1961). — RINGLER, R. L., S. MINAKAMI and T. P. SINGER: Biochem. Biophys. Res. Comm. 3, 417 (1960). — ROBBIE, W. A.: Arch. Biochem. 5, 49 (1944); — in Methods in Medical Research, ed. by V. R. POTTER, Vol. 1, p. 307. Chicago: The Year Book Publisher 1948. — RODKEY, L., and E. G. BALL: J. biol. Chem. 182, 17 (1950). — ROMBERGER, J. A.: Plant Physiol. 34, 589 (1959).
SEARLS, R. L., and D. R. SANADI: J. biol. Chem. 235, 2485 (1960). — SEKUZU, I., and K. OKUNUKI: J. Biochem. (Tokyo) 43, 107 (1956). — SEKUZU, I., Y. ORII and K. OKUNUKI: J. Biochem. (Tokyo) 48, 214 (1960). — SEKUZU, I., S. TAKEMORI, Y. ORII and K. OKUNUKI: Biochim. Biophys. Acta 37, 64 (1960). — SHIBATA, K.: J. Biochem. (Tokyo) 45, 599 (1958). — SIMON, E. W.: Biochem. J. 69, 67 (1958). — SINGER, T. P., and E. B. KEARNEY: J. biol. Chem. 183, 409 (1950). — SISLER, E. C., and H. J. EVANS: Tobacco Sci. 2, 132 (1958). — SLATER, E. C.: Biochem. J. 46, 484 (1950). — SMITH, F. G., and E. STOTZ: J. biol. Chem. 179, 865 (1949). — SMITH, J. L.: Fed. Proc. 20, 48 (1961). — SMITH, L.: In: Methods in Enzymology, ed. by S. P. COLOWICK and N. O. KAPLAN, Vol. 2, p. 732. New York: Academic Press, Inc. 1955; — in Methods of Biochemical Analysis, ed. by D. GLICK, Vol. 2, p. 427. New York: Interscience, 1955a; — J. biol. Chem. 215, 833 (1955b). — SMITH, L., and B. CHANCE: Ann. Rev. Plant Physiol. 9, 449 (1958). — SMITH, L., and H. CONRAD: Arch. Biochem. Biophys. 63, 403 (1956). — SMITH, L., and E. STOTZ: J. biol. Chem. 179, 891 (1953). — STEIN, A. M., N. O. KAPLAN and M. M. CIOTTI: J. biol. Chem. 234, 979 (1959). — STERN, H., and F. B. JOHNSTON: Plant Physiol. 32, 476 (1957). — STICKLAND, R. G.: Biochem. J. 77, 636 (1960). — STOTZ, E.: In: Methods in Enzymology, ed. by S. P. COLOWICK and N. O. KAPLAN, Vol. 2, p. 740. New York: Academic Press, Inc. 1955. — STRITTMATTER, P.: J. biol. Chem. 235, 2492 (1960). — STRITTMATTER, P., and S. F. VELICK: J. biol. Chem. 221, 253 (1956); 221, 277 (1956a); 228, 785 (1957). — STRONG, F. M.: Topics in Microbial Chemistry, p. 32. New York: Wiley 1958. — SWEDIN, B., and H. THEORELL: Nature (Lond.) 185, 71 (1940).
TAGAWA, K., and M. SHIN: J. Biochem. (Tokyo) 46, 865 (1959); 46, 875 (1959a). — TAKEMORI, S.: J. Biochem. (Tokyo) 47, 382 (1960). — THEORELL, H.: Arkiv Kemi, Mineral. Geol. A 16, 1 (1942). — THEORELL, H., and Å. ÅKESON: J. Amer. Chem. Soc. 63, 1804 (1941). — THIMANN, K. V., C. S. YOCUM and D. P. HACKETT: Arch. Biochem. Biophys. 53, 239 (1954).
UMBREIT, W. W., R. H. BURRIS and J. F. STAUFFER: Manometric Techniques. Minneapolis: Burgess Publishing Co., 1957.
VAN BRUGGEN, J. T., F. J. REITHEL, C. K. CAIN, P. A. KATZMAN, E. A. DOISY, R. D. MUIR, E. C. ROBERTS, W. L. GABY, D. M. HOMAN and L. R. JONES: J. biol. Chem. 148, 365 (1943). — VAN TAMELEN, E. E., J. P. DICKIE, M. E. LOOMANS, R. S. DEWEY and F. M. STRONG: J. Amer. Chem. Soc. 83, 1639 (1961).
WAINIO, W. W., and M. ARONOFF: Arch. Biochem. Biophys. 57, 115 (1955). — WAINIO, W. W., C. VAN DER WENDE and N. F. SHIMP: J. biol. Chem. 234, 2433 (1959). — WARBURG, O.: Biochem. Z. 189, 354 (1927). — WHATLEY, F. R.: In: Modern Methods of Plant Analysis, ed. by K. PAECH and M. V. TRACEY, Vol. I, p. 452. Berlin-Göttingen-Heidelberg: Springer-Verlag 1956. — WINZLER, R. J.: J. Cell. Comp. Physiol. 17, 263 (1941). — WISKICH, J. T. R. K. MORTON and R. N. ROBERTSON: Australian J. biol. Sci. 13, 109 (1960). — WOLF, G.

694   David P. Hackett: Enzymes of Terminal Respiration.

F. Kieffer and C. Martius: Fed. Proc. 18, 354 (1959). — Wosilait, W. D.: J. biol. Chem 235, 1196 (1960). — Wosilait, W. D., and A. Nason: J. biol. Chem. 208, 785 (1954); — in Methods in Enzymology, ed. by S. P. Colowick and N. O. Kaplan, Vol. 2, p. 725. New York: Academic Press 1955. — Wosilait, W. D., A. Nason and A. J. Terrell: J. biol. Chem. 206, 271 (1954).

Yamanaka, T., T. Horio and K. Okunuki: Biochim. Biophys. Acta 40, 349 (1960). — Yamashita, J., T. Higashi, T. Yamanaka, M. Nozaki, H. Mizushima, H. Matsubara, T. Horio and K. Okunuki: Nature (Lond.) 179, 959 (1957). — Yocum, C. S., and D. P. Hackett: Plant Physiol. 32, 186 (1957). — Yonetani, T.: J. biol. Chem. 235, 845 (1960). — Young, L. C. T., and E. E. Conn: Plant Physiol. 31, 205 (1956).

Zelitch, I.: J. biol. Chem. 201, 719 (1953); — in Methods in Enzymology, ed. by S. P. Colowick and N. O. Kaplan, Vol. 1, p. 528. New York: Academic Press, Inc. 1955; — J. biol. Chem. 224, 251 (1957); 233, 1299 (1958); 234, 3077 (1959). — Zelitch, I., and G. A. Barber: Plant Physiol. 35, 205 (1960); 35, 626 (1960a). — Zelitch, I., and S. Ochoa: J. biol. Chem. 201, 707 (1953). — Ziegler, D. M., D. E. Green and K. A. Doeg: J. biol. Chem. 234, 1916 (1959).

Summary of Recommendations on Enzyme Terminology.

by the

Commission on Enzymes of the International Union of Biochemistry, 1961*.

Enzyme units.

(1) One *unit* (U) of any enzyme should be defined as that amount which will catalyse the transformation of 1 micromole of substrate per minute, or, where more than one bond of each substrate molecule is attacked, 1 micro-equivalent of the group concerned per minute, under defined conditions. Where two identical molecules react together, the unit will be the amount which catalyses the transformation of 2 micromoles per minute. The temperature should be stated, and where practicable should be 25° C. The other conditions, including pH and substrate concentration, should be optimal. In order to avoid inconvenient numbers, terms such as milli-unit (mU), kilo-unit (kU), etc., may be used.

(2) *Enzyme assays* should be based wherever possible upon measurements of initial rates of reaction in order to avoid complications, and the substrate concentration should be sufficient for saturation of the enzyme, so that the kinetics approach zero order. Where a sub-optimal concentration of substrate must be used, the Michaelis constant should be determined so that the observed rate may be converted into that which would be obtained on saturation with substrate.

(3) *Specific activity* should be expressed as units of enzyme per milligram of protein.

(4) *Molecular activity* should be defined as units per micromole of enzyme at optimal substrate concentration, that is, as the number of molecules of substrate transformed per minute per molecule of enzyme.

(5) When the enzyme has a prosthetic group or catalytic centre whose concentration can be measured, the catalytic power can be expressed as *catalytic centre activity*, i.e. the number of molecules of substrate transformed per minute per catalytic centre. The term "turnover number," which has been employed with various meanings, should no longer be used.

(6) *Concentration* of an enzyme in solution should be expressed as units per millilitre.

Symbols of enzyme kinetics.

(7) In mathematical equations for enzyme kinetics, the symbols given in Appendix B (p. 699, *eds,*) should be used for velocity, saturation velocity, Michaelis constant, substrate and inhibitor constants, and velocity constants.

---

* Reprinted, with permission, from Report of the Commission on Enzymes of the International Union of Biochemistry 1961. London: Pergamon Press, 1961.

(8) All equilibria involving combinations of enzymes with substrates, inhibitors or products should be expressed in terms of dissociation constants rather than association constants.

(9) The term "Michaelis constant" and the symbol K_m should be used only to denote the substrate concentration at which the velocity is equal to half the saturation velocity.

(10) The terms "substrate constant" (K_s) and "inhibitor constant" (K_i) should be used to denote the equilibrium (dissociation) constants of the reactions $E + S = ES$ and $E + I = EI$ respectively.

(11) The velocity constants of the individual steps involved in an enzyme reaction should be numbered as in the following example:

$$E + S \underset{k_{-1}}{\overset{k_{+1}}{\rightleftharpoons}} ES \underset{k_{-2}}{\overset{k_{+2}}{\rightleftharpoons}} EP \underset{k_{-3}}{\overset{k_{+3}}{\rightleftharpoons}} E + P$$

Thus k_{+n} will denote the velocity constant of the nth step in the forward direction, i.e. proceeding from substrate to product, while k_{-n} will denote that of the reverse reaction of the same step.

(12) The velocity of an enzyme reaction should be denoted by v, and the value of v corresponding to saturation of the enzyme with substrate should be denoted by V.

The nomenclature of coenzymes.

(13) The nicotinamide nucleotide coenzymes should in future be known by their chemical names "nicotinamide-adenine dinucleotide" (NAD) and "nicotin-amide-adenine dinucleotide phosphate" (NADP) respectively. The names "cozymase," "phosphocozymase," "coenzyme I" (CoI), "coenzyme II" (CoII), "diphosphopyridine nucleotide" (DPN), "triphosphopyridine nucleotide" (TPN), "codehydrogenase I," "codehydrogenase II" should no longer be used. The mononucleotide should continue to be known as "nicotinamide mono-nucleotide" (NMN).

(14) The names "flavin-adenine dinucleotide" (FAD) and "flavin mononucleotide" (FMN) should be retained.

(15) For the reduced form of NAD, two alternative abbreviations should be permitted, namely $NADH_2$ (corresponding to $FADH_2$) or, where it is desired to show the release of a H^+ ion in the reduction, $NADH + H^+$. However, when the latter form is used, the oxidized form should always be written as NAD^+; under no circumstances should the reduction be shown as a change from NAD to NADH. Similar forms should be permitted for NADP.

(16) The name "coenzyme Q" should be dropped and the name "ubiquinone" used instead. This may be abbreviated as "UQ," and when it is desired to indicate the number of isoprene units in the side-chain a numerical suffix may be added thus "UQ_{10}."

(17) Although names of the form "coenzyme X" are not recommended, the name "coenzyme A" should be retained, in the absence of any practicable alternative. Two abbreviations should be permissible, namely CoA for normal use or, where it is desired to indicate the thiol group, CoASH.

Classification and nomenclature of cytochromes.

(18) Cytochromes should be defined as haemoproteins whose principal biological function is electron and/or hydrogen transport by virtue of a reversible valence change of their haem iron.

(19) The name "cytochrome" implies a single haemoprotein entity; the term "cytochrome system" should be used to denote any wider system in which one or more cytochromes, apart from cytochrome oxidase, are involved.

(20) Cytochromes should be classified at present in four groups, according to the nature of their prosthetic haem groups, namely Cytochromes A, B, C and D, containing formylporphyrin-iron, protoporphyrin-iron, a substituted meso-porphyrin-iron with covalent porphyrin-protein linkages, and dihydropor-phyrin-iron respectively. The criteria for assignment to groups should be those given in Chapter 5 (of the Report; *eds.*).

(21) When a cytochrome contains haem groups of two different kinds attached to one specific protein, both groups should be shown in the name, e.g. "cytochrome CD."

(22) Haemoproteins closely related to cytochromes, but with a haemoglobin-like spectrum and a reactivity with ligands which do not react with cytochrome *c*, should be called "cytochromoids."

(23) There should be no sub-classification of groups A—D at present.

(24) A newly discovered haemoprotein should not be classed as a cytochrome until it has been shown to come within the definition given in (18) above. It may temporarily be named on the pattern "haemoprotein 560 (*Bacterium X*)," where 560 is the wavelength in mμ of the α band of its spectrum.

(25) When a haemoprotein is first established as a cytochrome, it should receive a *provisional name* of the type "cytochrome 560 (*Bacterium X*)." Efforts should then be made to establish its group; when this has been done provisionally, an *interim name* of the form "cytochrome B (560, *Bacterium X*)" should be given, in which the group is indicated by a *capital* letter. Finally, when the allocation has been clearly determined and its individuality properly established, it should be allocated an official *final name*, based on an italic *small* letter and a subscript number, e.g. "cytochrome b_1," and be included in the list of cytochromes (see Appendix C) (p. 699; *eds.*).

(26) The names of the majority of already well-established cytochromes, including those used in the names of enzymes, should remain unchanged, since they will rank as final names.

(27) Final names should only be allotted by authority, preferably by a Standing Committee, and not by individual workers.

(28) It is recommended that the International Union of Pure and Applied Chemistry should be consulted about setting up a Standing Committee on Cytochromes, to decide on the inclusion of cytochromes within the different groups, to allot final names to cytochromes when necessary, to consider the establishment of new groups if needed, and to keep the list of cytochromes up to date.

Classification and nomenclature of enzymes.

(29) Names purporting to be names of enzymes, especially those ending in "-ase," should only be used for single enzymes. When it is desired to name a system containing more than one enzyme on the basis of the overall reaction catalysed by it, the word "system" should be included in the name, e.g. "the succinate oxidase system."

(30) The basis for classification and naming should be the overall reaction catalysed, as expressed by the formal equation; this means that the intimate mechanism of the reaction, and the formation of intermediate complexes with the enzyme, will not be taken into account.

(31) Systematic names cannot be given to enzymes until it is known what reactions they catalyse. This applies for example to any enzyme that is only known to catalyse an isotopic exchange.

(32) The Commission recommends that the system of classification shown in Appendix D (p. 700; *eds.*) be approved; this divides enzymes into six main classes, each of which is divided into a number of sub-classes and sub-sub-classes, according to the nature of the reaction catalysed.

(33) It also recommends that the enzymes should be coded on a four-number system intimately connected with the system of classification, as shown in Appendix D and Appendix E (refer to Report; *eds.*). On this system the first number indicates the main class, the second and third show the sub-class and sub-sub-class respectively, thus defining the type of reaction, and the fourth is the number of the enzyme within its sub-sub-class.

(34) Once given, the number of an enzyme should remain attached to it as a permanent means of identification (unless it has been wrongly classified and has to be moved to another group). New enzymes should be placed at the end of the appropriate sub-sub-classes, so that the numbering of the existing enzymes therein will not be disturbed.

(35) New enzyme numbers should be allotted only by authority (e.g. by future Enzyme Commissions or by a Standing Committee) and not by individual workers.

(36) It is recommended that there shall be both systematic and trivial nomenclatures for enzymes; the systematic name will be formed in accordance with definite rules, and will identify the enzyme and indicate its action as precisely as possible; the trivial name will be sufficiently short for general use, and in a great many cases will be the name already in current use.

(37) The systematic and trivial nomenclatures should be in accordance with the Rules (1) to (31) set out in Chapter 6 of this report.

(38) The systematic and trivial enzyme names given in the list in Appendix E (refer to Report; *eds.*) should be used henceforth.

(39) Where an enzyme is the main subject of a paper or abstract, it is recommended that its code number, systematic name (where a satisfactory one exists) and source should be given at its first mention in the text; thereafter the trivial name may be used.

(40) Enzymes which are not the main subject should be identified at their first mention by their code numbers.

(41) When the paper deals with an enzyme which is not yet in the Commission's list, the author may introduce a new systematic name and/or a new trivial name, both formed only according to the recommended rules.

(42) A Standing Committee should be set up with power to approve the names of new enzymes, to allot enzyme numbers, to alter or delete existing names or numbers, and generally to keep the list of enzymes up to date.

The terminology of enzyme formation.

(43) In discussing enzyme formation evoked by the presence of chemical substances, the term "induction" should be used, rather than "adaptation," the enzyme-forming system should be described as "inducible" and the enzyme formed as "induced." "Sequential induction" should be used instead of "simultaneous adaptation" or "successive adaptation."

(44) The names of enzyme precursors should no longer be formed by the use of the suffix "-ogen"; the prefix "pre" should be used instead.

Recommended Symbols for Enzyme Kinetics.

v — Velocity of reaction catalysed by an enzyme.

V — Value of v when the enzyme is saturated with substrate, as given by the Michaelis equation.

K_m — "Michaelis constant." Concentration of substrate at which $v = V/2$.

K_s — "Substrate constant." Equilibrium (dissociation) constant of the reaction $E + S \rightleftharpoons ES$.

K_i — "Inhibitor constant." Equilibrium (dissociation) constant of the reaction $E + I \rightleftharpoons EI$.

k_{+n}, k_{-n} Velocity constants of the forward and backward reactions in the nth step of an enzyme reaction.

List of Cytochromes.

Cytochrome a		Widely distributed
	a_1	E. coli, etc.
	a_3	Distribution similar to a
		There is no general agreement about the separate existence of a and a_3
	b	Widely distributed
	b_1	E. coli
	b_2	Yeast (yeast lactate dehydrogenase, 1.1.2.3.)
	b_3	Plants
	b_5	Microsomes
	b_6	Plant chloroplasts
	b_7	Arum spadix
	B	(559, Streptomyces)
	c	Widely distributed
	c_1	Widely distributed
	c_2	Rhodospirillum rubrum
	c_3	Desulphovibrio
	c_4	Azotobacter, etc.
	c_5	Azotobacter
	C	(555, plant chloroplasts). Also known as Cytochrome f
	C	(556, Helix pomatia). Also known as Cytochrome h
	C	(554, halotolerant bacteria). Also known as Cytochrome b_4
	C	(552, Chromatium)
	C	(554, Chlorobium)
	C	(550, Rhodospheroides)
	C	(550, Rhodopseudomonas)
	CD	(625, 553, 548, Pseudomonas aeruginosa) Also known as Cytochrome GB
	d	E. coli, etc. Also known as Cytochrome a_2
	d_1	Acetobacter peroxidans Also known as Cytochrome a_4

"Appendix D".

Key to Numbering and Classification of Enzymes.

1. Oxidoreductases

1.1 Acting on the CH—OH group of donors

    1.1.1 With NAD or NADP as acceptor
    1.1.2 With a cytochrome as an acceptor
    1.1.3 With O_2 as acceptor
    1.1.99 With other acceptors

1.2 Acting on the aldehyde or keto-group of donors

    1.2.1 With NAD or NADP as acceptor
    1.2.2 With a cytochrome as an acceptor
    1.2.3 With O_2 as acceptor
    1.2.4 With lipoate as acceptor
    1.2.99 With other acceptors

1.3 Acting on the CH—CH group of donors

    1.3.1 With NAD or NADP as acceptor
    1.3.2 With a cytochrome as an acceptor
    1.3.3 With O_2 as acceptor
    1.3.99 With other acceptors

1.4 Acting on the CH—NH_2 group of donors

    1.4.1 With NAD or NADP as acceptor
    1.4.3 With O_2 as acceptor

1.5 Acting on the C—NH group of donors

    1.5.1 With NAD or NADP as acceptor
    1.5.3 With O_2 as acceptor

1.6 Acting on $NADH_2$ or $NADPH_2$ as donor

    1.6.1 With NAD or NADP as acceptor
    1.6.2 With a cytochrome as an acceptor
    1.6.4 With a disulphide compound as acceptor
    1.6.5 With a quinone or related compound as acceptor
    1.6.6 With a nitrogenous group as acceptor
    1.6.99 With other acceptors

1.7 Acting on other nitrogenous compounds as donors

    1.7.3 With O_2 as acceptor
    1.7.99 With other acceptors

1.8 Acting on sulphur groups of donors

    1.8.1 With NAD or NADP as acceptor
    1.8.3 With O_2 as acceptor
    1.8.4 With a disulphide compound as acceptor
    1.8.5 With a quinone or related compound as acceptor
    1.8.6 With a nitrogenous group as acceptor

1.9 Acting on haem groups of donors

    1.9.3 With O_2 as acceptor
    1.9.6 With a nitrogenous group as acceptor

1.10 Acting on diphenols and related substances as donors

    1.10.3 With O_2 as acceptor

1.11 Acting on H_2O_2 as acceptor

1.98 Enzymes using H_2 as reductant

1.99 Other enzymes using O_2 as oxidant

    1.99.1 Hydroxylases
    1.99.2 Oxygenases

2. Transferases

2.1 Transferring one-carbon-groups

    2.1.1 Methyltransferases
    2.1.2 Hydroxymethyl-, formyl- and related transferases
    2.1.3 Carboxyl- and carbamoyltransferases

2.2 Transferring aldehydic or ketonic residues

2.3 Acyltransferases

    2.3.1 Acyltransferases
    2.3.2 Aminoacyltransferases

2.4 Glycosyltransferases

    2.4.1 Hexosyltransferases
    2.4.2 Pentosyltransferases

2.5 Transferring alkyl or related groups

2.6 Transferring nitrogenous groups

    2.6.1 Aminotransferases
    2.6.2 Amidinotransferases
    2.6.3 Oximinotransferases

2.7 Transferring phosphorus-containing groups

    2.7.1 Phosphotransferases with an alcohol group as acceptor
    2.7.2 Phosphotransferases with a carboxyl group as acceptor
    2.7.3 Phosphotransferases with a nitrogenous group as acceptor
    2.7.4 Phosphotransferases with a phospho-group as acceptor
    2.7.5 Phosphotransferases, apparently intramolecular
    2.7.6 Pyrophosphotransferases
    2.7.7 Nucleotidyltransferases
    2.7.8 Transferases for other substituted phospho-groups

2.8 Transferring sulphur-containing groups

    2.8.1 Sulphurtransferases
    2.8.2 Sulphotransferases
    2.8.3 CoA-transferases

3. Hydrolases

3.1 Acting on ester bonds

    3.1.1 Carboxylic ester hydrolases
    3.1.2 Thiolester hydrolases
    3.1.3 Phosphoric monoester hydrolases
    3.1.4 Phosphoric diester hydrolases
    3.1.5 Triphosphoric monoester hydrolases
    3.1.6 Sulphuric ester hydrolases

3.2 Acting on glycosyl compounds

3.2.1 Glycoside hydrolases
3.2.2 Hydrolysing N-glycosyl compounds
3.2.3 Hydrolysing S-glycosyl compounds

3.3 Acting on ether bonds

3.3.1 Thioether hydrolases

3.4 Acting on peptide bonds (peptide hydrolases)

3.4.1 α-aminopeptide aminoacido-hydrolases
3.4.2 α-carboxypeptide aminoacido-hydrolases
3.4.3 Dipeptide hydrolases
3.4.4 Iminopeptide hydrolases
3.4.5 Prolinepeptide hydrolases
3.4.6 Peptide peptidohydrolases

3.5 Acting on C−N bonds other than peptide bonds

3.5.1 In linear amides
3.5.2 In cyclic amides
3.5.3 In linear amidines
3.5.4 In cyclic amidines
3.5.99 In other compounds

3.6 Acting on acid-anhydride bonds

3.6.1 In phosphoryl-containing anhydrides

3.7 Acting on C−C bonds

3.7.1 In ketonic substances

3.8 Acting on halide bonds

3.8.1 In C-halide compounds
3.8.2 In P-halide compounds

3.9 Acting on P−N bonds

4. Lyases

4.1 Carbon-carbon lyases

4.1.1 Carboxy-lyases
4.1.2 Aldehyde-lyases
4.1.3 Ketoacid-lyases

4.2 Carbon-oxygen lyases

4.2.1 Hydro-lyases
4.2.99 Other carbon-oxygen lyases

4.3 Carbon-nitrogen lyases

4.3.1 Ammonia-lyases
4.3.2 Amidine-lyases

4.4 Carbon-sulphur lyases

4.5 Carbon-halide lyases

5. Isomerases

5.1 Racemases and epimerases

5.1.1 Acting on aminoacids and derivatives
5.1.2 Acting on hydroxyacids and derivatives
5.1.3 Acting on carbohydrates and derivatives

5.2 *Cis-trans* isomerases

5.3 Intramolecular oxidoreductases

5.3.1 Interconverting aldoses and ketoses
5.3.2 Interconverting keto- and enol-groups
5.3.3 Transposing C=C bonds

5.4 Intramolecular transferases

5.4.1 Transferring acyl groups
5.4.2 Transferring phosphoryl groups
5.4.99 Transferring other groups

5.5 Intramolecular lyases

6. Ligases

6.1. Forming C−O bonds

6.1.1 Aminoacid-RNA ligases

6.2 Forming C−S bonds

6.2.1 Acid-thiol ligases

6.3 Forming C−N bonds

6.3.1 Acid-ammonia ligases (amide synthetases)
6.3.2 Acid-aminoacid ligases (peptide synthetases)
6.3.3 Cyclo-ligases
6.3.4 Other C−N ligases
6.3.5 C−N ligases with glutamine as N-donor

6.4 Forming C−C bonds

Sachverzeichnis.

(Deutsch – Englisch).

Wegen allgemeiner Stichworte wie Extraktion, Abtrennung, Reinigung usw. der einzelnen Stoffgruppen vergleiche man auch das Inhaltsverzeichnis am Anfang dieses Bandes.
cis-, trans-, n, D-, L- und ähnliche Isomere sind unter dem Anfangsbuchstaben der Verbindung und nicht unter dem Präfix eingeordnet.
Alle Iso-Verbindungen finden sich unter Iso-.
Ä, Ö, Ü sind wie Ae, Oe, Ue eingereiht.
Bei gleicher Schreibweise in beiden Sprachen sind die Verbindungen jeweils einfach aufgeführt.

Acetat-aktivierende Enzyme, Chloroplasten, *acetate activating enzymes, chloroplasts* 570.
Aceto-CoA-kinase 636
Acetoin, Synthese, *acetoin, synthesis* 631.
Acetolactatsynthetase, *acetolactate synthetase* 345—346
Acetyl CoA 616, 622
α-N-Acetyl-D-Glucosamin L-Phosphat, *α-N-acetyl-D-glucosamine L-phosphate* 478.
Acetylornithinase 336—337
Acetylornithin δ-Transaminase, *acetylornithine δ-transaminase* 336
Acetylphosphat, *acetyl phosphate* 193, 616, 617
—, Bestimmung, *determination* 193.
Aconitase 617, 618—620
—, Chloroplasten, *chloroplasts* 570
Aconitat, Bestimmung, *aconitate, determination* 619.
Acyloin-Reaktion, *acyloin reaction* 205.
Adenosin, *adenosine* 405.
Adenosylmethionin-Homocystein Transmethylase, *adenosylmethionine-homocysteinetransmethylase* 323—324.
Adenosylmethionin-Nicotinsäure Transmethylase, *adenosylmethionine-nicotinic acid transmethylase* 324—325.
S-Adenosylmethionin, *S-adenosylmethionine* 321
S-Adenosylmethionindethioadenosylase, *S-adenosylmethionine dethioadenosylase* 322
AISCAR (5-Amino-4-imidazol-N-succinocarboxamidribotid), *AISCAR (5-amino-4-imidazole-N-succino-carboxamide ribotide)*
—, Aktivitätsbestimmung, *assay* 437.
—, Darstellung, *preparation* 437.
—, enzymatische Spaltung, *enzymatic cleavage* 437.
—, Synthese, *synthesis* 436.
α-Alanin-β-Alanintransaminase, *α-alanine-β-alanine transaminase* 372—373.
β-Alaninglutamattransaminase, *β-alanine-glutamate transaminase* 371.
Alanin-Glutamattransaminase, Chloroplasten, *alanine-glutamate transaminase, chloroplasts* 570.

α-Alaninglycintransaminase, *α-alanine-glycine transaminase* 371—372.
Alanin-Ketomalonattransaminase, *alanine-ketomalonate transaminase* 380.
Alaninracemase, *alanine racemase* 381—383.
L-Alanindehydrogenase, *L-alanine dehydrogenase* 294—295.
Aldehyd Dehydrogenase, *aldehyde dehydrogenase* 467.
Aldehyde, langkettige, *aldehydes, long-chain* 467.
Aldehydoxydasen, Nitratreduktion, *aldehyde oxidases, nitrate reduction* 110.
Aldolase 530.
—, Aktivatoren und Hemmstoffe, *activators and inhibitors* 532.
—, Bestimmung der Aktivität, *assay* 530.
—, —, chemische, *chemical* 531.
—, —, enzymatische, *enzymatic* 531.
—, Vorkommen, *distribution* 533.
Aldolkondensation, *aldol condensation* 205.
Alkoholdehydrogenase, *alcohol dehydrogenase* 631.
Alliinase 315, 316.
D-Altronsäure-γ-Lakton, *D-altronic acid-γ-lactone* 207.
Ameisensäure, *formic acid* 687.
Ameisensäuredehydrogenase, *formic dehydrogenase* 643.
Amethopterin 190.
Amidasen, *amidases* 412, 413.
Amidsynthese, *amide synthesis* 393.
γ-Aminobutyrat-Glutamattransaminase, *γ-aminobutyrate-glutamate transaminase* 373—374.
5-Amino-4-Imidazolcarboxamidribotid-transformylase, Versuch und Darstellung, *5-amino-4-imidazolecarboxamide ribotide transformylase, assay and preparation* 438.
5-Aminoimidazolribotid, Synthese, *5-aminoimidazole ribotide, synthesis* 436.
5-Aminoimidazolribotidcarboxylase, *5-aminoimidazole ribotide carboxylase* 436.
5-Amino-4-Imidazol-N-Succinocarboxamid Ribotid s. AISCAR, *5-amino-4-imidazole-N-succinocarboxamide ribotide see AISCAR*

δ-Aminolaevulinsäuretransaminase, *δ-aminolaevulinic transaminase* 379.

2-Amino-4-oxy-6F-ormylpteridin, *2-amino-4-hydroxy-6-formylpteridine* 198.

Aminopterin 190.

Aminosäure abhängiger Pyrophosphat-ATP-Austausch, *amino acid dependent pyrophosphate-ATP exchange* 406.

— aktivierende Enzyme, Chloroplasten, *activating enzymes, chloroplasts* 571.

— C-S Spaltungsenzyme, *C-S cleaving enzymes* 313.

Aminosäuredeaminasen, *amino acid deaminases* 309.

Aminosäuredecarboxylasen, höhere Pflanzen, *amino acid decarboxylases, higher plants* 302.

—, Mikroorganismen, *microorganisms* 304.

Aminosäureoxydasen, *amino acid oxidases* 688.

Aminosäurereduktase, *amino acid reductase* 317, 318.

Aminosäure-sRNAsynthese, *amino acid-sRNA synthesis* 406.

Aminosäuren aktivierende Enzyme, *amino acid-activating enzymes* 405.

Aminosäuretransacetylase, *amino acid transacetylase* 336.

Aminosäuretransaminase, verzweigte Kette, *amino acid transaminase, branched-chain* 347.

D-Aminosäure-Apooxydase, *D-amino acid apooxidase* 185.

D-Aminosäureoxydase, *D-amino acid oxidase* 186

—, Mikroorganismen, *microorganisms* 297, 298.

DAminosäureoxydasen, *D-amino acid oxidases* 296.

D-Aminosäuretransaminasen, *D-amino acid transaminases* 362—363.

L-Aminosäureoxydase, *L-amino acid oxidase* 642.

—, Mikroorganismen, *microorganisms* 298—300.

L-Aminosäureoxydasen, *L-amino acid oxidases* 296.

L-Aminosäuretransaminase, *L-amino acid transaminases* 363.

Aminosäuren, verzweigte Kette, Abbau, *amino acids, branched-chain, degradation* 347.

—, —, Synthese, *synthesis* 343.

Aminotripeptidase 414, 415

Aminoxydase, Vorkommen in Erbsensämlingen, *amine oxidase, of pea seedlings* 300—302.

Aminoxydasen, *amine oxidases* 300.

Ammoniak-Bestimmung, CONWAY-Methoden *ammonia, determination*, CONWAY-methods 126.

—, —, NESSLERs Reagens, NESSLER *reagent* 127.

—, —, RUSSELL-Methode, RUSSELL *method* 127.

Ammoniak, Oxydation, *oxidation* 158—160.

—, Nachweismethoden, *assay methods* 126.

Amylopektin, *amylopectin* 510, 516.

Amylose 510.

Amytal (5-Äthyl-5-isoamyl-barbiturat) 655, 656, 661.

Ananasenzym, *pineapple enzyme* 255.

Antimycin A 588, 653, 655, 660.

Apfelsäuredehydrogenase,*malic dehydrogenase* 617, 628.

—, Chloroplasten, *chloroplasts* 570.

Apfelsäureenzym, *malic enzyme* 635.

L-Arabinokinase, aus Phaseolus aureus, Aktivitätsbestimmung, Darstellung und Eigenschaften, *L-arabinokinase, from Phaseolus aureus, assay, preparation and properties* 484.

β-L-Arabinose 1-phosphat, *β-L-arabinose 1-phosphate* 478, 485, 486.

D-Araboascorbinsäure, *D-arabo-ascorbic acid* 218.

Arginase 341—342.

Arginin, Abbau, *arginine, degradation* 342.

Arginindecarboxylase, *arginine decarboxylase* 307.

Arginindesiminase, *arginine desiminase* 342.

Argininosuccinase 340—341.

Argininosuccinatsynthetase,*argininosuccinate synthetase* 339—340.

Arsenolyse, *arsenolysis* 393.

Asclepain, Reinigung, *asclepain, purification* 417.

Ascorbase 218.

Ascorbicase 218.

Ascorbinase 218.

L-Ascorbinsäure, *L-ascorbic acid* 204.

Ascorbinsäure, oxydierende Enzyme, *ascorbic acid, oxidising enzymes* 226.

—, reduzierende Enzymsysteme, *reducing enzymes* 226.

Ascorbinsäuredehydrase, *ascorbic acid dehydrase* 218.

Ascorbinsäureoxydase, *ascorbic acid oxidase* 218, 680, 681

—, Aktivitätsbestimmungen, *activity determination* 224.

—, Eigenschaften, *properties* 221.

—, Kupferbestimmungsmethode, *copper determination method* 220.

—, Nachweis und Reinigung, *assay and purification* 684.

—, Sauerstoffaffinität, *oxygen affinity* 650.

—, Titration mit N-Bromsuccinimid, *titration with N-bromosuccinimide* 224.

—, Vorkommen, *occurence* 225.

Asparaginase 413, 414.

Asparaginsäure β-Semialdehyddehydrogenase *aspartic β-semialdehyde dehydrogenase* 326—327.

Asparaginsynthetase, *asparagine synthetase* 400.

Asparagintransaminasen, *asparagine transaminases* 377—378.

Aspartase 309.

Aspartat-Glutamattransaminase, Chloroplasten, *aspartate-glutamate transaminase,
chloroplasts* 570.
β-Aspartokinase 325—326.
„Assimilationskraft", *"assimilatory power"*
576.
Atmungskette, *respiratory chain* 647.
—, —, Fluorimetrie, *fluorimetry* 659.
—, —, Gebrauch von Cytochrom-Hemmstoffen, *use of cytochrome inhibitors* 654.
—, —, — von Flavoprotein-Hemmstoffe,
flavoprotein inhibitors 655.
—, —, — von Oxydase-Hemmstoffen, *oxidase inhibitors* 653.
—, —, Spektrophotometrie, *spectrophotometry* 656.
—, —, Spektrophotometrie, Instrumente,
spectrophotometry, instruments 658.
—, Untersuchung mittels Differenzspektra
studies, difference spectra 657, 659.
„Atmungskontrolle", *"respiratory control"*
652.
Atmungsintensität, Messung, manometrische,
respiratory rate, measurement, manometric
648.
—, —, polarographische, *polarographic* 648.
—, —, spektrophotometrische, *spectrophotometric* 649.
—, —, volumetrische, *volumetric* 648.
ATP (Adenosintriphosphat), *ATP (adenosine triphosphate)*.
ATP-Shikimisäuretransphosphorylase,
ATP-shikimic acid transphosphorylase
272.
Azid, *azide* 654.

Bakterien, Enzyme der Nitrifikation, *bacteria,
enzymes of nitrification* 161.
—, — —, Aktivitätsmessung, *measurement
of activity* 164.
—, — —, Cofaktoren, *co-factors* 164.
—, — —, Darstellung, *preparation* 161.
—, — —, Eigenschaften und Mechanismen,
properties and mechanisms 166.
—, — —, Extraktionsmethoden, *methods of
extraction* 162.
—, — —, Fraktionierung, *fractionation* 163.
Bakteriochlorophyll, Bestimmung, *bacteriochlorophyll, determination* 573.
—, Chromatophoren, *chromatophores* 572.
p-Benzochinonreduktase, *p-benzoquinone
reductase* 665.
Bernsteinsäure, Bildung aus Glutaminsäure,
*succinic acid, formation from glutamic
acid* 347—348.
Bernsteinsäuredehydrogenase, *succinic
dehydrogenase* 624—627.
Bernsteinsäuresemialdehyd-Dehydrogenase,
succinic semialdehyde dehydrogenase 348.
Bis-dehydroconiferylalkohol, *bis-dehydro-
coniferyl alcohol* 283, 284.
Biosynthese, aromatische, in Mikroorganismen, *biosynthesis, aromatic, in microorganisms* 260.
Biotin 199

D-Biotin Oxydase, *d-biotin oxidase* 199.
Brenztraubensäure-(de)carboxylase, *pyruvic
(de)carboxylase* 630—632.
Brenztraubensäureoxydase, *pyruvic oxidase*
621.
Butyryl-CoA 637.

Carbamatkinase, *carbamate kinase* 338.
Carbamylphosphat-Aspartattranscarba-
mylase, *carbamyl phosphate-aspartate
transcarbamylase* 441.
—, Nachweis und Darstellung, *assay and
preparation* 442.
Carbamylphosphat synthesierende Enzyme,
carbamyl phosphate synthesizing enzymes
337—338.
Carbamylphosphatsynthetase, Darstellung
*carbamyl phosphate synthetase, prepara-
tion* 440
—, Nachweis, *assay* 439.
Carboxydismutase 579.
—, Bestimmung der Aktivität, *assay* 579.
—, Eigenschaften, *properties* 581—582.
—, Reinigung, *purification* 579, 580.
Carboxypeptidase 414.
Carnosinase 414.
Cellulose 476.
Chinasäure, *quinic acid* 260, 261, 268.
Chinasäuredehydrogenase, *quinic dehydro-
genase* 268—270
Chinone, bei der Photosynthese, *quinones,
in photosynthesis* 605.
Chinonreduktase, *quinone reductase* 664, 665.
Chitin 508.
—, Synthese, *synthesis* 509.
Chlorogensäure, *Chlorogenic acid* 666, 682, 683.
Chloroplasten, *chloroplasts* 570, 571.
—, ganze, Isolierung, *whole, isolation* 571.
—, —, Reinigung, *purification* 571.
—, Bruchstücke, Darstellung, „*broken*", *pre-
paration* 572.
—, isolierte, Photosynthese, *isolated, photo-
synthesis* 570.
—, nichtzyklische Photophosphorylierung in,
noncyclic photophosphorylation in 587, 588.
—, Protein-Bestandteile, *protein constituents*
589.
—, TPN-Reduziersysteme, *TPN-reducing
system* 589.
—, zyklische Photophosphorylierung in,
cyclic photophosphorylation in 584.
Chloroplastenextrakt, *chloroplast extract* 572.
Chromatium, Chromatophoren, *chromatophores* 572.
Chromatophoren, *chromatophores* 572.
—, *Chromatium*, Bakterienkultur, *chromatium, bacterial culture* 572.
—, —, Darstellung aus Zellenextrakten, *preparation of cell extracts* 573.
—, —, Isolierung und Reinigung, *isolation
and purification* 572.
—, nichtzyklische Photophosphorylierung in,
noncyclic photophosphorylation in 588.
—, Protein-Bestandteile, *protein constituents*
589.

Chromatophoren, *Rhodospirillum rubrum*, Isolierung und Reinigung, *Rhodospirillum rubrum, isolation and purification* 574.
—, zyklische Photophosphorylierung in, *cyclic photophosphorylation in* 585.
Chymopapain, Reinigung, *purification* 417.
Citratbildung, *citrate, formation* 616.
Citrullinasesystem, *citrullinase system* 343.
Citrullinureidase, *citrulline ureidase* 343.
CoA 198.
—, Abbau, *degradation* 198.
CoA-3′-nucleotidase 198.
CoA-peptidase 199.
CoA-pyrophosphatase 199.
Coenzyme A s. CoA.
Coenzym Q s. Ubichinon, *coenzyme Q s. ubiquinone* 605, 612, 665.
Coenzyme, Nomenklatur, *coenzymes, nomenclature* 696.
$C^{14}O_2$ Fixierung, Produkte, $C^{14}O_2$ *fixation, products* 576.
Coniferylaldehyd, *coniferyl aldehyde* 283, 284.
Coniferylalkohol, *coniferyl alcohol* 283, 285.
—, Dehydrogenierung, *dehydrogenation* 285, 286.
—, gekoppelte Oxydation, *coupled oxidation* 287, 288.
$C^{14}P$, Darstellung, $C^{14}P$, *preparation* 442.
p-Cumarinsäure, *p-coumaric acid* 278.
Cyanid, *cyanide* 654.
Cystathionase 315.
D-Cysteindesulfhydrase, D-*cysteine desulfhydrase* 313.
L-Cysteindesulfhydrase, L-*cysteine desulfhydrase* 312.
Cysteinsulfinattransaminase, *Cysteinesulfinate transaminase* 369—371.
Cytochrom b, *cytochrome b* 671.
Cytochrom b_2, *cytochrome b_2* 671.
Cytochrom b_3, *cytochrome b_3* 671.
Cytochrom b_5, *cytochrome b_5* 670, 672.
Cytochrom b_6, *cytochrome b_6* 600, 602, 672.
Cytochrom b_7, *cytochrome b_7* 672.
Cytochrom c, *cytochrome c* 217, 672, 673.
Cytochrom c_1, *cytochrome c_1* 672.
Cytochrom c_2, *cytochrome c_2* 602.
—, Darstellung, *preparation* 603.
—, Eigenschaften, *properties* 603, 604.
Cytochrom f, *cytochrome f* 600, 672.
—, Darstellung, *preparation* 600.
—, Eigenschaften, *properties* 601, 602.
Cytochrom 552, *cytochrome 552* 605.
Cytochrom 553, Darstellung, *cytochrome 553, preparation* 600, 601.
Cytochrome, Bakterien, *cytochromes, bacterial* 602.
—, in Blättern und Algen, *in leaves and algae* 599, 600.
—, Chromatophoren, *chromatophores* 572.
—, Identifizierung, spektrophotometrische, *identification, spectrophotometric* 662.
—, Klassifikation und Nomenklatur, *classification and nomenclature* 696, 699.
Cytochrome b, *cytochromes b* 668—672.
—, Aktivität, *activity* 670.

Cytochrome, Eigenschaften, *properties* 669.
—, Reinigung, *purification* 670.
—, Bestimmung der Aktivität, *assay* 669, 670.
Cytochrome c, *cytochromes c* 672—675.
—, Eigenschaften, *properties* 674.
—, Reinigung, *purification* 673, 674.
—, Bestimmung der Aktivität, *assay* 673.
Cytochromoxydase (a—a_3), *cytochrome oxidase (a—a_3)* 675.
Cytochromoxydase, *cytochrome oxidase* 217, 650.
—, Azid-Hemmung, *azide inhibition* 654.
—, Cyanid-Hemmung, *cyanide inhibition* 654.
—, Eigenschaften, *properties* 677.
—, Kohlenmonoxyd-Hemmung, *carbon monoxide inhibition* 654.
—, Nachweis, *assay* 675.
—, Reinigung, *purification* 676.
Cytochrom c Oxydase, Sauerstoffaffinität, *cytochrome c oxidase, oxygen affinity* 650.
Cytochrom c Reduktase, *cytochrome c reductase* 665—667.
—, Hemmung, *inhibition* 667.
Cytochromsysteme, Hemmung, *cytochrome system, inhibition* 654.

Deaminase, dehydrierende, *deaminases, dehydrative* 310.
Decarboxylierung, Aminosäuren, *amino acid decarboxylation* 302.
Dehydrasen, *dehydrases* 310.
L-Dehydroascorbinsäure, L-*dehydroascorbic acid* 204, 229.
—, 2,4-Dinitrophenylhydrazin-Methode, *2,4-dinitrophenylhydrazine method* 230.
—, Vorkommen, *distribution* 233.
Dehydrochinasäure, *dehydroquinic acid* 260, 268.
5-Dehydrochinasäure, *5-dehydroquinic acid* 260, 261, 262, 265, 267.
5-Dehydrochinase, *5-dehydroquinase* 267, 268.
Dehydrodiconiferylalkohol, *dehydro-diconiferyl alcohol* 283, 284.
Dehydro-di-pinoresinol 283, 284.
Dehydroshikimisäure, *dehydroshikimic acid* 260.
5-Dehydroshikimisäure, *5-dehydroshikimic acid* 262, 265, 267.
Dehydroshikimisäuredehydrase, *dehydroshikimic dehydrase* 265, 266.
Dehydroshikimisäurereduktase, *dehydroshikimic reductase* 268.
5-Dehydroshikimisäurereduktase, *5-dehydroshikimic reductase* 270—272.
Dephospho-CoA 189, 199.
Dephospho-CoA kinase 194.
Dephospho-CoA pyrophosphorylase 192, 199.
DesamidoDPN 189.
Desamido-DPN-Bildung, *desamido-DPN formation* 188.
Desaminierung, nichtoxydative, *deamination, non-oxidative* 309.
—, oxydative, *oxidative* 291.
Desulfhydrasen, *desulfhydrases* 312.

45

Diaceton-2-keto-L-gulonsäure, *diacetone-2-keto-L-gulonic acid* 211.
Diaminopimelinsäuredecarboxylase, *diaminopimelic acid decarboxylase* 308.
α-ε-Diaminopimelinsäureracemase, *α-ε-diaminopimelic acid racemase* 388, 389.
Diaphorase 664.
Dihydrofolsäurereduktase, *dihydrofolic acid reductase* 190.
Dihydroliponsäuredehydrogenase, *dihydrolipoic dehydrogenase* 622.
Dihydroorotase 442.
—, Nachweis und Darstellung, *assay and preparation* 443.
Dihydroorotsäuredehydrogenase, Nachweis und Darstellung, *dihydroorotic acid dehydrogenase, assay and preparation* 444.
Dimethylpropionthetindethiomethylase 314.
6,7-dimethyl-8-ribityllumazine 176.
α, β-Dioxysäuredehydrase, *α, β-dihydroxy acid dehydrase* 346, 347.
Diphosphatase, saure, *Escherichia coli, diphosphatase, acid, Escherichia coli* 39.
C—l-Diphosphatasen, *C—1-diphosphatases* 36.
Diphosphopyridinnucleotid (DPN), *diphosphopyridinenucleotide (DPN)* 187, 196, 197.
Disaccharide, Synthese, *disaccharides, synthesis* 498.
DPN s. Diphosphopyridinnucleotid, *DPN s. diphosphopyridinenucleotide*
DPN-ase 197.
DPN-kinase 189.
DPN-pyrophosphorylase 188.
DPN-triosephosphat-dehydrogenase, *DPN triosephosphate dehydrogenase* 535.
—, Hemmstoffe, *inhibitors* 536.
—, Nachweis, *assay* 535.
—, Vorkommen, *distribution* 536.
DPN-synthetase 188, 189.
DPNH (reduziertes Diphosphopyridin nucleotid), *DPNH (reduced diphosphopyridine nucleotide)*
DPNH-dehydrogenase 663, 664.
DPNH-oxydase, Mitochondrien, *DPNH oxidase, mitochondria* 649.
—, *S. faecalis*, Sauerstoffaffinität, *S. faecalis, oxygen affinity* 650.
—, Nachweis, *assay* 649, 651.
DPNH-oxydierende Lipoflavoprotein, *DPNH-oxidising lipoflavoprotein* 666.

Endgruppenatmung, Enzyme der, *terminal respiration, enzyme of* 647.
Endgruppenoxydase, *terminal oxidase* 653.
—, Identifikation, *identification* 650.
Endonucleasen, *endonucleases* 62.
Enolase, 2-phosphoglycerat Dehydrase, *enolase, 2-phosphoglycerate dehydrase* 541.
—, —, Aktivatoren und Hemmstoffe, *activators and inhibitors* 542.
—, —, Nachweis, *assay* 541.
—, —, Vorkommen, *distribution* 542.
3-Enolpyruvylshikimat, *3-enolpyruvyl shikimate* 260, 262.

3-Enolpyruvylshikimat-5-phosphat (ESP)) *3-enolpyruvylshikimate-5-phosphate(ESP)* 261, 262.
3-Enolpyruvylshikimisäure, *3-enolpyruvyl shikimic acid* 261.
D-Enzyme, aus Kartoffeln, *D-enzyme, from potatoes* 515, 516.
Q-Enzym, aus Kartoffeln, *Q-enzyme, from potatoes* 514, 515.
Enzymbildung, Terminologie, *enzyme formation, terminology* 698.
Enzymeinheiten, *enzyme units* 695.
—, Aktivität katalytischer Zentren *catalytic centre activity* 695.
—, Molekularaktivität, *molecular activity* 695.
—, spezifische Aktivität, *specific activity* 695.
Enzymkinetik, *enzyme kinetics* 695, 699.
Enzymterminologie, *enzyme terminology* 695.
Enzyme, kondensierende, Chloroplasten, *enzymes, condensing, chloroplasts* 570.
—, intrazelluläre Lokalisierung, *intracellular localization* 29.
—, Klassifikation, *classification* 697, 700.
DL-Epipinoresinol 283, 284.
5-Epishikimisäure, *5-epishikimic acid* 271.
Erbsenzym, *pea enzyme* 253.
ESP (3-Enolpyruvylshikimat-5-phosphat, *ESP (3-enolpyruvyl shikimate-5-phosphate)*.
—, enzymatische Bildung, *enzymic formation* 273.
Esterasen, *esterases* 448.
Ethanolaminphospholipoprotein, *ethanolamine phospholipoprotein* 572.

FAD s. Flavinadenindinucleotid, *FAD see Flavin adenine dinucleotide*
FAD-ase 196, 199.
FAD-pyrophosphorylase 185.
Ferredoxin 592.
—, Bakterien, *bacterial* 593.
—, Identität mit PPNR, *identity with PPNR* 593.
—, Isolierung und Reinigung, *isolation and purification* 593—596.
—, physiologische Rolle, *physiological role* 593.
Ferredoxin-TPN-reduktase, *ferredoxin-TPN reductase* 596.
—, Aktivitätsnachweis, *assay* 598.
—, Eigenschaften, *properties* 598.
—, Isolierung und Reinigung, *isolation and purification* 597.
—, Krystallisationsmethode, *crystallization procedure* 599.
Ferricytochromoxydase, *ferricytochrome oxidase* 654.
Fettsäureperoxydase, *fatty acid peroxidase* 467.
Fettsäuren, enzymatische Synthese, *fatty acids, enzymatic synthesis* 472.
—, — —, Nachweis, *assay* 473.
—, — —, pH-Optima, *pH optima* 473.
—, — —, Stabilität, *stability* 473.
Ficin, Reinigung, *purification* 417.

Flavinadenindinucleotid (FAD), *flavin adenine dinucleotide (FAD)* 182, 196, 689.

Flavine und Nitratreduktase, *flavins and nitrate reductase* 93.

Flavinmononucleotid (FMN), *flavinmononucleotide (FMN)* 182, 184, 185, 196, 228, 688.

Flavokinase 183.

Flavoproteine,*flavoproteins* 662, 666.

Fluoroacetyl-CoA 637.

FMN s. Flavinmononucleotid, *s. flavinmononucleotide.*

FMN-phosphatase 196.

Folsäure, *folic acid* 179, 189, 198.

Folsäurereduktase, *folic acid reductase* 190.

Formiat, Umwandlung, *formate, transformation* 350.

Formylase 177, 178.

Formylglycinamidinribotidkinosynthase, Darstellung, *formylglycinamidine ribotide kinosynthase, preparation* 435.

Formylkynurenin, *formylkynurenine* 177.

N-Formyl-L-kynurenin, *N-formyl-L-kynurenine* 179.

Fructokinase 547.

Fructose-1,6-diphosphat, *fructose-1,6-diphosphate* 205.

Fructose-1,6-diphosphatase 528.

Fructosediphosphatase, alkalische, *fructose diphosphatase, alkaline* 39.

—, neutrale, *neutral* 39.

—, Pflanzen, *plants* 37.

—, saure, *acidic* 39.

—, Tier, *animal* 36.

Fructose-1,6-diphosphatase, alkalische, Spezifität, *fructose-1,6-diphosphatase, alkaline, specificity* 529.

—, Nachweis, *assay* 528.

—, neutrale, *neutral* 529

—, pH 5,5-Enzym, *pH 5.5 enzyme* 530.

—, pP 6,9-Enzym, *pH 6.9 enzyme* 530.

—, pH 8,5-Enzym, *pH 8.5 enzyme* 530,

—, Reinigung, *purification* 529.

—, Vorkommen, *distribution* 530.

D-Fructose 476.

D-Fructose-6-phosphat, D-*fructose 6-phosphate* 476.

D-Fructuronsäure, D-*fructuronic acid* 208

Fumarase 627, 636.

—, Chloroplasten, *chloroplasts* 570.

L-Galaktoascorbinsäure, L-*galactoascorbic acid* 218.

D-Galaktokinase 478.

— aus *Phaseolus aureus*, Nachweis, Darstellung und Eigenschaften, D-*galactokinase, from Phaseolus aureus, assay, preparation and properties* 484.

— aus Sacharomyces fragilis, Eigenschaften *from Sacharomyces fragilis' properties* 483

— aus Saccharomyces fragilis, Reinigung, *from Saccharomyces fragilis, purification* 482.

— —, Nachweis, *assay* 481.

Galaktolipide, *galactolipids* 571.

D-Galaktonsäure-γ-Lakton, D-*galactonic acid-γ-lactone* 210.

L-Galaktonsäure-γ-Lakton, L-*galactonic acid-γ-lactone* 207, 208, 210, 211.

L-Galaktonsäure-γ-Laktondehydrogenase, L-*galactonic acid-γ-lactone dehydrogenase* 208.

α-D-Galaktose L-Phosphat, α-D-*galactose L-phosphate* 478, 481, 485, 486, 492.

α-D-Galaktose-1-Phosphat-Uridyl-Transferase α-D-*galactose 1-phosphate uridyltransferase* 478.

α-D-Galaktose-1-Phosphaturidyl-Transferase, aus *Saccharomyces fragilis*, α-D-*galactose 1-phosphate uridyl transferase, from Saccharomyces fragilis* 492.

D-Galakturonsäure, D-*galacturonic acid* 208, 212.

D-Galakturonsäurelakton, D-*galacturonic acid lactone* 208.

D-Galakturonsäure-γ-Lakton, D-*galacturonic acid-γ-lactone* 211.

D-Galakturonsäuremethylester, D-*galacturonic acid methyl ester* 208.

α-D-Galakturonsäure-1-Phosphat, α-D-*galacturonic acid 1-phosphate* 478, 485, 486.

GDP (Guanosin-5′-diphosphat), *GDP (guanosine 5′-diphosphate)*.

GDP-D-Mannose 491.

GDP-D-Mannosepyrophosphorylase aus Bierhefe, *GDP-D-mannose pyrophosphorylase, from brewer's yeast* 491—492.

GDP-Zucker, *GDP-sugars* 478.

β-1,3-D-Glucan, 507.

L-Glucoascorbinsäure, L-*glucoascorbic acid* 218

Glucokinase 547.

D-Gluconsäure-γ-Lakton, D-*gluconic acid-γ-lactone* 210.

Glucose-L-C14 207.

Glucose-Cycloacetoacetat, *glucose-cycloacetoacetate* 206.

Glucoseoxydase, *glucose oxidase* 688, 689.

—, Sauerstoffaffinität, *oxygen affinity* 650

Glucose-6-Phosphat, *glucose-6-phosphate* 207.

α-D-Glucose-L-Phosphat, α-D-*glucose-1-phosphate* 474, 478, 485, 486, 492, 510, 511, 512, 513

Glucose-6-Phosphat-Dehydrogenase s. G-6-P-Dehydrogenase, *glucose-6-phosphate dehydrogenase see G-6-P dehydrogenase*

D-Glucuronsäure, D-*glucuronic acid* 208.

D-Glucuronsäurekinase aus *Phaseolus aureus*, Darstellung, D-*glucuronic acid kinase, from Phaseolus aureus, preparation* 483.

— — —, Eigenschaften, *properties* 484

— — —, Nachweis, *assay* 483.

D-Glucuronsäurelakton, D-*glucuronic acid lactone* 207, 211.

D-Glucuronsäure-γ-Lakton, D-*glucuronic acid-γ-lactone* 206, 208, 211, 212.

α-D-Glucuronsäure-1-Phosphat, α-D-*glucuronic acid 1-phosphate* 478, 483, 485, 486.

D-Glutamat, D-*glutamate* 399.
Glutamatalanintransaminase, *Glutamate-alanine transaminase* 368, 369.
Glutamat-Aspartattransaminase, *Glutamate-aspartate transaminase* 363.
—, Bestimmung der Aktivität, *assay* 363.
—, Eigenschaften, *properties* 367, 368.
—, Nachweis, chromatographischer, *assay, chromatographic* 364.
—, —, durch Messung von α-Ketoglutarat, *by measurement of α-ketoglutarate* 365.
— — — von Oxalacetat, *of oxalacetate* 364.
— — mit spezifischen Decarboxylasen, *with specific decarboxylases* 364.
—, Reinigung, *purification* 366, 367.
Glutamatphosphohistidinoltransaminase, *Glutamate-phosphohistidinol transaminase* 375, 376.
Glutamatracemase, *glutamate racemase* 383, 384.
Glutaminase 413.
Glutaminsäuredecarboxylase, *glutamic decarboxylase* 303, 304.
—, *Escherichia coli, glutamic acid decarboxylase, Escherichia coli* 305, 306.
Glutaminsäuredehydrogenase, *glutamic dehydrogenase* 158.
—, Chloroplasten, *chloroplasts* 570.
L-Glutaminsäuredehydrogenase, höhere Pflanzen, L-*glutamic dehydrogenase, higher plants* 291, 292.
—, Mikroorganismen, *microorganisms* 293, 294.
D-Glutaminsäureoxydase, D-*glutamic acid oxidase* 298.
Glutaminsynthetase, *glutamine synthetase* 393.
—, Chloroplasten, *chloroplasts* 570.
—, Elektrophorese, *electrophoresis* 397.
—, optische Spezifität der, *optical specifity of* 399.
—, Reinigung, *purification* 395, 396.
—, Stabilität, *stability* 399.
—, Sedimentationsanalyse, *ultracentrifugal analysis* 397.
Glutamintransaminasen, *glutamine transaminases* 377, 378.
Glutamylcysteinsynthetase, *glutamylcysteine synthetase* 401.
—, Darstellung, *preparation* 402.
γ-Glutamylhydroxamat, γ-*glutamyl hydroxamate* 393.
Glutamyl-Übertragung, *glutamyl transfer* 394.
Glutathionreduktase, *glutathione reductase* 680.
Glutathionsynthetase, *Glutathione synthetase* 403.
Glycerate-2,3-diphosphatase 40.
D-Glycerinaldehyd, D-*glyceraldehyde* 206.
DL-Glycerinaldehyd, DL-*glyceraldehyde* 206.
L-Glycerinaldehyd, L-*glyceraldehyde* 205.
Glycin, Synthese, *glycine, synthesis* 349.
Glycinamidribotidkinosynthase, *glycinamide ribotide kinosynthase* 429.
—, Aktivitätsbestimmung, *assay* 429.
—, Darstellung, *preparation* 431.

Glycinamidribotidtransformylase, *glycinamide ribotide transformylase* 432.
—, Darstellung, *preparation* 434.
—, Aktivitätsbestimmung, *assay* 433.
Glycin-Glutamattransaminase, *glycine-glutamate transaminase* 380.
Glycinoxydase, *glycine oxidase* 642, 643.
Glycinreduktase-Systeme, *glycine reductase system* 318, 319.
Glycolyse, *glycolysis* 520.
Glycolytisches System, Aktivatoren und Hemmstoffe, *glycolytic system, activators and inhibitros* 522.
— —, Darstellung, *preparation* 521.
— —, Nachweis, *assay* 520.
Glycylglycindipeptidase, *glycylglycine dipeptidase* 414.
Glykolaldehyd, *glycolaldehyde* 218.
Glykolaldehyddehydrogenase, *glycolaldehyde dehydrogenase* 642.
Glykolaldehyd, Transformation, *glycolaldehyde, transformation* 352.
Glykolsäureoxydase, *glycolic acid oxidase* 638—640, 686.
—, Aktivitätsbestimmung, *assay* 687.
—, Eigenschaften, *properties* 688.
—, Reinigung, *purification* 687.
Glykoside, Synthese, *glycosides, synthesis* 504.
Glyoxalatcyclus, *glyoxalate cycle* 636.
—, Enzyme des, *enzymes of* 616.
Glyoxalsäure, *glyoxylic acid* 686, 687.
Glyoxalsäurereduktase, *glyoxalic acid reductase* 641, 642.
G-6-P-Dehydrogenase, Reinigung, *G-6-P-dehydrogenase, purification* 554, 555.
—, Aktivitätsbestimmung, *assay* 552—554.
Guajacylglycerol-β-Coniferyläther, *guaiacyl-glycerol-β-coniferyl ether* 283, 284.
Guajacylglycerol-bis-Coniferyläther, *guaia-cylglycerol-bis-coniferyl ether* 283, 284.
Guajacylglycerol-α-Dehydrodiconiferyl-β-Coniferyläther, *guaiacylglycerol-α-de-hydrodiconiferyl-β-coniferyl ether* 283, 284.
Guajacylpropanderivate, *guaiacylpropane derivatives* 283.
Guajakol, *guaiacol* 685.
L-Gulonsäure, L-*gulonic acid* 207, 208, 211, 215.
L-Gulonsäuredehydrogenase, L-*gulonic acid dehydrogenase* 213, 214.
L-Gulonsäurelakton, L-*gulonic acid lactone* 213.
L-Gulonsäure-γ-Lakton, L-*gluconic acid-γ-lactone* 206, 207, 212, 215.

Harnstoffcyclus, Enzyme des, *Urea cycle, enzymes of* 337.
Hemicellulose 476.
Heteropyrithiamin, Bestimmung, *hetero-pyrithiamine, determination* 195.
Hexokinase 547, 525.
—, Aktivatoren und Hemmstoffe, *activators and inhibitors* 526.

Hexokinase, Reinigung, *purification* 551, 552.
—, Aktivitätsbestimmung, *assay* 525, 550, 551.
—, Vorbereitung, *preparation* 525.
Histidase 310.
L-Histidin, Biosynthese, L-*histidine, biosynthesis* 329.
Histidin-Alanin 380.
Histidin-Alanintransaminase 380.
L-Histidinoldehydrogenase 332, 333.
L-Histidinolphosphatphosphatase, L-*histidinol phosphate phosphatase* 332.
Homocysteindesulfhydrase, *homocysteine desulfhydrase* 313.
Homoserindehydrogenase, *homoserine dehydrogenase* 327.
L-Homoserinkinase, L-*homoserine kinase* 327, 328.
Homoserinphosphatmutaphosphatase, *homoserine phosphate mutaphosphatase* 328, 329.
Hydrochinon, *hydroquinone* 683.
α-Hydroxy-β-Ketosäurereduktoisomerase, α-*hydroxy-β-keto acid reductoisomerase* 346.
3-Hydroxykynurenin, *3-hydroxykynurenine* 178.
Hydroxylamin, Bestimmung, *hydroxylamine, estimation* 125.
—, —, CSÁKYs Methode, *determination,* CSÁKY *method* 128.
—, —, 8-Hydroxychinolinmethode, *8-hydroxyquinoline method* 128.
—, —, Nitrobenzolmethode, *nitrobenzene method* 129.
—, —, o-Phenanthrolinmethode, *o-phenanthroline-method* 129.
—, Oxydation, *oxidation* 158—160.
Hydroxylamin-„Oxydase", höhere Pflanzen, *hydroxylamine "oxidase", of higher plants* 160, 161.
Hydroxylaminoxydase, Aktivitätsbestimmung, *hydroxylamine oxidase, assay* 165.
Hydroxylaminoxydasesysteme, Cofaktoren, *hydroxylamine oxidase systems, co-factors* 167.
—, Hemmstoffe, *inhibitors* 168.
—, Stabilität, *stability* 168.
Hydroxylaminreduktase, Darstellung, *hydroxylamine reductase, preparation* 115.
—, Aktivitätsbestimmung, Farbstoffreduktion, *assay, dye reduction* 130—133.
—, Eigenschaften und Mechanismen, Bakterien, *properties and mechanisms, bacteria* 155—157.
—, —, Pilze und höhere Pflanzen, *properties and mechanisms, fungi and higher plants* 146, 147.
—, Elektronendonatoren und Cofaktoren, Bakterien, *electron donators and co-factors, bacteria* 145—146.
—, — —, Pilze und höhere Pflanzen, *fungi and higher plants* 138, 139, 142.
—, Extraktion aus Bakterien, *extraction from bacteria* 117.

Hydroxylaminreduktase, Hemmstoffe, Pilze und höhere Pflanzen, *inhibitors, fungi and higher plants* 140 bis 142.
—, physiologischeFaktoren, Pilze und höhere Pflanzen, *physiological factors, fungi and higher plants* 158.
—, Reinigung, *purification* 118, 120.
—, Nachweismethoden, Fehlerquellen, *assay methods, sources of error* 121—126.
Hydroxylaminreduktasen, *hydroxylamine reductases* 112.
Hydroxylaminreduktasesysteme, Hemmstoffe, *hydroxylamine reductase systems, inhibitors* 154, 155.
—, Stabilität, *stability* 154, 155.
Hyponitrit, Bestimmung, *hyponitrite, determination* 129, 130.
—, Darstellung, *preparation* 136.
Hyponitritreduktase, Darstellung, *hyponitrite reductase, preparation* 114.
—, Elektronendonatoren, Pilze und höhere Pflanzen, *electron donors, fungi and higher plants* 141.
—, Hemmstoffe, Pilze und höhere Pflanzen, *inhibitors, fungi and higher plants* 140 bis 142.
—, Nachweismethoden, Fehlerquellen, *assay methods, sources of error* 121—126.
Hyponitritreduktasen, *hyponitrite reductases* 112.

IAc (3-Indolacetaldehyd), Umbau in Indolessigsäure, *IAc (3-indoleacetaldehyde), conversion to IAA* 243.
Imidazolacetolphosphattransaminase, *Imidazoleacetol phosphate transaminase* 331.
Imidazolglycerolphosphatdehydrase, *Imidazoleglycerol phosphate dehydrase* 329, 331.
Indolessigsäure (IES), *IAA (Indole acetic acid)* 238, 685.
—, Abbau, *breakdown* 248.
—, biologischer Test, *bioassay* 238.
—, colorimetrische Nachweismethoden, *colorimetric assay methods* 238.
—, Oxydationsreaktion, *oxidation reaction* 249.
—, Synthese, *synthesis* 240.
—, Trennung von Vorstufen, *separation from precursors* 239.
Indolessigsäureoxydase, Vorkommen, *IAA oxidase, distribution* 248.
Indolessigsäureoxydasen, Darstellung und Eigenschaften, *IAA oxidases, preparation and properties* 251.
Indolessigsäure oxydierende Enzyme, Spezifität, *IAA oxidising enzymes, specificity* 252.
Indol-3-glycerolphosphatsynthetase, *indole-3-glycerol phosphate synthetase* 353, 354.
Inosinicase 438.
Inositol 200.
Isocitronensäuredehydrogenase, *isocitric dehydrogenase* 189.
—, Chloroplasten, *chloroplasts* 570.
—, DPN-spezifisch, *DPN specific* 621.

Isocitronensäuredehydrygenase, TPN-spezifische, *TPN specific* 620.
Isocitritase 617, 637—638.
Isomeraseprodukt, Darstellung, *isomerase
product, preparation* 558.

Kallose, *Callose* 507, 508.
Karotenoide, Chromatophoren, *carotenoids,
chromatophores* 572.
Karotin, *carotene* 201.
Karotinoxydase, *carotene oxidase* 201.
Kartoffelapyrase, *potato apyrase* 187.
KDHP (2-Keto-3-deoxy-D-Arabohepton-
säure-7-Phosphat), *KDHP (2-keto-3-de-
ɔxy-D-araboheptonic acid-7-phosphate)*
—, Darstellung, *preparation* 263.
—, Eigenschaften, *properties* 264.
—, enzymatischer Umbau in Dehydrochinat,
enzymic conversion to dehydroquinate 265.
—, Identifikation durch Papierchromatographie, *identification by paper chromatography* 263.
KDHP-Synthetase 262, 265.
—, Nachweis, *assay* 262.
KDPA (2-Keto-3-deoxy-7-Phosphohepton-
säure), *KDPA (2-keto-3-deoxy-7-phospho-
heptonic acid)* 262.
2-Keto-3-deoxy-D-Araboheptonsäure (KDA)
2-keto-3-deoxy-D-araboheptonic acid (KDA)
261, 262.
2-Keto-3-deoxy-D-Araboheptonsäure-7-Phos-
phat (KDHP), *2-keto-3-deoxy-D-arabohep-
tonic acid-7-phosphate (KDHP)* 260, 261.
2-Keto-L-Gulonsäure, *2-keto-L-gulonic acid*
208.
2-Keto-Gulonsäure, *2-keto-gulonic acid* 205,
211, 212.
3-Keto-L-Gulonsäure, *3-keto-L-gulonic acid*
212, 213.
α-Ketosäureoxydasen, *α-keto acid oxidases*
621—623.
Kohlendioxydfixierung, Messung, *carbon dioxide fixation, measurement* 576.
Kohlenhydratcyclus, reduzierender, *carbohydrate cycle, reductive* 577.
Kohlenhydratoxydase, *carbohydrate oxidase,*
689—690.
Kohlenhydratsynthese, Enzym der *carbohydrate synthesis, enzymes of* 474.
Kohlenmonoxyd, Cytochromoxydase-Hem-
mung, *carbon monoxide, cytochrome oxidase inhibition* 654.
—, Polyphenoloxydase-Hemmung, *polyphenol oxidase inhibition* 654.
Kondensierende Enzyme, *condensing enzyme,*
616.
KREBS-Cyclus, Enzyme des, KREBS *cycle,
enzymes of* 616.
Kupferoxydasen, *copper oxidases* 650, 681.
L-Kynurenin, L-*kynurenine* 179.
Kynureninase 179.
Kynureninformamidase, *kynurenine formami-
dase* 177, 178.
Kynurenin-Hydroxylase, *kynurenine hydroxy-
lase* 178.

Kynurenintransaminase, *kynurenine transaminase* 374, 375.

Laccase 218, 285, 287, 650, 681.
—, Nachweis und Reinigung. *assay and puri-
fication* 683.
Laktonase, *lactonase* 552.
—, Nachweis und Reinigung, *assay and puri-
fication* 556.
Laktonase I, *lactonase I* 214.
Laktonase II, *lactonase II* 214.
Lecithinase 454.
Leucinaminopeptidase, *leucine aminopepti-
dase* 414.
Leucindecarboxylase, *leucine decarboxylase*
307.
L-Leucindehydrogenase, L-*leucine dehydro-
genase* 295, 296.
Lignin, Bildung aus Shikimisäure, *lignin,
formation from shikimic acid* 278.
—, Coniferensynthese, *coniferous synthesis*
283.
Ligninsynthese, *lignin synthesis* 277, 278.
Lipasen, *lipases* 448.
—, Aktivatoren und Hemmstoffe, *activators
and inhibitors* 450, 451.
—, Definition, *definition* 448.
—, Enzymdarstellung, *enzyme preparation*
449.
—, Nachweismethoden, *assay procedures* 448
bis 450.
—, pH-Optima, *pH optima* 451.
—, Reinigung, *purification* 451, 452.
—, Spezifität, *specificity* 451.
—, Vorkommen, *distribution* 448.
Liponsäure, *lipoic acid* 622.
Lipoxydase, *lipoxidase* 469, 688.
—, Eigenschaften, *properties* 470.
—, Nachweis, *assay* 469.
—, manometrischer Nachweis, *assay, mano-
metric* 469.
—, —, spektrophitometrischer, *spectrophoto-
metric* 469.
—, Reinigung, *purification* 470.
Lipoyldehydrogenase 664.
Lupinenzym, *lupin enzyme* 251.
Lysindecarboxylase, *E. coli, lysine decarboxy-
lase, E. coli* 306.
C—S-Lyase 316, 317.
Lysinracemase, *lysine racemase* 386.
Lysolecithinisomerase, *lysolecithin isomerase*
462.
Lysophospholipase B 456.
Lyxonsäure, *lyxonic acid* 218.

Malatsynthetase, *malate synthetase* 636.
Malease 629, 630.
Maltodextrine, *maltodextrins* 516.
Maltopentaose 515.
Maltotriose 513 .
Malyl-CoA 637.
D-Mannonsäure-γ-Lakton, D-*mannonic acid-
γ-lactone* 207.
α-D-Mannose 1-phosphat, α-D-*mannose
1-phosphate* 478.

D-Mannuronsäure, D-*mannuronic acid* 211.
Melibiose 498.
Menadionreduktase, *menadione reductase* 665
Methämoglobin-Reduzierfaktor, *methemo-globin reducing factor* 590.
Methioninaktivierendes Enzym, *methionine activating enzyme* 321.
Methionindethionmethylase, *methionine dethiomethylase* 314.
Methioninracemase, *methionine racemase* 385 386.
S-Methylmethionin-Homocysteintransme-thylase, *S-methylmethionine-homocysteine transmethylase* 322, 323.
6-Methylaminopurin, *6-methylaminopurine* 405.
5-Methylcytosin, *5-methylcytosine* 405.
Milchsäuredehydrogenase, *lactic dehydro-genase* 623, 632—634.
Myokinase 624.

Naphthochinone, Photophosphorylierung, *naphthoquinones*, *photophosphorylation* 610
Nicotinsäure, *nicotinic acid* 176, 177, 187, 196.
Nicotinsäuremethylpherase, *nicotinic acid methylpherase* 324, 325.
Nitratassimilation, *nitrate assimilation* 67.
Nitratatmung, *nitrate respiration* 68.
Nitratreduktase, *nitrate reductase* 67.
—, Bakterien, *bacteria* 76.
—, —, Assimilations-Enzym, *assimilatory enzyme* 82.
—, —, dissimilierendes Enzym, *dissimilatory enzyme* 78.
—, —, Elektronendonatoren und Cofaktoren, *electron donors and co-factors* 95, 96.
—, —, Fraktionierung und Stabilität, *frac-tionation and stability* 78.
—, —, Stabilität und Hemmstoffe, *stability and inhibitors* 101—107.
—, —, Substrate und Cofaktoren, *substrates and co-factors* 98—101.
—, Darstellung, *preparation* 68.
—, Extraktionsmethoden, *methods of ex-traction* 68.
—, —, Bakterien, *bacteria* 74, 75, 76, 77.
—, —, Pilze, *fungi* 72.
—, —, höhere Pflanzen, *higher plants* 73.
—, Fraktionierung und Stabilität, *fractio-nation and stability* 71.
—, — — —, Elektronendonatoren und Co-faktoren, *electron donors and co-factors* 91—93.
—, — — —, Hemmstoffe, *inhibitors* 96, 97, 102.
—, — — —, Stabilität, *stability* 96, 102.
—, GRIESS-ILOSVAY colorimetrische Methode GRIESS-ILOSVAY *colorimetric method* 83 bis 87.
—, Metall- und Anionenbedarf, *metal and anion requirement* 94.
—, Nachweismethoden, *assay methods* 83.
—, —, Berechnung von Nitratverlust, *esti-mation of nitrate loss* 88.

Nitratreduktase, physiologische Fehlerquel-len, *sources of error* 89—91.
—, —, manometrische Methode, *manometric method* 85.
—, —, spektrophotometrische Methode, *spectrophotometric method* 88.
—, —, Faktoren, Bakterien, *physiological factors, bacteria* 109.
—, — —, Pilze und höhere Pflanzen, *fungi and higher plants* 108.
—, Schwermetalle, *heavy metals* 71.
—, thermische Inaktivierung, *thermal in-activation* 73.
—, Verteilung in zellfreien Extrakten, *distrib-ution in cell-free extracts* 79.
—, Wirkungsmechanismus, Pilze und höhere Pflanzen, *fungi and higher plants, mecha-nism of action* 97, 100, 101.
Nitratreduktion, dissimilatorische, *nitrate re-duction, dissimilatory* 68.
— in Bakterien, *in bacteria* 68.
—, Lichtabhängigkeit, *light dependence* 67.
Nitrit, Oxydation, *nitrite, oxidation* 158 bis 160.
Nitritoxydasesysteme, Co-faktoren, *nitrite oxidase systems, co-factors* 167.
—, Hemmstoffe, *inhibitors* 168.
—, Stabilität, *stability* 168.
Nitritreduktase, Cofaktoren und Hemmstoffe, Bakterien, *nitrite reductase, co-factors and inhibitors, bacteria* 142—144.
—, — —, Pilze und höhere Pflanzen, *fungi and higher plants* 137—139, 141.
—, Darstellung, *preparation* 113.
—, Eigenschaften und Mechanismen, Bak-terien, *properties and mechanisms, bacteria* 148—153.
—, — —, Pilze und höhere Pflanzen, *fungi and higher plants* 146, 147.
—, Elektronendonatoren, Bakterien, *electron donors, bacteria* 142—144.
—, —, Pilze und höhere Pflanzen, *fungi and higher plants* 137—139, 141.
—, Extraktion aus Bakterien, *extraction from bacteria* 116.
—, Nachweis, Farbstoffreduktion, *assay, dye reduction* 130—133.
—, —, Fehlerquellen, *sources of error* 121 bis 126.
—, —, manometrischer, *manometric* 124, 125, 133.
—, physiologische Faktoren, Pilze und höhere Pflanzen, *physiological factors, fungi and higher plants* 157, 158.
—, Reinigung, *purification* 114, 118, 119.
—, Stabilität, *stability* 150, 151.
Nitritreduktasen, *nitrite reductases* 112.
Nitroverbindungen, aromatische, Reduktion *Nitro-compounds, aromatic, reduction* 111, 112.
NMN (Nicotinsäureamidmononucleotid), *NMN (Nicotinamide mononucleotide)*
NMN-Phosphatase, *NMN-phosphatase* 197.
5'-Nucleotidasen, *5'-nucleotidases* 57.

Nucleotidpyrophosphatasen, *nucleotide pyrophosphatases* 55.

Omphalia-Enzym, *Omphalia enzyme* 256.
Ornithin, Synthese, *ornithine, synthesis* 335.
Ornithintransaminase, *ornithine transaminase* 376.
Ornithin-δ-transaminase, *ornithine δ-transaminase* 335.
Ornithintranscarbamylase, *ornithine transcarbamylase* 338, 339.
Orotidin-5'-phosphatdecarboxylase, *orotidine-5'-phosphate decarboxylase* 446.
Orotidin-5'-phosphatpyrophosphorylase, *orotidine-5'-phosphate pyrophosphorylase* 446.
Ortho phosphoserine phosphatase 54.
Oxaloacetat, *oxaloacetate* 617, 618, 628.
Oxalsäure, *oxalic acid* 218.
Oxanthranilsäure, *hydroxyanthranilic acid* 179.
Oxime, Bestimmung, *oximes, determination* 129.
α-Oxoglutaratoxydase, *α-oxoglutarate oxidase* 621.
γ-Oxybutyratdehydrogenase, *γ-hydroxybutyrate dehydrogenase* 348.
Oxychinoline, *hydroxyquinolines* 655.
Oxydase, Fichtensaft, *oxidase, spruce sap* 286, 287.
α-Oxydation, *α-oxidation* 467.
—, Hemmstoffe, *inhibitors* 468.
β-Oxydation, *β-oxidation* 465.
—, geradekettige Fettsäuren, *even-chain fatty acids* 465.
—, ungeradkettige Fettsäuren, *odd-chain fatty acids* 466.
Oxymethyltetrahydrofolatdehydrogenase, *hydroxymethyltetrahydrofolate dehydrogenase* 350.
p-Oxyphenylbrenztraubensäure, *p-hydroxyphenylpyruvic acid* 261, 274.
β-Oxypropionsäure, *β-hydroxypropionic acid* 466.
Oxythiamin, *oxythiamine* 181.

Pantethein, pantetheine 191.
Pantethein 4'-phosphat, *pantetheine 4'-phosphate* 199.
Pantothenatkinate, *pantothenate kinase* 191.
Pantothenatsynthetase, *pantothenate synthetase* 180.
Pantothensäure, *pantothenic acid* 180, 190, 191, 198, 207.
Pantothenylcystein, *pantothenylcysteine* 191.
Pantoyltaurin, *pantoyltaurine* 207.
Papain, Reinigung, *papain, purification* 417.
Pektin, *pectin* 476.
Pentabromoaceton, *pentabromoacetone* 617.
Pentosephosphatcyclus, Enzyme des, *pentose phosphate cycle, enzymes of* 546—549.
Peptidasen, *peptidases* 412, 414.
Peptidsynthese, *peptide synthesis* 401.
Peroxydase, *peroxidase* 216, 285.

Peroxydase, Aktivitätsbestimmung und Reinigung, *assay and purification* 685.
—, Bestimmung, *determination* 216.
—, Eigenschaften, *properties* 686.
Peroxydasen, *peroxidases* 684—686.
Phenazinmethosulfat, *phenazine methosulphate* 625.
Phenoloxydasen, Pilze, *phenoloxidases, mushroom* 285.
L-Phenylalanin, L-*phenylalanine* 281.
Phenylalanin, *phenylalanine* 261.
—, Biosynthese, *biosynthesis* 261.
Phenylalanin-Alanintransaminase, *phenylalanine-alanine transaminase* 380.
Phenylalanindeaminase, *phenylalanine deaminase* 281—283.
Phenylbrenztraubensäure, *phenylpyruvic acid* 261, 274.
Phenylcumaranderivat, *phenylcoumaran derivative* 283.
Phenylmilchsäure, *phenyllactic acid* 274.
Phosphatase, Aktivitätsbestimmung, *phosphatase, assay* 22.
—, —, Orthophosphatwirkung, *orthophosphate effect* 28.
Phosphatasen, alkalische, *phosphatases, alkaline* 51.
—, Lokalisation, *localization* 34.
—, saure, *acid* 51.
Phosphatidsäurephosphatase, *phosphatidic acid phosphatase* 462.
Phosphatstoffwechsel, Enzyme, *phosphate metabolism, enzymes* 21.
—, —, allgemeine Nachweismethoden, *general assay methods* 25—28.
—, —, Isolierung, *isolation* 21.
—, —, Kernfraktion, *nuclear raction* 32.
—, —, "large particles"-Fraktion, *larger particles fraction* 33.
—, —, Mitochondrienfraktion, *mitochondrial fraction* 33.
—, —, Plastidenfraktion, *plastid fraction* 32.
—, —, Reinigung, *purification* 21.
—, Stabilität, *stability* 22.
—, —, Zellwandfraktion, *cell wall fraction* 33.
Phosphatidylglycerol 571.
Phosphodiesterasen, *phosphodiesterases* 62.
Phosphoenolbrenztraubensäurecarboxylase, *phosphoenolpyruvic carboxylase* 634.
Phosphoenolpyruvatcarboxykinase, *phosphoenolpyruvate carboxykinase* 635.
Phosphoenolpyruvatcarboxylase, Chloroplasten, *phosphoenolpyruvate carboxylase, chloroplasts* 570.
Phosphoglucomutase 523.
—, Aktivatoren, *activators* 524.
—, Aktivitätsbestimmung, *assay* 524.
—, Darstellung, *preparation* 524.
6-Phosphogluconsäure, *6-phospho gluconic acid* 207.
6-Phosphogluconsäuredehydrogenase, Reinigung, *6-phosphogluconic dehydrogenase, purification* 554, 555.
—, Aktivitätsbestimmung, *assay* 552—554.

Phosphoglucose-Isomerase 526.
—, Hemmstoffe, *inhibitors* 527.
—, Nachweis, *assay* 526.
—, Darstellung, *preparation* 526.
3-Phosphoglycerat, Umbau, *3-phosphogly-cerate, transformation* 349.
Phosphoglyceratkinase, *phosphoglycerate kinase* 537.
—, Aktivatoren und Hemmstoffe. *activators and inhibitors* 539.
—, Aktivitätsbestimmung, *assay* 537.
—, Darstellung, *preparation* 538.
—, Vorkommen, *distribution* 539.
Phosphoglyceromutase 539.
—, Aktivatoren und Hemmstoffe, *activators and inhibitors* 540.
—, Aktivitätsbestimmung, *assay* 539.
—, Vorkommen, *distribution* 540.
Phosphohexoisomerase 566, 567.
Phosphohexokinase 527.
—, Aktivatoren und Hemmstoffe, *activators and inhibitors* 528.
—, Darstellung, *preparation* 528.
—, Nachweis, *assay* 527.
Phosphoinositidphosphorylase, *phosphoino-sitide phosphorylase* 462.
Phosphoketopentoepimerase 558.
—, Aktivitätsbestimmung, *assay* 562, 563.
—, Reinigung, *purification* 563, 564.
Phospholipase A, Co-faktoren und Hemm-stoffe, *phospholipase A, cofactors and in-hibitors* 455.
—, Nachweismethode, *assay method* 455.
—, pH-Optima, *pH optima* 456.
—, Reinigung, *purification* 456.
—, Spezifität, *specificity* 456.
—, Vorkommen, *distribution* 454.
Phospholipase B 456.
—, Aktivatoren und Hemmstoffe, *activators and inhibitors* 458.
—, Nachweismethode, *assay method* 456, 457.
—, pH-Optima, *pH optima* 458.
—, Reinigung, *purification* 459.
—, Spezifität, *specifity* 457.
—, Stabilität, *stability* 458.
Phospholipase C 459.
—, Aktivatoren, *activators* 460.
—, Nachweismethode, *assay method* 459.
—, pH-Optima, *pH-optimum* 460.
—, Reinigung, *purification* 460.
—, Spezifität der Substrate, *substrate specifi-city* 460.
—, Stabilität, *stability* 460.
Phospholipase D 461.
—, Aktivatoren, *activators* 462.
—, Nachweismethode, *assay method* 461.
—, pH-Optima, *pH optima* 462.
—, Reinigung, *purification* 462.
—, Spezifität der Substrate, *substrate speci-ficity* 462.
—, Stabilität, *stability* 462.
Phospholipasen, *phospholipases* 454.
—, Nomenklatur, *nomenclature* 454.
Phospholipide, *phospholipids* 462.

Phosphomonoesterasen, *phosphomonoesterases* 50.
4-Phosphopantothenat, *4-phosphopanto-thenate* 192.
Phosphopantothensäure, *phosphopantothenic acid* 191.
4-Phosphopantothenylcystein, *4-phospho-pantothenylcysteine* 192.
Phosphopantothenylcysteincarboxylase, *phosphopantothenylcysteine decarboxylase* 192.
Phosphoproteinphosphatase 52.
Phosphor, colorimetrische Bestimmung, *phosphorus, colorimetric estimation* 23.
Phosphoriboisomerase 556, 577.
—, Aktivitätsbestimmung, *assay* 556, 557.
—, Reinigung, *purification* 557, 558.
5-Phosphoribosylpyrophosphat s. PRPP-kinase, *5-phosphoribosylpyrophosphate kinase see PRPP-kinase*
Phosphoribosylpyrophosphat-Nicotinsäure-Transferase, *phosphoribosylpyrophosphate-nicotinic acid transferase* 188.
Phosphoribulokinase 577.
—, Eigenschaften, *properties* 578.
—, Nachweismethode, *assay method* 577.
—, Reinigung, *purification* 578.
Phosphorylase 522.
—, Aktivitätsbestimmung, *assay* 522.
—, Eigenschaften, *properties* 523.
—, Hemmstoffe, *inhibitors* 523.
— aus Kartoffeln, *from potatoes* 511—514.
—, Reinigung, *purification* 523.
Phosphorylierung, oxydative, Leistungs-fähigkeit, *phosphorylation, oxidative, effi-ciency* 651.
—, —, manometrische Methode, *manometric method* 652.
—, —, Sauerstoff-Elektrode, *oxygen electrode* 652.
—, photosynthetisch, *photosynthetic* 582.
3-Phosphoshikimisäure, *3-phosphoshikimic acid* 271.
5-Phosphoshikimisäure, *5-phosphoshikimic acid* 261, 262, 272.
Phosphotransacetylase 193, 194.
—, Bakterien, *bacterial* 622.
Photophosphorylierung, Bakterien, *photo-phosphorylation, bacterial* 612.
—, nichtcyclische, in Chloroplasten, *noncyclic, in chloroplasts* 587—588.
—, —, Mechanismus, *mechanism* 586—587.
—, cyclische, *cyclic* 583, 610.
—, —, in Chloroplasten, *in chloroplasts* 584, 585.
—, —, Mechanismus, *mechanism* 583.
Photosynthese, Chinone in, *photosynthesis, quinones in* 605, 610.
—, Enzymsysteme, *enzyme systems in* 569.
—, HILLs Reaktion, HILL *reaction* 570.
—, Kohlendioxydassimilation, *carbon dioxide assimilation* 574.
—, Teilung von Licht- und Dunkel-Phase, *separation of light and dark phases* 575.
Phytase 200.

DL-Pinoresinol 283, 284.
Plastocyanin 589.
Plastochinon, *plastoquinone* 605, 606, 611, 677, 678.
—, Eigenschaften, *properties* 609.
—, Extraktion und Reinigung, *extraction and purification* 607—609.
Plastochinon A, *plastoquinone A* 606, 612.
Plastochinon B, *plastoquinone B* 606, 612.
Plastochinon C, *plastoquinone C* 606.
Polyphenolase, Pflanzen, *polyphenolase, plant* 642.
Polyphenoloxydase, *polyphenol oxidase* 217, 680, 681—682.
—, chronometrische Bestimmung, *chronometric determination* 217.
—, „latente", *"latent"* 589.
—, Reinigung, *purification* 683.
—, Sauerstoffaffinität, *oxygen affinity* 650, 651.
Polyphosphat-ADP-Phosphotransferase, *polyphosphate-ADP-phosphotransferase* 50.
Polyphosphat-AMP-Phosphotransferase, *polyphosphate-AMP-phosphotransferase* 49.
Polyphosphatasen, *polyphosphatases* 47.
Polyphosphatdepolymerasen, *polyphosphate depolymerases* 47.
Polysaccharide, Chromatophoren, *polysaccharide, chromatophores* 572.
—, Synthese, *synthesis* 507.
PPNR (photosynthetisch wirksame Pyridinnucleotidreduktase), *PPNR (photosynthetic pyridine nucleotide reductase)* 590.
Prephensäure, *prephenic acid* 260, 261, 262, 274.
Prephensäurearomatase, *prephenic aromatase* 262, 274.
Prephensäuredehydrogenase, *prephenic dehydrogenase* 274, 275.
Prolidase 414.
L-Prolin, Biosynthese, L-*proline, biosynthesis* 333—334.
Prolinase 414.
Prolinracemase, *proline racemase* 387, 388.
Prolinreduktase, *proline reductase* 318.
Propionyl-CoA 637.
Protease-Aktivität, Messung, *protease activity. measurement* 416.
Proteasen, *proteases* 412, 416.
Proteinsynthese, *protein synthesis* 405, 408.
Protocatechusäure, *protocatechuic acid* 261, 265.
—, Nachweise, *assays* 266.
Protocatechusäureoxydase, *protocatechuic acid oxidase* 266.
Protoplasten, Isolierung, *protoplasts, isolation* 34.
PRPP-Amidotransferase 426.
—, Aktivitätsbestimmung, *assay* 426.
—, Darstellung, *preparations* 427.
PRPP-Kinase 422.
—, Aktivitätsbestimmung, *assay* 422.
—, Darstellung, *preparation* 424.

„Pseudohämoglobin" s. RHP, *"pseudohemoglobin" see RHP*. 603.
Pseudouridin, *pseudo-uridine* 405.
Purinnucleotide, Enzyme der Synthese, *purine nucleotides, enzymes of synthesis* 421, 422.
Pyridinnucleotid und Nitratreduktase, *pyridine nucleotide and nitrate reductase* 91.
Pyridinnucleotide, Oxydierung durch andere Mittel als Cytochromsysteme, *pyridine nucleotides, oxidation by means other than cytochrome system* 679, 680.
Pyridoxal (Vitamin B$_6$) 186.
Pyridoxalphosphokinase 186.
Pyrimidinkinase, *pyrimidine kinase* 174.
Pyrimidinnucleotide, Enzyme der Synthese, *pyrimidine nucleotides, enzymes of synthesis* 421, 439.
Pyrithiamin, *pyrithiamine* 181, 207.
Pyrogallol 685.
Pyrophosphat, Bestimmung, *pyrophosphate, determination* 193.
Pyrophosphatasen, anorganische, *pyrophosphatases, inorganic* 41.
O-5-P-Pyrophosphorylase, Darstellung, *O-5-P pyrophosphorylase, preparation* 425.
Δ'-Pyrrolin-5-Carboxylatreduktase,$_2$ Δ'-*pyrroline-5-carboxylate reductase* 334.
Pyruvatkinase, *pyruvate kinase* 543.
—, Aktivatoren und Hemmstoffe, *activators and inhibitors* 544.
—, Aktivitätsbestimmung, *assay* 543.
—, Vorkommen, *distribution* 544.

Racemasen, *racemases* 381.
Raffinose 476, 498.
RHP (*Rhodospirillium* heme protein) 603.
—, Darstellung, *preparation* 603.
—, Eigenschaften, *properties* 603, 604.
Riboflavin 175, 176, 182, 184, 196.
Riboflavin-5'-phosphat s. Flavinmononucleotid, *riboflavin 5'-phosphate see also flavinmononucleotide (FMN)*.
Ribonucleasen, *ribonucleases* 58.
Ribonucleinsäure s. sRNA, *ribonucleic acid see also sRNA* 405.
Ribosomen, *ribosomes* 408.
—, Aminosäurenzusammensetzung, *amino acid composition* 409.
sRNA, Darstellung, *sRNA, preparation of* 406.

Saccharide, Komplex, Vorstufenbildung, *saccharides, complex, precursor formation* 478.
Saccharose, *sucrose* 489, 500.
Saccharosephosphat, *sucrose phosphate* 498, 501.
Sauerstoffaffinität, scheinbare, *oxygen affinity, apparent* 650.
Sedoheptulose-1,7-Diphosphat, *sedoheptulose-1,7-diphosphate* 529.
Sedoheptulose-1,7-diphosphatase 40.
Sedoheptulose-7-Phosphat, Darstellung, *sedoheptulose-7-phosphate, preparation* 565.
Serin, Synthese, *serine, synthesis* 349.

Serinalanintransaminase, *serine-alanine transaminase* 376.

Serinaldolase, *serine aldolase* 351—352.

D-Serindehydrase, *D-serine dehydrase* 311.

L-Serindehydrase, *L-serine dehydrase* 310.

Shikimisäure, *shikimic acid* 260, 261, 267, 268.

— als Vorstufen in höheren Pflanzen, *as precursor in higher plants* 275.

Stärke, Synthese, *starch, synthesis* 510.

Stickoxydreduktase, Aktivitätsbestimmung, manometrisch, *assay, manometric* 134—136.

—, Cofaktoren und Hemmstoffe, Bakterien, *nitric oxide reductase, co-factors and inhibitors, bacteria* 144, 145.

—, Eigenschaften und Mechanismen, Bakterien, *properties and mechanisms, bacteria* 153.

—, Elektronendonatoren, Bakterien, *electron donors, bacteria* 144, 145.

—, Extraktion aus Bakterien, *extraction from bacteria* 117.

—, Reinigung, *purification* 120.

Stickoxydreduktasen, *nitric oxide reductases* 112.

Stickoxydreduktasesysteme, Hemmstoffe, *nitric oxide reductase systems, inhibitors* 154, 155.

—, Stabilität, *stability* 154, 155.

Succinyl-Coa-Synthetase 623, 624.

Sulfolipids, Pflanzen, *sulfolipid, plants,* 571.

Synthetasen, *synthetases* 393.

Tartronsäuresemialdehyd, *tartronic acid semialdehyde* 205.

5,6,7,8-Tetrahydrofolsäure, *5,6,7,8-tetrahydrofolic acid* 189.

2-Tetrahydroxybutyl-5-methyl-4-carboxyfuran 206.

Tetrametaphosphatasen, *tetrametaphosphatases* 47.

Tetrapolyphosphatasen, *tetrapolyphosphatases* 45.

Thetine, *thetins* 321.

Thiamin, *thiamine* 173, 180, 194, 207.

Thiaminase 194.

Thiaminmonophosphat, *thiamine monophosphate* 174, 195.

Thiamonmonophosphatase, *thiamine monophosphatase* 174.

Thiaminphosphatase, *thiamine phosphatase* 175, 195.

Thiaminphosphatsynthetase, *thiamine phosphate synthetase* 174, 175.

Thiaminpyrophosphat, *thiamine pyrophosphate* 180.

Thiaminpyrophosphatase, *thiamine pyrophosphatase* 195.

Thiaminpyrophosphatkinase, *thiamine pyrophosphatase kinase* 182.

Thiaminpyrophosphokinase, *thiamine pyrophosphokinase* 174, 180, 181.

Thiazolkinase, *thiazole kinase* 174, 175.

Thiochrom, *thiochrome* 182.

Trehalosephosphat, *trehalose phosphate* 501, 503, 504.

L-Threonin, Biosynthese, *L-threonine, biosynthesis* 325.

L-Threonindehydrase, *L-threonine dehydrase* 310.

D-Threonindehydrase, *D-threonine dehydrase* 311.

Threoninracemase, *threonine racemase* 384, 385.

Threonsäure, *threonic acid* 218.

Threoninsynthetase, *threonine synthetase,* 328, 329.

β- und γ-Tocopherol-Chinone, *β and γ tocopherol quinones* 612.

TPN s. Triphosphopyridinnucleotid, *TPN see triphosphopyridinenucleotide*

TPN-Reduktase, *TPN reductase* 591, 593.

TPN-Reduktionsfaktor, *TPN-reducing factor* 590.

TPN-Triosephosphat-Dehydrogenase, *TPN-triosephosphate dehydrogenase* 534.

—, Aktivatoren und Hemmstoffe, *activators and inhibitors* 535.

—, Aktivitätsbestimmung, *assay* 534.

—, Reinigung, *purification* 535.

—, Vorkommen, *distribution* 535.

TPNH₂-diaphorase 596.

TPNH₂-Plastocyaninreduktase, *TPNH₂-plastocyanin reductase* 596.

Transacetylase 616.

Transaldolase 564.

—, Nachweis, *assay* 564, 565.

—, Reinigung, *purification* 565, 566.

Transamidinierung, *transamidination* 380, 381.

Transaminasen, *transaminases* 361.

Transhydrogenase 667.

—, Eigenschaften und Reinigung, *properties and purification* 668.

—, Nachweis, *assay* 667.

Transketolase 558.

—, Nachweis, *assay* 559, 560.

—, Reinigung, *purification* 560—562.

Transmethylierung, Enzyme der, *transmethylation, enzymes of* 319.

Trimetaphosphatasen, *trimetaphosphatases* 45.

Triosephosphatisomerase, *triosephosphate isomerase* 533.

—, Nachweis, *assay* 533.

—, Vorkommen, *distribution* 534.

Triosephosphat-Phosphoglycerat-Dehydrogenase, *triosephosphate-phosphoglycerate dehydrogenase* 536.

—, Aktivatoren und Hemmstoffe, *activators and inhibitors* 537.

—, Nachweis, *assay* 536.

—, Vorkommen, *distribution* 537.

Triphosphopyridinnulceotid, *triphosphopyridinenucleotide* 187, 196, 228.

Tripolyphosphatasen, *tripolyphosphatases* 44.

Tryptamin, *tryptamine* 238.

—, Umbau in IAc und Indolessigsäure (IES), *conversion to IAc and IAA* 244.

Tryptophan, Synthese, *tryptophan, synthesis* 352.

Tryptophanalanintransaminase 379.
Tryptophanglutamattransaminase 379.
Tryptophanpyrrolase 177.
Tryptophansynthetase 354, 355.
L-Tryptophan 238.
L-Tryptophandecarboxylase 308.
—, enzymatischer Umbau in Indolessigsäure (IES), *L-tryptophan, enzymes converting to IAA* 240.
—, — — —, Vorkommen, *distribution* 240.
TTP (L-Tryptophan) 238.
Tyrase 278—281.
Tyrosin, *tyrosine* 261.
—, Biosynthese, *biosynthesis* 261.
L-Tyrosin, *L-tyrosine* 278.
Tyrosinapodecarboxylase, *tyrosine apodecarboxylase* 187.
Tyrosindecarboxylase, *tyrosine decarboxylase* 309.
Tyrosinglutamattransaminase, *tyrosineglutamate transaminase* 377.
Tyrosin aktivierende Enzyme, *tyrosine-activating enzyme* 407.

Ubichinon (Coenzym Q), *ubiquinone (coenzyme Q)* 605, 612, 665.
—, Eigenschaften und Reinigung, *properties and purification* 678—679.
—, Nachweis, *assay* 677.
Ubichinonreduktase, *ubiquinonereductase* 665.
UDP (Uridin-5′-diphosphat), *UDP (uridine 5′-diphosphate)* 474.
UDP-N-acetyl-D-glucosamin, *UDP-N-acetyl-D-glucosamine* 479, 508.
UDP-N-Acetyl-D-Glucosamin-Chitintransglucosylase aus *Neurospora crassa, UDP-N-acetyl-D-glucosamine-chitin transglucosylase, from Neurospora crassa* 508, 509.
UDP-N-Acetyl-D-Glucosaminpyrophosphorylase, aus Bäckerhefe, *UDP-N-acetyl-D-glucosamine pyrophosphorylase, from baker's yeast* 490.
— aus *Phaseolus aureus, UDP-N-acetyl-D-glucosamine pyrophosphorylase from Phaseolus aureus* 489, 490.
UDP-L-Arabinose 479, 485, 493, 497, 509.
UDP-L-Arabinose-4-Epimerase 495, 510.
— aus *Phaseolus aureus, from Phaseolus aureus* 495.
UDP-D-Galaktose, *UDP-D-galactose* 478, 481, 485, 492, 493.
UDP-D-Galaktose-4-epimerase, *UDP-D-galactose 4-epimerase* 478, 481, 493.
— aus *Phaseolus aureus, from Phaseolus aureus* 495.
— aus *Saccharomyces fragilis, from Saccharomyces fragilis* 493, 494.
UDP-D-Galaktose-Pyrophosphorylase, *UDP-D-galactose pyrophosphorylase* 485.
UDP-D-Galakturonsäure, *UDP-D-galacturonic acid* 479, 485, 493, 495.
— aus Rettich, *from radish* 495.

UDP-D-Galakturonsäure-4-Epimerase, *UDP-D-galacturonic acid 4-epimerase* 495.
UDP-D-Glucose 474, 476, 480, 481, 485, 486, 487, 493, 495, 498, 499, 501, 502, 503, 504, 506, 507, 508, 510.
UDP-D-Glucose-Dehydrogenase 478.
— aus Erbsen, *from peas* 495—497.
UDP-D-Glucose-Diphenoltransglucosylase, aus Weizenkeimlingen, *UDP-D-glucose-diphenol transglucosylase, from wheat germ* 504—506.
UDP-D-Glucose-D-Fructose-6-Phosphat-glucosylase aus Weizenkeimlingen, *UDP-D-glucose-D-fructose 6-phosphate transglucosylase from wheat germ* 500, 501.
UDP-D-Glucose-D-Fructosetransglycosylase, aus Weizenkeimlingen, *UDP-D-glucose-D-fructose transglucosylase from wheat germ* 499, 500.
UDP-D-Glucose-β-1,3-D-Glucantransglucosylase aus *Phaseolus aureus, UDP-D-glucose-β-1,3-D-glucan transglucosylase from Phaseolus aureus* 507, 508.
UDP-D-Glucose-D-Glucose-6-Phosphattransglucosylase aus Hefe, *UDP-D-glucose-D-glucose 6-phosphate transglucosylase from yeast* 501—504.
UDP-D-Glucose-Phenol-D-Glucosidtransglucosylase aus Weizenkeimlingen, *UDP-D-glucose-phenol-D-glucoside transglucosylase from wheat germ* 506.
UDP-D-Glucosepyrophosphorylase aus Bierhefe, *UDP-D-glucose pyrophosphorylase from brewer's yeast* 488, 489.
— aus *Phaseolus aureus, from Phaseolus aureus* 486—488.
UDP-D-Glucuronsäure, *UDP-D-glucuronic acid* 478, 479, 480, 481, 485, 493, 495, 497, 504, 509.
UDP-D-Glucuronsäuredecarboxylase, *UDP-D-glucuronic acid decarboxylase* 510.
— aus Weizenkeimlingen, *from wheat germ* 497.
UDP-D-Xylose 479, 481, 485, 492, 495, 497, 509.
UDP-D-Xylose-D-Xylodextrintransxylosylase aus Spargel, *UDP-D-xylose-D-xylodextrin transxylosylase from asparagus* 509, 510.
Umbau, enzymatischer, TTP in Indolessigsäure, Darstellung und Eigenschaften, *enzymes converting TTP to IAA, preparation and properties* 242.
—, —, von Indolacetonitril in Indolessigsäure, *enzymes converting indoleacetonitrile to IAA* 246.
—, —, Zwischenprodukte in Indolessigsäure, *enzymes converting to IAA, intermediates* 243.

Vitamin K-Reduktase, *vitamin K-reductase* 665.
Vitamin K_1 in Chloroplasten, *vitamin K_1 in chloroplasts* 605.

Vitamin K$_1$, Extraktion, *extraction* 607.
Vitamin K$_3$ 611.
Vitamin K$_5$ 611.

Weinsäuredehydrogenase, *tartaric dehydrogenase* 629.
Weizenblattenzym, *wheat leaf enzyme* 254.

Xanthinoxydasen, Nitratreduktion, *xanthine oxidases, nitrate reduction* 110.
Xylan 509.

Xylonsäure, *xylonic acid* 218.
α-D-Xylose-1-Phosphat, α-D-*xylose 1-phosphate* 478, 485, 486.

Zimtsäure, *cinnamic acid* 281.
Zuckernucleotidpyrophosphorylasen aus *Phaseolus aureus*, Eigenschaften, *sugar nucleotide pyrophosphorlyases, from Phaseolus aureus, properties* 486.
— — —, Nachweis und Darstellung, *assay and preparation* 485.

Subject Index.

(English—German).

For general terms such as extraction, isolation purification etc. of various groups of compounds cf. also the table of contents at the beginning of the volume.

cis-, trans-, n-, D-, L- and similar isomers are listed according to the first letter of the following word.

All iso-compounds are to be found under Iso-,

Ä, Ö, Ü are taken as Ae, Oe, Ue.

Where English and German spelling of a word is identical, the italicised (German) entry is omitted.

Acetate activating enzymes, chloroplasts, *Acetat-aktivierende Enzyme, Chloroplasten* 570.

Aceto-CoA-kinase 636.

Acetoin, synthesis, *Synthese* 631.

Acetolactate synthetase, *Acetolactatsynthetase* 345—346.

Acetyl CoA 616, 622.

α-N-acetyl-D-glucosamine 1-phosphate, *α-N-acetyl-D-Glucosamin 1-Phosphat* 478.

Acetylornithinase 336—337.

Acetylornithine δ-transaminase, *Acetylornithin δ-transaminase* 336.

Acetyl phosphate, *Acetylphosphat* 193, 616, 617.

— —, determination, *Bestimmung* 193.

Aconitase 617, 618—620.

—, chloroplasts, *Chloroplasten* 570.

Aconitate, determination, *Aconitat, Bestimmung* 619.

Acyloin reaction, *Acyloin-Reaktion* 205.

Adenosine, *Adenosin* 405.

S-Adenosylmethionine, *S-Adenosylmethionin* 321.

S-Adenosylmethionine dethioadenosylase, *S-Adenosylmethionindethioadenosylase* 322.

Adenosylmethionine-homocysteine transmethylase, *Adenosylmethionin-homocysteintransmethylase* 323—324.

Adenosylmethionine-nicotinic acid transmethylase, *Adenosylmethionin-nicotinsäuretransmethylase* 324—325.

AISCAR (5-amino-4-imidazole-N-succinocarboxamide ribotide), *(5-Amino-4-imidazol-N-succinocarboxamidribotid)*.

—, assay, *Nachweis* 437.

—, enzymatic cleavage, *enzymatische Spaltung* 437.

—, preparation, *Darstellung* 437.

—, synthesis, *Synthese* 436.

α-Alanine-β-alanine transaminase, *α-Alanin-β-alanintransaminase* 372—373.

L-Alanine dehydrogenase, *L-Alanindehydrogenase* 294—295.

β-Alanine-glutamate transaminase, *β-Analin-glutamattransaminase* 371.

α-Alanine-glycine transaminase, *α-Alaninglycintransaminase* 371—372.

Alanine-glutamate transaminase, chloroplasts, *Alanin-glutamattransaminase, Chloroplasten* 570

Alanine-ketomalonate transaminase, *Alaninketomalonattransaminase* 380.

Alanine racemase, *Alaninracemase* 381—383.

Alcohol dehydrogenase, *Alkoholdehydrogenase* 631.

Aldehyde dehydrogenase, *Aldehyddehydrogenase* 467.

Aldehyde oxidases, nitrate reduction, *Aldehydoxydasen, Nitratreduktion* 110.

Aldehydes, long-chain, *Aldehyde, langkettige* 467.

Aldol condensation, *Aldolkondensation* 205.

Aldolase 530

—, activators and inhibitors, *Aktivatoren und Hemmstoffe* 532.

—, assay, *Nachweis* 530.

—, —, chemical, *chemischer* 531

—, —, enzymatic, *enzymatischer* 531

—, distribution, *Vorkommen* 533

Alliinase 315, 316.

D-altronic acid-γ-lactone, *D-Altronsäure-γ-lakton* 207.

Amethopterin 190.

Amidases, *Amidasen* 412, 413

Amide synthesis, *Amidsynthese* 393.

Amine oxidase, of pea seedlings, *Aminooxydase, Erbsenkeimlinge* 300—302.

Amine oxidases, *Aminoxydasen* 300.

Amino acid-activating enzymes, *Aminosäuren aktivierende Enzyme* 405.

— — —, chloroplasts, *Chloroplasten* 571.

Amino acid C-S cleaving enzymes, *Aminosäure C-S Spaltungsenzyme* 313.

Amino acid deaminases, *Aminosäuredeaminasen* 309.

Amino acid decarboxylases, higher plants, *Aminosäuredecarboxylasen, höhere Pflanzen* 302.

— — —, microorganisms, *Mikroorganismen* 304.

Amino acid decarboxylation, *Decarboxylierung, Aminosäure* 302.

Amino acid-dependent pyrophosphate-ATP exchange, *Aminosäureabhängiger Pyrophosphat-ATP-Austausch* 406

Amino acid oxidases, *Aminosäureoxydasen* 688.

Amino acid reductases, *Aminosäurereduktasen* 317, 318.

Amino acid-sRNA synthesis, *Aminosäure-sRNA-Synthese* 406.

Amino acid transacetylase, *Aminosäuretransacetylase* 336.

Amino acid transaminase, branched-chain, *Aminosäuretransaminase, verzweigte Kette* 347.

D-amino acid apooxidase, D-*Aminosäureapooxydase* 185.

D-amino acid oxidase, D-*Aminosäureoxydase* 186.

— — —, microorganisms, *Mikroorganismen* 297—298.

D-amino acid oxidases, D-*Aminosäureoxydasen* 296.

D-amino acid transaminases, D-*Aminosäuretransaminasen* 362—363.

L-amino acid oxidase, L-*Aminosäureoxydase* 642.

— — —, microorganisms, *Mikroorganismen* 298—300.

L-amino acid oxidases, L-*Aminosäureoxydasen* 296.

L-amino acid transaminases, L-*Aminosäuretransaminasen* 363.

Amino acids, branched-chain, degradation, *Aminosäuren, verzweigte Ketten, Abbau* 347.

— —, —, synthesis, *Synthese* 343.

γ-aminobutyrate-glutamate transaminase, *γ-Aminobutyrat-glutamattransaminase* 373—374.

2-amino-4-hydroxy-6-formylpteridine, *2-Amino-4-oxy-6-formylpteridin* 198.

5-amino-4-imidazolecarboxamide ribotide transformylase, assay and preparation, *5-Amino-4-imidazolcarboxamidribotidtransformylase, Nachweis und Darstellung* 438.

5 aminoimidazole ribotide, synthesis, *5-Aminoimidazolribotid, Synthese* 436.

5-aminoimidazole ribotide carboxylase, *5-Aminoimidazolribotidcarboxylase* 436.

5-amino-4-imidazole-N-succinocarboxamide ribotide see: AISCAR, *5-Amino-4-imidazol-N-succinocarboxamidribotid s. AISCAR*

δ-aminolaevulinic transaminase, *δ-Aminolaevulinsäuretransaminase* 379.

Aminopterin 190.

Aminotripeptidase 414, 415.

Ammonia, assay methods, *Ammoniak, Nachweismethoden* 126.

—, determination, CONWAY methods, *Bestimmung, CONWAY-Methoden* 126.

—, —, RUSSELL method, RUSSELL-*Methode* 127.

—, —, NESSLER reagent, NESSLERs *Reagens* 127.

—, oxidation, *Oxydation* 158—160.

Amylopectin, *Amylopektin* 510, 516.

Amylose 510.

Amytal (5-ethyl-5-isoamyl-barbiturate) 655, 656, 661.

Antimycin A 588, 653, 655, 660.

L-arabinokinase, from *Phaseolus aureus*, assay, preparation and properties, L-*Arabinokinase, aus Phaseolus aureus, Nachweis, Darstellung und Eigenschaften* 484.

β-L-arabinose 1-phosphate, β-L-*Arabinose 1-Phosphat* 478, 485, 486.

D-arabo-ascorbic acid, D-*Araboascorbinsäure* 218.

Arginase 341—342.

Arginine decarboxylase, *Arginindecarboxylase* 307.

Arginine, degradation, *Arginin, Abbau* 342.

Arginine desiminase, *Arginindesiminase* 342.

Argininosuccinase 340—341.

Argininosuccinate synthetase, *Argininosuccinatsynthetase* 339—340.

Arsenolysis, *Arsenolyse* 393.

Asclepain, purification, *Reinigung* 417.

Ascorbase 218.

L-ascorbic acid, L-*Ascorbinsäure* 204.

Ascorbic acid, oxidising enzymes, *Ascorbinsäure, oxydierende Enzyme* 226.

— —, reducing enzymes, *reduzierende Enzymsysteme* 226.

Ascorbic acid dehydrase, *Ascorbinsäuredehydrase* 218.

Ascorbic acid oxidase, *Ascorbinsäureoxydase* 218, 680, 681.

— — —, activity determination, *Aktivitätsbestimmungen* 224.

— — —, assay and purification, *Nachweis und Reinigung* 684.

— — —, copper determination method, *Kupferbestimmungsmethode* 220.

— — —, occurence, *Vorkommen* 225.

— — —, oxygen affinity, *Sauerstoffaffinität* 650.

— — —, properties, *Eigenschaften* 221.

— — —, titration with N-bromosuccinimide, *Titration mit N-Bromsuccinimid* 224.

Ascorbicase 218.

Ascorbinase 218.

Asparaginase 413, 414.

Asparagine synthetase, *Asparaginsynthetase* 400.

Asparagine transaminases, *Asparagintransaminasen* 377—378.

Aspartase 309

Aspartate-glutamate transaminase, chloro-
plasts, *Aspartatglutamattransaminase,
Chloroplasten* 570.
Aspartic β-semialdehyde dehydrogenase,
*Asparaginsäure-β-semialdehyd-dehydro-
genase* 326—327.
β-Aspartokinase 325-326.
"Assimilatory power", *„Assimilationskraft"*
576.
ATP (adenosine triphosphate), *(Adenosin-
triphosphat)*.
ATP-shikimic acid transphosphorylase, *ATP-
Shikimisäuretransphosphorylase* 272.
Azide, *Azid* 654.

Bacteria, enzymes of nitrification, *Bakterien,
Enzyme der Nitrifikation* 161.
—, — —, co-factors, *Cofaktoren* 164.
—, — —, fractionation, *Fraktionierung* 163.
—, — —, measurement of activity, *Aktivitätsmessung* 164.
—, — —, methods of extraction, *Extraktionsmethoden* 162.
—, — —, properties and mechanisms, *Eigenschaften und Mechanismen* 166.
—, — —, preparation, *Darstellung* 161.
Bacteriochlorophyll, chromatophores, *Chromatophoren* 572.
—, determination, *Bestimmung* 573.
p-Benzoquinone reductase, *p-Benzochinonreduktase* 665.
Biosynthesis, aromatic, in microorganisms,
*Biosynthese, aromatische, in Mikroorganis-
men* 260.
Biotin 199.
D-Biotin oxidase, *D-Biotin Oxydase* 199.
Bis-dehydroconiferylalcohol, *Bis-dehydroconiferylalkohol* 283, 284.
Butyryl-CoA 637.

Callose, *Kallose* 507, 508.
Carbamate kinase, *Carbamatkinase* 338.
Carbamyl phosphate-aspartate transcarbamylase, *Carbamylphosphat-aspartat-
transcarbamylase* 441.
— — —, assay and preparation, *Nachweis
und Darstellung* 442.
Carbamyl phosphate synthesizing enzymes,
Carbamylphosphat synthesierende Enzyme
337—338.
Carbamyl phosphate synthetase, assay, *Carbamylphosphatsynthetase, Nachweis* 439.
— — —, preparation, *Darstellung* 440.
Carbohydrate cycle, reductive, *Kohlenhydratcyclus, reduzierender* 577.
Carbohydrate oxidase, *Kohlenhydratoxydase*
689—690.
Carbohydrate synthesis, enzymes of, *Kohlenhydratsynthese, Enzyme der* 474.
Carbon dioxide fixation, measurement,
Kohlendioxydfixierung, Messung 576.
Carbon monoxide, cytochrome oxidase inhibition, *Kohlenmonoxyd, Cytochrom-
oxydase-Hemmung* 654.
— —, polyphenol oxidase inhibition, *Polyphenoloxydase-Hemmung* 654.

Carboxydismutase 579.
—, properties, *Eigenschaften* 581—582.
—, assay, *Nachweis* 579.
—, purification, *Reinigung* 579—580.
Carboxypeptidase 414.
Carnosinase 414.
Carotene, *Karotin* 201.
Carotene oxidase, *Karotinoxydase* 201.
Carotenoids, chromatophores, *Karotenoide,
Chromatophoren* 572.
Cellulose 476.
Chitin 508.
—, synthesis, *Synthese* 509.
Chlorogenic acid, *Chlorogensäure* 666, 682,
683.
Chloroplasts, *Chloroplasten* 570, 571.
Chloroplast extract, *Chloroplasten-Extrakt*
572.
—, "broken", preparation, *„gebrochene",
Darstellung* 572.
—, cyclic photophosphorylation in, *zyklische
Photophosphorylierung in* 584.
—, isolated, photosynthesis by, *isolierte,
Photosynthese durch* 570.
—, noncyclic photophosphorylation in, *nichtzyklische Photophosphorylierung in* 587 to
588.
—, protein constituents, *Proteinbestandteile*
589.
—, TPN-reducing system, *TPN-Reduziersysteme* 589.
—, whole, isolation, *ganze, Isolierung* 571.
—, —, purification, *Reinigung* 571.
Chromatium, chromatophores, *Chromatophoren* 572.
Chromatophores, *Chromatophoren* 572.
—, *Chromatium*, bacterial culture, *Bakterienkultur* 572.
—, —, isolation and purification, *Isolierung
und Reinigung* 572.
—, —, preparation of cell extracts, *Darstellung aus Zellextrakten* 573.
—, cyclic photophosphorylation in, *zyklische
Photophosphorylierung in* 585.
—, noncyclic photo-phosphorylation in,
nichtzyklische Photophosphorylierung in
588.
—, protein constituents, *Proteinbestandteile*
589.
—, *Rhodospirillum rubrum*, isolation and
purification, *Isolierung und Reinigung*
574.
Chymopapain, purification, *Reinigung* 417.
Cinnamic acid, *Zimtsäure* 281.
Citrate, formation, *Citratbildung* 616.
Citrullinase system, *Citrullinasesystem* 343.
Citrulline ureidase, *Citrullinureidase* 343.
CoA 198.
—, degradation, *Abbau* 198.
CoA-3′-nucleotidase 198.
CoA-peptidase 199.
CoA-pyrophosphatase 199.
Coenzyme A s. CoA.
Coenzyme Q, see Ubiquinone, *Coenzym Q,
s. Ubichinon* 605, 612, 665.

Coenzymes, nomenclature, *Coenzyme, Nomenklatur* 696
$C^{14}O_2$ fixation, products, *$C^{14}O_2$-Fixierung, Produkte* 576.
Condensing enzyme, *kondensierendes Enzym* 616.
Coniferyl alcohol, *Coniferylalkohol* 283, 285.
— —, coupled oxidation, *gekuppelte Oxydation* 287—288.
— —, dehydrogenation, *Dehydrogenierung* 285—286.
Coniferyl aldehyde, *Coniferylaldehyd* 283, 284.
Copper oxidases, *Kupferoxydasen* 650, 681.
p-Coumaric acid, *p-Cumarinsäure* 278.
$C^{14}P$, preparation, *Darstellung* 442.
Cyanide, *Cyanid* 654.
Cystathionase 315.
D-Cysteine desulfhydrase, D-*Cysteinsulfhydrase* 313.
L-Cysteine desulfhydrase, L-*Cysteindesulfhydrase* 312.
Cysteinesulfinate transaminase, *Cysteinsulfinattransaminase* 369—371.
Cytochrome b, *Cytochrom b* 671.
Cytochrome b_2, *Cytochrom b_2* 671.
Cytochrome b_3, *Cytochrom b_3* 671.
Cytochrome b_5, *Cytochrom b_5* 670, 672.
Cytochrome b_6, *Cytochrom b_6* 600, 602, 672.
Cytochrome b_7, *Cytochrom b_7* 672.
Cytochrome c 217, 672, 673.
Cytochrome c_1, *Cytochrom c_1* 672.
Cytochrome c_2, *Cytochrom c_2* 602.
—, preparation, *Darstellung* 603
—, properties, *Eigenschaften* 603—604.
Cytochrome f, *Cytochrom f* 600, 672.
—, preparation, *Darstellung* 600
—, properties, *Eigenschaften*, 601—602
Cytochrome 552, *Cytochrom 552* 605.
Cytochrome 553, preparation, *Cytochrom 553, Darstellung* 600—601.
Cytochrome oxidase $(a — a_3)$, *Cytochromoxydase $(a — a_3)$* 217, 650, 675.
— —, assay, *Aktivitätsbestimmung* 675
— —, azide inhibition, *Azid-Hemmung* 654.
— —, carbon monoxide inhibition, *Kohlenmonoxyd-Hemmung* 654.
— —, cyanide inhibition, *Cyanid-Hemmung* 654.
— —, properties, *Eigenschaften* 677.
— —, purification, *Reinigung* 676.
Cytochrome c oxidase, oxygen affinity, *Cytochrom-c-Oxydase, Sauerstoffaffinität* 650.
Cytochrome c reductase, *Cytochrom-c-Reduktase* 665—667.
— —, inhibition, *Hemmung* 667.
Cytochrome system, inhibition, *Cytochromsystem, Hemmung* 654.
Cytochromes, bacterial, *Cytochrome, Bakterien* 602
—, classification and nomenclature, *Klassifikation und Nomenklatur* 696, 699.
—, chromatophores, *Chromatophoren* 572.

Cytochromes, identification, spectrophotometric, *Identifikation, spektrophotometrische* 662.
—, in leaves and algae, *in Blättern und Algen* 599—600.
Cytochromes b, *Cytochrome b* 668—672.
—, activity, *Aktivität* 670.
—, assay, *Nachweis* 669—670.
—, properties, *Eigenschaften* 669.
—, — and purification, *und Reinigung* 670.
Cytochromes c, *Cytochrome c* 672—675.
—, assay, *Nachweis* 673.
—, properties, *Eigenschaften* 674.
—, purification,. *Reinigung* 673—674.

Deaminases, dehydrative, *Deaminasen, dehydrierende* 310.
Deamination, non-oxidative, *Desaminierung, nichtoxydative* 309.
—, oxidative, *oxydative* 291.
Dehydrases, *Dehydrasen* 310.
L-dehydroascorbic acid, L-*Dehydroascorbinsäure* 204, 229.
—, 2,4-dinitrophenylhydrazine method, *2,4-Dinitrophenylhydrazin-Methode* 230.
—, distribution, *Vorkommen* 233.
Dehydro-diconiferyl alcohol, *Dehydrodiconiferylalkohol* 283, 284.
Dehydro-di-pinoresinol 283, 284.
5-Dehydroquinase, *5-Dehydrochinase* 267—268.
Dehydroquinic acid, *Dehydrochinasäure* 260, 268.
5-Dehydroquinic acid, *5-Dehydrochinasäure* 260, 261, 262, 265, 267.
Dehydroshikimic acid, *Dehydroshikimisäure* 260.
5-Dehydroshikimic acid, *5-Dehydroshikimisäure* 262, 265, 267.
Dehydroshikimic dehydrase, *Dehydroshikimisäuredehydrase* 265—266.
Dehydroshikimic reductase, *Dehydroshikimisäurereduktase* 268.
5-Dehydroshikimic reductase, *5-Dehydroshikimisäurereduktase* 270—272.
Dephospho-CoA 189, 199.
Dephospho-CoA kinase 194.
Dephospho-CoA pyrophosphorylase 192, 199.
Desamido-DPN 189.
Desamido-DPN formation, *Desamido-DPN-Bildung* 188.
Desulfhydrases, *Desulfhydrasen* 312.
Diacetone-2-keto-L-gulonic acid, *Diaceton-2-keto-L-gulonsäure* 211.
Diaminopimelic acid decarboxylase, *Diaminopimelinsäuredecarboxylase* 308.
α-ε-Diaminopimelic acid racemase, *α-ε-Diaminopimelinsäureracemase* 388—389.
Diaphorase 664
Dihydrofolic acid reductase, *Dihydrofolsäurereduktase* 190.
Dihydrolipoic dehydrogenase, *Dihydroliponsäuredehydrogenase* 622.
Dihydroorotase 442.

Dihydroorotase, assay and preparation, *Nach-weis und Darstellung* 443.
Dihydroorotic acid dehydrogenase, assay and preparation, *Dihydroorotsäuredehydrogenase, Nachweis und Darstellung* 444.
α, β-Dihydroxy acid dehydrase, *α, β-Dioxysäuredehydrase* 346—347.
Dimethylpropionthetin dethiomethylase 314.
6,7-dimethyl-8-ribityllumazine 176.
Diphosphatase, acid, *Escherichia coli, Diphosphatase, saure, Escherichia coli* 39.
C—1-diphosphatases, *C—1-Diphosphatasen* 36.
Diphosphopyridinenucleotide (DPN), *Diphosphopyridinnucleotid (DPN)* 187, 196, 197.
Disaccharides, synthesis, *Disaccharide, Synthese* 498.
DPN see: Diphosphopyridinenucleotide, *DPN s. Diphosphopyridinnucleotid*.
DPN-ase 197.
DPN-kinase 189.
DPN-pyrophosphorylase 188.
DPN-synthetase 188, 189.
DPN triosephosphate dehydrogenase, *DPN Triosephosphatdehydrogenase* 535.
— — —, assay, *Nachweis* 535.
— — —, distribution, *Vorkommen* 536.
— — —, inhibitors, *Hemmstoffe* 536.
DPNH (reduced diphosphopyridine nucleotide), *DPNH (reduziertes Diphophopyridinnucleotid)*.
DPNH dehydrogenase 663—664.
DPNH oxidase, assay, *DPNH-Oxydase, Nachweis* 649, 651.
— —, *S. faecalis*, oxygen affinity, *Sauerstoffaffinität* 650.
— —, mitochondria, *Mitochondrien* 649.
DPNH-oxidising lipoflavoprotein, *DPNH-oxydisierendes Lipoflavoprotein* 666.

Endonucleases, *Endonucleasen* 62.
Enolase, 2-phosphoglycerate dehydrase, *2-phosphoglycerat dehydrase* 541.
— —, activators and inhibitors, *Aktivatoren und Hemmstoffe* 542.
— —, assay, *Nachweis* 541.
— —, distribution, *Vorkommen* 542.
3-Enolpyruvyl shikimate, *3-Enolpyruvylshikimat* 260, 262.
3-Enolpyruvyl shikimate-5-phosphate (ESP), *3-Enolpyruvylshikimat-5-phosphat (ESP)* 261, 262.
3-Enolpyruvyl shikimic acid, *3-Enolpyruvylshikimisäure* 261.
Enzyme formation, terminology, *Enzymbildung, Terminologie* 698.
Enzyme kinetics, *Enzymkinetik* 695, 699.
Enzyme terminology, *Enzymterminologie* 695.
Enzyme units, *Enzymeinheiten* 695.
— —, molecular activity, *Molekularaktivität* 695.
— —, specific activity, *spezifische Aktivität* 695.

D-Enzyme, from potatoes, *D-Enzym, aus Kartoffeln* 515, 516.
Q-Enzyme, from potatoes, *Q-Enzym, aus Kartoffeln* 514, 515.
Enzymes, classification, *Enzyme, Klassifikation* 697, 700.
—, condensing, chloroplasts, *kondensierende, Chloroplasten* 570.
Enzymes converting to IAA, intermediates, *Umbau in Indolessigsäure (IES), Zwischenprodukte* 243.
Enzymes converting indoleacetonitrile to IAA, *Umbau von Indolacetonitril in Indolessigsäure (IES)* 246.
Enzymes converting TTP to IAA, preparation and properties, *Umwandlung von TTP in Indolessigsäure, Darstellung und Eigenschaften* 242.
Enzymes, intracellular localization, *Enzyme, intrazelluläre Lokalisierung* 29.
DL-Epipinoresinol 283, 284.
5-Epishikimic acid, *5-Epishikimisäure* 271.
ESP (3-enolpyruvyl shikimate-5-phosphate), *ESP (3-Enolpyruvylshikimat-5-phosphat)*.
—, enzymic formation, *enzymatische Bildung* 273.
Esterases, *Esterasen* 448.
Ethanolamine phospholipoprotein, chromatophores, *Ethanolaminphospholipoprotein, Chromatophoren* 572.

FAD see flavin adenine dinucleotide, *FAD s. Flavinadenindinucleotid*.
FAD-ase 196, 199.
FAD pyrophosphorylase 185.
Fatty acid peroxidase, *Fettsäure-Peroxydase* 467.
Fatty acids, enzymatic synthesis, *Fettsäuren, enzymatische Synthese* 472.
— — —, assay, *Nachweis* 473
— — —, pH optima, *pH Optima* 473.
— — —, stability, *Stabilität* 473.
Ferredoxin 592.
—, bacterial, *Bakterien* 593.
—, identity with PPNR, *Identität mit PPNR* 593.
—, isolation and purification, *Isolierung und Reinigung* 593—596.
—, physiological role, *physiologische Rolle* 593.
Ferredoxin-TPN reductase, *Ferredoxin-TPN-reduktase* 596.
— —, assay, *Nachweis* 598.
— —, crystallization procedure, *Kristallisationsmethode* 599.
— —, isolation and purification, *Isolierung und Reinigung* 597.
— —, properties, *Eigenschaften* 598.
Ferricytochrome oxidase, *Ferricytochromoxydase* 654.
Ficin, purification, *Reinigung* 417.
Flavin adenine dinucleotide, *Flavinadenindinucleotid* 182, 196, 689.
Flavinmononucleotide, *Flavinmononucleotid* 182, 184, 185, 196, 228, 688.

Flavins and nitrate reductase, *Flavine und Nitratreduktase* 93.

Flavokinase 183.

Flavoproteins, *Flavoproteine* 662, 666.

Fluoroacetyl-CoA 637.

FMN s. flavinmononucleotide, *FMN s. Flavinmononucleotid*.

FMN-phosphatase 196.

Folic acid, *Folsäure* 179, 189, 198.

Folic acid reductase, *Folsäurereduktase* 190.

Formate, transformation, *Formiat, Transformation* 350.

Formic acid, *Ameisensäure* 687.

Formic dehydrogenase, *Ameisensäuredehydrogenase* 643.

Formylglycinamidine ribotide kinosynthase, preparation, *Formylglycinamidinribotidkinosynthetase Darstellung* 435.

Formylase 177, 178.

Formylkynurenine, *Formylkynurenin* 177.

N-Formyl-L-kynurenine, *N-Formyl-L-kynurenin* 179.

Fructokinase 547.

D-Fructose 476.

Fructose diphosphatase, acidic, *Fructosediphosphatase, saure* 39.

—, —, alkaline, *alkalische* 39.

—, —, animal, *tierische* 36.

—, —, neutral, *neutrale* 39.

—, —, plants, *pflanzliche* 37.

Fructose-1,6-diphosphate, *Fructose-1,6-diphosphat* 205.

Fructose-1,6-diphosphatase 528.

—, assay, *Nachweis* 528.

—, alkaline, specificity, *alkalische, Spezifität* 529.

—, distribution, *Vorkommen* 530.

—, pH 5.5 enzyme 530.

—, pH 6.9 enzyme 530.

—, pH 8.5 enzyme 530.

—, neutral, *neutrale* 529.

—, purification, *Reinigung* 529.

D-Fructose 6-phosphate, *D-fructose 6-phosphat* 476.

D-fructuronic acid, *D-Fructuronsäure* 208.

Fumarase 627, 636.

—, chloroplasts, *Chloroplasten* 570.

L-galactoascorbic acid, *L-Galaktoascorbinsäure* 218.

D-galactokinase, *D-Galaktokinase* 478.

—, from *Phaseolus aureus*, assay, preparation and properties, *aus Phaseolus aureus, Nachweis, Darstellung und Eigenschaften* 484.

—, from *Saccharomyces fragilis*, assay, *Nachweis* 481.

—, — —, properties, *Eigenschaften* 483.

—, — —, purification, *Reinigung* 482.

Galactolipids, *Galaktolipide* 571.

D-galactonic acid-γ-lactone, *D-Galaktonsäure-γ-lakton* 210.

L-galactonic acid-γ-lactone, *L-Galaktonsäure-γ-lakton* 207, 208, 210, 211.

L-galactonic acid-γ-lactone dehydrogenase, *L-Galaktonsäure-γ-laktondehydrogenase* 208.

α-D-galactose 1-phosphate, *α-D-galactose 1-phosphat* 478, 481, 485, 486, 492.

— — uridyl transferase, 478.

— — — —, from *Saccharomyces fragilis*, 492.

D-galacturonic acid, *D-Galakturonsäure* 208, 212.

— — lactone, *D-Galakturonsäurelakton* 208.

— acid-γ-lactone, *D-Galakturonsäure-γ-lakton* 211.

— acid methyl ester, *D-Galakturonsäuremethylester* 208.

α-D-Galacturonic acid 1-phosphate, *α-D-Galakturonsäure 1-phosphat* 478, 485, 486.

GDP (guanosine 5'-diphosphate), *GDP (Guanosin 5'-diphosphat)*.

GDP-D-mannose 491.

— pyrophosphorylase, from brewer's yeast, *GDP-D-mannosepyrophosphorylase, aus Bierhefe* 491—492.

GDP-sugars, *GDP-Zucker* 478.

β-1,3-D-glucan, *β-1,3-D-Glucan* 507.

L-glucoascorbic acid, *L-Glucoascorbinsäure* 218.

Glucokinase 547

D-gluconic acid-γ-lactone, *D-Gluconsäure-γ-lakton* 210.

Glucose-1-C¹⁴ 207.

Glucose-cycloacetoacetate, *Glucose-cycloacetoacetat* 206.

Glucose oxidase, *Glucoseoxydase* 688—689.

— —, oxygen affinity, *Sauerstoffaffinität* 650.

α-D-glucose 1-phosphate, *α-D-glucose 1-phosphat* 474, 478, 485, 486, 492, 510, 511, 512, 513.

Glucose-6-phosphate, *Glucose-6-phosphat* 207

— dehydrogenase see G-6-P dehydrogenase

D-glucuronic acid, *D-Glucuronsäure* 208.

— — kinase, from *Phaseolus aureus*, assay,. *D-Glucuronsäurekinase, aus Phaseolus aureus, Nachweis* 483.

— — —, — —, preparation, *Darstellung* 483.

— — —, — —, properties, *Eigenschaften* 484.

— — lactone, *D-Glucuronsäurelakton* 207, 211.

— acid-γ-lactone, *D-Glucuronsäure-γ-lakton* 206, 208, 211, 212.

α-D-glucuronic acid 1-phosphate, *α-D-glucuronsäure 1-phosphat* 478, 483, 485, 486.

D-glutamate, *D-Glutamat* 399.

Glutamate-alanine transaminase, *Glutamatalanintransaminase* 368—369.

Glutamate-aspartate transaminase, *Glutamataspartattransaminase* 208.

— —, assay, *Nachweis* 363.

— —, —, chromatographic, *chromatographischer* 364.

Glutamate-aspartate, transaminase. assay, by measurement of α-ketoglutarate, *durch Messung von α-Ketoglutarat* 365.

— —, —, — of oxalacetate, *Oxalacetat* 364.

— —, —, with specific decarboxylases, *mit spezifischen Decarboxylasen* 364.

— —, properties, *Eigenschaften* 367—368.

— —, purification, *Reinigung* 366—367.

Glutamate-phosphohistidinol transaminase, *Glutamatphosphohistidinoltransaminase* 375—376.

Glutamate racemase, *Glutamatracemase* 383—384.

Glutamic acid decarboxylase, *Escherichia coli*, *Glutaminsäuredecarboxylase, Escherichia coli* 305—306.

L-Glutamic acid dehydrogenase, of microorganisms, L-*Glutaminsäuredehydrogenase, Mikroorganismen* 293—294.

D-Glutamic acid oxidase, D-*Glutaminsäureoxydase* 298.

Glutamic decarboxylase, *Glutaminsäuredecarboxylase* 303—304.

Glutamic dehydrogenase, *Glutaminsäuredehydrogenase* 158.

— —, chloroplasts, *Chloroplasten* 570.

L-Glutamic dehydrogenase, higher plants, L-*Glutaminsäuredehydrogenase, höhere Pflanzen* 291—292.

Glutaminase 413.

Glutamine synthetase, *Glutaminsynthetase* 393.

— —, chloroplasts, *Chloroplasten* 570.

— —, electrophoresis, *Elektrophorese* 397.

— —, optical specificity, *optische Spezifizität* 399.

— —, purification, *Reinigung* 395, 396.

— —, stability, *Stabilität* 399.

— —, ultracentrifugal analysis, *Sedimentationsanalyse* 397.

Glutamine transaminases, *Glutamintransaminasen* 377—378 .

γ-Glutamyl hydroxamate, γ-*Glutamylhydroxamat* 393.

Glutamyl transfer, *Glutamyl Übertragung* 394.

Glutamylcysteine synthetase, *Glutamylcysteinsynthetase* 401.

— —, preparation, *Darstellung* 402.

Glutathione reductase, *Glutathionreduktase* 680.

Glutathione synthetase, *Glutathionsynthetase* 403.

Glycerate-2,3-diphosphatase, *Glycerat-2,3-diphosphatase* 40.

D-glyceraldehyde, D-*Glycerinaldehyd* 206.

DL-glyceraldehyde, DL-*Glycerinaldehyd* 206.

L-glyceraldehyde, L-*Glycerinaldehyd* 205.

Glycinamide ribotide kinosynthase, *Glycinamidribotidkinosynthase* 429.

— — —, assay, *Nachweis* 429.

— — —, preparation, *Darstellung* 431.

Glycinamide ribotide transformylase, *Glycinamidribotidtransformylase* 432.

— — —, assay, *Nachweis* 433.

— — —, preparation, *Darstellung* 434.

Glycine, synthesis, *Glycin, Synthese* 349.

Glycine-glutamate transaminase, *Glycinglutamattransaminase* 380.

Glycine oxidase, *Glycinoxydase* 642—643.

Glycine reductase system, *Glycinreduktase Systeme* 318, 319.

Glycolaldehyde, *Glykolaldehyd* 218.

Glycolaldehyde dehydrogenase, *Glykolaldehyddehydrogenase* 642.

Glycoldehyde, transformation, *Glycoldehyd, Umwandlung* 352.

Glycolic acid oxidase, *Glykolsäureoxydase* 638—640, 686.

— — —, assay, *Nachweis* 687.

— — —, properties, *Eigenschaften* 688.

— — —, purification, *Reinigung* 687.

Glycolytic system, activators and inhibitors, *Glycolitisches System, Aktivatoren und Hemmstoffe* 522.

— —, assay, *Nachweis* 520.

— —, preparation, *Darstellung* 521.

Glycolysis, *Glycolyse* 520.

Glycosides, synthesis, *Glykoside, Synthese* 504,

Glycylglycine dipeptidase, *Glycylglycindipeptidase* 414.

Glyoxalate cycle, *Glyoxalatcyclus* 636.

— —, enzymes of, *Enzyme des* 616.

Glyoxalic acid reductase, *Glyoxalsäurereduktase* 641—642.

Glyoxalic acid, *Glyoxalsäure* 686, 687.

G-6-P dehydrogenase, assay, *G-6-P-dehydrogenase, Nachweis* 552—554.

—, purification, *Reinigung* 554, 555.

L-gulonic acid, L-*Gulonsäure* 207, 208, 211,215.

L-gulonic acid dehydrogenase, L-*Gulonsäuredehydrogenase* 213, 214.

L-gulonic acid lactone, L-*Gulonsäurelakton* 213.

D-gulonic acid-γ-lactone, D-*Gulonsäure-γ-lakton* 210.

L-gulonic acid-γ-lactone, L-*Gulonsäure-γ-lakton* 206, 207, 212, 215.

Guaiacol, *Guajakol* 685.

Guaiacylglycerol-bis-coniferyl ether, *Guajacylglycerol-bis-coniferyläther* 283, 284.

Guaiacylglycerol-β-coniferyl ether, *Guajacylglycerol-β-coniferyläther* 283, 284.

Guaiacylglycerol-α-dehydrodiconiferyl-β-coniferyl ether, *Guajacylglycerol-α-dehydrodiconiferyl-β-coniferyläther* 283, 284.

Guaiacylpropane derivatives, *Guajacylpropanderivaten* 283.

Hemicellulose 476.

Heteropyrithiamine, determination, *Heteropyrithiamin, Bestimmung* 195.

Hexokinase 525, 547.

—, activators and inhibitors, *Aktivatoren und Hemmstoffe* 526.

—, assay, *Nachweis* 525, 550, 551.

—, preparation, *Darstellung* 525.

—, purification, *Reinigung* 551, 552.

Histidase 310

L-Histidine, biosynthesis, L-*Histidin, Biosynthese* 329.

Histidine-alanine, transaminase, *Histidin-alanintransaminase* 380.
L-Histidinol dehydrogenase 332—333.
L-Histidinol phosphate phosphatase, L-*Histidinolphosphatphosphatase* 332.
Homocysteine desulfhydrase, *Homocysteindesulfhydrase* 313.
Homoserine dehydrogenase, *Homoserindehydrogenase* 327.
L-Homoserine kinase, L-*Homoserinkinase* 327—328.
Homoserine phosphate mutaphosphatase, *Homoserinphosphatmutaphosphatase* 328—329.
Hydroquinone, *Hydrochinon* 683.
Hydroxyanthranilic acid, *Oxyanthranilsäure* 179.
γ-Hydroxybutyrate dehydrogenase, *γ-Oxybutyratdehydrogenase* 348.
α-Hydroxy-β-keto acid reductoisomerase, α-*Hydroxy-β-ketosäurereductoisomerase* 346.
3-Hydroxykynurenine, *3-Hydroxykynurenin* 178.
Hydroxylamine, determination, CSÁKY method, *Hydroxylamin, Bestimmung*, CSÁKYs *Methode* 128.
—, —, 8-hydroxyquinoline method, *8-oxychinolinmethode* 128.
—, —, nitrobenzene method, *Nitrobenzolmethode* 129.
—, —, o-phenanthroline method, *o-phenanthrolinmethode* 129.
—, estimation, *Berechnung* 125.
Hydroxylamine oxidase, assay, *Hydroxylaminoxydase, Nachweis* 165.
Hydroxylamine "oxidase", of higher plants, *Hydroxylamin „Oxydase", höhere Pflanzen* 160—161.
Hydroxylamine oxidase systems, co-factors, *Hydroxylaminoxydasesysteme, Cofaktoren* 167.
— —, inhibitors, *Hemmstoffe* 168.
— —, stability, *Stabilität* 168.
—, oxidation 158—160.
— reductase, assay, dye reduction, *Hydroxylaminreduktase, Nachweis, Farbstoffreduktion* 130—133.
— —, assay methods, sources of error, *Nachweismethoden, Fehlerquellen* 121—126.
— —, electron donors and co-factors, bacteria, *Elektronendonatoren und Cofaktoren, Bakterien* 145—146.
— —, — —, fungi and higher plants, *Pilze und höhere Pflanzen* 138, 139, 142.
— —, extraction from bacteria, *Extraktion aus Bakterien* 117.
— —, inhibitors fungi and higher plants, *Hemmstoffe, Pilze und höhere Pflanzen* 140—142.
— —, physiological factors, fungi and higher plants, *physiologische Faktoren, Pilze und höhere Pflanzen* 158.
— —, properties and mechanisms, bacteria, *Eigenschaften und Mechanismen, Bakterien* 155—157.

Hydroxylamine, properties and mechanisms, fungi and higher plants, *Eigenschaften und Mechanismen, Pilze und höhere Pflanzen* 146, 147.
— —, preparation, *Darstellung* 115.
— —, purification, *Reinigung* 118, 120.
— —, systems, inhibitors, *Hydroxylaminreduktasesysteme, Hemmstoffe* 154, 155.
— — —, stability, *Stabilität* 154, 155.
Hydroxylamine reductases, *Hydroxylaminreduktasen* 112.
Hydroxymethyltetrahydrofolate dehydrogenase, *Oxymethyltetrahydrofolatdehydrogenase* 350.
β-Hydroxypropionic acid, *β-Oxypropionsäure* 466.
p-Hydroxyphenylpyruvic acid, *p-Oxyphenylbrenztraubensäure* 261, 274.
Hydroxyquinolines, *Oxychinoline* 655.
Hyponitrite, determination, *Hyponitrit, Bestimmung* 129, 130.
—, preparation, *Darstellung* 136.
Hyponitrite reductase, assay methods sources of error, *Hyponitritreduktase, Nachweismethoden, Fehlerquellen* 121—126.
— —, electron donors, fungi and higher plants, *Elektronendonatoren, Pilze und höhere Pflanzen* 141.
— —, inhibitors, fungi and higher plants, *Hemmstoffe, Pilze und höhere Pflanzen* 140—142.
— —, preparation, *Darstellung* 114.
Hyponitrite reductases, *Hyponitritreduktasen* 112.

IAA (Indole acetic acid), *Indolessigsäure (IES)* 238, 685.
—, breakdown, *Abbau* 248.
—, bioassay, *biologischer Test* 238.
—, colorimetric assay methods, *colorimetrische Nachweismethoden* 238.
—, oxidation reaction, *Oxydationsreaktion* 249.
—, oxidising enzymes, specificity, *oxydierende Enzyme, Spezifität* 252.
—, synthesis, *Synthese* 240.
—, separation from precursors, *Trennung von Vorstufen* 239.
IAA oxidase, distribution, *Indolessigsäureoxydase, Vorkommen* 248.
IAA oxidases, preparation and properties, *Indolessigsäureoxydasen, Darstellung und Eigenschaften* 251.
IAc (3-indoleacetaldehyde), conversion to IAA, *IAc (3-Indolacetaldehyd), Umbau in Indolessigsäure* 243.
Imidazoleacetol phosphate transaminase, *Imidazolacetolphosphattransaminase* 331.
Imidazoleglycerol phosphate dehydrase, *Imidazolglycerolphosphatdehydrase* 329, 331.
Indoleacetic acid (IAA), *Indolessigsäure (IES)* 238, 685.
Indole-3-glycerol phosphate synthetase, *Indol-3-glycerolphosphatsynthetase* 353—354.
Inosinicase 438.

Inositol 200.
Isocitric dehydrogenase, *Isocitronensäure-dehydrogenase* 189.
— —, chloroplasts, *Chloroplasten* 570.
— —, DPN specific, *DPN spezifische* 621.
— —, TPN specific, *TPN spezifische* 620.
Isocitritase 617, 637—638.
Isomerase product, preparation, *Isomeraseprodukt, Darstellung* 558.

KDHP (2-keto-3-deoxy-D-araboheptonic acid-7-phosphate, *KDHP (2-Keto-3-deoxy-D-araboheptonsäure-7-phosphat)*.
—, enzymic conversion to dehydroquinate, *enzymatischer Umbau in Dehydrochinat* 265.
—, identification by paper chromatography, *Identifikation durch Papierchromatographie* 263.
—, preparation, *Darstellung* 263.
—, properties, *Eigenschaften* 264.
— synthetase 262, 265.
— —, assay, *Nachweis* 262.
KDPA (2-keto-3-deoxy-7-phosphoheptonic acid), *KDPA (2-Keto-3-deoxy-7-phosphoheptonsäure)* 262.
α-Keto acid oxidases, *α-Ketosäureoxydasen* 621—623.
2-Keto-3-deoxy-D-araboheptonic acid(KDA), *2-Keto-3-deoxy-D-araboheptonsäure (KDA)* 261, 262.
2-Keto-3-deoxy-D-araboheptonic acid-7-phosphate (KDHP), *2-Keto-3-deoxy-D-araboheptonsäure-7-phosphat (KDHP)* 260, 261.
2-Keto-gulonic acid, *2-Keto-gulonsäure* 205, 211, 212.
2-Keto-L-gulonic acid, *2-Keto-L-gulonsäure* 208.
3-Keto-L-gulonic acid, *3-Keto-L-gulonsäure* 212, 213.
KREBS cycle, enzymes of, *Krebs Cyclus, Enzyme des* 616.
Kynureninase 179.
L-Kynurenine, *L-Kynurenin* 179.
Kynurenine formamidase, *Kynureninformamidase* 177, 178.
Kynurenine hydroxylase, *Kynurenin-hydroxylase* 178.
Kynurenine transaminase, *Kynurenintransaminase* 374—375.

Laccase 218, 285, 287, 650, 681.
—, assay and purification, *Nachweis und Reinigung* 683
Lactic dehydrogenase, *Milchsäuredehydrogenase* 623, 632—634.
Lactonase, *Laktonase* 552
—, assay and purification, *Nachweis und Reinigung* 556.
Lactonase I, *Laktonase I* 214.
Lactonase II, *Laktonase II* 214.
Lecithinase 454.
Leucine aminopeptidase, *Leucinaminopeptidase* 414.
Leucine decarboxylase *Leucindecarboxylase* 307.

L-Leucine dehydrogenase, *L-Leucindehydrogenase* 295—296
Lignin, coniferous synthesis, *Coniferensynthese* 283.
—, formation from shikimic acid, *Bildung aus Shikimisäure* 278.
Lignin synthesis, *Ligninsynthese* 277—278.
Lipases, *Lipasen* 448.
—, activators and inhibitors, *Aktivatoren und Hemmstoffe* 450, 451
—, assay procedures, *Nachweismethoden* 448—450.
—, — —, enzyme preparation, *Darstellung* 449.
—, definition, *Definition* 448.
—, distribution, *Vorkommen* 448.
—, pH optima, *pH Optima* 451.
—, purification, *Reinigung* 451, 452.
—, specificity, *Spezifität* 451.
Lipoic acid, *Liponsäure* 622.
Lipoxidase, *Lipoxydase* 469, 688.
—, assay, *Nachweis* 469.
—, —, manometric, *manometrischer* 469.
—, —, spectrophotometric, *spektrophotometrischer* 469.
—, properties, *Eigenschaften* 470.
—, purification, *Reinigung* 470.
Lipoyl dehydrogenase 664.
C—S-Lyase 316—317.
Lysine decarboxylase, E. coli, *Lysindecarboxylase, E. coli* 306.
Lysine racemase, *Lysinracemase* 386.
Lysolecithin isomerase, 462.
Lysophospholipase B 456.
Lyxonic acid, *Lyxonsäure* 218

Malate synthetase, *Malatsynthetase* 636.
Malease 629—630.
Malic dehydrogenase, *Apfelsäuredehydrogenase* 617, 628.
— —, chloroplasts, *Chloroplasten* 570.
Malic enzyme, *Apfelsäureenzym* 635.
Maltodextrins, *Malzdextrine* 516.
Maltotriose 513.
MalylCoA 637.
D-Mannonic acid-γ-lacton, *D-Mannonsäure-γ-lakton* 207.
α-D-Mannose 1-phosphate, *α-D-mannose 1-phosphat* 478.
D-Mannuronic acid, *D-Mannuronsäure* 211.
Melibiose 498.
Menadione reductase, *Menadionreduktase* 665.
Methemoglobin reducing factor, *Methämoglobin-reduzierender Faktor* 590.
Methionine activating enzyme, *Methionin aktivierendes Enzym* 321.
Methionine dethiomethylase, *Methionin-dethiomethylase* 314.
Methionine racemase, *Methioninracemase* 385—386.
6-methylaminopurine, *6-Methylaminopurin* 405.
5-methylcytosine, *5-Methylcytosin* 405.

S-Methylmethionine-homocysteine trans-methylase, *S-Methylmethionin-homo-cystein transmethylase* 322—323.
Myokinase 624.

Naphthoquinones, photophosphorylation, *Naphthochinone, Photophosphorylierung* 610.
Nicotinic acid, *Nicotinsäure* 176, 177, 187, 196.
Nicotinic acid methylpherase, *Nicotinsäuremethylferase* 324—325.
Nitrate assimilation, *Nitratassimilation* 67.
Nitrate reductase, *Nitratreduktase* 67.
— —, assay methods, *Bestimmungsmethoden* 83.
— —, — —, estimation of nitrate loss, *Berechnung von Nitratverlust* 88.
— —, — —, manometric method, *manometrische Methode* 85.
— —, — —, sources of error, *Fehlerquellen* 89—91.
— —, — —, spectrophotometric method, *spektrophotometrische Methode* 88.
— —, bacteria, *Bakterien* 76.
— —, —, assimilatory enzyme, *assimilierendes Enzym* 82.
— —, —, dissimilatory enzyme, *dissimilierendes Enzym* 78
— —, —, electron donors and co-factors, *Elektronendonatoren und Cofaktoren* 95, 96.
— —, —, fractionation and stability, *Fraktionierung und Stabilität* 78.
— —, —, stability and inhibitors, *Stabilität und Hemmstoffe* 101—107.
— —, —, substrates and co-factors, *Substrate und Cofaktoren* 98—101.
— —, distribution in cell-free extracts, *Verteilung in zellfreien Extrakten* 79.
— —, fractionation and stability, *Fraktionierung und Stabilität* 71.
— —, fungi and higher plants, electron donors and co-factors, *Pilze und höhere Pflanzen, Elektronendonatoren und Cofaktoren* 91—93.
— —, — — —, inhibitors, *Hemmstoffe* 96, 97, 102.
— —, — — —, stability, *Stabilität* 96, 102.
— —, — — —, mechanism of action, *Wirkungsmechanismus* 97, 100, 101.
— —, GRIESS-ILOSVAY colorimetric method *colorimetrische Methode nach* GRIESS-ILOSVAY 83—87.
— —, heavy metals, *Schwermetalle* 71
— —, metal and anion requirements, *Metall- und Anionenbedarf* 94.
— —, methods of extraction, *Extraktionsmethoden* 68.
— —, — —, bacteria, *Bakterien* 74, 75, 76, 77.
— —, — —, fungi, *Pilze* 72.
— —, — —, higher plants, *aus höheren Pflanzen* 69.

Nitrate reductase, physiological factors, bacteria, *Beeinflussung durch physiologische Faktoren, Bakterien* 109.
— —, — —, fungi and higher plants. *Pilze und höhere Pflanzen* 108.
— —, preparation, *Darstellung* 68.
— —, thermal inactivation, *Hitzeinaktivierung* 73.
Nitrate reduction, in bacteria, *Nitratreduzierung, in Bakterien* 68.
— —, dissimilatory, *dissimilierende* 68.
— —, light dependence, *Lichtabhängigkeit* 67.
— respiration, *Nitratatmung* 68.
Nitric oxide reductase, assay, manometric, *Stickoxydreduktase, Aktivitätsbestimmung, manometrische* 134—136.
— — —, co-factors and inhibitors, bacteria, *Cofaktoren und Hemmstoffe, Bakterien* 144, 145.
— —, electron donors, bacteria, *Elektronendonatoren, Bakterien* 144, 145.
— — —, extraction from bacteria, *Extraktion aus Bakterien* 117.
— — —, properties and mechanisms, bacteria, *Eigenschaften und Mechanismen, Bakterien* 153.
— — —, purification, *Reinigung* 120.
Nitric oxide reductase systems, inhibitors, *Stickoxdyreduktasesysteme, Hemmstoffe* 154, 155.
— — —, stability, *Stabilität* 154, 155.
Nitric oxide reductases, *Stickoxydreduktasen* 112.
Nitrite, oxidation, *Nitrit, Oxydation* 158—160.
Nitrite oxidase systems, co-factors, *Nitritoxydasesysteme, Co-faktoren* 167.
— — —, inhibitors, *Hemmstoffe* 168.
— — —, stability, *Stabilität* 168.
Nitrite reductase, assay, dye reduction, *Nitritreduktase, Nachweis, Farbstoffreduktion* 130—133.
— —, —, manometric, *manometrischer* 124, 125, 133.
— —, assay methods, sources of error, *Nachweismethoden, Fehlerquellen* 121—126.
— —, co-factors and inhibitors, bacteria, *Cofaktoren und Hemmstoffe, Bakterien* 142—144.
— —, — —, fungi and higher plants, *Pilze und höhere Pflanzen* 137—139, 141.
— —, electron donors, bacteria, *Elektronendonatoren, Bakterien* 142—144.
— —, — —, fungi and higher plants, *Pilze und höhere Pflanzen* 137—139, 141.
— —, extraction from bacteria, *Extraktion aus Bakterien* 116.
— —, physiological factors, fungi and higher plants, *physiologische Faktoren, Pilze und höhere Pflanzen* 157, 158.
— —, preparation, *Darstellung* 113.
— —, properties and mechanisms, bacteria, *Eigenschaften und Mechanismen, Bakterien* 148, 149—153.

728   Subject Index.

Nitrite reductase, assay dye, fungi and higher plants, *Pilze und höhere Pflanzen* 146, 147.
— —, purification, *Reinigung* 114, 118, 119.
— —, stability, *Stabilität* 150, 151.
Nitrite reductases, *Nitritreduktasen* 112.
Nitro-compounds, aromatic, reduction, *Nitroverbindungen, aromatische, Reduktion* 111, 112.
NMN (Nicotinamide mononucleotide), *NMN (Nicotinsäureamidmononucleotid)*
NMN-phosphatase, *NMN-Phosphatase* 197.
5'-Nucleotidases, *5'-Nucleotidasen* 57.
Nucleotide pyrophosphatases, *Nucleotidpyrophosphatasen* 55.

Omphalia enzyme, *Omphalia Enzym* 256.
Ornithine, synthesis, *Ornithin, Synthese* 335.
Ornithine transaminase, *Ornithintransaminase* 376.
Ornithine δ-transaminase, *Ornithin δ-transaminase* 335.
Ornithine transcarbamylase, *Ornithintranscarbamylase* 338—339.
Orotidine-5'-phosphate decarboxylase, *Orotidin-5'-phosphatdecarboxylase* 446.
Orotidine-5'-phosphate pyrophosphorylase, *Orotidin-5'-phosphatpyrophosphorylase* 446.
Ortho phosphoserine phosphatase 54.
Oxalic acid, *Oxalsäure* 218.
Oxaloacetate, *Oxalacetat* 617, 618, 628.
Oxidase, spruce sap, *Oxydase, Fichtensaft* 286 bis 287.
α-Oxihation, *α-Oxydation* 467.
—, inhibitors, *Hemmstoffe* 468.
β-Oxidation, *β-Oxydation* 465.
—, even-chain fatty acids, *geradekettige Fettsäuren* 465.
—, odd-chain fatty acids, *ungeradekettige Fettsäuren* 466.
Oximes, determination, *Oxime, Bestimmung* 129.
α-Oxoglutarate oxidase, *α-Oxoglutaratoxydase* 621.
Oxygen affinity, apparent. *Sauerstoffaffinität, scheinbare* 650.
Oxythiamine, *Oxythiamin* 181

Pantetheine, *Pantethein* 191.
Pantetheine 4'-phosphate, *Pantethein 4'-phosphat* 199.
Pantothenate kinase, *Pantothenkinatase* 191.
Pantothenate synthetase, *Pantothenatsynthetase* 180.
Pantothenic acid, *Pantothensäure* 180, 190, 191, 198, 207.
Pantothenylcysteine, *Pantothenylcystein* 191
Pantoyltaurine, *Pantoyltaurin* 207.
Papain, purification, *Reinigung* 417.
Pea enzyme, *Erbsenzym* 253.
Pectin, *Pektin* 476.
Pentabromoacetone, *Pentabromoaceton* 617.

Pentose phosphate cycle, enzymes of, *Pentosephosphatcyclus, Enzyme des* 546—549.
Peptidases, *Peptidasen* 412, 414.
Peptide synthesis, *Peptidsynthese* 401.
Peroxidase, *Peroxydase* 216, 285.
—, assay and purification, *Nachweis und Reinigung* 685.
—, determination, *Bestimmung* 216
—, properties, *Eigenschaften* 686.
Peroxidases, *Peroxydasen* 684—686.
Phenazine methosulphate, *Phenazinmethosulfat* 625
Phenoloxidases, mushroom, *Phenoloxydasen, Pilze* 285.
Phenylalanine, *Phenylalanin* 261
—, biosynthesis, *Biosynthese* 261.
Phenylalanine-alanine transaminase, *Phenylalanin-alanintransaminase* 380.
Phenylalanine deaminase, *Phenylalanindeaminase* 281—283.
L-Phenylalanine, L-*Phenylalanin* 281.
Phenylcoumaran derivative, *Phenylcumaranderivat* 283.
Phenyllactic acid, *Phenylmilchsäure* 274.
Phenylpyruvic acid, *Phenylbrenztraubensäure* 261, 274.
Phosphatase, assay, *Nachweis* 22.
— —, orthophosphate effect, *Orthophosphat-Wirkung* 28.
Phosphatases, acid, *Phosphatasen, saure* 51.
—, alkaline, *alkalische* 51.
—, localization, *Lokalisierung* 34.
Phosphate metabolism, enzymes, *Phosphatstoffwechsel, Enzyme des* 21.
— —, —, cell wall fraction, *Zellwandfraktion* 33.
— —, —, general assay methods, *allgemeine Versuchsmethoden* 25—28.
— —, —, isolation, *Isolierung* 21.
— —, —, larger particles fraction, *"large particles" Fraktion* 33.
— —, —, mitochondrial fraction, *Mitochondrienfraktion* 33.
— —, —, nuclear fraction, *Kern-Fraktion* 32.
— —, —, plastid fraction, *Plastiden-Fraktion* 32.
— —, —, purification, *Reinigung* 21.
— —, —, stability, *Stabilität* 22.
Phosphatidic acid phosphatase, *Phosphatidsäurephosphatase* 462.
Phosphatidylglycerol 571.
Phosphodiesterases, *Phosphodiesterasen* 62.
Phosphoenolpyruvate carboxykinase, *Phosphoenolpyruvatcarboxykinase* 635.
Phosphoenolpyruvate carboxylase, chloroplasts, *Phosphoenolpyruvatcarboxylase, Chloroplasten* 570
Phosphoenolpyruvic carboxylase, *Phosphoenolbrenztraubensäurecarboxylase* 634.
Phosphoglucomutase 523.
—, activators, *Aktivatoren* 524.
—, assay, *Nachweis* 524.
—, preparation, *Darstellung* 524.
6-phosphogluconic acid, *6-Phosphogluconsäure* 207.

6-Phosphogluconic dehydrogenase, assay, *6-Phosphogluconsäuredehydrogenase, Nachweis* 552—554.
— —, purification, *Reinigung* 554, 555
Phosphoglucose isomerase 526.
— —, assay, *Nachweis* 526.
— —, inhibitors, *Hemmstoffe* 527.
— —, preparation, *Darstellung* 526.
3-Phosphoglycerate, transformation, *3-Phosphoglycerat, Umwandlung* 349.
Phosphoglycerate kinase, *Phosphoglycerat kinase* 537.
— —, activators and inhibitors, *Aktivatoren und Hemmstoffe* 539.
— —, assay, *Nachweis* 537.
— —, distribution, *Vorkommen* 539.
— —, preparation, *Darstellung* 538.
Phosphoglyceric acid mutase, *Phosphoglyceromutase* 539
— —, activators and inhibitors, *Aktivatoren und Hemmstoffe* 540.
— — —, assay, *Aktivitätsbestimmung* 539.
— — —, distribution, *Vorkommen* 540.
Phosphohexoisomerase 566, 567.
Phosphohexokinase 527.
— activators and inhibitors, *Aktivatoren und Hemmstoffe* 528.
—, assay, *Nachweis* 527.
—, preparation, *Darstellung* 528.
Phosphoinositide phosphorylase, *Phosphinositidphosphorylase* 462.
Phosphoketopentoepimerase 558.
—, assay, *Nachweis* 562, 563.
—, purification, *Reinigung* 563, 564.
Phospholipase A, assay method, *Nachweismethode* 455.
—, cofactors and inhibitors, *Co-faktoren und Hemmstoffe* 455.
—, distribution, *Vorkommen* 454.
—, pH optima, *pH Optima* 456.
—, purification, *Reinigung* 456.
—, specificity, *Spezifität* 456.
Phospholipase B 456.
—, activators and inhibitors, *Aktivatoren und Hemmstoffe* 458.
—, assay method, *Nachweismethode* 456, 457.
—, pH optima, *pH Optima* 458.
—, purification, *Reinigung* 459.
—, specificity, *Spezifität* 457.
—, stability, *Stabilität* 458.
Phospholipase C 459
—, activators, *Aktivatoren* 460.
—, assay method, *Nachweismethode* 459.
—, pH optima, *pH Optima* 460.
—, purification, *Reinigung* 460.
—, stability, *Stabilität* 460.
—, substrate specificity, *Spezifität der Substrat* 460.
Phospholipase D 461.
—, activators, *Aktivatoren* 462.
—, assay method, *Nachweismethode* 461.
—, pH optima, *pH Optima* 462.
—, purification, *Reinigung* 462.
—, stability, *Stabilität* 462.

Phospholipase, substrate specificity, *Spezifität der Substrat* 462.
Phospholipases, *Phospholipasen* 454.
—, nomenclature, *Nomenklatur* 454.
Phospholipids, *Phospholipide* 462.
Phosphomonoesterases, *Phosphomonoesterasen* 50
4-Phosphopantothenate, *4-Phosphopantothenat* 192.
Phosphopantothenic acid, *Phosphopantothensäure* 191.
4-Phosphopantothenylcysteine, *4-Phosphopantothenylcystein* 192.
Phosphopantothenylcysteine decarboxylase, *Phosphopantothenylcysteindecarboxylase* 192.
Phosphoprotein phosphatase 52.
Phosphoriboisomerase 577.
—, assay, *Nachweis* 556, 557.
—, purification, *Reinigung* 557, 558.
5-Phosphoribosylpyrophosphate kinase see PRPP-kinase, *5-Phosphoribosylpyrophosphat kinase s. PRPP-kinase.*
Phosphoribosylpyrophosphate-nicotinic acid transferase, *Phosphoribosylpyrophosphat-nicotinsäure-transferase* 188.
Phosphoribulokinase 577.
—, assay method, *Nachweismethode* 577.
—, properties, *Eigenschaften* 578.
—, purification, *Reinigung* 578.
Phosphorus, colorimetric estimation, *Phosphor, colorimetrische Bestimmung* 23.
Phosphorylase 522.
—, assay, *Nachweis* 522.
—, inhibitors, *Hemmstoffe* 523.
—, from potatoes, *aus Kartoffeln* 511—514.
—, properties, *Eigenschaften* 523.
—, purification, *Reinigung* 523.
Phosphorylation, oxidative, efficiency, *Phosphorylierung, oxydative, Leistungsfähigkeit* 651.
—, —, manometric method, *manometrische Methode* 652.
—, —, oxygen electrode method, *Sauerstoff-Elektroden-Methode* 652.
—, photosynthetic, *photosynthetische* 582.
3-Phosphoshikimic acid, *3-Phosphoshikimisäure* 271.
5-Phosphoshikimic acid, *5-Phosphoshikimisäure* 261, 262, 272.
Phosphotransacetylase 193, 194.
—, bacterial, *Bakterien* 622.
Photophosphorylation, bacterial, *Photophosphorylierung, Bakterien* 612.
—, cyclic, *zyklische* 583, 610.
—, —, in chloroplasts, *in Chloroplasten* 584.
—, —, chromatophores, *Chromatophoren* 585.
—, —, mechanism, *Mechanismus* 583.
—, noncyclic, in chloroplasts, *nichtzyklische, in Chloroplasten* 587, 588.
—, —, chromatophores, *Chromatophoren* 588.
—, —, mechanism, *Mechanismus* 586, 587.
Photosynthesis, carbon dioxide assimilation, *Photosynthese, Kohlendioxydassimilation* 574.

Photosynthesis, enzyme systems, *Enzym-systeme* 569.
—, Hill reaction, *Hill-Reaktion* 570.
—, quinones in, *Chinone in* 605, 610.
—, separation of light and dark phases, *Teilung von Licht- und Dunkel-Phasen* 575.
Phytase 200.
Pineapple enzyme, *Ananasenzym* 255.
dl-Pinoresinol 283, 284.
Plastocyanin 589.
Plastoquinone, *Plastochinon* 605, 606, 611, 677, 678.
—, extraction and purification, *Extraktion und Reinigung* 607—609.
—, properties, *Eigenschaften* 609.
Plastoquinone A, *Plastochinon A* 606, 612.
Plastoquinone B, *Plastochinon B* 606, 612.
Plastoquinone C, *Plastochinon C* 606.
Polyphenolase, plants, *Polyphenolase, Pflanzen* 642.
Polyphenol oxidase, *Polyphenoloxydase* 217, 680, 681, 682.
— —, chronometric determination, *chronometrische Bestimmung* 217.
— —, "latent", „*latente*" 589.
— —, oxygen affinity, *Sauerstoffaffinität* 650, 651.
— —, purification, *Reinigung* 683.
Polyphosphate-ADP-phosphotransferase, *Polyphosphat-ADP-phosphotransferase* 50
Polyphosphate-AMP-phosphotransferase, *Polyphosphat-AMP-phosphotransferase* 49.
Polyphosphate depolymerases, *Polyphosphatdepolymerasen* 47.
Polyphosphatases, *Polyphosphatasen* 47.
Polysaccharide, chromatophores, *Polysaccharid, Chromatophoren* 572.
Polysaccharides, synthesis, *Polysaccharide, Synthese* 507.
Potato apyrase, *Kartoffelapyrase* 187.
PPNR (photosynthetic pyridine nucleotide reductase), *PPNR (photosynthetische Pyridinnucleotidreduktase)* 590.
Prephenic acid, *Prephensäure* 260, 261, 262, 274.
Prephenic aromatase, *Prephensäurearomatase* 262, 274.
Prephenic dehydrogenase, *Prephensäuredehydrogenase* 274, 275.
Prolidase 414.
Prolinase 414.
l-Proline, biosynthesis, l-*Prolin, Biosynthese* 333, 334.
Proline racemase, *Prolinracemase* 387, 388.
Proline reductase, *Prolinreduktase* 318.
Propionyl CoA 637.
Protease activity, measurement, *Proteaseaktivität, Messung* 416.
Proteases, *Proteasen* 412, 416.
Protein synthesis, *Proteinsynthese* 405, 408.
Protocatechuic acid, *Protocatechusäure* 261, 265.
— —, assays, *Nachweismethoden* 266.
— — oxidase, *Protocatechusäureoxydase* 266.

Protoplasts, isolation, *Protoplasten, Isolierung* 34.
PRPP-amidotransferase 426.
—, assay, *Nachweis* 426.
—, preparations, *Darstellung* 427.
PRPP-kinase 422.
—, assay, *Nachweis* 422.
—, preparation, *Darstellung* 424.
"Pseudohemoglobin" see RHP, „*Pseudohämoglobin*" *s. RHP* 603.
Pseudo-uridine, *Pseudouridin* 405.
Purine nucleotides, enzymes of synthesis, *Purinnucleotide, Enzyme de Synthese* 421, 422.
Pyridine nucleotide and nitrate reductase, *Pyrimidinnucleotid und Nitratreduktase* 91.
— nucleotides, oxidation by means other than cytochrome system, *Pyridinnucleotide, Oxydation durch andere Mittel als Cytochromsysteme* 679, 680.
Pyrimidine kinase, *Pyrimidinkinase* 174.
Pyrimidine nucleotides, enzymes of synthesis, *Pyrimidinnucleotide, Enzyme der Synthese* 421, 439.
Pyridoxal (Vitamin B$_6$) 186.
Pyridoxal phosphokinase 186.
Pyrithiamine, *Pyrithiamin* 181, 207.
Pyrogallol 685.
Pyrophosphate, determination, *Pyrophosphat, Bestimmung* 193.
Pyrophosphatases, inorganic, *Pyrophosphatasen, anorganische* 41.
0-5-P Pyrophosphorylase, preparation, *0-5-P pyrophosphorylase, Darstellung* 425.
Δ'-Pyrroline-5-carboxylate reductase, *Δ'-Pyrrolin-5-carboxylatreduktase* 334.
Pyruvate kinase, *Pyruvatkinase* 543.
— —, activators and inhibitors, *Aktivatoren und Hemmstoffe* 544
— —, assay, *Nachweis* 543.
— —, distribution, *Vorkommen* 544.
Pyruvic (de)carboxylase, *Brenztraubensäure-(de)carboxylase* 630—632.
Pyruvic oxidase, *Brenztraubensäureoxydase* 621.

Quinic acid, *Chinasäure* 260, 261, 268.
Quinic dehydrogenase, *Chinasäuredehydrogenase* 268—270.
Quinone reductase, *Chinonreduktase* 664, 665.
Quinones, in photosynthesis, *Chinone, in der Photosynthese* 605.

Racemases, *Racemasen* 381.
Raffinose 476, 498.
Respiratory chain, *Atmungskette* 647.
— — studies, use of cytochrome inhibitors, *Anwendung von Cytochrom-Hemmstoffen* 654.
— — —, use of flavoprotein inhibitors, *Anwendung von Flavoprotein Hemmstoffe* 655.
— — —, use of oxidase inhibitors, *Anwendung von Oxydase-Hemmstoffe* 653.

Respiratory chain studies, difference spectra, *Differenzspektra* 657, 659.
— — —, fluorimetry, *Fluorimetrie* 659.
— — —, spectrophotometry, *Spektrophotometrie* 656.
— — —, —, instruments, *Instrumente* 658.
"Respiratory control", „*Atmungskontrolle*" 652.
Respiratory rate, measurement, manometric, *Atmungsintensität, Messung, manometrische* 648.
— —, —, polarographic, *polarographische* 648.
— —, —, spectrophotometric, *spektrophotometrische* 649.
— —, —, volumetric, *volumetrische* 648.
RHP (*Rhodospirillum* heme protein) 603.
—, preparation, *Darstellung* 603.
—, properties, *Eigenschaften* 603, 604.
Riboflavin 175, 176, 182, 184, 196.
Riboflavin-5′-phosphate see also flavinmononucleotide (FMN), *Riboflavin-5′-phosphat s. Flavinmononucleotid.*
Ribonucleases, *Ribonucleasen* 58.
Ribonucleic acid see also: sRNA, *Ribonucleinsäure s. sRNA* 405.
Ribosomes, *Ribosomen* 408.
—, amino acid composition, *Aminosäuren-Zusammensetzung* 409.
sRNA, preparation of, *sRNA, Darstellung* 406

Saccharides, complex, precursor formation, *Saccharide, komplexe, Bildung von Vorstufen* 478.
Sedoheptulose-1,7-diphosphatase 40.
Sedoheptulose-1,7-diphosphate, *Sedoheptulose-1,7-diphosphat* 529.
Sedoheptulose-7-phosphate, preparation, *Sedoheptulose-7-phosphat, Darstellung* 565.
Serine, synthesis, *Serin, Synthese* 349.
Serine-alanine transaminase, *Serinalanintransaminase* 376.
Serine aldolase, *Serinaldolase* 351—352.
D-Serine dehydrase, D-*Serindehydrase* 311.
L-Serine dehydrase, L-*Serindehydrase* 310.
Shikimic acid, *Shikimisäure* 260, 261, 267, 268.
— —, as precursor in higher plants, *als Vorstufen in höheren Pflanzen* 275.
Starch, synthesis, *Stärke, Synthese* 510.
Succinic acid, formation from glutamic acid, *Bernsteinsäure, Bildung aus Glutaminsäure* 347—348.
Succinic dehydrogenase, *Bernsteinsäuredehydrogenase* 624—627.
Succinic semialdehyde dehydrogenase, *Bernsteinsäuresemialdehyd-dehydrogenase* 348.
Succinyl CoA synthetase 623—624
Sucrose, *Saccharose* 498, 500.
Sucrose phosphate, *Saccharosephosphat* 498, 501.
Sugar nucleotide pyrophosphorylases, from *Phaseolus aureus*, assay and preparation, *Zuckernucleotidpyrophosphorylasen, aus Phaseolus aureus, Bestimmung der Aktivität und Darstellung* 485.

Sugar nucleotide pyrophosphorylases, from *Phaseolus aureus*, properties, *Eigenschaften* 486.
Synthetases, *Synthetasen* 393.

Tartaric dehydrogenase, *Weinsäuredehydrogenase* 629.
Tartronic acid semialdehyde, *Tartronsäuresemialdehyd* 205.
Terminal oxidase, *Endgruppenoxydase* 653.
— —, identification, *Identifizierung* 650.
Terminal respiration, enzymes of, *Endgruppenatmung, Enzyme der* 647.
5,6,7,8-tetrahydrofolic acid, *5,6,7,8-Tetrahydrofolsäure* 189.
2-Tetrahydroxybutyl-5-methyl-4-carboxyfuran 206.
Tetrametaphosphatases, *Tetrametaphosphatasen* 47.
Tetrapolyphosphatases, *Tetrapolyphosphatasen* 45.
Thetins, *Thetine* 321.
Thiaminase 194.
Thiamine, *Thiamin* 173, 180, 194, 207.
Thiamine monophosphatase, *Thiaminmonophosphatase* 174.
Thiamine monophosphate, *Thiaminmonophosphat* 174, 195.
Thiamine phosphatase, *Thiaminphosphatase* 175, 195.
Thiamine phosphate synthetase, *Thiaminphosphatsynthetase* 174, 175.
Thiamine pyrophosphatase, *Thiaminpyrophosphatase* 195.
Thiamine pyrophosphate, *Thiaminpyrophosphat* 180.
Thiamine pyrophosphate kinase, *Thiaminpyrophosphatkinase* 182.
Thiamine pyrophosphokinase, *Thiaminpyrophosphokinase* 174, 180, 181.
Thiazole kinase, *Thiazolkinase* 174, 175.
Thiochrome, *Thiochrom* 182.
Threonic acid, *Threonsäure* 218.
D-Threonine dehydrase, D-*Threonindehydrase* 311.
L-Threonine, biosynthesis, L-*Threonin, Biosynthese* 325.
L-Threonine dehydrase, L-*Threonindehydrase* 310.
Threonine racemase, *Threoninracemase* 384, 385.
Threonine synthetase, *Threoninsynthetase* 328, 329.
β and γ Tocopherol quinones, *β- und γ-Tocopherolchinone* 612.
TPN see: Triphosphopyridinenucleotide, *TPN s. Triphosphopyridinnucleotid.*
TPN-reducing factor, *TPN-reduzierender Faktor* 590.
TPN reductase, *TPN-Reduktase* 591, 593.
TPN triosephosphate dehydrogenase, *TPN triosephosphatdehydrogenase* 534.
— — —, activators and inhibitors, *Aktivatoren und Hemmstoffe* 535.

TPN triosephosphate dehydrogenase, assay, *Bestimmung der Aktivität* 534.
— — —, purification, *Reinigung* 535.
— — —, distribution, *Vorkommen* 535.
TPNH$_2$-diaphorase 596.
TPNH$_2$-plastocyanin reductase, *TPNH$_2$-plastocyaninreduktase* 596.
Transacetylase 616.
Transaldolase 564.
—, assay, *Bestimmung der Aktivität* 564, 565.
—, purification, *Reinigung* 565, 566.
Transamidination, *Transamidinierung* 380, 381.
Transaminases, *Transaminasen* 361.
Transhydrogenase 667.
—, assay, *Bestimmung der Aktivität* 667.
—, properties and purification, *Eigenschaften und Reinigung* 668.
Transketolase 558.
—, assay, *Nachweis* 559, 560.
—, purification, *Reinigung* 560—562.
Transmethylation, enzymes of, *Transmethylierung, Enzyme der* 319.
Trehalose phosphate, *Trehalosephosphat* 501, 503, 504.
Trimetaphosphatases, *Trimetaphosphatasen* 45.
Triosephosphate isomerase, *Triosephosphatisomerase* 533.
— —, assay, *Nachweis* 533.
— —, distribution, *Vorkommen* 534.
Triosephosphate-phosphoglycerate dehydrogenase, *Triosephosphat-phosphoglyceratdehydrogenase* 536.
— — —, activators and inhibitors, *Aktivatoren und Hemmstoffe* 537.
— — —, assay, *Nachweis* 536.
— — —, distribution, *Vorkommen* 537.
Triphosphopyridinenucleotide, *Triphosphopyridinnucleotid* 187, 196, 228.
Tripolyphosphatases, *Tripolyphosphatasen* 44.
Tryptamine, *Tryptamin* 238.
—, conversion to IAc and IAA, *Umwandlung in IAc und Indolessigsäure (IES)* 244.
Tryptophan, synthesis, *Tryptophan, Synthese* 352.
Tryptophan-alanine transaminase 379.
Tryptophan-glutamate transaminase 379.
Tryptophan pyrrolase 177.
Tryptophan synthetase 354, 355.
L-Tryptophan 238.
—, enzymes converting to IAA, *enzymatische Umwandlung in Indolessigsäure (IES)* 240.
— — — —, distribution, *Vorkommen* 240.
L-Tryptophan decarboxylase 308.
TTP (L-tryptophan).
Tyrase 278—281.
Tyrosine, *Tyrosin* 261.
Tyrosine-activating enzyme, *Tyrosin aktivierende Enzyme* 407.
Tyrosine apodecarboxylase, *Tyrosinapodecarboxylase* 187.

Tyrosine, biosynthesis, *Tyrosin, Biosynthese* 261.
Tyrosine decarboxylase, *Tyrosindecarboxylase* 309.
Tyrosine-glutamate transaminase, *Tyrosinglutamattransaminase* 377.
L-Tyrosine, L-*Tyrosin* 278.

Ubiquinone (Coenzyme Q), *Ubichinon (Coenzym Q)* 605, 612, 665.
—, assay, *Nachweis* 677.
—, properties and purification, *Eigenschaften und Reinigung* 678, 679.
Ubiquinone reductase, *Ubichinonreduktase* 665.
UDP (uridine 5'-diphosphate), *UDP (Uridin 5'-diphosphat)* 474.
UDP-N-acetyl-D-glucosamine, *UDP-N-acetyl-D-glucosamin* 479, 508.
UDP-N-acetyl-D-glucosamine-chitin transglucosylase, from *Neurospora crassa*, *UDP-N-acetyl-D-glucosamin-chitin transglucosylase, aus Neurospora crassa* 508, 509.
UDP-N-acetyl-D-glucosamine pyrophosphorylase, from baker's yeast, *UDP-N-acetyl-D-glucosaminpyrophosphorylase, aus Bäckerhefe* 490.
— — —, — *Phaseolus aureus* 489, 490.
UDP-L-arabinose 479, 485, 493, 497, 509.
— — 4-epimerase 495, 510.
— — —, from *Phaseolus aureus*, *aus Phaseolus aureus* 495.
UDP-D-galactose, *UDP-D-galaktose* 478, 481, 485, 492, 493.
— — 4-epimerase 478, 481, 493.
— — —, from *Phaseolus aureus*, *aus Phaseolus aureus* 495.
— — —, — *Saccharomyces fragilis* 493, 494.
— — pyrophosphorylase 485.
UDP-D-galacturonic acid, *UDP-D-Galakturonsäure* 479, 485, 493, 495.
— — —, from radish, *aus Rettich* 495.
— — — 4-epimerase 495.
UDP-D-glucose 474, 476, 480, 481, 485, 486, 487, 493, 495, 498, 499, 501, 502, 503, 504, 506, 507, 508, 510.
UDP-D-glucose dehydrogenase 478.
— — —, from peas, *aus Erbsen* 495—497.
UDP-D-glucose-diphenol transglucosylase, from wheat germ, *UDP-D-glucose-diphenoltransglucosylase, aus Weizenkeimlingen* 504—506.
UDP-D-glucose-D-fructose 6-phosphate transglucosylase, from wheat germ, *UDP-D-glucose-D-fructose-6-phosphattransglucosylase, aus Weizenkeimlingen* 500, 501.
UDP D glucose-D-fructose transglucosylase, from wheat germ, *UDP-D-glucose-D-fructosetransglucosylase, aus Weizenkeimlingen* 499, 500.
UDP-D-glucose-β-1,3-D-glucan transglucosylase, from *Phaseolus aureus*, *UDP-D-glucose-β-1,3-D-glucantransglucosylase, aus Phaseolus aureus* 507, 508.

UDP-D-glucose-D-glucose 6-phosphate trans-glucosylase, from yeast, *UDP-D-glucose-D-glucose 6-phosphattransglucosylase, aus Hefe* 501—504.

UDP-D-glucose-phenol-D-glucoside trans glucosylase, from wheat germ, *UDP-D-glucose-phenol-D-glucosid-transglucosylase, aus Weizenkeimlingen* 506.

UDP-D-glucose pyrophophorylase, from brewer's yeast, *UDP-D-glucosepyrophosphorylase, aus Bierhefe* 488, 489.

— — —, from *Phaseolus aureus, aus Phaseolus aureus* 486—488.

UDP-D-glucuronic acid, *UDP-D-glucuronsäure* 478, 479, 480, 481, 485, 493, 495, 497, 504, 509.

UDP-D-glucuronic acid decarboxylase, *UDP-D-glucuronsäuredecarboxylase* 510.

— — — —, from wheat germ, *aus Weizenkeimlingen* 497.

UDP-D-xylose 479, 481, 485, 492, 495, 497, 509.

UDP-D-xylose-D-xylodextrin trans-xylosylase, from asparagus, *UDP-D-xylose-D-xylodextrintransxylosylase, aus Spargel* 509, 510.

Urea cycle, enzymes of, *Harnstoffcyclus, Enzyme des* 337.

Vitamin K-reductase, *Vitamin K-reduktase* 665.

Vitamin K_1, in chloroplasts, *Vitamin K_1, in Chloroplasten* 605.

—, extraction, *Extraktion* 607.

Vitamin K_3 611.

Vitamin K_5 611.

Wheat leaf enzyme, *Weizenblattenzym* 254.

Xanthine oxidases, nitrate reduction, *Xanthinoxydasen, Nitratreduktion* 110.

Xylan 509.

Xylonic acid, *Xylonsäure* 218.

α-D-xylose 1-phosphate, *α-D-xylose-1-phosphat* 478, 485, 486.

Table des Matières.

Pour la contribution écrite en français:

F. CHAPEVILLE et P. FROMAGEOT, Enzymes du Métabolisme du Soufre.

(français-allemand)

Acide aminophénylsulfurique, *m-Amino-phenylschwefelsäure* 1.
ADP-sulfurylase 2, 4.
—, préparation, *ADP-sulfurylase, Darstellung* 4.
APS-kinase 2, 4.
—, préparation, *APS-kinase, Darstellung* 4.
—, purification, *APS-kinase, Reinigung* 6.
Arylsulfatases, *Arylsulfatasen* 11.
ATP-sulfurylase 2, 3.
—, préparation, *ATP-sulfurylase, Darstellung* 4.
—, purification, *ATP-sulfurylase, Reinigung* 5.
Cholinesulfatase, *Cholinsulfatase* 12.
Chondroitine-sulfokinase, *Chondroitinsulfokinase* 2.
Chondrosulfatases, *Chondrosulfatasen* 12.
Enzymes d'activation du sulfate, *Sulfataktivierung, Enzyme der* 3.
— de transfert du sulfate, *Sulfattransferasen* 6
Glucosulfatases, *Glucosulfatasen* 12.
Levure de boulangerie, *Hefe* 2.
C—S-lyase 17.
Méthylviologène, *Methylviologen* 10.
Myrosulfatases, *Myrosulfatasen* 12.
Phénol-sulfokinase, *Phenolsulfokinase* 2.

Phénol-sulfokinase, préparation, *Phenolsulfokinase, Darstellung* 7.
Phénylsulfate, *Phenylsulfat* 1.
Phosphate de pyridoxal, *Pyridoxalphosphat* 19.
3'-Phosphoadénosine-5'-phosphosulfate, *3'-Phosphoadenosin-5'-phosphosulfat* 1.
Rhodanèse, *Rhodanese* 14.
Sérine-sulfhydrase, *Serinsulfhydrase* 18.
Stéroïdesulfatases, *Steroidsulfatasen* 12.
Sulfatases, *Sulfatasen* 10.
Sulfate, *Sulfat* 1.
— «actif», *Sulfat „aktives"* 1.
— de p. nitrophényle, *p-Nitrophenylsulfat* 2.
—, systèmes réducteurs, *Sulfat, Reduktionssysteme* 8.
—, —, mesure de l'activité, *Sulfat, Reduktionssysteme, Aktivitätsmessung* 8.
—, —, préparation, *Sulfat, Reduktionssysteme, Darstellung* 9.
Sulfite-oxydase, *Sulfitoxydase* 16.
Sulfite-réductase, *Sulfitreduktase* 15.
Sulfokinase des stéroïdes, *Steroidsulfokinase* 2.
Sulfokinases, *Sulfokinasen* 2.
β-Thioglucosidase 12.
Thiosulfate-réductase, *Thiosulfatreduktase* 13, 14.

Sachverzeichnis.

Zu dem französischen Beitrag.

F. CHAPEVILLE et P. FROMAGEOT, Enzymes du Métabolisme du Soufre.

(deutsch-französisch)

ADP-sulfurylase 2,4.
—, Darstellung, *ADP-sulfurylase, préparation* 4.
m-Aminophenylschwefelsäure, *Acide aminophénylsulfurique* 1.
APS-kinase 2, 4.
—, Darstellung, *APS-kinase, préparation* 4.
—, Reinigung, *APS-kinase, purification* 6.
Arylsulfatasen, *Arylsulfatases* 11.
ATP-sulfurylase 2, 3.
—, Darstellung, *ATP-sulfurylase, préparation* 4.

ATP-sulfurylase, Reinigung, *ATP-sulfurylase, purification* 5.
Cholinsulfatase, *Cholinesulfatase* 12.
Chondroitinsulfokinase, *Chondroitine-sulfokinase* 2.
Chondrosulfatasen, *Chondrosulfatases* 12.
Glucosulfatasen, *Glucosulfatases* 12.
Hefe, *Levure de boulangerie* 2.
C—S-lyase 17.
Methylviologen, *Méthylviologène* 10.
Myrosulfatasen, *Myrosulfatases* 12.
p-Nitrophenylsulfat, *Sulfate de p. nitrophényle* 2.

Phenolsulfokinase, *Phénol-sulfokinase* 2.
—, Darstellung, *Phénol-sulfokinase, préparation* 7.
Phenylsulfat, *Phénylsulfate* 1.
3′-Phosphoadenosin-5′-phosphosulfat, *3′-Phosphoadénosine-5′-phosphosulfate* 1.
Pyridoxalphosphat, *Phosphate de pyridoxal* 19.
Rhodanese, *Rhodanèse* 14.
Serinsulfhydrase, *Sérine-sulfhydrase* 18.
Steroidsulfatasen, *Stéroïdesulfatases* 12.
Steroidsulfokinase, *Sulfokinase des stéroïdes* 2.
Sulfat, *Sulfate* 1.
— „aktives", *Sulfate «actif»* 1.
—, Reduktionssysteme, *Sulfate, systèmes réducteurs* 8.

Sulfat, „aktives", Aktivitätsmessung, *Sulfate, systèmes réducteurs, mesure de l'activité* 8.
—, —, Darstellung, *Sulfate, systèmes réducteurs, préparation* 9.
Sulfatasen, *Sulfatases* 10.
Sulfataktivierung, Enzyme der, *Enzymes d'activation du sulfate* 3.
Sulfattransferasen, *Enzymes de transfert du sulfate* 6.
Sulfitoxydase, *Sulfite-oxydase* 16.
Sulfitreduktase, *Sulfite-réductase* 15.
Sulfokinasen, *Sulfokinases* 2.
β-Thioglucosidase 12.
Thiosulfatreduktase, *Thiosulfate-réductase* 13, 14.